The SAGE Handbook of
Geomorphology

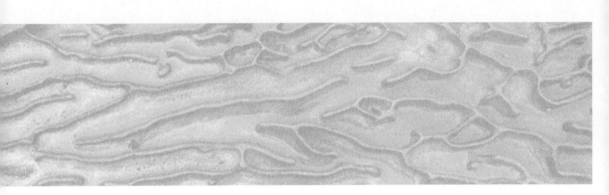

SAGE has been part of the global academic community since 1965, supporting high quality research and learning that transforms society and our understanding of individuals, groups, and cultures. SAGE is the independent, innovative, natural home for authors, editors and societies who share our commitment and passion for the social sciences.

Find out more at: **www.sagepublications.com**

The SAGE Handbook of
Geomorphology

Edited by

Kenneth J. Gregory
and Andrew S. Goudie

Los Angeles | London | New Delhi
Singapore | Washington DC

First published 2011
This paperback edition first published 2014

SAGE Publications Ltd
1 Oliver's Yard
55 City Road
London EC1Y 1SP

SAGE Publications Inc.
2455 Teller Road
Thousand Oaks, California 91320

SAGE Publications India Pvt Ltd
B 1/I 1 Mohan Cooperative Industrial Area
Mathura Road, Post Bag 7
New Delhi 110 044

SAGE Publications Asia-Pacific Pte Ltd
33 Pekin Street #02-01 Far East Square
Singapore 048763

Library of Congress Control Number 2010939890

British Library Cataloguing in Publication data
A catalogue record for this book is available from the British Library

ISBN 978-1-4129-2905-9
ISBN 978-1-4462-9583-0 (pb)

Typeset by Cenveo Publisher Services
Printed in Great Britain by Henry Ling Limited, Dorchester
Printed on paper from sustainable resources

Contents

Acknowledgements

This Handbook, initiated by members of the International Association of Geomorphologists (IAG), has evolved considerably in content during the preparation of the chapters. Basil Gomez assisted by André Roy and Vic Baker contributed to the establishment of the basis for the volume during the early stages. I was brought in at a later stage and have been most grateful to Andrew Goudie for the many excellent ways in which he has provided help and support to get the Handbook to its final conclusion. Thanks are also expressed to the publishers especially Robert Rojek for the initial discussions and to Sarah-Jayne Boyd who has fended my many queries.

We are extremely grateful to all the contributors not only for preparing their chapters, but for contributing royalties to the IAG, and for their patience in awaiting the culmination of such an extensive project.

Kenneth J. Gregory

COMMENT FOR PAPERBACK EDITION

Since publication of this Handbook in 2011 the discipline of geomorphology has continued to be extremely active, as illustrated by the 8th International Conference on Geomorphology (IAG) held in Paris from 27–31 August 2013. That conference attracted more than 1400 delegates, involving the presentation of more than 900 scientific papers. As explained in the original Acknowledgements above, this Handbook was initiated by members of the International Association of Geomorphologists and we are delighted that the volume will now be more widely available as a paperback edition. Since 2010 other geomorphological societies (Table 1.3) have continued to be extremely active, with many adapting to the new means of communication which are becoming increasingly significant (Gregory et al., 2013).

Recent years have reinforced the opportunities available to the science of geomorphology, and the Conclusion includes 11 examples (Table 33.1) of research investigations that demonstrate impacts of significance of global warming for the land surface and land surface processes. Opportunities for geomorphology are now even more acute and could be the subject for further research investigations, so that we hope that this Handbook will continue to provide the basis for further research enquiries benefitting society.

Kenneth J. Gregory and Andrew S. Goudie

References

Gregory, K.J., Lane, S.N., Lewin, J., Ashworth, P.J., Downs, P.W., Kirkby, M.J. and Viles, H.A. (2013) Communicating geomorphology: global challenges for the twenty-first century. *Earth Surface Processes and Landforms*, doi: 10.1002/esp.3461

List of Contributors

Victor R. Baker is Regents' Professor of Hydrology and Water Resources, Geosciences and Planetary Sciences, University of Arizona. His 350 research papers and chapters concern palaeohydrology, planetary geomorphology and history/philosophy. His 16 books include *Catastrophic Flooding* (1981), *The Channels of Mars* (1982), *Flood Geomorphology* (1988, co-edited), *Ancient Floods, Modern Hazards* (2002, co-edited) and *Megaflooding on Earth and Mars* (2009, co-edited). He was the 1998 President of the Geological Society of America (GSA) and his honours include GSA Distinguished Scientist and Distinguished Career Awards, Foreign Membership in the Polish Academy of Sciences, Honorary Fellowship in the European Union of Geosciences and the David Linton Award of the BSG.

Paul Bierman is a geomorphologist with wide-ranging interests including environmental geology, hydrology, isotope geochemistry, glacial geology, surface process and rates of weathering and denudation. He works at the interface between active research, education and science literacy at all levels. He received his degrees from Williams College and the University of Washington. He directs UVM's Cosmogenic Nuclide Extraction Lab - one of only a handful of laboratories in the country dedicated to the preparation of samples for analysis of 10-Be and 26-Al from pure quartz (uvm.edu/cosmolab). He is the recipient (1996) of the Donath Medal as the most promising young geologist in the country. He was also the recipient of NSF's highest award, the Director's Award for Distinguished Teaching Scholars, in 2005.

Paul Bishop has been Professor of Physical Geography in the University of Glasgow since 1998. His undergraduate degree is from the School of Earth Sciences at Macquarie University in Sydney, as are his PhD and recently awarded DSc. His research focuses principally on long-term landscape evolution, with a key interest being the ways in which tectonics and surface processes interact, as mediated by bedrock rivers. Low-temperature thermochronology and cosmogenic nuclide analysis are key techniques for that research, which has always been undertaken with teams of post-doctoral researchers and PhD students.

Derek Booth is Affiliate Professor in the departments of Earth and Space Sciences and Civil and Environmental Engineering at the University of Washington, where he was previously a research professor until 2006. He has studied geomorphology, hydrology and watershed management for the past 30 years with the US Geological Survey, with the Basin Planning Program for King County (Washington), in academia and now in private practice. He is presently the president and senior geologist at Stillwater Sciences, an environmental consulting firm active in stream and watershed analysis in California, Oregon and Washington; and the senior editor of *Quaternary Research*. His interests are river dynamics and deposits, urban watershed management and stormwater, landscape processes and geologic hazards. His work emphasizes field-based collection and analysis of basic physical data crucial to understanding landscape conditions and watershed processes and their likely responses to human disturbance.

Tony G. Brown was originally appointed as a lecturer in Soils at the University of Leicester having completed a PhD on Holocene fluvial geomorphology and palaeoecology at the University of Southampton. Since the 1980s his research has focussed on Quaternary and particularly Holocene fluvial

geomorphology floodplain environments with a particular interest in geochronology and human environ-
mental impact. In 1991, and after moving to the University of Exeter in 1993, he directed the Quaternary
Palaeoenvironments Research Group. In 2009, he took up his present post as Director of the
Palaeoenvironment Laboratory at the University of Southampton (PLUS). He has published over 150
refereed research articles, edited five books and is author of the standard research monograph *Alluvial
Geoarchaeology* CUP (1997) and a recent review of floodplain geoarchaeology (*Aggregate-related
Archaeology: Past, Present and Future*. Oxbow) commissioned by English Heritage. Over the last
20 years he has received research project funding from NERC, English Heritage, AHRC, Leverhulme,
HLF and industry.

Joanna Bullard is a Reader in Aeolian Geomorphology at Loughborough University. She completed her
undergraduate degree at the University of Edinburgh and her PhD at the Sheffield Centre for International
Drylands Research, Sheffield University, investigating the geomorphological variability of linear sand
dunes in the southwest Kalahari, southern Africa. Her current research focuses on aeolian processes and
landforms and includes determining geomorphological controls on dust emissions from the Simpson-
Strzelecki Desert, Australia and the development of dunefields in the Atacama Desert, South America.
Her research has taken her to environments ranging from the forests of Central America to the proglacial
sandur plains of Iceland and Greenland. A particular interest is in the interaction between different
geomorphological process systems and she is currently exploring the interaction between glacifluvial and
aeolian processes in West Greenland. She is the physical geography editor of the RGS-IBG Book Series
and an associate editor for the journal *Earth Surface Processes and Landforms*.

Michael Church is Emeritus Professor at the University of British Columbia, where he taught courses in
environment and resources, hydrology, geomorphology, field studies and research methods for 38 years.
His research interests focus on fluvial sediment transport and the interpretation of river channel changes.
He has, throughout his career, also participated in practical work on water management in forestry, mining
and fisheries, and has worked extensively on management issues on large rivers. He is a registered
Professional Geoscientist in British Columbia.

Nicholas J. Clifford is Professor of Physical Geography at King's College London and completed his
undergraduate and postgraduate degrees at the University of Cambridge. He is a fluvial geomorphologist,
with interests in numerical simulation of flows and habitats in rivers; in river restoration and management;
and in the broader field of environmental time series analysis. He has held appointments at University
College London and the University of Nottingham, and has published widely in the fields of geomorphology, physical geography and geographical philosophy and methods. He is co-editor of Sage's *Key
Concepts* and *Key Methods in Geography*, and is Managing Editor of the journal *Progress in Physical
Geography*.

Peter Cowell is head of school in the School of Geosciences at the University of Sydney and Director of
the University of Sydney Institute of Marine Science. His has undertaken more than 25 years research on
coastal and continental-shelf sediment deposits. This work has included how these deposits were formed
under the influence of long-term changes in sea level and other environmental factors. The results of these
studies have been used to develop computer models that simulate the formation of known deposits over
past centuries and millennia, and these models have been adapted to forecast future coastal impacts of
climate change over coming decades and centuries based on geological calibration. The work has been
widely published internationally, and has involved collaboration with leading coastal scientists in
Australia, New Zealand, The Netherlands, the UK, Italy, the USA and Brazil, where the models have been
trialed under a wide range of environmental conditions.

Michael Crozier is Professor Emeritus at Victoria University of Wellington holding a personal Chair in
Geomorphology from 1998. He is President of the International Association of Geomorphologists, past
president of the New Zealand Geographical Society and former member of the New Zealand Conservation
Authority. Academic awards include a Fulbright Scholarship to USA, Leverhulme fellowship to

University of Bristol and a Distinguished Visiting Professorship at the University of Durham. He has held academic positions at the Universities of Otago, Alberta, Trent and Vienna. His research focus includes landslides, natural hazards, and human impact on natural systems.

Mark Dickson is a coastal geomorphologist at the University of Auckland. He obtained a BSc from Massey University and a PhD from the University of Wollongong. He worked as a postdoctoral researcher at the University of Bristol and a FRST postdoctoral research fellow at the National Institute of Water and Atmospheric Research (NZ). His research is focused on eroding cliffed coastlines, employing both field methods and numerical modelling.

Peter Downs was until recently a consulting fluvial geomorphologist with Stillwater Sciences in Berkeley, California before joining the University of Plymouth as a Senior Lecturer in Physical Geography. He was previously a Lecturer in Geography at the University of Nottingham after obtaining his BSc and PhD degrees at the universities of Leicester and Southampton respectively. His research interests include catchment- and reach-scale sediment budgets, the dynamics and sensitivity of river channels, applications of geomorphology to river basin management, and river restoration planning, design and evaluation. The research has occurred in widely varying fluvial environments in the UK, California, Mississippi and New Zealand, and has frequently been interdisciplinary, linking riverine dynamics to aquatic habitat provision, biological response, flood control, and river instability management. He has published widely including the co-authored text *River Channel Management: Towards sustainable catchment hydrosystems* (2004). He was awarded Chartered Geographer status in 2005, and currently sits on the editorial board of the Institution of Civil Engineers' *Water Management* journal.

Tom Farr received BS and MS degrees from Caltech, and a PhD from the University of Washington, all in Geology. After a short time as an engineering geologist, he joined the Radar Sciences Group at the Jet Propulsion Laboratory, where he has been since 1975. At JPL, he helped develop the first geologic applications of imaging radar using aircraft, satellites and the Space Shuttle. He has also been a science investigator on European and Japanese satellite programs and has assisted in the interpretation of radar images from Venus and from Saturn's moon Titan. His scientific research includes the use of remote sensing and digital topographic data for study of landscapes on Earth and other planets, including how they are formed and modified by climate and tectonic or volcanic activity.

Derek Ford grew up in the limestone city of Bath, England, and began caving and climbing in the nearby Mendip Hills at age 12. He graduated BA Hons, DPhil in geography at Oxford University, taught in geography and geology at McMaster University, Ontario, Canada, and is now Emeritus Professor. His area of research interest has been karst, including hydrogeology, cave genesis, landforms, palaeokarst, clastic and precipitate deposits, dating and palaeoenvironmental studies of speleothems. He has undertaken field research or directed graduate research in Canada, USA, Caribbean and Latin America, and collaborative research throughout Europe and in China; 50 MSc and PhD students supervised to completion, 20 post-doctoral fellows and foreign visiting scientists; over 200 refereed journal papers or chapters, 12 books written or edited, many consulting reports (particularly for Parks Canada), one coffee table book, producer of one movie for the National Film Board of Canada. President, Canadian Association of Geographers International Spelelogical Union. Honours include FRS Canada, medals, career awards and honorary fellowships from 14 nations.

Hugh French obtained his PhD degree from the University of Southampton, UK, in 1967. He taught at the University of Ottawa, Canada, in the departments of geography, geology and earth sciences and was dean of the Faculty of Science, president of the International Permafrost Association and founder and editor-in-chief of the international journal *Permafrost and Periglacial Processes*. He received the Roger Brown Award of the Canadian Geotechnical Society for 'outstanding contributions to permafrost science and engineering' in 1989 and the Canadian Association of Geographers Award 'for scholarly distinction in geography' in 1995. He is now Emeritus Professor, University of Ottawa. His undergraduate text, *The Periglacial Environment, Third Edition*, 2007, was first published in 1976 and subsequently revised in 1996; it has been the international standard for more than 30 years. He has conducted field research in most parts of northern Canada, the Canadian High Arctic islands, central and northern Alaska, Svalbard, northern Siberia, the Qinghai-Xizang (Tibet) Plateau, and Antarctica, and Pleistocene periglacial investigations in southern England, central Poland and the New Jersey Pine Barrens, eastern USA.

Andrew Goudie was at the School of Geography, Oxford University, became fellow of Hertford College, and was appointed Professor of Geography and head of department. In October 2003, he became Master of St Cross College, Oxford. He has been an honorary secretary and vice Pesident of the Royal Geographical Society, executive secretary and chairman of the British Geomorphological Research Group, a member of the Council of the Institute of British Geographers, and president of the Geographical Association, and president of the International Association of Geomorphologists. In 1991, he was awarded the Founders' Medal of the Royal Geographical Society and the Mungo Park Medal by the Royal Scottish Geographical Society; in 2002 a medal from the Royal Academy of Belgium and a DSc by Oxford University. In 2007, he received the Geological Society of America's Farouk El-Baz Award for Desert Research. His research interests include the geomorphology of deserts, climatic change, environmental archaeology and the impact of humans on the environment. He has worked extensively in South Africa, Botswana, Swaziland and Namibia and in Oman, Jordan, Bahrain and the United Arab Emirates. He is the author or co-author of the following books *Geomorphological Techniques* (1981, 1990), *Chemical Sediments and Geomorphology* (1983), *The Geomorphology of England and Wales* (1990), *Salt Weathering Hazards* (1997), *Aeolian Environments, Sediments and Landforms* (1999), *Great Warm Deserts of the World* (2002), *Encyclopedia of Geomorphology* (2004), *Desert Dust in the Global System* (2006), *The History of the Study of Landforms* (volume 4) (2008), *Geomorphological Hazards and Disaster Prevention* (2010) and *Landscapes and Geomorphology* (2010).

Kenneth J. Gregory was Warden of Goldsmiths College University of London and is now Emeritus Professor of the University of London and Visiting Professor of the University of Southampton. He obtained his BSc, PhD and DSc from the University of London. He was made CBE in 2007 for services to geography and higher education, is a fellow of University College London, a former secretary and chairman of the British Geomorphological Research Group, and currently president of the British Society for Geomorphology. His research interests are in river channel change and management, palaeohydrology and the development of physical geography. Recent publications include *The Changing Nature of Physical Geography* (2000), *River Channel Management* (with Peter Downs, 2004), and *The Earths Land Surface* (2010). He edited *Palaeohydrology: Understanding Global Change* with Gerardo Benito (2003), *Physical Geography* (2005) and was lead editor for *Environmental Sciences: A Companion* (2008). He has three Honorary degrees, awards received include the Founder's Medal of the Royal Geographical Society (1993), the Linton award of the BGRG (1999) and the Geographical medal of the Royal Scottish Geographical Society (2000).

Richard Huggett is a Reader in physical geography in the University of Manchester. He obtained his BSc and PhD from University of London. His recent publications include *Topography and the Environment* (with Joanne E. Cheesman, 2002), *Fundamentals of Biogeography* (2nd edn, 2004), *Physical Geography: A Human Perspective* (with Sarah Lindley, Helen Gavin, and Kate Richardson, 2004), *The Natural History of the Earth: Debating Long-term Change in the Geosphere and Biosphere* (2006), *Physical Geography: The Key Concepts* (2010), and *Fundamentals of Geomorphology* (3rd edn, 2011).

Vishwas Kale is a Professor in the department of geography, University of Pune, India. He obtained his MS in geography from the University of Pune and his PhD in archaeology at the Deccan College, Pune. His research interests include fluvial and flood geomorphology, palaeohydrology, Quaternary geomorphology and landscape evolution. He has published over 20 research papers in these topics. With Hervé Piégay, he peer-reviewed international journals and jointly authored a textbook on *Introduction to Geomorphology*. He is on the editorial board of the journal *Geomorphology*.

G. Mathias (Matt) Kondolf is a fluvial geomorphologist and Professor of Environmental Planning at the University of California, Berkeley. He holds a PhD in geography and environmental engineering from the Johns Hopkins University, a MS in earth sciences from the University of California Santa Cruz and an AB in geology *cum laude* from Princeton University. He teaches courses in hydrology, river restoration, and environmental science. He conducts research on aspects of environmental river management, including sediment management in regulated rivers, geomorphic and ecological processes in river restoration, and resolving conflicts among restoration goals, and he frequently advises government and international agencies in the United States and abroad on these topics. With Hervé Piégay, he is co-editor of *Tools in Fluvial Geomorphology* (2003).

Hélène Lamarre is a research assistant at the Canadian Research Chair in Fluvial Dynamics, University of Montreal. She obtained her PhD degree from the University of Montréal in 2006, her thesis investigating sediment transport, sedimentary structures and bed stability in step-pool channels. She contributed to the development of a new approach to track individual particles in order to quantify bed load transport in gravel-bed rivers. The method is based on the insertion of passive integrated transponders in clasts of different sizes and shapes. This cost-effective approach is now used by many geomorphologists in Québec and Europe to link sediment transport to channel hydraulics and morphology. Her current research interests focus on the interaction between flow velocity, bed stability and bed load transport in gravel-bed rivers with a wide range of slope and channel morphology. Her research has been in the south of Québec and in the French Alps, and recent work has been published in *Geomorphology*, *Earth Surface Processes and Landforms* and *Journal of Sedimentary Research*. Since 2002, she lectures to bachelor students on hydrologic processes that operate at or near Earth's surface.

Stuart Lane is Professor of Physical Geography and Director of the Institute of Hazard, Risk and Resilience in Durham University. His research is concerned with the mathematical modelling and analysis of geomorphological and hydrological processes, including river flows, sediment and solute transport, river morphology, river habitat and flood risk. His work is unusual in that it has focused upon innovative collaborations across disciplines and he has worked with engineers, earth scientists, mathematicians, biologists and social scientists, amongst others. He has published over 170 papers, edited three research monographs and is currently Managing Editor of the journal *Earth Surface Processes and Landforms*. He has been awarded two best paper prizes, the Jan de Ploey award from the International Association of Geomorphologists, a Philip Leverhulme Prize for Earth and Environmental Science and an award from the Association of Rivers Trusts for contributions to science.

Andreas Lang has been Chair of Physical Geography at the University of Liverpool since 2003 and Head of the Department of Geography since 2007. He obtained his PhD at the University of Heidelberg, Germany in 1995. Before taking on his current position he was employed as postdoctoral researcher at the Max-Planck Institute for Nuclear Physics in Heidelberg (1995–98) and at the University of Bonn (1998–2001), and as Senior Lecturer at the University Leuven (Belgium, 2001–03). Currently he is the Chairman of the British Society for Geomorphology. His research interests are focused on longer-term landscape evolution – especially the changing sediment fluxes in river systems in response to climate change and human impact. He is also involved with developing and applying geochronological techniques and has been influential in introducing dosimetric dating in geomorphological research.

Dénes Lóczy was research fellow at the Geographical Research Institute of the Hungarian Academy of Sciences for 18 years until 1997, when he moved to the University of Pécs where he is Director of the Institute of Environmental Sciences. He studied Geography and English at Eötvös Loránd University, Budapest, and habilitated at the University of Pécs. He published a handbook *Land Evaluation* (2002) and a two-volume university text-book *Geomorphology* (2005, 2008), both in Hungarian, co-edited collections of papers for Hungarian publishers, Gebrüder Borntraeger and Springer Verlag and wrote 120 academic papers, partly in Hungarian, partly in English. In 1992 he was Alexander von Humboldt scholarship holder in Darmstadt, Germany. Between 2001 and 2005 he was Secretary of the International Association of Geomorphologists. He is a member of the Editorial Boards of *Zeitschrift für Geomorphologie* (Stuttgart) and *Studia Geomorphologica Carpatho-Balkanica* (Kraków).

Anne Mather is Reader in Earth Sciences in the School of Geography, Earth and Environmental Sciences at the University of Plymouth. She completed her undergraduate degree at the University of Hull and her PhD at the University of Liverpool. Her current research focuses on long-term environmental change in tectonically active, dryland areas, involving the direct and indirect responses of alluvial and fluvial systems to regional tectonics and embracing landform and environmental landscape reconstruction on 10^3 to 10^6 year time scales and catchment/basin spatial scales. Her main geographical areas of research include Spain, Turkey, Morocco and Chile.

John Menzies is currently Professor of Earth Sciences and Geography at Brock University. He received his BSc (Hons) degree in geography from the University of Aberdeen and his PhD from the University of Edinburgh. His research interests are in the mechanics of drumlin formation, in subglacial environments, in the rheology of subglacial sediments and in glacial micromorphology in Quaternary sediments and pre-Quaternary sedimentary rocks. He is the editor/author of *Modern Glacial Environments*, *Past Glacial Environments* and the revised student edition of *Modern and Past Glacial Environments*. He is on the editorial board of *Sedimentary Geology*.

Cherith Moses is a senior lecturer in physical geography at the University of Sussex. She graduated with a BSc in geography and a PhD in geomorphology from Queen's University Belfast during which time she was also a research fellow at the Australian National University. Her research focuses on field and laboratory monitoring and measuring of rock weathering and surface change to further understanding of environmental and climatic impacts on landform development. She has a particular interest in limestone landscapes, interactions of weathering processes and weathering with erosion, and applies rock weathering research to the study of building stone decay and the development of rock coasts. She has conducted field research in Australia, Asia, Europe and North Africa.

Nicholas A. Odoni is a researcher at the Institute of Hazard, Risk and Resilience in Durham University. His research is concerned with the development of innovative landscape models in both geomorphology and hydrology. He is also interested in more general problems in geomorphological modelling, in particular the application of experiment design and metamodelling techniques to clarify model emergent behaviour and uncertainty. His PhD research at the University of Southampton was concerned with models of long-term catchment evolution, and addressed the problem of exploring model equifinality in an exemplar landscape evolution model. His more recent work in Durham has been concerned with interactive modelling of diffuse interventions in river catchments to reduce flood risk.

Takashi Oguchi is Vice-Director and Professor at the Centre for Spatial Information Science, The University of Tokyo, Japan. He received his PhD in Geography from The University of Tokyo, and broadened his experience at the University of Arizona, Colorado State University, and the Centre for Ecology and Hydrology (UK). He has participated in research projects on fluvial/hillslope geomorphology, geomorphometry, geoarchaeology, water quality and spatial databases. He has been one of the three editors-in-chief of *Geomorphology* (Elsevier) since 2003, and on the editorial board of *Catena* (Elsevier), *Geographical Research* (Blackwell), *Geography Compass* (Blackwell) and *Open Geology Journal* (Bentham Science).

Hervé Piégay is Research Director at the National Centre of Scientific Research, working at École Normale Supérieure of Lyon where he is a co-director of a research unit of 200 persons. He received his PhD at the University Paris IV – Sorbonne focusing on interactions between riparian vegetation and channel geomorphology. He has led research on the contemporary history of rivers and their catchments, underlining human controls on environmental changes, wood in rivers, floodplain and former channel sedimentation, geomorphology and society, with specific research on human perceptions of river environments. He is involved in integrated sciences and interacts with practitioners providing knowledge for river management, planning and restoration and methodological frameworks and tools using GIS and remote sensing. He is mainly involved in the Rhône basin but also on the Rhine, in Indonesia (Progo), Vietnam (Mekong), USA (Sacramento), and Italy (Magra). He has edited books including *Tools in Fluvial Geomorphology* (with M.G. Kondolf, 2003) and *Gravel-bed Rivers 6: From Process Understanding to*

River Restoration (with H. Habersack and M. Rinaldi, 2007). He is associate editor of *Geodinamica Acta* (Lavoisier) and is on the editorial board of *Geomorphology* (Elsevier).

Colin Pain recently retired from Geoscience Australia where he led the team that developed a system of regolith mapping used widely in Australia. He received BA and MA degrees in Geography at the University of Auckland and a PhD in Geomorphology at the Australian National University. He has worked on landforms and landscape evolution in Papua New Guinea, the Philippines and Arizona. His research interests include the relationships between regolith and landforms and the application of digital elevation models and satellite imagery to landform mapping and research. He is on the editorial board of the *Journal of Maps*. He is currently based in Dubai where he is team leader of the Soil Survey of the Northern Emirates, and holds an honorary position in the MED-Soil Research Group at the University of Sevilla.

David Petley is the Wilson Professor of Hazard and Risk at Durham University. His research focuses upon landslide and rockfall mechanics, using a combination of state-of-the-art field monitoring techniques and bespoke laboratory testing. He is the director of the International Landslide Centre at Durham, with research interests in the UK, Europe, South and East Asia, and Australasia, and vice president of the Natural Hazards Division of the European Geosciences Union. He has written over 80 academic papers, and is co-author with Keith Smith of the leading textbook on natural hazards, *Environmental Hazards – Assessing Risk and Reducing Disaster*.

Jonathan Phillips is Professor of Earth Surface Systems in the department of geography, University of Kentucky, and Chief Scientist for Copperhead Road Geosciences, LLC. His research interests are in fluvial and soil geomorphology, biogeomorphology and nonlinear dynamics of earth surface systems. After receiving a BA from Virginia Tech and an MA from East Carolina, he earned his PhD in Geography at Rutgers. He has held previous academic appointments at Arizona State, East Carolina and Texas A&M Universities. He has published extensively in the earth and environmental sciences.

Jim Pizzuto is a fluvial geomorphologist in the department of geological sciences, University of Delaware who has published papers in the last 25 years on a variety of topics, including river bank erosion, the evolution of sediment pulses, floodplain evolution, the influence of land use and climate change on rivers, dam removal, and the transport and storage of sediment and contaminants in fluvial systems.

Bruce L. Rhoads is Professor and head of the Department of Geography at the University of Illinois. His research programme consists of three distinct, but interrelated areas of interest: integrated watershed science in human-dominated environments, including stream channelization and naturalization; basic research on the dynamics of meandering rivers and river confluences; and the philosophical and methodological underpinnings of geomorphology and the relation of these to those of geography, geology and other sciences.

Keith Richards is Professor of Geography at the University of Cambridge, and a fellow of Emmanuel College. He is a fluvial geomorphologist whose research interests focus on river channel forms and processes in a wide range of environments; hydrological processes and sediment production and transfer processes in river basins; and the modelling of fluvial and hydrological systems. He has had interests in river and floodplain restoration, inter-relationships between hydrological and ecological processes in floodplain environments, and glacial hydrology. He is a former secretary and chairman of the British Geomorphological Research Group, and a former vice president (research) of the Royal Geographical Society/Institute of British Geographers. He has published some 200 papers, is the author or co-author of *Rivers: Form and Process in Alluvial Channels* (1982; 2004 facsimile) and *Arsenic Pollution: A Global Synthesis* (2009); and editor or co-editor of *Geomorphology and Soils* (1985); *Slope Stability: Geotechnical Engineering and Geomorphology* (1987); *River Channels: Environment and Process* (1987); *Landform Monitoring, Modelling and Analysis* (1998); *Glacier Hydrology and Hydrochemistry* (1998); and themed issues of the *Philosophical Transactions, Royal Society Series A* (2004), *Geomorphology* (2007) and *Geoforum* (2007).

David Robinson graduated with a BSc in geography and a PhD in geomorphology from Kings College, University of London, before moving to Sussex where he is currently Reader in Physical Geography. He has researched and published extensively on rock weathering in a variety of environments, particularly in Europe and North Africa, on experimental weathering, and on the measurement of rock-surface change. He has a particular interest in sandstone weathering and the evolution of sandstone landscapes. More recently he has become heavily involved with European colleagues in studies of the weathering and erosion of rock coasts in macro-tidal, temperate environments. He is currently the Honorary Treasurer and serves on the Executive Committee of the British Society for Geomorphology.

André G. Roy is Professeur Titulaire at the Université de Montréal, where he also holds the Canada Research Chair in Fluvial Dynamics. He served as president of the *Canadian Geomorphological Research Group*, and of the *Canadian Association of Geographers*, and as Chair of the Grant Selection Committee of the Environmental Earth Sciences Committee of the *Natural Sciences and Engineering Research Council (Canada)*, and as Group Chair of the Earth Sciences Grant Selection Committees of NSERC. His research focuses on the study of flow turbulence and sediment transport in rivers and on the detailed quantification of fluvial processes. Recently he has worked on the response of rivers to environmental change. He has published more than 110 peer reviewed articles in a broad range of scientific journals in geomorphology, water resources and the earth sciences. He has received both national and international recognition for excellence in research and teaching, including the award for Scholarly Distinction from the Canadian Association of Geographers.

Mike J. Smith is currently a Senior Lecturer in GIS at the School of Geography, Geology and the Environment at Kingston University. He received a BSc (Hons) degree in geography from the University of Wales, Aberystwyth, an M.Sc. degree in geography from the University of British Columbia and a PhD degree in palaeo-glaciology from the University of Sheffield. Mike principally lectures to bachelor and masters' programmes on the application of remote sensing in the geosciences. His research interests are based around the application of digital elevation models in geomorphology and specifically focused upon the visualisation and gemorphometric modelling of glacial landscapes. Recent interests also include field spectroscopy of loess. He is the founder and Editor of the *Journal of Maps*.

László Sütő, was a PhD student of the Earth Sciences Doctoral School at the University of Debrecen, and has undertaken research on anthropogenic geomorphology and landscape evaluation for 10 years until 2007. He has been research fellow at interdisciplinary environmental projects of the Environmental Management and Law Association between 1998 and 2001. At present he is Senior Lecturer at the Institute of Tourism and Geography at College of Nyíregyháza. He is co-author of the book *Anthropogenic Geomorphology*, published by Springer Verlag in 2010, author of 52 articles, partly in Hungarian, partly in English on anthropogenic geomorphology, landscape evaluation, tourism and education in geography. He is a member of the Hungarian Geological Society and the Hungarian Geographical Society.

Graham Taylor has worked on regolith geology in Australia for the last 40 years, publishing more than 100 refereed papers in international journals and a book with Tony Eggleton. He has worked at ANU, HKU and the University of Canberra and continues in retirement to work on regolith, particularly bauxite.

Michael Thomas is Professor Emeritus in Environmental Science at the University of Stirling, Scotland. He holds degrees from Reading and London Universities and began his career in tropical geomorphology as a lecturer at the University of Ibadan in Nigeria from 1960–64, later based at St. Andrews and then Stirling, Scotland. Research into tropical weathering and geomorphology later branched out to include studies of Quaternary sedimentation and climate change. His book, *Geomorphology in the Tropics* (1994) continues to be a standard reference. His research was recognised by election to FRSE in 1988 and the award of the Centenary Medal (2000) of the RSGS and the David Linton Award (2001) of the BSG. He was Joint Editor of *Catena* 1996–2006.

Colin E. Thorn recently retired from the department of geography at the University of Illinois where he has spent thirty years as a geomorphologist, following brief spells at the Universities of Montana and Maryland in the same role. His research interests are focused upon periglacial and theoretical geomorphology. In the realm of periglacial geomorphology he has worked primarily in alpine contexts including the USA, Canada, Norway, Sweden and Finland. He has also worked briefly in a number of other periglacial contexts. His research on theoretical matters is centred primarily on the conceptual underpinnings of primary geomorphological principles.

Heather Viles is Professor of Biogeomorphology and Heritage Conservation at the University of Oxford. She grew up in Essex and studied at the Universities of Cambridge and Oxford before post-doctoral research at University College London on acid rain impacts on English cathedrals. Her research focuses on weathering of rocks in extreme environments (including on Mars) and the deterioration and conservation of building stones, with particular emphasis on the role of organisms. She is currently Vice-President (expeditions and fieldwork) of the Royal Geographical Society with IBG. Her most recent book, with Andrew Goudie, is *Landscape and Geomorphology* (2010).

Jeff Warburton is Reader in Geomorphology at Durham University. He completed his undergraduate studies at the University of Wales Aberystwyth before moving to the University of Colorado to carry out Masters Research. His PhD on glacio-fluvial sediment transfer was awarded by Southampton University prior to postdoctoral research in Natural Resource Engineering at Lincoln University in New Zealand. His research is concerned primarily with understanding the geomorphology and sediment transfer processes operating in upland and mountain catchments. Particular emphasis is placed on upland peat erosion, peat mass movements, debris flows/shallow landslides, glacio-fluvial sediment transfer and the geomorphology of mountain streams. This work is underpinned by intensive field monitoring programmes and geomorphological laboratory experiments. He also has a long-term interest in geocryology and the development of frost-sorted patterned ground.

Thad Wasklewicz is Associate Professor in the Department of Geography, East Carolina University. He received a PhD in geography from Arizona State University. His research interests include debris flow process-form interactions in alpine and arid settings of the western United States and Japan. The research has also expanded to encompass debris flow development after wildfires. Much of this work involves analyses of high-resolution digital terrain models developed from terrestrial laser scanning techniques.

Martin Williams is Emeritus Professor in Geographical and Environmental Studies, University of Adelaide, Australia. He holds a PhD degree from the Australian National University and a Doctor of Science (ScD) degree from the University of Cambridge. He has conducted extensive fieldwork in Africa, Australia, India and China, and is author of over 200 research papers on landscape evolution, climatic change and prehistoric environments in those regions. His two most recent books are *Quaternary Environments* (2nd ed., 1998) and *Interactions of Desertification and Climate* (1996).

Paul Williams was born in Bristol, England, and began caving in the nearby Mendip Hills as a schoolboy. He graduated BA Hons from Durham University and PhD and ScD from Cambridge University. He is a senior fellow of the International Association of Geomorphologists. He taught from 1964 at the University of Dublin (Trinity College) and later was a research fellow at the Australian National University, Canberra. He has been Professor at the University of Auckland (School of Environment) since 1972, where he supervised numerous PhD and Masters students. He has published on land use hydrology, coastal geomorphology and the Quaternary, although his chief area of research interest is karst, including landform evolution, hydrogeology, palaeoenvironmental studies of speleothems, and applied work. He has undertaken field research in New Zealand, Australia, Papua New Guinea, Niue, Ireland, France, USA, Vietnam, Russia and China. He is a member of the editorial boards of *Zeitschrift für Geomorphologie* and *Progress in Physical Geography* and a former member of the board of *Earth Surface Processes and Landforms*. He is a member of the World Commission on Protected Areas of IUCN and a member of their Caves and Karst Task Force. He has served on the executive of the International Association of

Geomorphologists and is currently a member of the executive (Bureau) of the International Speleological Union.

Colin Woodroffe is a coastal geomorphologist in the School of Earth and Environmental Sciences at the University of Wollongong. He has a PhD and ScD from the University of Cambridge, and was a lead author on the coastal chapter in the 2007 Intergovernmental Panel on Climate Change (IPCC) Fourth Assessment report. He has studied the stratigraphy and development of coasts in Australia and New Zealand, as well as islands in the West Indies, and Indian and Pacific Oceans. He has written a comprehensive book on *Coasts, form, process and evolution*, and recently co-authored a book on *The Coast of Australia*.

List of Figures

List of Colour Plates

List of Tables

Introduction to the Discipline of Geomorphology

Kenneth J. Gregory and Andrew Goudie

The word geomorphology, which means literally 'to write about (Greek *logos*) the shape or form (*morphe*) of the earth (*ge*)', first appeared in 1858 in the German literature (Laumann,1858; see Roglic, 1972, Tinkler,1985). The term was referred to in 1866 by Emmanuel de Margerie as 'la géomorpholgie'; it first appeared in English in 1888 (McGee, 1888a,b) and was used at the International Geological Congress in 1891 in papers by McGee and Powell. The term came into general use, including by the US Geological Survey, after about 1890, and it received wide currency in Mackinder's lecture to the British Association meeting in Ipswich in 1895 when he referred to 'what we now call geomorphology, the causal description of the earth's present relief' or the 'half artistic, half genetic consideration of the form of the lithosphere' (Mackinder, 1895: 367–379). The International Geographical Congress in London in 1895 had a section entitled Geomorphology, and A. Penck used the term in his paper to the meeting (Penck, 1895; Stoddart, 1986).

Although the late 19th century was when geomorphology was defined, the subject of study was recognizable much earlier (Tinkler, 1985) being significantly influenced by developments such as those in stratigraphy and uniformitarianism in geology and evolution in biology. Although there were many origins in geology, it became more geographically based with the contributions of W.M. Davis (1850–1934) who developed a normal cycle of erosion, suggested that it developed through stages of youth, maturity and old age, conceived other cycles including the arid, coastal and glacial cycles, and proposed that landscape was a function of structure, process and stage or time. His attractive ideas dominated geomorphology for the first half of the 20th century and arguably provided a foundation for later work and also encouraged debate by stimulating contrary views. Although alternative approaches, such as those by G.K.Gilbert, were firmly based upon the study of processes rather than on landform evolution, for the first half of the 20th century the influence of Davisian ideas ensured that geomorphology emphasized the historical development of landforms because the cycle of erosion introduced by Davis was appealing in its simplicity.

From its 19th century foundations geomorphology has developed enormously as reflected in the chapters in Section 1 of this book, including the history of geomorphology (Chapter 2), explanation (Chapter 3) and theory (Chapter 4), followed by geomorphology and environmental management (Chapter 5), and society (Chapter 6). This introduction focuses on the emergence and growth of the discipline, and then outlines the context of tensions, debates and issues that have arisen, many of which are elaborated in subsequent chapters.

EMERGENCE OF THE DISCIPLINE

It is possible to see how definitions of, or comments on, the scope of geomorphology (Table 1.1) reflect the way in which the discipline of geomorphology emerged and was established. Until 1900 many early definitions saw geomorphology as being concerned with description of the Earth's relief or with the form of the land (Table 1.1). However, according to Kirk Bryan (1941), Davis attempted to subdivide geomorphology as the genetic description of landforms into geomorphogeny, concerned with the history, development and changes of landforms, and geomorphography, concerned with their description. This distinction was not adopted (see Beckinsale and Chorley, 1991: 107) and indeed Davis preferred the term physiography,

although he used the term geomorphology in an identical way. However, the distinction between the description of landforms and their development was identified by Russell (1949, 1958) who contrasted geographical with geological geomorphology (Table 1.1). In 1958 Russell commented that

geomorphology has not developed as substantially as Gilbert forecast in 1890. Physiography concentrated on problems of erosion, almost to the exclusion

Table 1.1 Some definitions of geomorphology

Definition/quotation	Source
Qu'on peut appeler, a l'exemple de plusieurs savants américains, la géomorphologie'.	De Margerie, 1886, 315
Such genetic study of topographic forms (which has been denominated geomorphology) is specifically applicable in the investigation of the Cenozoic phenomena of the eastern United States.	McGee, 1888a, 547
These two ideas (drainage system and base level) gradually developed by a younger generation of students, are the fundamental principles of a new subscience of geology sometimes called geomorphology or physical geography.	Gilbert, 1902, 638
What we now call geomorphology, the causal description of the earth's present relief or the 'half artistic, half genetic consideration of the form of the lithosphere'.	Mackinder, 1895, 373
Die Geomorphologie.	Penck, 1895
Further illustration of the growing recognition of form as the chief object of the physiographic study of the lands is seen in the use of the term 'geomorphology' by some American writers.	Davis, 1900, 161
The geomorphologist may concern himself deeply with questions of structures, process and time, but the geographer wants specific information along the lines of what, where and how much.	Russell, 1949
Flood-plain deposits, deltas and deltaic plains are considered with reference to the sedimentary, structural and morphological processes under which they originated, as examples to illustrate the value of a more geological geomorphology.	Russell, 1958
To place geomorphology upon sound foundations for quantitative research into fundamental principles, it is proposed that geomorphic processes be treated as gravitational or molecular shear stresses acting upon elastic, plastic or fluid earth materials to produce the characteristic varieties of strain, or failure, that constitute weathering, erosion, transportation and deposition.	Strahler, 1952
Geomorphology is primarily concerned with the exogenous processes as they mould the surface of the earth, but the internal forces cannot be disregarded when one considers fundamental concepts of the origin and development of landforms.	Leopold, Wolman and Miller, 1964, 3
Whenever anyone mentions theory to a geomorphologist, he instinctively reaches for his soil auger.... Geomorphology is that science which has for its *objects* of study the geometrical features of the earth's terrain, an understanding of which has been attempted in the past within clearly definable, but not always clearly defined spatial and temporal scales and in terms of the processes which produced, sustain and transform them within those scales.	Chorley, 1978

Continued

Table 1.1   Cont'd

Definition/quotation	Source
Although the term is commonly restricted to those landforms that have developed at or above sea level, geomorphology includes all aspects of the interface between the solid earth, the hydrosphere and the atmosphere. Therefore not only are the landforms of the continents and their margins of concern but also the morphology of the sea floor. In addition the close look at the Moon, Mars and other planets, provided by space craft has created an extra-terrestrial aspect to geomorphology.	Chorley, Schumm and Sugden, 1984
Geomorphology may be defined as the science which studies the nature and history of landforms and the processes of weathering, erosion and deposition which created them. As such it has attracted, and overlapped with, the work of geologists, geographers, soil scientists and hydrologists.	Selby, 1985, 8
We see geomorphology as an holistic, chronological, integrative field-based science, that is integral to the study of a dynamically vibrant planet.	Baker and Twidale, 1991
Geomorphology is now a discipline that has major research frontiers ranging in scale from the transport paths of individual particles over a river bed to the combined tectonic and surface processes responsible for the 100 million year history of sub-continental scale landscapes.	Summerfield, 2005a
Geomorphology could...find itself at the centre of a group of new kinds of science, consistent with its traditional embracing of both geology and geography, the long term and the short, the global and the local, and using the tools of landscape-scale modelling to integrate both phenomena and scales.	Richards and Clifford, 2008

of other parts of the discipline, and developed a terminology which became elaborated beyond usefulness. Disregard of the third dimension and inadequate geophysical backgrounds led to unrealistic results by physiographers.

Despite the approach of G.K.Gilbert, it was not until the mid 20th century that processes shaping the land and landforms gained prominence, as articulated by Strahler (1952) in dynamic geomorphology, and reinforced by Leopold, Wolman and Miller (1964) in their book *Fluvial Processes in Geomorphology*. These two contributions were outstanding stimuli for the instigation of process geomorphology. Strahler (1952) suggested that there were two quite different viewpoints of geomorphology, namely dynamic (analytical) and historical (regional) geomorphology which became associated with timeless and timebound perspectives. Also in the second half of the 20th century investigations into the Quaternary became more prominent. Whereas, previously, historical geomorphology had been focused on denudation chronology and especially on the morphogenesis of tertiary landscapes, investigation of contemporary glacial, deglacial and periglacial processes gave a significant stimulus to the investigation of Quaternary glacial systems, and this was further stimulated once investigations could be anchored to improved Quaternary dating using isotope stages.

A further strand was the inception of theory in geomorphology, crystallized by Chorley (1978), including his frequently cited comment that 'whenever anyone mentions theory to a geomorphologist, he instinctively reaches for his soil auger' (Table 1.1). Definitions used by two important texts in the 1980s reflected the multi-disciplinary nature of geomorphology (Selby, 1985) and the expanding horizons, to include non-terrestrial parts of the earth's surface and the potential inclusion of other planets (Chorley, Schumm and Sugden, 1984) (Table 1.1).

Although Baker and Twidale (1991) regretted the dominance of process geomorphology and pressed for a more holistic view (Table 1.1), further adjustment was achieved as a consequence of the advent of plate tectonics in the 1960s. Continental drift had long been acknowledged as a tectonic basis for geomorphology, and endogenic processes had been shown to be significant in studies over the past century (Haschenburger and Souch, 2004). However, the fundamental importance of plate tectonics was subsequently reinforced by geochronological techniques including cosmogenic dating methods. This reinvigorated geomorphology, so that Summerfield (2005a) saw two scales of geomorphology: small-scale process geomorphology contrasting with macroscale geomorphology reflecting advances made by researchers outside the traditional geomorphological community. However, Summerfield (2005a) sees the potential for links between these two groups of researchers, countering (Summerfield, 2005b) the view that the growing role of geophysicists could lead to a reduced role for geographical geomorphologists as visualized by Church (2005). Summerfield (2005b) believes that 'there is enormous scope to advance geomorphology as a whole probably at its most exciting time since it emerged as a discipline'. This has to cope with the advocation of earth system science (see Clifford and Richards, 2005) from which geomorphology could emerge at the centre of a group of new kinds of science (Richards and Clifford, 2008, see Table 1.1).

This therefore presents a paradox. On the one hand throughout the gradual emergence and broadening definition of geomorphology it is understandable why 'geomorphology is, and always has been, the most accessible earth science to the ordinary person: we see scenery as we sit, walk, ride or fly. It is part of our daily visual imagery ...' (Tinkler, 1985: 239). On the other hand, despite this accessibility and centrality, the discipline of geomorphology 'remains little known and little understood, certainly in relation to other academic disciplines, and especially outside university circles' (Tooth, 2009). Thus a discipline that should be very familiar is insufficiently known, although there is considerable potential to build upon the heritage and focus which differs from, and complements, other geosciences.

The discipline that emerged over more than a century, as expressed in the definitions and comments in Table 1.1, is now poised to develop further as a result of the techniques now available. Such emergence has effectively taken a century or so and during that time books published (Table 1.2) have reinforced the construction of the discipline, many aspects of which are discussed in a recent international encyclopedia of geomorphology (Goudie, 2004). Four substantial histories of the study of landforms have been published (Chorley et al., 1964, 1973; Beckinsale and Chorley, 1991; Burt et al., 2008) together with other perspectives including those of Davies (1968) and Tinkler (1985). Other books are referred to in Table 1.1 but particular ones collected in Table 1.2 exemplify seven themes. First were reactions to Davisian ideas: although credited with over 600 publications, Davis did not publish a book with geomorphology in the title so that *Geographical Essays* (1909) and his book in 1912 were his major works, and his ideas were conveyed in numerous articles. Although Davisian ideas were arguably the single most influential theory in geomorphology from the mid 20th century onwards (Tinkler, 1985: 147), they were challenged, amended and resisted by proffered

alternatives (e.g. Hettner, 1921; Gregory, 2000). This second group included Penck's *Die Morphologische Analyse* published in 1924, a significant alternative to the Davisian approach to geomorphology which did not become widely known in the English-speaking world until it was available in translation in 1953 (Czech and Boswell, 1953). Third were textbooks necessary to establish the foundations, and books with geomorphology in the title included those by Wooldridge and Morgan (1937), Worcester (1939), Lobeck (1939), von Engeln (1942) and Thornbury (1954), although others that were influential had alternative titles such as *Physiography* (e.g. Salisbury, 1907). Whereas such texts endeavoured to present a perception of geomorphology as a whole, a fourth group offered a particular approach which may have been regional (e.g. Cotton, 1922), climatic (e.g. Tricart and Cailleux, 1955, 1965, 1972) or founded upon an alternative cyclic approach (e.g. King, 1962). A fifth group was made up of books which concentrated upon techniques (e.g. King, 1966) or upon processes (e.g. Ritter, 1978), whereas a sixth group comprises more recent texts (e.g. Chorley, Schumm and Sugden, 1984; Selby, 1985; Summerfield, 1991). In addition to these, a seventh group comprises those which cover the history of geomorphology (Chorley et al., 1964, 1973; Beckinsale and Chorley, 1991; Burt et al., 2008).

GROWTH OF THE DISCIPLINE

The growth of the present discipline over the last century was considerable, and is reflected in the creation of organized societies, the inauguration of journals and the proliferation of sub-branches of the subject.

Although geomorphology was represented in existing geographical and geological societies, a move towards its separate identification was exemplified by the creation of the Quaternary Geology and Geomorphology Division of the Geological Society of America in 1955 (Table 1.3), a division which has made very significant awards to recognize contributions made by distinguished geomorphologists. Inevitably individual countries created their own geomorphological societies, with the Swiss Geomorphological Society established in 1946 and succeeded by the British Geomorphological Research Group founded in 1959–1960. Such national societies engendered international contacts through their publications, often as collections of papers arising from meetings. It was from an international meeting organized by the British

Table 1.2 Some key publications establishing the subject, emphasizing books with geomorphology in their title

Date	Author(s)	Title
1909	Davis, W.M.	*Geographical Essays*
1912	Davis, W.M.	*Die Erklarende Beschreibung der Landformen*
1921	Hettner, A.	*Die oberflachenformen des Festlandes, ihre Untersuchung und Darstellung; Probleme und Methoden der Morphologie*
1922	Cotton, C.A.	*Geomorphology of New Zealand*
1924	Penck, W.	*Die Morphologische Analyse*
1937	Wooldridge, S.W. and Morgan, R.S.	*The Physical Basis of Geography: An outline of Geomorphology*
1939	Worcester, P.G.	*A Textbook of Geomorphology*
1939	Lobeck, A.K.	*Geomorphology: An Introduction to the Study of Landscapes*
1942	von Engeln, O.D.	*Geomorphology*
1953	Czech, H. and Boswell, K.C.	*Morphological Analysis of Landforms* (translation of Penck, 1924)
1954	Thornbury, W.D.	*Principles of Geomorphology*
1955	Tricart, J. and Cailleux, A.	*Introduction à la géomorphologie climatique*
1962	King, L.C.	*The Morphology of the Earth*
1964	Chorley, R.J., Dunn, A.J. and Beckinsale, R.P.	*The History of the Study of Landforms, Vol. I, Geomorphology before Davis*
1966	King, C.A.M.	*Techniques in Geomorphology*
1968	Davies, G.L.	*The Earth in Decay: A History of British Geomorphology 1578–1878*
1972	Tilley, P.	*The Surface Features of the Land* (translation of Hettner, 1928)
1973	Chorley, R.J., Beckinsale, R.P and Dunn, A.J.	*The History of the Study of Landforms, Vol. II, The Life and Work of William Morris Davis*
1978	Ritter, D.F.	*Process Geomorphology*
1984	Chorley, R.J., Schumm, S.A. and Sugden, D.A.	*Geomorphology*
1985	Tinkler, K.J.	*A Short History of Geomorphology*
1985	Selby, M.J.	*Earth's Changing Surface: An Introduction to Geomorphology*
1991	Beckinsale, R.P. and Chorley, R.J.	*The History of the Study of Landforms or The Development of Geomorphology, Vol. III: Historical and Regional Geomorphology 1890–1950*
1991	Summerfield, M.A.	*Global Geomorphology*
2008	Burt, T.P., Chorley, R.J., Brunsden, D., Cox, N.J. and Goudie, A.S.	*The History of the Study of Landforms or The Development of Geomorphology, Vol. IV: Quaternary and Recent Processes and Forms (1890–1965) and the Mid-Century Revolutions*

Geomorphological Research Group (now the British Society for Geomorphology) that the International Association of Geomorphology arose, (http://www.geomorph.org/main.html) now having successfully held congresses every 4 years in seven locations (Manchester, Frankfurt, Hamilton, Bologna, Tokyo, Zaragosa and Melbourne). Geomorphological societies have continued to be created and 22 are listed in Table 1.3.

Publication of research is essential for the growth of any discipline and, although in the first half of the 20th century many important geomorphological papers were published in geological and geographical journals, the growth of research activity and the increasing number of publications required establishment of dedicated geomorphological journals. A *Journal of Geomorphology* was inaugurated in 1938 but survived for just 4 years, probably affected by World War 2. Many journals were inaugurated due to the enthusiasm and vision of a single individual: Professor Jean Tricart was the inspiration for *Revue de Géomorphologie Dynamique* (1950–) and later the inception of *Earth Surface Processes* (1977–), which became *Earth Surface Processes and Landforms* in 1979, edited from its inception by Professor Mike Kirkby, under whose editorship it

Table 1.3 Examples of geomorphological societies

Society	Date of foundation	Number of members	Objectives	Web site
Swiss Geomorphological Society (SGS)	1946	200	Founded by a Geomorphological Working Group at the University of Basel	—
Quaternary Geology and Geomorphology Division of the Geological Society of America	1955		To bring together scientists interested in Quaternary geology and geomorphology, to facilitate presentation and discussion of their problems and ideas, to promote research and publication of results in those fields of geology	http://rock.geosociety.org/qgg/index.htm
British Geomorphological Research Group (BGRG)	1960		Established from a group focused on morphological mapping from 1958, first AGM of BGRG held in October 1960. Became BSG in 2006	—
Deutscher Arbeitkreis fuer Geomorphologie (German Geomorphologists Group)	1974		The professional organization for German Geomorphologists, organized within the German Association of Geography (DGfG)	http://gidimap.giub.uni-bonn.de:9080/geomorph/
Japanese Geomorphological Union	1979		Founded to expand interdisciplinary communication among sciences concerned with landform changes and related environmental aspects. Publishes *Transactions Japanese Geomorphological Union*	http://wwwsoc.nii.ac.jp/jgu/index.html
The Geomorphology Speciality Group (GSG)	1979	533	A component of the Association of American Geographers, to foster better communication among those working in the geomorphic sciences, especially geography. *Geomorphorum* is issued twice a year	—

Continued

Table 1.3   Cont'd

Society	Date of foundation	Number of members	Objectives	Web site
Australia New Zealand Geomorphology Group (ANZGG)	1982		Established primarily to organize conferences on geomorphic themes for the benefit of the community of geomorphologists in Australia and New Zealand	http://www.anzgg.org/
Sociedad Espanola de Geomorfologia (SEG), (Spanish Geomorphological Society)	1987	340	The development and promotion of geomorphology through cooperation and national and international exchange. Among its aims is the promotion and dissemination of knowledge of geomorphology, from different fields of knowledge such as geography, geology, engineering and biology and various fields of academia, government, business and industry. Edits the journal *Cuaternario & Geomorfología* in collaboration with the Spanish Association for Quaternary Studies (AEQUA) since 1987	http://www. geomorfologia.es/
Commission on Geomorphology of the Austrian Geographical Society (2000–)	1987		Informal grouping within the scope of the Vienna Institute for Geography; since 2000 is the Austrian Research Group Geomorphology and Environmental Change (Austrian Research Association on Geomorphology and Environmental Change) and also Austria	http://www.geomorph.at/
The Czech Association of Geomorphologists	1988	50	Founded as the Geomorphological Commission by the Physical Geography section of the Czech Geographical Society	http://www.kge.zcu.cz/ geomorf/index.html
International Association of Geomorphologists	1989		IAG/AIG was founded at the Second International Conference on Geomorphology in Frankfurt/Main (Germany) in 1989 to strengthen international geomorphology. Principal objectives are development and promotion of geomorphology as a science through international cooperation and dissemination of knowledge of geomorphology	http://www.geomorph. org/main.html
Stowarzyszenie Geomorfologow Polskich (Association of Polish Geomorphologists)	1991	246	Dedicated to the advancement of the science of geomorphology as well as representative Polish geomorphologists in the country and abroad. Its primary activities are the publication of scientific literature, the organization of scientific	http://www.sgp.org.pl/

Continued

Table 1.3 Cont'd

Society	Date of foundation	Number of members	Objectives	Web site
			conferences, the creation of research grants, awarding of medals and awards, operation of task commissions, and other special activities like the protection of unique landforms. Headquarters office in Poznan	
Canadian Geomorphology Research Group (CGRG)	1993	250	To advance the science of geomorphology in Canada by (1) organizing and sponsoring technical sessions, workshops, and field trips; (2) publishing newsletters twice a year; (3) operating a listserver (CANGEORG) which maintains a comprehensive bibliography of Canadian geomorphological, Quaternary, and environmental geoscience publications; (4) supporting publication of technical reports and field guides; (5) presenting the J. Ross Mackay Award in recognition of a significant achievement by a young geomorphologist in Canada; and (6) cooperating with related earth science associations within Canada	http://cgrg.geog.uvic.ca/
Uniao da Geomorfologia Brasileira (UGB)	1996		Objectives are to (1) bring all those in Brazil or abroad to engage in Brazilian geomorphology and related fields; (2) promote the progress of Brazilian geomorphology; (3) encourage scientific and technological research related to the geomorphological context; (4) maintain exchange with professionals from related areas and national and foreign counterparts; (5) conduct regular meetings; (6) promote the expertise of scientists and technicians in various fields of geomorphology; (7) promote scientific and technical meetings that discuss matters of interest to the development of geomorphology; (8) disseminate technical and scientific information of interest to the policyholder; and (9) keep journals of members' work and news of interest to those involved in geomorphology in Brazil	http://www.ugb.org.br/home/?pg=1
Asociacia Slovenskych Geomorfologov (Association of Slovak geomorphology)	1996		A voluntary association of scientists and selection professionals in the field of geomorphology and its related disciplines. Based in Bratislava	http://www.asg.sav.sk/stanovy.htm

Continued

Table 1.3 Cont'd

Society	Date of foundation	Number of members	Objectives	Web site
Associazione Italiana di Geografia Fisica e Geomorfologia (Italian Association of Physical Geography and Geomorphology) (AIGeo)	2000		Established by the former National Group of Physical Geography and Geomorphology, for promoting, encouraging and coordinating research in the fields of physical geography and geomorphology. It also intends to promote educational ventures for physical geographers and geomorphologists and to facilitate the diffusion of environmental and territory knowledge	http://www.aigeo.it/
Associaçao Portuguesa de Geomorfólogos (Portuguese Association of Geomorphology) (APGeom)	2000		Dedicated to the interdisciplinary study and systematic forms of surface and the processes that create and transform. Founded to promote scientific knowledge in the context of geomorphology and its application in various areas of national interest	http://www.apgeom.pt/Apres/apres.htm
Mexican Society of Geomorphology (MSG)	2003		Involved in the organization of IAG Regional Geomorphology Conference, October–November 2003 in Mexico City	
British Society for Geomorphology (BSG)	2006		Professional organization for British geomorphologists, provides a community and services for those involved in teaching or research in geomorphology, both in the UK and overseas	http://www.geomorphology.org.uk/
Groupe Français de Géomorphologie (French Geomorphology Group) (GFG)			The GFG is an association (1901 Act) of people whose work directly or indirectly affects geomorphology. Geomorphology, as an environmental science, participates in the understanding and management of the environment and security of goods and people. Since 1995 GFG has published *Géomorphologie: Relief, Processus, Environnement*	http://www.gfg.cnrs.fr/spip.php?article18

Other organizations include The Geographical Society of China, Geomorfolosko drustvo Slovenije, The Southern African Association of Geomorphologists.

has become a leading international journal. Between these dates was the foundation of *Zeitschrift für Geomorphologie* (1956–) and *Geomorphological Abstracts* (1960–). By collecting together abstracts of papers published worldwide, *Geomorphological Abstracts* enabled wider knowledge of research outputs, catalysing greater dissemination and understanding of geomorphological activity. In 1989 the creation of the journal *Geomorphology* (Table 1.4) established an international serial that has become extremely important for the publication of research papers. Other journals listed in Table 1.4 are of three types: some are devoted to aspects of the earth's surface and its processes (e.g. hydrology, glacial and periglacial, coastal or arid environments, Quaternary morphogenesis); some are primarily geographical or geological but contain important geomorphology papers; and an environmental group emphasizes the fact that many papers now reflect research by multi-disciplinary teams of researchers.

As any discipline grows and expands it naturally fragments with the distinction of branches: such sub-divisions arise as groupings of active researchers, as a means of interfacing with other disciplines, providing communities which are easier to convene than those of the entire growing discipline, so that, naturally, publications tend to concentrate in certain disciplinary areas (e.g. journals included in Table 1.4). It is impossible to compile a list of all the branches of geomorphology that have been suggested but many of them are collected in Table 1.5: some are major branches with much activity and many adherents, such as fluvial geomorphology which attracted a large

Table 1.4   Examples of journals publishing papers on geomorphology (developed from Gregory, 2010)

Year initiated	Journal	Comments
1938	Journal of Geomorphology	Discontinued after several years of publication
1950	Revue de Géomorphologie Dynamique	Journal edited and inspired by Professor Jean Tricart
1956	Zeitschrift für Geomorphologie	Publishes papers from the entire field of geomorphological research, both applied and theoretical. Since 1960 has published 153 Supplementbände (Supplementary volumes) which cover specific important topics
1960	Geomorphological Abstracts	At first published abstracts of papers in geomorphology but later expanded to Geo Abstracts covering related disciplines
1977	Earth Surface Processes and Landforms	From 1977 to 1979 was Earth Surface Processes but then expanded its name. Described as an international journal of geomorphology publishing in all aspects of earth surface science
1989	Geomorphology	Publishes peer-reviewed works across the full spectrum of the discipline from fundamental theory and science to applied research of relevance to sustainable management of the environment
2002	Journal of Geophysical Research – Earth Surface	Focuses on the physical, chemical and biological processes that affect the form and function of the surface of the solid Earth
Hydrological		
1963	Journal of Hydrology	
1970	Nordic Hydrology	
1971	Water, Air and Soil Pollution	
1984	Regulated Rivers	
1987	Hydrological Processes	
Glacial and Periglacial		
1947	Journal of Glaciology	
1969	Arctic and Alpine Research (called Arctic, Antarctic and Alpine Research from 1999)	
1977	Polar Geography and Geology	
1980	Annals of Glaciology	

Continued

Table 1.4   Cont'd

Year initiated	Journal	Comments
1990	*Permafrost and Periglacial Processes*	
1990	*Polar and Glaciological Abstracts*	

Coastal

1973	*Coastal Zone Management*	
1984	*Journal of Coastal Research*	

Arid

1978	*Journal of Arid Environments*	
2009	*Aeolian Research*	

Quaternary

1970	*Quaternary Research, Quaternary Newsletter*	
1972	*Boreas*	
1982	*Quaternary Science Reviews*	
1985	*Journal of Quaternary Science*	
1990	*Quaternary Perspectives, Quaternary International*	
1991	*The Holocene*	

Physical Geology

1973	*Geology*	
1975	*Environmental Geology*	

Physical Geography

1965	*Geografiska Annaler Series*	
1977	*Progress in Physical Geography*	
1980	*Physical Geography*	

Environment

1972	*Science of the Total Environment*	
1973	*Catena*	
1976	*Geo Journal, Environmental Management*	
1990	*Global Environmental Change*	
1997	*Global Environmental Outlook*	

Geomorphological journals are followed by examples of other categories. Many geographical and geological journals such as *Geographical Journal* (1831–) and *Bulletin Geological Society of America* (1890–) contain important geomorphological papers.

proportion of geomorphological research activity. In Britain, whereas less than 20 per cent of publications were fluvial before 1960, this increased towards 30 per cent by 1975 (Gregory, 1978). The branches (Table 1.5) can be envisaged according to purpose, including quantitative research, which was much apparent after the 1960s during the quantitative revolution; applied research, which

increased after the 1970s with the impact of the environmental revolution; and engineering geomorphology with a particular focus on applied aspects. A related group is defined according to analysis including process, climatic, historical, human activity or structural-based; the difficulty of recognizing separate fields is illustrated by the way in which karst geomorphology could relate to

Table 1.5   Some branches of geomorphology (developed from Gregory, 2009 in Gregory et al., 2009)

Branch of geomorphology	Objective (links to other disciplines and sub-disciplines)
According to purpose	
Quantitative	Use of quantitative, mathematical and statistical methods for the investigation of landforms, geomorphological processes and form process relationships requiring modelling.
Applied	Application of geomorphology to the solution of problems especially relating to resource development and mitigation of environmental hazards.
Engineering geomorphology	Provides a spatial context for explaining the nature and distribution of particular ground-related problems and resources, and also concerned with evaluating the implications of landform changes for society and the environment. The focus is particularly on the risks from surface processes (geohazards) and the effects of development on the environment, particularly the operation of surface processes and the resulting changes to landforms or the level of risks (see Fookes et al., 2005).
According to analysis	
Process	Exogenetic and endogenetic processes and the landforms produced.
Climatic	The way in which assemblages of process domains are associated with particular climatic zones. Sometimes extended with crude parameters to define morphoclimatic zones. Three levels of investigation recognized as:
	Dynamic – the investigation of processes (as above).
	Climatic – the way in which contemporary processes are associated with contemporary climatic zones.
	Climatogenetic – allowing for the fact that many landforms are the product of past climates and are not consistent with the climatic conditions under which they now occur.
Historical	Analysis of processes and landform evolution in past conditions. Sometimes referred to as palaeogeomorphology and interacting with fields such as palaeohydrology.
Structural/tectonic	Study of landforms resulting from the structures of the lithosphere and the associated processes of faulting, folding and warping.
Karst	The processes and landforms of limestone areas which have solution as a dominant process and give rise to distinctive suites of landforms. Special landforms and drainage above and below ground are due to solubility of calcareous rocks including limestone, marble and dolomite (carbonates), and gypsum, anhydrite and salt (evaporites) in natural waters. Derived from the geographical name of part of Slovenia.
Anthropogeomorphology	Study of human activity as a geomorphological agent.

Continued

Table 1.5   Cont'd

Branch of geomorphology	Objective (links to other disciplines and sub-disciplines)
According to process domains	
Aeolian	Wind-dominated processes in hot and cold deserts and other areas such as some coastal zones.
Coastal	Assemblage of processes and landforms that occur on coastal margins.
Fluvial	Investigates the fluvial system at a range of spatial scales from the basin to specific within-channel locations; at time scales ranging from processes during a single flow event to long-term Quaternary change; undertaking studies which involve explanation of the relations among physical flow properties, sediment transport and channel forms; of the changes that occur both within and between rivers. Results can contribute in the sustainable solution of river channel management problems.
Glacial	Concerned with landscapes occupied by glaciers, and with landscapes which have been glaciated because they were covered by glaciers in the past.
Periglacial	Non-glacial processes and features of cold climates, including freeze–thaw processes and frost action typical of the processes in the periglacial zone and in some cases the processes associated with permafrost, but also found in high altitude, alpine, areas of temperate regions.
Hillslope	The characteristic slope forms and the governing processes including processes of mass wasting.
Tropical	Processes, morphology and landscape development in tropical systems associated with chemical weathering, mass movement and surface water flow.
Urban	Processes and morphology in urban environments (urban hydrology, urban ecology).
Weathering	Processes involving the gradual breakdown and alteration of materials through a combination of physical, chemical and biological processes.
Soil geomorphology or pedogeomorphology	Attempts to describe and explain relationships between soils and landforms, including study of the evolution (temporal aspects) and distribution (spatial aspects) of soil and soil materials, and the landscapes in which they are formed and altered.
Mountain geomorphology	Dynamics of earth-surface processes and the formation of landforms in high mountains, with particular reference to the interactions between tectonics, climate, vegetation, hydrology and geomorphological processes.
Extra-terrestrial geomorphology	The origins of landforms and landscapes on planets other than Earth because geomorphological systems cannot be studied solely on the terrestrial land surface (see Baker, 2008).
Seafloor engineering geomorphology	As new mapping technology reveals that ocean floors exhibit a wide variety of relief, sediment properties and active geologic processes such as erosion, faulting, fluid expulsion and landslides, detailed surveys of sea floor geomorphology combine with other disciplines to contribute to solution of engineering problems (see Prior and Hooper, 1999).
Multi-disciplinary hybrids	
Hydrogeomorphology	The geomorphological study of water and its effects (fluvial geomorphology; geographical hydrology; hydrology).
Biogeomorphology	The influence of animals and plants on earth surface processes and landform development (ecology).

structural- or to process-based classification. A major grouping is based on particular processes or groups of processes, and includes the 13 categories in Table 1.5. A final grouping is of multi-disciplinary groups which include bio-geomorphology and hydrogeomorphology, branches which have been established to foster links to research in other disciplines, in these two cases in ecology and hydrology.

Although it is now possible to visualize three broad types of approach, namely process, macro geomorphology and historical/Quaternary, many other sub-divisions of the subject continue and yet more are created such as ice sheet geomorphology (e.g. Fleisher et al., 2006) or seafloor geomorphology (Table 1.5). Such fragmentation of the discipline into many branches could dishearten the new student of the discipline but two implications arise. First, as this fragmentation has been characterized as investigating more and more about less and less, the so-called fissiparist or reductionist trend, there has been a growing awareness of the need to return to the 'big picture' with pleas for a more holistic view. In very general terms the first part of the 20th century saw the emergence of some branches of geomorphology, the second part of the 20th century witnessed the fissiparist creation of many more branches and sub-divisions, so that the 21st century has seen concerted efforts to realize a more holistic approach, a trend facilitated by new techniques available and required by the holistic nature of many problems demanding solution.

A second implication relates to how the chapters in this Handbook should be organized in view of the breadth and diversity of the branches available. If all branches of geomorphology were accorded a chapter, the volume could become too thick to hold together in one binding – perhaps a reason for the *Treatise of Geomorphology* being developed by Elsevier as an online publication. The method adopted here is to have an initial group of chapters dealing with foundation and relevance indicating how geomorphology developed (Chapter 2), evolved explanation (Chapter 3) and employed theory (Chapter 4) as well as demonstrating the relevance of environmental management (Chapter 5) and the importance to society (Chapter 6). In order to achieve the aims of the subject, approaches have included observations and experiments (Chapter 7), geomorphological mapping (Chapter 8), remote sensing (Chapter 12), geographical information systems (Chapter 13), and have required dating methods (Chapter 11). Many approaches are associated with processes and environments but some such as biogeomorphology (Chapter 14), human activity (Chapter 15) and extra-terrestrial geomorphology (Chapters 16 and 35) are clearly defined

approaches which merit separate treatment. A major section of 13 chapters (Chapters 17–27) covers processes and environments, explaining how geomorphology has progressed through the investigation of specific groups of process assemblages and their impact on environments. A final group of chapters on environmental change shows how landscape evolution is now dependent upon our understanding of tectonics (Chapter 28), on how environments (Chapter 29) and environmental change can be interpreted (Chapter 30) and how approaches to landscape change and response can be visualized (Chapter 31). The conclusion includes short statements on challenges and perspectives from key leaders of several geomorphological organizations (Chapter 32) and a final chapter emphasizing the relevance of geomorphology to global climate change (Chapter 33).

THE CONTEXT OF DEBATES

As geomorphology has grown, accompanied by the profileration of sub-branches and ideas, it is inevitable that a number of tensions have arisen between the different branches, debates have ensued as a consequence, and discussion of certain issues has occurred, all conditioning the nature of geomorphology. Some individuals have been particularly influential, with Davis, Gilbert, Strahler and Chorley et al. recognized as fashion dudes (Sherman, 1996), and specific articles and books have been equally seminal: analysis of the references published in articles published in *Geomorphology* (1995–2004) showed that, of the 31,696 works cited, only 22 were referenced at least 20 times (Doyle and Julian, 2005). Debates are healthy drivers contributing to the progress of any discipline. A sequence of stages of development for remote sensing as suggested by Curran (1985, 6–7, following Jensen and Dahlberg (1983), was applied to physical geography (Gregory, 2000) and is tentatively adapted for geomorphology as shown in Table 1.6. The chapters of this Handbook amplify, illustrate and illuminate many of the discipline's debates but several are introduced here, not in order to steal the thunder of the subsequent chapters but to provide a context for those chapters.

Any discipline is limited in the subject matter that it can encompass and by its interfaces with other disciplines. So one debate concerns the *spatial limits of the discipline*, often referred to recently as the closure that restricts the spatial and temporal extent of the subject (Lane, 2000, 432). Analysis of the literature (e.g. Kondolf and Piegay,

Table 1.6 Discipline growth applied to geomorphology

Stage of growth of discipline (adapted after Jensen and Dahlberg, 1983)	Application to geomorphology
Preliminary growth period with small absolute increments of literature and little or no social organization	**Youth:** pre-1900, with origins in geology as well as in geography
	1900–1960: a period when geomorphology grew so that by the 1960s although some believed maturity to have been accomplished, the emphasis upon long-term landscape evolution meant that insufficient attention had been given to processes, to other branches of geomorphology and the relations of geomorphology to other disciplines
A period of exponential growth when the number of publications double at regular intervals and specialist research units are established	**Maturity:** 1960–2000, substantial growth achieved as illustrated by many new journals (Table 1.4) and books reflecting the branches of geomorphology (Table 1.5), with the influence of systems, models, quantitative and statistical methods and remote sensing
A subsequent period when the growth rate begins to decline and although annual increments remain constant, specialization and controversy increase	**Old age:** 2000– , growth rate may have declined but multi-disciplinary research has progressed involving geomorphologists, being exemplified by hybrid branches of the subject
A final period when the rate of growth approaches zero, specialist research units and social organization break down, and the subject reaches maturity	**Rejuvenation:** 2010– , a new phase where the role of geomorphology is redefined, enhanced by new techniques and potentially a more vibrant holistic and resilient discipline

2003; Doyle and Julian, 2005) can indicate the spread of geomorphological research activity, showing how geomorphology has come to be dominantly associated with the land surface of the Earth (Gregory, 2010), although it could encompass the sea floor, certainly in terms of seafloor engineering geomorphology (Prior and Hooper, 1999). A further extension can be in planetary terms because Baker (2008) has suggested that, to be a complete science of landforms and landscapes, geomorphology should not be restricted to the terrestrial portions of the Earth's surface because systems of landforms and their generative processes are best understood in a planetary context, so that to exclude extraterrestrial landscapes from geomorphology is illogical. If geomorphology includes planetary geomorphology as the study of the geomorphology of planets other than Earth then branches of geomorphology such as coastal can also be visualized in an extraterrestrial context (Parker and Currey, 2001).

Spatial limits are complemented by the debate about *temporal limits for geomorphology*. Indeed time pervades all fields of geomorphology (Thornes and Brunsden, 1977). Although prior to 1971 time had not been given explicit attention, a

useful distinction between timebound (known periods of time) and timeless changes was highlighted by Chorley and Kennedy (1971: 251). Studies of change benefitted from the distinction between steady, graded and cyclic timescales suggested by Schumm and Lichty (1965), and by greatly refined Quaternary timescales, together with growing awareness that some land-forming events have occurred very rapidly. It is now accepted that geomorphological research is analogous to different levels of microscope magnification: some investigations relate to short periods of days or weeks; others may be concerned with change over hundreds or thousands of years; and yet others could be concerned with developments over millions of years. The concept of landscape evolution space has been introduced (Phillips, 2009a,b) as a tool for assessing landscapes and geomorphic systems, providing a systematic means for assessing the various factors that contribute to the potential for change in geomorphic systems (see also Chapter 33).

As geomorphological analysis can apply at a range of spatial and temporal scales, a further debate has concerned *relating space and time*. This has involved models of landscape evolution and change, consideration of themes such as thresholds and complex response (e.g. Schumm, 1979), coupled with the problem of transferring understanding from one timescale to another. Attempts to use spatial variations as a model for change over time periods, often referred to as space–time substitution or the ergodic hypothesis (e.g. Paine, 1985) have proved fruitful in certain situations and there is further scope for their development and also for relating ecological and geomorphological systems (Viles et al., 2008).

A further debate centres on *gradualism and catastrophism.* Whereas early interpretations saw many features of the Earth's surface as a consequence of catastrophic events, ideas developed during the 19th century gradually led to the notion that in uniformitarian terms the present was the key to the past with many processes and environments seen as the consequence of gradual and progressive change. However, geomorphological hazards and extreme events prompted the view that certain features of the Earth's surface can only be explained as a consequence of catastrophic events. Geomorphological systems cannot be explained entirely as the result of continuing processes, so that catastrophism has played a greater role than previously thought.

However, as discipline is limited, there are *contrasting approaches* with at least three alternative foci now perceived for geomorphology: (1) geographical, interpreting morphology and processes; (2) geophysical, concentrating upon the broad structural outlines (see Church, 2005; Summerfield, 2005b); and (3) chronological, focused on the history of change. A more evolutionary geomorphology involving global structural geomorphology can be seen as counteracting the emphasis placed upon the investigation of processes but may not always be clearly differentiated from the disciplines of geology and tectonics. Baker (1988) suggested that from 1888 to 1938, there were separate approaches, one grounded in geology, and a separate one with its roots in geography, but by the 1960s, geomorphology, led by fluvial studies, changed its emphasis from historical studies to process studies, so that the geology/geography dispute became irrelevant. The implications of plate tectonics for the earth's surface certainly produced a shift in the focus of geomorphology. A further recent approach is complexity which to some extent succeeds the realization that uncertainty exists in environmental systems meaning that it is easier to predict than to explain.

There has also been debate about the degree to which geomorphology should include consideration of *human impact.* Prior to the mid 20th century comparatively little geomorphological attention was given to human impact (though the work of Marsh, 1864, is a notable exception), but

it was then increasingly recognized that it was impossible to investigate the surface of the Earth, and especially Earth surface processes, without reference to anthropogenic impact (see Gregory, 2000: Chapter 7; Goudie, 2005) and numerous implications are introduced in Chapter 15. However this can be extended further by considering whether geomorphology should include a greater cultural component. Just as a more society-oriented climatology or cultural climatology is envisioned so we can visualize a cultural geomorphology (e.g. Gregory, 2006). This does not detract from existing investigations of form, process and change but progresses by allowing for differences in human impact and legislative control according to culture and affecting future change.

To counter the increasingly specialized, fissiparist investigations, mentioned previously, the advantages of a more *holistic view* have become apparent. A holistic approach has arisen in at least two senses. First, within geomorphology, it has arisen to counter the greater specialist emphasis upon components of the land surface without sufficiently acknowledging the links between them. Thus linkages between components (e.g. Brierley et al., 2006) can emphasize ways in which nested hierarchical relationships between compartments in a catchment demonstrate both connectivity and disconnectivity in relation to geomorphic applications to environmental management. Second, as holism applies literally to the whole as more than the sum of the parts, it is the basis for greater links developed in multi-disciplinary investigations between geomorphology, and sub-disciplines of physical geography (e.g. Gregory et al., 2002), and also with other environmental and earth sciences such as the interface of geomorphology and ecosystems ecology (e.g. Renschler et al., 2007). Hybrid disciplines have been fostered, including ecogeomorphology and hydrogeomorphology, and multi-disciplinary investigations have been encouraged. Holistic approaches, countering the fissiparist trend which characterizes many of the sub-branches, have been in keeping with greater general awareness of environment, and hence with applications of geomorphology.

Potential *applications of geomorphology* have become more evident, with the ways in which applications may be communicated including applicable outputs embracing publications ranging from review papers, book chapters and books; and applied outputs which include interdisciplinary problem-solving, educational outreach, protocols and direct involvement (Gregory et al., 2008). Awareness of the potential effects of global warming becomes more urgent, with greater frequency of high magnitude events, possibly catastrophic ones, giving opportunities for geomorphic research as shown in the concluding chapter (Chapter 33), including investigation of potential implications of global change for coasts, flooding, glaciers and ground ice. There is also an increasing concern with other impacts of global change, such as deforestation, desertification and soil erosion (Slaymaker et al. 2009) and also with the geomorphological significance of hazards and disasters (Alcantara-Ayala and Goudie, 2010).

Emphasis upon process studies may have led to a relative neglect of landforms (Goudie, 2002), so that greater awareness of landforms (Gregory, 2010) and of visually attractive landscapes needs to be re-developed by geomorphologists, as an exemplar of proselytization of the discipline. The debate about whether *geomorphology is sufficiently visible* (Tooth, 2009) reminds us that there is a need not only for internal understanding but also of dissemination of the nature of the discipline and of the way in which the discipline can contribute in environmental problems. Subsequent chapters offer many examples of ways in which

geomorphology is becoming increasingly relevant, although as the land surface is of increasingly wider interest its study may be subsumed within other disciplines. A recent report offering new horizons for research in Earth surface processes (NRC, 2009) identifies nine grand challenges and proposes four high-priority research initiatives (Table 1.7), which resound with the issues identified above and merit consideration against the background of the subsequent chapters.

The establishment of geomorphology (Table 1.7) might be thought of, slightly tongue in cheek, according to headings borrowed from Davis' geographical cycle of landscape development which dominated much of the early growth of the discipline. Using this interpretation the initial origins pre-1900, reminiscent of the way in which the cycle was associated with initial uplift, were followed by Youth up to 1960, and then by Maturity to at least 2000. However, is Old Age an appropriate appellation for the current state of the discipline? One possibility is that any symptoms of old

Table 1.7 Nine grand challenges and four high-priority research initiatives for research on earth surface processes as proposed by NRC (2009)

Challenges facing earth surface processes	Comments
What does our planet's past tell us about its future?	New tools and techniques to analyse the extensive natural record of Earth's landscape evolution will help scientists understand the processes that shaped Earth and predict how changing earth surface processes will shape the landscapes of the future.
How do geopatterns on the Earth's surface arise and what do they tell us about processes?	New observational tools and powerful ways to present spatial data, as in geographic information systems, can help scientists to understand how geopatterns form.
How do landscapes record climate and tectonics?	Some of the most intriguing research questions centre on the relative sensitivity and rates of the numerous feedback mechanisms among climate, topography, ecosystems, physical and chemical denudation, sedimentary deposition and the deformation of rocks in active mountain belts.
How do biogeochemical reactions at Earth's surface respond to and shape landscapes?	Chemical erosion and weathering of bedrock creates soil, essential for anchoring and nourishing life, and also contributes to landscape evolution and nutrient cycles.
What transport laws govern the evolution of the Earth's surface?	Mathematical laws to define fundamental rates of processes such as landslides, glacial erosion and chemical erosion are required to allow researchers to understand the mechanics and rate of landscape change.
How do ecosystems and landscapes co-evolve?	Understanding the linkages among living ecosystems, earth surface processes and landscapes, needed to fully understand Earth's changing surface.
What controls landscape resilience to change?	Changes under the influence of drivers such as climate, plate tectonics, volcanism and human activities – and when conditions change with sufficient magnitude and duration.

Continued

Table 1.7   Cont'd

Challenges facing earth surface processes	Comments
How will Earth's surface evolve in the Anthropocene?	Understanding, predicting and adapting to changing landscapes increasingly altered by humans is a pressing challenge which falls squarely within the purview of earth surface science. Research on the interactions between humans and landscapes needed to meet this challenge.
How can earth surface science contribute to a sustainable earth surface?	Some disrupted and degraded landscapes should be restored or redesigned.
Research initiatives	
Interacting landscapes and climate	Quantitative understanding of climatic controls on earth surface processes, and the influence of landscape on climate over time scales from individual storm events to the evolution of landscapes, will shed light on the connection between landscapes and climate.
Quantitative reconstruction of landscape dynamics across time scales	Developing detailed reconstructions of the evolution of Earth's surface, based on information recorded in landscapes and in sedimentary records, will provide information on how Earth's surface has changed over various time scales.
Co-evolution of ecosystems and landscapes	Forge a new understanding of the co-evolution of ecosystems and landscapes to address pressing problems of future environmental change.
Future of landscapes in the Anthropocene	How can we predict and respond to rapidly changing landscapes that are increasingly altered by humans?

age which appeared in the first decade of the 21st century are now poised to be followed by Rejuvenation, akin to the reasons for the instigation of a new cycle of erosion. Many of the subsequent chapters demonstrate how geomorphology is poised, after more than a century of development, to enter a new revitalized stage which characterizes a vibrant, holistic and resilient discipline. Readers can reach their own conclusions before this theme is returned to in the conclusion in Chapter 33.

REFERENCES

Alcantara-Ayala, I. and Goudie, A.S. (eds) (2010) *Geomorphological Hazards and Disaster Prevention*. Cambridge, Cambridge University Press.

Baker, V.R. and Twidale, C.R. (1991) The re-enchantment of geomorphology. *Geomorphology* 4, 73–100.

Baker, V.R. (1988) Geological fluvial geomorphology. *Bulletin Geological Society of America* 100, 1157–1167.

Baker, V.R. (2008) Planetary landscape systems: A limitless frontier. *Earth Surface Processes and Landforms* 33, 1341–1353.

Beckinsale, R.P. and Chorley, R.J. (1991) *The History of the Study of Landforms or The Development of Geomorphology Vol. 3: Historical and Regional Geomorphology 1890–1950*. Routledge, London.

Brierley, G.J., Fryirs, K. and Jain, V. (2006) Landscape connectivity: The geographic basis of geomorphic applications. *Area* 38, 165–174.

Bryan, K. (1941) Physiography. *Geological Society of America*, 50th Ann. Vol. 3–15.

Burt, T.P., Chorley, R.J., Brunsden, D., Cox, N.J. and Goudie, A.S. (2008) *The History of the Study of Landforms or the Development of Geomorphology Vol. 4: Quaternary and Recent Processes and Forms (1890–1965) and the Mid-Century Revolutions*. Geological Society, London.

Chorley, R.J. (1978) Bases for theory in geomorphology. In C. Embleton, C., Brunsden, D. and Jones D.K.C. (eds). *Geomorphology: Present Problems and Future Prospects*. Oxford University Press, Oxford, UK.

Chorley, R.J. and Kennedy, B.A. (1971) *Physical Geography: A Systems Approach*. Prentice Hall, London.

Chorley, R.J., Dunn, A.J. and Beckinsale, R.P. (1964) *The History of the Study of Landforms, Vol. I. Geomorphology before Davis*. Methuen, London.

Chorley, R.J., Beckinsale, R.P and Dunn, A.J. (1973) *The History of the Study of Landforms Vol. II. The Life and Work of William Morris Davis*. Methuen, London.

Chorley, R.J., Schumm, S.A. and Sugden, D.A. (1984) *Geomorphology*. Methuen, London and New York.

Church, M. (2005) Continental drift. *Earth Surface Processes and Landforms* 30, 129–30.

Clifford, N.J. and Richards, K.S. (2005) Earth system science: An oxymoron? *Earth Surface Processes and Landforms* 30, 379–83.

Cotton, C.A. (1922) Geomorphology of New Zealand: An introduction to the study of land-forms. Dominion museum, Wellington: Dominion museum.

Curran, P.J. (1985) *Principles of Remote Sensing*. Longman, Harlow: Longman.

Czech, H. And Boswell, K.C. (1953) *Morphological Analysis of Landforms* (a translation of Penck, 1924). Macmillan, London.

Davies, G.L. (1968) *The Earth in Decay: A History of British Geomorphology 1578–1878*. MacDonald Technical & Scientific, London.

Davis, W.M. (1895) Bearing of physiography on uniformitarianism. *Bulletin of the Geological Society of America* 7, 8–11.

Davis, W.M. (1900) The physical geography of the lands. *Popular Science Monthly* 57, 157–170.

Davis, W.M. (1909) *Geographical Essays*. Ginn and Co, Boston.

Davis, W.M. (1912) *Die Erklarende Beschreibung der Landformen*. Teubner, Leipzig.

De Margerie, E. (1886) Géologie. *Polybiblion Revue Bibliographique Universelle, Partie littéraire* 24, 310–330.

Doyle, M.W. and Julian, J.P. (2005) The most-cited works in Geomorphology. *Geomorphology* 72, 238–249.

Engeln, von O.D. (1942) *Geomorphology*. Macmillan, New York.

Fleisher, P.J., Knuepfer, P.L.K. and Butler, D.R. (2006) Introduction to the special issue: Ice sheet geomorphology. *Geomorphology* 75, 1–3.

Fookes, P.G., Lee, E.M. and Milligan, G. (eds) (2005) *Geomorphology for Engineers*. Whittles Publishing, Dunbeath, UK.

Gilbert, G.K. (1902) John Wesley Powell. *Annual Report of the Smithsonian Institution for 1902*, pp. 633–640.

Goudie, A.S. (2005) *The Human Impact on the Natural Environment*, 6th edn. Blackwell, Oxford.

Goudie, A.S. (2002) Aesthetics and relevance in geomorphological outreach. *Geomorphology* 47, 245–249.

Goudie, A.S. (ed.) (2004) *Encyclopedia of Geomorphology*. International Association of Geomorphologists, and Routledge, London and New York.

Gregory, K.J. (1978) Fluvial processes in British basins. In C. Embleton, D. Brunsden and D.K.C. Jones, (eds.) *Geomorphology Present Problems and Future Prospects*. Oxford University Press, Oxford, pp. 40–72.

Gregory, K.J. (2000) *The Changing Nature of Physical Geography*. Arnold, London.

Gregory, K.J. (2006) The human role in changing river channels. *Geomorphology* 79, 172–191.

Gregory, K.J. (2010) *The Earth's Land Surface*. Sage, London.

Gregory, K.J., Gurnell, A.M. and Pettts, G.E. (2002) Restructuring physical geography. *Transactions of the Institute of British Geographers* 27, 136–154.

Gregory, K.J. Benito, G. Downs, P.W. (2008) Applying fluvial geomorphology to river channel management: Background for progress towards a palaeohydrology protocol. *Geomorphology* 98, 153–172.

Gregory, K.J., Simmons, I.G., Brazel, A.J., Day, J.W., Keller, E.A., Sylvester, A.G. and Yanez-Arancibia, Y. (2009) *Environmental Sciences. A Student's Companion*. Sage, London.

Haschenburger, J.K. and Souch, C. (2004) Contributions to the understanding of geomorphic landscape published in the Annals. *Annals of the Association of American Geographers* 94, 771–793.

Hettner, A. (1921) *Die oberflachenformen des Festlandes, ihre Untersuchung und Darstellung;Probleme und Methoden der Morphologie*. Teubner, Leipzig (2nd edition 1928).

Jensen, J. R. and Dahlberg, R. E. (1983) Status and Content of Remote Sensing Education in the United States, *International Journal of Remote Sensing* , 4, 235–245.

King, C.A.M. (1966) *Techniques in Geomorphology*. Arnold, London.

King, L.C. (1962) *The Morphology of the Earth*. Oliver and Boyd, Edinburgh.

Kondolf, G.M. and Piegay, H. (eds) (2003) *Tools in Geomorphology*. Wiley, Chichester.

Lane, S.N. (2000) Review of J.D. Phillips 'Earth surface systems: Complexity, order and scale'. *Annals Association of American Geographers* 90, 432–434.

Leopold, L.B., Wolman, M.G. and Miller, J.P. (1964) *Fluvial Processes in Geomorphology*. Freeman, San Francisco.

Lobeck, A.K. (1939) *Geomorphology: An Introduction to the Study of Landscapes*. McGraw Hill, New York.

Mackinder, H.J. (1895) Modern geography, German and English. *Geographical Journal* 6, 367–379.

Marsh G.P. (1864) *Man and Nature*. Scribner, New York.

McGee, W.J. (1888a) The geology at the head of Chesapeake Bay. *US Geological Survey, 7th Annual Report (1885–1886)*, 537–646.

McGee, W.J. (1888b) The classification of geographic form by genesis. *National Geographic Magazine* 1, 27–36.

NRC (National Research Council) (2009) *Landscapes on the Edge: New Horizons for Research in Earth Surface Processes*. Committee on Challenges and Opportunities in Earth Surface Processes. National Research Council, Washington.

Paine, A.D.M. (1985) Ergodic reasoning in geomorphology – time for review of the term? *Progress in Physical Geography* 9, 1–15.

Parker, T.J. and Currey, D.R. (2001) Extraterrestrial coastal geomorphology. *Geomorphology* 37, 303–328.

Penck, A. (1895) Die geomorphologie als genetische Wissenschaft: Eine Einleitung zur Diskussion uber geomorphologische Nomenklatur. *Report of the 6th International Geographical Congress* 737–747.

Penck, W. (1924) *Die Morphologische Analyse*. Engelhom, Stuttgart.

Phillips, J.D. (2009a) Landscape evolution space and the relative importance of geomorphic processes and controls. *Geomorphology* 109, 79–85.

Phillips, J.D. (2009b) Changes, perturbations, and responses in geomorphic systems. *Progress in Physical Geography* 33, 1–14.

Prior, D.B. and Hooper, J.R. (1999) Sea floor engineering geomorphology: Recent achievements and future directions. *Geomorphology* 31, 411–439.

Renschler, C.S. Doyle, M.W. and Thoms, M. (2007) Geomorphology and ecosystems: Challenges and keys for success in bridging disciplines. *Geomorphology*, 89, 1–8.

Richards, K.S. and Clifford, N. (2008) Science, systems and geomorphologies: Why LESS may be more. *Earth Surface Processes and Landforms* 33, 1323–1340.

Ritter, D.F. (1978) *Process Geomorphology*. McGraw Hill, Boston.

Roglic, J. (1972) Historical review of morphologic concepts. In M. Herak and V.T. Springfield (eds) *Karst: Important karst regions of the northern hemisphere*. Elsevier, Amsterdam.

Russell, R.J. (1949) Geographical geomorphology. *Annals of the Association of American Geographers* 39, 1–11.

Russell, R.J. (1958) Geological geomorphology. *Geological Society of America Bulletin* 69, 1–22.

Salisbury, R.D. (1907) *Physiography*. Holt, New York.

Schumm, S.A. (1979) Geomorphic thresholds: The concept and its applications. *Transactions Institute of British Geographers* NS4, 485–515.

Schumm, S.A. and Lichty, R.W. (1965) Time, space and causality in geomorphology. *American Journal of Science* 263, 110–119.

Selby, M.J. (1985) *Earth's Changing Surface. An Introduction to Geomorphology*. Clarendon Press, Oxford.

Sherman, D.J. (1996) Fashion in geomorphology. In B.L. Rhoads and C.E. Thorn (eds). *The Scientific Nature of Geomorphology*. Wiley, Chichester, pp. 87–114.

Slaymaker O., Spencer T. and Embleton-Hamann (eds) (2009) *Geomorphology and Global Environmental Change*. Cambridge University Press, Cambridge.

Stoddart, D.R. (1986) *On Geography and its History*. Blackwell, Oxford.

Strahler, A.N. (1952) Dynamic basis of geomorphology. *Geological Society of America Bulletin* 62, 923–938.

Summerfield, M.A. (1991) *Global Geomorphology*. Longman, Harlow.

Summerfield, M.A. (2005a) A tale of two scales, or the two geomorphologies. *Transactions of the Institute of British Geographers* 30, 402–415.

Summerfield, M.A. (2005b) The changing landscape of geomorphology. *Earth Surface Processes and Landforms* 30, 779–781.

Thornbury, W.D. (1954) *Principles of Geomorphology*. Wiley, New York.

Thornes, J.B. and Brunsden, D. (1977) *Geomorphology and Time*. Methuen, London.

Tilley, P. (1972) *The Surface Features of the Land*. Macmillan, London (translation of Hettner, 1928).

Tinkler, K.J. (1985) *A Short History of Geomorphology*. Croom Helm, London and Sydney.

Tooth, S. (2009) Invisible geomorphology. *Earth Surface Processes and Landforms* 34, 752–754.

Tricart, J. and Cailleux, A. (1965) *Introduction à la géomorphologie climatique*. Sedes, Paris.

Tricart, J. and Cailleux, A. (1972) *Introduction to Climatic Geomorphology*, translated by De Jonge, C.J.K. Longman, London.

Viles, H.A., Naylor, L.A., Carter, N.E.A. and Chaput, D. (2008) Biogeomorphological disturbance regimes: Progress in linking ecological and geomorphological systems. *Earth Surface Processes and Landforms* 33, 1419–1435.

Wooldridge, S.W. and Morgan, R.S. (1937) *The Physical Basis of Geography*. Longman, London.

Worcester, P.G. (1939) *A Textbook of Geomorphology*. Van Nostrand, New York.

PART I

Foundation and Relevance

2

Geomorphology: Its Early History

Andrew Goudie

The subject matter of geomorphology – landscapes and the processes that mould them – has been something that has fascinated the human race for thousands of years (Goudie and Viles, 2010). Written documents relating to geomorphological knowledge developed during the European Renaissance, but much of this work was hugely influenced by biblical concerns, especially by the belief that Earth was created by Divine Intervention only 6000 years ago and had been moulded subsequently by catastrophes like Noah's flood (Bauer, 2004). The time span for geomorphological processes to operate and for forms to develop was very brief. However, towards the end of the 18th century ideas began to change (Chorley et al., 1964; Tinkler, 1985), notably in Edinburgh. James Hutton (1788), often seen as the founder (albeit unreadable) of modern geomorphology, his more lucid disciple, John Playfair (1802), and Charles Lyell (1830) argued for the importance of gradual sub-aerial denudation over millennia (Werritty, 1993). Gradualist and uniformitarian ideas took hold, and the concept that Earth was old and had a long history was appreciated. Additionally, the fluvialists argued for the dominance of rivers in denuding the landscape through slow, long-continued action (Kennedy, 2006). In effect, the real foundations of modern geomorphology were established in the early 19th century, although the term itself was not to be coined and adopted until decades later.

However, these radical ideas did not go unchallenged and the diluvialists, who included Buckland and Sedgwick, still pursued the view that catastrophic flooding had caused many surface features. There were also some structuralists who believed that valleys were essentially clefts or rents in the ground surface rather than the product of stream erosion as had been maintained by Playfair. Even Lyell argued that many phenomena, including erratic blocks in unexpected places, could be due to marine rather than sub-aerial action. However, a possible explanation for erratic blocks and other mysterious phenomena shortly became available – glacial agency.

THE GLACIAL THEORY

So, as we have seen, in the early years of the 19th century, the diluvial theory, which arose from a belief in the Biblical Flood (Noachian Deluge), was usually invoked to explain many geomorphological phenomena. However, in the 1820s and 1830s some scientists started to suggest that glaciers had once been much more extensive than today and could account for much of what was then called 'drift'. Notable was the work of Esmark in Norway, and Jean-Pierre Perraudin, Ignatz Venetz and Jean de Charpentier in the Alps (Wright, 1896; Imbrie and Imbrie, 1979; Teller, 1983). In Germany, Bernhardi (1832) proposed that glacier ice had once extended across Europe as far south as Germany.

The most famous exponent of the glacial theory was, however, Louis Agassiz. In 1836, following a tour around Switzerland with Venetz and de Charpentier, he became an enthusiast for the idea that in the past glaciers had been much more extensive than now. Agassiz developed this theory as his *Discours de Neuchâtel* (Agassiz, 1840). In 1840 he visited Scotland and recognized evidence for former glaciations. He managed to convert Oxford's William Buckland to the acceptance of his views, even though Buckland has been an arch diluvialist and ardent catastrophist (Oldroyd, 1999). Agassiz also visited Ireland and recognized the evidence for glaciations there (Davies, 1968).

The Glacial Theory was not well received by some members of the geological establishment in Britain, most notably Charles Lyell and Roderick Murchison, though the latter's opposition eventually thawed (Gilbert and Goudie, 1971). Lyell found that the glacial theory was incompatible with his uniformitarian ideas, as in a sense it was, and attributed many of the allegedly glacial phenomena to marine submergence and wave action (Dott, 1998). He formulated the theory that drift was the product of deposition by icebergs at times of high sea-level, and the presence of marine shells in some drift deposits at high elevations supported this notion. Even towards the end of the 19th century some opposition still remained. In 1893, for instance, H.H. Howorth produced his massive *neocatastrophist The Glacial Nightmare and the Flood – a second appeal to common sense from the extravagance of some recent geology*, and tried to return to a fundamentalist–catastrophic interpretation of the evidence. Moreover, well into the 20th century British geomorphologists continued to argue that glaciers protected rather than eroded the landscape, with, for example, Gregory (1913) denying the role of glacier excavation in fjord formation. This chapter in the history of British glacier studies is well reviewed by Evans (2008).

The significance of the Ice Age beyond Europe was soon recognized. In New Zealand, F. von Hochstetter and J. von Haast were impressed by ancient moraines, lakes, fjords and the massive gravel plains of Canterbury (see Haast, 1879). Haast's work stimulated comparable researches in the Australian Alps by R. Von Lendenfeld (1886). In India, Sir Joseph Hooker remarked that he had met with ancient moraines in each valley he had ascended at about 7000–8000 feet (2134–2439 m) (Hooker, 1854: vol. ii, 103–4). Other observations from Kashmir and the Karakorams in the west to Sikkim in the east are described by Godwin-Austen (1864) and many subsequent workers.

Agassiz's views were adopted in the USA by Hitchcock (1841), who argued that the drift of Massachusetts was a glacial deposit. However, full appreciation of glaciation in North America partly resulted from Agassiz's visit in 1846 and it was Dana (1849) who was probably the first to suggest the former extensive glaciation of the Canadian Cordillera. During the 1850s and 1860s survey parties found evidence for a great Cordilleran Ice

Sheet. More detailed investigations were carried out in the 1870s and 1880s by G.M. Dawson (1878) and T.C. Chamberlin (Jackson and Clague, 1991).

Various expeditions demonstrated that glaciers had formerly been more extensive in high mountains of lower latitudes as along the Andes of South America, the Atlas Mountains, Lebanon and the Caucasus and northern China (Geikie, 1874: 379). Finally, when J.W. Gregory ascended Mount Kenya (Gregory, 1894) he discovered abundant evidence that proved that glaciers had once extended over 1600 m below their present level.

A major development in glacial ideas occurred in the 1870s when it became recognized that there had been more than one glacial advance and that these had been separated by warm phases, called interglacials (Hamlin, 1982). People such as A. Geikie (1882, 1893) began to appreciate the complexity of drift stratigraphy in Scotland. In addition, Croll recognized that orbital fluctuations could have caused multiple alternations of glacial and interglacials (Croll, 1875). These trends led to the work by J. Geikie, who in *The Great Ice Age* (1874) appreciated the importance of interglacial periods. In its turn the work by the Geikies was extremely influential in the subsequent development of the classic and durable Penck and Brückner model of glacial chronology in the Alps (1909).

Scientists also started to be intrigued by other sorts of climatic change that might have occurred in non-glaciated regions. J.S. Newberry, who explored the Colorado Plateau in the 1850s, recognized these classic landscapes as having been 'formerly much better watered than they are today' (1861: 47). Lake basins, of the type that abound in the Basin and Range Province of the American West, with their spectacular abandoned shorelines, gave particularly clear evidence of hydrological change. Subsequently, other American scientists, like Gilbert and Russell, examined these same lake basins in greater depth. The travels of J.W. Gregory (1894) in the newly discovered East African rift valley revealed the former greater extent of many of the lakes that occurred within it. By World War 1 a picture was emerging of the scale of climatic change that had taken place in lower latitudes and of the very substantial alterations that had taken place in climatic belts as made evident not only by desiccated or shrunken lakes, but also by old river systems and ancient sand dunes (Penck, 1914).

RIVER VALLEYS AND THE POWER OF FLUVIAL DENUDATION

Although Hutton, Playfair and Lyell had made clear the role of rivers in landscape development, the acceptance of fluvialism was not a straightforward matter for, as Kennedy (2006, 4) pointed out, there were many phenomena that appeared to cast doubt upon whether valleys had actually been produced by rivers: the non-accordance of valley junctions (hanging valleys), especially in the Alps and other mountain ranges; the widespread occurrence of deep lakes in the upper courses of valleys which could not have been excavated by the 'normal' action of rivers; the existence of cases – such as those of the Cotswold Hills – where the stream was minuscule (misfit) compared with the size of the valley; the widespread occurrence of valleys with no streams in them at all (dry valleys); the existence of valleys – including the fjords of Scandinavia – which patently continued out under the sea; widespread deposits of non-local sands, gravels and erratic boulders – Buckland's Diluvium; and the fact that rivers sometimes ran into valleys which cut dramatically through high ground, as in the Weald of southeastern England or in southern Ireland.

Related to the question of the origin of valleys was the question of the origin of planed off strata and of planation surfaces. In 1846, Ramsay had proposed that the roughly height-accordant summits of South Wales were a series of relicts that had been cut by wave action. Mackintosh argued (1869) that most facets of the British landscape, including escarpments (cliffs) and tors (stacks), had a marine origin.

In the 1860s, however, geomorphologists, as they soon came to be known, began to appreciate once again that rivers moulded valleys and were capable of achieving a great deal of geomorphological work and planation (Tinkler, 1985: 94 et seq.). There were various reasons for this. First, increasing acceptance of the power of former glaciers to cause wholesale transformation of the landscape and to produce features such as lake basins (Ramsay, 1862), explained away some drainage anomalies. Second, catastrophic/structural views on valley development were viewed with less favour. Third, when geomorphologists moved away from the relatively stable landscape of the British Isles to places like the Pacific islands, Assam or the mountains of Ethiopia, they encountered strong evidence of the power of rivers. Fourth, data on sediment loads of rivers demonstrated that they could indeed achieve a great deal of work. Fifth, some of the older and less progressive pioneers of the discipline were gradually passing from the scene (Davies, 1969: 317).

Croll (1875) made an early attempt to quantify rates of geomorphological change and used data on the amount of material being transported. Croll's fellow Scot, A. Geikie (1868), was equally concerned to demonstrate the power of sub-aerial erosion in comparison with that of the sea, and provided data on suspended loads for a range of the world's rivers, expressing them as a rate of surface lowering. The findings of Croll and Geikie were substantiated and strengthened by those of Ewing (1885) and Reade (1885).

Among the ardent fluvialists was Greenwood, who in 1857 produced *Rain and Rivers; or Hutton and Playfair against Lyell and all comers* (Stoddart, 1960). In it he championed the power of rainwash. More influential was Jukes (1862, 1866), who worked on the rivers of southern Ireland and showed that they had not only excavated their valleys but had also adjusted their courses to the underlying geological structures. Scrope (1866) was another exponent of fluvialism who pointed to the speed with which floods could transform landscapes.

A major figure in the revival of fluvialism (Chorley, et al., 1964: Chapter 20) was J.D. Dana (1850a, 1850b) who had travelled around the heavily dissected Pacific Islands. As Natland (1997, 326) wrote 'To become a fluvialist, all one has to do is ascend a large Tahitian Valley and get caught in a rainstorm'. One important convert to fluvialism was Ramsay, who as we saw earlier, had regarded the sea as the cause of planation in highland Wales. He recognized the role that rivers had played in developing the drainage of the Weald (Ramsay, 1872). The power of fluvialism, however, became sealed as a fundamental concept in geomorphology because of the impact that the landscapes of the American West, including the Grand Canyon, had on American geomorphologists such as J.W. Powell, C. Dutton, G.K. Gilbert and W.J. McGee (Orme, 2007a). Here there was abundant and dramatic evidence for the power of rivers. This, together with that from French hydraulic engineers, was used in France to good effect by La Noë and Margerie (see Broc, 1975). Their espousal of fluvialism transformed French geomorphology at the end of the 19th century. Another important figure in Europe was Rütimeyer (1869) who demonstrated that in the Alps valleys were not cracks in the crust but had been excavated by rivers.

ROCK DECAY

During the 19th century great strides were made in the understanding of physical, chemical and biological weathering processes, and these are well summarised by G.P. Merrill in his *A Treatise on Rocks, Rock-weathering and Soils* (1897) (Goudie and Viles, 2008). Knowledge of weathering phenomena owed a great deal to the growth of an independent science of pedology, or soil science, most notably by scholars like Dokuchayev (1883) in Russia.

The possible power of thermal fatigue weathering to cause rock disintegration was known to some early investigators. Merrill (1897: 180–3) summarises such views, which were adopted by many of the early desert geomorphologists such as Walther (1900), W. Penck (1924) and Hume (1925). Also in the early 19th century a great deal was learned about salt weathering because of its simulation in the laboratory as an analogue of frost weathering of building stones. Nineteenth century geologists were also well aware of the power of frost in producing angular debris (e.g. De la Beche, 1839) and recognized that one mechanism was the 9 percent volume expansion that accompanies the phase change of water to ice (e.g. Ansted, 1871).

With regard to chemical weathering, 19th century scientists carried out a wide range of chemical and mineralogical studies of weathering products and solutes, including laboratory simulations. There were also important studies of rates of chemical denudation, most notably by Bischof (1854). Various other studies hinted at the importance of organic acids to mineral decomposition (Goudie and Viles, 2008). Awareness of laterite, an enigmatic product of tropical weathering, goes back to Buchanan's work in south India in the early 1800s (see Goudie, 1973, for a discussion of early work on laterite and other duricrusts). By the end of the century laterite had also been recognized in the Seychelles, West Africa and Brazil (Prescott and Pendleton, 1952).

One particular aspect of weathering-related studies was the science of limestone (Karstic) relief and solution processes (see Rogliæ, 1972; Jakucs, 1977). Prime importance must be accorded to work on the Dinaric Karst and in particular to the extensive studies of one of A. Penck's students, Jovan Cvijić. His *Das Karstphänomen* (1893) and many subsequent works laid the theoretical foundations of many of our current ideas, though Serbian scholars had made some important studies before him (Ćalić, 2007).

MOUNTAIN BUILDING

During the 19th century there was considerable interest in how mountains formed (Adams, 1938) and in motions of Earth's crust (Chorley, 1963). E. Suess in Austria and Dana (1873) in the USA proposed that mountains formed through compressive stresses generated by a gradual thermal contraction of the whole earth (Oreskes, 1999: 10). Suess argued that, on a contracting Earth, mountains resulted from a wrinkling of the crust to accommodate a diminishing surface area.

The belief in the power of secular cooling was something that had been promulgated earlier in the 19th century by geologists such as Eliede Beaumont (1852) and De La Beche (1834). Indeed the contraction theory was the dominant paradigm for most of the 19th century (Oldroyd, 1996: 171). Dana (1873) also believed in the secular cooling model, but believed that as Earth contracted its rocks would be squeezed to the greatest degree on continental margins. Dana developed his geosynclinal theory (Knopf, 1948) of sedimentary accumulation, compression and uplift. His idea that the earth's and ocean basins had always occupied the positions that they do now ('permanentism') came under attack in the early 20th century when ideas on continental drift appeared (Le Grand, 1988).

The contraction theory had its limitations, not least for explaining the shear amount of folding in the Alps and elsewhere as exemplified by nappe structures (Heim, 1878). It became evident that mountains were not always caused by vertical movements of the crust, as contraction theory tended to suggest, but by horizontal shortening (Penck, 1909). An opponent of the contraction theory was Fisher (1881), who proposed the idea of convection currents within Earth's interior. In addition, severe reservations with respect to the contraction theory arose because of the recognition of the importance of isostasy (Watts, 2001). Contractionism also suffered in the 1890s when radioactivity was discovered. Radioactive decay generated heat and this meant that Earth was not cooling down and contracting as rapidly as one in which the only heat source was its initial accretion (Rogers and Santosh, 2004: 4).

The importance of isostasy was also made evident by studies in formerly glaciated terrains which would have been affected by downwarping and upheaval in response to ice cap advance and recession respectively. This was the birth of the theory of glacio-isostasy (Jamieson, 1865, 1882). Jamieson's work was followed by that of De Geer (1888, 1892) in Fennoscandia. Early proponents of glacio-isostasy in America were Whittlesey (1868) and Shaler (a pupil of Agassiz) (1874). During his classic study of pluvial Lake Bonneville G.K. Gilbert (1890) found a dome-like pattern of uplift of former shorelines and inferred that this indicated hydro-isostatic recovery following the desiccation of the lake.

Building upon the work of people such as Dana and Fisher, and using his experience from the American West, where many mountains appeared to be composed of igneous rocks intruded into sedimentary sequences, Dutton, who invented the term 'isostasy' in 1882 (Orme, 2007a), argued that crustal deformation could be understood as a response to isostatic compensation (1889).

His model, in simple form, was that uplifted portions of the continent are eroded, that material is transported to coastal regions, that the weight of this material causes subsidence along the continental margins, which causes displacement of materials at depth, with this material moving laterally and producing igneous intrusions and further uplift of the continent (Orestes, 1999: Figure 2.5, p. 31). Gilbert (1890) built upon Dutton's ideas and noted that in the Basin and Range Province mountain building was associated with many faults and with crustal extension rather that crustal contraction (Haller, 1982). The significance of crustal tension was also recognized by Suess, as it was by Gregory (1894) who, working in the context of East Africa, was the first to use the term 'rift valley' (Dawson, 2008).

The hypothesis that mountain building could result from continental drift, though hinted at by Antonio Snider-Pellegrini in 1858 (Hallam, 1973: 1), was not developed in a concerted way until the early 20th century through the work of Taylor (1910) and Wegener (1912).

DAVIS AND THE CYCLE

William Morris Davis has been described as an Everest among geomorphologists (Chorley et al., 1973). He was the leading American geomorphologist of the late 19th and early 20th centuries. He spent most of his career at Harvard where he was an exacting but skilful teacher. Above all he was a very prolific author, writing more than 500 articles and books, many of them beautifully illustrated with his own line drawings.

His great contribution was to produce a deductive model of landscape evolution, called the Cycle of Erosion or the Geographical Cycle (Davis, 1899). This was developed during the 1880s and 1890s (Orme, 2004, 2007b) during a time when, following Darwin, evolutionary concepts were in vogue. His theory of landscape development was the dominant paradigm in American geomorphology from the late 19th to the mid-20th century (Sack, 1992). Davis believed that landscapes were the product of three factors: structure (geological setting, rock character, etc.), process (weathering, erosion, etc.) and time (stage) in an evolutionary sequence. Stage was what most interested him. He suggested that the starting point of the cycle was the uplift of a broadly, flat, low-lying surface. This is followed by a phase he termed *youth*, when streams become established and start to cut down and to develop networks. Much of the original flat surface remains. In the phase he termed *maturity* the valleys have

widened so that the original flat surface has been largely eroded away and streams drain the entire landscape. The streams begin to meander across wide floodplains and the hillslopes become gradually less steep. In *old age* the landscape becomes so denuded that a low relief surface close to sea level develops, with only low hills (monadnocks) rising above it. This surface is then called a peneplain.

Initially, the Davisian model was postulated in the context of development under humid temperate ('normal') conditions, but it was then extended by Davis and successors to other environments, including arid, glacial, coastal, savanna, limestone and periglacial landscapes (Birot, 1968). His model was immensely influential and dominated much thinking in Anglo-Saxon geomorphology in the first half of the last century. The model was, however, largely deductive and theoretical and suffered from a rather vague understanding of surface processes, from a paucity of data on rates of operation of processes, from a neglect of climate change and from assumptions he made about the rates and occurrence of tectonic uplift. However, it was elegant, simple and tied in with broad, evolutionary concerns in science at the time.

In France the Davisian model was popularised by de Lapparent (1896) (see Giusti, 2004). Chorley et al. (1973) argued that Davis's cyclic model was not very successful in Germany, where it was opposed by such figures as Hettner and the Pencks (Tilley, 1968), though this may be something of an exaggeration (Wardenga, 2004). W. Penck's model of slope evolution (1924), often seen as the antithesis of Davis, involved more complex tectonic changes than that of Davis, and he regarded slopes as evolving in a different manner (slope replacement rather than slope decline) through time. An alternative model of slope development by parallel retreat leading to *pediplanation* was put forward by L.C. King in southern Africa. His model (1963) represents an amalgam of the views of Davis and Penck; episodic uplift resulting in both downwearing and backwearing, with the parallel retreat of slopes leading to the formation of low angle rock cut surfaces (pediments) which coalesced to form pediplains through the process of pediplanation. Thorn (1988) provides a useful comparative analysis of the Davis, Penck and King models of slope evolution (see also Chapter 4 in this Handbook).

By the mid 20th century the Davisian model was becoming less dominant and was the subject of a penetrating assault by Chorley (1965). This was partly because there was a growing awareness of crustal mobility that could not sustain notions of initial uplift followed by prolonged structural quiescence (Orme, 2007b).

DENUDATION CHRONOLOGY AND LONG-TERM EVOLUTION

The explanation of how landscapes came to attain their present form has always been a major objective of geomorphologists. Up to the 1960s, many workers adopted an historical approach to landscape evolution. Their aim was to identify the sequence of stages of erosional development that demonstrated how contemporary landscapes had been sculptured from hypothetical initial fairly uniform and featureless topographies. This sequential approach, with its focus on denudation, came to be known as 'denudation chronology' (Jones, 2004). During the first half of the 20th century, this became a major preoccupation of geomorphological studies in America, under the influence of D.W. Johnson, in Britain, where S.W. Wooldridge was a dominant figure, and in France, where H. Baulig's study of the Massif Central established a blueprint for subsequent work.

Classical denudation chronology sought to identify evidence of past planation surfaces and erosional levels in a landscape, in whatever way they formed, and to place them in a time sequence. To this end, two key concepts were employed. The first was that topographic 'flats', bevels and benches, together with accordant ridge and summit levels, represented the remnants of marine platforms, peneplains, pediplains produced during past periods of relatively stable base level. Often the studies that were undertaken focussed on a debate as to whether or not the identified erosional remnants were of sub-aerial or marine origin. A second concept was that there had been a progressive but episodic fall in base level through time, so that the most elevated features were the oldest. The resulting 'geomorphological staircases' often rose via terraces and benches to the more fragmentary remains of 'summit surfaces' preserved on ridges and escarpments. The identification and delimitation of such surfaces was usually based on visual observation, augmented by field mapping, profiling and various kinds of cartographic analysis, including the use of superimposed and projected profiles. Relatively little emphasis was placed on the study of surficial deposits.

Since the 1960s there has been less interest in classical denudation chronology. The 1960s witnessed the onset of radical changes to prevailing views of the past arising from growing knowledge about global tectonics and Quaternary climate change. Moreover, many geomorphologists concentrated on understanding the role of present day processes rather than trying to establish a long-term evolutionary history based on often small fragments of ancient landscapes preserved in the landscape at the present day.

CLIMATIC GEOMORPHOLOGY

In the 20th century, particularly in Germany and France, climatic geomorphology was a major approach. However, ideas about the importance of climate in determining processes and landforms germinated in the 19th century as more and more scientists carried out investigations outside Europe and more and more professional earth scientists became involved in scientific expeditions to areas that had previously been little known or had been impossible to access for logistical or political reasons. One strand of the development of climatic geomorphology was the study of periglacial and permafrost processes by European explorers of the vast sub-arctic regions of North America and Eurasia, though it was Lozinski who provided the first unifying concepts of periglacial geomorphology just before World War 1 (French, 2003). Other distinctive cold climate phenomena were also recognized. Nivation was a term introduced by Matthes (1900) to describe and explain the processes associated with late-lying seasonal snow patches and landforms derived from them (nivation benches and nivation hollows), while solifluction, the slow downslope movement of a saturated soil mass usually associated with freeze–thaw cycles and frost heave, was identified in the Falkland Islands by Andersson (1906).

Among the phenomena that scientists studied in lower latitudes were loess, desert dunes, desert weathering, coral reefs, deep weathering, laterites and inselbergs. Loess, a largely non-stratified and non-consolidated silt, containing some clay, sand and carbonate is a widespread and geomorphologically important deposit. During the 19th century many theories were advanced concerning its origin, including fluvial, marine, lacustrine and pedological ones. It is the subject of an enormous literature that developed after Lyell (1834) had drawn attention to the loamy deposits of the Rhine valley. It was, however, Von Richthofen (1882), working in China, who cogently argued that these intriguing deposits probably had an aeolian origin and were produced by dust storms transporting silts from deserts and depositing them on desert margins.

The colonization of the Sahara by the French from the 1880s onwards led to some of the first serious work on desert sand dunes (Goudie, 1999). However, dunes were not the only field of interest of desert travellers, for the exploration of deserts in the 19th century gradually led to the emergence of studies that established the nature of desert processes and their differences from those in other environments. French scientists were very active in the Western Sahara and accumulated a great deal of vital information on the full range of desert landforms (see Chudeau, 1909; Gautier, 1908).

Also notable was the work of Walther, who worked in the deserts of North Africa, Sinai the USA and Australia. His *Das Gesetz der Wüstenbildung in Gegenwart und Vorzeit* (1900) was the first full-scale book devoted to desert geomorphological processes and he championed the role of such mechanisms as thermal fatigue weathering, salt weathering and deflation.

American scientists also contributed greatly to the development of knowledge on desert land-forms and processes (Udden, 1894; Free, 1911). Especially remarkable was the work of W.P. Blake on stone pavements, desert varnish, old lake basins, calcretes (caliche) and wind grooving of rock surfaces (e.g. Blake, 1855, 1904). It was also in the American West that W.J. McGee (1897) drew attention to the role of sheetfloods on pedi-ment surfaces. Also notable were Gilbert's studies in the Colorado Plateau on rates of denudation in arid regions (Gilbert, 1876). The development of ideas on the role of wind in drylands is discussed by Goudie (2008a, 2008b).

THE TROPICS

During the voyage of the *Beagle* Charles Darwin saw many coral reefs. In 1842, he summarized his subsidence theory to explain the sequence of fringing reefs, barriers reefs and atolls (Spencer et al., 2008). As Davis (1913: 173) remarked:

> for forty years the scientific world accepted it as demonstrated. Darwin's diagram of a subsiding island and an up growing reef have been repro-duced over and over again on countless black-boards, as representing one of the great discoveries of geological science.

Dana (1851, 1872) was a strong supporter of this theory and did much to make coral reefs a legiti-mate object of scientific enquiry in North America (Spencer et al., 2008: 870). The other key figure was Jukes (1847), but unlike Darwin and Dana he worked not on open-ocean atolls but on the Great Barrier Reef of Australia. However, like Dana, he wholeheartedly accepted Darwin's subsidence theory (Stoddart, 1988, 1989). Apart from coral reefs themselves, there was a recognition of some other features of lower latitude coastlines, includ-ing aeolianites (Rathbun, 1879; Branner, 1905) and beachrock (Beaufort, 1817; Darwin, 1841).

Geomorphologists gradually came to see the distinctive nature of humid tropical landforms and processes. Deep weathering was described from eastern China by Kingsmill (1864), and Russell

(1889) appreciated the extent of deep weathering in the tropics in comparison with higher latitudes. However, the most important early paper on deep weathering was by Branner (1896) who stressed the importance of such factors as rank vegetation, termites, lichens, bacteria and lightning-generated nitric acid in assisting the role of tepid tropical rain. Pumpelly (1879) believed that the rock sur-face beneath the deeply weathered layer would be highly irregular and that if stripped off this uneven surface of weathering would be exposed. We have here the germs of an idea that developed in the 20th century to account for such phenomena as tors, inselbergs and etchplains (e.g. Falconer, 1911: Wayland, 1933; Büdel, 1957).

Passarge's work in the Kalahari (Passarge, 1904) had an influence in Davis's formulation of the arid cycle of erosion (Davis, 1905), while in Poland, Romer (1899) introduced the idea that the main morphological zones of Earth coincided with climatic zones and may have been affected by them (Kozarski, 1993: 348). The development of climatic geomorphology by A. Penck (1905) and von Richthofen in Germany and by E. de Martonne (1909) in France was facilitated and stimulated by the first global syntheses and clas-sifications of soils (e.g. by Dokuchayev), plants (e.g. by Schimper) and climates (e.g. by Köppen). De Martonne's *Traité de Géographie Physique*, which was translated into English, Polish and Spanish 'directly or indirectly fuelled a full cen-tury of studies in physical geography across con-tinental Europe' (Broc and Giusti, 2007).

In the USA, Davis recognized 'accidents', whereby non-temperate and non-humid climatic regions were seen as deviants from his normal cycle of erosion and he introduced, as we saw earlier, his arid cycle (Davis, 1905). Some (see Derbyshire, 1973) regard Davis as one of the founders of climatic geomorphology, although the leading French climatic geomorphologists, Tricart and Cailleux (1972), criticized Davis for his neglect of the climatic factor in landform develop-ment. Much important work was undertaken on dividing the world into morphoclimatic regions with distinctive landform assemblages in France (e.g. Birot, 1968), Germany (e.g. Büdel, 1982) and New Zealand (Cotton, 1942).

In the later years of the 20th century the popu-larity of climatic geomorphology became less as certain limitations became apparent (see Stoddart, 1969):

1   Much climatic geomorphology was based on inadequate knowledge of rates of processes and on inadequate measurement of process and form. Assumptions were made that, for example, rates of chemical weathering were high in the humid tropics and low in cold regions, whereas

subsequent empirical studies have shown that this is far from inevitable.

2   Some of the climatic parameters used for morphoclimatic regionalization (e.g. mean annual air temperature) were meaningless or crude from a process viewpoint. Macro-scale regionalization was seen as having little inherent merit and ceased to be a major goal of geographers, who eschewed 'placing lines that do not exist around areas that do not matter'.

3   Conversely, and paradoxically, climatic geomorphology had a tendency to concentrate on bizarre forms found in some 'extreme' environments rather than on the overall features of such areas.

4   Many landforms that were supposedly diagnostic of climate (e.g. pediments in arid regions or inselbergs in the tropics) are either very ancient relict features that are the product of a range of past climates or they have a form that gives an ambiguous guide to origin.

5   The impact of the large, frequent and abrupt climatic changes of the Late Cainozoic has disguised any simple climate–landform relationship. For this reason, Büdel (1982) attempted to explain landforms in terms of fossil as well as present day climatic influences. He recognized that landscape were composed of various 'relief generations' and saw the task of what he termed 'climato-genetic geomorphology' as being to recognize, order and distinguish these relief generations, so as to understand today's highly complex relief.

Although these tendencies have tended to reduce the relative importance of traditional climatic geomorphology, notable studies still appear that look at the nature of landforms and processes in different climatic settings (e.g. M. Thomas, 1994, and Wirthmann, 1999, on the humid tropics; D. Thomas, 1998, on dry lands and French, 1999, on periglacial regions).

G.K. GILBERT AND DYNAMIC EQUILIBRIUM

G.K. Gilbert was a remarkable American geomorphologist who, in many respects, was ahead of his time (Baker and Pyne, 1978). Although he died over 90 years ago, par excellence his career exemplifies many of the concerns of modern geomorphology. Working for much of his career in the American West, he made diverse and impressive contributions to the discipline. He helped to explain and name the structure and topography of the Basin ånd Range province with its many

alternations of mountains and playas, he explained and classified the igneous intrusions that had created the Henry Mountains of the Colorado Plateau, he studied the greatest pluvial lake of the American West – Lake Bonneville – and recorded the evidence of its fluctuating levels, he established that large lakes could depress Earth's crust and so contributed to the growth of ideas about crustal mobility, and he helped to demonstrate that the craters on the Moon were the result of meteorite impact. However, the name of Gilbert is most often associated with that approach which is often termed dynamic geomorphology (see Chapter 3 in this Handbook).

This blossomed in the second half of the 20th century, and was defined by Strahler (1952) as an approach which treats geomorphic processes as 'gravitational or molecular shear stresses, acting on elastic, plastic or fluid earth materials to produce the characteristic varieties of strain or failure which we recognize as the processes of weathering, erosion, transportation and deposition'. As Slaymaker (2004: 307) remarked, 'the work of G.K.Gilbert is the first seminal antecedent of the study of geomorphic process or dynamic geomorphology'. This is exemplified in Gilbert's report on the *Geology of the Henry Mountains* (1877), his study of the convexity of hill tops (1909) and his work on the transportation of debris by running water (1914). Compared to Davis, Gilbert 'eschewed long-term cyclic interpretations in favour of an open-systems framework whereby landforms sought equilibrium shapes in response to changing fluxes of energy and mass' (Orme, 1989: 78).

In some areas of geomorphology, studies based on an analysis of force and resistance occurred earlier than in others (e.g. Terzaghi's work on slopes and rock mechanics in the 1920s; Bagnold's work on aeolian forms and processes in the 1930s; Hjulström's studies of processes in gravel rivers in the 1930s and the work of various physicists, such as Nye, Glen and Perutz, on glacier dynamics in the 1950s). Moreover, some geomorphologists, while they were great exponents of the Davisian model, were also greatly interested in processes. This is, for example, the case with D.W. Johnson's study of *Shore Processes and Shoreline Development* (1918).

CONCLUSION

The purpose of this chapter has been to present some of the main ideas that developed in geomorphology between the end of the 18th century and the second half of the 20th century. Among the main ideas considered have been diluvialism and catastrophism, fluvialism and uniformitarianism, the ice age, the weathering of rocks, the growth of mountains, the cycle of erosion, denudation chronology, climatic geomorphology and dynamic equilibrium. It has not been possible to cover the history of all aspects of geomorphology – the discipline is too diverse for that – and so much more could have been said about such approaches as anthropogeomorphology, biogeomorphology, coastal geomorphology and so on. There are also dangers of over simplifying the story of the evolution of geomorphology and of placing individual geomorphologists into convenient camps. As Ahnert (1998: 326) pointed out, for example, Albrecht Penck 'included a large number of quantitative functional equations for geomorphological processes in his *Morphologie der Erdoberfläche* (1894)', while his son, Walther, known predominantly for his model of slope development (1924), was a process-oriented geomorphologist who 'emphasised the principle that the key to the explanation of the development of forms lay in an understanding of the process mechanisms involved'.

Since the 1960s there have been at least three major developments – plate tectonics, the revolution in Quaternary science and quantitative process-oriented geomorphology – that have taken place. These have been reviewed by Burt et al. (2008) and are discussed in depth in many of the chapters of the present volume. The shift towards physically based models and process studies is also described in Chapter 4 in this Handbook.

REFERENCES

Adams, F.D. (1938) *The Birth and Development of the Geological Sciences*. New York: Dover Publications.

Agassiz, L. (1840) *Etudes sur les Glaciers*. Privately published, Neuchâtel.

Ahnert, F. (1998) *Introduction to Geomorphology*. London: Arnold.

Andersson, J.G. (1906) Solifluction, a component of subaerial denudation. *Journal of Geology* 14: 91–112.

Ansted, D.T. (1871) On some phenomena of the weathering of rocks, illustrating the nature and extent of subaerial denudation. *Transactions of the Cambridge Philosophical Society* 5: 328–39.

Baker, V.R. and Pyne, S. (1978) G.K. Gilbert and modern geomorphology. *American Journal of Science* 278: 97–123.

Bauer, B. (2004) Geomorphology, in A.S. Goudie (ed.), *Encyclopedia of Geomorphology*. London: Routledge, pp. 428–35.

Beaufort, F. (1817) *Karamania*. London.

Bernhardi, R. (1832) An hypothesis of extensive glaciation in prehistoric time, in K.T. Mather and S.L. Mason (eds) *Source Book in Geology*. New York: McGraw Hill, 1939, pp. 327–28.

Birot, P. (1968) *The Cycle of Erosion in Different Climates*. London: Batsford.

Bischof, G. (1854) *Elements of Chemical and Physical Geology*. London: Cavendish Society.

Blake, W.P. (1855) On the grooving and polishing of land rocks and minerals by the sand. *American Journal of Science and Arts* 20: 178–82.

Blake, W.P. (1904) Origin of pebble-covered plains in desert regions. *Transactions of the American Institute of Mining Engineers* 34: 161–2.

Branner, J.C. (1896) Decomposition of rocks in Brazil. *Bulletin of the Geological Society of America* 7: 255–314.

Branner, J.C. (1905) The stone reefs of Brazil, their geological and geographical, with a chapter on the coral reefs. *Bulletin of the Museum of Comparative Zoology* 44.

Broc, N. (1975) Les débuts de la géomorphologie en France; le tournant des années 1890. *Revue d'histoire des Sciences* 28: 31–60.

Broc, N. and Giusti, C. (2007) Autour du Traité de Géographie Physique d'Emmanuel de Martonne: du vocabulaire géographique aux théories en géomorphologie. *Géomorphologie* 2: 125–44.

Büdel, J. (1957) Die doppelten Einebnugsflächen in den feuchten Tropen, *Zeitschrift für Geomorphologie* NF 1: 201–88.

Büdel, J. (1982) *Climatic Geomorphology*. Princeton, NJ: Princeton University Press.

Burt, T.P., Chorley, R.J., Brunsden, D., Cox, N.J. and Goudie, A.S. (eds) (2008) *The History of the Study of Landforms or the Development of Geomorphology, Col. 4. Quaternary and recent processes and forms (1890–1965) and the mid-century revolutions*. London: Geological Society.

Ćalić, J. (2007) Karst research in Serbia before the time of Jovan Cvijić. *Acta Carsologica* 36: 315–9.

Chorley, R.J. (1963) Diastrophic background to twentieth-century geomorphological thought. *Bulletin of the Geological Society of America* 74: 953–70.

Chorley, R.J. (1965) A re-evaluation of the geomorphic system of W.M. Davis, in R.J. Chorley and P. Haggett (eds), *Frontiers in Geographical Teaching*. London: Methuen, pp. 21–38.

Chorley, R.J., Dunn, A.J. and Beckinsale, R.P. (1964) *History of the Study of Landforms, Vol. 1*. London: Methuen.

Chorley, R.J., Beckinsale, R.P. and Dunn, A.J. (1973) *History of the Study of Landforms, Vol. 2*. London: Methuen.

Chudeau, R. (1909) *Sahara Soudanais*. Paris: Armand Colin.

Cotton, C.A. (1942) *Climatic Accidents in Landscape-making*. Christchurch: Whitcombe and Tombs.

Croll, J. (1875) *Climate and Time*. London: Daldy, Isbister and Co.

Cvijic, J. (1893) Das Karstphänomen. *Geographischen Abhandlungen Wien*, V: 218–329.

Dana, J.D. (1849) *Geology. United States Exploring Expedition, During the Years 1838, 1839, 1840, 1841, 1842 under the Command of Charles Wilkes U.S.N., Vol. XC*. Philadelphia: Sherman.

Dana, J.D. (1850a) On the degradation of the rocks of New South Wales and formation of valleys. *American Journal of Science* 59: 289–94.

Dana, J.D. (1850b) On denudation in the Pacific. *American Journal of Science.* 2nd series, 9: 48–62.

Dana, J.D. (1851) On the coral reefs and islands. *American Journal of Science and Arts.* 2nd series, 12: 25–51, 165–86, 329–38.

Dana, J.D. (1872) *Corals and Coral Islands.* New York: Dodd, Mead and Co.

Dana, J.D. (1873) On some results of the Earth's contraction from cooling. *American Journal of Science* 5: 423–43.

Darwin, C. (1841) On the remarkable bar of sandstone off Pernambuco on the coast of Brazil. *The London, Edinburgh and Dublin Philosophical Magazine* 19, 257–60.

Darwin, C.R. (1842) *The Structure and Distribution of Coral Islands.* London: Smith Elder.

Davies, G.L. (1968) The tour of the British Isles made by Louis Agassiz in 1840. *Annals of Science* 24: 131–46.

Davies, G.L. (1969) *The Earth in Decay: A History of British Geomorphology, 1578–1978.* London: Macmillan.

Davis, W.M. (1899) The geographical cycle. *Geographical Journal* 14: 481–504.

Davis, W.M. (1905) The geographical cycle in an arid climate. *Journal of Geology* 13: 381–407.

Davis, W.M. (1913) Dana's confirmation of Darwin's Theory of Coral Reefs. *American Journal of Science* 35: 173–88.

Dawson, J.B. (2008) Discovery of the African rift valleys: early work on the Gregory Rift Valley and volcanoes in northern Tanzania. *Geological Society Memoir* 33: 3–7.

Dawson, M.C. (1878) On the superficial geology of British Columbia. *Quarterly Journal of the Geological Society* 34: 89–123.

De Geer, G. (1888) Om skandinaviens nivåförändringar under quatärperioden. *Geol. Foren. Stockh. Förh.* 10: 366–79.

De Geer, G. (1892) On Pleistocene changes of level in eastern North America. *Proceedings Boston Society of Natural History* 25: 454–477.

De La Beche, H.T. (1834) *Researches in Theoretical Geology.* London: C Knight.

De La Beche, H.T. (1839) *Report on the Geology of Cornwall, Dorset and West Somerset.* Memoir of the Geological Survey of Britain.

De Martonne, E. (1909) *Traité de Geographie Physique.* Paris: A. Colin.

Derbyshire, E. (ed.) (1973) *Climatic Geomorphology.* London: Macmillan.

Dokuchayev, V.V. (1883) *Russian Chernozem* (in Russian). St Petersburg: Free Economic Society Press.

Dott, R.H. (1998) Charles Lyell's debt to North America: his lectures and travel from 1841–1853, in D.J. Blundell and A.C. Scott (eds) *Lyell: The Past is the Key to the Present.* Geological Society London, Special Publications 143, 53–69.

Dutton, C.E. (1889) Isostasy. *Bulletin of the Philosophical Society of Washington* 11: 51–64

Elie de Beaumont, L. (1852) *Notice Sur les Systèmes de Montagnes.* Paris: Bertrand.

Evans, I.S. (2008) Glacial erosional processes and forms: mountain glaciations and glacier geography, in T.P. Burt, R.J. Chorley, D. Brunsden, N.J. Cox and A.S. Goudie (eds), *The History of the Study of Landforms or the Development of Geomorphology, vol 4. Quaternary and recent processes and forms (1890–1965) and the mid-century revolutions.* London: Geological Society, pp. 413–94.

Ewing, A.L. (1885) An attempt to determine the amount and rate of chemical erosion taking place in the limestone (calciferous to Trenton) valley of Centre County, Pa., and hence applicable to similar regions throughout the Appalachian region. *American Journal of Science* 29: 29–31.

Falconer, J.D. (1911) *The Geology and Geography of Northern Nigeria.* London: Macmillan.

Fisher, O. (1881) *The Physics of the Earth's Crust.* London: Macmillan.

Free, E.E. (1911) The movement of soil material by the wind. *Bulletin of the Department of Agriculture Bureau of Soils* 68: 272.

French, H.M. (1999) *The Periglacial Environment,* 2nd edn. London: Longman.

French, H.M. (2003) The development of periglacial geomorphology; 1– up to 1965. *Permafrost and Periglacial Processes* 14: 29–60.

Gautier, E.F. (1908) *Sahara Algérien.* Paris: Armand Colin.

Geikie, A. (1868) On denudation now in progress. *Geological Magazine* 5: 249–54.

Geikie, A. (1882) *Text Book of Geology.* London: Macmillan.

Geikie, A. (1893) *Text Book of Geology,* 3rd edn. London: Macmillan.

Geikie, J. (1874) *The Great Ice Age.* London: Isbister.

Gilbert, G.K. (1876) The Colorado plateau province as a field for geological study. *American Journal of Sciences and Arts* 12, 85–104.

Gilbert, G.K. (1877) *Report on the Geology of the Henry Mountains.* Washington, DC: US Geological and Geographical Survey.

Gilbert, G.K. (1890) *Lake Bonneville.* US Geological Survey Monograph, 1: 43.

Gilbert, G.K. (1909) The convexity of hill tops. *Journal of Geology* 17: 344–50.

Gilbert, G.K. (1914) The transportation of debris by running water. *US Geological Survey, Professional Paper* 86.

Gilbert, E.W. and Goudie A.S. (1971) Sir Roderick Impey Murchison, Bart, KBE, 1792–1871. *Geographical Journal* 137: 505–11.

Giusti, C. (2004) Géologues et géographes français face a la théorie davisienne (1896–1909): retour sur "l'intrusion" de la géomorphologie dans le géographie. *Géomorphologie* 3: 241–54.

Godwin-Austen, H.H. (1864) On the glaciers of the Mustagh Range (trans-Indus). *Proceedings of the Royal Geographical Society* 34: 19–56.

Goudie, A.S. (1973) *Duricrusts in Tropical and Subtropical Landscapes.* Oxford: Clarendon Press.

Goudie, A.S. (1999) The history of desert dune studies over the last 100 years, in A.S. Goudie, I. Livingstone and

S. Stokes (eds). *Aeolian Environments, Sediments and Landforms.* Chichester: Wiley. pp. 1–13.

Goudie, A.S. (2008a) The history and nature of wind erosion in deserts. *Annual Review of Earth and Planetary Science* 36: 97–119.

Goudie, A.S. (2008b) Aeolian processes and forms in T.P. Burt, R.J. Chorley, D. Brunsden, N.J. Cox and A.S. Goudie (eds). *The History of the Study of Landforms or the Development of Geomorphology, vol 4. Quaternary and recent processes and forms (1890–1965) and the mid-century revolutions.* London: Geological Society, pp. 767–804.

Goudie, A.S. and Viles H.A. (2008) Weathering processes and forms, in T.P. Burt, R.J. Chorley, D. Brunsden, N.J. Cox and A.S. Goudie (eds). *The History of the Study of Landforms or the Development of Geomorphology, vol 4. Quaternary and recent processes and forms (1890–1965) and the mid-century revolutions.* London: Geological Society, pp. 129–64.

Goudie, A.S. and Viles, H.A. (2010) *Geomorphology and Landscape: A Very Short Introduction.* Oxford: Oxford University Press.

Greenwood, G. (1857) *Rain and Rivers, or Hutton and Playfair Against Lyell and All Comers.* London: Longman.

Gregory, J.W. (1894) Contributions to the geology of British East Africa – Part 1. The glacial geology of Mount Kenya. *Quarterly Journal of the Geological Society* 50: 515–30.

Gregory, J.W. (1913) *Nature and Origin of Fjords.* London: John Murray.

Haast, J. (1879) *Geology of the Provinces of Canterbury and Westland.* Christchurch: Lyttleton Times.

Hallam, A. (1973) *A Revolution in the Earth Sciences.* Oxford: Clarendon Press.

Haller, J. (1982) Heretical views on mountain building in Europe and North America; harbingers of modern tectonics. *International Journal of Earth Sciences* 71: 427–40.

Hamlin, C. (1982) James Geikie, James Croll and the eventful Ice Age. *Annals of Science* 39: 565–83.

Heim, A. (1878) *Untersuchungen Über den Mechanismus der Gebirgsbildung.* Basle: B. Schwabe.

Hitchcock, E. (1841) *Final Report on the Geology of Massachusetts.* Northampton: JH Butler.

Hjulström, F. (1935) Studies on the morphological activity of rivers as illustrated by the river Fryis. *Bulletin of the Geological Institute, University of Uppsala* 25: 221–527.

Hooker, J. (1854) *Himalayan Journals.* London: John Murray.

Howorth, H.H. (1893) *The Glacial Nightmare and the Flood – a Second Appeal to Common Sense from the Extravagance of Some Recent Geology.* London: Sampson Low, Marston and Co.

Hume, W.F. (1925) *Geology of Egypt, Vol. 1.* Cairo: Government Press.

Hutton, J. (1788) Theory of the Earth. *Transactions of the Royal Society of Edinburgh* 1: 209–304.

Imbrie, J. and Imbrie, K.P. (1979) *Ice Ages. Solving the Mystery.* London: Macmillan.

Jackson, L.E. and Clague, J. (1991) The Cordilleran le Sheet: One hundred and fifty years of exploration and discovery. *Géographie Physique et Quaternaire* 45: 269–80.

Jakucs, L. (1977) *Morphgenetics of Karst Regions: Variants of Karst Evolution.* Bristol: Adam Hilger.

Jamieson, T.F. (1865) On the history of the latest geological changes in Scotland. *Quarterly Journal of the Geological Society of London* 21: 161–203.

Jamieson, T.F. (1882) On the cause of the depression and re-elevation of the land during the glacial period. *Geological Magazine* NS, 9: 400–7.

Johnson, D.W. (1918) *Shore Processes and Shoreline Development.* New York: John Wiley.

Jones, D.K.C. (2004) Denudation chronology, in A.S. Goudie (ed.), *Encyclopedia of Geomorphology.* London: Routledge, pp. 244–8.

Jukes, J.B. (1847) *Narrative of the Surveying Voyage of the H.M.S Fly, Commanded by Captain F.P. Blackwood, R.N., in Torres Strait, New Guinea and Other Islands of the Eastern Archipelago, During the Years 1842–1846.* London: T. and W. Brone.

Jukes, J.B. (1862) On the mode of formation of some of the river valleys in the south of Ireland. *Quarterly Journal of the Geological Society of London* 18: 378–403.

Jukes, J.B. (1866) Atmospheric V. Marine denudation. *Geological Magazine* 3: 232–5.

Kennedy, B.A. (2006) *Inventing the Earth: Ideas on Landscape Development since 1740.* Oxford: Blackwell.

King, L.C. (1963) Canons of landscape evolution. *Bulletin of the Geological Society of America* 64: 721–51.

Kingsmill, T.W. (1864) Notes on the geology of the east coast of China. *Journal of the Geological Society of Dublin* 10: 1–6.

Knopf, A. (1948) The geosynclinal theory. *Bulletin of the Geological Society of America* 59: 649–70.

Kozarski, S. (1993) Geomorphology in Poland, in H.J. Walker and W.E. Grabau (eds), *The Evolution of Geomorphology.* Chichester: John Wiley, pp. 347–61.

de Lapparent, A. (1896) *Leçons de Géograhie Physique.* Paris: Masson.

Le Grand, H.E. (1988) *Drifting Continents and Shifting Theories.* Cambridge: Cambridge University Press.

Lyell, C. (1830) *Principles of Geology.* London: John Murray.

Lyell, C. (1834) Observations on the loamy deposit called 'loess' of the basin on the Rhine. *Edinburgh New Philosophical Journal* 17: 110–3, 118–20.

Mackintosh, D. (1869) *The Scenery of England and Wales.* London: Longman Green.

Matthes, F.E. (1900) Glacial sculpture of the Bighorn Mountains, Wyoming. *United States Geological Survey, 21st Annual Report 1899–1900,* pp. 167–90.

McGee, W.J. (1897) Sheetflood erosion. *Bulletin of the Geological Society of America* 8: 87–112.

Merrill, G. (1897) *A Treatise on Rocks, Rock-weathering and Soils.* New York: Macmillan.

Natland, J.H. (1997) At Vulcan's shoulder: James Dwight Dana and the beginnings of planetary volcanology. *American Journal of Science* 297: 312–42.

Newberry, J.S. (1861) *Report Upon the Colorado River of the West, Part III.* Washington, DC: Geological Report.

Oldroyd, D.R. (1996) *Thinking About the Earth. A History of Ideas in Geology.* London: Athlone Press.

Oldroyd, D. (1999) Early ideas about glaciation in the English Lake District: The problem of making sense of glaciation in a glaciated region. *Annals of Science* 56: 175–203.

Oreskes, N. (1999) *The Rejection of Continental Drift.* New York: Oxford University Press.

Orme, A.R. (1989) The twin foundations of geomorphology, in G.L.H. Davies and A.R. Orme (eds), *Two Centuries of Earth Science.* Los Angeles: University of California, pp. 31–90.

Orme, A.R. (2004) American geomorphology at the dawn of the 20th century. *Physical Geography* 25: 361–81.

Orme, A.R. (2007a) Clarence Edward Dutton (1841–1912): soldier, polymath and aesthete, in P.N. Wyse Jackson (ed.), *Four Centuries of Geological Travel: The Search for Knowledge on Foot, Bicycle, Sledge and Camel.* Geological Society of London Special Publications, 287: 271–86.

Orme, A.R. (2007b) The rise and fall of the Davisian cycle of erosion: prelude, fugue, coda and sequel. *Physical Geography* 28: 474–506.

Passarge, S. (1904) *Die Kalahari.* Berlin: D. Reimer.

Penck, A. (1894) *Morphologie der Erdoberfläche.* Stuttgart: Engelhorn.

Penck, A. (1905) Climatic features in the land surfaces. *American Journal of Science* 19: 165–74.

Penck, A. (1909) The origin of the Alps. *Bulletin of the American Geographical Society* 41: 65–71.

Penck, A. (1914) The shifting of the climatic belts. *Scottish Geographical Magazine* 30: 281–93.

Penck, W. (1924) *Morphological Analysis of Landforms.* 1953 English translation by H. Czech and K.C. Boswell. London: Macmillan.

Penck, A. and Brückner E. (1909) *Die Alpen in Eiszeizalter.* Leipzig: C.H. Tauchnitz.

Playfair, J. (1802) *Illustrations of the Huttonian System of the Earth.* Edinburgh: William Creech.

Prescott, J.A. and Pendleton, R.L. (1952) *Laterite and Lateritic Soils.* Farnham Royal: Commonwealth Agricultural Bureau.

Pumpelly, R. (1879) The relation of secular rock-disintegration to loess, glacial drift and rock basins. *American Journal of Science,* 3rd series, 17: 137.

Ramsay, A.C. (1846) The denudation of South Wales. *Memoir Geological Survey of Great Britain* 1.

Ramsay, A.C. (1862) On the glacial origin of certain lakes in Switzerland, the Black Forest, Great Britain, Sweden, North America and elsewhere. *Quarterly Journal of the Geological Society* 18: 185–204.

Ramsay, A.C. (1872) On the river-courses of England and Wales. *Quarterly Journal of the Geological Society* 30: 81–95.

Rathbun, R. (1879) Prof. Hartt on the Brazilian sandstone reefs. *The American Naturalist* 13: 347–58.

Reade, T.M. (1885) Denudation of the two Americas. *America Journal of Science* 29: 290–300.

Rogers, J.J.W and Santosh, M. (2004) *Continents and Supercontinents.* New York: Oxford University Press.

Rogliæ, J. (1972) Historical review of morphological concepts, in M. Herak and V.T. Stringfield (eds), *Karst: Important Karst Regions of the Northern Hemisphere.* Amsterdam: Elsevier, pp. 1–17.

Romer, E. (1899) Wplyw Klimatu na formy powierzchini ziemi. *Kosmos* 24: 243–71.

Russell, I.C. (1889) Subaerial decay of rocks. *US Geological Survey, Bulletin* 52.

Rütimeyer, L. (1869) *Uber Thal und Seebildung in den Alpen.* Schweighauser: Basel.

Sack, D. (1992) New wine in old bottles: the historiography of a paradigm change. *Geomorphology* 5: 251–63.

Scrope, G.P. (1866) On the origin of valleys. *Geological Magazine* 3: 193–9.

Shaler, N.S. (1874) Preliminary report on the recent changes of level on the coast of Maine. *Memoir of the Boston Society of Natural History* 2: 320–40.

Slaymaker, O. (2004) Dynamic Geomorphology, in A.S. Goudie (ed.), *Encyclopedia of Geomorphology.* London: Routledge, pp. 307–10.

Spencer, T., Stoddart, D.R. and McLean, R.F. (2008) Coral reefs, in T.P. Burt, R.J. Chorley, D. Brunsden, N.J. Cox and A.S. Goudie (eds), *The History of the Study of Landforms or the Development of Geomorphology, vol 4. Quaternary and Recent Processes and Forms (1890–1965) and the Mid-century Revolutions.* London: Geological Society. pp. 863–922.

Stoddart, D.R. (1960) Colonel George Greenwood: the father of modern subaerialism. *Scottish Geographical Magazine* 76: 108–10.

Stoddart, D.R. (1969) Climatic geomorphology, in R.J. Chorley (ed.), *Water, Earth and Man.* London: Methuen, pp. 473–85.

Stoddart, D.R. (1988) Joseph Beete Jukes, the 'Cambridge Connection', and the theory of reef development in Australia in the nineteenth century. *Earth Sciences History* 7, 99–110.

Stoddart, D.R. (1989) From colonial science to scientific independence: Australian reef geomorphology in the nineteenth century, in K.J. Tinkler (ed.), *History of Geomorphology from Hutton to Hack.* Boston: Unwin Hyman, pp. 151–63.

Strahler, A.N. (1952) Dynamic basis of geomorphology. *Geological Society of America Bulletin* 63: 923–38.

Suess, E. (1883–1908) *Das Antlitz der Erde.* Vienna: F. Tempsky.

Taylor, F.B. (1910) Bearing of the Tertiary mountain belt on the origin of the Earth's plan. *Bulletin of the Geological Society of America* 21: 179–226.

Teller, J.T. (1983) Jean de Charpentier 1788–1855. *Geographers: Biobibliographical Studies* 7: 17–22.

Thomas, D.S.G (ed.) (1998) *Arid Zone Geomorphology,* 2nd edn. Chichester: Wiley.

Thomas, M.F. (1994) *Geomorphology in the Tropics: A Study of Weathering and Denudation in Low Latitudes.* Chichester: Wiley.

Thorn, C. (1988) *An Introduction to Theoretical Geomorphology.* Boston: Unwin Hyman.

Tilley, P. (1968) Early challenges to Davis' concept of the cycle of erosion. *Professional Geographer* 20: 265–9.

Tinkler, K.J. (1985) *A Short History of Geomorphology.* London: Croom Helm.

Tricart, J. and Cailleux A. (1972) *Introduction to Climatic Geomorphology*. London: Longman.

Udden, J.A. (1894) Erosion, transportation, and sedimentation performed by the atmosphere. *Journal of Geology* 2: 318–31.

Von, L.R. (1886) A glacial period in Australia. *Nature* 34: 522–4.

Von, R. F. (1882) On the mode of origin of the loess. *Geological Magazine* NS, 9: 293–305.

Walther, J. 1900 *Das Gesetz der Wüstenbildung in Gegenwart und Vorzeit*. Berlin: Quelle and Meyer.

Wardenga, U. (2004) The influence of William Morris Davis on geographical research in Germany. *GeoJournal* 59: 23–6.

Watts, A. (2001) *Isostasy and Flexure of the Lithosphere*. Cambridge: Cambridge University Press.

Wayland, E.J. (1933) Peneplains and some other erosional platforms. *Annual Report and Bulletin, Protectorate of Uganda Geological Survey, Department of Mines*, Note 1: 77–9.

Wegener, A. (1912) Die Entstehung der Kontinente. *Petermann's Mitteïlungen*, pp. 185–95, 253–6, 305–9.

Werritty, A. (1993) Geomorphology in the UK, in H.J. Walker and W.E. Grabau (eds), *The Evolution of Geomorphology*. Chichester: Wiley, 458–67.

Whittlesey, C. (1868) Depression of the ocean during the ice period. *Proceedings of the American Association for the Advancement of Science* 16: 92–7.

Wirthmann, A. (1999) *Geomorphology of the Tropics*. Heidelberg: Springer.

Wright, G. (1896) Agassiz and the Ice Age. *The American Naturalist* 32: 165–71.

3

The Nature of Explanation in Geomorphology

Keith Richards and Nicholas J. Clifford

Geomorphology emerged as a scientific discipline in the latter part of the 19th century, and rapidly developed from the early 20th century under the dual influence of its parent disciplines, geology and geography. Geomorphology, as a science, has moved beyond description to yield explanations and predictions of landform change, but as a relatively young discipline marked by rapid development from different origins, it is perhaps not surprising that the subject has been characterized by multiple, often competing, methods and perspectives. In this chapter, the nature of explanation in geomorphology as it has evolved over the 20th century is examined, and connections are made between the objects and objectives of geomorphological enquiry, the manner in which this enquiry takes place and the kind of explanation which is ultimately achieved. For much of its history, geomorphological enquiry was directed to explaining long-term, broad-scale forms and their changes; but in the second half of the 20th century, emphasis moved to focus on smaller-scale studies of landform-changing processes. Latterly, attention has focused on integrating these perspectives, aided by advances in data collection, data handling and modelling and by the availability of a suite of environmental reconstruction techniques.

THE WHAT, WHERE, WHEN, HOW AND WHY OF GEOMORPHOLOGY

Geomorphology can readily be defined as the scientific study of landforms, but this is a skeleton which clearly needs a little flesh. Like many sciences, it is essentially concerned with answering the questions 'What?, Where?, When?, How? and Why?', but in the specific context of the earth's surface forms and the processes which maintain or change them. These questions have purposes which can variously be regarded as descriptive or explanatory.

Description is initially concerned with rigorous identification of the objects of enquiry ('What?'). This can imply qualitative 'naming' of earth surface features – that is, through the classification of topographic elements, such as 'tors' or 'inselbergs'. A philosophical issue that arises here is whether these categories and the classificatory structures of which they are a part are 'real' (i.e. are 'natural kinds'; see Rhoads and Thorn, 1996), or simply convenient mental constructs to impose some degree of regularity on the apparently diverse character of surface forms. Such philosophical questions also have implications for the manner of enquiry and type of explanatory process which follows initial description (Harrison, 2001). Is the landscape *naturally* constructed of discrete entities for which we *require* names – drumlins, cirques, barchans, yardangs, inselbergs, etc – or is it simply a continuous three-dimensional surface, to some of whose topographic attributes we *arbitrarily* assign these names? A specifically classificatory approach to landforms defines nominal-scale variables (Panel 3.1), which can then be associated with other nominal-scale independent variables, such as lithological or climatic types, to provide a descriptive association which yields the first stages of explanation. For example, inselbergs are commonly associated with seasonal tropical climates, an association which once identified invites further consideration of the 'How?' and 'Why?' questions. However, as Haines-Young and Petch (1986) have argued, vague, qualitative and generalized classes of landform make it impossible to identify one-to-one associations with climates or lithologies, and equifinality (Chorley, 1962) is almost inevitable. The notion of equifinality implies that a given landform may result from more than one process regime, or process history. This clearly makes explanation difficult, since the clues to its origin contained within a landform are necessarily ambiguous if equifinality really does occur. Indeed, it invites cessation of critical testing, since to invoke multiple causes is a convenient excuse for terminating further enquiry. To avoid this dilemma, description and classification of landforms must be more detailed, rigorous and genetically based. Since this presupposes explanation of their origins, it is evident that continual recycling and revision must occur amongst description, classification, analysis and explanation. This can be seen in the evolving approaches to types of sub-glacial deposit. These have been classed as lodgement, melt-out, flow, deformation and waterlain tills on the basis of clast fabric, but can be re-interpreted as deforming beds with spatially varying geotechnical character from the evidence of their micro-morphology (Van der Meer et al., 2003). A similar re-interpretation of A-tents or pop-ups (fractured slabs with triangular apex juxtapositions, which span a wide range of sizes, locations and lithologies) results in a common attribution as neo-tectonic features of recent (geological) origin (Twidale and Bourne, 2009).

Given these difficulties, description in a modern, scientific geomorphology has become increasingly quantitative and based on measurement, since numerical representation of topography in the form of interval- or ratio-scale variables (Panel 3.1) can be relatively consistent and reproducible. It cannot, of course, be considered to be objective, since the selection of attributes for measurement is dependent on the aims, assumptions and theoretical perspectives of the geomorphologist *qua* scientist. Neither will it always be the case that a numerical index provides a representative, precise and accurate summary of the concepts which it is intended to represent. Panel 3.1 defines these requirements more fully, and illustrates the problem of representation via the example of indexes of drainage basin shape, which variously compare the basin outline with a circle (Horton, 1945) or a lemniscate loop (Chorley, 1957). However, as long as a rigorous operational definition is provided for the representation and measurement of surface form, a given landscape should generate near-identical numerical morphological attributes whoever effects the measurement. While *different* operational definitions (which include statements about data sources and measurement devices) are used, there can be no expectation that different researchers will generate comparable data. As an example, Panel 3.2 illustrates the detail required in an operational definition for drainage density, and the variability that can occur in measurements of this morphometric property for the same drainage basins, especially as represented in differing map editions, and at different map scales.

Quantitative description in geomorphology enables representation of two kinds of topographic attribute: landforms and land form. The former are

---

Panel 3.1 Aspects of variable definition

Measurement scales

Variables may be defined which are measurable using one of the following measurement scales:

1 *Nominal* – in this scale, items are simply classified and named (e.g. rock types – chalk, clay, granite)
2 *Ordinal* – in this scale, items are ranked in sequence according to some criterion (e.g. the rock hardness sequence clay < chalk < granite)
3 *Interval* – in this scale a quantitative, measurable attribute is defined which, however, has no true zero (the most familiar example is temperature, measured on scales with arbitrary zeros, and 'freezing' conditions defined at 0°C and 32°F)
4 *Ratio* – this scale is also numerical, but there is a true zero (so that, for example, if a rock has an unconfined compressive strength twice as large as that of another rock, this is true whether the strength is measured in lb ft^{-2} or kN m^{-2})

Representation of concepts

The object of measurement on quantitative (interval and ratio) scales is to provide a clear, unambiguous, reproducible and unique representation of a given theoretical concept.

An example of the problems created by this requirement is the representation of the rather complex concept of 'drainage basin shape', which is difficult to encapsulate in a single number. Horton (1945) compared the planimetric form of basins to a circle, using the form ratio, $F = A/L^2$ (A is area and L is basin length). This takes the value 0.79 in a circular basin. Drainage basins are not often circular, however, and Chorley (1957) introduced the teardrop-shaped lemniscate loop as an alternative index. This is represented by the parameter $k = \pi L^2/4A$. Although the loop is a more realistic representation of a basin shape, the index itself is merely a constant multiplied by the reciprocal of the form ratio.

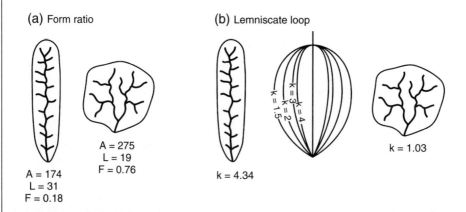

Accuracy and precision

The object of measurement is to obtain the true value of an index, but this is impossible given the practical limitations of measurement and sampling. The operational definition to be used must be capable of generating both accurate and precise values.

An accurate measurement (a in the diagram below) is unbiased; the average of a large sample of repeat measurements is close to the true value (in the example, however, the variance of this sample is large; although accurate, it is an imprecise estimate). A precise measurement (b) is one which is repeated with low variance (although this can be biased, as in the example in the diagram).

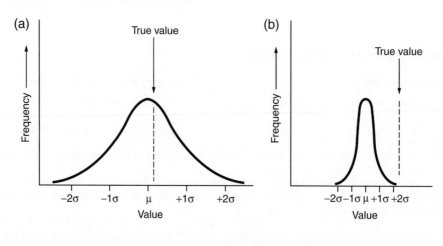

Panel 3.2 Operational definition, measurement and operator variance

An example of the range of decisions to be made in constructing an operational definition for a geomorphological variable is provided in this panel, the variable in question being drainage density (Dd). This is defined as $Dd = \Sigma L/A$, where ΣL is the total length of streams in a basin of area A.

   As illustrated in the diagram below, area may be estimated by counting grid squares or using a planimeter, and length by using dividers or a chartometer. These traditional, low-technology methods are now replaced by digitizers (which generate vectors of point data) or scanners (which generate raster data) coupled to computer programs capable of calculating areas and lengths. It is unlikely that measurements of an area or length by different methods will be identical, so an operational definition must state the method of measurement and cross-calibration between methods is necessary.

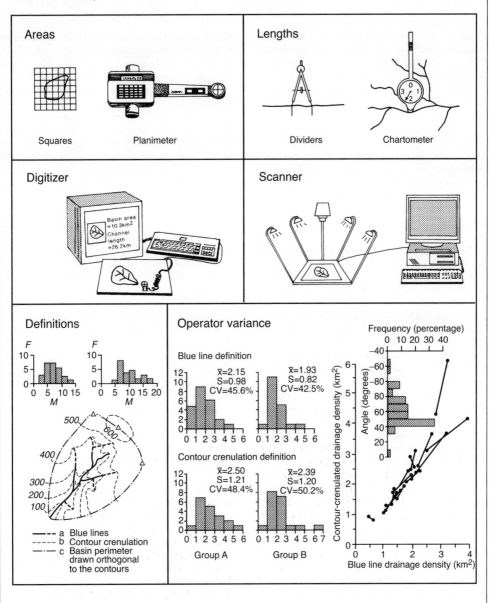

Before measurement can be undertaken, the items to be measured must be defined; these include the basin perimeter (watershed) and network extent. The latter may be represented by the mapped blue lines on a large-scale topographic map, or by extending stream heads to the last significant contour indentation (bottom left diagram). The histograms of stream magnitude (*M*) show that, without clear guidelines for stream head definition, operators (here in two groups) can disagree markedly on the network extent. This is the problem of operator variance (Chorley, 1958). The final diagram illustrates that mean drainage densities obtained by groups may be similar, but that estimates by paired individuals may diverge markedly, and more so for the subjective contour crenulation method (Richards, 1979). The inset histogram is of the angles connecting measurements made by two operators on the same basin. If the same proportional difference occurs in the 'blue line' and 'contour crenulation' methods, the angle is 45°. That it is often >45° indicates greater proportional differences in the more subjective contour-based method.

discrete, identifiable (classified) features such as cirques, drumlins or barchan dunes. The numerical description of such features is the task of specific morphometry (Goudie, 1990). By contrast, the land form of a particular, arbitrarily bounded region defines its general topographic characteristics. Land form can be summarized by the general morphometry derived from altitude matrices (Evans, 1972), otherwise known as digital elevation models or digital terrain models – high-resolution regular grids imposed on the topography, with altitudes identified for each grid point, for which successive differentiation of the resulting matrix generates a slope matrix (rate of change of altitude) and slope curvature matrix (rate of change of slope), with a slope aspect matrix being constructed from the azimuths of the steepest slopes. Such matrices can be used to derive frequency distributions of topographic variables, to examine inter-relationships between the properties of land form, or to form the basis for Geographical Information Systems which can be used to map the spatial pattern of slope- and aspect-controlled surface energy receipt, evaporation, vegetation cover, soil moisture status and runoff (Moore et al., 1988; Lane et al., 1998: Figure 3.1).

The increasing use of remotely sensed data to provide basic information on both form and process has renewed the long-standing issue of the interaction between data collection and quality, representation of form and inference of process. Automatic extraction of drainage networks in GIS packages, for example, is only as good as the rule-base on which the program operates. Basic properties such as basin shape, area and relief are all sensitive to original data resolution and subsequent interpolation, refinement and smoothing (Yang et al., 2001; McMaster, 2002), although derived properties may be more or less robust (Hurtrez et al., 1999). These representation problems are compounded when the forms generated are used as input to models of process. For example, the hydrological model TOPMODEL uses a topographic index (a/tan B) to assess the distribution of the saturated contributing area; where a is the area drained per unit contour length, and B is the local gradient (Beven and Kirkby, 1979). This index can be calculated for each cell in a digital terrain model, and subsequently cumulated into a frequency distribution, or mapped on a cell-by-cell basis over the catchment. Since the 1990s, such process-based applications have become increasingly sophisticated, and with higher-resolution data becoming available, and where prior knowledge of potential problems is used to create sensible strategies for their remediation, coupling of hydrological models to geomorphological process models using automated data collection and extraction algorithms is now viable (Lane et al., 2004). A good example is the SHALSTAB model of Montgomery and Dietrich (1994), in which a model of shallow translational landslides is coupled with a soil hydrology model and a digital elevation model in order to predict the spatial occurrence of shallow landslides.

A scientific geomorphology must also explain the forms that it describes and its explanations commonly – and as we shall see later in this chapter, necessarily – involve two levels of analysis. The first is concerned with answering the questions 'Where?' and 'When?', while the second addresses the questions 'How?' and 'Why?' These two levels deal, respectively, with configurational or contextual aspects of explanation, and immanent aspects. The contextual aspects are geographical and historical/geological (Simpson, 1963). Answers to the question 'Where?' deal with the geographical background to the creation of the land surface, and with understanding the spatial variation of earth surface process types and intensities and their relationship to spatial variations of climate and lithology. Answers to the question 'When?' deal with the geological dimension – temporal variations of climate during the Tertiary and Quaternary periods which have controlled the complex evolution of physical landscapes by altering the balance of process

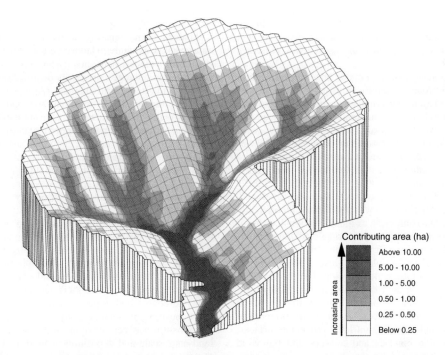

Contributing area (ha)

Above 10.00

5.00 - 10.00

1.00 - 5.00

0.50 - 1.00

0.25 - 0.50

Below 0.25

Increasing area

Figure 3.1　A digital terrain model of a small Australian catchment, with an 'overlay' of information on upslope contributing area classes (in hectares), which relate to the likelihood of soil saturation (after Moore et al., 1988)

types and intensities. Within the framework of this contextual information, answers are sought to questions about the fundamental nature of earth surface processes – the unchanging, immanent aspects of explanation which are independent of time and place. These questions are concerned with the nature of endogenic (internal, tectonic) processes and exogenic (external, surface processes such as weathering, erosion, transport and deposition), with the underlying physics and chemistry of the process mechanisms, and with the relationships between those processes and the forms that they produce.

THE GEOMORPHOLOGICAL BANDWAGON PARADE

It has been well known for many years that a full understanding of the nature and causes of landforms requires appraisal of form, process, materials and time. The influential American physical geographer W.M. Davis constructed a theoretical model of landform evolution – the cycle of erosion (1899) – in which landforms were functions of 'structure, process and stage', and

more recently Gregory (1978) has summarized this symbolically as:

$$F = \int_0^t (P, M)\, dt \qquad (3.1)$$

which is short-hand for 'landforms (F) represent the integrated effect of changing processes (P) acting on materials (M) over time (t)'. Notwithstanding the undeniable necessity of this overall appraisal of the elements of geomorphology, the relative emphases on form, process and materials have changed over the years. This may be interpreted as reflecting shifts of fashion as bandwagons roll by (Jennings, 1973), and as influential figures advocate particular approaches, and educate succeeding generations to further these (Yatsu, 1992; Sherman, 1996); or, more meaningfully, as paradigms shift (Kuhn, 1970) following revolutions in scientific thought (Sack, 1992). For example, the Davisian evolutionary scheme has been interpreted as a reflection of the scientific influence of Darwinian thinking in the late 19th century (Stoddart, 1966), whereas Gilbert's geomorphological approach (almost contemporary with Davis's, see p. 9) has been related to 19th century developments in chemistry, particularly thermodynamics, and hence to later formalisations of systems approaches (Chorley, 1962).

All such historiographies are, however, themselves prone to simplification and bias, while their construction and criticism is one way in which geomorphological methodologies and explanations evolve (Rhoads, 1999). In reality, what historically appear as major changes of emphasis are more likely to reflect a series of pragmatic and gradual adjustments as knowledge and understanding improve, and, increasingly, scientific methods of enquiry and techniques are developed. 'Time' for example, was represented theoretically in the Davisian cycle as a unidirectional scale for landform evolution, in an organic analogy involving the sequence from a 'youthful' stage through 'maturity' to an 'old age' stage. The empirical geomorphology which developed in the Davisian tradition took the form of denudation chronology. This involved reconstruction of the erosional history of the landscape through measurement of the altitudes of summit-level erosional surfaces, valley-side benches and terraces and the wind gaps abandoned by the drainage network after river capture. However, erosional processes clearly have the unfortunate tendency of destroying the evidence of prior conditions and earlier features, and the empirical basis of denudation chronology gave way to the evidence provided by sedimentary deposits and weathering residues which preserve evidence of former conditions, as the methods were progressively developed which enabled recovery of that evidence. Coupled with the developing knowledge of the climatic variations which have characterized the last 2–3M years and have influenced landform development, these trends naturally led to the replacement of denudation chronology by what is now known as Quaternary science.

Quaternary science is a multi-disciplinary field that underpins the interpretation of long-term landform evolution. It embodies a range of techniques of environmental reconstruction applicable to various forms of sedimentary deposit that enable the identification of environmental conditions at stages in that evolution; including sedimentology, palynology and the analysis of various floral and faunal remains (Lowe and Walker, 1984). It also includes several dating techniques that permit the ordering of events in time, and even the construction of absolute chronologies; in particular, radiocarbon and other radiometric techniques, and luminescence dating. These represent an essential basis for correlating the discontinuous history of terrestrial landscape change with the continuous record of climate change revealed by marine sediments and the

oxygen-isotope stratigraphy (Shackleton, 2000); and indeed, for recognizing the existence of a neo-catastrophist signature in landscape change, where periods of statis are separated by periods of rapid activity and change, driven both by external events and internal dynamics (Ager, 1995). A recent addition to the technical apparatus that helps explain landform change is the suite of cosmogenic isotopic dating techniques that permit the direct estimation of erosion amounts and rates (Cockburn and Summerfield, 2004), and that allow the testing of hypotheses about modes of erosional development, implying reversion to an albeit new form of denudation chronology.

The importance of absolute dates for estimating process rates and reconstructing landform history is underscored by recognition of the 'tintinnabulations' that may arise from inherent oscillations in coupled geomorphic systems (Humphrey and Heller, 1995) that occur independently of climate changes (see p. 13). Thus, there remains a need either to couple both types of dating (of erosion and deposition) in a single study, or to model the relationship between erosion, transport at the landscape scale and deposition or there is a risk that these potential oscillatory interactions will be ignored and the history over-simplified. While such integration may be necessary for improved scientific explanation, it may not be sufficient. In the 1960s, for example, the growing appreciation of climate change was poorly deployed in geomorphological explanation, in part, because of a continued emphasis on process-form equilibrium (Grove, 2008; and see p. 11), but also because of cultural and disciplinary divisions between researchers from the differing scientific backgrounds of geography, geology and engineering. These have essentially partitioned specialist knowledge amongst various disciplinary groups. This may inhibit further integration of geomorphology and Quaternary science (Summerfield, 2005), and may also help to explain the differing reactions from geomorphologists to the emerging project of Earth System Science (Clifford and Richards, 2005; Richards and Clifford, 2008).

The changing nature of the analysis of 'process' in geomorphology also shows that interpretation of trends in emphasis as being merely bandwagons or revolutions is unrealistic. Rather, there is a history of pseudo-parallel development of different kinds of process geomorphology. One example, particularly associated with a French school (e.g. Tricart and Cailleux, 1972), is the generalized interpretation of configurational climatic control of process regimes. This involves the identification of process intensity domains in terms of climatic parameters, as illustrated (Figure 3.2) by the analysis conducted by Peltier (1950) and summarized in the form of the maps

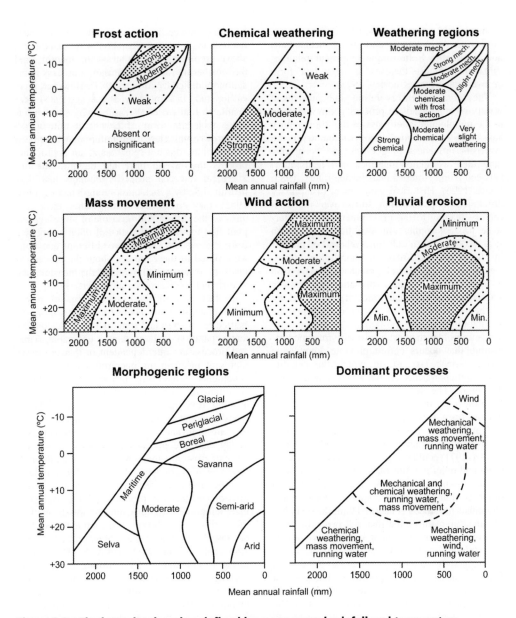

Figure 3.2   The intensity domains, defined by mean annual rainfall and temperature, of certain earth surface processes (after Peltier, 1950)

of morphoclimatic regions which characterize 'climatic geomorphology'. More detailed interpretation of geomorphological processes focuses on their immanent properties, and their interpretation in terms of underlying physical and chemical mechanisms, and this can be broadly classed as 'dynamic geomorphology'. However, two broad strands exist in this enquiry. The first is the 'functional' approach which characterized the period from 1950 to 1975, approximately. This is based

on positivist assumptions concerning the primacy of observation and measurement, on the identification of empirical relationships between morphological and process variables, and on the structuring of those relationships via systems analysis (Chorley and Kennedy, 1973). It is essentially a 'black-box' method, in that it fails to identify the mechanisms underlying the empirically observed linkages. The second strand is the truly dynamic geomorphology envisaged by

Strahler (1952), which is based on realist attempts to uncover the nature of underlying mechanisms not identifiable from simple empirical relationships, and which is therefore a 'white-box' form of investigation (see Chorley, 1978).

The more general evolution and subsequent variety of approaches to process geomorphology in the middle and latter parts of the 20th century has been more broadly associated with the divergent influences of geology and geography (and their associated methods and techniques) between North America and Europe (Dury, 1983; Bauer, 1996). Historical experience, disciplinary differences and paradigm shifts over the 20th century leading to the pre-eminence of smaller-scale process studies are also linked by Smith et al. (2002) in their call for a renewal of interest in long-term, broad-scale geomorphology. Here, they argue that the development and progression of methodologies and explanations follows a spiral, rather than linear trajectory. The analogy of the spiral is used to justify not simply the inevitability, but also the benefits of periodically re-visiting older themes; a critically informed re-visiting that occurs before previous knowledge and experience is lost, in which the benefits of current approaches enable a new synthesis for the future. The degree to which this is itself a simplification or parody of a subject history remains open, however (see p. 18).

Finally, the nature of geomorphological approaches to 'structure' has also evolved; where Sparks (1971) could study the relationship between rocks and relief in terms of qualitative assessment of the association between topography and lithological and structural controls, later geomorphologists have been concerned with the quantitative measurement of the material properties that determine resistance to shear, compressive or tensile stresses, or to a variety of chemical and mechanical weathering processes. A critical requirement here is that processes and material properties are analysed as interdependent phenomena; it is difficult to select which of the potential multitude of material properties to measure without some *a priori* theoretical consideration of the processes whose balance and intensity are modulated by the nature of the materials. This is nicely illustrated by two studies of inselbergs in east Africa. The specific locations of inselbergs as positive relief features were studied by Pye et al. (1986), and subtle geochemical differences were found to distinguish the exposed inselberg rocks from those underlying the surrounding plains. These material differences are thus 'activated' as a significant property by the chemical weathering process hypothesized as the main influence on the evolution of an inselberg-dominated landscape (the regolith–bedrock

interface descending less rapidly in the more potassic rock, until unweathered bedrock emerges above ground as erosion strips the regolith, and the weathering rate slows further; a process interaction modelled by Ahnert, 1987). However, in an earlier study in Zimbabwe, Pye et al. (1984) were unable to find material reasons why some inselbergs took the form of bornhardt domes and others of kopjes. Here, their failure may have been because they did not hypothesize an appropriate process that would have led them to identify a critical material property. Perhaps had they measured Selby's rock mass strength index (Selby, 1982), they could have distinguished these features as the differential outcome of stress-release joint development and rock-fall *after* emergence from the regolith cover. Two different questions were posed in these studies about the process of landform origin, which required the measurement of different, critical material properties.

In fact, one of the consequences of recent developments in the nature of investigations of 'structure, process and time' is that their convenient separation is increasingly untenable. Time and process are interdependent in that processes represent mechanisms operating over time, and statements concerning process rates demand simultaneous measurement of processes and absolute timescales; processes and materials are interdependent in that particular material properties are relevant in specific process contexts. Today, therefore, generalized models – such as the cycle of erosion – are eschewed by a geomorphology which recognizes the need for problem-specific, and scientific, hypothesis testing of the inter-dependence of process and materials and their temporal variations, in attempts to explain the earth's surface forms. Thus, while this section has largely dealt with the progression of the late 19th century contributions of W.M. Davis, it is fitting to conclude it by returning to the 'rival paradigm' retrospectively attributed to his contemporary, G.K. Gilbert, which provides a model for the physically based, scientific and field-oriented approach that has underpinned geomorphology from the mid 20th century (Yochelson, 1980). Gilbert's research ranged across phenomena as varied as lake shore features, meteor craters and bedload transport in rivers (Gilbert, 1885, 1893, 1914), revealing an awareness of the instability of terrestrial processes in his studies of isostatic effects on Lake Bonneville shorelines (Gilbert, 1890) and the rhythmic sedimentary consequences of astronomic cycles (Gilbert, 1900), Gilbert's work illustrates a method quite different from the over-generalized and untestable (Bishop, 1980) theorizing of Davis. Although a truly critical rationalist geomorphology (Haines-Young and Petch, 1986) may be elusive, Gilbert's process-based, geomorphological problem-solving is at least a goal to which geomorphologists might aspire in order that their attempts at explanation are rigorous.

EVOLUTION AND EQUILIBRIUM: THE EXAMPLE OF THE APPALACHIANS

The development of different emphases and approaches in geomorphology is particularly evident in the distinction that exists between evolutionary and equilibrium interpretations of landforms (Ritter, 1988; Montgomery, 1989). While Davis's cycle of erosion emphasized the time dependence of landforms, Gilbert's classic paper for the embryo United States Geological Survey, 'The Report on the Geology of the Henry Mountains' (Gilbert, 1877), introduced the concept of time-independent equilibrium to the interpretation of landforms. This dichotomy can be well illustrated by using a case study of a landscape whose peculiar characteristics appear to have encouraged generations of geomorphologists to exercise their explanatory skills: the Appalachians in the eastern United States (Johnson, 1931; Gardner and Sevon, 1989). The Ridge and Valley province of Appalachia is associated with a series of NE–SW trending ridges and valleys (Figure 3.3a), but the rivers drain to the SE across the structural and topographic grain. Summit levels display marked accordance of altitude, the drainage system is markedly asymmetric as it crosses the main axis of tectonic uplift, and the rivers appear discordant to structural influences. As underlying theoretical perspectives have changed during the 20th century, there have been significant alterations in the interpretation of this landscape.

Long-term evolutionary interpretations

The Davisian model of a cycle of erosion following initial rapid uplift, characterized by an irreversible and directed sequence of landscapes from youthful, through mature to the old-age peneplain (Davis, 1899; King and Schumm, 1980), provided an early theoretical framework for analysis of the long-term denudational history of Appalachia. It was necessarily modified to accommodate partial peneplanations occurring during periods of base-level stillstand, and a series of such partial peneplains has been identified in Appalachia, of which the Fall Zone, Schooley, Harrisburg and Somerville are the most important (Johnson, 1931). These were reconstructed

from summit accordances (see Figure 3.3a, for example), valley-side benches, knick points in river-long profiles and wind gaps in ridges. The Davisian cycle assumed that processes of river incision create slopes, on which weathering forms soils, in which creep occurs to result in slope angle decline, with progressive comminution of debris also allowing river-long profiles to decline in gradient; until the low-angle peneplain is formed. The partial surfaces identified in Appalachia, eating back into higher-level and older surfaces, possibly imply a rather different model of landform development, however – that of parallel retreat of slopes and pedimentation. This indicates the importance of understanding processes and their palaeoclimatic implications before interpreting the purely morphological evidence of landform evolution; if surfaces do exist in Appalachia, they are likely to reflect Tertiary climatic conditions and process regimes very different from those of the generally cold Quaternary.

The Appalachian drainage patterns are generally discordant with the underlying structure, and this has been interpreted as the result of superimposition (Meyerhoff and Olmstead, 1936), particularly from assumed Cretaceous cover strata (Figure 3.3b). However, examination of the offshore sedimentary record suggests three periods of rapid sedimentation on the continental shelf (and therefore terrestrial erosion) in the early Jurassic, late Cretaceous and the Miocene onwards (Poag and Sevon, 1989). This is not consistent with a denudational chronology inferred rom terrestrial morphological evidence interpreted in a Davisian theoretical framework, but rather is a reflection of the interaction of complex tectonic, palaeoclimatic and sea level histories.

Equilibrium interpretation

Certain elements of the Appalachian landscape are undoubtedly well adjusted to rock structure and lithology (e.g. river gradients are steep on more resistant rocks). Thus Hack (1960) queried the emphasis of evolutionary geomorphologists on the discordances and argued instead that the landscape displayed evidence of dynamic equilibria. These involved mutual adjustment of landform components to rock resistance and types and intensities of process, with weathering and erosion processes being balanced. For example, an equilibrium soil thickness is defined by the balance between surface removal by erosion and production of soil material by bedrock weathering. This balance varies spatially as the outcropping rock types vary; a silty limestone produces an

(a)

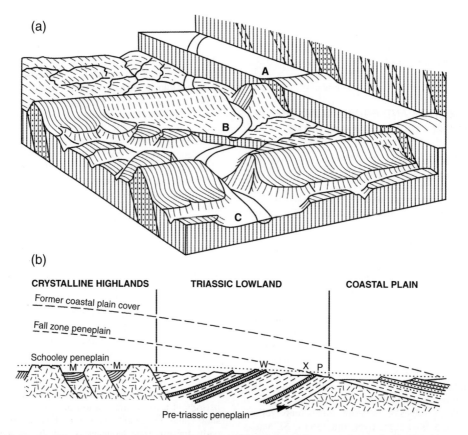

(b)

CRYSTALLINE HIGHLANDS   TRIASSIC LOWLAND   COASTAL PLAIN

Former coastal plain cover

Fall zone peneplain

Schooley peneplain

Pre-triassic peneplain

Figure 3.3   (a) A typical ridge-and-valley Appalachian landscape (ridges aligned SW-NE, left-to-right), illustrating a 'three-cycle' development in which A represents the first-stage lowland, dissected to form B and then the present landscape C, with remnants of the A and B surfaces on summits (after King and Schumm, 1980). (b) Johnson's (1931) diagram illustrating the relationship between the supposed Jurassic Fall-Zone peneplain (note his spelling) and the Mio-Pliocene Schooley surface in northern New Jersey; the former is hypothesized to have been buried by Cretaceous marine sediments, the latter to have been formed after superimposition of the drainage system from this cover (M is the Musconetcong River and its tributaries, W and P the Watchung and Palisades trap ridges, and x the intersection of the two surfaces)

erodible soil even on gentle slopes, so a thin soil under which rapid weathering occurs balances the surface removal. Cherty limestone, however, produces a self-armouring soil with a surface veneer of chert fragments which result in slow surface removal on steep slopes, balanced by slow weathering beneath a thick soil.

This equilibrium interpretation is pervasive at all scales, from the local to the regional; for example, the migration of the main NE–SW trending divide of Appalachia to the NW of the main axis of uplift, with resultant drainage discordance as the rivers cross the structural grain, is attributable to variation in rock resistance and rate of

erosion. A belt of resistant metamorphic rocks outcropping to the NW of the Atlantic–Gulf Coast watershed is picked out by a high gradient zone on maps of stream gradient (Hack, 1973), and is associated with slow rates of erosion which enabled the river heads, and the divide, to migrate in a north-westerly direction until the presently observable asymmetry was created. At an intermediate sub-regional scale, Battiau-Queney (1989) has argued that block tectonic processes control topographic development. Crustal blocks behave independently of one another, controlled by structural discontinuities in the middle and deep crust, and the landforms of each block reflect a dynamic equilibrium between surface processes, rock resistance, and rates of uplift which develops within the block.

Perhaps even more significant is the undermining of the denudation chronologist's reconstruction of planation surfaces from visible summit accordance. This arises from the suggestion that accordant summits reflect the fact that, in areas of spatially roughly uniform climate, hydrology and erosional regime, roughly uniformly spaced streams and common slope angles would result in watershed geometry that renders similarity of watershed elevations inevitable. An equilibrium interpretation is therefore possible for summit accordance (Hack, 1960), and the notion of

accordance as evidence of earlier planation and subsequent dissection is firmly rejected.

A resolution: scale and the interpretation of landforms

A resolution of these apparently conflicting interpretations is, however, possible, and depends largely on the scale at which explanation is attempted. Costa and Cleaves (1984) demonstrate this via an appraisal of the landforms of parts of Maryland. Figure 3.4a shows that there is a thick saprolite beneath a dissected low-relief upland

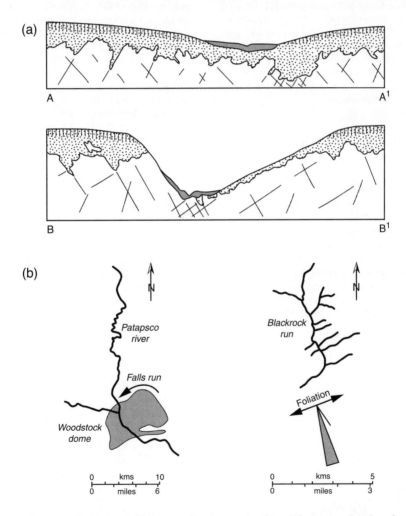

Figure 3.4   (a) The existence of thick saprolite beneath a late Miocene upland surface in Maryland; and (b) adjustment of the river system to rock character, including gneiss outcrops such as the Woodstock Dome, and the orientation of joints and foliation (after Costa and Cleaves, 1984)

surface, and this is testimony to an episodic erosional history involving long periods of stability and weathering. The dissected surface truncates rock outcrops, and is clearly older than the present river valleys, since only the higher order streams have been able to cut through it. There is evidence that the main river alignments had been established by the Cretaceous, and a deeply weathered low-relief surface was probably formed by the late Miocene; its dissection probably results in the most recent sedimentation episode offshore (see above). However, Figure 3.4b shows that there are also equilibrium features in the landscape, particularly in the form of river network patterns adjusted to rock resistance. Small headwater streams skirt around gneiss domes, while larger and more powerful rivers cut across them indiscriminately. Stream alignments also frequently show adjustment to joint and foliation orientations. The section below on time and area scales explores in greater detail the reasons why this resolution of equilibrium and evolutionary interpretations is possible; but there remain problems for the resolution.

Remaining difficulties: thresholds and complex response

The simplest form of mutual adjustment between landforms and their process environment is one in which a continuous response occurs. For example, if the time axis of Figure 3.5a is replaced by 'mass removed', and the ordinate is 'channel gradient', the graph could represent the continuous reduction of long profile gradient associated with the progressive lowering of relief during the cycle of erosion. However, many changes in the landscape require that a threshold is exceeded; change is therefore not continuous, and any equilibrium

must be viewed as metastable. On hillslopes, mass movements occur when the shear stress acting on slope materials exceeds their shear resistance, and in the bedload transport process that controls river channel form a threshold fluid force must be exceeded before transport is initiated. These are 'extrinsic thresholds', triggered by an external agency (usually related to a heavy rainfall), according to Schumm's (1979) definitions. This is, however, somewhat confusing because the threshold is actually an intrinsic material property.

More subtle thresholds are Schumm's 'intrinsic' or 'geomorphic' thresholds. These are the result of cycles of sediment storage and release during the denudation of the land surface, giving rise to a 'complex response' (the 'tintinnabulations' mentioned above). When a fluvial landscape is affected by base level lowering, a wave of erosion passes headwards through the river system. Headwards locations are affected some time after the initial stimulus, and sediments released by tributary erosion accumulate in the major valleys until their surface gradient is such that renewed valley fill erosion is triggered (the intrinsic threshold is exceeded). This results in a damped oscillatory response to the initial stimulus, as illustrated in Figure 3.6. Such behaviour has been observed in small-scale laboratory as early as Lewis (1944) and in semi-arid drainage basins, but it remains to be demonstrated in large, humid-climate, vegetated basins, and may not be a consequence of altered runoff and sediment yield over a whole catchment in the absence of base level change. Nevertheless, the existence of complex response indicates that geomorphologists cannot assume a one-to-one relationship between environmental change and landscape change, because of the temporally lagged spatial transmission of change. Thus time and space relations are crucial to an understanding of the evolution of landforms.

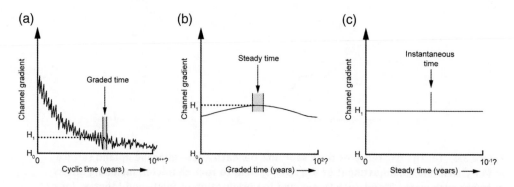

Figure 3.5   Cyclic (a), graded (b) and steady (c) timescales in the variation of river gradient

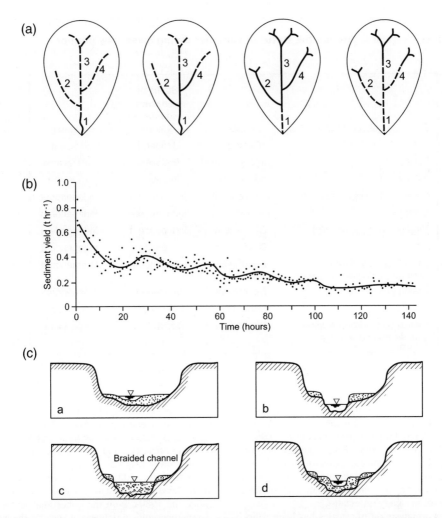

Figure 3.6   (a) A sequence in which base level lowering triggers stream incision (shown by a continuous line), which extends headwards until eroding tributaries cause valley filling downstream (pecked lines) (after Schumm and Hadley, 1957). (b) The damped oscillatory sediment yield response to base level lowering of a laboratory basin, and (c) the consequences for the formation of a series of inset alluvial fills and terraces in the main valley (after Schumm, 1977)

TIME AND AREA SCALES

The reason that Costa and Cleaves (1984) are able to reconcile evolutionary and equilibrium interpretations is that the form of geomorphological analysis required is often dependent on the time and area scales being investigated. Schumm and Lichty (1965) have illustrated this by distinguishing between three timescales: long-term, or cyclic time; medium-term, or graded time; and short-term, or steady time (Figure 3.5). Cyclic time covers 10^{4++} years, and is the timescale over which secular evolution of the landscape occurs. Graded time (10^{0} to 10^{2} years) is the timescale of equilibrium, for example, that in which river channel gradient and cross-section properties are adjusted to the prevailing climate and relief and are maintained by negative feedbacks. On steady timescales (10^{-1} to 10^{0} years), landforms are static and short-term variations of process actually adjust to them (see the next section). In addition to these different modes of process–form relationship on different timescales, the status of variables changes, being irrelevant, independent,

Panel 3.3 Time scales and the status of variables during given time periods

Drainage basin variables	Cyclic	Status of variables during Graded	Steady
	(Geologic)	(Modern)	(Present)
Time	Independent	Irrelevant	Irrelevant
Initial relief	Independent	Irrelevant	Irrelevant
Geology (lithology, structure)	Independent	Independent	Independent
Climate	Independent	Independent	Independent
Vegetation (type, density)	Dependent	Independent	Independent
Relief	Dependent	Independent	Independent
Hydrology (runoff and sediment yield per unit area)	Dependent	Independent	Independent
Drainage network form	Dependent	Dependent	Independent
Hillslope form	Dependent	Dependent	Independent
Hydrology (short-term water and sediment discharge)	Dependent	Dependent	Dependent

dependent or indeterminate according to the timescale (Panel 3.3). On the cyclic scale, time and initial relief are independent controls; they are irrelevant in steady time. River channel forms are dependent on catchment properties over graded times, but are themselves among the independent controls of the behaviour of flow and sediment transport on steady timescales.

This emphasis on timescales has been echoed by Church and Mark (1980), who classify the variable set invoked by a geomorphological investigation as 'frozen' and 'relaxed'. If a system is being studied over time T, there are certain variables that change over timescales t where $t \gg T$. These variables are therefore essentially 'frozen' in the context of the problem in hand; plate tectonic processes, for example, are irrelevant to studies of splash and wash erosion and local slope development. Other phenomena change over timescales t where $t \ll T$, and these can only be represented by their mean or their cumulative effect, and the system is too damped to respond continuously to their changes. These are the 'relaxed' variables, and an example is the viscosity of water, which varies minute-by-minute in relation to temperature, but whose variations are irrelevant to studies over timescales in which T is greater than, say, a year. Such an interpretation is mirrored by hierarchical notions of ecosystem theory (O'Neill et al., 1986), in which systems are

viewed as consisting of relatively isolated levels of process operation: at 'lower' levels processes operate at faster rates and are represented as averages of background variation, while at 'higher' levels slower response rates permit definition of the boundary conditions within which the system at the level of interest is perceived as operating. Since the 1980s, both the conceptual as well as the methodological richness of systems theory has radically increased. Seminal works in geomorphology on abstract representation of correlation structures (Melton, 1958) and non-linearity in cause–effect (Culling, 1988) are now part of a wider movement to develop and deploy complexity science. This brings additional explanatory potential in describing and explaining scale dependency (and independency) in explanation, and in recognising multiple thresholds and complex causation (Phillips, 1999), and has contributed substantively in incorporating bio-physical processes (especially the role of vegetation) as an important component in mediating feebacks between land forms and environmental processes (Stallins, 2006).

Areal scales also influence the appropriate framework for explanation, and their effects inevitably tend to be correlated to timescale effects. For example, Slaymaker (1972) undertook a series of experimental studies of sediment transport in small Welsh catchments, and showed that locally, it was possible to demonstrate a nice (equilibrium) adjustment of form and process: for example, correlation exists between stone creep on scree slopes and scree angle and aspect. At the larger catchment scale, however, sediment supply and removal were often imbalanced as the landforms continued to readjust to the effects of Quaternary glaciation – for example, exporting from Quaternary valley fills more than is supplied from adjacent hillslopes. More recently, Slaymaker (1987) has contrasted investigations at the scale of individual slopes and channel reaches, which emphasize landform dynamics (stress fields and strain responses), with those at the basin and sub-regional scales which deal with landform kinematics (their rates and modes of evolution).

Although the dynamics and kinematics of landforms must be combined in a complete process geomorphology, different modes of explanation are appropriate to particular time and space scales. The evolution–equilibrium axis is paralleled by additional dichotomies, such as historical–functional, and configurational–immanent. Large-scale landscapes viewed over long time periods are investigated in terms of their history and development (involving the methods of denudation chronology and Quaternary science), while small-scale landforms or landform components are analysed in terms of how they work, and how their forms and processes are mutually adjusted (dynamic geomorphology). Investigation of the specific locational and historical/geological circumstances within which landform evolution takes place demands detailed reconstruction of the configurational aspects of explanation (see for example, a rare, opportunistic example to infer a long temporal evolutionary sequence from an analysis of a landform at differing points in space: so-called ergodic reasoning; Savigear, 1962; Paine, 1985). By contrast, the functioning of a landform is addressed via analysis of the fundamental and time–space invariant nature of natural geomorphological processes, and their underlying physico-chemical controls. However, dichotomies like these must be reconstructed and reconciled for complete understanding to be possible, and the following section considers how this occurs.

REQUIREMENTS FOR INTEGRATION OF CONTRASTING APPROACHES

There are four important determinants of the integration that is possible between approaches that seem irreconcilable because of their contrasts in scale and method. These are: (1) the principle of uniformitarianism; (2) the covering law model of deductive explanation; (3) the existence of spatially distributed feedbacks between form and process; and (4) the development of models that can capture these feedbacks.

Uniformitarianism is often caricatured (Gould, 1987) as the statement that 'the present is the key to the past', and is associated in particular with James Hutton's *Theory of the Earth* (1788). It is a principle enshrined in 19th century geology, but as Gould (1965) has argued, for the wrong reasons. One interpretation of uniformitarianism is that present-day observations of geomorphological processes can be extrapolated into the past on the assumption that they are uniform in time and space, to provide information on the age of landforms. This is what Gould called 'substantive uniformitarianism', and it is untenable because process rates are demonstrably not uniform, neither in time nor in space. More fundamental is 'methodological uniformitarianism'; this is the assumption that the underlying immanent physical and chemical phenomena responsible for earth surface processes are time- and space-invariant. This is not a geological assumption *per se*; it underpins all of science. It is the reason that Baker (1983) can utilize relationships between the properties of bedforms created by sediment transport in aqueous flows and the power of those flows to estimate the power, and thence the depth, velocity, and discharge, of the catastrophic discharges resulting from the sudden Late-Glacial drainage of Lake Missoula in Washington State. That discharge (perhaps 1×10^7 m^3 s^{-1}, or 30 times an average annual maximum flow in the Amazon) resulted in the transport of gravel which formed huge bedforms in the Channelled Scablands analogous to those created in sand by more humble flows. Substantive uniformitarianism involving extrapolation of measured rates cannot cope with such extreme events or their products; methodological uniformitarianism, however, provides the basis for understanding those events on the grounds that their underlying physics are 'normal'.

The 'covering law model' provides a basis for deductive explanation of particular events or phenomena (Haines-Young and Petch, 1986). It involves the combination of a statement (or statements) that is general, time- and space-invariant and is concerned with the immanent properties of systems (a general law), with statements that are case-specific, and define the configurational geographical and historical circumstances (the initial conditions) within which the general law operates to generate the event to be explained. Panel 3.4 illustrates how an event – the observed change of channel pattern from braided to meandering in the South Platte River in

Panel 3.4 The 'covering law' model of explanation

The event: channel change in the South Platte

As the figure below illustrates, the South Platte River experienced a radical change in its channel pattern during the late 19th and mid 20th century.

A deductive explanation for this 'event' is based on the combination of two statements, or sets of statements.

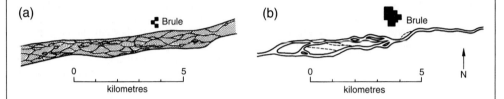

(a) Brule

(b) Brule

0 ____ 5 kilometres

0 ____ 5 kilometres

N

The covering law

This statement is general, space–time invariant and is concerned with immanent properties of the phenomenon to be explained.

For example, unconfined alluvial rivers exhibit a multi-thread braided channel pattern if stream power per unit bed area exceeds 50 W m^{-2} at the mean annual flood, but a single-thread meandering channel pattern if the power is less than 50 W m^{-2} at this discharge.

The initial conditions

This statement, or set of statements, relates to the specific case being studied. It is therefore time- and place-specific, and defines the geographical and historical context – the configurational properties – of the problem to be explained.

For example, before 1850, as the sketch maps above show, the South Platte River was braided. However, in the late 19th century a series of reservoirs impounded headwater flows, and the magnitude of the mean annual flood was accordingly decreased. In the mid 20th century, the power per unit bed area at the mean annual flood was only about 2.5 W m^{-2}. The South Platte River has been a single-thread meandering river during most of the 20th century.

The deduced explanation

Based on a combination of the general and the specific statements of I and II, an explanation for an event can be formulated.

Thus, the South Platte River changed its channel pattern from a braided to a meandering form because of a decline in its stream power at the mean annual flood following reservoir impoundment in its headwaters.

The continuing problem of explanation

However, this explanation may only be a partial one, and this illustrates that the covering law model must be used critically, and in a spirit of continual enquiry. Nadler and Schumm (1981) have shown that the irrigation water supplied by the reservoirs impounding the South Platte River gradually raised downstream floodplain water tables, encouraging the spread of vegetation such as tamarisk. This helped to stabilize the floodplain and channel-bank sediments, reducing their erodibility and contributing to conversion of the channel pattern from an unstable, shifting multi-thread form, to a more stable meandering character. Thus not only did the actual stream power decrease (as implied by the initial conditions), but the threshold power required to destabilize a channel and create a multi-thread form increased (the stream power threshold defined in the 'covering law' therefore needs to be a variable threshold).

the 19th century – can be explained deductively by combining a covering law-like statement about channel pattern (in this example a very simple one), with a specific statement about impoundment and reduced flow magnitudes in the river in question. The specific event being considered cannot adequately be explained in the absence of either the covering law or the initial conditions. This scientific method is the essential basis of critical rationalism (Haines-Young and Petch, 1986); this is a rational approach in which hypotheses are formulated and critically tested through the design of rigorous experiments which seek falsification. Empirical evidence is always partial, and therefore cannot permit verification of a hypothesis. However, critical experiments designed to probe the weaknesses in hypotheses, and cause them to be falsified, provide a rational basis for their modification and improvement (Popper, 1959). As Bishop (1980) notes, the Davisian scheme is incapable of being tested experimentally, and of being falsified, and is therefore 'non-scientific' in these terms. Nevertheless, Panel 3.4 draws attention to the difficulties of falsification in experimentation involving 'open' environmental systems, where the interaction of multiple uncontrolled variables leads to difficulties in explaining observations at variance with hypotheses. Furthermore, the experimental method may generate regularities that are no more than the product of the closure created by the experiment itself, so that a truly critical, rational and realist geomorphology (Richards, 1990) must recognize the need to interpret and understand the nature of the mechanisms generating observable regularities, and not to assume that those mechanisms are necessarily identifiable through conventional experimental activity. Advocacy of purposive sampling in geomorphology (Pitty, 1971) provides one aspect of realist experimentation (Pawson and Tilley, 1997) where the scientific method is designed to produce the outcomes which may be postulated a priori as evidence of causal connections; abductive reasoning (Inkpen and Wilson, 2009) provides another; and concentration on 'thick description' in case-intensive research (Richards et al., 1997) provides a third.

The concept of spatially distributed feedback provides an essential link between the long-term and short-term study of geomorphological problems. To separate timescales, as do Schumm and Lichty (1965), is arbitrary and artificial; the past obviously influences the present at all scales. It does so because inherited landforms influence the nature and distribution of present processes. For example, clay slopes in southern Britain often have very low-gradient shear surfaces which reflect periglacial mass movement on those slopes in the

Late Glacial. Present rates of processes such as soil creep on these slopes are constrained by the inherited low angles. Ashworth and Ferguson (1986) illustrate measurements of bed velocity and current direction in a braided river reach, the bed topography of which at time t_0 determines the pattern of these flow properties as a specific discharge enters at the head of the reach, and this in turn determines the transport of bed material from point to point. Locally, continuity of bedload transport then dictates whether erosion or deposition occurs at each point, and these processes then change the topography. The discharge (and input bed material) entering the reach at time t_1 then encounters a new form, and the spatial pattern of velocity and bed load transport is different. The reach may be of constant average width, and therefore in one sense at equilibrium. However, within that width its topography is continually evolving because of the feedback (spatially distributed) between form and process, as summarized in Figure 3.7. This model of inheritance of form controlling the subsequent process distribution has also been used by Williams (1987) in a discussion of the evolution of Chinese tower karst landforms. It is also central to the account by Owen et al. (1998) of the changing style of glaciation and glacial deposition in Himalayan valleys in successive glacial periods; as the valleys deepen, so successive glacial advances take different forms as glacial processes respond differently to reflect the changing topography. These examples show that the key to geomorphological explanation is the recognition that landscape evolution is a recursive process.

This conclusion gives the lie to arguments that have been made at various times about the failure of geomorphology to integrate its process-oriented and historical dimensions; Douglas made this case in 1982, and it was repeated by Smith et al. in 2002. In fact, the research agenda to effect this integration has been in progress for nearly half a century, and involves the development of models that capture form–process feedbacks. In 1971, Kirkby developed a mathematical model which could be used to explore how characteristic slope forms evolved in response to particular surface process regimes (soil creep producing convex slopes, wash and rill erosion producing concave slopes). Ahnert (1976) generalized this to three (spatial) dimensions, and was able to apply his SLOP3D landform evolution model to simulate the Quaternary evolution of slopes in the northern Eifel region in Germany (Ahnert, 1987). These models embodied understanding about process, but the spatial variation of process rate depended on local gradient and the convergence/divergence of transport paths (represented by the upslope area draining to a specific point; see Figure 3.1). At any

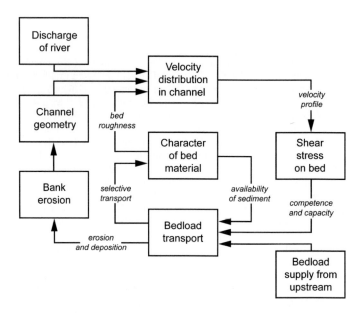

Figure 3.7 The feedback between form and process that results in continual evolution of channel form (based on Ashworth and Ferguson, 1986; and Richards, 1988)

time step, therefore, the topography presents a boundary condition that determines how the process distribution effects further change; form and process are entwined. The latest incarnations of this tradition are perhaps the CHILD model (Tucker, 2001) and CAESAR (Coulthard et al., 2002) in which simplified representations of the erosion process combine with a capability to simulate deposition, so that although valleys may incise and slopes retreat, valley fill sediment accumulation can also be modelled, together with the cut and fill processes that create river terraces. While issues remain concerning the formulation of the 'laws' that represent fluvial and hillslope processes, as well as the implementation of the various algorithms on numerical grids, model parameterization and calibration, and model testing (Codilean et al., 2006), this is the modelling framework required to integrate process-based geomorphology (with its short-term, small-scale focus on equilibrium and functional and dynamic behaviour) with historical geomorphology (with its long-term, large-scale emphasis on historical change associated with Tertiary and Quaternary climatic change and tectonism). It is also the modelling framework required to connect the chronology of complex, threshold-related, quasi-oscillatory behaviour of terrestrial landscapes to the continuous, smoothed record of global climatic change provided by the oxygen isotope record and recorded in the smoothed, deep-ocean

sedimentary record. Such modelling may also represent the most productive way to engage geomorphological enquiry with the wider project of earth system science.

Panel 3.5 therefore demonstrates the ultimate unity of geomorphology. It is concerned with explaining the nature of landforms, which it can seek to achieve using something comparable to the covering law model. This demands, first, general 'laws' concerned with the functional nature of landforms, the immanent properties of earth surface processes, and the adjustment of form to process. As suggested above, it may be difficult to identify those 'laws' by conventional experimental activity in the natural, uncontrolled environment, but in any event what is required is an understanding of the nature of earth surface processes, rather than a set of empirical regularities. The second requirement is detailed geographical and geological configurational information, including a history of process variation and landform development. The history can only be constructed, however, if process understanding is brought to bear on past erosional and depositional features: if the present is the key to the past via methodological uniformitarianism. Furthermore, inherited forms also influence the types, intensities and spatial distributions of present processes, so that the past must also be regarded as a key to the present. Ultimately, therefore, 'evolutionary' and 'equilibrium'

Panel 3.5 The structure of integrated geomorphological explanation

This summary diagram suggests that geomorphological explanations require both functional and historical elements. The former provides general understanding of present process and the basis for reconstructing the past; the latter provides the specific initial conditions which contextualize this understanding (see text).

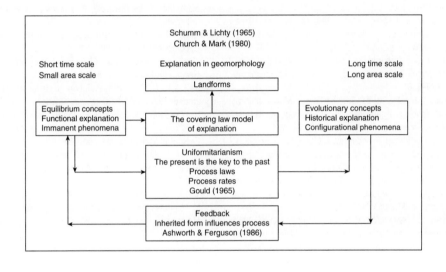

geomorphology each make their contributions to the overall goal, which is the explanation of the nature, origin and development of landforms.

REFERENCES

Ager, D. (1995) *The New Catastrophism*. Cambridge University Press, Cambridge, p. 252.

Ahnert, F. (1976) Brief description of a comprehensive three-dimensional process–response model of landform development. *Zeitschrift fur Geomorphologie, Supplementband* 25: 29–49.

Ahnert, F. (1987) Approaches to dynamic equilibrium in theoretical simulations of slope development. *Earth Surface Processes and Landforms* 12: 3–15.

Ahnert, F. (1998) *Introduction to Geomorphology*. Arnold, London, p. 352.

Ashworth, P.J. and Ferguson, R.I. (1986) Interrelationships of channel processes, changes and sediments in a proglacial braided river. *Geografiska Annaler, Series A* 68: 361–371.

Baker, V.R. (1983) Large-scale fluvial palaeohydrology, in K.J. Gregory (ed.). *Background to Palaeohydrology*. John Wiley & Sons, Chichester, pp. 453–478.

Battiau-Queney, Y. (1989) Constraints from deep crustal structure on long-term landform development of the British

Isles and Eastern United States, in T.W. Gardner and W.D. Sevon (eds). *Appalachian Geomorphology*. Elsevier, Amsterdam, pp. 53–70.

Bauer, B.O. (1996) Geomorphology, geography and science, in B.L. Rhoads and C.E. Thorn (eds). *The Scientific Nature of Geomorphology: Proceedings of the 27th Binghamton Symposium in Geomorphology (27–29 September 1996)*. John Wiley & Sons, New York, pp. 381–413.

Beven, K.J. and Kirkby, M.J. (1979) A physically-based variable contributing area model of basin hydrology. *Hydrological Sciences Bulletin* 24: 43–69.

Bishop, P. (1980) Popper's principle of falsifiability and the irrefutability of the Davisian cycle. *Professional Geographer* 32: 310–315.

Chorley, R.J., Schumm, S.A. and Sugden, D. (1984) *Geomorphology*. Methuen, London, p. 606.

Chorley, R.J. (1957) Climate and morphometry. *Journal of Geology* 65: 628–638.

Chorley, R.J. (1962) *Geomorphology and General Systems Theory*. United States Geological Survey, Professional Paper 500-B.

Chorley, R.J. (1978) Bases for theory in geomorphology, in C. Embleton, D. Brunsden and D.K.C. Jones (eds). *Geomorphology: Present Problems and Future Prospects*. Oxford University Press, Oxford, pp. 1–13.

Chorley, R.J. and Kennedy, B.A. (1973) *Physical Geography: A Systems Approach*. Prentice-Hall, London.

Chorley, R.J., Dunn, A.L. and Beckinsale, R.P. (1964) *The History of the Study of Landforms, or the Development of Geomorphology. Vol. 1, Geomorphology Before Davis* (esp. Part IV, The Western Explorations, pp. 467–648). Methuen, London.

Chorley, R.J., Beckinsale, R.P. and Dunn, A.L. (1973) *The History of the Study of Landforms, or the Development of Geomorphology. Vol. 2, The Life and Work of William Morris Davis.* Methuen, London.

Church, M.A. and Mark, D.M. (1980) On size and scale in geomorphology. *Progress in Physical Geography* 4: 324–391.

Clifford, N. and Richards, K. (2005) Earth system science: an oxymoron? *Earth Surface Processes and Landforms* 30: 379–383.

Cockburn, H.A.P. and Summerfield, M.A. (2004) Geomorphological applications of cosmogenic isotope analysis. *Progress in Physical Geography* 28: 1–42.

Codilean, A.T., Bishop, P. and Hoey, T. (2006) Surface process models and the links between tectonics and topography. *Progress in Physical Geography* 30: 307–333.

Costa, J.E. and Cleaves, E.T. (1984) The Piedmont landscape of Maryland: a new look at an old problem. *Earth Surface Processes and Landforms* 9: 59–74.

Coulthard, T.J., Macklin, M.G. and Kirkby, M.J. (2002) Simulating upland river catchment and alluvial fan evolution. *Earth Surface Processes and Landforms* 27: 269–288.

Culling, W.E.H. (1988) A new view of the landscape. *Transactions of the Institute of British Geographers* NS 13: 345–360.

Davis, W.M. (1899) The geographical cycle. *Geographical Journal* 14: 481–504.

Douglas, I. (1982) The unfulfilled promise: Earth surface processes as a key to landform evolution. *Earth·Surface Processes and Landforms* 7: 101.

Dury, G.H. (1983) Geography and geomorphology: the last 50 years. *Transactions of the Institute of British Geographers* NS 8: 90–99.

Evans, I.S. (1972) General geomorphometry, derivatives of altitude and descriptive statistics, in R.J. Chorley (ed.), *Spatial Analysis in Geomorphology.* Methuen, London, pp. 17–90.

Gardner, T.W. and Sevon, W.D. (1989) *Appalachian Geomorphology.* Amsterdam, Elsevier, p. 318.

Gilbert, G.K. (1877) *Report on the Geology of the Henry Mountains.* United States Geographical and Geological Survey of the Rocky Mountain Region, p. 160.

Gilbert, G.K. (1885) *The Topographic Features of Lake Shores.* United States Geological Survey, 5th Annual Report 1883–84, pp. 69–123.

Gilbert, G.K. (1890) *Lake Bonneville.* United States Geological Survey Monograph 1, p. 438.

Gilbert, G.K. (1893) The moon's face: a study of the origin of its features. *Philosophical Society of Washington, Bulletin* 12: 241–292.

Gilbert, G.K. (1900) Rhythms and geologic time. *American Association for the Advancement of Science, Proceedings* 49: 1–19.

Gilbert, G.K. (1914) *The Transportation of Debris by Running Water.* United States Geological Survey, Professional Paper 86, p. 263.

Goudie, A.S. (ed.) (1990) *Geomorphological Techniques,* 2nd edn. Unwin Hyman, London, p. 570.

Gould, S.J. (1965) Is uniformitarianism necessary? *American Journal of Science* 263: 223–228.

Gould, S.J. (1987) *Time's Arrow, Time's Cycle – Myth and Metaphor in the Discovery of Geological Time.* Harvard University Press, Cambridge, MA, p. 222.

Gregory, K.J. (1978) A physical geography equation. *National Geographer* 12: 137–141.

Gregory, K.J. (2000) *The Changing Nature of Physical Geography.* Edward Arnold, London, p. 368.

Grove, A.T. (2008) The revolution in palaeoclimatology around 1970, in T.P. Burt, R.J. Chorley, D. Brunsden, N.J. Cox, and A.S. Goudie (eds). *The History of the Study of Landforms or the Development of Landforms.* The Geological Society, London, pp. 961–1004.

Hack, J.T. (1960) Interpretation of erosional topography in humid temperate regions. *American Journal of Science* 258-A: 80–97.

Hack, J.T. (1973) Drainage adjustment in the Appalachians, in M. Morisawa (ed.), *Fluvial Geomorphology.* State University of New York, Binghamton, NY, pp. 51–69.

Haines-Young, R. and Petch, J. (1986) *Physical Geography: Its Nature and Method.* Harper and Row, London, p. 230.

Harrison, S.J. (2001) On reductionism and emergence in geomorphology. *Transactions of the Institute of British Geographers* NS 26: 327–339.

Horton, R.E. (1945) Erosional development of streams and their drainage basins: hydrophysical approach to quantitative morphology. *Geological Society of America Bulletin* 56: 275–370.

Humphrey, N.F. and Heller, P.L. (1995) Natural oscillations in coupled geomorphic systems: an alternative origin for cyclic sedimentation. *Geology* 23: 499–502.

Hurtrez, J.E., Sol, C. and Lacazeau, F. (1999) Effect of drainage area on hypsometry from an analysis of small-scale drainage basins in the Siwalik Hills (Central Nepal). *Earth Surface Processes and Landforms* 24: 799–808.

Inkpen, R. and Wilson, G. (2009) Explaining the past: abductive and Bayesian reasoning. *The Holocene* 19: 329–334.

Jennings, J.N. (1973) 'Any milleniums today, Lady?' The geomorphic bandwaggon parade. *Australian Geographical Studies* 11: 115–133.

Johnson, D.W. (1931) *Stream Sculpture on the Atlantic Slope: A Study in the Evolution of Appalachian Rivers.* Columbia University Press, New York, p. 142.

King, P.B. and Schumm, S.A. (1980) *The Physical Geography (Geomorphology) of William Morris Davis.* Geo Books, Norwich, p. 174.

Kirkby, M.J. (1971) Hillslope process–response models based on the continuity equation, in D. Brunsden (ed.), *Slopes: Form and Process.* Institute of British Geography Special Publications, vol. 3, pp. 15–30.

Kuhn, T.S. (1970) *The Structure of Scientific Revolutions*, 2nd edn. University of Chicago Press, Chicago.

Lane, S.N., Richards, K.S. and Chandler, J. (1998) *Landform Monitoring, Modelling and Analysis*. Wiley & Sons, Chichester, p. 452.

Lane, S.N., Brookes, C.J., Kirkby, A.J. and Holden, J. (2004) A network-index based version of TOPMODEL for use with high-resolution digital topographic data. *Hydrological Processes* 18: 191–201.

Lewis, W.V. (1944) Stream trough experiments and terrace formation. *Geological Magazine* 81: 241–253.

Lowe, J.J. and Walker, M.J.C. (1984) *Reconstructing Quaternary Environments*. Longman, London, p. 389.

McMaster, K.J. (2002) Effects of digital elevation modle resolution on derived stream network positions. *Water Resources Research* 38: 13-1–13-8.

Melhorn, W.N. and Flemal, R.C. (1975) *Theories of Landform Development*. George Allen & Unwin, London, p. 306.

Melton, M.A. (1958) Geometric properties of mature drainage systems and their representation in an E_4 phase space. *The Journal of Geology* 66: 35–54.

Meyerhoff, H.A. and Olmstead, E.W. (1936) The origins of Appalachian drainage. *American Journal of Science* 232: 21–46.

Montgomery, K. (1989) Concepts of equilibrium and evolution in geomorphology: the model of branch systems. *Progress in Physical Geography* 13: 47–66.

Montgomery, D.R. and Dietrich, W.E. (1994) A physically-based model for the topographic control on shallow landsliding. *Water Resources Research* 30: 1153–1171.

Moore, I.D., O'Loughlin, E.M. and Burch, G.J. (1988) A contour-based topographic model for hydrological and ecological applications. *Earth Surface Processes and Landforms* 13: 305–320.

Nadler, C.T. and Schumm, S.A. (1981) Metamorphosis of the South Platte and Arkansas Rivers, eastern Colorado. *Physical Geography* 2, 95–115.

O'Neill, R.V., DeAngelis, D.L., Waide, J.B. and Allen, T.F.H. (1986) *A Hierarchical Concept of Ecosystems*. Princeton University Press, Princeton.

Owen, L.A., Derbyshire, E.A. and Fort, M. (1998) The Quaternary glacial history of the Himalaya. *Quaternary Proceedings* 6: 91–120.

Paine, A.D.M. (1985) 'Ergodic' reasoning in geomorphology: time for a review of the term? *Progress in Physical Geography* 9: 1–15.

Pawson, R. and Tilley, N. (1997) *Realistic Evaluation*. Sage, London.

Peltier, L.C. (1950) The geographical cycle in periglacial regions as it is related to climatic geomorphology. *Annals Association of American Geographers* 40: 214–236.

Phillips, J.D. (1999) *Earth Surface Systems: Complexity, Order, and Scale*. Blackwell Publishers, Oxford.

Pitty, H.F. (1971) *Introduction to Geomorphology*. Methuen, London.

Pitty, A.F. (1982) *TheNature of Geomorphology*. Methuen, London, p. 161.

Poag, C.W. and Sevon, W.D. (1989) A record of Appalachian denudation in postdrift Mesozoic and Cenozoic sedimentary deposits of the US Middle Atlantic continental margin, in T.W. Gardner and W.D. Sevon (eds), *Appalachian Geomorphology*. Elsevier, Amsterdam, 119–157.

Popper, K. (1959) *The Logic of Scientific Discovery*. Routledge, London.

Pye, K., Goudie, A.S. and Thomas, D.S.G. (1984) A test of petrological control in the development of bornhardts and koppies on the Matopos Batholith, Zimbabwe. *Earth Surface Processes and Landforms* 9: 455–467.

Pye, K., Goudie, A.S. and Watson, A. (1986) Petrological influence on differential weathering and inselberg development in the Kora area of central Kenya. *Earth Surface Processes and Landforms* 11: 41–52.

Rhoads, B.L. (1999) Beyond pragmatism: The value of philosophical discourse for physical geography. *Annals of the Association of American Geographers* 89: 760–771.

Rhoads, B.L. and Thorn, C.E. (1996) Toward a philosophy of geomorphology, in B.L. Rhoads and C.E. Thorn (eds). *The Scientific Nature of Geomorphology: Proceedings of the 27th Binghamton Symposium in Geomorphology (27–29 September 1996)*. Wiley & Sons, New York, pp. 115–143.

Rice, J. (1977) *Introduction to Geomorphology*. Routledge, London, p. 387.

Richards, K. S. (1979) Prediction of drainage density from surrogate measures. *Water Resources Research* 15, 435–42.

Richards, K.S. (1990) 'Real' geomorphology. *Earth Surface Processes and Landforms* 15: 195–197.

Richards, K.S. and Clifford, N.J. (2008) Science, systems and geomorphologies: why LESS may be more. *Earth Surface Processes and Landforms* 33: 1323–1340.

Richards, K.S., Brookes, S.M., Clifford, N., Harris, T. and Lane, S. (1997) Theory, measurement and testing in 'real' geomorphology and physical geography, in D.R. Stoddart, (ed.), *Process and Form in Geomorphology*. Routledge, London, pp. 265–292.

Ritter, D.F. (1988) Landscape analysis and the search for geomorphic unity. *Geological Society of America Bulletin* 100: 160–171.

Sack, D. (1992) New wine in old bottles: the historiography of a paradigm change. *Geomorphology* 5: 251–263.

Savigear, R.A.G. (1962) Some observations on slope development in north Devon and north Cornwall. *Transactions and Papers (Institute of British Geographers)* 31: 23–42.

Schumm, S.A. (1979) Geomorphic thresholds: the concept and its applications. *Transactions Institute of British Geographers* NS 4: 485–515.

Schumm, S.A. and Lichty, R.W. (1965) Time, space and causality in geomorphology. *American Journal of Science* 263: 110–119.

Selby, M.J. (1982) Rock mass strength and the form of some inselbergs in the central Namib desert. *Earth Surface Processes and Landforms* 7: 489–497.

Shackleton, N.J. (2000) The 100,000-year Ice-Age cycle identified and found to lag temperature, carbon dioxide, and orbital eccentricity. *Science* 289, 1897–1902.

Sherman, D. (1996) Fashion in geomorphology, in B.L. Rhoads and C.E. Thorn (eds), *The Scientific Nature of Geomorphology: Proceedings of the 27th Binghamton Symposium in Geomorphology* (*27–29 September 1996*). Wiley & Sons, New York.

Simpson, G.G. (1963) Historical science, in C.C. Albritton (ed.), *The Fabric of Geology*. Freeman, San Francisco, pp. 24–48.

Slaymaker, O. (1972) Patterns of present subaerial erosion and landforms in mid-Wales. *Transactions of the Institute of British Geographers* 55: 47–68.

Slaymaker, O. (1987) The process geomorphology of British Columbia Pacific Ranges (coast mountains) – spatial scale consideration, in A. Godard and A. Rapp (eds). *Processus et Mesure de L'erosion*. CNRS, Paris, pp. 23–32.

Smith, B.J., Warke, P.A. and Whalley, W.B. (2002) Landscape development, collective amnesia and the need for integration in geomorphological research. *Area* 34: 409–418.

Sparks, B.W. (1971) *Rocks and Relief*. Longman, London, p. 404.

Stallins, J.A. (2006) Geomorphology and ecology: Unifying themes for complex systems in biogeomorphology. *Geomorphology* 77: 207–216.

Stoddart, D.R. (1966) Darwin's impact on geography. *Annals of the Association of American Geographers* 56: 683–698.

Strahler, A.N. (1952) The dynamic basis of geomorphology. *Geological Society of America Bulletin* 63, 923–938.

Summerfield, M.A. (2005) A tale of two scales, or the two geomorphologies. *Transactions of the Institute of British Geographers* NS 30: 402–415.

Tricart, J. and Cailleux, A. (1972) *Introduction to Climatic Geomorphology*. Longman, London, p. 295.

Tucker, G.E., Lancaster, S.T., Gasparini, N.M. and Bras, R.L. (2001) The channel–hillslope integrated landscape development (CHILD) model, in R.S. Harmon, and W.W. Doe III (eds), *Landscape Erosion and Evolution Modeling*. Kluwer Academic Plenum Publishers, New York pp. 349–388.

Twidale, C.R. and Bourne, J.A. (2009) On the origin of A-tents (pop-ups), sheet structures, and associated forms. *Progress in Physical Geography* 33: 1–16.

Van der Meer, J., Menzies, J. and Rose, J. (2003) Subglacial till: the deforming glacier bed. *Quaternary Science Reviews* 22: 1659–1685.

Williams, P.W. (1987) Geomorphic inheritance and the development of tower karst. *Earth Surface Processes and Landforms* 12: 453–466.

Yang, D., Hereth, S. and Musiake, K. (2001) Spatial resolution sensitivity of catchment geomorphologicl properties and the effect on hydrological simulation. *Hydrological Processes* 15: 2085–2099.

Yatsu, E. (1992) To make geomorphology more scientific. *Transactions Japanese Geomorphological Union* 13: 87–124.

Yochelson, E.L. (ed) (1980) *The Scientific Ideal of G.K. Gilbert*. US Geological Survey Special Paper 183, 129–142.

4

The Role and Character of Theory in Geomorphology

Bruce L. Rhoads and Colin E. Thorn

In the earth sciences, …, the most notable advances are almost invariably associated with the construction of a theoretical model which, in a particularly symmetrical and harmonious manner, seems to embrace a large part of reality… the establishment of an appropriate theoretical model stimulates the imagination to such an extent that the scope of what is looked upon as "observed reality" is so enlarged that a new wealth of factual information becomes readily available.

(Chorley, 1963: 953)

The only true prisoners of theory are those that do not know it.

(Chorley, 1978: 1)

In a brief survey of a topic as profound and pervasive as theory, it is necessary to be expedient. Our review is constrained in a number of ways, it is: (1) largely chronological in sequence; (2) overwhelmingly English-language in origin; (3) emphasizes developments in fluvial geomorphology; and (4) focuses primarily on ideas rather than the people involved in theory development. These constraints must be recognized, but it seems likely that given the centrality of fluvial geomorphology to the body of geomorphological theory, it may be possible to extend ideas expressed here to other areas of the field. We leave it for others with appropriate expertise to do so. Also, although interpersonal relations, such as student-mentor associations and research groups, are important in theory development (Kennedy, 2006), detailing such relations is not central to the task at hand. The purposes of the chapter are to lay out a framework from philosophy of science, the model–theoretic view, that we believe captures the essence of how theory is viewed (albeit often implicitly) in geomorphology and to use a review of theoretical developments since the mid 1850s to illustrate this argument.

THE NATURE OF THEORY

The word theory is commonly used in two ways. First, as in the sequence: conjecture, hypothesis, theory where increasing surety is implied. Second, as in the expression 'a scientific theory' where a set of linked ideas is intended to provide a general, explanatory framework for knowledge spanning a significant portion of a science, or perhaps its entirety. We will ignore the first use, justifiable though it be, and consider exclusively the second one. Theory is the superstructure of science. Science without theory would be shorn of its explanatory and predictive capacities. It is these two great attributes of science that provide understanding of the world around us and allow knowledge to be used in reliable ways.

Scientists, including geomorphologists, have not devoted much attention to the question: what exactly is a scientific theory? Instead, they have simply gone about the business of developing and applying theories in scientific practice. The study of scientific theories has largely been conducted by philosophers of science. Until recently, the Received View, developed by the logical positivists during the first half of the 20th century, was the most prominent perspective on the structure of scientific theories, including geomorphological theories (Harvey, 1969; Thornes, 1978). According to this perspective, theories are axiomatic systems consisting of observation statements (statements confirmed by empirical testing) linked through correspondence rules (explicit or partial definitions) to theoretical statements (statements about abstract unobservable entities) (Brown, 1977; Suppe, 1977) (Figure 4.1). In other words, theories are formal logico-linguistic structures. *Representation* is often viewed as a key function of scientific theories and in the Received View representation occurs through a connection between statements and the world. Observation statements are compared against data via empirical testing to determine their validity. Verified observation statements then provide empirical support for the theoretical statements through the linking correspondence rules (Figure 4.1). Through this process, a theory, as a set of statements, becomes connected representationally to relevant aspects of real-world phenomena.

Logical positivism, along with its epistemic heart – the Received View of scientific theories – was severely criticized during the 1960s and is now viewed as a deeply flawed account of science and of scientific theories. Over the past 40 years,

an alternative perspective on scientific theories known as the model–theoretic view (MTV) has been developed to try to account for the role of theories in scientific practice (Giere, 1988; Rhoads and Thorn, 1996a). In particular, MTV emphasizes the important representational role that models play in the application of theoretical principles to scientific problems (Da Costa and French, 2000; Suppe, 2000; Bailer-Jones, 2002). In MTV, models, rather than statements, become the primary constituents of theory structure. Models can be used to connect theoretical principles to testable hypotheses about real-world phenomena (theoretical models) and also to provide data-based representations of these phenomena (data models). The comparison of hypotheses derived from theoretical models with evidence embedded in data models serves as the basis for empirical testing (Figure 4.1).

Theoretical models are discrete specifications of general theoretical statements or principles. In this sense, such models embody theoretical principles and stand in an unproblematic truth relation to these principles. Because many different discrete specifications of general theoretical principles are possible, an interconnected set, or family, of theoretical models can be developed from these principles (Figure 4.1). The formulation of theoretical models should not be viewed as a product

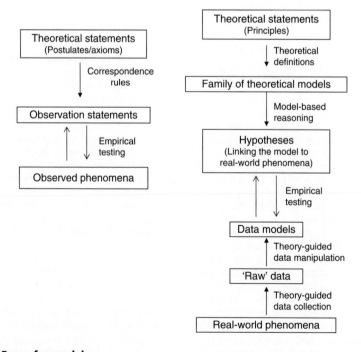

Figure 4.1   Bases for models

of simple deduction; constructing such models is often a complex process that cannot be reduced to formal logic and involves analogic and heuristic reasoning (Cartwright, 1999; Giere, 2004).

The primary purpose of theoretical models is not merely to embody theoretical principles, but to *represent* real-world phenomena of scientific interest (Giere, 2004). In other words, scientists use models as representational tools to facilitate intellectual access to phenomena. The method of representation is varied, including mathematical equations, diagrams, pictorial displays, images, descriptions and physical artifacts. Although a model itself commonly is not cast in linguistic form, through its representational nature it entails propositions, or hypotheses, about the phenomena it is meant to represent (Bailer-Jones, 2003). These hypotheses provide the basis for empirical testing of the extent to which the model, as a representative device, denotes salient aspects of the phenomena of interest (Giere, 1988).

Data models serve as the other major representational device in science. Like theoretical models, these models also are meant to represent phenomena, but in this case the representation derives from information collected about the phenomena, rather than from abstract theoretical statements or principles. Scientists rarely use 'raw' data in empirical testing. Instead, raw data are subjected to various manipulation procedures to produce processed data, or data models (Harris, 2003). The processing of data is theory-guided and conducted with a theoretical objective in mind. Often the theoretical objective is defined by theoretical hypotheses derived from a theoretical model. Thus, a data model can provide the evidential basis for evaluations of theoretical hypotheses. In other cases, a data model may be relatively autonomous in the sense that, while it is valued for its capacity to represent phenomena of interest, it is not yet entirely clear how the model connects to theoretical principles (Da Costa and French, 2000). However, over time, scientists usually work diligently at connecting such a data model to established theoretical principles through the development of suitable theoretical models, thereby diminishing the conceptual autonomy of the data model.

What then according to MTV is a scientific theory? In contrast to the Received View, MTV suggests that a scientific theory is not a well-defined entity that can be reduced to a formal system of statements. Generally, scientific theories can be conceived as theoretical principles and a relevant family (group) of theoretical models that embody these principles (Giere, 1988). Data models also are important, however, and in many cases the development of theory proceeds not from the top down (theoretical models formulated from theoretical principles), but from the bottom

up (the development of relatively autonomous data models that accurately represent real-word phenomena leads to sustained scientific efforts to link these models to theoretical principles and models). In this sense, models and theories are intimately intertwined at all levels of scientific practice.

How does the model–theoretic view of theory relate to geomorphology? One purpose of this chapter is to explore this question. What emerges is that MTV seems to be at least generally consistent both with past and present perspectives on theory in geomorphology, even though these perspectives have been implicit rather than explicit.

THEORY IN GEOMORPHOLOGY

While MTV is the product of a concerted, and we believe compelling, disciplinary effort by philosophers of science to create a generic model of scientific research, we must not ignore the efforts within geomorphology itself to assess its own theoretical frameworks. An important pioneering attempt to ground physical geography firmly in a coherent philosophy of science is that of Haines-Young and Petch (1986) who invoked critical rationalism. While this work has been undermined somewhat by critiques of critical rationalism, the aim of Haines-Young and Petch (1986) remains noteworthy. Three other examples are more useful in the present context: Chorley (1978); Rhoads and Thorn (1993) and Church (1996). These efforts are more focused than MTV, but help to illuminate the role of theory in geomorphology. Chorley (1978, 2) enumerated seven main categories of explanation, the most relevant of which are teleological, historical, functional and realist, and discussed the development of geomorphic theory within these categories. Rhoads and Thorn (1993) examined how the character of geomorphological explanations varies with scale. As temporal and spatial scales increase, theory becomes increasingly conjectural as information density providing the basis for theory decreases and as complications imposed by historical contingencies increase. Testing of theories over geological time scales often is based on limited and sometimes equivocal evidence. Quantification is less likely because the information required to meaningfully calibrate or validate mathematical and statistical models is generally unavailable. Thus, qualitative, conceptual models are prevalent when looking into the past over geological time scales. Church (1996) undertook a similar task recognizing four distinctive modes of theoretical development: (1) small scales that are approached statistically;

(2) the scale of classical mechanics where deterministic theories are sustainable; (3) large scales where the system is viewed deterministically, but contingent endogenous effects cannot be predicted (now approached using nonlinear dynamic models); and, finally, (4) the largest spatial and temporal scales where narrative, particularistic models are invoked. As Church (1996: 153) notes 'Our theoretical construction of order in nature is bound by the tyranny of the scales.' However, it is important to appreciate that the four scales of theoretical development earmarked by Church are not truly tied to absolute time or space scales, rather, similar to the point emphasized by Rhoads and Thorn (1993), the approaches are constrained by the resolution of information. Finally, while it is important that the concepts and findings among modes of theoretical development be consistent, it does not follow that products generated by one mode will be derivable from other modes.

THE NATURE OF GEOMORPHOLOGY

At root geomorphology is the scientific investigation of the surface forms and processes of the terrestrial portion of earth; while submarine features may be included, as may other planets, those undertaking such research may not necessarily consider themselves geomorphologists (Rhoads and Thorn, 1996a). Like many disciplines geomorphology has a readily defined core or *domain*, but the periphery is uncertain. Our concern in this chapter is with central issues. Consequently, we address theory in geomorphology as it has sought to explain sub-aerial landforms, landscapes (assemblages of landforms), and the processes responsible for their initiation, development and destruction. Theory defined in this broad sense is relevant not only to basic knowledge of geomorphological systems, but also to efforts to understand human impacts on these systems, including the effects of pervasive global environmental change (e.g. Slaymaker et al., 2009).

Geomorphologists pursue two primary goals: first, exploration of general principles that explain geomorphological processes, the ensuing landforms, and their interaction; and, second, explanation of the unique histories of individual landforms and landscapes. The salient component of the latter exercise missing from the former is an effective treatment of 'contingency' (Gould, 1989) – in essence the unique history or sequence of events that actually happened, but did not have to happen. Identification of these two important threads (e.g. Thornes, 1983) suggests a dichotomy in the discipline, and unfortunately there is some practical

truth to this observation (Rhoads and Thorn, 1996a). However, we believe the two issues to be merely two sides of the same coin, and would follow Schumm (1977: 10) in his observation about the fluvial system, '…it is possible to view the [fluvial] system either as a physical system or a historical system. In actuality it is a physical system with a history.' Such an integrated view complicates theoretical needs, but clearly captures reality.

THE EARLY DEVELOPMENT OF BASIC GEOMORPHOLOGICAL PRINCIPLES

During the 19th century many concepts emerged in geology that laid the foundation for a formal science of geomorphology. These concepts can be thought of as 'regulative principles', defined by von Engelhardt and Zimmermann (1988) as a hierarchical set of guidelines that are not testable scientific statements, but instead guide the development of models and testable hypotheses. In this sense, they are roughly analogous to the theoretical principles of the MTV, and in the earth sciences often permit researchers to grapple with the large range of time and space scales under consideration. Developments in geology that influenced geomorphology appear in Chapter 2 of this volume and are the subject of a seminal treatise by Chorley et al. (1964, 1973) and Beckinsale and Chorley (1991), and a much briefer, but highly informative, volume by Tinkler (1985). Here our strategy is simply to present a set of thumbnail sketches of salient mid 19th century issues impinging upon the embryonic discipline of geomorphology.

Uniformitarianism

The victory of uniformitarianism over catastrophism after a convoluted and protracted debate is clearly a huge theoretical shift in the earth sciences that took much of the 1800s to be achieved. Central to the conflict were the opposing notions of diluvialism (the creation of landforms and landscapes initially by the Mosaic flood or Noachian deluge, and subsequently, by a series of floods) and fluvialism (the creation of valleys by long-term fluvial action). Much of the debate was carried out within a geologic framework that lacked a Quaternary period and the associated concept of glaciation (Tinkler, 1985). A major contribution to resolution of the diluvial versus fluvial conflict was the increasing acceptance of glaciation after about 1840, but again not without conflict. For example, Lyell at one point accepted glaciation in full-blown form, only to end his days believing only in mountain glaciation (Tinkler, 1985). Eventually glaciation represented a more plausible 'catastrophe', than regional, or indeed global, flooding.

Marine planation

Lacking unequivocal consensus on either fluvial erosion or even the existence of glaciation it is hardly surprising that marine erosion, as opposed to any form of sub-aerial erosion, was initially widely invoked as a/the primary agent of regional scale or larger erosion as uniformitarian concepts first encroached upon catastrophism. Lyell (1830) favored marine erosion of landmasses during gradual emergence, while Ramsay (1846) invoked planation during submergence and marine transgressions. Both versions were uniformitarian in approach and Ramsay's laid the foundation for a concept that lasted nearly a hundred years in British geomorphology (Tinkler, 1985).

Glaciation

Establishment of glaciation is one of the great benchmarks in the history of earth sciences. Accepted by Hutton and Playfair, it was essentially lost for several decades, sputtered when reintroduced, but became firmly established when Agassiz (1840) became its champion. Agassiz in many ways usurped the glory due his mentor, Charpentier, and certainly was guilty of exaggerating the scope of ice sheets. Resistance to his ideas was widespread in Great Britain where the presence of marine shells in drift at quite high elevations provided a genuine intellectual challenge. However, Ramsay (1862) extended the scope of glaciers from transport to erosion, while Geikie (1874) established the 'The Great Ice Age' and the concept of an interglacial. In general, increased extent of alpine glaciation in the past received more ready acceptance than did continental glaciation in presently ice-free regions.

Diastrophism

As the 19th century ended, the sea, in the form of eustasy, retained a central role in constraining geomorphological thought. Eustasy is most closely identified with Suess and his text *Das Autlitz der Erde* (1883–1908). Suess promoted the view that the continents, excepting their margins, are essentially stable and changes in ocean basin capacity produce globally synchronous changes in sea level. Lengthy periods of rising sea level were associated with sediment infilling and short periods of sea level drop were believed to be produced by abrupt sea bottom subsidence. Chorley (1963) flags Chamberlain's (1909) paper as the apogee of the eustatic theory. The fundamental theoretical significance of Suess' concept was that global correlation of sea levels, and hence of many landscapes, was feasible.

In reviewing this period of geomorphology, Chorley (1963) emphasizes the preoccupation at the time with diastrophism – the relative or absolute changes in the position, level or altitude of rocks forming Earth's crust (Chorley et al., 1984), while Beckinsale and Beckinsale (1989) view the issues somewhat more broadly. Diastrophism incorporates more specific concepts such as orogenic, epeirogenic, isostatic, igneous and eustatic changes; two of which isostasy (Dutton, 1889) and epeirogeny (Gilbert, 1890) emerged in quick succession. Davis' (1889) initial version of the *cycle of erosion* is synchronous with these other momentous concepts, portions of which were initially invoked in support of eustasy, although Davis did not support the concept (Chorley, 1963). Eventually, these same concepts were reinterpreted and provided the evidence for the demise of Suess' global model. Nothing could point to the importance of theory more convincingly than reinterpretation of the same set of observations to both support and undermine an idea.

Basic fluvial principles

The exploration of the American West, with its dramatic exposed landscapes, yielded several basic principles related to the action of running water on the landscape. Observations of canyon systems led John Wesley Powell to propose the concept of base level, which refers to elevational control on erosion of a stream into a landscape (Ritter, 1978). Sea level represents the base level for large rivers flowing into oceans, whereas the elevation at which a tributary joins a main river is the base level for the tributary. Temporary base levels can be associated with bedrock controls that lead to discontinuities in river longitudinal profiles, such as waterfalls. During his exploration of the Henry Mountains, Grove Karl Gilbert (1877) developed several key ideas about fluvial action. A major contribution was the concept of the graded stream, in which a river adjusts its slope so that it is able to transport the load supplied to it without aggrading or degrading its bed. His background in mechanics led him to think in terms of mutual adjustments between force and resistance in the form of dynamic equilibrium. He postulated several laws of geomorphic adjustment, including the laws of structure, divides and declivities.

It would be fair to characterize the 19th century as an era of development of foundational concepts and regulative principles for a nascent geomorphology. Generally speaking, simple types of observations, mainly visual, generated concepts which, in turn, led to attempts to generate supporting data. Because most ideas were geological in scope and were only weakly connected to or supported by evidence, they tended to be qualitative in nature (Rhoads and Thorn, 1993) and invoked narrative, particularist explanations (Church, 1996).

MODELS OF LANDSCAPE EVOLUTION

Geomorphology, so named in English in the 1880s (Tinkler, 1985), served a central role in geology for much of the 19th century. However, as the century turned and geology went 'underground', geomorphology was freed to become focused, not on quintessential geologic issues such as the existence of glaciation, but truly on landform and landscape development. However, the emergence of geomorphology from the historical science of geology and from principles such as uniformitarianism, planation, diastrophism and fluvial action led to an initial emphasis on models of regional landscape evolution over geological timescales (millions of years). The most influential of these models, the cycle of erosion developed by William Morris Davis, dominated geomorphological research for the next fifty years, the teaching of geomorphology for even longer, and in both spheres served as the touchstone against which all would-be competitors were compared.

The cycle of erosion

The cycle of erosion proposes a cyclical, that is repetitive, model of landscape development at a regional scale. It derives from a fundamental analogy with the life cycle of organisms by proposing that landscapes progress through successive, sequential changes through time, much like the process of aging. At a time when Darwin's work cast a long shadow across much of science it is interesting to note how the concept of randomness over time promoted by Darwin became a matter of elapsed time intrinsically producing change for Davis and followers. As both Chorley (1965) and Stoddart (1966) have pointed out 'evolution' for Darwin was an undirected process, but for Davis

it was a developmental scheme akin to ecological succession. As noted by Harris and Twidale (1968), the Davisian model really incorporates three models: one of landscape development, one of river development (both channel form and networks), and one of slope development. The cycle of erosion is usually discussed in terms of structure, process, and stage. Davis (1899) actually used the word 'time', not stage, initially because at the time there was no widely understood scale of geologic time. Davis (1899) 'guessed' that the span of his geographical cycle might end up being comparable to 'Cretaceous or Tertiary time'. He then discussed this undefined, but enormous, time span in relative terms, identifying anthropomorphic 'stages' (e.g. youth, maturity, old age). Beckinsale (1976) points out that Davis actually emphasized five factors, adding relief and texture of dissection to the above three. He also originally emphasized rate of uplift as an important variable, but this issue has been reduced merely to the notion that periods of uplift are brief and the intervening periods of stability are dominated by erosion. The climatic norm for the cycle of erosion was mid-latitude temperate conditions. However, this quickly became recognized as a shortcoming, and, describing deviations from the 'normal cycle' as 'climatic accidents', Davis (1905) himself published an 'arid cycle'. Other climatic variants followed with Peltier's (1950) 'periglacial cycle' seemingly the last.

For Davis, structure was a regional concept – the geological framework upon which the cycle operated. Process was an umbrella term embracing the various weathering, erosional, and depositional mechanisms geomorphologists invoke today. Stage, a relative concept of time, was expressed in the form of sequential changes in landscape form that could be expected throughout the course of an entire 'geographical cycle'. The most publicized 'stages' are those of 'youth', 'maturity' and 'old age'. Upon cessation of uplift (presumably rapid) a landscape enters a brief period of youth during which relative relief increases, attains a maximum when the summits of the original interfluves are eliminated, at which time the landscape enters maturity. During maturity, fluvial dissection is extensive, the landscape reaches its maximum ruggedness before it starts to become increasingly rounded and floodplains become extensive. The period of maturity passes sequentially into one of old age when a rolling landscape of reduced relative and absolute relief prevails – ultimately a 'plain without relief' (Davis, 1899: 497) is produced. However, Davis chose to emphasize the period immediately prior to this, hence the famous 'peneplain' interrupted by 'monadnocks', or upstanding erosional remnants.

Cycles elapse during periods of stability between uplift; consequently, their duration is variable and it is possible to see the imprint of multiple cycles in the landscape at any one time. As landscapes develop so too do drainage networks with floodplains expanding, meandering becomes widespread, and the drainage network becomes well integrated. According to Davis individual rivers over time may become 'graded', especially during the stage of old age. The sense in which Davis used this term is similar to that of G.K. Gilbert, except he modified it by suggesting that grade is attained over long periods of time, rather than constituting a continual mutual adjustment between force and resistance [see Dury (1966) and Knox (1975) for comprehensive discussions]. Davis applied the concept of grade to slopes as well, invoking a balance between the debris to be moved and the ability to move it. While the concept of grade is essentially similar for both rivers and slopes it was not anticipated that the state would be reached in both contexts at the same time in any given landscape.

Davis' main contribution to slope concepts was to present an explanation of the convexo-concave profile found on many hillslopes (Werritty, 2008). He attributed the upper convexity to soil creep (Davis, 1892), an idea later elaborated on by G.K. Gilbert (1909). The lower concavity was ascribed to decreasing grain size under the influence of increasing surface wash. Davis' explanation of the upper convexity has held up well, but his views about the lower concavity have been less successful. Carson and Kirkby (1972) and Young (1972) provide comprehensive discussions of the shortcomings of Davisian slope concepts. A basic assumption underlying the stages of landscape evolution in Davis' model is the notion that hillslopes always reduce their gradients over time – an assumption that subsequent research demonstrated does not always hold (Chorley, 1965).

Die Morphologische Analyse

Resistance to Davisian thought was always present in some quarters; however, in the English-speaking world there were very few who provided serious resistance for perhaps nearly forty years. Davis' contemporary, G.K. Gilbert, a scientist with the U.S. Geological Survey, was largely uninfluenced by Davis, pursued his own path, and sowed seeds in the discipline of geomorphology that provide foundations for contemporary research. However, during Davis' lifetime Gilbert's ideas were overwhelmed by the popularity of the cycle of erosion. Gilbert received considerable professional recognition during his career, but by virtue of

personality and career positions Gilbert seems to have been both disinterested and/or unable to counter the enormous impact of Davis on geomorphology. Gilbert's approach to geomorphology may be characterized as an engineering and empirical one. The elegance of his work is self-evident and it seems quite remarkable that he was so highly recognized in his time and yet his approach took so long to bear fruit. Perhaps the explanation simply rests with Sherman's (1996) notion of 'fashion dudes' in science; alternatively, it may simply be that Davis' biologically founded model was easier to understand and therefore more appealing than Gilbert's overtly engineering (quantitative) approach. The happy fit between the Davisian model and development of denudation chronologies created considerable support in England, especially as S.W. Wooldridge was not only a prime mover in denudation studies, but also an extremely influential geomorphologist within British academia.

Alternatives to the cycle of erosion did emerge beyond the shores of the United States and England. Walther Penck, son of Albrecht Penck and family friend of W.M. Davis, developed a very precisely crafted model of landscape development. His *Die Morphologische Analyse* was published posthumously in 1924, but was widely known to the English-speaking world through Davis' 1932 translation. Unfortunately, Davis' rendition of Walther Penck's convoluted German misrepresented it badly, and a sound translation did not emerge until Czech and Boswell's (1972) edition. As Bremer (1983) has noted, Walther Penck's work attracted little attention in Germany, so we are confronted with the quirky situation of a text written in convoluted German having impact primarily in a misrepresented form in the English-speaking world. Only essential details are presented here. A lengthy appraisal may be found in Chorley et al. (1973) and a short one in Thorn (1988).

The model of landscape evolution developed by Walther Penck viewed the shape of landforms as the product of the ratio between the rate of endogenous movement and the rate of erosion. As the last two could be examined, he believed the first could be calculated. Unlike Davis his conceptualization was profoundly precise, so much so that Young (1972) was able to create a canonical diagram of it and essentially produce a modern process–response model. Penck challenged Davis' self-admitted simplification of producing an uplifted landscape that was then eroded during a tectonically stable period. Penck did not see landforms following predictable sequences, but rather following possible pathways that reflected the changing interplay between uplift and erosional intensity (rates). His model focused primarily on slope development with uniform rates of downcutting by rivers into an uplifting surface producing straight slopes, accelerating rates of downcutting during accelerated uplift (waxing development) producing slopes of convex profile, and decreasing rates of downcutting during periods of decreasing uplift (waning development) producing slopes of concave profile.

While the internal logic of Penck's model is exemplary, the external logic has some flaws, primarily stemming from the fact that he emphasized mainly the interplay between tectonics and erosional incision while ignoring many other important factors such as climate change, lithologic variability, and weathering rates – all of which can strongly influence landform development (see Burbank et al., 1996 for an interesting modern integration of such matters). The critical imprint on the development of geomorphology is that Davis tagged Penck with the concept of the 'parallel retreat of slopes' and this is how he was long remembered. However, Penck actually promoted the concept of 'slope replacement' – each slope segment retreating upslope parallel to itself, but being replaced by one at a lower angle advancing upslope. Despite the differences in their models, Penck and Davis both made visual observations of landforms and used their models to explain the genesis of these forms, but failed to conduct rigorous empirical testing to confirm or refute the basic tenets of their models. Neither model was constrained by what would be considered today as a realistic estimate of the time available for landscape development.

Pediplanation

Lester King was a classically trained Davisian geomorphologist whose most important work was undertaken in South Africa. Frustrated by what he perceived as the inadequacy of the Davisian model to explain South African landscapes and fueled with exposure to the slope work of T.J.D. Fair, the tectonic ideas of A.L. du Toit, and drawing upon selected portions of Davisian and Penckian thought, King (1963, 1967, 1983) created a sweeping model of landscape development that spanned the details of slope processes, although without process measurements, and eventually reached deep into the realm of crustal geophysics (Thorn, 1988).

King's model is cyclical but, unlike Davis, he invokes the parallel retreat of slopes. This process produces time-transgressive upland surfaces which, subsequent to uplift, are largely unmodified further save by reduction in areal extent as younger time-transgressive surfaces at lower elevations attack their margins. The ultimate landscape is a pediplain produced by coalescence of pediments and dominated by concavities with residual landforms referred to as inselbergs or bornhardts. Ultimately, King (1983) went on to explain the cyclicity in his landscape model by a combination of onshore unloading due to erosion and offshore loading due to sedimentation. These processes become the fundamental components of cymatogeny – large-scale crustal deformation dominated by vertical movement that does not greatly disturb rock structures. Cymatogeny then becomes associated with an expanding earth in King's model. Neither cymatogeny, nor an expanding earth, have been widely accepted ideas. King also invoked semi-arid environments as both the historical and contemporary norm. Like Davis, King essentially treated his model as a set of regulative principles by invoking so many ad hoc modifications that it was effectively unfalsifiable. The model of pediplanation may be seen as the last hurrah of English-language attempts to develop an overarching qualitative model of landscape evolution in geomorphology. However, the work of Büdel, which was not translated comprehensively until 1982 (despite his original papers in German dating from as early as the 1930s), should be noted. Clearly, Büdel attempted to work on a global scale with his 'double etchplain' concept being perhaps the most famous one to enter the English-language world.

The demise of the cycle of erosion

All three models reviewed here are intimately associated with their progenitors – those who followed in the footsteps of Davis, Penck and King were disciples of their approaches, rather than major innovators. Neither Penck's model nor King's model enjoyed the widespread popularity of the cycle of erosion and the downfall of the latter greatly changed the character of geomorphology. Substantive weaknesses are reviewed in Chorley (1965), Flemal (1971), Chorley et al. (1973), and Higgins (1975). Single cause–single effect relationships, a preoccupation with stage at the expense of structure and process, invocation of a 'normal' climate, misinterpretation of Appalachian climatic history (critical because the Appalachians were the core area for Davis' explanations), and a lack of quantification all emerged as shortcomings and helped undermine the Davisian approach to geomorphology.

Rhoads and Thorn (1996b) have examined the Davisian approach to geomorphology in terms of its explanatory methodology. Davis and his followers commonly invoked morphogenetic terms, that is terms that both describe a landform while simultaneously explaining its origin (e.g. peneplain). They then proceeded to explain the landscape using the morphogenetic terms rather than test the actual validity of the model against field observations, most of which involved simple visual inspection of landscapes or of topographic maps. Because morphogenetic landscape terminology was deeply embedded in, or derived from, the cycle of erosion, descriptive explanations involved circular reasoning. The model served essentially as a grand regulative principle that was assumed to be true and therefore could guide explanation of all possible cases. It was not presented as a model to be rigorously tested through comparative evaluation of hypotheses derived from it and from competing models. In this sense the cycle of erosion contained elements of a 'ruling' theory, a methodological approach disparaged by Chamberlain (1897). Although Davisian geomorphology continued to dominate university teaching of geomorphology well into the 1970s, it lost its pre-eminence with those who saw themselves as cutting-edge researchers.

Clearly the models of Davis, Penck and King organized underlying theoretical principles about diastrophism, fluvial activity, and uniformitarianism into coherent constructs that could be used in geomorphological explanations. In this sense, they conform to the MTV of theoretical models. However, as applied in explanatory fashion, the Davisian model, to a large degree King's model, and to some extent Penck's model each became a grand regulative principle – essentially a lens through which to interpret the landscape, rather than an entity to be tested. The virtual absence of hypothesis testing against quantitative field data and the lack of absolute temporal control (save that provided by stratigraphic principles) fostered such tendencies, but they are not inherent to the nature of the models. Thus, the inherent high levels of uncertainty associated with models applied over geological time spans was compounded by circular modes of explanatory reasoning (Rhoads and Thorn, 1993, 1996b). From the perspective of geomorphology, all three models fall into Chorley's (1978) historical theoretical category and Church's (1996) landscape history category. The era of explanatory landscape-evolution models, which according to Church (1996) ended in about 1950, was dominated by theory at geological time scales and regional spatial scales, and is a period when geomorphic theory was constructed and applied essentially with little regard to the standards and constraints prevailing in the physical sciences. This situation changed abruptly in the 1940s and 1950s with Robert Horton sowing the seed and Arthur Strahler becoming the chief promoter: theory in geomorphology became fundamentally and permanently re-orientated.

THE SHIFT TOWARD PHYSICALLY BASED MODELS

Robert Horton (1945), a hydrologist by training, developed a deterministic model of channel initiation and drainage network development that, in stark contrast to the cycle of erosion, drew upon quantitative mechanistic principles from physics as its foundation. Although many details of the model are not supported by subsequent empirical findings, it was not the details, but the mechanistic approach that had an impact on a new generation of geomorphologists, who were beginning to seriously question the utility of the cycle of erosion. Horton clearly articulates that he is addressing the same problem as Davis and (Horton, 1945, 367) draws the analogy between his own work and that of Davis as 'two pictures of the same object taken in different lights'; however, he also claims his own approach to be the more fundamental of the two. Horton's paper may be viewed as a modern quantitative process study, and as such stands as an important turning point in geomorphology – the salient difference between the Davisian and Hortonian approaches is that a descriptive explanatory model based on an organic analogy of a 'life cycle' of landscape change is challenged by a quantitative model of erosional landscape development deeply embedded in fundamental physical reasoning. It is worth noting here the latent significance of G.K. Gilbert's approach to geomorphology, one that resonates strongly with Horton's view, but that at the time was languishing in the shadow of Davisian geomorphology.

Another seminal study that departed substantially from Davisian geomorphology was the work by Ralph Bagnold on the physics of sand transport and desert dunes (Bagnold, 1941). In this work, Bagnold applied the basic principles of turbulent boundary-layer theory to develop and empirically evaluate models of the shear stress exerted by wind on sand surfaces and the fluid threshold required to initiate motion of sand grains on these surfaces. He then applied this understanding to explain the geometry and sedimentology of desert ripples and dunes. In the 1950s, Bagnold began to explore the movement of cohesionless sediment in flowing water (Bagnold, 1951, 1956). His work caught the attention of Luna Leopold at the United States Geological Survey, one of the main figures in the rise of process-based geomorphology, who had Bagnold turn his considerable intellectual talents towards problems of meandering rivers

(Bagnold, 1960) and sediment transport in rivers (Bagnold, 1966).

THE 1950S AND 1960S: THE EMERGENCE OF PROCESS GEOMORPHOLOGY

The 1950s and 1960s was not a period when theoretical introspection was overt, but the clear shift of geomorphologists to an alignment with mainstream scientific methodologies, combined with individual conceptual breakthroughs, such as magnitude and frequency, clearly took English-speaking geomorphology out of the Davisian world and into the realm of the process geomorphology of today, as well as into the world of reductionist science. Such statements are not valid for much of the French and German geomorphology, which continued to cling, at least in part, to climatic geomorphology that is essentially a byproduct of Davisian thought. As this trend did not lead to the clear reduction in scale of research associated with process geomorphology, climatic geomorphology generally retained narrative presentations and remained particularist in style.

In the early 1950s a siren call for change came in the form of Strahler's (1952a) paper *Dynamic Basis of Geomorphology*. In this paper, Strahler argued for the study of geomorphological processes 'as manifestations of various types of shear stresses that act upon earth materials to produce processes of erosion, weathering, transportation, and deposition'. In other words, geomorphology should be grounded in fundamental principles from physics, particularly Newtonian mechanics, and from chemistry – a position consistent with the work by Horton and Bagnold. Moreover, this approach should focus on quantitative determinations of landform characteristics and causative factors using methods of mathematical statistics, should develop concepts of open systems and steady states for geomorphological phenomena, and should derive general mathematical models from physical reasoning to serve as quantitative geomorphological laws.

The new approach resulted in a veritable explosion of geomorphological research during the 1950s and 1960s, but its reach extends beyond that period to the present. Early work by Strahler and his protégés emphasized empirical models developed through statistical analysis of relations among geomorphological variables, many of which were metrics of form (Strahler, 1952b, 1957; Melton, 1958; Morisawa, 1962). At the same time, the hydraulic geometry of streams was introduced by scientists at the U.S. Geological Survey as an empirical statistical model relating discharge to its component properties of width, depth and velocity both at a station and in the downstream direction (Leopold and Maddock, 1953). Geomorphological theory had now taken a radical step into the world of functionalism: the use of empirical data and statistical analysis to explore relationships between process(es) and form (Chorley, 1978). Geomorphology no longer operated within a quasi-independent scientific framework, but rather sought to ally itself with the mainstream world of science in which physics, engineering, and chemistry provided the fundamental frameworks. It is a period that emphasized empiricism, and attempted primarily to determine repeated and predictable regularities in form and function.

The invocation of a systems approach to geomorphology, which is apparent in Strahler's early papers (e.g. 1952a, 1952b, 1958), provided his group with an opportunity to employ statistical analysis to probe the structure of geomorphological systems. A prime example of this approach is the work by Mark Melton (1958) on the correlation structure of drainage basins (Keylock, 2003). Early attempts to infuse geomorphology with systems thinking emphasized an equilibrium perspective, stimulating a completely different way of conceptualizing regional landscape development. John T. Hack (1960, 1975) addressed Davis' classic fieldwork area, the Appalachians, invoking an alternative conceptual model in which the entire landscape is in 'dynamic equilibrium'. In such a framework topographic variability is derived from spatial variability in rock resistance and process intensity and not from long-term cyclical development. Hack appealed to the work of G.K. Gilbert (e.g. 1877, 1886) to authenticate his approach, in the process resurrecting Gilbert's style of investigation, which combined geological and mechanistic reasoning, as a foundation for contemporary process-based studies. In a related study, Wolman and Miller (1960) introduced geomorphologists to the concept of geomorphic work – one of the most influential ideas in modern geomorphology – through the paired notions of magnitude and frequency of geomorphic events. An outcome of this study was the proposition that river channel morphology is adjusted to a 'dominant' discharge, which presumably reflects an equilibrium adjustment of this morphology to prevailing flow conditions.

Strahler's (1952a) call for systems thinking was not new: as Smalley and Vita-Finzi (1969) were at pains to point out systems thinking is an old approach in earth sciences. However, it was Chorley (1962) who captured the imagination of geomorphologists. His paper highlights the general system(s) theory (GST) of von Bertalanffy (1956, 1962). Chorley and Kennedy (1971) became the essential textbook illustrating the integration of GST and geomorphology. A symbolic language for GST is laid out and models in the form of canonical diagrams showing the interaction (correlation) among morphometric and process measurements pervade the text. As such Strahler's plea for rational equations (i.e. physically based relations that are dimensionally balanced) is unanswered through this approach.

The treatise *Theoretical Geomorphology* (Scheidegger, 1961) was the first book to tackle the study of geomorphology within a framework of the physical (Newtonian) mechanics of earth-surface processes and the relation of these mechanistic processes to the physical resistance of earth materials – an approach Chorley (1978) refers to as realist theory. Davis' cycle of erosion is mentioned briefly at the beginning of Scheidegger's volume, but the remainder of the book draws upon a body of literature largely international in flavor (e.g. German and Russian) to develop a suite of physically based mathematical models for describing all manner of geomorphological phenomena. The reaction within the geomorphological community was delayed because few geomorphologists trained in the Davisian tradition had the necessary background in mathematics and physics to master Scheidegger's approach.

On a related note, geomorphological research at the U.S. Geological Survey adopted physical concepts such as entropy (Leopold and Langbein, 1962) and minimization principles (Langbein and Leopold, 1964, 1966) to develop models of process-form adjustment in 'quasi-equilibrium' fluvial systems. The use of theory in these studies had a distinctly teleological emphasis, that is geomorphic systems exhibit goal-directed behavior and adjust toward attractor states (Chorley, 1978). Another fundamental conceptual breakthrough was provided by Schumm and Lichty (1965), who examined the roles of independent and dependent variables and their inter-relationships over varying timescales. Their paper attempted to account for differences in investigative approaches across the range of scales of geomorphological inquiry, and in this sense, laid the foundation for subsequent efforts to address change in theory, methodology and styles of reasoning over scale in geomorphology (e.g. Rhoads and Thorn, 1993; Church, 1996).

THE 1970s AND 1980s: QUANTITATIVE MODELING, EXPERIMENTATION AND DISEQUILIBRIUM

During the 1970s and 1980s geomorphology became infused with detailed field measurements of processes and morphological attributes. Also, a gradual transition occurred from the use of statistical models at the beginning of this period to increasing development of mathematical physically based models toward the end. Underlying this transition was an emphasis on training in mathematics, physics and chemistry to provide the theoretical background knowledge required to study geomorphological problems using the process approach. Modern geomorphology invokes theoretical principles that pervade the hard sciences. This great theoretical change in geomorphology, initiated in the 1950s and 1960s, became broadly entrenched in the 1970s and 1980s. The massive sea change that swept through geomorphology after the demise of the cycle of erosion was fulfilled during these two decades.

A major conceptual development during the 1970s was the introduction of the concept of geomorphic thresholds and the idea that episodes of abrupt change can occur in geomorphic systems as thresholds are transcended (Schumm, 1979). The notion of thresholds was not entirely new – thresholds of sediment movement in the form of the Hjulstrom curve or Shields diagram were widely recognized – yet Schumm's perspective was novel because it extended the idea to the scale of entire geomorphological systems. Thresholds represent distinct nonlinearities in system response, thereby frustrating attempts to describe system dynamics in the differential equations of continuum mechanics. Moreover, the crossing of thresholds can often trigger spatial variations in watershed-scale response of fluvial systems in the form of 'complex response' – a distinctly non-equilibrium mode of system dynamics.

The turn toward the possibility of disequilibrium dynamics in geomorphological systems was further advanced through the development of event-based conceptual models highlighting the manner in which landscapes are changed by particular formative events and how these landscapes then recover from the influence of formative events (Wolman and Gerson, 1978; Brunsden and Thornes, 1979). The concept of geomorphic effectiveness holds that the impact of a formative event on shaping the landscape depends not only on how much an event reconfigures landscape morphology, but also on the capacity of post-event processes to 'undo' these morphological changes. Thus, in this framework magnitude and sequencing of events in relation to recovery times (Graf, 1977) becomes critically important in understanding landscape dynamics. This kind of thinking continues under the rubric of 'landscape sensitivity' (e.g. Brunsden, 2001).

As noted by Kirkby (1989), quantitative modeling in geomorphology was initially dominated by regression techniques, which had the advantage of fueling the need for field observations, but the disadvantage of providing a relative poor link between observation and theory. However mathematical models began to emerge during the 1970s (Ahnert, 1976, 1977; Kirkby, 1976) and by the end of the 1980s a variety of quantitative models was being exploited in geomorphological inquiry, from those based on classical statistical and deterministic mathematics to those that might be called *avant garde,* such as catastrophe theory (Graf, 1979), fractals (Tarboton et al., 1988) and chaos theory (Culling, 1987; Slingerland, 1989). At the same time, these changes in theoretical modeling were matched by rapid improvements in the sophistication of data acquisition and analysis, be it by satellites or micro-sensors attached to data loggers. In some, but not all, areas of geomorphic research problems stemming from data shortages were replaced by those stemming from data overload.

Experimental approaches to the evaluation of geomorphological models also began to emerge during the 1970s, especially through the work by Stanley Schumm and his students on the dynamics of watersheds, drainage networks, and river channels (Schumm et al., 1987). Another example was the development of field experiments, exemplified by the work of William Dietrich and colleagues on the dynamics of meandering rivers (Dietrich and Smith, 1983; Dietrich, 1987). The extent to which such studies constitute 'true' experiments is debatable (Church, 1984; Slaymaker, 1991), but this approach was new in the sense that an explicit effort was made to use physically based theory to guide a systematic programme of data collection in the field. The resulting field-based information was then used to directly evaluate the accuracy of predictions of the theoretical model.

THE CONTEMPORARY SCENE

Since the 1990s, geomorphology has been influenced strongly by several trends: (1) an increasing concern with complexity and nonlinear dynamics, (2) rapid advances in measurement technology, (3) increasing computational and information-processing capabilities, (4) enhanced collaborations with other disciplines, especially engineering and the life sciences, (5) interest in philosophical issues, (6) concern about practical aspects of human impacts of geomorphological systems, and (7) a renewed focus on landform development over geological time scales (Rhoads, 2004). The first trend has fueled the development of alternative models that represent geomorphological phenomena as nonlinear dynamic systems (NDS) (Phillips,

1999). While at the disciplinary-wide level NDS approaches to landscape modeling remain embryonic, progress among a group of aficionados has been impressive. As Phillips (1992a,1992b) has pointed out it is possible to embrace equilibrium (specifically equilibria) concepts, nonlinear behavior, feedback, bifurcations, chaos, and dissipative structures within NDS, while also making clear links to traditional concepts in geomorphology. Regardless of whether NDS becomes a pre-eminent overarching theoretical framework in geomorphology, it has already contributed to a substantive shift in the domain of theory underpinning geomorphology. The potential represented by NDS is reflected in Church's (1996) identification of it as one of his four categories. A conceptual advantage of the NDS approach is that it captures the short-term predictability of geomorphological responses as well as long-term unpredictability. We begin, perhaps, to see contingency embedded within 'scientific laws', if we allow ourselves a little terminological latitude.

The capacity to obtain new types of information is leading to innovative insights into the dynamics of geomorphological systems, but the voluminous amounts of data that can now be generated through new measurement technologies is placing increasing demands on data-processing capabilities to transform data into interpretable results. Sensor technology and cyber infrastructure have become key buzzwords and promise to radically transform the type, amount, and information content of data collected in the field. However, like most field-oriented scientists, geomorphologists must struggle with critical concerns about research design and how appropriate arrays of sensors can be deployed to address in meaningful ways theoretical questions of fundamental importance.

Theoretical modeling has been greatly affected by the tremendous advances in computational capacity over the past 15 years. Physically based numerical modeling has become a standard tool in the theoretical arsenal of geomorphologists. Often, a primary objective of such modeling is prediction. As noted by Wilcock and Iverson (2003: 3) 'predictive models are playing an increasingly prominent and sometimes controversial role in both basic and applied studies in geomorphology'. A key issue in such modeling concerns the extent to which numerical models based on Newtonian continuum mechanics can be used to accurately predict landscape dynamics over a range of spatial and temporal scales versus the advantages offered by other modeling approaches, such as ruled-based, heuristic mathematical models (e.g. cellular automata). Important questions that continue to be addressed in the world of theoretical modeling include: what do we model and to what purpose, how do we formulate geomorphic models, and how do we evaluate model predictions (Wilcock and Iverson, 2003)?

The theme of complexity overlaps with the interest in collaborations with other disciplines. The influence of biota on geomorphological processes (and vice versa) has been relatively neglected. Over the past two decades, a strong appreciation of the role of vegetation in landform and landscape development has emerged. The essentially unvegetated landscapes of Davisian geomorphology have given way to models in which vegetation may play a central role, such as that of woody debris in some rivers (Brooks and Brierley, 2002; Daniels and Rhoads, 2004). In certain circumstances animals may now be seen as playing important roles in particular aspects of landscape development (e.g. Butler, 1995). Some landforms, such as the river systems of the Okavango Inland delta, appear to be almost completely dominated by the dynamics of biota (McCarthy et al., 1992).

The increasing concern about human impacts on geomorphological systems reflects a societal need to embrace what geomorphologists have to say in the realm of planning and management. The enormous strength of engineering as a discipline is its site-specific capabilities while concern about problems in larger spatial contexts (e.g. the watershed scale) and/or for longer time periods often invokes the need for geomorphological expertise. Another important facet of this issue is an appreciation of just how profound and pervasive is the human impact at a global scale (Hooke, 1994). Whereas, traditionally, geomorphologists shied away from applied issues and sought 'natural' landscapes to study, such an approach is no longer feasible. Furthermore, funding opportunities increasingly drive research geomorphologists into the applied domain. Thus, over the past decade in particular, geomorphologists have become important players in the realm of environmental restoration, including stream restoration (Wade et al., 2002; Wohl et al., 2005).

A prominent development over the past 20 years has been the emergence of an interest in philosophical aspects of geomorphology. The role of theory in geomorphology was addressed by Rhoads and Thorn (1993), at least partly in response to a perceived concern about overemphasis on reductionist, mathematical-modeling approaches to inquiry raised by Baker and Twidale (1991). Rhoads and Thorn (1993) noted that the character of theory changes with scale; in particular, quantification is less likely when dealing with large (geologic) space and timescales because the data required to calibrate or meaningfully validate mathematical or statistical models generally are unavailable. Also, theory tends to become increasingly conjectural as the basis for empirical support diminishes – a characteristic that is not 'bad', but

simply an artifact of limits imposed on theory by the spatial and temporal structure of the world.

Geomorphologists also have delved into the realm of ontological and epistemological issues, considering how various philosophical perspectives on these issues, such as realism, pragmatism, empiricism and positivism, shed light on the aims of geomorphological inquiry (Richards, 1990, 1996; Rhoads, 1994, 1999; Rhoads and Thorn, 1994; Baker, 1996). More recently, Rhoads (2006) has called into question the mechanistic materialist underpinnings of the dynamic basis of geomorphology as advocated by Strahler (1952a) and has called for a more holistic view of geomorphological dynamism grounded in process philosophy. Styles of reasoning have also received attention by geomorphologists with perhaps the most important development being recognition of the distinction between abductive and deductive reasoning (Baker and Twidale, 1991; Rhoads and Thorn 1993).

The scope of research continues to expand in geomorphology, in part because of the prominence of collaboration, particularly with scientists from other disciplines. Inter-disciplinary collaboration creates new frontiers, especially at the boundaries of disciplines, and hence inherently at the interface of two or more disciplines. Not only does such work open up new themes, it also introduces new perspectives and techniques directly to the matter at hand. It would seem that at the present time biology and engineering are the two richest sources of external inputs to geomorphology.

Another trend worth emphasizing is the development of mathematical models of landscape evolution. While reductionist science has greatly expanded our knowledge base, it is important to remember that a fundamental goal of geomorphologists is the explanation of landform and landscape development, presumably at scales that are visually meaningful. If we accept this premise, it is important that our most powerful techniques, of which numerical modeling must certainly be one, are applied to problems at the appropriate scale. One of Richard Chorley's (1978: 10) most prescient and penetrating observations was

> geomorphology can only continue to make a unique contribution to the earth sciences if, in the study of process, physical truth is sufficiently coarsened in both space and time as to accord with the scales on which it is profitable to study geomorphologically viable landform objects.

Today we see landscape modeling moving ever upward in scale and even reaching into the watershed scale (e.g. Willgoose et al., 1992). Numerical have shed light on how the style of landscape evolution depends critically on the timescale of tectonic uplift (t_p) in relation to the response time of erosional processes (τ). Thus, dynamic equilibrium in the sense of Hack (1960) is associated with short response times and slow tectonic forcing ($t_p \gg \tau$), Davisian-style evolution corresponds to episodic forcing and slow response times ($\tau \gg t_p$), Penck's model of waxing and waning landscape development holds when $t_p \approx \tau$, and King's ideas about backwearing and pedimentation are a variant of the Davisian model under conditions of an asymmetric uplift geometry (Kooi and Beaumont, 1996). Steady-state landscapes, those where topography remains constant through time, develop when rates of fluvial erosion across a landscape balance rates of tectonic uplift (Montgomery, 2001).

After a period of concern about the role of geohistorical inquiry in geomorphology in the early 1990s (Baker and Twidale, 1991), such studies now are flourishing. Geohistorical investigations generally apply theoretical principles within abductive frameworks of reasoning to infer how various erosional and depositional processes shaped a contemporary landform or set of landforms from some prior configuration (Baker and Twidale, 1991; Rhoads and Thorn, 1993). A critical aspect of this retrodictive, reconstructive approach is the capacity to establish accurate chronologies. Enrichment in this branch of geomorphological inquiry has occurred through the development of new dating techniques, which provide empirical evidence to test numerical models of landscape evolution (Bishop, 2007). A dramatic example is the possibilities being created by terrestrial *in situ* cosmogenic nuclide dating (Grosse and Phillips, 2001). The fundamental innovation here is that it is now possible to determine how long a bedrock surface has been exposed at the earth's surface. Previously, such information had to be determined indirectly. The new technique has already shed light on the age of glaciated surfaces, revealing many in Scandinavia to be very old and to have experienced multiple glaciation essentially without modification (e.g. Kleman and Stroeven, 1997).

A PERSPECTIVE ON CONTEMPORARY THEORY

We return now to the question of whether the model–theoretic view (MTV) of theory is relevant for characterizing theory in geomorphology. It seems to us the answer generally is yes, with perhaps a few qualifications. The 19th century was largely a period when regulative (theoretical) principles were developed which subsequently served as the foundation for the formulation of theoretical models in the 20th century, including those by Davis, Penck and King. Although these models were qualitative, that characteristic was largely a product of the scale of analysis to which they applied and the type of information (mainly visual) that could be generated at geological time scales using available technology. Nonetheless, these models were largely aids to interpretation of observations (mainly visual) of relations among geomorphological features within landscapes, rather than sources of theoretical hypotheses that then became the focus of rigorous empirical testing to ascertain the overall validity of the model. Instead, the models were assumed valid and immune from empirical disconfirmation.

The movement to ground geomorphology in the background knowledge of the basic sciences in the mid 20th century has provided a sustained foundation for contemporary geomorphology. The theoretical principles from the 19th century, when geomorphology was tied closely to historical and physical geology, are now complemented by theoretical principles from physics, chemistry, and biology. The initial emphasis on statistical empiricism in the 1950s and 1960s led to the emergence of models, such as hydraulic geometry and correlation structures that could be characterized within MTV as data models. Theoretical principles (e.g. conservation of mass, momentum and energy) infused into geomorphology from the basic sciences led to the development of new theoretical models, particularly deterministic physically based mathematical models. The interplay between data models and theoretical models has been active in geomorphology and coincides with a division of labor in geomorphology into those who engage in the formulation of theoretical models based on general theoretical principles and those who focus on the development of data models through collection and processing of field and laboratory information. Although a division can be identified, it does not mean that membership is mutually exclusive; research groups as well as individuals often participate in both activities. The point is that both types of approaches have contributed substantially to the theoretical content of geomorphology. As one example, a data model of the dynamics of river confluences derived from laboratory experiments (Best, 1987, 1988) has served as a foundation both for field studies of confluence dynamics (Rhoads and Sukhodolov, 2001) and as the impetus for the development of theoretical numerical models of these dynamics (Bradbrook et al., 2001). In turn, the simulation results of the theoretical models, which serve as sophisticated theoretical hypotheses, have been compared with patterns of measured data (data models) to assess the validity of the theoretical model.

The MTV notion that theory in contemporary science consists of 'families of models' derived from basic principles seems to be consistent with how geomorphology operates. No longer are we under the influence of one dominant model as in the era of Davisian geomorphology. Today models abound, in statistical, stochastic, deterministic, quantitative, qualitative, conceptual, mathematical and narrative forms. MTV certainly seems more consistent with how theory is utilized in geomorphology than the logical positivist view of theory as a formal, hierarchically structure set of axiomatic statements. As noted by Kirkby (1990: 63), 'models have always been needed to organize our views of how the natural landscape functions'. Conversely, little, if any, evidence of axiomatized, logico-linguistic forms of theory is apparent in geomorphology. One implication of MTV is that it provides a less formal view of theory than does the Received View. It is hard to say 'this right here is the theory – its content, structure and domain of applicability'. Nor may it be possible (or even useful) to reach widespread consensus on which theoretical principles, family of models, theoretical hypotheses and data models are part of a specific geomorphological theory. Instead, it may be MTV's pliant, somewhat fuzzy perspective on theory that captures the true essence of the success of theory in science.

The notion of theory as 'families of models' also helps to accommodate differences in the characteristics of theory across an expansive domain of time and space scales. These characteristics are primarily shaped by the regulative principles employed; thus, the structure of theory is not necessarily uniform across scale. Increasing spatial–temporal scale generally produces an inherent loss in precision (density of information per unit of time and space). Increasing scale also usually means that quantitative models, which often require information commensurate with their scale of application to evaluate rigorously, become, at best, less constrained and at worst, highly speculative. This claim then brings us back to the arguments by Chorley (1978), Rhoads and Thorn (1993) and Church (1996) that the type of explanation (theoretical model) appropriate at one scale may not be entirely appropriate at another. Indeed, it emphasizes the centrality of scale-linkage problems within geomorphic theory and the need to wrestle with these problems directly (e.g. Kennedy, 1977; Haigh, 1987; de Boer, 1992). We truly need a 'family of models' of various types to accommodate the enormous scope, both temporal and spatial, of geomorphological inquiry.

CONCLUSION

It seems fair to say that theory has always pervaded or infused geomorphology, as it has most sciences, but that the notion of exactly what constitutes a theory is somewhat ambiguous. While the logical positivist view that a theory is an axiomatized logico-linguistic structure has dominated formal perspectives on theory both in science and philosophy, we believe the model–theoretic view provides a perspective that accords better with the important role that models play in geomorphological inquiry. We have tried to show how models seem to pervade all of what we do in geomorphology, including much work that is overtly 'theoretical', but also, through the development of data models, that which is highly empirical. The intersection of data models and theoretical models within the context of hypothesis testing is an important aspect of theoretical inquiry in geomorphology. It is also important to note that the characteristics of theory, expressed in the details of a model deemed appropriate at that scale by guiding theoretical considerations, vary with scale. Different types of models may be more appropriate than others at specific scales. Such variation seems to be inherent in how nature itself is structured in relation to our capacity both to know it and to obtain information about it (time–space scales that are much less than or much greater than human life spans). Nevertheless, our capacity to overcome time–space limitations is ever increasing through the development of new technologies that extend our capacity to explore the world both theoretically (through computational simulations) and empirically (through new data-generating techniques). Such advances may never overcome the limitations completely but they almost certainly will lead to new ways of exploring the world through models. Technology alone, however, will not replace the need for conceptual ingenuity to provide new theoretical insights, to develop inventive theoretical models, and to synthesize and integrate new data into innovative data models. Nevertheless, in seeking to order both space and time, we should remember that the search for order may be a human need, rather than a natural reality. Perhaps indeed it is a 'naughty world' (Kennedy, 1980) and we do well to remember two admonitions from Goodman:

> Whatever made the world and whatever makes it go, the scientist writes the laws. And whether or not nature behaves according to the law depends entirely upon whether we succeed in writing laws that describe its behavior. ... The scientist demands little of nature but much of himself. Where there is change, he looks for constancy in the rate of change; and failing that, for constancy in the rate of change of the rate of change. And where he finds no constancies, he settles for approximations.

> (Goodman, 1967: 96–9)

REFERENCES

Agassiz L.J.R. (1840) On the polished and striated surfaces of the rocks which form the beds of the glaciers in the Alps. *Proceedings of the Geological Society* 3: 321–2.

Ahnert F. (1971) A general and comprehensive theoretical model of slope profile development. *University of Maryland Occasional Papers, Geography No. 1.*

Ahnert F. (1976) Brief description of a comprehensive three-dimensional process–response model of landform development. *Geocom Bulletin* 6: 99–122.

Ahnert F. (1977). Some comments on the quantitative formulation of geomorphological processes in a theoretical model. *Earth Surface Processes* 2: 191–201.

Anderson M.G. (1988) (ed.) *Modelling Geomorphological Systems.* Chichester: Wiley.

Bagnold R.A. (1941) *The Physics of Blown Sand and Desert Dunes.* London: Methuen.

Bagnold R.A. (1951) The movement of a cohesionless granular bed by fluid flow over it. *British Journal of Applied Physics* 2: 29–34.

Bagnold R.A. (1956) The flow of cohesionless grains in fluids. *Philosophical Transactions of the Royal Society of London A* 249: 235–97.

Bagnold R.A. (1960) *Some Aspects of the Shape of Meanders,* U.S. Geological Survey Professional Paper 282-E. Washington, DC, pp. 134–44.

Bagnold R.A. (1966) *An Approach to the Sediment Transport Problems from General Physics,* U.S. Geological Survey Professional Paper 422-I. Washington, DC.

Bailer-Jones D.M. (2002) Scientists' thoughts on scientific models. *Perspectives on Science* 10: 275–301.

Bailer-Jones D.M. (2003) What scientific models represent. *International Studies in the Philosophy of Science* 17: 59–74.

Baker V.C. (1996) The pragmatic roots of American quaternary geology and geomorphology. *Geomorphology* 16: 197–215.

Baker V.C and Twidale C.R. (1991) The reenchantment of geomorphology. *Geomorphology* 4: 73–100.

Beckinsale R.P. (1976) The international influence of William Morris Davis. *Geographical Review* 66: 448–66.

Beckinsale R.P. and Beckinsale R.D. (1989) Eustasy to plate tectonics: unifying ideas on the evolution of the major features of the earth's surface, in K.J. Tinkler (ed.), *History of Geomorphology from Hutton to Hack.* Boston: Unwin Hyman, pp. 205–21.

Beckinsale R.P. and Chorley R.J. (1991) *The History of the Study of Landforms or the Development of Geomorphology. 3. Historical and Regional Geomorphology 1890–1950.* London: Routledge.

Best J.L. (1987) Flow dynamics at river channel confluences: implications for sediment transport and bed morphology, in F.G. Ethridge, R.M. Flores and M.D. Harvey (eds), *Recent Developments in Fluvial Sedimentology,* Special Publication 39. Tulsa, OK: Society of Economic Paleontologists and Mineralogists, pp. 27–35.

Best J.L. (1988) Sediment transport and bed morphology at river channel confluences. *Sedimentology* 35: 481–98.

Bishop P. (2001) Long-term landscape evolution: linking tectonics and surface processes. *Progress in Physical Geography* 32: 329–65.

Bishop P. (2007) Long-term landscape evolution: linking tectonics and surface processes. *Earth Surface Processes and Landforms* 32: 329–365.

Bradbrook K.F., Lane S.N. and Richards K.S. (2000) Numerical simulation of three-dimensional time-averaged flow structure at river channel confluences. *Water Resources Research* 26: 2731–46.

Bremer H. (1983), Penck A. (1858–1945) and Penck W. (1888–1923) Two German geomorphologists. *Zeitschrift für Geomorpologie* NF 27: 129–38.

Brooks A.P. and Brierley G.J. (2002) Mediated equilibrium: the influence of riparian vegetation and wood on the long-term evolution and behavior of a near-pristine river. *Earth Surface Processes and Landforms* 27: 343–67.

Brown H.I. (1977) *Perception, Theory and Commitment: the New Philosophy of Science.* Chicago: University of Chicago Press.

Brunsden D. (2001) A critical assessment of the sensitivity concept in geomorphology. *Catena* 42: 99–123.

Brunsden D. and Thornes J.B. (1979) Landscape sensitivity and change. *Transactions of the Institute of British Geographers* NS 4: 463–84.

Büdel J. (1982) *Climatic Geomorphology,* translated by L. Fischer and D. Busche. Princeton: Princeton University Press.

Burbank D.W., Leland J., Fielding E., Anderson R.S., Brozovic N., Reid M.R. and Duncan C. (1996) Bedrock incision, rock uplift and threshold hillslopes in the northwestern Himalayas. *Nature* 379: 505–10.

Butler D.R. (1995) *Zoogeomorphology: animals as Geomorphic Agents.* Cambridge: Cambridge University Press.

Carson M.A. and Kirkby M.J. (1972) *Hillslope Form and Process.* Cambridge: Cambridge University Press.

Cartwright N. (1999) *The Dappled World: a Study of the Boundaries of Science.* New York: Cambridge University Press.

Chamberlain T.C. (1987) The method of multiple working hypotheses. *Journal of Geology* 5: 837–48.

Chamberlain T.C. (1909) Diastrophism as the ultimate basis of correlation. *Journal of Geology* 17: 685–93.

Chorley R.J. (1962) *Geomorphology and General Systems Theory,* U.S. Geological Survey Professional Paper 500-B. Washington, DC.

Chorley R.J. (1963) Diastrophic background to twentieth-century geomorphological thought. *Geological Society of America Bulletin* 74: 953–70.

Chorley R.J. (1965) A re-evaluation of the geomorphic system of W.M. Davis, in R.J. Chorley and P. Haggett (eds), *Frontiers in Geographical Teaching.* London: Methuen, pp. 21–38.

Chorley R.J. (1978) Bases for theory in geomorphology, in C. Embleton, D. Brunsden and D.K.C. Jones (eds), *Geomorphology Present Problems and Future Prospects.* Oxford: Oxford University Press, pp. 1–13.

Chorley R.J., Beckinsale R.P. and Dunn A.J. (1973) *The History of the Study of Landforms. 2. The Life and Work of W.M. Davis.* London: Methuen.

Chorley R.J., Dunn A.J. and Beckinsale R.P. (1964) *The History of the Study of Landforms. 1. Geomorphology Before Davis.* London: Methuen.

Chorley R.J. and Kennedy B.A. (1971) *Physical Geography: a Systems Approach.* London: Prentice Hall.

Chorley R.J., Schumm S.A. and Sugden D.E. (1984) *Geomorphology.* London: Methuen .

Church M. (1984) On experimental method in geomorphology, in T.P. Burt and D.E. Walling (eds), *Catchment Experiments in Fluvial Geomorphology.* Norwich: Geo Books, pp. 563–80.

Church M. (1996) Space, time and the mountain: how do we order what we see? in B.L. Rhoads and C.E. Thorn (eds), *The Scientific Nature of Geomorphology.* Chichester: Wiley, pp. 147–70.

Culling W.E.H. (1987) Equifinality: modern approaches to dynamical systems and their potential for geographical thought. *Transactions of the Institute of British Geographers* NS 12: 57–72.

Czech H. and Boswell K.C. (translation) (1972) *Morphological Analysis of Land Forms.* New York: Hafner. (See Penck, 1924, below for the original version.)

Da Costa N. and French S. (2000) Models, theories, and structures: thirty years on. *Proceedings of the Philosophy of Science* 67: S116–27.

Daniels M.D. and Rhoads B.L. (2004) Effect of LWD configuration on three-dimensional flow structure in two low-energy meander bends at varying stages. *Water Resources Research* 40: W11302, doi: 10 1029/2004WR003181.

Davis W.M. (1889) The rivers and valleys of Pennsylvania. *National Geographical Magazine* 1: 183–253.

Davis W.M. (1892) The convex profile of bad-land divides. *Science* 20 245.

Davis W.M. (1899) The geographical cycle. *Geographical Journal* 14: 481–504.

Davis W.M. (1905) The geographical cycle in an arid climate. *The Journal of Geology* 13: 381–407.

Davis W.M. (1909) *Geographical Essays*, in D.W. Johnson (ed.). Boston: Ginn.

de Boer D.H. (1992) Hierarchies and spatial scale in process geomorphology: a review. *Geomorphology* 4: 303–18.

Dietrich W.E. (1987) Mechanics of flow and sediment transport in river bends, in K. Richards (ed.), *River Channels: form and Process.* Oxford: Blackwell, pp. 179–224.

Dietrich W.E. and Smith J.D. (1983) Influence of the point bar on flow through curved channels. *Water Resources Research* 19: 1173–92.

Dury, G.H. (1966). The concept of grade, in G.H. Dury (ed) *Essays in Geomorphology.* Heinemann, London, 211–233.

Dutton C. (1889) On some of the greater problems of physical geology. *Bulletin of the Philosophical Society of Washington* 11: 51–64.

Flemal R.C. (1971) The attack on the Davisian system of geomorphology: a synopsis. *Journal of Geologic Education* 19: 3–13.

Geikie J. (1874) *The Great Ice Age.* London: Dalby and Ibister.

Giere R.N. (1988) *Explaining Science: a Cognitive Approach.* Chicago: University of Chicago Press.

Giere R.N. (2004) How models are used to represent reality. *Philosophy of Science* 71: 742–52.

Gilbert G.K. (1877) Report on the geology of the Henry Mountains, United States, in *Geographical and Geological Survey of the Rocky Mountain Region.* Washington: US Geographical and Geological Survey, p. 160.

Gilbert G.K. (1886) The inculcation of the scientific method by example. *American Journal of Science* (*Third Series*) 31: 284–99.

Gilbert G.K. (1890) *Lake Bonneville.* U.S. Geological Survey Monograph 1, pp. 23–65.

Gilbert G.K. (1909) The convexity of hillslopes. *Journal of Geology* 17: 344–50.

Goodman N. (1967) Uniformity and simplicity. *Geological Society of America Special Paper* 89: 93–9.

Gould S.J. (1989) Response by Stephen Jay Gould. *Bulletin Geological Society of America* 101: 998–1000.

Graf W.L. (1977) The rate law in fluvial geomorphology. *American Journal of Science* 277: 178–91.

Graf W.L. (1979) Catastrophe theory as a model for change in fluvial systems, in D.D. Rhodes and E.J. Williams (eds), *Adjustments of the Fluvial System.* London: Allen & Unwin, pp. 13–32.

Grosse J.C. and Phillips F.M. (2001) Terrestrial *in situ* osmogenic nuclides: theory and application. *Quaternary Science Reviews* 20: 1475–560.

Hack J.T. (1960) Interpretation of erosional topography in humid temperate regions. *American Journal of Science* (*Bradley volume*) 258A: 80–97.

Hack J.T. (1975) Dynamic equilibrium and landscape evolution, in W.N. Melhorn and R.C. Flemal (eds), *Theories of Landform Development.* Boston: Allen & Unwin, pp. 87–102.

Haigh M.J. (1987) The holon: hierarchy theory and landscape research. *Catena* 10(Supplement): 181–92.

Haines-Young R.H. and Petch J.R. (1986) *Physical Geography: its Nature and Methods.* London: Harper and Row.

Harris T. (2003) Data models and the acquisition and manipulation of data. *Philosophy of Science* 70: 1508–17.

Harris S.A. and Twidale C.R. (1968) Cycles geomorphic, in R.W. Fairbridge (ed.), *The Encyclopedia of Geomorphology.* New York: Reinhold, pp. 237–40.

Harvey D. (1969) *Explanation in Geography.* London: Edward Arnold.

Higgins C.G. (1975) Theories of landscape development: a perspective, in W.N. Melhorn and R.C. Flemal (eds), *Theories of Landform Development.* London: Allen & Unwin, pp. 1–28.

Hooke RLeB. (1994) On the efficacy of humans as geomorphic agents. *GSA Today* 4: 217, 224–5.

Horton R.E. (1945) Erosional development of streams and their drainage basins, hydrophysical approach to quantitative morphology. *Bulletin of the Geological Society of America* 56: 275–370.

Kennedy B.A. (1977) A question of scale? *Progress in Physical Geography* 1: 154–7.

Kennedy B.A. (1980) A naughty world. *Institute of British Geographers Transactions* 4: 550–8.

Kennedy B.A. (2006) *Inventing the Earth*. Malden, MA: Blackwell.

Keylock C.J. (2003) Mark Melton's geomorphology and geography's quantitative revolution. *Transactions of the Institute of British Geographers* 28: 142–57.

King L.C. (1963) Canons of landscape evolution. *Bulletin of the Geological Society of America* 64: 721–51.

King L.C. (1967) *The Morphology of the Earth*. New York: Hafner.

King L.C. (1983) *Wandering Continents and Spreading Sea Floors on an Expanding Earth*. Chichester: Wiley.

Kirkby M.J. (1976) Deterministic continuous slope models. *Zeitschrift für Geomorphologie Band* 25 (NF Supplement): 1–19.

Kirkby M.J. (1989) The future of modeling in physical geography, in B. Macmillan (ed.), *Remodelling Geography*. Oxford: Blackwell.

Kirkby M.J. (1990) The landscape viewed through models. *Zeitschrift für Geomorphologie Band* 79 (NF Supplement): 63–81.

Kleman J. and Stroeven A.P. (1997) Preglacial surface remnants and Quaternary glacial regimes in northwestern Sweden. *Geomorphology* 19: 35–54.

Knox J.C. (1975) Concept of the graded stream, in W.N. Melhorn and R.C. Flemal (eds), *Theories of Landform Development*. London: Allen & Unwin, pp. 169–98.

Kooi H. and Beaumont C. (1996) Large-scale geomorphology: classical concepts reconciled and integrated with contemporary ideas via a surface processes model. *Journal of Geophysical Research* 101: 3361–86.

Langbein W.B. and Leopold L.B. (1964) Quasi-equilibrium states in channel morphology. *American Journal of Science* 262: 782–94.

Langbein W.B. and Leopold L.B. (1966) *River Meanders – theory of Minimum Variance*, U.S. Geological Survey Professional Paper No. 422-H. Washington, DC.

Leopold L.B. and Langbein W.B. (1962) *The Concept of Entropy in Landscape Evolution*, U.S. Geological Survey Professional Paper 500-A. Washington, DC.

Leopold L.B. and Maddock Jr. T. (1953) *The Hydraulic Geometry of Stream Channels and Some Physiographic Implications*, U.S. Geological Survey Professional Paper 252. Washington, DC.

Lyell C. (1830) *Principles of Geology*, 1st edn, vol. 1. London: John Murray.

McCarthy T.S., Ellery W.N. and Stanistreet I.G. (1992) Avulsion mechanisms on the Okavango Fan, Botswana – the control of a fluvial system by vegetation. *Sedimentology* 39: 779–95.

Melton M.A. (1958) Correlation structure of morphometric properties of drainage systems and their controlling agents. *Journal of Geology* 66: 422–60.

Montgomery D. (2001) Slope distributions, threshold hillslopes, and steady-state topography. *American Journal of Science* 301: 432–54.

Morisawa M. (1962) Quantitative geomorphology of some watersheds in the Appalachian Plateau. *Geological Society of American Bulletin* 73: 1025.

Peltier L.C. (1950) The geographic cycle in periglacial regions as it is related to climatic geomorphology. *Annals of the Association of American Geographers* 40: 214–36.

Penck W. (1924) *Die morphologische Analyse: ein Kapitel der physikalischen Geologie*. Geographische Abhandlungen 2, Reihe Heft 2. Stuttgart: Engelhorn.

Phillips J.D. (1992a) The end of equilibrium. *Geomorphology* 5: 195–201.

Phillips J.D. (1992b) Nonlinear dynamical systems in geomorphology: revolution or evolution? *Geomorphology* 5: 219–29.

Phillips J.D. (1999) *Earth Surface Systems: complexity, Order and Scale*. Oxford: Blackwell.

Ramsay A.C. (1846) *The Denudation of South Wales Memoir of the Geological Survey of Great Britain, 1*. London: Her Majesty's Stationery Office, pp. 297–335.

Ramsay A.C. (1862) On the glacial origin of certain lakes in Switzerland, the Black forest, Great Britain, Sweden, North America, and elsewhere. *Quarterly Journal of the Geological Society* 18: 185–204.

Rhoads B.L. (1994) On being a 'real' geomorphologist. *Earth Surface Processes and Landforms* 19: 269–72.

Rhoads B.L. (1999) Beyond pragmatism: The value of philosophical discourse in physical geography. *Annals of the Association of American Geographers* 89: 760–71.

Rhoads B.L. (2004) Whither physical geography? *Annals of the Association of American Geographers* 94: 748–55.

Rhoads B.L. (2006) The dynamic basis re-envisioned: a process perspective on geomorphology. *Annals of the Association of American Geographers* 96: 14–30.

Rhoads B.L. and Sukhodolov A.N. (2001) Field investigation of three-dimensional flow structure at stream confluences: 1. Thermal mixing and time-averaged velocities. *Water Resources Research* 37: 2393–410.

Rhoads B.L. and Thorn C.E. (1993) Geomorphology as science: the role of theory. *Geomorphology* 6: 287–307.

Rhoads B.L. and Thorn C.E. (1994) Contemporary philosophical perspectives on physical geography with emphasis on geomorphology. *Geographical Review* 84: 90–101.

Rhoads B.L. and Thorn C.E. (1996a) Toward a philosophy of geomorphology, in B.L. Rhoads and C.E. Thorn (eds), *The Scientific Nature of Geomorphology*. Chichester: Wiley, pp. 115–43.

Rhoads B.L. and Thorn C.E. (1996b) Observation in geomorphology, in B.L. Rhoads and C.E. Thorn (eds), *The Scientific Nature of Geomorphology*. Chichester: Wiley, pp. 21–56.

Richards K.S. (1990) 'Real' geomorphology. *Earth Surface Processes and Landforms* 15: 195–7.

Richards K.S. (1996) Samples and cases: generalization and explanation in geomorphology, in B.L. Rhoads and C.E. Thorn (eds), *The Scientific Nature of Geomorphology*. Chichester: Wiley, pp. 171–90.

Ritter D.F. (1978) *Process Geomorphology*. Dubuque, IA: WC Brown.

Scheidegger A.E. (1961) *Theoretical Geomorphology*. Berlin: Springer-Verlag.

Schumm S.A. (1977) *The Fluvial System*. New York: Wiley.

Schumm S.A. (1979) Geomorphic thresholds: the concept and its applications. *Transactions of the Institute of British Geographers* NS 4: 485–515.

Schumm S.A. and Lichty R.W. (1965) Time, space and causality in geomorphology. *American Journal of Science* 263: 110–19.

Schumm S.A., Mosley M.P. and Weaver W.E. (1987) *Experimental Fluvial Geomorphology*. New York: Wiley.

Sherman D.J. (1996) Fashion in geomorphology, in B.L. Rhoads and C.E. Thorn (eds), *The Scientific Nature of Geomorphology*. Chichester: Wiley 87–114.

Slaymaker O. (ed.) (1991) *Field Experiments and Measurement Programs in Geomorphology*. Rotterdam: Balkema.

Slaymaker O., Spencer T. and Embleton-Hamann C. (eds) (2009) *Geomorphology and Global Environmental Change*. Cambridge: Cambridge University Press.

Slingerland R. (1989) Predictability and chaos in quantitative dynamic stratigraphy, in T.A. Cross (ed.), *Quantitative Dynamic Stratigraphy*. Englewood Cliffs, NJ: Prentice-Hall, pp. 43–53.

Smalley I.J. and Vita-Finzi C. (1969) The concept of 'system' in the earth sciences, particularly geomorphology. *Bulletin Geological Society of America* 80: 1591–4.

Stoddart D.R. (1966) Darwin's impact on geography. *Annals of the Association of American Geographers* 56: 683–98.

Strahler A.N. (1952a) Dynamic basis of geomorphology. *Bulletin Geological Society of America* 63: 923–38.

Strahler A.N. (1952b) Hypsometric (area–altitude) analysis of erosional topography. *Geological Society of America* 63: 1117–42.

Strahler A.N. (1957) Quantitative analysis of watershed geomorphology. *American Geophysical Union Transactions* 38: 913–20.

Strahler, A.N. (1958) Dimensional analysis applied to fluvially eroded landforms. *Bulletin Geological Society of America* 69: 279–300.

Suess E. (1883–1908) *Das Antlitz der Erde*, three volumes. Vienna: Tempsky.

Suppe F. (1977) The search for philosophic understanding of scientific theories, in F. Suppe (ed.), *The Structure of Scientific Theories*. Urbana, IL: University of Illinois Press.

Suppe F. (2000) Understanding scientific theories: an assessment of developments, 1969–1998. *Philosophy of Science* 67(Proceedings): S102–15.

Tarboton D.G., Bras R.L. and Rodriguez-Iturbe I. (1988) The fractal nature of river networks. *Water Resources Research* 24: 1317–22.

Thorn C.E. (1988) *An Introduction to Theoretical Geomorphology*. Boston: Unwin Hyman.

Thornes J.B. (1978) The character and problems of theory in contemporary geomorphology, in C. Embleton, D. Brunsden and D.K.C. Jones (eds), *Geomorphology: present Problems and Future Prospects*. London: Oxford University Press, pp. 14–24.

Thornes J.B. (1983) Evolutionary geomorphology. *Geography* 68: 225–35.

Tinkler K.J. (1985) *A Short History of Geomorphology*. London: Croom Helm.

von Bertalanffy L. (1956) General systems theory. *General Systems Yearbook* 1: 1–10.

von Bertalanffy L. (1962) General systems theory: a critical review. *General Systems Yearbook* 7: 1–20.

von Engelhardt W. and Zimmermann J. (1988) *Theory of Earth Science* (translated by L. Fischer). Cambridge: Cambridge University Press.

Wade R.J., Rhoads B.L., Newell M.D., Wilson D., Garcia M. and Herricks E.E. (2002) Integrating science and technology to support stream naturalization near Chicago, Illinois. *Journal of American Water Resources Association* 38: 931–44.

Werritty A. (2008) Valley-side slopes and drainage basins. II: Geometry and evolution, in T.P. Burt, R.J. Chorley, D. Brunsden, N.J. Cox and A.S. Goudie (eds), *The History of the Study of Landforms. Volume 4: Quaternary and Recent Processes and Forms (1890–1965) and the Mid Century Revolutions*. Bath, UK: The Geological Society of London, pp. 351–93.

Wilcock P.R. and Iverson R.M. (eds) (2003) *Prediction in Geomorphology*. Washington, DC: American Geophysical Union.

Willgoose G., Bras R.L. and Rodriguez-Iturbe I. (1992) The relationship between catchment and hillslope properties: implications of a catchment evolution model. *Geomorphology* 5: 21–37.

Wohl E., Angermeier P.L., Bledsoe B., Kondolf G.M., MacDonnell L., Merritt D.M., Palmer, M.A., Poff, N.L. and Tarboton, D. (2005) River restoration. *Water Resources Research* 41: W10301 doi:10.1029/2005WR003985.

Wolman M.G. and Gerson R. (1978) Relative scales of time and effectiveness of climate in watershed geomorphology. *Earth Surface Processes* 3: 189–208.

Wolman M.G. and Miller J.P. (1960) Magnitude and frequency of forces in geomorphic processes. *Journal of Geology* 68: 54–74.

Young A. (1972) *Slopes*. Edinburgh: Oliver and Boyd.

5

Geomorphology in Environmental Management

Peter W. Downs and Derek B. Booth

'Environmental management' is both a multi-layered social construct, in which environmental managers interact with the environment and each other, and a field of study emphasizing the need for interdisciplinary understanding of human–environment interactions (Wilson and Bryant, 1997). Environmental managers are those

> whose livelihood is primarily dependent on the application of skill in the active and self-conscious, direct or in-direct, manipulation of the environment with the aim of enhancing predictability in a context of social and environmental uncertainty.
>
> (Wilson and Bryant, 1997: 7)

Thus the goal of environmental management is '...to harmonize and balance the various enterprises which man has imposed on natural environments for his own benefit' (Goudie, 1994: 181). The perception of 'benefit', however, depends on the prevailing management vision and objectives for a particular environmental facet (water, rivers, beaches, deserts, etc.) or location. These variables lead to at least five fundamental dimensions to environmental management: the importance of place, implications of scale, situation in time, the cultural context and political framework (Downs and Gregory, 2004).

These dimensions result in *considerable* scope for social and environmental uncertainty that is largely beyond the realm of this chapter. Instead, we concern ourselves primarily with the 'application of skill' in enhancing the predictability of environmental manipulation from a geomorphological perspective. Geomorphology, as the study of the origins and evolution of Earth's landforms and the processes that shape them, is clearly part of the disciplinary scientific basis of environmental management, and there is a long-standing concern with 'applicable' geomorphology (see Gregory, 1979) that investigates geomorphological processes and resulting landforms under the influence of human activity. Graf (2005) argues that geomorphology has been closely associated with public policy and land-use management in the United States since the late 19th century, especially through Grove Karl Gilbert's studies for the US Geological Survey (Gilbert, 1890, 1914, 1917). While few applicable geomorphology studies occurred through the first half of the 20th century (Cooke and Doornkamp, 1974), geomorphology has both re-connected and increasingly strengthened its association with management applications since that time (Graf, 2005) through process-based studies beginning with Horton (1945) and chapters by Leopold (Leopold, 1956) and Strahler (Strahler, 1956) in the seminal publication *Man's Role in Changing the Face of the Earth* (Thomas, 1956). By the century's end, Graf (1996: 443) contended that

> Geomorphology as a natural science is returning to its roots of a close association with environmental resource management and public policy ... there is a new emphasis on application of established theory to address issues of social concern.

Despite an extensive literature on the applicability of geomorphology to problems in environmental management, we find few descriptions of direct geomorphological application. On inspection, the majority of texts regarding 'applied geomorphology' (see Table 5.1) actually involve *applicable* studies rather than the true '…application of geomorphological techniques and analysis to a planning, conservation, resource evaluation, engineering or environmental problem' (Brunsden et al., 1978: 251). Perhaps this phenomenon can be explained because true 'applied geomorphologists' are environmental professionals whose priorities lay primarily with their clients rather than with writing for publication, leaving reports on applied geomorphology mostly to a small cadre of academics who consult part time and whose career success is defined mainly by published manuscripts, chapters and books.

Table 5.1 Example texts in 'applied' geomorphology

Authors/Editors	Year	Title	Publisher
Steers, J.A.	1971	*Applied Coastal Geomorphology*	M.I.T. Press, Cambridge, MA
Coates, D.R. (ed)	1971	*Environmental Geomorphology*	State University of New York Publications in Geomorphology, Binghampton
Coates, D.R. (ed)	1972	*Environmental Geomorphology and Landscape Conservation. Volume I. Prior to 1900*	Dowden, Hutchinson & Ross, Stroudsburg
Coates, D.R. (ed)	1973	*Environmental Geomorphology and Landscape Conservation. Volume III. Non-urban regions*	Dowden, Hutchinson & Ross, Stroudsburg
Cooke, R.U. and Doornkamp, J.C.	1974	*Geomorphology in Environmental Management*	Oxford University Press, Oxford
Coates, D.R. (ed)	1976	*Geomorphology and Engineering*	Allen & Unwin, London
Hails, J.R. (ed)	1977	*Applied Geomorphology: A Perspective of the Contribution of Geomorphology to Interdisciplinary Studies and Environmental Management*	Elsevier Scientific Pub. Co., New York
Dunne, T. and Leopold, L.B.	1978	*Water in Environmental Planning*	W.H. Freeman, San Francisco
Cooke, R.U., Brunsden, D., Doornkamp, J.C. and Jones, D.K.C.	1982	*Urban Geomorphology in Drylands*	Clarendon Press, Oxford
Craig, R.G. and Craft, J.L.	1982	*Applied Geomorphology*	Allen & Unwin, London
Verstappen, H. Th.	1983	*Applied Geomorphology: Geomorphological Surveys for Environmental Development*	Elsevier, Amsterdam
Cooke, R.U.	1984	*Geomorphological Hazards in Los Angeles*	Allen & Unwin, London
Costa, J.E. and Fleischer, P.J. (eds)	1984	*Developments and Applications of Geomorphology*	Springer-Verlag, Berlin
Petts, G.E.	1984	*Impounded Rivers: Perspectives for Ecological Management*	John Wiley and Sons, Chichester
Doornkamp, J.C.	1985	*The Earth Sciences and Planning in the Third World*	Liverpool University Press, Liverpool
Graf, W.L.	1985	*The Colorado River – Instability and Basin Management*	Association of American Geographers, Washington, DC
Fookes, P.G. and Vaughan, P.R	1986	*Handbook of Engineering Geomorphology*	Blackie, Glasgow
Hart, M.G	1986	*Geomorphology, Pure and Applied*	Allen & Unwin, London

Continued

Table 5.1 Cont'd

Authors/Editors	Year	Title	Publisher
Toy, T.J. and Hadley, R.F.	1987	Geomorphology and Reclamation of Disturbed Lands	Academic Press, Orlando
Brookes, A.	1988	Channelized Rivers: Perspectives for Environmental Management	John Wiley and Sons, Chichester
Hooke, J.M. (ed)	1988	Geomorphology in Environmental Planning	John Wiley and Sons, Chichester
Cooke, R.U. and Doornkamp, J.C.	1990	Geomorphology in Environmental Management: A New Introduction	Clarendon Press, Oxford
Morisawa, M.	1994	Geomorphology and Natural Hazards	Elsevier, Amsterdam
McGregor, D.M and Thompson, D.A. (eds)	1995	Geomorphology and Land Management in a Changing Environment	John Wiley and Sons, Chichester
Brookes, A. and Shields Jr. F.D. (eds)	1996	River Channel Restoration: Guiding Principles for Sustainable Projects	John Wiley and Sons, Chichester
Bird, E.C.F.	1996	Beach Management	John Wiley and Sons, Chichester
Viles, H. and Spencer, T.	1996	Coastal Problems: Geomorphology, Ecology and Society at the Coast	Arnold, London
Thorne, C.R., Hey, R.D. and Newson, M.D. (eds)	1997	Applied Fluvial Geomorphology for River Engineering and Management	John Wiley and Sons, Chichester
Hooke, J.M.	1998	Coastal Defence and Earth Science Conservation	Geological Society, Bath
Giardino, J.R. and Marston, R.A.	1999	Engineering Geomorphology	Elsevier, Amsterdam
Oya, M.	2001	Applied Geomorphology for Mitigation of Natural Hazards	Kluwer Academic Publishers, Boston
Marchetti, M. and Rivas, V.	2001	Geomorphology and Environmental Impact Assessment	A.A. Balkema, Lisse
Allison, R.J. (ed)	2002	Applied Geomorphology: Theory and Practice	John Wiley and Sons, Chichester
Kneupfer, P.L.K. and Petersen, J.F. (eds)	2002	Geomorphology in the Public Eye	Elsevier, Amsterdam
Downs, P.W. and Gregory, K.J.	2004	River Channel Management: Towards Sustainable Catchment Hydrosystems	Arnold, London
Bremer, H. and Dieter Burger, D.	2004	Karst and Applied Geomorphology: Concepts and Developments	Gebruder Borntraeger, Berlin
Brierley, G.J. and Fryirs, K.A.	2005	Geomorphology and River Management: Applications of the River Styles Framework	Blackwell, Oxford
Fookes, P.G. Lee, E.M. and Milligan, G. (eds)	2005	Geomorphology for Engineers	Whittles, Caithness
Anthony, D.J., Harvey, M.D., Laronne, J.B. and Mosley, M.P. (eds)	2006	Applying Geomorphology to Environmental Management	Water Resources Publications, Highlands Ranch, CO
James, L.A. and Marcus, W.M. (eds)	2006	The Human Role in Changing Fluvial Systems	Elsevier, Amsterdam
Fookes, P.G. Lee, E. M. and Griffith, J.S.	2007	Engineering Geomorphology: Theory and Practice	Whittles, Caithness

In this review, we will first consider the evolving role for geomorphology in environmental problem-solving and the ways in which geomorphological services are provided to environmental managers. Three specific roles for geomorphology in environmental management are illustrated, namely applications to natural hazard avoidance and diminution, environmental restoration and conservation, and the sustainable development of natural resources. Evidence from these contributions provides the basis for suggesting some core skills and standards required by applied geomorphologists. Finally, we assess the potential future role for geomorphology in environmental management and argue that truly successful applications will demand reconceptualizing management problems from a geomorphological perspective, and not simply applying geomorphology within the constraints of traditional management practice.

INCREASING OPPORTUNITY FOR APPLICATION

The substance and style of geomorphology applications in environmental management have evolved over time. Scientifically, the discipline of geomorphology has continued to add technical expertise and tools that allow its applied scientists to better tackle environmental issues. Notable in this regard has been the greater availability of predictive tools, frequently stemming from GIS-based terrain modelling, with which to evaluate alternative management scenarios, thus increasing the visibility and transparency of geomorphology-centred solutions. Socially, environmental managers and the public have gradually recognized the relevance of geomorphology in environmental problem-solving, leading to greater numbers of geomorphologists interacting with public policy (Kneupfer and Petersen, 2002). In tandem with a growing public awareness that environmental conditions are important in determining human quality of life (since at least Rachel Carson's 1962 book *Silent Spring*), scientific studies of the impacts of humans on ecosystem processes have increasingly highlighted their geomorphological underpinnings. Consequently, geomorphologists have been asked to tackle an ever-broadening variety of environmental management problems.

The evolution of contributions of fluvial and coastal geomorphology to environmental management in the United Kingdom is illustrative. Original engineering solutions based on dominating and controlling nature had no place for geomorphology, but a subsequent shift from 'hard' to 'soft' engineering solutions ushered in what Hooke (1999) calls the 'first phase' of geomorphology application, based on the recognition that landforms change naturally during the lifespan of an engineering project and that projects disrupting natural geomorphic processes were frequently causing deleterious effects elsewhere. A second phase occurred once strategic geomorphological questions were asked during early project planning phases, where local geomorphological baseline information was collected and utilized to answer specific landform questions ahead of implementation, and where geomorphologists were actively involved in the project design (and ultimately its appraisal). Hooke's predicted third phase (which, a decade later, we argue is now the present) emphasizes our understanding the conditions governing geomorphological variability, instability and equilibrium, and the widespread use of enhanced modelling and remote data to predict the effects and the risks of different management scenarios.

There are now several different drivers promoting geomorphological contributions to environmental management. First and most commonly, geomorphology is 'part of the solution' (Gardiner, 1994; FISRWG, 1998; King et al., 2003) contributing within a multi- and interdisciplinary framework to solve specific environmental problems (Jewitt and Görgens, 2000; Rogers, 2006). For many years, especially in the lowlands, landscapes were imagined to be largely static during the engineering time-frame of a project, and thus geomorphological processes were ignored. However, both research and empiricism have showed the fallacy of this assumption, leading to the progressive integration of geomorphology in management. Now, because environmental management objectives usually involve either reducing the risk posed to the built environment by landform change, minimizing the impact of floods or mass movements, or restoring charismatic native aquatic and terrestrial species, geomorphology is frequently a tool in the service of end-points in engineering, land-use planning, and biology (respectively). Examples are provided later in this chapter.

A second driver is the widely recognized connection between environmental degradation and socio-economic deterioration (Kasperson and Kasperson, 2001), with the most vulnerable communities existing in areas of high environmental sensitivity and low social resilience (Fraser et al., 2003). Included in this category are contributions in reducing the impact of natural disasters, a topic which demands an understanding of the coupled nature of human and natural disaster vulnerability (Alcántara-Ayala, 2002). Geomorphology has become implicitly recognized as a strategic component of environmental and social justice related to sustainability and global change. Understanding the mutual vulnerability (and dependence) of society and landscape may be one of the most demanding challenges to geomorphology (Slaymaker, 2009) if it wants to be seen as a key element of global environmental management.

A third driver is more intrinsic and involves a growing effort to support and retain 'geodiversity', the earth science counterpoint to biodiversity. Geodiversity is defined as 'the natural range (diversity) of geological (rocks, minerals, fossils), geomorphological (landform, processes) and soil features. It includes their assemblages, relationships, properties, interpretations and systems' (Gray, 2004: 8). At a process level, geomorphological contributions to this goal involve the restoration and maintenance of physical integrity in environmental systems (e.g. Graf, 2001), for instance in trying to maintain functional river processes and forms even under a modified hydrological regime. At the level of landforms, geodiversity involves the preservation and conservation management of parkland environments (Gordon et al., 1998, 2002) or unique landforms (Downs and Gregory, 1994). In England, this has taken the form of a network of Regionally Important Geological/Geomorphological Sites (RIGS) (see McEwan, 1996) and progress towards a series of Local Geodiversity Action Plans as the mechanism for delivering geoconservation. Thus in contrast to the other drivers, in this application geomorphology may be the solution and endpoint.

In response to these increased demands for geomorphological services, employment opportunities for 'professional' geomorphologists have increased. Whereas geomorphology expertise was once provided almost solely by geomorphologists contracted from academia, there is now a far wider client base (Table 5.2). One result is greater collaborative problem-solving between professional and academic geomorphologists, frequently through academic involvement as 'expert advisors' on projects. A second result, however, is greater ambiguity in defining professionalism in geomorphology. This is a problem of long standing (see Brunsden et al., 1978) and sees the title 'geomorphologist' liberally and sometimes disingenuously applied: we return to this theme later after reviewing a suite of recent contributions of geomorphology in environmental management.

GEOMORPHOLOGICAL CONTRIBUTIONS: SERVICES FOR ENVIRONMENTAL MANAGEMENT

Geomorphological application in environmental management arises primarily as proposed changes

Table 5.2 Client groups for geomorphological services

Client groups

- Individuals
- Developers
- Various tiers of government: local, state, federal (and tribes in the United States)
- Consulting engineers, planners, landscape architects and biologists
- Conservation-focused NGOs and environmental advocacies
- Natural resource managers (energy utilities, irrigation districts, forest managers, resource conservation districts, aggregate miners)
- Reinsurance officers
- Lawyers
- Universities
- International agencies/organizations

Source: Updated from Brunsden, 1996.

in land use, development of natural resources, restoration initiatives, and measures to reduce natural hazards are filtered through numerous protective policies and regulations imposed at a local, national or supra-national scale. In the Pacific North west of the United States, where a maze of state and federal entities each have jurisdiction over some element of a proposed project, Shannon (1998) reports that there are normally 17 tribal, state and federal agencies involved as water passes through a drainage basin, creating multiple opportunities for geomorphology applications with somewhat different objectives.

Many geomorphology services in the United States stem from the 1973 Endangered Species Act and the 1977 Clean Water Act while, in Europe, the 1992 European Union Habitats Directive, the 2000 Water Framework Directive (WFD) and now the 2007 Floods Directive are starting to profoundly influence approaches to environmental management (Clarke et al., 2003; Manariotis and Yannopoulos, 2004; Wharton and Gilvear, 2006). Therefore, the focus and scale of geomorphology services can vary widely, for instance, from helping protect and restore summer breeding habitats of the federally endangered riparian bird, the south-western willow flycatcher (*Empidonax traillii extimus*) (Graf et al., 2002), to a strategic role in the WFD's goal of good 'hydromorphological' quality across all rivers of the member states.

The geomorphological tools applicable to a particular problem will depend on the perceived problem, the management context, available funding and the geomorphological environment of concern, resulting in at least seven categories of geomorphological service including:

1   Project orientation – designed to provide initial insights into the problem or issue;
2   Determination of current site conditions – using desk study, field work, monitoring, and analysis, and usually designed, at least implicitly, to understand the sensitivity to change of the landform;
3   Interpretative analytical investigation of past site conditions – using conceptual models, historic databases and various landform dating techniques to inform probable historical conditions;
4   Prediction of future site conditions – interpretative or using numerical modelling tools applied to predict landform sensitivity to various potential management scenarios;
5   Problem solution and design – almost always as part of a multi-disciplinary team;
6   Post-project appraisal monitoring and evaluation – ideally including implementation, effectiveness and validation monitoring to inform an evaluation of project sustainability;
7   Expert advisory – frequently related to litigation, insurance claims and expert witness testimony.

A list of potential contributions under each type of service is provided in Table 5.3. With the possible exception of expert advisory, project involvement generally implies involvement in more than one service. For example, land-use planning projects frequently require at the least the first four services, and project implementation supports the first five. The sixth service, post-project appraisal, is something of an enigma: while, logically, every project should be evaluated (and thus monitored) as the basis for informing future practice (Downs and Kondolf, 2002), funding is rarely set aside. Assessment of appraisal practice, notably the National River Restoration Science Synthesis in the United States, has revealed a far more ad hoc basis for project appraisal (Bernhardt et al., 2005, 2007). This is particularly unfortunate because a critical requirement of post-project appraisal is a suitable baseline data set (service 2), meaning that land-use planning and engineering projects should integrally involve appraisal design at their outset.

As illustration of the range of services geomorphology can provide to environmental management, we provide a series of examples below organized according to three reasonably distinct management requirements, namely: hazard avoidance and diminution, sustainable development of natural resources, and environmental restoration and preservation. They are drawn from recently published journal articles and book chapters, and our personal experiences in fluvial and hillslope geomorphology.

Hazard avoidance and diminution: geomorphology services in support of engineering

Perhaps the arena of longest standing interest (see Cooke and Doornkamp, 1974; Coates, 1976) is the application of geomorphology to the diminution of hazards associated with landscape change, usually related to erosion and sedimentation. More broadly, this reflects geomorphology's technical contribution to reducing the impact of natural disasters (see Alcántara-Ayala, 2002). Such application relates primarily to site-specific geomorphology studies in the service of river, coastal, geotechnical and agricultural engineering, but it can also include risk assessment services to the insurance industry (Doornkamp, 1995). These services have spawned the sub-discipline of 'engineering geomorphology' (e.g. Fookes and Vaughan, 1986; Fookes et al., 2005, 2007); that is, geomorphology applied in assessing the risks to construction associated with surface processes and landform change (Fookes et al., 2007). Examples of the application to river engineering and geotechnical engineering are provided below: numerous other examples are contained in a special issue of *Geomorphology* edited by Giardino and Marston (1999).

River bed and bank protection: geomorphology and river engineering

With increasing settlement and agriculture of floodplain lands has come a need for channel 'stability', a sub-set of channelization (see Brookes, 1988) wherein the river's natural tendency for lateral migration is perceived as a hazard and forcibly resisted by structural reinforcement of river banks to prevent erosion and fix the channel planform. Planned bank protection frequently involves symptomatic and piecemeal application of rip-rap, gabion baskets, concrete walls and sheet steel piling. Likewise, erosion of the channel bed has frequently been arrested by concrete or structural grade controls which, together with bank protection, can result in a fully immobile channel. However, such schemes are invariably detrimental to instream habitat for native aquatic species, have high failure rates, and have frequently exacerbated channel erosion problems downstream or upstream requiring additional protection measures that are also environmentally deleterious. As a consequence, there is now a suite

Table 5.3 Geomorphological services in environmental management

Service	Environmental management purpose
Project orientation	Hypothesis generation about probable field conditions
	Development of initial conceptual model of system functioning
Determination of current conditions	Undertake point-in-time (baseline) inventory, as basis for project design and post-project monitoring and evaluation
	Determine likely compliance with regulations
	Determine necessity of additional studies
	Contribute catchment- or network-based studies to river basin and floodplain plans
	Determine opportunities for environmental improvement/restoration
	Identify critical locations: at-risk habitats, extent of current erosion/deposition risk
	Quantify impact of past human activities
	Determine whether apparent risk warrants remedial action
	Early phase advice on whether proposed management approach will succeed
	Initial indication of measurable success criteria
Investigation of past conditions	Indicate significance of evolutionary trajectory of the landform for planned activities
	Indicate whether the proposed development is likely to cause impact
	Guidance on whether proposed restoration will have the desired beneficial impact
	Indicate similarity to historical conditions
	Basis for developing sustainable management options
	Basis for conceptual project design
Prediction of future conditions	Assist in option selection: judge likely success of proposed management actions
	Provide guidelines for necessary project design
Problem solution/design	Advise on sustainable approaches to management solution
	Advise on siting infrastructure (roads, bridges, houses, etc.)
	Contribute bounding parameters to project design
	Impact analysis of likely compliance with environmental regulations
	Risk assessment related to policies for river-basin management, land-use zoning, flood management
	Collaborative contribution as part of project solution team
Post-project monitoring and evaluation	Monitor effectiveness of implemented solution
	Evaluate project success and contribution to knowledge
	Identify unanticipated actions that require remedial attention
	Evaluate efficacy of management method in comparison with others
	Evaluate completion of commitments by project proponents
Expert advisory	Supply analysis for use in legal cases
	Provision of expert witness testimony and expert opinion
	Advocacy for client

of protective regulations in most countries that amplify the need for far more *strategic* applications that minimize both the environmental impact and the economic cost to taxpayers. These changes have provided numerous opportunities for geomorphology to contribute to river engineering and management (e.g. Thorne et al., 1997; Skidmore et al., 2009).

Breaking with the tradition of always utilizing river bank protection near to floodplain development, geomorphologists are now frequently required to analyse the risk posed to the development by the river's natural evolutionary tendency. Techniques include the use of field reconnaissance surveys (Downs and Thorne, 1996; Thorne, 1998), overlays of historical aerial photographs and large-scale topographic map (Graf, 1984, 2000), and other techniques (Lawler et al., 1997) including procedures for large rivers (Thorne, 2002). Management approaches are then proposed based on the cause, severity, extent and mode of bank failure in the vicinity of the perceived need (Thorne et al., 1996). Likewise, because research has shown that channel-bed erosion frequently occurs via the upstream migration of an incising knickpoint in channelized rivers (Schumm et al., 1984; Simon, 1989), a regional geomorphological assessment of river bed and bank conditions (Figure 5.1) can determine the most effective site (and minimum requirement) for grade-control structures (see examples in Darby and Simon, 1999) or be used to propose maintenance activities in low-energy river environments prone to sedimentation (Sear et al., 1995; Newson et al., 1997; Landwehr and Rhoads, 2003). Again strategic approaches such as use of multiple low structures to manage a large knickpoint by mimicking step-pool channel morphology to provide better habitat characteristics (e.g. Chin et al., 2008) show the influence of geomorphology input.

Reducing hillslope and landslide hazards: geomorphology and geotechnical engineering

Slide-prone areas can offer some of the most attractive sites for new development for many reasons, ranging from the economic value associated with views from the hillside to the simple fact of being some of the last remaining undeveloped land in areas of otherwise dense population. The value of geomorphology in reducing the risk of landslides has almost always been recognized for many decades, but the importance of that role relative to other hazard-mitigation strategies continues to vary through history and by locality. Spectacular examples of previously unrecognized landslides demonstrate the cost of ignorance (see Leighton et al., 1984; Linden, 1989), although the human response to instabilities once recognized can range from complete avoidance to massive hillslope reconstruction.

Geomorphology input often consists of landslide hazard mapping as the basis for minimizing slope instability risk in populated areas. Specific sites or entire regions are assigned a relative

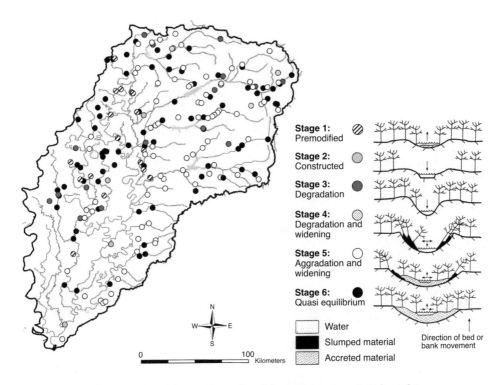

Figure 5.1   Using reconnaissance survey and rapid assessment protocols and to characterize sites in the Yazoo River catchment, Mississippi, according to their stages in river bed and bank erosion (see legend) following the passage of multiple knickpoints (adapted from Simon et al., 2007b)

hazard rating, based typically on one or more of the factors understood to determine stability – past landslides, slope angle, surficial and underlying geologic material(s), hillslope hydrology, vegetation, prior engineering works, active geomorphic process (such as wave action at the base of a slope). The choice of 'relevant' factors and the assignment of hazard levels are based on some combination of local (empirical) knowledge and geomorphic principles (e.g. Tubbs, 1974). At its simplest, 'steep slopes' are deemed 'hazardous', and the only geomorphic principle being used is the gross importance of hillslope gradient in driving downslope processes. More recent and increasingly sophisticated approaches continue to

make use of basic topographic information but now include high-resolution data, additional types of information on slope conditions and material properties, and predictive techniques adapted from the field of artificial intelligence (Ayalew and Yamagishi, 2005; Chacón et al., 2006; van Westen et al., 2008).

Beyond landslide mapping, a geomorphologist's interpretational skills can be used, for instance, in recognizing the presence of prior landslides which is often one of the best predictors of actual or potential instability (Figure 5.2). Geomorphic features used to identify past landslides commonly include hummocky topography, bent trees, springs and seeps, and arcuate

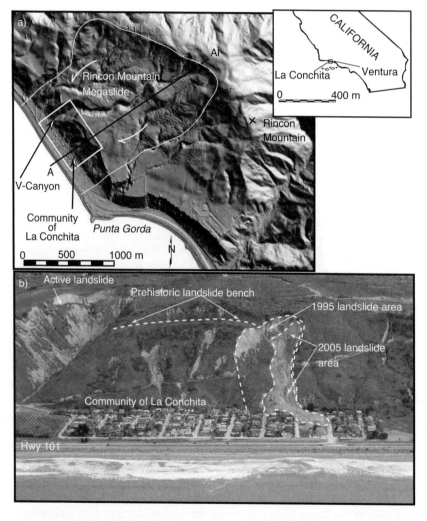

Figure 5.2 Interpretation of LiDAR/aerial imagery to identify multiple ages of landslides above La Conchita, California, including a prehistoric landslide that lay unrecognized during the development of the community of La Conchita (from Gurrola et al., 2010)

scarps (Dunne and Leopold, 1978; Sidle et al., 1985). This approach can be combined with more rigorous, mechanistic modelling of driving and resisting forces in a soil or rock mass in critical areas identified by the geomorphologists. In some cases such analyses are devoid of additional geomorphic input, which may be justifiable insofar as the land mass often behaves as predicted by engineering analysis. In many cases, however, the heterogeneity of hillslope deposits, the presence of groundwater or surface water, and the varied influences of human infrastructure and human activities require an integrative analysis geomorphology is well-suited to implement.

Sustainable development of natural resources: geomorphology services in support of natural resources management and land-use planning

A second arena for service is in the sustainable management of natural resources, including agriculture and land-use planning. In this sense, sustainable management can be defined in relation to the preservation or enhancement of the total stock of natural capital, zero or minimum net negative impact of management operations, and zero or minimum requirement for ongoing management intervention to uphold system values (Clark, 2002). Relative to engineering geomorphology, this arena has been less clearly articulated as a focus area for geomorphology in environmental management, but services generally relate to the assessment of geomorphological impacts to assist in the development of plans for habitat conservation, soil conservation, urban development, water supply, water quality, forest management, river basin management or beach management. Three examples are provided related to clean water provision, urban planning and the management of large dams.

Assessing fine sediment sources and pathways: geomorphology and clean water protection

Many protective policies for clean water focus on minimizing the deleterious impact of land management activities, and this frequently involves trying to reduce excess fine sediment delivery to potable water supplies or aquatic habitat: more than ten federal laws in the United States allow federal, state and tribal agencies to govern sediment quality (Owens et al., 2005). For aquatic habitat, especially for salmonid species, research has indicated the detrimental impact of too much fine sediment on spawning, rearing and shelter habitat (Beschta and Jackson, 1979; Sear, 1993;

Anderson et al., 1996). Such 'excess' fine sediment is usually derived from land-surface disturbance caused by agriculture (Collins et al., 1997; Walling et al., 1999), forestry clearance and management (Luce and Wemple, 2001; Owens et al., 2005), or early-phase urban development (Wolman, 1967). Such impacts also leave a significant legacy impact on the channel morphology processes that can affect fine sediment loads for decades to hundreds of years after land-use change (Trimble, 1997; Fitzpatrick and Knox, 2000; Prosser et al., 2001).

The geomorphologist's goal is frequently to identify the likely source and pathways of excess sediment by determining sediment sources or a catchment sediment budget. This information allows regulators to promote best management practices or take punitive actions, as necessary. The approach to defining a sediment budget depends partly on the size of the catchment but usually involves a collation and analysis of available data, catchment modelling and field surveys (Reid and Dunne, 1996; Gregory and Downs, 2008). Various catchment models can be used; one popular approach is to overlay readily available digital information regarding the channel network, geology, hillslope gradients and vegetation data on a digital elevation model to produce a discrete set of landscape units or process domains (Montgomery and Foufoula-Georgiou, 1993; Montgomery, 1999) which, by virtue of their coherence should produce similar unit-area sediment production rates. The resulting map of units subsequently provides a way of organizing field survey to cover a representative range of different units, verifying mapped erosion sources and estimating their dimensions, and estimating the delivery ratio of hillslope material to the channel network (Figure 5.3). Field survey also records channel erosion processes including the dimensions of bank failure and the extent of vertical incision relative to evidence of adjustment from channel morphology, vegetation and structures such as bridges. Data that can help corroborate or constrain rates of sediment yield are extremely useful, for example using dated rates of sedimentation from reservoirs, bays and floodplains, dated structures, dendrochronology and 'finger printing' of sediment sources to determine provenance (Oldfield et al., 1979; Walling, 1999; Walling et al., 1999).

Hydromodification and river channel change: geomorphology and urban planning

The geomorphic study of stream channels altered by human disruptions to hydrology has a long history since the first systematic discussion by Leopold (1968). A quarter of a century later, the term 'hydromodification' was coined in the engineering and regulatory literature (Frederick and Dressing, 1993), and in many parts of the

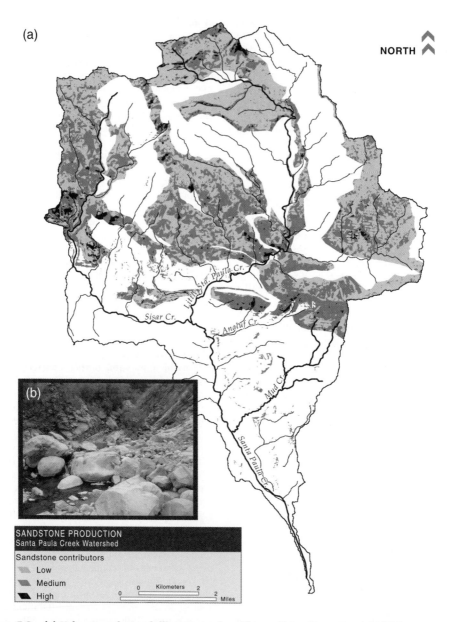

Figure 5.3 **(a) Using terrain modelling to explore the sediment source and yield characteristics of a mountainous watershed in southern California. Overlays of geology, land cover and hillslope gradient are used to characterize coarse sediment production for analysing potential influences on salmonid habitat. While many habitat concerns focus on excess fine sediment production, in this watershed coarse sediment derived largely from sandstone sources (inset photograph) (b) provides both the overarching structure for fish habitat and natural barriers to fish passage and so is critical**

United States it is now used as the shorthand term for all manners of land-use change, particularly urbanization, affecting downstream channels.

The dimensions of channels in an urban stream network generally follow the overall pattern of discharge changes across that network, with larger flow peaks resulting in greater channel adjustment. However, the application of geomorphic understanding leads to a more complex interpretation of existing data and suggests that any locally observed correlations between channel size, rate of channel change, and watershed characteristics are likely to be non-universal (Booth and Henshaw, 2001). Geomorphologists should be aware that local channel gradient and the pattern of gradient changes across a channel network are particularly important factors but are rarely incorporated into case-study analyses, so a geomorphic perspective is needed to determine, for example, whether the measurements were taken in reaches that are more or less susceptible to change (Montgomery and Buffington, 1997). The location of urban development relative to the channel network is also important: developments that concentrate urban effects in only a few areas tend to have less impact on the channel network as a whole than equivalent development spread across the watershed (Ebisemiju, 1989; Alberti et al., 2007). Flow increases introduced at one point in the channel network may be far more effective at eroding sediment than at another, because of the spatial variability of watershed soils and the distribution of alluvial and bedrock (or other non-alluvial) reaches (e.g. Booth, 1990).

Observed channel stability may reflect true re-establishment of fluvial equilibrium (as anticipated, for example, by Hammer, 1972, Neller, 1988, Ebisemiju, 1989, Henshaw and Booth, 2000), but alternatively it may simply represent the product of flushing all mobile sediment from the system to produce a relatively static, non-alluvial channel (e.g., Tinkler and Parish, 1998). In either case, the re-attainment of 'channel stability' can express the condition of a substantially altered flow regime with negligible physical impacts but potentially catastrophic (and unrecognized) biological consequences. This outcome reinforces the need for geomorphologists to be involved in integrated assessments of the urban impacts on river systems (Nilsson et al., 2003).

Water supply impact assessments: geomorphology and the management of large dams

Following the golden age of multi purpose large dam building in the mid 20th century (Beaumont, 1978), a plethora of research including that from geomorphologists has indicated the deleterious impact of regulated rivers on fluvial ecosystems (Petts, 1984; Ligon et al., 1995, Collier et al., 1996, Graf, 2001). Dam operators are now frequently required to develop revised flow release schedules to minimize further impacts to downstream fluvial ecosystems or, in the case of hydropower dams in the United States regulated by the Federal Energy Regulatory Commission (FERC), to consider impacts on fish and wildlife equally with power generation and flood control before a new licence is issued (see Masonis and Bodi, 1998). Geomorphologists are therefore now providing services to assess the future downstream impacts of large dams, modify flow release schedules or evaluate the potential impacts of dam removal.

Geomorphological studies in relation to the re-licensing of hydropower dams in the United States generally occurs as an integral part of a suite of studies that also encompass water quality, aquatic species, special-status plants and wildlife, recreation, aesthetics and cultural resources. Analyses consider the character and changes to hydrology and sediment supply dynamics caused by the dam's operation, the morphological condition of the downstream channel and the characteristics of downstream transport and storage of sediment and large wood (Stillwater Sciences et al., 2006). Where operational changes are proposed, they frequently involve altering the flow release schedule to better suit the downstream ecology. Increasingly this involves the prescription of 'flushing flows' (Reiser et al., 1989) designed to partially restore the flood pulse advantage (Bayley, 1991) by flushing fine sediment, stimulating coarse sediment transport, and facilitating floodplain inundation (Downs et al., 2002). Designing such flow releases to have a sustainable impact is difficult (Kondolf et al., 1993; Kondolf and Wilcock, 1996; Schmidt et al., 2001), primarily because promoting sediment transport in regulated rivers may result in further channel incision and bed armouring. Geomorphologists may therefore need also to design a programme of coarse sediment augmentation to parallel the prescribed high-flow releases.

When dam removal is proposed, geomorphologists are generally involved in assessing the dynamics of sediment re-distribution that follows the removal of the dam. Concerns usually exist for the upstream dynamics of the resulting knickpoint; the rate of fluvial sediment excavation and likely channel morphology within the former reservoir site; and the ecological impact of sediment released downstream (Pizzuto, 2002; Doyle et al., 2003). Emphasis on predicting future conditions under a variety of dam-removal scenarios has resulted in the development of sediment transport models that can accommodate pulsed sediment supply (see Cui and Wilcox, 2008) (Figure 5.4) and the use of scaled (e.g. Bromley and Thorne, 2005) or generic physical models (Cui et al., 2008; Wooster et al., 2008) to provide guidelines for sediment management during dam removal (Randle et al., 2008; Downs et al., 2009).

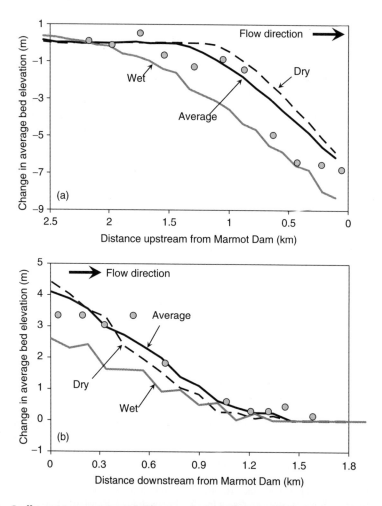

Figure 5.4   Sediment transport modelling used to predict the likely impact of the removal of Marmot Dam (Sandy River, Oregon) for the year following dam removal, under average, wet and dry year scenarios (exceedance probability of peak flow and annual runoff of 50 per cent, 10 per cent and 90 per cent, respectively). The 14-m high dam was removed in July 2007 and the cofferdam breached in October 2007. Plots show predictions from (a) the former reservoir area and (b) the depositional wedge immediately downstream of Marmot Dam. Data points are from post-project surveys undertaken 1 year later: 2008 had an annual runoff exceedance probability of approximately 29 per cent (adapted from Downs et al., 2009)

Environmental restoration and preservation: geomorphology services in support of conservation management and landscape design

A third service area relates to conservation management practices and landscape design, including planning for restoration, preservation, and recreation. Stemming from the concept of 'restoration ecology' (Jordan et al., 1987), '… the process of repairing damage caused by humans to the diversity and dynamics of indigenous ecosystems' (Jackson et al., 1995, 71), environmental restoration is now a big business with expenditure on river restoration alone estimated at over $1 billion annually since 1990 (Bernhardt et al., 2005). Growing recognition of the importance for ecology of habitat structure and function has greatly increased the visibility and relevance of geomorphology as a contributing discipline.

It also marks a sharp departure from geomorphology's role in environmental management as a discipline helping to reduce or minimize environmental impacts associated with human activity (previous examples) towards applications directly involved with environmental reconstruction and repair, translating biological objectives into implementation practicalities and elevating the geomorphologist's need for numerical simulation modelling. Examples related to river restoration design and parkland planning are provided below.

River restoration assessment and design: geomorphology and restoration ecology

A plethora of quantitative scientific research since the 1970s demonstrated unequivocally both the deleterious impact of human activities on fluvial ecosystems and substantial declines in populations of native aquatic species. This caused a perception shift in river management away from minimizing and mitigating human impacts towards active restoration involving undoing the impact of past actions. A generally acknowledged preference for process-based restoration where possible (NRC, 1992; SRAC, 2000), and increasing understanding of the critical role of hydro-geomorphological processes in maintaining aquatic habitat diversity has served to position geomorphology as a central discipline in river restoration, with geomorphologists frequently providing the functional link between achieving species-based restoration goals and managing on-going risks to floodplain inhabitants related to river flooding and erosion. The geomorphologist's role in river restoration can encompass project orientation including assessing uncertainties inherent to the restoration design (e.g. Wheaton et al., 2008); determining past, present and future conditions for baseline purposes including the use of rapid assessment protocols and baseline data surveys (Downs and Gregory, 2004; Brierley and Fryirs, 2005), understanding the extent of hydrosystem modification, specifying conceptual models including the recovery potential for the existing channel (Fryirs and Brierley, 2000); project design including relationships between geomorphology and valued flora and biota; and project monitoring and evaluation.

Process-based restoration may involve the geomorphologist in prescribing environmental high flows to restore the ecological benefits related to high-flow events, determining the feasibility of weir or large dam removal (see preceding section for both) or reconnecting floodplains and backwater channels through the removal of embankments (Toth et al., 1998; Buijse et al., 2002). In all cases, a key element is balancing the sediment supply and transport to result in a sustainable design. Where prompted recovery is chosen, the geomorphologist will likely be involved in the siting and design of instream structures constructed of logs or boulders, often to mimic natural channel features (e.g. Lenzi, 2002; Chin et al., 2008), ensuring that they promote the required beneficial processes and are structurally stable (see Hey, 1994; Downs and Thorne, 2000). Where restoration necessitates reconstructing the channel morphology (e.g. reinstating a meandering planform), the geomorphologist should be involved in ensuring the restoration project works in its catchment context (Kondolf and Downs, 1996; Kondolf et al., 2001) and that the design functions with the river's contemporary flow regime and sediment supply (Shields, 1996; Soar and Thorne, 2001). In the case of regulated rivers, this can involve re-scaling the channel (Figure 5.5). Geomorphologists should also help develop a monitoring and evaluation programme to ensure that adequate learning results from the restoration 'experiment' (Downs and Kondolf, 2002; Rhoads et al., 2008; Skinner et al., 2008), ensuring the long-term sustainability of the implemented solution (Gregory and Downs, 2008; Newson and Clark, 2008).

Conservation and recreation provision: geomorphology and parkland planning

Development pressures increasingly make parkland settings a focal point for both conservation and recreation planning, whether the parkland is in a pristine isolated setting or on the urban fringe. There is an inherent tension implicit between conservation and recreational land uses that make it critical to understand the geomorphological sensitivity of the landscape (Brunsden and Thornes, 1979) in addition to its floristic and wildlife values. In urban settings, for instance, river corridors are increasingly viewed by municipalities and regional planners as critical multipurpose open spaces (e.g. Barker, 1997) where there is pressure not only to provide recreational facilities (river access points, trails, etc.), but also to provide the urban population accessible exposure to 'natural ecosystems'. At the same time, where habitat for threatened and endangered native species exists in the river corridor there may be regulatory pressure to designate exclusion zones for the public. From the other extreme, many montane habitats support dynamic and fragile habitats where the valued biodiversity is linked closely to geomorphological processes (past and present) that should be understood in developing management plans (Gordon et al., 1998, 2002).

The geomorphologist's role in such situations is to help parameterize the dynamics and sensitivity of the landscape unit so that planners can better

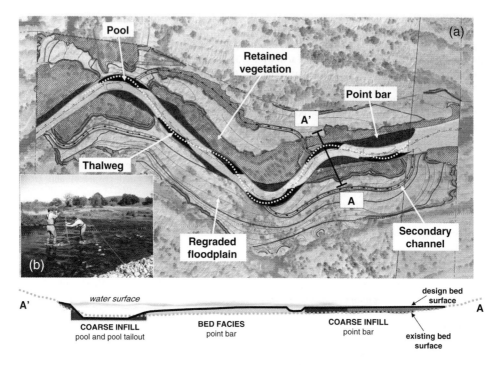

Figure 5.5 (a) Restoration design for reconstructing an incised reach of the Merced River, California. The channel design was based on optimizing channel width, depth and bed sediment to ensure sediment transport continuity, provide suitable velocities of flow for salmonids, and acceptable flood inundation frequencies for re-establishing native floodplain vegetation. The design was based on bedload transport equations from Parker (1990), channel meander characteristics from Soar and Thorne (2001), and 2 years of baseline physical and biological monitoring data. (b) Inset photograph shows scour chains being installed adjacent to surface sediment tracers (bright bed sediments)

understand the processes that maintain and shape the mosaic of habitat patches that form the parkland extent and achieve an acceptable coexistence for the various desired uses. For river corridors, this may involve determining reach-scale channel adjustments in the context of the catchment conditions and so requires an assessment of the evolutionary dynamics of the corridor reach and the likely cumulative effects of other activities within the catchment. Similar to other geomorphological applications, there is an emphasis on understanding current conditions in the context of recent historical changes and projecting such understanding into the future, cognizant of the legacy impacts of natural changes and human impacts in the catchment. Future conditions can be predicted using hydrological models to predict likely changes to channel hydraulic geometry or stream power as an indirect measure of possible erosion threat, and direct monitoring or sediment transport model simulations can determine the

potential impact of a sediment pulse moving through a channel network (e.g. Gomez et al., 2009). In other cases, the geomorphologist's knowledge of established conceptual models from prior research may provide the basis for determining likely future conditions. Relating this understanding back to the planning group can assist in determining a riparian buffer zone for preservation purposes, areas at risk of erosion where facilities and structures should be avoided, and the best routes for trails.

SKILLS AND STANDARDS FOR THE EARLY 21ST CENTURY APPLIED GEOMORPHOLOGIST

The examples above suggest that while a wide range of geomorphological capabilities are relevant, applied geomorphological investigations frequently require similar generic skills. These skills derive from the various project services outlined in Table 5.3. First, most projects require the geomorphologist to determine how the landform functions at present, involving not only the specific dynamics of the project site but also its spatial context and regional setting. Second, the practising geomorphologist needs then to discern how the landscape unit functioned during some reference period from the recent historical past, typically before a series of human-related constraints were placed on landscape function (Ebersole et al., 1997) and third, predict how the landscape unit is likely to function in the future under the same, additional, or fewer constraints. Predicting future conditions, although critical for environmental management, is a relatively new topic for geomorphology and the specific role it should play in geomorphological applications to environmental management is still being debated (e.g., Wilcock et al., 2003; Lancaster and Grant, 2003).

With these three basic analyses complete, a sustainable project solution is likely to involve the geomorphologist in determining how and why changes occurred between past and present conditions. This requires an assessment of the integrated impacts of legacy factors on contemporary geomorphological processes, which can be very challenging (Slaymaker, 2009). Furthermore, the geomorphologist should be capable of contributing as part of the project team to discussions about how the landscape should function in the future (Montgomery et al., 1995). Such discussions must acknowledge management desires and constraints, the impact of legacy factors (including earlier management episodes) and, increasingly, expected landscape response under changes to fundamental drivers such as climate. This process also requires a high level of interdisciplinary appreciation in order to work alongside other scientists and engineers, because project goals are far more likely to be focused on imperilled native species, water quality concerns, or reducing natural hazards than on sustaining geodiversity, *per se*. As a related factor, the applied geomorphologist should understand and be able to communicate how to manipulate site conditions to make the landscape unit function in the preferred way, sustainably and at an acceptable level of risk. This step is increasingly being translated to involve non-structural as well as structural methods of management, and to involve the economic valuation of non-use and recreational management benefits in addition to traditional economic-use valuation (Downs and Gregory, 2004).

Finally, geomorphologists need to be able to design and implement geomorphological monitoring and evaluation so that a significant learning experience can be achieved from the implemented project and so applied to future projects. This step has implications for baseline data collection and so should be considered at the project outset, a hallmark of what is broadly known as 'adaptive management' (Holling, 1978; Ralph and Poole, 2003).

In Table 5.4, the seven generic skills outlined above are subdivided according to a series of techniques that imply technical training. Besides those listed, there are of course other skills, such as clear and concise report and proposal writing capabilities, attention to detail with data collection and analyses, and good interpersonal communication and networking capacity that will also shape an individual's success as a professional. As few geomorphologists are likely to be trained in all of the techniques identified in Table 5.4 (and others no doubt equally essential, but beyond the authors' experience), there is an implication that geomorphological contributions to environmental management will likely occur as part of a team.

The required skills identified in Table 5.4 also reflect a breadth of techniques that are certainly beyond the range of training provided under any one degree course or single training class. Their multiplicity also frames the vexing issue of professionalism, standards and ethics in geomorphology (see Brunsden et al., 1978; Brunsden, 1996; Leopold, 2004). Unlike related disciplines involved with environmental management such as engineering and biology, there are no undergraduate degrees awarded in 'geomorphology' and there is no professional institution that accredits professional geomorphology training. Geomorphologists seeking independent recognition of their skills can aspire only towards accreditation from a neighbouring discipline either by way of an examination that is rigorous but may have only marginal linkage with their daily activities, such as Professional Geologist in the United States as accredited by the American Institute for Professional Geologists or individual state licensing boards; or by peer review that lacks the rigor of examination but may relate more specifically to the skills of the applicant, such as the status of Chartered Geographer (Geomorphology) conferred by the United Kingdom's Royal Geographical Society (with the Institute of British Geographers).

The result has been the creation of a new and unregulated market for training 'professional geomorphologists', frequently via unexamined short courses. In particular, training offered by Wildland Hydrology (www.wildlandhydrology. com) has been tremendously popular in the United States with the course's founder, David Rosgen, lauded as the 'river doctor' by *Science* magazine (Malakoff, 2004). While Rosgen has probably

Table 5.4 Core skills and techniques required by the early 21st century applied geomorphologist

Service	Skills	Example methods
Project orientation	1 Background information assembly	• Assimilation of geology, soils, vegetation, precipitation, population history, land management history, prior geomorphology reports
	2 Terrain modelling using GIS	• Empirical modelling for expected conditions • Development of process domains/landscape units
	3 Development of conceptual models	• Expert interpretation using known site details and accepted process–form linkages in academic literature
Determination of current conditions	4 Mapping and inventorying	• Survey, aerial photographic interpretation • Field reconnaissance using rapid assessment protocols
	5 Baseline data collection	• Collation of existing data records • Collection of additional data to supplement existing records (e.g. transects and cross-sections, grain size determination)
	6 Field interpretation of current morphology and process	• Field reconnaissance using rapid assessment protocols • Expert judgment
	7 Classification and characterization of landscape units	• Application of *a priori* classification hierarchy • Characterization via statistical analysis of attributes
	8 Monitoring of site dynamics	• Repeat measurements (tracer studies, repeat transects/cross-sections) over a designated interval or following large forcing events such as high intensity rainfall, floods, storm surges)
	9 Determine regional sediment flux (also for past conditions)	• Estimate of sediment yield, budget for watershed or coastal zone
Investigation of past conditions	10 Reconstruction of historical data series	• Air photo/map/survey overlay • Reconstruction of sediment flux from historical records • Use of narrative accounts, ground photographs • Vegetation composition and age
	11 Palaeo-environmental reconstruction for pre-historical conditions	• Stratigraphic analysis and interpretation of sedimentary deposits • Geochronology dating methods, e.g. radio carbon, lead 210, 237 • Erosion estimates using short-lived radio nuclides
Prediction of future conditions	12 Sensitivity analysis of potential for change	• According to measured potential for changes related to threshold: e.g. stream power • Interpretation of departure from 'expected' conditions: e.g. using hydraulic geometry comparisons, discriminant bi-variate plots • Positioning of units in expected sequence of change: e.g. channel evolution model • Statistical deterministic or probabilistic analysis • Using hydrological and sediment transport models (see below)

Continued

Table 5.4 Cont'd

Service	Skills	Example methods
	13 Computer and physical model simulations	• Computer modelling of hillslope stability • Computer modelling of river bank stability • Computer modelling of sediment transport in rivers, and near shore • Modelling of planform change • Physical modelling using scale models or generic experiment in flume
Problem solution/ design	14 Expert interpretation and integration	• Based on the geomorphologist's experience and mental models, project perception • Ability to determine and contextualize the historical legacy on contemporary geomorphological processes
	15 Contribution to project objectives for sustainable/minimum maintenance/ impact	• Contribution via problem-solving forum of technical specialists, government agency representatives, other stakeholders
	16 Project siting	• Interpretation or risk analyses to determine minimum conflict point or maximum benefit between natural process and project requirements
	17 Project design	• Use of empirical and numerical models to propose process-based dimensions suitable to contemporary forcing mechanisms • Experience with implementation methods and techniques • Design of adaptive monitoring and evaluation programmes, experience in hypotheses setting
	18 Project implementation oversight	• In assistance to project engineer
Post-project monitoring and evaluation	19 Determination of measurable success criteria	• Expert knowledge of geomorphological system relationships (analytical references)
	20 Development of monitoring and evaluation plan	• Identification of primary variables, methods, locations and frequency of monitoring • Suggestions for suitable analyses
	21 Adaptive management response to outcomes of post-project appraisal	• Ability to interpret evaluation in context of implemented project to determine success and next steps
Expert advisory	22 Data provision	• Analytical expertise to provide data for open use or to bolster case
	23 Cross-examination capability	• Expert knowledge of specific geomorphological system and related systems to answer questions in deposition and in court

done more to popularize the discipline of fluvial geomorphology than anyone, many academic geomorphologists have concerns with the adequacy of the course content in its applied context (Juracek and Fitzpatrick, 2003; Simon et al., 2007a; Roper et al., 2008), and argue that the courses lack the rigor of academic geomorphology training and have no examination to identify the proficiency of the taker. This debate is of more than academic interest because, in some parts of the United States,

it is now implied, if not explicitly stated, that geomorphology professionals *must* have extensive training in Wildland Hydrology classes, effectively usurping university training as the basis for defining professional geomorphology status, at least for fluvial geomorphology. Belatedly, several universities are responding with certificate programmes in river and stream restoration (e.g. University of Minnesota, Portland State University and North Carolina State University).

FUTURE: APPLIED
GEOMORPHOLOGY RE-CAST

While the lack of a 'professional geomorphologist' accreditation is a clear weakness and probable hindrance to geomorphology's future role in environmental management, we suggest that the most important issue for geomorphologists to consider is the *way* in which geomorphology contributes to problem solving. Because, historically, engineers have been charged with solving environmental management issues and have achieved this largely through implementing structural solutions, applied geomorphologists have commonly adopted this prevailing paradigm by finding site-based remedies to an undesirable problem, and using structural solutions to achieve an immediate fix using short-term funding. Such 'engineering geomorphology' has undoubtedly heightened the problem-solving emphasis of geomorphology but does not reflect a true geomorphological approach to environmental management, being more limited and narrowly focused than might reasonably be derived from geomorphology's academic heritage. Several illustrations are provided below of overarching issues that would help re-cast applied geomorphology to the benefit of both geomorphology and

environmental management. Other issues and a selection of inherent strengths and exciting opportunities that geomorphologists could, or should, bring to environmental problem-solving are outlined in Table 5.5.

Regional-scale approaches and longer-term planning horizons to offset cumulative impacts

Geomorphologists have long argued that site-based, structural approaches to environmental management should be discarded in favour of seeing environmental management in a broader landscape context (e.g. the catchment, littoral cell) and in terms of cause-and-effect linkages (Sear, 1994; Hooke et al., 1996; Kondolf and Downs, 1996). While a movement towards catchment-based approaches began several decades ago with integrated river basin management plans, cause-and-effect-based solutions are less frequently adopted, especially over long-term planning horizons. This is at odds with geomorphological research that frequently points to decades-to-centuries legacy effects (e.g., for the movement of disturbed sediments through a watershed), and

Table 5.5 Prospects for geomorphology in environmental management: strengths, weaknesses, opportunities, threats

Strengths	Weaknesses
• Directly concerned with the surface of the earth • Directly concerned with regional (e.g. catchment) functions that are the basis for maintaining healthy ecosystems and valued native biological populations • Long history of studying the role of human impact in system functioning • Well positioned to integrate biology, engineering and planning into practical solutions • Well positioned to practise design and management with nature to achieve truly sustainable designs	• Poor representation at policy levels • Poor representation on funding bodies to ensure adequate research funds • Lack of standard methods • Lack of routine monitoring of geomorphic systems • Viewed as a sub-set of engineering, especially in more quiescent landscapes • Lack of a professional group and professional accreditation
Opportunities	Threats
• 'Ecosystem services' a natural processes-based spin on conservation and management • Well placed to integrate human activities as part of a process–form–habitat–biota–culture link • New dating techniques for process rates that work well within timeframes of human occupation • New forms of digital data and processing (e.g. LiDAR and terrain modelling) as the basis for better approaches to cumulative impacts • Improved predictive models as the basis for strong representation under conditions of global change	• Geomorphology practised by others with little or insufficient training, and so lacking in broad areas of necessary skills (Table 5.4) • Perception of simplistic geomorphological descriptions of system functioning (e.g. the 'bankfull' paradigm) that do not apply in all cases

which suggest that geomorphology approaches to environmental problem-solving should be based on the same time frame to avoid costly management errors involved with over-engineered and wrongly positioned structures (Gregory and Downs, 2008). Three challenges for geomorphology research are to improve our analytical approaches for setting contemporary processes of landscape change into the context of historical legacy factors (which, inherently requires reconciliation of process geomorphology studies with historical approaches; Slaymaker, 2009), to better predict the impacts of cumulative effects at a project site, especially into the future (Reid, 1998) and, not least, to re-cast public perception of landscape change.

Acknowledging environmental management in terms of uncertainty, risk and probability

Long-term, larger spatial scale approaches to environmental management challenge geomorphologists to cast their predictions probabilistically, so that risk analyses are possible. Deterministic, reductionist approaches to geomorphological science needs to be allied to probabilistic outcomes to allow this potential to be tapped. This 'uncertainty' challenge is also not unique to geomorphology but applies generally across the environmental sciences and into civil engineering (e.g. Johnson and Rinaldi, 1998; Brookes et al., 1998). It implies the use, as suggested frequently, of adaptive environmental assessment and management (Holling, 1978) approaches that are risk-tolerant rather than risk-adverse (Clark, 2002), and that maximize the chance of surprise discoveries (McLain and Lee, 1996) as a function of 'learning by doing' (Haney and Power, 1996). It also implies a change towards using a series of moderate interventions with moderate risk as the basis for environmental management, rather than society's current preference for one large intervention that frequently represents a huge risk (Cairns, 2002).

Providing the scientific basis for an ecosystem service-based approach

Another significant opportunity for geomorphology in the coming decades will result from efforts worldwide to redefine environmental management in terms of the services to humans provided by functioning ecosystems. A groundswell of support

has developed for considering the economic advantages of functioning ecosystems as part of environmental problem-solving, motivated in part by a stark appraisal of the extent of degradation and unsustainable use of ecosystem services by authors of the Millennium Ecosystem Assessment (MEA, 2005). Geomorphology is ideally based to exploit a paradigm of environmental management based on 'what do you want your landscape to do for you?' rather than 'this activity is forbidden'. In this regard, ecosystem services can be seen as moving the popular concept of 'design with nature' (McHarg, 1969) towards an overall and more pluralistic concept of 'management with nature' (Downs and Gregory, 2004). Critically, however, this will require reconceptualizing human agency and geomorphological systems to consider human influence not *on* the environment, but *within* it through a combined approach that encompasses sociocultural and biophysical processes (Urban, 2002). This might entail, for example, applied geomorphologists adopting socially based concepts such as 'ecohealth', which sees human health as a primary outcome of effective ecosystem management (Parkes et al., 2008).

Applying regional-scale, risk-based, ecosystem services-oriented approaches to environmental management would significantly enhance the potential for geomorphology application. It would also require that geomorphologists are well-represented in policy formulation, so that the discipline is not just relevant to the problem-solving milieu but central to how environmental managers define both their 'problems' and 'solutions'. Such development will require (1) educational outreach by academic and professional geomorphologists to non-geomorphology audiences and (2) employment of geomorphologists in government, non-governmental and consulting organizations to reinforce relevance. Within geomorphology, practitioners and academics need to foster a better symbiosis between basic and applied research that facilitates the application of cutting-edge skills to the interdisciplinary, value-laden questions of environmental management. In particular, there is the need for (3) better dissemination to practitioners of basic and applicable geomorphology research and (4) a steady supply of new 'tools' that increase the technical capabilities of professional geomorphologists (adapted from Gregory et al., 2008: Table 3).

CONCLUSION

This review has focused on issues surrounding the geomorphologist's application of skill to environmental problem-solving. Despite applied geomorphology's long but largely unrepresented history within the geomorphology discipline, escalating environmental awareness and better technical expertise have brought increasing opportunities to contribute to environmental management. Most frequently, geomorphologists are part of a project team seeking to navigate a suite of protective policies and regulations in projects focused on hazard avoidance and diminution, the sustainable development of natural resources, or environmental restoration and preservation. Multiple service areas, with their attendant need for technical analyses, have opened new venues for geomorphological training beyond the traditional basis in academia.

Despite potential threats to the meaningful application of geomorphology to environmental management, there are tremendous opportunities, especially if geomorphologists demonstrate to environmental managers the value of regional and long-term approaches cast in terms of uncertainty, risk and probability of outcome. Realizing these opportunities requires geomorphologists to re-cast 'applied geomorphology', wherein the goal for environmental management that is not the control or manipulation of the natural environment, but rather the maximizing of ecosystem services. An approach to environmental management based on ecosystem services would seek to maximize beneficial outcomes rather than simply to minimize the infringement of regulations, making it inherently more integrated with natural processes and landscape evolution rather than preoccupied with perceived risk and static morphology. This approach also implies funding environmental management to be far less construction-oriented, and so far more amendable to geomorphological approaches based on adaptive management.

Undoubtedly, thankfully, applied geomorphology has made great strides since the first edition of Cooke and Doornkamp's *Geomorphology in Environmental Management* was published in 1974. The future seems to offer enormous opportunity for continuing to expand the role of geomorphology in environmental problem-solving, particularly if geomorphologists can embrace temporal and spatial scales of problem solving more closely allied to geomorphology's scientific origins, and better integrate concerns for environmental conservation and social justice to gain improved understanding of the mutual vulnerability and dependence of society and landscape (Slaymaker, 2009). The challenges, however, are technical, conceptual and ethical, and all three must be considered equally in advancing the role of geomorphology in environmental management.

ACKNOWLEDGEMENTS

We are indebted to Lauren Klimetz and Natasha Bankhead-Pollen for provision of graphics in Figure 5.1, to Larry Gurrola and Ed Keller for Figure 5.2, and to Sapna Khandwala and Sebastian Araya for cartographic assistance and expertise.

REFERENCES

Alberti, M., Booth, D.B., Hill, K., Coburn, B., Avolio, C., Coe, S., et al. (2007) The impact of urban patterns on aquatic ecosystems: an empirical analysis in Puget lowland sub-basins. *Landscape and Urban Planning* 80: 345–436.

Alcantara-Ayala, I. (2002) Geomorphology, natural hazards, vulnerability and prevention of natural disasters in developing countries. *Geomorphology* 47: 107–24.

Anderson, P.G., Taylor, B.R. and Balch, G.C. (1996) *Quantifying the Effects of Sediment Release on Fish and Their Habitats.* Canadian Manuscript Report of Fisheries and Aquatic Sciences No. 2346. Vancouver, British Columbia.

Ayalew, L. and Yamagishi, H. (2005) The application of GIS-based logistic regression for landslide susceptibility mapping in the Kakuda-Yahiko Mountains, Central Japan. *Geomorphology* 65: 15–31.

Barker, G. (1997) *A Framework for the Future: Green Networks with Multiple Uses in and Around Towns and Cities,* English Nature Research Reports Number 256. Peterborough, UK: English Nature.

Bayley, P.B. (1991) The flood-pulse advantage and restoration of river–floodplain systems. *Regulated Rivers: Research and Management* 6: 75–86.

Beaumont, P. (1978) Man's impact of river systems: a worldwide review. *Area* 10: 38–41.

Bernhardt, E.S., Palmer, M.A., Allan, J.D., Alexander, G., Barnas K., Brooks S., et al. (2005) Synthesizing US river restoration efforts. *Science* 308: 636–7.

Bernhardt, E.S., Sudduth, E.B., Palmer, M.A., Allan, J.D., Meyer, J.L., Alexander, G., et al. (2007) Restoring rivers one reach at a time: results from a survey of US river restoration practitioners. *Restoration Ecology* 15: 482–93.

Beschta, R.L. and Jackson, W.L. (1979) The intrusion of fine sediments into a stable gravel bed. *Journal of the Fisheries Research Board of Canada* 36: 204–10.

Booth, D.B. (1990) Stream channel incision following drainage basin urbanization. *Water Resources Bulletin* 26: 407–18.

Booth, D.B. and Henshaw, P.C. (2001) Rates of channel erosion in small urban streams, in M. Wigmosta and S. Burges (eds), *Land Use and Watersheds: Human Influence on Hydrology and Geomorphology in Urban and Forest Areas.* Vol. 2. AGU Monograph Series, Water Science and Applications. Washington, DC, pp. 17–38.

Brierley, G.J. and Fryirs, K. (2005) *Geomorphology and River Management: Applications of the River Styles Framework*. Oxford: Blackwell.

Bromley, J.C. and Thorne, C.R. (2005) A scaled physical modelling investigation of the potential response of the Lake Mills Delta to different magnitudes and rates of removal of Glines Canyon Dam from the Elwha River W.A., in G.E. Moglen (ed.), *Managing Watersheds For Human and Natural Impacts: Engineering, Ecological, and Economic Challenges: Proceedings of the 2005 Watershed Management Conference July 19–22 2005 Williamsburg VA*. Reston VA: American Society of Civil Engineers.

Brookes, A. (1988) *Channelized Rivers: Perspectives for Environmental Management*. Chichester: John Wiley.

Brookes, A., Downs, P.W. and Skinner, K.S. (1998) Uncertainty in the engineering of wildlife habitats. *Journal of the Institution of Water and Environmental Management* 12: 25–9.

Brunsden, D. (1996) Geomorphologie sans frontiers, in S.B. McCann and D.C. Ford (eds), *Geomorphologie Sans Frontiers*. Chichester: John Wiley and Sons, pp. 1–29.

Brunsden, D., Doornkamp, J.C. and Jones, D.K.C. (1978) Applied geomorphology: a British view, in C. Embleton, D. Brunsden and D.K.C. Jones (eds), *Geomorphology: Present Problems and Future Prospects*. Oxford: Oxford University Press, pp. 251–62.

Brunsden, D. and Thornes, J.B. (1979) Landscape sensitivity and change, *Transactions of the Institute of British Geographers* NS 4: 463–84.

Buijse, A.D., Coops, H., Staras, M., Jans, L.H., Van Geest, G.J., Grift, R.E., et al. (2002) Restoration strategies for river floodplains along large lowlands in Europe. *Freshwater Biology* 47: 889–907.

Cairns, Jr. J. (2002) Rationale for restoration, in M.R. Perrow and A.J. Davy (eds), *Handbook of Ecological Restoration. Vol. 1: Principles of Restoration*. Cambridge: Cambridge University Press, pp. 10–23.

Carson, R. (1962) *Silent Spring*. Boston: Houghton Mifflin.

Chacón, J., Irigaray C., Fernández T. and El Hamdouni R. (2006) Engineering geology maps: landslides and geographical information systems. *Bulletin of Engineering Geology and the Environment* 65: 341–411.

Chin, A., Anderson, S., Collison, A., Ellis-Sugai, B.J., Haltiner, J.P., Hogervorst J.B. (2008) Linking theory and practice for restoration of step-pool streams. *Environmental Management* 43: 645–61.

Clark, M.J. (2002) Dealing with uncertainty: adaptive approaches to sustainable river management. *Aquatic Conservation: Marine and Freshwater Ecosystems* 12: 347–63.

Clarke, S.J., Bruce-Burgess, L. and Wharton, G. (2003) Linking form and function: towards an eco-hydromorphic approach to sustainable river restoration. *Aquatic Conservation: Marine and Freshwater Ecosystems* 13: 439–50.

Coates, D.R. (ed.) (1976) *Geomorphology and Engineering*. London: Allen and Unwin.

Collier, M., Webb, R.H. and Schmidt, J.C. (1996) *Dams and Rivers: Primer on the Downstream Effects of Dams*, Circular No. 1126. Washington, DC: United States Geological Survey.

Collins, A.L., Walling, D.E. and Leeks, G.J.L. (1997) Fingerprinting the origin of fluvial suspended sediment in larger river basins: combining assessment of spatial provenance and source type. *Geografiska Annaler* 79A: 239–54.

Cooke, R.U. and Doornkamp, J.C. (1974) *Geomorphology in Environmental Management*. Oxford: Oxford University Press.

Cui, Y. and Wilcox, A. (2008) Development and application of numerical models of sediment transport associated with dam removal, in M.H. Garcia (ed.), *Sedimentation Engineering: Processes, Measurements, Modeling, and Practice*, ASCE Manuals and Reports on Engineering Practice No. 110. Reston, VA: American Society of Civil Engineers, pp. 995–1010.

Cui, Y., Wooster, J.K., Venditti, J., Dusterhoff, S.R., Dietrich, W.E. and Sklar L. (2008) Simulating sediment transport in a flume with forced pool-riffle morphology: examinations of two one-dimensional numerical models. *Journal of Hydraulic Engineering* 134: 892–904.

Darby, S.E. and Simon, A. (eds) (1999) *Incised River Channels: Processes, Forms, Engineering and Management*. Chichester: John Wiley and Sons.

Doornkamp, J.C. (1995) Perception and reality in the provision of insurance against natural perils in the UK. *Transactions of the Institute of British Geographers* 20: 68–80.

Downs, P.W. and Gregory, K.J. (1994) Sensitivity analysis and river conservation sites: a drainage basin approach, in D. O'Halloran, C. Green, M. Harley, M. Stanley and J. Knill (eds) *Geological and Landscape Conservation*. London: Geological Society, pp. 139–43.

Downs, P.W. and Gregory, K.J. (2004) *River Channel Management: Towards Sustainable Catchment Hydrosystems*. London: Arnold.

Downs, P.W. and Kondolf, G.M. (2002) Post-project appraisals in adaptive management of river channel restoration. *Environmental Management* 29: 477–96.

Downs, P.W. and Thorne, C.R. (1996) A geomorphological justification of river channel reconnaissance surveys. *Transactions of the Institute of British Geographers* 21: 455–68.

Downs, P.W. and Thorne, C.R. (2000) Rehabilitation of a lowland river: reconciling flood defence with habitat diversity and geomorphological sustainability. *Journal of Environmental Management* 58: 249–68.

Downs, P.W., Cui Y., Wooster, J.K., Dusterhoff, S.R., Booth, D.B., Dietrich, W.E., and Sklar, L.S. (2009) Managing reservoir sediment release in dam removal projects: an approach informed by physical and numerical modelling of non-cohesive sediment. *International Journal of River Basin Management* 7: 1–20.

Downs, P.W., Skinner, K.S. and Kondolf, G.M. (2002) Rivers and streams, in M.R. Perrow and A.J. Davy (eds), *Handbook of Ecological Restoration. Vol. 2: Restoration in Practice*. Cambridge: Cambridge University Press, pp. 267–96.

Doyle, M.W., Stanley, E.H. and Harbor, J.M. (2003) Channel adjustments following two dam removals in Wisconsin. *Water Resources Research* 39: 1011. doi:10.1029/2002WR001714.

Dunne, T. and Leopold, L.B. (1978) *Water in Environmental Planning*. San Francisco: W.H. Freeman.

Ebersole, J.L., Liss, W.J. and Frissell, C.A. (1997) Restoration of stream habitats in the western United States: restoration as reexpression of habitat capacity. *Environmental Management* 21: 1–14.

Ebisemiju, F.S. (1989) Patterns of stream channel response to urbanization in the humid tropics and their implications for urban land use planning: a case study from southwestern Nigeria. *Applied Geography* 9: 273–86.

Federal Interagency Stream Restoration Working Group (FISRWG) (1998) *Stream Corridor Restoration: Principles, Processes and Practices*, United States National Engineering Handbook Part 653. Washington, DC.

Fitzpatrick, F.A. and Knox, J.C. (2000) Spatial and temporal sensitivity of hydrogeomorphic response and recovery to deforestation, agriculture, and floods. *Physical Geography* 21: 89–108.

Fookes, P.G. and Vaughan, P.R. (1986) *Handbook of Engineering Geomorphology*. Glasgow: Blackie.

Fookes, P.G., Lee, E.M. and Griffith, J.S. (2007) *Engineering Geomorphology: Theory and Practice*. Caithness: Whittles.

Fookes, P.G., Lee, E.M. and Milligan, G. (2005) *Geomorphology for Engineers*. Caithness: Whittles.

Fraser, E.D.G., Mabee, W. and Slaymaker, O. (2003) Mutual vulnerability, mutual dependence: the reflexive nature of human society and the environment. *Global Environmental Change* 13: 137–44.

Frederick, R.E. and Dressing, S.A. (1993) Technical guidance for implementing BMPs in the coastal zone. *Water Science and Technology* 28: 129–35.

Fryirs, K. and Brierley, G.J. (2000) A geomorphic approach to the identification of river recovery potential. *Physical Geography* 21: 244–77.

Gardiner, J.L. (1994) Sustainable development for river catchments. *Journal of the Institution of Water and Environmental Management* 8: 308–19.

Giardino, J.R. and Marston, R.A. (1999) Engineering geomorphology: an overview of changing the face of the earth. *Geomorphology* 31: 1–11.

Gilbert, G.K. (1890) *Lake Bonneville*, Monograph 1, United States Geological Survey. Washington, DC: US Government Printing Office, p. 438.

Gilbert, G.K. (1914) *Transportation of Debris by Running Water*, Professional Paper 86, United States Geological Survey. Washington, DC: US Government Printing Office.

Gilbert, G.K. (1917) *Hydraulic Mining Debris by Running Water*, Professional Paper 105, United States Geological Survey. Washington, DC: US Government Printing Office.

Gomez, B., Cui, Y., Kettner, A.J., Peacock, D.H. and Syvitski, J.P.M. (2009) Simulating changes to the sediment transport regime of the Waipaoa River, New Zealand, driven by climate change in the twenty-first century. *Global and Planetary Change* 67: 153–66.

Gordon, J.E., Dvorák, I.J., Jonasson, C., Josefsson, M., Kociánová, M. and Thompson, D.B.A. (2002) Geo-ecology and management of sensitive montane landscapes. *Geografiska Annaler* 84A: 193–203.

Gordon, J.E., Thompson, D.B.A., Haynes, V.M., Brazier, V. and Macdonald, R. (1998) Environmental sensitivity and conservation management in the Cairngorm Mountains, Scotland. *Ambio* 27: 335–44.

Goudie, A.S. (1994) Environmental management, in Goudie A.S. and Thomas D.S.G. (eds), *The Encyclopedic Dictionary of Physical Geography*, 2nd edn. Oxford: Blackwell.

Graf, W.L. (1984) A probabilistic approach to the spatial assessment of river channel instability. *Water Resources Research* 20: 953–62.

Graf, W.L. (1996) Geomorphology and policy for restoration of impounded American rivers: what is 'natural'? in B.L. Rhoads and C.E. Thorn (eds), *The Scientific Nature of Geomorphology*. New York: John Wiley and Sons, pp. 443–73.

Graf, W.L. (2000) Locational probability for a dammed, urbanizing stream: Salt River, Arizona, USA. *Environmental Management* 25: 321–35.

Graf, W.L. (2001) Damage control: restoring the physical integrity of America's rivers. *Annals of the Association of American Geographers* 91: 1–27.

Graf, W.L. (2005) Geomorphology and American dams: the scientific, social, and economic context. *Geomorphology* 71: 3–26.

Graf, W.L., Stromberg, J. and Valentine, B. (2002) Rivers, dams, and willow flycatchers: a summary of their science and policy connections. *Geomorphology* 47: 169–88.

Gray, J.M. (2004) *Geodiversity: Valuing and Conserving Abiotic Nature*. Chichester: John Wiley and Sons.

Gregory, K.J. (1979) Hydrogeomorphology: how applied should we become? *Progress in Physical Geography* 3: 84–101.

Gregory, K.J. and Downs, P.W. (2008) The sustainability of restored rivers: catchment-scale perspectives on long term response, in S.E. Darby and D.A. Sear (eds), *River Restoration: Managing the Uncertainty in Restoring Physical Habitat*. Chichester: John Wiley and Sons, pp. 253–86.

Gregory, K.J., Benito, G. and Downs, P.W. (2008) Applying fluvial geomorphology to river channel management: background for progress towards a palaeohydrology protocol. *Geomorphology* 98: 153–72.

Gurrola, L.D., DeVecchio, D.E. and Keller, E.A. (2010) Rincon Mountain megaslide: La Conchita, Ventura County, California. *Geomorphology* 114: 311–18..

Hammer, T.R. (1972) Stream channel enlargement due to urbanization. *Water Resources Research* 8: 1530–46.

Haney, A. and Power, R.L. (1996) Adaptive management for sound ecosystem management. *Environmental Management* 20: 879–86.

Henshaw, P.C. and Booth, D.B. (2000) Natural restabilization of stream channels in urban watersheds. *Journal of the American Water Resources Association* 36: 1219–36.

Hey, R.D. (1994) Environmentally sensitive river engineering, in P. Calow and G.E. Petts (eds), *The Rivers Handbook: Hydrological and Ecological Principles*. Oxford: Blackwell Scientific. pp. 337–62.

Holling, C.S. (ed.) (1978) *Adaptive Environmental Assessment and Management*, International Institute for Applied Systems, Analysis 3. New York: John Wiley and Sons.

Hooke, J.M. (1999) Decades of change: contributions of geomorphology to fluvial and coastal engineering and management. *Geomorphology* 31: 373–89.

Hooke, J.M., Bray, M.J. and Carter, D.J. (1996) Coastal groups, littoral cells, policies and plans in the UK. *Area* 27: 358–67.

Horton, R.E. (1945) Erosional development of streams and their drainage basins: hydrophysical approach to quantitative morphology. *Bulletin of the Geological Society of America* 263: 303–12.

Jackson, L.L., Lopoukline, N. and Hillyard, D. (1995) Ecological restoration – a definition and comments – commentary. *Restoration Ecology* 3: 71–5.

Jewitt, G.P.W. and Gorgens, A.H.M. (2000) Facilitation of interdisciplinary collaboration in research: lessons from a Kruger National Park Rivers Research Programme project. *South African Journal of Science* 96: 410–4.

Johnson, P.A. and Rinaldi, M. (1998) Uncertainty in the design of stream channel restorations, in B.M. Ayyub (ed.), *Uncertainty Modeling and Analysis in Civil Engineering*. London: CRC Press, pp. 425–37.

Jordan, W.R., Gilpin, M.E. and Aber, J.D. (1987) Restoration ecology: ecological restoration as a technique for basic research, in W.R. Jordan, M.E. Gilpin and J.D. Aber (eds), *Restoration Ecology: A Synthetic Approach to Ecological Research*. Cambridge: Cambridge University Press, pp. 3–21.

Juracek, K.E. and Fitzpatrick, F.A. (2003) Limitations and implications of stream classification. *Journal of the American Water Resources Association* 39: 659–70.

Kasperson, J.X. and Kasperson, R.E. (2001) *Global Environmental Risk*. Tokyo: UN University Press.

King, J., Brown, C. and Sabet, H. (2003) A scenario-based holistic approach to environmental flow assessments for rivers. *River Research and Applications* 19: 619–39.

Kneupfer, P.L.K and Petersen, J.F. (2002) Geomorphology in the public eye: policy issues, education, and the public. *Geomorphology* 47: 95–105.

Kondolf, G.M. and Downs, P.W. (1996) Catchment approach to planning channel restoration, in A. Brookes and F.D. Shields Jr. (eds), *River Channel Restoration: Guiding Principles for Sustainable Projects*. Chichester: John Wiley and Sons, pp. 129–48.

Kondolf, G.M. and Wilcock, P.R. (1996) The flushing flow problem: defining and evaluating objectives. *Water Resources Research* 32: 2589–99.

Kondolf, G.M., Sale, M.J. and Wolman, M.G. (1993) Modification of fluvial gravel size by spawning salmonids. *Water Resources Research* 29: 2265–74.

Kondolf, G.M., Smeltzer, M.W. and Railsback, S.F. (2001) Design and performance of a channel reconstruction project in a coastal California gravel-bed stream. *Environmental Management* 28: 761–76.

Lancaster, S.T. and Grant, G.E. (2003) You want me to predict what?, in P.R. Wilcock and R.M. Iverson (eds), *Prediction in Geomorphology*, Geophysical Monograph 135. Washington, DC: American Geophysical Union, pp. 41–50.

Landwehr, K. and Rhoads, B.L. (2003) Depositional response of a headwater stream to channelization, east central Illinois, USA. *River Research and Applications* 19: 77–100.

Lawler, D.M., Thorne, C.R. and Hooke, J.M. (1997) Bank erosion and instability, in C.R. Thorne, R.D. Hey and M.D. Newson (eds), *Applied Fluvial Geomorphology for River Engineering and Management*. Chichester: John Wiley and Sons, pp. 137–72.

Leighton, F.B., Cann, L. and Poormand, I. (1984) History of Verde Canyon landslide, San Clemente, Orange County, California. *California Geology* 37: 173–6.

Lenzi, M.A. (2002) Stream bed stabilization using boulder check dams that mimic step-pool morphology features in Northern Italy. *Geomorphology* 45: 243–60.

Leopold, L.B. (1956) Land use and sediment yield, in Thomas WL (ed.), *Man's Role in Changing the Face of the Earth*. Chicago: University of Chicago Press, pp. 639–47.

Leopold, L.B. (1968) *Hydrology for Urban Land Planning – ? a Guidebook on the Hydrologic Effects of Urban Land Use*, Circular 554. Washington, DC: United States Geological Survey.

Leopold, L.B. (2004) Geomorphology: a sliver off the corpus of science. *Annual Review of Earth and Planetary Science* 32: 1–12.

Ligon, F.K., Dietrich, W.E. and Trush, W.J. (1995) Downstream ecological effects of dams: a geomorphic perspective. *BioScience* 45: 183–92.

Linden, K.V. (1989) The Portuguese bend landslide. *Engineering Geology* 27: 301–73.

Luce, C.H. and Wemple, B.C. (2001) Introduction to special issue of hydrologic and geomorphic effects of forest roads. *Earth Surface Processes and Landforms* 26: 111–3.

Malakoff, D. (2004) The River Doctor. *Science* 305: 937–9.

Manariotis, I.D. and Yannopoulos, P.C. (2004) Adverse effects on Alfeios River Basin and an integrated management framework based on sustainability. *Environmental Management* 34: 261–9.

Masonis, R.J. and Bodi, F.L. (1998) River law, in R.J. Naiman and R.E. Bilby (eds), *River Ecology and Management: Lessons from the Pacific Coastal Ecoregion*. New York: Springer, pp. 553–71.

McEwan, L. (1996) Student involvement with the regionally important geomorphological site (RIGS) scheme: an opportunity to learn geomorphology and gain transferable skills. *Journal of Geography in Higher Education* 20: 367–78.

McHarg, I.L. (1969) *Design with Nature*. New York: Doubleday/Natural History Press.

McLain, R.J. and Lee R.G. (1996) Adaptive management, promises and pitfalls. *Environmental Management* 20: 437–48.

Millenium Ecosystem Assessment (MEA) (2005) *Ecosystems and Human Well-being: Synthesis*. Washington, DC: Island Press.

Montgomery, D.R. (1999) Process domains and the river continuum. *Journal of the American Water Resources Association* 35: 397–410.

Montgomery, D.R. and Buffington, J.R. (1997) Channel-reach morphology in mountain drainage basins. *Geological Society of America Bulletin* 109: 596–611.

Montgomery, D.R. and Foufoula-Georgiou, E. (1993) Channel network source representation used digital elevation models. *Water Resources Research* 29: 3925–34.

Montgomery, D.R., Grant, G.E. and Sullivan, K. (1995) Watershed analysis as a framework for implementing ecosystem management. *Water Resources Bulletin* 31: 369–85.

National Research Council (NRC) (1992) *Restoration of Aquatic Ecosystems: Science, Technology and Public Policy*. Washington DC: National Academy Press.

Neller, R.J. (1988) Complex channel response to urbanisation in the Dumaresq Creek drainage basin, New South Wales, in R.F. Warner (ed.), *Fluvial Geomorphology of Australia*. Sydney: Academic Press, pp. 323–41.

Newson, M.D. and Clark, M.J. (2008) Uncertainty and the sustainable management of restored rivers, in S.E. Darby and D.A. Sear (eds), *River Restoration: Managing the Uncertainty in Restoring Physical Habitat*. Chichester: John Wiley and Sons, pp. 287–301.

Newson, M.D., Hey, R.D., Bathurst, J.C., Brookes, A., Carling, P.A., Petts, G.E., et al. (1997) Case studies in the application of geomorphology to river management, in C.R. Thorne, R.D. Hey and M.D. Newson (eds), *Applied Fluvial Geomorphology for River Engineering and Management*. Chichester: John Wiley and Sons, pp. 311–63.

Nilsson, C., Pizzuto, J.E., Moglen, G.E., Palmer, M.A., Stanley, E.H., Bockstael, N.E., et al. (2003) Ecological forecasting and the urbanization of stream ecosystems: challenges for economists, hydrologists, geomorphologists, and ecologists. *Ecosystems* 6: 659–74.

Oldfield, F., Rummery, T.A., Thompson, R. and Walling, D.E. (1979) Identification of suspended sediment sources by means of magnetic measurements: some preliminary results. *Water Resources Research* 15: 211–8.

Owens, P.N., Batalla, R.J., Collins, A.J., Gomez, B., Hicks, D.M., Horowitz, A.J., et al. (2005) Fine-grained sediment in river systems: environmental significance and management issues. *River Research and Applications* 21: 693–717.

Parker, G. (1990) Surface-based bedload transport relation for gravel rivers. *Journal of Hydraulic Research* 28: 417–36.

Parkes, M.J., Morrison, K.E., Bunch, M.J. and Venema, H.D. (2008) *Ecohealth and Watersheds: Ecosystem Approaches to Re-integrate Water Resources Management with Health and Well-being*, Publication series No. 2. Winnipeg: Network for Ecosystem Sustainability and Health

and the International Institute for Sustainable Development.

Petts, G.E. (1984) *Impounded Rivers: Perspectives for Ecological Management*. Chichester: John Wiley and Sons.

Pizzuto, J. (2002) Effects of dam removal of river form and process. *Bioscience* 52: 683–91.

Prosser, I.P., Rutherfurd, I.D., Olley, J.M., Young, W.J., Wallbrink, P.J. and Moran, C.J. (2001) Large-scale patterns of erosion and sediment transport in river networks, with examples from Australia. *Marine and Freshwater Research* 52: 81–99.

Ralph, S.C. and Poole, G.C. (2003) Putting monitoring first: designing accountable ecosystem restoration and management plans, in D.R. Montgomery, S. Bolton and D.B. Booth (eds), *Restoration of Puget Sound Rivers*. Seattle, WA: University of Washington Press.

Randle, T.J., Greimann, B.P. and Bountry, J.A. (2008) Guidelines for assessing dam removal impacts, in *EOS Transactions of the American Geophysical Union*, Fall Meeting 2008, abstract H41I-02.

Reid, L.M. (1998) Cumulative watershed effects and watershed analysis, in R.J. Naiman and R.E. Bilby (eds), *River Ecology and Management: Lessons for the Pacific Coastal Ecosystem*. New York: Springer, pp. 476–501.

Reid, L.M. and Dunne, T. (1996) *Rapid Evaluation of Sediment Budgets*. Reiskirchen: Catena Verlag.

Reiser, D.W., Ramey, M.P. and Wesche, T.A. (1989) Flushing flows, in J.A. Gore and G.E. Petts (eds), *Alternatives in Regulated River Management*. Baton Rouge: CRC Press, pp. 91–135.

Rhoads, B.L., Garcia, M.H., Rodriguez, J., Bombardelli, F., Abad, J. and Daniels, M. (2008) Methods for evaluating the geomorphological performance of naturalized rivers: examples from the Chicago Metropolitan area, in S.E. Darby and D.A. Sear (eds), *River Restoration: Managing the Uncertainty in Restoring Physical Habitat*. Chichester: John Wiley and Sons, pp. 209–28.

Rogers, K.H. (2006) The real river management challenge: integrating scientists, stakeholders and service agencies. *River Research and Applications* 22: 269–80.

Roper, B.B., Buffington, J.M., Archer, E., Moyer, C. and Ward, M. (2008) The role of observer variation in determining Rosgen stream types in northeastern Oregon mountain streams. *Journal of the American Water Resources Association* 44: 417–27.

Sacramento River Advisory Council (SRAC) (2000) *Sacramento River Conservation Area Handbook*. Sacramento, CA: California Department of Water Resources.

Schmidt, J.C., Parnell, R.A., Grams, P.E., Hazel, J.E., Kaplinski, M.A., Stevens, L.E. and Hoffnagle, T.L. (2001) The 1996 controlled flood in Grand Canyon: flow, sediment transport, and geomorphic change. *Ecological Applications* 11: 657–71.

Schumm, S.A., Harvey, M.D. and Watson, C.C. (1984) *Incised Channels: Morphology, Dynamics and Control*. Littleton, CO: Water Resources Publications.

Sear, D.A. (1993) Fine sediment infiltration into gravel spawning beds within a regulated river experiencing floods: ecological implications for salmonids. *Regulated Rivers: Research and Management* 8: 373–90.

Sear, D.A. (1994) River restoration and geomorphology. *Aquatic Conservation: Marine and Freshwater Ecosystems* 4: 169–77.

Sear, D.A., Newson, M.D. and Brookes, A. (1995) Sediment-related river maintenance: the role of fluvial geomorphology. *Earth Surface Processes and Landforms* 20: 629–47.

Shannon, M.A. (1998) Social organizations and institutions, in Naiman R.J. and Bilby R.E. (eds), *River Ecology and Management: Lessons for the Pacific Coastal Ecosystem*. New York: Springer, pp. 529–52.

Shields, Jr. F.D. (1996) Hydraulic and hydrologic stability, in A. Brookes and F.D. Shields Jr. (eds), *River Channel Restoration: Guiding Principles for Sustainable Projects*. Chichester: John Wiley and Sons, pp. 24–74.

Sidle, R.C., Pearce, A.J. and O'Loughlin, C.L. (1985) *Hillslope Stability and Land Use*, Water Resources Monograph 11. Washington, DC: American Geophysical Union.

Simon, A. (1989) A model of channel response in disturbed alluvial channels. *Earth Surface Processes and Landforms* 14: 11–26.

Simon, A., Doyle, M., Kondolf, G.M., Shields, Jr. F.D., Rhoads, B.L. and McPhillips, M. (2007a) Critical evaluation of how the Rosgen classification and associated "natural channel design" methods fail to integrate and quantify fluvial processes and channel response. *Journal of the American Water Resources Association* 43: 1–15.

Simon, A., Klimetz, L. and Klimetz, D. (2007b) *Characterization of Channel Morphology and Sediment Loads for the Yazoo River Basin, Mississippi*, USDA-ARS National Laboratory Technical Report No. 59.

Skidmore, P.B., Thorne, C.R., Cluer, B., Pess, G.R., Castro, J., Beechie, T.J. and Shea, C.C. (2009) *Science Base and Tools for Evaluating Stream Engineering, Management, and Restoration Proposals*, NOAA Technical Memorandum NMFS-NWFSC. US Department of Commerce.

Skinner, K.S., Shields, Jr. F.D. and Harrison, S. (2008) Measures of success: uncertainty and defining the outcomes or river restoration schemes, in S.E. Darby and D.A. Sear (eds), *River Restoration: Managing the Uncertainty in Restoring Physical Habitat*. Chichester: John Wiley and Sons, pp. 187–208.

Slaymaker, O. (2009) The future of geomorphology. *Geography Compass* 3: 329–49.

Soar, P.J. and Thorne, C.R. (2001) *Channel Restoration Design for Meandering Rivers Coastal and Hydraulics Laboratory ERDC/CHL CR-01-1*. Vicksburg, MS: US Army Engineer Research and Development Center.

Strahler, A.N. (1956) The nature of induced erosion and aggradation, in W.L. Thomas (ed.), *Man's Role in Changing the Face of the Earth*. Chicago: University of Chicago Press, pp. 621–38.

Stillwater Sciences, Confluence Research and Consulting, and Heritage Research Associates Inc. (2006) *Scientific Approaches for Evaluating Hydroelectric Project Effects*. Arcata: Stillwater Sciences.

Thomas, W.L. (ed.) (1956) *Man's Role in Changing the Face of the Earth*. Chicago: University of Chicago Press.

Thorne, C.R. (1998) *Stream Reconnaissance Handbook: Geomorphological Investigation and Analysis of River Channels*. Chichester: John Wiley and Sons.

Thorne, C.R. (2002) Geomorphic analysis of large alluvial rivers. *Geomorphology* 44: 203–19.

Thorne, C.R., Allen, R.G. and Simon, A. (1996) Geomorphological river channel reconnaissance for river analysis, engineering and management. *Transactions of the Institute of the British Geographers* 21: 469–83.

Thorne, C.R., Newson, M.D. and Hey, R.D. (1997) Application of applied fluvial geomorphology: problems and potential, in C.R. Thorne, R.D. Hey and M.D. Newson (eds), *Applied Fluvial Geomorphology in River Engineering Management*. Chichester: John Wiley and Sons, pp. 365–70.

Tinkler, K.J. and Parrish, J. (1998) Recent adjustments to the long profile of Cook Creek, an urbanized bedrock channel in Mississauga, Ontario, in K.J. Tinkler and E.E. Wohl (eds), *Rivers Over Rock: Fluvial Processes in Bedrock Channels*, Geophysical Monograph 107. Washington, DC: American Geophysical Union, pp. 167–88.

Toth, L.A., Melvin, S.L., Arrington, D.A. and Chamberlain, J. (1998) Managed hydrologic manipulations of the channelised Kissimmee River: implications for restoration. *BioScience* 8: 757–64.

Trimble, S.W. (1997) Contribution of stream channel erosion to sediment yield from an urbanizing watershed. *Science* 278: 1442–4.

Tubbs, D.W. (1974) *Landslides in Seattle*, Information Circular 52. Washington, DC: Washington Division of Geology and Earth Resources.

Urban, M.A. (2002) Conceptualizing anthropogenic change in fluvial systems: drainage development on the Upper Embarras River, Illinois. *Professional Geographer* 54: 204–17.

van Westen, C.J., Castellanos, E. and Kuriakose, S.L. (2008) Spatial data for landslide susceptibility, hazard, and vulnerability assessment: an overview. *Engineering Geology* 102: 112–31.

Walling, D.E. (1999) Linking land use, erosion and sediment yields in river basins. *Hydrobiologia* 410: 223–40.

Walling, D.E., Owens, P.N. and Leeks, G.J.L. (1999) Fingerprinting suspended sediment sources in the catchment of the River Ouse, Yorkshire, UK. *Hydrological Processes* 13: 955–75.

Wharton, G. and Gilvear, D.J. (2006) River restoration in the UK: meeting the dual needs of the European Union Water Framework Directive and flood defence? *International Journal of River Basin Management* 4: 1–12.

Wheaton, J.M., Darby, S.E. and Sear, D.A. (2008) The scope of uncertainties in river restoration, in S.E. Darby and D.A. Sear (eds), *River Restoration: Managing the Uncertainty in Restoring Physical Habitat*. Chichester: John Wiley and Sons, pp. 21–39.

Wilcock, P.R., Schmidt, J.C., Wolman, M.G., Dietrich, W.E., Dominick, D., Doyle, M.W., et al. (2003) When models meet managers: examples from geomorphology, in P.R. Wilcock and R.M. Iverson (eds), *Prediction in Geomorphology*, Geophysical Monograph 135. Washington, DC: American Geophysical Union, pp. 27–40.

Wilson, G.A. and Bryant, R.L. (1997) *Environmental Management: New Directions for the Twenty-First Century.* London: UCL Press.

Wolman, M.G. (1967) A cycle of sedimentation and erosion in urban river channels. *Geografiska Annaler* 49A: 385–95.

Wooster, J.K., Dusterhoff, S.R., Cui, Y., Dietrich, W.E., Sklar, L. and Malko, M. (2008) Sediment supply and relative size distribution effects on fine sediment infiltration into immobile deposits. *Water Resources Research* 44: W03424. doi: 10 1029/2006WR005815.

6

Geomorphology and Society

Mathias Kondolf and Hervé Piégay

Humans evolved in the context of geomorphic processes, and their survival and prosperity depended in no small measure on their understanding of these processes, the threats posed to their survival and implications for food production. Human settlements are often vulnerable to geomorphic hazards by virtue of their location in the path of geomorphic processes such as flooding, landslides or debris flows. In the next level of interaction, human settlement alters geomorphic processes, such as exacerbating the severity of flooding by reducing infiltration and increasing stormwater runoff (for a given unit of precipitation), increasing erosion and sediment yields through land clearance, triggering mass movements by altering drainage patterns or undercutting the toes of slopes and inducing seismic activity through crustal loading under large reservoirs (Gupta, 1985, 2002; Chao, 1995). Moreover by managing or regulating geomorphic processes, humans have altered their environments, often simplifying natural systems to improve conditions for human survival, such as increasing habitat for game (Andersen, 2005) or enhancing water availability for agriculture (Evenari et al., 1961).

We cannot speak about geomorphology and society without considering human perception and perceived opposition between nature and culture. Human perception of geomorphological processes underlies the choice of management actions, but differs among cultures. Sometimes processes are viewed negatively and humans fight against them, whereas in other contexts, they are considered beneficial. Increasingly, the discourse is shifting from an attitude of 'development

against nature' to one of 'live with nature'. So long as the scale of human intervention does not undermine the geomorphic processes (and the biodiversity and resources they support), humans benefit from the healthy functioning of geomorphic processes. For example, fluvial corridors that have not incised can improve downstream water quality through the self-epuration of nitrates that occurs in a floodplain forest well-connected to the groundwater (Naiman et al., 2005). The scientific community has an increasingly important role to play in providing 'reality checks' and innovative solutions to advance these societal goals.

In this chapter, we explore some of the many interfaces between human society and geomorphic processes, considering natural processes and settlements, pressures and impacts on earth systems, natural hazards and risks, benefits of environmental infrastructures, and lastly, perceptions and policy.

HUMAN SOCIETY

Human settlements in relation to geomorphic processes

When considering interactions between human settlement and geomorphic processes, we can identify two levels of interaction. First, human settlements as affected by natural geomorphic processes; and second, human modifications to geomorphic processes and rates. The first level of interaction is illustrated by the Quechua village destroyed by the lahar on the flank of an Andean volcano, or the floodplain settlement inundated by large floods. Here humans have placed themselves in the way of geomorphic processes that would have occurred whether the humans were there or not. If the processes were sufficiently frequent, we often find that indigenous settlements were placed out of harm's way. For example, the local Ababda tribesmen living in the valley floor near the shrine of Sheik El Shazli, in the Red Sea Range of Egypt, located their dwellings just above the level reached by periodic flash floods in the Wadi Um Smarah, while recent settlers have built structures in the path of flash floods and debris flows, which in turn have backed up floodwaters and caused flooding of structures previously above floods (Figure 6.1) (Gohar and Kondolf, 2007). Medieval settlements in river valleys near Stuttgart, Germany, were optimally placed with respect to several resources and hazards. These primarily agricultural communities located their villages on bedrock where possible, so as not to displace prime agricultural land, here windblown loess and floodplain silts forming fertile loams. Village sites also had a spring or surface water source, and were located above the level of floods. It was only in the 19th century that floodplains, heretofore reserved for agriculture, became extensively occupied by dwellings, resulting in loss of arable land and creation of flood hazards.

Historically, human settlements have typically been located along rivers, to take advantage of water supply, navigation and mill power.

Where high ground was available nearby, it would traditionally be preferred for settlement, but in some cases the advantages of the location outweighed the disadvantages of periodic flooding, and cities were developed where there was nothing but low floodplain. However, even subtle differences in topography of floodplains were recognized and exploited by early settlers. Natural levees along large meandering rivers constitute a good example. Natural levees result when flood waters overflow the channel onto the floodplain. As these waters leave the channel, their velocity slows and coarser sediments in suspension (sand) abruptly settle out, building up a berm of sand along the channel margin. Over time, this berm can build up, so that higher and higher floods are required to overtop it, and it acts as a 'natural levee'. Typically, natural levees drop off steeply along the channel side, sloping more gently downwards away from the channel, into the back swamp. Along the Sacramento River, early settlements, such as Colusa, were restricted to the natural levee (Kelley, 1989). Likewise, in New Orleans, the older neighborhoods of the French Quarter and Garden District were built on the natural levee of the Mississippi, which was not only higher, but also conveniently adjacent to the channel, for docking, etc. With greater distance from the river's edge, the land sloped down into the back swamps, which were annually flooded and recognized as entirely ill-suited for human settlement. However, the advent of powerful pumps in the 20th century made possible an unsustainable occupation of these low-lying lands. Once drained and their soils

PRE-DEVELOPMENT

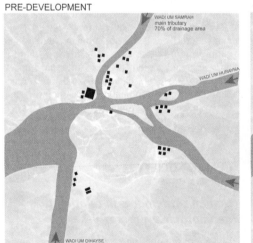

1996 FLOOD

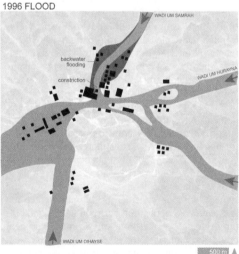

Figure 6.1   Development in floor of flash-flood prone wadi floor, El Sheik El-Shazli, Red Sea Governate, Egypt, showing increased flood levels upstream resulting from construction in floodway. (From Gohar and Kondolf, 2007)

were exposed to the atmosphere, vast areas subsided below sea level (Campanella, 2002). The pattern of inundation depths during Hurricane Katrina in New Orleans reflected the fluvial landforms, with natural levees either dry or only shallowly inundated, while many backswamp areas were over 4 m deep. The natural levees above flood level were not only along the modern river, but also along ancient distributary channels running through the center of the city, such as the Metarie–Gentilly Ridge (Plate 1; see colour plate section, page 587). Along the Mekong River in Cambodia, settlements and permanent agriculture (e.g. fruit trees) are located on natural levees, with back-swamp areas used for annual crops (Campbell, 2009). In Bangladesh, artificial flood-control levees have typically failed to control floods, but ironically serve as 'high ground' onto which residents take refuge during inundations (Sklar, 1992).

Where no high ground is available, societies have often taken refuge from floods by migrating from flood-prone areas during the wet season or building structures above the anticipated flood levels. In the Mekong River Delta, dwellings built on frequently inundated floodplains are commonly built on stilts, such that the living spaces remain above flood waters during the prolonged inundation of the wet season, and the houses are accessed by small boats. In Dadun, a 'water village' of the Pearl River Delta in China, the flood of 1962 inspired many residents to build second and third stories on their houses in anticipation of future floods (Bosselmann et al., 2009).

Human effects on geomorphic processes

The next level of interaction of humans with geomorphic processes involves human activity changing the nature or rate of the geomorphic process. As described by Downs and Gregory (2004) with respect to river use, the magnitude and extent of human alterations has increased over time, from hydraulic civilizations such as Egypt, to the massive dam construction of the 20th century. Examples include deforestation and road construction increasing erosion rates and triggering landslides by removing protective vegetation cover, decreasing slope strength, altering infiltration and drainage patterns and undercutting the toes of slopes. Land-use changes can increase runoff from a given rainfall, resulting in higher peak flows, which in turn leads to channel incision and destabilization. Levees prevent floodwaters from spreading out onto the floodplain and thereby increase the magnitude of floods and the speed with which they propagate downstream. These are

all natural geomorphic processes (soil erosion, mass wasting, infiltration and runoff, flood flows, overbank flooding), but they are altered (often intensified) by human-induced changes.

Human actions have a wide range of effects on geomorphic processes. Anthropic increases in erosion may involve either increased strength of the geomorphic process acting on the landscape (e.g. increased flood heights and velocities) or reduced strength of the soil, hillslope or river bank being acted upon (e.g. by devegetation, toe undercutting). At the extreme are human activities that directly alter the landscape, such as massive grading for subdivisions, open-cast mining, highway construction and dredging for harbors. Certainly, one of the most striking illustrations of such large-scale landscape modification, with profound ecological implications, is the 'mountaintop removal' practice of coal mining, in which the tops of mountains are literally removed and dumped in intervening valleys, obliterating the pre-existing topography, disrupting (and typically contaminating) surface runoff and groundwater flow (Palmer et al., 2010). More pervasively, human activities change the rate of geomorphic processes that lead to effects downstream, such as changes in vegetation cover resulting in increases or decreases in infiltration or erosion rates. The Sierra Nevada range in California is illustrative. Mining (for gold in the 19th century and for construction aggregate in the 20th century) directly modified large areas of the landscape. Hydraulic mining for gold literally washed away mountains and delivered over 1.3B m^3 of sediment to rivers in the Central Valley of California, causing a five-fold increase in sediment loads over the period 1860–1885 over their pre-disturbance value (Gilbert, 1917). Dam construction in the 20th century reduced the yield of sediment from mountains to the Valley down to about 20 per cent of its pre-disturbance level, and mining of sand and gravel from channels and floodplains resulted in an annual deficit of about 40 M m^3 (Plate 2, page 588).

Examining the movement of coarse sediment through the river network from eroding uplands to the coast is illustrative (Figure 6.2). River-derived sediment nourishes beaches, which protect coastal cliffs from wave erosion. Where dams trap sediment, preventing its transport to the coast, and where downstream of the dams, sand and gravel has been removed from the river channel for construction aggregate (e.g. northern Italy, southern California, Japan), the coastal sediment deficit has resulted in sand-starved beaches that have narrowed or disappeared, accelerating erosion of coastal cliffs (Inman, 1985; Gaillot and Piégay, 1999). Ironically, one purpose of many of the dams on these coastal rivers was to protect houses

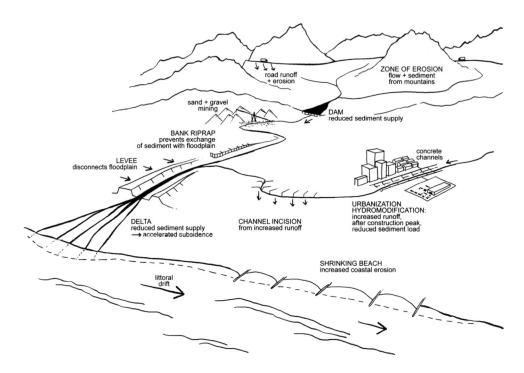

Figure 6.2   Coarse sediment movement from eroding uplands to sea level, showing human alterations to sediment transport continuity

on floodplains adjacent to the river from damage from floods, but the end result has been to damage or destroy houses built on coastal cliffs, whose erosion has been accelerated. To mitigate for such sand starvation, the US Army Corps of Engineers annually places millions of tonnes of sand on beaches around the US, in what is termed 'coastal nourishment'. Elsewhere, eroding coastlines are protected by boulders, concrete jacks or concrete rubble. In the Ministry of Works in Japan, some scientists joke that humans do what was formerly the river's job: instead of allowing the river to carry its sand to the coast, humans remove the sand from the channel to make concrete, which is removed from the system for the life of a structure, but ultimately winds up along the coast as rubble to protect against coastal erosion.

In these examples, we see human effects as most often changing the rate and magnitude, and sometimes the direction, but not the essential nature of geomorphic processes. In addition, we often see that site-specific reactive management can often set off a vicious cycle, in which actions taken to 'fight nature' conceived in a simple cause-consequence context have unintended consequences downstream (both in space and time). As geomorphic systems consist of spatially nested cascading sub-systems (hillslope ⇒ river ⇒ coast)

(Piegay and Schumm, 2003), channel rectification to improve drainage for agriculture concentrates flow in-channel, and typically induces bed incision. This, in turn, creates unstable, over-steepened banks, whose collapse causes erosion of agricultural fields that the original project was to benefit. More broadly, land-use changes at the basin scale can be viewed in terms of the alteration of discharge and/or sediment load, which will inevitably affect alluvial channels and sediment supply to downstream (Gregory, 2006). The increased impermeable surface that accompanies urbanization results in higher peak flows with increased transport capacity, incision of river beds and propagation of sediment 'slugs' downstream. In the urban area of Lyon, downstream stream reaches are covered by sand deposits, with pronounced effects on fish species and fishers (Schmitt et al., 2008). Moreover, floods downstream are flashier than before, increasing flood hazards to human safety.

Humans, a key component of the earth system

When we consider fluvial geomorphology, it is safe to say that the vast majority of geomorphic studies have been on rivers in human-altered environments, but the human impacts have not always been fully appreciated. As demonstrated by Jacobson and Coleman (1986), the floodplains along rivers and streams of the eastern seaboard of the US typically contain a meter or more of sediment aggraded during periods of intense agricultural land use from the late 18th century through the early 20th century. This 'legacy sediment' is easily eroded by the modern stream, contributing to the sediment yield, even if the uplands have healed and no longer produce significant sediment. As a result, rivers along the eastern US did not manifest decreases in sediment load over the 20th century, despite dramatic reductions in upland erosion rates (Meade, 1982). Whereas the braided rivers in Europe are considered as valuable ecosystems (Ward et al., 1999), many of the braided channels in southern Europe resulted from high loads of coarse sediment from Alpine areas that were heavily populated during the 19th century, deforested and overgrazed. The post-19th century channel narrowing observed in many rivers responding to reduced sediment loads (Kondolf et al., 2002) raises questions about what kind of policy should be promoted to preserve or restore river forms that may be based on an idealized conception of nature (Dufour and Piegay, 2009).

The selection of study sites in fluvial geomorphology is hardly random, and this per force must have influenced our notions of what is normal or 'natural'. Most geomorphic study sites have been either in preferred geographical locales or near major research centers (Graf, 2001). Geomorphologists have often been drawn to study sites that offer conditions that are flume-like or otherwise permit a focus on certain variables of interest. The famous bedload trap on the East Fork of the Green River, in Wyoming, was the site of repeated bedload measurements and calibration of the Helley–Smith bedload sampler (Leopold and Emmett, 1976). A sand-dominated bedload passed over a pavement of gravel, such that the sand was mobile over a wide range of flows, even low velocities, while the gravel bed was mobilized only rarely. Most of the measurements were of the sandy bedload passing over the stable gravel bed. Yet the bedload transport regime at this site could be viewed as an artifact of human impacts. The sand was supplied by Muddy Creek, a tributary entering the East Fork a short distance upstream of the bedload trap. Its name notwithstanding, Muddy Creek had an unnaturally high sand load because it received irrigation return flow (from another basin) that artificially increased its flows during much of the year, which induced bank erosion and increased its transport capacity (Dietrich et al., 1979). As a result, its transport mode, a large sand load moving as bedload over a stable gravel bed, is somewhat unusual and artificial. Yet observations in the East Fork have been generalized to rivers worldwide. Similarly, much geomorphic research has been conducted on river reaches that displayed attributes deemed especially desirable, such as the meander bends on the Popo Agie River, Wyoming, used as a model for meander bend geometry by Leopold and Langbein (1966). When viewed in the larger context of the longer length of river, the studied reach can be seen as limited, and indeed, exceptional.

Much fluvial geomorphic research has focused on effects of changes in climate, in some cases ignoring human influences. For example, some of the river incision in the UK attributed by geographers to climatic changes has almost certainly been the result of massive aggregate mining from river channels in the 19th and 20th centuries (Sear and Archer, 1998). Thus it is an important question – and not always easily answered – how to distinguish the effect of changing climate from human impacts, one that demands a thoughtful answer lest we erroneously attribute the effects to the only cause we have thought of a priori. The multiple working hypotheses approach advocated by Chamberlin (1897) is basic to good fluvial geomorphic research.

Geomorphologists often try to study rivers in their 'natural' character, trying to avoid human impacts, such as the 'reference reach' so often sought in the design of river restoration. Typically, however, the channels that need restoration are lowland streams that have been heavily modified by human activity, while the less disturbed 'reference reaches' differ in some fundamental way from the streams to be restored, either because they occur in mountainous uplands, enjoy bedrock control or otherwise may be misleading as models for restoration of the disturbed reach.

The 'reference reach' can be misleading in other respects. The riparian forest along the Eygues River in south-eastern France has been designated as a site of special interest under the Natura 2000 program. Yet this forest did not exist in the 19th century. Rather it is an artifact of rural depopulation and land-use changes in the late 19th and early 20th centuries (Kondolf et al., 2007). In North America, the goals of restoration are often stated as a return to 'pre-disturbance' conditions, with the disturbance implicitly being the European transformation of the landscape in 18th to 19th centuries. But this too, is an arbitrary goal, given that Native Americans were managing the landscape for game and forage, notably by setting periodic fires (Anderson, 2005).

The whole question of what we mean by 'restoration' of rivers and other landscape features or ecosystems presumably must involve some underlying geomorphology and ecology. But concept itself is culturally determined. Downs and Gregory (2004, Table 9.2) review terms such as restoration, rehabilitation and enhancement, making clear that the social context is key. While to successfully implement a restoration project requires some understanding of the underlying geomorphic processes, the very decision to restore and the specific goals set for the project are largely socially determined.

More practically, we can say that scientists must understand that rivers worldwide are reacting to different human pressures, sometimes over long time scales, and their actual structure – even if 'natural looking' – can be largely the result of previous human pressures. Field-based studies must consider this point when designing experiments and hypothesis testing.

NATURAL HAZARDS, MEMORY AND TIME SCALES

We can view many geomorphic processes as broadly predictable, even though we cannot predict precisely where and when the next occurrence will be. For example, available evidence indicates that there is a 30 per cent chance that an earthquake of magnitude 7.5 will occur on the Hayward Fault near Oakland and Berkeley, California, within the next three decades (Petersen et al., 2008), but we cannot predict the date when the earthquake will occur. Most of the actions that we can take to reduce loss of life and property in an earthquake require timescales of months to years, for example, retrofitting buildings so that they do not collapse from shaking, quake-proofing homes by fixing bookshelves to walls, reducing potential sources of fire, etc. The tremendous interest in being able to predict the precise date/time of an earthquake may be misplaced, because it is not clear how much better prepared we could be simply from knowing an exact date when a large earthquake would occur. In effect, we already know enough to guide our actions so that we are prepared, yet the long recurrence intervals between earthquakes means that most people have not experienced a large earthquake themselves, so they are not motivated to prepare for a large earthquake that will occur.

The New Madrid Fault in the Mississippi lowlands near St Louis, Missouri, experienced a major earthquake in 1810, when extensive subsidence caused entire towns to disappear and the river changed its course, as described by Mark Twain (1883). The US Geological Survey had mapped the fault and issued warnings about the likelihood of a repeat performance by this fault, but these warnings did not capture public imagination. Then a local scientist without a background in geology predicted that a large earthquake would occur on or about 2 December 1990 (Fowler, 1991; Spence et al., 1993). Despite the US Geological Survey stating that the probability of an earthquake on that day was the same as any other, the response was massive: schools were closed, many businesses shut down, shelves were stripped bare of earthquake supplies like flashlight batteries and matches, and millions stayed home to await the earthquake. When the predicted earthquake did not occur, many members of the public then became skeptical of the legitimate warnings (Spence et al., 1993).

Natural processes that threaten human settlement, such as earthquakes, large floods and hurricanes, tend to have return intervals (for any given place) of typically 50 years or more – longer than the time scales on which humans usually plan and prioritize (Table 6.1). For example, a politician whose term is 2 to 6 years will be motivated more by risks on a comparable time scale than by risks of long-term hazards whose likelihood in any single year may be a few percent or less. Chances are pretty good that a real-estate developer can sell all the lots on his subdivision long before the property is inundated by a large flood. More generally, we can posit a mismatch between the long intervals separating these events and human perception, such that people tend to ignore the threat if the process does not recur frequently enough that they experience it within their lifetimes.

Some societies have developed mechanisms to remind their members of the risks posed by such geomorphic processes, even if current residents may have no direct experience with floods, earthquakes, etc. The indigenous Tlingit people of the Pacific Northwest of North America inhabit an environment with deep subduction-zone earthquakes occurring with a return period of over a century (Emmons, 1991). Although many members of the tribe would never experience an earthquake themselves, the collective memory of the experience was traditionally maintained by annual earthquake rituals in which seismic shaking was simulated. According to traditional Tlingit beliefs, the earth is held up by 'old woman underneath'. Earthquakes occur when Raven tries to dislodge her from her stance.

In France, a recent law (July 2003) aims to promote a 'new risk culture', in which people become aware of the risks in their environments. Government agencies must now inform the population every two years of the risks in each commune, take steps in advance to reduce

Table 6.1 Time scales of natural hazards and human perception

Natural hazard	Time scale (y)	Human perception	Time scale (y)
Earthquake, California	50–100	Human life	70
Hurricane, Florida (at any one point on coast)	50	Planning horizon for real estate development	5–10
Major fire	5	Term of office, elected official	2–4
Cold weather in northeast/Midwest US	2	Government budget	1
		Attention span of media	0.02

exposure, and devise emergency response plans to take effect once the event occurs.

SOCIETY AND PERCEPTION

How we perceive geomorphic features and processes affect the way we act on them

Many perception studies have shown that landscape features perceived as natural are more positively appreciated by respondents than those that appear to be artificial, built by humans (Gregory and Davies, 1993; Kaplan et al., 1998; Van den Berg et al., 2003). Nevertheless, the perception of nature varies among our largely urbanized population, as a function of individual attitudes depending on socio-cultural settings, and economic development (Figure 6.3c). Moreover, even if there is some consensus in perceptions of nature, many examples illustrate culturally varied reactions such as preferences for single-thread meandering channels or against wood in rivers.

In North America, stream restoration projects became popular in the 1990s, with many of the projects aiming to stabilize dynamic channels. Purely from an ecological perspective, the greatest habitat diversity is found along a dynamically migrating channel, with its active erosion and deposition, recruitment of large wood from eroding banks, establishment of pioneer riparian stands on fresh sandbar surfaces and a complex of undulating bedforms, backwater habitats behind bars and floodplain water bodies (Stanford and Ward, 1993; Ward and Stanford, 1995). Yet a large number of North American 'restoration' projects sought to stabilize dynamic channels by installing large wood and boulders in the outside banks of meanders, and building boulder weirs across the channel. An understanding of riverine ecology would suggest that stabilizing such channels would result in a decrease in ecological value for native species (Ward and Stanford, 1995), but the

goal of stability is rarely questioned in such projects, and appears to reflect a deep-seated cultural preference (Kondolf, 2006). In many cases, the objectives of these restoration projects were to create single-thread meandering channels, whether such channels would have existed on the site naturally or not. These projects reflected what appears to be a well-developed cultural preference for stable, meandering channels, which can also be seen in the landscapes designed by Capability Brown on English estates of the late 18th century (Kondolf, 2006). The ideal landscape was considered to be the meander bend of the River Thames visible from Richmond terrace, and the dimensions of this bend were reproduced almost exactly in the 'Serpentine', a constructed crescent-shaped lake in Hyde Park, reflecting the theory of beauty articulated by Hogarth (1753), who identified the serpentine line as an element appearing in great works of art from classical times, as well as in pleasing landscapes:

> The eye hath this sort of enjoyment in winding walks, and serpentine rivers, and all sorts of objects, whose forms, as we shall see hereafter, are composed principally of what I call waving and serpentine lines ... that leads the eye in a wanton kind of chase, and from the pleasure that gives the mind, entitles it to the name of beautiful.

The preference for the meandering form found expression in landscape paintings of the 19th century, and more recently in advertisements for automobiles, which often depict sinuous highways (and river channels) (Figure 6.4) (Kondolf, 2006).

In North America, many 'restoration' projects have sought to create stable, single-thread meandering channels on reaches that were considered unstable (and thus 'bad'). In many cases, these idealized channels were built on rivers whose flow regime and sediment load would naturally support a more dynamic (and thus messier) channel form, and many of these imposed idealized channel forms have washed out within months or years of construction (e.g. Smith and Prestegaard, 2005).

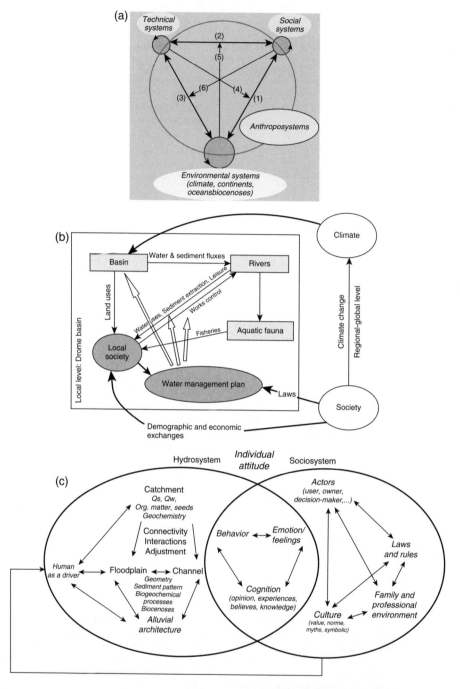

Figure 6.3   From a physically based conceptual approach to the 'anthroposystem'.
(a) The 'anthroposystem' as defined by Lévêque et al. (2000), a complex system where the 'environmental' components (physics, chemistry, biology) interact with the social and the technical components. Interactions are not only considered as human pressures on environment but also as technical developments and social reactions to the environment, its characters and evolution. Modified from Lévêque et al., 2000. (b) Example of the conceptual framework of natural and anthropic factors influencing the fluvial dynamics of the Drôme River, France (Pont et al., 2009). (c) Perception of the river by an individual strongly depends on his social environment and his own characters (e.g. feeling, knowledge, experience). (Adapted from Le Lay and Piegay, 2007)

While the justifications for such projects were usually stated in terms of width–depth ratios and channel types under the Rosgen (1994) stream classification system, the fact that the meandering channel has been the goal of so many projects (even in cases where such an idealized form would be out of keeping with the channel forms that would naturally occur) suggests that these designs were motivated by a desire to create beauty – even though the designers themselves are unaware of their aesthetic biases and unexamined bias towards the ideal meandering channel.

Wood in rivers has long been viewed negatively because of perceived impacts to fish passage, navigation and risks to infrastructure. Historical

analysis of journals of early settlers indicates that wood was widespread in channels along the Pacific coast of North America, with debris jams spanning the entire channel of even large rivers (Maser and Sedell, 1994). Indeed, the US Army Corps of Engineers (the Corps) devoted much of its efforts in the mid 19th century to 'snagging' (i.e. removing large wood) from the Mississippi River (Barry, 1997). In the Apalachicola River, Florida (as typical of rivers used for navigation) large wood was extensively removed: the Corps recorded removing over 334 000 logs, stumps and overhanging trees from the 1870s through the 1990s (USACE, 2001). Increasingly, the scientific community has begun to reconsider this

Figure 6.4 Magazine advertisement for Jeep Cherokee sport utility vehicles, featuring a sinuously bending highway that transitions into a meandering river. (From Kondolf (2009), used by permission of BBD&O, New York)

element in the historical river landscape and its process role in ecosystem and habitat (Maser and Sedell, 1994).

A recent international study on wood perception demonstrated cross-cultural differences in how university students perceived wood in streams (Piégay et al., 2005, Chin et al., 2008; Le Lay et al., 2008). In regions with more advanced environmental consciousness (Sweden, Germany, Oregon), wood in channels was mostly viewed positively, but in other regions (mainly other European countries) wood was considered as a natural element to be fought against or controlled, and in others (China and areas of India and Russia) wood in the channel was considered to be 'out of nature'. Moreover, one's view of wood in channels can be strongly coloured by one's position and responsibilities: if you are concerned primarily with controlling floods you are more likely to lean towards channel clearing and reducing roughness, whereas if your mission is to restore native fish populations, you are more likely to value wood in channels and seek ways to allow it to remain there (Le Lay, 2006).

Naturalness and wilderness, or natural and pristine states, constitute the ends of a gradient on which the most appreciated nature is plotted. Many surveys have shown that the most appreciated natural-looking landscapes must be harmonious, organized, '… and in fact maintained by human actions so that gardened natural landscapes are more positively appreciated than landscapes ordered by natural processes' (Mosley, 1989; Gregory and Davis, 1993). Moreover, if this nature provides activities and is secure, it gets an additional value (Akasawa et al., 2004). Whether a river must be cleaned or not is a question not only of fish habitat or flood conveyance. The public may expect the river to be cleaned because it is perceived as 'dirty'. Again, interactions between nature and culture must be considered. A naturalistic but still organized landscape seems to be more appreciated than a 'messy' landscape (Nassauer, 1995), but we can view the latter as landscapes whose underlying organization is not obvious to our culturally trained eyes.

Lessons: reconsidering the way approaching geomorphology and society

Although most models of human interactions with natural systems tend to treat the two as separate systems that interact in specific ways, such as return interval of flooding, in many cases the two systems are directly coupled. Examples include building levees in coastal Louisiana, which encourage development in the 'protected' areas behind the levees but also contribute to loss of freshwater wetlands and thus reduction of their dampening effects on storm surge, which ultimately exacerbates coastal flooding and thus drives up economic damages from hurricanes. Levees can be seen as having the effect of filtering frequent smaller floods while exacerbating the damages caused by larger disasters (Werner and McNamara, 2007). Similarly, 'beach nourishment' programmes (in which sand is mechanically added to eroding beaches) increase the value of coastal properties thereby maintained, and thus create greater incentive to continue (and even increase) beach nourishment programs (McNamara and Werner, 2008).

This new paradigm is illustrated by the conceptual frameworks in which the scientists have worked since the 1960s. After the emergence of 'systems' thinking focused on disciplinary issues (e.g. the so-called fluvial system of Schumm (1977) for the geomorphologists), systems thinking progressively incorporated all the environmental components of the earth system (e.g. physics, ecology and then chemistry) notably in the 'hydrosystem concept' which emerged in the 1980s (Amoros and Petts, 1993). New paradigms are wider and consider society as well, and the studied earth system is becoming 'anthropocentred' (Figure 6.3a). Following the emergence of the 'anthroposystem concept' in the early 2000s, interdisciplinary teams developed conceptual and integrative models (Figure 6.3). For example, in the Drôme River basin, fluvial processes, determined by sediment and water fluxes, were viewed as key drivers of aquatic and riparian habitat diversity in a catchment framework where the residents interact with natural factors through land uses and modification of vegetative cover, economic activities directly linked to the river, management strategies, and cultural heritage. In the long term, the main driving forces affecting the evolution of both the natural and the human components of the system are climate variability, demographic and economic exchanges, and legislation. When we consider individual attitudes (Figure 6.3c), the system is becoming more complex. The motivation of a given person to act is influenced not only by how the system works but also what perceptions and representations he has of it, again depending on his social framework and his own characteristics (behaviour, beliefs, knowledge, experience, feeling).

Geomorphic studies must widen their spatial and temporal perspective and combine retrospective and prospective approaches. There is a need to highlight the trajectory of earth sub-systems and how geomorphic changes are transferred to each of them in a cascading framework with its own timeframe and thresholds. Moreover, this question is not only a problem of human pressure and impact assessment but also a problem of development and human well-being. Geomorphic understanding and its controlled factors must also consider the consequences of geomorphic characteristics or evolution related to the multiple stakes in terms of uses, resource exploitation, security, and wider benefits. The fluvial/earth system must be now studied as an anthroposystem.

From a philosophical point of view, one can question why the 'natural landscape' remains such a compelling reference for the scientific community. To what degree may the attraction to study 'natural' systems imply a negative perception of 'artificial' ecosystems. Might this explain why 'artificial' systems have been understudied in comparison to 'natural'? Can we find artificial ecosystems from whose structure and functioning we can draw lessons for sustainable design? What role should humans play on earth? They are definitely a part of Nature and have influenced the earth's evolution. After centuries of fighting to extract culture from nature, perhaps it is time to reintroduce humans in their environment.

Finally, there is a need to reconsider our objectives in the context of emerging ethical questions. For example, why do we attempt to restore ecosystems and to what extent is this possible? Must we restore because we have been 'bad', we have destroyed the living communities, or because it is necessary for human survival in the long term? This new, anthropocentred, question is a challenging issue. What do we mean by a 'healthy ecosystem'? Can we really expect from it multiple benefits? With the emergence of the concepts of natural infrastructures and ecosystem services, a new paradigm is based on the concept of 'win–win', in which humans must balance the beneficial uses of natural resources whilst minimizing negative consequences. To succeed will require that we connect earth sciences with social sciences.

CONCLUSIONS

For most of the 20th century, most academic and many research geomorphologists focused on fundamental processes and landscape evolution, leaving applied problems to be addressed by engineers. The engineers, when faced with a badly behaving river, would tend to assume that problems could be solved with installation of structures and rarely looked beyond the reach scale for solutions. With societal demand for improved ecosystem health, there is a recognized

need to look at a larger scale, and to manage river basins rather than focusing on a short 'problem reach'. As the field of river restoration evolved in North America, many of the early efforts were attempts to create habitat, stabilize channels with structures and/or create idealized meandering forms based on simplified ideas drawn from geomorphic research of the 1960s, boiled down to 'cookbook' approaches. As noted above, these projects reflected a cultural bias towards stable meandering channels. They also echoed the reach-scale engineering approaches that have prevailed in the past (albeit with somewhat different objectives). With the widespread failure of these projects (e.g. Smith and Prestegaard, 2005; Kondolf, 2006; Miller and Kochel, 2010), there is increasing demand that restoration consider process on larger temporal and spatial scales, which requires application of 'real geomorphology' (Sear, 1994).

The techniques used in geomorphic analysis can be viewed in terms of costs and benefits. For questions more important to society, greater investment is justified, such that multiple working hypotheses, and sufficient data collection and analysis are warranted. A less expensive option is an expert-based, qualitative approach. Even when giving lip service to the need to understand the underlying problem before throwing engineering fixes in the river, managers tend to be attracted to approaches that promise an easy, cheap answer. In France, managers are often unwilling to pay more than 10 000 euros for a geomorphic study, while they might not think twice about paying over 100 000 euros for an engineering study. In North America, planning and monitoring studies are expected to constitute only a small fraction of the total budget of restoration projects, which implicitly assumes that the projects will involve a lot of construction. However, in many cases the most effective restoration strategy may be to simply set aside land for a riparian corridor, or breach levees that disconnect floodplain from overbank flows, which involve low construction costs but may warrant substantial investment in planning and evaluation to optimize interventions and draw lessons from performance of projects.

REFERENCES

Amoros, C. and Petts, GE. (1993) *Hydrosystemes fluviaux*. Masson, Paris, p. 300.

Anderson, M.K. (2005) *Tending the Wild: Native American Knowledge and Management of California Natural Resources*. University of California Press, Berkeley, p. 555.

Asakawa S., Yoshida K. and Yabe K. (2004) Perceptions of urban stream corridors within the greenway system of

Sapporo, Japan. *Landscape and Urban Planning* 68: 167–182.

Barry, J. (1997) *The Rising Tide: The Great Mississippi Flood of 1927 and How it Changed America.* Simon & Schuster, New York.

Bosselmann, P.C., Kondolf, G.M., Feng, J., Bao, G., Zhang, Z. and Liu, M. (2010) The future of a Chinese water village: alternative design practices aimed to provide new life for traditional water villages in the Pearl River Delta, *Journal of Urban Design* 15: 243–267.

Campanella, R. (2002) *Time and Place in New Orleans: Past Geographies in the Present Day.* Pelican Publishing, Gretna, Louisiana.

Campbell, I.C. (2009) *The Mekong: Biophysical Environment of an International River Basin.* Elsevier, Amsterdam, p. 432.

Chamberlin, T.C. (1897) Studies for students, *Journal of Geology* V: 837–848.

Chao, B.F. (1995) Anthropogenic impact on global geodynamics due to reservoir water impoundment, *Geophysical Research Letters* 22: 3529–3532.

Chin, A., Daniels, M.D., Urban, M.A., Piégay, H., Gregory, K.J., Bigler, W., Butt, A.Z., Grable, J.L., Gregory, S.V., Lafrenz, M., Laurencio, L.R. and Wohl, E. (2008) Perceptions of Wood in rivers and challenges for stream restoration in the United States. *Environmental Management,* 41: 893–903.

Dietrich, W.E., Smith, J.D. and Dunne, T. (1979) Flow and sediment transport in a sand-bedded meander, *Journal of Geology* 87: 305–315.

Downs, P. and Gregory, K. (2004) *River Channel Management: Towards Sustainable Catchment Hydrosystems.* Arnold, London, p. 395.

Dufour, S. and Piégay, H. (2009) The myth of the lost paradise to target river restoration: forget natural reference, focus on human benefits. *River Research and Applications* 25: 568–581.

Emmons, G.T. (1991) *The Tlingit Indians.* University of Washington Press, Seattle.

Evenari, M., Shanan, L. and Tadmor, N. (1961) Ancient agriculture in the Negev. *Science* 133: 979–996.

Fowler, G. (1991) Iben Browning, 73: Researcher studied climate and quakes. (Obituary) *New York Times* 20 July 1991.

Gaillot, S. and Piégay, H. (1999) Impact of gravel-mining on stream channel and coastal sediment supply, example of the Calvi Bay in Corsica (France). *Journal of Coastal Research* 15: 774–788.

Gilbert, G.K. (1917) *Hydraulic Mining Debris in the Sierra Nevada.* U.S. Geological Survey Professional Paper 105.

Gohar, A. and Kondolf, G.M. (2007) *Flooding Risks in El-Sheikh el-Shazli,* Report to US Agency for International Development, Cairo, September 2007. Available online at: http://landscape.ced.berkeley.edu/~kondolf/LivingCoast/.

Graf, W.L. (2001) Damage control: restoring the physical integrity of America's rivers. *Annals of the Association of American Geographers* 91: 1–27.

Gregory, K. (2006) The human role in changing river channels. *Geomorphology* 79: 172–191.

Gregory, K.J. and Davis, R.J. (1993) The perception of riverscape aesthetics: an example from two Hampshire rivers. *Journal of Environmental Management* 39: 171–185.

Gupta, H. (1985) The present status of reservoir induced seismicity investigations with special emphasis on Koyna earthquakes. *Tectonophysics* 118: 257–279.

Gupta, H.K. (2002) A review of recent studies of triggered earthquakes by artificial water reservoirs with special emphasis on earthquakes in Koyna, India. *Earth-Science Reviews* 58: 279–310.

Inman, D.L. (1985) Budget of sand in southern California: river discharge vs. cliff erosion, in J. McGrath (ed.), *California's Battered Coast,* Proceedings from a Conference on Coastal Erosion. California Coastal Commission, pp. 10–15.

Jacobson, R.B. and Coleman, D.J. (1986) Stratigraphy and recent evolution of Maryland Piedmont flood plains. *American Journal of Science* 286: 617–637.

Kaplan, R., Kaplan, S. and Ryan, R.L. (1998) *With People in Mind. Design and Management for Everyday Nature.* Washington, Island Press, p. 239.

Kelley, R. (1989) *Battling the Inland Sea: American Political Culture, Public Policy, and the Sacramento Valley, 1850–1896.* University of California Press, Berkeley, p. 395.

Kondolf, G.M. (2001) Historical changes to the San Francisco Bay-Delta watershed: Implications for ecosystem restoration, in H.J. Nijland and M.J.R. Cals (eds), *River Restoration in Europe: Practical Approaches,* Proceedings of the Conference on River Restoration, Wageningen, Netherlands, May 2000, pp. 327–338.

Kondolf, G.M. (2006) River restoration and meanders. *Ecology and Society* [online] URL: http://www.ecologyandsociety.org/vol11/iss2/art42/.

Kondolf, M. (2009) Rivers, meanders, and memory, in M. Treib (ed.), *Spatial Recall.* London, Taylor & Francis (Routledge), pp. 106–119.

Kondolf, G.M., Piégay, H. and Landon, N. (2002) Channel response to increased and decreased bedload supply from land-use change: Contrasts between two catchments, *Geomorphology* 45: 35–51.

Kondolf, G.M., Piégay, H. and Landon, N. (2007) Changes since 1830 in the riparian zone of the lower Eygues River, France. *Landscape Ecology* 22: 367–384.

Le Lay, Y. (2006) L'évaluation environnementale du bois en rivière par les gestionnaires des cours d'eau français. *Géocarrefour* 84: 265–75.

Le Lay, Y. and Piégay, H. (2007) Le bois mort dans les paysages fluviaux français: éléments pour une gestion renouvelée. *L'Espace Géographique* 1: 51–64.

Le Lay, Y., Piégay, H., Gregory, K.J., Dolédec, S., Chin, A. and Elosegi, A. (2008) Variations in cross-cultural perception of riverscapes in relation to in-channel wood. *Transactions of the Institute of British Geographers* 33: 268–287.

Leopold, L.B. and Langbein, W.B. (1966) River meanders. *Scientific American* June: 60–70.

Leopold, L.B. and Emmett, W.W. (1976) Bedload measurements, East Fork River, Wyoming. *Proceedings of the National Academy of Sciences USA* 73: 1000–1004.

Levêque, C., Pavé, A., Abbadie, L., Weill A. and Vivien, F.D. (2000) Les zones ateliers, des dispositifs pour la recherche sur l'environnement et les anthroposystèmes. Une action du programme 'Environnement, vie et sociétés' du CNRS. *Natures Sciences et Sociétés* 8: 43–52.

Lewis, P. (2003) *New Orleans: Making of an Urban Landscape.* Center for American Places.

Maser, C. and Sedell, J.R. (1994) *From the Forest to the Sea: The Ecology of Wood in Streams, Rivers, Estuaries and Oceans.* CRC Press, New York.

McNamara, D.E. and Werner, B.T. (2008) Coupled barrier island–resort model: 1. Emergent instabilities induced by strong human-landscape interactions. *Journal of Geophysical Research* 113, F01016, doi:10.1029/2007JF000840.

Meade, R.H. (1982) Sources, sinks, and storage of river sediment in the Atlantic drainage. *Journal of Geology* 90: 235–252.

Miller, J.R. and Kochel, R.C. (2010) Assessment of channel dynamics, instream structures and post-project channel adjustments in North Carolina and its implications to effective stream restoration. *Environmental Earth Science* 59: 1681–1692.

Mosley, M.P. (1989) Perceptions of New Zealand river scenery. *New Zealand Geographer* 45: 2–13.

Naiman, J., Décamps, H. and McClain M.E. (eds) (2005) *Riparia: Ecology, Conservation, and Management of Streamside Communities.* Elsevier.

Nassauer, J.I. (1995) Messy ecosystems, orderly frames. *Landscape Journal* 14: 161–170.

Palmer, M.A., Bernhardt, E.S., Schlesinger, W.H., Eshleman, K.N., Foufoula- Georgiou, E., Hendryx, M.S. Lemly, A.D. and Likens, G.E. (2010) Mountaintop mining consequences. *Science* 327: 148–149. doi: 10.1126/science.1180543.

Petersen, M.D., Frankel, A.D., Harmsen, S.C., Mueller, C.S., Haller, K.M., Wheeler, R.L. , Leyendecker, E.V.,. Wesson, R.L., Harmsen, S.C., Cramer, C.H., Perkins, D.M. and Rukstales, K.S. (2008) Documentation for the 2008 Update of the United States National Seismic Hazard Maps, *U.S. Geological Survey Open-File Report* 2008–1128, p. 61.

Piégay, H. and Schumm, S.A. (2003) System approach in fluvial geomorphology, in M.G. Kondolf and H. Piégay (eds), *Tools in Fluvial Geomorphology.* J. Wiley and Sons, Chichester, pp. 105–134.

Piégay, H., Gregory, K.J., Bondarev, V., Chin, A., Dalhstrom, N., Elosegi, A. , Gregory, V., Joshi, V., Mutz, M., Rinaldi, M., Wyzga, B. and Zawiejska, J. (2005) Public perception as a barrier to introducing wood in rivers for restoration purposes. *Environmental Management* 36: 665–674.

Pont, D., Piégay, H., Farinetti, A., Allain, S., Landon, N. and Liébault, F. (2009) Conceptual framework and interdisciplinary approach for the sustainable management of gravel-bed rivers: the case of the Drôme River basin (SE France). *Aquatic Sciences* 71: 356–370.

Rosgen, D.L., (1994) A classification of rivers. *Catena,* 22, 169–199.

Schmitt, L., Grosprêtre, L., Breil, P., Lafont, M., Vivier, A., Perrin, J.F., Valette, L., Valin, K., Cordier, R. and Cottet, M.

(2008) Préconisations de gestion physique de petits hydrosystèmes périurbains: l'exemple du bassin de l'Yzeron (France), in Verniers, G. and Petit, F. (eds), *Actes du Colloque 'La gestion physique des cours d'eau: bilan d'une décennie d'ingénierie écologique'.* Namur, 10–12 October 2007. Groupe Interuniversitaire de Recherches en Ecologie Appliquée, Laboratoire d'Hydrographie et de Géomorphologie Fluviatile, Direction des Cours d'Eau Non Navigables, Direction Générale des Ressources Naturelles et de l'Environnement – Ministère de Région wallonne, pp. 177–186.

Schumm, S.A. (1977) *The Fluvial System.* Wiley-Interscience, New York.

Sear, D.A. (1994) River restoration and geomorphology. *Aquatic Conservation: Marine and Freshwater Ecosystems* 4: 169–177.

Sear, D.A. and Archer, D. (1998) The geomorphological impacts of gravel mining: Case study of the Wooler Water, Northumberland UK, in P. Klingeman, P.D. Komar and R.D. Hey (eds), *Gravel-bed Rivers in the Environment.* Water Resources Press, Boulder, Colorado, pp. 415–432.

Sklar, L. (1992) *Technical Review of the Bangladesh Flood Action Plan.* Berkeley, California, International Rivers Network.

Smith, S.M. and Prestegaard, K.L. (2005) Hydraulic performance of a morphology-based stream channel design. *Water Resources Research* 41: doi: 10.1029/2004WR003926.

Spence, W.J., Herrmann, R.B., Johnston, A.C. and Reagor, B.G. (1993) *Responses to Iben Browning's Prediction of a 1990 New Madrid, Missouri, Earthquake,* US Geological Survey Circular 1083. US Government Printing Office, Washington DC.

Stanford, J.A. and Ward, J.V. (1993) An ecosystem perspective of alluvial rivers, connectivity and the hyporheic corridor. *Journal North American Benthological Society* 12: 48–60.

Twain, M. (1883) *Life on the Mississippi.* James R. Osgood & Co., New York.

USACE (US Army Corps of Engineers) (2001) *The 2001 Annual Maintenance Report and Five-year Analysis Report for the Apalachicola–Chattahoochee–Flint Waterway.* US Army Corps of Engineers, Mobile District, Mobile, Alabama.

Van den Berg, A.E., Koole, S.L. and Van der Wulp, N.Y. (2003) Environmental preference and restoration: (how) are they related? *Journal of Environmental Psychology* 23: 135–146.

Ward, J. and Stanford, J. (1995) Ecological connectivity in alluvial river ecosystems and its disruption by flow regulation. *Regulated Rivers: Research and Management* 10: 1–15.

Ward, J.V., Tockner, K., Edwards, P.J., Kollmann, J., Bretschko, G., Gurnell, A.M. , Edwards, P.J., Petts, G.E. and Ward, J.V. (1999) A reference system for the Alps: The 'Fiume Tagliamento'. *Regulated Rivers: Research & Management* 15: 63–75.

Werner, B.T. and McNamara, D.E. (2007) Dynamics of coupled human-landscape systems. *Geomorphology* 91: 393–407.

PART 2

Techniques and Approaches

7

Observations and Experiments

Michael Church

Observation and experiment are fundamental procedures in science. Science is said to be derived from the facts of observation – observations preferably, although by no means always, made experimentally. Observations are, then, the fundamental building blocks of scientific knowledge. In a naïve view, observations are supposed to be directly acquired by careful, unprejudiced observers, independent of any theory. They constitute the firm reference points against which theory is tested. Observations made under similar conditions by different observers ought to be similar, hence verifiable, and therein is supposed to lie the authority of science. Observations and experiments are closely connected, the latter being a formally arranged instance of the former designed, first of all, to ensure that the observations will sharply discriminate between alternative possible ideas (theories) about some phenomenon and, second, to ensure that an independent observer can indeed replicate the observations.

There are serious problems with this naïve view. First of all, there is more to seeing than meets the eye. Two observers, regarding the same phenomenon, may not 'see' the same thing at all because the visual (or other sensory) stimulus is interpreted differently by each observer in light of her/his (different) prior experience. How one first regards a *trompe l'oeil* image is perhaps a trivial confirmation of this circumstance. This is a difficult issue to analyse because it is, of course, impossible for any one of us to appropriate someone else's perceptions and compare them directly with our own. The problem makes clear, however, that scientific facts are public facts: an observation must be publicly expressed as a statement before it can be compared with a view to verification – before it can gain scientific value – and this necessarily entails interpretation by the expositor. Exposition that will qualify a fact as scientifically interesting, in turn, presupposes a learned context, most obviously concerning what are currently the significant scientific questions and how various other observers have approached them. But effective context extends all the way to one's general social and cultural conditioning. What appear to be relevant and, therefore, scientifically interesting facts is guided by the current state of scientific knowledge and by the larger social context of the observer.

A more widely analysed objection to the naïve view is the awkward circumstance that the 'facts of observation' are not prior artifacts at all – they are themselves thoroughly theory laden. 'Scientific' observations (in fact, all our observations of the world around us) are informed by and interpreted via the theories we have formed about the world on the basis of our prior experience and formal learning. Observations have no scientific relevance apart from reference to an appropriate theoretical context. Even more, in science, observations often are mediated by instrumental readings, the interpretation of which depends on more or less elaborate theories about the relation between the primary stimulus and the response of the instrument. The notion that one must learn to be a competent scientific observer effectively gives the lie to the naïvely empiricist view expounded above.

None of this undermines the view that science is based on the 'facts' of observation, but it acknowledges that observations must be recognized as perceptions interpreted in a particular conceptual context. Testing the concepts is the real task of science.

The foregoing paragraphs, intended to introduce the reader to some of the issues surrounding scientific observation, represent only a very tentative toe dipped into the deep, deep waters of the philosophy of science. We will pursue philosophical themes no further, it being sufficient for our purpose to recognize the contingent nature of observations and the contrived nature of experimentation. Readers who wish to delve farther into the fundamental nature of observation and theory in science, and the relations between them, should read Chalmers's (1999) introduction to recent arguments in the philosophy of science that opens with a discussion about the status of observation and experiment. Geomorphologists have not been given to reflecting on these matters (at least, not publicly) but interested readers will find useful discussions in Haines-Young and Petch (1986) and in Rhoads and Thorn (1996). For a view of science in its social context, Latour (1987) is a fundamental, but approachable work.

OBSERVATION

We have decided that a scientifically meaningful observation – a 'fact of observation' – is compounded of a perception and some interpretation of it. Scientifically relevant observations are systematic ones rather than merely casual ones. That is, the observations are made in accordance with some programme that renders them relevant and some protocols that render them interpretable and verifiable within the context of the program. (A 'programme', as referred to here, is a structured sequence of activities that can vary from a general paradigm within which it is intended to describe and comprehend the phenomena of geomorphology to an exact prescription for testing a specific concept.) This is not to say that casual observations are worthless – very often the germ of an important new idea or perspective arises from an observation casually made. Indeed, a hallmark of superior scientific judgment is the capacity to recognize and to follow up casual and, especially, unexpected observations that seem to be at variance with one's current understanding.

Scientific observations are selective; in geomorphology, some phenomenon of interest is picked out for attention amidst its environmental surroundings. In classical bench science, selection is often achieved by isolating the phenomenon of interest in an experimental setting that eliminates all extraneous phenomena. In field science, this is usually not possible, so the observer must deliberately focus attention upon the interesting phenomenon. This circumstance is well illustrated through the almost lost art of sketching, a near-universal tool in geomorphology until the mid 20th century (Figure 7.1). Technical sketches serve to emphasize the phenomenon of interest in a way that remains faithful to the field situation – what is deleted is the supposed irrelevant detail of the surrounding scene – or they may idealize the element of central interest. In the latter case, there is already a conscious element of deliberate interpretation present in the recorded observation.

Classification

Selective observation not only moves observation from the casual to the systematic, it represents the first step toward classification. Before classification, observations represent singular experiences. Insofar as science is about generalizing the experiences of the world around us in order to facilitate understanding, singular events have little relevance. (The alert reader will detect in the foregoing statement an apparent contradiction of the notion that narrative earth history – a description of the historical development of a particular landscape, for example – might constitute significant science. On this issue, see Simpson, 1963.) Classification implies some generalizing principle – a theory – and is useful only within that context because science is about testing principles, not about pigeonholing observations. Hence, a worthwhile classification is one that establishes classes according to the properties of theoretical interest that elements of the class possess. A class is a set of events or objects identified by certain definitive properties. This reverses the usual process of definition. For example, 'open channels containing unidirectionally flowing water, self-formed in the sediments deposited from transport by the flow' defines the class of alluvial stream channels – a definition that is customarily presented the other way around ('alluvial channels are ...'). The properties, in particular the property of self-formation in its own deposits, are the elements of theoretical interest and the basis in this case for a worthwhile class. To be useful, classes must also have definable limits and the properties of the class must apply equally to all recognized members of the class.

A problem with some classifications, in particular ones based on the morphological properties of landforms, is the possibility that the same end

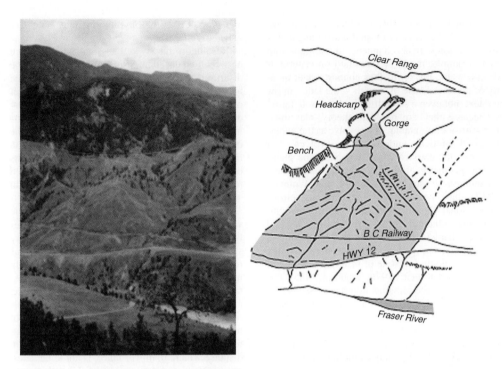

Figure 7.1   Photograph and sketch of landscape to emphasize selectivity in geomorphological observation. The shaded principal feature in the sketch is a large earthflow, which does not stand out in the photograph. The distance across the photograph at railway level is approximately 1 km

form has arisen from quite different processes. This is the problem of equifinality (see Chorley, 1962). So, for example, it is commonly supposed that sand waves on the bed of a stream arise from the initial entrainment of sand by a turbulent eddy, creating a defect in the streambed that then propagates into a wave train (Best, 1992), but it has been established that they also develop spontaneously as a fluid mechanical instability at the interface between water and mobile sediment (Venditti et al., 2005). The possibility for equifinality obfuscates what we might otherwise be able to learn about the genesis of the particular class of landforms merely from their classification.

Another difficulty is that, whilst some classifications may appear to arise 'naturally' as the consequence of the appearance that the phenomena in question consist of a number of discrete types, other exercises seek to categorize what is, in truth, a continuum of phenomena, such that boundaries are essentially arbitrary. It is difficult to think of an entirely natural classification in geomorphology because most landform features exhibit a more or less continual range of variation.

Consider, for example, the variety of river channel patterns (Schumm, 1985), or the arbitrary divisions of Wentworth's (1922) universally accepted classification of sedimentary grain sizes. (There are some minor variations in class limits in various engineering and soil science classifications, but the principle of arbitrary size grades remains the same.) On the other hand, classifications of the basic morphological units that make up river channels – riffles, pools, rapids, steps, cascades, etc., might qualify as a natural classification, though their metric properties are not entirely distinct (see Grant et al., 1990; Montgomery and Buffington, 1997, for two views on such units). Compare, with these examples, such celebrated natural classifications as the periodic table of the elements, or the species concept in life science.

A somewhat related issue arises from the circumstance that many needs for classification arise in the context of specific problems, often applied problems. The criteria for classification will then be ones that illuminate the problem at hand that may or may not follow currently established science. This is especially true, for example, in the context of land management. For such purposes river channels, to continue our main example, might be classified on criteria as diverse as the life-forms they contain, water quality or perceived state of 'health' – the latter, in the context, not even a scientifically defensible term.

Because they are based in theory, classifications are not immutable. They evolve as the underlying theoretical understanding evolves. For example, the classical classification of the planimetric geometry of stream channels as braided, meandered or straight (Leopold and Wolman, 1957) has been elaborated into a multidimensional classification based on the recognition that channel form encompasses several distinct characteristics, including talweg pattern, channel division (presence or absence of islands), bar morphology, and style of instability (Mollard, 1973; Kellerhals et al., 1976; Schumm, 1985). The elaboration of the classification has proceeded in pace with developing theoretical understanding of the factors governing river channel morphology. Kondolf et al. (2003) give an extended discussion of river and stream classification that explores further some of the principles outlined here.

Measurement

Classification is the first level of measurement. Measurement is a fundamental property of scientific observation because it establishes the degree of precision with which we can be said to know the phenomenon under scrutiny. There are four levels of measurement (Table 7.1), each level carrying more specific information than the last. Mathematical operations are conducted on only the highest two orders of measurement. For these measures, the distance between measurement units has specific meaning, hence the resolution

Table 7.1 Measurement scales

Scale	Property	Example
Nominal	Identity	Soil type classification, landform
Ordinal	Order	Schmidt hammer data on rock strength, stream channel order
Interval	Additivity of difference (relative additivity)	Time, temperature, stream link magnitude
Ratio	Absolute additivity	Length, mass, velocity

and precision of the observations may be established exactly.

Resolution refers to the smallest difference between two observations that may in principle be discriminated by the measurement system – that is, it refers to the smallest unit of difference that one may discriminate on an instrument scale or by direct observation. It is a property of the measuring system, not of the measured object. The increasing pervasiveness of digital electronic data ties the concept of resolution to the number of bits of information (in the computational sense) used to represent an observation, or to the elemental scale of pixelated information.

The precision is the range of non-systematic variation of the measurements, which includes both resolution and sampling variation. A very straightforward example of resolution versus realizable precision is provided by erosion pin measurements. Pins are established at some distance behind an eroding streambank, for example, and the distance from the pin to the bank is periodically measured. A resulting distribution of measurements is shown in Figure 7.2. The resolution is set by the minimum resolvable variation in distance which, in this instance, is apt to be determined by uncertainty in establishing the exact position of the irregular edge of the eroding bank rather than by the divisions of the measuring tape. The realized precision is well indicated by the presence of negative measurements. Negative measurements are certainly possible if a segment of a cohesive bank 'leans' outward before failing. In the case illustrated, the banks were noncohesive and such a situation is unlikely to be common. The negative measurements are apt to represent error arising from inconsistency in placing the tape between the pin and the bank edge (which could be overcome by using two pins to establish a line of measurement).

A different kind of resolution problem arises when measurements are grouped. For example, in the size grading of sediments, it is usual to pass the grains through square mesh screens in order to divide them into classes based on the mesh openings. This leads to a number of issues related both to the measurements and to the 'binning' of the results. The relation between the square openings and grain shape is such that the classification process does not yield an unequivocal measure of grain size (Figure 7.3). Hence, there is bound to be some misclassification of individual grains and some distortion of the size gradation curve. Furthermore, what we learn from this procedure is not the distribution of sizes but the cumulative distribution of sizes at certain reference points (the mesh openings). The derivative function – the actual grain size distribution – is only approximated. The degree of approximation

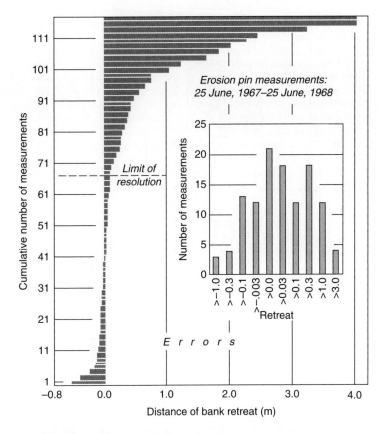

Figure 7.2   Distribution of erosion pin measurements along an eroding terrace edge of a gravel outwash deposit in Arctic Canada (Ekalugad Fjord, Baffin Island). The negative measurements imply a measurement system precision of ±0.1 m, the small number of more negative measurements indicating either outward leaning blocks or blunders

is not easily specified: ideally (i.e. absent imperfect classification), it will depend on the relation between grain size and shape (Figure 7.3).

There are further lessons in this case. The measure of grain size is, in any case, arbitrary. It is usually accepted as the length of the second of the three principal axes of the grain (Figure 7.3, inset), whereas a measure based on the cube root of volume might be regarded as more definitive for many purposes. The result would not usually be quite the same. In a common alternative procedure, grain size is measured as the 'equivalent settling diameter' computed from a theoretical relation between grain size and its settling behaviour in water (see Ferguson and Church, 2004). Grain shape affects this relation as well, but so, now, does grain density. This is nevertheless a preferred measure in many applications of grain size statistics, most obviously to study sedimentation problems in fluid transport. For the present,

its most interesting aspect is the explicit demonstration of the relation between observation and theory.

Scales

Measurements are intrinsically tied to dimension and they depend upon reference scales. The standard reference scales in science collectively make up the Système Internationale (SI) system of units. Only dimensionless measures are meaningfully comparable and the SI system really represents a means to make measures non-dimensional in a comparable way. Hence, a distance, L, of 4.2 m measured with a ruler should really be recognized as $L/L_m = 4.2$, in which L_m is a standard metre, a platinum–iridium bar held at the International Bureau of Weights and Measures in Paris. Any other measurement made with a

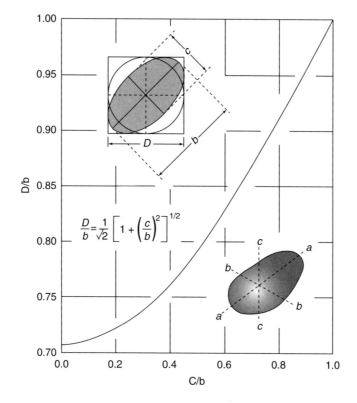

$$\frac{D}{b} = \frac{1}{\sqrt{2}}\left[1 + \left(\frac{c}{b}\right)^2\right]^{1/2}$$

Figure 7.3   The relation between grain 'size', as measured by passage through a square-mesh sieve, and grain shape: inset, principal grain axes. (From Church, 2003)

similar ruler should be directly comparable with the standard because the scale of the ruler should ultimately be traceable to the standard, hence the measurements are comparable with each other. [In fact, the physical bar was abandoned as the ultimate standard in 1960. Today, the standard metre is defined as the length of the path travelled by light in a vacuum during 1/299 792 458 second. The second, in turn, is defined by 9 129 631 770 oscillations of a caesium-133 atom (Taylor, 1995). These apparently curious definitions are the consequence of attempting to establish standards that are directly related to the fundamental physical properties of the universe, hence truly universal and immutable (and so always, in principle, recoverable). The bizarre numbers derive from the attempt to preserve the magnitudes of the earlier, more straightforward definitions.]

SI scales and other conventional systems of measures are arbitrary. Natural phenomena, however, possess intrinsic scales. Such scales may be guessed intuitively or they may be constructed from a systematic consideration of the phenomenon. Consider, for example, the transformation of

information by gravitationally driven waves in a fluid. The waves propagate with velocity $U = \sqrt{gd}$, in which g is the acceleration of gravity and d is the depth of the medium. A universal scale for wave propagation can be constructed by comparison with the velocity of the fluid itself, u. Hence we consider U/u, the magnitude of which determines whether or not wave information may be propagated against the fluid flow. The inverse of this ratio is the well-known Froude number, $Fr = u/\sqrt{gd}$. The most desirable intrinsic scales are canonical scales – ones that yield universal scales of order 1 and must, therefore, indicate the characteristic magnitude of a phenomenon. The Froude number provides an example of such a scale.

Dimensional analysis provides a means to derive the intrinsic scales of a problem systematically. Dimensional analysis converts a problem from the arbitrary scales of standard measurement to the intrinsic scales. In the process, the number of variables in the problem is reduced to $m - n$, wherein m is the number of original variables and n is the number of standard dimensions involved. Consider, for example, an idealized simple water flow (more generally, the flow of a practically incompressible fluid). Relevant variables include the properties of the fluid, fluid density, ρ, with standard dimensions [M L^{-3}], and fluid viscosity, μ [M L^{-1} T^{-1}]; forces to which the flow is subject, represented by g, the acceleration of gravity [L T^{-2}]; and some resulting velocity, u [L T^{-1}], and length scale, l [L] (usually flow depth). There are five quantities and three standard dimensions, so that we seek two intrinsic scales. These are readily discovered to be $\sqrt{gl}$ and $\mu/\rho l$, both characteristic velocities. They lead to the Froude number, $u/\sqrt{gl}$, and the Reynolds number, $\rho u l/\mu$. These natural scales are valid anywhere on Earth (and on any other planet, provided that we adopt the appropriate value of g). The former represents the ratio of inertial forces to gravitational forces and is a measure of macroscopic dynamic effects in the flow, while the latter represents the ratio of inertial forces to viscous forces and is a measure of the importance of microscopic – in fact, molecular – dynamic effects. An interesting outcome is that the characteristic domain of Froude numbers is near 1, the limit value for propagation of information against the flow by gravity waves (i.e. for upstream wave travel). This value defines an important transition in the flow from 'subcritical' to 'supercritical' regime. Accordingly, it may be accepted as a canonical scale for water flows. Characteristic Reynolds numbers, on the other hand, vary with the fluid because viscosity is a property of the fluid. For water, $10^3 < Re < 10^6$ characteristically, which seems not to provide a satisfactory natural scale. An important transition indexed by the Reynolds number (Re) is the transition from laminar to turbulent flow. Unfortunately, the transition is asymmetrical: the upward transition occurs at $Re_u \approx 2500$, while the downward transition occurs near $Re_d \approx 500$. Re_u signals the transition from the dominance of molecular forces to that of inertial forces. A useful canonical scale might be $Re_n = Re/Re_u$, but it is not used.

For more complex problems, a formal means is available to conduct dimensional analysis, known as the Buckingham Π theorem (or, simply, Pi theorem) (see Barenblatt, 2003, or the introductory pages of many physics texts).

Natural scales play a fundamental role in designing scientific investigations. When phenomena observed in one set of standard scales are to be used to interpret or infer the behaviour of a natural system of a different magnitude, the natural scales must be preserved. Hence, the entire issue of scale modelling is subject to natural scales. We will take up this topic under scaled experimentation.

Characteristic scales are depicted informally for many natural processes in diagrams that show scales of length and time (Figure 7.4).

Linear trajectories in this diagram indicate characteristic system velocities that may specify the transfer of matter or energy or information, according to the circumstances. For example, characteristic velocities in Figure 7.4 define distinctively different phenomena that inform river channel and drainage basin development, and may be the starting point for seeking phenomena that determine the natural scales.

Scaling

It is a characteristic of many natural phenomena that they appear to possess no natural canonical scale. Such phenomena exhibit similar physical appearances over a wide range of standard scales, hence preserve characteristic scale relations. To continue our recent example, turbulent eddies in fluid flows display self-similarity, even across different media and occurrence (e.g. atmosphere, ocean and river flows). Without some external reference scale, it is not possible to determine the scale of the flow by reference to the internal dynamics of the flow. Figure 7.4 defines similarity ranges for phenomena related to channeled flows and to the fluvial landscape. Such phenomena cannot, in reality, exhibit infinite similarity ranges, however. At the lower end, one bumps into some limit for the controlling dynamic phenomenon (such as the turbulent–laminar transition in fluids) or an indivisible elementary unit (such as individual grains of sediment); at the upper end, one approaches energetic limits to drive the phenomenon or geometrical limits set by the landscape within which the phenomenon is occurring. These limits, the points of scaling transitions, may become the most interesting feature of such phenomena, for they identify their underlying limits.

Scaling relations are represented as power functions, that is:

$$\varphi = Cx_1^\alpha x_2^\beta \tag{7.1}$$

in which φ is some observable of interest, x_i are scaling variables of the system (typically, other observables) upon which the scale of φ depends, and α and β are powers. C is a constant, dimensionless in rationally constructed relations and of order 1 if the variates are all canonical forms. Scaling relations appear as power laws because they preserve dimensional homogeneity. If φ and the x_i are all naturally scaled (dimensionless), then α, β, etc., may take any values. It is possible to construct 'incomplete' scaling relations. The well-known downstream hydraulic geometry is, in fact, a set of empirical scaling relations for channel form through the drainage system, but

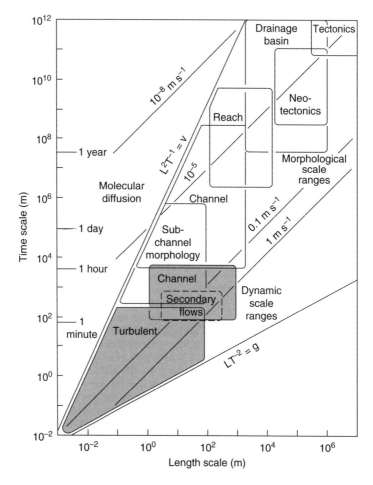

Figure 7.4 Characteristic scales for turbulent flow of water and for fluvial sediment yield in the landscape. Limit velocities are defined for diffusional processes and for gravitational wave propagation, while characteristic velocities are defined for various phenomena associated with fluvial systems. Trajectories for various virtual velocities are superimposed. Scales for channel processes are shaded

they are incomplete. This much is obvious from their dimensional inhomogeneity, but it can also be inferred from recognition that they do not explicitly include the properties of the materials that bound the channels, which must be important in determining channel dimensions. Scaling relations sometimes defy dimensional homogeneity because they express some systematic distorting factor in the system. Such relations are described as 'allometric' when they express real systemic distortion of form or function, and as 'fractals' when they represent distortions that are associated with the scale of representation itself, that is, such that self-similarity of appearance is maintained

across scales of representation under corresponding transformation of the measurements.

Errors

Observations are subject to error. 'Error' in scientific parlance refers not to mistakes, but to the range of variation of observations about the expected or exact value of the subject. The realizable precision of observations (see above) is a significant contributor to the total error of observations, but the error can also include local bias in the observing system and real variation in objects being measured which have been classified as being similar, but yet exhibit variations of detail. Error is the measure of uncertainty in an observation. No observation is without error, so its presence is no reflection upon the usefulness of the observation or the competence of the observer. But if we are to rationally interpret measurements, the error statistics must be known and must be propagated correctly through derivative, calculated quantities.

The precision of observations can be determined by repeated observations (x_i, $i = 1 \ldots n$) of the same object or process in order to determine the variation in results. The precision is summarized by the variance of those observations, or $s^2 = \Sigma(x_i - <x>)^2/(n - 1)$, in which $<x>$ denotes the mean, or by the standard deviation, s. Relative variance is expressed by the coefficient of variation, $s/<x>$ (which is not a stable statistic if $<x>$ is near zero). 'Repeated' observations commonly are observations on different members supposed to belong to the same class. In this case, precision and real residual variation (i.e. within-class variation) are confounded together.

Bias (or lack of accuracy) is systemic. It arises from fixed errors in the observing system, such as the calibration error of an instrument. Bias need not be constant: it may vary over the range of observations (see Figure 7.5a), so that it may be compounded of some constant error (mean bias) plus a variable systemic error. Biases can be dealt with only by comparing the observing system with a reference system previously determined to be accurate (or, at the least, to have a known bias).

Bias is often considered in relation to the performance of a measuring instrument. A measuring instrument is a physical (or sometimes chemical) system designed to respond to some environmental stimulus in a systematic manner to reveal the magnitude or intensity of the stimulus. It transposes some physical stimulus or effect into terms that our senses can appreciate. In modern science, such instruments are usually electronic and characteristically consist of electromechanical, electro-optical or electro-piezometric systems by which the phenomenon of interest is transduced into an instrument signal. Bias may arise through

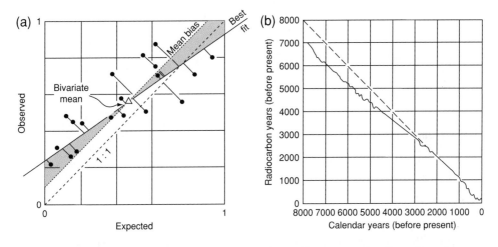

Figure 7.5 Examples of bias in scientific observations: (a) An example of an observing system subject to errors of precision, of real residual variation, and of both local and constant bias. Mean bias is indicated by the displacement of the bivariate mean of the data from the 1:1 line; local bias arises from the varying offset between the best-fit relation and the 1:1 line (shaded); errors of precision are indicated by the displacement of the data from the line of best fit (lines subtended from data points). (b) Dating bias of the radiocarbon assay for the age of organic materials. Before about 2500 years before present, the radiocarbon-derived ages drift away from calendar ages at a rate of about 225 years/ millennium, though the rate is not constant, in the direction of being too young. There are further variations for recent dates, not resolved in the diagram. 'Present' is conventionally interpreted to be ad 1950, the approximate date when the technique was developed. Adoption of a base year prevents published dates from becoming immediately obsolete

the instrument calibration – the relation between the phenomenon of interest and the signal. Instruments possess a number of general properties that might affect both precision and the occurrence of bias. The 'static response' (total response to a signal) of an instrument includes offset (constant bias induced by the system, which can often be 'dialed out' when the instrument is placed in 'calibration mode', so that it measures an internal standard), range of response (which affects the possibility for gross bias at the limits of the range), resolution (which bears on realisable precision), sensitivity (which may contribute to local or constant bias), linearity (which affects resolution across the range) and hysteresis (does it respond in the same way on the way up the scale as it does on the way down?). 'Dynamic response' gives the time to respond to a step change in the signal and commonly varies as $e^{-t/\delta}$, in which t is the response time of the instrument. When $t = \tau$, the response is $e^{-1} = 63.2\%$ complete. When $t = 4.6\tau$, the response of the instrument is 99% complete. For most modern electronic instruments, τ is a small fraction of a second.

Biases may arise when calibration relations must be used to infer some information of geomorphological interest from a 'proxy' relation – from an indirect measure of the relevant information. The most prominent example, perhaps, is the well-known dating bias of the radiocarbon method (Figure 7.5b). It arises because the flux of carbon-14 in the upper atmosphere is not constant, contrary to an assumption of the underlying elementary theory. Since the bias has been determined by the comparison of radiocarbon ages with ages determined by other means (principally, by dendrochronology), a correction exists for it, but the correction does not always lead to an unambiguous date because of fluctuations in the radiocarbon record. Many other calibrated geochronological methods possess biases that arise from one or another necessary assumption of the method.

Biases can also arise because the observer is predisposed toward some underlying concept of the way things should be – possibly toward some dominating ('ruling') theory. The observer may consequently be inclined to disbelieve and to reject observations that are contrary to the expected outcome. This obviously is bad practice; done consciously, it is dishonest practice. But this matter closely parallels one of the most important exercises of scientific judgment; that is, the detection of blunders (errors in the common sense of the term). Sometimes an observation should be rejected because it evidently is subject to some gross error (misreading an instrument; conditions exceeding the range of the instrument; instrument failure; incorrect booking ...). It is important to identify such blunders and to discard them, preferably after having determined why they are blunders. Determining the cause of the blunder is, in turn, important because sometimes a contrary measurement that is in fact not blundered turns out to be more important for broadening understanding than all the observations that fall close to their expected outcomes. Such measurements, when they can be confirmed not to be blunders, lead to revision of the science. An example, surprising and disputed at the time, was F. Law's (1956) discovery, from studies in northern England, that the water yield from forests was smaller than the yield from open land, the reason being the large evapo-transpiration from the forest.

Geomorphologically significant observations are often constructed from the combination of a number of elementary measurements as, for example, in summing elementary volumes to determine the total volume of a sediment deposit. Each elementary measurement is subject to some error. The error of summed independent elements, each with some error, s_i, and weight, a_i, is $s = \sqrt{\sum a_i^2 s_i^2}$, from which one may deduce that the error of a scaled result, with scale transform, a, is $s_t = a s_m$, in which s_m is associated with the actual measurement. For summed measurements, each of which has the same elementary (but independent) expected error, $s = \sqrt{n s_i^2}$. The error of an average of a sequence of similar measurements is $s/n = s_i/\sqrt{n}$; that is, the average is known to better precision than the individual observations. This makes sense since it is a summary of substantially more information.

Often, we require the product of measurements, each of which is subject to error, as the individual dimensions of a sedimentary volume. In the case $v = xyz$, $s_v/v = [s_x^2/x^2 + s_y^2/y^2 + s_z^2/z^2]^{1/2}$. Note that the errors are expressed in relative form. These are the most elementary error calculations, and they all assume that individual errors are independent of each other; for more complicated functional expressions, error calculations involve combinations of the above results, or more elaborate calculations that lie beyond the scope of this introductory discussion. Readers should consult a text on error analysis, such as Barford (1985). Such texts also treat the expected distribution of errors, which provides additional diagnostic tools for the analyst.

Sampling

Many programmes of geomorphological observations entail sampling to determine an areal or temporal representative value of some phenomenon of interest, or to study variability.

Classical sampling criteria are reviewed by Cochran (1963) and by Griffiths (1967) and are discussed in some textbooks of statistics for environmental studies. One may wish to sample for purposes of determining the areal distribution of landscape or landform properties, or one may sample in order to determine a summary measure of the entire surface. If there is no predisposition to expect any particular pattern in the data, then either random or systematic sampling schemes might be adopted. In the case of determining areal distribution, the latter certainly is more efficient, but is considered more risky because the sampling arrangement may become confounded with undetected regular variation on the surface. Systematic sampling often is chosen for convenience of navigation.

Geomorphological surfaces are, however, often zoned in more or less well-known ways. In the case of sedimentary deposits, for example, one encounters more or less well defined surficial facies. Similarly, time has regular divisions and weather exhibits episodic variations that influence the observations to be made. Surprisingly little attention has been paid to sampling in these circumstances. In such cases, schemes must be constructed to sample fairly over the known divisions of the sample space. Stratified, systematic unaligned sampling, a technique in which a study area is divided into an arbitrary number of cells, each of which is sampled at a random location, preserves the efficiency of systematic sampling, but avoids confounding with an underlying pattern. It is, furthermore, a 'universal method' – that is, it avoids the need for areal weighting associated with purely stratified schemes (Smartt and Grainger, 1974; Wolcott and Church, 1991).

Standard observing programs

Some branches of environmental science, in particular climatology and hydrology, are dominated by standard programmes of observations (of weather and of water flows and levels in the cited cases), directed by state agencies. These programmes reflect the daily synoptic nature of significant fluctuations in the subject phenomena. Such programmes provide significant information to geomorphologists but it is not obvious that similar programmes would capture significant geomorphological information (unless observations of sediment transport in rivers are accepted as peculiarly 'geomorphological' measurements). Significant changes of landscape occur on timescales much longer than the synoptic regime of weather, even though individual events – such as mass failures – may be sudden. The question arises, then, whether standard observing programmes have any place in geomorphology.

Such programmes have been attempted. Examples include programmes of benchmark and long-term monitoring sites directed by the Water Resources Division of the United States Geological Survey (Osterkamp et al., 1991; Osterkamp and Emmett, 1992), the purpose of which is to determine long-term changes in sediment yield and in both hillslope and stream channel landforms in response to variations in climate, land cover and land use. More recently, some units of the American Long Term Ecological Research network have incorporated geomorphologically significant observations. Elsewhere, various organizations have established research observatories, usually drainage basins, for similar purposes. Maintaining such programmes is difficult because of the extended timescales involved, which may cover decades before cumulatively significant results are revealed. In such cases, problems are not merely financial; they also entail continuity of personnel and, therefore, of purpose, and they must deal with changes in observing technology that may inadvertently introduce biases into long term sequences of observations. The concept of geomorphological observatories continues to be discussed, but has yet to be highly elaborated.

An alternative that is often used to obtain coarse-grained measurements of geomorphological change is to use historical documents to describe or to measure the state of the landscape at different times. Maps, terrestrial photographs (see Bierman et al., 2005) and aerial photographs have all been used to this end. Aerial photographs are incomparably the most useful because of their metric properties and high ground resolution. Increasingly, measurements made from satellite imagery and from various forms of airborne remote sensing (cf. lidar) are becoming available as well and will, in the future, become important sources of retrospective information.

EXPERIMENTS

Ideally, experiments are specially constructed programmes of observations directed at critically testing theories or generalizations about nature. An experiment is designed to acquire unequivocal evidence for the effect of one phenomenon upon another or, at least, for the pattern of co-variation of the subject phenomena. Accordingly, experiments are contrived so as to control all sources of variation that may affect the phenomena under study save those the effect of which it is desired to determine.

Experiments exhibit four fundamental characteristics (Church, 1984):

1 There must be a theory or conceptual model of the processes or relations of interest that will be supported or refuted by the experiment. Usually, the theory is not directly tested.
2 The theory gives rise to specific hypotheses that will be falsified or supported by the experiment. If the underlying theory is well developed, the hypotheses will be exact predictions (but sometimes the hypothesized outcomes will be qualitative).
3 To test the hypotheses we require definition of explicit phenomena of interest and operational statements of the measurements that will be made of them (sufficiently complete that replicate measurements may be made elsewhere by other investigators, desirably in a different context).
4 We also need a formalized schedule of measurements made under conditions controlled insofar as possible to ensure that the remaining variability be predictable under the experimental hypotheses.

Two less fundamental but no less important features must also be present: first, a specified scheme for analysis of the measurements that will unequivocally discriminate the possibilities in point 2, above; and, second, a data collection and management system designed for the purposes of the analysis. These last two requirements are often given little consideration in the planning stage of an experimental programme, whence observations that turn out to be critical for the success of the analysis are neglected.

It is easy to see how the requirements for experimental work may be met in classical laboratory contexts, but the rubric poses an immediate problem for field science. It is impossible to control all the extraneous sources of variability in the landscape. Relatively little geomorphological work qualifies as properly experimental (the increasingly common practice of nominating almost any programme of geophysical observations an 'experiment' notwithstanding). Are experiments possible at all in geomorphology?

The question is perhaps best addressed by considering the time and space scales of environmental phenomena to which a landform or landscape of interest is sensitive. The essential factor that seems most difficult to control in field experimentation is weather, the ubiquitous driving force in geomorphology. But do landforms respond distinctively to the driving forces of weather on all

scales? It appears at least possible that significant geomorphological changes occur, in many situations, over a sufficiently long time that most or all of the range of normally occurring weather might be sampled within the necessary time frame of the 'experiment' (although, of course, the landform may respond chiefly to a restricted range of extreme weather events). The landform is not specially affected by short-term variations of weather. At the other extreme, very long-term climate change may lie beyond the response time of the landform – it certainly lies outside the practical bounds of experimentation.

Space and timescales are linked because of the relatively restricted range of virtual velocities of sedimentary material associated with each mode of sediment flux in the landscape. (Virtual velocity is the rate of progression of material through the landscape, including resting time.) A landform with characteristic dimension L and typical virtual velocity u will have a response time L/u. For small L/u, the landform will be affected by relatively high-frequency features of climate, while for large L/u a landform or an entire landscape will reflect low frequency effects of climate. Figure 7.4 illustrates these conjectures for fluvial landscapes. Time and space scales may, of course, be quite different for different modes of sediment flux. For example, there appears to be a mismatch between hillslope and fluvial response times, upon which hinges some of the most interesting features of landscape.

A second factor that has high variability in natural landscapes is material properties. In addition to the intrinsic physical properties of earth materials, these may include water content of surficial materials and the extent and character of surface vegetative cover, or the occurrence of a distinctive surface material. It is conceivable that these factors would be included in a worthwhile theory about landscape features, so they may present no insuperable problem. However, the scale of their spatial variability may be important. The local scale of spatial variability of soil hydrological properties, for example, may be particularly difficult to overcome. If the characteristic scale of variability of some landform properties is much smaller than the scale of the experimental unit, it may be possible to conclude that one is dealing with an essentially homogeneous material. Similarly, if the scale is far larger than the scale of the experimental unit, then one certainly is dealing with an effectively homogeneous unit.

Similar problems must be faced in varying degree in other sciences, so that a number of experimental strategies have been developed in order to cope with them. All of them might be useful in geomorphology.

Classical experiments in the field

These entail intentional, direct interference with the natural conditions of the landscape in order to obtain controlled, and therefore unequivocal, results about a limited set of the processes that change the landscape, or of the conditions that govern landscape change. Such experiments are practical at small scale to study processes and conditions that have local expression and, usually, synoptic to seasonal forcing. Such experiments are directly analogous to classical laboratory procedures and are best illustrated by example.

A simple experiment was described by Mackay (1978) to confirm that the depth of winter snow cover influences the propensity of periglacial ice wedges to crack (open) in winter. Snow fences were erected at a site where cracking had regularly been observed for some years. Snow depth at the site was essentially doubled and, over 3 years, observation of 25 ice wedges yielded six cracking events (frequency 6/75, or 8 per cent) versus an average of 20 to 50 per cent of ice wedges cracking over 9 years in an adjacent network of 100 wedge sites with natural snow cover. The year-to-year variation was induced by variation in snow accumulation. In this case, the phenomenon of interest is seen to depend upon a seasonal meteorological phenomenon that was susceptible to sufficiently controlled manipulation to confirm a result inferred from ordinary observations.

A second example of a simple experiment is a study of beach face dynamics in which the water table under the beach was controlled by pumping (Chappell et al., 1979). It was shown that deliberately lowering the water table under the beach augmented sand deposition. Water table variations are controlled by the tide and, in turn, control the propensity for sand to be deposited on the beach face or eroded from it. On the rising tide, the relatively low water table allows some of the water moving onshore with the wave wash to be absorbed into the beach so that the backwash is comparatively weak and sand is accreted onto the beach whereas, on the falling tide, the relatively high water table and wave wash induce high pore water pressures immediately under the surface so that the sand fluidizes and is drawn off the beach by the backwash. In this experiment, individual waves provided the forcing for the flux of sand, but the summary geomorphological effect was observed over the tidal cycle. The investigators had only to ensure that wave conditions were sufficiently energetic to produce the sediment exchange.

A more elaborate experiment is described by Mackay (1997), who drained a 600 × 350 m lake on the western Arctic coast of Canada in order to observe permafrost aggradation into the thaw bulb present under the former lake and the associated geomorphological effects. In the long term, ground temperatures are in equilibrium with the mean annual air temperature at the surface, so the effect of removing the lake was to reduce the effective mean annual temperature of the surface from a value near 0°C to −10.5°C. Permafrost aggradation began in the 32 m deep thaw bulb. Multiple effects were observed, including the development and early growth of ice wedge polygons, surface heave reflecting the development of aggradation ice, thinning of the active layer as surface vegetation developed, and – perhaps most important – definitive evidence for groundwater flow of the intrapermafrost water, promoting convective heat transfer. The experiment showed that a three-dimensional conductive–convective approach to heat transfer is necessary to understand permafrost aggradation in this circumstance. The studied effects depend mainly on mean climate rather than short-term or even seasonal weather, so that an experimental interpretation of the landscape manipulation is feasible because secular climate was effectively fixed throughout the experiment, even though there is a significant long-term warming trend in the region. Some of the observed effects were quantitatively studied for the first time. Their particular expression was, however, conditioned by the physiography and materials of the site.

Paired experiments

By far the most common means that has been employed in the attempt to overcome the constraint of time varying forcing by weather in landscape studies is a 'paired' experiment, in which two (or more) ostensibly similar landscape units are prepared or selected for study. While one or more units are deliberately manipulated, one or more others remain untreated and act as 'reference units' that can be used to discount the time-varying effects of weather. Today, such designs often arise in the context of environmental impact investigations. However, paired catchment studies represent the most common example, having been established many times in order to study the effects of land use on water and sediment production, most often in the applied context of forestry or agriculture. A historical perspective on paired drainage basin experiments was provided by Whitehead and Robinson (1993).

The earliest such experiment, conducted at Wagon Wheel Gap, Colorado, between 1910 and 1926 (Bates and Henry, 1928), remains instructive. The purpose was to study the effect on runoff volume of forest cutting and burning (not a geomorphological purpose, but no less instructive for that). The criteria for choice of the experimental drainage basins were that:

- the selected basins be practically contiguous in order to minimize differences in experienced weather, particularly the amount and timing of precipitation;
- the basins be situated on identical geological structures and be as nearly alike as possible in topography and aspect so that water retention and drainage conditions would be similar;
- the vegetation on all units be regionally representative.

This amounts to a set of requirements for the establishment of experimental control, but they have been extensively criticized because their application relies on judgments. It is possible to apply objective criteria to the comparison of landscape units (see Church, 2003a) although it has almost never been done in practice because the large variability in landscape characteristics requires that a very large data set be gathered in order to achieve well discriminated results.

A paired study is conducted by observing all units for some period before any treatment is undertaken so that pre-treatment correlations can be established between phenomena in the various units (e.g. mean annual runoff; annual sediment yield; summer water temperature; nutrient efflux). Then treatment is undertaken and the post-treatment phenomena in the treated units are compared with results predicted by the prior correlation – in effect, a calibration relation.

A variation on this strategy is to select two landscape units that present a strong contrast in some particular property and to study the comparative behaviour of the two units over time. This has frequently been done in geomorphology (e.g. studies of comparative sediment yield based on land use variation), but it lends no assurance that experimental control has satisfactorily been established. A further variation is to embed treatment units within the reference unit. In this case, one may obtain a confounded result (the treatment affects the putative reference unit), or one may confront significant scaling problems (of which more below).

Paired experiments have been extensively criticized, while Hewlett et al. (1969) gave a classic defence of experimental catchment studies. Only two criticisms appear to be fundamental – that is, to go beyond the question whether the requirements for adequate experimental control have been met. These are that the results are not transferable and that they cannot be extrapolated to different scale. The first criticism seems to arise from the view that study of the landscape unit is

an attempt to characterize the behaviour of the complete environmental system, and that individual environmental systems are always, in some sense, unique. But this is not usually the experimental purpose and should not be an objection to a properly defined experiment. The second objection is, however, fundamental, because processes entailing mass or energy transfer – in particular, water and sediment transfers in drainage basins – are subject to scaling effects (see Dooge, 1986; Church et al., 1999). Scaling arises from storage effects, which become more significant in larger landscape units. Hence, transfer of information from one landscape unit to another should be undertaken by use of appropriate scaling functions rather than by direct comparison.

A basic criticism that can be levelled at most paired experiments involving drainage basins, however, is that the experimental units are rarely replicated, except in plot scale experiments. This leaves the investigator with no measure of sampling variance in either the reference or experimental unit, no means to ensure that observed variations in either the reference or experimental unit are indeed due to the experimental manipulation, hence with no basis for a rigorous comparison of the outcomes. Underwood (1992) has described this problem in detail, and has proposed statistical means to overcome it in an ecological context.

Statistical experiments

Plot scale experiments are a logical extension of paired comparisons designed to overcome the problem of replication identified above. Since small plots often may be relatively easily established, multiple plot studies are common for site-scale effects such as soil erodibility, with either systematic or random variation designed to remove the variability that is not of central concern to the experiment. Four distinct strategies are used to select the experimental units:

1 *Replication* involves observation of some experimental treatment or effect applied to many different units (plots, hillslopes, ...) in the attempt to control extraneous natural variability by distributing it over all of the outcomes of interest in a large sample. Additional variability is reduced as $\sqrt{n}$, the sample size. However, if there are a number of extraneous factors, each one must be proportionally distributed across the experiment and, as the number of them increases, the probability for this condition decreases. The problem may be costly to evaluate and difficult to eliminate.

2 *Prescreening* of experimental units is an attempt to select ones that are homogeneous with respect to extraneous factors. This is the strategy that guides the selection of paired catchments (see above), but it often involves difficult judgments or substantial expense.

3 *Blocking* is the grouping of experimental units on the basis of prior information so that they are substantially more homogenous within groups than between them. Experimental variation is sought between the groups. The subject of statistical experimental design arises from this approach and offers numerous strategies for overcoming extraneous variability by assigning experiment units proportionally (a classic reference is Cochran and Cox, 1957; there are many books on the subject).

4 *Adjustment of experimental results* is an *ex post facto* attempt to isolate the experimental effect using additional information gathered during the experiment. The additional information must be independent of the experimental effects. Bovis (1978) presented an interesting example of this strategy. He studied rates of surface erosion on a montane transect, using major vegetation cover types for the initial stratification. No discrimination was achieved, but when the data were restratified (adjusted) according to locally observed stand types and species associations, the expected significant differences in soil erodibility emerged. Formal statistical methods of adjustment are provided by the analysis of covariance.

A problem that arises in many field experiments is pseudoreplication of experimental units (Hurlbert, 1984; there is a substantial subsequent literature on this topic). Pseudoreplication occurs if one supposes that repeated samples represent legitimate replications of an experiment when in fact they are not independent of each other. This often arises in field contexts when the samples are all drawn from one site. Pseudoreplication further occurs when one supposes that replicate samples represent replicate treatments. In either case, one ends up with a mistaken estimate of sampling variance and invalid inferences.

Inadvertent experiments

'Inadvertent experiments' are events or developments or processes that occur in the landscape that are not designed to be geomorphological experiments at all, but that might fortuitously be interpreted as if they were experiments in order to make observations and to draw conclusions that are geomorphologically significant. Such opportunities are not common but, when they occur, they are unusually important because they may reveal effects on time and space scales that are well beyond the scope of deliberately contrived experiments. One simply is observing an uncontrolled process. Only in rare circumstances will the extraneous conditions be interpretable as sufficiently constant to permit experimental interpretation of the observed effects, and it may be an even rarer circumstance that the geomorphologist is able to compile sufficiently comprehensive information about the case to permit the interpretation. It is therefore critical for geomorphologists to have a clear understanding of the criteria for an experiment so that, if an opportunity arises to document an inadvertent experiment, it will not be missed.

Most inadvertent experiments of relevance to geomorphology have to do with manipulation of the land surface, such as land conversion from forest to uses that expose soil to accelerated erosion, or with engineering modification of hillslopes or stream courses. In many natural resource management situations – but especially in forestry – the environmental impacts of resource development are coming to be regarded as so significant that there is increasing conviction that such activities should always be conducted experimentally. Adaptive management (Holling, 1978) is a concept that requires managers to deliberately review cumulative management experience, to draw lessons from experience concerning desirable or undesirable outcomes and to adjust (adapt) actions accordingly. In effect, then, all activities must be regarded as experiments, with provision for monitoring and reporting the effects of the activities that are sufficient to enable adaptive decisions to be taken. There is, in this paradigm, substantial scope for geomorphologically significant observations to be made. But adaptive management is expensive (because the reporting requirements for effective learning are as stringent as those of a deliberately contrived experiment), and so properly designed programmes remain, in commercial resource development contexts, rare.

Inadvertent experiments are so diverse in their character that the quotation of an example is the best way to elaborate the concept farther. Lewin (1976) describes a reach of a gravel-bed river (the River Ystwyth in central Wales) that has been repeatedly straightened and returned to a rectilinear, plane-bed state. High flows rapidly initiated meander development, which could be studied *ab initio*. The channel was originally rectified in 1864, permitting a century long view of development recorded on maps. Restraightened in 1969, the process was followed once more, in synoptic detail, using modern survey methods.

Scaled physical experiments

The use of scaled physical models, usually in a laboratory setting, is an important means to overcome the time and energy costs of mounting experiments in the field. It is also the closest one might come to emulating the highly successful experimental methods of the basic sciences (though it is by no means clear that that is a particular virtue). For rigorous results, scaling considerations are critical.

Similarity under scale change may follow geometrical, kinematic or dynamical criteria. The subject is highly developed, especially in hydraulics (Yalin, 1982) and reference here will be limited to some complications of direct geomorphological interest. In a river channel, major criteria for dynamical similarity are that the Froude and Reynolds scalings be preserved, that is, $u_m/\sqrt{gl_m} = u_p/\sqrt{gl_p}$ and $u_m l_m/\nu = u_p l_p/\nu$ ($\nu = \mu/\rho$ is the kinematic viscosity of water, and m and p denote model and prototype, respectively). In fact, both scalings cannot be preserved simultaneously, unless the viscosity of the fluid can be arbitrarily adjusted. It is usual to preserve Froude similarity and to ensure that the Reynolds number remains within the turbulent domain (which is actually an advantage of a non-canonical scale). This limits scale adjustments, without distortion, in most work to a practical range of about 20:1. Froude scaling imposes the conditions $Q_m/Q_p = (l_m/l_p)^{2/5}$ and $t_m/t_p = (l_m/l_p)^{1/2}$ – that is, timescale varies as the square root of the geometric scale – which may also present practical problems unless large fractions of prototype time can be dismissed as non-critical (i.e. no significant processes of change occur). On this consideration, for example, the observing period in an experimental alluvial stream channel might be foreshortened by discounting the large fraction of time when prototype flows are below the threshold for significant sediment movement.

Scaling considerations affect the way in which observations may be interpreted. From the first Froude condition quoted above, we find $l \propto Q^{2/5}$. But in the downstream hydraulic geometry of river channels, we find $d \propto Q^f$, $0.3 < f < 0.4$, usually, while w (channel width) $\propto Q^b$; $b \approx 0.5$. Hence the aspect ratio of river channels, w/d systematically changes under scale change, and river channels in similar materials are *not* simple scale models of each other. The reason is that the channel bounding materials have strength limitations that do not scale in the same way, hence rivers in similar materials form 'regime groups' rather than similarity groups. Significant scale transformations require a change of material dimensions as well as gross dimensions but this eventually introduces new difficulties since, as

material texture become fine the physics of their transport changes, even in scaled-down flows, so that the essential processes that form the prototype river channel may not be preserved.

The tradition of laboratory experimentation in geomorphology commenced in the 19th century, even though it has always remained a very subordinate thread of enquiry in the discipline. An impressive list of classical experiments is given by Goudie (1990; Table 1.2). Nearly all have entailed the use of physical models – often called 'hardware models' in geomorphological literature. Nearly all except certain weathering experiments have been conducted on a reduced scale, but very few have given formal attention to scales. Most have had only a poorly developed theoretical context and very few indeed have been designed to discriminate well-formulated hypotheses. Most have, in fact, been informal experiments (q.v.). Physical modelling of processes in the landscape is by no means the preserve of geomorphology and much may be learned from the experience of other disciplines, particularly civil engineering, where the tradition is much more highly developed.

Achieving appropriate scaling of landscape processes remains a significant problem. Only a subset of the land-forming processes in which geomorphologists are interested is amenable to such treatment. A major concern, already alluded to above, is the difficulty to deal appropriately with time for processes that proceed slowly in comparison with human timescales. So, for example, there is a rich tradition of scaled experimentation in hydraulics and, by extension, fluvial geomorphology, because fluids, and sediments carried by fluids, have relatively high virtual velocities, but there is little scope for experimental study of hillslope development. On the other hand, there is substantial scope for laboratory study of earth materials and their behaviour, and much has been done in the laboratory in the tradition of geotechnical testing. No explicit scaling is required to study the weathering properties of sample materials, or the strength and yield characteristics of earth materials, but scaling problems return in the form of the difficulty to transfer results from the scale of laboratory samples to the scale of material behaviour in the field.

There is a significant stochastic element in many geomorphological processes and a further measure of variation is introduced by the variable forcing in the field. These elements are difficult to capture in physical experiments. They imply that experiments that seek to reproduce the variation seen in field examples must be repeated, possibly many times. But for many experiments, involving elaborate equipment and considerable lapse of time, replication is scarcely feasible.

Indeed, experiments – beyond very simple and usually highly reduced ones – have rarely been replicated. (An example of a radically reduced experiment, easy to replicate, is mass failure in a granular pile, intended to reveal failures on scree slopes or on the slip face of sand dunes.) Lack of replication severely hampers comparisons with prototype behaviour, or even comparisons between experiments under varied conditions, for there is no means to determine just what a significantly changed outcome might be. Carefully contrived, however, scaled physical experiments can confirm or controvert the significance of particular processes in geomorphology and they can uncover certain parametric relations amongst elements of geomorphological systems. They deserve further emphasis and development.

Informal experiments

Geomorphologists have more commonly used models as suggestive analogues of full-sale systems or landscapes without invoking formally scaled similarity (hence the contrast of 'analogue model' versus 'scaled model'). S.A. Schumm and his co-workers have been notable exponents of this approach (Schumm et al., 1987). A difficulty noted above with formal scale transformation is the changing physical behaviour of earth materials under large transformations. Paola quantifies this problem by considering the 'granularity' of a landscape system, defined as L/D, the ratio of system size/grain size. For a landscape, granularity approaches infinity and a scaled transformation requires impractically small grains. Quite beyond this conundrum, earth materials incorporate cohesive effects and effects of surface vegetation cover that render exact transformations unachievable for most systems.

Models that are at least analogues of their prototype have been achieved in some cases by making a radical change of materials. Accordingly, Schumm et al. (1987) report the use of a sand–kaolinite mixture to mimic the behaviour of bedrock in stream tray experiments on river erosion into rock. More dramatic substitutions include a change even of physical principles, so that the parallel structure of physical laws allows electrical analogues to be constructed for the flow of water or ice.

Paola offers a more general rationalization for informal experiments based on the scale-free range in behaviour of many earth systems. He introduces the concepts of 'external similarity', in which a small system looks and behaves like a large system (as a scaled model), and 'internal similarity', in which a small part of a system looks and behaves like the entire system (e.g. the drainage network). The former may be thought of as an 'imposed' similarity in as much as it can be designed (or confirmed) by scale transformations. The latter can be thought of as 'spontaneous' similarity in the sense that it arises naturally and is based on the self-similar (or fractal) appearance hence, one assumes, behaviour of many earth systems. Many natural systems exhibit both external and internal similarity; others may exhibit internal similarity even though formative processes may change across the scale range (i.e. dynamical similarity is not maintained).

These ideas lead to a strategy for developing an analogue model. A model of some field situation may be constructed using materials (often different than those of the prototype) that are expected to yield phenomena similar to those in the prototype that one wishes to study. Verification by trial is pursued by comparing prototype measurements with corresponding measurements made in the model. In short, one attempts to create a situation in which spontaneous similarity emerges.

Such considerations provide a basis for believing that the behaviour of small sedimentary systems developed in the laboratory (Schumm et al., 1987) or small examples observed in nature (e.g. Schumm's (1956) classical study of drainage system development on a filled site; Rachocki's (1981) observations of small alluvial fans) may reveal principles that govern the behaviour of much larger units in the landscape. Recently, Lague et al. (2008) have attempted to study the geomorphological response to tectonic processes in a physical model with overall length scale of order 1 metre. One way to consider the results is to regard them as hypotheses about the way in which larger systems may work.

Numerical experiments

The advent of large-scale computation has made possible the algorithmic, numerical simulation of many geomorphological processes (see Kirkby et al., 1987 for a general introduction to numerical modelling in physical geography). The first question is whether such exercises represent experimentation in any real sense. The question may be answered by reference to the conditions for an experiment given above. If the basis for the model is a candidate theory, if the operation of the model leads to more or less exact predictions that may be compared against field evidence by the specification of appropriate measurements, and if an appropriate schedule of simulations may be undertaken to effect a reasonably comprehensive comparison, then it appears that the conditions for an experimental investigation may properly be fulfilled. Indeed, numerical experimentation possesses one decided advantage over almost any other experimental arrangement one may make in geomorphology: one may control the extraneous conditions absolutely (in the numerical model) by holding them constant. This opens the way for truly controlled sensitivity tests amongst the experimental variables that are often difficult or impossible to achieve in any other way.

The comparison with field measurements is, in fact, not an intrinsic part of the numerical experiment, but it is the critical test of the relevance of the experiment and, presumably, of the underlying theory to geomorphological explanation and understanding.

How is a numerical experiment in geomorphology (actually, one may substitute for 'geomorphology' any environmental science) different from a calculation one might make from any physical theory, a calculation that one scarcely would think of as an experiment? The distinction lies in the circumstance that environmental systems are complex systems: a theory about an environmental system combines elements of many more fundamental physical and/or chemical and/or biological processes, so that the outcome may be by no means obvious. The usual outcome of a numerical experiment is a more or less comprehensive test of the model, hence of the underlying theory, which is usually a system theory. Does it, or does it not, satisfactorily simulate the observed field conditions? The experiment is accordingly seen to be a test of theoretical understanding, which conforms with the purpose of any scientific experiment. It appears that numerical experiments are perfectly admissible experiments.

Another possible objection to numerical experiments as rigorous tests of theory in geomorphology is that, because of the complexity of the environment, theories about it are almost never complete theories. At some point in the argument, some arbitrary parameterization or empirical statement appears. In the minds of many investigators, this circumstance defines the difference between a theory and a model, and it might appear to disqualify a numerical investigation of the model as an experiment in the sense that experiments are designed to test true generalizations about nature. The distinction appears to be pedantic: in fact, the circumstance that the model is incomplete may be a compelling reason to conduct experimental trials. The outcome of numerical experimentation is much more often a test of our understanding of the environmental system (as exemplified in the construction of the numerical model) than a definitive test of theory.

Experimental strategy

In many geomorphological studies there is an essentially historical element, or a particular dependence upon place, that precludes the adoption of an experimental approach. In many other cases, experimental control cannot satisfactorily be achieved. For problems that are amenable to experimental treatment, we may recognize two stages of experimentation. An initial experiment will be 'exploratory'; an attempt to secure acceptable scientific evidence to support or controvert the hypothetical proposal (see Flueck, 1978). A 'confirmatory' experiment is an attempt to secure critical evidence for or against a result that has already been reported. In an exploratory experiment, the design may be altered as the experiment proceeds (cf. Bovis, 1978). However, the evidence marshalled in support of the result must meet experimental criteria:

- data must have been collected and analysed under satisfactory experimental control;
- the data must be sufficient to discriminate the possible outcomes and the analyses must be accompanied by statements of the probability of observing the actual outcome;
- the experiment must be publicly reported and the data available for scrutiny.

These requirements establish confidence in the results. The second criterion is often difficult to achieve since, in the face of environmental variability, it is often difficult to obtain a sufficient number of observations to produce a discriminating result.

Confirmatory experiments begin from a well articulated context, so that they should be stringently designed and they should not be altered during their execution. Successfully completed, they lead to strong inferences (Platt, 1964). Exploratory and confirmatory experiments fit different circumstances of investigation. A characterization of different experimental circumstances is given in Figure 7.6.

The discretionary power of inferential statistics can be applied in the context of exploratory and confirmatory experiments. In an exploratory experiment, one wishes to be reasonably certain not to ignore a relation that may prove interesting. Hence, we set a relatively low a-level for the detection of a significant result (say, $a < 0.10$) to minimize the chance of making a Type II error (falsely rejecting a hypothesis that is in fact true). But in a confirmatory experiment we are rather concerned with avoiding the acceptance of weak evidence to sustain a claim, so we set a stringent a-level (say $a < 0.01$) to minimize the chance of committing a Type I error (accepting as true a hypothesis that is in fact false).

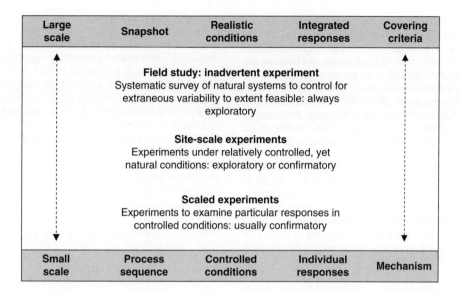

Figure 7.6 **Contextual characterization of exploratory and confirmatory experiments in geomorphology. (After a personal communication from J. Kane, 2006)**

Most experiments in geomorphology are exploratory experiments. They are attempts to find support for and to sharpen hypotheses that are not, initially, very well articulated. This is neither unreasonable nor unusual in view of the formative stage in which the science still finds itself. Indeed, it follows a well-worn path in the development of any science in which most experimentation is as much an exploration of nature as a means of critically testing well-articulated theories.

Of course, the resolving power of any experiment depends upon the quantity of evidence that can be marshalled. In some instances – in particular, cases in which the total population of potential subjects is limited by regional considerations – one may have too few available cases to obtain high resolution. In many cases, geomorphologists may be able to sample substantially the entire population of landforms. One no longer has merely a sample or an experimental test, but a statement about the way the world is, complete with a statement of population variance.

CONCLUSIONS

In this chapter we have explored aspects of observation and experiment, giving some emphasis to circumstances that are particularly significant in the conduct of geomorphological research.

Most geomorphological processes – geomorphological in the proper sense that they entail significant changes in the landscape – happen on long timescales, and many occur at large spatial scales as well. Individual events that are directly accessible to observation – a landslide; a flood; a windstorm – contribute to the production of geomorphological change but rarely, by themselves, constitute significant changes. The facts of observation in geomorphology – other than the current configuration of the landscape – largely have to do with the environmental physics and chemistry that underlie the evolution of the landscape. Aspects of this underlying science may often be accessible to at least informal experiment at reduced scale or in the laboratory. In the laboratory, however, one must always be mindful of the question whether – with full experimental control – one still has a functioning environmental system for study. A pervasive characteristic of the environment is variability at all spatial and temporal scales, in both boundary and forcing conditions. Absent a significant portion of that variability – eliminated in the interest of establishing experimental control over the observations – one may wonder whether one still has a functional system – one that is able to reveal the real pattern of cause and effect in the environment. This remains a matter for skilled judgment about what are the effective timescales on which the system is forced, and the effective spatial scales over which the material response occurs.

To be sure, the remarkable developments in sensors, in data recording systems, and in computational power that we are witnessing today promise to overcome at least some of the practical difficulties that have faced geomorphologists in the past, both in passive observation and in active experimentation. Most of these developments stem from dramatic improvements in electronically mediated data acquisition, transmission, storage and analysis. They enable us to sharpen observation, to probe for spatial and temporal patterns where none is immediately apparent, at frequencies formerly inaccessible and to begin to construct time sequences of environmental change on a global scale. But these improvements also remove the phenomena of study farther from direct visual observation: we construct an increasingly complex picture of the world using increasingly complex instruments. Elementary transparency is lost. In this respect, geomorphology is increasingly moving toward the mainstream of geophysical science. For geomorphologists, this dramatically widens the range of available techniques for observation and experiment, but it deepens the essential issue of interpretation.

ACKNOWLEDGEMENTS

The paper has benefited substantially from the insights of Professor Rob Ferguson of Durham University and Professor Olav Slaymaker in my own university. It owes many of its themes to the awkward questions posed by generations of students at the University of British Columbia, where I have masqueraded as an authority on these matters for some forty years.

REFERENCES

Barenblatt, G.I. (2003) *Scaling*. Cambridge, The Cambridge University Press.

Barford, N.C. (1985) *Experimental Measurements: Precision, Error and Truth*. John Wiley and Sons, Chichester.

Bates, C.G. and Henry, A.J. (1928) Forest and stream-flow experiment at Wagon Wheel Gap, Colorado: Final report on completion of the second phase of the experiment, *Monthly Weather Review* Supplement 30.

Best, J. (1992) On the entrainment of sediment and initiation of bed defects: insights from recent developments within turbulent boundary layer research. *Sedimentology* 39: 797–811.

Bierman, P.R., Howe, J., Stanley-Mann, E., Peabody, M., Hilke, J. and Massey, C.A. (2005) Old images record landscape changes through time. *GSA Today* 15: 4–10.

Bovis, M.J. (1978) Soil loss in the Colorado Front Range: sampling design and a real variation. *Zeitschrift für Geomorphologie* Suppl. Bd. 29: 10–21.

Chalmers, A.F. (1999) *What Is This Thing Called Science?* Open University Press, Buckingham.

Chappell, I., Eliot, I.G., Bradshaw, M.P. and Lonsdale, E. (1979) Experimental control of beach face dynamics by water-table pumping. *Engineering Geology* 14: 29–41.

Chorley, R.J. (1962) *Geomorphology and General Systems Theory*, United States Geological Survey, Professional Paper 500B. Washington, DC.

Church, M. (1984) On experimental method in geomorphology, in T.P. Burt and D.E. Walling (eds), *Catchment Experiment in Fluvial Geomorphology*. Geobooks, Norwich, pp. 563–580.

Church, M. (2003a) What is a geomorphological prediction?, in P.R. Wilcock and R.M. Iverson (eds), *Prediction in Geomorphology*, American Geophysical Union Monograph 135, pp. 183–194. doi: 10.1029/135GM013.

Church, M. (2003b) Grain size and shape, in G.V. Middleton (ed.), *Encyclopedia of Sediments and Sedimentary Rocks*. Kluwer, Dordrecht, pp. 338–345.

Church, M., Ham, D., Hassan, M. and Slaymaker, O. (1999) Fluvial sediment yield in Canada: scaled analysis. *Canadian Journal of Earth Sciences* 36: 1267–1280.

Cochran, W.G. (1963) *Sampling Techniques*, 2nd edn. John Wiley, New York.

Cochran, W.G. and Cox, G.M. (1957) *Experimental Designs*, 2nd edn. John Wiley, New York.

Dooge, J.C.I. (1986) Looking for hydrologic laws. *Water Resources Research* 22: 46S–58S.

Ferguson, R.I. and Church, M. (2004) Grain settling. A simple universal equation for grain settling velocity. *Journal of Sedimentary Research* 74: 933–947.

Flueck, J.A. (1978) The role of statistics in weather modification experiments. *Atmosphere–Ocean* 16: 60–80.

Goudie, A. (ed.) (1990) *Geomorphological Techniques*. Unwin Hyman, London.

Grant, G.E., Swanson, F.J. and Wolman, M.G. (1990) Pattern and origin of stepped-bed morphology in high gradient streams, Western Cascades, Oregon. *Geological Society of America Bulletin* 102: 340–352.

Griffiths, J.C. (1967) *Scientific Method in the Analysis of Sediments*. McGraw-Hill, New York.

Haines-Young, R. and Petch, J. (1986) *Physical Geography: Its Nature and Methods*. Harper and Row, London.

Hewlett, J.D., Lull, H.W. and Reinhart, K.G. (1969) In defence of experimental watersheds. *Water Resources Research* 5: 306–316.

Holling, C.S. (1978) *Adaptive Environmental Assessment and Management*. John Wiley, Chichester.

Hurlbert, S.H. (1984) Pseudoreplication and the design of ecological field experiments. *Ecological Monographs* 54: 187–211.

Kellerhals, R., Church, M. and Bray, D.I. (1976) Classification and analysis of river processes, American Society of Civil Engineers. *Journal of the Hydraulics Division* 102: 813–829.

Kirkby, M.J., Naden, P.S., Burt, T.P. and Butcher, D.P. (1987) *Computer Simulation in Physical Geography*. John Wiley, Chichester.

Kondolf, G.M., Montgomery, D.R., Piégay, H. and Schmitt, L. (2003) Geomorphic classification of rivers and streams, in G.M. Kondolf and H. Piégay (eds), *Tools in Fluvial Geomorphology*. John Wiley, Chichester, pp. 171–204.

Lague, D., Crave, A. and Davy, P. (2008) Laboratory experiments simulating the geomorphic response to tectonic uplift. *Journal of Geophysical Research* 113: 20. doi: 10.1029/2002JB001785.

Latour, B. (1987) *Science in Action*. Harvard University Press, Cambridge, Massachusetts, p. 274.

Law, F. (1956) The effect of afforestation upon the yield of catchment areas. *Journal of the British Waterworks Association*. 38: 484–494.

Leopold, L.B. and Wolman, M.G. (1957) River Channel Patterns – Braided, Meandering and Straight. *United States Geological Survey Professional Paper 282B*, Washington, DC.

Lewin, J. (1976) Initiation of bed forms and meanders in coarse-grained sediment. *Geological Society of America Bulletin* 87: 281–285.

Mackay, J.R. (1978) *The Use of Snow Fences to Reduce Ice-wedge Cracking, Garry Island, Northwest Territories*. Geological Survey of Canada, Current Research, Part A, Paper 78-1A, Ottawa, Ontario. pp. 523–524.

Mackay, J.R. (1997) A full-scale field experiment (1978-1995) on the growth of permafrost by means of lake drainage, western Arctic coast: a discussion of the method and some results. *Canadian Journal of Earth Sciences* 34, 17–33.

Montgomery, D.R. and Buffington, J.M. (1997) Channel-reach morphology in mountain drainage basins. *Geological Society of America Bulletin* 109: 596–611.

Osterkamp, W.R. and Emmett, W.W. (1992) The Vigil Network: long-term monitoring to assess landscape changes, in J.D. Bogen, D.E. Walling and T.J. Day (eds), *Erosion and Sediment Transport Monitoring Programs in River Basins*, Proceedings of the Oslo Symposium. International Association of Hydrological Sciences publication 210, pp. 397–404.

Osterkamp, W.R., Emmett, W.W. and Leopold, L.B. (1991) The Vigil Network: a means of observing landscape change in drainage basins. *Hydrological Sciences Journal* 36: 331–344.

Platt, J.R. (1964) Strong inference. *Science* 146: 347–353.

Rachocki, A.H. (1981) *Alluvial Fans*. John Wiley, Chichester.

Rhoads, B.L. and Thorn, C.E. (1996) *The Scientific Nature of Geomorphology*. John Wiley, Chichester.

Schumm, S.A. (1956) Evolution of drainage systems and slopes in badlands at Perth Amboy, New Jersey. *Geological Society of America Bulletin* 67: 597–646.

Schumm, S.A. (1985) Patterns of alluvial rivers. *Annual Review of Earth and Planetary Science* 13: 5–27.

Schumm, S.A., Mosley, M.P. and Weaver, W.E. (1987) *Experimental Fluvial Geomorphology*. Wiley-Interscience, New York.

Simpson, G.G. (1963) Historical science, in C.C. Albritton (ed.), *The Fabric of Geology*. Freeman Cooper, Stanford, pp. 24–43.

Smartt, P.F.M. and Grainger, J.E.A. (1974) Sampling for vegetation survey: some aspects of the behaviour of unrestricted, restricted and stratified techniques. *Journal of Biogeography* 1: 193–206.

Taylor, B.N. (1995) *Guide for the Use of the Standard International System of Units*. United States National Institute of Standards and Technology, Gaithersburg, Maryland.

Underwood, A.J. (1992) Beyond BACI: the detection of environmental impacts on populations in the real, but variable, world. *Journal of Experimental Marine Biology and Ecology* 161: 145–178.

Venditti, J.G., Church, M. and Bennett, S.J. (2005) Bed form initiation from a flat sand bed. *Journal of Geophysical Research* 110, F01009: p. 19. doi: 10.1029/2004JF000149.

Wentworth, C.K. (1922) A scale of grade and class terms for clastic sediments. *Journal of Geology*, 30, 377–392.

Whitehead, P.G. and Robinson, M. (1993) Experimental basin studies – an international and historical perspective of forest impacts. *Journal of Hydrology* 145: 217–230.

Wolcott, J. and Church, M. (1991) Strategies for sampling spatially heterogeneous phenomena: the example of river gravels. *Journal of Sedimentary Petrology* 61: 534–543.

Yalin, M.S. (1982) On the similarity of physical models. *Hydraulic Modelling in Maritime Engineering*, Proceedings of a conference held on 13–14 October, 1981, by the Institution of Civil Engineers. Thomas Telford, London, pp. 1–14.

8

Geomorphological Mapping

Mike J. Smith and Colin F. Pain

Mapping of landforms is probably as old as the making of maps. Mountain ranges, volcanoes and plains all appear on early representations of land. Often the mountains were represented as hachures or 'hairy caterpillars', the depiction of which reached a high art form with the maps of Erwin J. Raisz (e.g. Raisz, 1951, see also Robinson, 1970) and A.K. Lobeck (e.g. 1957) (Plate 3a, page 589). These maps were compiled according to set standards, and specified symbols were used. Subsequently there have been several attempts to set standardized symbols and methods, especially by the International Geographical Union (IGU) (e.g. Leszczycki, 1963; International Geographical Union, 1968). This was driven, in part, by the complexity of information that can be presented on one or more of the following key topics: *morphometry* (shape/location), *morphogenesis* (evolution, including geological control), *morphochronology* (relative and absolute age) and *morphodynamics* (genesis and processes).

More recently there have been several developments related to geomorphic mapping using polygons rather than symbols (Plate 3b and c, page 589). Land systems mapping (Christian and Stewart, 1953) and regolith landform mapping (Pain et al., 2001) provide examples. However, over the past two or three decades geomorphic maps as such have not been a major part of the study of landforms (Gustavsson, 2006; Gustavsson et al., 2006), a circumstance underlined by the formation of the International Association for Geomorphology (IAG) Working Group on Applied Geomorphic Mapping (AppGeMa – Pain et al.,

2008) and publication of a technical handbook on the topic (Smith et al., in press).

This chapter briefly reviews the development of geomorphic mapping, outlines its present status, considers the main issues and looks at likely future developments.

RETROSPECTIVE

There is a considerable literature on geomorphic maps, and reviews can be found in Fairbridge (1968), Cooke and Doornkamp (1990), Mitchell (1991), Goudie (1981, 2004) and Verstappen (in press). We can make a very broad distinction between maps that (1) use *symbols, shading and colours* to depict landforms and (2) those that use *polygons* that represent specific types of landforms. The former were developed before geographic information systems (GIS) came into widespread use, and the maps of Raisz and Lobeck have already been mentioned. The latter were also developed before GIS, but are more amenable to GIS manipulation and therefore may have become more common over the past two or three decades. Certainly, in Australia, polygon maps of landforms and related phenomena are routinely converted into digital GIS coverages as part of the collection of legacy data for new projects (e.g. Fitzpatrick et al., 2004).

The IGU Subcommission on Geomorphological Mapping sponsored a series of conferences and publications in the 1960s and 1970s. These set out a recommended set of symbols and colours for mapping at mainly detailed scales. Examples of maps and mapping systems are Bondesan et al. (1989), Castiglioni et al. (1990), Gullentops (1964), Joly (1962), Tricart and Usselmann (1967) and van Zuidam (1982). Morphological mapping (Waters, 1958) played an important part during this period; Savigear (1965) was perhaps the best known example, with Rose and Smith (2008) providing historical context. Brunsden et al. (1975) provide several fine examples of the application of morphological maps to site investigations for highway construction. During this period geomorphic maps using symbols were a common feature of many papers, even where the main topic was not mapping but some other systematic aspect of geomorphology. For example, de Dapper et al. (1989) used a geomorphic map to illustrate a paper on using SPOT imagery to map land degradation.

The previous example also illustrates the changing emphasis of data sources used for geomorphological mapping (see Smith and Pain, 2009). Initial work was field based (e.g. Wright, 1912), but the time and cost savings of utilizing remotely sensed data has meant a shift towards aerial photography (e.g. Prest et al., 1968), satellite imagery (e.g. Levine and Kaufman, 2008) and digital elevation models (DEMs). One of the first uses of DEMs was that of Moore and Simpson (1982) who used elevations of a national gravity station network in Australia to produce a national DEM and to briefly demonstrate digital manipulations that emphasized the main landform features of the continent. Thelin and Pike (1991) produced similar relief shaded DEM visualisations from 1:250,000 topographic data of the United States, together with a description of the main landform features. This visualization was the digital equivalent of maps such as those produced by Raisz (1951), and marked an important milestone in digital geomorphic mapping (see also Pike, 2000). More recently Guzzetti and Reichenbach (1994), using 1:25,000 topographic data, produced a similar DEM and landform unit map of Italy.

Whilst mapping landforms using polygons may be more 'GIS-friendly', and therefore now more common, polygon mapping as implemented through land systems mapping was developed shortly after World War 2 (Christian and Stewart, 1953). It was used extensively in Australia and Africa (Ollier, 1977) during the second half of the 20th century. More recently, regolith landform mapping has been developed in Australia as a means of mapping the pervasive regolith that covers much of the continent and, in particular, makes mineral exploration difficult. Chan (1988) explained the rationale behind the approach, whilst Pain et al. (1991) have modified and expanded the Australian application of regolith landform mapping (see also Pain et al., 2001).

One aspect of the development of geomorphic mapping is that, unlike geology and soils, geomorphic maps have rarely been part of the routine work of government agencies. For example, Australia has long had a complete coverage of geology maps at 1:250,000 compiled during mapping campaigns undertaken by state and Commonwealth government geological agencies. Similarly, New Zealand has a complete coverage of soil maps compiled by the New Zealand Soil Bureau and its successors. In contrast, Australia has only an incomplete coverage of land system maps compiled by CSIRO between the 1950s and 1980s. In the 1980s Geoscience Australia and some state geological surveys began regolith landform mapping (see http://crcleme.org.au/Pubs/regmaps.html), but these programmes halted as other priorities took over.

In 1973 van Dorsser and Salomé (1973) commented on five methods of detailed geomorphological mapping being carried out at the time, in Switzerland, Czechoslovakia, Poland, the Netherlands and France; in 1985 they extended this list to include Belgium and Germany (Salomé and van Dorsser 1985). Contemporaneously, geomorphic mapping was also being undertaken in Italy (e.g. Bondesan et al., 1989) (Plate 4, page 590). More recently geomorphological maps have been implemented through government funding to universities such as in Germany with the GMK25 1:25,000 series of maps (German Working Group on Geomorphology; http://gidimap.giub.uni-bonn.de/gmk.digital/downloads.htm) and in Brazil (e.g. Santos et al., 2009).

CURRENT STATE OF THE ART

Of the two approaches mentioned above, symbols, colours and patterns, and polygons, each of which serve different although related purposes, the use of polygons has tended to dominate over the last decade or so. Gustavsson et al. (2006) provide a useful summary, pointing out that thematic and applied maps now tend to dominate over 'scientific' maps that present a holistic view. They highlight a perceived general decline in geomorphic mapping up to about 2005, noting that it is surprising given that it coincides with the rise in the use of GIS by practitioners. More recently there has been renewed interest in academic geomorphic mapping, exemplified by the variety consistently being published by the *Journal of Maps* (http://www.journalofmaps.com). Landforms are also still being mapped as part of applied projects; for example, they are an important part of many engineering (Fookes et al., 2007) and land-use planning (e.g. Clarke et al., 2010) projects.

Whilst the 1960s and 1970s saw an explosion in mapping and legend systems (Verstappen, in press) as noted above, subsequently there has been both rationalisation and fragmentation. Otto et al. (in press) outline the most common legend systems, concluding with the work of Gustavsson et al. (2006). Their legend combines the data acquisition and mapping stages through the use of a template within a GIS and attempts to present complex geomorphological information without the requirement for overly complex legends. As such, whilst it is designed to operate at scales from 1:5000 to 1:50,000, it is not intended to replace highly detailed legends. It is also simplified by separating elements pertaining to materials, processes, genesis and morphography enabling users to identify descriptive and interpretive aspects.

The biggest driver of geomorphic mapping has been technology: the availability of new data sources has allowed new insights and rapid mapping to be performed, organized within the framework of a GIS. Whilst more mundane, a digital data framework for the organization of spatial data has perhaps had the single greatest impact; researchers have been able to take the paradigm of 'layers' of input data and produce layers of thematic, mapped, output. The co-registration of data in to a single geodetic reference system is simple, yet profound. However, the addition of new sources of digital spatial data has opened up vast regions of Earth's surface (and indeed other planets) for study. Satellite imagery remains an important ongoing data source with an increasing trend towards longer archives, higher spatial resolution and greater data volumes (Figure 8.1); however, it has been the availability of DEMs that has been at the forefront of much recent research (see Smith and Pain, 2009 for a detailed review). Whilst DEMs have been generated from contours and aerial photos for some considerable time, the advent of routine space-based data collection through photogrammetric processing using dedicated fore/aft sensors (e.g. SPOT 5, ASTER) and interferometric synthetic aperture radar (InSAR; Rosen, et al., 2000) has been significant. This period has also seen the release of two sub-100 m spatial resolution global DEMs in the form of the Shuttle Radar Topography Mission (http://www2.jpl.nasa.gov/srtm/) and ASTER Global-DEM (http://www.gdem.aster.ersdac.or.jp/) (see Hirt et al., 2010 for a comparison of the two DEMs).

The move towards DEMs of higher spatial resolutions and vertical accuracies, demanded by applications such as flood modelling (Casas et al., 2006), has been met largely by the development of laser ranging technologies and, in particular LiDAR (light detection and ranging; Baltsavias, 1999). The relatively high cost of LiDAR surveys has limited its application, with the biggest success reserved for airborne InSAR, particularly Intermap (http://www.intermap.com/). Whilst not offering the same spatial resolution or vertical accuracy of LiDAR, the data is of significantly better quality than space-borne products yet remains cost effective, remarkably enabling the acquisition of western Europe and the conterminous United States. Smith et al. (2006) review a range of nationally available datasets for geomorphological mapping and assess their completeness against independent field mapping. Of final note is the rapid emergence of terrestrial LiDAR which offers the quantitative acquisition of highly dense ranging measurements. These data are being used for small-scale mapping and modelling (e.g. Hodge et al., 2009) and their increased application is readily apparent.

As a footnote to this section on spatial data, it should be noted that fieldwork remains an important data collection technique and largely unchanged (e.g. Griffiths and Abraham, 2008), although methods such as plane tabling (e.g. Rose, 1989) have been largely replaced by traditional surveying, the use of global positioning system hand-held receivers (e.g. Dykes, 2008) or, as described above, terrestrial LiDAR systems.

With the availability of spatial data and a digital organizational framework, the next stage in geomorphic mapping involves the identification of pertinent landforms. This is most commonly an interpretative stage performed by an experienced observer; however, the objectiveness of such interpretations has been questioned (e.g. Siegel, 1977) and as a result there is considerable research into automated and semi-automated techniques for landform identification or *feature extraction* (see Seijmonsbergen et al., in press for further discussion). The primary inputs to these techniques are DEMs and morphography is therefore the context through which landforms are identified. This topic has evolved in to a research area in its own right, termed *geomorphometry*, or quantitative land surface analysis (see Chapter 13, Oguchi and Wasklewicz). Hengl et al. (2008) provide a detailed outline of geomorphometry and its applications. In addition to elevation, DEMs can also be used to calculate parameters for landforms, including gradient, aspect, profile/plan curvature and roughness (Evans, 1972), although care must be taken to use DEMs with an appropriate resolution (Pain, 2005). More recent work (Camargo et al., 2009) has combined these parameters with surface reflectance data in order to refine the methodologies.

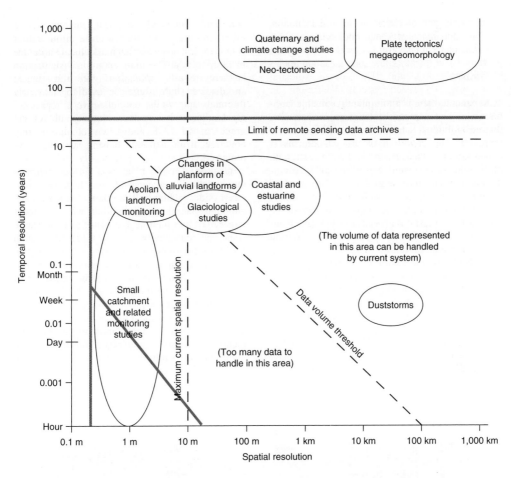

Figure 8.1 **Constraints of spatial and temporal resolutions of satellite sensors upon geomorphological research. (This figure has been redrafted from Millington and Townshend (1987), reflecting current sensors)**

The actual feature extraction methods employed borrow heavily from remote sensing and can be grouped as either *supervised* or *unsupervised*. Supervised techniques include object-oriented, clustering (e.g. maximum likelihood) and regression (e.g. multinomial). Unsupervised techniques include object-oriented, clustering (e.g. fuzzy k-means) and machine learning (e.g. decision tree). A detailed review of these techniques is beyond the scope of this chapter.

Whilst it is recognized that manual mapping is not repeatable and open to interpretation, it remains commonly employed due to simplicity and speed of deployment (Smith, in press). With digital satellite imagery a variety of enhancements can easily be applied to improve the visual acuity and aid identification and mapping of landforms. The most common techniques include false colour

composites, band ratios, convolution filtering (high pass filter) and contrast stretches (Lillesand et al., 2008); Clark (1997) and Jansson and Glasser (2005) provide advice from glacial geomorphology that is applicable to other domains. Smith and Wise (2007) highlight three primary controls on the representation of landforms on satellite imagery:

- *Relative size*: the spatial resolution of imagery in relation to the size of the landform.
- *Azimuth biasing*: the orientation of a landform with respect to the angle of solar illumination. Increasing linearity of the landform will cause visual changes to the planform shape.
- *Landform signal strength*: the tonal and textural differentiation of the landform with respect to its surroundings, affected by the variation in surface

materials and elevation of solar illumination. The appearance of shadowing maximizes the morphological response ('relief effect') from the terrain and therefore low solar elevation angles are best.

In general, there is a minimum resolvable landform size and a range of landform orientations that an individual satellite image will be able to represent. The *definition* of these landforms is dependent upon the surface cover and the strength of the relief effect. Smith and Wise (2007) recommend solar elevation angles <20°; however, as polar orbiting satellites have fixed overpass times, a solar elevation during acquisition cannot be specified. As a result it is necessary to take advantage of seasonal variations, with this variation greater at higher latitudes.

With DEMs, processing might initially appear to be far simpler as there is a single consistent data source that is representative of morphology and therefore, in principle, not open to bias. However, the availability of 'altitude' as a parameter has led to the computation of a variety of derivative products as noted above. Some of these are particularly pertinent for manual mapping and have been reviewed by Smith and Clark (2005) and Hillier

and Smith (2008). The simplest visualization is a *greyscale* image of the raw elevation values (Figure 8.2a); however, this masks local variations in morphology. The most common visualization is therefore *relief shading* (Figure 8.2b) where an idealized light source is used to illuminate the landscape; whilst very effective at representing landforms it introduces the azimuth biasing noted above. Alternative visualizations that are free from azimuth biasing are therefore preferred.

Gradient (Figure 8.3a) is a further common manipulation as many landforms have relatively steep sides which allow their identification, although interpreters unfamiliar with the technique may find the image difficult to understand. Improved presentation can be achieved through inverting the greyscale colouring scheme so that flatter areas appear darker. Curvature measures the rate of change of slope and is comprised of three elements (Schmidt et al., 2003): profile, planform and tangential curvature. *Profile curvature* is the most pertinent for geomorphic mapping as it measures downslope curvature and therefore identifies breaks-of-slope (Figure 8.3b), a common identifier for landforms. Like gradient, it can also be difficult to interpret. Other techniques that

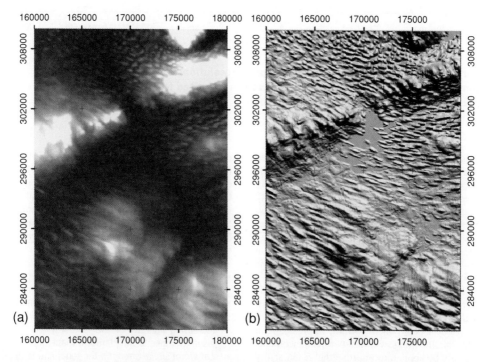

Figure 8.2   DEM visualisation using (a) greyscaling and (b) relief shading (illumination angle 20°). (Reproduced from Ordnance Survey Ireland, Copyright Permit MP001904)

have been employed include local contrast stretch (Smith and Clark, 2005), residual relief separation (Hillier and Smith, 2008) and spatial wavelet transform (Hillier, 2008). The local contrast stretch performs a standard linear contrast stretch over a user-specified region (or kernel) and can therefore highlight variations in terrain (Figure 8.4a). Residual relief separation uses a coarse median filter to estimate the underlying regional relief which is then subtracted from the original DEM to leave the 'residual' relief. The latter are localized topographic variations which are primarily landforms (Figure 8.4b). As with the residual relief separation, the component of the landscape containing the landforms of interest is extracted with a spatial wavelet transform. Using appropriate coefficients, a wavelet transform is computed along a profile (or a mesh of profiles) and the location and scale of each feature is determined. The approach has the benefit of being scale invariant and can therefore identify multiple landforms at different sizes.

In summary, simple greyscale viewing is not appropriate, with the most common technique relief shading. This is easy to interpret and highlight subtle topographic features; however, it suffers from azimuth biasing. The use of multiple illumination angles can attempt to offset this disadvantage. Non-azimuth biased visualizations are therefore preferred, such as gradient and profile curvature. The last three techniques use alternative processing techniques to identify landforms from the underlying regional relief, offering the advantage of non- azimuth bias. There is no single visualization technique that is ideally suited to geomorphological mapping, although non-azimuth biased techniques are methodologically preferred. Best practice therefore involves initial mapping using a bias-free visualization technique and, once complete, supplementing the work with mapping from multiple relief shaded images (at least two, but ideally three or four illumination azimuths).

With visualization and mapping complete, the final stage in the creation of a geomorphic map is dissemination. The last decade has seen a dramatic change in the role and scope of the internet, so much so that this is driving much of the public interest in the outputs of academic research and therefore the need for good output documents for consumption by non-specialists is of increasing importance. Perhaps the single biggest disruptor in this area has been Google Earth; the idea of a virtual globe allowing access to high resolution

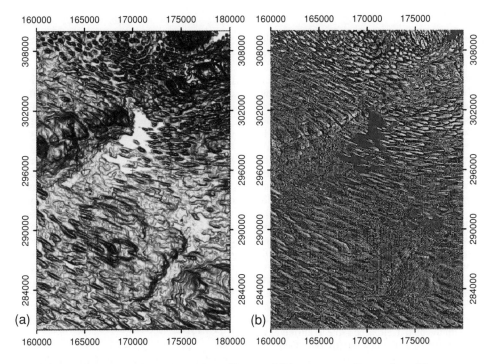

Figure 8.3   DEM visualisation using (a) gradient and (b) curvature. (Reproduced from Ordnance Survey Ireland, Copyright Permit MP001904)

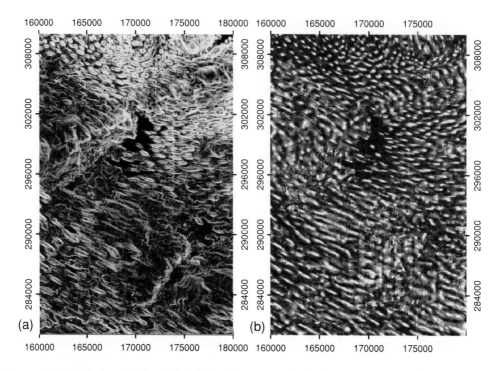

Figure 8.4   DEM visualisation using (a) local contrast stretching and (b) residual relief separation. (Reproduced from Ordnance Survey Ireland, Copyright Permit MP001904)

aerial imagery for virtually any part of the globe was unthinkable even a short time ago. Yet this paradigm for interaction with spatial data, infrastructure and access to the actual data itself has been achieved in a very short space of time. The added ability for overlaying vector and raster datasets makes it a useful 'platform' for data dissemination although there are limitations to what can be achieved cartographically (Otto et al., in press). An alternative is to use a bespoke web mapping system that allows users to view and query your data, as well as access the actual content using web mapping services (e.g. WMS, WCS, WFS). Finally, electronic versions of maps can be disseminated as GeoPDFs; the standard portable document format (PDF) has been extended to incorporate spatial reference systems that allow features to be identified and measurements taken. It is also possible to incorporate layers allowing a simple 'GIS-style' interface where different datasets can be switched on and off. Finally, the raw data can be included in the PDF with the map itself, accessible as attributes to map features (e.g. Gurney, 2010). The GeoPDF has gained considerable support amongst both GIS vendors for their creation, as well as major data suppliers (e.g. US Army, US Geological

Survey, Australian Hydrographic Service) for map dissemination.

Something that ties the above three dissemination methods together is the application of data *standards*. Standards are vital to the effective dissemination of data rather than the reliance and 'tie-in' to commercial systems and, within the geospatial domain, this has been effectively tackled by the Open Geospatial Consortium (OGC; http://www.opengeospatial.org). In this regard, both KML (used by Google Earth) and GeoPDF are OGC standards. Considerable work has been completed in relation to web mapping systems with WMS, WCS and WFS all being OGC standards; indeed support for WMS is near-ubiquitous and with the availability of open-source web mapping systems such as MapServer (http://mapserver.org), web mapping has never been more accessible (Mitchell, 2005).

KEY ISSUES

There are two main reasons for compiling a geomorphic map. First, academic geomorphologists compile *general maps* of landforms as a means of understanding the sites or locations they are working on. Second, *geomorphic maps* may be compiled for *specific purposes* such as road building (e.g. Jones et al., 1983). The special purpose map will take less time to compile, but may leave out features that turn out to be important later on. General maps, on the other hand, may take more resources to compile, but will usually have a much wider application. Experience suggests that a general map should be compiled first, perhaps at a smaller scale than the specific maps, which can be compiled at a later date. Thus in work carried out on floodplains in Australia, the specific application has to do with water availability, quality and distribution, both surface and under ground. However, the best approach is usually to compile a general map at a regional scale, and then to produce specific maps with much more detail for selected areas (e.g. Kernich et al., 2009). In Australia at present, general geomorphic maps are produced to provide context for detailed maps that use landforms supported by field observations as surrogates for surface materials. The latter are important in predicting, for example, infiltration and recharge.

When developing the scope for geomorphic maps it is important to consider the four primary elements: morphometry, morphogenesis, morphochronology and morphodynamics. Whilst it is almost impossible (and, indeed, usually not necessary) to combine all these elements into a single cartographic document, the principal focus of the mapping should be considered prior to the setting out of data collection methods and compilation of materials. For many applications, the inclusion of information about processes (genesis) and landform evolution (chronology) is important for interpreting landscapes because they allow predictions of materials (e.g. Clarke et al., 2010).

Of particular pertinence to the two preceding points is the issue of *scale*; at what scale do you map geomorphic features for inclusion in to general maps and how are these landforms, or their component parts, included in specific maps? The use of an appropriate legend system is required in order to represent the landforms included and take into account the methods by which the maps are compiled.

There continues to be a need for standards, in spite of many attempts starting with the IGU in the 1960s and 1970s. The earlier section has outlined the renewed interest in geomorphological mapping and the revision of legend systems to meet the current increase in demand, yet symbolization and data structure remain inextricably tied together (Mentlik et al., 2006). Whilst the physical model implemented is linked to the underlying GIS selected for the mapping, this will have been derived from an abstract logical model. At the lowest level this will simply be the informal deployment of a conceptual model through the creation of different data layers, but may also involve the rigorous, structured, ingestion and creation of data leading to the production of final data layers related to map features.

Standards for regolith landform mapping have also been proposed (Pain et al., 1991; Pain, 2008a, 2008b 2008c), but uptake has been slow. However, geologists have been particularly successful in being able to agree upon a basic set of standards for mapping, nomenclature and digital data (OneGeology, 2010). OneGeology provides a good model for geomorphic mapping to follow – in fact there have been some attempts to include geomorphic standards within OneGeology. Examples of national standards include those implemented by the British Geological Survey (Ambrose, 2000) and the US Geological Survey (FGDC, 2006); the latter has also been implemented within ArcGIS (http://resources.esri.com/mapTemplates/index.cfm?fa=codeGalleryDetails&scriptID=16317).

The applications of geomorphic maps are many. Road alignment and road building (Jones et al., 1983) are common examples from the construction sector, whilst Fookes et al. (2007) provide many examples related to engineering. Elsewhere, landform mapping, as part of a series of surface maps, continues to be important for mineral exploration (e.g. McFarlane, 1981; Chan, 1988; Pain, 2008b), whilst academic resurgence is demonstrated through a range of maps published by the *Journal of Maps* (http://www.journalofmaps.com).

THE FUTURE

Although geomorphic mapping has been recognized as being useful for a large number of applications for some considerable time, future work can only see these applications increase in number and importance. The re-emergence of geomorphic mapping as the core inter-disciplinary interface on commercial, governmental and academic projects has developed a role central to the synthesis of diverse data. This central role has, in part, been driven by technological developments and subsequent access to a variety of new and innovative datasets. Geomorphic mapping is not new in, and of, itself, but it has been the greatest beneficiary, and exploiter, of remotely sensed data (Smith and Pain, 2009). This has enabled mapping remote regions, in greater (topographic) detail, over increasingly shorter time periods. The acquisition of topographic data has been particularly transformative and whilst we have had nearly 15 years of archive topographic data from satellite platforms, it is the emergence of aerial and terrestrial datasets that is particularly compelling for new applications. To date there has been considerable application of high-resolution regional and national DEMs, with the exploitation of terrestrial LiDAR systems becoming particularly evident at the field-scale. Complementary data sets are also being explored, including passive airborne systems such as radiometrics (composition of the upper 50 cm of the surface) and aeromagnetics (imaging subsurface features), and active systems such as airborne electromagnetics which can provide three-dimensional detail on conductivity down to as deep as 100 m. Hyperspectral remote sensing has been available from airborne platforms for over 20 years, but has had limited, experimental, use from space. This is set to change with a number of satellites due for launch over the next 5 years, further adding to the data available to geomorphologists.

These data are now firmly rooted within the paradigm of data 'layers' that are fully exploited by GIS. Whilst the underlying technology will remain the same, the ability to handle, manipulate and analyse large, parallel, datasets will begin to be realized. Dataset size remains a significant issue with considerable work focused upon compression, manipulation and generalisation. Raw processing power and storage space will always remain a potential constraint with current computing paradigms allowing the virtualisation of entire computing ecosystems; this improves stability but also allows the system to be scalable. Many GIS vendors are now focused upon making these environments as flexible, and reliable, as possible. As a result there has been a move towards 'cloud' computing for intensive processing tasks. If a project requires greater processing, it is simply a case of purchasing greater capacity that can automatically be increased. This kind of approach is already in progress through the Berkeley Open Infrastructure for Network Computing (BOINC), where projects such as Seti@Home use free computer time provided by users worldwide (Korpela et al., 2001; Anderson, 2004). A similar although more limited approach is used at Geoscience Australia; individual work station computers are used after hours to run, in parallel, routine manipulations of AEM data that would take weeks on a single computer.

As a result of these changes a range of application areas have been exploited by geomorphic mapping. Some of these areas are now highlighted.

The prediction of soil attributes from terrain attributes has long been a part of soil science, especially pedology and soil mapping. In recent times this has become emphasized with the development of a new sub-discipline of soil science, pedometrics, and the study of pedotransfer functions. This is based on the well-established idea that soil catenas develop in many landscapes in response to the way water moves through and over the landscape (e.g. Moore et al., 1993; McSweeney et al., 1994; Børgesen et al., 2008). It follows that terrain attributes can be used to characterize soil attributes. The need for this kind of work arises because of the high cost of field soil surveys, and the increasing availability of digital data at a high enough resolution to make it possible to compile maps of similar utility.

In a similar approach Wilford (in press) has developed an index of weathering that combines geological, radiometric and DEM data to predict depth and intensity of weathering. This index uses changes in radiometric signature with intensity of weathering as calibrated by field and laboratory observations, and use DEM data to perform predictions for different geological units. It works at both local and regional scales, and provides fundamental information about soil parent materials and other landscape attributes.

Geomorphological mapping is now seeing something of a renaissance; from an initial period in the 1960s and 1970s where field-based morphological mapping was initially developed, through to the 1980s and 1990s which were characterized by increasing niche interest. The last decade has seen a dramatic increase in activity, a result of digital management workflows (principally through the deployment of GIS) and a dramatic increase in remotely sensed data (both satellite imagery and DEMs). Perhaps of greater importance is the realization that geomorphology really is an 'interface' discipline, not just physically, in the sense of studying Earth's surface, but also between pure and applied sciences that seek to derive greater benefit from integrating Earth surface processes and landforms in to their analyses. Geomorphology finds itself to be a key integrating discipline, as well as a major user and developer of massive datasets and processing algorithms. In the same way that geological mapping is a key underpinning resource to societal development, so geomorphology is increasingly being seen as a necessity. Future trends will undoubtedly see geomorphological mapping become more ubiquitous and the products made widely available. Distribution will make use of distribution channels and, in particular, digital routes including virtual globes, web mapping and GeoPDFs. The sophistication of managing and modelling data will increase, particularly within the realm of elevation point clouds and derived DEMs. The next decade should be viewed with anticipation as the discipline is significantly enhanced.

REFERENCES

Ambrose, K. (2000) *Specifications for the Preparation of 1:10 000 Scale Geological Maps*, 2nd edn. British Geological Survey, Nottingham.

Anderson, D.P. 2004. BOINC: a system for public-resource computing and storage. IEEE/ACM International Workshop on Grid Computing (GRID'04), 5: 4–10.

Baltsavias, E.P. (1999) Airborne laser scanning: basic relations and formulas. *ISPRS Journal of Photogrammetry & Remote Sensing* 54: 199–214.

Bondesan, M., Castiglioni, G.B. and Gasperi, G. (1989) *Geomorphological Map of the Po Plain*, Progress report of the working group. Materiali – Departimento di Geografia, Universita di Padova. 8/1989, p. 23.

Børgesen, C.D., Iversen, B.V., Jacobsen, O.H. and Schaap, M.G. (2008) Pedotransfer functions estimating soil hydraulic properties using different soil parameters. *Hydrological Processes* 22: 1630–39.

Brunsden, D., Doornkamp, J.C., Fookes, P.G., Jones, D.K.C. and Kelly, J.M.H. (1975) Large scale geomorphological mapping and highway engineering design. *Quarterly Journal of Engineering Geology* 8: 227–53.

Camargo, F.F., Florenzano, T.G., de Almeida, C.M. and de Oliveira, C.G. (2009) Geomorphological mapping using object-based analysis and ASTER DEM in the Paraíba do Sul Valley, Brazil. *International Journal of Remote Sensing* 30: 6613–20.

Casas, A., Benito, G., Thorndycraft, V.R. and Rico, M. (2006) The topographic data source of digital terrain models as a key element in the accuracy of hydraulic flood modelling. *Earth Surface Processes and Landforms* 31: 444–56.

Castiglioni, G.B., Biancotti, A., Bondesan, M., Cortemiglia, G.C., Elmi, C., Favero, V., Gasperi, G., Marchetti, G., Orombelli, G., Pellegrini, G.B. and Tellini, C. (1999) Geomorphological map of the Po Plain, Italy, at a scale of 1:250,000. *Earth Surface Processes and Landforms* 24: 1115–20.

Castiglioni, G.B., Bondesan, M. and Elmi, C. (1990) Geomorphological mapping of the Po Plain (Italy), with an example in the area of Ravenna. *Zeitschrift fur Geomorphologie Supplement Band* 80: 35–44.

Chan, R.A. (1988) Regolith terrain mapping for mineral exploration in Western Australia. *Zeitschrift für Geomorphologie Supplement Band* 80: 205–21.

Christian, C.S. and Stewart, G.A. (1953) *General Report on Survey of Katherine–Darwin Region, 1946*, CSIRO Land Research Series 1: 150.

Clark, C.D. (1997) Reconstructing the evolutionary dynamics of palaeo-ice sheets using multi-temporal evidence, remote sensing and GIS. *Quaternary Science Reviews* 16: 1067–92.

Clarke, J.D.A., Gibson, D. and Apps, H. (2010) The use of LiDAR in applied interpretive landform mapping for natural resource management, Murray River alluvial plain, Australia. *International Journal of Remote Sensing* 31: 6275–96.

Cooke, R.U. and Doornkamp, J.C. (1990) *Geomorphology in Environmental Management, a New Introduction*, 2nd edn. Clarendon Press, Oxford, p. 410.

de Dapper, M., Goossens, R. and Ongena, T. (1989) The use of SPOT images for the assessment and mapping of geomorphology and land degradation by savanisation in a wet-and-dry tropical forested environment (Lubumbashi, Shaba, Zaire. *Supplementi di Geografia Fisica e Dinamica Quaternaria* II: 87–91.

Dykes, A.P. (2008) Geomorphological maps of Irish peat landslides created using hand-held GPS. *Journal of Maps* v2008: 258–76.

Evans, I.S. (1972) General geomorphometry, derivatives of altitude and descriptive statistics, in R.J. Chorley (ed.), *Spatial Analysis in Geomorphology*. Harper and Row, New York, pp. 17–90.

Fairbridge, R.W. (ed.) (1968) *The Encyclopedia of Geomorphology, Encyclopedia of Earth Sciences 3*. Reinhold, New York, p. 1295.

Federal Geographic Data Committee (2006) *FGDC Digital Cartographic Standard for Geologic Map Symbolization*. Reston, VA, Federal Geographic Data Committee Document Number FGDC-STD-013-2006. p. 290.

Fitzpatrick, A., Wilkinson, K., Kellett, J., Pain, C.F. and Claridge, J. (2004) Datasets, in T. Chamberlain and K.E. Wilkinson (eds), *Salinity Investigations Using Airborne Geophysics in the Lower Balonne Area, Southern Queensland*. Queensland Department of Natural Resources and Mines, pp. 19–46.

Fookes, P.G., Lee, E.M. and Griffiths, J.S. (2007) *Engineering Geomorphology: Theory and Practice*. Whittles Publishing, p. 288.

Goudie, A.S. (ed.) (1981) *Geomorphological Techniques*. George Allen & Unwin, London, p. 395.

Goudie, A.S. (ed.) (2004) *Encyclopedia of Geomorphology*. Routledge, London, p. 1200.

Griffiths, J.S. and Abraham, J.K. (2008) Factors affecting the use of applied geomorphology maps to communicate with different end-users. *Journal of Maps* v2008: 201–10.

Gullentops, F. (1964) Trois examples de cartes geomorphologiques detaillees. *Acta Geographica Lovaniensia* 3: 425–30.

Gurney, S.D. (2010) Contemporary (2001) and 'Little Ice Age' glacier extents in the Buordakh Massif, Cherskiy Range, north east Siberia. *Journal of Maps* v2010: 7–13.

Gustavsson, M. (2006) Development of a detailed geomorphological mapping system and GIS geodatabase in Sweden. Acta Universitatis Upsaliensis. *Digital Comprehensive Summaries of Uppsala Dissertations from the Faculty of Science and Technology* 236: 36.

Gustavsson, M., Kolstrup, E. and Seijmonsbergen, A.C. (2006) A new symbol-and-GIS based detailed geomorphological mapping system: Renewal of a scientific discipline for understanding landscape development. *Geomorphology* 77: 90–111.

Guzzetti, F. and Reichenbach, P. (1994) Towards a definition of topographic divisions for Italy. *Geomorphology* 11: 57–74.

Hengl, T., Hannes, I. and Reuter, H.I. (2008) *Geomorphometry: Concepts, Software, Applications.* Elsevier, London, p. 796.

Hillier, J.K. (2008) Seamount detection and isolation with a modified wavelet transform. *Basin Resesarch* 20: 555–73.

Hillier, J. and Smith, M.J. (2008) Residual relief separation: DEM enhancement for geomorphological mapping. *Earth Surface Processes and Landforms* 33: 2266–76.

Hirt, C., Filmer, M.S. and Featherstone, W.E. (2010) Comparison and validation of the recent freely available ASTER-GDEM ver1, SRTM ver4.1 and GEODATA DEM-9S ver3 digital elevation models over Australia. *Australian Journal of Earth Sciences* 57: 337–47.

Hodge, R., Brasington, J. and Richards, K.S. (2009) In situ characterization of grain-scale fluvial morphology using terrestrial laser scanning. *Earth Surface Processes and Landforms* 34: 954–68.

International Geographical Union, Commission on Applied Geomorphology, Subcommission on Geomorphological Mapping (1968) The unified key to the detailed geomorphological map of the World 1:25,000 – 1:50,000 in Klimaszewski, M. Part I. Problems of the detailed geomorphological map. Bashenina, N.V., Gellert, J. Jolly, F. et al. Part II. Project of the unified key to the detailed geomorphological map of the World. Folia Geographica, Series Geographica–Physica 2, Krakow.

Jansson, K.N. and Glasser, N.F. (2005) Using Landsat 7 ETM+ imagery and digital terrain models for mapping lineations on former ice sheet beds. *International Journal of Remote Sensing* 26: 3931–41.

Joly, M.F. (1962) Principes pour une méthode de cartographie géomorphologique. *Bulletin de l'Association de Géographes Français* 39: 271–8.

Jones, D.K.C., Brunsden, D. and Goudie, A.S. (1983) A preliminary geomorphological assessment of part of the Karakoram Highway. *Quarterly Journal of Engineering Geology* 16: 331–55.

Kernich, A.L., Pain, C.F., Clarke, J.D.A. and Fitzpatrick, A.D. (2009) Geomorphology of a dryland fluvial system: the Lower Balonne River, southern Queensland. *Australian Journal of Earth Sciences* 56: S139–S153.

Korpela, E., Werthimer, D., Anderson, D., Cobb, J. and Lebofsky, M. (2001) SETI@HOME – massively distributed computing for SETI. *Computing in Science and Engineering* 3: 78–83.

Leszczycki, S. (ed.) (1963) *Problems of Geomorphological Mapping.* Geographical Studies 46, Institute of Geography of the Polish Academy of Sciences. p. 178.

Levine, N.S. and Kaufman, C.C. (2008) Land use, erosion, and habitat mapping on an Atlantic Barrier Island, Sullivan's Island, South Carolina, *Journal of Maps* v2008: 161–74.

Lillesand, T.M., Kiefer, R.W. and Chipman, J.W. (2008) *Remote Sensing And Image Interpretation.* John Wiley and Sons, New York.

Lobeck, A.K. (1951) *Physiographic Diagram of Australia.* The Geographical Press, Columbia University, New York.

Lobeck, A.K. (1957) *Physiographic Diagram of the United States.* The Geographical Press, C.S. Hammond and Co., Maplewood, New Jersey.

Löffler, E. and Ruxton, B.P. (1969) Relief and land form map of Australia (1:5,000,000 scale), AWRC, The representative basin concept in Australia (A Progress Report). Australian Water Resources Council Hydrological Series 2.

McFarlane, M.J. (1981) *Morphological Mapping in Laterite Areas and its Relevance to the Location of Economic Minerals in Laterite. Lateritisation Processes.* Proceedings of the International Seminar on Lateritisation Processes, Trivandrum, India. A.A. Balkama, Rotterdam. pp. 308–317.

McSweeney, K.K., Gessler, P.E., Hammer, D., Bell, J. and Petersen, G.W. (1994) Towards a new framework for modeling the soil-landscape continuum. Factors of Soil Formation: A Fiftieth Anniversary Retrospective. *Soil Science Society of America Special Publication* 33: 127–45.

Mentlik, P., Jedlicka, K., Minar, J. and Barka, I. (2006) Geomorphological information system: physical model and options of geomorphological analysis. *Geografie – Sbornik Cesk Geograficke Spolecnosti* 1: 15–32.

Millington, A.C. and Townshend, J.R.G. (1987) The potential of satellite remote sensing for geomorphological investigations, an overview, in V. Gardiner (ed.), *International Geomorphology.* Wiley, Chichester, pp. 331–42.

Mitchell, C.W. (1991) *Terrain Evaluation*, 2nd edn. Longman, UK, p. 441.

Mitchell, T. (2005) *Web Mapping Illustrated.* O'Reilly Media, Sebastopol, CA.

Moore, R.F. and Simpson, C.J. (1982) Computer manipulation of a digital terrain model (DTM) of Australia. Bureau Mineral Resources. *Journal of Australian Geology and Geophysics* 7: 63–7.

Moore, I.D., Gessler, P.E., Nielson, G.A. and Peterson, G.A. (1993) Soil attribute prediction using terrain analysis. *Soil Science Society of America Journal* 57: 443–52.

Ollier, C.D. (1977) Terrain classification: methods, applications and principles, in J.R. Hails (ed.), *Applied Geomorphology.* Elsevier, Amsterdam, pp. 277–316.

One Geology (2010) *Making Geological Map Data for the Earth Accessible.* http://www.onegeology.org.

Otto, J.-C., Gustavsson, M. and Geilhausen, M. (in press) Cartography: design, symbolisation and visualisation of geomorphological maps, in M.J. Smith, P. Paron and J. Griffiths (eds), *Geomorphological Mapping: A Handbook of Techniques and Applications.* Elsevier, London.

Pain, C.F. Chan, R., Craig, M., Gibson, D., Ursem, P. and Wilford, J. (1991) *RTMAP BMR Regolith Database Field Handbook.* BMR Record 1991/29, p. 125.

Pain, C.F. (2005) Size does matter: relationships between image pixel size and landscape process scale, in A. Zerger and R.M. Argent (eds), *MODSIM 2005 International Congress on Modelling and Simulation.* Modelling and Simulation Society of Australia and New Zealand, December 2005, pp. 1430–36.

Pain, C.F. (2008a) *Field Guide for Describing Regolith and Landforms CRC LEME.* CSIRO Exploration and Mining, PO Box 1130, Bentley WA 6102, Australia, p. 94.

Pain, C.F. (2008b) Regolith description and mapping, in K. Scott and C.F. Pain (eds), *Regolith Science*. CSIRO Publishing, Melbourne, pp. 263–67.

Pain, C.F. (2008c) A revised map of Australia's physiographic regions: a hierarchical background for digital soil mapping. *Geophysical Research Abstracts* 10: EGU2008-A-05870.

Pain, C.F., Paron, P. and Smith, M.J. (2008) Applied geomorphological mapping (AppGeMa): a working group of the International Association of Geomorphologists. *Geophysical Research Abstracts* 10: EGU2008-A-05888.

Pain, C.F., Craig, M.A., Gibson, D.L. and Wilford, J.R. (2001) Regolith–landform mapping: an Australian approach, in P.T. Bobrowsky (ed.), *Geoenvironmental Mapping, Method, Theory and Practice*. A.A. Balkema, Swets and Zeitlinger, The Netherlands, pp. 29–56.

Pike, R.J. (2000) Geomorphometry – diversity in quantitative surface analysis. *Progress in Physical Geography* 24: 1–20.

Prest, V.K., Grant, D.R. and Rampton, V.N. (1968) *The Glacial Map of Canada*. Geological Survey of Canada.

Raisz, E. (1951) The use of air photos for landform maps. *Annals of the Association of American Geographers* 41: 324–30.

Robinson, A.H. (1970) Erwin Josephus Raisz 1893–1968. *Annals of the Association of American Geographers* 60: 189–93.

Rose, J. (1989) Glacial stress patterns and sediment transfer associated with the formation of superimposed flutes. *Sedimentary Geology* 62: 151–76.

Rose, J. and Smith, M.J. (2008) Glacial geomorphological maps of the Glasgow region, western central Scotland. *Journal of Maps* v2008: 399–416.

Rosen, P.A., Hensley, S., Joughin, I.R., Li. F.K., Madsen, S.N., Rodríguez, E. and Goldstein, R.M. (2000) Synthetic aperture radar interferometry. *Proceedings of the IEEE* 88: 333–82.

Salomé, A.I. and van Dorsser, H.J. (1985) Some reflections on geomorphological mapping systems. *Zeitschrift für Geomorphologie* 29: 375–80.

Santos, L.J.C., Chisato, O.-F., Canali, N.E.-S., Fiori, A.P., Da Silveira, C.T. and Da SILVA, J.M.F. (2009) Morphostructural mapping of Parana State, Brazil. *Journal of Maps* v2009: 170–78.

Savigear, R.A.G. (1965) A technique of morphological mapping. *Annals of the Association of American Geographers* 55: 514–38.

Schmidt, J., Evans, I.S. and Brinkmann, J. (2003) Comparison of polynomial models for land surface curvature calculation. *International Journal of Geographic Information Science* 17: 797–814.

Seijmonsbergen, A.C., Hengl, T. and Anders, N.S. (in press) Semi-automated identification and extraction of geomorphological features using digital elevation data, in M.J. Smith, P. Paron and J. Griffiths (eds), *Geomorphological Mapping: A Handbook of Techniques and Applications*. Elsevier, London.

Siegal, B.S. (1977) Significance of operator variation and the angle of illumination in lineament analysis of synoptic images. *Modern Geology* 6: 75–85.

Smith, M.J. (in press) Digital mapping: visualisation, interpretation and quantification of landforms, in M.J. Smith, P. Paron and J. Griffiths (eds), *Geomorphological Mapping: A Handbook of Techniques and Applications*. Elsevier, London.

Smith, M.J. and Clark, C.D. (2005) Methods for the visualisation of digital elevation models for landform mapping. *Earth Surface Processes and Landforms* 30: 885–900.

Smith, M.J. and Wise, S.M. (2007) Problems of bias in mapping linear landforms from satellite imagery. *International Journal of Applied Earth Observation and Geoinformation* 9: 65–78.

Smith, M.J. and Pain, C.F. (2009) Applications of remote sensing in geomorphology. *Progress in Physical Geography* 33: 568–82.

Smith, M.J., Rose, J. and Booth, S. (2006) Geomorphological mapping of glacial landforms from remotely sensed data: an evaluation of the principal data sources and an assessment of their quality. *Geomorphology* 76: 148–65.

Smith, M.J., Paron, P. and Griffiths, J. (eds) (in press). *Geomorphological Mapping: A Handbook of Techniques and Applications*. Elsevier, London.

Thelin, G.P. and Pike, R.J. (1991) Landforms of the conterminous United States – a digital shaded-relief portrayal. US Geological Survey Miscellaneous Investigations Map 2206, 1: 3,500,000.

Tricart, J. and Usselmann, P. (1967) Feuille géomorphologique Neuf-Brisach 1/2, 1/50.000, Recherche Cartographique sur Programme No 77, C.N.R.S., t. XVII. *Revue de Géomorphologie Dynamique* 17: 10–21.

van Dorsser, H.J. and Salomé, A.I. (1973) Different methods of detailed geomorphological mapping. *K.N.A.G. Geografisch Tijdschrift* 7: 71–4.

van Zuidam, R.A. (1982) Considerations on systematic medium-scale geomorphological mapping. *Zeitschrift fur Geomorphologie,* 26: 473–80.

Verstappen, H.T. (in press) Old and new trends in geomorphological and landform mapping, in M.J. Smith, P. Paron and J. Griffiths (eds), *Geomorphological Mapping: A Handbook of Techniques and Applications*. Elsevier, London.

Waters, R.S. (1958) Morphological mapping. *Geography* 43: 10–18.

Wilford, J. (in press) A weathering intensity index for the Australian continent using airborne gamma-ray spectrometry and digital terrain analysis, *Geoderma* in press.

Wright, W.B. 1912. The drumlin topography of south Donegal. *Geological Magazine* 9: 153–59.

9

The Significance of Models in Geomorphology: From Concepts to Experiments

Nicholas A. Odoni and Stuart N. Lane

Amid all the revolutions of the globe the economy of Nature has been uniform, and her laws are the only things that have resisted the general movement. The rivers and the rocks, the seas and the continents have been changed in all their parts; but the laws which direct those changes, and the rules to which they are subject, have remained invariably the same

(Playfair, 1802, *Illustrations of the Huttonian Theory*, p. 374)

THE PERVASIVE PRESENCE OF MODELS IN GEOMORPHOLOGY: UNIVERSALITY VERSUS SINGULARITY

Playfair's quote summarizes the essential presumption of a geomorphological modeller: that there are things – in Playfair's view, *laws* – that can be taken as given; and that these laws operate to produce the diversity of landforms as we see them. Whilst the term 'law' is somewhat unfortunate, as most landforms are the consequence of particular realizations of one or a combination of laws, it emphasizes the notion that there are things of universal importance that we can identify and use to explain, perhaps even predict the future of,

the landforms that we observe on earth's surface. Modelling in geomorphology is about working with these universalities, whether finding them or using them as part of geomorphological enquiry. The commitment of geomorphological science, over the last 200 years, to this universality is why models are a pervasive component of geomorphological enquiry. The nature of this commitment, the kind of universality sought, has changed markedly. In 1967, in a benchmark paper concerned with modelling in geomorphology, Chorley saw the formalization of modelling as a practice as central to developing the discipline 'from a subjective catalogue of phenomena into a coherent and rational discipline' (p. 90). This formalization mapped all types of geomorphological activity onto different types of modelling, with the resultant models subject to assessment, rejection and reformulation.

This chapter takes a similar perspective: all geomorphological enquiry can be cast as a modelling activity; and then thinks through the different implications of such a casting. But, and in contrast to Chorley's wider remit, it then focuses on models in their own right, reflecting how the use of geomorphological models has come to refer to a specific set of activities through which we seek to emulate in either laboratories or computers the behaviour of landforms. (Note that here the words 'emulate' and 'emulation' are used in their usual sense, and should not be confused with the more specialist use, also commented upon later, with respect to derived models, metamodels and emulators.) This emulation is the primary focus of this chapter but before we consider emulation we will consider the wider and all-pervasive practice of modelling in geomorphological enquiry through our perceptual and conceptual modelling activities.

PERCEPTUAL MODELS, CONCEPTUAL MODELS AND SYSTEMS

Perceptualization

The pervasive presence of models in geomorphological research arises from our belief that there are elements of the geomorphological world that 'travel' in space and time and so can be valuable elements of geomorphological explanation. For instance, whilst every landform might appear to be unique when explored in detail, there are generalizations of different kinds that can be invoked as means of understanding, explaining or even predicting landscape form and process.

Consider river meanders. It is possible to: (1) generalize their properties statistically (e.g. Leopold and Wolman, 1960); (2) describe their appearance mathematically, whether regular (e.g. Langbein and Leopold, 1966), irregular (e.g. Ferguson, 1976) or random (e.g. Speight, 1965); or (3) represent them as a morphodynamic system arising from the interaction between river morphology, sediment transport and river hydraulics (e.g. Dietrich, 1987). These are very different types of generalizations, underpinned by very different approaches and methods, but they all share a search for generalization that has, within them, whether explicit or implicit, an underlying perceptual model (Kirkby, 1996; Lawrence, 1996). This model is experiential in the sense that it may be informed by field visits, reading material including journal articles, attending academic (and other) workshops and conferences, chance discussions and debates, previous modelling activities, public engagement and a multitude of other experiences, the very muddiness of being a geomorphologist. Whilst Kirkby (1996) identifies the perceptual model as a property of the mind, it may also be a property of a group of researchers or even a whole community (e.g. see Shermann, 1996, on 'fashion' in geomorphology). Thus, the perceptual model is not necessarily a tightly defined or traced construct, or even a fully formulated conception, but such models underlie and precede all, more formal, model building. Indeed, the dependence of modelling upon perceptual models may be one reason why geomorphological modellers pay so much attention to establishing the validity of the models that they develop: geomorphologists wish to be certain that their models are not simply a product of their personal or a particular community's imagination. This is something we will consider below.

Conceptualization and systems

The translation of a perceptual model to a form that can be worked with, whether for population with data, or expressed as a series of equations and hence encoded in a mathematical model, is a process of formalization called model conceptualization. Sometimes, this may be left as implicit, as little more than a series of hypotheses to be tested, either individually or as a whole, and being conceptual statements of what the geomorphologist thinks. Figure 9.1 shows an early conceptualization of the river meander (Lyell, 1831), in which observation of river form is used to dismiss the hypothesis that rivers could have been created by a biblical flood. For hypothesis, read process. In other cases, these hypotheses may be used to design geomorphological simulations of one of two broad forms: (1) hardware models, in which the hypotheses are converted into scaled-down physical representations of geomorphological systems, such as studying river processes in a laboratory flume; or (2) mathematical models, including statistical models, in which the hypotheses are treated as processes and represented or inferred as sets of equations which are then either analysed mathematically, solved numerically and/ or applied to particular data.

Behind both of these approaches to simulation is an important recognition. It is rare for an individual hypothesis to be sufficient to address a geomorphological question. For instance, in a landscape evolution model there will be a suite of processes that need to be represented. Each may be based upon mass conservation and momentum conservation rules, but manifest in different forms, according to the process being considered. Further, these processes may interact with each other, leading to feedbacks that may be negative (cancelling) or positive (reinforcing): thus we need to think of geomorphological *systems*. Hardware models recognize these interactions by aiming to study whole landforms, albeit where space and time are scaled to allow laboratory experimentation, with each model run allowing testing of a particular hypothesis (e.g. changing sediment feed rate in a laboratory river model). Simulation models more generally, whether of the hardware or mathematical–numerical kind, recognize the

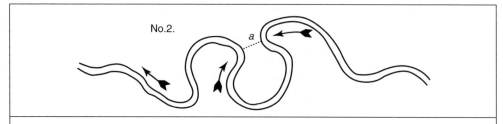

When the tortuous flexures of a river are extremely great, the aberration from the direct line of descent is often restored by the river cutting through the isthmus which separates two neighbouring curves. Thus, in the annexed diagram, the extreme sinuosity of the river has caused it to return for a brief space in a contrary direction to its main course, so that a peninsula is formed, and the isthmus (at a) is consumed on both sides by currents flowing in opposite directions. In this case an island is soon formed, on either side of which a portion of the stream usually remains. These windings occur not only in the channels of rivers flowing through flat alluvial plains, but large valleys also are excavated to a great depth through solid rocks in this serpentine form. In the valley of the Moselle, between Berncastle and Roarn, which is sunk to a depth of from six to eight hundred feet through an elevated platform of transition rocks, the curves are so considerable that the river returns, after a course of seventeen miles in one instance, and nearly as much in two others, to within a distance of a few hundred yards of the spot it passed before. The valley of the Meuse, near Givet, and many others in different countries, offer similar windings. Mr. Scrope has remarked, that these tortuous flexures are decisively opposed to the hypothesis, that any violent and transient rush of water suddenly swept out such valleys; for great floods would produce straight channels in the direction of the current, not sinuous excavations, wherein rivers flow back again in an opposite direction to their general line of descent.

Figure 9.1   A simple conceptual geomorphological model from Lyell (1831: 170–1)

importance and scope of interactions by coupling the actions of different processes together in the form of a system.

Although systems theory was formalized in the 20th century by von Bertalanffy (e.g. 1950, 1968), it is probably fair to say that most earth scientists have always thought in systems terms, even though they did not always use systems terminology (for a good discussion, see Thorn, 1988: 164–92). For example, in Hutton's theory of the rock cycle, from his *Theory of the Earth* (Hutton, 1788, 1795), the planet Earth is in effect being described as a system, with different system components, materials and processes, and through which matter is transported and recycled. Later workers formalized adoption of system principles in geomorphology, such that it is now commonplace (e.g. Strahler, 1950, 1952; Chorley, 1962; Leopold et al., 1964; Gregory and Walling, 1973).

A system can be imagined as having the following properties: (1) objects (e.g. a grain of sediment); (2) processes that act on objects (e.g. momentum transfer whether from a fluid or other grains, that makes the grain move) and which connect objects together, and often specified in the form of rules; (iii) boundaries, often introduced to make the modelling problem tractable (e.g. defining the spatial extent of the

deposit over which sediment movement will be simulated); (4) boundary conditions, necessary to recognize that when boundaries are set, additional or auxiliary information is required (e.g. the sediment feed rate); and (5) exogenous drivers that cause change in the boundary conditions (e.g. a change in sediment feed rate). In simulation models, this kind of systems analysis – that is, conceptualizing how the real system may be broken down into these different components – is critical to understanding the kinds of modelling approaches required. However, it also reminds us that a critical element of both mathematical–numerical simulation models and hardware models is the specification of the system, in which decisions have to be taken over what kinds of objects to include (e.g. individual particles, or volumes of particles), what processes to include and how to couple them (e.g. should river bank erosion be modelled, and how, or will it be treated as a boundary condition) and what the boundaries are for the model. The latter may be geographical (such as defining the spatial extent of the model and its associated spatial resolution) or temporal (such as defining the time period the model will represent and its temporal resolution). As the spatial extent and time period are chosen, so important decisions arise as to how to represent processes that often reflect Schumm and Lichty's

(1965) classification of how variables switch from being dependent, to independent, to irrelevant as spatial and temporal scales are changed. Choosing what to model, how to model it and, above all, what system to work with, are important stages in model closure, and taken together, these mean that most models are not 'general' but developed to be relevant to particular places, over certain spatial scales and for particular time periods. This is a theme that we will return to when we think critically about the use of modelling in geomorphology.

TYPES OF MODEL IN GEOMORPHOLOGY

A survey of text books and review papers shows the multitude of hardware and simulation models used throughout geomorphology (e.g. Chorley and Kennedy, 1971; Carson and Kirkby, 1972; Young, 1972; Haines-Young and Petch, 1986; Thorn, 1988; Scheidegger, 1991; Selby, 1993; Kirkby, 1989, 1996 and 2000; Rhoads and Thorn, 1996; Dietrich et al., 2003; Martin and Church, 2004; Willgoose, 2005). Such reviews together indicate that some model types are dependent upon others, that is, some types of model are necessary components of other types of model. With this in mind, Figure 9.2 presents a model typology, with the main types of model used in geomorphology arranged as a hierarchy, each level being itself divided between two main model types. The reader should refer to this typology as each model type is discussed.

Before discussing the model types in detail, it should be noted that the models in the upper levels of the hierarchy are considered to be more basic than those at the lower. In this respect, by terming a model 'basic', we do not imply it is crude or simplistic; rather, we allude to the sequences of thought and structuring that go into the model's construction. For example, we assume that a researcher generally devises a qualitative model of a system before a quantitative one. Similarly, although a numerical computer model may appear to be more sophisticated than its qualitative counterpart, it is also less complete, as some of the qualitative model's properties may not be expressible as equations, or only so after much simplification and loss of their richness of detail.

Theory, data and models

The division between the two topmost model types – data models and theoretical models – derives from a particular branch of the philosophy of science and encompassed within the 'model–theoretic view' (MTV) (see, for example, Giere,

1988, 1999; Baker, 1996). Odoni and Lane (2010) refine this concept to identify two sorts of models, or rather modelling approaches: (1) model–theoretic and (2) data–theoretic. Following the general MTV split between model types, 'model–theoretic' approaches to modelling place theory central to model development: a model's perceptualization and conceptualization is driven via *a priori* theoretical arguments. Data–theoretic approaches to models, by contrast, begin with the aim of presenting data but implicitly or explicitly accept that some theory has to be applied in forming any data model (e.g. Baker, 1996). The distinction between model–theoretic and data–theoretic therefore provides an important reminder that data do not exist without theories, such as those governing sampling, observation, processing, analysis and presentation (Baker, 1996). Odoni and Lane (2010) argue that in the context of over-arching theory, field-based disciplines like hydrology and geomorphology require continually the interface of models and data. For example, if a flow hydrograph taken during a large flood includes discharges that occurred when the river was spilling well over its banks the discharges may have been inferred from the gauging station's rating curve. In turn, the rating curve will have been inferred from sampled readings taken at the station, supported by extrapolation to the more difficult to measure higher flow estimates. There is therefore a complex mixture of data and theory in the flow hydrograph, and this mixture is typically encountered throughout geomorphology. More specifically, it means that those who work with data have no choice but to use models and those who work with models have no choice but to use data. Often we portray data as being the ultimate arbiter of model performance, failing to recognize the extent to which data themselves are underpinned by particular sets of theories and assumptions and necessitating a much more complex understanding of the relationship between models and data (Lane, 2010).

Hardware models

Continuing with Figure 9.2, the next broad distinction in modelling terms is between hardware and simulation models (also sometimes called 'software models'). Hardware models (e.g. Haines-Young and Petch, 1986; Knighton, 1998) are based upon hard, physical parts as opposed to abstractions, which simulation models rely on. Hardware models are also usually representations of systems scaled simultaneously in space and time, simulating a system's behaviour through the use of real objects and materials (e.g. Haines-Young and Petch, 1986). Figure 9.3 shows a

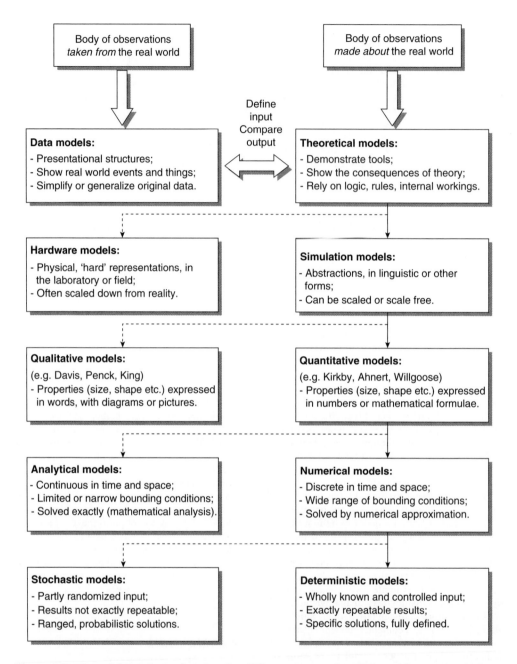

Figure 9.2   A model typology, showing the different types of model used in geomorphology, presented as a hierarchical structure

hardware model of a meandering river channel and patterns of erosion and sedimentation such a model can simulate; Figure 9.4 by contrast shows of a hardware model applied to problems over much grander scales of time and space, in this instance demonstrating sedimentation patterns over a fold–thrust belt under different conditions of sediment supply.

The critical issue associated with hardware models relates to the scaling required to allow the model to be applied over the spatial and temporal scales of interest to the modeller. Scaling can be an issue because properties of the behaviour of materials do change as they are scaled: for instance, when simulating a river in the laboratory, the grain size is commonly reduced but, as the

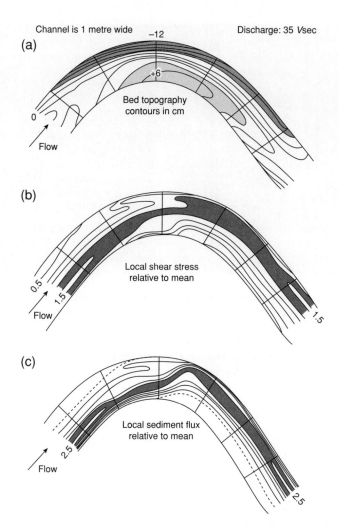

Figure 9.3   An experimental meandering channel: (a) bed topography in a laboratory meander bend; (b) distribution of relative shear stress in the meander bend in (a); region where tau ≥ 1.5 tau_bar is shaded, contour interval 0.5 with supplementary contour at 1.25; (c) distribution of relative sediment transport in the meander bend in (a); region with 2.0 × *Qs* ≤ *Qs* ≤ 2.5 × *Qs* is shaded; contour interval as in (b). (After Hooke (1975), reprinted with permission of the University of Chicago Press)

grain size is reduced so the importance of turbulence in maintaining suspension changes, and the turbulent properties of the flow may not scale in the same proportion to the sediment. Similarly, at very small grain sizes, new processes may be introduced such as cohesion. The hardware modeller normally considers this problem through two approaches. The first minimizes the amount of scaling required by making the facility as large as possible as compared with the spatial scale of the river system. In some situations, it may be possible to avoid having to scale at all such as in

Bagnold's use of a wind tunnel is his study of the saltation of sand and the formation of desert dunes (Bagnold, 1941). The second identifies dimensionless descriptors of key processes in the real system (e.g. in a river flow, the Froude number, the flow Reynolds number, the particle Reynolds number) and then makes sure that the hardware model maintains the most important dimensionless descriptors. The simplest descriptors secure geometrical similarity (e.g. if a square centimetre in the model represents a 10 × 10 m square at the real scale, then a 10 × 10 km catchment can be

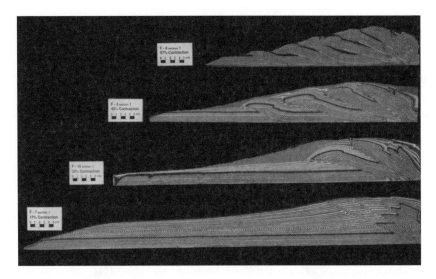

Figure 9.4   Sedimentation patterns generated in a series of sandbox experiments, devised to demonstrate evolution of a fold-thrust belt and its associated sedimentation patterns under different conditions of sediment supply (from Storti and McClay, 1995, and also discussed generically in Beaumont et al., 2000). Initial conditions are the same throughout; no sand is added in the top panel experiment, whereas increased amounts are added in the lower three, the bottom panel showing the effects of the highest amount of added sand

represented by a 10 × 10 m model). With kinematic or dynamic similarity, the scaling seeks to reproduce known dynamical quantities.

Even where sufficient scaling is achieved, and as with simulation models, hardware models may not include all processes that could matter in a system. For example, Hancock and Willgoose (2001) report impressive simulations of fluvial network evolution, but their model does not include a means to simulate creep. This is not a problem if creep is unimportant in the real system, but means that the generalization of conclusions from the model must be undertaken with caution when being considered for situations where creep is important. By contrast, Roering et al. (2001) agitated the bed of their model to include the grain settlement processes characteristic of creep-type behaviour, making their model appear to be the more complete of the two. However, this means of representing creep itself needs to be assessed. Both models avoid other processes that may be important, for example representing solution weathering or attrition of fluvially transported sediment. This does not mean that the models are wrong, just that their relevance is restricted to particular conditions. They are easy to criticize unfairly because, as with all models, they are not complete. A particular reason to be cautious regarding such criticism is that models are often

designed to address questions at particular scales. A model's assumptions may be quite acceptable in the context of the purpose of the research for which it is being used not least because the purpose of a model inevitably frames the scales over which the model is to be applied, and so changes the importance of the variables to be included (Schumm and Lichty, 1965). A particular challenge in modelling arises when information is transferred across scales, in space and time (e.g. Lane and Richards, 1997), often necessitating either explicit capturing of the full scale range that matters or inclusion if additional models to represent processes that impact on the model but which operate at scales not explicitly included.

Simulation models

In our typology (Figure 9.2), simulation models are divided between qualitative and quantitative models. Although qualitative models underpin hardware models, we introduce them specifically under the label 'simulation' because they share a key quality – abstraction – and qualitative models can be used very powerfully to *simulate* system behaviour. A good example of this is provided in Phillips (1995) who shows how the analysis of the interactions between objects in a system can be analysed to search for the tendency for systems to display deterministic chaos.

Qualitative models

Qualitative models are ubiquitous in geomorphology, and are usually expressed diagrammatically. They also have a long history of use in the science, Davis's theory of the geographical cycle, which was as much a simulation model as a theory, being a good example in this respect (e.g. Young, 1972; Thorn, 1988). Qualitative models may appear to be little more than 'conceptual models', but they differ because they allow an element of prediction: if **A** happens, then **B** may follow. Such models can then assist understanding of a problem in its early stages, or in illustrating hypotheses about phenomena yet to be measured. A common difficulty, however, will be in understanding more deeply *why* the predictions are justified, and hence to be trusted (or not). In particular, the response of **B** (e.g. increasing or decreasing) may depend on the value of **A** or **B** as a result of non-linear

response. Table 9.1 shows a highly simplified model, lacking in treatment of feedbacks, which aims to predict the equilibrium response of river channel morphology to changes in water and sediment discharge. A discussion of the problems of equilibrium treatments is beyond the scope of this chapter, but what Table 9.1 shows is that the river response is sensitive to how changes in the water and sediment discharges relate to each other, as well as dependent upon the properties of the channel being predicted, which themselves will feed back on one another. Thus, most qualitative models are based upon the treatment of interactions within the system, endogenous feedbacks, as well as exogenous drivers, where the distinction between endogenous and exogenous is a function of the scale of consideration: a process that is exogenous at one scale may need to be represented as endogenous if either the spatial or temporal scale is increased. This is why the thinking contained in Schumm and Lichty (1965) is so important to model building in geomorphology. Figure 9.5 shows an example of qualitative model set up to include feedbacks, here for the

Table 9.1 Response of channel variables (e.g. slope _s_, grain size _D_, depth _d_ and width _w_), to changes in water (_Qw_) or sediment (_Qs_) discharge (modified from Schumm, 1969)

Driver		Adjustment	Example of change
Qs., Qw++	→	s−, D+, d+, w++	Long-term effect of urbanization leading to an increased frequency and magnitude of runoff, generally leading to channel erosion.
Qs., Qw+	→	s−, D+, d+, w+	Initial response to increase in catchment vegetation cover through afforestation. Land preparation generally increases water discharge but may either increase or decrease sediment delivery according to nature of management. May lead to widening of channel as mean discharge and/or frequency of bankfull channel increases. No necessary increase in sediment delivery, which is why only a small increase in _Qw_ may increase _w_.
Qs−−, Qw−	→	s−, D+, d+, w*	Longer-term response to increase in catchment vegetation is a net negative effect on water balance and a major reduction in sediment delivery. Effect on width determined by the relative effects of reduction in sediment discharge and water discharge.
Qs., Qw+	→	s−, D+, d+, w+	Increase in runoff due to more frequent extreme events. If the system is supply limited, the amount of sediment delivered does not change excessively. In this case, the increase in _Qw_ will dominate and both _d_ and _w_ will increase.
Qs+ Qw+	→	s−, D*, d−, w*	Increase in runoff due to more frequent extreme events. Overland flow increases erosion of surface sediment leading to a rapid increase in sediment delivery above a critical surface erosion threshold). Thus, system is transport limited and an increase in _Qs_ will dominate over the increase in _Qw_. Generally, _d_ will decrease. _w_ may increase or decrease depending upon the nature of the river bank. Feedback effects of _D_ and _s_ may make both of these responses difficult to predict. The response of _D_ is also difficult to predict as it depends on the type of sediment delivered to the reach of river under consideration.

+ indicates increase, ++ a strong increase, − a decrease, −− a strong decrease, . a negligible change and * an unpredictable response.

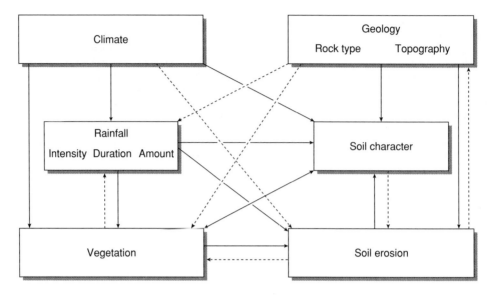

Figure 9.5   Model of the main factors affecting soil formation and erosion, presented as an influence diagram, indicating stronger and weaker factor influences (solid and hatched arrows respectively) (Selby, 1993, adapted from Morisawa, 1968). Note also how the model indicates feedback effects between model components

relationships between soil formation and erosion. Such models are purely schematic, but sometimes, by using arrows of different thickness or style, major or minor influences can be distinguished from each other. Likewise, by attributing '+' or '−' signs we see whether influence and response (i.e. cause and effect) are expected to be positively or negatively correlated. Taken as a whole, feedbacks can be seen: topography, via slope, may influence soil erosion; erosion will lead, via elevation changes, to changes in topography. Phillips (1995) analyses models of this kind to explore how systems are likely to behave.

Quantitative models

Model quantification is generally held to be useful for the following main reasons (e.g. Thorn, 1988; Scheidegger, 1991; Kirkby, 1996): (1) it regulates model output, ensuring that model predictions are correctly derived from the model's structure, theory and rules; (2) it usually involves making models more physically based, thus potentially improving their trustworthiness, and their applicability to a wider range of conditions and locations; (3) it allows the testability of predictions against data, and clarifies whether the model passes or fails any such test; and (4) it allows model output to be represented in many different ways, thereby

widening the range of observations that can be used to compare model with reality.

An important task in quantifying a model is to find a way to express model relationships mathematically. In geomorphology, since there are so many objects and processes that could be modelled but which, for practical reasons, cannot be measured or wholly known, system properties tend to be simplified and/or aggregated in space and time, lumped, with the assumption that the lumped components will behave consistently within each space or time unit, for all space and time units. This is why gemorphological models may be physically based but also require semi-empirical equations (e.g. Thorn, 1988; Beven, 1996), sometimes called auxiliary relationships, to represent those processes not modelled explicitly. Sometimes, the relationships introduced are themselves models (e.g. use of the Manning equation to parameterize energy losses in hydrodynamic models) and often the relationships require data and/or parameters of uncertain value (e.g. Manning's n). The introduction of parameters is at once both a challenge and an opportunity. The challenge arises from two aspects. First, some parameters are not readily determined from measurements. Second, the meaning of parameters measured in the field is not always the same as in models. For example (Lane, 2005), in one-dimensional hydraulic models of river flow, Manning's *n* is used to scale energy losses, associated with friction, secondary circulation and turbulence at the spatial scale at which the model is applied. Field estimates of Manning's *n* may be estimated from grain size, from inverting the Manning's equation for velocity by measuring section-averaged velocity, wetted perimeter and channel slope or from pictures. These methods of estimating *n* from field data each make different assumptions about what is controlling *n* and are of uncertain resemblance to the meaning of *n* in hydrodynamic models. Indeed, the value of *n* required in a one-dimensional model will be different from that required in a two-dimensional model, because the process content of the two models is different, regardless of what *n* value might be measured in the field.

The opportunity arises from the challenge. Parameters represent one of the ways in which models can be forced to agree with measurements, a process that is generally referred to as calibration. This provides the modeller with an opportunity to compensate for the inadequacies in their model by 'forcing empirical adequacy' (Oreskes et al., 1994), something that can risk the modeller getting the 'right results for the wrong reasons' (Beven, 1989). This kind of approach gives primacy to data, but we often overlook the fact that many of our data provide incomplete or incorrect representations of the system being modelled, risking that the model is getting the 'wrong results for the wrong reasons' (Lane and Richards, 2001; Lane, 2010). A critical additional issue arises here: if models need to be made to work in particular places through calibration, the modeller has to ask whether or not the parameterizations developed for particular places also transfer. In other words, the supposed generality of a quantitative simulation model may become restricted emphasizing that data for boundary conditions may not be enough: data commensurate with model predictions may be required to assist with parameterization. This is why model development in geomorphology has been highly constrained by innovation in measurements: recent abilities to estimate long-term erosion rates (e.g. Bishop, 2007), for instance, have been central to revitalizing our focus on long-term landform evolution models.

One of the interesting elements of modelling practice in geomorphology is that the researcher may be more interested in situations where the model does not work rather than in situations where it does. This is in marked contrast to applications by policy makers or consultants for whom the logic of a model that is shown not to work is quite illogical to the delivery of their professional responsibilities (Lane et al., 2006). However, for the researcher, model failings provide an opportunity to develop a model. Such development has to proceed cautiously, as all too often the temptation is to improve a model by making it more complex. To counter this, Kirkby (1990, 1996) identifies four aspects of modelling to concentrate upon so as to improve their applicability and predictive power: (1) they should be as physically based as possible, so as to make them more generic, and easier to link to other models based on the same principles; (2) they should be as simple as possible, thus making them easier both to build and to understand, and increasing the likelihood that they are parsimonious with the data needed to run them and to parameterize them; (3) they should represent the richness of model behaviour through devising expressive formulations with few factors, rather than through many different sub-processes and rules; and (4) they should allow scaling up or down as broadly as possible, without the necessity either to change the model's structure or to recalibrate it by adjusting parameter values.

Analytical models

Mathematical formulation drives quantification and how it is achieved. Analytical models are interesting in that they are purely mathematical and approach geomorphological questions by solving equations exactly, with minimal or even zero data assumptions, using mathematical analysis. An analytical model is therefore the most tightly regulated of all the forms of theoretical model in the hierarchy, and generates exact solutions directly attributable to the modelled process relationships. Obtaining such solutions, however, is only possible after great simplification of both the model's components and the possible boundary and driving conditions. This is a strong limitation, and accordingly few such models have been devised in geomorphology, although some mathematicians have recently considered processes like river braiding using this approach (e.g. Hall, 2005, 2006). A good example of a geomorphologically led application is provided by Kirkby (1971) for slope profile development. Each slope profile was considered to be bounded at the slope crest, a drainage divide, and the toe, where slope material was removed according to specified base level conditions. The hillslope was assumed to be composed of a homogeneous, infinitely divisible, non-cohesive regolith, and mass conservation at any point was applied in the form of a partial differential equation devised to express mass continuity:

$$\frac{\partial z}{\partial t} + k \frac{\partial q}{\partial x} = 0 \qquad (9.1)$$

where z is the elevation, t is the time, x is the horizontal distance from the divide, k is a constant, and q is the local rate of mass flux resulting from modelled transport processes. Erosion or deposition was therefore directly related to the net mass flux at any point, causing z to increase or reduce as appropriate. The mass flux term, q, was then assumed to include quantities moved according to various different transport laws, specified by their capacity, C. Kirkby proposed two basic forms of the law, one for threshold controlled processes, such as shallow landslides, and another for slow mass movements (mainly creep), surface wash and stream transport, given by:

$$C = f(a)\left(\frac{-\partial z}{\partial x}\right)^n \tag{9.2}$$

where: $f(a)$ is a function of the upslope drainage area and increases with distance from the divide, and n is a positive number, assumed to be constant and spatially uniform. Guided by observational data Kirkby recast equation (9.2) to represent water-driven processes using:

$$C \propto a^m (\text{slope})^n \tag{9.3}$$

where m and n can be estimated from field observations. After Culling (1963, 1965) and Kirkby (1967), creep was then implemented by recasting equation (9.2) as:

$$C \propto \left(-\frac{\partial z}{\partial x}\right) \tag{9.4}$$

By solving equation (9.1) for different process combinations and uplift, the latter represented through a base-level condition, the model simulated a range of equilibrium and characteristic declining forms. The characteristic profile predicted under dominant creep processes confirmed earlier qualitative predictions made by Davis (1892) and Gilbert (1909). Similarly, the model predicted that river long profiles would be concave. Kirkby's analytical model therefore clearly demonstrated that, for a particular level of detail, simple, generalized equations such as equations (9.2), (9.3) and (9.4) can begin to provide both adequate explanation of the main causes of slope development and a means of predicting their likely form. Extending this approach, Luke (1972, 1974) applied similar equations to the evolution of two-dimensional forms. Similarly, Smith and Bretherton (1972),

simulated channel formation, predicting that channels should develop if:

$$F < q \frac{\partial F}{\partial q} \tag{9.5}$$

where F is a generic sediment transport function and q is the discharge of water per unit width. Depending on the form of transport law used to represent F, this result has been shown to hold under some conditions (Dietrich and Dunne, 1993). The models by Kirkby, Smith et al. and Luke are highly simplified and make critical assumptions regarding boundary conditions and exogenous drivers (e.g. constant runoff, no variation in bedrock or sediment, and so on). To overcome these limitations, a different approach is required, and this is achieved by the next major model type.

Numerical models

In a numerical model, the partial differential equations used to represent the process relationships are solved by numerical approximation rather than exact mathematical analysis. As numerical solution is complex and requires many calculations, such models have to be run on computers. Two solution methods, 'finite element' and 'finite difference', are commonly used in geomorphological models to achieve this. In general terms, finite element methods are considered generally to be the more mathematically correct and are more easily applied to irregular grids, whereas finite difference methods are more easily programmed and usually faster to run, but have often to be limited to orthogonal or regular grids.

A key feature of numerical models is that the spatial domain is divided into discrete units, each rather like an individual sub-model in its own right, and time is divided into small increments, called 'time steps', a whole series of them comprising the complete simulation period. Thus, in a one-dimensional slope profile model, during each time step, each hillslope segment might receive mass, at a certain rate, from its immediate upslope neighbour; it may also pass mass downslope, also at a certain rate, to its immediate downslope neighbour. At the end of the time step, an accounting of the mass balance is made for every segment in the profile, all the segment elevations are updated, and the simulation continues to the next time step, and so on. The equivalent procedure applies to two-dimensional models. Note also here that the mass transfer is usually accounted for by considering the detailed calculations for each cell in relation to all of its neighbours, a scheme we term here 'nearest-neighbour' transfer. By contrast with this scheme, numerical models may also allow for direct removal of material from the spatial domain (e.g. Ahnert, 1976, to simulate solution weathering), or dispersal of material over many cells at once, such as would be required, for example, to simulate shallow landslides (e.g. Tucker and Slingerland, 1997), a method we term 'extended dispersal'. A feature of extended dispersal is that it is often threshold related and used to simulate episodic processes, whereas nearest-neighbour implementation is most commonly used for continuous processes, such as creep.

Numerical approximation and discretization may introduce numerical artefacts into the solution, biasing the output in certain directions (e.g. Howard, 1997; Braun and Sambridge, 1997). Thus, numerical modellers have to give substantial attention to whether or not the behaviour of their model is a real behaviour of the system being considered or a product of the way the equations are being solved. For this reason, validating a numerical model may not be enough and standards in other areas of science where numerical modelling is applied recognize that validation should only be attempted (indeed, even may be optional) once a model is fully verified (Lane and Richards, 2001). Verification involves a range of activities, such as looking at the sensitivity of model predictions to changes in the resolution of discretization (e.g. Hardy et al., 2003) or to parameterization as a means of checking for sensible model behaviour (Lane et al., 1994). The recognition of the importance of verification in geomorphological modelling has resulted in the development of standards for certain types of geomorphological models, such as for the application of fluid dynamics methods to rivers research (e.g. Lane et al., 2005).

Both spatial discretization and temporal discretization are directly linked and given their importance they need special mention. Imagine a unit of the landscape covering 100 m × 100 m. If material is being moved over this landscape at the rate of 200 m per year, and the model is being run over long timescales, such as thousands of years, then for the simplest discretization, the model time step has to be 0.5 year or shorter, if each 100 m × 100 m unit is to influence the solution. Two broad solutions exist to deal with this issue. The first allows for a time step that scales with process rate (e.g. a Courant Number criterion; Bates and Lane, 2000). The second uses higher-order solutions in which the number of cells used to evaluate the discretization, and hence influence a particular cell, is increased. These higher-order solutions have benefits, but are normally at the expense of greater computational time.

In geomorphological terms, the critical role played by numerical models is the possibility for introducing dynamic feedbacks between processes with more than a single target object or state variable. Thus, using landscape evolution modelling as an example, whilst Kirkby's (1971) analysis had, in effect, a single state variable (homogeneous regolith) there may also be state variables for bedrock, vegetation, water, different classes of sediment and so on. Likewise, climate may also be varied from warm to cold, or wet to dry, and back again, and boundary conditions can include not only uplift, but tilting and warping, isostatic rebound, wedge elevation or subduction, and other variants (e.g. Ahnert, 1987; Tucker and Slingerland, 1997; Bogaart et al., 2003; Fischer et al., 2004). Similar flexibility is afforded in all types of numerical model in this respect, whatever the system being modelled. The disadvantage of such sophistication is that it greatly increases the computational burden, so that complex models may take many hours, days or weeks to complete a single simulation. This is an especially difficult problem if the model has to be calibrated rigorously before it can be used for the intended research purpose, as the number of simulations required to conduct the calibration increases exponentially with the number of parameters in the model. Various techniques are available to reduce the burden and Beven and co-workers (e.g. Binley et al., 1991) in particular have proposed a methodology based on Monte Carlo sampling, developed initially for hydrological models, that has potential application to most types of numerical model. Factor space and uncertainty exploration can also be made computationally much less burdensome through the use of 'metamodels' and 'emulators', something which we also discuss briefly below.

Deterministic models versus stochastic models

In Figure 9.2, we make one final distinction between models: deterministic versus stochastic. This is a subtle distinction and relates to a continuum between those models where it is assumed that a particular input always produces the same output (deterministic), and those models that have a randomizing element (such as a random-number generator) in their calculating procedures, or in which certain variables are varied randomly (within a prescribed range) during the course of the simulation (stochastic). The reason for seeing this as a continuum is that the extent to which stochasticity is introduced does vary. For instance, one of the responses to the problem of parameter determination is to see a parameter value as stochastic and to sample randomly from an *a priori* distribution of possible parameter values.

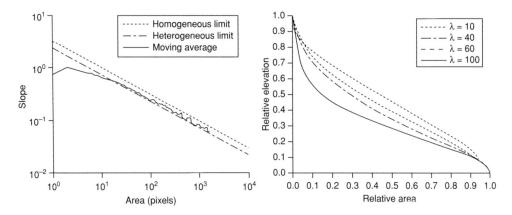

Figure 9.6   The effect of spatial heterogeneity in the erosivity of sediment on the evolution of a modelled landscape, evidenced here by differences in the derived area–slope relationship and hypsometric curves. (After Moglen and Bras, 1995)

Where there is no *a priori* distribution, then the analysis may go one step further and sample from all possible parameter values, conditioning the possible parameter distribution *a posteriori* by comparison of model predictions with observations (e.g. Binley et al., 1991). This latter approach is crucial in geomorphological modelling, but much under-used, because it provides a means of assigning estimates of uncertainty to model predictions. In practice, the uncertainty estimates assigned are a restricted form of uncertainty, associated with what is allowed to be uncertain, commonly only parameter values (see below). But, in principle, this example provides a means by which model predictions, through stochastic analysis, can be allowed to capture elements of the provisional nature of geomorphological models. The results of stochastic analysis are not always reassuring. For example Moglen and Bras (1995; also see Figure 9.6) applied a stochastic variation in sediment erosivity properties to a landscape evolution model, to test its effects on channel formation, routing and extension. This was a test of the landscape's initial conditions and these conditions were shown to have a major influence upon model predictions, in particular the sinuosity of the channel network and how this was evidenced by other metrics, for example slope–area relationship and the hypsometric distribution.

BEYOND THE PRIMARY MODEL: DERIVED MODELS, METAMODELS AND EMULATORS

We now digress from the model typology to consider different types of models and modelling,

beginning with derived models. A growing difficulty that has accompanied the increasing sophistication and functionality of numerical models is how to understand what drives the output. This somewhat counter-intuitive problem occurs because the output generated by a complex model is *emergent behaviour* of a complex (modelled) system. Since the model will have been devised to try to improve our understanding of that system (or at least of its real counterpart), then the conclusion that the cause of the emergent behaviour in the model is still obscure may well defeat the aim the modelling was intended to serve. A further problem is that the number of input variables also often increases with the model's sophistication and functionality. This in turn increases the modelling uncertainties and the computational burden, sometimes enormously, because there are so many factor value combinations that might be tested and the model takes a finite time (hours or longer) to complete a single simulation. [The run size required for a fully factored numerical experiment of k factors (variables) tested at N levels each (values) is N^k simulations. Thus, a model with 20 uncertain factors to be tested at only three levels would require about 3.5×10^9 simulations, a huge burden even if each simulation only takes 1 s.] In these circumstances, modest simulation run sizes may be unfeasible even with many powerful computers available for the task.

These problems – understanding what drives the output and dealing with the computational burden needed to explore the model uncertainties – can be approached and their effects lessened through the use of derived or secondary models, generically called 'metamodels', or, in their more specialist forms, 'emulators' (e.g. Law and Kelton, 1991; Kleijnen et al., 2005; Kennedy, et al. 2006). For simplicity here, a metamodel (or emulator) can be understood as a 'model of the model'. Although hardly used in the geosciences to date, they have been widely applied for 30 or more years in other disciplines, particularly in engineering and operations research (e.g. Barton, 1992; Kleijnen et al., 2005). (Note that Challenor et al., 2006, use an emulator approach in relation to understanding climate change, and Odoni, 2007, derives metamodels for a range of output metrics in a landscape evolution modelling study.) The guiding concept is to treat the primary model (the 'simulator') as a complicated mathematical function, whose variables are assumed to drive the simulator's output according to potentially discernible patterns. To help to identify these patterns, it is essential that the simulation run is conducted as a designed experiment, sampling across the reaches of the factor space using structured or semi-randomized, Latin–hypercube type designs (e.g. Sacks et al., 1989; Law and Kelton, 1991; Wu and Hamada, 2000; Sanchez, 2005). The metamodel is then derived by statistical inference, such as through regression analysis where the dependent variable is a metric derived from the simulator's output, and the explanatory variables are the factors varied in the simulations. This can be represented generically by:

$$Y_X = G(X) + \varepsilon \qquad (9.6)$$

where Y_X is the output metric, X is the source model's set of input variables (factors) x_1, x_2, x_3 and so on, $G(X)$ is the overall function of those variables (comprising log functions, quadratics, exponentials etc. of the various x_i terms, and also of interaction effect $x_i x_j$ terms as necessary) and ε is the error term.

Note that since the metamodel is derived by statistical modelling, the researcher has the option to drop the non-significant and least significant variables from the metamodel. This leads to a much clearer understanding of what drives the simulated output (which of the x_i terms are important and which not), thereby helping to solve the first problem noted above. This dropping of non-significant terms also indicates any scope for increased model parsimony, and hence the potential to develop 'reduced complexity models' in future (see also discussion below). Therefore, a metamodel which has a good fit (high R^2, well-balanced residuals and so on), can be used to predict the simulator's output for the metric in question at any point in the factor space (but always within the ranges applied in the initial numerical experiment). Very large output sets (10^6 points and larger) can be generated very quickly this way, considerably reducing the computational burden. The same applies with

regard to the use of emulators, which are more sophisticated versions of metamodels, but include a separate error function.

Much work is being conducted into how both metamodels and emulators can improve our understanding and use of numerical models more generally (see, in particular, the work of the Modelling Uncertainty group at Manchester, http://mucm.group.shef.ac.uk/index.html). Given their usefulness in other disciplines, it will be interesting to see what take-up there is of these methods in the geosciences, and how our use and understanding of models is thereby affected.

MODELS AT THE EXTREMES

There are two additional types of models that we did not want to classify into the above typology, not only because they do not fit, but because, we argue, they make the mental images – the perceptualizations – of the modellers' dominant in the modelling process potentially dominant over theory. As far as we know, they have never been compared in this way before.

The first extreme type of model is based upon artificial neural networks (ANNs). The general ideas behind ANN models derive from advances in biology and psychology, relating to our understanding of the human brain, and from developments in electronics and computing, relating to the operation of switching and signalling devices. The mimicry of the way that the human brain works is made permissible under the idea that we learn, through observation, how to make predictions, at least over the short term and in the immediate 'locality' of our experience. Thus, ANNs mimic how the brain works by using observations to learn, in a black box, what the system does. There is no *a priori* specification of what the network should look like, as in a regression model. They have the potential, therefore, to capture more of the richness of system behaviour shown in perceptual or qualitative models but not captured by the usual system models using mathematics and numerical computation. They are also, as parallel processing devices, often quick to run once the training has been completed (although large complex networks may be just as slow to run as a conventional numerical model of the same system). The main disadvantages, however, are that model training may take many days or weeks to complete, and this is assuming that suitable training sets are available for the purpose in the first place, which will not always be the case. Also, an ANN model itself cannot be opened up to scrutiny and assessed against whether or not it has learnt the accepted physical principles that we bring to bear on geomorphological questions. However, Licznar and Nearing (2003) compared an ANN model of runoff and soil erosion with a physically based hydrology and soil erosion model and found that the ANN gave substantially better predictions. This conclusion was reached through what is called a 'split test' in modelling: that is, the data used to train the ANN were not the same data used to judge the performance of the ANN.

The second type of model that has emerged over the last decade or so is the 'reduced complexity' model (RCM). The notion of an RCM is a pleonasm: all models have reduced complexity, so linking a model with the reduced complexity label is not really necessary. Brasington and Richards (2007) refer to this as the 'paradox' of RCMs. However, it has come to describe a class of simulation models whose process content is informed as much by the modeller's mental image of the landform being modelled as it is the modeller's mental image of the fundamental processes that matter in geomorphological modelling. Murray and Paola's (1994) seminal model of river braiding illustrates this. Their model shows that: (1) river braiding is a consequence of flow contraction and expansion, coupled to (2) a non-linear sediment transport law, which leads to (3) sediment flux over-reacting to variation in flow strength, and never managing to optimize the bed topography (Paola, 2001). The model reproduces the geometric scaling of planform inundation observed in field braided rivers (Murray and Paola, 1997), a critical emergent element of river braiding. In this sense it provides a fundamental

high level explanation of the braiding process. At the same time, mathematical analysis (Lane, 2006) shows that the model does not conserve flow momentum and that the model has some odd properties (such as greater sediment divergence than flow divergence in response to topographic forcing). The point about RCMs is that this does not matter in the sense that RCMs are 'synthetic' as opposed to 'reductionist' (Paola, 2001). The validity of the model rests in its ability to show the emergent properties of the system and not reproduce the underlying detailed physics of the system, the process content of the model (Murray, 2007). Opening up the model to detailed scrutiny is of little worth, in the same way that opening up an ANN model is largely meaningless. The worth of the model, as perhaps with *any* geomorphological model as applied in a research environment, is more about whether or not it helps us to think differently, to challenge our prior assumptions (Lane, 2010) and hence change our mental image of the landform, than whether or not it is true to our long-held underlying principles or predicts the detailed evolution of the landform being studied.

UNCERTAINTIES AND GEOMORPHOLOGICAL MODELLING

Perhaps one area where geomorphological modellers have been less forthcoming than in other disciplines, notably hydrology, is in relation to uncertainties. Table 9.2 summarizes the types

Table 9.2 Model uncertainties (Modified from Lane, 2003)

Type of uncertainty	Explanation
Closure uncertainties	This relate to uncertainties that arise because certain processes have been included or excluded during the perceptualization and conceptualization stages of model development. A common strategy to deal with this is to include processes, test their effects, and to ignore them or to simplify them if they seem to have only a small effect upon model predictions. The problem with this is that the effect of a process can often change as the effects of other processes (i.e. system state) changes. Thus, it is impossible to be certain that a given process will always be unimportant. Note that problems of closure are not just associated with models, but are an inherent characteristic of all science.
Structural uncertainties	These also arise in the perceptualization and conceptualization stages, and relate to a lack of certainty in exactly how objects and relationships in a model link together. A good example of this is whether or not a component is an active or a passive component of a model. If it is active, it is allowed to evolve as dictated by the model. If it is passive, then it is treated as a fixed boundary condition. A good example is topography in an erosion model: the topography may be taken as given; or, as erosion occurs, so the denudational unloading may be fed back into uplift and hence topography. Which of these two treatments is acceptable depends upon the timescale over which the model is applied and so, as with closure uncertainties, structural uncertainties limit a model's applicability and remind us of the importance of evaluating a model in relation to the use to which it will be put. Any model can be criticized on the grounds of structural uncertainties without a purpose-specific evaluation.

Continued

Table 9.2 Cont'd

Type of uncertainty	Explanation
Solution uncertainties	These uncertainties arise in numerical simulation, because most numerical models are approximate rather than exact solutions of the governing equations. Commonly, numerical solution involves an initial guess, operation of the model upon that initial guess, and its subsequent correction. This continues until operation of the model on the previous corrected guess does not alter by the time of the next guess. This process can result in severe numerical instability in some situations, which is normally easy to detect. However, more subtle consequences, such as numerical diffusion associated with the actual operation of the solver, can be more difficult to detect. Following guidelines in relation to good practice can help in this respect.
Process uncertainties	These arise where there is poor knowledge over the exact form of the process representation used within a model. A good example of this is the treatment of turbulence in models of river flow. Turbulence can have an important effect upon flow processes as it extracts momentum from larger scales of flow and dissipates it at smaller scales. Most river models average turbulence out of the solution, but then have to model the effects of turbulence upon time-averaged flow properties. Turbulence models vary from the simple to the highly and it can be shown that different turbulence treatments are more or less suitable to different model applications. Thus, the process representation in a model, as with structural aspects of a model, needs careful evaluation with reference to the specific application for which the model is being used.
Parameter uncertainties	These arise when the right form of the process representation is used but there are uncertainties over the value of parameters that define relationships within the model. Particular problems can arise when parameters have a poor meaning in relation to measurement. This can arise in two ways. First, some parameters are difficult to measure in the field as they have no simple field equivalent. Second, during model optimization, parameters can acquire values that optimize agreement with measurements, but which are different to the value that they actually take on the basis of field measurements. A good example of this is the bed roughness parameter used in one-dimensional flood routing models. It is common to have to increase this quite significantly at tributary junctions, to values much greater than might be suggested by the shape of the river or the bed grain size. In this case, there is a good justification for it, as one-dimensional models represent not only bed roughness effects but also two- and three-dimensional flow processes and turbulence through the friction equation. Roughness is therefore representing the effects of these other processes in order to achieve what Beven (1989) labels 'the right results but for the wrong reasons'.
Initialisation uncertainties	These are associated with the initial conditions required for the model to operate. They might include the geometry of the problem (e.g. the morphology of the river and floodplain system that is being used to drive the model) or boundary conditions (e.g. the flux of nutrients to a lake in a eutrophication model).
Validation uncertainties	Given the above six uncertainties, a model is unlikely to reproduce reality exactly, and validation is required to assess the extent to which there is a reasonable level of agreement. However, validation data themselves have an uncertainty attached to them because they themselves constitute a data model. This is not simply due to possible measurement error, but also when the nature of model predictions (their spatial and temporal scale, the parameter being predicted) differs from the nature of a measurement. A commonly cited example of this is validation of predictions of soil moisture status in hillslope hydrological models, when point measurements of soil moisture status (in space and time) are used to validate areally integrated predictions. This creates problems for modelling in two senses. First, apparent model error may actually be validation data error. Second, if validation data are then used for model optimisation (and note that data used for model optimization should not then be used for validation), uncertainty will be introduced into model predictions as the data that the model are optimised to may be incorrect. Following Beven (1989) this means that we may get the wrong results for the wrong reasons

of uncertainties that can emerge in geomorphological modelling: not all of these uncertainties apply to all of the types of models above; but they do allow us to make three general comments about uncertainty in geomorphological modelling.

First, as with any area of science, there is a sense in which geomorphological models produce understanding that is provisional because of uncertainties that we cannot avoid. We tend to assume that the role of scientific research is to reduce uncertainty when actually, as Wynne (1992) demonstrates, the role of scientific research is to generate more uncertainty. Indeed, uncertainty may play a very valuable role in preventing our prevailing view of the world from becoming too settled, and so forcing further innovation in geomorphological modelling and hence understanding. As an example, Lane and Richards (1998) argue that three-dimensional models of river flow are required in tributary junctions in order to represent properly the effects of secondary circulation upon flow processes. Bradbrook et al. (2001), having used a three-dimensional model for this purpose, demonstrate the significant new uncertainties that have emerged from: (1) the difficulty of specifying inlet conditions in each of the tributaries in three-dimensions; (2) problems of designing a numerical mesh that provides a stable numerical solution that minimizes numerical diffusion; (3) uncertainties over the performance of a roughness treatment in three dimensions; and (4) problems with finding an appropriate turbulence model. Uncertainties of this kind help innovation in geomorphological research rather than hindering it.

Second, the primary focus of uncertainty in geomorphological modelling tends to be upon parameter uncertainties. Although establishing such uncertainties may be severely demanding in computational terms, they are relatively easy to do, simply by exploring model response to changes in parameter values, something that can be done to varying levels of sophistication (e.g. Binley et al., 1991). Unfortunately, some of these other uncertainties are much harder to handle. Few geomorphologists are formally trained in numerical solution methods, for instance, and hence the range of tools that might be used to assess numerical performance. Identifying closure, structural and process uncertainties requires geomorphological models to be viewed critically, both by the modellers and others. This is not always easy, especially when geomorphological communities emerge locked into particularly fashionable views of what is thought to matter or deemed admissible (Shermann, 1996). Unlike other areas of modelling (e.g. climate modelling; Winsberg, 2003), we know surprisingly little about the ethnographical dimensions of geomorphological practice in general and geomorphological modelling in particular, even though such understanding may be central to rethinking scientific practices themselves.

Third, and perhaps most importantly, many of the uncertainties listed in Table 9.2 will be constructed through the mental images that the geomorphological modeller or community brings to bear upon the models that they develop. It is hard to shape these mental images through simply reading text books or working through equations. They are also shaped through the multitude of ways in which we can experience landforms, whether through remotely sensed imagery or fieldwork. Landforms, whether their geometry, their sedimentological history, their erosional history etc., provide an inordinate amount of detail regarding why and how they form. Letting the landform and its history steer the development of geomorphological models implies a much less formal role for 'data' in geomorphological enquiry, in which the critical role of field experience in shaping model development is recognized explicitly. Chorley (1978) bemoaned that the immediate response of a geomorphologist to the word 'theory' was to reach for their soil auger. Perhaps reaching for a soil auger, in response to theory, is no bad thing.

CONCLUSION: THE SIGNIFICANCE OF MODELS IN GEOMORPHOLOGY

In this chapter we have sought to demonstrate that models underpin all of geomorphological enquiry and have done since well before the arrival of the computer as a simulation tool. In turn, models are underpinned by the ways that we, as geomorphologists, imagine landforms and their associated processes. The significance of models is best realized when they are developed through a close exchange between observation, of both form and process, and theory. This exchange can be bidirectional. Theory and models can be very helpful in the first instance in steering our approach to field data collection: what to look for, what to measure or where to sample. They also generate hypotheses for testing. However, models without observations, whether informal or formal, primary through fieldwork or secondary through tools like remote sensing and archival records, are highly likely to be very poor models indeed. As such, innovations in our ability to measure geomorphological forms and processes, and to reconstruct geomorphological history, are central to informing the ways that geomorphological models will develop in the future.

REFERENCES

Ahnert, F. (1976) Brief description of a comprehensive three-dimensional process–response model of landform development, *Zeitschrift für Geomorphologie* N.F. Supplementband 25: 29–29.

Ahnert, F. (1987) Approaches to dynamic equilibrium in theoretical simulations of slope development, *Earth Surface Processes and Landforms* 12: 3–15.

Bagnold, R.A. (1941) *The Physics of Blown Sand and Desert Dunes*, Methuen, London, p. 265.

Baker, V.R. (1996) Hypotheses and geomorphological reasoning (Chapter 3), in B.L. Rhoads and C.E. Thorn (eds), *The Scientific Nature of Geomorphology*, John Wiley, Chichester.

Barton, R.R. (1992) Metamodels for simulation input–output relations, in J.J. Swain, D. Goldsman, R.C. Crain and J.R. Wilson (eds), *Proceedings of the 1992 Winter Simulation Conference* SCS, San Diego. pp. 289–99.

Bates, P.D. and Lane, S.N. (2000) Hydraulic modelling in hydrology and geomorphology: a review of high resolution approaches, (Chapter 1) in P.D. Bates and S.N. Lane (eds), *High Resolution Flow Modeling in Hydrology and Geomorphology*. John Wiley, Chichester.

Beaumont, C., Kooi, H. and Willett, S. (2000) Coupled tectonic-surface process models with applications to rifted margins and collisional orogens, (Chapter 3), in M.A. Summerfield (ed.), *Geomorphology and Global Tectonics*. John Wiley, Chichester.

Beven, K. (1989) Changing ideas in hydrology – the case of physically-based models, *Journal of Hydrology* 105: 157–72.

Beven, K. (1996) Equifinality and uncertainty in geomorphological modelling. (Chapter 12), in B.L. Rhoads and C.E. Thorn (eds), *The Scientific Nature of Geomorphology*. John Wiley, Chichester.

Binley, A.M., Beven, K.J., Calver, A. and Watts, L.G. (1991) Changing responses in hydrology: assessing the uncertainty in physically-based model predictions, *Water Resources Research* 27: 1253–61.

Bishop, P. (2007) Long-term landscape evolution: linking tectonics and surface processes, *Earth Surface Processes and Landforms* 32: 329–65.

Bogaart, P.W., Tucker, G.E. and de Vries, J.J. (2003) Channel network morphology and sediment dynamics under alternating periglacial and temperate regimes: a numerical simulation study, *Geomorphology* 54: 257–77.

Bradbrook, K.F., Lane, S.N., Richards, K.S., Biron, P.M. and Roy, A.G. (2001) Flow structures and mixing at an asymmetrical open-channel confluence: a numerical study, *ASCE Journal of Hydraulic Engineering* 127: 351–68.

Brasington, J. and Richards, K. (2007) Reduced-complexity, physically-based geomorphological modelling for catchment and river management, *Geomorphology* 90: 171–7.

Braun, J. and Sambridge, M. (1997) Modelling landscape evolution on geological time scales: a new method based on irregular spatial discretization, *Basin Research* 9: 27–52.

Carson, M.A. and Kirkby, M.J. (1972) *Hillslope Form and Process*. Cambridge University Press, Cambridge. p. 476.

Challenor, P.G., Hankin, R.K.S. and Marsh, R. (2006) Towards the probability of rapid climate change, in H.J. Schellnhuber, W. Cramer, N. Nakicenovic, T. Wigley and G. Yohe (eds), *Avoiding Dangerous Climate Change*. Cambridge University Press, Cambridge. pp. 55–63.

Chorley, R.J. (1962) Geomorphology and general systems theory. U.S. Geological Survey Professional Paper 500-B. p. 10.

Chorley, R.J (1967) Models in geomorphology. (Ch. 3), in R.J. Chorley and P. Haggett (eds), *Models in Geography*. Methuen, London. pp. 50–96.

Chorley, R.J. (1978) Bases for theory in geomorphology, in C. Embleton, D. Brunsden and D.K.C. Jones (eds), *Geomorphology: Present Problems and Future Prospects*. Oxford University Press, Oxford.

Chorley, R.J. and Kennedy B.A. (1971) *Physical Geography: A Systems Approach*. Prentice-Hall International, London. p. 370.

Culling, W.E.H. (1963) Soil creep and the development of hillside slopes, *Journal of Geology* 71: 127–61.

Culling, W.E.H. (1965) Theory of erosion on soil-covered slopes, *Journal of Geology* 73: 230–54.

Davis, W.M. (1892) The convex profile of bad-land divides, *Science* 20: 245.

Dietrich, W.E. (1987) Mechanics of flow and sediment transport in river bends, in K.S. Richards (ed.), *River Channels*. IBG Special Publication 18, Blackwell, Oxford. pp. 279–327.

Dietrich, W.E. and Dunne, T. (1993) The channel head. (Chapter 7), in K. Beven and M.J. Kirkby (eds), *Channel Network Hydrology*, John Wiley, Chichester.

Dietrich, W.E., Bellugi, D.G., Sklar, L.S., Stock, J.D. and Heimsath, A.M. (2003) Geomorphic transport laws for predicting landscape form and dynamics, in P.R. Wilcock and R.M. Iverson (eds), *Prediction in Geomorphology*. Geophysical Monograph Series, number 135, American Geophysical Union, Washington D.C. pp. 103–32.

Ferguson, R.I. (1976) Disturbed periodic model for river meanders, *Earth Surface Processes and Landforms* 1: 337–47.

Fischer, K.D., Jahr, T. and Jentzsch, G. (2004) Evolution of the Variscan foreland-basin: modelling the interactions between tectonics and surface processes, *Physics and Chemistry of the Earth* 29: 665–71.

Giere, R.N. (1988) *Explaining Science: A Cognitive Approach*. University of Chicago Press, Chicago. p. 321.

Giere, R.N. (1999) *Science Without Laws*. The University of Chicago Press, Chicago. p. 285.

Gilbert, G.K. (1909) The convexity of hilltops, *Journal of Geology* 17: 344–50.

Gregory, K.J. and Walling, D.E. (1973) *Drainage Basin Form and Process: A Geomorphological Approach*. Edward Arnold, London, p. 456.

Haines-Young, R.H. and Petch, J.R. (1986) *Physical Geography: Its Nature and Methods*. Paul Chapman Publishing, London. p. 230.

Hall, P. (2005) On the non-parallel instability of sediment-carrying channels of slowly varying width, *Journal of Fluid Mechanics* 529: 1–32.

Hall, P. (2006) Nonlinear evolution equations and the braiding of weakly transporting flows over gravel beds, *Studies in Applied Mathematics* 117: 27–69.

Hancock, G. and Willgoose, G. (2001) The interaction between hydrology and geomorphology in a landscape simulator experiment, *Hydrological Processes* 15: 115–33.

Hardy, R.J., Lane, S.N., Ferguson, R.I. and Parsons, D.R. (2003) Assessing the credibility of a series of computational fluid dynamic simulations of open channel flow, *Hydrological Processes* 17: 1539–60.

Hooke, R., and Le, B. (1975) Distribution of sediment and shear stress in a meander bend, *Journal of Geology* 83: 543–65.

Howard, A.D. (1997) Badland morphology and evolution: interpretation using a simulation model, *Earth Surface Processes and Landforms* 22: 211–27.

Hutton, J. (1788) Theory of the Earth; or an investigation of the laws observable in the composition, dissolution, and restoration of land upon the Globe, *Transactions of the Royal Society of Edinburgh* 1: 290–304.

Hutton, J. (1795) *Theory of the Earth*. William Creech, Edinburgh, 2 volumes.

Kennedy, M.C., Anderson C.W., Conti S. and O'Hagan A. (2006) Case studies in Gaussian process modelling of computer codes, *Reliability Engineering and System Safety* 91: 1301–9.

Kirkby, M.J. (1967) Measurement and theory of soil creep, *Journal of Geology* 75: 359–78.

Kirkby, M.J. (1971) Hillslope process-response models based on the continuity equation, (Chapter 2), in C. Embleton and J.T. Coppock (eds), *Slopes - Form and Process*. Special Publication 3. Institute of British Geographers, London.

Kirkby, M.J. (1989) A model to estimate the impact of climatic change on hillslope and regolith form, *Catena* 16: 321–41.

Kirkby, M.J. (1990) The landscape viewed through models. *Zeitschrift fur Geomorphologie* 79, 63–81.

Kirkby, M.J. (1996) A role for theoretical models in geomorphology? (Chapter 10), in B.L. Rhoads and C.E. Thorn (eds), *The Scientific Nature of Geomorphology*. John Wiley and Sons, Chichester.

Kirkby, M.J. (2000) Limits to modelling in the Earth and environmental sciences, (Chapter 15), in S. Openshaw and R.J. Abrahart (eds), *Geocomputation*. Taylor and Francis, London and New York.

Kleijnen, J.P.C., Sanchez, S.M., Lucas, T.W. and Cioppa, T.M. (2005) A user's guide to the brave new world of designing simulation experiments, *INFORMS Journal on Computing* 17: 263–89.

Knighton, D. (1998) *Fluvial Forms and Processes: A New Perspective*. Hodder Headline Group. London. p. 383.

Lane, S.N. (2003) Numerical modelling in physical geography: understanding, explanation and prediction, (Ch. 17), in N.J. Clifford and G. Valentine, *Key Methods in Geography* Sage, London. pp. 263–90.

Lane, S.N. (2005) Roughness: time for a re-evaluation? *Earth Surface Processes and Landforms* 30: 251–3.

Lane, S.N. (2006) Approaching the system-scale understanding of braided river behaviour, in G.H. Sambrook Smith, J.L. Best, C.S. Bristow and G.E. Petts (eds), *Braided Rivers: Process, Deposits, Ecology and Management*. IAS Special Publication 36, Blackwell Publishing.

Lane, S.N. (2010) Making mathematical models perform in geographical spaces, in J. Agnew and D. Livingstone (eds), *Handbook of Geographical Knowledge*. Sage, London.

Lane, S.N. and Richards, K.S. (1997) Linking river channel form and process: time, space and causality revisited, *Earth Surface Processes and Landforms* 22: 249–60.

Lane, S.N. and Richards, K.S. (1998) Two-dimensional modelling of flow processes in a multi-thread channel, *Hydrological Processes* 12: 1279–98.

Lane, S.N. and Richards, K.S. (2001) The 'validation' of hydrodynamic models: some critical perspectives, in P.D. Bates and M.G. Anderson (eds), *Model Validation for Hydrological and Hydraulic Research*. John Wiley and Sons, Chichester. pp. 413–38.

Lane, S.N., Richards, K.S. and Chandler, J.H. (1994) Distributed sensitivity analysis in modelling environmental systems, *Proceedings of the Royal Society, Series A* 447: 49–63.

Lane, S.N., Hardy, R.J, Ferguson, R.I. and Parsons, D.R. (2005) A framework for model verification and validation of CFD schemes in natural open channel flows, in P.D. Bates, S.N. Lane and R.I. Ferguson (eds), *Computational Fluid Dynamics: Applications in Environmental Hydraulics*. Wiley, Chichester. pp. 169–92.

Lane, S.N., Brookes, C.J., Heathwaite, A.L. and Reaney, S.M. (2006) Surveillant science: challenges for the management of rural environments emerging from the new generation diffuse pollution models, *Journal of Agricultural Economics* 57: 239–57.

Langbein, W.B. and Leopold, L.B. (1966) River meanders theory of minimum variance. USGS, Professional paper, 422H. Washington DC, 1966. pp. 1–15.

Law, A.M. and Kelton, W.D. (1991) *Simulation Modeling and Analysis*, 2nd edn. McGraw-Hill, New York. p. 759.

Lawrence, D.S.L. (1996) Physically based modelling and the analysis of landscape development. (Chapter 11), in B.L. Rhoads and C.E. Thorn (eds), *The Scientific Nature of Geomorphology*. John Wiley and Sons, Chichester.

Leopold, L.B. and Wolman, M.G. (1960) River meanders, *Geological Society of America Bulletin* 71: 769–93.

Leopold, L.B., Wolman, M.G. and Miller, J.P. (1964) *Fluvial Processes in Geomorphology*. Freeman, San Francisco, p. 522.

Licznar, P. and Nearing, M.A. (2003) Artificial neural networks of soil erosion and runoff prediction at the plot scale, *Catena* 51: 89–114.

Luke, J.C. (1972) Mathematical models for landform evolution. *Journal of Geophysical Research* 77: 2460–4.

Luke, J.C. (1974) Special solutions for nonlinear erosion problems. *Journal of Geophysical Research* 79: 4035–40.

Lyell, C. (1831) *Principles of Geology,* 1st edn. (Two volumes). John Murray, London.

Martin, Y. and Church, M. (2004) Numerical modelling of landscape evolution: geomorphological perspectives, *Progress in Physical Geography* 28: 317–39.

Moglen, G.E. and Bras, R.L. (1995) The effect of spatial heterogeneities on geomorphic expression in a model of basin development, *Water Resources Research* 31: 2613–23.

Morisawa, M. (1968) *Streams: Their Dynamics and Morphology.* McGraw-Hill, New York. p. 175.

Murray, A.B. (2007) Reducing model complexity for explanation and prediction, *Geomorphology* 90: 178–91.

Murray, A.B and Paola, C. (1994) A cellular model of braided rivers, *Nature* 371: 54–7.

Murray, A.B and Paola, C. (1997) Properties of a cellular braided stream model, *Earth Surface Processes and Landforms* 22: 1001–25.

Odoni, N.A. (2007) Exploring equifinality in a landscape evolution model. Unpublished PhD thesis, School of Geography, University of Southampton, U.K.

Odoni, N.A and Lane, S.N. (2010) Knowledge-theoretic models in hydrology, *Progress in Physical Geography* (in press).

Oreskes, N., Shrader-Frechette, K. and Belitz, K. (1994) Verification, validation and confirmation of numerical models in the earth sciences, *Science* 26: 641–6.

Paola, C. (2001) Modelling stream braiding over a range of scales, in M.P. Mosley (ed.), *Gravel Bed Rivers V.* New Zealand Hydrological Society, New Zealand. pp. 11–46.

Phillips, J.D. (1995) Self-organization and landscape evolution, *Progress in Physical Geography* 19: 309–21.

Playfair, J. (1802) *Illustrations of the Huttonian Theory of the Earth.* Creech, London. p. 528.

Rhoads, B.L. and Thorn, C.E. (1996) Towards a philosophy of geomorphology (Chapter 5), in B.L. Rhoads and C.E. Thorn (eds), *The Scientific Nature of Geomorphology.* John Wiley and Sons, Chichester.

Roering, J.J., Kirchner, J.W., Sklar, L.S. and Dietrich, W.E. (2001) Hillslope evolution by nonlinear creep and landsliding: an experimental study, *Geological Society of America Bulletin* 29: 143–6.

Sacks, J., Welch, W.J., Mitchell, T.J. and Wynn, H.P. (1989) Design and analysis of computer experiments, *Statistical Science* 4: 409–35.

Sanchez, S.M. (2005) Work smarter, not harder: guidelines for designing simulation experiments, in M.E. Kuhl, N.M. Steiger, F.B. Armstrong and J.A. Joines (eds), *Proceedings of the 2005 Winter Simulation Conference.* pp. 69–82.

Scheidegger, A.E. (1991) *Theoretical Geomorphology,* 3rd edn. Springer Verlag, Berlin, p. 434.

Schumm, S.A. (1969) River metamorphosis, *ASCE Journal of the Hydraulics Division* 95: 255–73.

Schumm, S.A. and Lichty, R.W. (1965) Time, Space, and causality in geomorphology. *American Journal of Science* 263: 110–9.

Selby, M.J. (1993) *Hillslope Materials and Processes,* 2nd edn. Oxford University Press, Oxford, p. 451.

Shermann, D.J. (1996) Fashion in geomorphology, (Chapter 4), in B.L. Rhoads and C.E. Thorn (eds), *The Scientific Nature of Geomorphology.* John Wiley and Sons, Chichester.

Smith, T.R and Bretherton, F.P. (1972) Stability and the conservation of mass in drainage basin evolution, *Water Resources Research* 8: 1506–29.

Speight, J.G. (1965) Meander spectra of the Angabunga River, *Journal of Hydrology* 3: 1–15.

Storti, F. and McClay, K. (1995) Influence of syntectonic sedimentation on thrust wedges in analogue models, *Geology* 23: 999–1002.

Strahler, A.N. (1950) Equilibrium theory of erosional slopes approached by frequency distribution analyses, *American Journal of Science* 248: 673–96 and 800–14.

Strahler, A.N. (1952) Dynamic basis of geomorphology, *Geological Society of America Bulletin* 63: 923–38.

Thorn, C.E. (1988) *An Introduction to Theoretical Geomorphology.* Unwin Hyman, Boston, p. 247.

Tucker, G.E. and Slingerland, R.L. (1997) Drainage basin response to climate change, *Water Resources Research* 33: 2031–47.

Tucker, G.E. and Bras, R.L. (1998) Hillslope processes, drainage density, and landscape morphology, *Water Resources Research* 34: 275164.

von Bertalanffy, L. (1950) The theory of open systems in physics and biology, *Science* 3: 23–8.

von Bertalanffy, L. (1968) *General System Theory: Foundations, Development, Applications.* George Braziller, New York. p. 295.

Willgoose, G. (2005) Mathematical modeling (sic) of whole landscape evolution, *Annual Review of Earth and Planetary Sciences* 33: 443–59.

Winsberg, E. (2003) Model-based reasoning: technology, science, values, *Philosophy of Science* 70: 442–4.

Wu, C.F.J. and Hamada, M. (2000) *Experiments: Planning, Analysis and Parameter Design Optimization.* John Wiley and Sons, New York. p. 630.

Wynne, B. (1992) Uncertainty and environmental learning: reconceiving science and policy in the preventive paradigm, *Global Environmental Change* 2: 111–27.

Young, A. (1972) *Slopes. Geomorphology Texts 3.* K.M. Clayton (general editor), Longman, London and New York, p. 288.

10

Process and Form

Richard Huggett

Landforms, the subject matter of geomorphology, are physical features of earth's surface. They are omnipresent, plain to see and occur at a variety of geographical scales, ranging from mima mounds to mountains to major tectonic plates; and they have 'lifespans' lasting from days to millennia to aeons.

Process, form and the inter-relationships between them are basic to understanding the origin and development of landforms. It is perhaps worth briefly considering what form and process are, as most geomorphologists use the terms without defining them. In geomorphology, form or morphology has three facets: (1) constitution (chemical and physical properties), (2) configuration (size, shape and other geometric properties) and (3) mass–flow characteristics (e.g. discharge, precipitation rate and evaporation rate) (Strahler, 1980). These form variables contrast with dynamic variables (chemical and mechanical properties representing the expenditure of energy and the doing of work) associated with geomorphic processes; they include power, energy flux, force, stress and momentum. Geomorphic processes are the multifarious chemical and physical means by which Earth's surface undergoes modification (cf. Thornbury, 1954: 19). They are driven by geological forces emanating from inside the earth (endogene processes), by forces originating at or near Earth's surface and in the atmosphere (exogene processes), and by forces coming from outside Earth (extra-terrestrial processes, such as asteroid impacts). They include processes of transformation and transfer associated with weathering, gravity, water, wind and ice.

Mutual interactions between form and process are the core of geomorphic investigation – form affects process and process affects form. In a wider setting, atmospheric processes, ecological processes and geological processes influence, and in turn are influenced by, geomorphic process–form interactions (Figure 10.1).

The nature of the two-way connection between Earth surface process and Earth surface form has lain at the heart of geomorphic discourse. The language in which geomorphologists have expressed these connections has altered with changing cultural, social and scientific contexts. In broad terms, and simplifying a very complex history of the subject, a qualitative approach begun by classical thinkers and traceable through to the mid 20th century preceded a quantitative approach. Early writers pondered the origin of Earth's surface features, linking the forms they saw, such as mountains, to assumed processes, such as catastrophic floods. An excellent example is the work of Nicolaus Steno (alias Niels Steensen, 1638–1686). While carrying out his duties as court physician to Grand Duke Ferdinand II at Florence, Steno explored the Tuscan landscape and devised a six-stage sequence of events to explain the current plains and hills (Steno, 1916 edn). The first true geomorphologists, such as William Morris Davis and Grove Karl Gilbert, also tried to infer how the landforms they saw in the field were fashioned by geomorphic processes.

Currently, geomorphologists study landforms in at least four ways (Slaymaker, 2009; see also Baker and Twidale, 1991). First is a process–response

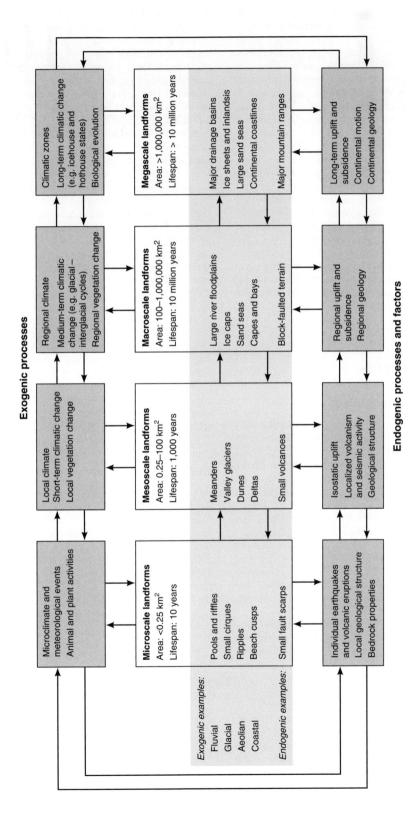

Figure 10.1 Landforms at different scales and their interactions with exogenic and endogenic processes (after Huggett (2007a))

(process–form) or functional approach that builds upon chemistry and physics, utilizes a systems methodology, and is referred to as surface process, or simply process geomorphology (e.g. Embleton and Thornes, 1979). Second is the landform evolution approach that has its roots in historical geological science (geohistory), and sometimes called historical geomorphology. Third is an approach that focuses on characterizing landforms and landform systems and that stems from geographical spatial science. Fourth, is an environmentally sensitive approach to landforms, systems of landforms, and landscape at regional to global scales. The historical and process approaches dominate modern geomorphology (Summerfield, 2005), with the latter predominating, at least in Anglo-American and Japanese geomorphology.

This chapter will explore process and historical studies, which largely have their own perspectives on process–form interactions, and the status and prospects of marrying process and historical explanations.

PROCESS STUDIES

Process geomorphology

The first process (or dynamic) geomorphologist was Grove Karl Gilbert. However, such thinkers as Herodotus, Aristotle and Strabo in the classical period and Leonardo da Vinci, Nicholas Steno, Giovanni Targioni-Tozetti, Jean-Étienne Guettard and James Hutton in the modern period presaged Gilbert's approach (see Chorley et al., 1964). Since the 1950s, process geomorphology has bloomed into a major area of enquiry. Its objects of investigation are typically small and medium landforms that last for a millennium or less and the processes that fashion them. Investigation takes place in the field, in the laboratory and on computers using numerical simulation models.

A shift of emphasis in process studies came with Strahler's (1952: 923) proposal of a 'system of geomorphology grounded in basic principles of mechanics and fluid dynamics'. He hoped that this system would

> enable geomorphic processes to be treated as manifestations of various types of shear stresses, both gravitational and molecular, acting upon any type of earth material to produce varieties of strain, or failure, which we recognize as the manifold processes of weathering, erosion, transportation and deposition.

In fact, Strahler and his students, as well as Leopold, Wolman and Miller (1964), were largely empirical, involving a statistical treatment of form variables (such as width, depth and meander wavelength) and surrogates for variables that controlled them (such as discharge) (see Lane and Richards, 1997). The challenge of characterizing the geomorphic processes themselves was eventually taken up by Culling (1960, 1963, 1965) and Kirkby (1971), and it was not until the 1980s that geomorphologists, particularly those at the Universities of Washington and Berkeley, USA (e.g. Dietrich and Smith, 1983), developed Strahler's vision of a truly dynamic geomorphology (see Lane and Richards, 1997). Having said that, it should be noted that there were early attempts to tackle the mechanics and dynamics of geomorphic systems, including Nye's (1951) application of plasticity theory to the flow of ice sheets and glaciers, Bagnold's classic study of the physics of blown sand and desert dunes (Bagnold, 1954), his classic paper on the bedload stresses set up during the transport of cohesionless grains in fluids (Bagnold, 1956), and his approach to the problem of sediment transport from the viewpoint of general physics (Bagnold, 1966). There is no doubt that Strahler's groundbreaking ideas spawned a generation of Anglo-American geomorphologists who researched the small-scale erosion, transport and deposition of sediments in a mechanistic and fluid dynamic framework (cf. Martin and Church, 2004).

Geomorphic systems

Part of the success of process geomorphology was the adoption of a systems approach, which provided a standard way of portraying systems and, to some extent, a common language for discussing static and changing conditions (see Huggett, 2007b, 2010). Strahler (1950, 1952; see also 1980) introduced open systems theory to geomorphology, though Grove Karl Gilbert first mooted the idea of a system in the subject, and in doing so took an open systems approach in his concept of dynamic equilibrium:

> The tendency to equality of action, or the establishment of a dynamic equilibrium, has already been pointed out ... but one of its most important results has not been noticed ... in each basin all lines of drainage unite in a main line, and a disturbance upon any line is communicated through it to the other main line and thence to every tributary. And as any member of the system may influence all others, so each member is influenced by every other.
>
> (Gilbert, 1877: 123–4)

Thus, Gilbert saw equilibrium landforms adjusting to geomorphic processes (Chorley, 1965b).

Strahler's low-key comments on geomorphic systems ushered in a revival of Gilbertian thinking in geomorphology. Melton (1958), Hack (1960), Chorley (1962), Howard (1965), Schumm (1977) and many others took up his call to action. For example, Hack (1960) abandoned the cyclic theory of landform development (as proposed by Davis) and instead adopted Gilbert's concept of dynamic equilibrium as a philosophical base for interpreting erosional topography in the Central Appalachians, USA. In this conception,

> The landscape and the processes molding it are considered a part of an open system in a steady state of balance in which every slope and every form is adjusted to every other. Changes in topographic form take place as equilibrium conditions change, but it is not necessary to assume that the kind of evolutionary changes envisaged by Davis ever occur.
>
> (Hack, 1960: 81)

Open systems thinking led to a new typology of systems, as first proposed by Chorley and Kennedy (1971), and adopted and adapted by Strahler (1980). According to these authors, there are four levels of systems: morphological (form) systems, cascading (flow) systems, process–response (process–form) systems (the terms in parenthesis are Strahler's) (Figure 10.2) and control systems, in which human intervention leads to changes in process–response systems.

Form systems

Morphological or form systems are sets of morphological variables that are thought to inter-relate in a meaningful way in terms of system origin or system function. An example is a hillslope represented by variables pertaining to hillslope geometry and to hillslope composition, all of which form an inter-related set (Figure 10.2a). Another example is the interacting components of a drainage basin: divides, hillslopes, floodplains and channels. Work on form systems led to a careful consideration of morphological variables and to the defining of appropriate quantitative descriptors of landscape form. Correlation sets between system variables, sometimes expressed as interaction matrices, may represent form systems, and indeed flow and process–form systems (e.g. Phillips, 1999, 73–80).

Flow systems

Cascading systems or flow systems are 'interconnected pathways of transport of energy or matter or both, together with such storages of energy and matter as may be required' (Strahler, 1980, 10). An example is a hillslope represented as a store of material linked by flow associated with erosional processes (Figure 10.2a), or more generally the energy and material flows in a drainage basin, including water and sediment movement. Other examples of flow systems include the water cycle, the biogeochemical cycle and the sedimentary cycle, all of which occur at scales ranging from minor cascades in small segments of a landscape, through medium-scale cascades in drainage basins and seas, to mighty circulations involving the entire globe. The idea of flow systems went hand-in-hand with the quantification of material cycles and fluxes in the landscape. Admittedly, attempts to quantify solid and dissolved sediment fluxes in the world's major rivers predated the geomorphic systems era. However, the physical process approach in geomorphology, started by Horton (1945) and Strahler (1950, 1952) in the United States, was largely responsible for the measuring of process rates in different environments.

Process–form systems

Process–response systems or process–form systems are conceived as an energy flow system linked to a morphological system in such a way that system processes may alter the system form and, in turn, the changed system form alters the system processes. (Figure 10.2c) depicts a hillslope as a process–form system, which forms the basis of a mathematical process–response model describing the evolution of soil-mantled slopes:

$$\frac{\mathrm{d}z}{\mathrm{d}t} = U - P - \nabla \cdot q_s$$

(see Figure 10.2c for definition of terms). Another case is the morphological components of a drainage basin linked to energy and mass flow variables; for instance, channel width, depth, and drainage density interacting with water and sediment movement, shear forces, and so forth. Process–form systems generated a sizeable literature focussed on the connections between landform and the processes that create it, mostly at small and medium scales (see, for instance, Stoddart 1997 and back issues of *Earth Surface Process and Landforms* and *Geomorphology*). Some process–form models, such as hillslope development models (Chapter 20), have wider applicability to large-scale features and to very long timescales.

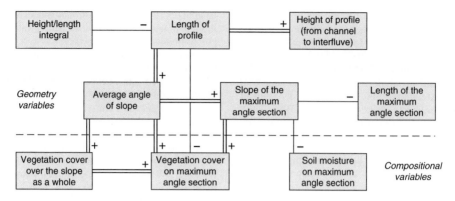

(a) Form system

(b) Flow system

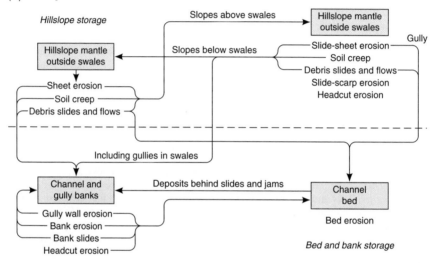

(c) Process – form system

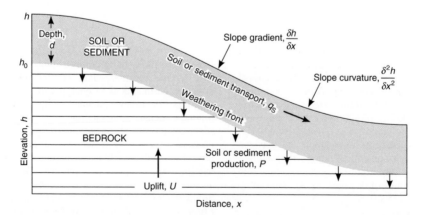

Figure 10.2   Types of geomorphic systems. (a) Valley-side slopes in Manitoba, Canada, depicted as a form system (adapted from Chorley and Kennedy (1971)). (b) Sediment stores and erosional processes in steepland drainage basins of the California coastal range, USA, depicted as a flow system (adapted from Lehre (1982)). (c) A hillslope as a process–form system

Control systems

These process–form systems interact with humans. They include managed rivers, coasts with sea defences and some caves.

A recent line of enquiry in geomorphology is role of life in landform development, which, in effect, extends the notion of a control system to include organisms other than humans. To be sure, from the 1980s onward, some geomorphologists have emphasized the importance of biotic processes to landscape development (e.g. Viles, 1988; Butler, 1995). Such work has culminated in the contention that life has a 'topographic signature' (Dietrich and Perron, 2006). The argument runs that over short timescales, biotic processes mediate chemical reactions, disrupt the ground surface, expand soil and strengthen soil by weaving a network of roots, which changes affect weathering, soil formation and erosion, slope stability and river dynamics. Over geological timescales, biotic effects are less patent but no less significant. Animals and plants help to shape climate, and in turn, climate dictates the mechanisms and rates of erosion that constrain topographic evolution (Dietrich and Perron, 2006). Future studies should reveal just how clear the topographic signature of life is.

Chaos and complexity

From the 1960s onward, some geomorphologists began questioning simplistic notions of equilibrium and steady state. Three seminal ideas emerged: (1) thresholds, (2) bifurcation theory, and (3) chaos (see Huggett, 2007b).

Howard (1965) noted that geomorphic systems might possess thresholds that separate two rather different system economies. Schumm (1973, 1977) introduced the notions of metastable equilibrium and dynamic metastable equilibrium, showing that thresholds within a fluvial system cause a shift in its mean state. The thresholds, which may be intrinsic or extrinsic, are not part of a change continuum, but show up as dramatic changes resulting from minor shifts in system dynamics, such as caused by a small disturbance. In metastable equilibrium, static states episodically shift when thresholds are crossed. In dynamic metastable equilibrium, thresholds trigger episodic changes in states of dynamic equilibrium.

Some geomorphologists applied bifurcation theory to geomorphic systems in the late 1970s and early 1980s. They based their arguments on catastrophe theory, which is a special branch of bifurcation theory developed by René Thom (1975), and tried to use Thom's ideas (his cusp catastrophe proved a favourite) to explain certain processes at the Earth's surface. Examples included Chappell's (1978) cusp catastrophe model expressing relationships between wave energy, water table height relative to a beach surface, and erosion and accretion; Graf's (1979, 1982) model of the condition at stream junctions in the northern Henry Mountains of Utah; and Thornes's (1983) model of sediment transport in a river. The cusp catastrophe model still has currency, being used, for example, to explain the instability of a slip-buckling slope (Qin et al., 2001).

All these intimations of complex dynamics and non-equilibrium within systems found a firm theoretical footing with the theory of nonlinear dynamics and chaotic systems that scientists from a range of disciplines developed, including geomorphology itself. Classical open systems research characteristically deals with linear relationships in systems near equilibrium. A fresh direction in thought and a deeper understanding came with the discovery of deterministic chaos by Edward Lorenz in the 1963. The key change was the recognition of nonlinear relationships in systems. In geomorphology, nonlinearity means that system outputs (or responses) are not proportional to systems inputs (or forcings) across the full gamut of inputs (cf. Phillips, 2006).

Nonlinear relationships produce rich and complex dynamics in systems far removed from equilibrium, which display periodic and chaotic behaviour. The most surprising feature of such systems is the generation of 'order out of chaos', with systems states unexpectedly moving to higher levels of organization under the driving power of internal entropy production and entropy dissipation. Systems of this kind, which dissipate energy in maintaining order in states removed from equilibrium, are dissipative systems. The theory of complex dynamics predicts a new order of order, an order arising out of and poised perilously at the edge of, chaos. It is a fractal order that evolves to form a hierarchy of spatial systems whose properties are holistic and irreducible to the laws of physics and chemistry. Geomorphic examples are flat or irregular beds of sand on streambeds or in deserts that self-organize themselves into regularly spaced forms – ripples and dunes – that are rather similar in size and shape (e.g. Baas, 2002; see Murray et al., 2009 for other examples). Conversely, some systems display the opposite tendency – that of non-self-organization – as when relief reduces to a plain. A central implication of chaotic dynamics for the natural world is that all Nature may contain fundamentally erratic, discontinuous and inherently unpredictable elements. Nonetheless, nonlinear Nature is not all complex and chaotic. Phillips (2006) astutely noted that 'Nonlinear systems are not all, or always, complex, and even those which can be chaotic are not chaotic under all circumstances.

Conversely, complexity can arise due to factors other than nonlinear dynamics'.

Phillips (2006) has proposed methods for detecting chaos in geomorphic systems. He argued that convergence versus divergence of a suitable system metric (elevation or regolith thickness for instance) is an immensely significant indicator of stability behaviour in a geomorphic system. In landscape evolution, convergence associates with downwasting and a reduction of relief, whilst divergence relates to dissection and an increase of relief. More fundamentally, convergence and divergence underpin developmental, 'equilibrium' conceptual frameworks, with a monotonic move to a unique endpoint (peneplain or other steady-state landform), as well as evolutionary, 'non-equilibrium' frameworks that engender historical happenstance, multiple potential pathways and end-states, and unstable states. The distinction between instability and new equilibria is critical to understanding the dynamics of actual geomorphic systems, and for a given scale of observation or investigation, it separates two conditions. On the one hand sits a new steady-state equilibrium governed by stable equilibrium dynamics that develops after a change in boundary conditions or in external forcings. On the other hand sits a persistence of the disproportionate impacts of small disturbances associated with dynamic instability in a non-equilibrium system (or a system governed by unstable equilibrium dynamics) (Phillips, 2006). The distinction is critical because the establishment of a new, steady-state equilibrium implies a consistent and predictable response throughout the system, predictable in the sense that the same changes in boundary conditions affecting the same system at a different place or time would produce the same outcome. In contrast, a dynamically unstable system possesses variable modes of system adjustment and inconsistent responses, with different outcomes possible for identical or similar changes or disturbances.

HISTORICAL STUDIES

Landforms at all scales have a history. However, such landforms as ripples in riverbeds and terracettes on hillslopes tend to be short lived, so that their history tends to pass unrecorded unless burial by sediments ensures their survival in the stratigraphic record. However, some landforms occupying small areas form slowly and persist for millions of years, flared slopes are a case in point (see Ollier and Bourman, 2002). For this reason, geomorphologists with a prime interest in geohistorical changes usually deal with relatively more persistent landforms at scales ranging from coastal features and landslides, and river terraces, through plains and plateaux, to regional and continental drainage systems.

This section will consider three areas of geohistorical enquiry. First, it will explore the 'traditional' use of morphological and stratigraphical evidence to study regional landforms. Second, it will probe the use of numerical landscape modelling in exploring landscape evolution. Third, it will examine evolutionary geomorphology, which takes a very long-term perspective, adopting the position that Earth's landforms have evolved as a whole through a non-repeated set of states, rather than set of never-ending erosion cycles, with vestiges of earlier states surviving in present landscape as exhumed and relict features.

Historical geomorphology and regional landforms

William Morris Davis fathered modern historical geomorphology at the turn of the 20th century (see Davis, 1909). He provided a means of interpreting long-term change in landforms by considering structure, process and stage. By 'stage', Davis meant the state of progression through a predictable geographical cycle of change from youth, through maturity, to senility. Davis's work, and the studies that it engendered, used a 'particularistic narrative model of explanation' (Harrison, 2001: 327), or what might be termed a storytelling approach, where historical contingency plays a key explanatory role (cf. Church, 1996).

Historical geomorphology has developed since Davis's time, and the interpretation of long-term changes of landscape no longer relies on the straightjacket of the geographical cycle (e.g. Chorley, 1965a). It rests now on various chronological analyses, particularly those based on stratigraphic studies of Quaternary sediments, and upon a much fuller appreciation of geomorphic processes (e.g. Brown, 1980). Observed stratigraphic relationships furnish relative chronologies, whilst absolute chronologies derive from sequences dated using historical records, radiocarbon analysis, dendrochronology, luminescence and palaeomagnetism. Such quantitative chronologies offer a means for calculating long-term rates of change in the landscape; the dates of events may allow the estimation of process rates.

The study of Tertiary landscape evolution in southern Britain nicely shows how emphasis in historical geomorphology has changed from land-surface morphology being the key to interpretation to a more careful examination of stratigraphical and morphological evidence for past geomorphic processes in a firmer timeframe. As Jones (1981: 4–5) put it, this

> radical transformation has in large part resulted from a major shift in methodology, the heavily morphologically-biased approach of the first half of the twentieth century having given way to studies that have concentrated on the detailed examination of superficial deposits, including their faunal and floral content, and thereby provided a sounder basis for the dating of geomorphological events.

The key to Wooldridge and Linton's (1939, 1955) classic model of landscape evolution in Tertiary south-east England was three basic surfaces, each strongly developed on the chalkland flanks of the London Basin (Figure 10.3). First is

an inclined, recently exhumed, marine-trimmed surface that fringes on the present outcrop of Palaeogene sediments. Wooldridge and Linton called this the sub-Eocene surface (but now more accurately termed the sub-Palaeogene surface). Second is an undulating summit surface lying above about 210 m, mantled with thick residual deposits of 'clay-with-flints', and interpreted by Wooldridge and Linton as the remnants of a region-wide subaerial peneplain, as originally suggested by Davis in 1895, rather than a high-level marine plain lying not far above the present summits. Third is a prominent, gently inclined, erosional platform, lying between about 150 and 200 m and cutting into the summit surface and seemingly truncating the sub-Eocene surface. As it bears sedimentary evidence of marine activity, Wooldridge and Linton

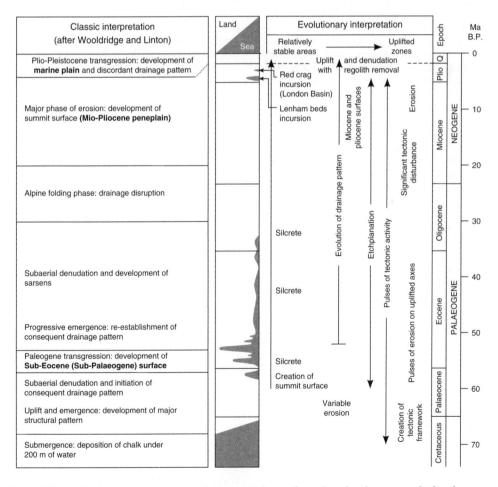

Figure 10.3 **Classic and evolutionary interpretations of Tertiary landscape evolution in southern England (adapted from Jones (1999))**

interpreted it as a Pliocene marine plain. Wooldridge and Linton believed that the two higher surfaces – the summit surface and the marine platform – were not warped. They argued, therefore, that these surfaces must have formed after the tectonic episode that deformed the sub-Eocene surface, and that the summit plain had to be a peneplain fashioned during the Miocene and Pliocene.

The Wooldridge and Linton model of Tertiary landscape evolution was the ruling theory until at least the early 1960s and perhaps as late as the early 1970s. Following Wooldridge's death in 1963, interest in the long-term landform evolution or denudation chronology of Britain waned. Critics accused denudation chronologists of letting their eyes deceive them: most purely morphological evidence is 'so ambiguous that theory feeds readily on preconception' (Chorley, 1965b: 151). However, alongside the denigration of and declining interest in denudation chronology, some geomorphologists reappraised the evidence for long-term landscape changes. This fresh work led in the early 1980s to the destruction of Wooldridge and Linton's 'grand design' and to the creation of a new evolutionary framework that discarded the obsession with morphological evidence in favour of the careful examination of Quaternary deposits (Jones, 1999: 5–6). The reappraisal was in part inspired by Pinchemel's (1954) alternative idea that the gross form of the chalk backslopes in southern England and northern France results from intersecting Palaeogene erosion surfaces that suffered exhumation and modification during the Neogene and Quaternary. Foremost among the architects of the new evolutionary model of Tertiary landscape evolution in southern England were Jones (1981) and Green (1985). Jones (1999) confirmed this model in a region-wide synthesis that presented a subtle interplay between sea-level changes, tectonic deformation and denudation involving phases of etchplanation under tropical conditions. Jones (1999) owned that this evolutionary model needs substantiating in a number of important regards. Uncertainty surrounds the true nature of the structural foundations of the area and its tectonic evolution, which fuels the continuing controversy over the temporal and spatial dimensions of uplift and the relative importance of mid-Tertiary (Miocene) tectonic activity. Likewise, the recent suggestions of Pleistocene uplift and warping need confirming, elaborating and accurately dating. Moreover, the nature and palaeo-environmental significance of residual soils, including the varied types of silcrete, and the number, age and geographical extent of Neogene marine incursions, most especially the baffling Lenham Beds incursion, still demand much investigation.

Numerical modelling of landscape evolution

Earth surface processes fashion landscapes over hundreds of thousands to millions of years. It is therefore impossible to study landscape evolution directly, but mathematical models offer a means of probing long-term changes in landscape form. Early researchers into this line of investigation were Kirkby (1971) and Ahnert (1976). Kirkby used the continuity equation of debris moving on hillslopes and in rivers as a basis for a hillslope model (Kirkby, 1971). An example of his work includes climatic change in a simulation to reproduce Savigear's (1952) observed sequence of cliffs in various stages of retreat in South Wales. Kirkby ran the model for three phases. First, for a period, starting 500,000 years ago and ending 50,000 years ago corresponding roughly to inland valley development with a fixed base level under mainly periglacial conditions; second, for a period of cliff retreat from 50,000 to 10,000 years ago; and third, for a period of basal removal covering the last 10,000 years. The observed upper convexities of the slope profiles as surveyed by Savigear can, according to the model, only be formed during the periglacial phase and require at least 100,000 years to form. They are today relict features.

Developments in the mathematical modelling of long-term landscape change associated with huge advances in computational technology, coupled with a set of process equations designated 'geomorphic transport laws' (Dietrich et al., 2003) (Table 10.1), have greatly aided the understanding of some aspects of landscape evolution. As Martin and Church (2004: 334) put it,

> The modelling of landscape evolution has been made quantitatively feasible by the advent of high speed computers that permit the effects of multiple processes to be integrated together over complex topographic surfaces and extended periods of time.

A recent example is the use of a numerical landscape to predict the spacing of ridges and valleys in hilly terrain (Perron et al., 2009). What the models cannot do is predict unique events that may play a starring role in shaping the evolution of many landscapes, which leads the discussion to evolutionary geomorphology.

Evolutionary geomorphology and large-scale landforms

Ollier (1981, 1992), the leading proponent of evolutionary geomorphology, argued that the land surface has changed in a definite direction

Table 10.1 Geomorphic transport laws

Process	Geomorphic transport law	Mechanisms	Definition of terms
Soil production rate	$P = P_0^{-ad}$	Salt and freeze–thaw weathering, atmospheric dust input, mineral alteration leading to loss of physical strength; bioturbation (animal burrowing, root growth, and tree throw), geochemical reactions mediated by microbes	P, soil production rate from bedrock; d, soil thickness; a, constant
Slope-dependent downslope movement (creep)	$q_s = -K\nabla h$	Wetting and drying, freezing and thawing, shear flow, bioturbation	q_s, volumetric sediment transport rate per unit width; K, constant; h, elevation
Landsliding	None available, but important starts for earthflows, deep-seated landslides, and landslide dynamics	Stress exceeds material strength owing to earthquakes, elevated pore pressures derived from precipitation or from undermining of toe; released sediment travels downslope	—
Surface wash and splash	Many short-term empirical and mechanistic expressions, but no geomorphic transport law available	Rainsplash and overland flow displace and remove particles; rill and gully incision	—
River incision into bedrock	$E = k_b A^m S^n$	Plucking and particle wear due to river flow and sediment transport	E, incision rate; k_b, constant that may depend on uplift rate and rock strength; A, drainage area; S, local slope; m, n, constants
Debris flow incision into bedrock	$E = k_d f\left[\rho_s D^2 \left(\dfrac{\partial u}{\partial y}\right)^a L_s\right]^p$	Particle impact and sliding wear of bedrock during mass transport	D, representative grain diameter; k_d, constant that depends on bedrock properties; f, frequency of debris flows; ρ_s, bulk density of debris flow; u, debris flow velocity as a function of distance y above the bed; L_s, length of debris flow 'snout'; a, p, constants
Glacial scour	$E = cU_b$	Sediment-rich basal sliding wears bedrock	U_b, basal ice velocity; and c, constant
Wind transport and scout	Extensive theory for sediment transport by wind; some theory for rock abrasion	Abrasion by wind-suspended particles	—

Source: Adapted from Dietrich and Perron (2006).

through time, and it has not suffered an 'endless' progression of erosion cycles. In other words, Earth's landscapes have evolved as a whole. In doing so, they have been through several geomorphological 'revolutions' that have led to distinct and essentially irreversible changes of process regimes. These revolutions probably occurred during the Archaean, when the atmosphere was reducing rather than oxidizing, during the Devonian, when a cover of terrestrial vegetation appeared, and during the Cretaceous, when grassland appeared and spread. The breakup and coalescence of continents would also alter landscapes. The geomorphology of Pangaea was, in several respects, unlike present geomorphology (Ollier 1991: 212). Vast inland areas lay at great distances from the oceans, many rivers were longer by far than any present river, and terrestrial sedimentation was more widespread. When Pangaea broke up, rivers became shorter, new continental edges were rejuvenated and eroded continental margins warped tectonically. Once split from the supercontinent, each Pangaean fragment followed its own history: each experienced its own unique events, including the creation of new plate edges and changes of latitude and climate. In Ollier's opinion, geomorphologists should view the landscape evolution of each continental fragment from this very long-term perspective, in which the current fads and fashions of geomorphology – process studies, dynamic equilibrium, and cyclical theories – have limited application (Ollier 1991, 212). The morphotectonic evolution of south-east Australian landscapes supports this view as it seems to represent a response to unique, non-cyclical events (Ollier and Pain, 1994; Ollier, 1995). Lidmar-Bergström (1996) came to a similar conclusion having studied the evolution of relief in Sweden, which has produced long-lasting covers of sediment and several exhumed surfaces, including a sub-Cambrian peneplain and a sub-Mesozoic etchplain.

UNIFYING PROCESS AND HISTORICAL EXPLANATIONS

Big challenges face research into the interaction of geomorphic forms and processes, the biggest of which is probably the integration of process and historical approaches. Several geomorphologists have recently identified this challenge. For instance, after reviewing numerical models of landscape evolution, Martin and Church (2004) concluded that advances in such modelling would demand a reintegration of process studies and historical investigations because suitable data sets to calibrate models of even a high temporal are almost non-existent. The unification issue centres on the vexed but crucial debate surrounding what Simpson (1963) called 'immanence' (processes that may always occur under the right historical conditions) and 'configuration' (the state or succession of states created by the interaction of immanent process with historical circumstances). The contrast is between a 'what happens' approach (timeless knowledge – immanence) and a 'what happened' approach (timebound knowledge – configuration). In simple terms, geomorphologists may study geomorphic systems in action today, but such studies are necessarily short term, lasting

for a few years or decades and principally investigate immanent properties. Yet geomorphic systems have histories that goes back centuries, millennia, or millions of years. Using the results of short-term studies to predict how geomorphic systems will change over long periods is difficult owing to environmental changes and the occurrence of singular events (configuration in Simpson's parlance). Schumm, (1991; see also Schumm and Lichty, 1965) tried to resolve this problem, and in doing so established some links between process studies and historical studies. He argued that, as the size and age of a landform increase, so present conditions can explain fewer of its properties and geomorphologists must infer more about its past. Evidently, such small-scale landforms and processes as sediment movement and river bedforms are explicable with recent historical information. River channel morphology may have a considerable historical component, as when rivers flow on alluvial plain surfaces that events during the Pleistocene determined. Explanations for large-scale landforms, such as structurally controlled drainage networks and mountain ranges, require mainly historical information. A corollary of this idea is that the older and bigger a landform, the less accurate will be predictions and postdictions about it based upon present conditions. It also shows that an understanding of landforms requires a variable mix of process geomorphology and historical geomorphology; and that the two subjects should work together rather than stand in polar opposition.

The reconciliation of process and historical explanations remains a challenging issue in geomorphology, but progress in this endeavour is evident in at least four lines of enquiry. The first is space–time substitution, the second is process philosophy, the third is tectonic geomorphology, and the fourth is the application of nonlinear dynamics to geomorphic systems.

Space–time substitution

A common practice in geomorphology is to study change through time by identifying similar landforms of differing age at different locations, and then arranging them chronologically to create a time sequence or topographic chronosequence. Such space–time substitution, based on the ergodic hypothesis (see Chorley et al., 1984; Paine, 1985, Gregory, 2000), has proved salutary in understanding landform development; and to an extent, it provides a link between process studies and historical reconstructions. Two broad types of space–time substitution are used (Paine, 1985). In the first category, which looks at equilibrium or 'characteristic' landforms, the assumption is that the geomorphic processes and forms under consideration are in equilibrium with landforms and environmental factors. For instance, modern rivers on the Great Plains display relationships between their width–depth ratio, sinuosity and suspended load, which aid the understanding of channel change through time (Schumm, 1963).

In the second category of space–time substitution, which looks at developing or 'relaxation' landforms, the argument is that similar landforms of different ages occur in different places. A developmental sequence emerges by arranging the landforms in chronological order. The reliability of such space–time substitution depends upon the accuracy of the landform chronology. Least reliable are studies that simply assume a time sequence. Charles Darwin, investigating coral-reef formation, thought that barrier reefs, fringing reefs and atolls occurring at different places represented different evolutionary stages of island development applicable to any subsiding volcanic peak in tropical waters. Davis applied this evolutionary schema to landforms in different places and derived what he deemed was a time sequence of landform development: the geographical cycle. This seductively simple approach is open to misuse. The temptation is to fit the landforms into some preconceived view of landscape change, even though other sequences might be constructed. More useful are situations where, although an absolute chronology is unavailable, field observations enable geomorphologists to place the landforms in the correct order. This occasionally happens when, for instance, adjacent hillslopes become progressively cut off from the action of fluvial or marine processes at their bases. A classic example is a segment of the South Wales coast, in the British Isles, where a sand spit growing from west to east has affected the Old Red Sandstone cliffs between Gilman Point and the Taf estuary (Savigear, 1952). In consequence, the western-most cliffs have been subject to sub-aerial denudation without waves cutting their bases the longest, while the cliffs to the east are progressively younger. The most informative examples of space–time substitution arise where absolute landform chronologies exist. Historical evidence of slope profiles along Port Hudson bluff, on the Mississippi River in Louisiana, southern United States, revealed a dated chronosequence (Brunsden and Kesel, 1973). The Mississippi River was undercutting the entire bluff segment in 1722. Since then, the channel has shifted about 3 km downstream with a concomitant stopping of undercutting. The changing conditions at the slope bases have reduced the mean slope angle from 40° to 22°.

Several problems beset space–time substitution. First, not all spatial differences are temporal differences. Second, landforms of the same age might differ through historical accidents. Third, different sets of processes may produce the same landform. Fourth, process rates and their controls may have changed in the past, with human impacts presenting particular problems. Fifth, equilibrium conditions are unlikely to have endured for the timescales over which the locational data substitute for time. And sixth, some ancient landforms are relicts of past environmental conditions.

Process philosophy

A recent attempt by Rhoads (2006) to meld process and historical approaches appealed to process philosophy, and especially to the process philosophy developed by Arthur North Whitehead, the British mathematician and philosopher. To understand what process philosophy is, it may help to consider these quotations from Whitehead's cannon as picked out by Rhoads: 'Every scheme for the analysis of nature has to face these two facts, *change* and *endurance*. ... The mountain endures. But when after the ages it has been worn away, it has gone' (Whitehead, 1925: 86–87, italics in original). However, both change and endurance are dynamic, because endurance is 'the process of continuously inheriting a certain identity of character transmitted throughout a historical route of events' (Whitehead, 1925: 108). Environmental contingencies help to determine endurance: 'a favourable environment is essential to the maintenance of a physical object' (Whitehead, 1925: 109). However, in the fullness of time, all geomorphic features are fated to change: 'One all pervasive fact, inherent in the very character of what is real is the transition of things, the passage one to another' (Whitehead, 1925: 93). In other words, everything in geomorphology (and physical geography for that matter) takes part in an overall process of becoming, being and fading away. From this process perspective, timeless and timebound views – immanence and configuration – fuse in the dual notions of change and endurance.

Tectonic geomorphology and continental landforms

Great strides in tectonic geomorphology over the last couple of decades have shown how the application of geomorphic process modelling to long-term landform change, and specifically large-scale tectonic processes, can help with historical explanations (Burbank and Anderson, 2001).

Important interactions between endogenic factors and exogenic processes produce macroscale and megascale landforms (Figure 10.1). Since the 1990s, geomorphologists have come to realize that the global tectonic system and the world climate system interact in complex ways (see Chapter 28). The interactions give rise to fundamental changes in atmospheric circulation patterns, in precipitation, in climate, in the rate of uplift and denudation, in chemical weathering, and in sedimentation (Raymo and Ruddiman, 1992; Small and Anderson, 1995; Rea et al., 1998; Montgomery et al., 2001; Peizhen et al., 2001; Codilean et al., 2006). The interaction of large-scale landforms, climate and geomorphic processes occurs in at least three ways. First, plate tectonic processes directly affect topography (uplift). Second, topography may cause local and regional climate change, which in its turn may promote uplift (e.g. Molnar and England, 1990; Hodges, 2006). Third, topography may indirectly influence chemical weathering rates and the concentration of atmosphere carbon dioxide. Only the first of these is relevant to the discussion in this chapter.

Plate tectonics explains some major features of Earth's topography. An example is the striking connection between mountain belts and processes of tectonic plate convergence. However, the nature of the relationship between mountain belts (orogens) and plate tectonics is far from clear, with several questions remaining unsettled (Summerfield, 2007). What factors, for example, control the elevation of orogens? Why do the world's two highest orogens – the Himalaya–Tibetan Plateau and the Andes – include large plateaux with extensive areas of internal drainage? Does denudation shape mountain belts at the large scale, and are its effects more fundamental than the minor modification of landforms that are essentially a product of tectonic processes? Since the 1990s, researchers have addressed such questions as these by treating orogens, and landscapes more generally, as products of a coupled tectonic–climatic system with the potential for feedbacks between climatically influenced surface processes and crustal deformation (Beaumont et al., 2000; Hodges et al., 2004; Pinter and Brandon, 1997; Willett, 1999; Willett, et al., 2006; Wobus et al., 2003).

The elevation of orogens appears crucially to depend upon the crustal strength of rocks. Where crustal convergence rates are high, surface uplift soon creates (in geological terms) an elevation of around 6 to 7 km that the crustal strength of rocks can just about sustain, although individual mountain peaks may stand higher where the strength of the surrounding crust supports them.

However, in most mountain belts, the effects of denudation prevent elevations from attaining this upper ceiling. As tectonic uplift occurs and elevation increases, river gradients become steeper so raising denudation rates. The growth of topography is also likely to increase precipitation (through the orographic effect) and therefore runoff, which will also tend to enhance denudation (Summerfield and Hulton, 1994). In parts of such highly active mountain ranges as the Southern Alps of New Zealand, rivers actively incise and maintain, through frequent landslides, the adjacent valley-side slopes at their threshold angle of stability. In consequence, an increase in the tectonic uplift rate produces a speedy response in denudation rate as river channels cut down and trigger landslides on adjacent slopes (Montgomery and Brandon, 2002). Where changes in tectonic uplift rate are (geologically speaking) rapidly matched by adjustments in denudation rates, orogens seem to maintain a roughly steady-state topography (Summerfield, 2007). The actual steady-state elevation is a function of climatic and lithological factors, higher overall elevations being attained where rocks are resistant and where dry climates produce little runoff. Such orogens never achieve a perfect steady state because there is always a delay in the response of topography to changing controlling variables such as climate, and especially to changing tectonic uplift rates because the resulting fall in baselevel propagates along drainage systems to the axis of the range. Work with simulation models suggests that variations in denudation rates across orogens appear to affect patterns of crustal deformation (Beaumont et al., 2000; Willett, 1999). For relatively simple orogens, the prevailing direction of rain-bearing winds seems significant. On the windward side of the orogen, higher runoff generated by higher precipitation totals leads to higher rates of denudation than on the drier, leeward side. As a result, crustal rocks rise more rapidly on the windward flank than on the leeward flank, so creating a patent asymmetry in depths of denudation across the orogen and producing a characteristic pattern of crustal deformation. Such modelling studies indicate that a reversal of prevailing rain-bearing winds will produce a change in topography, spatial patterns of denudation, and the form of crustal deformation (Summerfield, 2007). In addition, they show that the topographic and deformational evolution of orogens result from a complex interplay between tectonic processes and geomorphic processes driven by climate.

To test the idea that continental land form, as expressed in the spatial distribution of elevations and slopes, is a primary expression of the complex interplay of tectonic and climatic systems,

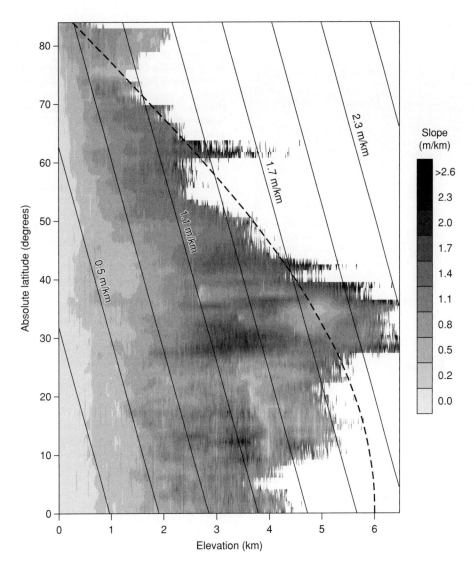

Figure 10.4 Average slopes of continental surfaces as a function of elevation and absolute latitude. Open areas at higher elevations are elevation–latitude coordinates not presently represented by continental surfaces on the modern Earth. Diagonal lines are contours of a first-order trend surface through slope values. Slope is expressed as rise/run in m/km. The average slope over all continental surfaces is about 3 m/km. The heavy black and white line is the best-fit cosine of maximum elevation at each latitude (r^2 = 0.49). The Tibetan Plateau is visible as the grouping of low-slope values around 5 km and 350 (after McElroy and Wilkinson (2005))

McElroy and Wilkinson (2005) used height of land and steepness of slope data from a global 30-arc second digital elevation model. They found that continental topography, as manifest in area–elevation relationships, is largely insensitive to latitude, and therefore to climate. However, their analysis revealed that land surface slope does vary with absolute latitude, with steeper slopes associated with higher latitudes (Figure 10.4). McElroy and Wilkinson argued that the poleward influence of cryogenically mediated processes of continental denudation explains this finding.

Nonlinear dynamics

Advances in the application of nonlinear dynamics to geomorphic systems promise to cast new light on the process–history problem. Phillips (2007a, 2007b) contends that geomorphic systems have manifold environmental controls and forcings, which can produce many different landscapes. Moreover, some controls and forcings are causally contingent and specific to time and place. Dynamical instability creates and enhances some of this contingency by encouraging the effects of small initial variations and local disturbances to persist and grow disproportionately big. Phillips then reasons that the combined probability of any particular set of global controls is low, and the probability of any set of local, contingent controls is even lower. In consequence, the likelihood of any landscape or geomorphic system existing at a particular place and time is negligibly small; all landscapes are perfect, in the sense that they are an improbable coincidence of several different forces or factors. This fascinating notion dispenses with the view that all landscapes and landforms are the inevitable outcome of deterministic laws. In its stead, it offers a powerful and integrative new view that sees landscapes and landforms as circumstantial and contingent outcomes of deterministic laws operating in a specific environmental and historical context, with several outcomes possible for each set of processes and boundary conditions. This view may help to reconcile different process and historical geomorphological traditions.

REFERENCES

Ahnert, F. (1976) Brief description of a comprehensive three-dimensional process–response model of landform development, *Zeitschrift für Geomorphologie* 25: 29–49.

Baas, A.C.W. (2002) Chaos, fractals and self-organization in coastal geomorphology: simulating dune landscapes in vegetated environments, *Geomorphology* 48: 309–28.

Bagnold, R.A. (1954) *The Physics of Blown Sand and Desert Dunes*. 2nd edn. Methuen, London.

Bagnold, R.A. (1956) The flow of cohesionless grains in fluids, *Philosophical Transactions of the Royal Society, London* A249: 235–97.

Bagnold, R.A. (1966) An approach to the sediment transport problem from general physics. *United States Geological Survey, Professional Paper* 282-E. pp. 135–44.

Baker, V.R. and Twidale, C.R. (1991) The reenchantment of geomorphology, *Geomorphology* 4: 73–100.

Beaumont, C., Kooi, H. and Willett, S. (2000) Coupled tectonic–surface process models with applications to rifted margins and collisional orogens, in M.A. Summerfield (ed.), *Geomorphology and Global Tectonics*. John Wiley & Sons, Chichester. pp. 29–55.

Brown, E.H. (1980) Historical geomorphology – principles and practice, *Zeitschrift fur Geomorphologie, Supplementband* 36: 9–5.

Brunsden, D. and Kesel, R.H. (1973) The evolution of the Mississippi River bluff in historic time, *Journal of Geology* 81: 576–97.

Burbank, D.W. and Anderson, R.S. (2001) *Tectonic Geomorphology: A Frontier in Earth Science*. Blackwell Science. Malden, Massachusetts.

Butler, D.R. (1995) *Zoogeomorphology: Animals as Geomorphic Agents*. Cambridge University Press, Cambridge.

Chappell, J. (1978) On process–landform models from Papua New Guinea and elsewhere, in J.L. Davies and M.J. Williams (eds), *Landform Evolution in Australia*. Australian National University Press, Canberra. pp. 348–61.

Chorley, R.J. (1962) Geomorphology and General Systems Theory. *US Geological Survey Professional Paper 500-B*.

Chorley, R.J. (1965a) A re-evaluation of the geomorphic systems of W.M. Davis, in R.J. Chorley and P. Haggett (eds), *Frontiers in Geographical Teaching: The Madingley Lectures for 1963*. Methuen, London. pp. 21–38.

Chorley, R.J. (1965b) The application of quantitative methods to geomorphology, in R.J. Chorley and P. Haggett (eds), *Frontiers in Geographical Teaching; The Madingley Lectures for 1963*. Methuen, London. pp. 147–63.

Chorley, R.J. and Kennedy, B.A. (1971) *Physical Geography: A Systems Approach*. Prentice-Hall, London.

Chorley, R.J., Dunn, A.J. and Beckinsale, R.P. (1964) *The History of the Study of Landforms, or the Development of Geomorphology, vol. 1: Geomorphology before Davis*. Methuen, London.

Chorley, R.J., Schumm, S.A. and Sugden, D.A. (1984) *Geomorphology*. Methuen, London.

Church, M. (1996) Space, time and the mountain – how do we order what we see?, in B.L. Rhoads and C.E. Thornes (eds), *The Scientific Nature of Geomorphology*. John Wiley & Sons, New York. pp. 147–50.

Codilean, A.T., Bishop, P. and Hoey, T.B. (2006) Surface process models and the links between tectonics and topography, *Progress in Physical Geography* 30: 307–33.

Culling, W.E.H. (1960) Analytical theory of erosion, *Journal of Geology* 68: 336–44.

Culling, W.E.H. (1963) Soil creep and the development of hillside slopes, *Journal of Geology* 71: 127–61.

Culling, W.E.H. (1965) Theory of erosion on soil-covered slopes, *Journal of Geology* 73: 230–54.

Davis, W.M. (1895) On the origin of certain English rivers, *Geographical Journal* 5: 128–46.

Davis, W.M. (1909) *Geographical Essays*. Ginn, Boston, Massachusetts.

Dietrich, W.E. and Smith, J.D. (1983) Influence of the point bar on flow through curved channels, *Water Resources Research* 9: 1173–92.

Dietrich, W.E. and Perron, J.T. (2006) The search for a topographic signature of life, *Nature* 439: 411–8.

Dietrich, W.E., Bellugi, D.G., Sklar, L.S., Stock, J.D., Heimsath, A.M. and Roering, J.J. (2003) Geomorphic transport laws for predicting landscape form and dynamics, in P.R. Wilcock and R.M. Iverson (eds), *Prediction in Geomorphology.* Geophysical Monograph 135. Washington DC: American Geophysical Union. pp. 103–32.

Embleton, C. and Thornes, J. (eds) (1979) *Process in Geomorphology.* Edward Arnold, London.

Gilbert, G.K. (1877) *Geology of the Henry Mountains (Utah).* United States Geographical and Geological Survey of the Rocky Mountain Region. United States Government Printing Office, Washington DC.

Graf, W.L. (1979) Catastrophe theory as a model for changes in fluvial systems, in D.D. Rhodes and G.P. Williams (eds), *Adjustments of the Fluvial System.* Kendall Hunt, Dubuque, Iowa. pp. 13–32.

Graf, W.L. (1982) Spatial variation of fluvial processes in semi-arid lands, in C.E. Thorn (ed.), *Space and Time in Geomorphology.* Allen & Unwin, London. pp. 192–217.

Green, C.P. (1985) Pre-Quaternary weathering residues, sediments and landform development: examples from southern Britain, in K.S. Richards, R.R. Arnett and S. Ellis (eds), *Geomorphology and Soils.* Allen & Unwin, London. pp. 58–77.

Gregory, K.J. (2000) *The Changing Nature of Physical Geography.* Arnold, London.

Hack, J.T. (1960) Interpretation of erosional topography in humid temperate regions, *American Journal of Science* 258A: 80–97.

Harrison, S. (2001) On reductionism and emergence in geomorphology, *Transactions of the Institute of British Geographers* (New Series) 26: 327–39.

Hodges, K.V. (2006) Climate and the evolution of mountains. *Scientific American* 295: 54–61.

Hodges, K.V., Wobus, C.W., Ruhl, K., Schildgen, T. and Whipple, K. (2004) Quaternary deformation, river steepening, and heavy precipitation at the front of the Higher Himalayan ranges, *Earth and Planetary Science Letters* 220: 379–89.

Horton, R.E. (1945) Erosional development of streams and their drainage basins: hydrophysical approach to quantitative morphology. *Bulletin of the Geological Society of America* 56: 275–370.

Howard, A.D. (1965) Geomorphological systems – equilibrium and dynamics, *American Journal of Science* 263: 302–12.

Huggett, R.J. (2007a) *Fundamentals of Geomorphology.* 2nd edn. Routledge, London.

Huggett, R.J. (2007b) A history of the systems approach in geomorphology (3rd edition, 2011), *Géomorphologie: Relief, Processus, Environnement* 2007, 2: 145–58.

Huggett, R.J. (2010) *Physical Geography: The Key Concepts.* Routledge, Abingdon.

Jones, D.K.C. (1981) *Southeast and Southern England. The Geomorphology of the British Isles.* Methuen, London and New York.

Jones, D.K.C. (1999) Evolving models of the Tertiary evolutionary geomorphology of southern England, with special reference to the Chalklands, in B.J. Smith, W.B. Whalley and P.A. Warke (eds), *Uplift, Erosion and Stability: Perspectives on Long-Term Landscape Development.* Geological Society Special Publication 162. The Geological Society, London. pp. 1–23.

Kirkby, M.J. (1971) Hillslope process: response models based on the continuity equation, in D. Brunsden (ed.), *Slope: Form and Process.* Institute of British Geographers Special Publication No. 3. pp. 15–30.

Kirkby, M.J. (1984) Modelling cliff development in South Wales: Savigear re-viewed, *Zeitschrift für Geomorphologie* 28: 405–26.

Lane, S.N. and Richards, K.S. (1997) Linking river channel form and process: time, space and causality revisited, *Earth Surface Processes and Landforms* 22: 249–60.

Lehre, A.K. 1982. Sediment budget of a small coastal range drainage basin in north-central California, in F.J. Swanson, R.J. Janda, T. Dunne and D.N. Swanson (eds), *Sediment Budgets and Routing in Forested Drainage Basins.* US Forest Service, Pacific Northwest Forest Range Experimental Station. General Technical Report PNW-141. pp. 66–77.

Leopold, L.B., Wolman M.G. and Miller J.P. (1964) *Fluvial Processes in Geomorphology.* WH Freeman, San Francisco and London.

Lidmar-Bergström, K. (1996) Long term morphotectonic evolution in Sweden, *Geomorphology* 16: 33–59.

Martin, Y. and Church, M. (2004) Numerical modelling of landscape evolution: geomorphological perspectives, *Progress in Physical Geography* 28: 317–39.

McElroy, B. and Wilkinson, B. (2005) Climatic control of continental physiography, *The Journal of Geology* 113: 47–58.

Melton, M.A. (1958) Geometric properties of mature drainage systems and their representation in E_4 phase space, *Journal of Geology* 66: 35–54.

Molnar, P. and England, P. (1990) Late Cenozoic uplift of mountain ranges and global climate change: chicken or egg? *Nature* 346: 29–34.

Montgomery, D.R. and Brandon, M.T. (2002) Topographic controls on erosion rates in tectonically active mountain ranges, *Earth and Planetary Science Letters* 201: 481–9.

Montgomery, D.R., Balco, G. and Willett, S.D. (2001) Climate, tectonics, and the morphology of the Andes, *Geology* 29: 579–82.

Murray, A.B., Lazarus, E., Ashton, A., Baas, A., Coco, G., Coulthard, T. et al., (2009) Geomorphology, complexity, and the emerging science of the Earth's surface, *Geomorphology* 103: 496–505.

Nye, J.F. (1951) The flow of glaciers and ice sheets as a problem in plasticity, *Proceedings of the Royal Society of London,* Series A, 207: 554–72.

Ollier, C.D. (1981) *Tectonics and Landforms.* Geomorphology Texts 6. Longman, London and New York.

Ollier, C.D. (1991) *Ancient Landforms.* Belhaven Press, London and New York.

Ollier, C.D. (1992) Global change and long-term geomorphology, *Terra Nova* 4: 312–9.

Ollier, C.D. (1995) Tectonics and landscape evolution in southeast Australia, *Geomorphology* 12: 37–44.

Ollier, C.D. and Pain, C.F. (1994) Landscape evolution and tectonics in southeastern Australia. *AGSO Journal of Australian Geology and Geophysics* 15: 335–45.

Ollier, C.D. and Bourman, R.P. (2002) Flared slopes, footslopes, and the retreat of overhanging slopes: examples of convergent landform development, *Physical Geography* 23: 321–34.

Paine, A.D.M. (1985) 'Ergodic' reasoning in geomorphology: time for a review of the term? *Progress in Physical Geography* 9: 1–15.

Peizhen, Z., Molnar, P. and Downs, W.R. (2001) Increased sedimentation rates and grain sizes 2–4 Myr ago due to the influence of climate change on erosion rates, *Nature* 410: 891–7.

Perron, J.T., Kirchner, J.W. and Dietrich, W.E. (2009) Formation of evenly spaced ridges and valleys, *Nature* 460: 502–5.

Phillips, J.D. (1999) *Earth Surface Systems: Complexity, Order, and Scale*. Blackwell, Oxford.

Phillips, J.D. (2006) Deterministic chaos and historical geomorphology: a review and look forward, *Geomorphology* 76: 109–21.

Phillips, J.D. (2007a) The perfect landscape, *Geomorphology* 84: 159–69.

Phillips, J.D. (2007b) Perfection and complexity in the lower Brazos River, *Geomorphology* 91: 364–77.

Pinchemel, P. (1954) *Les Plaines de Craie du Nord-Ouest du Bassin Parisien et du Sud-est du Bassin de Londres et Leurs Bordures*. Armand Colin, Paris.

Pinter, N. and Brandon, M.T. (1997) How erosion builds mountains, *Scientific American* 276 (April): 60–5.

Qin, S.Q., Jiao, J.J. and Wang, S. (2001) A cusp catastrophe model of instability of slip-buckling slope, *Rock Mechanics and Rock Engineering* 34: 119–34.

Raymo, M.E. and Ruddiman, W.F. (1992) Tectonic forcing of late Cenozoic climate, *Nature* 359: 117–22.

Rea, D.K., Snoeckx, H. and Joseph, L.H. (1998) Late Cenozoic eolian deposition in the North Pacific: Asian drying, Tibetan uplift, and cooling of the northern hemisphere, *Paleoceanography* 13: 215–24.

Rhoads, B.L. (2006) The dynamic basis of geomorphology reenvisioned. *Annals of the Association of American Geographers* 96: 14–30.

Savigear, R.A.G. (1952) Some observations on slope development in South Wales, *Transactions of the Institute of British Geographers* 18: 31–51.

Schumm, S.A. (1963) Sinuosity of alluvial rivers on the Great Plains, *Geological Society of America Bulletin* 79: 1573–88.

Schumm, S.A. (1973) Geomorphic thresholds and complex response of drainage systems, in M. Morisawa (ed.), *Fluvial Geomorphology*. State University of New York, Binghamton, Publications in Geomorphology. pp. 299–310.

Schumm, S.A. (1977) *The Fluvial System*. John Wiley & Sons, New York.

Schumm, S.A. (1991) *To Interpret the Earth: Ten Ways to be Wrong*. Cambridge University Press, Cambridge.

Schumm, S.A. and Lichty, R.W. (1965) Time, space, and causality in geomorphology, *American Journal of Science* 263: 110–9.

Simpson, G.G. (1963) Historical science, in C.C. Albritton (ed.), *The Fabric of Geology*. Addison-Wesley, Reading, MA. pp. 24–48.

Slaymaker, O. (2009) The future of geomorphology, *Geography Compass* 3: 329–49.

Small, E.E. and Anderson, B.S. (1995) Geomorphically driven late Cenozoic rock uplift in the Sierra Nevada, California, *Science*, 270: 277–80.

Steno, N. (1916) *The prodomus of Nicolaus Steno's dissertation concerning a solid body enclosed by process of Nature within a solid*. An English version with an introduction and explanatory notes by John Garrett Winter, with a foreword by William H. Hobbs. University of Michigan Humanistic Series, vol. XI. *Contributions to the History of Science, Part II*. Macmillan, New York and London.

Stoddart, D.R. (ed.) (1997) *Process and Form in Geomorphology*, Routledge, London.

Strahler, A.N. (1950) Equilibrium theory of erosional slopes, approached by frequency distribution analysis. *American Journal of Science*, 248: 673–96 and 800–14.

Strahler, A.N. (1952) Dynamic basis of geomorphology, *Bulletin of the Geological Society of America* 63: 923–38.

Strahler, A.N. (1980) Systems theory in physical geography, *Physical Geography* 1: 1–27.

Summerfield, M.A. (2005) A tale of two scales, or the two geomorphologies, *Transactions of the Institute of British Geographers* NS 30: 402–15.

Summerfield, M.A. (2007) Internal–external interactions in the Earth system, in I. Douglas, R.J. Huggett and C.R. Perkins (eds), *Companion Encyclopedia of Geography: From Local to Global*. 2nd edn. Routledge, London. pp. 93–107.

Summerfield, M.A. and Hulton, N.J. (1994) Natural controls of fluvial denudation rates in major world drainage basins, *Journal of Geophysical Research* 99: 13871–83.

Thom, R. (1975) *Structural Stability and Morphogenesis*. Benjamin, New York.

Thornbury, W.D. (1954) *Principles of Geomorphology*. 1st edn. John Wiley & Sons, New York.

Thornes, J.B. (1983) Evolutionary geomorphology, *Geography* 68: 225–35.

Viles, H.A. (ed.) (1988) *Biogeomorphology*. Basil Blackwell, Oxford.

Whitehead, A.N. (1925) *Science and the Modern World*. The Free Press, New York.

Willett, S.D. (1999) Orography and orogeny: the effects of erosion on the structure of mountain belts, *Journal of Geophysical Research* 104: 28957–81.

Willett, S.D., Hovius, N., Brandon, M.T. and Fisher, D.M. (eds) (2006) *Tectonics, climate, and landscape evolution*.

Geological Society of America, Special Paper 398. The Geological Society of America, Boulder, Colorado.

Wobus, C.W., Hodges, K.V. and Whipple, K.X. (2003) Has focused denudation sustained active thrusting at the Himalayan topographic front?, *Geology* 31: 861–4.

Wooldridge, S.W. and Linton, D.L. (1939) *Structure, surface and drainage in south-east England*. Publication 10. Institute of British Geographers, London.

Wooldridge, S.W. and Linton, D.L. (1955) *Structure, Surface and Drainage in South-East England*. George Philip & Sons, London.

11

Dating Surfaces and Sediments

Tony G. Brown

Geomorphology, as a sub-discipline of earth sciences and concerned with the formation of landforms and earth surface processes at all scales, is a historical science. Although it has been more apparent in stratigraphic geology, methods of dating surfaces and sediments continue to have a profound influence on the development of geomorphology. Indeed it is arguable that one of the reasons that traditional geology departments in Europe rather neglected physical geology, or geomorphology as it was to become, was that it was seen as being far too recent to be amenable to standard stratigraphic and biostratigraphic methods of correlation and dating. Before the advent of chronometric dating including radiocarbon (which, although discovered in the 1940s was not extensively used until the last 40 years of the 20th century), the study of landforms was trapped in a circular argument in which climate and time were reciprocal; that is, either an entity was very old, or had developed very rapidly in response to the prevailing climate. This promoted intuitive, but untestable, hypotheses such as superposition, exhumation and rejuvenation and was an underlying contributory factor in the abandonment of denudation chronology in the 1960s in favour of detailed studies of geomorphological processes. The increasing availability of radiocarbon dating in the 1970s was also responsible for a partial return to studies of denudation chronology, albeit in the form of Late Pleistocene and Holocene alluvial and glacial chronologies (Brown, 1997). However, it was not until the development of direct sediment dating techniques and surface dating that the story has turned a full circle and

earth scientists are able to tackle multi-temporal problems of landform evolution under changing forcing conditions.

Before discussing the different techniques that can be used to estimate dates, it is worth thinking rather more about what we mean by sediment or surface dating. It appears simple – the estimation of an age of a sediment body or a rock or soil surface. However, there are a number of hidden assumptions built into the concept, and therefore, into our interpretation of any dates. All sediments and surfaces have a number of dates of formation, continuation and, in some cases, burial. What is normally regarded as a sediment date by geomorphologists is the age measured in some system, relatable to calendrical years, since the last period of grain deposition. However, this may be one of two, three or even more episodes in a complex denudational grain history subsequent to a number of geological cycles of sedimentary origin, or after crystallization from a melt of igneous origin as illustrated in Figure 11.1. This is becoming of critical importance for two reasons. First, the sources of error and applicability of many methods of direct sediment dating is intimately connected to grain history, and secondly methods of direct sediment dating can also be used to reveal grain history and therefore have a relationship to denudation flux rates. A similar question arises with geomorphological surfaces as geomorphologists generally attempt to date the emergence or creation of landforms, such as a tor, inselberg or roche mountoneé, but this is only the latest in a number of possible surface ages. Most rock surfaces that make up landforms are probably

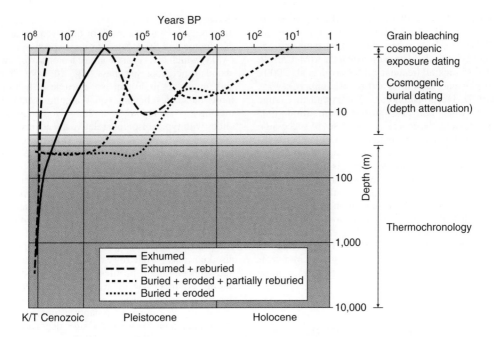

Figure 11.1   Grain history chronology with application ranges for grain dating techniques

previous discontinuities in the rock mass, or sedimentary boundaries. Far from being an exception, exhumation is a normal state. So, for example, a tor forming at the apex or outer edge of a granitic pluton has an intrusion age, cooling period, subduction and burial period and an unroofing period or age, and finally a development age as a result of differential weathering and erosion. The problems that this poses for landform interpretation, including modelling, will be discussed later in this chapter. The technical developments in surface and sediment dating also feed into our increasingly interconnected view of earth surface processes over time and towards an integration of tectonics, climate, sediment and biogeochemical cycling incidentally resulting in a critical challenge to the concept of the homeostatic control over earth surface conditions (Leeder, 2007).

DATING EROSIONAL SURFACES

Where erosional surfaces are not buried there is no stratigraphic method of dating except that the surface must post-date the depositional age of the rock or sediment that constitutes it. In the case of bedrock this is generally pre-Tertiary and probably millions of years before any geomorphological processes that were responsible for the

creation of the surface. The result is that erosional landforms have been difficult to date directly. For very young (Late Holocene) rock surfaces biological proxies such as lichenometry, or if trees are growing on the surface, dendrochronology have been used.

Lichenometry and dendrochronology

The slow growth rates of lichen on rock surfaces (some may be over 5000 years old in the Arctic) makes lichen theoretically suitable for rock surface dating using lichenometry (Innes, 1985). Likewise the growth of trees with annual growth rings can be used to date surfaces using dendrochronology. In both methods growth is a variable that is environmentally dependent and this environmental effect will be included in a lichen or tree growth curve derived from the local area. Due to the problems of lichen thallus coalescence, surface weathering and the limited age range of dated surfaces available any growth curve, lichenometry is only reliable and generally applicable for the last few hundred years. However, this has proved valuable in studies of both Late Holocene flood history (Merrett and Macklin, 1999) and glacial retreat, especially in relation to the Little Ice Age both in Europe (Small and

Clarke, 1974; Matthews and Matthews, 2005) and in South America (Winchester and Harrison, 2000; Harrison et al., 2008). Lichenometry has also been used to date snow avalanches (McCarroll, 1993), rockfall frequency (McCarroll et al., 2000), for channel definition and flood events (Gregory, 1977). Dendrochronology can extend this range somewhat but also suffers from limitations imposed by the ecological processes of succession and disturbance in what are generally rather high-stress environments. An additional methodology most used in semi-arid zones is soil formation (Birkeland, 1974). The rate of formation of horizons, especially the B horizon and clay, Fe and in some cases $CaCO_3$ accumulation can be used to assign both rank order and approximate ages to soils, even allowing for the differential rate of soil formation caused by climatic change (Johnson et al., 1990). This methodology grades into sediment dating and so will be discussed again in the next section.

Weathering and soil formation

Both weathering and soil formation are clearly products of both time and climate (Birkeland, 1974). Weathering has tended to be used as a dating method in cool temperate regions, whilst soil formation has been used both in temperate chronosequences such as studies of glacial retreat (Huggett, 1998; Bernasconi et al., 2008) and more rarely alluvial terraces (see Brown, 1997) and arid to semi-arid environments, particularly under desert pavements (McFadden et al., 1987). The most common approach to the quantification of weathering has been the use of apparatus such as the Schmidt Hammer for the measurement of rock hardness (Matthews and Shakesby, 1984; McCarroll, 1991). Our understanding of the formation of desert pavements has been greatly increased by the aeolian model of pavement formation proposed by Wells and now tested using exposure dating (Wells et al., 1995). This provides a mechanism by which the A horizon accumulates and develops a vesicular character (Av horizon) which is a function of time under continuously arid conditions. The advent of optically stimulated luminescence (OSL) and cosmogenic dating (see below) has resurrected these techniques by supplying an additional test of their reliability and the limits to their application.

Cosmogenic surface dating

Cosmogenic dating is the use of isotopes generated at the surface of rocks as a result of the exposure of the rock surface to solar radiation

(Stuart, 2001). A major advance has occurred in this research field as a result of the invention and development of the accelerator mass spectrometer (AMS) which can separate a rare isotope from an abundant neighbouring mass (e.g. ^{14}C from ^{12}C) and suppresses molecular isobars (one of two or more molecules having equal atomic weights) making it possible to measure long-lived radionuclides with typical isotopic relative abundances of 10^{-12} to 10^{-18}. In 1977 Richard Muller showed how accelerators could be used for the detection of ^3H (tritium), ^{14}C and ^{10}Be and reported the first date using ^3H (Muller, 1977). Since then the most commonly used of these cosmogenic nuclides have been ^{10}Be and ^{26}Al on siliceous rocks such as granites, ^{36}Cl on limestones and it is also theoretically feasible to use ^{53}Mn on haematite. There are two principal approaches: surface exposure dating also known as cosmic ray exposure (CRE) dating, and concentration profiles. Surface exposure dating assumes that there has either been no erosion of the rock surface over the time period in question or that it can be reliably estimated. The concentration profile has to consider depth attenuation and nature of the material. The cosmogenic technique will be introduced only in outline here, and more details are given in reviews by Gosse and Phillips (2001), Bierman and Nichols (2004) and by Cockburn and Summerfield (2004). The fundamental principle is that cosmogenic isotopes (e.g. ^{10}Be and ^{26}Al) are created when elements (target elements) in the atmosphere or on Earth are bombarded by cosmic rays that penetrate the atmosphere from outer space. The accumulation of these isotopes within a rock surface can be used to establish how long it has been exposed to cosmic radiation, rather than being shielded by ice or sediment. In the case of ^{10}Be and ^{26}Al, the target mineral is quartz due to nuclear interactions between ^{28}Si and ^{16}O, principally due to bombardment with cosmic ray nucleons (neutrons and protons) and by slow and fast muons (elementary particles with 200 times the mass of electrons created only by cosmic rays). To date, cosmogenic isotopic dating has been applied to glacial advances and retreats (Matthews et al., 2008), fault scarps (Benedetti et al., 2002), tors (Phillips et al., 2006; Hagg, 2009), strath terraces, bedrock channels the linkage of long-term landform evolution with earth surface processes and tectonics (Bishop, 2007). The case study presented here is the dating of megaflood deposits in Siberia. A particularly innovative use of cosmogenic exposure dating has been on the Cap Canaille cliff in southern France with the aim of estimating the age of a major cliff collapse which resulted in a coastal tsunami (Recorbet et al., 2009).

The Altai Mountains, in southern Siberia have become well known in recent years for the preservation of landforms and sediments associated with some of the largest floods in Earth's history (Baker et al., 1993). These floods may have had significant effects on global climate through the reorganization of the drainage of Asia (Grosswald, 1980; Velichko et al., 1984; QUEEN Project, Thiede and Bauch, 1999). However, the significance of these events is difficult to assess until they are dated. The geomorphic features which include giant current ripples (Rudoy and Baker, 1993; Carling et al., 2001), flood berms, lake sediments, dropstones, spillways and palaeochannels (Rudoy and Baker, 1993), were caused by the failure of ice-dammed lakes in the Kuray and Chuya Basins. It is believed that the failure of these dams released billions of cubic metres of water with maximum estimated discharges reaching 18×10^6 m^3 s^{-1} (Rudoy, 1998). This water and associated sediment discharged into the Ob River system and then, according to Grosswald (1980, 1998), into Lake Mansi. As a result Lake Mansi may have overflowed through a series of gaps and spillways into the Aral, Caspian and eventually Black and Mediterranean Seas (Grosswald, 1980). However, more recent research by the QUEEN Group (Svendsen et al., 1999) has suggested a far smaller extent of the Weichselian (MIS 5d-2) ice sheets east of the Ural Mountains. They propose a maximum mid-Weichselian extent when there would have been no ice-free corridor to the Arctic Ocean and a glacial lake (Lake Komi) extending up to 100 m asl, whilst in the late Weichselian (western European LGM), ice was even more limited and the Ob could have discharged into the Arctic Ocean. The Lake Komi shorelines have been OSL-dated at 93 ± 13 ka, 7 ± 12 ka and 88 ± 11 ka (Mangerud et al., 1999). It is therefore important to try and correlate the Altai megafloods with this northern ice-sheet history. If overflowing to the south did occur then it would represent a massive transfer of water and energy from Central Asia to the mid latitudes. Alternatively, the impact would have been on the North Atlantic and Arctic Oceans (Velichko et al., 1984, 1997). They were believed to be Late Pleistocene on the basis of their superficial location in the sedimentary sequence of the region but they are likely to predate 15.6 K BP on the basis of recent radiocarbon dating of overlying lake sediments (Blyakharchuk et al., 2004). The flood deposits were well suited for cosmogenic isotope dating of the surface few centimetres from horizontal faces of the blocks for the following reasons:

- all the samples contain quartz (metamorphics/schists and granites);
- the high altitude (700–2100 m asl);

- the lack of overhanging vegetation as most of the samples come from a desert–steppe environment;
- the large block size (types a and b) reducing the possibility of subsequent burial and exhumation or overturning.

The results displayed in Figure 11.2 show a high degree of clustering around the date of 15,000–16,000 BP and this agrees with an independent study using a very similar methodology by Reuther et al. (2006). There is also evidence from one deposit of a previous flood c. 30,000 BP.

Cosmogenic burial dating

It is also possible to use cosmogenic isotopes for burial dating from a depth profile of cosmogenic measurements and in effect a sediment dating technique. This method is almost the inverse of surface dating as it uses the slow decay of ^{10}Be, ^{26}Al (Table 11.1) to determine the time since exposure. The reason that a ^{10}Be/^{26}Al ratio depth profile is required is that for shallow burial the post-burial nuclide production by cosmic ray muons cannot be ignored (Fabel, 2004). In fact, to shield a sample completely from cosmic rays requires at least 27 m of rock or 37 m of sediment. Therefore, the first applications were to caves (Granger and Muzikar, 2001; Anthony and Granger, 2004) but subsequent applications have been made to open-air sediments using an estimate of the post-burial production. In many cases this should also include an estimate of the changing depth of burial and this is again problematic. In general, these errors will produce ages that are too young, especially if post-burial production is ignored (a so-called naïve date). Additionally, the conversion of this profile into an age for terrace deposition is complicated by pre-depositional inheritance and post-depositional disturbance including superficial deposition and/or erosion including wind ablation (Brocard et al., 2003). The older the terraces the greater the scatter of terrace surface dates and it is generally accepted that the oldest dates are closest to the probable age of terraces but all estimates have large error ranges up to 20 per cent (Siame et al., 1997). This is an exciting area of future research with the potential to not only extend sediment budgets and denudation rate estimation over longer time periods but also date terrace gravels which may contain important archaeological remains in the form of stone tools but which lack faunal remains or carbonates. Combined studies with OSL dating are clearly an immediate next step. Attempts are currently being made to date the terraces of the Lower Rhine in Germany using

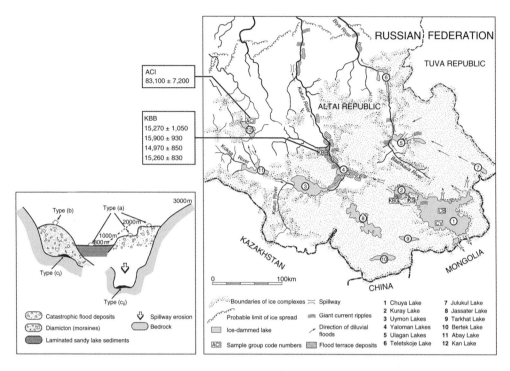

Figure 11.2   Location and sampling structure for the cosmogenic dating of flood deposits in the Altai Mountains, southern Siberia. Results shown are the ¹⁰Be exposure ages only

T_1 type (C_i)   striations created by advancing glacier (predating ice-dammed lake).

T_2 type (b)   blocks on moraines marking maximum extent of the glacier slightly predating/contemporaneous with ice-dammed lake.

T_3 type (a)   dropstones deposited during the short period of lake existence (from floating ice).

T_4 type (a)   blocks on flood berms immediately postdating the ice-dammed lake and dating the flood(s).

T_4 type (C_{ii})   bedrock scour immediately post-dating the ice-dammed lake and dating the flood(s).

burial dating along with palaeomagnetism and both pollen and mammal biostratigraphy (Boenigk and Frechen, 2006; Dehnert et al., 2006). The method has huge potential as a result of the rather convenient range of 0.1–5 Ma and the fact that it is in theory applicable to nearly all earth surface sediments.

A particularly innovative use of cosmogenic isotopes, in conjunction with OSL, has been worked by Matmon (2009) dating the evolution of drainage of the Makhtesh Hazera drainage system on the western flank of the Dead Sea rift. This study attempted to relate the sub-catchment denudation rates to the evolution of the macro-scale landform which in this case is a spectacular crater-form basin developed on an uplifted anti-cline (Zilberman, 2009).

DATING SEDIMENTS

The dating of sediments and therefore depositional landforms such as river terraces, slope deposits, alluvial fans as well as tectonic movements has developed from relative stratigraphic dating, through biostratigraphic correlation to independent or chronometric dating, the earliest of which was radiocarbon dating. Radiocarbon dating was the first of the radionuclide methods widely applied in geomorphology dating almost back to the first development of the method by Libby in the 1950s. Its pre-eminence as a dating method is illustrated by the fact that it alone of all dating methods has a dedicated journal, *Radiocarbon*.

Table 11.1 The half lives of short- to long-lived isotopes used in earth sciences

Radionuclide	Half-life	Environmental origin	Principal applications
^{7}Be	≈54 days	Natural; galactic cosmic rays and solar energetic particles in Earth's atmosphere	Natural tracer
^{228}Th	1.9 years	Natural	Sedimentation rate in estuaries
^{132}Cs	2.0 years	Chernobyl fallout, now unmeasurable	Lake and fluvial sediments
^{210}Pb	22.8 years	Natural	Lake and fluvial sediments
^{137}Cs	30.3 years	Thermonuclear weapons testing and Chernobyl fallout	Lake and fluvial sediments and erosion rates (directly)
^{32}Si	≈140 years	Natural atmospheric precipitation and nuclear testing	Lake sediments and erosion rates
^{39}Ar	269 years	Natural noble gas	Neotectonics thermochronology
^{241}Am	432.2 years	Thermonuclear weapons testing	Lake sediments and erosion rates
^{14}C	5730 years	High-altitude production through cosmic ray impacts on nitrogen and uptake by living organisms	Post-death/burial dates, lakes, fluvial, glacial and mire sediments
^{36}Cl	300,000 years	Natural cosmogenic isotope	Exposure and burial of fluvial deposits, denudation rates
^{10}Be	1,500,000 years	Natural cosmogenic isotope	Exposure and burial of fluvial deposits, denudation rates
^{53}Mn	3,700,000 years	Natural cosmogenic isotope found in meteorites	Terrestrial weathering and landform evolution
^{26}Al	7,100,000 years	Natural cosmogenic isotope	Exposure and burial of fluvial deposits, denudation rates

Source: Data are from various sources including the Smithsonian/NASA Astrophysics Data System.

Longer-lived radionuclides: Radiocarbon

Cosmogenic nuclides are formed within Earth's atmosphere and surface as a result of a steady flux of cosmic rays. One of the most important atmospherically formed cosmogenic nuclides is ^{14}C, which can be found in any organic material. Just as with other cosmogenic radionuclides the estimation of the production rate is critical. Fortunately, from the matching of long tree ring chronologies and most recently long varve sequences from Lake Suigetsu which is located near the coast of the Sea of Japan, a calibration curve now exists back well beyond 10,000 BP (Kitagawa and van der Plicht, 1998, 2000; Bronk-Ramsey, 2009). Its establishment as a method over 40 years ago, and its universal use, has yielded a vast literature on ^{14}C dating including recent developments of the accelerator mass spectrometry, improved processing to remove contaminants (e.g. the A-box method; Higham et al., 2006) which have led to a dramatic reduction in the size of sample required (by a factor of 100) and extension of the maximum reliable dating to c. 50 ka BP.

When many ^{14}C measurements can be made, such as on organic lake and ombrotrophic mire sequences the age precision can be greatly improved by wiggle-matching with the calibration curve (Bronk-Ramsey et al., 2001). Radiocarbon can be used on fluvial sediments in different ways: to give maximum ages, for example from soils, to give activity dates such as fragile organic matter (e.g. a leaf) incorporated into a single-episode deposit or minimum ages such as a peat developed over a flood deposit (Brown, 1997; Lewin et al., 2005; Chiverrell et al., 2007). When non-fragile organic matter, such as peat or wood, is incorporated into sediments, only a maximum sedimentation age can be determined, as a result of an unknown time-lapse or 'window' between the formation of the organic matter and its incorporation into the sediment. This is a particular problem with colluvial sedimentation and can cause unavoidable uncertainty in the depositional timeframe although this can be reduced using frequency analyses (Lang, 2003). Soils also present considerable problems stemming from the potentially long and variable residence time of the different forms of C compounds in soils, which is related to the soil-forming environment and which can change over time. Several studies have shown a systematic variation with the humin and humic fraction in alluvial sediments (Brown et al., 2005, 2007). One approach to these problems in fluvial sequences has been to mass the dates and plot them as probability density functions (Johnston et al., 2006) after removal of the artefacts of calibration. This implicitly accepts that there is considerable noise in the system, and it is inevitably biased towards sequences with high amounts of organic materials. As a geomorphological methodology it relies upon correlation with other proxy series; this may suggest rather than imply causality.

Long-lived radionuclides: Uranium series

This group of dating methods comprises the radioactive decay series of the uranium isotopes ^{238}U, ^{235}U and ^{232}U as based on the measurement of radiometric disequilibria of $^{230}Th-^{234}U$. The event which can be dated with these dating methods is the closing of a geochemical system, which was previously disturbed by chemical fractionation. Uranium series can be applied to pedogenic carbonates within or sealing sediment bodies (Sharp et al., 2003) and to carbonate fossils. Pedogenic ages are minimum ages, as a result of the unknown time elapsed between the sediment formation and carbonate precipitation which can be at least as long as c. 40,000 years (Birkeland, 1974). Another challenging problem is the polycyclic formation of pedogenic carbonates and the possible re-opening of the system which may result in recrystallisation. Microsampling techniques may be used to solve this problem (Mallick and Frank, 2002). Recent uses of uranium series for dating in geomorphology have used isochron analysis (extraction of multiple sub-samples from a horizon) to correct for detrital contamination and have shown that reliable dates can then be obtained for geomorphic events such as river terrace aggradation and abandonment (Candy et al., 2005).

One of the most important attributes of uranium series dating is its extension beyond the range of ^{14}C. This has been used in Israel to extend the Lake Lisan record from sedimentary carbonates. Stein and Goldstein (2006) used uranium–thorium dating for the correlation of this environmental chronology with global events, seismic events, palaeomagnetic events and to extend the calibration of the ^{14}C timescale.

Radiation dosimetry

Radiation dosimetry is almost the opposite of cosmogenic dating as it relies on changes to mineral crystals that have accumulated since the burial of the sediment grains. The applicability of radiocarbon is still limited to very Late Pleistocene and Holocene landforms where suitable organic materials have been incorporated. These limitations do not exist with methods based on radiation dosimetry, and this is why thermoluminescence (TL) and optically stimulated luminescence (OSL) have so much potential in geomorphology. The principle of this dating method is the time-dependent accumulation of energy in minerals, primarily quartz and feldspar, resulting from omnipresent low-level radiation from natural radioactivity and cosmic radiation. An excellent summary of the method with information on sampling methodology can be found in Duller (2008) with far more detail in (Krbetschek, 2009) but at its simplest the resetting of the accumulated energy is caused by heat or daylight, the latter of which is of crucial importance for the dating of sediments. Consequently, the last exposure of the mineral grains to daylight is determined, and thus the age of sediment deposition. Insufficient resetting of the accumulated energy would result in an age overestimation. Three different dosimetric dating techniques are currently used in geomorphology: thermoluminescence (TL), optically stimulated luminescence (OSL) and electron spin resonance (ESR). The basic difference between these methods is the way the accumulated energy is exploited and detected. Furthermore, their sensitivity to the resetting process by daylight is different. OSL shows the fastest bleaching characteristics by daylight, making this method the most important for dating fluvial sediments (Jain et al., 2004), although there are problems. These include some quartz which is frequently not fully bleached or zeroed, feldspar contamination and dose rate disequilibrium caused by high uranium levels. Even colluvial sediments with a short transport distance may be successfully dated by OSL (Fuchs et al., 2004). Two intrinsic tests on OSL date reliability are stratigraphic order in a vertical sequence and the exploitation of divergent environmental dosimetry within a single level or member (Brown et al., 2010). In the case of TL, bleaching behaviour is less reliable, as is also the case for ESR. Nevertheless, recent developments in single quartz grain analysis allow the selective extraction of only well-bleached grains (Beerten and Stesmans, 2005).

One of the first and most successful geomorphological applications of OSL has been to the dating of aeolian landforms in arid and semi-arid environments. The dating of longitudinal and source-bordering dunes within the Lake Eyre Basin of Australia by Nanson (Nanson et al., 1992) and Maroulis et al. (2007), has shown that although dune formation is climatically driven, there are important relationships and feedbacks between the aeolian and fluvial systems. Similar work on beach ridges in the Kalahari Basin by Burrough et al. (2009), has provided an improved chronology for mega-lake expansion and contraction.

OSL is the preferred technique for the dating fluvial-transported sediments because of their bleaching characteristics being faster than TL. However, when dealing with fluvial sediments the total resetting of the former luminescence signal often does not occur, resulting in an age overestimation, which has a stronger impact on young samples than on older ones (Jain et al., 2004). To deal with the problem of insufficient bleaching, various approaches are being followed in ongoing research studies (Arnold et al., 2007; Fuchs et al., 2007). Depending on the fluvial process of sediment transportation, certain mineral grain sizes are prone to be better bleached than others. In a case study on recent river flood sediments from the Elbe River (Germany), Fuchs et al. (2005) showed that fine-grained quartz seems to be better bleached than fine-grained feldspar. The residual luminescence signal from fine-grain quartz resulted in an estimated residual age of c. 0.1 ka, which would be problematic only for very young samples. Another approach is the reduction of the analysed sample material to small aliquots or even single grains (e.g. Fuchs and Wagner, 2005; Thomas et al., 2005). Based on the assumption that insufficiently bleached samples consist of a mixture of well and incomplete bleached mineral grains, the measurement of several sub-samples would result in a specific distribution. The part of the distribution with the lowest values represents the sub-samples with the best bleaching characteristics, thus supposed to be closest to the true deposition age (Olley et al., 1999). A very promising approach to dating insufficiently bleached samples is the analyses of specific OSL components, each with certain bleaching characteristics. Analysing only the fast bleaching component, Singarayer et al. (2005) achieved promising results for insufficiently bleached samples from various environments, applying the linear modulation technique.

The normal age range of the technique is limited by saturation of the quartz fast component signal, but experiments by Wang et al. (2007) have demonstrated the potential of a slower thermally transferred OSL (TT-OSL) signal from quartz which could considerably increase the maximum age range of the method. There is now a broadly applicable protocol for SAR TT-OSL (Stevens et al., 2009) which, under favourable conditions, can date sediments up to, and even beyond, the Brunhes–Matuyama palaeomagnetic boundary (0.78 Ma).

Short-lived radionuclides

There are several radionuclides with short half-lives of the order of decades, which have been used in geomorphology (Table 11.1). The criteria for a radioisotope to be a candidate for geomorphological dating are:

- The chemistry of the isotope (element) is known.
- The half-life of the isotope is known.
- The initial amount of the isotope per unit substrate is known or accurately estimated.
- The substrate adsorbs and incorporates an adequate amount of the isotope (in sedimentary systems, this is the finest, usually clay- or colloidal-size material).
- Once the isotope is attached to the substrate, the only change in concentration resulting from radioactive decay.
- In order to be useful, it is relatively easy to measure.
- The isotope has an effective range for the scale of time investigated (about 8 half-lives).

With these criteria met, the age of a substance can be calculated by the following formula

$$T_{age} = \ln(A_0/A_s) \times 1/\lambda \qquad (11.1)$$

where A_0 is the isotopic activity at time zero (the present) and A_s is the activity of the unknown, λ is the decay constant for the isotope.

The isotope ^{210}Pb is suitable for most recent sediments and ecosystem studies, where changes in deposition rates have occurred within the last century. A member of the ^{238}U series, ^{210}Pb is subject to disequilibria with its distant relative ^{226}Ra (radium) due to the physio-chemical activity of the intermediate gaseous progenitor ^{222}Rn (radon). Radioactive disequilibrium arises when the gaseous ^{222}Rn escapes into the atmosphere. With a half-life of 3.8 days, the ^{222}Rn decays through a series of isotopes of very short half-life to ^{210}Pb. This process produces excess ^{210}Pb in the atmosphere and subsequently the hydrosphere. Like Be (beryllium), the highly reactive Pb is rapidly adsorbed by, or incorporated on to, particulate material. Precipitation of this material produces excess ^{210}Pb over ^{210}Pb in equilibrium with ambient [226]Ra already in sediments. This excess ^{210}Pb provides the mechanism for age and depositional assessment. A modification of this technique can also be used to assess soil erosion of agricultural land (Walling and He, 1999). The most common use of ^{210}Pb in geomorphology is the assessment of recent erosion rates from lake or reservoir (dam) sediments.

There are also the so-called 'transuranic' elements that have been released into the atmosphere on a global scale since the early 1950s, the main source of which has been nuclear weapons testing. The first of these to be used to date lake sediments was ^{137}Cs which was measured in five lakes in the Lake District, UK in 1971 by Pennington et al. (1973). More recently, other transuranic radionuclides such as [238]Pu, [239]Pu, [240]Pu (plutonium) and [241]Am (americium) have been measured in lakes (Olsson, 1986) including Lake Ontario (Joshi, 1988) and Blelham Tarn in Cumbria, UK (Michel et al., 2001). The depth profile of these radionuclides in Blelham Tarn was compared with ^{137}Cs which also had a more recent peak resulting from the Chernobyl fallout in 1986. The same radionuclides have also been measured from fluvial sediments from the Danube and the Black Sea coast (Mihai and Hurtgen, 1997) and this opens the possibility of their use for the tracing and estimation of fine sediment flux over the last 50 years.

Applications generally concern transport rates of fine sediment down slopes, through floodplains or into lakes and reservoirs. In particular, ^{137}Cs has been particularly widely used for estimating both the rate of sediment deposition on floodplains and the rate of soil erosion from the truncation of soil profiles with reference to non-eroding soils (Zhang and Walling, 2005). Further development can be expected in this area with the difference between dating methods and sediment tracing techniques becoming increasingly blurred.

Amino acid racemization

Amino acid racemization (AAR) is a method which dates mollusc shells and is based upon the rate at which amino acids change after the death of the organism. Quaternary molluscs commonly contain six to ten amino acids including aspartic acid, glutamic acid, glycine, alanine, valine, leucine, isoleucine, proline and phenylalanine. These represent the residuum of the original calcification proteins. During diagenesis the proteins hydrolyse to release low molecular weight polypeptides and free amino acids. Loss of amino acids from fossils can occur by decomposition,

diffusion or leaching and the process of racemisation. Racemisation is the conversion of the original L-amino acids (laevo or left-handed) into a mixture of D (dextro or right-handed) and L-amino acid forms. The D/L ratio (enantiomeric ratio) is therefore a measure of time since the organism's death, subject to other variables which affect racemisation reactions. These are:

- Temperature (the rate increases with increasing temperatures)
- Genus used
- Amino acid types present
- Contamination and other physical–chemical diagenetic effects

In 1954 Abelson, separated amino acids from sub-fossil shell and suggested the possible use of the kinetics of the degradation of amino acids as the basis for a dating method (Abelson, 1955; O'Donnell et al., 2007). During the 1970s and 1980s hopes were entertained that it would provide a much improved chronology for Quaternary raised beaches (Andrews et al., 1979; Miller et al., 1981; Keen et al., 1979). However, inconsistencies with other dating methods at key sites led to doubts about the validity of the method and reduced its use. Despite these problems there has been considerable refinement of the method and recent studies have used exhaustive bleach treatment of shell carbonate and a reverse-phase high pressure liquid chromatography method on single species samples (Penckman et al., 2007). This has both decreased sample size and increased reliability, helping to resurrect AAR as a technique of major importance. Recent AAR work in Europe is of immediate importance in archaeology but also of importance in geomorphology, as it can provide a chronology for early Pleistocene fluvial systems and palaeohydrology (Parfitt et al., 2005). There remain two essential requirements: first, that we have a closed carbonate system; and second, that the temperatures are, or have not been, high enough to exceed the range of the method, and this limits the technique to temperate environments. If it can be shown that amino acids isolated from silicates provide a closed system then the technique may become applicable to siliceous microfossils such as diatoms (Penckman et al., 2008).

STRATIGRAPHIC METHODS AND CONSIDERATIONS

Just as in the rest of earth sciences stratigraphy and stratigraphic methods of correlation and sequential dating remain important. Stratigraphic integrity and interpretation is fundamental to reliable dating using any of the methods mentioned in this chapter, and checks of consistency, such as divergent dosimetry (deliberately assaying samples from the same bed or member with different environmental dose rates) in OSL dating, are based upon stratigraphic integrity. This becomes particularly evident in the modelling of sequences using methods such as Bayesian modelling. Bayesian modelling can be used to refine and improve the resolution of sedimentary sequences but it is entirely dependent upon the stratigraphic model used for date phasing. Where it is used in single sequence, such as a lake or floodplain core this is not a problem, but where dates from different locations such as channel and floodplain facies are being modelled then problems can arise from the synchroneity of deposition at different places and the lateral diachrony of single sedimentary members. In a study of a floodplain convergence zone of the River Trent in central England (Howard et al., 2008) a model developed from the radiocarbon dates rejected most of the OSL dates when they were tested against the model. Likewise a model based upon the members dated using OSL rejected many of the radiocarbon dates. Despite these problems, which at least cause geomorphologists to think about their morphostratigraphic models, Bayesian modelling is a powerful tool in increasing geomorphological precision (Chiverell et al., 2008). The two remaining stratigraphical methods that will be discussed here are tephrochronology and biostratigraphy.

Tephrochronology

Technically tephrochronology is a correlation tool whereby volcanic ashes within any sedimentary body can be chemically typed and related to a volcanic eruption that is either of known historical date (e.g. Krakatoa in 1883), or has been dated by radionuclide methods, most commonly potassium–argon dating. It is therefore dependent upon the spatial and temporal database of tephra and considerable research effort is currently being expended on improving these databases for several parts of the globe including the 'Tephrabase' for Europe (Newton et al., 2007). The RESET Project is constructing a European database and eventually in theory therefore, any ash will be able to be correlated with these known events and a date assigned. However, in practice there are a number of problems as listed below:

- *Reworking.* Ash is highly mobile due to its low bulk density and lack of cohesion and can be reworked into sediments at a later date.

- *Lack of unique chemistry.* Although, in theory, all eruptions will vary chemically the limits of detection and eruptions with very similar chemistries means that not all be identifiable to a single event.
- *Detection.* Ash is not always detectable within sediments, particularly where mixing has blurred any depositional boundaries and sampling for microtephra is expensive and speculative.
- *Occurrence.* There are areas of the world where ashes are very rare, due to the prevailing wind regime and distance from volcanic centres.

Despite these reservations tephrochronology has much to offer geomorphology as a result of its potential resolution and wide applicability, and although generally used to date lake and bog sequences tephra had been used to date fluvial terraces and sea level change (Kubo, 1997).

Biostratigraphy

Certain areas of biostratigraphy have remained, or even emerged, as powerful dating tools. Given the geologically short timescale that geomorphologists tend to be concerned with (10^1 to 10^5 years), many groups of organisms show insufficient evolutionary change to permit a formal indicator fossil approach, although for the Tertiary and Pleistocene as a whole, first- and last appearance biostratigraphy (FAD and LAD) has been extremely important (Rose, 2009). Some groups with high turnover rates do evolve enough, and this includes some small mammals and perhaps even some large mammals such as the mammoth lineage (Lister and Sher, 2001). This is the principle behind the 'vole clock' by which the rapid rate of extinction and speciation of voles is utilized, through the identification of voles species largely from dental morphology (Piertney et al., 2005). Vole species are also good indicators of environmental conditions at the time of deposition. For example, the short-tailed grass vole and the bank vole appear both now and in sites dating back to the warm interglacial Middle and Late Pleistocene period. During colder periods, species like the narrow-skulled vole are common. The northern vole is not usually present in the warmest periods and often indicates seasonally flooded grassland. However, its main value lies in providing a method of dating Pleistocene sites more precisely. More frequently, the selective and directional change of an assemblage can be used to provide a stratigraphic position within the Pleistocene and the Tertiary. This is the case with large herbivores, which in the case of the British Isles diverged from mainland European sequences through successive marine isotope stages (MIS) becoming an increasingly distinctive assemblage (Schreve, 2001). Vertebrate remains also have the advantage that they can potentially be used to correlate fluvial sequences back into the Tertiary, especially in the case of large rivers entering tectonic depressions with high subsidence rates (Mörs, 2002). The weakness of this methodology remains that new site discoveries will inevitably change FADs and LADs for some species.

The biostratigraphic approach used to be far more widely used and, in the case of pollen analysis of warm marine isotope stages, was the main method of assigning age through correlation. The procedure was clearly problematic with some chronologically separated stages not being distinctive enough to be separable, and conversely single chronological stages appearing distinctive because of local factors such as wetter or dryer localities or relative position in the interglacial cycle. An example is the likely inseparability of 'Hoxnian'-type pollen sequences from MIS 11 and MIS 9 (Thomas, 2001). This has led to the technique being largely abandoned as a biostratigraphic tool. Yet all palynologists know that, as with large herbivores or small mammals, there is pattern and order in the Pleistocene history of vegetation at any point on Earth's surface above and beyond local climatic variability. For example, if an organic sample from a north-west European sedimentary sequence contains more than a single, or a few, pollen grains of *Pterocarya* (wingnut) type; then we know that it is an early Pleistocene interglacial (O'Brien and Jones, 2003). What should now be done is that securely dated sequences, particularly lacustrine ones, could be used for statistical comparison of pollen assemblages as part of a more palaeoecological approach incidentally providing a statistical basis for assemblage dating which can be used in a multi-method dating approach.

Archaeology

If the date of manufacture of a human (or hominin) artefact or structure is known with certainty, then it is clear that its incorporation in or over sediments can be used to provide a minimum age of deposition. This can be the case for historical artefacts such as coins of a known date of issue or, in some cases, pottery. But there are many problems, even with dated artefacts, including reworking and bioturbation, which can, under certain circumstances even work small artefacts up a soil column. With early artefacts, especially stone tools, the situation is far more problematical and there has long been recognized a serious problem of circular reasoning (or reinforcement syndrome) with a pattern of occurrence in

sedimentary units being regarded as a chronology and then occurrence being used to date other sediments. However, prior to the availability of radiocarbon dating, this approach was inevitably used to provide some chronology for Pleistocene events (cf. Zeuner, 1946). For these reasons, and because of the much greater availability of independent chronometric dating, the use of archaeology will not be discussed in this chapter but more details can be found in Brown (1997) and Brown et al. (2003). One area of overlap with provenancing is the use of the by-products of human activity as spatio-temporal tracers. This has been relatively successful, especially in the area of ancient mining (Brown et al., 2003, 2009) with the additional benefit that such events can be used as long-term experiments in the flux of point sources through catchments over hundreds or even thousands of years. This area now overlaps with studies of grain provenance and history as discussed further in the chapter.

DATES AS RATES

The use of longer-lived cosmogenic isotopes, particularly ^{10}Be and ^{26}Al, for the estimation of medium- to long-term erosion rates is a relatively new methodology in research concerning the feedbacks between tectonics, erosion and climate (Bierman and Steig, 1996; Granger et al., 1996; Scaller et al., 2001; Vance et al., 2003; von Blanckenburg, 2005). Total erosion rates (cf. denudation rates) can be measured from cosmogenic nuclides produced *in situ* because the concentration of a steadily eroding bedrock is inversely proportional to the rate of erosion (Lal, 1991). A conceptually similar use of uranium series disequilibrium and ^{10}Be was applied to estimate erosion rates from catchments over 30 years ago (Plater et al., 1988; Brown et al., 1988). However, in order to use these (cosmogenic) methods in catchment sediment budgeting, several fundamental assumptions have to be made. In particular the assumption of a steady-state erosion rate (referring to the timescale of cosmogenic production in the parent rock material) is a potential drawback of the approach (Vance et al., 2003). Depending on the sampling strategy, the measurement of cosmogenic nuclides allows the estimation of: (1) mean erosion rates of catchments from fluvial sediment, (2) hillslope erosion rates from bedrock outcrops, or (3) bedrock exposure ages (if the erosion rate is zero). However, using cosmogenic inventories of river sediments to estimate mean erosion rates of catchments, all studies still make the assumption that the erosion rate is at steady state for the timescale of cosmogenic production in soil/bedrock, and sediment storage is minor/short in comparison to cosmogenic nuclide decay times.

In theory the surface nuclide concentration, C, of river sediments is inversely proportional to the erosion rate, ε:

$$C = P_0 / (\lambda + \rho\varepsilon/\Lambda) \qquad (11.2)$$

where λ is the decay constant of the nuclide, *rho;* the density of material and Λ is the attenuation length (the depth at which the cosmic rays are reduced by a factor of $1/e$ (von Blanckenburg, 2005). The surface production rate, P_0, depends on the intensity of secondary cosmic rays (nucleons and muons) and therefore is scaled for altitude, magnetic inclination and latitude [equation (11.2)]. Because production rates are still not well constrained, it is estimated that cosmogenic production decreases with depth with an exponential length scale of approximately 0.6 m. This includes some impreciseness and yet the effects of: (1) non-steady state of erosion rates, (2) long-term sediment storage, (3) landslides, (4) progressive concentration of quartz in soil through weathering and soil erosion, and (5) processes on grain-size selection are rarely known. However, major advances to solve some of these uncertainties have been achieved by numerical modelling studies (Niemi et al., 2005; Reinhardt et al., 2007; Schaller and Ehlers, 2006). One of the most exciting future applications of terrestrial nuclides in combination with other dating evidences (e.g. OSL dating), is the estimation of palaeo-erosion histories as shown by Schaller et al. (2002) in the case of the Meuse river. New developments to measure ^{14}C in sediments promise much for the reconstruction of long-term catchment-wide sediment fluxes. This methodology needs to be tested in an integrated study of a medium-sized catchment budget with independent dating controls.

GRAIN PROVENANCE AND HISTORY

Low-temperature thermochronometers such as apatite or zircon accumulate a fission-track age resulting from the spontaneous fission decay of ^{238}U allowing the determination of time elapsed since the grains cooled below 110°C or 200°C, respectively. This has been widely used to estimate the thermotectonic evolution or fission-track stratigraphy (Lisker et al., 2009) including apatite-fission track (APFT) stratigraphy, and this can include late Tertiary and Pleistocene uplift which is of obvious geomorphological significance (Green, 2002; Anell et al., 2009). This subject is now a matter of lively debate and

we are close to an overlap between thermochronometer methods such as APFT and OSL or AAR which all suggest significant uplift of active margins, including parts of north-west Europe, which may have driven both denudation and river terrace deposition in combination with the quasi-cyclical Milankovitch climatic forcing (Bridgland and Westaway, 2007). An alternative method is the uranium–lead dating of zircon grains and this can also be used as a provenancing technique. This has now become feasible as a result of the use of relatively low-resolution but also low-cost LA-ICP-MS, allowing a statistical sample of grain ages to be determined. In areas with heterogenous ages of primary zircon grains such as the Himalayas this allows the spatial differentiation of sediment source areas. This approach has been used by Clift (2009) to unravel changing courses of the Indus River which may have adversely affected the Harappan civilization during the late Neolithic in modern Pakistan. The advent of single-grain OSL dating also has a provenancing aspect. The frequency distribution of grain age/bleaching is dependent upon many factors one of which is the source of sediments; overland flow/soils, floodplain sediments or bedrock.

A MULTI-METHOD DATING CASE STUDY: SOUTH-WEST ENGLAND

As a result of the acidic nature of the bedrock nearly all the river gravels west of the Blackdown Hills in the south-west peninsular of the British Isles are entirely decalcified, with no preservation of shells or mammal bones. This has led to the lack of an alluvial chronology for this area. In an attempt to resolve this, a combination of OSL and ^{14}C has been used to provide at least a preliminary chronology, with the aim of relating this to climate change and to a new conceptual model for terrace formation in unglaciated basins with wider applicability to terrace staircase sequences elsewhere (Brown et al., 2010). The Exe catchment lay beyond the maximum extent of the British Pleistocene ice sheets and the drainage pattern evolved from the Tertiary to the Middle Pleistocene, by which time the major valley systems were in place and downcutting began to create a staircase of paired strath terraces (Brown et al., 2010). The higher terraces (8–6) typically exhibit altitudinal overlap or appear to be draped over the landscape, whilst the middle terraces show greater altitudinal separation, and the lowest terraces are of a cut-and-fill form (Figure 11.3). The terrace deposits investigated in this study were deposited in cold-transitional phases of the glacial–interglacial

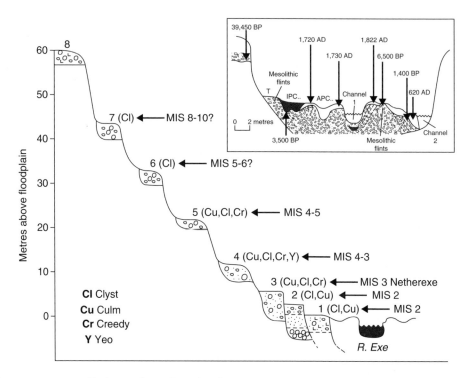

Figure 11.3 A preliminary chronology for the Exe terraces based upon OSL dating from Brown et al. (2010) with additional ^{14}C dates from Fyfe et al. (2004) with an inset of OSL dates from the Five Fords reach of the river Culm. All OSL dates have a ±10% error term associated with them

Milankovitch climatic cycles, with the OSL and ^{14}C dates revealing that the lowest four terrace gravel terraces were deposited in the Devensian Marine Isotope Stage (MIS) 4–2. This conforms to a cascade process–response model of basin terrace evolution in the Exe valley, which emphasizes the role of lateral erosion in the creation of strath terraces and the reworking of inherited resistant lithological components down through the staircase (Brown et al., 2009). Elements of this evolution can even be seen in the Holocene with the dissection of the Lateglacial and early Holocene braidplain by multiple (anastomosing) channels, with later Holocene evolution being dominated by avulsion into former channels.

CONCLUSIONS AND INTERFACES WITH OTHER DISCIPLINES

The increasing ability to date or at least propose estimates of dates (see earlier discussion) has implications well beyond geomorphology. The distinguished English Palaeolithic archaeologist

John Wymer remarked in his later years that during his lifetime archaeology had progressed from a situation where lithics (or lithic typology) was used to date river gravels to the converse with river gravels dating lithics, a far preferable situation in his opinion (Wymer, personal communication, 1987). Indeed this has allowed critical questions in Palaeolithic archaeology to be tackled afresh, such as: is there an ordered typological progression of stone tools or when was Europe colonized by hominins? This is just as true in geomorphology; the dead end that was denudation chronology becomes a fertile avenue, and we can propose evolutionary landscape models which integrate tectonics, erosion and rock properties. This will involve developing a more complete understanding of long-term grain history. There are also three large questions that this new array of dating methods have the potential to answer. The first is, To what extent are any landforms in equilibrium with geophysical factors including both climate and tectonics? The second is, To what extent are landforms inherited and from when? And the third is, Have humans significantly altered the geomorphic and related cycles and to what extent does this mitigate or exaggerate climatic forcing?

ACKNOWLEDGEMENTS

The author would like to thank members of LUCIFS and particularly G. Erkens, M. Fuchs, T. Hoffman, J.-J. Macaire, K.-M. Moldenhauer and D. E. Walling for their discussions and assistance. Others who have greatly assisted with this review are P. Toms, K. Penckman and L. Kay. The author gratefully acknowledges assistance with the diagrams from P. Baldaro of the Cartographic Unit, University of Southampton.

REFERENCES

Abelson, P.H. (1955) *Organic Constituents of Fossils.* Carnegie Washington Yearbook 54: 107–9.

Andrews, J.T., Bowen, D.Q. and Kidson, C. (1979) Amino acid ratios and the correlation of raised beach deposits in southwest England and Wales, *Nature* 281: 566–8.

Anell, I., Thybo, H. and Artemieva, I.M. (2009) Cenozoic uplift and subsidence in the North Atlantic region: Geological evidence revisited, *Tectonophysics* 474: 78–105.

Anthony, D.M. and Granger, D.E. (2004) A late Tertiary origin for multilevel caves along the western escarpment of the Cumberland Plateau, Tennessee and Kentucky established by cosmogenic ^{26}Al and ^{10}Be. *Journal of Cave and Karst Studies* 66: 46–55.

Arnold, L.J., Bailey, R.M. and Tucker, G.E. (2007) Statistical treatment of fluvial dose distributions from southern Colorado arroyo deposits, *Quaternary Geochronology* 2: 162–7.

Baker, V.R., Benito, G. and Rudoy, A. (1993) Palaeohydrology of Late Pleistocene superflooding, Altai Mountains, Siberia, *Science* 259: 348–50.

Beerten, K. and Stesmans, A. (2005) Single quartz grain electron spin resonance (ESR) dating of a contemporary desert surface deposit, Eastern Desert, Egypt, *Quaternary Science Reviews* 24: 223–31.

Benedetti, L., Finkel, R., Papanastassiou, D., King, G., Armijo, R., Ryerson, F.J., Farber, D. and Flerit, F. (2002) Post-glacial slip history of the Sparta fault (Greece) determined by ^{36}Cl cosmogenic dating: Evidence for non-periodic earthquakes. *Geophysics Research Letters* 29: 1246.

Bernasconi, S.M. and BIGLINK Project Members (2008) Weathering, soil formation and initial ecosystem evolution on a glacier forefield: A case study from the Damma Glacier, Switzerland, *Mineralogical Magazine* 72: 19–22.

Bierman, P. and Steig, E.J. (1996) Estimating rates of denudation using cosmogenic isotope abundances in sediment, *Earth Surface Processes and Landforms* 21: 125–39.

Bierman, P.R. and Nichols, K.K. (2004) Rock to sediment – Slope to sea with Be-10 – Rates of landscape change, *Annual Review of Earth and Planetary Sciences* 32: 215–55.

Birkeland, P.W. (1974) *Pedology, Weathering, and Geomorphological Research.* Oxford University Press, Oxford.

Bishop, P. (2007) Long-term landscape evolution: Linking tectonics and surface processes, *Earth Surface Processes and Landforms* 32: 329–65.

Blyakharchuk, T.A., Wright, H.E., Borodavko, P.S., van der Knaap, W.O. and Ammann, B. (2004) Late Glacial and Holocene vegetational changes on the Ulagan high-mountain plateau, Altai Mountains, southern Siberia, *Palaeogeography, Palaeclimatology, Palaeoecocology* 209: 259–79.

Boenigk, W. and Frechen, M. (2006) The Pliocene and Quaternary fluvial archives of the Rhine system, *Quaternary Science Reviews* 25: 550–74.

Bridgland, D.R. and Westaway, R. (2007) Preservation patterns of late cenozoic fluvial deposits and their implications: results from IGCP 449, *Quaternary International* 189: 5–38.

Brocard, G.Y., van der Beek, P.A., Bourlès, D.L., Siame, L.L. and Mugnier, J.-L. (2003) Long-term fluvial incision rates and postglacial river relaxation time in the French Western Alps from ^{10}Be dating of alluvial terraces with assessment of inheritance, soil development and wind ablation effects, *Earth and Planetary Science Letters* 209: 197–214.

Bronk-Ramsey, C.J. (2009) Suigetsu 2006: A wholly terrestrial radiocarbon calibration curve. 20th International Radiocarbon Conference 2009, Big Island, Hawaii, presentation.

Bronk-Ramsey, C.J., van der Plicht, J. and Weninger, B. (2001) 'Wiggle matching' radiocarbon dates, *Radiocarbon* 43: 381–9.

Brown, A.G. (1997) *Alluvial Environments: Geoarchaeology and Environmental Change.* Cambridge University Press, Cambridge.

Brown, A.G., Petit, F. and James, A. (2003) Archaeology and Human Artefacts, in H. Piegay and M. Kondolf (eds.), *Tools in Fluvial Geomorphology.* Wiley, Chichester. pp. 59–76.

Brown, A.G., Hatton, J., Pearson, L., Roseff, R. and Jackson, R. (2005) The Severn-Wye Revisited: Lateglacial-Holocene Floodplain Palaeoenvironments in the Lugg Valley, in D.N. Smith, M.B. Brickley and W. Smith (eds.), *Fertile Ground.* Papers in Honour of Susan Limbrey. Symposia for the Association of Environmental Archaeology No. 22. Oxbow Books, Oxford. pp. 16–29.

Brown, A.G., Bennett, J. and Rhodes, E. (2009a) Roman Mining on Exmoor: A Geomorphological Approach at Anstey's Combe, Dulverton, *Environmental Archaeology* 14: 50–61.

Brown, A.G., Basell, L.S., Toms, P.S. and Scrivener, R.C. (2009b) Towards a budget approach to Pleistocene terraces: Preliminary studies using the River Exe in South West England, UK, *Proceedings of the Geologists Association* 120: 275–81.

Brown, A.G., Basell, L.S, Toms, P.S., Bennett, J., Hosfield, R.T. and Scrivener, R.C. (2010) Late Pleistocene evolution of the Exe Valley: A chronstratigraphic model of terrace formation and its implications for Palaeolithic Archaeology, *Quaternary Science Reviews* 29: 897–912.

Brown, A.G., Carey, C., Challis, K., Howard, A.M., Kincey, M., Tetlow, E. and Cooper, L. (2007) Predictive modelling of multi-period geoarchaeological resources at a river confluence. Phase II Report (PNUM 3357). ALSF Report. English Heritage, London.

Brown, A.G., Carey. C., Erkens, G., Fuchs, M., Hoffman, T., Macaire, J-J., Moldenhauer, K.M. and Walling, D.E. (2009) From sedimentary records to sediment budgets: Multiple approaches to catchment sediment flux. *Geomorphology* 108: 35–47.

Brown, L., Pavich, M.J., Hickman, R.E., Klein, J. and Middleton, R. (1988) Erosion of the eastern United States observed with 10Be, *Earth Surface Processes and Landforms* 13: 441–57.

Burrough, S.L., Thomas, D.S.G. and Bailey, R.M. (2009) Mega-Lake in the Kalahari: A Late Pleistocene record of the Palaeolake Makgadikgadi system, *Quaternary Science Reviews* 28: 1392–411.

Candy, I., Black, S. and Sellwood, B. (2005) U-series isochron dating of immature and mature calcretes as a basis for constructing Quaternary landform chronologies for the Sorbas basin, southeast Spain, *Quaternary Research* 64: 100–11.

Carling, P.A., Kirkbride, A.D., Parnachov, S., Borodavko, P.S. and Berger, G.B. (2001) Late-glacial catastrophic flooding in the Altai Mountains of south-central Siberia: a synoptic overview and introduction to flood deposit sedimentology, in I.P. Martini, V.R. Baker and G. Garzon (eds.), *Flood and Megaflood Processes and Deposits: Recent and Ancient Examples*. Special Publication 32 of the IAS, Blackwell Science, Oxford.

Chiverrell, R.C., Harvey, A.M., Hunter, S.Y., Millington, J. and Richardson, N.J. (2007) Late Holocene environmental change in the Howgill Fells, Northwest England, *Geomorphology* 100: 41–69.

Chiverrell, R.C., Foster, G.C., Thomas, G.P.S., Marshall, P. and Hamilton, D. (2008) Robust chronologies for landform development, *Earth Surface Processes and Landforms* 34: 319–28.

Clift, P. (2009) Harappan collapse, *Geoscientist,* 19: 18–22.

Cockburn, H.A.P. and Summerfield, M.A. (2004) Geomorphological applications of cosmogenic isotope analysis, *Progress in Physical Geography* 28: 1–42.

Dehnert, A., Akçar, N., Kubik, P., Kasper, H.U., Preusser, F. and Schlüchter, Ch. (2006) Burial dating of Rhine River terrace sediments, Lower Rhine Embayment, Germany. 4th. Swiss Geosciences Meeting, Bern.

Duller, G. (2008) *Luminescence Dating: Guidelines on Using Luminescence Dating in Archaeology*. English Heritage, London.

Fabel, D. (2004) Cosmogenic burial dating of shallow deposits – a preliminary study, in I.C. Roach (ed.) *Regolith 2004*. CRC Leme. pp. 86–7.

Fuchs, M. and Wagner, G.A. (2005) The chronostratigraphy and geoarchaeological significance of an alluvial geoarchive: comparative OSL and AMS ^{14}C dating from Greece, *Archaeometry* 47: 849–60.

Fuchs, M., Lang, A. and Wagner, G.A. (2004) The history of Holocene soil erosion in the Phlious Basin, NE-Peloponnese, Greece, provided by optical dating, *The Holocene* 14: 334–45.

Fuchs, M., Straub, J. and Zöller, L. (2005) Residual luminescence signals of recent river flood sediments: A comparison between quartz and feldspar of fine- and coarse-grain sediments, *Ancient TL* 23: 1.

Fuchs, M., Woda, L. and Bürkert, A. (2007) Chronostratigraphy of a sediment record from the Hajar mountain range in north Oman: Implications for optical dating of insufficiently bleached sediments, *Quaternary Geochronology* 2: 202–7.

Fyfe, R.M., Brown, A.G. and Coles, B.J. (2004) Vegetational change and human activity in the Exe Valley, Devon, UK, *Proceedings of the Prehistoric Society* 69: 161–82.

Gosse, J.C. and Phillips, F.M. (2001) Terrestrial in situ cosmogenic nuclides: Theory and application, *Quaternary Science Reviews* 20: 1475–560.

Granger, D.E. and Muzikar, P.F. (2001) Dating sediment burial with in situ-produced cosmogenic nuclides: theory, techniques, and limitations, *Earth and Planetary Science Letters* 188: 269–81.

Granger, D.E., Kirchner., J.W. and Kinkel, R. (1996) Spatially-averaged long-term erosion rates measured from in situ-produced cosmogenic nuclides in alluvial sediment, *Journal of Geology* 104: 249–57.

Green, P.F. (2002) Early Tertiary paleo-thermal effects in Northern England: reconciling results from apatite fission track analysis with geological evidence, *Tectonophysics* 349: 131–44.

Gregory, K.J. (1977) Lichens and the determination of river channel capacity, *Earth Surface Processes* 1: 273–85.

Grosswald, M.G. (1980) Late Wieichselian ice sheet of northern Eurasia, *Quaternary Research,* 13: 1–32.

Grosswald, M.G. (1998) New approach to the Ice Age palaeohydrology of Northern Eurasia, in G. Benito, V.R. Baker and K.J. Gregory (eds.), *Palaeohydrology and Environmental Change*. Wiley, Chichester. pp. 199–214.

Hagg, J. (2009) Application of *in-situ* cosmogenic nuclide analysis to landform evolution in Dartmoor, southwest Britain. Unpublished PhD Thesis, University of Edinburgh.

Harrison, S., Glasser, N.F., Winchester, V., Haresign, E., Warren, C.R., Duller, G. and Kubik, P.W. (2008) Glaciar León, Chilean Patagonia: late Holocene chronology and geomorphology, *Holocene* 18: 343–52.

Higham, T.F.G., Jacobi, R.M., Ramsey, C. and Bronk, (2006) AMS Radiocarbon dating of ancient bone using ultrafiltration, *Radiocarbon* 48: 179–95.

Howard, A.J., Brown, A.G., Carey, C.J., Challis, K., Cooper, L.P., Kincey, M. and Tom, P. (2008) Archaeological resource modelling in temperate river valleys: a case study from the Trent Valley, UK, *Antiquity* 82: 1040–54.

Huggett, R.J. (1998) Soil chronosequences, soil development, and soil evolution: a critical review, *Catena* 32: 155–72.

Innes, J.L. (1985) Lichenometry, *Progress in Physical Geography* 9: 187–254.

Jain, M., Murray, A.S. and Bøtter-Jensen, L. (2004) Optically stimulated luminescence dating: How significant is incomplete light exposure in fluvial environments?, *Quaternaire* 15: 143–57.

Johnson, D.L., Keller, E.A. and Rockwell, T.K. (1990) Dynamic pedogenesis: new views on some key soil concepts, and a model for interpreting Quaternary soils, *Quaternary Research* 33: 306–19.

Johnstone, E., Macklin, M.G. and Lewin, J. (2006) The development and application of a database of radiocarbon-dated Holocene fluvial deposits in Great Britain, *Catena* 66: 14–23.

Joshi, S.R. (1988) West Valley plutonium and americium-241 in Lake Ontario sediments off the mouth of Niagara River, *Water, Air and Soil Pollution* 42: 159–68.

Keen, D.H., Harmon, R.S. and Andrews, J.T. (1981) U-series and amino acid dates from Jersey, *Nature* 289: 162–4.

Kitagawa, H. and van der Plicht, J. (1998) Atmospheric radiocarbon calibration to 45,000 yr BP: Late glacial fluctuations and cosmogenic isotope production. *Science* 279: 1187–90.

Kitagawa, H. and van der Plicht, J. (2000) Atmospheric radiocarbon calibration beyond 11,900 cal BP from Lake Suigetsu laminated sediments, *Radiocarbon* 42: 369–380.

Krbetschek, M. (ed.) (2009) *Luminescence Dating: An Introduction and Handbook.* Springer, Berlin.

Kubo, S. (1997) Reconstruction of palaeo sea-level and landform changes since the Marine Isotope stage 5e using buried terraces in the lower Sagami plain central Japan, *Quaternary Research* 36: 147–63.

Lal, D. (1991) Cosmic labelling of erosion surfaces: in-situ nuclide production rates and erosion models, *Earth and Planetary Science Letters* 104: 424–39.

Lang, A. (2003) Phases of soil erosion-caused colluviation in the loess hills of South Germany, *Catena* 51: 209–21.

Leeder, M. (2007) Cybertectonic Earth and Gaia's weak hand: sedimentary geology, sediment cycling and the Earth system, *Journal of the Geological Society of London* 164: 277–96.

Lewin, J., Macklin, M.G. and Johnstone, E. (2005) Interpreting alluvial archives: sedimentological factors in the British Holocene fluvial record, *Quaternary Science Reviews* 24: 1873–89.

Lisker, F., Ventura, B. and Glasmacher, U.A. (2009) Apatite thermochronology in modern geology. *Geological Society, London, Special Publications* 324: 1–23.

Lister, A.M. and Sher, A.V. (2001) The origin and evolution of the woolly mammoth. *Science* 294: 1094–7.

Mallick, R. and Frank, N. (2002) A new technique for precise uranium-series dating of travertine micro-samples, *Geochimica et Cosmochimica Acta* 66: 4261–72.

Mangerud, J., Svendsen, J.I. and Astakhov, V.I. (1999) Age and extent of the Barents and Kara icea sheets in Northern Russia, *Boreas* 28: 46–80.

Maroulis, J.C., Nanson, G.C., Price, D.M. and Pietsch, T. (2007) Aeolian-fluvial interaction and climate change: source-bordering dune development over the past ~100 ka on Cooper Creek, central Australia. *Quaternary Science Reviews* 26: 386–404.

Matmon, A. (2009) The evolution of Makhtesh Hazera and its drainage system since the Middle Pleistocene, in Y. Enzel, N. Greenbaum and M. Laskow (eds.), *GLOCOPH Israel 2009: Field Trip Guidebook,* pp. 187–90.

Matthews, J.A. and Shakesby, R.A. (1984) The status of the 'Little Ice Age' in southern Norway: Relative-age dating of Neoglacial moraines with Schmidt hammer and lichenometry, *Boreas* 13: 333–46.

Matthews, J.A. and Matthews, J.A. (2005) Little Ice Age glacier variations in Jotunheimen, southern Norway: A study in regionally controlled lichenometric dating of recessional moraines with implications for climate and lichen growth rates. *The Holocene* 15: 1–19.

Matthews, J.A., Shakesby, R.A., Schnabel, C. and Freeman, S. (2008) Cosmogenic ^{10}Be and ^{26}Al ages of Holocene moraines in southern Norway I: Testing the method and confirmation of the date of the Erdalen Event (c. 10 ka) at its type site, *The Holocene* 18: 1155–64.

McCarroll, D. (1991) The Schmidt hammer, weathering and rock surface roughness, *Earth Surface Processes and Landforms* 16: 477–80.

McCarroll, D. (1993) Lichens: Lichenometric dating of diachronous surfaces, *Earth Surface Processes and Landforms* 18: 527–39.

McCarroll, D., Shakesby, R.A. and Matthews, J.A. (2000) Enhanced rockfall activity during the Little Ice Age: Further lichenometric evidence from a Norwegian talus. *Permafrost and Periglacial Processes* 12: 157–64.

McFadden, L.D., Wells, S.G. and Jercinovich, M.J. (1987) Influences of eolian and pedogenic processes on the origin and evolution of desert pavements, *Geology* 15: 504–08.

Merrett, S.P. and Macklin, M.G. (1999) Historic river response to extreme flooding in the Yorkshire Dales, northern England, in A.G. Brown and T.M. Quine (eds.), *Fluvial Processes and Environmental Change.* Wiley, Chichester. pp. 345–60.

Michel, H., Barci-Funel, G., Dalmasso, J., Ardisson, G., Appleby, P.G., Haworth, E. and El-Daoushy, F. (2001) Plutonium, americium and cesium records in sediment cores from Blelham Tarn, Cumbria (UK), *Journal of Radioanalytical and Nuclear Chemistry* 247: 107–10.

Mihai, A.A. and Hurtgen, Ch. (1997) Plutonium and americium in sediment samples along the Romanian sector of the Danube river and the Black Sea coast, *Journal of Radioanalytical and Nuclear Chemistry* 222: 275–8.

Miller, G.H., Hollin, J.T. and Andrews, J.T. (1979) Aminostratigraphy of the UK Pleistocene, *Nature* 281: 539–43.

Mörs, Th. (2002) Biostratigraphy and palaeoecology of continental Tertiary vertebrate faunas in the Lower Rhine Embayment (NW Germany), *Netherlands Journal of Geosciences* 81: 177–83.

Muller, R.A. (1977) Radioisotope dating with a cyclotron, *Science* 196: 489–94.

Nanson, G.C., Chen, X.Y. and Price, D.M. (1992) Lateral migration, thermoluminescence chronology and colour variation of longitudinal dunes near Birdsville in the Simpson Desert, Central Australia. *Earth Surface Processes and Landforms* 17: 807–19.

Niemi, N.A., Oskin, M., Burbank, D.W., Heimsath, A.M. and Gabet, E.J. (2005) Effects of bedrock landslides on cosmogenically determined erosion rates, *Earth and Planetary Science Letters* 237: 480–98.

O'Brien, C.E. and Jones, R.L. (2003) Early and middle pleistocene vegetation history of the Medoc region, southwest France, *Journal of Quaternary Science* 18: 557–79.

O'Donnell, T.H., Macko, S.A. and Wehmiller, A.F. (2007) Stable carbon isotope composition of amino acids in modern and fossil Mercenaria, *Organic Geochemistry* 38: 485–98.

Olley, J.M., Caitcheon G.G. and Roberts R.G. (1999) The origin of dose distributions in fluvial sediments, and the prospect of dating single grains from fluvial deposits using optically stimulated luminescence, *Radiation Measurements* 30: 207–17.

Olsson, (1986) Radiometric dating, in B.E. Berglund, H.J.B. Birks, M. Ralska-Jasiewiczowa and H.E. Wright (eds.), *Palaeoecological Events During the Last 15,000 Years.* Wilet and Sons, Chichester. pp. 273–97.

Parfitt, S.A., Barendregt, R.W., Breda, M., Candy, I., Collins, M.J., Coope, G.R., Durbidge, P., Field, M.H., Lee, J.R., Lister, A.M., Mutch, R., Penkman, K.E.H., Preece, R.C., Rose, J., Stringer, C.B., Symmons, R., Whittaker, J.E., Wymer, J.J., and Stuart, A.J. (2005) The earliest humans in Northern Europe: Artefacts from the Cromer forest-bed formation at Pakefield, Suffolk, UK, *Nature* 438: 1008–12.

Penckman, K.E.H., Kaufman, D.S., Maddy, D. and Collins, M.J. (2008) Closed-system behaviour of the intra-crystalline fraction of amino acids in mollusc shells, *Quaternary Geochronology* 3: 2–25.

Penkman, K.E.H., Preece, R.C., Keen, D.H., Maddy, D., Schreve, D.C. and Collins, M. (2007) Amino acids from the intra-crystalline fraction of mollusc shells: Applications to geochronology. *Quaternary Science Reviews* 26: 2958–69.

Pennington, W., Cambray, R.S. and Fisher, E.M. (1973) Observations on lake sediments using fallout ^{137}Cs as a tracer, *Nature* 242: 324–6.

Phillips, W.M., Hall, A.M., Mottram, R., Fifield, L.K. and Sugden, D.E. (2006) Cosmogenic [10]Be and [26]Al exposure ages of tors and erratics, Cairngorm Mountains, Scotland: Timescales for the development of a classic landscape of selective linear glacial erosion, *Geomorphology* 73: 222–45.

Piertney, S., Stewart, W., Lambin, X., Telfer, S., Aars, J. and Dallas, J. (2005) Phylogeographic structure and postglacial evolutionary history of water voles (Arvicola terrestris) in the United Kingdom, *Molecular Ecology* 14: 1435–44.

Plater, A.J., Dugdale, R.E. and Ivanovicj, I. (1988) The application of uranium disequilibrium concepts to sediment yield determinations, *Earth Surface Processes and Landforms* 13: 171–82.

Recorbet, F., Rochette, P., Braucher, R., Bourlés, D., Benedetti, L., Hantz, D. and Finkel, R.C. (2009) Evidence for major collapse of a coastal cliff 2 ka ago in Cassis (South East France), *Geomorphology* in press.

Reinhardt, L.J., Hoey, T.B., Barrows, T.T., Dempster, T.J., Bishop, P. and Fifield, L.K. (2007) Interpreting erosion rates from cosmogenic radionuclide concentrations measured in rapidly eroding terrain, *Earth Surface Processes and Landforms* 32: 390–406.

Reuther, A.U, Herget, J., Ivy-Ochs, S., Borodavko, P., Kubik, P.W. and Heine, K. (2006) Constraining the timing of the most recent cataclysmic flood event from ice-dammed lakes in the Russian Altai Mountains, Siberia, using cosmogenic in situ [10]Be, *Geology* 34: 913–6.

Rose, J. (2009) Early and Middle Pleistocene landscapes of eastern England, *Proceedings of the Geologists' Association* 120: 3–33.

Rudoy, A.N. (1998) Mountain ice-dammed lakes of southern Siberia and their influence on the development and regime of the intracontinental runoff systems of North Asia in the Late Pleistocene, in G. Benito, V.R. Baker and K.J. Gregory (eds.), *Palaeohydrology and Environmental Change.* Wiley, Chichester pp. 215–34.

Rudoy, A.N. and Baker, V.R. (1993) Sedimentary effects of cataclysmic Late Pleistocene glacial outburst flooding, Altai Mountains, Siberia, *Sedimentary Geology* 85: 53–62.

Schaller, M. and Ehlers, T.A. (2006) Limits to quantifying climate driven changes in denudation rates with cosmogenic radionuclides, *Earth and Planetary Science Letters* 248: 153–67.

Schaller, M., von Blanckenburg, F., Hovius, N. and Kubik, P.W. (2001) Large-scale erosion rates fro in situ-produced cosmogenic nuclides in European river sediments, *Earth and Planetary Science Letters* 188: 441–58.

Schaller, M., von Blanckenburg, F., Hovius, N. and Kubik, P.W. (2002) Paleo-erosion rate record in a 1.6 Ma terrace sequence of the Meuse river, *Geochimica et Cosmochimica Acta* 66: A673–4.

Schreve, D.C. (2001) Differentiation of the British late Middle Pleistocene interglacials: the evidence from mammalian biostratigraphy, *Quaternary Science Reviews*, 20: 1693–705.

Sharp, W.D., Ludwig, K.R., Chadwick, O.A., Amundson, R. and Glaser, L.L. (2003) Dating fluvial terraces by 230Th/U on pedogenic carbonate, Wind River Basin, Wyoming, *Quaternary Research* 59: 139–50.

Siame, L.L., Bourlès, D.L., Sébrier, M., Bellier, O., Casano, J.C., Araujo, M., Perez, M., Raisbeck, G.M. and Yiou, F. (1997) Cosmogenic dating ranging from 20 to 700 ka of a series of alluvial fan surfaces affected by the El Tigre fault, Argentina. *Geology* 25: 975–8.

Singarayer, J.S., Bailey, R.M., Ward, S. and Stokes S. (2005) Assessing the completeness of optical resetting of quartz OSL in the natural environment, *Radiation Measurements* 40: 13–25.

Small, R.J. and Clark, M.J. (1974) The medial moraines of lower Glacier de Tsidjiore Nouve, Valais, Switzerland, *Journal of Glaciology*, 13: 255–64.

Stein, M. and Goldstein, S.L. (2006) U-Th and radiocarbon chronologies of late Quaternary lacustrine records of the Dead Sea basin: Methods and applications, in Y. Enzel, A. Agnon and M. Stein (eds.), *New Frontiers in Dead Sea Palaeoenvironmental Research*. Geological Society of America, Special Paper 401. pp. 141–54.

Stevens, T., Buylaert, J-P. and Murray, A.S. (2009) Towards development of a broadly-applicable SAR TT-OSL dating protocol for quartz. Radiation Measurements. doi:10.1016/j.radmeas.2009.02.015.

Stuart, F.M. (2001) In situ cosmogenic isotopes: principles and potential for archaeology, in D.R. Brothwell and A.M. Pollard (eds.), *Handbook of Archaeologiccal Sciences*. Wiley and Sons, Chichester. pp. 93–100.

Svendsen, J.I., Astakov, V.I., Bolshiyanov, D.Yu., Demidov, I., Dowdeswell, J.A., Gataullin, V., Hjort, C., Hubberten, H.W., Larsen, E., Mangerud, J., Melles, M., Möller, P., Saarnisto, M. and Siegert, M.J. (1999) Maximum extent of the Eurasian ice sheets in the Barents and Kara Sea region during the Weichselian, *Boreas,* 28: 234–42.

Thiede, J. and Bauch, H. (1999) The Late Quaternary history of northern Eurasia and the adjacent Arctic Ocean: An introduction to QUEEN. *Boreas* 28: 3–5.

Thomas, G.N. (2001) Late Middle Pleistocene pollen biostratigraphy in Britain; pitfalls and possibilities in the separation of interglacial sequences, *Quaternary Science Reviews* 20: 21–1630.

Thomas, P.J., Jain, M., Juyal, N. and Singhvi, A.K. (2005) Comparison of single-grain and small-aliquot OSL dose estimates in <3000 years old river sediments from South India. *Radiation Measurements* 39: 457–69.

Vance, D., Bickle, M., Ivy-Ochs, S. and Kubik, P.W. (2003) Erosion and exhumation in the Himalaya from cosmogenic isotope inventories of river sediments, *Earth and Planetary Science Letters* 206: 273–88.

Velichko, A.A., Kononov, Yu.M. and Faustova, M.A. (1997) The last glaciation of earth: size and volume of ice sheets, *Quaternary International* 41/42: 43–51.

Velichko, A., Isayeva, L., Makeyev, V., Matishov, G. and Faustova, M. (1984) Late Pleistocene glaciation of the Arctic Shelf, and the reconstruction of Eurasian Ice Sheets, in A.A. Velichko (ed.), *Late Quaternary Environments of the Soviet Union*. Longman, London. pp. 35–44.

von Blanckenburg, F. (2005) The control mechanisms of erosion and weathering at basin scale from cosmogenic nuclides in river sediment, *Earth and Planetary Science Letters* 237: 462–79.

Walling, D.E. and He, Q. (1999) Using fallout Lead-210 measurements to estimate soil erosion on cultivated land. *Soil Science Society American Journal* 63: 1404–12.

Wang, X.L., Wintle, A.G. and Lu, Y.C. (2007) Testing a single-aliquot protocol for recuperated OSL dating, *Radiation Measurements* 42: 380–91.

Wells, S.G., McFadden, L.D., Poths, J. and Olinger C.T. (1995) Cosmogenic ^{3}He surface-exposure dating of stone pavements: Implications for landscape evolution in deserts, *Geology* 23: 613–16.

Winchester, V. and Harrison, S. (2000) Dendrochronology and lichenometry: colonisation, growth rates and dating of geomorphological events on the east side of the North Patagonian Icefield, *Geomorphology* 34: 181–94.

Yamskikh, A.F., Yamskikh, A.A. and Brown. A.G. (1999) Siberian-type Quaternary floodplain sedimentation: the example of the Yenisei River, in A.G. Brown and T.A. Quine (eds.), *Fluvial Processes and Environmental Change*. Wiley, Chichester. pp. 241–52.

Zhang, X. and Walling, D.E. (2005) Characterizing land surface erosion from Cesium-137 profiles in lake and reservoir sediments, *J. Environmental Quality* 34: 514–23.

Zeuner, F. (1946) *Dating the Past: An Introduction to Geochronology*. Methuen, London.

Zilberman, E. (2009) The formation of the Makhteshim: Deep erosional craters in anticline ridges of the Negev desert, southern Israel, in Y. Enzel, N. Greenbaum and M. Laskow (eds.), *GLOCOPH Israel 2009: Field Trip Guidebook*. pp. 192–86.

12

Remote Sensing in Geomorphology

Tom G. Farr

Remote sensing techniques have evolved considerably since the days of aerial photography and color–infrared film. Imaging spectrometers produce images in over 200 spectral bands, thermal infrared sensors pick up temperature and emissivity variations, and imaging radars map surfaces in multiple wavelengths and polarizations (Lillesand et al., 2008). High-resolution sensors provide images from space at better than 1 m resolution of nearly anywhere on the planet. At the same time, new techniques for measuring the third dimension, topography, at moderate-to-high resolution have been developed: lidar or laser ranging, and interferometric imaging radars have joined the time-honored technique of stereophotogrammetry to provide geomorphologists with unprecedented accuracy and coverage.

The development of these techniques has coincided with a vast improvement in computer capabilities and the development of geographic information systems: software that places the remote sensing data into a geographic context. Thus, an ordinary geologist can derive information from these sometimes esoteric technologies without knowing the details of the physics of interactions, calibration, geometric rectification, etc. (Butler and Walsh, 1998).

Remote sensing techniques are virtually the only way we have to study other planets. The techniques discussed here are being used to map and interpret the surfaces of the other terrestrial planets (Mercury, Venus, Mars) and asteroids as well as the moons of the outer planets, notably the Galilean satellites of Jupiter and Saturn's Titan.

Details of individual systems will be given as tables in each section.

IMAGING

Imaging sensors operate in virtually all of the spectral 'windows' in which Earth's atmosphere is transparent (Figure 12.1). At the shorter wavelengths of the visible to near infrared (VNIR) part of the electromagnetic spectrum, the signal received is reflected sunlight; in the thermal IR range, emitted radiation from the warm Earth is collected; and in the microwave part of the spectrum, radars provide their own illumination in order to achieve useful spatial resolution.

Visible–near infrared

The wavelength range of reflected sunlight, about 0.4–2.5 µm, contains information on the composition of surfaces. Reflectance spectroscopy is a well-developed field which has sought to explain spectrum absorptions in terms of atomic and molecular vibrations within mineral structures (Clark, 1999). These spectral signatures are fairly unique and can often be used to identify mineral components. Libraries of laboratory reflectance spectra have been collected and made available for comparison to unknown spectra (Clark et al., 2007; Baldridge et al., 2009). Due to their short wavelength, VNIR sensors collect information

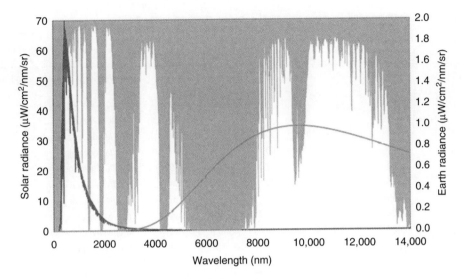

Figure 12.1   Energy available for remote sensing and atmospheric transmittance. Left curve shows solar radiance at Earth's surface. Note the peak in the visible part of the spectrum. Right curve shows emission spectrum of Earth, assuming it operates as a black body at about 37°C. Note that the scale for the emission curve is different and much lower than that for solar radiance. The white background of plot depicts wavelengths at which the atmosphere is relatively transparent. Note the deep absorptions at about 1400 and 1900 nm due to water vapor and the broad thermal infrared 'window' between about 8000 and 14,000 nm

from only the upper few micrometers of a surface. Thus weathered veneers and coatings will affect the interpretation of the spectra.

Types of visible–near infrared sensors

VNIR instruments have been flown for decades in aircraft and in orbit (Table 12.1). Most collect images in several wavelength ranges or spectral bands. This multi-spectral technique yields moderate to high-resolution images which can be combined in a variety of ways to discriminate or identify compositional units. Examples of these types of imagers include the Landsat series in orbit since 1972; ASTER and MODIS, which also include thermal IR bands; and several other systems, listed in Table 12.1.

A more recent development is imaging spectrometry or hyperspectral imaging, in which the spectrum is sampled nearly continuously at high spectral resolution, typically yielding over 200 image bands (Schaepman et al., 2009) (Plate 5, see the colour plate section, page 591). AVIRIS, an airborne imaging spectrometer, has been flying since 1987 and EO-1, a spaceborne demonstration is currently in orbit. Full sampling of the reflectance spectrum significantly improves the ability

to identify compositional units, but obviously increases the amount of data to be stored or downlinked and the computer processing load.

Another type of VNIR sensor is also very useful in geomorphology, the high-resolution panchromatic camera. This sacrifices spectral discrimination in order to capture more light, producing images in some cases with better than 1 m spatial resolution from orbit. Thus air-photo quality images may be obtained of nearly anywhere on earth.

Finally, VNIR images from the above systems or weather satellites can be used to track atmospheric phenomena responsible for aeolian sediment movement (Herman et al., 1997; Baddock et al., 2009). Plumes of sediment from north Africa have been tracked west into the Atlantic Ocean and dust from central China has also been detected (Goudie, 2008).

Calibration and corrections

Because the sunlight VNIR instruments collect passes through Earth's atmosphere, its effects must be taken into account if the resulting images will be compared with laboratory spectra or models. The area of atmospheric correction of

Table 12.1 Common visible–near infrared remote-sensing systems

Acronym/ abbreviation	Dates	Resolution (m)	Swath (km)	Wavelengh range (μm)	Number of bands	URL
MSS	1972–1992	80	100	0.5–1.1	4	http://edc.usgs.gov/guides/landsat_mss.html
TM	1982–	30	100	0.4–2.4	6	http://edc.usgs.gov/products/satellite/tm.html
MODIS	1999–	250–1000	2330	0.4–2.2	19	http://modis.gsfc.nasa.gov/
ASTER	1999–	15, 30	60	0.5–2.4	9	http://asterweb.jpl.nasa.gov/
EO-1	2000–	30	7.7	0.4–2.5	220	http://eo1.usgs.gov/
SPOT	1986–	2.5–20	60	0.5–1.7	Pan+4	http://www.spot.com/
Ikonos	1999–	1, 4		0.45–0.9	Pan+4	http://www.geoeye.com/
GeoEye	2008–	0.4, 1.6	15	0.45–0.9	Pan+4	http://www.geoeye.com/
QuickBird	2001–	0.6, 2.4	16	0.45–0.9	Pan+4	http://www.digitalglobe.com/
AVIRIS	1987–	~20	~10	0.4–2.5	224	http://aviris.jpl.nasa.gov/
MASTER	1998–	~10	~10	0.4–2.4	25	http://master.jpl.nasa.gov/

ASTER, Advanced Spaceborne Thermal Emission and Reflection Radiometer; AVIRIS, Airborne Visible/Infrared Imaging Spectrometer; EO-1, Earth Observing Mission 1; MASTER, MODIS/ASTER airborne simulator; MODIS, Moderate Resolution Imaging Spectroradiometer; MSS, Multispectral Scanner System; SPOT, Satellite Pour l'Observation de la Terre; TM, Thematic Mapper.

VNIR data is continuously developing (Gao et al., 2009). The most difficult effects to be dealt with are water vapor absorption and scattering due to dust. These effects vary temporally and spatially, so are difficult to counter *a priori*. Some software packages include atmospheric corrections based on the Lowtran or Modtran standard atmospheric models which incorporate crude spatial and temporal variability and provide reasonably good results for average conditions (Gao et al., 2009). Greater accuracy may be obtained by sampling the atmosphere during data acquisition with a balloon sonde or by measuring the spectral reflectance of patches of ground with a field spectrometer and then forcing the remotely sensed images to match the field spectra (Milton et al., 2009).

Older imaging systems were not very geodetically accurate, that is, they were not well registered to map coordinates. This is still somewhat true for airborne systems as their platform is inherently less stable than orbital platforms, but improvements in GPS and inertial measurement units have increased airborne image geometric quality as well. Software packages commonly include interactive programs for image registration so that multiple images may be co-registered or images may be registered to a map base.

Discrimination and identification

A wide range of processing techniques have been developed over the years to extract information from VNIR multispectral or hyperspectral data. As mentioned above, a fully calibrated and atmospherically corrected VNIR data set may be compared directly with available laboratory spectra of known materials allowing identification of the mapped units. However, Nature often is more complex. Lacking an accurate atmospheric correction, band ratioing or statistical methods are usually employed. Computing ratios of multispectral bands reduces the effects of illumination and the atmosphere and enhances certain compositions. For example, the ratio of Landsat Thematic Mapper band 3 (0.66 μm) to band 1 (0.48 μm) is proportional to Fe^{3+} content (Mustard and Sunshine, 1999; Short, 2009).

A number of multi-dimensional statistical analysis techniques have been developed over the years to allow discrimination of units in VNIR multi-spectral and hyperspectral data (Landgrebe, 2005; Landgrebe and Biehl, 2009). N-dimensional clustering (where N is the number of bands) can be accomplished either in an unsupervised or supervised manner, the latter by 'training' the program on known sites in the image. The clusters represent spectrally distinct units which must be characterized with other methods. A variation of unsupervised clustering interprets clusters at the edge of the distribution as end-members and the other points as mixtures between the end-members (Adams and Gillespie, 2006). The result is a map of end-members and their proportions;

the characterization of the end-members is accomplished with field work or other data.

The above statistical techniques work well on multi-spectral VNIR data, but bog down with the hundreds of bands contained in an hyperspectral data set. However, experiments in dimensional reduction have been reasonably successful. To accomplish this, a principal component transformation is performed on the imaging spectrometer data, yielding images which represent decreasing amounts of statistical variance (Lillesand et al., 2008). In this way, the first 10 or so principal component images can represent nearly 100 per cent of the total spectral variety of the scene and can then be used in clustering or other analyses. However, the spectral signatures of minor components of the scene may be lost in this manner.

Thermal infrared

Emitted thermal radiation from Earth at ambient temperature peaks around 9 μm and Earth's atmosphere has a 'window' from about 8 to 14 μm (Figure 12.1). The flux is much less than reflected sunlight so that, typically, these sensors produce images with lower spatial and spectral resolution than VNIR sensors. Due to their longer wavelength, thermal infrared (TIR) signals are emitted from deeper within a material and thus are indicative of the bulk properties and less sensitive to weathered surfaces and coatings. TIR remote sensing techniques are not as well-developed as VNIR, mostly because laboratory spectra are more difficult to obtain and atmospheric corrections are more difficult. The physical condition of the target is also important, with large variations in signatures caused by smooth, pitted, or powdered surfaces (e.g. Kirkland et al., 2002; Vaughan et al., 2005).

Thermal infrared sensors

Few TIR sensors are in operation (Table 12.2). Multi-spectral systems that are available include TM band 6, ASTER and MODIS, in orbit since 1999, and the airborne MASTER (Hook et al., 2001). TIMS was an airborne system in use from about 1982 to 1999 (Hook et al., 1999). SEBASS, an airborne hyperspectral TIR imager, with 256 bands extending from 2.4 to 13.5 μm has been flown for some projects (Kirkland et al., 2002).

Calibration and corrections

As mentioned, correction for atmospheric effects is more difficult at TIR wavelengths. This is because there are more effects that vary more, both spatially and temporally, than at VNIR wavelengths. Water vapor and other atmospheric species are particularly active as are aerosols. The effects are not only absorptive, but the warm atmosphere also contributes to the desired signal radiating from the surface.

As with the VNIR part of the spectrum, atmospheric models like MODTRAN can be used to correct for atmospheric effects (Hook et al., 1999) and field spectrometers (Hook and Kahle, 1996) can be used to characterize large homogenous areas for calibration via the empirical-line technique.

Information extraction

Even with atmospheric effects, TIR images contain spectral information, much like VNIR images. However, the surface temperature

Table 12.2 Common thermal–infrared remote-sensing systems

Acronym/ abbreviation	Dates	Resolution (m)	Swath (km)	Wavelength range (μm)	Number of bands	URL
TM	1982–	120	100	10.4–12.5	1	http://edc.usgs.gov/products/satellite/tm.html
MODIS	1999–	1000	2330	3.7–14.4	16	http://modis.gsfc.nasa.gov/
ASTER	1999–	90	60	8.1–11.6	5	http://asterweb.jpl.nasa.gov/
TIMS	1982–1999*	~10	~10	8.2–12.6	6	—
MASTER	1998–	~10	~10	3.1–12.9	25	http://masterweb.jpl.nasa.gov/
SEBASS	1999–	2	0.25	3.0–13.5	256	http://www.aero.org

ASTER, Advanced Spaceborne Thermal Emission and Reflection Radiometer; MASTER, MODIS/ASTER airborne simulator; MODIS, Moderate Resolution Imaging Spectroradiometer; SEBASS, Spatially Enhanced Broadband Array Spectrograph System; TIMS, Thermal Infrared Multispectral Scanner; TM, Thematic Mapper.

produces a strong masking effect, with sunlit sides of hills having a much different appearance than cooler, shaded sides. For this reason, one of the most widely used processing techniques for multi-spectral TIR data is the decorrelation stretch (Gillespie et al., 1984, 1986). This effectively uses a modification of principal components to decrease the temperature effect and emphasize variations in emissivity, the spectral component of TIR data. Emissivity may also be more quantitatively derived by estimating the temperature of the surface. This can be done by assuming a value for emissivity for one band of a multi-spectral set and solving for temperature which is then used to solve for emissivity in the other bands (Gillespie et al., 1998; Hook et al., 1999).

An improvement on this technique is the 'alpha-residual' technique which simplifies the temperature-emissivity equations to allow an 'alpha' spectrum to be derived, which is similar to the emissivity spectrum and can be compared with ground or laboratory spectra (e.g. Salisbury and D'Aria, 1992; Hook et al., 1999).

Another important quantity which can be derived from TIR images is thermal inertia. This requires thermal images, which can be single broadband images, acquired at times of maximum heating and cooling, typically early afternoon and pre-dawn. The difference, often corrected for visible albedo differences, is a measure of the heat capacity of the surficial materials. In its simplest interpretation, dense rocky areas have high thermal inertia while less dense materials, such as dust, heat up and cool quickly, thus having a low thermal inertia. Water has a high thermal inertia, but springs seeping through sediments evaporate and cause anomalous cooling, yielding a distinctive signature (Kahle, 1987).

Synthetic aperture radar

Radars are active sensors that operate in the microwave part of the spectrum, roughly centimeters to meters wavelength (Henderson and Lewis, 1998). Not only is Earth's atmosphere transparent in these wavelengths, but also are clouds; thus, combined with the fact that they supply their own illumination, radars can provide images day or night and in all weather conditions. This makes them particularly useful in monitoring surface changes occurring at night or beneath rain or volcanic eruption clouds. In contrast to the sensitivity of the shorter wavelengths to surface composition, radars are most sensitive to the physical roughness of a surface at scales near the observing wavelength. Secondarily, they are sensitive to the near-surface dielectric constant, which is a function of density (rock versus loose soil) and moisture content. Very low dielectric constants typical of loose, dry materials may be penetrated many wavelengths and if a solid interface is present at a shallow enough depth, an image may be obtained of the subsurface (Plate 6, page 591) (Schaber et al., 1986).

Radars are more sensitive to the imaging geometry than shorter wavelength sensors. For a typical surface of moderate roughness or vegetation cover, slight topographic variations will be enhanced, especially at near-nadir look angles. Thus, radar images obtained at small look angles (typically 20–30° off nadir) are very useful for delineating subtle topographic variations, similar to low sun-angle air photos.

Radar sensors

Imaging radars using a variety of wavelengths have been flown on aircraft and spacecraft (Table 12.3). Multiple wavelengths allow sensitivity to a variety of roughness scales in the centimeter to meter range, typical of desert landscapes. Radars are also sensitive to vegetation structure at these scales (i.e. leaves, branches, trunks). As radars produce their own illumination, polarization of the transmitted beam may also be selected. Most imaging radars transmit and receive linear polarizations; some use only one combination, often vertical transmit and vertical receive (VV), but some interlace H and V transmission and reception to allow all four combinations to be obtained: HH, VV, HV and VH. This polarimetric mode allows more complete characterization of the surface roughness and electrical properties (Elachi, 1988; Henderson and Lewis, 1998; Campbell, 2002).

Imaging radars are sometimes called side-looking radars as they must image to the side of a moving platform. In the direction away from the platform, the range direction, resolution is obtained by sending short pulses. In the along-track or azimuth direction, resolution is obtained by adding pulses. This synthetic aperture technique allows high resolution to be achieved independent of platform altitude.

The side-looking geometry of imaging radars produces geometric distortions which cause difficulties in interpretation and can destroy some of the information in the scene. The fact that imaging radars divide the range into small intervals to create range pixels yields a projected geometry when the slant-range pixels are laid down in a grid. A slant-range image appears compressed in the near range because of this projection. This effect is more pronounced in airborne radar images because the range of angles is greater than in spaceborne images. A simple transformation produces a ground-range image from the slant-range.

Table 12.3 Common radar remote-sensing systems

Acronym/ abbreviation	Dates	Resolution (m)	Swath (km)	Band/pol*	URL
Seasat	1978	25	100	LHH	http://southport.jpl.nasa.gov/
SIR-A	1981	25	50	LHH	—
SIR-B	1984	25	~50	LHH	—
SIR-C/X-SAR	1994	25	~50	XVV, Cquad, Lquad	http://edc2.usgs.gov/sir-c/sir-c.php//
ERS	1991–	25	100	CVV	http://earth.esa.int/ers/
Envisat	1998–	25	100	CVV, CHH	http://envisat.esa.int/
JERS	1992–1998	20	75	LHH	http://www.eorc.jaxa.jp/JERS-1/
PALSAR	2006–	10–100	40–350	Lquad	http://www.eorc.jaxa.jp/ALOS/en/ about/palsar.htm
Radarsat	1995–	10–100	45–500	CHH	http://ccrs.nrcan.gc.ca/radar/ index_e.php
SRTM	2000	30	N/A	CHH, CVV	http://jpl.nasa.gov/srtm
TerraSAR-X	2007–	1–16	5–100	XHH, HV, VV, VH	http://www.terrasar.de/
Airsar	1988–2004	10	~10	Cquad, Lquad, Pquad	http://southport.jpl.nasa.gov/
Geosar	2000–	3	25	Xquad, Pquad	http://www.geosar.com/
C/X-SAR	1988–	~10	~20	X and C quad	http://www.ccrs.nrcan.gc.ca/ glossary/index_e.php?id=3169
Star-3i	2001–	2	10	XHH	http://www.intermap.com/

*X ~ 3 cm, C ~ 6 cm, L ~ 25 cm, P ~ 70 cm wavelength.

Envisat, environmental satellite; ERS, Earth Resources Satellite; JERS, Japanese Earth Resources Satellite; PALSAR, phased array L-band synthetic aperture radar; SIR, Shuttle Imaging Radar; SRTM, Shuttle Radar Topography Mission.

Note, however, that topographic variations are not accounted for in this transformation.

Topographic variations are distorted in the same way as the overall image: Their near-range points are compressed relative to their far-range points (Figure 12.2). The general case of radar foreshortening compresses the image of a mountain's near-range slope and extends the image of its back slope (Figure 12.2a). The effect is obviously exacerbated by small look angles and steep slopes to the extreme case of layover, in which the top of the mountain is imaged before the bottom of the near-range slope (Figure 12.2b). Image data in the laid-over area are lost. Image data can also be lost on the backslope if the slope is steep enough and the look angle large enough to put the slope in radar shadow (Figure 12.2c).

These distortions, if not extreme like layover or shadow, can be rectified using separate topographic data, yielding an orthorectified map-view which can be overlaid on other images and maps.

As a generalization, since most slope angles on Earth are less than about 35°, imaging radars with small look angles, such as Seasat, enhance the topography at the expense of surface roughness information. Conversely, larger look angles, such

as SIR-A's 47°, reduce the effect of topography and enhance the sensitivity to surface roughness.

Synthetic aperture radars are coherent sources, like lasers, and the phase of the radar returns is recorded as well as the amplitude. If two radar images are obtained of the same area, but the antennas are displaced, the phase signals from the two antennas may be combined as an interferogram. The phase differences are related to the surface topography. This technique of interferometric SAR (InSAR) will be discussed later in the chapter.

Calibration and interpretation

Most imaging radars have been calibrated so that brightness in their images represents values of the backscatter coefficient or cross-section, a way of expressing quantitatively the fraction of the transmitted power scattered back to the sensor. While radar images can be interpreted qualitatively based on relative brightness, backscatter cross-section can be inverted to predict quantitatively the surface roughness, near the scale of the radar wavelength used (Dobson and Ulaby, 1998).

Radar waves impinging on a surface scatter from the surface and near-surface volume.

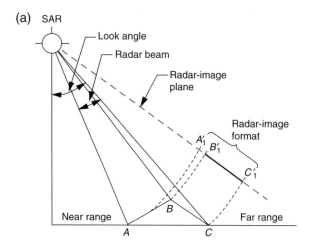

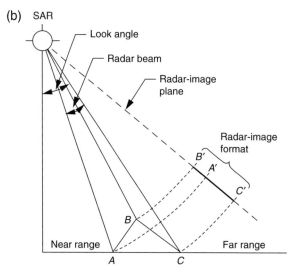

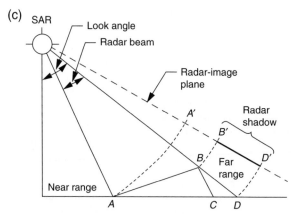

Figure 12.2 Geometry of side-looking radar. (a) Because radar uses time-delay to discriminate between objects, the radar-facing slope (AB) appears fore-shortened as opposed to slope BC. (b) The extreme case of foreshortening, layover, places the top of the mountain (B) in front of its base (A). Data from AB is lost. This situation is more common for radars with small look (incidence) angles and areas of high relief. (c) The opposite situation, typical for radars with large look angles, is shadowing on the far slope (BC). Data here is also lost. (From Ford et al., 1993)

Very smooth surfaces at the scale of the radar wavelength scatter the incident radiation in the specular direction (Figure 12.3a); this scattering is a mirror-like reflection often called 'specular reflection'. Rougher surfaces cause more of the radar energy to be scattered randomly until it reaches a diffuse, Lambertian distribution. Virtually all imaging radars are monostatic, that is, the same antenna is used for transmission and reception. Comparison of the cases in Figure 12.3a shows that the smooth surface has no component of its return in the direction of the transmission arrow, so the resulting image tone would be black. With increasing roughness, more energy is scattered back to the antenna, resulting in lighter image tones for rougher surfaces.

Changes in the local incidence angle for the surfaces shown in Figure 12.3a produce characteristic curves of image brightness, quantified as radar backscatter versus incidence angle (Figure 12.3b). Clearly, a smooth surface produces a sharp peak when the radar illumination is perpendicular to it (incidence angle = 0°), while a very rough surface scatters the signal equally in all directions with little dependence on incidence angle. The curves in Figure 12.3b show that smooth surfaces can be brighter than rough surfaces at small incidence angles (typically less than 20–25°); an erroneous interpretation of roughness can result if incidence angle is not taken into account.

Surface roughness, while not a characteristic that most geomorphologists consider routinely (Shepard et al., 2002), is affected by several surficial processes including mantling by aeolian dust (Farr, 1992), rock spalling, desert pavement formation, and disintegration due to salt weathering (Farr and Chadwick, 1996).

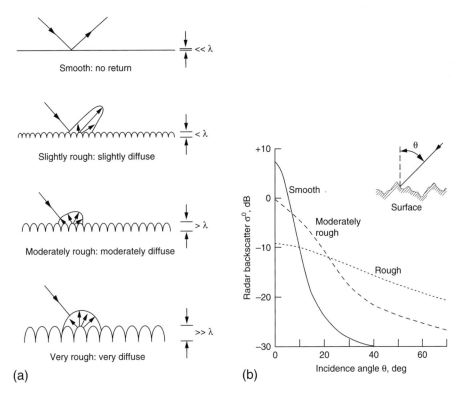

(a)   (b)

Figure 12.3   Radar response to roughness. (a) Smooth areas act like mirrors and scatter the radar beam away from the receiver. Rougher surfaces (at the scale of the wavelength) scatter more and more radiation randomly. (b) Radar backscatter (image brightness) as a function of incidence angle for different roughnesses. Smooth surfaces reflect directly back only for normal incidence while rough surfaces scatter relatively consistently through a wide range of angles. Note that, for very small angles, smooth surfaces may appear brighter than rough surfaces. (From Ford et al., 1993)

Sonar

Sonar (sound navigation and ranging) is an active remote-sensing technique (like radar) using sound waves to make maps of the sea floor. As this is a relatively specialized area of remote sensing in geomorphology, this section will provide only a brief overview of the technique and its application. Sonar has advanced over the past couple of decades due to advances in navigation, sub-sea vehicles, and data processing (Blondel and Murton, 1997; Jonsson et al., 2009). In particular, shallower waters of the continental slopes have been mapped extensively with multi-beam sonar techniques showing in great detail a variety of geologic landforms and processes (Prior and Hooper, 1999; Brown and Blondel, 2009). Deeper parts of the ocean have also been imaged with deep-tow vehicles (Prior and Hooper, 1999).

Interestingly, the wavelength of sound waves typically used for sonar is about the same as that of electromagnetic waves typically used for radar (i.e. centimeters to meters). Because of this, some of the same interpretation techniques may be used. For example, backscatter of sound waves from the sea floor may be used to estimate the surface roughness near the wavelength. In addition, because sea floor sediments are often more transparent to acoustic radiation, subsurface sounding can be accomplished (Prior and Hooper, 1999).

TOPOGRAPHIC MAPPING

Topographic mapping has developed rapidly since the days of stereo air photography. Stereo air photography still plays a major role in topographic mapping: stereo photographs from space are often used to map remote or hostile areas. In addition, interferometric radar (InSAR) and laser ranging (lidar) have come into their own, providing new sources of high-quality digital elevation models (DEM) (Table 12.4; Maune, 2007).

Stereo photography

Stereo photography has been used ever since photography was invented to provide a three-dimensional view of subjects (Greve, 1996; Lillesand et al., 2008). From the air, calibrated stereo photographs are obtained which are converted to topographic maps. This capability was extended to space with orbital sensors operating in the VNIR either by combining images acquired on subsequent orbits, with side-lap, or by designing the sensor to acquire either a forward or aft view in addition to the nadir view on a single orbit. Both ASTER and the Japanese ALOS have fore-and-aft stereo imagers. ASTER has been used operationally to produce DEMs which are available from the USGS (Hirano et al., 2003). The Japanese space agency has produced a global DEM using ASTER stereo images (Langenberg, 2009) (see Resources section).

InSAR

Interferometric synthetic aperture radar (InSAR) exploits the fact that radar sensors collect phase as well as amplitude of the backscattered signal. Thus two radar images, obtained at different vantage points, may be combined as an interferogram

Table 12.4 Common digital topography systems

Acronym/ abbreviation	Dates	Posting (m)	Accuracy (m)	Technique	URL
ASTER	1999–	30	25	Stereo	http://asterweb.jpl.nasa.gov/gdem.asp
SRTM	2000	30, 90	~5	InSAR	http://jpl.nasa.gov/srtm
Airsar	1988–2004	10	~2	InSAR	http://southport.jpl.nasa.gov/
Geosar	2000–	3	1	InSAR	http://www.geosar.com/
Star-3i	2001–	2	~1	InSAR	http://www.intermap.com/
Lidar	N/A	~2	~0.5	Lidar	http://www.ncalm.org/
IceSAT	2003–	170	~0.1	Lidar	http://icesat.gsfc.nasa.gov/
NED	N/A	10–30	5–10	Stereo	http://seamless.usgs.gov/

ASTER, Advanced Spaceborne Thermal Emission and Reflection Radiometer; NED, National Elevation Dataset; SRTM, Shuttle Radar Topography Mission.

in which phase differences are proportional to surface heights (Massonnet, 1997; Madsen and Zebker, 1998; Burgmann et al., 2000). The two radar images may be acquired simultaneously with two antennas (single-pass InSAR), or with a single radar system flying nearly the same flight path twice (repeat-pass InSAR). As the geometry of the system, particularly the baseline between the two acquisitions, needs to be known with high accuracy, the fixed baseline single-pass technique is preferable for producing DEMs. This technique was used to create a near-global DEM with the Space Shuttle in 2000 (Farr et al., 2007) and is proposed for a measurement of water surfaces (Alsdorf et al., 2007). In addition, changes in water vapor content of the atmosphere between repeat-pass acquisitions can cause phase delays not related to topography, and vegetation movement between acquisitions can decorrelate the phase signals, yielding no data (Massonnet and Feigl, 1995; Zebker et al., 1997). However, the simplicity of a single radar system and the fact that many of its disadvantages can be at least partially overcome, has led to a large amount of repeat-pass InSAR data being acquired. It should be noted that a 'normal' amplitude image is automatically produced in addition to the DEM and the image and DEM are inherently registered, allowing ortho-rectification of the image to be accomplished easily.

As radar waves will only partially penetrate into a vegetation canopy, the DEM produced by InSAR will not be a 'bald-earth' model. The amount of penetration is proportional to the wavelength and inversely proportional to the density of the canopy. This is an area of continuing research (Carabajal and Harding, 2006; Hofton et al., 2006a; Sexton et al., 2009).

An additional aspect of repeat-pass InSAR relevant to geomorphology is its ability to measure sub-centimeter changes in topography over large areas (Madsen and Zebker, 1998; Massonnet and Feigl, 1998; Smith, 2002). As the repeat-pass spatial baseline separation approaches zero, the system becomes less sensitive to topography and more sensitive to changes. At a baseline of zero, the only phase differences observed are changes in the topography between the two observations at scales of a fraction of a wavelength. Thus, a typical orbital radar operating at a wavelength of 5.5 cm, is sensitive to changes of a few millimeters. The sensitivity is only to changes along the radar line of sight, typically 20–40° off nadir. If the baseline separation is non-zero, as it is in almost all cases, the contribution of static topography can be eliminated through the use of a pre-existing DEM. An extensive literature exists on the application of this technique to study of earthquake deformation (e.g. Massonnet et al., 1993), volcano inflation and deflation (e.g. Mouginis-Mark et al., 2000; Zebker et al., 2000; Masterlark and Lu, 2004; Lundgren and Lu, 2006), landslides (e.g. Roering et al., 2009), and subsidence (e.g. Fielding et al., 1998; Bell et al., 2008). Note that the technique will not work on surfaces that have changed too much at the sub-resolution scale such as sediment deposition and mass-wasting processes which would decorrelate the phase returns much like moving vegetation noted above. The decorrelation phenomenon can be turned to advantage, however, if one is only interested in a sensitivity to change rather than its measurement. Areas of temporal phase decorrelation are easily detected and have been used to map small landslides and damaged buildings after an earthquake (Fielding et al., 2005).

Ground-based InSAR is a relatively new application in which a fixed station is directed at a target, typically the side of a mountain, and repeated images are obtained (Tarchi et al., 2003). At a distance of over 1 km, the sensitivity to line-of-sight changes is on the order of millimeters. The main application of this technique has been landslide monitoring.

Lidar

Pulsed lasers may be used as altimeters and are known as (in analogy to radar) light detection and ranging, or lidar. If the location and attitude of the platform are accurately known, then the measurement may be converted to topography (Lillesand et al., 2008; Wehr, 2008). Lidars have been flown on aircraft, helicopters and spacecraft, and have also been used on the ground for making very high-resolution three-dimensional models of buildings and terrain (e.g. Morris et al., 2008). When flown in spacecraft, the limited dwell time for the lidar generally precludes scanning the beam, producing only a profile. Aircraft and helicopter systems allow scanning, producing broad area coverage. When combined with highly accurate differential GPS positioning, a rapidly pulsed airborne scanning lidar produces topographic measurements of unprecedented resolution and accuracy. After post-processing, typical DEMs have approximately 1 m postings with height accuracy of a few centimeter.

A unique characteristic of lidar is its ability to penetrate through vegetation canopies to reveal the topography beneath (e.g. Haugerud et al., 2003) (Plate 7, page 592). Some lidar systems have the additional capability of recording the full waveform returning from each pulse, allowing the vertical structure of the vegetation to be inferred (Lefsky et al., 1999). Another measurement which is possible with some lidar systems is the reflectance of the surface at the laser wavelength, typically in the near IR or blue–green wavelengths. The latter wavelengths have also been used for shallow bathymetric mapping (Guenther et al., 2000; Storlazzi et al., 2003; Guenther, 2007).

Applications of lidar in geomorphology have been growing rapidly (Luthcke et al., 2002), especially as a result of the recent efforts of the National Center for Airborne Laser Mapping (NCALM) consortium (see Resources section). Data have been made available on the GEON web site as well as the USGS Seamless Server (see Resources section). Studies are being conducted on fault morphology (Hudnut et al., 2002; Sherrod et al., 2004; Frankel et al., 2007), sediment transport on alluvial fans (Staley et al., 2006; Frankel and Dolan, 2007), volcano morphology (Garvin, 1996; Hofton et al., 2006b), and coastal change mapping (Shrestha et al., 2005; Hapke et al., 2006).

FUTURE DIRECTIONS

Geographic information systems

Older remote sensing data are generally presented in instrument-specific formats and geometries (so-called level 0 or 1 products). However, with the advent of better location and attitude information for remote sensing platforms, further processing can yield geo-coded data (level 2) which are geodetically registered. Thus, the data may be compared directly to maps and other geo-coded data sets. This is most often accomplished in software packages called geographic information systems or GIS. These packages allow multiple layers of image data, topography, or other thematic data (e.g. geologic maps) to be viewed and analyzed with a variety of tools such as cross-correlation between map units and multi-spectral signatures, drainage analysis, models, etc.

Most GIS are capable of handling several different types and formats of data. The two main forms of data are raster and vector. Raster data, usually images, are composed of an evenly spaced grid of pixels while vector data are described by a series of points with associated attributes, such as a boundary within which lies a body of water. As there are a variety of GIS available, a factor to keep in mind when considering which package

to use is its capabilities in both image processing and vector-based cartographic applications. A more complete description of the application of GIS to geomorphology can be found in the chapter on GIS.

Web-based mapping

Google Earth is the most well-known example of the ease of access of high-resolution spatial data over the internet. Other providers include Microsoft's Bing Maps and Wikimapia (see Resources section). In its simplest form, these services allow the user to view any place on Earth in high resolution, with a topographic base. The databases are geo-coded, and user data may be imported into them and overlain onto the base maps. In addition, numerous web-based user groups have sprung up, adding content, interpretations, discussions, etc. For geomorphologists, access to high-resolution satellite images of anywhere on Earth provides an unprecedented way to compare landforms. New applications are being invented daily, so that it is often rewarding to peruse these and other sites for innovative uses of web-based spatial data (Lyon et al., 2006; Whitmeyer et al., 2010).

Signatures of geomorphic processes

Comparison of results from a variety of remote sensing studies of landscapes and geomorphic processes begins to yield a suggestion that geomorphic processes leave 'signatures' on the landscape which are often fairly unique. For example, the processes of desert varnish development and the formation of desert pavements on alluvial fans in Death Valley cause a darkening in VNIR and a smoothing in radar images which is indicative of older surfaces (Daily et al., 1979). In contrast, salt weathering on fans in north-west China causes a smoothing in radar images, but a brightening in VNIR because desert varnish can not form in the erosive environment (Farr and Chadwick, 1996). The winnowing process by which shales and limestones are removed and quartzites retained on the fans of Death Valley shows up clearly in TIR images (Gillespie et al., 1984). Lava flow surfaces in the Mojave Desert are subject to mantling by aeolian silt which attenuates the roughness and brightens the surface (Farr, 1992). The amount of smoothing has a direct relation to the thickness of aeolian cover and the age of the flows. The ages of lava flows in Hawaii may also be estimated based on the development of silica-rich coatings which incorporate oxidized tephra, which affect TIR and VNIR reflectance respectively (Farr and Adams, 1984; Kahle et al., 1988).

Topographic signatures are also commonly used to deduce geomorphic processes and the high resolution of lidar DEMs is allowing new understanding of small-scale processes such as channel initiation and overland flow (Volker et al., 2007). Some more speculative work has been done in a search for topographic signatures of biologic activity as well (Dietrich and Perron, 2006). More work needs to be done to establish the uniqueness of these possible signatures.

Ground-based remote sensing

Remote sensing need not be as remote as an aircraft or spacecraft; ground-based techniques are providing very high-resolution views at the outcrop scale for applications in data calibration, geomorphology, archaeology, etc. Tripod-based or terrestrial lidars have been used to produce microtopographic maps of small areas, typically about 10 m, with resolutions of a few centimeters. These have been used to interpret emplacement of lava flows (Morris et al., 2008) and to study other surficial processes (Collins and Kayen, 2006).

A concept called the Virtual Outcrop combines close-range lidar with imaging (Xu et al., 2000; Waggott et al., 2005; Lambers et al., 2007). This produces a very high-resolution three-dimensional model of an area typically on the order of 10 m in size. Applications include mapping trenches, cliffs and caves, and archiving the models for future reference. The image data may be a simple photograph or it may be a multi-spectral dataset which can be used to identify compositions, etc.

Hand-held spectrometers are available commercially that provide reflectance spectra for surfaces which may be used to identify minerals and weathering products (Milton et al., 2009).

These field data may also be used to calibrate more-remote data by characterizing large homogeneous areas.

Future instruments and technology

In 2007 the National Research Council released a Decadal Survey on Earth Science and Applications from Space (NRC, 2007). This report outlined a series of orbital remote sensing missions for study of the planet. Among these missions are several that have direct application to geomorphology (Table 12.5). In the near-term, geomorphologists may be interested in SMAP, designed to measure soil moisture and freeze/thaw; ICESat-II, to continue lidar measurements of ice-sheet heights; and DESDynI, a radar optimized for repeat-pass interferometry for measurement of surface and ice-sheet changes and vegetation structure. In a later timeframe, a hyperspectral imager is recommended for compositional and mineral mapping; an orbiting lidar for topographic mapping; and a mission to measure snow accumulation. Recently, it was decided to continue the Landsat series with the Landsat Data Continuity Mission (LDCM) which will be essentially identical to previous Landsat TM, possibly including a broadband TIR channel.

Meanwhile, other nations and space agencies are proceeding to develop new capabilities. In particular, the Canadian and European Space Agencies will continue their series of orbiting imaging radars and the German Space Agency recently lofted a second TerraSAR-X which will be capable of working with its twin, already in orbit, to produce topography at about 10 m resolution.

Looking farther out, it is likely that resolution from space will continue to improve, with the only impediment being government regulations. Global DEMs at better than 10 m resolution will likely be available in the next one to two decades and complete image coverage of Earth at better than 50 cm

Table 12.5 Proposed future remote-sensing systems

Acronym/ abbreviation	Launch date	Resolution (m)	Swath (km)	Wavelength range (μm)	Number of bands	URL
LDCM	2012	15–30	25	0.44–1.37	9	http://ldcm.nasa.gov/about.html
SMAP	2013	1–40 (km)	1000	LHH, LVV, LHV	6	http://smap.jpl.nasa.gov/
DESDynI SAR	2015	35	350	Lquad	1	http://desdyni.jpl.nasa.gov/
DESDynI Lidar	2015	25	2	1.064	1	—

DESDynI, Deformation, Ecosystem Structure, and Dynamics of Ice; LDCM, Landsat Data Continuity Mission; SMAP, Soil Moisture Active-Passive.

could also become available. Geomorphologists will have much to work with during the coming years.

ACKNOWLEDGEMENTS

This work was carried out at the Jet Propulsion Laboratory, California Institute of Technology, under contract with NASA.

RESOURCES: USEFUL WEB SITES

Tutorials and reports

- http://rst.gsfc.nasa.gov/
- http://www.r-s-c-c.org/index.html
- http://dynamo.phy.ohiou.edu/tutorial/tutorial_files/frame.htm
- http://disc.sci.gsfc.nasa.gov//geomorphology/

NASA data gateways

- https://wist.echo.nasa.gov/~wist/api/imswelcome/
- http://gcmd.gsfc.nasa.gov/index.html
- http://www.asf.alaska.edu/

USGS data gateways

- http://edc.usgs.gov/
- http://earthexplorer.usgs.gov
- http://glovis.usgs.gov/ImgViewer/Java2Img Viewer.html
- http://gisdata.usgs.net/index.php

European Space Agency data gateway

- http://earthnet.esrin.esa.it/

Data of Canada

- http://geogratis.cgdi.gc.ca/

Japanese Aerospace Exploration Agency data gateway

- http://www.eorc.jaxa.jp/en/index.html

Global Land Cover Facility

- http://glcf.umiacs.umd.edu/index.shtml

Lidar

- http://www.ncalm.org/
- http://calm.geo.berkeley.edu/ncalm/ddc.html
- http://lidar.cr.usgs.gov/
- http://pugetsoundlidar.ess.washington.edu/
- http://core2.gsfc.nasa.gov/lidar/terrapoint
- http://core2.gsfc.nasa.gov/Topo/topo.html
- http://www.geongrid.org/index.php/topography/
- http://www.ngs.noaa.gov/RESEARCH/RSD/main/lidar/lidar.shtml
- http://www.csc.noaa.gov/lidar/
- http://www.jalbtcx.org/
- http://atm.wff.nasa.gov/

Web mapping

- http://earth.google.com/
- http://www.bing.com/maps/
- http://wikimapia.org/

Links to other remote-sensing web sites

- https://engineering.purdue.edu/~biehl/MultiSpec/
- http://keck.library.unr.edu/default.htm
- http://www.techexpo.com/WWW/opto-knowledge/IS_resources.html

REFERENCES

Adams, J.B. and Gillespie, A.R. (2006) *Remote Sensing of Landscapes with Spectral Images: A Physical Modeling Approach.* Cambridge University Press, Cambridge. p. 362.

Alsdorf, D.E., Rodriguez, E. and Lettenmaier, D.P. (2007) Measuring surface water from space. *Reviews of Geophysics* 45. doi: 10.1029/2006RG000197.

Baddock, M.C., Bullard, J.E. and Bryant, R.G. (2009) Dust source identification using MODIS: A comparison of techniques applied to the Lake Eyre Basin, Australia. *Remote Sensing of Environment* 113: 1511–28. doi:10.1016/j.rse.2009.03.002.

Baldridge, A.M., Hook, S.J., Grove, C.I. and Rivera, G. (2009) The ASTER Spectral Library, Version 2.0. *Remote Sensing of Environment* 113: 711–5. http://speclib.jpl.nasa.gov/. doi:10.1016/j.rse.2008.11.007.

Bell, J.W., Amelung, F., Ferretti, A., Bianchi, M. and Novali, F. (2008) Permanent scatterer InSAR reveals seasonal and long-term aquifer-system response to groundwater pumping and artificial recharge. *Water Resources Research* 44: doi:10.1029/2007WR006152.

Blondel, P. and Murton, B.J. (1997) *Handbook of Seafloor Sonar Imagery.* John Wiley, New York. p. 308.

Brown, C.J. and Blondel, P. (2009) Developments in the application of multibeam sonar backscatter for seafloor habitat mapping. *Applied Acoustics* 70: 1242–7. doi:10.1016/j.apacoust.2008.08.004.

Burgmann, R., Rosen, P.A. and Fielding, E.J. (2000) Synthetic aperture radar interferometry to measure Earth's surface topography and its deformation. *Annual Review of Earth and Planetary Sciences* 28: 169–209.

Butler, D.R. and Walsh, S.J. (1998) The application of remote sensing and geographic information systems in the study of geomorphology: An introduction. *Geomorphology* 21: 179–81.

Campbell, B.A. (2002) *Radar Remote Sensing of Planetary Surfaces.* Cambridge University Press, Cambridge. p. 342.

Carabajal, C.C. and Harding, D.J. (2006) SRTM C-band and ICESat laser altimetry elevation comparisons as a function of tree cover and relief. *Photogrammetric Engineering and Remote Sensing* 72: 287–98.

Clark, R.N. (1999) Spectroscopy of Rocks and Minerals, and Principles of Spectroscopy, in A.N. Rencz (ed.), *Manual of Remote Sensing, vol. 3. Remote Sensing for the Earth Sciences.* John Wiley and Sons, New York. pp. 3–58.

Clark, R.N., Swayze, G.A., Wise, R., Livo, E., Hoefen, T., Kokaly, R. and Sutley, S.J. (2007) USGS digital spectral library splib06a: U.S. Geological Survey, Digital Data Series 231. http://speclab.cr.usgs.gov/spectral.lib06/.

Collins, B.D. and Kayen, R. (2006) Applicability of terrestrial LIDAR scanning for scientific studies in Grand Canyon National Park, Arizona. USGS Open File Report, 2006-1198. p. 27.

Daily, M., Farr, T., Elachi, C. and Schaber, G. (1979) Geologic interpretation from composited radar and Landsat imagery. *Photogrammetric Engineering and Remote Sensing* 45: 1109–16.

Dietrich, W.E. and Perron, J.T. (2006) The search for a topographic signature of life. *Nature* 439: doi:10.1038/nature04452.

Dobson, M.C. and Ulaby, F.T. (1998) Mapping soil moisture distribution with imaging radar, in F.M. Henderson and A.J. Lewis (eds), Principles and Applications of Imaging Radar, vol. 2: *Manual of Remote Sensing.* Wiley, New York, pp. 407–434.

Elachi, C. (1988) *Spaceborne Radar Remote Sensing: Applications and Techniques,* IEEE Press, New York. p. 254.

Farr, T.G. (1992) Microtopographic evolution of lava flows at Cima volcanic field, Mojave Desert, California. *Journal of Geophysical Research* 97: 15171–9.

Farr, T.G. and Adams, J.B. (1984) Rock coatings in Hawaii. *Geological Society of America Bulletin* 95: 1077–83.

Farr, T.G. and Chadwick, O.A. (1996) Geomorphic processes and remote sensing signatures of alluvial fans in the Kun Lun Mountains, China. *Journal of Geophysical Research* 101: 23091–100.

Farr, T.G., Caro, E., Crippen, R., Duren, R., Hensley, S., Kobrick, M., Paller, M., Rodriguez, E., Rosen, P., Roth, L., Seal, D., Shaffer, S., Shimada, J., Umland, J., Werner, M., Oskin, M., Burbank and Alsdorf, D. (2007) The Shuttle Radar Topography Mission. *Reviews of Geophysics* 45: doi: 1029/2005RG000183.

Fielding, E.J., Blom, R.G. and Goldstein, R.M. (1998) Rapid subsidence over oil fields, measured by SAR interferometry. *Geophysical Research Letters* 25: 3215–8.

Fielding, E.J., Talebian, M., Rosen, P.A., Nazari, H., Jackson, J.A., Ghorashi, M. and Walker, R. (2005) Surface ruptures and building damage of the 2003 Bam, Iran, earthquake mapped satellite synthetic aperture radar interferometric correlation. *Journal of Geophysical Research* 110: doi:10.1029/2004JB003299.

Ford, J.P., Plaut, J.J., Weitz, C.M., Farr, T.G., Senske, C.A., Stofan, E.R., Michaels, G. and Parken, T.J. (1993) Guide to Magellan Image Interpretation. JPL Publication 93-24. p. 148. Available at: http://history.nasa.gov/JPL-93-24/jpl_93-24.htm.

Frankel, K.L. and Dolan, J.F. (2007) Characterizing arid region alluvial fan surface roughness with airborne laser swath mapping digital topographic data. *Journal of Geophysical Research* 112: doi: 10.1029/2006JF000644.

Frankel, K.L., Brantley, K.S., Dolan, J.F., Finkel, R.C., Klinger, R.E., Knott, J.R., Machette, M., Owen, L.A., Phillips, F.M., Slate, J.L, and Wernicke, B.P. (2007) Cosmogenic [10]Be and [36]Cl geochronology of offset alluvial fans along the northern Death Valley fault zone: Implications for transient strain in the eastern California shear zone. *Journal of Geophysical Research* 112: doi: 10.1029/2006JB004350.

Gao, B.C., Montes, M.J., Davis, C.O. and Goetz, A.F.H. (2009) Atmospheric correction algorithms for hyperspectral remote sensing data of land and ocean. *Remote Sensing of Environment* 113: S17–S24. doi: 10.1016/j.rse.2007.12.015.

Garvin, J.B. (1996) Topographic characterization and monitoring of volcanoes via airborne laser altimetry. *Geological Society of London, Special Publication* 110: 137–52.

Gillespie, A., Kahle, A.B. and Palluconi, F.D. (1984) Mapping alluvial fans in Death Valley, California using multichannel thermal infrared images. *Geophysical Research Letters* 11: 1153–6.

Gillespie, A., Kahle, A.B. and Walker, R.E. (1986) Color enhancement of highly correlated images. I. Decorrelation and HIS contrast stretches. *Remote Sensing of Environment* 20: 209–35.

Gillespie, A., Rokugawa, S., Matsunaga, T., Cothern, J.S., Hook, S. and Kahle, A.B. (1998) A temperature and emissivity separation algorithm for Advanced Spaceborne Thermal Emission and Reflection Radiometer

(ASTER) images. *IEEE Transactions on Geoscience and Remote Sensing* 36: 1113–26.

Goudie, A.S. (2008) The history and nature of wind erosion in deserts. *Annual Review of Earth and Planetary Sciences* 36: 97–119.

Greve, C. (1996) *Digital Photogrammetry, an Addendum to the Manual of Photogrammetry.* ASPRS. p. 247.

Guenther, G.C. (2007) Airborne lidar bathymetry, in D. Maune (ed.), *Digital Elevation Model Technologies and Applications: The DEM Users Manual.* 2nd edn. American Society for Photogrammetry and Remote Sensing. pp. 253–320. Available at: http://www.jalbtcx.org/downloads/Publications/DEM_Chapter08.pdf.

Guenther, G.C., Brooks, M.W. and LaRocque, P.E. (2000) New capabilities of the SHOALS airborne lidar bathymeter. *Remote Sensing of Environment* 73: 247–55.

Hapke, C.J., Reid, D., Richmond, B.M., Ruggiero, P. and List, J. (2006) National assessment of shoreline change Part 3: Historical shoreline change and associated coastal land loss along sandy shorelines of the California coast. USGS Open-File Report 2006-1219. p. 72.

Haugerud, R.A., Harding, D.J., Johnson, S.Y., Harless, J.L. and Weaver, C.S. (2003) High-resolution lidar topography of the Puget Lowland, Washington. *Geological Society of America Today* June: 4–10.

Henderson, F.M. and Lewis, A.J. (1998) Principles and Applications of Imaging Radar, vol. 2: *Manual of Remote Sensing.* Wiley, New York. p. 866.

Herman, J.R., Bhartia, P.K., Torres, O., Hsu, C., Seftor, C. and Celarier, E. (1997) Global distribution of UV-absorbing aerosols from Nimbus 7/TOMS data. *Journal of Geophysical Research* 102: 16911–22.

Hirano, A., Welch, R. and Lang, H. (2003) Mapping from ASTER stereo image data: DEM validation and accuracy assessment. *ISPRS Journal of Photogrammetry and Remote Sensing* 57: 356–70. doi:10.1016/S0924-2716(02)00164-8.

Hofton, M., Dubayah, R., Blair, J.B. and Rabine, C. (2006a) Validation of SRTM elevations over vegetated and non-vegetated terrain using medium footprint lidar. *Photogrammetric Engineering and Remote Sensing* 72: 279–85.

Hofton, M.A., Malavassi, E. and Blair, J.B. (2006b) Quantifying recent pyroclastics and lava flows at Arenal Volcano, Costa Rica, using medium-footprint lidar. *Geophysical Research Letters* 33: doi: 10.1029/2006GL027822.

Hook, S.J. and Kahle, A.B. (1996) The micro Fourier transform interferometer (mFTIR): A new field spectrometer for acquisition of infrared data of natural surfaces. *Remote Sensing of Environment* 56: 172–81.

Hook, S.J., Abbott, E.A., Grove, C., Kahle, A.B. and Palluconi, F. (1999) Use of multispectral thermal infrared data in geological studies, in A.N. Rencz, (ed.), 1999. Remote Sensing for the Earth Sciences, vol. 3. *Manual of Remote Sensing.* Wiley, New York, pp. 59–110.

Hook, S.J., Myers, J.J., Thome, K.J., Fitzgerald, M. and Kahle, A.B. (2001) The MODIS/ASTER airborne simulator (MASTER) – a new instrument for earth science studies. *Remote Sensing of Environment* 76: 93–102.

Hudnut, K.W., Borsa, A., Glennie, C. and Minster, J.B. (2002) High-resolution topography along surface rupture of the 16 October 1999 Hector Mine, California, earthquake (Mw 7.1) from airborne laser swath mapping. *Bulletin of the Seismological Society of America* 92: 1570–6.

Jonsson, P., Sillitoe, I., Dushaw, B., Nystuen, J. and Heltne, J. (2009) Observing using sound and light– a short review of underwater acoustic and video-based methods. *Ocean Science Discussions* 6: 819–70.

Kahle, A.B. (1987) Surface emittance, temperature, and thermal inertia derived from Thermal Infrared Multispectral Scanner (TIMS) data for Death Valley, California. *Geophysics* 52: 858–74.

Kahle, A.B., Gillespie, A.R., Abbott, E.A., Abrams, M.J., Walker, R.E., Hoover, G. and Lockwood, J.P. (1988) Relative dating of Hawaiian lava Flows using multi-spectral thermal infrared images: A new tool for geologic mapping of young volcanic terranes. *Journal of Geophysical Research,* 93: 15239–51.

Kirkland, L.E., Herr, K.C., Keim, E.R., Adams, P.M., Salisbury, J.W. and Hackwell, J.A. (2002) First use of an airborne thermal infrared hyper-spectral scanner for compositional mapping. *Remote Sensing of Environment* 80: 447–59.

Lambers, K., Eisenbeiss, H., Sauerbier, M., Kupferschmidt, D., Gaisecker, T. and Sotoodeh, S. (2007) Combining photogrammetry and laser scanning for the recording and modelling of the Late Intermediate Period site of Pinchango Alto, Palpa, Peru. *Journal of Archaeological Science.* doi: 10.1016/j.jas.2006.12.008.

Landgrebe, D. (2005) Multispectral land sensing: Where from, where to? *IEEE Transactions on Geoscience and Remote Sensing* 43: 414–21. doi: 10.1109/TGRS.2004.837327.

Landgrebe, D. and Biehl, L. (2009) Multispec, http://cobweb.ecn.purdue.edu/%7Ebiehl/MultiSpec/.

Langenberg, H. (2009) Cartography: Terra cognita. *Nature Geoscience* 2: 542. doi: 10.1038/ngeo603.

Lefsky, M.A., Cohen, W.B., Acker, S.A., Parker, G.G., Spies, T.A. and Harding, D. (1999) Lidar remote sensing of the canopy structure and biophysical properties of Douglas-Fir Western Hemlock forests. *Remote Sensing of Environment* 70: 339–61.

Lillesand, T.M., Kiefer, R.W. and Chipman, J.W. (2008) *Remote Sensing and Image Interpretation.* 6th edn. Wiley, New Jersey. p. 756.

Lundgren, P. and Lu, Z. (2006) Inflation model of Uzon caldera, Kamchatka, constrained by satellite radar interferometry observations. *Geophysical Research Letters* 33: doi: 10.1029/2005GL025181.

Luthcke, S.B., Ekholm, S. and Blair, J.B. (2002) Introduction to the special issue on Laser Altimetry. *Journal of Geodynamics* 334: 343–5.

Lyon, S.W., Lembo, A.J., Walter, M.T. and Steenhuis, T.S. (2006) Internet mapping tools make scientific applications easy, Eos. *Transactions of the American Geophysical Union* 87: 386.

Madsen, S.N. and Zebker, H.A. (1998) Imaging radar interferometry, in Henderson, F.M. and Lewis, A.J. (eds) Principles and Applications of Imaging Radar, vol. 2: *Manual of Remote Sensing*. Wiley, New York, pp. 359–380.

Massonnet, D. (1997) Satellite radar interferometry, *Scientific American* 276(Feb.): 46–53.

Massonnet, D. and Feigl, K.L. (1995) Discrimination of geophysical phenomena in satellite radar interferogram., *Geophysical Research Letters* 22: 1537–40.

Massonnet, D. and Feigl, K.L. (1998) Radar interferometry and its application to changes in the Earth's surface. *Reviews of Geophysics* 36: 441–500.

Massonnet, D., Rossi, M., Carmona, C., Adragna, F., Peltzer, G., Feigl, K. and Rte, T. (1993) The displacement field of the Landers earthquake mapped by radar interferometry. *Nature* 364: 138–42.

Masterlark, T. and Lu, Z. (2004) Transient volcano deformation sources imaged with interferometric synthetic aperture radar: Application to Seguam Island, Alaska. *Journal of Geophysical Research* 109: doi: 10.1029/2003JB002568.

Maune, D.F. (ed.) (2007) *Digital Elevation Model Technologies and Applications: The DEM Users Manual*, 2nd edn. ASPRS, p. 620.

Milton, E.J., Schaepman, M.E., Anderson, K. Kneubuhler, M. and Fox, N. (2009) Progress in field spectroscopy. *Remote Sensing of Environment* 113: S92–S109. doi: 10.1016/j.rse.2007.08.001.

Morris, A.R., Anderson, F.S., Mouginis-Mark, P.J., Haldemann, A.F.C., Brooks, B.A. and Foster, J. (2008) Roughness of Hawaiian volcanic terrains. *Journal of Geophysical Research* 113: doi: 10.1029/2008JE003079.

Mouginis-Mark, P.J., Crisp, J.A. and Fink, J.H. (eds) (2000) Remote Sensing of Active Volcanism. *AGU Geophysics Monograph* 116: 272.

Mustard, J.F. and Sunshine, J.M. (1999) Spectral analysis for Earth science: Investigations using remote sensing data, in A.N. Rencz (ed.), Remote Sensing for the Earth Sciences, vol. 3. *Manual of Remote Sensing*. Wiley, New York, pp. 251–306.

NRC (2007) *Earth Science and Applications from Space, National Imperatives for the Next Decade and Beyond*. National Academies Press, Washington DC. p. 428.

Prior, D.B. and Hooper, J.R. (1999) Sea floor engineering geomorphology: recent achievements and future directions, *Geomorphology* 31: 411–39.

Roering, J.J., Stimely, L.L., Mackey, B.H. and Schmidt, D.A. (2009) Using DInSAR, airborne LiDAR, and archival air photos to quantify landsliding and sediment transport. *Geophysical Research Letters* 36: doi: 10.1029/2009GL040374.

Salisbury, J.W. and D'Aria, D. (1992) Emissivity of terrestrial materials in the 8-14 mm atmospheric window. *Remote Sensing of Environment* 42: 83–106.

Schaber, G.G., McCauley, J.F., Breed, C.S. and Olhoeft, G.R. (1986) Shuttle imaging radar: Physical controls on signal penetration and subsurface scattering in the eastern Sahara. *IEEE Transactions on Geoscience and Remote Sensing* GE-24: 603–23.

Schaepman, M.E., Ustin, S.L., Plaza, A.J., Painter, T.H., Verrelst, J. and Liang, S. (2009) Earth system science related imaging spectroscopy – An assessment. *Remote Sensing of Environment* 113: S123–37. doi:10.1016/j.rse.2009.03.001.

Sexton, J.O., Bax, T., Siqueira, P., Swenson, J.J. and Hensley, S. (2009) A comparison of lidar, radar, and field measurements of canopy height in pine and hardwood forests of southeastern North America. *Forest Ecology Management* 257: 1136–47. doi:10.1016/j.foreco.2008. 11.022.

Shepard, M.K., Campbell, B.A., Bulmer, M.H., Farr, T.G., Gaddis, L.R. and Plaut, J.J. (2002) The roughness of natural terrain: A planetary and remote sensing perspective. *Journal of Geophysical Research* 106: 32777–95.

Sherrod, B.L., Brocher, T.M., Weaver, C.S., Bucknam, R.C., Blakely, R.J., Kelsey, H.M., Nelson, A.R. and Haugerud, R. (2004) Holocene fault scarps near Tacoma, Washington, USA. *Geology*, 332: 9–12. doi: 10.1130/G19914.1.

Short, N.M. (2009) *The Remote Sensing Tutorial*. http://rst.gsfc.nasa.gov/.

Shrestha, R.L., Carter, W.E., Sartori, M., Luzum, B.J. and Slatton, K.C. (2005) Airborne Laser Swath Mapping: Quantifying changes in sandy beaches over time scales of weeks to years. *ISPRS Journal of Photogrammetry and Remote Sensing* 59: 222–32. doi: 10.1016/j.isprsjprs.2005.02.009.

Smith, L.C. (2002) Emerging applications of interferometric synthetic aperture radar (InSAR) in geomorphology and hydrology. *Annals of the Association of American Geographers* 92: 385–98.

Staley, D.M., Wasklewicz, T.A. and Blaszczynski, J.S. (2006) Surficial patterns of debris flow deposition on alluvial fans in Death Valley, CA using airborne laser swath mapping data, *Geomorphology* 74: 152–63. doi: 10.1016/j.geomorph.2005.07.014.

Storlazzi, C.D., Logan, J.B. and Field, M.E. (2003) Quantitative morphology of a fringing reef tract from high-resolution laser bathymetry: Southern Molokai, Hawaii. *Geological Society of America Bulletin* 115: 1344–55.

Tarchi, D., Casagli, N., Moretti, S., Leva, D. and Sieber, A.J. (2003) Monitoring landslide displacements by using ground-based synthetic aperture radar interferometry: Application to the Ruinon landslide in the Italian Alps. *Journal of Geophysical Research* 108: doi: 10.1029/2002JB002204.

Vaughan, R.G., Hook, S.J., Calvin, W.M. and Taranik, J.V. (2005) Surface mineral mapping at Steamboat Springs, Nevada, USA, with multi-wavelength thermal infrared images. *Remote Sensing of Environment* 99: 140–58.

Volker, H.X., Wasklewicz, T.A. and Ellis, M.A. (2007) A topographic fingerprint to distinguish alluvial fan formative processes. *Geomorphology* 88: 34–45. doi: 10.1016/j.geomorph.2006.10.008.

Waggott, S. Clegg, P., Trinks, I., McCaffrey, K., Holdsworth, B., Jones, R. and Hobbs, R. (2005) Towards the virtual outcrop. *Geomatics World* Jan./Feb.: 32–3.

Wehr, A. (2008) LIDAR: Airborne and terrestrial sensors, in Z. Li, J. Chen and E. Baltzavias (eds), *Advances in*

*Photogrammetry, Remote Sensing and Spatial Information Science*s. ISPRS Congress Book. pp. 73–84.

Whitmeyer, S.J., Nicoletti, J. and DePaor, D.G. (eds) (2010) The digital revolution in geologic mapping. *Geological Society of America Today* 20: 4–10. doi: 10.1130/GSATG70A1.

Xu, X., Aiken, C.L.V., Bhattacharya, J.P., Corbeanu, R.M., Nielsen, K.C., McMechan, G.A. and Abdelsalam, M.G. (2000) Creating virtual 3-D outcrop. *The Leading Edge* February: 197–202.

Zebker, H.A., Rosen, P. and Hensley, S. (1997) Atmospheric effect in interferometric synthetic aperture radar surface deformation and topographic maps. *Journal of Geophysical Research* 102: 7547–63.

Zebker, H.A., Amelung, F. and Jonsson, S. (2000) Remote sensing of volcano surface and internal processes using radar interferometry, in P.J. Mouginis-Mark, J.A. Crisp and J.H. Fink (eds), Remote Sensing of Active Volcanism. *AGU Geophysics Monograph* 116, pp. 179–206.

13

Geographic Information Systems in Geomorphology

Takashi Oguchi and Thad A. Wasklewicz

Geographic information systems (GIS) are systems of hardware and software used for storage, retrieval, mapping and analysis of geographic data. The history of GIS is relatively short, and the applications in physical geography and earth sciences, including geomorphology, became frequent only around the end of the 20th century. However, the propagation of GIS in geomorphology during the last 10 years has been rapid. GIS combined with digital elevation models (DEMs) have already become one of the most common approaches of geomorphological research, and many recent studies in geomorphology have utilized GIS and DEMs (Figure 13.1). GIS are strongly linked with methodology and concepts of traditional geomorphology established before the advent of GIS. For example, mapping is very frequently required for research in geography and earth science including geomorphology, and GIS can efficiently provide sophisticated cartographic products. A more essential contribution of GIS to geomorphology is the capability of quantitative topographic analysis, which has expanded the methodology of manual morphometric measurements that developed mainly in the mid 20th century. In contrast, the relationships between geomorphological methods based on GIS/DEM and conventional qualitative methods such as air-photo interpretation and field descriptions are less clear, although the purposes of both approaches are often similar, such as the classification of landform units, understanding of process–form relationships, and detection of topographic changes. This paper reviews GIS applications in geomorphology with emphasis on historical

development and discussion about future perspectives. It also includes the description of application examples in seven fields of geomorphology: (1) visualization of topography, (2) basic morphometric analysis, (3) analysis of streamnets and watersheds, (4) automated landform classification, (5) soil erosion modelling, (6) landslide susceptibility modelling and (7) detection and analysis of topographic change. These fields were selected because of a relatively large number of studies conducted so far.

HISTORICAL BACKGROUND

One of the most important methods and concepts in GIS is the overlay of map layers, which permits the analysis of relationships among different elements in the geographical space, and the evaluation of the effects of some elements on the others. Manual methods of map layer overlay and visual interpretation of the results were known to be effective for understanding cause and effect relationships even in the 19th century. For example, John Snow, a British medical doctor, investigated a cholera outbreak in London and drew a map by overlaying three layers: the distribution of deaths, locations of pumps for obtaining drinking water and streets. He detected a problematic pump that had supplied polluted water, and it proved the communication of cholera through water (Snow, 1855). Such manual approaches of overlay were often used for geographical and environmental studies in the early to

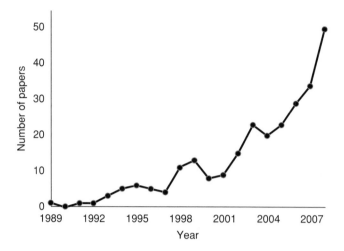

Figure 13.1   Change in the annual number of papers including 'GIS' in the title, abstract, or key words, published in four major international journals of geomorphology (*Catena, Earth Surface Processes and Landforms, Geomorphology*, and *Zeitschrift für Geomorphologie*), during 1989–2008

mid 20th century (e.g. McHarg, 1969). However, manual map overlay is tedious and inefficient, and the obtained results may be subjective or speculative because of the lack of quantitative/statistical analysis, although the methods are surely helpful for hypothetical reasoning. Therefore, the use of digital spatial data and computers to perform map overlay and subsequent quantitative analysis became important.

The first GIS (CGIS: Canada Geographic Information System) was invented by Roger Tomlinson in the 1960s for land-use management and resource monitoring in Canada. He noticed the necessity of a computer system which could replace the manual processing of paper maps, and was successful in obtaining support from the Canadian government and private companies including IBM. CGIS was designed to include innovative methods and ideas common to present-day GIS, such as raster/vector conversion, polygon overlay, and spatial queries using polygons. Although CGIS was operated with one of the most powerful commercial computer systems at that time, the hardware did not permit efficient data processing. The Laboratory of Computer Graphics and Spatial Analysis at Harvard University, USA, also provided some useful mapping software packages from the 1960s. Their efforts led to a GIS software package called ODYSSEY, developed during the late 1970s and the early 1980s. It can be regarded as a prototype of currently available commercial GIS software,

although the number of its users was very small (Foresman, 1998).

A significant change in GIS occurred in the 1980s when many GIS-related private companies were established. The resultant development of commercial GIS software was coupled with new hardware including engineering workstations (EWS) with fast computation capability but compact in size, leading to the propagation of GIS especially in governmental organizations such as census and environmental bureaus. However, applications of GIS in geomorphology were still limited. It should be noted that even before the 1980s some basic methods to handle raster DEMs were developed for topographic analyses and visualization (Sharpnack and Akin, 1969; Evans, 1972), and the automated delineation of streams and watersheds from a raster DEM became a relatively major topic in the 1980s (e.g. Mark, 1984; Jenson and Domingue, 1988). Nevertheless, connections between such DEM-related techniques and GIS were still ambiguous at that time. One possible reason is ESRI Arc/Info, a successor of ODESSEY and the most common GIS software package at that time, was designed mainly for vector data, not for raster data including raster DEMs. However, the first comprehensive textbook of GIS, written by a physical geographer (Burrough, 1986), dealt with spatial analyses both in vector and raster, and it predicted the future progress of GIS applications to earth and environmental sciences including geomorphology.

In the late 1980s and the early 1990s, some researchers stressed the importance of GIS for the future development of geomorphology, introducing the basic concepts of GIS to geomorphologists (e.g. De Roo et al., 1989; Dikau, 1989, 1993). Close relationships between DEM-based geomorphometry and GIS were also pointed out (e.g. Gardner et al., 1990). Applications of GIS to geomorphological topics such as landslides, soil erosion and coastal processes started appearing in scientific journals as well as proceedings of GIS-related conferences. For example, more than ten papers were presented at the Remote Sensing and GIS session in the IAG 3rd International Geomorphology Conference in Hamilton, Canada, in 1993. In such studies during the early 1990s, the methods of handling topographic data tended not to be straightforward because, as noted, available GIS software was often vector-based. For example, Flacke et al. (1990) used vector triangulated irregular networks (TINs) to handle topographic parameters for soil erosion modelling. Although Carrara et al. (1991) and Keller (1992) used raster DEMs to obtain topographic derivatives, the statistical values computed for each vector polygon unit were used for final overlay analyses with Arc/Info. Only in some limited cases, originally developed raster-based GIS software was applied to raster DEMs (e.g. Dikau, 1993). This situation, however, did not last long because of the propagation of software that can handle both vector and raster data, such as IDRISI from Clark University, USA. The ArcGrid extension for raster data analyses within Arc/Info was also released and updated in the 1990s.

The use of GIS in geomorphology was markedly accelerated in the mid to late 1990s, in response to the release of windows-based GIS software for personal computers (PCs) such as ESRI ArcView and IDRISI for Windows. PCs with fast processors, large memory and powerful graphic accelerators became widely available with reasonable prices, resulting in the era of 'desktop GIS'. The availability of basic GIS data such as DEMs also increased dramatically, because of the governmental preparation and compilation of spatial data, as well as the development of the Internet as a way of efficient data distribution. For example, the GTOPO30, a global DEM with a horizontal grid spacing of 30 arc seconds (ca. 1 km in the mid latitude regions), was released in 1996 and has been freely downloadable from the USGS web site. DEMs with ca. 30 to 50 m horizontal resolutions, such as the USGS 30-m DEM, were compiled from existing contour maps in many countries as part of governmental infrastructure. The availability and price of such medium-resolution DEMs significantly differ among countries. For instance, DEMs for the USA can be freely downloaded by anyone, while those for Europe are often sold, and those for some developing countries are classified for political and/or military reasons.

More detailed DEMs with 1 to 10 m horizontal resolutions can be obtained from photogrammetry and laser/radar remote sensing, and geomorphological applications using such fine-scaled DEMs have been increasing since the late 1990s. Advanced remote sensing also yielded the Shuttle Radar Topography Mission (SRTM) DEM, covering most of the earth's surface with ca. 30 or 90 m horizontal resolutions. The SRTM DEM is freely downloadable and thus useful for research in areas where other medium-resolution DEMs are unavailable. The data processing capability of PCs has also been increasing steadily, enabling rapid analysis of large topographic data sets (points, grids, TINs and vector).

Recent geomorphological journals carry many papers related to GIS and/or DEMs, and the number of articles has drastically increased during the last few years (Figure 13.1). Books devoted to the theory and applications of GIS in geomorphology were also published in recent years (Wilson and Gallant, 2000; Bishop and Shroder, 2004; Hengl and Reuter, 2009). Major textbooks about GIS in general, such as Burrough and McDonnell (1998) and Longley et al. (2005), also introduce various geomorphological topics including derivation of morphometric parameters from a DEM, soil erosion modelling, and landslide risk assessment.

MAJOR APPLICATIONS OF GEOGRAPHIC INFORMATION SYSTEMS IN GEOMORPHOLOGY

GIS combined with DEMs have been used to address various issues in geomorphology, and as noted above, there are several particular topics to which GIS have been frequently applied. We review these topics with representative examples, including both relatively early studies and recent ones.

Visualization of topography

Currently, most thematic maps and images for geographical and geoscientific research are constructed using digital data and computers. Although cartographic products generated from commercial graphic software such as Adobe Illustrator and Freehand have been used, the application of GIS software is a common exercise in today's geomorphic research. The implementation of GIS techniques in geographical research was a logical transition, because data with real geographic coordinates are more advantageous than merely geometric data used in graphic software. Visual representation of topography using a map is essential in geomorphology, and various efforts were made to do it effectively with computers (Vitek et al., 1996). In the 1960s and 1970s, some computer software packages for drawing or 'typing' topographic maps were invented (e.g. SYMAP from the Laboratory of Computer Graphics and Spatial Analysis at Harvard University), but resultant maps did not look superior to conventional printed maps because of the limitation in distributing and printing the digital representations from these software packages. A good example of effective digital topographic representation developed with GIS is shaded-relief maps constructed from DEM-based hill shading. The technique of digital hill shading was already known in the 1960s (Yoeli, 1965), but its strength in topographic visualization came to be widely known through the poster-size shaded-relief map of the conterminous United States created by Thelin and Pike (1991). Because of its superb landform representation, the map became a best-seller of the USGS, and was widely used in schools and universities. Then modules for DEM-based hill shading were added to commercial GIS software packages such as IDRISI and ArcView. Accordingly, shaded-relief images frequently appear in recent geomorphological publications, mainly as the background of thematic maps to let readers understand the general topographic characteristics of the study area. Although shaded-relief images themselves only permit visual topographic interpretation, even recent papers show the importance of such images in investigating various geomorphological features including glacial landforms (Smith and Clark, 2005), landslides (Van Den Eeckhaut et al., 2005) and active faults (Oguchi et al., 2003).

Another effective way of topographic representation is the use of three-dimensional (3-D) images such as block diagrams and bird's-eye views, as well as dynamic animated cartography including fly-through movies (Mitas et al., 1997). In the early 1990s, the production of 3-D images required the map designer to develop original algorithms (McLaren and Kennie, 1989); whereas commercial GIS and remote sensing software packages released after the 1990s often included a module for 3-D cartography based on DEMs. Topographic visualization in 3-D and its animation may involve hill shading and the draping/superimposing of satellite remote sensing or air-photo images to make results more realistic. Although such complex data processing used to take a long time,

the problem has been reduced by enhanced computational and graphic capability of computers. A recent notable progression of 3-D applications in geomorphology is the emergence of the Google Earth (http://earth.google.com), which enables fast creation of 3-D topographic images and fly-through animations for any place on the earth without data preparation on user's side. Although Google Earth is essentially a tool for visualization without analytical functions, its usefulness in science has been recognized (Butler, 2006).

Basic morphometric analysis

DEMs can provide a variety of information about landform structure. The most basic information is height as recorded in a DEM, and it can be directly used to derive area–altitude relationships, that is, hypsographs or hypsometric curves. GIS enable the fast derivation of the relationships without tedious manual techniques once required by users. The relationships are effective in detecting flat surfaces such as peneplains and in analyzing the general structure of watersheds (Fielding et al., 1994; Hurtrez et al., 1999), as was already recognized before the GIS era through classic morphometric studies (e.g. Gardiner, 1978).

Basic derivatives from DEMs include slope, aspect and curvature. Techniques to obtain these parameters from a raster DEM based on a finite difference method and a moving window became available in the late 1960s (Sharpnack and Akin, 1969); then various options of calculation were proposed and compared (Evans, 1972; Zevenbergen and Thorne, 1987; Guth, 1995). The three parameters play important roles in determining earth surface processes. Slope and curvature affect the rate of surface sediment transport and slope stability, while aspect exerts strong influence on soil moisture and in turn hydrological and pedogenic processes. Therefore, modules for computing these parameters from a DEM are often integrated into GIS software. Other simple topographic parameters, such as relative relief within a window of certain size, can also be obtained from a DEM (e.g. Johansson, 1998).

The previously described DEM derivatives are prone to errors and this issue has received special attention because these values have been widely used in geomorphic research (Heuvelink, 1998). The error reflects the cell size of the original DEM (Kienzle, 2004), methods of DEM construction (Bolstad and Stowe, 1994), algorithms of parameter calculation (Florinsky, 1998), and the size of the moving window used for calculation (Albani et al., 2004). Some researchers show contour-based topographic models give more accurate parameter values than raster DEMs (e.g. Mizukoshi and Aniya, 2002). However, recent applications mostly depend on raster DEMs rather than contour-based data because of their much higher availability and simpler data structure. In addition, the spatially equal data density of a raster DEM also supplies statistically meaningful results.

Relatively simple analyses of these basic parameters may give various geomorphological insights. For example, the frequency distribution and statistical values of slope obtained from DEMs have been related to dominant hillslope processes and the presence/absence of an equilibrium condition within mountains (Katsube and Oguchi, 1999; Montgomery, 2001). Lithological and structural control on landforms can also be discussed quantitatively using basic topographic parameters and GIS, if digital geological data are also available (Kühni and Pfiffner, 2001). The increasing availability of very-fine-resolution DEMs facilitates the analysis of fine-scale topographic configurations such as debris flow lobes on alluvial fans (Staley et al., 2006) and complex lava surface topography (Marsella et al., 2009) based on basic DEM derivatives.

Analysis of stream-nets and watersheds

Methods to delineate stream-nets and watersheds from a raster DEM have been proposed since the 1970s (Collins, 1975), and as noted, the 1980s saw significant development in this field. This fact reflects the large demand for the methods not only in geomorphology, but also in other disciplines including hydrology, environmental sciences and civil engineering. In most cases, stream-net and watershed delineation depend on DEM-derived flow direction and flow accumulation. Although this methodology is relatively simple and has already been established, applications of the method tend to face some common problems such as the occurrence of spurious pits, reduced accuracy in low-relief areas especially plains, and the limited number (typically eight) of possible stream flow directions. To tackle these problems, various researchers proposed revised methods (e.g. Martz and Garbrecht, 1998). The concurrent use of vector data showing the location of streams and divides may improve the quality of delineation results, especially when a high accuracy DEM is unavailable (Mayorga et al., 2005).

The delineation of watershed boundaries is relatively easy in a technical sense because they are almost uniquely determined from height distribution. In contrast, deriving stream-nets is more complex and different methodologies may lead to different results. The appropriate determination of the channel head location along a topographically determined valley line is a particularly difficult task. The most basic method to locate channel heads is the use of a threshold drainage area (Jenson and Domingue, 1988), and it is often employed in the hydrological modules of commercial GIS software. Another objective method is to select cells with relatively high Horton/Strahler stream orders after assigning orders to all DEM cells by assuming streams starts from ridges, which may be effective for highly dissected terrain including badlands (Lin and Oguchi, 2004). However, in many cases the location of a channel head varies depending on several local conditions such as slope angle, land use, and drainage area (e.g. Dietrich et al., 1993). Vogt et al. (2003) have proposed a new method that combines several environmental variables to account for varying landscape conditions in determining the location of a channel head.

The automated delineation of stream-nets and watersheds facilitate relevant geomorphological analysis such as the numerical classification of watersheds based on abundant data (Cheng et al., 2001). A large sample number is highly important to obtain statistically meaningful results, but classical manual methods of watershed analysis, as often performed in the mid 20th century, were inefficient to provide large data sets. Increasing availability of relatively high-resolution DEMs for broad areas such as the SRTM data also helps comprehensive studies with large samples, although the resolution of the DEMs may be still insufficient for precise stream-net and watershed delineation (Hancock et al., 2006). At the same time, very fine DEMs with one to a few meters resolution, constructed from photogrammetry and radar/laser remote sensing, permit the analysis of complex drainage systems in a small area (Lin and Oguchi, 2004).

The automated delineation of stream-nets and watersheds also facilitates the applications of quantitative models in fluvial geomorphology to the real world. As a result of this technique, hydro-geomorphological parameters such as stream power can be analyzed not only for selected points, but also successively along the river (Reinfelds et al., 2004; Jain et al., 2006; Figure 13.2). Specific fluvial features including knickzones can also be determined objectively from DEMs and stream-net data (Hayakawa and Oguchi, 2009).

Automated landform classification

Landform classification has a long-established tradition as a basic method of geomorphological

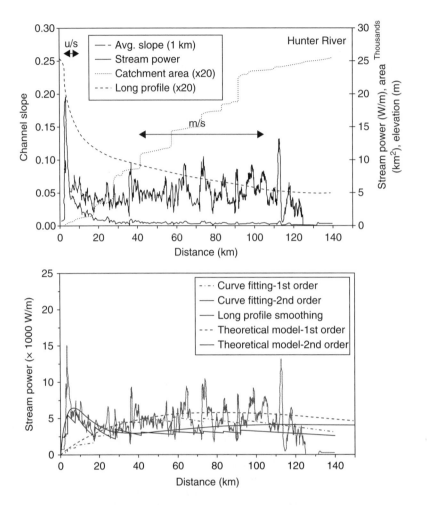

Figure 13.2 Downstream variation in stream power and its components for the Hunter River, Australia, based on GIS and DEM analyses. In the upper figure stream power is based on the long profile smoothing method. In the lower figure stream power based on theoretical models and curve fitting are also shown. (Modified after Jain et al., 2006)

studies. It traditionally depended on visual interpretation of maps, air photos and satellite images as well as field surveys. Such approaches, based on expert's judgment, tend to be qualitative and subjective because different observers may yield different results. Automated methods based on DEMs and GIS have been proposed to increase the objectiveness and reproducibility of landform classification. The possibility of using basic topographic parameters such as slope and curvature for landform classification was already discussed in the late 1980s (Dikau, 1988; Pike, 1988). Geometric criteria were used to establish the categories of the classification in these studies, which differed from conventional classification

systems. The conventional classifications, on the other hand, took into account the genesis of each topographic unit. Another early example of DEM-based landform classification (Chorowicz et al., 1989) delineated many topographic profiles and measured their shape; then the results were extended to an area for mapping landform objects. This seems to reflect difficulties in directly analyzing areal units, related to limitations in the software and hardware available at that time. In addition, the detected landform objects are simply geometric like convex slopes, concave slopes, and crest lines.

Although studies of automated landform classification in the early 1990s were still relatively simple, more origin-based topographic units such as valley heads and flood plains were detected from DEMs (Tribe, 1991; Lee et al., 1992). Furthermore, the combination of altitude and slope derived from DEMs was found to be effective to differentiate large-scale landform units such as mountains, hills and lowlands (Graff and Usery, 1993; Guzzetti and Reichenbach, 1994). More complex origin-based classification of small scale landforms such as moraines and solifluction slopes, became possible in the late 1990s (Giles and Franklin, 1996; Brown et al., 1998) in response to the enhanced capability of GIS and the increased availability of high-quality DEMs. Such studies often depended on methods developed in the field of remote sensing including the maximum likelihood and artificial neural network. The accuracy of classifications in these studies is 50 to 90 per cent, implying that in many cases the conventional manual landform classification based on expert's knowledge is still more reliable and useful than automated classification. However, recent studies of DEM-based automated landform classification have tested new parameters such as topographic openness (Prima et al., 2006) and third-order partial derivatives (Minár and Evans, 2008; Figure 13.3), as well as new methodology including object-oriented approaches (Van Asselen and Seijmonsbergen, 2006). These approaches provide potential for increased accuracy in landform classification. In addition to higher objectiveness and increased accuracy, the automated methods may permit faster construction of landform classification maps and make subsequent analysis easier, because classification results are available in a digital format from the beginning.

Recently, the increased availability of DEMs and advances in computational processing rates have allowed landform classification to occur at regional (Miliaresis, 2001) and global (Iwahashi and Pike, 2007) scales. Previous work classifying landforms at broad scales, either manually or digitally, could only represent large topographic units (i.e. a mountain range or a broad plain). The recent studies have been able to provide more detailed landform classification over a broad area.

Soil erosion modelling

Most application examples mentioned above can be performed if a suitable DEM is available. If other spatial data are also combined with a DEM, the variation of geomorphological applications increases. An example of such approaches is soil erosion modelling using the universal soil loss equation (USLE) or similar types of equations such as the revised USLE (RUSLE). Although the

USLE was originally developed before the GIS era to predict the rate of soil erosion, its structure fits well to the capability of layer overlay in raster GIS. Therefore, some papers in the 1980s already featured USLE-based studies using GIS (Bork and Hensel, 1988), and similar methods have been applied to various regions of the world in recent publications (e.g. Da Silva, 2004; Ozcan et al., 2008). Some GIS textbooks and the manuals of major GIS software packages such as ArcGIS and IDRISI have also introduced the USLE-style soil erosion modelling as a typical contribution of GIS to environmental problems. Data layers required for this type of application include rainfall, soil properties, soil management practice, vegetation and topography. Information about vegetation is often collected using satellite remote sensing, while climatological and pedological data are usually obtained from existing analog/digital maps and literature. Field observations and laboratory analyses of soil samples may also be required.

More hydrology-oriented models of soil erosion such as areal non point source watershed environment response simulation (ANSWERS) have also been linked to GIS (Desmet and Govers, 1995). This type of approach owes much to the development of distributed hydrological models in GIS. Soil erosion modelling also permits the estimation of sediment yields from watersheds as well as the simulation of landform development (Peeters et al., 2006; Bhattarai and Dutta, 2007), if the contribution of other erosional processes such as mass movement can be assumed to be negligible. Regional-scale erosion modelling is also becoming realistic with increasing availability of data for broad-scale areas (e.g. Verstraeten, 2006).

Landslide risk/susceptibility modelling

Like soil erosion modelling, evaluation of landslide risk/susceptibility has been one of the most common GIS/DEM applications in geomorphology. Other than numerous recent publications, early literature in the 1980s already addressed this issue (Okimura and Kawatani, 1987). Both shallow failures and deep-seated landslides have been research targets. Data required for landslide risk/susceptibility modelling are similar to those for soil erosion modelling (e.g. topography, rainfall, geology/soil and land use/vegetation; Figure 13.4) because both geomorphic processes reflect surface instability, and in most cases, water-related processes. Distribution of artificial infrastructure such as roads has also been considered in some research

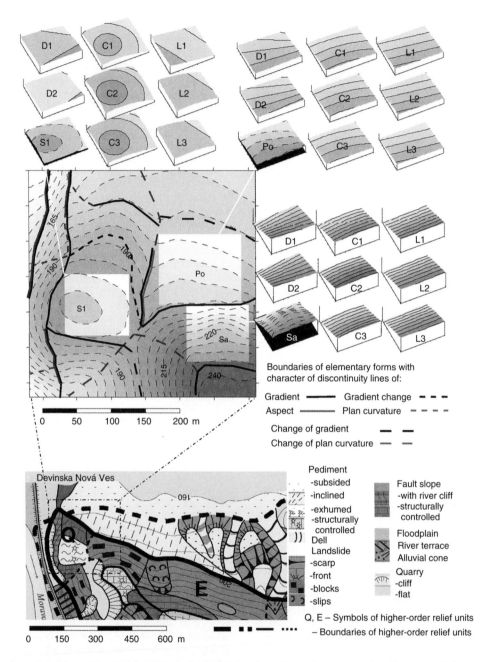

Figure 13.3 Automated landform classification based on elementary landforms and their boundaries derived from a DEM, for Devínska Kobyla Mountain, Slovakia. (After Minár and Evans, 2008)

(Larsen and Parks, 1997). In addition, the inventories of past landslides constructed from air photos, satellite images and literature are also often used to construct, calibrate or validate landslide risk/susceptibility models. One distinct difference between soil erosion modelling and landslide modelling is that the result of the former is usually presented as potential erosion rates, while that of the latter as the likelihood of landslide occurrence.

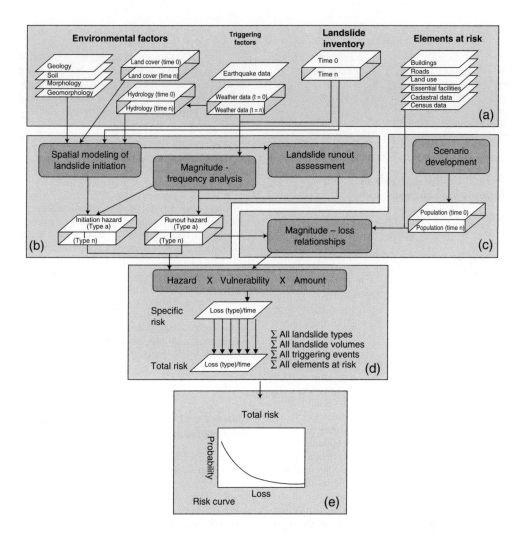

Figure 13.4 Schematic representation of the landslide risk assessment procedure. (a) Basic data sets required, both of static, as well as dynamic (indicated with 'time...') nature, (b) Susceptibility and hazard modelling component, (c) Vulnerability assessment component, (d) Risk assessment component, (e) Total risk calculation in the form of a risk curve. (After Van Westen et al., 2008)

Various methods to assess landslide risk/susceptibility using GIS were developed during the 1990s. Empirical types of approaches are based on statistical models such as linear overlay of weighted factors (Shu-Quiang and Unwin, 1992) and discriminant analysis (Carrara et al., 1991). GIS are also linked with the deterministic geotechnical models of slope stability which calculate the factor of safety based on topography, soil properties and hillslope hydrology (Van Western and Terlien, 1996). Delineation of watersheds and stream-nets are sometimes combined

with such models, because upstream area can be used as a surrogate of the amount of water input to a certain part of a hillslope. Final landslide susceptibility maps usually show categorical levels of landslide potential such as very high, high, low and very low, to make maps easier to understand for practical and management purposes.

Some recent examples intended to offer improved and more complex methods of landslide susceptibility assessment including the use of soil wetness modelling (Gritzner et al., 2001), stochastic hydrological modelling (Iida, 1999), logistic regression of different types (Dai et al., 2001; Van Den Eeckhaut et al., 2006) and artificial neural network (Ermini et al., 2005). The use of such methods became possible with the increased availability of fast computers, relevant software packages and high-resolution spatial data (Van Werstern et al., 2008).

Detection and analysis of topographic change

A set of topographic data for different periods enables the detection and analysis of topographic change. GIS facilitate this type of research because they allow the overlay of data for different periods and subsequent quantitative analysis using multiple layers. Data with different projection and distortion can be overlaid at the same coordinate system for reliable comparisons, using the functions of geometric correction, interpolation, and ortho-rectification contained within a GIS.

One of the earliest examples of this type of GIS application is the analysis of coastal erosion (Hill and Turnipseed, 1989). Later applications in coastal geomorphology analyzed height data as well as horizontal locations of landforms to discuss 3-D topographic changes (Rogers et al., 2004). Analytical or digital photogrammetry based on air photos or stereo satellite images often plays an important role in obtaining 3-D topographic data for different periods. The detected topographic changes at coastal zones permit the estimation of dominant geomorphic processes and sediment movement patterns along the coast (Mitsova et al., 2009). The success of beach management practices (i.e. beach renourishment after a storm) has also been considered through assessing topographic changes of beach–dune settings (Gares et al., 2006).

Similar methods have been applied to changes in fluvial landforms. Channel location data for different periods can be obtained from maps and remote sensing images including air photos. The data permit the detection and analysis of lateral channel migration. If a meandering river underwent complex lateral movement, GIS techniques can be employed to estimate channel locational probabilities (Wasklewicz et al., 2004; Figure 13.5). The locational probabilities can be further evaluated with the aid of GIS-based cellular statistics (Staley and Wasklewicz, 2006). Measures of spatial statistics such as spatial autocorrelation can also be employed to analyze highly complex temporal change in land cover with a braided river (Takagi et al., 2007). Height data for different

periods along a river are useful to analyze the progress of fluvial erosion and deposition (Ham and Church, 2000; Otto et al., 2008). A series of DEMs for different periods have also been used for analyzing other types of topographic changes such as gullying (De Rose et al., 1998) and periglacial processes (Wangensteen et al., 2006).

PROBLEMS AND FUTURE PERSPECTIVES

The increase in GIS/DEM applications in various fields of geomorphology has been a recent definitive trend, indicating that some major problems in the past such as expensive software and hardware, limited availability of digital data and limited computational power of computers have already been solved at least to some extent. However, there are still some common problems with GIS/DEM applications in geomorphology. This section describes the examples of such problems, and makes proposals to overcome the problems for developing future research.

Characteristics and quality of digital elevation models

As noted, DEMs are the most frequently used data for GIS applications in geomorphology, and their availability has significantly increased in recent years. Some widely used DEMs are arranged using the longitude/latitude (geographic) projection; for instance, the SRTM DEM has a grid interval of 1 or 3 arc seconds. The longitude/latitude projection is basic and very often used in GIS, especially when the area of data coverage is relatively large. However, the exact metric grid interval of a DEM with that projection differs from place to place, which cannot be neglected if the DEM covers a relatively wide range of latitude. In contrast, if a relatively small area is the target of study, it is possible to neglect the change in the grid interval and assume that each DEM grid is rectangular with a fixed size. However, some GIS software packages including ESRI products do not accept rectangular raster data and only completely square data can be handled. A typical solution of this problem is the application of an interpolation method to alter the side lengths of a DEM cell. Interpolation is also employed to change the projection system, for example from the longitude/latitude (geographic) system to Universal Transverse Mercator (UTM) to enable metric calculations within GIS.

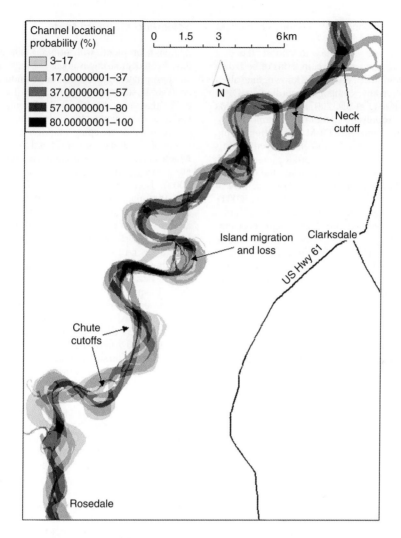

Figure 13.5   Locational probability of a segment in the Lower Mississippi River, USA. (Modified after Wasklewicz et al., 2004)

It should be noted that interpolated DEMs are secondary products and tend to include certain errors (Hu et al., 2009). Therefore, the choice of an appropriate interpolation method is highly important for successful conversion of a DEM (Ahmadzadeh and Petrou, 2001). For example, when Kriging is employed as an interpolation method, it is necessary to use a function model (linear, exponential, Gaussian, etc.) and a certain level of the nugget effect corresponding to those in the actual semi-variogram. Nevertheless, most interpolation modules of GIS software have only limited choices of models and parameters, and users tend to apply a default setting without paying sufficient attention to the characteristics of

data. Therefore, interpolation performed may significantly lower the quality of the resultant DEM and its derivatives (Ahmadzadeh and Petrou, 2001).

As noted, errors including measurement, digitization and airborne scanning errors in original DEMs should also be taken into account, although the problem itself is usually beyond the control of users. For example, DEMs from remote sensing data such as the SRTM DEM often contain local spikes and pits, and DEMs from contour data tend to have 'terraced' altitude distribution especially in lowlands. Geomorphological analyses using DEMs with large errors may be useless. For example, the quality of automated landform classification strongly depends on the level of DEM error (Lee et al., 1992).

It is therefore necessary to choose the best DEM for a research purpose, in terms of both data error and resolution. As widely known, the values of DEM derivatives change according to DEM resolution (Deng et al., 2007) and coarse DEMs can only represent generalized topography. Increase in apparent DEM resolution using interpolation may not be effective in settling this problem even if an appropriate interpolation method is employed, because the optimal DEM resolution is proportional to the density of original height measurements (Kienzle, 2004). The increasing availability of high-resolution DEMs, however, will settle the major problems related to data resolution. Simple contraction of high-resolution DEMs provides medium- to low-resolution DEMs, and thus a DEM having the best resolution for a certain purpose can be obtained readily. Therefore, choice of the best DEM based on error characteristics will be more important. For example, photogrametrically derived DEMs may be more suitable to analyze slope distribution in lowland than DEMs derived from vector contour data with the terracing artifacts (Hashimoto et al., 2007).

Although the use of high-quality data with small error is desirable, it is also important to do something meaningful using currently available data, because no spatial data including DEMs are perfectly accurate. If the detected geomorphological differences are sufficiently larger than the error range of the data used, it is acceptable to make geomorphological inferences about the differences, even if the error range is relatively large. Such judgment about the credibility of the results can be made only when concrete information about data error is available. If both DEMs and other data such as geology are used for analysis, the quality of the other data including completeness and precision also needs to be considered. As a result of increased availability of spatial data in recent years, it is not rare to have more than two available datasets for the same theme and area, but detailed information about the quality of these datasets may be unavailable. In these cases, individuals must consider means to verify the DEM uncertainty from field-based data quality assessments (i.e. Wechsler, 2007). In summary, more attention needs to be directed toward errors of DEMs and other spatial data, and sharing information about errors of available data will support this direction. Several recent books provide further information on reducing errors in DEM production for the enhancement of accurate GIS analyses in geomorphology (Li et al., 2005; El-Sheimy et al., 2005; Hengl and Reuter, 2009).

Scale and mapping units

Although problems with original data resolution have been decreasing as noted above, determining an appropriate size of units for geomorphological analysis is still an important issue. The smallest units that can be mapped depend on data resolution or map scales. For example, the size of the smallest mapping units on 1:10,000 geomorphological maps is considered to be 0.25 cm², which corresponds to a landform dimension of 50 × 50 m (Van Asselen and Seijmonsbergen, 2006). Therefore, the combination of data from 1:10,000 maps and a 50-m DEM is acceptable if the altitudinal distribution of units is a target of analysis. However, if a DEM derivative for each unit such as slope needs to be analyzed, a DEM with a finer resolution is necessary because the meaningful values of the DEM derivative for a unit is derived from a certain number of DEM cells. Despite the widespread concern with scale in geomorphology (Dikau, 1990), many studies still only consider a particular scale of analysis (Evans, 2003) or a very narrow range of scales (Schmidt and Andrew, 2005; Shary et al., 2005). Furthermore, these approaches are often done without consideration of the spatial scale [defined as the extent a feature extends (occupies) in space and is commonly referred to as length scale in disciplines such as physics, geology and biology] of the surface roughness (Atkinson and Tate, 2000; Reuter et al., 2009). This is an area of research where GIS and geomorphological mapping can play a significant role in the issue of scale that plagues many scientific fields.

The choice of data resolution also depends on the general purpose of geomorphological analysis. To investigate the difference in the distribution of small landform units according to large-scale topographic divisions, it may be acceptable to derive the divisions using a relatively coarse DEM even if it is unsuitable for deriving the small units. Various combinations of data resolution and the size of mapping units are possible, provided the researcher supplies a logical or quantifiable justification for the combined scales of analysis. However, many existent publications do not present such justification to readers, even when data with different spatial resolutions are used simultaneously.

Effective spatial scales used also vary depending on models used for research. Landslide hazard assessment based on a geotechnical model including a safety factor calculation is suitable for a small area with detailed information about soil and topography, but it may not work well for a large area, where statistical approaches are more effective. It may be also unrealistic to apply a single statistical model to a very broad area, because of substantial differences in the environment within the area. The choice of suitable mapping units is also related to analytical methods. For example, the units of landform partitioning for landslide hazard assessment can be divided into grid cells, terrain units, unique-condition units, slope units and topographic units and their effectiveness differs according to the type of analytical approaches including statistical, physically based, and heuristic (Guzzetti et al., 1999). However, different combinations of topographic partitioning and analytical approaches have been used, because methods for selecting the best combinations have not been established.

It should be also noted that different DEM scales or resolutions can result in dissimilar findings despite conducting the identical analyses within the same study area. For example, the inferred effects of variables on the occurrence of patterned ground may change markedly according to the cell size of the raster data (Luoto and Hjort, 2006). This problem, the so-called modifiable areal unit problem (MAUP), has been receiving significant attention in the field of spatial statistics, and the concept was introduced to ecology and remote sensing during the 1990s (Jelinski and Wu, 1996; Marceau and Hay, 1999). Although the MAUP is a complex issue and a simple solution is unavailable, geomorphologists need to pay more attention to it because the MAUP in geomorphology may appear differently according to the type of landforms. A variety of case studies are needed to address the problem. It is also necessary for geomorphologists to understand what has been discussed in the relevant fields such as GIScience, photogrammetry and remote sensing, particularly in relation to data resolution, data error and the methodology of spatial analysis. Not only the MAUP, but also recent discussion by GIScientists about the choice of window size in deriving morphometric parameters from a DEM (Albani et al., 2004; Ehsani et al., 2010) is worthy of attention, because in geomorphology, a 3×3 moving window has been unconsciously used in most applications.

Calibration and validation of models

GIS technology helps geomorphologists apply physical, mathematical and statistical models to geomorphological issues such as soil erosion and landslides. Calibration and validation of the applied model are highly important, because numerical models are usually flexible and various models can be candidates for the target phenomenon. However, as indicated by Guzzetti et al.

(1999, 2006), most previous studies provided little information on the quality of the proposed models.

The quality of a geomorphological model can be assessed using either the same data used for model construction or independent information. Assessments of the first type include the matching level evaluation of obtained results such as the calculation of user's accuracy, producer's accuracy and the kappa statistic. Sensitivity analysis of input variables is another assessment method of the first type, which assumes that results from a robust statistical model do not change significantly with a relatively small change in the input variables. These assessments, however, may be regarded as the evaluation of model performance rather than strict model validation or confirmation.

Validation of a model using independent data is considered to be more effective (Chung and Fabbri, 2003). A way of such validation is to divide the original data into two subsets: one for training or modelling, and the other for validation. The division can be based on space (two different areas), time (old and new) or random selection (Chung and Fabbri, 2003). This approach reflects the relatively infrequent occurrence of some geomorphological events such as landslides. For a frequently occurring phenomenon, validation using a newly obtained different dataset is easy, because a new event will take place soon after the construction of a model. Despite the strength of validation based on data partitioning, there is a common problem – it is often difficult to ensure that the partitioned datasets are comparable with almost the same environmental conditions. For instance, geology may be significantly different even between adjacent areas (Guzzetti et al., 2006), and some environmental conditions such as vegetation cover changes with time. Although random partitioning seems to solve the problem, its effectiveness has not been fully explored because the method has rarely been employed in previous studies.

The lower availability of data for a low-frequency geomorphological phenomenon limits not only validation procedures, but also original model construction. Therefore, developing modeling methods suitable for rare events, as attempted by Van Den Eeckhaut et al. (2006), is highly important. In contrast, data for high-frequency phenomena such as soil erosion tend to have higher availability, facilitating the construction and validation of models. Indeed, some researchers performed relatively robust calibration of soil erosion models based on field data (e.g. Van Oost et al., 2005; Warren et al., 2005). However, even in the case of soil erosion, availability of useful field data tends to be limited. For instance, direct observation of catchment sediment yield is usually performed only at a few locations typically at the outlet of a whole catchment. It should be noted that different parameter sets may give similar gross erosion rates for a watershed, although the modelled distribution of erosion rates within the watershed may be significantly different. Selecting the best parameter sets under such a circumstance is crucial for not only distributed erosion assessment but also understanding the effect of each environmental factor. Distributed data such as gully dimensions and point erosion rates estimated from ^{137}Cs and rare-earth element tracers (Van Oost et al., 2005; Warren et al., 2005; Zhang et al., 2005) are effective in solving this problem, although collecting such data with a high spatial resolution also requires tedious field surveys and laboratory analyses. Furthermore, the cost associated with producing high-resolution geochronology data can rapidly become prohibitive. A possible future direction is to combine an erosion model with the detection of topographic change based on a set of high-resolution DEMs, to directly link the model to distributed validation data within GIS. As noted, recent development of digital photogrammetry and radar remote sensing has facilitated the acquisition of high-resolution DEMs for different periods, which should be useful for solid geomorphological modelling.

The problem of model calibration and validation is directly related to uncertainty in original data and modelling results. Methods to deal with uncertainty could be helpful for the production of more reliable modelling outputs. For example, the introduction of fuzzy logic may improve the performance of a GIS model if the amount of available data is small, or if a factor used is expressed as a categorical variable but its actual spatial change is gradual rather than abrupt. Mitra et al. (1998) indicated that the performance of a USLE-based model of soil erosion, for a wide range of spatial scale, can improve if fuzzy logic is applied to variables used. However, complex data manipulation with some assumptions required for fuzzy logic applications may make the final modelling result ambiguous and more difficult to evaluate. Indeed, geomorphological studies based on fuzzy logic has been limited to a few cases (e.g. Schneevoigt et al., 2008), although its basic concept has been widely known. On the one hand, this situation could reflect the rather serious limitation of fuzzy logic in solving geomorphological problems. On the other hand, it may be worthwhile to evaluate the potential of fuzzy logic for geomorphological research in detail. The same seems to apply to other methods addressing uncertainty in data and its effects on modelling, including Monte Carlo simulation (Davis and

Keller, 1997) and artificial neural network (Ermini et al., 2005).

Roles of expert knowledge

Traditional geomorphology heavily depended on qualitative and heuristic methods such as field observations and visual interpretation of topographic maps and air photos. The reliability of such a method depends on how well and how much the investigator understands the origin of landforms and geomorphological processes acting on them. Therefore, maps obtained by this method cannot be readily evaluated in terms of reliability and certainty (Guzzetti et al., 1999). Indeed, the result of visual detection of landform units may significantly vary among researchers (Van Den Eeckhaut et al., 2005). However, geomorphology has historically developed with such heuristic approaches including descriptions of landforms and surface deposits using diagrams and text as well as manual landform classification. Despite the more quantitative and objective landform analyses provided by GIS/DEM approaches, it is not constructive to neglect the classic and important achievements that have stimulated geomorphologists into current research themes. Indeed, landforms are complex objects affected by various environmental factors, and it is impossible to quantify all their characteristics and all factors affecting their development. Therefore, even today, a certain level of heuristic reasoning is often required to better understand landforms.

A relatively common way to add the traditional heuristic reasoning to GIS-oriented geomorphological studies is to discuss the implication of obtained results based on broad knowledge from conventional geomorphology, which may ensure that the studies contribute to the progress of geomorphology in general. Traditional geomorphology includes not only descriptive studies, but also process-oriented studies. Knowledge from the latter is also useful to strengthen the implication of studies based on GIS and DEMs, because such studies usually deal with static structure of landforms rather than the movement of material, although certain processes may be inferred from topographic structure alone (Lane, 1998). To enable such discussion with a wide scope, researchers are expected to have sufficient knowledge about geomorphology including traditional and qualitative approaches. In other words, mere GIS specialists have difficulties in undertaking good geomorphological research, despite their prowess in computer sciences, statistics, programming and handling software.

Among the major subfields of GIS-related studies, automated landform classification is strongly related to qualitative expert knowledge, especially in establishing the classification method and evaluating classification results. Commonly used landform units such as alluvial fans, talus cones, and river terraces are usually detected based on a heuristic approach, but the detection of such landform units based solely on an automated method is still difficult. Automated landform classification has often been combined or compared with an existing landform classification map. However, as previously noted, existing maps based on experts' knowledge may have limited reliability. In addition, the results of automated landform classification have often been compared with published maps only visually, because existing maps may not be available in a digital format.

Many challenges remain and there are various possibilities for future methodological development. Indeed, different approaches toward automated landform classification have been proposed in recent years (Van Asselen and Seijmonsbergen, 2006; Prima et al., 2006; Iwahashi and Pike, 2007; Stepinski and Bagaria, 2009). This situation differs from some other applications such as the modelling of landslides or soil erosion for which similar methods tend to be applied to various case studies. A realistic direction of automated landform classification is to reduce the dependency on expert knowledge compared to traditional geomorphological mapping (Van Asselen and Seijmonsbergen, 2006), not trying to provide a fully automated method. The balance between automated and traditional methods in such studies is flexible, but should be determined according to the purpose of landform classification.

CONCLUSIONS

This chapter has provided a review of GIS/DEM applications in geomorphology with emphasis on their historical development, and has indicated some current problems and future directions. GIS were initially used just as mapping tools or to perform simple computations, but currently they have become a powerful, integrated technology useful for both scientific and applied purposes. GIS can be applied to analyzing and understanding landforms in an objective, qualitative and efficient way. GIS and DEMs permit statistical studies for a broad area based on abundant data, and thus support us to establish a general theory of landform structure. Detailed studies for small areas are also enhanced by GIS and high-resolution DEMs. Because of such advantages over traditional geomorphological methods

including visual map reading and field surveys on limited accessible sites, GIS/DEM applications in geomorphology have significantly increased during the past several decades.

Although GIS and DEMs have been playing an important role in various subfields of geomorphology, they have some common problems, and their solution will make GIS/DEM applications more meaningful. As noted, the solutions often require broad knowledge on relevant fields including spatial statistics, photogrammetry, remote sensing and traditional qualitative geomorphology. It is not easy for a single person to deal with all these aspects along with GIS/DEM analyses. Handling GIS and DEMs tends to take a much longer time than people usually think for several reasons: (1) careful data arrangement including the conversion of format and projection as well as corrections of errors are often required before starting analysis; (2) available software is not always user-friendly; (3) additional tools such as macros, modules and external programs often need to be used along with basic software packages; and (4) various personal 'tips' are necessary for efficient data handling and avoiding possible computer errors. Therefore, it is desirable to have a good team in which researchers with different types of expertise can work together to solve the major problems with GIS/DEM applications in geomorphology.

It is also necessary to increase the fields in geomorphology to which GIS analyses can be applied. For example, although modelling of erosional processes such as landslides and soil erosion using GIS and DEMs has been very common, only a limited number of studies have provided a model of depositional processes (Warren et al., 2005). This situation may result in a significant problem when we apply a GIS model to a broad area, because erosion in certain parts of a watershed leads to sediment supply, which tends to cause deposition in downstream areas. This example indicates that, although numerous case studies of GIS/DEM applications in geomorphology have already been made, there are still many important issues left unexamined. In any case, GIS have opened a wide window for geomorphological research, and their use will contribute significantly to future development of geomorphology, at least during the next decade.

ACKNOWLEDGEMENTS

Part of this material is based upon work supported by the National Science Foundation under Grant No. 0239749, entitled '*CAREER: Alluvial Fan Form Quantification to Advance Geographic*

Science and Education'. Further support was provided through a Short-term Fellowship No. 04004 and Grant-in-Aids for Scientific Research (16089205 and 19300306) from the Japan Society for the Promotion of Science.

REFERENCES

Ahmadzadeh, M.R. and Petrou, M. (2001) Error statistics for slope and aspect when derived from interpolated data. *IEEE Transactions on Geoscience and Remote Sensing* 39, 1823–33.

Albani, M., Klinkenberg, B., Andison, D.W. and Kimmins, J.P. (2004) The choice of window size in approximating topographic surfaces from digital elevation models. *International Journal of Geographical Information Science* 18, 577–93.

Atkinson, P.M. and Tate, N.J. (2000) Spatial scale problems and geostatistical solutions: a review. *Professional Geographer* 54, 607–23.

Bhattarai, R. and Dutta, D. (2007) Estimation of soil erosion and sediment yield using GIS at catchment scale. *Water Resources Management* 21, 1635–47.

Bishop, M. and Shroder, J. (eds) (2004) *Geographic Information Science and Mountain Geomorphology*. Springer-Verlag, Berlin/Heidelberg.

Bolstad, P.V. and Stowe, T. (1994) An evaluation of DEM accuracy: Elevation, slope and aspect. *Photogrammetric Engineering and Remote Sensing* 60, 1327–32.

Bork, H.-R. and Hensel, H. (1988) Computer-aided construction of soil erosion and deposition maps. *Geologisches Jahrbuch* A104, 357–71.

Brown, D.G., Lusch, D.P. and Duda, K.A. (1998) Supervised classification of types of glaciated landscapes using digital elevation data. *Geomorphology* 21, 233–50.

Burrough, P.A. (1986) *Principles of Geographical Information Systems for Land Resources Assessment*. Clarendon Press, Oxford.

Burrough, P. A. and McDonnell, R. (1998) *Principles of Geographical Information Systems* (2nd edn.). Oxford University Press, Oxford.

Butler, D. (2006) Virtual globes: the web-wide world. *Nature* 439, 776–8.

Carrara, A., Cardinali, M., Detti, R., Guzzetti, G., Pasqui, V. and Reichenbach, P. (1991) GIS techniques and statistical models in evaluating landslide hazard. *Earth Surface Processes and Landforms* 16, 427–45.

Cheng, Q., Russell, H., Sharpe, D., Kenny, F. and Qin, P. (2001) GIS-based statistical and fractal/multifractal analysis of surface stream patterns in the Oak Ridges Moraine. *Computers and Geosciences* 27, 513–26.

Chorowicz, J., Kim, J., Manousiis, S., Rudant, J.-P., Foin, P. and Veillet, I. (1989) A new technique for recognition of geological and geomorphological patterns in digital terrain models. *Remote Sensing of Environment* 29, 229–39.

Chung, C.-J.-F. and Fabbri, A.G. (2003) Validation of spatial prediction models for landslide hazard mapping. *Natural Hazards* 30, 451–72.

Collins, S.H. (1975) Terrain parameters directly from a digital terrain model. *The Canadian Surveyor* 29, 507–18.

Da Silva, A.M. (2004) Rainfall erosivity map for Brazil. *Catena* 57, 251–9.

Dai, F.C., Lee, C.F. and Zhang, X.H. (2001) GIS-based geo-environmental evaluation for urban land-use planning: a case study. *Engineering Geology* 61, 257–71.

Davis, T.J. and Keller, C.P. (1997) Modelling uncertainty in natural resource analysis using fuzzy sets and Monte Carlo simulation: slope stability prediction. *International Journal of Geographical Information Science* 11, 409–34.

Deng, Y., Wilson, J.P. and Bauer, B.O. (2007) DEM resolution dependencies of terrain attributes across a landscape. *International Journal of Geographical Information Science* 21, 187–213.

De Roo, A.P.J., Hazelhoff, L. and Burrough, P.A. (1989) Soil erosion modelling using "ANSWERS' and Geographical Information Systems. *Earth Surface Processes and Landforms* 14, 517–32.

De Rose, R.C., Gomez, B., Marden, M. and Trustrum, N.A. (1998) Gully erosion in Mangatu forest, New Zealand, estimated from digital elevation models. *Earth Surface Processes and Landforms* 23, 1045–53.

Desmet, P.J.J. and Govers, G. (1995) GIS-based simulation of erosion and deposition patterns in an agricultural landscape: a comparison of model results with soil map information. *Catena* 25, 389–401.

Dietrich, W.E., Wilson, C.J., Montgomery, D.R. and McKean, J. (1993) Analysis of erosion thresholds, channel networks, and landscape morphology using a digital terrain model. *Journal of Geology* 101, 259–78.

Dikau, R. (1988) Case studies in the development of derived geomorphic maps. *Geologisches Jahrbuch* A104, 329–38.

Dikau, R. (1989) The application of a digital relief model to landform analysis in geomorphology, in J. Raper, (ed.), *Three Dimensional Applications in Geographical Information Systems*. Taylor & Francis, London. pp. 51–77.

Dikau, R. (1990) Geomorphic landform modelling based on hierarchy theory, in K. Brassel and H. Kishimoto (eds.), *Proceedings 4th International Symposium on Spatial Data Handling, 23rd–27th July*. Zürich. pp. 230–9.

Dikau, R. (1993) Geographical information systems as tools in geomorphology. *Zeitschr. Geomorph. N.F.* 92, 231–9.

Ehsani, A.H., Quiel, F. and Malekian, A. (2010) Effect of SRTM resolution on morphometric feature identification using neural network-self organizing map. *GeoInformatica* *GeoInformatica* 14, 405–424.

El-Sheimy N., Valeo C. and Habio A. (2005) *Digital Terrain Modeling: Acquisition, Manipulation and Applications*. Artech House, London.

Ermini, L., Catani, F. and Casagli, N. (2005) Artificial Neural Networks applied to landslide susceptibility assessment. *Geomorphology* 66, 327–43.

Evans, I.S. (1972) General geomorphometry, derivatives of altitude and descriptive statistics, in R.J. Chorley (ed.),

Spatial Analysis in Geomorphology. Methuen, London. pp. 17–90.

Evans, I.S. (2003) Scale-specific landforms and aspects of the land surface, in I.S. Evans, R.Dikau, E. Tokunaga, H. Ohmori and M. Hirano (eds), *Concepts and Modelling in Geomorphology: International Perspectives*. Terrapub, Tokyo. pp. 61–84.

Fielding, E., Isacks, B., Barazangi, M. and Duncan, C. (1994) How flat is Tibet? *Geology* 22, 163–7.

Flacke, W., Auerswald, K. and Neufang, L. (1990) Combining a modified universal soil loss equation with a digital terrain model for computing high resolution maps of soil loss resulting from rain wash. *Catena* 17, 383–97.

Florinsky, I.V. (1998) Accuracy of local topographic variables derived from digital elevation models. *International Journal of Geographical Information Science* 12, 47–61.

Foresman, T. (ed.) (1998) *The History of Geographic Information Systems: Perspectives from the Pioneers*. Prentice Hall, Upper Saddle River.

Gardiner, V. (1978) Redundancy and spatial organization of drainage basin form indices: an empirical investigation of data from north-west Devon. *Transactions of the Institute of British Geographers* 3, 416–31.

Gardner, T.W., Sasowsky, K.C. and Day, R.L. (1990) Automated extraction of geomorphometiric properties from digital elevation data. *Zeitschr. Geomorph. N.F. Suppl. Bd.* 80, 57–68.

Gares, P.A., Wang, Y. and White, S.A. (2006) Using LiDAR to monitor a beach nourishment project at Wrightsville Beach, North Carolina, USA. *Journal of Coastal Research* 22, 1206–19.

Giles, P.T. and Franklin, S.E. (1996) Comparison of derivative topographic surfaces of a DEM generated from stereoscopic SPOT images with field measurements. *Photogrammetric Engineering and Remote Sensing* 62, 1165–71.

Graff, L.H. and Usery, E.L. (1993) Automated classification of generic terrain features in digital elevation models. *Photogrammetric Engineering and Remote Sensing* 59, 1409–17.

Gritzner, M.L., Marcus, W.A., Aspinall, R. and Custer, G. (2001) Assessing landslide potential using GIS,soil wetness modeling and topographic attributes, Payette River, Idaho. *Geomorphology* 37, 149–65.

Guth, P.L. (1995) Slope and aspect calculations on gridded digital elevation models: Examples from a geomorphometric toolbox for personal computers. *Zeitschr Geomorph NF Suppl Bd.* 101, 31–52.

Guzzetti, F. and Reichenbach, P. (1994) Towards a definition of topographic divisions for Italy. *Geomorphology* 11, 57–74.

Guzzetti, F., Carrara, A., Cardinali, M. and Reichenbach, P. (1999) Landslide hazard evaluation: a review of current techniques and their application in a multi-scale study, Central Italy. *Geomorphology* 31, 181–216.

Guzzetti, F., Reichenbach, P., Ardizzone, F., Cardinali, M. and Galli, M. (2006) Estimating the quality of landslide supceptibility models. *Geomorphology* 81, 166–84.

Ham, D.G. and Church, M. (2000) Bed-material transport estimated from channel morphodynamics: Chilliwack River, British Columbia. *Earth Surface Processes and Landforms* 25, 1123–42.

Hancock, G.R., Martinez, C., Evans, K.G. and Moliere, D.R. (2006) A comparison of SRTM and high-resolution digital elevation models and their use in catchment geomorphology and hydrology: Australian examples. *Earth Surface Processes and Landforms* 31, 1394–412.

Hashimoto, A., Oguchi, T., Hayakawa, Y., Lin, Z., Saito, K. and Wasklewicz, T.A. (2007) GIS analysis of depositional slope change at alluvial-fan toes in Japan and the American Southwest. *Geomorphology* 100, 120–30.

Hayakawa, Y.S. and Oguchi, T. (2009) GIS analysis of fluvial knickzone distribution in Japanese mountain watersheds. *Geomorphology* 111, 27–37.

Hengl, T. and Reuter, H.I. (eds) (2009) *Geomorphometry: Concepts, Software, Applications*. Elsevier, Amsterdam.

Heuvelink, G.B.M. (1998) *Error propagation in environmental modelling with GIS*. Taylor & Francis, London.

Hill, J.M. and Turnispeed, P. (1989) Spatial analysis of coastal land loss by soil type. *Journal of Coastal Research* 5, 83–91.

Hu, P., Liu, X. and Hu, H. (2009) Accuracy assessment of digital elevation models based on approximation theory. *Photogrammetric Engineering and Remote Sensing* 75, 49–56.

Hurtrez, J.E., Sol, C. and Lucazeau, F. (1999) Effect of drainage area on hypsometry from an analysis of small-scale drainage basins in the Siwalik Hills (Central Nepal). *Earth Surface Processes and Landforms* 24, 799–808.

Iida T. (1999) A stochastic hydro-geomorphological model for shallow landsliding due to rainstorm. *Catena* 34, 293–313.

Iwahashi, J. and Pike, R.J. (2007) Automated classification of topography from DEMs by an unsupervised nested-means algorithm and a three-part geometric signature. *Geomorphology* 86, 409–40.

Jain, V., Preston, N., Fryirs, K. and Brierley, G. (2006) Comparative assessment of three approaches for deriving stream power plots along long profiles in the upper Hunter River catchment, New South Wales, Australia. *Geomorphology* 74, 297–317.

Jelinski, D.E. and Wu, J. (1996) The modifiable areal unit problem and implications for landscape ecology. *Landscape Ecology* 11, 129–40.

Jenson, S.K. and Domingue, J.O. (1988) Extracting topographic structure from digital elevation data for geographic information system analysis. *Photogrammetric Engineering and Remote Sensing* 54, 1593–1600.

Johansson, M. (1998) Analysis of digital elevation data for palaeosurfaces in south-western Sweden. *Geomorphology* 26, 279–95.

Katsube, K. and Oguchi, T. (1999) Altitudinal changes in slope angle and profile curvature in the Japan Alps: A hypothesis regarding a characteristic slope angle. *Geographical Review of Japan* 72B, 63–72.

Keller, F. (1992) Automated mapping of mountain permafrost using the program PERMAKART within the geographic information system ARC/INFO. *Permafrost and Periglacial Processes* 3, 133–8.

Kienzle, S. (2004) The effect of DEM raster resolution on first order, second order and compound terrain derivatives. *Transactions in GIS* 8, 83–111.

Kühni, A. and Pfiffner, O.A. (2001) The relief of the Swiss Alps and adjacent areas and its relation to lithology and structure: topographic analysis from a 250-m DEM. *Geomorphology* 41, 285–307.

Lane, S.N. (1998) The use of digital terrain modelling in the understanding of dynamic river systems, in S.N. Lane, K.S. Richards and J.H. Chandler, (eds), *Landform Monitoring, Modelling and Analysis*, Wiley, Chichester. pp. 311–42.

Larsen, M.C. and Parks, J.E. (1997) How wide is a road? The association of roads and mass-wasting in a forested montane environment. *Earth Surface Processes and Landforms* 22, 835–48.

Lee, J., Snyder, P.K. and Fisher, P.F. (1992) Modeling the effect of data errors on feature extraction from digital elevation models. *Photogrammetric Engineering and Remote Sensing* 58, 1461–7.

Li, Z., Zhu, Q. and Gold, C. (2005) *Digital Terrain Modeling. principles and methodology*. CRC Press, Boca Raton.

Lin, Z. and Oguchi, T. (2004) Drainage density, slope angle, and relative basin position in Japanese bare lands from high-resolution DEMs. *Geomorphology* 63, 159–73.

Longley, P.A., Goodchild, M.F., Maguire, D.J. and Rhind, D.W. (2005) *Geographic Information Systems and Science* (2nd edn.). Wiley, Chichester.

Luoto, M. and Hjort, J. (2006) Scale matters-A multi-resolution study of the determinants of patterned ground activity in subarctic Finland. *Geomorphology* 80, 282–94.

Marceau, D.J. and Hay, G.J. (1999) Remote sensing contributions to the scale issue. *Canadian Journal of Remote Sensing* 25, 357–66.

Mark, D.M. (1984) Automated detection of drainage networks from digital elevation models. *Cartographica* 21, 168–78.

Marsella, M., Proietti, C., Sonnessa, A., Coltelli, M., Tommasi, P. and Bernardo, E. (2009) The evolution of the Sciara del Fuoco subaerial slope during the 2007 Stromboli eruption: Relation between deformation processes and effusive activity. *Journal of Volcanology and Geothermal Research* 182, 201–13.

Martz, L.W. and Garbrecht, J. (1998) The treatment of flat areas and depressions in automated drainage analysis of raster digital elevation models. *Hydrological Processes* 12, 843–55.

Mayorga, E., Logsdon, M.G., Ballester, M.V.R. and Richey, J.E. (2005) Estimating cell-to-cell land surface drainage paths from digital channel networks, with an application to the Amazon basin. *Journal of Hydrology* 315, 167–82.

McHarg, I. (1969) *Design with Nature*. The Natural History Press, New York.

McLaren, R.A. and Kennie, T.J.M. (1989) Visualization of digital terrain models: techniques and applications, in J. Raper (ed.), *Three Dimensional Applications in Geographical Information Systems*. Taylor & Francis, London. pp. 79–90.

Miliaresis, G.C. (2001) Geomorphometric mapping of Zagros Ranges at regional scale. *Computers & Geosciences* 27, 775–86.

Minár, J. and Evans, I.S. (2008) Elementary forms for land surface segmentation: The theoretical basis of terrain analysis and geomorphological mapping. *Geomorphology* 95, 236–59.

Mitas, L., Brown, W.M. and Mitasova, H. (1997) Role of dynamic cartography in simulations of landscape processes based on multivariate fields. *Computers & Geosciences* 23, 437–46.

Mitasova, H., Overton, M.F., Recalde, J., Freeman, D.J. and Freeman, C.W. (2009) Raster-based analysis of coastal terrain dynamics from multitemporal LiDAR data. *Journal of Coastal Research* 25, 507–14.

Mitra, B., Scott, H.D., Dixon, J.C. and McKimmey, J.M. (1998) Application of fuzzy logic to the prediction of soil erosion in a large watershed. *Geoderma* 86, 183–209.

Mizukoshi H. and Aniya M. (2002) Use of Contour-based DEMs for deriving and mapping topographic attributes. *Photogrammetric Engineering and Remote Sensing* 68, 83–93.

Montgomery, D.R. (2001) Slope distributions, threshold hillslopes, and steady-state topography. *American Journal of Science* 301, 432–54.

Oguchi, T., Aoki, T. and Matsuta, N. (2003) Identification of an active fault in the Japanese Alps from DEM-based hill shading. *Computers & Geosciences* 29, 885–91.

Okimura, T. and Kawatani, T. (1987) Mapping of the potential surface failure sites on granite mountain slopes, in V. Gardiner (ed.), *International Geomorphology, Part 1*. Wiley, Chichester. pp. 121–38.

Otto, J.-C., Goetz, J. and Schrott, L. (2008) Sediment storage in Alpine sedimentary systems – quantification and scaling issues. *IAHS-AISH Publication* 325, 258–65.

Ozcan, A.U., Erpul, G., Basaran, M. and Erdogan, H.E. (2008) Use of USLE/GIS technology integrated with geostatistics to assess soil erosion risk in different land uses of Indagi Mountain Pass, Çankýrý, Turkey. *Environmental Geology* 53, 1731–41.

Peeters, I., Rommens, T., Verstraeten, G., Govers, G., Van Rompaey, A., Poesen, J. and Van Oost, K. (2006) Reconstructing ancient topography through erosion modeling. *Geomorphology* 78, 250–64.

Pike, R.J. (1988) The geometric signature: quantifying landslide-terrain types from digital elevation models. *Mathematical Geology* 20, 491–511.

Prima, O.D.A., Echigo, A., Yokoyama, R. and Yoshida, T. (2006) Supervised landform classification of Northeast Honshu from DEM-derived thematic maps. *Geomorphology* 78, 373–86.

Reinfelds, I., Cohen, T., Batten, P. and Brierley, G. (2004) Assessment of downstream trends in channel gradient, total and specific stream power: a GIS approach. *Geomorphology* 60, 303–416.

Reuter, H.I., Hengl, T. Gessler, P. and Soille, P. (2009) Preparation of DEMs for geomorphometric analyses, in T. Hengl and H.I. Reuter (eds.), *Geomorphometry: Concepts, Software, and Applications*. Elsevier, Amsterdam. pp. 87–120.

Rogers, S.S., Sandweiss, D.H., Maasch, K.A., Belknap, D.F. and Agouris, P. (2004) Coastal change and beach ridges along the Northwest Coast of Peru: Image and GIS analysis of the Chira, Piura and Colán Beach-Ridge Plains. *Journal of Coastal Research* 20, 1102–25.

Schmidt, J. and Andrew, R. (2005) Multi-scale landform characterization. *Area* 37, 341–50.

Schneevoigt, N.J., van der Linden, S., Thamm, H.-P. and Schrott, L. (2008) Detecting Alpine landforms from remotely sensed imagery. A pilot study in the Bavarian Alps. *Geomorphology* 93, 104–19.

Sharpnack, D.A. and Akin, G. (1969) An algorithm for computing slope and aspect from elevations. *Photogrammetric Engineering and Remote Sensing* 35, 247–8.

Shary, P.A., Sharaya, L.S. and Mitusov, A.V. (2005) The problem of scale-specific and scale-free approaches in geomorphometry. *Geografia Fisica e Dinamica Quaternaria* 28, 81–101.

Shu-Quiang, W. and Unwin, D.J. (1992) Modelling landslide distribution on loess soils in China: an investigation. *International Journal of Geographical Information Systems* 6, 391–405.

Smith, M.J. and Clark, C.D. (2005) Methods for the visualization of digital elevation models for landform mapping. *Earth Surface Processes and Landforms* 30, 885–900.

Snow, J. (1855) *On the Mode of Communication of Cholera.* John Churchill, London.

Staley, D.M. and Wasklewicz, T.A. (2006) Expanding the analytical potential of channel locational probability maps using GIS and cellular statistics. *River Research and Applications* 22, 27–37.

Staley, D.M., Wasklewicz, T.A. and Blaszczynsk, J.S. (2006) Surficial patterns of debris flow deposition on alluvial fans in Death Valley, CA using airborne laser swath mapping data. *Geomorphology* 74, 152–63.

Stepinski, T.F. and Bagaria, C. (2009) Segmentation-based unsupervised terrain classification for generation of physiographic maps. *IEEE Geoscience and Remote Sensing Letters* 6, 733–737.

Takagi, T., Oguchi, T., Matsumoto, J., Grossman, M.J., Sarker, M.H. and Matin, M.A. (2007) Channel braiding and stability of the Brahmaputra River, Bangladesh, since 1967: GIS and remote sensing analyses. *Geomorphology* 85, 294–305.

Thelin, G.P. and Pike R.J. (1991) Landforms of the conterminous United States – a digital shaded-reflect portrayal. Misc. Investigation Series Map I-2206, U.S.G.S., U.S. Dept. of Interior.

Tribe, A. (1991) Automated recognition of valley heads from digital elevation models. *Earth Surface Processes and Landforms* 16, 33–49.

Van Asselen, S. and Seijmonsbergen, A.C. (2006) Expert-driven semi-automated geomorphological mapping for a mountainous area using a laser DTM. *Geomorphology* 78, 309–20.

Van Den Eeckhaut, M., Poesen, J., Verstraeten, G., Vanacker, V., Moeyersons, J., Nyssen, J. and van Beek, L.P.H. (2005) The effectiveness of hillshade maps and expert knowledge in mapping old, deep-seated landslides. *Geomorphology* 67, 351–63.

Van Den Eeckhaut, M., Vanwalleghem, T., Poesen, J., Govers, G., Verstraeten, G. and Vandekerckhove, L. (2006) Prediction of landslide susceptibility using rare events logistic regression: A case-study in the Flemish Ardennes (Belgium). *Geomorphology* 76, 392–410.

Van Oost, K., Govers, G., Cerdan, O., Thauré, D., Van Rompaey, A., Steegen, A., Nachtergaele, J., Takken, I. and Poesen, J. (2005) Spatially distributed data for erosion model calibration and validation: The Ganspoel and Kinderveld datasets. *Catena* 61, 105–21.

Van Western, C.J. and Terlien, M.T.J. (1996) An approach towards deterministic landslide hazard analysis in GIS: A case study from Manizales (Colombia). *Earth Surface Processes and Landforms* 21, 853–68.

Van Westen, C.J., Castellanos, E. and Kuriakose, S.L. (2008) Spatial data for landslide susceptibility, hazard, and vulnerability assessment: An overview. *Engineering Geology* 102, 112–31.

Verstraeten, G. (2006) Regional scale modelling of hillslope sediment delivery with SRTM elevation data. *Geomorphology* 81, 128–40.

Vitek, J.D., Giardino, J.R. and Fitzgerald, J.W. (1996) Mapping geomorphology: a journey from paper maps, through computer mapping to GIS and Virtual Reality. *Geomorphology* 16, 233–49.

Vogt, J.V., Colombo, R. and Bertolo, F. (2003) Deriving drainage networks and catchment boundaries: a new methodology combining digital elevation data and environmental characteristics. *Geomorphology* 53, 281–98.

Wangensteen, B., Guðmundsson, Á., Eiken, T., Kääb, A., Farbrot, H. and Etzelmüller, B. (2006) Surface displacements and surface age estimates for creeping slope landforms in Northern and Eastern Iceland using digital photogrammetry. *Geomorphology* 80, 59–79.

Warren, S.D., Mitasova, H., Hohmann, M.G., Landsberer, F.Y., Ruzycki, T.S. and Senseman, G.M. (2005) Validation of a 3-D enhancement of the Universal Soil Loss Equation for prediction of soil erosion and sediment deposition. *Catena* 64, 281–96.

Wasklewicz, T.A., Anderson, S. and Liu, P.-S. (2004) Geomorphic context of channel locational probabilities along the Lower Mississippi River, USA. *Geomorphology* 63, 145–158.

Wechsler, S.P. (2007) Uncertainties associated with digital elevation models for hydrologic applications: a review. *Hydrology and Earth System Sciences* 11, 1481–500.

Wilson J.P. and Gallant J.C. (eds) (2000) *Terrain Analysis: Principles and Applications.* Wiley, New York.

Yoeli, P. (1965) Analytical hill shading (a cartographic experiment). *Surveying and Mapping* 25, 573–9.

Zevenbergen, L.W. and Thorne, C.R. (1987) Quantitative analysis of land surface topography. *Earth Surface Processes and Landforms* 12, 47–56.

Zhang, X.-C., Li, Z.-B. and Ding, W.-F. (2005) Validation of WEPP sediment feedback relationships using spatially distributed rill erosion data. *Journal of Soil Science Society of America* 69, 1440–7.

14

Biogeomorphology

Heather Viles

WHAT IS IT AND WHY IS IT IMPORTANT?

Biogeomorphology (also commonly referred to as ecogeomorphology) is the branch of geomorphology which focuses on the interactions between ecological and geomorphic processes. As such, it covers a very wide range of subject matter, from the mutual interdependence of microorganisms and weathering on bare rock surfaces to the interactions between forest cover and fluvial dynamics within whole catchments. Biogeomorphic interactions occur in all terrestrial environments, from the hyper-arid zone to the wet tropics, and are also heavily affected by human impacts on the environment. Thus, biogeomorphology has a very broad canvas, one which has global reach and covers many different scales. Biogeomorphology also has many practical applications in terms of environmental management. Geomorphic theory has recently started to take on board ideas about biogeomorphic interactions, following the earlier incorporation by ecological theory of ideas such as ecosystem engineering. Biogeomorphic interactions play key roles in the overall Earth System over a range of timescales and are thus of importance more widely than just to geomorphology. For example, changes in plant and animal communities on Earth over tectonic timescales have been associated with alterations in soils and geomorphology as well as with changing climate.

Because of the wide scope of biogeomorphology, and the difficulty of observing, quantifying and modelling many biogeomorphic interactions, this branch of geomorphology is extremely diverse and disparate. Unlike many other sub-fields of geomorphology, for example, it does not have its own journal. Most people who research on biogeomorphic interactions would not call themselves biogeomorphologists. Furthermore, research in this area is increasingly carried out by inter-disciplinary teams involving hydrologists, geochemists, geomorphologists and ecologists thus contributing to the eclectic nature of much biogeomorphological research. Such eclecticism is reflected in extensive cross-fertilization of methods and ideas.

HISTORY OF BIOGEOMORPHOLOGICAL RESEARCH AND THEORY

Like much of today's geomorphology, the origins of biogeomorphic research can be traced back to the 19th century. Natural historians such as Charles Darwin, Charles Lyell and Archibald and James Geikie wrote widely and passionately about a range of interactions between the living world and the underlying soils and rocks. However, their work built upon much earlier observations of potential links between the organic and inorganic worlds which have become largely forgotten. During the 19th century many detailed studies were made of the role of individual organisms in various earth surface processes. For example, Julius Sachs made experimental studies of the role of plant roots in chemical weathering of minerals in the 1860s and 1870s (Mottershead and Viles, 2004). Perhaps the most famous work of this sort was the intensive studies made by Charles Darwin into the role of earthworms in denudation, in which he attempted to quantify their role in bioturbation and erosion of sediment. As Darwin expressed it:

> Worms have played a more important part in the history of the world than most persons would at first suppose.... In many parts of England a weight of more than ten tons (10,516 kilogrammes) of dry earth annually passes through their bodies and is brought to the surface on each acre of land.
>
> (Darwin, 1881: 308)

He based these estimates on his own measurements of worm casts at Down House, Kent, as well as other people's observations at sites in England and France. Further, more detailed, calculations of the erosion of the worm-cast material at Down House produced the following results:

> It was found by measurements and calculations that on a surface with a mean inclination of 9° 26′, 2.4 cubic inches of earth which had been ejected by worms crossed, in the course of a year, a horizontal line one yard in length; so that 240 cubic inches would cross a line 100 yards in length. This latter amount in a damp state would weigh 11½ pounds. Thus a considerable weight of earth is continually moving down each side of every valley, and will in time reach its bed. Finally this earth will be transported by the streams flowing in the valleys into the ocean, the great receptacle for all matter denuded from the land.
>
> (Darwin, 1881: 311)

These sorts of observations and calculations of the roles of a wide range of organisms made their way into the geological texts of the time in the form of general statements about the importance of biotic controls on geomorphology. Charles Lyell, in his book *Principles of Geology*, for example, stated that:

> The powers of the organic creation in modifying the form and structure of the earth's crust are most conspicuously displayed in the labours of the coral animals. We may compare the operation of these zoophytes in the ocean to the effects produced on a smaller scale upon the land by the plants which generate peat.
>
> (Lyell, 1875: 587)

James Geikie, in his book *Earth Sculpture*, went into more detail and summarized the contribution of organisms to denudation as follows:

> The various humous acids...are powerful agents of chemical change. Without their aid rain-water would be a less effective worker. The living plants themselves, however, attack rocks, and by means of the acids in their roots dissolve out the mineral matters required by the organisms. Further, their roots penetrate the natural division-planes of rocks and wedge these asunder; and thus, by allowing freer percolation of water, they prepare the way for more rapid disintegration. Nor can we neglect the action of tunnelling and burrowing animals, some of which aid considerably in the work of destruction.
>
> (Geikie, 1898: 23–4)

However, perhaps the fullest 19th century treatment of the role of plants and animals in geomorphology came in Archibald Geikie's *Text-book of Geology* in which he devoted almost 25 pages to considering the roles they played in terms of destructive, conservative and reproductive impacts. His discussion ranges widely, bringing together observations from around the world on organic impacts such as beaver dams, mole rat and crayfish burrowing, mollusc boring, bison wallowing, protective role of calcareous nullipores, and the formation of biological deposits such as coral reefs, tufas and marls. A wide range of landforms created dominantly by biological effects has been recognized since Geikie's insightful discussion, some of which are illustrated in Figure 14.1.

Another full discussion of the role of biota in earth surface processes came in George Perkins Marsh's visionary book *Man and Nature* first published in 1864. In this book, Marsh aimed to draw the reader's attention to the ways in which humans were damaging the earth's surface, often through human disruption of biogeomorphic interactions and how successful environmental management might overcome this. For example, Marsh writes that

> every plant, every animal, is a geographical agency, man a destructive, vegetables and even wild beasts, restorative powers. The rushing waters sweep down earth from the uplands; in the first moment of repose, vegetation seeks to re-establish itself on the bared surface and, by the slow deposit of its decaying products, to raise again the soil which the torrent had lowered.
>
> (Marsh, 1965: 53–4)

Marsh writes clearly and a touch romantically about the ways in which deforestation can upset such processes, as illustrated in the following quotation:

> The soil is bared of its covering of leaves, broken and loosened by the plough, deprived of the fibrous rootlets which held it together, dried and pulverised by sun and wind, and at last exhausted by new combinations. The face of the earth is no

Figure 14.1   Examples of biogenically produced landforms at a wide range of scales: (top left) Part of the southern section of the Great Barrier Reef, Australia; (top right) small tufa barrage on a stream at Cwm Nash, South Wales, anchored with LWD; (middle left) a nebkha on Agate Beach, near Luderitz, southern Namibia; (middle right) termite mounds in the Kimberley area, north-west Australia; (bottom left) badger mounding in Wytham Wood, near Oxford (image courtesy of John Crouch); and (bottom right) depressions in sandstones at Golden Gate Highlands National Park, South Africa, inhabited and developed by the lichen *Lecidea aff. sarcogynoides*

longer a sponge, but a dust heap, and the floods which the waters of the sky pour over if hurry swiftly along its slopes, carrying in suspension cast quantities of earth particles

(Marsh, 1965: 186–7)

Over the course of the 19th century, as evidenced by the works referred to above, it is clear that biotic influences on geomorphology were a central part of the dominant understanding

of how the earth's surface functioned and, in turn, how it could be managed more successfully. These ideas were based on some surprisingly detailed monitoring as well as a whole suite of common sense observations.

Things changed, at least in the Anglo-American world, greatly over the early years of the 20th century. Whilst there continued to be a series of interesting, and sometimes bizarre, observations on biogeomorphological topics in the literature (such as J Harlen Bretz's observations in 1946 on the role that bison played in eroding boulders in mid-west USA), biotic influences became neglected by geomorphic theory. Whilst a number of observations were made of biogenic processes, the dominant theme during this period can be identified as a concern for identifying biogenically produced landforms. Examples include Gregory (1911) who wrote about 'constructive waterfalls' which were produced by tufa accumulating around breaks in a stream profile.

Despite the many observations of biogeomorphological processes and biologically produced landforms in the late 19th/early 20th centuries they were not included in the Anglo-American geomorphological theories of the time. Geomorphologists such as G.K. Gilbert and W.M. Davis were interested in the evolution of landscapes over long timespans, and the major factors on which their theories were based were tectonics and denudation. Plant and animal life and their impact on processes and landforms were not of interest to the development of geomorphic theory at this time. W.M. Davis's 'cycle of erosion', for example, seems to be occurring in a landscape largely free of plants and animals, according to his block diagrams. Most geomorphology texts in the early part of the 20th century ignored any biotic influences on landscape evolution. There were some exceptions. Shaler's *Aspects of the Earth* (1889) does acknowledge the role of forests in geomorphology, for example, the role of riparian forests in protecting river banks in alluvial streams. Lobeck's book *Geomorphology*, published in 1939, went even further as it contained a whole chapter on organisms and geomorphology, largely focusing on major biogenic landforms such as coral reefs and peat bogs. Lobeck also summarized some interesting examples of the extensive fluvial landscape modifications caused by log jams and beaver activity as well as some smaller biogenic landforms such as termite and ant mounds, and flamingo nests. Other books promised much, for example, Isaiah Bowman's *Forest Physiography* published in 1911, whose title suggested a biogeomorphic focus. The book itself, however, focused largely on a discussion of the geomorphic characteristics of the USA.

For ecologists, however, the start of the 20th century marked the widespread uptake of the succession theory of F.E. Clements and brought at least some aspects of geomorphology into ecological thought. Clements's theory acknowledged the important interplay between plants and soil characteristics through the fundamental process of facilitation. This involved the idea that plants can modify the nature of the soil on which they grow, making it more suitable for other plants to grow in. As ideas of succession developed over

the 20th century, so ideas over the relationship between plants and their soil environment became more complex. The development of the concept of the biosphere by Vernadsky in 1923 and others is also relevant (see Vernadsky, 1998), as it linked organisms together at a global scale and illustrated how the biosphere was linked to the hydrosphere, atmosphere and lithosphere.

Within mainland Europe the dominant geomorphological theory in the early part of the 20th century was that of climatic geomorphology. In this theoretical framework, the characteristics of landforms and landscapes in any one area were seen to be conditioned by the climate. Clearly, the influence of climate on landforms is both direct (more rainfall = more erosion, for example) and also mediated through plants, as more rain produces lusher vegetation which, in turn, enhances weathering and reduces erosion. Thus, vegetation plays a more central role in the basic functioning of geomorphic systems according to this theoretical framework, as is shown clearly in the work of Julius Büdel, Jean Tricart and others (Birot, 1968; Büdel, 1982; Tricart, 1957; Tricart and Cailleux, 1965). Birot uses vegetation as a way of structuring his whole theory of climatic geomorphology, as he outlines

> More significant still is the indirect action of climate, through the intermediary of vegetation, and as a result vegetation will be used as the basis of the classification of systems of erosion. Firstly, the vegetation is an important agent in the process of rock decomposition (whether living, that is, through the roots or dead) and secondly it controls the conditions under which detritus is carried on the slopes.
>
> (Birot, 1968: 56)

Within Birot's climatic geomorphology he recognized a fundamental distinction between the geomorphology of areas with continuous vegetation and those with discontinuous vegetation. In the first case all transport processes were obstructed, whereas in the second case all transport processes, including aeolian, could occur.

Walther Penck (1953) wrote extensively about the complex role of vegetation in geomorphology. He mentions for example that

> Credit is due to K v Terzagi for having pointed out the extraordinary – and usually much underrated – significance of vegetation in the solution of rocks, especially of limestone. In the Limestone Alps, for example, on reaching the upper limit of the forest, one becomes aware of how cushions of vegetation are, as it were, sunk into the limestone
>
> (Penck: 44).

He also wrote provocatively about whether or not the root systems of plants reduced denudation and discouraged mass movements. Whilst deduction might suggest that forest cover should reduce downslope movement of material through the binding role of tree roots, Penck maintained that the lack of any clear morphological signature above and below the tree-line, and the correlation between slope angle and soil thickness precluded such a simple relationship. Thus, within the European tradition of geomorphology biotic influences (at least those of vegetation) remained centre stage throughout the 20th century, with healthy debates about the nature of such influences.

In the Anglo-American geomorphological community the ongoing marginalization of biogeomorphology was enhanced in the 1960s and 1970s, when a concern for small-scale quantification of processes within catchments led to a decline in studies focusing on biotic influences, as things such as plant influences on soil erosion were much harder to measure than inorganic processes such as rainsplash, sheet wash etc. Whilst some of the earliest catchment experiments were inspired by comparisons between forested and unforested areas and aimed to investigate the impacts of changing land use cover, hydrologists seemed to pick up on biological influences more readily than geomorphologists. The dominance of linear mathematical treatments of geomorphological systems further prevented the inclusion of biological influences which often cause notable nonlinearities in systems behaviour. There were notable exceptions of geomorphologists, especially fluvial geomorphologists, tackling the nonlinear impacts of vegetation such as the work by Langbein and Schumm (1958) on the influence of vegetation on denudation at the catchment scale across different rainfall regimes producing a nonlinear relation between runoff and erosion rate). Schumm (1968) went further with such ideas, looking at the influences of past vegetation assemblages on palaeohydrology and erosion. However, within ecology in the 1960s and 1970s the focus on the ecosystem concept strengthened the engagement of ecologists with geomorphology, as landforms and earth surface processes make up at least part of the inorganic component of the ecosystem.

Starting in the 1980s there was a large increase in the number of biogeomorphological studies in the Anglo-American geomorphic tradition, inspired by a number of key researchers, notably John Thornes in the UK and Geoff Humphreys in Australia. These studies generally focused on measurements of individual biogenic processes, with a view to providing quantitative evidence of their importance in a wide range of geomorphic processes and environments. John Thornes, for example, kick-started a reassessment of the role of vegetation in erosion, whilst Geoff Humphries began a re-awakening of geomorphological work on animal bioturbation and its contribution to erosion (Humphreys and Mitchell, 1983; Thornes, 1985). Work also began on quantifying the influence of micro-organisms on geomorphology in the 1980s (Viles, 1984), an area largely neglected by the 19th century biogeomorphic pioneers who lacked the techniques to study them in adequate detail. These studies, with their focus on quantification, helped to reposition biogeomorphology more centrally within geomorphology. The compilation of work edited by Viles (1988) illustrates the state of the art at that time in terms of the types of studies being carried out on identifying and measuring the inter-relationships between organisms and geomorphology. John Thornes's (1990) edited volume on vegetation and erosion similarly captures the state of the art on plants and their role in reducing (or even enhancing) soil erosion. The breadth and depth of coverage of animal impacts on geomorphology in 1995 of Butler's *Zoogeomorphology* indicates the growth of interest in this area. Over the 1980s and 1990s developments in computer modelling within geomorphology (e.g. the growth of cellular automata methods) encouraged the inclusion of vegetation and animal effects within geomorphological models (e.g. Thornes et al., 1996).

CURRENT DEVELOPMENTS IN BIOGEOMORPHOLOGICAL THEORY

Recently, there have been several papers which reflect an increased interest in bringing together insights from geomorphology, ecology and evolutionary biology in order to develop an articulated biogeomorphological theory. Three conceptual frameworks have been proposed, all of which have some overlapping elements. One key starting point was the development of the concept of 'ecosystem engineering'. This concept was proposed by Jones et al. (1994) who noted that organisms can, through their own structures and/ or through their activities, modify their habitat and change their local environment resulting in clear impacts on the ecosystem. Examples are the role of beavers in modifying fluvial habitats through the production of beaver dams, and the action of lichens growing on soils in arid areas which promote the formation of biological soil crusts. Ecosystem engineering has some overlaps and links with the concept of ecological engineering, which can be defined as the utilization of ecological processes for sustainable environmental management (Odum, 1962; Mitsch and Jorgensen, 1989). Ecosystem engineering as a concept formalizes the two-way mutual interactions of geomorphology and ecology which are the focus of biogeomorphology.

Another way of trying to conceptualize the mutual interdependence of geomorphic and ecological systems has been to start from the perspective of complexity and nonlinear systems dynamics (e.g. Phillips, 2006; Stallins, 2006). Phillips (2006) illustrates how the non-linear nature of geomorphic systems makes complex behaviour (e.g. deterministic chaos and dynamical instability) common. The presence of chaotic dynamics indicates the importance of both global laws and local factors to explaining geomorphic systems. Local factors include specific details of timing and geography, and can involve organic influences which themselves vary greatly over time and space. Stallins (2006) develops these ideas, and introduces the notion of 'ecological memory' whereby the ongoing interactions between ecology and geomorphology become 'encoded' in a landscape. He uses the example of vegetated coastal dunes to explain this concept. Along dune systems of coastal barrier islands periodic storms cause extensive disturbances, which a variety of plant and sediment interactions adapt to in different ways.

Stallins (2006) sees ecosystem engineers as providing a link between microscale process–form interactions and the larger landscape scale.

Viles et al. (2008) built on these ideas, and the geomorphological disturbance model of Bull (1991), to consider how disturbances affect biogeomorphic systems in non-linear ways, as a result of the complex web of interactions between the organic and inorganic parts of these systems. Understanding the inter-relationships that are at the heart of these non-linear systems is a vital step in trying to predict their future behaviour and how they react to human-induced and natural disturbances. Initial conceptual modelling by Viles et al. (2008), as summarized in Figure 14.2, should be built upon and developed, as such knowledge is crucial to attempts to manage these systems, and improve environmental management. Corenblit et al. (2008) and Corenblit and Steiger (2009) take a broader perspective and build on ideas from evolutionary biology (especially the niche construction concept of Odling-Smee et al., 2003). They develop the framework of an evolutionary geomorphology based on the notion that evolving biological communities are closely entwined with evolving geomorphological systems over long timespans. Thus, the development of land plants in the Silurian would have been associated with critical changes

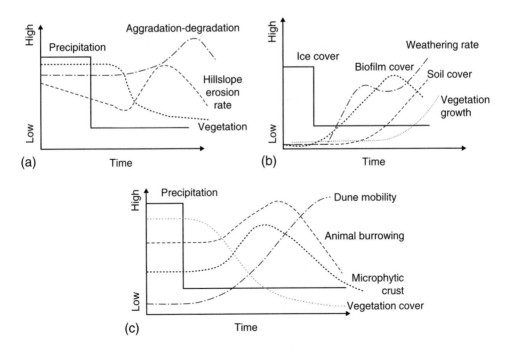

Figure 14.2 **Conceptual diagrams showing the biogeomorphological consequences of disturbance: (a) on arid hillslopes (after Bull, 1991); (b) on weathering systems in deglaciating areas; and (c) on vegetated dunes in drylands**

in soils and geomorphology, and global scale alterations in sediment budgets (Corenblit and Steiger, 2009). Over a wide range of timescales geomorphic patterns may be a result of organic evolution. Similar ideas have been developed by Phillips (2009) who proposes that soils are 'extended composite phenotypes' which represent the cumulative evolutionary imprint of organisms.

CURRENT STATE OF THE ART

A vast amount of research which fits under the broad umbrella of biogeomorphology has been carried out in recent years. Within the first decade of the 21st century there have been several special issues of journals bringing together papers from conferences and meetings on biogeomorphic topics (e.g. Naylor et al., 2002; Urban and Daniels, 2006, Renschler et al., 2007), as well as a whole host of papers in disparate journals covering a wide range of ecological, geomorphological, environmental and earth science fields. Some important hydrological contributions to the development of biogeomorphological ideas have also been published recently (e.g. Gurnell et al., 2000; Tockner et al., 2002; Gregory, 2003). More and more examples of the interactions between organisms and geomorphology at a wide range of scales are being researched in ever-increasing detail. This makes it an almost impossible task to present a meaningful summary of the state of the art, so I adopt a two-fold approach here. First, Table 14.1

Table 14.1 Studies in biogeomorphology that focus on single organism/geomorphological interactions

Environment	Biogeomorphic interaction under study	Reference
Temperate environments (a) hillslopes	Aphaenogaster ants and bioturbation	Richards, 2009
	Pocket gophers and ecosystem engineering	Reichman and Seabloom, 2002
	Earthworms and beetle larvae and mineral weathering	Suzuki et al., 2003
	Grass roots and water-borne erosion	De Baets et al., 2006
	Badger mounds and erosion	Eldridge, 2004
	Treethrow and soils	Phillips et al., 2008
Temperate environments (b) river channels	Large woody debris and channel morphology	Kail, 2003
	Lower plants and tufa deposition	
	Beaver dams, fluvial geomorphology and riparian hydrochemistry	Butler and Malanson, 2005 Hill and Duval, 2009
Alpine and arctic environments	Grizzly bears and erosion	Butler, 1992
	Fungi and rock weathering	Etienne, 2002
	Biological soil crusts and soils/biodiversity	Breen and Levésque, 2008
Arid and semi-arid environments	Cape porcupines and soils	Bragg et al., 2005
	Vegetation and dune stability	Hugenholtz and Wolfe, 2005
	Mesquite shrub and aeolian erosion	King et al., 2006
	Eucalyptus trees and soil infiltration	Eldridge and Freudenberger, 2005
	Sheep, goats and soil microtopography	Stavi et al., 2009
	Termites and wetland island development	Gumbricht et al., 2004
Humid tropical and subtropical environments	Cyanobacteria and rock weathering	Büdel, 1999
	Caddis fly larvae and tufa deposition	Drysdale, 1999
	Forests and slope stability	Genet et al., 2010
Coastal environments	Limpets and erosion of chalk	Andrews and Williams, 2000
	Snow geese and salt marsh erosion	Jefferies et al., 2006
	Vegetation and dune stability	Baas, 2002
	Intertidal seagrass and sediment stabilization	Bos et al., 2007
	Vegetation, snails and bistable salt marsh states	Paramor and Hughes, 2004

gives a sense of the diversity and breadth of research, in terms of the organisms involved, as well as the geomorphic and ecological consequences. Table 14.1 illustrates that many studies in biogeomorphology still focus on single organism/geomorphological process interactions often observed over a relatively short timespan. Relatively few studies tackle multiple, complex process interactions and the change in systems over longer timespans, although such an approach is becoming more common as biogeomorphological research matures. Second, the rest of this section focuses on three particular contexts to illustrate the depth and emerging coherence of biogeomorphological research. The examples chosen are not aimed to be comprehensive, but rather to illustrate contrasting situations and, in particular, the different roles of microorganisms, plants and animals. These examples all illustrate, in different ways, the complexity of the multiple interactions between organisms and geomorphic processes.

Biological crusts in arid and semi-arid environments: the influence of microorganisms and lower plants

A range of lower plants and microorganisms produce notable biological crusts on many soil and rock surfaces within arid and semi-arid environments. These environments are known to be highly variable over time and space in terms of both geomorphology and ecology, with disequilibrium common. Belnap and Lange (2003) provide a full account of the distribution, characteristics and impacts of biological crusts. These crusts, which may be millimetres or centimetres in thickness, make many contributions to geomorphic and ecological processes some of which are reviewed by Viles (2008). Much recent research has focused on testing the nature and importance of these contributions within different field areas. Looking first at geomorphological roles, on bare rock surfaces biological crusts can either provide a bioprotective role or can be important agents of rock weathering (as recorded in the Atlas Mountains, Morocco, by Robinson and Williams, 2000). Similarly, on soil surfaces biological crusts can act as an important agent of surface stabilization, minimizing both aeolian and water-borne erosion (as recorded by McKenna Neuman et al., 1996 and Uchida et al., 2000). Biological crusts can have a key influence on hydrological pathways on rock and soil surfaces, concentrating flow and reducing (or increasing) infiltration depending on their roughness characteristics.

Biological crusts also have impacts on mineral weathering rates (as evidenced in a recent study in the Gurbantunggut Desert, China, by Chen et al., 2009) as well as on surface albedo, microtopography and microclimates. Furthermore, and linked to many of the above processes, biological soil crusts have been reported to influence underlying topsoil characteristics in Inner Mongolia (e.g. Guo et al., 2008) whilst, in turn, at larger scales the nature of biological soil crusts has been shown to be heavily influenced by underlying soil characteristics across a central Mexican drylands transect (e.g. Rivera-Aguilar et al., 2009).

Turning now to the influences of biological crusts on ecology, it is clear that several of these are intimately linked to their geomorphological impacts. Biological crusts can encourage or discourage the growth of vascular plants. For example, crust influences on hydrological pathways can help shrub growth downslope, by concentrating runoff into the shrub patch (Gilad et al., 2004). Furthermore, seed germination can be enhanced or retarded by biological crusts, depending on their roughness and other physical and chemical characteristics. Su et al. (2009) illustrate the role of moss and algal crusts of different dates on the south east fringes of the Tengger Desert, China, in aiding germination of two plant species. A third species remained unaffected. Biological crusts can also contribute to local biodiversity in terms of arthropods and a range of other organisms, as illustrated by Lalley et al. (2006) from lichen-dominated soil crusts in the Namib Desert which were found to provide a home for some arthropods not found in adjacent shrub patches. Linked with many of these influences is the role that biological crusts have been found to play in nutrient cycling, particularly of nitrogen and carbon (as reported, for example, by Thomas and Dougill, 2006, in the Kalahari Desert).

Research on biological crusts (especially those on soil rather than rock surfaces) is now well advanced and illustrates the participation of ecologists and geomorphologists looking at different aspects of the iterative, complex inter-relations between biotic and abiotic processes across a very wide range of dryland environments. Biological crusts are ecosystem engineers and also key components of the geomorphological functioning of many dryland areas. The challenge now lies in linking the types of research findings listed above to general models of landscape dynamics within disequilibrium dryland environments. Research on biological crusts also has relevance to environmental management, and there have been many attempts to look both at the impact of disturbance on these crusts and the services they provide (see review in Viles, 2008) as well as kick-starting biological crust growth to aid surface protection and stabilization and ecological restoration (e.g. Bowker, 2007). Little research effort has so far been expended on understanding the longer term, evolutionary aspects of biological crusts and their inter-relations with geomorphic and ecological systems, and how the systems have adapted to changes in climate etc. over century to millennial and longer timespans.

Riparian vegetation and bank stability

There is a vast literature on the many influences of vegetation on riverine environments, illustrating the fertile nature of this research field and its many practical applications for river management schemes. As with the recent research on biological crusts, work in this field has focused on empirical testing of the many geomorphic and ecological roles of riparian vegetation, but there has been much more progress here in terms of modelling these, and applying results to practical environmental management. Furthermore, there has been more research effort looking at the longer term development of these linked ecological and geomorphological systems.

As with biological crusts, riparian vegetation has been found to have a number of different impacts on its surrounding environment, which can be divided into geomorphological and ecological (although there are complex links between the two). Looking first at the geomorphic impacts, riparian vegetation has been found to have two main effects, that is, mechanical and hydrological (Simon and Collinson, 2002). Mechanical effects include the stabilizing effect of root reinforcement and the destabilizing effect of weight loading provided by mature trees (Pollen, 2007). Hydrological effects include the stabilizing effect of canopy interception and transpiration and the destabilizing effect of increasing infiltration rate and capacity (Simon et al., 2006). Further geomorphic impacts of vegetation that have been identified are influences on sub-aerial processes (e.g. freeze–thaw and soil desiccation) where herbaceous vegetation can be particularly important (Wynn and Mostaghimi, 2006). Root reinforcement has been shown in many studies to contribute to bank stability (e.g. Abernethy and Rutherfurd, 2000), although some modelling exercises have suggested that such impacts have been overstated relative to hydrological ones (Van de Wiel and Darby, 2007). Docker and Hubble (2009) provide empirical evidence for the variability of the root reinforcement effect depending on tree species from south-eastern Australia. Here, two eucalyptus species showed greater

reinforcing effects than two other tree species studied. There were also highly variable impacts on soil strength across a single tree's root system. Simon et al. (2006) studied both mechanical and hydrological impacts of two riparian tree species on the Upper Truckee River in California and found Lemmon's willow to exert significant control on bank strength through root reinforcement and impacts on soil hydrology, whereas the impacts of Lodgepole pine were less marked. The enhancement of bank strength should lead to fewer bank failures and a decline in fine sediment delivery to the river system, and thus the variable stabilizing role of different species can have environmental management implications.

Riparian vegetation also has impacts on flow regimes, especially in small channels, where above-ground biomass acts as a further agent of surface stabilization under flood conditions when flexible stems and branches adopt streamlined profiles along the banks. Gurnell and Petts (2006) show the complex interactions between these impacts along the River Tagliamento in Italy where *Populus* and *Salix* species play highly important roles in establishing vegetated islands. Many riparian plant species exhibit adaptations to life in these dynamic environments, including ability to survive flooding, and seeds which can germinate and sprout quickly on newly deposited sediment. Riparian vegetation also has important impacts on ecological dynamics and is, in turn, affected by ecological processes such as succession and the impacts of episodic fires, floods and other disturbances. Introduced, non-native species often have very different characteristics and are not as effective in these linked ecological and geomorphological roles. Hubble et al. (2010), for example, illustrate the fact that many Australian riparian tree species have developed particularly deep root systems to tap low-lying water tables, and these root system designs in turn make them more effective than introduced species in providing reinforcement to stream banks.

Dead riparian vegetation also contributes to fluvial biogeomorphology, through the production of large woody debris (LWD) which can exert a strong influence on river flow regimes, sediment deposition, channel form and also provides a substrate for seed germination. A huge amount of research in recent years has been aimed at quantifying LWD in various stream systems around the world, both managed and natural, and assessing its impact on fluvial dynamics. Brooks et al. (2003) for example, examine the differences in LWD production and its impacts on fluvial systems between two catchments, one managed and one not, in south-east Australia. As a range of studies have shown, these various impacts of riparian vegetation can last for centuries as they affect the evolution of channel systems and islands (e.g. O'Connor et al., 2003). Corenblit et al. (2009) go further and develop a conceptual framework for considering riparian systems as emergent and evolving systems over timescales from ecological succession to organism evolution. In a similar way to Phillip's (2009) view of soils as 'extended composite phenotypes', Corenblit et al. (2009) regard fluvial landforms within systems affected by riparian vegetation as extended products of genetic, phenotypic and life-history adaptations of the organisms involved. These many diverse contributions to research on fluvial biogeomorphology have direct relevance and applicability to many riverine management and restoration schemes, and the findings of the research are being used to provide better management solutions.

Animal bioturbation on hillslopes

Since the 1980s there has been a plethora of studies on bioturbation by animals and its influence on soils and sediment movement. These papers follow the pioneering work of Charles Darwin and others which were neglected during the intervening century as pedology and geomorphology focused on the influence of inorganic processes. In recent years, the view of the soil as a 'biomantle' (Johnson, 1990) has become much more accepted and the contribution of organisms to soil formation and movement acknowledged. A wide range of animal species is involved in bioturbation, ranging from the small (termites, ants, earthworms, burrowing spiders) to the relatively large (badgers, Arctic foxes, wombats). Paton et al. (1995) illustrate nicely that the dominance of different types of organisms in bioturbation broadly follows climate with, for example, vertebrates dominating in arid environments, earthworms in temperate environments and ants, earthworms and vertebrates showing joint dominance in the humid subtropics. Extensive monitoring studies have been carried out to try and measure the amount of soil moved in various directions by animal bioturbation, and some attempts have been made to include the data in models of slope processes (e.g. Yoo et al., 2005). Furthermore, there has been some speculation over the role that animal bioturbation plays in soil carbon turnover.

Animals play several geomorphological, pedological and ecological roles in soils, and in turn, have adapted to the conditions within soils and the disturbances, such as fire, which affect life on vegetated hillslopes (see the reviews of Gabet et al., 2003; Wilkinson et al., 2009). Burrowing animals (such as earthworms, beetles, woodlice, mole rats and ground squirrels) move sediment up and down within the soil profile, and on the surface move sediment up- and downslope as they remove material from the openings of these burrow systems. Animals making surface scrapes also move material upslope and downslope on the surface, such as lyre birds and wombats in Australia. Some animals construct mounds (e.g. termites, pocket gophers, moles and badgers) which represent a short- to medium-term store of sediment on the surface. A range of other biotic and abiotic processes can then move the material brought to the surface by burrowing, scraping and mounding downslope. Of course, such animal bioturbation is not the only way in which soil moves vertically and downslope; plant effects such as treethrow, and abiotic processes are also important. Quantification of the amount of sediment moved by this range of processes has involved a range of different techniques, from direct monitoring, and monitoring using tracers to the use of optical stimulated luminescence (OSL) dating (see, for example, Gabet, 2000; Richards, 2009; Wilkinson and Humphreys, 2005).

Little work has as yet been done on the longer term dimensions of animal bioturbation and its interaction with pedological, ecological and geomorphological dynamics, although some speculations have been made about the alterations in soils that would have occurred when termites, earthworms and other prodigious bioturbators emerged in the fossil record (Retallack, 2001). Over shorter, successional timespans, changes in climate and other disturbances affecting the soil organism communities will have had pervasive impacts on the tempo and nature of bioturbation and biologically enhanced soil erosion. Increased knowledge on such interactions between bioturbators and hillslope processes are of undoubted importance for environmental management efforts.

CURRENT DEBATES AND AN AGENDA FOR THE FUTURE

Despite the many examples presented above of the role of organic agency in shaping geomorphology (and vice versa) doubts persist over the importance of such biotic processes. Phillips (1995), for example, following the logic of Schumm and Lichty (1965), but approaching the problem from a non-linear systems dynamics framework, noted that biogeomorphological processes were usually short term and small scale relative to the key landscape shaping influences. More recently,

Dietrich and Perron (2006) have proposed that there is no clear topographic signature of life at the large scale, that is, that the lineaments of the landscape would not look any different in the absence of life. However, discussion rages over this topic with many clear examples of small-scale landscapes diagnostic of biotic processes (some of which have been reviewed above, for example, channel systems within streams affected by riparian vegetation) and clear evidence that over evolutionary timescales major differences in geomorphology have been linked with emergence of new species and ecological communities.

Despite the excellent progress reviewed above in biogeomorphological research, there is still a need for a more integrated view of the interlinkages between inorganic processes and microorganisms, plant and animal activity in any one area. Modelling has proved to be a fertile tool (see, for example, Baas, 2002; Fonstad, 2006), and one which could be exploited more (Viles et al., 2008). Research has started to focus on longer timespans, and the practical implications of biogeomorphological interactions for environmental management have begun to be put in practice, but there is much more work that could be done in this area. Biogeomorphological interactions are one component of the earth system, and further research could usefully be targeted at investigating how biogeomorphological interactions influence global carbon and nitrogen cycling. As Corenblit and Steiger (2009) propose, the time is right to bring together ecological, geomorphological and evolutionary biological perspectives to produce a firm conceptual framework for future biogeomorphological research.

ACKNOWLEDGEMENTS

I would like to thank Mark Page for helping with bibliographic searching and to the participants of the biogeomorphology session at the International Association of Geomorphologists Congress in Melbourne, Australia, 2009 for providing inspiration and examples for this chapter.

REFERENCES

Abernethy, B. and Rutherfurd, I.D. (2000) The effect of riparian tree roots on the mass stability of riverbanks. *Earth Surface Processes and Landforms* 25, 921–37.

Andrews, C. and Williams, R.B.G. (2000) Limpet erosion of chalk shore platforms in southest England. *Earth Surface Processes and Landforms* 25, 1371–81.

Baas, A.C.W. (2002) Chaos, fractals and self-organization in coastal geomorphology: simulating dune landscapes in vegetated environments. *Geomorphology* 48, 309–28.

Belnap, J. and Lange, O. (eds) (2003) *Biological Soil Crusts: Structure, Function and Management.* Springer-Verlag: Berlin.

Birot, P. (1968) *The Cycle of Erosion in Different Climates.* B.T. Batsford: London.

Bos, A.R., Bouma, T.J., de Kort, G.L.J. and van Katwijk, M.M. (2007) Ecosystem engineering by annual intertidal seagrass beds: sediment accretion and modification. *Estuarine, Coastal and Shelf Science* 74, 344–8.

Bowker, M.A. (2007) Biological soil crust rehabilitation in theory and practice: an underexploited opportunity. *Restoration Ecology* 15, 13–23.

Bowman, I. (1911) *Forest Physiography.* John Wiley and Sons: New York.

Bragg, C.J., Donaldson, J.D. and Ryan, P.G. (2005) Density of cape porcupines in a semi-arid environment and their impact on soil turnover and related ecosystem processes. *Journal of Arid Environments* 61, 261–75.

Breen, K. and Levésque, E. (2008) The influence of biological soil crusts on soil characteristics along a high arctic glacier foreland, Nunavut, Canada. *Arctic, Antarctic and Alpine Research* 40, 287–97.

Bretz, J.H. (1946) Bison-polished boulders on the Alberta Great Plains. *Journal of Geology* 54, 262–2.

Brooks, A.P., Brierley, G.J. and Millar, R.G. (2003) The long-term control of vegetation and woody debris on channel and flood-plain evolution: insights from a paired catchment study in southeastern Australia. *Geomorphology* 51, 7–29.

Büdel, B. (1999) Ecology and diversity of rock-inhabiting cyanobacteria in tropical regions. *European Journal of Phycology* 34, 361–70.

Büdel, J. (1982) *Climatic Geomorphology.* Princeton University Press: Princeton, New Jersey.

Bull, W.B. (1991) *Geomorphic Responses to Climate Change.* Oxford University Press: Oxford.

Butler, D.R. (1992) The grizzly bear as an erosional agent in mountainous terrain. *Zeitschrift für Geomorphologie* 36, 179–89.

Butler, D.R. (1995) *Zoogeomorphology: Animals as Geomorphic Agents.* Cambridge University Press: Cambridge.

Butler, D.R. and Malanson, G.P. (2005) The geomorphic influences of beaver dams and failures of beaver dams. *Geomorphology* 71, 48–60.

Chen, R., Zhang, Y., Li, Y., Wei, W., Zhang, J. and Wu, N. (2009) The variation of morphological features and mineral components of biological soil crusts in the Gurbantunggut Desert of northwestern China. *Environmental Geology* 57, 1135–43.

Corenblit, D. and Steiger, J. (2009) Vegetation as a major conductor of geomorphic changes on the Earth surface: toward evolutionary geomorphology. *Earth Surface Processes and Landforms* 34, 891–6.

Corenblit, D., Gurnell, A.M., Steiger, J. and Tabacchi, E. (2008) Reciprocal adjustments between landforms and living organisms: extended geomorphic evolutionary insights. *Catena* 73, 261–73.

Corenblit, D., Steiger, J., Gurnell, A.M. and Naiman, R.J. (2009) Plants intertwine fluvial landform dynamics with ecological succession and natural selection: a niche construction perspective for riparian systems. *Global Ecology and Biogeography* 18, 507–20.

Darwin, C. (1881) *The Formation of Vegetable Mould Through the Action of Worms with Observations on Their Habits.* John Murray: London.

De Baets, S., Poesen, J., Gyseels, G. and Knapen, A. (2006) Effects of grass roots on the erodibility of topsoils during concentrated flow. *Geomorphology* 76, 54–67.

Dietrich, W.E. and Perron, J.T. (2006) The search for a topographic signature of life. *Nature* 439, 411–8.

Docker, B.B. and Hubble, T.C.T. (2009) Modelling the distribution of enhanced soil shear strength beneath riparian trees of south eastern Australia. *Ecological Engineering* 35, 921–34.

Drysdale, R.H. (1999) The sedimentological significance of hydropsychid caddis-fly larvae (Order: Trichoptera) in a travertine-depositing stream: Louis Creek, NW Queensland, Australia. *Journal of Sedimentary Research* 69, 145–50.

Eldridge, D.J. (2004) Mounds of the American badger (*Taxidea taxus*): significant features of north American shrub-steppe ecosystems. *Journal of Mammology* 85, 1060–7.

Eldridge, D.J. and Freudenberger, D. (2005) Ecological wicks: Woodland trees enhance water infiltration in a fragmented agricultural landscape in eastern Australia. *Austral Ecology* 30, 336–47.

Etienne, S. (2002) The role of biological weathering in periglacial areas: a study of weathering rinds in south Iceland. *Geomorphology* 47, 75–86.

Fonstad, M.A. (2006) Cellular automata as analysis and synthesis engines at the geomorphology–ecology interface. *Geomorphology* 77, 217–34.

Gabet, E.J. (2000) Gopher bioturbation: evidence for non-linear hillslope diffusion. *Earth Surface Processes and Landforms* 25, 1419–28.

Gabet, E.J., Reichman, O.J. and Seabloom, E.W. (2003) The effects of bioturbation on soil processes and sediment transport. *Annual Review of Earth and Planetary Sciences* 31, 249–73.

Geikie, A. (1893) *Text-book of Geology.* 3rd edn. Macmillan and Co: London.

Geikie, J. (1898) *Earth Sculpture.* John Murray: London.

Genet, M., Stokes, A., Fourcaud, T. and Norris, J.E. (2010) The influence of plant diversity on slope stability in a moist evergreen deciduous forest. *Ecological Engineering* 36, 265–275.

Gilad, E., von Hardenberg, J., Provenzale, A., Shachak, M. and Meron, E. (2004) Ecosystem engineers: from pattern formation to habitat creation. *Physical Review Letters* 93, doi: 10.1103/PhysRevLett.93.098105.

Gregory, J.W. (1911) Constructive waterfalls. *Scottish Geographical Magazine* 27, 537–46.

Gregory, S.V. (ed.) (2003) *The Ecology and Management of Wood in World Rivers.* American Fisheries Society: Bethesda, Maryland.

Gumbricht, T., McCarty, J. and McCarty, T.S. (2004) Channels, wetlands and islands in the Okavango delta, Botswana and their relation to hydrological and sedimentological processes. *Earth Surface Processes and Landforms* 29, 15–29.

Guo, Y., Zhao, H., Zuo, Z., Srake, S. and Zhao, X. (2008) Biological soil crust development and its topsoil properties in the process of dune stabilization, Inner Mongolia, China. *Environmental Geology* 54, 653–62.

Gurnell, A.M. and Petts, G.E. (2006) Trees as riparian engineers: the Tagliamento River, Italy. *Earth Surface Processes and Landforms* 31, 1558–74.

Gurnell, A.M., Hupp, C.R. and Gregory, S.V. (2000) Linking hydrology and ecology: special issue. *Hydrological Processes* 14, 2813–3179.

Hill, A.R. and Duval, T.P. (2009) Beaver dams along an agricultural stream in southern Ontario, Canada: their impact on riparian zone hydrology and nitrogen chemistry. *Hydrological Processes* 23, 1324–36.

Hubble, T.C.T., Docker, B.B. and Rutherfurd, I.D. (2010) The role of riparian trees in maintaining riverbank stability: a review of Australian experience and practice. *Ecological Engineering* 36, 292–304.

Hugenholtz, C.H. and Wolfe, S.A. (2005) Biogeomorphic model of dunefield activation and stabilization on the northern Great Plains. *Geomorphology* 70, 53–70.

Humphreys, G.S. and Mitchell, P.B. (1983) A preliminary assessment of the role of bioturbation and rainwash on sandstone hillslopes in the Sydney Basin, in R.W. Young and G.C. Nanson (eds), *Aspects of Australian Sandstone Landscapes.* Australia and New Zealand Geomorphology Group, pp. 66–80.

Jefferies, R.L., Jano, A.P. and Abraham, K.F. (2006) A biotic agent promotes large-scale catastrophic change in the coastal marshes of Hudson Bay. *Journal of Ecology* 94, 234–42.

Johnson, D.L. (1990) Biomantle evolution and the redistribution of earth materials and artefacts. *Soil Science* 149, 84–102.

Jones, C.G., Lawton, J.H. and Shachak, M. (1994) Organisms as ecosystem engineers. *Oikos* 69, 373–86.

Kail, J. (2003) Influence of large woody debris on the morphology of six central European streams. *Geomorphology* 51, 207–23.

King, J., Nickling, W.E. and Gillies, J.A. (2006) Aeolian shear stress ratio measurements within mesquite-dominated landscapes of the Chihuahuan Desert, New Mexico, USA. *Geomorphology* 82, 229–44.

Lalley, J.S., Viles, H.A., Henschel, J.R. and Lalley, V. (2006) Lichen-dominated soil crusts as arthropod habitat in warm deserts. *Journal of Arid Environments* 67, 579–93.

Langbein, W.B. and Schumm, S.A. (1958) Yield of sediment in relation to mean annual precipitation. *Transactions of the American Geophysical Union* 39, 1076–84.

Lobeck, A.K. (1939) *Geomorphology.* McGraw-Hill: New York.

Lyell, C. (1875) *Principles of Geology.* 12th. edn. (2 vols.). John Murray: London.

Marsh, G.P. (1965) *Man and Nature.* D. Lowenthal (ed.). Belknap Press of Harvard University Press: Cambridge, Mass.

McKenna-Neuman, C., Maxwell, C.D., and Boulton, J.W. (1996) Wind transport of sand surfaces crusted with photoautotrophic microorganisms. *Catena* 27, 229–47.

Mitsch, W.J. and Jorgensen, S.E. (1989) *Ecological Engineering: An Introduction to Ecotechnology*. New York: Wiley.

Mottershead, D.N. and Viles, H.A. (2004) Experimental studies of rock weathering by plant roots: updating the work of Julius Sachs (1832–1897), in D.J. Mitchell and D.E. Searle (eds) *Stone Deterioration in Polluted Urban Environments*. Science Publishers Inc: Plymouth, pp. 61–72.

Naylor, L.A., Viles, H.A. and Carter, N.E.A. (2002) Biogeomorphology revisited: looking towards the future. *Geomorphology*, 47, 3–14.

O'Connor, J.E., Jones. M.A. and Haluska, T.L. (2003) Flood plain and channel dynamics of the Quinault and Queets Rivers, Washington, USA. *Geomorphology* 51, 31–59.

Odling-Smee, F.J., Laland, K.N. and Feldman, M.W. (2003) *Niche Construction: The Neglected Process in Evolution*. Princeton University Press: Princeton, NJ.

Odum, H.T. (1962) Man in the ecosystem. Proceedings, Lockwood Conference on the Suburban Forest and Ecology, *Bulletin Connecticut Agricultural Station* 652, 57–75.

Paramor, O.A.L. and Hughes, R.G. (2004) The effects of bioturbation and herbivory by the polychaete Nereis diversicolor on loss of saltmarsh in south east England, *Journal of Applied Ecology* 31, 449–63.

Paton, T.R., Humphreys, G.S. and Mitchell, P.B. (1995) *Soils: A New Global View*. UCL Press: London.

Penck, W. (1953) *Morphological Analysis of Land Forms: A Contribution to Physical Geology*. Translated by Hella Czech and Katharine Cumming Boswell. New York: St. Martin's Press.

Phillips, J.D. (1995) Biogeomorphology and landscape evolution: the problem of scale. *Geomorphology*, 13, 337–47.

Phillips, J.D. (2006) Evolutionary geomorphology: thresholds and nonlinearity in landform response to environmental change. *Hydrology and Earth System Science* 10, 731–42.

Phillips, J.D. (2009) Soils as extended composite phenotypes. *Geoderma* 149, 143–51.

Phillips, J.D., Marion, D.A. and Turkington, A.V. (2008) Pedologic and geomorphic impacts of a tornado blowdown event in a mixed pine-hardwood forest. *Catena* 75, 278–87.

Pollen, N. (2007) Temporal and spatial variability in root reinforcement of streambanks: accounting for soil shear strength and moisture. *Catena*, 69, 197–205.

Reichman, O.J. and Seabloom, E.W. (2002) The role of pocket gophers as subterranean ecosystem engineers. *Trends in Ecology and Evolution* 17, 44–9.

Renschler, C.S., Doyle, M.W. and Thoms, M. (2007) Geomorphology and ecosystems: challenges and keys for success in bridging disciplines. *Geomorphology* 89, 1–8.

Retallack, G.J. (2001) *Soils of the Past: An Introduction to Palaeopedology*. Blackwell Science: Oxford.

Richards, P.J. (2009) Aphaenogaster ants as bioturbators: impacts on soil and slope processes. *Earth-Science Reviews* 96, 92–106.

Rivera-Aguilar, V., Godínez-Alvarez, H., Moreno-Torres, R. and Rodríguez-Zaragoza, S. (2009) Soil physic-chemical properties affecting the distribution of biological soil crusts along an environmental transect at Zapotitlán drylands, Mexico. *Journal of Arid Environments* 73, 1023–8.

Robinson, D.A. and Williams, R.B.G. (2000) Accelerated weathering of a sandstone in the high Atlas Mountains of Morocco by an epilithic lichen. *Zeitschrift für Geomorphologie* 44, 513–28.

Schumm, S.A. (1968) Speculations concerning palaeohydrologic controls of terrestrial sedimentation. *Bulletin Geological Society of America* 79, 1573–88.

Schumm, S.A. and Lichty, R.W. (1965) Time, space and causality in geomorphology. *American Journal of Science* 262, 110–9.

Shaler, N.S. (1889) *Aspects of the Earth*. Scribner: New York.

Simon, A. and Collinson, A.J.C. (2002) Quantifying the mechanical and hydrologic effects of riparian vegetation on streambank stability. *Earth Surface Processes and Landforms* 24, 527–46.

Simon, A., Pollen, N. and Langendoen, E. (2006) Influence of two woody riparian species on critical conditions for streambank stability: Upper Truckee River, California. *Journal American Water Resources Association* 42, 99–113.

Stallins, J.A. (2006) Geomorphology and ecology: unifying themes for complex systems in biogeomorphology. *Geomorphology* 77, 207–16.

Stavi, I., Ungar, E.D., Lavee, H. and Sarah, P. (2009) Livestock modify ground surface microtopography and penetration resistance in a semi-arid shrubland. *Arid Land Resources and Management* 23, 237–47.

Su, Y.G., Li, X.R., Zheng, J.-G. and Huang, G. (2009) The effect of biological soil crusts of different successional stages and conditions on the germination of seeds of three desert plants. *Journal of Arid Environments* 73, 931–6.

Suzuki, Y., Matsubara, T. and Hoshino, M. (2003) Breakdown of mineral grains by earthworms and beetle larvae. *Geoderma* 112, 131–42.

Thomas, A.D. and Dougill, A.J. (2006) Distribution and characteristics of cyanobacterial soil crusts in the Molopo Basin, South Africa. *Journal of Arid Environments* 64, 270–83.

Thornes, J.B. (1985) The ecology of erosion. *Geography* 70, 222–36.

Thornes, J.B. (1990) *Vegetation and Erosion*. John Wiley and Sons: Chichester.

Thornes, J.B., Shao, J.X., Diaz, E., Roldan, A., McMahon, M. and Hawkes, J.C. (1996) Testing the MEDALUS hillslope model. *Catena* 26, 137–60.

Tockner, K., Ward, J.V., Edwards, P.J. and Kollmann, J. (2002) Riverine landscape: an introduction. *Freshwater Biology* 47, 497–500.

Tricart, J. (1957) *Application du Concept de Zonalite á la Géomorphlgie*. Tijdschrift van het Koninklijk Nederlandsch Aardrijkskundig Geomootschap, Amsterdam. pp. 422–4.

Tricart, J. and Cailleux, A. (1965) *Introduction á la Géomorphologie Climatique*. Sedes: Paris.

Uchida, T., Ohte, N., Kimoto, A., Mizuyama, T. and Changhua, L. (2000) Sediment yield on a devastated hill in southern China: effects of microbiotic crust on surface erosion processes. *Geomorphology* 32, 129–45.

Urban, M.A. and Daniels, M. (2006) Introduction: exploring the links between geomorphology and ecology. *Geomorphology* 77, 203–6.

Van der Wiel, M.J. and Darby, S.E. (2007) A new model to analyse the impact of woody riparian vegetation on the geotechnical stability of riverbanks. *Earth Surface Processes and Landforms* 32, 2185–98.

Vernadsky, V.I. (1998) *The Biosphere.* Nevramont Publishing Group: New York.

Viles, H.A. (1984) Biokarst: review and prospect. *Progress in Physical Geography* 8, 523–42.

Viles, H.A. (ed.) (1988) *Biogeomorphology.* Blackwell: Oxford.

Viles, H.A. (2008) Understanding dryland landscape dynamics: Do biological crusts hold the key? *Geography Compass,* 2/3, 899–919.

Viles, H.A., Naylor, L.A., Carter, N.E.A. and Chaput, D. (2008) Biogeomorphological disturbance regimes: progress in linking ecological and geomorphological systems. *Earth Surface Processes and Landforms* 33, 1419–35.

Wilkinson, M.T. and Humphreys, G.S. (2005) Exploring pedogenesis via nuclide-based soil production rates and OSL-bioturbation rates. *Australian Journal of Soil Research,* 43, 767–79.

Wilkinson, M.T., Richards, P. and Humphreys, G.S. (2009) Breaking ground: pedological, geological and ecological implications of soil bioturbation. *Earth-Science reviews* 97, 257–272.

Wynn, T.M. and Mostaghimi, S. (2006) Effects of riparian vegetation on stream bank subaerial processes in south-western Virginia, USA. *Earth Surface Processes and Landforms* 31, 399–413.

Yoo, K., Amundson, R., Heimsath, A.M. and Deitrich, W.E. (2005) Process-based model linking pocket gopher (Thomomys bottae) activity to sediment transport and soil thickness. *Geology* 33, 917–20.

15

Human Activity and Geomorphology

Dénes Lóczy and László Sütő

THE EMERGENCE OF ANTHROPOGEOMORPHOLOGY: RECOGNIZING THE SIGNIFICANCE OF HUMAN INTERVENTION

Landscapes are the products of interactions between abiotic and biotic factors. Ever since its appearance, even before the beginning of recorded history, human society has been involved in the system of geomorphic processes. The intensity of involvement has been proportional to the size of the human population, to its demands upon the environment and to the level of technological progress achieved to satisfy growing demands.

The significance of human activities in the evolution of the Earth's surface was recognized as early as the mid-19th century. In the United States George Perkins Marsh published a book entitled *Man and Nature* in 1864 (reprinted as *The Earth as Modified by Human Action*, Marsh, 1965). The Italian geologist Antonio Stoppani (1873) found human action comparable to other landscape-shaping forces and was the first to speak of an 'Anthropozoic era' (Crutzen, 2006). Elisée Reclus (1869) and Karl Ritter's students (e.g. Fischer, 1915) investigated this issue in France and Germany, respectively. After World War 2 the first comprehensive descriptions of topographic and hydrographic changes in the various sectors of the economy (e.g. Fels, 1935, 1965) and the first, although brief, chapters on anthropogeomorphology appeared in geomorphology and physical geography text-books worldwide (Bulla, 1954; Tricart, 1978; Louis and Fischer, 1979). In the

1960s urban geomorphology became an established discipline in several countries (in the Soviet Union: Kotlov, 1961; in the USA: Coates, 1976; in Great Britain: Douglas, 1983).

The above publications, however, are more the exception than the rule. Mainstream research still focused on geological structure and then on climatic conditions in long-term relief evolution. Interest in contemporary processes, including the impact of humans, only became widespread with Quaternary studies gaining greater prominence. A famous recognition in the geomorphology of the 20th century was made by Sherlock (1922), who claimed that the amount of earth moved by humans exceeds the amount dislocated by any other geomorphic agent. It was also claimed that economic activities do not coincide in space with natural landscapes (Erdősi, 1987). The combination of biophysical processes and human action complicates the analysis of anthropogenic landscapes (Goudie and Viles, 1997a).

In the 20th century, total mastery over the environment became a manifested goal (Davydov, 1951). Human impact on the environment was delimited in space by the Russian geochemist Vladimir Ivanovitch Vernadsky (1924), who coined the term 'noösphere' (Greek: 'globe of thought') to denote the extent of influence of humankind as part of the biosphere. Subsequent research aimed to estimate human impact spatially. For instance, the total area transformed by human activities was estimated at 30–50 per cent of Earth's land surface (Vitousek et al., 1997; Hooke, 2000). In order to delimit the period when human action prevailed, Paul J. Crutzen (2006) introduced a new term 'to emphasize the central role of humankind in geology and ecology by proposing to use the term "Anthropocene" for the current geological epoch' (Crutzen, 2006: 13). The discussion on the issue whether the Holocene/Anthropocene boundary can be established on stratigraphic grounds has begun (Zalasiewicz et al., 2008).

It is claimed by Peter K. Haff (2003) that the Anthropocene offers unique opportunities to geomorphology. As opposed to 'classical geomorphology', which studies the biophysical forces that drive landscape change, in Haff's concept, a new geomorphology or 'neogeomorphology' must anticipate landscape change driven by a combination of biophysical and social forces. Meant to influence the course geomorphic evolution takes ('the future trajectory of the global landscape'), neogeomorphology is engaged with prediction with practical implications for landscape engineering and management.

In the 1960s, the term 'anthropogeomorphology' was created (Golomb and Eder, 1964) to denote investigations of the direct human creation of landforms and human modification of natural (biophysical) processes (mainly through land use changes). Based on the traditional elements of the landscape (material, process and landform) used in geomorphology, the place of anthropomorphology can be established (Table 15.1). The table shows that natural processes are sometimes accompanied by human-induced processes (N+A) and, on other occasions, the processes planned and controlled by human society are influenced by natural processes (A+N). The 'end product' of landscape rehabilitation is ideally an assemblage of landforms which may be similar to, but not entirely identical with, the initial natural landform (indicated as 'N').

A useful grouping of human influences seems to be according to the purpose of the intervention: intentional (e.g. earth removal and accumulation) and unintentional changes (accelerated soil erosion) (Brown, 1970; Williams, 2008).

Human action, as summarized most comprehensively by Goudie (2006), takes place at micro-, meso- and macroscales and influences geomorphological processes (Gregory and Walling, 1987). It can be extremely localized (a road cut) with limited spatial impact, or of regional (e.g. land drainage) or even continental scale (e.g. navigation canals across catchments like the Danube–Main–Rhine Canal). Timescales may also vary from short term, which is typical of human activity, to geological timescales. Thus, anthropogeomorphology has to employ a unique combination of totally different timescales. This partly explains why human impact on the landscape has long been regarded as an 'external disturbance' (see later). Long-term denudation estimates put the formation of planated surfaces to a time interval of some ten or hundred million years. Since the survival of human society as it exists at present is far from being secured, the extrapolation of the rate of surface change caused by human activities for a million years (for instance, by Huggett, 2003) is hardly justified. In addition, natural (biophysical) and man-induced geomorphic processes may also differ in magnitude, frequency and direction. Anthropogeomorphology can be approached both from the natural and social sciences. Geomorphologist focus on the magnitude and rate of anthropogeomorphological processes (Thomas, 1956; Turner et al., 1990), while the human geographers' priority is the socio-economic impulses contributing to geomorphological dynamics as well as the historical overview of human impact, a common approach to anthropogeomorphology. In addition, planners merge the above approaches for practical purposes (Nir, 1983).

Another subdiscipline concerned with the social aspects of geomorphology is *applied* or *environmental geomorphology* (Hails, 1977). It serves the purposes of human society, particularly through developing resources (through terrain mapping and evaluation, monitoring landscape change and reducing hazards (contributing to environmental impact assessment and environmental management; Goudie, 2001). One of its branches, *engineering geomorphology* directly contributes to urban development (Cooke and Doornkamp, 1990). The particular benefits of geomorphological knowledge in these tasks can best be illustrated by case studies (Goudie, 2006). Márton Pécsi (1971) lists engineering geomorphological tasks at various timescales: temporary

Table 15.1 Grades of transition from natural to human-induced geomorphic processes with the disciplines which study them (Lóczy, 2008)

Material	Process	Landform	Discipline
N	N	N	'Classical' geomorphology
N	A	A	Anthropogeomorphology
N	N+A	A	Engineering geomorphology
A	A+N	'N'	Land rehabilitation research
A	A	A	Engineering

N = natural; A = 'artificial' (produced, redeposited, triggered, operated or transformed by human action). The disciplines in *italics* are covered by applied geomorphology.

changes (e.g. hollows caving in); cumulative serial events (slumps repeated over decades); periodic changes (collapses, slides caused by undermining) and episodic changes (rockfalls).

Recognizing the significance of anthropogeomorphology, at the 6th International Conference on Geomorphology in 2005, organized in Zaragoza, the International Association of Geomorphologists (IAG/AIG) launched a Working Group entitled 'Human Impact on the Landscape' (HILS) for the investigation of the role of the human factor in geomorphological processes.

FROM HUMAN IMPACT TO HUMAN AGENCY

Applied geographical investigations concerned with human impact on environmental processes have long remained within the confines of sub-disciplines like soil erosion studies, fluvial or coastal geomorphology. Each of the geomorphological branches has included consideration of human action among the traditional geomorphic processes (Gregory, 2000).

Martin Haigh (1978) introduced a systematic classification of human impact. In a simplified form, his system distinguishes the following groups of human impact:

- directly human-induced processes – like construction; excavation or hydrological interventions;
- indirectly human-induced processes – like acceleration of erosion and sedimentation; ground subsidence; slope failure and earthquakes triggered.

The processes in the second group can be conceived as system responses to human impact, usually with a time delay and negative consequences to society (Zepp, 2008). In practice, however, the distinction between the two groups is not always clear.

Although apparently a unifying concept, environmental management could not integrate biophysical and sociocultural processes into a single conceptual framework. The term 'human impact' reflects the attitude that human interventions come 'from outside' into the geomorphological system, much like external disturbances. In this concept (Figure 15.1) human society exerts an influence on geomorphic evolution in three ways:

- directly, producing landforms (e.g. by excavation – identical with Haigh's first group);

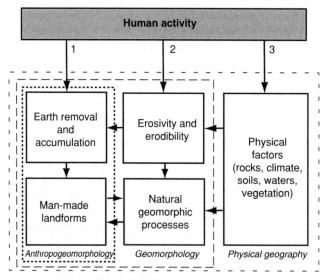

1 = Direct human impact on landform
2 = Human impact on geomorphic process (and indirectly on landform)
3 = Human impact on conditions influencing process

Figure 15.1   The 'human impact' model (Lóczy, 2008). 1 = direct human impact on landform; 2 = human impact on geomorphic processes; 3 = human impact on conditions influencing processes

- effects on some geomorphic processes (e.g. accelerating soil erosion; Trimble and Crosson, 2000);
- indirectly, influencing the biophysical conditions which control geomorphic processes (e.g. climate change induced by human activities; Slaymaker et al., 2009).

The first impact falls in the scope of anthropogeomorphology; the second impact relates to many branches of geomorphology and the third impact belongs to physical geography.

In the Hungarian system of anthropogenic geomorphology (Table 15.2), primary landforms serve the goals of production, while secondary landforms are necessary corollaries. It is notable, however, that in József Szabó's opinion (Szabó, et al., 2010) investigations should focus on processes which operate on anthropogenic landforms after direct human intervention stops.

The above approach of direct and indirect human impact on the landscape has prevailed in anthropogeomorphology until now (e.g. environmental impact statements). Recently, however, the philosophical foundations of anthropogeomorphology have been challenged (Urban, 2002). The traditional assumption that human activities are external to the biophysical environment is questioned on the grounds of the reinterpretation of the principles formulated in phenomenological philosophy (Abram, 1996) and human geography (Cosgrove, 1989). Since biophysical landscapes and human society has co-evolved over many millenia and anthropic forces are guided by human perception and valuation, humans can be integrated together with all other geomorphological agents (Urban, 2002). This concept is reflected in the term 'human agency', which is increasingly common in the academic language. Since human and biophysical influences on the system are arranged on a continuum, the above concept makes the delimitation of relevant research even more difficult. For heavily modified landscapes, a more complex model with feedbacks between biophysical processes and man-made landforms and their human perception is suggested (Figure 15.2).

Mining: excavation and accumulation

Analysing anthropogenic landforms according to economic activities, the most spectacular examples of direct and intentional landscape transformation are probably found in mining landscapes (Bell, 2010). The extent of human action is often described by the amount of earth moved or by the area affected by mining. (According to Erdősi, 1987, in the environs of Mecsek Mountains, south-west Hungary, 135 million cubic metres of earth was moved by mining activities between World War 2 and the 1980s.) Together with collieries, ore mines (particularly iron and copper ore mines) also occupy and transform comparable areas (pits and disposal sites of tailings).

Underground mining of coal is characterized by three types of primary features: *mine workings* as excavation landforms, *levelled terrains* as planated landforms and *waste tips* as accumulation landforms, all with various further evolution stages. Secondary landforms are mostly created by undermining. The most common is ground subsidence, which is controlled by geological (stratigraphic, tectonic, hydrogeological) conditions, soil mechanics and the dimensions of mining activities (size of mining area, duration of activity, thickness and depth of coal seams). The above parameters determine the rate at which the settling movement extends upwards. (In the hard-coal mining area at Pécs it is estimated at 3 mm day^{-1}; Erdősi, 1987). Andrew Goudie (2006) mentions further examples of ground subsidence caused by gold and salt mining, groundwater and oil abstraction.

It is clear from a generalized sketch of *ground subsidence* (Kratzsch, 1983) (Figure 15.3) that undermining affects a much larger area of impact than that immediately above voids created by underground mine workings. (Erdősi, 1970, found the subsidence trough of the Pécs–Somogy area almost seven times larger – 13.5 km^2 – on the surface than the orthogonal projection of the extraction void at depth.) There are numerous studies on the rates of subsidence in mining areas employing various methods, ranging from levelling to remote sensing techniques. The new sensors allow measurement at the required level of precision (Wright and Stow, 1999). Examples include Stoke-on-Trent in Britain, where 52 individual coal seams ranging in thickness from 1 cm to 3.5 m have been worked. SAR interferometry was used in the mapping of areas affected and a differential interferogram pojected on a satellite image as isopleths of subsidence shows changes in surface elevation (Haynes et al., 1997; Baek et al., 2008).

Underground mines present a series of ecological problems both during operation and after closure. Near-surface underground voids are a potential immediate geomorphological hazard because they may open up to the surface as sinkholes (crown holes). Examples are abundant in the coal-mining areas of Germany, particularly in the Ruhr district (Niese, 2007). The type of hazard depends on overburden thickness (or depth of deposit) and quality, excavation height and the

Table 15.2 A genetic classification of man-made landforms (Szabó, in Szabó et al., 2010)

Type of intervention	Landform type	Direct product		Indirect product	
		Primary	Secondary	Modified in quality	Modified in quantity
Montanogenic	E	—	Open-cast pits	Subsidence bowls, sinkholes	Fluvial features caused by mine water inflow
	P	—	Waste-filled valleys	Accumulation in pits	
	A	—	Spoil heaps	Bulges around tips	
Industrogenic	E	Cooling lake basins	Quarries for landfill	Mass movements on industrial raw material disposal sites	Accelerated erosion by sewage inflow
	P	Industrial parks	Sludge reservoirs		
	A	Sockels for windmills	Slag disposal sites		
Urbanogenic	E	Cave dwellings	Loam pits	Cellar collapses	Erosion by runoff from sealed surfaces
	P	Levelled construction sites	Garbage disposal sites, sanitary landfills		
	A	Tells, burial mounds, tumuli	Rubble mounds		
Traffic	E	Road cuts	Hollow roads	Slumps on embankments	Increased piping
	P	Airfields	Mounds removed		
	A	Embankment	Roadside A		A in culverts
Water management	E	Artificial channels	Navvy pits	Shore erosion because of impoundment	Rapid incision
	P	Polders	Cutoffs		
	A	Flood-control dykes, dams	A from dredging channels		A behind dams
Agrogenic	E	Waterholes	Excavation pits	Rapid gullying	Deflation forms
	P	Terraces	Pseudoterraces	Sheet wash	Silt spreading
	A	Lynchets	Stone ridges, walls	Alluvial fans	Delta expansion
Warfare	E	Moats	Bomb craters	Avalanches triggered by explosions	Erosion by water-courses modified for defence puposes
	P	Glacis, airfields	Destroying settlements		
	A	Earthworks	'Trümmelberge' (rubble heaps)		
Tourism, sports	E	Recreation lake basins	Field sports (riding, moto-cross, quad) landscapes	Abrasion along recreation lake shores	Accelerated erosion along hiking paths
	P	Sports tracks			
	A	Ski-jumping ramps			

Landform types: E = excavation; P = planation (levelling); A = accumulation.

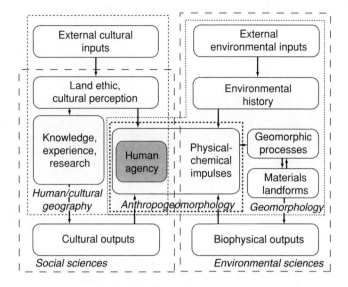

Figure 15.2   The 'human agency' model of heavily modified landscapes with the integration of biophysical and sociocultural processes. (Modified after Urban, 2002, with the author's permission)

dip of the worked deposits (Hollmann and Nürenberg, 1972). Superimposed on one another, the three zones in the hanging wall above the coal seams worked are:

- zone of complete loss of structure;
- zone of shattering;
- loosened zone and
- shelter zone (no jointing).

The mechanical processes in these zones are projected as surface deformation, called (secondary) *montanogenic* landforms and arranged in increasing size by Erdősi (1987) as:

- crown holes (sites of very localized collapse);
- collapse pits (maximum 15–20 m across);
- collapse fields (with multiple ruptures and holes);
- subsidence troughs (cumulative products of periodic subsidence events).

Scarps produced by collapse can be distinguished from agricultural terraces on the basis that their alignment is independent of contour lines (Erdősi, 1987). During mineral extraction the groundwater table is usually kept at a low level and this also generates subsidence. A hydrological simulation of the optimal spatial allocation of pumping wells is helpful in avoiding damage (Bayer et al., 2008).

Open-cast mining creates large-scale excavation (mine pits) and accumulation features (spoil heaps). On heaps, steep slopes, saturation by infiltrating water or by impounded groundwater, marginal undercutting and slope overloading all

increase landslide hazard (GWP Consultants, 2007). In developing countries under various climatic conditions, thick overburden is a particularly serious problem, studied by engineers applying geotechnical modelling: sensitivity analyses based on long-term monitoring, for instance, in Turkey (Kasmer and Ulusay, 2006) and in Nigeria (Okagbue, 2007). Occasional spontaneous combustion which loosens up spoil and vertical air motion on slopes may also modify geomorphic processes.

Stone and aggregate *quarrying* generates typical anthropogenic processes on quarry walls, floors and debris cones or aprons (Dávid, 2007). Weathering can be studied on quarry faces (Petersen, 2002). Quarries are also investigated for conservation purposes as part of the geological heritage (Prosser, 2003) and as habitats of particular value (Cottle, 2004). The problems of limestone quarries are integral to karst research (Gunn and Bailey, 1993).

Land cultivation: soil erosion and sedimentation

It is assumed that the impact of farming on soil erosion was discernible as early as Neolithic times. Recently, intensive geoarchaeological investigations using state-of-the-art dating techniques are directed at identifying such impacts and their significance in the decline of ancient civilizations. Degraded slopes and coarse alluvia accumulated on valley floors in Greece point to intensive land use (Dufaure, 1976).

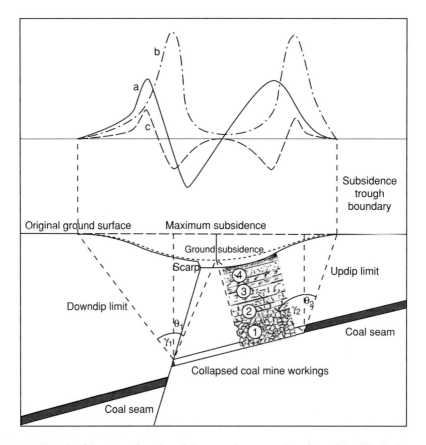

Figure 15.3   Sketch of impact of undermining on the ground surface. (Modified after Brady and Brown 1993: *Rock Mechanics: For Underground Mining*, 2nd edn. Springer Verlag, Fig. 15 – With kind permission of Springer Science and Business Media)

a = strain (faulted); b = displacement (unfaulted); c = strain (unfaulted). Zones above void: 1 = collapse zone with complete loss of structure; 2 = zone of shattering; 3 = loosened zone; 4 = shelter zone.

Impacts of Greek colonization survive in the landscapes of Sicily (Neboit, 1980). From the now abandoned agricultural lands of the Middle Ages lynchets and ridge-and-furrow fields have been described (summarized in Rackham, 2000). In central Europe, land use and climate changes jointly resulted in reworking of deposits in sediment cascades (Lang and Hönscheidt, 1999).

Terraces are common practices of erosion control on cultivated hillslopes steeper than 12 per cent. Today in the Mediterranean region, terraced vineyards and olive, citrus and chestnut plantations are often neglected and the lack of maintenance causes serious erosion problems (Douglas et al., 1994; Seeger and Ries, 2008). In tropical landscapes, in addition to rice, extensive terrace systems were established for coffee,

cocoa, pepper and tea plantations. In Hungary, the vineyard terraces of Tokaj–Hegyalja date back at least to the 1620s (Boros, 1996). Typical local features are 'obalas', that is, stones heaped up along plot boundaries or on terrace risers. In central-European landscapes, ploughland terraces (German: *Ackerterrassen*) have developed along long-standing field boundaries through tillage (Schaeffer, 1957). With different land use types (arable/grassland or arable/forest) different rates of soil erosion can produce scarps, another type of 'pseudoterrace' (German: *Nutzwechselstufen*), called 'taluds' in the United States (Soil Survey Staff, 1951). Taluds are farmland steps resultant from soil accumulation beyond hedgerows and stone walls running transverse to the slope in the 'bocage' landscapes of western Europe (Neboit-Guilhot, 1999).

There are numerous studies on land use changes leading to serious damage on hillslopes by sheet wash and *gully erosion*. In the Pyrenees, G. Soutadé (1980) related the formation of *chalades* (short and deep gullies on gneissic slopes) to large-scale deforestation by shepherds and charcoal burners in the 18th century. The triggering influences of gully formation, however, were exceptionally intense rainfall events. More ambiguous is the origin of erosion features on tropical lands. The origin of extensive gully systems (in Malagasy: *lavaka*) on the heavily deforested island of Madagascar has led to debates among geomorphologists. As opposed to earlier views claiming that they result from deforestation and overgrazing, their polygenetic nature, and the combination of natural and cultural factors in their development, is now widely accepted (Wells and Andriamihaja, 1993). Elsewhere, evidence for the damage done by forest clearance (reduced rainfall interception, more intense storms and runoff, soil desiccation) is easier to find (Pimental, 1993). In his classic work Stanley Trimble (1983) set up a sediment budget for the Coon Creek (Wisconsin) and pointed out that unregulated agricultural practices in the late 19th century and early 20th century accelerated erosion and resulted in valley floor sedimentation.

Overgrazing may also upset the equilibrium of soil formation and erosion. In the environmentally sensitive region of the Sahel, desertification caused by the North Atlantic and El Niño/Southern Oscillation is aggravated by the traditionally stock-keeping people's overgrazing (Geist, 2005).

In some parts of the world, sedimentation associated with human *settlement* is well documented (Walling and Probst, 1997). In the United States, catchment erosion rates and stream suspended sediment yields are reconstructed from the rate of delta progradation (Pasternack et al., 2001) or from changes in the rate of floodplain sedimentation (Knox, 2006).

Whether human actions are the sole culprit in the modification of geomorphic processes or natural climatic fluctuations have to be equally seriously considered remains to be decided. The debates on the origin of arroyos in North America (Leopold, 1951) and on the valley fills in the Mediterranean (Vita Finzi, 1969) illustrate how opposing views have emerged (Goudie, 2006).

Water management: river channelization, water transfer and canal construction

Water management was a basic condition for the survival of the first civilizations in Egypt,

Mesopotamia, the Indus and Ganga valleys and in China and has never ceased to be a necessity ever since. Its impacts have been intensively studied by geomorphologists, particularly over the past half a century (e.g. Thomas, 1956). Ranging from dam construction to urbanization, Gregory (2006) lists 36 distinct types of human influence on rivers, all worthy of geomorphological investigation.

Dams are spectacular engineering structures and, at the same time, remarkable landscape elements. The first dams were built in Egypt more than 6000 years ago. For the construction of the famous dam of Sadd el Kafara near Heluan (Egypt), during the period between 2950 and 2750 BC, more than 100,000 tonnes of building material were used (Hill, 1984). The 3700-year-old Ma'rib dam in Yemen, 3.2 km long, 36.5 m high and 152 m wide, was breached several times (Hehmayer, 1989). Among the largest of modern structures is the double dam on the River Vahsh of Tajikistan, higher than 300 m and double the size of the Grand Coulee Dam on the Columbia River. Major dams, which are higher than 15 m or those of 5–15 m height which impound a reservoir of more than 3 million cubic metres capacity, number more than 45,000 and half of them are found in China (Pottinger, 2000).

River flow regulation involves topographic changes, as flood control calls for the construction of embankments (primary accumulational landforms), which rise above their lowland environments as landmarks and the driest habitats. The navvy pits, that is, the quarries on the floodplains from where the material was obtained for the dams (secondary excavational features), are usually less spectacular but act as 'seminatural' wetlands and are of great ecological value today (Szabó, in Szabó et al., 2010).

Since 1955 a range of approaches have been proposed to interpret the indirect consequences of human interventions into the fluvial system (Table 15.3) (Gregory, 2006). The concept of river metamorphosis as a holistic approach (Schumm, 1977) has been successfully applied to rivers under human impact on several continents.

The downstream effects of dams were found to be highly variable, ranging from minimum to considerable (7.5 m) scour along different downstream distances, by a survey of 21 dams on alluvial rivers in the semiarid western United States (Williams and Wolman, 1984). With less fluctuation in water regime, the rivers of humid areas may show entirely different temporal and spatial scales in adjustment.

Dams, particularly those of great dimensions, are increasingly regarded as questionable solutions in water management (Downs and Gregory, 2004). In 2000 the World Commission on Dams raised doubts about the usefulness of

Table 15.3 Approaches to describe human impact on rivers (Modified after Gregory, 2006)

Concept	Indicators	Source
Typology of channel change	Six types identified from the balance between bedload × particle diameter (q_sd) and water discharge × stream slope (q_ws)	Lane (1955)
Induced aggradation	Gullying or channel extension upstream + aggradation downstream	Strahler (1956)
Cycle of erosion in urban channels	Change in channel parameters induced by deforestation, cultivation, urbanization	Wolman (1967)
River metamorphosis, thresholds and complex response	Sediment and water discharge related to slope, median diameter of bed material, flow depth and width	Schumm (1969)
Rate law	Response time and reaction time to change in equilibrium	Graf (1977)

dams, also from the point of view of geomorphology (referring, among others, to the transformation of water regime and sediment transport, turbulence of flow and local scour or fill). While large-scale construction works are under way in some countries, such as in China and India, in others, in the USA, the decommissioning of old dams has begun (Graf, 1996, 2003).

Luna B. Leopold (1998) summarized his geomorphological objections to the world's largest water management project, the Three Gorges Scheme on the Yangtze (Chen et al., 2007), on the basis of:

- The rate of reservoir sedimentation is underestimated.
- The floor slope of the reservoir will not form in less than 70–150 years time and no reliable prediction for such a long time period can be made.
- Coarse sediments will deposit in the upper section of the reservoir.
- When floodwater is retained in the reservoir, suspended load will also deposit.
- The wear of plant equipment, scour below the dam and erosion of reservoir banks will be rapid.

Large-scale river flow regulation measures induce fundamental changes in river mechanism (Brookes, 1988). The channelization of the Tisza River (a tributary of the Danube in the Carpathian Basin) of 1419 km original length in the 19th century is often cited as an example of an intervention with far-reaching indirect consequences. Under natural conditions, the extensive floodplain of this low gradient lowland river remained waterlogged for several months almost every year. Channelization increased the gradient through cutting off 114 meanders, thus reducing the total river length to 955 km. The intervention brought about a lowering of the groundwater table and the alteration of soils (locally alkalization).

The absence of inundation resulted in desiccation of the cultivated floodplain beyond the dykes. On the other hand, significant accumulation began in the active floodplain (exact rates are currently being studied; Nagy et al., 2001). Geomorphologists keep emphasizing how important floodplains are as geoarchives, both for natural and human-induced environmental changes (e.g. Buch and Heine, 1995; Gábris and Nádor, 2007).

Probably related to global climate change as well as human intervention in drainage basins (Slaymaker et al., 2009), peak flood waves tend to increase on many rivers and flood control dykes cannot provide adequate protection against them any more. The most obvious solution, the constant raising of dykes, is debated because of the increased risks and alternative ways of flood control are sought. (Along the Tisza River the construction of 10–14 emergency reservoirs is proposed; Schweitzer et al., 2008.)

Because of the complexity of channel response regulated by thresholds (Schumm, 1979), it is difficult to predict the long-term adjustment of channelized rivers. Some general types of channel change within the straightened reaches, however, can be established (Brookes, 1987). The local history of channel evolution explains cases of hypersensitivity or undersensitivity (Gregory, 2006). On the Hungarian section of the Maros River, Tímea Kiss and co-workers (2008) found a sensitive response. As a consequence of a number of 19th-century cutoffs, the slope of the river bed doubled and braided units developed along the originally meandering–anastomosing reach.

Drainage canals and irrigation ditches are also striking landscape elements in many parts of the world since the Antiquity. Navigation canals, such as the Main–Danube Canal (171 km in length), which cuts through the European continental divide at 406 m elevation, are even more complicated engineering structures.

Water transfers are human interventions into the water budget of a region at the largest scale.

In China three routes for water transfer from the humid south to the dry north were decided as early as the 1950s (Zou, 1983). The construction of the central canal began in December 2002, but was suspended in 2009. The greatest challenge for engineers will be the 3.5 km long tunnel under the Huanghe (Yellow River). The eastern canal of 18,000,000,000 m^3 $year^{-1}$ capacity, along the ancient Grand Canal, is under construction and is likely to be built until 2012.

One of the largest engineering projects on Earth is the so-called 'Great Manmade River' in Libya. The huge, pipeline network, 3380 km in length, supplies 6,500,000 m^3 of freshwater per day from the Nubian Sandstone Aquifer through 1300 wells for the coastal cities. The project apparently does not involve major relief changes but the earthworks and the amounts of building stones, sand and cement employed are remarkable (As of March 2011, the political situation in Lybia makes the future of the project uncertain.)

Spreading irrigation or water extraction changes groundwater levels and cause rock decay and geomorphological change, particularly in coastal and desert environments, through accelerated salt weathering (Goudie and Viles, 1997b).

Coastal development: defence and land reclamation

The history of coastal engineering began with the construction of harbours in ancient times. Major interventions also had indirect impacts and sometimes made it impossible to operate the harbour, which had to be abandoned. Since coastal areas belong to the most densely inhabited zones on Earth and are among the most popular destinations of world tourism, human action combines with action of waves and currents to contribute to coastal development. Human impact on coastal processes may derive from three sources (Eurosion, 2004):

- from human activities on the mainland (changes in the water and sediment regimes of rivers);
- on-coast interventions (various engineering structures) and
- off-coast influences (dredging, sand extraction).

The engineering works on rivers (channelization, impoundment – see above) influence sediment supply to the river mouths. Typical examples are cited from Africa (Komar, 2000): the Nile delta 'starving' of silts after the construction of the Aswan High Dam in 1964 and the Volta delta after the construction of the Akossombo Dam (Ly, 1980).

Most of the beaches are maintained by 'hard engineering' structures: protected by sea-walls, riprap revetments and beach fills against flooding and erosion is prevented by groynes, jetties or breakwaters, which modify the movement of sediment along the coast (USACE, 2001). The identification of coastal cells (Carter, 1989) is a useful tool for the characterization of coastal processes and also for the estimation of the relative significance of human agency. Human modification of sediment transport along coasts, including beach nourishment is a major research topic worldwide (Bird, 1996).

In addition to the 'classic' sites where human action led to the transformation of coasts (for instance, in Britain, the Humber Estuary and Coast, Hull University, 1994), the intensive investigations which started along the eastern and southern coasts of Asia deserve particular attention. In Taiwan, 80 per cent of beaches have been significantly eroded over the past three decades (Hsu et al., 2007). It was only partly due to the tectonic environment (intensive uplift) but also to channelization (Tanshui River) and impoundment of rivers (Tsengwen River), sand mining from riverbeds (Choushui and Lanyang Rivers) and beaches, construction of coastal structures at right angles (Taichung Harbour) and parallel to coastline (Tainan City), reclamation of seafront lands for harbours, recreation and industrial purposes, and natural gas and groundwater extraction causing land subsidence (Wenfong coast). Increasingly more precise beach cross-section measurements and satellite observations allow the estimation of coastal retreat. Reduced sediment influx hinders the recovery of beach profile after storm surges along the Bohai Sea coast (Xue et al., 2009).

Given their crowdedness and lack of potential area for urban expansion, coastal land reclamation has been crucial in the development of two Asian metropolis areas, Hong Kong and Singapore. In Hong Kong the impacts on the groundwater system are remarkable (Guo and Jiao, 2009). In Singapore coastal land reclamation increased the land area from 581.5 km^2 in 1960 to 692.7 km^2 in 2007 through connecting 54 offshore islands. The impact on near-shore currents is considerable. In addition to the Jurong industrial area, Terminal 3 of Changi International Airport was built on reclaimed land (Douglas and Lawson, 2003). Filling the 870 ha area began in 1994 using more than 52 million cubic metres of sands transported from the nearby islands of Indonesia. The runway construction itself, however, could only start after dewatering and compaction completed in 1999. The terminal opened in January 2008.

Sometimes land reclamation is only a means of reparation for historical land loss. The recent plans of connecting the rock island of Heligoland on the North Sea with the flat sands of 'The Dune' involves the gaining of 60 ha dry land, in compensation for the land washed away by the sea during the 1721 storm tide (Zand-Vakili, 2008).

Urbanization: residential and infrastructural developments

Topography and human settlement are in a two-way interaction. Consequently, geomorphologists either evaluate the suitability of land for urban development or monitor process–response systems (Cooke, 1977). The accumulation of waste is typical of all settlements, from prehistoric tells to modern cities, where the centre is built on several metres of cultural layers. In cities of medieval history cellar and tunnel systems often form a complex of underground cavities, liable to collapse (Fodor and Kleb, 1986).

Residential development on hillslopes poses risks (Sidle et al., 1985). In cities of high relief, locally major elevations are levelled. In Rio de Janeiro, excavation activities in areas of unregulated development were the reasons for catastrophic landslides triggered by exceptional rainfall events. Subsidence areas within some cities are threatened by inundation. In New Orleans, in the aftermath of Hurricane Katrina of 2005, extensive areas were inundated. The extent of this hazard is now studied by GIS techniques (Farris et al., 2007).

Under the socialist system in eastern European countries, urban planning focused on the establishment of large prefabricated housing estates in cities, which involved major modifications of relief (Gyenizse and Szabó-Kovács, 2008). The consequences of urban planning and changing land use on the stability of urban terrain are analysed in the cities of Germany (Genske, 2003).

In the study of urban rivers a holistic approach is becoming prevalent and Chin and Gregory (2009) emphasize the following requirements in urban planning: consideration of impacts of land use on the catchment affected; retention of precipitation; delay of runoff; management of the effects of runoff within urban areas and planning for downstream consequences. They propose a conceptual model of predictable impacts.

In the urban environment, environmental hazards are manifest in particular manners. A system-based approach had to be developed for urban flood management (Zevenbergen et al., 2008) or surface instability (Szabó et al., 2010). Human agency influences landsliding in an intricate manner. A black box model attempting to relate the input (GDP increase and the increased level of technological capabilities of society involved) to the output (the increasing frequency of landslides) was suggested by Remondo et al. (2005), without revealing causal relationships.

Atmospheric pollution (especially sulphates and nitrates) centred on urban and industrial regions is an increasingly important factor in rock decay. Weathering crusts are intensively studied by both geomorphologists (Goudie, 1995) and civil engineers (e.g. Török, 2003).

Military activities: fortification and warfare

Land transformation for military purposes may be either constructive (defence lines, fortifications as well as rubble heaps accumulated after war) or destructive (bomb craters) (Illyés, in Szabó et al., 2010). The world's largest engineering structure, the 7240 km-long Great Wall of China began to be built 1100 BC and united into a single stronghold under the rule of the Qin Dynasty between 221 and 207 BC. Near Beijing the stone-and-brick wall is 4–8 m wide and 7–8 m high, but sections of the Great Wall have subsequently been quarried for housing and road construction.

The defence structures of the Roman Empire typically consist of fortified camps (castrae) and watch-towers along the *limes*, a major *defence line* of the Empire. In Britain the 120 km-long Hadrian's Wall was erected between 122 and 128 AD against the Picts of Scotland (English Heritage, 2007) and the Antonine Wall (Historic Scotland, 2007) to the north was built of stones and peat in 142 AD. The environmental impacts are studied by geoarcheologists and conservationists.

In the Carpathian Basin, the earth *motte* of Szabolcs, a former administrative centre, is almost 800 m long and up to 30 m high. Vast amounts of timber must have been used for its construction. Its present volume is estimated by A. Prékopa to be 326,000 m³ (Frisnyák, 1990).

In World War 1, along the stagnating front lines, military operations (artillery fire, land mine explosions) created a military landscape densely spotted with *craters*. In north-eastern France, a multitude of examples are found. The plateau of Chemin des Dames along the Aisne River was the scene of three battles in World War 1 and converted into a field of craters. The Butte de Vauquois, a mound of 308 m elevation in the Argonne Mountains, is a monument of underground warfare. In its various installations (shafts, galleries and ramifications) of 17 km length over 1.5 km distance, 218 German and 321 French mines exploded and shattered part of the mound. Trench warfare survives along the line of control in Kashmir where Pakistani and Indian troops face each other. The impacts of World War 2 were naturally even more widespread and of larger scale. On 18 April 1945, the Royal Air Force bombarded the lonely North Sea fortress island of Heligoland and transformed the surface of the island into a single crater field (Podjacki et al., 2003). The largest craters are 50 m across and 20 m deep. The area called Mittelland suffered a surface lowering of ca. 40 m exactly 2 years later, when 6700 tonnes of ammunition was blown up underground. Apart from nuclear bomb experiments, this event caused the largest-scale geomorphological transformation related to military operations.

During the Vietnam war between 1965 and 1971, it is estimated that 26 million craters were created by bombing in Indochina (Westing and Pfeiffer, 1972). In this war, however, indirect influences were equally significant. The large-scale deforestation by US Army operations resulted in soil erosion, floods and floodplain sedimentation in the rainy season and desiccation and deflation of lateritic soils in the dry season.

The most accelerated and greatest geomorphologic transformations of modern warfare were caused by the testing of nuclear weapons. The first nuclear bomb of 19 kilotonnes was blown up on 16 July 1945 in Alamogordo (New Mexico) and created a crater 800 m wide and 3 m deep. The desert sand melted into a greenish-coloured glass. Between 1945 and 1998, 2057 nuclear weapons tests were carried out (PBS, 2011).

Recreation: tourism and sports activities

Tourism as a rapidly expanding branch of the economy, has major local impacts on the environment (Holden, 2008). Outdoor sports activities affect large areas in some European regions. In high mountains, the land degradation and mass movements induced by ski tracks are increasingly studied (Mosimann, 1985). Snow compaction by caterpillars can cause sliding. Snowboarding is a typical human trigger of avalanches. The number of human-induced landslide events has grown in the Himalayas, too (Barnard et al., 2001). In addition to hiking, horse riding and, increasingly, the use of off-road vehicles causes significant environmental damage (Taylor, 2003).

CASE STUDY: HUMAN IMPACT OF MINING IN HUNGARY

Although Hungary is a small country, not particularly rich in mineral resources, there were 1217 officially listed mining sites in northern Hungary in 2005 (Farkas, 2006). In Hungary, the mining of copper, iron, lead, zinc, gold, uranium and bauxite induced significant landscape changes in the past. Part of the abandoned open pits are reclaimed, while elsewhere rehabilitation takes place spontaneously.

Major landscape transformation resulted from 250 years of *brown coal* mining (Figure 15.4). In the Borsod Coal Mining Area, near Arló village (south of the town of Ózd), a slide-dammed pond was formed in 1863, just 11 years after the beginning of mining activity (Peja, 1956; Leél-Őssy, 1973). In Borsod, human-induced landscape elements and areas affected by mining were mapped at 1:25,000 scale and a disturbance index was proposed as a measure of relief transformation (Sütő, 2000, 2007). The removal of material related to mining extended over 27 per cent of the study area and involved the reworking of a 5.2 m thick layer of earth over the past 250 years.

In the area of the Lyukóbánya and Pereces shafts (near Miskolc), the total length of mine galleries is almost 50 km, equal to that of the underground tunnels in Budapest. Similarly, in the Salgótarján–Medves Hill area (northern part of the Nógrád Coal Mining Area) 268 depressions are attributed to coal mining (Karancsi, 2002). Prolonged subsidence is typical in the foothills of the North-Hungarian, the Transdanubian and Mecsek Mountains, where mines were abandoned after the change of the political system in 1989. In the vicinity of the huge open pits of lignite extraction in the Mátra and Bükk forelands, approximately 200 mm surface subsidence was recorded to the early 1980s (Fodor and Kleb, 1986). Similar instances of subsidence in an area of some hundreds of square metres are mapped in the mining regions of the Transdanubian Mountains (Juhász, 1974). Above the northern mining field at Nagyegyháza, 99.9 per cent of total subsidence occurred within 8 years after mine closure. In some cases the surface may *rise* 2 to 12 cm locally after mine closure due to the hydration of near-surface clays and replenishment of confined water reservoirs (Somosvári, 2002). The radius of the area under the influence of undermining (r) is calculated from the product of average cover thickness (H) and the cotangent of the critical angle (b) (Martos, 1965) using the formula $r = H \times \cot b$. The surface projection of the mine workings is mapped using GeoMedia software. Mass movements developed towards the inner parts of the subsidence, while trench-like faults emerged in the inflection zone (Sütő, 2007).

The most extensive, almost 200 m deep *opencast pits* are now being reclaimed at Pécs. Between 1950 and 1980, 248 *spoil heaps* of ca. 200 million cubic metres total volume were built in Hungary.

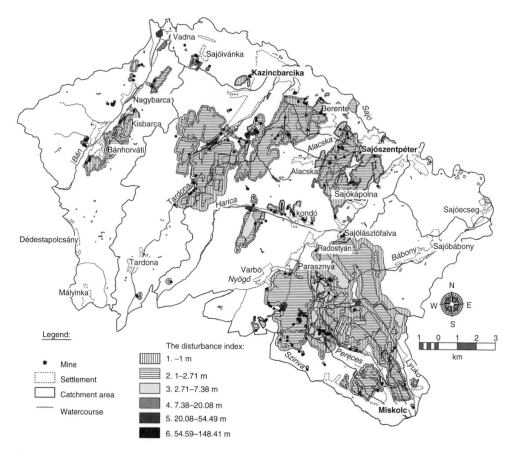

Figure 15.4   Disturbance map of the Borsod Mining Area, North-Hungary. (Sütő, 2007)

They are affected by physical and chemical weathering, burning, deflation, intensive mass movements and gully erosion. Based on experience from field work at Pécs between 1966 and 1974, Ferenc Erdősi (1970, 1987) found analogies between the further evolution of mining spoil heaps and that of residual hills in a natural topography. His typology of spoil heaps is based on morphological position: tips either fill valleys or raise slopes and exceptionally also occur on flat surfaces. Each exerts different kinds of influence on drainage and microclimate. One of the largest heaps in the Borsod Coal Mining Area is located in the Ádám Valley near Kazincbarcika. It is more than 100 m high and accommodates 1.2 million cubic metres of spoil (Sütő, 2000, 2007). Slides cause 8 m high failures on its slopes (Homoki et al. 2000).

Uranium ore mining has particularly great impacts on the landscape. At Kővágószőlős (Mecsek Mountains), to obtain 20.3 thousand tonnes of uranium, 46 million tonnes of rock had

to be moved and subsidence affects an area of almost 42 km².

Stone quarrying is practised in the vicinity of almost every settlement in the mountains of Hungary. In the Mátra Mountains 57 major quarries of 240 ha are listed, and fewer than 10 of them are reclaimed (Dávid, 2007). The geomorphic evolution of quarry walls is similar to that of the natural cliffs: debris fans, gorges, block fields develop.

Lakes in abandoned quarries are often spectacular tourist attractions (like the 'Tarn' formed in a former millstone quarry in the Megyer Hill near Sárospatak town; Rózsa and Kozák, 1995). Some quarries expose remarkable geological features along nature trails and are under protection (like the volcanic cone of Tokaj Hill of 20 km² area). Some quarries have almost demolished entire hills. The height of Esztramos Hill in North-Hungary was reduced from 380 to 340 m; the basalt hill near Zalahaláp village was totally removed and 7 million cubic metres of limestone was quarried from the Bél-kő within a century.

'HARD' VERSUS 'SOFT' (ECOLOGICAL) ENGINEERING

Until very recently, engineering structures like earthworks, bank revetments and similar structures were favoured to 'steer' geomorphic processes. The idea was to use durable materials (first of all, concrete) to resist the action of exogenic processes. In addition to high costs, such solutions receive negative reactions from the population affected but once introduced, structures are difficult to remove. For a long time, 'environmentally friendly' (or 'green') methods of mitigation for hazards (like grassed waterways) coexisted with 'hard' engineering but no combined techniques are found in literature before the 1960s. Becoming really widespread only in the 21st century, *ecological engineering* integrates biophysical and human-operated systems and strives to restore ecosystems disturbed by human activities, relying on the self-regulating capacities of ecosystems (Mitsch and Jørgensen, 2004). It not only applies to land reclamation activities but all fields where human-induced and biophysical processes interact. For instance, in urban planning the allocation of green areas and the operation of hydrological processes (stormwater management) can be designed observing ecological principles (Bergen et al., 2001). The participation of geomorphologists in river restoration projects promotes the application of the principles of 'soft' engineering (Brookes and Shields Jr, 1996; Chin and Gregory, 2009).

Coastal management (including optimization of sediment transport along coast) and river restoration efforts (like dam removal) are good examples of the opportunities offered by ecological engineering (Silvester and Hsu, 1993). Techniques for establishing halophilous *Spartina* marshes on dredged material deposits along the south Atlantic coast were pioneered by researchers at North Carolina State University in the late 1960s to early 1970s (Seneca et al., 1976). The objective of these early studies was to provide stabilization of shorelines and dredged materials, and to restore coastal habitats deteriorated in the wake of human population growth in coastal regions. The establishment of large-scale salt marsh began on several US coastlines in the 1970s. Ecological engineering with *Spartina* plantations dates back 40 years in China (Chung, 2006). It is mentioned in the paper that coastal defence by riprap was replaced by *Spartina*

plantation in a section of the Zhejiang coast. Within 2 years a salt-marsh developed and successfully protected the coast against successive typhoons of exceptional power by slowing down tidal currents and retaining large amounts of sediment. It is a promising solution against a predicted rise in sea-level, too (Li et al., 2002). The 2004 Christmas tsunami disaster on the Indian Ocean provided further evidence to the effectiveness of mangrove forests in coast protection ('bioshield function'). The extreme shore retreat rates measured on coasts from India to Indonesia (Ongkosongo, 1979) and China (for instance, 85 m year^{-1} in Jiangsu Province, China, between 1949 and 1990) are usually due to cutting coastal mangrove forests for firewood and settlement and since the 1970s replacing them with aquaculture, holiday resorts or harbours (Prasetya, 2007). Separate diagnostic parameters are established to assess the efficiency of coastal vegetation as a bioshield.

The guiding principle of resilience engineering (Hollnagel et al., 2006) is that the resilience of a system is inversely proportional to the amplitude of reaction to external influence, and directly proportional to the graduality of increase of reaction and to the recovery rate. The comparison between the traditional and resilient approach in flood management is tackled by Zevenbergen et al. (2008).

OUTLOOK

Anthropogeomorphology is one of the fastest developing branches of modern geomorphology. It is already integrated in the procedure of Environmental Impact Assessments (Marchetti and Rivas, 2001) and serves environmental management. It is increasingly recognized that, as vegetation cover changes, the geomorphic processes triggered reshape the ground surface which affects human life. Consequently, Murray et al. (2008) claim that geomorphology, with an emphasis of human agency, has a central place in the emerging science of Earth's surface dynamics. There are ever stronger connections between the study of human-induced processes and nature conservation.

Regarding both its subject and methodological approach, anthropogeomorphology is poised for further development. Taking advantage of the quantitative techniques offered by geophysics, GIS (USGS, 2008) and geoengineering, the discipline is ready to provide a more substantial contribution to the solution of environmental problems at all scales, from local to global.

REFERENCES

Abram, D. (1996) Merleau-Ponty and the voice of the Earth. In D. Macauley (ed.), *Minding Nature: The Philosophers of Ecology*. Guilford Press, New York. 82–101.

Baek, J., Kim, S.-W., Park, H.-J., Jung, H.-S., Kim, K.-D. and Kim, J.-W. (2008) Analysis of ground subsidence in coal mining area using SAR interferometry. *Geosciences Journal* 12, 277–84.

Barnard, P.L., Owen, L.A., Sharma, M.C. and Finkel, R.C. (2001) Natural and human-induced landsliding in the Gharwal Himalaya of northern India. *Geomorphology* 40, 21–35.

Bayer, P., Duran, E., Baumann, R. and Finkel, M. (2008) Optimized groundwater drawdown in a subsiding urban mining area. *Journal of Hydrology* 365, 95–104.

Bell, F.G. (2001) Some environmental impacts of mining. In P.G. Marinos, G.C. Koukis, G.C. Tsiambaos, and G.C. Stournaras (eds), *Engineering Geology and the Environment*. Swets and Zeltlinger, Lisse, The Netherlands. 3519–60.

Bergen, S.D., Bolton, S.M. and Fridley, J.L. (2001) Design principles for ecological engineering. *Ecological Engineering* 18, 201–10.

Bird, E.C.F. (1996) *Beach Management*. John Wiley and Sons, Chichester. 281 p.

Boros, L. (1996) *Tokaj-Hegyalja szőlő- és borgazdaságának földrajzi alapjai és jellemzői (Geographical background to and characteristics of viticulture in the Tokaj-Hegyalja Region)*. Észak- és Kelet-Magyarországi Földrajzi Évkönyv 3, Miskolc-Nyíregyháza. P. 322.

Brady, B.H.G. and Brown, E.T. (1993) *Rock Mechanics: For Underground Mining*. 2nd Edition. Springer Verlag, Berlin. p. 571.

Brookes, A. (1987) The distribution and management of channelized streams in Denmark. *Regulated Rivers* 1, 1, 3–16.

Brookes, A. (1988) *Channelized Rivers: Perspectives for Environmental Management*. John Wiley and Sons, New York. p. 326.

Brookes, A. and Shields Jr., F.D. (eds) (1996) *River Channel Restoration: Guiding Principles for Sustainable Projects*. Wiley, Chichester. p. 433.

Brown, E.H. (1970) Man shapes the earth. *Geographical Journal* 136, 74–85.

Buch, M.W. and Heine, K. (1995) Fluvial geomorphodynamics in the Danube River valley and tributary river systems near Regensburg during the Upper Quaternary – theses, questions and conclusions. *Zeitschrift für Geomorphologie*, Supplement-Band 100, 53–64.

Bulla, B. (1954) *Általános természeti földrajz (General Physical Geography) II*. Tankönyvkiadó, Budapest. p. 549.

Carter, R.W.G. (1989) *Coastal Environments*. Academic Press, New York. p. 617.

Chen, Zh.Y., Xu, K.Q. and Watanabe, M. (2007) Dynamic hydrology and geomorphology of the Yangtze River. In A. Gupta (ed.), *Large Rivers: Geomorphology and Management*. John Wiley and Sons, Chichester. pp. 457–70.

Chin, A. and Gregory, K.J. (2009) From research to application: management implications from studies of urban river channel adjustment. *Geography Compass* 3, 297–328.

Chung, Ch-Hs. (2006) Forty years of ecological engineering with *Spartina* plantations in China. *Ecological Engineering* 27, 49–57.

Coates, D.R. (ed.) (1976) *Geomorphology and Engineering*. Dowden, Hutchinson & Ross, Stroudsburg. p. 360.

Cooke, R.U. (1977) Urban geomorphology. *The Geographical Journal* 142, 59–65.

Cooke, R.U. and Doornkamp, J.C. (1990) *Geomorphology in Environmental Management: A New Introduction*. 2nd Edition. Clarendon Press, Oxford University Press, Oxford, New York. pp. xxiv, 410.

Cosgrove, D. (1989) *Social Formation and Symbolic Landscape*. University of Wisconsin Press, Madison, WI. p. 320.

Cottle, R. (2004) *Linking Geology and Biodiversity*. English Nature, Peterborough. p. 91. (Research Report 562).

Crutzen, P. (2006) The anthropocene. In E. Ehlers and Th. Krafft (eds), *Earth System Science in the Anthropocene: Emerging Issues and Problems*. Springer Verlag, Berlin. pp. 13–18.

Dávid, L. (2007) Anthropogenic geomorphological and after-use problems of quarrying: case studies from the UK and Hungary. *Geografia Fisica e Dinamica Quaternaria* 30, 161–5.

Davydov, M.M. (1951) Velikie gidrotekhnicheskiye sooruzheniya stalinskoy epokhi (Major hydrotechnological achievements in the Stalinic era). *Bol'shevik* 20, 24–36.

Douglas, I. (1983) *The Urban Environment*. Edward Arnold, London. p. 240.

Douglas, I. and Lawson, N. (2003) Airport construction: materials use and geomorphic change. *Journal of Air Transport Management* 9, 177–85.

Douglas, T.D., Kirkby, S.J., Critchley, R.W. and Park, G.J. (1994) Agricultural terrace abandonment in Alpujarra, Andalucia, Spain. *Land Degradation and Rehabilitation* 5, 281–91.

Downs, P.W. and Gregory, K.J. (2004) *River Channel Management: Towards Sustainable Catchment Hydrosystems*. Hodder Arnold, London. p. 408.

Dufaure, J.-J. (1976) La terrasse holocène d'Olympie et ses équivalents méditerranéen. *Bulletin de l'Association des géographes français* 433, 85–94.

English Heritage. Hadrian's Wall. Frontiers of the Roman Empire (2007) World Heritage Site Management Plan 2008–2014. English Heritage, Swindon. http://www.hadrians-wall.org/ResourceManager/Documents (Accessed 6 February 2008).

Erdősi, F. (1970) A szénbányászat által okozott felszínváltozás Pécs környékén (Topographic change by coal mining around Pécs). In: *Földrajzi tanulmányok a Dél-Dunántúl területéről (Geographical Studies from South-Transdanubia)*. MTA Dunántúli Tudományos Intézet, Pécs. pp. 85–107. (Dunántúli Tudományos Gyûjtemény 92, Series geographica 40)

Erdősi, F. (1987) *A társadalom hatása a felszínre, a vizekre és az éghajlatra a Mecsek tágabb környezetében (The Impact of Society on the Surface, Waters and Climate in the*

Broader Environs of the Mecsek Mountains). Akadémiai Kiadó, Budapest. pp. 228 app.

Eurosion (2004) *Living with coastal erosion in Europe: Sedimentation and space for sustainability.* European Commission, Brussels. http://www.euorsion.org/reports-online/index.html. (Accessed on 20 December 2008)

Farkas, J. (2006) *A Magyar Geológiai Szolgálat 2005. évi működési jelentése (Report of the Hungarian Geological Survey for 2005).* Hungarian Geological Survey, Budapest. p. 38.

Farris, G.S., Smith, G.J., Crane, M.P., Demas, C.R., Robbins, L.L. and Lavoie, D.L. (eds) (2007) *Science and Storms: The USGS Response to the Hurricanes of 2005.* US Geological Survey, Reston, VA. (Circular 1306). http://pubs.usgs.gov/circ/1306/pdf/. (Accessed 6 February 2009)

Fels, E. (1935) *Der wirtschaftende Mensch als Gestalter der Erde. Ein Beitrag zur allgemeinen Wirtschafts- und Verkehrsgeographie.* Bibliographisches Institut AG, Leipzig. p. 206.

Fels, E. (1965) Nochmals von der Anthropogene Geomorphologie. *Petermanns Geographische Mitteilungen* 1, 9–11.

Fischer, E. (1915) Der Mensch als geologischer Faktor. *Zeitschrift der Deutschen Geologischen Gesellschaft,* 67, 106–48.

Fodor, T. and Kleb, B. (1986) *Magyarország mérnökgeológiai áttekintése (An engineering geological overview of Hungary).* Hungarian State Geological Institute, Budapest. p. 199.

Frisnyák, S. (1990) *Magyarország történeti földrajza (An historical geography of Hungary).* Tankönyvkiadó, Budapest. p. 213.

Gábris, Gy. and Nádor, A. (2007) Long-term fluvial archives in Hungary: response of the Danube and Tisza rivers to tectonic movements and climatic changes during the Quaternary: a review and new synthesis. *Quaternary Science Reviews* 26, 2758–82.

Geist, H. (2005) *The Causes and Progression of Desertification.* Ashgate Publishing, Aldershot, UK. p. 258.

Genske, D.D. (2003) *Urban Land.* Springer Verlag, Berlin. p. 331.

GMR Project Info (2007) *Great Manmade River Project Info.* Libyan Ministry of Water Management, Tripoli. http://water-technology.net/projects/gmr/. (Accessed on 22 January 2007)

Golomb, B. and Eder, H.M. (1964) Landforms made by man. *Landscape* 14, 4–7.

Goudie, A.S. (1995) *The Changing Earth: Rates of Geomorphological Processes.* Blackwell, Oxford, UK, and Cambridge, MA. p. 302.

Goudie, A.S. (2001) Applied geomorphology: an introduction. *Zeitschrift für Geomorphologie* Supplementband 124, 101–10.

Goudie, A.S. (2006) *Human Impact on the Natural Environment.* 6th Edition. Blackwell Publishing, Oxford. p. 357.

Goudie, A.S. and Viles, H.A. (1997a) *The Earth Transformed.* Blackwell Publishing, Oxford. p. 276.

Goudie, A.S. and Viles, H.A. (1997b) *Salt Weathering Hazards.* John Wiley and Sons, Chichester. p. 256.

Graf, W.L. (1977) The rate law in fluvial geomorphology. *American Journal of Science,* 277, 178–191.

Graf, W.L. (1996) Geomorphology and policy for restoration of impounded American rivers: what is 'natural'? In B.L. Rhoads and C.E. Thorn (eds), *The Scientific Nature of Geomorphology.* John Wiley and Sons, New York. pp. 443–73.

Graf, W.L. (ed.) (2003) *Dam Removal Research: Status and Prospects.* The H. John Heinz III Center for Science, Economics and the Environment, Washington DC. p.165. http://www.heinzctr.org/publications/PDF/Dam_Research_Full_Report.pdf. (Accessed 6 August 2009)

Gregory, K.J. (2000) *The Changing Nature of Physical Geography.* Edward Arnold, London. p. 368.

Gregory, K.J. (2006) The human role in changing river channels. *Geomorphology* 79, 172–91.

Gregory, K.J. and Walling, D. (eds) (1987) *Human Activity and Environmental Processes.* John Wiley and Sons, Chichester.

Gunn, J. and Bailey, D. (1993) Limestone quarrying and quarry reclamation in Britain. *Geology* 21, 167–72.

Guo, H.P. and Jiao, J.J. (2009) *Coastal Groundwater System Changes in Response to Large-Scale Land Reclamation.* Nova Science Publishers, Hauppauge, NY. p. 89.

GWP Consultants (2007) Appendix 4-4 Slope design. In: *Quarry Design Handbook.* Pre-publication draft. GWP Consultants, Charlbury, Oxfordshire. p. 24. http://www.gwp.uk.com/pdfs/Appendix_4-4_Slope_design.pdf. (Accessed 11 February 2009)

Gyenizse, P. and Szabó-Kovács, B. (2008) A természeti környezet és a település kölcsönhatása Komló példáján (Interactions between the physical environment and the settlement: example of Komló, Hungary). *Földrajzi Értesítő* 57, 273–88.

Haff, P.K. (2003) Neogeomorphology: prediction and the anthropogenic landscapes. In P.R. Wilcock and R.M. Iverson (eds), *Prediction in Geomorphology.* AGU, Geophysical Monograph Series 135, 15–26.

Haigh, M. (1978) *Evolution of Slopes on Artificial Landforms.* Department of Geography, University of Chicago, Chicago. (Research Paper 183)

Hails, J.R. (ed.) (1977) *Applied Geomorphology: A Perspective of the Contribution of Geomorphology to Interdisciplinary Studies and Environmental Management.* Elsevier Science Publishers, Amsterdam. p. 418.

Haynes, M., Capes, R., Lawrence, G., Smith, A., Shilston, D. and Nicholls, G. (1997) Major urban subsidence mapped by differential SAR interferometry. *Proceedings of the 3rd ERS Symposium,* Florence. 1, 573–7.

Hehmayer, I. (1989) Irrigation farming in the ancient oasis of Ma'rib. *Proceedings of the Semiar for Arabian Studies* 19, 33–44.

Hill, D. (1984) *A History of Engineering in Classical and Medieval Times.* Routledge, London. p. 263.

Historic Scotland (2007) *The Antonine Wall. Frontiers of the Roman Empire.* World Heritage Site Management Plan 2007–2012. Historic Scotland, Edinburgh. p. 80. http://www.historic-scotland.gov.uk/antonin-wall-management-plan.pdf. (Accessed 6 February 2008)

Holden, A. (2008) *Environment and Tourism.* Routledge, London. p. 274.

Hollmann, F. and Nürenberg, R. (1972) Der 'Tagesnahe Bergbau' als technisches Problem bei der Durchführung von Baumaßnahmen im Niederrheinisch-Westfälischen Steinkohlengebiet. In *Mitteilungen der Westfälischen Berggewerkschaftskasse 30,* Bochum. p. 39.

Hollnagel, E., Woods, D.D. and Leveson, N. (2006) *Resilience Engineering, Concepts and Precepts.* Ashgate Publishing, Aldershot, UK. p. 397.

Homoki, E., Juhász, Cs., Baros, Z. and Sütő, L. (2000) Antropogenic geomorphological research on waste heaps in the East-Borsod coal basin (NE-Hungary). In M. Rzętały, (ed.), *Z Badań nad wpływen antropopresji na środowisko.* Studenckie Koło Naukowe Geografów UŚ, Sosnowiec. pp. 24–31.

Hooke, R. LeB. (2000) On the history of humans as geomorphic agents. *Geology* 28, 843–6.

Hsu, T.-W., Lin, Ts.-Y. and Tseng, I-F. (2007) Human Impact on coastal erosion in Taiwan. *Journal of Coastal Research* 32, 961–73.

Huggett, R.J. (2003) *Fundamentals of Geomorphology.* Routledge, London. p. 386.

Hull University. *Humber Estuary and Coast: Management Issues* 1994. The University of Hull Institute of Estuarine and Coastal Studies – Humberside County Council, Hull. p. 50. http://www.hull.ac.uk/coastalobs/media/pdf/humberestuaryandcoast.pdf. (Accessed 20 January 2009)

Juhász, Á. (1974) Anthropogene Einwirkungen und Geoprozesse in der Umgebung von Komló. *Földrajzi Értesítő* 23, 2, 223–4.

Karancsi, Z. (2002) Természetes és antropogén eredetű környezetváltozás a Medves térség területén. (Environmental changes of natural and human origin on the Medves region). PhD thesis, Department of Physical Geography and Geoinformatics, University of Szeged. p. 117.

Kasmer, O. and Ulusay, R. (2006) Stability of spoil piles at two coal mines of Turkey: geotechnical characterization and design considerations. *Environmental and Engineering Geoscience* 12, 337–52.

Kiss, T., Fiala, K. and Sipos Gy. (2008) Altered meander parameters due to river regulation works, Lower Tisza, Hungary. *Geomorphology* 98, 96–110.

Knox, J.C. (2006) Floodplain sedimentation in the Upper Mississippi Valley: natural versus human accelerated. *Geomorphology* 79, 286–310.

Komar, P.D. (2000) Coastal erosion – underlying factors and human impacts. *Shore and Beach* 68, 3–16.

Kotlov, F.V. (1961) Antropogennye izmeneniya reliefa na primere goroda Moskvy (Anthropogenic relief changes on the example of the city of Moscow). *Sbornik* 52, 134–50.

Kratzsch, H. (1983) *Mining Subsidence Engineering.* Springer Verlag, Berlin. p. 543.

Lane, E.W. (1955) The importance of fluvial morphology in hydraulic engineering. Proceedings of the American Society of Civil Engineers 81, 1–17.

Lang, A. and Hönscheidt, S. (1999) Age and source of colluvial sediments at Vaihingen-Enz, Germany. *Catena* 38, 89–107.

Leél-Őssy, S. (1973) Természeti és antropogén folyamatok és line 6: variation. *AGU Transactions* 32, 347–57.

Leopold, L.B. (1951) Rainfall intensity: an aspect of climatic variation. *AGU Transactions* 32, 347–57.

Leopold, L.B. (1998) *Sediment problems at Three Gorges Dam.* International Rivers Network, Berkeley, CA. p. 4. (Accessed 12 January 2001)

Li, C.X., Fan, D.D., Deng, B. and Wang, D.J. (2002) Some problems of vulnerability assessment in the coastal zone of China. In *Synthesis and Upscaling of Sea-Level Rise Vulnerability Assessment Studies.* Flood Hazard Research Centre, London. p. 8. http://www.survas.mdx.ac.uk/pdfs/3licongx.pdf. (Accessed 22 January 2009)

Lóczy, D. (2008) Az emberi társadalom és a geomorfológia. Alkalmazott geomorfológia (Human society and geomorphology: applied geomorphology). In D. Lóczy (ed.), *Geomorfológia (Geomorphology) II.* Dialóg Campus Kiadó, Budapest – Pécs. 395–431.

Louis, H. and Fischer, K. (1979) Allgemeine Geomorphologie. 4. Auflage. De Gruyter, Berlin. p. 814.

Ly, C.K. (1980) The role of Akossombo dam on the Volta River in causing erosion in central and eastern Ghana (West Africa). *Marine Geology* 35, 323–32.

Marchetti, M. and Rivas, V. (eds) (2001) *Geomorphology and Environmental Impact Statement.* Taylor and Francis, London. p. 221.

Marsh, G.P. (1864) *Man and Nature.* Reprinted in 1965: *The Earth as modified by human action.* Belknap Press – Harvard University Press, Cambridge, MA. p. 560.

Martos, F. (1965) *Bányakártan (Mine Damage Survey).* Tankönyvkiadó, Budapest. p. 145.

Mitsch, W.J. and Jørgensen, S.E. (2004) *Ecological Engineering and Ecosystem Restoration.* John Wiley and Sons, New York. p. 424.

Mosimann, T. (1985) Geo-ecological impacts of ski runs construction in the Swiss Alps. *Applied Geography* 5, 29–37.

Murray, A.B., Lazarus, E., Ashton, A., Baas, A., Coco, G., Coulthard, T., Fonstad, M., Haff, P., McNamara, D., Paola, C., Pelletier, J. and Reinhardt, L. (2008) Geomorphology, complexity and the emerging science of the Earth's surface. *Geomorphology* 103, 496–505. doi: 10.1016/geomorph. 2008.08.013.

Nagy, I., Schweitzer, F. and Alföldi, L. (2001) A hullámtéri hordalék-lerakódás (Accumulation on the active floodplain). *Vízügyi Közlemények* 83, 4, 539–64.

Neboit, R. (1980) Morphogenèse et occupation humaine dans l'Antiquité. *Bulletin de l'Association Géographique Française* 466, 21–27.

Neboit-Guilhot, R. (1999) Autour d'un concept d'érosion accélérée. L'homme, le temps et la morphogenèse. *Géomorphologie* 5, 159–72.

Niese, M. (2007) Tagesbrüche – Der Umgang mit Bergbaufolgeschäden im südlichen Ruhrgebiet. In H. Zepp (ed.), *Ökologische Problemräume Deutschlands.* Wissenschaftliche Buchgesellschaft, Darmstadt. pp. 57–68.

Nir, D. (1983) *Man as Geomorphological Agent.* Keter Publishing, Jerusalem. p. 165.

Okagbue, C.O. (2007) Stability of waste spoils in an area strip mine – geological and geotechnical considerations. *Earth Surface Processes and Landforms* 12, 289–300.

Ongkosongo, O.S.R. (1979) Human activities and their environmental impacts on the coast of Indonesia. In: *Proceedings of the Workshop on Coastal Area Development and Management in Asia and the Pacific,* Manila. pp. 67–75.

Pasternack, G.B., Grace, S., Brush, G.S. and Hilgartner, W.B. (2001) Impact of historic land-use change on sediment delivery to a Chesapeake Bay subestuarine delta. *Earth Surface Processes and Landforms* 25, 409–27.

PBS (2011) Nuclear weapons test map. Public Broadcasting Service http://www.pbs.org/wgbh/amex/bomb/maps/index.html. (Accessed 12 January 2011)

Pécsi, M. (1970) A mérnöki geomorfológia problematikája (Problems of engineering geomorphology). *Földrajzi Értesítő* 19, 4, 369–80.

Pécsi, M. (1971) A domborzati egyensúly megváltozása az ember műszaki-gazdasági tevékenysége következtében (Changes in relief equilibrium as a result engineering and economic activities by man). *MTA Biológiai Osztály Közleményei,* Budapest, 14, 29–37.

Peja, Gy. (1956) Suvadástípusok a Bükk É-i (harmadkori) előterében (Slump types in the northern [Tertiary] foreland of the Bükk Mountains). *Földrajzi Közlemények* 4(80), 217–40.

Petersen, J.F. (2002) The role of roadcuts, quarries and other artificial exposures in geomorphology education. *Geomorphology* 47, 289–301.

Pimental, D. (ed.) (1993) *World Soil Erosion and Conservation.* Cambridge University Press, Cambridge. p. 349.

Podjacki, O., Müller, P. and Steinecke, K. (2003) Nordseeküste: Helgoland – physio- und kulturgeographische Facetten. *Petermanns Geographische Mitteilungen* 147, 34–9.

Pottinger, L. (2000) *The Environmental Impacts of Large Dams.* International River Networks, Berkeley, CA. pp. 1–3. http://irn.org/basics/impacts.html. (Accessed 15 January 2001)

Prasetya, G. (2007) The role of coastal forests and trees in protecting against coastal erosion. In S. Braatz, S. Fortuna, J. Broadhead and R. Leslie (eds), *Coastal Protection in the Aftermath of the Indian Ocean Tsunami: What Role for Forests and Trees?* UN FAO Regional Office for Asia and the Pacific, Bangkok. pp. 103–80. http://www.fao.org/forestry/media/13191/1/0/. (Accessed 5 January 2009)

Prosser, C. (2003) Geology and quarries: some new opportunities. *Geology Today* 19, 65–70.

Rackham, O. (2000) *The History of the Countryside: The Classic History of Britain's Landscape, Flora and Fauna.* Phoenix, San Francisco. p. 445.

Remondo, J., Soto, J., González-Diéz, A., Díaz de Terán, J.R. and Cendrero, A. (2005) Human impact on geomorphic processes and hazards in mountainous areas in northern Spain. *Geomorphology* 66, 69–84.

Rózsa, P. and Kozák, M. (1995) Protection of volcanological natural monuments on the Miocene paleovolcano of Tokaj-Nagyhegy (NE-Hungary). *Geological Society of Greece, Special Publications 4/3.* 1069–73.

Schaeffer, I. (1957) Zur Terminologie der Kleinformen unseres Ackerlandes. *Petermanns Geographische Mitteilungen* 3, 194–200.

Schumm, S.A. (1977) *The Fluvial System.* Wiley Interscience, New York. p. 338.

Schumm, S.A. (1979) Geomorphic thresholds: the concept and its applications. *Transactions Institute of British Geographers,* New Series 4, 485–515.

Schumm, S.A. (1969) River metamorphosis. *Journal of the Hydraulics Division,* ASCE, HY1, 255–63.

Schweitzer, F., Lóczy, D. and Kis, É. (2008) Flood control strategies along the Tisza River in Hungary. In S.R. Basu and S.K. De (eds), *Issues in Geomorphology and Environment.* acb publishers, Kolkata. pp. 1–8.

Seeger, M. and Ries, J.B. (2008) Soil degradation and soil surface process intensities on abandoned fields in Mediterranean mountain environments. *Land Degradation and Development* 16, 488–501.

Seneca, E.D., Broome, S.W., Woodhouse Jr, W.W., Cammen, L.M., and Lyon III, J.T. (1976) Establishing *Spartina alterniflora* marsh in North Carolina. *Environmental Conservation* 3, 185–8.

Sherlock, R.L. (1922) *Man as a Geological Agent – an Account of his Action on Inanimate Nature.* H.F. and G. Witherby, London. p. 372.

Sidle, R.C., Pearce, A.J. and O'Loughlin, C.L. (1985) *Hillslope Stability and Land Use.* American Geophysical Union, Washington, DC. pp. vii, 140 (Water Resources Monograph Series 11).

Silvester, R. and Hsu, J.R.C. (1993) *Coastal Stabilization: Innovative Concepts.* Prentice-Hall, NJ p. 578.

Slaymaker, O., Spencer, T. and Embleton-Hamann, Ch. (eds) (2009) *Geomorphology and Global Environmental Change.* Cambridge University Press, Cambridge, p. 450.

Soil Survey Staff (1951) *Soil Survey Manual.* USDA Agricultural Handbook No 18. Washington, DC. p. 503.

Somosvári, Zs. (2002) Felszínemelkedés jelenségek okai (Reasons for ground rise). *Communications from the University of Miskolc.* Series A: Mining 62, 35–46.

Soutadé, G. (1980) *Modelé et dynamique actuelle des versant supra-forestiers des Pyrénées orientales.* Manuscript. Thése Etat. Albi. p. 452.

Stoppani, A. (1873) *Corso di Geologia* II. Cap. xxxi. Sec. 1327. Milano.

Strahler, A.N. (1956) The nature of induced erosion and aggradation. In W.L. Thomas (ed.), *Man's Role in Changing the Face of the Earth.* University of Chicago Press, Chicago, pp. 621–638.

Sütő, L. (2000) Mining agency in the East Borsod Basin, North-East Hungary. In Jankowski, A.T. and Pirozhnik I. I. (eds), *Nature Use in the Different Conditions of Human Impact.* Studenckie Koło Naukowe Geografów US, Sosnowiec. pp. 116–23.

Sütő, L. (2007) A szénbányászat geomorfológiára és területhasználatra gyakorolt hatásainak vizsgálata a Kelet-borsodi-szénmedencében (Effects of coal mining on the geomorphology and land use in the East Borsod Coal Basin). PhD thesis, Institute of Earth Sciences, University of Debrecen. p. 177.

Szabó, J., Dávid, L. and Lóczy, D. (eds) (2010) *Handbook of Anthropogenic Geomorphology.* Springer Verlag, Berlin. p. 298.

Taylor, R.B. (2003) *The Effects of Off-road Vehicles on Ecosystems.* Texas Parks and Wildlife Department, Austin, TX. http://www.tpwd.state.tx.us/publications. (Accessed 15 January 2009)

Thomas, W. (ed.) (1956) *Man's Role in Changing the Face of the Earth* I–II. University of Chicago Press, Chicago. p. 1236.

Török, Á. (2003) Surface strength and mineralogy of weathering crusts on limestone buildings in Budapest. *Building and Environment* 38, 1185–92.

Tricart, J. (1978) *Géomorphologie applicable.* Masson, Paris. p. 204.

Trimble, S.W. (1983) A sedimentary budget for Coon Creek basin in the Driftless Area, Wisconsin, 1853–1977. *American Journal of Science* 277, 876–87.

Trimble, S.W. and Crosson, P. (2000) U.S. Soil Erosion Rates – Myth and Reality. *Science* 289, 248–50.

Turner, B.L., Clark, W.C., Kates, R.W., Richards, J.F., Mathews, J.T. and Meyer, W.B. (1990) *The Earth as Transformed by Human Action.* Cambridge University Press, Cambridge. p. 729.

Urban, M.A. (2002) Conceptualizing anthropogenic change in fluvial systems: drainage development on the Upper Embarras River, Illinois. *The Professional Geographer* 54, 204–17.

USACE (2001) *Coastal Engineering Manual.* US Army Corps of Engineers. http://bigfoot.wes.army.mil/cem026.html. (Accessed 3 February 2009).

USGS (2008) *Significant Topographic Changes in the United States.* US Geological Survey, Reston, VA. http://topochange.usgs.gov/index.php. (Accessed 27 January 2009).

Vernadsky, V.I. (1924) La Geochimie. Felix Alcan, Paris. p. 404.

Vita-Finzi, C. (1969) The Mediterranean Valleys: geological changes in historical times. In *Agricultural History,* 46, 1; *American Agriculture 1790–1840,* pp. 254–5.

Vitousek, P.M., Mooney, H.A., Lubchenco, J. and Melillo, J.M. (1997) Human domination of earth's ecosystems. *Science* 277, 494–9.

Walling, D.E. and Probst, J.-L. (eds) (1997) Human impact on erosion and sedimentation. *Proceedings of an International Symposium* (Symposium S6) *of the International Association of Hydrological Sciences (IAHS),* Rabat, 1997. IAHS. p. 320.

Wells, N.A. and Andriamihaja, B. (1993) The initiation and growth of gullies in Madagascar: are humans to blame? *Geomorphology* 8, 1–46.

Westing, A. and Pfeiffer, E.W. (1972) The cratering of Indochina. *Scientific American* 226, 21–9.

Williams, B.K. (2008) The influence of human activity on geomorphological processes. *Scienceray,* p. 3. http://www.scienceray.com/Biology/Human-Biology/The-Influence-of-Human-Activity-on-Geomorphological-Processes.330421. (Accessed 20 December 2008)

Williams, G.P. and Wolman, M.G. (1984) *Downstream Effects of Dams on Alluvial Rivers.* US Geological Survey, Washington DC, Professional Paper 1286.

Wolman, M.G. (1967) A Cycle of Sedimentation and Erosion in Urban River Channels. Geografiska Annaler 49A, 385–395.

Wright, P. and Stow, R. (1999) Detecting mining subsidence from space. *International Journal of Remote Sensing* 20, 1183–8.

Xue, Z., Feng, A.-P., Yin, P. and Xia, D.-X. (2009) Coastal Erosion Induced by Human Activities: A Northwest Bohai Sea Case Study. *Journal of Coastal Research* 25, 3, 723–733.

Zalasiewicz, J., Williams, M., Smith, A., Barry, T.L., Coe, A.L., Bown, P.R., Brenchley, P., Cantrill, D., Gale, A., Gibbard, P., Gregory, F.J., Hounslow, M.W., Kerr, A.C., Pearson, P., Knox, R., Powell, J., Waters, C., Marshall, J., Oates, M., Rawson, P. and Stone. P. (2008) Are we now living in the Anthropocene? *GSA Today* 18, 4–8.

Zand-Vakili, A. (2008) Politiker plädieren für Vergrößerung Helgolands. *Die Welt on-line,* 3. Juli 2008. p. 2. http://www.welt.de/hamburg/article2174986/Politiker_plaedieren_fuer_Vergroesserung_Helgolands.html. (Accessed 23 January 2009)

Zepp, H. (2008) Der Mensch als geomorphologischer Faktor. In H. Zepp (ed) *Geomorphologie.* 4. Auflage. Ferdinand Schöningh, Paderborn. pp. 301–7.

Zevenbergen, C., Veerbeek, W., Gersonius, B. and van Herk, S. (2008) Challenges in urban flood management: travelling across spatial and temporal scales. *Journal of Flood Risk Management* 1, 81–8.

Zou, D.K. (1983) China's south-to-north water transfer proposals. In A.K. Biswas, D.K. Zou, J.E. Nickum, and Ch.M. Liu (eds) *Long-Distance Water Transfer: A Chinese Case Study and International Experiences.* United Nations University Press, Tokyo. p. 4. http://www.unu.edu/unupress/unubooks/8157e/8157E.htm. (Accessed 15 February 2007)

PART 3

Process and Environments

16

The Evolution of Regolith

Graham Taylor

Regolith is humanity's interface with planet Earth. It makes up the uppermost skin of the terrestrial parts of the planet and consists of weathered rock, sediments, water, gases, biota and, more rarely, materials like lavas and ashes. Regolith covers all terrestrial landscapes with the exception of those where fresh rock crops out. The word 'regolith' derives from the Greek 'blanket and rock' and was coined by Merrill in 1897.

Regolith provides many of the resources used by people regardless of their economic status. Most obviously the regolith supports and nurtures plants, which in turn provide fodder and shelter for wildlife. Humans eat both, use timber and plants to build shelter, burn them for warmth and cooking and harvest animals to provide protein. In many societies other resources are harvested from regolith including resources such as bauxite, iron ore, gold, diamonds, tin and many others. Many of these are essential to the economic growth of communities and provide valued trade goods for countries like Australia and Brazil. During 2004–2005 Australia produced about A\$9700 million of metallic resources from regolith materials and probably a similar value of industrial minerals.

The formation and nature of regolith depends, to a large degree, on the landscapes it covers; on the nature of the rocks from which it is formed, whether *in situ* or transported; on the composition and volume of water in it and moving through it; and, on the biota living within and on it. To a lesser extent the climate is important, as it ultimately determines the volume and seasonality of water in and regolith. Many argue that temperature is an important variable in the formation of regolith (e.g. Butt and Zeegers, 1992), while others (e.g. Taylor et al., 1990) believe it is

not as important. This chapter will examine examples of how regolith formation is influenced by these variables.

PARENT MATERIALS

Regolith is derived predominantly from rocks. It consists of detritus and chemically altered residues of physically and/or chemically altered components of rocks as well as many new minerals precipitated during the chemical alteration of parent rocks. The following chapter (Chapter 17) discusses rock and mineral weathering but it is necessary here to mention some aspects of weathering in order to understand the fundamental concepts of regolith evolution. Most rocks simultaneously break down (weather) by physical and chemical processes. These weathering processes are essentially controlled by the mineralogical composition of the rocks and the volume and composition of fluids with which the rocks come in contact. For example, a granite will contain cracks and joints as it moves closer to the Earth's surface than it did at the depth at which it originally crystallized. These joints open progressively as the overlying rocks are eroded and the granite pluton unloads. This is a physical response to the progressive reduction of lithostatic load on the granite. As these cracks in the granite open, groundwater is able to access them and react with the minerals bordering the cracks. This alteration in turn opens micro-pores in the chemically altered rim around the crack providing deeper access to water and deeper chemical weathering. The granite now consists of large fragments of relatively fresh rock (often referred to as core-stones or, if they are exposed at the surface, tors) and decomposed or freed granite minerals. Quartz in the original granite is not chemically altered significantly and is freed from its host. Feldspars on the other hand are readily altered to clay minerals (smectite, kaolinite) and during this alteration they release ions of Na^+, K^+, Ca^{2+} and Si^{4+} to the weathering solution. Feldspar weathering therefore adds new-formed minerals to the regolith and solutes to the regolith water. Some granites contain biotite, which weathers to a clay mineral plus Fe^{2+} in solution, amongst other solutes. The ferrous iron becomes oxidized on contact with oxygen which may be present in regolith water or regolith air. As soon as it oxidizes it precipitates as a yellow–brown to red–brown precipitate of goethite. It is this ferric oxyhydroxide that colours weathered granite and other weathered rocks that contain ferruginous minerals. So it is clear that physical and chemical weathering occur simultaneously but sometimes one predominates.

Some rocks (e.g. quartzite, chert, aplite) are relatively chemically stable at the Earth's surface and mainly break down by comminution. Scree slope deposits below quartz sandstone cliffs for example are generally composed of very large to small fragments of the sandstone forming a coarse regolith. This regolith accumulates almost entirely by physical weathering even though rainfall is moderate and chemical weathering minimal. There are circumstances where rocks that are more labile form regolith derived entirely from physically weathered fragments accumulated in scree deposits. In these cases where more labile rocks are physically weathered water is scarce and temperature extremes enhance physical weathering while the lack of water suppresses chemical weathering. Glacial deposits for regolith are almost entirely derived from physically weathered debris eroded by ice.

Other types of regolith originating essentially from physical weathering are uncommon, but some processes do add fresh rock fragments to existing regolith. Intense bushfires can raise the ground temperatures sufficiently high to significantly alter the uppermost regolith mineralogy (Eggleton and Taylor, 2008b) and to comminute the surface of tors (fresh granitic rocks).

Most regolith parent materials weather chemically even under relatively dry and cold climates. Chemically weathered parent material is the more significant type of weathering in building regolith materials that blanket landscapes. Chapter 17 discusses chemical weathering in detail so here I will only discuss those residues and products that are important in considering regolith evolution.

Chemical weathering produces three components important in forming regolith: (1) residues

that do not weather under the local conditions of weathering; (2) new-formed minerals and (3) solutes that move in the regolith water. Under most weathering conditions quartz, feldspar and muscovite are the most common residual products. Trace, but important, amounts of heavy minerals also commonly remain relatively inert during chemical weathering and remain as residues in the regolith (e.g. zircon, ilmenite, anatase). Providing these residual minerals are not removed from the weathering site by erosion, they will continue to weather, provided there is continual removal of the solutes formed during their breakdown. Quartz will simply dissolve; feldspars will lose cations and leave new-formed clay minerals. The suite of residual heavy minerals in regolith can prove important in determining the provenance of the regolith materials.

New-formed minerals are predominantly clay minerals and iron oxyhydroxides and oxides. Clays form from the decomposition of the silicate minerals in the parent rocks. The type of clay mineral formed depends to a large degree on the intensity of leaching or the length of time the parent rocks and regolith experience leaching. Ultimately, clay minerals are destroyed as leaching continues leaving gibbsite [$Al(OH)_3$]. The oxides and oxyhydroxides form from the oxidation of Fe^{2+} (and other readily oxidized cations) released to the regolith water system mainly from those same rock-forming silicate minerals and on contact with O_2 it oxidizes to one of the many forms of Fe^{3+}.

The soluble products of weathering are many and they derive from the decomposition of the parent rocks. They include Ca, Mg, K, Na and Si most commonly, but almost any of the cations found in rock-forming minerals can be found in regolith waters. The concentration of these weathering products in solution depends mainly on the volume of water moving through the regolith. If very little water moves then concentrations will rise and weathering will cease or become 'retrograde'. These solutes move through the regolith or off into deeper groundwater. In the regolith they may precipitate (e.g. to form calcrete or silcrete) or be taken up by regolith minerals to form other new-formed minerals.

Chemical weathering generally alters the parent rocks from the surface downwards. In so doing it results in more intensely weathered materials nearer the surface and less weathered materials at depth. Although many factors may disrupt this pattern (e.g. bioturbation) the layers or zones of more or less equal degree of weathering may be referred to as weathering facies. In the past many workers have used these layered facies to perform lithostratigraphic correlation (e.g. Firman, 1994), a technique that is fundamentally flawed as this weathering is imposed on materials already in place by weathering from the surface, and has nothing to do with stratigraphy.

REGOLITH AND LANDSCAPE

There are many different landscapes, but in terms of the evolution of regolith it is convenient to simplify them as shown in Figure 16.1. Basically there are three main regolith/landscape regimes that encompass all the main concepts necessary to understand regolith evolution. These are landscapes that are stable in terms of erosion, or *in situ* regimes; erosional regimes and, depositional regimes. In Figure 16.1 these correspond to a flat hill-top, upper slopes and lower slopes and alluvial plains respectively.

The *in situ* regime weathering exceeds erosion so most of the solid weathering products are retained. Solutes are leached downward and removed laterally by the groundwater. The residual weathered components are typically as shown in Figure 16.2a,b. The weathering front is the boundary between fresh and weathered rock. It may be sharp as in Figure 16.2a or gradual as in Figure 16.2b. The weathering front is usually irregular reflecting preferential weathering down

and along joints, bedding or other openings in the bedrock. Above the weathering front is *saprock*, a partially weathered material where some labile minerals are wholly or partially weathered and others not. It retains the original fabric of the underlying rock. Core-stones or large portions of the parent rock may remain unweathered.

The saprock grades upward into *saprolite* where essentially all the labile minerals are weathering and replaced by new-formed minerals; non-labile minerals such as quartz will be preserved in the saprolite. Smaller core-stones may occur and the original rock fabric is maintained despite volume changes induced by weathering.

The facies above the saprolite is more complex and is referred to in many ways. It is generally a facies or zone in which the mineralogy of the saprolite is intact but the parent rock fabric is no longer preserved. There are a number of possible reasons for this loss of fabric. They are: that too much of the parent material has been leached by weathering solutions to maintain the fabric so it collapses (*collapsed saprolite*); that it has been bioturbated by roots or fauna, or by physical breakup due to wetting and drying and the fabric is destroyed; or, that this facies has moved laterally which is possible on very minimal slopes (*transported regolith*).

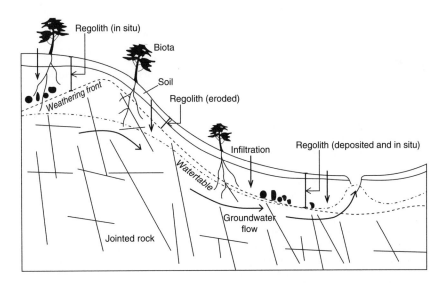

Figure 16.1 Hypothetical landscape showing three major regolith/landscape regimes from left to right: *in situ*, erosional and depositional. The weathering front (the boundary between regolith and bedrock) is shown, as is the watertable (the boundary between saturated and unsaturated ground, as well as directions of water flow for infiltration and for groundwater

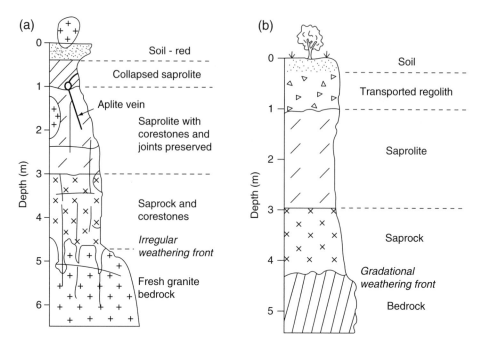

Figure 16.2   Two idealized *in situ* weathering profiles. (a) Developed on a granitic bedrock and (b) formed on deformed clastic sedimentary rocks

In regions toward the edge of flat elevated landscapes the lateral migration of weathering solutions (groundwater) solutes may precipitate forming a duricrusts or cemented regolith. If Fe^{2+} is a common ion being moved, it will oxidize to a ferricrete and Si^{4+} which is almost ubiquitous in groundwater and may precipitate to form a silcrete.

Above the collapsed saprolite occurs a *soil*. On hill tops or plateaux, these soils are generally derived from a pedogenetic reorganization of the existing regolith materials into soil horizons. These soils can vary enormously depending on the nature of the regolith from sandy earths with an organic-rich A horizon to deep chocolate clay-rich soils over basaltic regolith. These soils generally do not accumulate mineral material additional to that in the upper regolith although Eggleton and Taylor (2008a) show termites can add additional mineral matter from deep with the regolith profile and of course aeolian accessions to hill top and plateau soils is common (Chartres and Walker, 1988). Basalt-derived soils the world over contain measurable quantities of quartz, a substantial proportion of which arrives as wind-borne mineral matter (Mitzoa and Tahahashi, 1977).

The nomenclature for such profiles is somewhat confused and different workers use different terms for the same facies. Figure 16.3 shows some of the common terms used to describe *in situ* weathering profile facies.

Moving downslope from the hilltop (Figure 16.1) a convex slope (or cliff, in some cases) is encountered. This is generally a part of the landscape where the rate of erosion exceeds the rate of weathering so the regolith is generally thinner than on the tops. Weathering profiles similar to those on hill tops will be truncated or removed down to saprock and outcrops of slightly weathered to fresh rock may be exposed. Soils in this part of the landscape will be thin and vary from lithosols to minimal earths or podzolic soils. Because of their position on the slope they are well drained and therefore more likely to be redder than those lower on the slope.

The regolith as a whole on the upper slopes moves downslope under the influence of gravity and after heavy rain as a result of rotational slips and landslides.

The lower concave slopes are a zone of regolith accumulation. The regolith here progressively thickens downslope from the point at which deposition of materials from upslope begin and it merges on the downslope edge with alluvial valley fills. It can be a very complex body of materials

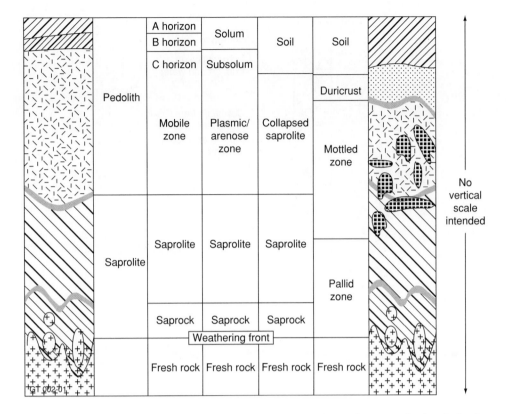

Figure 16.3   Some of the common terms used to describe the various weathering facies in an *in situ* regolith profile

containing components from sediment delivered by mud/land slides, gravity deposits from free faces above, colluvial material and sheet wash, so it may contain debris that is very coarse-grained to fine sands and mud.

Soils on the lower slopes can be complex. Where deposition is slow they can be well developed (e.g. red podzolic), but where deposition is more rapid soils may be minimally developed. Where coarse-grained materials form the regolith, no soil or lithosols occur. Paleosols are common as formerly extant soils are buried by new deposition as the locus of deposition moves across the slope. This is particularly common in alluvial fans deposited in this part of the landscape. Soils toward the top of this zone tend to be redder than those at its lower limits which are often yellow-coloured. Figure 16.4 shows a typical regolith profile formed on the convex lower slopes over a granitic bedrock.

Along valley axes there is generally relatively flat terrain or terraces that merge up the side slopes into the slope mantles. The nature of valley deposits and alluvium depends on many factors

but here we will just consider relative distance down the valley. In the higher reaches the alluvial plains will contain more gravel with some sand and further downstream the amount of sand increases, as does mud, and if the stream reaches low enough gradients sand and mud will dominate with little or no gravel present.

Because valleys form where there are rocks that have a lower resistance to weathering and hills where more resistant rocks occur the erosion of the valley may remove regolith that predates it. Sometimes remnants of the original weathering profile may be preserved below the alluvium (Figure 16.5). Aspandiar et al. (1997) discuss this with regard to the evolution of ancient regolith in north Queensland.

The nature of soils in valley situations depends on the nature of the alluvium and distance from the locus of sedimentation. In steep tracts lithosols are common in gravelly sediments but in more fine-grained sediments the soils may be well developed marginal to the major locus of contemporary sedimentation but very minimal close to the river where alluvial soils will occur.

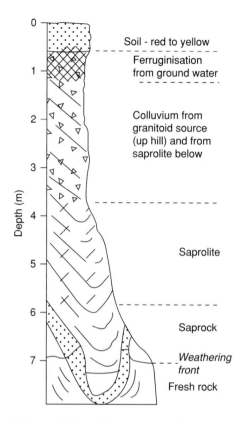

Figure 16.4 A typical regolith profile from the lower slopes of a regolith formed over- and downslope from a granitic parent material

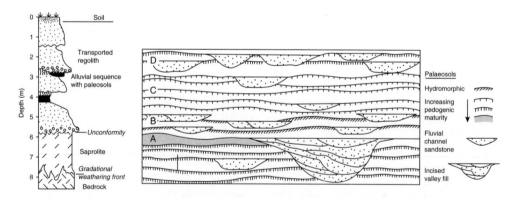

Figure 16.5 Left: Profile of regolith in a valley regolith sequence. Note the stripped regolith below the alluvium and the paleosols in the alluvial sequence. Right: A cross-section through an alluvial valley fill showing the distribution of channel deposits and overbank deposits as well as soils that may get preserved as the sequence accumulates and the locus of deposition shifts across the valley. A to D refer to progressively fining alluvial cycles as the valley fills and regional gradients decrease resulting in finer-grained deposits overall

REGOLITH AND CLIMATE

Climate is often viewed as a major determinant of the nature of the regolith (e.g. Butt and Zeegers, 1992). There is no doubt that the amount of available water has a major impact on regolith because of its ability to chemically weather rocks and transport solutes as well as in erosion to remove and deposit the physical particles produced by all weathering. But, the amount of water needs to be viewed over time. Where the regolith is old, then the total water available may well be much greater than that which has had time to work on newly exposed surfaces of rock. Also climate does not necessarily determine the total volume of water moving through rocks and causing weathering, groundwater may also be important in certain circumstances. So even in arid environments weathering may proceed, albeit more slowly than in wet ones, but if the regolith is stable (not being eroded) it may develop significantly over long periods. An example of this is the formation of gibbsite [$Al(OH)_3$] in basalt weathering under rainfalls of around 500 mm near Cooma in southern New South Wales where annual maximum temperatures are 20°C and minima are 4°C (Moore, 1996). Gibbsite also forms readily at Weipa in the far north of Queensland (Eggleton et al., 2008) where it is mined as bauxite, but here it forms under contemporary climates of >2000 mm of strongly seasonal rainfall and average annual temperatures of 33°C. This suggests that climate is perhaps not as important as many believe, but the amounts of time-integrated water are important.

High temperatures certainly increase chemical reaction rates, but does this translate directly to understanding rates of chemical weathering and regolith formation? Taylor et al. (1992) and Bird and Chivas (1993) show clearly that deep weathering occurred across parts of Australia under cool wet climates. Ambient temperatures only affect the regolith for a few centimetres depth, longer-term seasonal temperatures impact to about 10 cm and long-term climates that vary over 10^3 year scale perhaps a depth of several metres. Much of the chemical activity that forms regolith materials occurs at some depth below the surface and as a consequence it is much less influenced by surface temperatures and the surface climatic data we have.

Rainfall seasonality is another climatic variable of importance in regolith formation. Strongly seasonal climates provide water over a short period and this enables more weathering during the wet season and provides high water tables with fresh water. During this time elements such as Fe can be moved but as water tables fall during the dry season the Fe becomes oxidized and fixed within the regolith. Over longer cycles associated with long-term climates, changes of this nature can be formed also during periods when water tables fall, but with the rising water tables of wetter climates previously formed, regolith features such as ferruginization can be obliterated (e.g. Taylor and Eggleton, 2002, Figure 15.10, pp. 318–319).

REGOLITH AND TECTONICS

Uplift and subsidence impact mostly on landscape processes by basically causing either incision of drainage or accumulation of sediment/regolith. On incision by rivers hillsides are liable to be stripped of regolith or, if incision is relatively rapid, bare slopes and free-faces are all that remain (e.g. Grand Canyon). Episodic uplift (or climate-induced stream discharges) can leave terraces in valley bottoms and sides usually composed of alluvial regolith.

SOME SPECIAL REGOLITH FEATURES

Transported or in situ?

It is not always easy to determine whether a particular regolith or part thereof is *in situ* or transported. In many cases regolith profiles are composite and contain a lower *in situ* component and an upper transported component. Profiles that are entirely *in situ* are rare.

It is not always possible in the field to demonstrate whether a profile is *in situ* or transported, but requires mineralogical or geochemical analysis of the profile to establish whether there are any disconformities in vertical trends of mineralogy or chemistry. If present these indicate the presence of both transported and *in situ* regolith (e.g. Taylor and Eggleton, 2008a, 2008b) (Figure 16.6). In these cases profiles must extend to at least the depth of saprock so the nature of the bedrock and regolith can be seen to vary gradually without sharp break in mineral or chemical trends.

Duricrusts

Duricrusts are 'regolith material indurated by a cement, or the cement only, occurring at or near the surface or as a layer in the upper part of the regolith'. (Eggleton, 2001). There is a large variety of duricrusts generally named for their

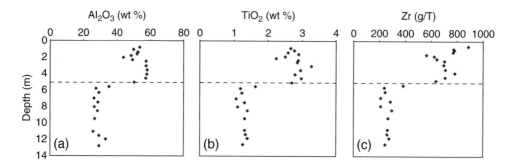

Figure 16.6   Change in amount of immobile elements across the Mottled Zone–Bauxite boundary (dashed line) at the Jacaranda pit, Andoom near Weipa in far northern Queensland. (a) Al$_2$O$_3$, (b) TiO$_2$, (c) Zr. (From Taylor and Eggleton, 2008)

cementing material: ferricrete has a ferruginous cement; silcrete a silica one, calcrete a carbonate one and a manganocrete a manganese cement.

Duricrusts are relatively common in regions of dynamically stable landscapes globally. There is a huge literature on duricrusts of various types, especially ferricretes as the Fe minerals adsorb many elements, some of which are very useful in geochemical exploration. Indeed, reworked ferricretes for some of the world's largest iron-ore deposits and are known to host secondary gold deposits (Butt and Zeegers, 1992; Taylor and Eggleton, 2001). Most are thought to form by the precipitation of the cementing mineral near the surface where geochemical conditions favour it, often at springs or seeps, sometimes at the top of an oscillating watertable and more rarely by replacement of regolith by the cement.

Ferricrete (laterite to some researchers) is formed by cementation of the regolith by Fe-oxides and oxyhydroxides. Theories on their genesis are almost as widespread as the number of researchers who work on them. To canvas all the ideas here would take too long, but good summaries can be found in Butt and Zeegers (1992), Bourman (1995), Taylor and Eggleton 2002, Ollier and Pain (1996) and Tardy (1993).

Calcrete is a druicrust cemented by carbonate-mineral cements, most commonly calcite, aragonite and Mg-calcite with some minor dolomite also precipitating in some calcretes. Their occurrence is usually restricted to arid and semi-arid climates as these carbonate cements are readily soluble and cannot persist in more humid climates.

The detail of their origin is also a matter of some dispute between researchers, but there is no shortage of CO$_3^{-}$ or Ca^{2+} or Mg^{2+} in most groundwater, particularly in more arid climates. Precipitation is usually due to either evaporation

at the groundwater surface or because of an increase in pH causing precipitation. The form of calcrete developed is generally either layered or nodular. It is hypothesised that nodules form in the pedogenic zone whilst layered calcretes are from the watertable evaporation (Plate 8, see the colour plate section, page 593).

Silcrete is as enigmatic as the duricrusts. They are regolith depleted of Al, enriched in Ti and cemented by silica in one of its various forms. Most are extremely hard and tough, white to grey in colour unless they contain some Fe in which care they are shades of red. They are widespread in Australia, Africa, southern Europe and England. They have been used by natives in Australia and Africa to manufacture tools and in Europe and England (where they are known as Saracen stones) by ancient and more modern civiliations to build memorials and buildings.

They have again been studied by many researchers with good modern summaries provided by Taylor and Eggleton (2001) and Thiry (1997). They are almost certainly precipitated in regolith material from soils or groundwater saturated in SiO$_2$. This is well demonstrated by Thiry and Milnes (1991) and by McNally and Wilson (1995) (Figure 16.7). McNally and Wilson show that the silcrete along the former Mirackina Channel was formed along the margins of the channel where water evaporated cementing the sediments of the channel. Subsequent erosion has caused relief inversion leaving hills where silcrete capped the channels sediment.

There are a number of features of silcrete that are not understood. The mechanism for the removal of Al from the regolith prior to the cementation with silica is unknown as Al is very insoluble, certainly much less so than silica. The meaning of the presence, almost ubiquitously, of geopetally arranged anatase is also unknown,

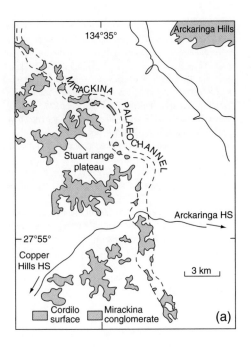

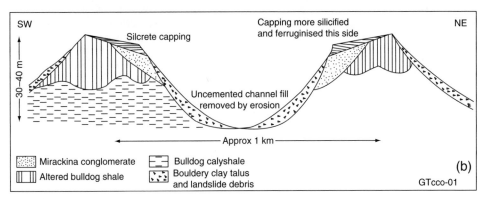

Figure 16.7 The evolution of the silcrete bounded Mirackina Palaeochannel in central Australia. (From McNally and Wilson, 1995)

and what determines whether the cement is microcrystalline quartz or quartz overgrowths on the regolith quartz grains is also a problem, although the former is a much more common cement.

ACKNOWLEDGEMENT

I would like to thank Ken Gregory for the effort he has put in to reduce my original manuscript to the size you now see.

REFERENCES

Aspandiar, M.F., Eggleton, R.A., Orr, T., van Eck, M. and Taylor, G. (1997) An understanding of regolith and landscape evolution as an aid to mineral exploration – the Charters Towers experience. *Resourcing the 21st Century.* Australian Institute of Mining and Metallurgy 1997 Annual Conference, Ballarat. 25–129.

Bird, M.I. and Chivas, A.R. (1993) Geomorphic and palaeoclimatic implications of an oxygen-isotope chronology for Australian deeply weathered profiles. *Australian Journal of Earth Sciences* 40, 345–58.

Bourman, R.P. (1995) A review of laterite studies in southern South Australia. *Transactions of the Royal Society of South Australia* 119, 1–28.

Butt, C.R.M. and Zeegers, H. (1992) Climate, geomorphological environment and geochemical dispersion models. In C.R.M. Butt and H. Zeegers (eds), *Regolith Exploration in Tropical and Subtropical Terrains.* Elsevier, Amsterdam, 3–24.

Chartres, C.J. and Walker, P.H. (1988) The effect of aeolian accessions on soil development on granitic rocks in south-eastern Australia; III, Micromorphological and geochemical evidence of weathering and soil development. *Australian Journal of Soil Research* 26, 33–53.

Eggleton, R.A. (2001) *The Regolith Glossary.* CRCLEME Perth, p. 144.

Eggleton, R.A. and Taylor, G. (2008a) Effects of some macrobiota on the Weipa Bauxite, Northern Australia. *Australian Journal of Earth Sciences* 55(Suppl), S71–S82.

Eggleton R.A. and Taylor G. (2008b) Impact of fire on the Weipa Bauxite, Northern Australia. *Australian Journal of Earth Sciences* 55(Suppl) S83–6.

Eggleton, R.A., Taylor, G., Le Gleuher, M., Foster, LD., Tilley, D.B. and Morgan C.M. (2008) Regolith profile mineralogy and geochemistry of the Weipa Bauxite, northern Australia. *Australian Journal of Earth Sciences* 55(Suppl), S17–S44.

Firman J.B. (1994) Paleosols in laterite and silcrete profiles; evidence from the south east margin of the Australian Precambrian shield. *Earth-Science Reviews* 36, 149–79.

McNally, G.H. and Wilson, I.R. (1995) Silcretes of the Mirackina Palaeochannel, Arckaringa, South Australia. *Australian Geological Survey Organisation Journal of Australian Geology and Geophysics* 16, 295–301.

Merrill, G.P. (1897) *A Treatise on Rocks, Rock Weathering and Soils.* Macmillan, New York. p. 411.

Mizota, C and Takahashi, Y. (1977) Eolian origin of quartz and mica in soils developed on basalts in north-western Kyushu and San-in, Japan. *Soil Science and Plant Nutrition* 28, 369–78.

Moore, C.L. (1996) Processes of chemical weathering of selected Cainozoic Eastern Australian basalts. PhD thesis, Australian National University, Canberra.

Ollier, C.D. and Pain, C.F. (1996) *Regolith, Soils and Landforms.* John Wiley and Sons, Chichester. p. 316.

Tardy, Y. (1993) *Petrologie des laterites et des sols tropicaux.* Masson, Paris. p. 459.

Taylor, G. and Eggleton R.A. (2001) *Regolith Geology and Geomorphology.* John Wiley and Sons, Chichester. p. 375.

Taylor, G. and Eggleton, R.A. (2008) Genesis of pisoliths and of the Weipa Bauxite deposit, northern Queensland. *Australian Journal of Earth Sciences* 55(Suppl), S87–S103.

Taylor, G., Truswell, E.M., Eggleton, R.A. and Musgrave, R. (1990) Cool climate bauxites. *Chemical Geology* 84, 183–4.

Taylor, G., Eggleton, R.A., Holzhauer, C.C., Maconachiee, L.L., Gordon, M., Brown, M.C. and McQueen, K.G. (1992) Cool climate lateritic and bauxitic weathreing. *Journal of Geology* 100, 669–77.

Thiry, M. (1997) Continental silicifications: A review. In H. Paquet and N. Clauer (eds) *Soils and Sediments: Mineralogy and Geochemistry.* Springer, Berlin. 191–222.

Thiry, M. and Milnes, A.R. (1991) Pedogenic and groundwater silcretes at Stuart Creek Opal Field, South Australia. *Journal of Sedimentary Petrology* 61, 111–27.

Wright, V.P. and Tucker, M.E. (1991) Calcretes: an introduction. In V.P. Wright and M.E. Tucker (eds), *Calcretes.* International Association of Sedimentologists Reprint Series, 2, 1–22.

17

Rock Surface Weathering: Process and Form

David A. Robinson and Cherith A. Moses

Weathering processes involve a range of physical, chemical and biological processes that generally weaken and break down rock materials, although some may, in the short term at least, strengthen rocks through, for example, the development of surface weathering crusts or rinds (Day, 1980; Conca and Rossman, 1982, 1985; Robinson and Williams, 1987). Although it is possible to identify distinct weathering processes, in most natural environments rocks are attacked simultaneously by more than one process often acting synergistically.

Weathering processes are fundamental to geomorphology. Occuring *in situ*, they release compounds in solution and prepare rock materials for removal and transportation by erosion and are, in themselves, directly responsible for the creation of characteristic landform features and landform evolution. Weathering processes are fundamental also for the creation of soils, determining many of their physical and chemical properties. Because weathering influences soil fertility, slope stability and the durability of engineering structures and building materials, it is a multidisciplinary field of study and there are varied approaches to research and the advancement of knowledge (e.g. Birkeland, 1984; Ollier, 1984; Ollier and Pain, 1996; Bland and Rolls, 1998; Whalley and Turkington, 2001; Lee and Fookes, 2005).

This review focuses on the weathering processes that act on exposed rock surfaces, on the retreat of rock surfaces and on the evolution and development of the resulting surface weathering features. The emphasis is on surface and immediate sub-surface change rather than on the geochemical alteration of minerals at depth as a result of chemical disequilibria between the conditions under which many minerals are formed and those encountered at Earth's surface (see Chapter 16).

Early interest in rock surface weathering focused on the description and documentation of weathering features, and to a lesser extent the products of weathering. This was particularly widespread from the early 1800s when scientists began to explore and map areas of the world very different from their home environments (Turkington and Paradise, 2005). Description of weathering features was accompanied by speculation as to their origin, often through deductive reasoning from field evidence. This remains a significant approach in weathering research, especially for the explanation of large-scale weathering features and landforms (Twidale, 1982; Young and Young, 1992; Robinson and Williams, 1994).

Attempts to understand the origin of features led to experimental simulations of weathering designed to test the efficacy of suggested processes. Salt, often very visible on the surface of disintegrating rocks, was the first, with the simulation of salt weathering attempted in 1828, followed later by tests of insolation and wetting and drying (Turkington and Paradise, 2005). Experimental weathering has become a major research field, greatly aided by continually improving environmental control and monitoring equipment linked to greatly enhanced tools for studying the results. For example, the application of electron microscopy has helped to advance understanding of the operation of biological weathering processes (Moses and Smith, 1993; Viles, 1995; Seaward, 1997; Robinson and Williams, 2000a; Carter and Viles, 2004, 2005). Significant developments in experimental weathering have resulted from studies of the durability of building stones (Schaffer, 1932; Smith et al., 2008) as well as geomorphology. Partly as a result of this, but also because of the distinctive suites of landforms often associated with them, weathering research has tended to focus on a limited range of rocks. Limestones, sandstones, gritstones and granites, have all been studied intensively, shales, schists and some other igneous rocks rather more rarely whilst many other rock types have been largely ignored.

Because rock surface weathering is strongly influenced by atmospheric conditions, especially moisture and temperature regimes, weathering studies frequently involve detailed climate monitoring, often at the micro-climatic scale, and climate changes impact significantly on weathering rates and processes. As a consequence, in many parts of the world weathered surfaces and products are polygenetic in origin making it potentially difficult to differentiate the impacts and results of present day processes from those of the past. The influence of Quaternary climatic changes are particularly widespread and influential. On shorter timescales, impacts associated with anthropogenic induced climate changes and atmospheric-pollution are also important and have been a significant focus of much recent research (Brimblecombe and Grossi, 2008, 2009; Smith et al., 2008) and conversely, the contribution of weathering to the carbon flux and temperature changes (Goudie, 2004a).

SURFACE WEATHERING PROCESSES

The most common approach to investigating rock weathering processes has been to study individual processes, often on more than one rock type or in contrasting environmental conditions, in order to minimize the number of variables that may influence the results. In this way, significant advances have been made in understanding particular process mechanisms.

Thermal shock and stress

Thermal shock is the impact of temperature changes on rocks which cause stresses sufficient to weaken and break down rocks or their constituent minerals. Historically, the process has most frequently been associated with the large diurnal variations in temperature characteristic of desert environments where cloudless skies result in very high levels of insolation during the day and radiation results in rapid cooling at night. However, it has been shown also that significant diurnal temperature changes occur in other climatic environments, most notably in polar regions (Hall, 1999; Hall and Andre, 2001), and that similar rates of rapid temperature change can occur in cold as in hot deserts, despite the large difference in the surface temperature of rocks in the two environments (Hall and Hall, 1991; McKay et al., 2009). In tropical and temperate environments, significant temperature changes can occur during the day as a result of shading by passing clouds (Jenkins and Smith, 1990). Additionally, nearly all rock faces vary in their direct exposure to insolation during the course of a day, with many being in the shade for 50% or more of daylight hours. Clasts on alluvial fans have been shown to have a distribution of crack orientations related to differential levels of insolation (Moores et al., 2008). Exposure to cooling winds has been shown also to cause significant differential chilling to the surface of masonry stone (Jenkins and Smith, 1990). Thermal stress is likely to be particularly large in course-grained igneous rocks composed of minerals of different colours and thermal characteristics (Hall and André, 2003; Gomez-Heras et al., 2006; Hall et al., 2008b). Likewise they may be large where rocks are differentially heated or cooled because of varying levels of insolation, shade and shadow or where varying levels of soiling darken the rock surface (Warke et al., 1996).

Doubts over the efficacy of weathering by temperature changes were raised by early experiments by Blackwelder (1933) and Griggs (1936) who found that disintegration occurred only when water was present in the cooling phase. This suggested that perhaps swelling and contraction due to wetting and drying or through chemical changes such as hydration were the real causes, but it is now recognized that the early experiments had serious limitations (Rice, 1976). Rates of temperature change on rock surfaces can exceed the critical threshold of $2°C$ min^{-1} at which ceramics develop cracks, reaching $> 8°C$ min^{-1} (Goudie, 2004b; McKay et al., 2009) and experiments have shown that fracture patterns similar to those found on rocks in cold regions can be simulated in the laboratory (Hall, 1999).

A particular form of thermal shock weathering occurs during wildfires which are a common occurrence in many seasonally dry environments (Blackwelder, 1927). During fires, rocks are frequently subjected to temperatures $> 500°C$ and

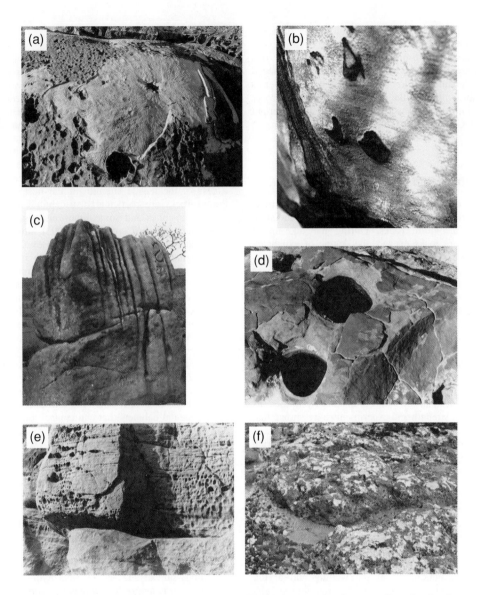

Figure 17.1   (a) Fire-weathered boulder Lawn Hill N.P. Queensland, Australia. The fire has resulted in explosive exfoliation of the exterior of the boulder. (b) Granular disintegration of a weakened subsurface layer exposed by the loss of a crust by surface scaling, Sussex, UK. (c) Pseudo-rillenkarren on an upstanding mass of gritstone, Derbyshire, UK. (d) Weather pits developed in a polygonally cracked sandstone, High Atlas, Morocco. Note the loss of some of the cracked outer crust. (e) Alveolar weathering in a sandstone, Sussex, UK. (f) Spalled upper surface of a chalk shore platform as a result of frost action, Sussex, UK

possibly > 1000°C (Figure 17.1a) and experiments have shown that such high temperatures can significantly damage rocks although the impact varies with rock type, boulder size and water content (Goudie et al., 1992; Allison and Goudie, 1994). The high temperatures achieved can evaporate water within rocks to steam which may not only stress rocks but also accelerate chemical weathering processes. This may help to explain the fact that fire appears to accelerate subsequent rates of weathering by other mechanisms such as salt and frost (Dorn, 2003; McCabe et al., 2007a,b).

Experiments also show that the impact of fire is different to simply heating rocks in a furnace (Warke and Smith, 1998). Exactly why this should be is uncertain but may relate to the speed of heating by fire and the high stresses this induces, though the deposition of soot and waxy residues on rock surfaces may also be important (McCabe et al., 2007a).

Wetting and drying

Exposure of rocks to atmospheric conditions results in significant fluctuations in moisture content within rocks. During the take-up of moisture rocks expand, the degree of expansion varying according to mineral composition, porosity and cementation. On subsequent drying rocks tend to contract, but often to a lesser degree than the expansion. Some rocks contain minerals which hydrate and once hydrated do not easily give up all the water they have absorbed. Others with surface-negative charges may hold onto water molecules with quasi-crystalline bonds. The frequency and intensity of wetting and drying are greatest on and in the near surface, decreasing into the rock creating differential stresses between the layers. Expansion is accompanied by reductions in rock strength and experiments have shown that this can lead to rock breakdown (Pissart and Lautridou, 1984; Hall and Hall, 1996). Clay-rich mudstones, shales and some limestones seem particularly vulnerable to weathering through wetting and drying although it is often difficult to discriminate wetting and drying from salt weathering as water in rocks is very rarely free of salts and many rocks that weather rapidly under wetting and drying often contain minerals such as gypsum (Cantón et al., 2001; Goudie, 2004c). There is considerable discussion of the relative importance of the wetting and drying as opposed to salt on the weathering and downwearing of shore platforms, which recent measurements have been shown to expand and contract considerably during tidal wetting and drying cycles (Stephenson and Kirk, 2001; Kanyana and Trenhaile, 2005; Trenhaile, 2006; Porter and Trenhaile, 2007).

Wetting and drying tends to increase rock porosity and water absorption capacity of rocks. Even in the absence of breakdown this may make rocks more susceptible to breakdown by other processes such as frost or salt weathering. Sumner and Loubser (2008) found no significant difference in weathering by wetting and drying with the level of saturation achieved and suggested that the form of wetting might be more important than the level of saturation.

Salt weathering

It has long been recognized that the growth of salt crystals within and on the surface of rocks can cause rock breakdown and disintegration (Evans, 1970; Doehne, 2002). Salt crystals initially develop and grow when a salt solution either evaporates or its temperature is reduced. Subsequently they may expand due to hydration or changes in temperature. A vast array of salts occurs within rocks (Goudie, 1974; Goudie and Viles, 1997), many of which can exert forces in excess of the tensile strength of the host rock. Porous rocks are particularly susceptible, especially sandstones and some limestones, igneous rocks much less so (Cardell et al., 2003), although significant damage can still occur (Ulusoy, 2007). Salt weathering is particularly common in arid environments where salts tend to accumulate because they are rarely flushed away and in coastal environments where there is a plentiful supply of salts. However, it is widespread also in more humid inland environments and especially in some urban environments where, due to atmospheric pollution and the widespread use of road de-icing chemicals many building stones are very susceptible to damage.

Salt weathering has been the subject of many experiments, especially since the introduction of environmental cabinets that can control temperature and humidity (Cooke, 1979). The efficacy of different salts varies significantly in ways that remain only partly understood. Variations in the solubility of different salts that occur within many rocks appear particularly important. Sodium sulfate, carbonate and nitrate, and magnesium sulfate, for example, all vary considerably in solubility and in particular, each decrease in solubility within the ambient temperature ranges that occur in many rocks (Goudie and Viles, 1997). Humidity is also important. Salts crystallize when saturated conditions prevail and several salts hydrate and dehydrate in response to changes in temperature and humidity. Hydration of some salts can occur very rapidly (Mortensen, 1933) and cause a doubling or trebling in crystal volume. The resulting pressures can be very considerable and are sufficient to break rocks (Winkler and Wilhelm, 1970). Goudie (2004c) estimates that in desert environments diurnal temperature and humidity changes suitable for hydration of salts such as sodium sulphate and sodium carbonate may occur on 150–270 days y^{-1}.

Sodium sulfate appears to be a particularly destructive salt (Goudie and Viles, 1997; Goudie, 2004c). It is highly soluble, suffers phase changes over commonly occurring ambient rock temperatures that involve large changes in volume and in addition has a needle-shaped crystal form that might also be important in its destructiveness. In contrast, experimental evidence suggests that calcium sulphate (gypsum) is very ineffective at causing rock breakdown as a result of crystallization within the pores of rocks (Goudie, 1974), most probably because of its low solubility. However, in polluted urban environments where it is a very widely occurring salt, crystallization of gypsum from the chemical reaction between atmospheric sulphur dioxide and calcium in limestones is believed to be the cause of widespread surface blistering and weathering of many building stones.

Most experiments have studied the weathering effects of salts independently of each other. However, in nature, most rocks contain more than one salt and, in some cases at least, combinations of two or more salts seems to increase the severity of damage over that done by any of the salts independently (Robinson and Williams, 2000b; Turkington and Smith, 2000; Warke and Smith, 2000). The study of the impacts on weathering rates of interactions between different salts is an area that needs much further study.

Although much salt weathering is believed to result from crystal growth, expansion and contraction of crystals due to temperature changes within rocks may also be significant (Sperling and Cooke, 1980). This is because some commonly occurring salts, such as halite and gypsum, have coefficients of expansion significantly greater than those of the minerals of many rocks in which they occur. However, experimental support for the importance of this process remains limited (Goudie 1974, 2004c).

There is a wealth of experimental evidence of the efficacy of salt weathering but the precise role of the various processes in the creation of weathering features remains elusive. This is largely because the scale of most experiments is too small to adequately model the impacts of salt weathering processes on actual rock masses and rock faces found in nature. However, salt weathering is generally thought to result in granular disintegration, spalling and flaking of rocks and create various alveolar or cavernous weathering forms, although the precise mechanisms whereby salt weathering creates such forms remain elusive.

Frost weathering

In climates where frosts occur, weathering by freeze–thaw is a widespread phenomenon (see Chapters 23 and 24) that occasionally produces strikingly visible levels of rock breakage (Robinson and Jerwood, 1987a,b). Traditionally, frost weathering was attributed to the 9 per cent

expansion in volume that accompanies the phase change of liquid water to ice. This can theoretically generate pressures of 207 MPa which far exceeds the maximum tensile strength of many rocks. However, the confined conditions necessary for this expansion to exert sufficient stress within rocks to cause extensive breakage to occur is probably rare. In recent years there has been increased scepticism as to the efficacy of the explanation for all but small scale surface flaking and grain detachment from rocks undergoing rapid freezing. Hydrofracturing, caused by high water pressures developed during freezing, may also contribute to frost weathering particularly of isolated boulders and some rock faces (Matsuoka and Murton, 2008).

The ice segregation model of frost weathering has been developed from the fundamental principles of the thermodynamics of water in porous media (Walder and Hallet, 1985) and relates to the migration of unfrozen water towards centres of ice formation and the growth of ice lenses (Walder and Hallet, 1986). This results from suction pressures that develop in the unfrozen water within rock pores as freezing removes liquid close to where ice crystallization occurs. A growing body of experimental evidence supports the fracture of porous rocks by ice lens growth (Hallet et al., 1991; Murton et al., 2000, 2001) under a variety of conditions. Experiments indicate that for individual blocks of rock, freezing intensity is less important than the number of freeze–thaw cycles and the rate of freezing. Ice segregation seems to occur most effectively under relatively mild freezing conditions between −3 and −8°C (Hallet et al., 1991). However, field evidence suggests freezing intensity and duration may be important, especially in the weathering of bedrock masses, probably because the penetration depth of cracking by frost increases with decreasing temperatures (Anderson, 1998).

The impact of the duration of freezing varies according to whether there is an open or closed system within a rock mass. If the system is closed and no additional water can enter, then once freezing is complete no further damage will occur until melting commences. In contrast where the system is open to the ingress of water, the volume of ice will continue to expand and cause further damage. This is the main explanation, for example, of why downward melting of ice in spring and summer in periglacial regions is associated with ice growth, ground heave and rock breakdown (Matsuoka and Murton, 2008).

The direction of freezing is also important and seems, along with rock type, to determine the form of the resultant breakdown. Freezing from all directions, as can occur on upstanding boulders, can cause multiple directions of fracturing. In contrast, uni-directional freezing of bedrock tends to produce linear cracking parallel to the direction of freezing. Under frozen ground conditions considerable ice growth may occur during summer thawing when water is supplied to the freezing front above the perennially frozen ground by meltwater from above. This has been successfully simulated by Murton et al. (2000, 2001) and shown to produce fracture patterns very similar to those found in both present and past periglacial regions.

Frost weathering can have significant impacts on porous rock used as building stone. In temperate environments freeze–thaw cycles that occur infrequently only during intense cold periods may be very important in opening pores and fractures that are then exploited by salt weathering events associated with far more frequent wetting and drying cycles (Warke et al., 2004, 2006). This is just one example of the importance of a rock's environmental past to its response to the present day weathering environment in which it is found. In the context of building stone research, this 'memory' or 'inheritance effect' has been shown to significantly vary the subsequent response of rocks to identical later weathering conditions (McCabe et al., 2007a,b).

Experimental studies of frost weathering have focused on soft porous rocks such as chalk and sandstones. There is a paucity of work on hard rock and in particular on the impact of frost on intact masses of hard rock with low porosity. Measured damage on such rocks has been restricted to very small changes in physical characteristics such as Young's modulus (Goudie et al., 1992; Allison and Goudie, 1994). A key question that remains to be answered is whether frost weathering of such rocks can actually cause breakdown of intact rock or whether damage to such rocks is restricted to crack and joint systems, including microcracks and fissures (Matsuoka and Murton, 2008).

Frost and salt weathering

In cold environments salts and frost can act together to cause damage much greater than either of the processes acting separately (Williams and Robinson, 1981, 2001), This can be particularly destructive of some shore platforms (Figure 17.1f) (Trenhaile and Rudakas, 1981; Robinson and Jerwood, 1987a,b). However, experiments have shown also that the presence of salts can sometimes reduce weathering by frost (McGreevy, 1982). The salt or salts present, their concentration and the severity of freezing seem to be critical

controls, because of variation in the depression of freezing point caused by different salts and the complex crystallization behaviour of salts and water at sub-zero temperatures. Sodium chloride, for example, is less destructive than sodium sulphate at temperatures above zero, but can be much more destructive under sub-zero conditions. More work is required to elucidate the weathering importance and precise mechanisms of these combined processes at sub-zero temperatures (Williams and Robinson, 1991, 2001).

Solution weathering

Of the range of chemical weathering processes, solutional weathering is held responsible for the formation of particular small-scale features, medium-scale landforms and whole landscapes. There is a long history of interest in the operation of solution based upon field evidence of its effects. These are most evident on soluble, monominerallic rock types, particularly carbonates (limestones, marble and dolomite) and evaporites (gypsum, anhydrite and salt) and are associated with the development of karst landscapes (Chapter 28) (Ford and Lunberg, 1987). On sandstones, karst-like features and landforms, known as pseudokarst, are produced by a combination of solutional preparation, followed by physical erosion of the residual material. This is usually loose sand in the case of quartz sandstones, but can also include clay minerals (Grimes, 1997; Wray, 1997). Solution also occurs, in combination with a range of other chemical weathering processes, on igneous and metamorphic rocks where it contributes to deep weathering and regolith and soil development (Chapter 18) but can also exhibit small scale surface weathering features such as solution basins (Emery, 1946; Campbell and Twidale, 1995; Twidale and Vidal-Romani, 2005). Evidence of solution is most often linked with particular landforms and associated small-scale features and the chemistry is well understood (Sweeting, 1972; Jennings, 1985; Trudgill, 1985; Ford and Williams, 2007).

There is increasing interest in the influence of biological processes on surface solution (Fornós and Ginés, 1996; Smith et al., 2000; Duane, 2006). Because of the influence of temperature and carbon dioxide on solution rates, there has been much interest in regional and global variations and the influence of climate. Such variations have been measured indirectly by analysis of the carbonate content of natural waters (e.g. Smith and Atkinson, 1976; Gunn, 1981) and rock exposure experiments (e.g. Gams, 1985; Reddy, 1988; Moses, 1996) and directly using microerosion meters and laser scanners (e.g. Smith et al., 1995, Trudgill et al., 2001; Swantesson, 2005). More recently, linkages between solution weathering, global carbon dioxide levels and climate change have been investigated (e.g. Kump et al., 2000; Liu and Zhao, 2000; Gombert, 2002; Navarre-Sitchler and Thyne, 2007). There is also a growing interest in the potential consequences of predicted climate change on rates of solution, particularly with respect to the potential impacts on buildings constructed from carbonate stones (Viles, 2002; Thornbush and Viles, 2007; Brimblecombe and Grossi, 2008, 2009; Grossi et al., 2008; Bonazza et al., 2009).

Advances in microscopy techniques have helped to facilitate an enhanced understanding of the mechanisms of solutional weathering as evidenced by high magnification views of rock and individual mineral surfaces. Individual grains may be removed from rock surfaces by a process known as solutional disintegration when the surrounding cement is more readily dissolved (Moses et al., 1995). A range of nanomorphologies are associated with the surfaces of individual minerals (Moses et al., 1995; Lee et al., 1998; Viles and Moses, 1998; Thornbush and Viles, 2007; Ruiz-Agudo et al., 2009). High magnification microscopy has also facilitated insights into the role of biological weathering processes in the genesis of features that have been traditionally ascribed entirely to solution such, for example solution basins (e.g. Moses and Smith, 1993) and rillenkarren (e.g. Fiol et al., 1996). One of the difficulties encountered has been in understanding the contribution that nanomorphologies make to the formation of the larger scale features associated with natural rock outcrops.

Biological weathering

Biologically mediated rock breakdown is associated with microorganisms, higher plants and, less directly, animals (Chapter 14) (Viles et al., 2008). Interest in biological weathering has focused on biophysical and biochemical processes associated with microorganisms. It is an area of research that has benefited greatly from advances in microscopy and from collaborations between biologists and geomorphologists. Biologists have largely been concerned with studying the metabolic activities of microorganisms whilst geomorphologists are more interested in their impact on the underlying rock. There is still much to be learned about the interaction of microorganisms with different rock types in a range of environments; in particular: understanding what happens at the interface between microorganisms and mineral surfaces, examining the interaction of biological with other weathering processes and evaluating the contribution that biological processes make to the evolution of rock surfaces and landscapes over time.

Lithobiontic communities occupy a range of niches on and within rocks. Epiliths live entirely on the surface, euendoliths actively bore tunnels into the rock, chasmoendoliths inhabit fissures and cracks in the work and cryptoendoliths live in the pore spaces of the rock (Golubic et al., 1981). The organisms involved include bacteria, fungi, archaea, algae and lichens. They are globally widespread and are influenced by the nature of the underlying substrate together with microenvironmental conditions, including oxygen and water availability, temperature, pH and nutrient availability (Brock et al., 1994; Viles, 1995). Lithobiontic communities consist of complex mixes of organisms and a variety of terms are used to refer to the community as a whole. Biofilms are surface films of microbial cells embedded in extracellular polymeric substances and are areas of high metabolic activity where the digestive enzymes excreted by microorganisms are concentrated. Not only do they interact with the minerals of the rock in which they live but they may also trap airborne particles such as aerosols, minerals and organic material. In the later stages of development they may contain small animals, such as mites, and lower and higher plants may develop (Gorbushina, 2007; Scheerer et al., 2009). They vary in thickness and are commonly referred to as biofilms, <~1 mm, biorinds, 1–5 mm and biocrusts, > 5 mm (Viles, 1995).

It is now well established that microorganisms can cause physical breakdown of rock surfaces as a result of their expansion and contraction, via the uptake of moisture, fatiguing the rock (Fry, 1927; Danin and Caneva, 1990; Moses and Smith, 1993). They also interact chemically, via their chemical exudates, causing solution of the rock surface and creating salts such as calcium oxalate. Studies of the interface between microorganisms and rock surfaces have been greatly assisted by advances in microscopy techniques and a wide range of studies have been conducted to examine the chemical interactions of bacteria, fungi, archaea, algae and lichens with individual rock types and minerals. Considerable progress has been made in understanding these interactions on natural rock outcrops (e.g. Viles, 1995; Chen et al., 2000) and on building stones and other forms of cultural heritage (e.g. Seaward, 1997; Warscheid and Braams, 2000; Gaylarde and Morton, 2002; Liscia et al., 2003; Crispim and Gaylarde, 2004; St. Clair and Seaward, 2004; Gaylarde et al., 2007; Scheerer et al., 2009).

Of the organisms that make up lithobiontic communities, the least well understood are bacteria and archaea (Scheerer et al., 2009).

The interaction of biological weathering with other weathering processes is a more recently evolving area of research that falls within the remit of biogeomorphology (Chapter 14) (Viles, 1988; Cox, 1989; Naylor et al., 2002; Corenblit et al., 2008). Of particular interest is the extent to which biofilms actively weather or protect rock surfaces, biodeterioration and bioprotection respectively (Carter and Viles, 2005). Bio-construction is also recognised (Naylor et al., 2002). Biophysical processes may assist chemical processes, such as solution, by plucking mineral grains from the surface (Moses and Smith, 1993). The decay of lichen thalli may expose a weakened rock surface to other processes such as salt weathering (Mottershead and Lucas, 2000a). The extent to which biofilms may protect rock surfaces, for example, from aggressive solution (Moses et al., 1995) or salt weathering (Mottershead et al., 2003; Mustoe, 2010) is far from being understood. It has been suggested that their contribution to weathering should be greatest in arid environments (Viles, 1995) but it seems this may not be the case (Stretch and Viles, 2002).

Field and laboratory experiments show that biofilms can also influence rock surface temperature fluctuations, affecting thermally controlled processes such as thermal fatigue and shock and frost action, as well as moisture uptake (Bjelland and Thorseth, 2002; Carter and Viles, 2003, 2004; Hall et al., 2005; Hall et al., 2008a,b). There has been much interest in biological weathering in the area of cultural heritage conservation. Most of this has been concerned with biodeterioration but there is a growing interest in the use of microorganisms not only to protect stonework (May, 2003) but also to bioremediate, that is, to ameliorate the effects of stone deterioration caused by other weathering processes (Webster and May, 2006).

A particular challenge lies in understanding the contribution that biological processes make to the evolution of rock surfaces over time. Folk et al. (1973) identified phytokarst as a unique medium-scale limestone landscape formed directly by cyanobacteria. Subsequent research indicates that the process–form relationship is more complex (Viles, 1995) and that a clear topographic influence is difficult to identify at anything above submetre scales (Deitrich and Perron, 2006). The role of microorganisms is thought to be not only specific to particular environments but also species specific (Carter and Viles, 2005). Some models of longer term, larger-scale landscape evolution neglect the impact of biological weathering (e.g. Phillips, 2005) but others include it (Viles et al., 2008).

WEATHERING IMPACTS ON ROCK SURFACES

Weathering may cause alterations to rock surfaces or result in the total breakdown and collapse of rock masses. The alterations result in a range of characteristic surface weathering features.

Weathering rinds, crusts and coating and sub-surface weakening

Weathering rinds, crusts and case-hardened layers are zones of chemical alteration on the outer portions of rocks (Gordon and Dorn, 2005). The alteration often results in a colour change and usually involves the loss or accretion of chemical constituents. They may develop entirely within the original rock or may involve also accretion on the outer surface when they are usually referred to as a rock coating. Chemicals included within a coating may be derived entirely from within the parent rock, or may include additions from external sources such as the atmosphere or drainage waters. Some may be derived entirely from external sources in which case they are correctly referred to as rock varnish and are not a true weathering feature.

Wetting and drying of rock surfaces is important in the development of weathering rinds, crusts and case-hardened layers. Penetration of water and atmospheric oxygen results in chemical alteration of minerals in rock surface layers. On limestones with high water absorption capacities for example, case hardening can develop to depths of more than 2 m (Day, 1980). Some highly mobile cations such as sodium and calcium and ferrous iron may be lost altogether resulting in the bleaching of depleted surface layers, but more commonly migration of water during drying carries the soluble products of chemical weathering towards the outer surface of rocks where they are deposited during drying. In some rocks, notably sandstones, this leaves behind a weak subsurface layer relatively depleted in iron and silica and a strengthened surface rind or crust enriched with these minerals. Deposition can be within the surface pores, as overgrowths on crystals in the surface layer or as a glaze on the outer surface. Crusts have been shown to help to waterproof the rock surface, which sheds rather than absorbs water (Robinson and Williams, 1989, 1992) and help to protect the surface from further weathering and erosion. They are denser and harder than the parent rock and the process is often referred to as case hardening (Day, 1980; Conca and Rossman, 1982). Fresh stone extracted from quarries for masonry use is known to harden within months as the so-called 'quarry sap' dries out.

Crust development is not always a continual process. Over time many crusts, particularly on sandstones, tend to deteriorate and when breached the weakened, poorly cemented sub-surface layer can be rapidly lost, undermining the crust which falls away as spalls. The cause of crust deterioration is poorly understood and requires further work but biological weathering processes by microorganisms may well be very important, particularly in the dissolution and mobility of silica.

Crusts and coatings may develop not only on rock surfaces exposed to the air but also along joint planes and at the rock–soil junction. These crusts are often unstable when exposed to full atmospheric conditions and begin to break down and be lost from the surface. Thus some rocks may exhibit more than one crust, some formed by sub-aerial weathering, others underground and in the process of destruction (Robinson and Williams, 1992).

Blistering, scaling, flaking and granular disintegration

Many rock surfaces exhibit a variety of surface spalling phenomena frequently referred to as scaling or flaking where thin layers of coherent rock, aligned parallel to the surface become loosened from the main body of the rock and eventually become detached and fall away. Scaling is the term generally used for the larger, thicker layers, ≥ 5 mm, flaking for smaller more fragile layers < 5 mm thick (Smith et al., 1994). Scaling is often the result of a weakened sub-surface, developed during case hardening, failing and as a result the crust becoming detached. Differences in porosity and density between the outer crust and the layers beneath will accentuate the build up of differential stresses during cycles of expansion and contraction due to wetting and drying or heating and cooling. Blistering and the surface detachment of scales and flakes can result also from salt weathering. The accumulation of salt within the surface pores can block the pore spaces and may cause internal pressures to increase, creating blisters which eventually burst and result in the detachment of the outer layer which falls away (Warke and Smith, 2000; Smith et al., 2002b). Flaking and scaling are particularly common on sandstones and many masonry stones. Extensive surface scaling can sometimes occur as a result of frost action, for example, on shore platforms (Robinson and Jerwood, 1987a, 1987b) on cliff faces and occasionally on masonary. Blistering is

a particular feature associated gypsum weathering of limestones in polluted urban environments (Smith et al., 1994).

Where the loss of a crust exposes a weakened sub-surface layer this frequently undergoes loss of individual crystals or groups of crystals by granular disintegration (Figure 17.1b) sometimes aided by salt or ice crystallization, until surface loss reaches the sound, unweakened rock below. Other surfaces may undergo granular disintegration due to the loss of cement by solutional processes.

Solutional sculpting

Surface features that form as the result of solutional sculpting are most evident on readily soluble rocks and are discussed in the section on solutional weathering. They are typically small-scale features, often referred to as karren (Sweeting, 1972) and are thought to form relatively quickly. Field tests suggest that rillenkarren may form over 10^{-3} years on limestone, 10^{-2} on gypsum and 10^{-1} years on salt (Mottershead and Lucas, 2000b). The development of karren may be fracture-controlled or hydraulically controlled and they are best developed on massively bedded, fine-grained and homogenous limestones (Ford and Williams, 2007). Vertical and steeply sloping surfaces that shed water quickly exhibit a range of solution rills whose development is thought to depend upon lithological constraints as well as runoff availability (Ford and Lundberg, 1987; Mottershead et al., 2000). Whilst rainfall intensity, drop size, water temperature and viscosity play significant roles in the growth of rillenkarren it has been suggested that their development may be assisted by biological weathering processes that mediate a physical removal of small particles of limestone (Fiol et al., 1996). Biological processes are thought to also play a key role in the development of karren, such as pits and pans, that form on horizontal and gently sloping surfaces (Fornos and Gines, 1996).

Alveoli, cavernous weathering and weather pits

Cavernous weathering is found in a wide variety of environments, including cold and humid regions (Goudie and Viles, 1997), but is particularly common in salt-rich locations such as deserts and coasts. The resulting weathering forms (Figure 17.1e) can vary in scale from massed hollows measured in tens of millimetres, usually termed alveolar or honeycomb weathering to large individual hollows known as tafoni that can reach several cubic metres in volume. Examples can be found at all sizes in between. They form in a variety of relatively massive rock types including granites, sandstones and limestones, but are rare or absent from rocks with closely spaced joints or discontinuities, such as schists and shales. Typically spherical or elliptical in form, the individual features often widen inwards and one of the continuing debates is how the debris is removed from within these 'cavernous' interiors to create the hollows. However, examples abound of weathering hollows across the entire spectrum of sizes that appear either to be in the process of formation and have not yet developed a 'cavernous' character or in decay with only the hollow remnant of the rear wall surviving.

It is widely agreed that cavernous weathering results from differential spalling, flaking and/or granular disintegration of the rock surface, but an active debate exists over the relative roles of chemical and physical weathering in creating the hollows, and especially in the creation of large tafoni. Weathering by salt crystallization is frequently indicated by the presence of salt crystals both on rock walls and in weathered debris. However, several workers suggest that chemical weathering, possibly enhanced in the presence of saline solutions may be at least as important, if not more important than physical effects (Martini, 1978; Young, 1987; Young and Young, 1992). Detailed examination of the microweathering features of sandstones displaying cavernous weathering found evidence of both physical and chemical weathering forms (Turkington, 1998).

One of the greatest problems is to explain the initial spatial distribution of hollows originating on a rock surface. Some hollows preferentially develop along joint or bedding planes where the rock is damper due local concentration of moisture movement. However, in many cases the hollows appear to develop randomly across seemingly homogenous rock surfaces. The patterns may reflect random variations in cementation or grain packing that may, for example, influence moisture distribution within the rock surface, but as the initial surface is destroyed by the creation of the hollow or hollows it is impossible to ascertain such possible origins. It has been suggested that once initiated hollows self-organize morphologically to minimize surface area (Turkington and Phillips, 2004) and once initiated microclimatological differences between the interior and exterior of cavernous features accentuate differences in weathering processes and rates between the hollow and the external surface of the rock through a positive feedback mechanism (Smith and McAllister, 1986). The interior of the hollow will be more sheltered and consequently undergo lesser temperature variation, greater, but less varied humidity, and the interior will be sheltered from direct insolation. This, in turn, will influence moisture movement and evaporation, salt crystallization and moisture-related chemical weathering reactions. This positive feedback further accentuates development (Dragovich, 1981). This continues until such time as environmental conditions change, when the features may become fossilized or, in the case of honeycomb weathering so destabilizes the rock that it begins to suffer mass breakdown and surface collapse that destroys the features (Robinson, 2007). Evidence from masonary stone indicates that honeycomb weathering can form in decades (Mottershead, 1994, 1997; Robinson and Williams, 1996) but tafoni appear to enlarge over tens of thousands of years (Norwick and Dexter, 2002).

Cavernous weathering is often associated with case hardening and some researchers have concluded that a hardened crust with a weakened subsurface layer may be a prerequisite for its development (Conca and Rossman, 1982, 1985). This is especially true of large tafoni where the outer rock often forms an overhanging 'visor' along the upper margins of the cavern. Models of the weathering of sandstone surfaces through case hardening and cavernous weathering have been proposed by Turkington and Paradise (2005) and Robinson (2007).

Approximately horizontal upper rock surfaces may also develop hollows. Variously known as weather pits, rock basins or gnammas, they are particularly characteristic of hard, massive rocks including limestones, granites, basalts, gritstones and some sandstones (Figure 17.1d). Predominantly circular in planform they vary from a few centimetres to more than a metre across and can be of similarly variable depths, although the diameter is usually much greater than the depth. More irregular linear forms can also be found. In wet weather they fill with water and although some are saucer shaped, with gently sloping side-walls, many exhibit an overlapping lip. Rainwater collects in the basins and it is generally agreed that water-based weathering processes are primarily responsible for their initiation and development, but on non-calcareous rocks the precise mechanisms remain uncertain. Overflow at the lowest point of the rim often creates a spillway that incises into the rock over time lowering the level of water collection until in some cases the basins become totally drained (Robinson and Williams, 1994). Rates of growth seem to vary, some seemingly inactive and unchanging over time whilst others enlarge at measurable rates of several centimetres a century (Twidale, 1982).

As weathering proceeds basins may widen and coalesce to produce compound forms.

Surface cracking

Weathering processes frequently generate stresses within rocks that are sufficient to cause them to crack or break. Crack patterns of various scales and shapes are often visible on rock surfaces. Of particular interest is polygonal cracking, also referred to as turtleback, tortoise shell (Williams and Robinson, 1989) or pachydermal weathering (Thomas et al., 2005). This is usually explained as a consequence of stresses resulting from expansion and contraction of rock surfaces caused by heating and cooling and/or wetting and drying. It most commonly develops on flat or, gently sloping exposed upper surfaces, or on rounded boulders of relatively porous sandstones in semi-arid to arid climates, although it is quite common also in more humid regions with significant temperature and moisture fluctuations, and on other rock types, including granites (Twidale, 1982; Williams and Robinson, 1989). It comprises shallow cracks that penetrate <10–30 cm perpendicular in to the rock below. The polygons are predominantly five- or six-sided and meet at 100–120° junctions. Most polygons average 100–200 mm in diameter but they can be as small as 10 mm and giant polygons greater than 1 m in diameter have also been described (Netoff, 1971). On the sides of rocks the polygons tend to become rectangular in form. In a review of polygonal weathering, Williams and Robinson (1989) concluded that the most common attribute of rocks on which polygons developed was some sort of weathering crust or rind, but this is not a universal characteristic of all rock surfaces on which they occur and their precise origin remains problematic as they have never been satisfactorily simulated. Interestingly similar features have recently been identified on rocks on Mars where their creation has been attributed to one or more of thermal contraction, moisture change, possibly involving partial dehydration of minerals or salt weathering processes that create uniform, near-surface tensile stress (Chan et al., 2008).

Over time the cracks often act as conduits for water to enter the rock, or flow over the surface, As a consequence the polygonanised surface can develop a considerable relief or they may often become detached (Figure 17.1d). In other cases, for reasons that are not understood, the centres of the polygons begin to spall, but the edges remain stable and coherent (Robinson and Williams, 1992).

MEASURING SURFACE CHANGE AND EROSION RATES

Accurate measurement of rates of recession of bare rock surfaces and the mapping of micromorphological change caused by weathering processes are two of the most important and challenging requirements of rock weathering studies. A number of techniques and methodologies have been developed to measure rates or degrees of change either directly or indirectly. Key indicators that are measured include surface recession, surface roughness, rock strength and small scale physical and chemical changes to the rock surface and subsurface. Measurements may be absolute, for example rates measured relative to a fixed surface (Cooke et al., 1995; André, 2002; Goldie, 2005; Häusselman, 2008; Nicholson, 2008), using microcatchments (Reddy, 1988; Halsey, 2000) or using micro-erosion meters, laser scanners and photogrammetry (Pentecost, 1991; Inkpen et al., 2000; Williams et al., 2000; Thornbush and Viles, 2004; Lim et al., 2005, 2010; Swantesson et al., 2006; Stephenson and Finlayson, 2009). Other approaches provide more indicative measures, for example, relative weight loss of exposure blocks (Moses, 2000; Trudgill, 2000; Dixon et al., 2006).

Many weathering processes effect change not only on the surface but also within the rock. For example, weathering rinds often have a zone of weakened material directly underneath them (e.g. Robinson and Willliams, 1987) and algae may weaken the rock just beneath the surface (Viles, 1987, Moses et al., 1995; Viles et al., 2000). The changes that take place beneath the surface may simply be physical, such as an increase in porosity or micro-fracturing; chemical either through the addition of compounds such as salts by precipitation or the selective removal of compounds by leaching. The simplest way to examine subsurface changes is to fracture or core the rock perpendicular to the surface so that changes can then be identified by a combination of visual and laboratory techniques including microscopy and a range of standard mechanical and chemical analyses. Unfortunately, these techniques damage the rock surface and destroy the sample. This imposes restrictions on their application for monitoring change over time, or where rock surfaces are subject to conservation. A range of non-destructive techniques are used to enhance our understanding of subsurface changes. These include the Schmidt hammer (Goudie, 2006), ultrasonic testing apparatus, such as the Grindosonic (Allison, 1988, 1990; Viles and Goudie, 2007) or PUNDIT (Warke et al., 2006) and mercury porosimity (e.g. Nicholson, 2001) and gas permeability testing (e.g. Warke et al., 2006). Enhanced understanding of the operation of weathering process mechanisms on and within rocks has been facilitated by technological developments in microscopic techniques, particularly those that provide chemical analyses and/or three-dimensional imagery (Taylor and Viles, 2000; Doehne et al., 2005; Cnudde et al., 2006; Dewanckele et al., 2009).

CURRENT ISSUES AND FUTURE DEVELOPMENTS

Rock weathering studies have been conducted at a wide range of spatial and temporal scales and using a variety of approaches. Short-term laboratory experiments, high magnification microscopy studies, rock exposure trials, studies that measure weathering rates directly and indirectly and studies that quantify and examine surface form have all contributed to our present understanding of rock surface and subsurface weathering. Improvements to environmental monitoring equipment for use both in the field and in the laboratory have greatly improved our knowledge of how conditions vary on and within rocks over a range of timescales. This has greatly improved our understanding of weathering processes and the accuracy with which we can simulate weathering cycles within laboratory. The techniques/ approaches used are usually appropriate for a particular study and there are often difficulties in drawing comparisons between studies, for example either because different sampling strategies have been employed or rates/processes have been measured over different time periods.

A major problem in weathering simulations is the lack of standardization in sample shape and experimental design. It has been shown, for example, that sample size and shape can significantly influence the results of salt weathering experiments (Goudie, 1974; Robinson and Williams, 1982). It is most probable that the same is true also of frost weathering. This makes direct comparison of the results of different experiments and the understanding of contrasting results between experiments difficult to interpret, especially as there remains uncertainty whether, for example, it is the length of edge to face, or surface area to volume that is the critical control. Most laboratory experimental weathering has used small samples and a major challenge for the future lies in increasing the scale of experiments to more nearly simulate real-life conditions and of upscaling the results of experiments to the field scale and integrating them into models of landscape evolution.

Measurement of the rates and nature of rock surface changes has been facilitated by the use and development of a range of increasingly sophisticated techniques and in particular the development of non-contact techniques using lasers, photogrammetry and imaging techniques to develop detailed digital terrain models of weathering surfaces. However, most techniques are best suited for use on relatively smooth horizontal, or near horizontal, surfaces and our ability to accurately measure the development of vertical, deeply pitted or hollowed surfaces remains very limited. Most measurement techniques cover very limited areas or numbers of points and the need to develop accurate techniques capable of measuring larger, more representative areas remains. Under the influence in part of research-funding mechanisms, there is an increased emphasis on applied research on cliffs, shore platforms and building stone, on rates of surface change and geotechnical impacts of weathering on rock surface stability. There is also an increasing interest in the products of weathering and in particular, the role of weathering in the production of quartz silt and loess (Smith et al., 2002b; Wright, 2007).

A particular difficulty lies in assessing the contribution of small scale weathering to larger scale landform and landscape evolution. This is exacerbated by the presently limited understanding of how weathering processes interact and how weathering and erosion interact at a range of scales. Viles (2001) identifies four key issues in rock weathering studies: first, whether there are characteristic spatio-temporal scales of landforms and processes; second, whether scales of process observation are the same as the scales of process operation; third, how to up- and down-scale observations (e.g. between microscopic scale, < 1 mm, to weathering landform scale, centimetre to metre); fourth, how and if different scales of processes and events interact to produce the geomorphology we see around us. Such questions provide the basis for ongoing discussions in rock weathering research. For example, Smith et al. (2002a) caution against upscaling from process to landscape scale suggesting that it may be more appropriate to first identify the key issues associated with the explanation of landscape change, before drawing upon process studies as required, that is, working from landscape to process. Although efforts have been made to understand geographical variations in weathering by focusing on the microscale boundary layer between the rock–atmosphere–hydrosphere–biosphere (Pope et al., 1995), it is suggested that clear linkages still need to be established between microscale and landscape scale enquiries (Turkington et al., 2005). Key issues that remain to be addressed include more detailed specifications of spatial and temporal scales and of the rates, durations and frequencies of weathering and related processes, forms and relationships (Phillips, 2005). Significant contributions are likely to be made through the use of numerical modelling of weathering processes (e.g. Walder and Hallet, 1985, 1986; Hallet, 2006; Murton et al., 2006; Trenhaile, 2008) and will be aided by recent developments in, for example, three-dimensional monitoring of rock surface topography and internal rock strength and relationships (Phillips, 2005).

REFERENCES

Allison, R.J. (1988) A non-destructive method of determining rock strength. *Earth Surface Processes and Landforms* 13, 729–36.

Allison, R.J. (1990) Developments in a non-destructive method of determining rock strength. *Earth Surface Processes and Landforms* 15, 571–7.

Allison, R.J. and Goudie, A.S. (1994) The effects of fire on rock weathering: An experimental study, in Robinson D.A. and Williams, R.B.G. (eds) *Rock Weathering and Landform Evolution.* Wiley, Chichester. pp. 41–56.

Anderson R.S. (1998) Near-surface thermal profiles in Alpine bedrock: Implications for the frost weathering of rock. *Arctic and Alpine Research* 30, 362–372.

André, M.-F. (2002) Rates of postglacial rock weathering on glacially scoured outcrops (Abisko-Riksgränsen area, 68°N). *Geografiska Annaler* 84A, 3–4, 139–50.

Birkeland, P.W. (1984) *Pedology, Weathering and Geomorphological Research.* Oxford University Press, New York.

Bjelland, T., and Thorseth, I.H. (2002) Comparative studies of the lichen–rock interface of four lichens in Vingen, western Norway. *Chemical Geology* 192, 81–98.

Blackwelder, E. (1927) Fire as an agent in rock weathering. *Journal of Geology* 35, 134–40.

Blackwelder, E. (1933) The insolation hypothesis of rock weathering. *American Journal of Science* 21, 140–44.

Bland, W. and Rolls, D. (1998) *Weathering.* Oxford University Press, New York.

Bonazza, A., Sabbioni, C., Messina, P., Guaraldi, C. and De Nuntiis, P. (2009) Climate change impact: Mapping thermal stress on Carrara marble in Europe. *Science of the Total Environment* 407, 4506–4512.

Brimblecombe, P. and Grossi, C.M. (2008) Millennium-long recession of limestone facades in London. *Environmental Geology* 56, 463–71.

Brimblecombe, P. and Grossi, C.M. (2009) Millennium-long damage to building materials in London. *Science of the Total Environment* 407, 1354–61.

Brock, T.D., Madigan, V.E.T., Martinko, J.M. and Parker, J. (1994) *Biology of Microorganisms.* 7th edn. Prentice-Hall, London.

Campbell, E. M. and Twidale, C. R. (1995) The various origins of minor granite landforms. *Caderno Lab. Xeolóxico de Laxe Coruña* 20, 281–306.

Cantón, Y., Solé-Benet, A., Queralt, I. and Pini, R. (2001) Weathering of a gypsum-calcareous mudstone under a semi-arid environment at Tabernas, SE Spain: Laboratory and field-based approaches. *Catena* 44, 111–32.

Cardell, C., Rivas, T., Mosquera, M.J., Birginie, J.M., Moropoulou, A., and Prieto, B. (2003) Patterns of damage in igneous and sedimentary rocks under conditions simulating sea-salt weathering. *Earth Surface Processes and Landforms* 28, 1–14.

Carter, N.E.A., and Viles, H.A. (2003) Experimental investigations into the interactions between moisture, rock surface temperatures and an epilithic lichen cover in the bioprotection of limestone. *Building and Environment* 38, 1225–1234.

Carter, N.E.A. and Viles, H.A. (2004) Lichen hotspots: Raised rock temperatures beneath Verrucaria nigrescens on limestone. *Geomorphology* 62, 1–16.

Carter, N.E.A. and Viles, H.A. (2005) Bioprotection explored: The story of a little known earth surface process. *Geomorphology* 67, 273–81.

Chan, M.A., Yonkee, W.A., Netoff, D., Seiler, W.M. and Ford, R.L. (2008) Polygonal cracks in bedrock on Earth and Mars: Implications for weathering. *Icarus* 194, 65–71.

Chen, J., Blume, H.P. and Beyer, L. (2000) Weathering of rocks induced by lichen colonization – a review. *Catena* 39, 121–46.

Cnudde, V., Masschaele, B., Dierick, M., Vlassenbroeck, J., Van Hoorebeke, L. and Jacobs, P. (2006) Recent progress in X-ray CT as a geosciences tool. *Applied Geochemistry* 21, 826–32.

Conca, J.L. and Rossman, G.R. (1982) Case hardening of sandstone. *Geology* 10, 520–3.

Conca, J.L. and Rossman, G.R. (1985) Core softening of cavernously weathered tonalite. *Journal of Geology* 93, 59–73.

Cooke, R.U. (1979) Laboratory simulation of salt weathering processes in arid environments. *Earth Surface Processes and Landforms* 4, 347–59.

Cooke, R.U., Inkpen, R.J. and Wiggs, G.F.S. (1995) Using gravestones to assess changing rates of weathering in the United Kingdom, *Earth Surface Processes and Landforms* 20, 531–46.

Corenblit, D., Gurnell, A.M., Steiger, J. and Tabacchi, E. (2008) Reciprocal adjustments between landforms and living organisms: Extended geomorphic evolutionary insights. *Catena* 73, 261–73.

Cox, N.J. (1989) Review of Biogeomorphology. *Progress in Physical Geography* 13, 620–4.

Crispim, C.A. and Gaylarde, C.C. (2004) Cyanobacteria and biodeterioration of cultural heritage: A review. *Microbial Ecology* 49, 1–9.

Danin, A. and Caneva, G. (1990) Deterioration of limestone walls in Jerusalem and marble monuments in Rome caused by cyanobacteria and cyanophilous lichens. *International Biodeterioration* 26, 397–417.

Day, M.J. (1980) Rock hardness: Field assessment and geomorphic importance. *The Professional Geographer* 32, 1, 72–81.

Deitrich, W.E. and Perron, J.T. (2006) The search for a topographic signature of life. *Nature* 439, 411–418.

Dewanckele, J., Cnudde, V., Boone, M., Van Loo, D., De Witte, Y., Pieters, K., et al. (2009) Integration of X-ray micro tomography and fluorescence for applications on natural building stones. *Journal of Physics: Conference Series* 186, 012082, 3pp.

Dixon, J.C., Campbell, S.W., Thorn, C.E. and Darmody, R.G. (2006) Incipient weathering rind development on introduced machine-polished granite disks in an Arctic environment, northern Scandinavia. *Earth Surface Processes and Landforms* 31, 111–21.

Doehne, E. (2002) *Salt Weathering: A Selective Review.* Geological Society, London, Special Publications, 205, 51–64.

Doehne, E., Carson, D. and Pasini, A. (2005) Combined ESEM and CT scan: The process of salt weathering. *Microscopy and Microanalysis* 11/I 2, 416–7.

Dorn, R.I. (2003) Boulder weathering and erosion associated with a wildfire, Sierra Ancha Mountains, Arizona. *Geomorphology* 55, 155–71.

Dragovich, D. (1981) Cavern microclimates in relation to preservation of rock art. *Studies in Conservation* 26, 143–9.

Duane, M.J. (2006) Coeval biochemical and biophysical weathering processes on Quaternary sandstone terraces south of Rabat (Temara), northwest Morocco. *Earth Surface Processes and Landforms* 31, 1115–28.

Emery, K. (1946) Marine solution basins. *The Journal of Geology* 54, 209–28.

Evans, I.S. (1970) Salt crystallization and rock weathering: A review. *Revue de Géomorphologie Dynamique* 19, 153–77.

Fiol, L., Fornós, J.J. and Ginés, A. (1996) Effects of biokarstic processes on the development of solutional rillenkarren in limestone rocks. *Earth Surface Processes and Landforms* 21, 447–52.

Folk, R.L., Roberts, H.H. and Moore, C.H. (1973) Black phytokarst from Hell, Cayman Islands, British West Indies. *Geological Society of America Bulletin* 84, 2351–60.

Ford, D.C. and Lundberg, J. (1987) A review of dissolution rills in limestone and other soluble rocks. *Catena Supplement* 8, 119–40.

Ford, D. and Williams, P. (2007) *Karst Hydrogeology and Geomorphology.* Wiley.

Fornós, J.J. and Ginés A. (eds) (1996) *Karren Landforms.* Servei de Publicacions, Universitat de les Illes Balears, Palma de Mallorca.

Fry, E.J. (1927) The mechanical action of crustaceous lichens on substrata of shale, schist, gneiss, limestone, and obsidian. *Annals of Botany* XLI, 437–60.

Gams, I. (1985) International comparative measurements of surface solution by means of standard limestone tablets. *Razprave iv. Razreda Sazu, Zbornik Ivana Rakovca/ Ivan Rakovec* Volume, XXVI, 1 sl., Ljubljana, 361–86.

Gaylarde, C. and Morton, G. (2002) Biodeterioration of mineral materials, in Britton G (ed.) *Environmental Microbiology* 1. Wiley, New York. pp. 516–28.

Gaylarde, C.C., Ortega-Morales, B.O. and Bartolo-Perez, P. (2007) Biogenic black crusts on buildings in unpolluted environments. *Current Microbiology* 54, 162–6.

Goldie, H.S. (2005) Erratic judgements: Re-evaluating solutional erosion rates of limestones using erratic-pedestal sites, including Norber, Yorkshire. *Area* 37, 433–42.

Golubic, S., Friedmann, E.I. and Schneider, J. (1981) The lithobiontic ecological niche, with special reference to microorganisms. *Journal of Sedimentary Research* 51, 475–478.

Gombert, P. (2002) Role of karstic dissolution in global carbon cycle. *Global and Planetary Change* 33, 177–84.

Gomez-Heras, M., Smith, B.J. and Fort, R. (2006) Surface temperature differences between minerals in crystalline rock: Implications for granular disintegration of granites through thermal fatigue. *Geomorphology* 78, 236–49.

Gorbushina, A.A. (2007) Minireview: Life on the rocks. *Environmental Microbiology* 9, 1613–31.

Gordon, S.J. and Dorn, R.I. (2005) Rind Weathering, in Goudie, A.S. (ed.) *Encyclopedia of Geomorphology*, vol. 2. Routledge, London and New York. pp. 853–5.

Goudie, A.S. (1974) Further experimental investigation of rock weathering by salt and other mechanical processes. *Zeitschrift für Geomorphologie* 21, 1–12.

Goudie, A.S. (2004a) Weathering and climate change, in Goudie, A.S. (ed.) *Encyclopedia of Geomorphology*. Routledge, London and New York. pp. 1112–3.

Goudie, A.S. (2004b) Insolation weathering, in Goudie, A.S. (ed.) *Encyclopedia ofGeomorphology*, vol. 1. Routledge, London and New York. pp. 566–7.

Goudie, A.S. (2004c) Salt weathering, in Goudie, A.S. (ed.) *Encyclopedia of Geomorphology*, vol. 2. Routledge, London and New York. pp. 894–7.

Goudie, A.S. (2006) The Schmidt Hammer in geomorphological research. *Progress in Physical Geography* 30, 703–18.

Goudie, A.S. and Viles, H.A. (1997) *Salt Weathering.* Wiley, Chichester.

Goudie, A.S., Allison, R.J. and McClaren, S.J. (1992) The relations between modulus of elasticity and temperature in the context of the experimental simulation of rock weathering by fire. *Earth Surface Processes and Landforms* 17, 605–15.

Griggs, D.T. (1936) The factor of fatigue in rock exfoliation. *Journal of Geology* 9, 783–96.

Grimes, K.G. (1997) Redefining the boundary between karst and pseudokarst: A discussion. *Cave and Karst Science* 24, 87–90.

Grossi, C.M., Bonazza, A., Brimblecombe, P., Harris, I. and Sabbioni, C. (2008) Predicting twenty-first century recession of architectural limestone in European cities. *Environmental Geology* 56, 3–4, 455–61.

Gunn, J. (1981) Prediction of limestone solution rates from rainfall and runoff data: Some comments. *Earth Surface Processes and Landforms* 6, 595–7.

Hall, K. (1999) The role of thermal stress fatigue in the breakdown of rock in cold regions. *Geomorphology* 31, 47–63.

Hall, K. and Hall, A. (1991) Thermal gradients and rock weathering at low temperatures: Some simulation data. *Permafrost and Periglacial Processes* 2, 103–12.

Hall, K. and Hall, A. (1996) Weathering by wetting and drying: some experimental results. *Earth Surface Processes and Landforms* 21, 365–76.

Hall, K. and André, M.-F. (2001) New insights into rock weathering from high-frequency rock temperature data: an Antarctic study of weathering by thermal stress. *Geomorphology* 41, 23–35.

Hall, K. and André, M.-F. (2003) Rock thermal data at the grain scale: applicability to granular disintegration in cold environments. *Earth Surface Processes and Landforms* 28, 823–36.

Hall, K., Staffan Lindgren, B. and Jackson, P. (2005) Rock albedo and monitoring of thermal conditions in respect of weathering: some expected and some unexpected results. *Earth Surface Processes and Landforms* 30, 801–11.

Hall, K., Guglielmin, M. and Strini, A. (2008a) Weathering of granite in Antarctica: I. Light penetration into rock and implications for rock weathering and endolithic communities. *Earth Surface Processes and Landforms* 33, 295–307.

Hall, K., Guglielmin, M. and Stini, A. (2008b) Weathering of granite in Antarctica. II Thermal stress at the grain scale. *Earth Surface Processes and Landforms* 33, 475–93.

Hallet, B. (2006) Why do freezing rocks break? *Science* 314, 1092–3.

Hallet, B., Walder, J.S. and Stubbs, J.W. (1991) Weathering by segregation ice growth in microcracks at sustained sub-zero temperatures: verification from an experimental study using acoustic emissions. *Permafrost and Periglacial Processes* 2, 283–300.

Halsey, D. (2000) Studying rock weathering with micro-catchment experiments. *Zeitschrift fur Geomorphologie* 120, 23–32.

Häuselmann, P. (2008) Surface corrosion of an alpine karren field: recent measurements at Innerbergli (Siebenhengste, Switzerland). *International Journal of Speleology* 37, 107–111.

Inkpen, R.J., Collier, P. and Fontana, D.J.L. (2000) Close-range photogrammetric analysis of rock surfaces. *Zeitschrift für Geomorphologie* 120, 67–81.

Jenkins, K.A. and Smith. B.J. (1990) Daytime rock surface temperature variability and its implications for mechanical rock weathering: Tenerife, Canary Islands. *Catena* 17, 449–59.

Jennings, J.N. (1985) *Karst Geomorphology.* 2nd edn. Blackwell.

Kanyana, J.I. and Trenhaile, A.S. (2005) Tidal wetting and drying on shore platforms: an experimental assessment. *Geomorphology* 70, 129–46.

Kump, L.R., Brantley, S.L. and Arthur, M.A. (2000) Chemical weathering, tmospheric CO_2 and climate. *Annual Review Earth and Planetary Sciences* 28, 611–67.

Lee, M. and Fookes, P. (2005) Climate and weathering, in Fookes, P.G., Lee, E.M. and Milligan, G. (eds) *Geomorphology for Engineers.* Whittles Publishing, CRC Press, Dunbeath. pp. 31–56.

Lee, R.R., Hodson, M.E. and Parsons, I. (1998) The role of intragranular microtextures and microstructures in chemical and mechanical weathering: Direct comparisons of experimentally and naturally weathered alkali feldspars. *Geochimica et Cosmochimica Acta* 62, 2771–88.

Lim, M., Rosser, N.J., Allison, R.J. and Petley, D.N. (2010) Erosional processes in the hard rock coastal cliffs at Staithes, North Yorkshire. *Geomorphology*, 114, 12–21.

Lim, M., Petley, D.N., Rosser, N.J., Allison, R.J., Long, A.J. and Pybus, D. (2005) Combined digital photogrammetry and time-of-flight laser scanning for monitoring cliff evolution. *Photogrammetric Record* 20, 109–29.

Liscia, M., Monteb, M. and Pacini, E. (2003) Lichens and higher plants on stone: a review. *International Biodeterioration & Biodegradation* 51, 1–17.

Liu, Z. and Zhao, J. (2000) Contribution of carbonate rock weathering to the atmospheric CO_2 sink. *Environmental Geology* 39, 1053–8.

Martini I.P. (1978) Tafoni weathering, with examples from Tuscany, Italy. *Zeitschrift für Geomorphologie* NF 22, 44–67.

Matsuoka, N. and Murton, J. (2008) Frost weathering: Recent advances and future directions. *Permafrost and Periglacial Processes* 19, 195–210.

May, E. (2003) Microbes on building stone—For good or ill? *Culture* 24, 5–8.

McCabe, S., Smith, B.J. and Warke, P.A. (2007a) Sandstone response to salt weathering following simulated fire damage: comparison of effects of furnace heating and fire. *Earth Surface Processes and Landforms* 32, 1874–83.

McCabe, S., Smith, B.J. and Warke, P.A. (2007b) Preliminary observations on the impact of complex stress histories on sandstone response to salt weathering: laboratory simulations of process combinations. *Environmental Geology* 52, 251–8.

McGreevy, J.P. (1982) Frost and salt weathering: Further experimental results. *Earth Surface Processes and Landforms* 7, 475–88.

McKay, C.P., Molaro, J.L. and Marinova, M.M. (2009) High frequency rock temperature data from hyper-arid desert environments in the Atacama and Antarctic dry valleys and implications for rock weathering. *Geomorphology* 110, 182–7.

Moores, J.E., Pelletier, J.D. and Smith, P.H. (2008) Crack propagation by differential insolation in desert surface clasts. *Geomorphology* 102, 472–81.

Mortensen, H. (1933) Die Salzprengung und irhe Bedcutung für die regional klimatische Gliederung der Wüsten. *Petermanns Geographische Mitteilungen* 79, 130–5.

Moses, C.A. (1996) Methods for investigating stone decay in polluted and 'clean' environments, Northern Ireland, in Smith, B.J. and Warke, P.A. (eds) *Processes of Urban Stone Decay.* Donhead Publishing Ltd., Donhead St Mary. pp. 212–27.

Moses, C.A. (2000) Field rock block exposure trials. *Zeitschrift für Geomorphologie* 120, 33–50.

Moses, C.A. and Smith, B.J. (1993) A note of the role of the lichen Colleme auriforma in solution basin development on a Carboniferous limestone. *Earth Surface Processes and Landforms* 18, 363–8.

Moses, C., Spate, A.P., Smith, D.I. and Greenaway, M.A. (1995) Limestone weathering in eastern Australia. Part 2: Surface micromorphology study. *Earth Surface Processes and Landforms* 20, 501–14.

Mottershead, D.N. (1994) Spatial variations in intensity of alveolar weathering of a dated sandstone structure in a coastal environment, Weston-Super Mare, UK, in Robinson D.A. and Williams, R.B.G. (eds) *Rock Weathering and Landform Evolution*. Wiley, Chichester. pp. 151–74.

Mottershead, D.N. (1997) A morphological study of coastal rock weathering on dated structures, south Devon, UK. *Earth Surface Processes and Landforms* 22, 491–506.

Mottershead, D. and Lucas, G. (2000a) The role of lichens in inhibiting erosion of a soluble rock. *Lichenologist* 32, 601–609.

Mottershead, D. and Lucas, G. (2000b) Field testing of Glew and Ford's model of solution flute evolution. *Earth Surface Processes and Landforms* 26, 839–46.

Mottershead, D.N., Moses, C.A. and Lucas, G.R. (2000) Lithological control of solution flute form: a comparative study. *Zeitschrift für Geomorphologie* 44, 491–512.

Mottershead, D., Gorbushina, A., Lucas, G., Wright, J. (2003) The Influence of Marine Salts, Aspect and Microbes in the Weathering of Sandstone in Two Historic Structures. *Building and Environment* 38, 1193–1204.

Murton, J.B., Peterson, R and Ozouf, J-C. (2006) Bedrock fracture by ice segregation in cold regions. *Science* 314, 1127–9.

Murton J.B., Coutard J.-P., Ozouf J.-C., Lautridou J.-P., Robinson D.A, Williams R.B.G, Guillemet G. and Simmons P. (2000) Experimental design for a pilot study on bedrock weathering near the permafrost table. *Earth Surface Processes and Landforms*, 25, 1281–1294.

Murton, J.B., Coutard, J.-P., Lautridou, J.-P., Ozouf, J.-C., Robinson, D.A. and Williams, R.B.G. (2001) Physical modelling of bedrock brecciation by ice segregation in permafrost. *Permafrost and Periglacial Processes* 12, 255–266.

Mustoe, G.E. (2010) Biogenic origin of coastal honeycomb weathering. *Earth Surface Processes and Landforms* 35, 424–34.

Navarre-Sitchler, A. and Thyne, G. (2007) Effects of carbon dioxide on mineral weathering rates at earth surface conditions. *Chemical Geology* 243, 53–63.

Naylor, L.A., Viles, H.A. and Carter, N.E.A. (2002) Biogeomorphology revisited: looking towards the future. *Geomorphology* 47, 3–14.

Netoff, D.I. (1971) Polygonal jointing in sandstone near Boulder, Colorado. *The Mountain Geologist* 8, 17–24.

Nicholson, D.T. (2001) Pore properties as indicators of breakdown mechanisms in experimentally weathered limestones. *Earth Surface Processes and Landforms* 26, 819–38.

Nicholson, D.T. (2008) Rock control on microweathering of bedrock surfaces in a periglacial environment. *Geomorphology* 101, 655–65.

Norwick, S.A. and Dexter, L.R. (2002) Rates of development of tafoni in the Moenkopi and Kaibab formations in Meteor Crater and on the Colorado Plateau, northeastern Arizona. *Earth Surface Processes and Landforms* 27, 11–26.

Ollier, C.D. (1984) *Weathering*. Longman, London.

Ollier, C. and Pain, C. (1996) *Regolith, Soils and Landforms*. Wiley, Chichester.

Pentecost, A. (1991) The weathering rates of some sandstone cliffs, Central Weald, England. *Earth Surface Processes and Landforms* 16, 83–91.

Phillips, J.D. (2005) Weathering instability and landscape evolution. *Geomorphology* 67, 255–72.

Pissart, A. and Lautridou, J.P. (1984) Variations de longuer de cylinders de Pierre de Caen (Calcaire bathonien) sous l'effet de séchage et d'humidification. *Zeitschrift fur Geomorphologie* 29, 111–6.

Pope, G.A., Dorn, R.I. and Dixon, J.C. (1995) A new conceptual model for the understanding of geographical variations in weathering. *Annals of the Association of American Geographers* 85, 38–64.

Porter, N.J. and Trenhaile, A.S. (2007) Short-term rock surface expansion and contraction in the inter-tidal zone. *Earth Surface Processes and Landforms* 32, 1379–97.

Reddy, M.M. (1988) Acid rain damage to carbonate stone: a quantitative assessment based on the aqueous geochemistry of rainfall runoff from stone. *Earth Surface Processes and Landforms* 13, 335–54.

Rice, A. (1976) Insolation warmed over. *Geology* 4, 61–2.

Robinson, D.A. (2007) Geomorphology of the inland sandstone cliffs of Southeast England, in Hartel, H., Cilek, V., Herben, T., Jackson, A. and Williams, R.B.G. (eds) *Sandstone Landscapes*. Academic Press, Praha.

Robinson, D.A. and Williams, R.B.G. (1982) Salt weathering of rock specimens of varying shape. *Area* 14, 293–9.

Robinson, D.A. and Jerwood, L.C. (1987a) Frost and salt weathering of chalk shore platforms near Brighton, Sussex, U.K. *Transactions Institute British Geographers* 12, 217–26.

Robinson, D.A. and Jerwood, L.C. (1987b) Weathering of chalk shore platforms during harsh winters in South-East England. *Marine Geology* 77, 1–14.

Robinson, D.A. and Williams, R.B.G. (1987) Surface crusting of sandstones in southern England and northern France, in Gardner, V. (ed.) *International Geomorphology*. Wiley, Chichester. pp. 623–35.

Robinson, D.A. and Williams, R.B.G. (1989) Polygonal cracking of sandstone at Fontainebleau, France. *Zeitschrift für Geomorphologie* 33, 59–72.

Robinson, D.A. and Williams, R.B.G. (1992) Sandstone Weathering in the High Atlas, Morocco. *Zeitschrift für Geomorphologie* 36, 413–29.

Robinson, D.A. and Williams, R.B.G. (1994) Sandstone weathering and landforms in Britain and Europe, in Robinson D.A. and Williams, R.B.G. (eds) *Rock Weathering and Landform Evolution*. Wiley, Chichester. pp. 371–92.

Robinson, D.A. and Williams, R.B.G. (1996) An analysis of the weathering of Wealden sandstone churches, in Smith, B.J. and Warke, P.A. (eds) *Processes of Urban Stone Decay*. Donhead, London, 133–49.

Robinson, D.A. and Williams, R.B.G. (2000a) Accelerated weathering of a sandstone in the High Atlas Mountains of Morocco by an epilithic lichen. *Zeitschrift für Geomorphologie* 44, 513–28.

Robinson, D.A. and Williams, R.B.G. (2000b) Experimental weathering of sandstone by combinations of salts. *Earth Surface Processes and Landforms* 25, 1309–15.

Ruiz-Agudo, E., Putnis, C.V., Jiménez-López, C. and Rodriguez-Navarro, C. (2009) An atomic force microscopy study of calcite dissolution in saline solutions: The role of magnesium ions. *Geochimica et Cosmochimica Acta* 73, 3201–17.

Schaffer, R.J. (1932) *The Weathering of Natural Building Stones*. Special Report 18, Building Research Establishment. Reprinted HMSO London. 1972.

Scheerer, S., Ortega-Morales, O. and Gaylarde, C. (2009) Microbial Deterioration of Stone Monuments–An Updated Overview, in Laskin, A.I., Gadd, G.M. and Sariaslani, S. (eds) *Advances in Applied Microbiology* 66, 97–139.

Seaward, M.R.D. (1997) Major impacts made by lichens in biodeterioration processes. *International Biodeterioration and Biodegradation* 40, 269–73.

Smith, B.J. and McAllister, J.J. (1986) Observations on the occurrence and origins of salt weathering phenomena near Lake Magadi, Southern Kenya. *Zeitschrift für Geomorphologie* 27, 11–26.

Smith, B.J, Whalley, W.B. and Magee, R.W. (1992) Assessment of building stone decay: a geomorphological approach, in Webster, R.G.M. (ed.) *Stone Cleaning and the Nature, Soiling, and Decay Mechanisms of Stone*. Donhead Publishing Ltd., London. pp. 249–57.

Smith, B.J., Magee, R.W. and Whalley, W.B. (1994) Breakdown patterns of quartz sandstone in a polluted urban environment: Belfast, N. Ireland, in Robinson D.A. and Williams, R.B.G. (eds) *Rock Weathering and Landform Evolution*. Wiley, Chichester. pp. 131–50.

Smith, B.J., Warke, P.A. and Moses, C.A. (2000) Limestone weathering in contemporary arid environments: a case study from southern Tunisia. *Earth Surface Processes and Landforms* 25, 1343–54.

Smith, B.J., Warke, P.A. and Whalley, W.B. (2002a) Landscape development, collective amnesia and the need for integration in geomorphological research. *Area* 33, 409–18.

Smith, B.J. Wright, J.S. and Whalley, W.B. (2002b) Sources of non-glacial loess-size quartz silt and the origins of 'desert loess'. *Earth Science Reviews* 59, 1–26.

Smith, B.J., Gomez-Heras, M. and McCabe, S. (2008) Understanding the decay of stone-built cultural heritage. *Progress in Physical Geography* 32, 439–61.

Smith, B.J., Turkington, A.V., Warke, P.A., Basheer, P.A.M., McAlister, J.J., Meneely, J., et al. (2002c) Modelling the rapid retreat of building sandstones: a case study from a polluted maritime environment, in Siegesmund S, Weiss T, Volbrecht A (eds) *Natural Stone, Weathering Phenomena,*

Conservation Strategies and Case Studies. Geological Society of London Special Publication, The Geological Society, London. 205, 347–62.

Smith, D.I. and Atkinson, T.C. (1976) Process, landforms and climate in limestone regions, in Derbyshire, E. (ed.) *Geomorphology and Climate*. Wiley, Chichester. pp. 367–409.

Smith D.I., Greenaway, M.A., Moses, C. and Spate, A.P. (1995) Limestone weathering in Eastern Australia. Part 1: Erosion rates. *Earth Surface Processes and Landforms* 20, 451–463.

Sperling, C.H.B. and Cooke, R.U. (1980) *Salt Weathering in Arid Environments. 1. Theoretical Considerations*. Papers in Geography, 9. Bedford College, London.

St. Clair, L.L. and Seaward, M.R.D. (eds) (2004) *Biodeterioration of Stone Surfaces*. Kluwer, Dordrecht.

Stephenson, W.J. and Kirk, R.M. (2001) Surface swelling of coastal bedrock on inter-titadal shore platforms, Kaikoura Peninsula, South Island, New Zealand. *Geomorphology* 41, 5–21.

Stephenson, W.J. and Finlayson, B.L. (2009) Measuring erosion with the microerosion meter – contributions to understanding landform evolution. *Earth Science Reviews* 95, 53–62.

Stretch, R.C. and Viles, H.A. (2002) The nature and rate of weathering by lichens on lava flows on Lanzarote. *Geomorphology* 47, 87–94.

Sumner, P.D. and Loubser, M.J. (2008) Experimental sandstone weathering using different wetting and drying moisture amplitudes. *Earth Surface Processes and Landforms* 33, 985–90.

Swantesson, J.O.H. (2005) Weathering and erosion of rock carvings in Sweden during the period 1994–2003. *Micro-mapping with laser scanner for assessment of breakdown rates*. Karlstad Univ. Stud. 29. Karlstad, Sweden.

Swantesson, J.O.H., Moses, C.A., Berg, G.E. and Jansson, K.M. (2006) Methods for measuring shore platform micro-erosion : a comparison of the micro-erosion meter and laser scanner. *Zeitschrift fur Geomorphologie*, 144, 1–17.

Sweeting, M.M. (1972) *Karst Landforms*. MacMillan, London.

Taylor, M.P. and Viles, H.A. (2000) Improving the use of microscopy in the study of weathering: sampling issues. *Zeitschrift fur Geomorphologie* 120, 145–58.

Thornbush, M.J, and Viles, H. A. (2004) Integrated digital photography and image processing for the quantification of colouration on soiled limestone surfaces in Oxford, England. *Journal of Cultural Heritage* 5, 285–290.

Thornbush, M.J and Viles, H.A. (2007) Simulation of the dissolution of weathered versus unweathered limestone in carbonic acid solutions of varying strength. *Earth Surface Processes and Landforms* 32, 841–52.

Trenhaile, A.S. (2006) Tidal wetting and drying on shore platforms: an experimental study of surface expansion and contraction. *Geomorphology* 76, 316–31.

Trenhaile, A.S. (2008) Modeling the role of weathering in shore platform development. *Geomorphology* 94, 24–39.

Trenhaile, A.S. and Rudakas, P.A. (1981) Freeze-thaw and shore platform development in Gaspé, Quebec. *Géographie Physique et Quaternaire* 35, 171–81.

Trudgill, S.T. (1985) *Limestone Geomorphology*. Longman.

Trudgill, S.T. (2000) Weathering overview–Measurement and modelling. *Zeitschrift fur Geomorphologie* 120, 187–93.

Trudgill, S.T., Viles, H.A., Inkpen, R.J., Moses, C., Gosling, W., Yates, T., et al. (2001) Twenty-year weathering remeasurements at St. Paul's Cathedral, London. *Earth Surface Processes and Landforms* 26, 1129–42.

Turkington, A.V. (1998) Cavernous weathering in sandstone: lessons to be learned from natural exposure. *Quarterly Journal of the Geological Society* 31, 375–83.

Turkington, A.V. and Smith, B.J. (2000) Observations of three-dimensional salt distribution in building sandstone. *Earth Surface Processes and Landforms* 25, 1317–32.

Turkington, A.V. and Phillips, J.D. (2004) Cavernous weathering, dynamical instability and self organization. *Earth Surface Processes and Landforms* 29, 665–75.

Turkington, A.V. and Paradise, T.R. (2005) Sandstone weathering: a century of research and innovation. *Geomorphology* 67, 229–53.

Turkington, A.V., Phillips, J.D. and Campbell, S.W. (2005) Weathering and landscape evolution. *Geomorphology* 67, 1–6.

Twidale, C.R. (1982) *Granite Landforms*. Elsevier, Amsterdam and New York.

Twidale, C.R. and Vidal-Romani, J.R. (2005) *Landforms and Geology of Granite Terrains*. A.A. Balkema Publishers, Leiden, The Netherlands.

Ulusoy, M. (2007) Different igneous masonary blocks and salt crystal weathering rates in architecture of historical city of Konya. *Building and Environment* 42, 3014–24.

Viles, H.A. (1987) Blue-green algae and terrestrial limestone weathering on Aldabra Atoll: An SEM and light microscope study. *Earth Surface Processes and Landforms* 12, 319–30.

Viles, H.A. (ed.) (1988) *Biogeomorphology*. Blackwell, Oxford.

Viles, H.A. (1995) Ecological perspectives on rock surface weathering: towards a conceptual model. *Geomorphology* 13, 21–35.

Viles, H.A. (2001) Scale issues in weathering studies. *Geomorphology* 41, 63–72.

Viles, H.A. (2002) Implications of future climate change for stone deterioration, in Siegesmund, S., Vollbrecht, A. and Weiss, T. (eds) *Natural Stone, Weathering Phenomena, Conservation Strategies and Case Studies*, Geological Society London, Special Publication, 205, 407–18.

Viles, H.A. and Moses, C.A. (1998) Experimental production of weathering nanomorphologies on carbonate stone. *Quarterly Journal of Engineering Geology* 31, 347–57.

Viles, H.A. and Goudie, A.S. (2007) Rapid salt weathering in the coastal Namib desert: Implications for landscape development. *Geomorphology* 85, 49–62.

Viles, H.A., Spencer, T., Telek, K. and Cox, C. (2000) Observations on 16 years of microfloral recolonisation data from limestone surfaces, Aldabra Atoll, Indian Ocean: implications for biological weathering. *Earth Surface Processes and Landforms* 25, 1355–70.

Viles, H.A., Naylor, L.A., Carter, N.E.A. and Chaput, D. (2008) Biogeomorphological disturbance regimes: progress in linking ecological and geomorphological systems. *Earth Surface Processes and Landforms* 33, 1419–35.

Walder, J.S. and Hallet, B. (1985) A theoretical model of the fracture of rock during freezing. *Geological Society of America Bulletin* 96, 336–46.

Walder, J.S. and Hallet, B. (1986) The physical basis of frost weathering: toward a more fundamental and unified perspective. *Arctic and Alpine Research* 18, 27–32.

Warke, P.A. and Smith, B.J. (1998) The effects of direct and indirect heating on the validity of rock weathering simulation studies and durability tests. *Geomorphology* 22, 347–57.

Warke, P.A. and Smith, B.J. (2000) Salt distribution in clay rich weathered sandstone. *Earth Surface Processes and Landforms* 25, 1333–42.

Warke, P.A., Smith, B.J. and Magee, R.W. (1996) Thermal response characteristics of stone: Implications for weathering of soiled surfaces in urban environments. *Earth Surface Processes and Landforms* 21, 295–306.

Warke, P.A., Smith, B.J. and McKinley, J. (2004) Complex weathering effects on the durability of building sandstone, in Prikryl, R. (ed) *Dimension Stone*, Balkema, Leiden, 229–236.

Warke, P.A., McKinley, J. and Smith, B. J. (2006) Variable weathering response in sandstone: Factors controlling decay sequences. *Earth Surface Processes and Landforms* 31, 715–35.

Warscheid, Th. and Braams, J. (2000) Biodeterioration of stone: a review. *International Biodeterioration and Biodegradation* 46, 343–68.

Webster, A and Maya, E. (2006) Bioremediation of weathered-building stone surfaces. *Trends in Biotechnology* 24, 255–260.

Whalley, W.B. and Turkington, A.V. (2001) Weathering and Geomorphology. *Geomorphology* 41 (Special Volume).

Williams, R.B.G. and Robinson, D.A. (1981) Weathering of sandstone by the combined action of frost and salt. *Earth Surface Processes and Landforms* 6, 1–9.

Williams, R.B.G. and Robinson, D.A. (1989) Origin and distribution of polygonal cracking of rock surfaces. *Geografisker Annalar, Series A* 71, 145–159.

Williams, R.B.G. and Robinson, D.A. (1991) Frost weathering of rocks in the presence of salts-a review. *Permafrost and Periglacial Processes* 2, 347–53.

Williams, R.B.G. and Robinson, D.A. (2001) Experimental frost weathering of sandstone by various combinations of salts. *Earth Surface Processes and Landforms* 26, 811–18.

Williams, R.G.B., Swantesson, J.O.H. and Robinson, D.A. (2000) Measuring rates of surface downwearing and mapping microtopography: the use of micro-erosion meters and laser scanners in rock weathering studies. *Zeitschrift für Geomorphologie* 120, 51–66.

Winkler, E.M. and Wilhelm, E.J. (1970) Saltburst by hydration pressures in architectural stone in urban atmosphere. Bulletin of the *Geological Society of America* 81, 567–72.

Wray, R.A.L. (1997) A global review of solutional weathering forms on quartz sandstones. *Earth-Science Reviews* 42, 137–60.

Wright, J.S. (2007) An overview of the role of weathering in the production of quartz silt, *Sedimentary Geology* 202, 337–51.

Young, R.W. (1987) Sandstone landforms of the tropical east Kimberley region, northwestern Australia. *Journal of Geology* 95, 205–18.

Young, R.W. and Young, A.R.M. (1992) *Sandstone Landforms.* Springer-Verlag, Heidelberg and New York.

18

Fluids, Flows and Fluxes in Geomorphology

André G. Roy and Hélène Lamarre

The surface of the Earth is largely shaped by the action of fluids at a range of temporal and spatial scales. Understanding fluid dynamics is critical to explain and predict changes in the landscape and such an understanding represents in many ways the essence of process geomorphology. The geomorphological tradition includes fundamental contributions to the study of fluid flows and of their effects on the Earth surface. For example, the seminal contributions of Gilbert (1914), Bagnold (1941) and Leopold et al. (1964) have had a lasting influence on the discipline. The movement of fluid is also one of the dynamic characteristics of our environment and it plays a major role in creating diversity over the surface of the Earth. In this chapter, we will present the fundamentals that pertain to fluids, flows and fluxes as they relate to the transport and deposition of sediments and to the development and maintenance of landforms. Fluid flow is the driver that is responsible for most continental denudation and for the transfer of material from the continents to the oceanic sedimentary basins. Its importance is enhanced by the effects of anthropogenic activities as they change the nature, quality and quantity of fluid flows and of their associated sediment fluxes.

Fluids, flows and fluxes present a great diversity in nature and are a vast topic. As a result, choices in the coverage of the material had to be made in order to emphasize the most important and useful aspects of the theme for geomorphologists. We have focused on the nature of flows and on their interactions with sediments. Many landforms result from the stress applied by fluid

flow on particles, thus generating a displacement of matter once thresholds for the initiation of movement have been reached. In this chapter, we are concerned with flows that produce flux of matter and that govern processes of sediment transport and of erosion and deposition of sediments leading to the development and changes in landform. The emphasis will be on water flows, overland and in river channels. Parallels with air flow will also be drawn. The scope of the chapter is largely inspired by the trinity proposed by Leeder (1983) and re-examined by Best (1993). The trinity describes the interactions between flow, sediments and forms with a particular emphasis on the role of flow turbulence (Figure 18.1). This will be an important part of the subject matter presented here.

Covering the basics for understanding fluid flow–sediment interactions as they apply to continental environments is the main thrust of this chapter. This approach builds upon classical work in various fields, including sedimentology (Allen, 1970), geomorphology (Carson, 1971) and engineering (Williams, 1982), but also and perhaps more importantly incorporates advances in the study of turbulent flows. Our focus in the treatment of the subject may appear incomplete or scanty for some geomorphological applications. For instance, we will neglect other types of flows and fluxes like those related with heat transfer and glacial processes. Even with such a restriction, the environmental conditions where flows and fluxes are encountered remain vast and diverse.

The chapter is divided into four parts. It begins with definitions of fluids, flows and fluxes and

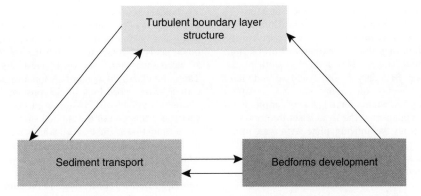

Figure 18.1 Leeder's trinity (Best, 1993). (Reproduced with the permission of Wiley Publishers)

with a characterization of the forces acting on fluids and of the nature and properties of flows in nature. The second part deals with flow turbulence and with the physical stresses generated by fluid flows on sediments. Some of the feedbacks between form and flow will be described, especially as these feedbacks are critical for the development and maintenance of bedforms. In the third part, we address the relation between fluid stresses and the motion of particles. Here, we discuss sediment transport events and their relations with the frequency and magnitude of fluid flow events at a range of temporal and spatial scales. Finally, global fluxes are discussed with an emphasis on the changes witnessed in the sediment fluxes with anthropogenic activities.

DEFINING FLUIDS, FLOWS AND FLUXES

What do we mean when we speak of fluid, flow and flux. The term fluid refers to a particular state of matter that deforms continuously and permanently under an applied stress. The term applies to gases and to most liquids (Middleton and Southard, 1984). As far as geomorphological work is concerned, air and water are the main fluids as they are involved in the displacement of solid matter and in sediment transport. Fluids are opposite to solids but a clear distinction between the two states may be difficult to determine in some cases. A fluid flow is a quantity of fluid in motion that can be defined by boundary conditions, pattern, rate and changeability (Allen, 1994). Flux is defined as the amount of matter that flows through a unit area per unit time.

Properties of fluids: air and water

The most important fluid properties are density and viscosity which are essential for understanding fluid motion and the interaction with sediments.

Density

Mass density (ρ) is defined as the mass per unit volume of a substance or $\rho = \dfrac{M}{V}$ where M is the mass and V is the volume. Because gravitational force and momentum exchange are both proportional to mass, mass density appears in numerous equations describing fluid motion and sediment transport. In some cases, density is expressed as the dimensionless quantity specific gravity. For many practical purposes, the metric system was designed so that water will have a mass density of one gram per unit cubic centimeter or 1000 kilograms per cubic meter. Water density is affected by temperature and its densest state is at 4°C. The density of air, under ordinary conditions, is three orders of magnitude smaller than that of water. Air density varies with pressure and temperature. On the other hand, weight density (γ) is the weight per unit volume of a substance. The properties of γ and ρ are related via Newton's law (i.e. $F = Ma$) as follows: $\gamma = \rho g$ where g is gravitational acceleration and has a standard value of 9.80665 m/s^2.

Viscosity

Viscosity can be expressed in two forms: dynamic (μ) and kinematic (ν). Dynamic viscosity is the internal friction of a fluid that offers a resistance to forces tending to cause the fluid to flow. It is a measure of fluid friction expressed in Nsm^{-2} in SI units. The dynamic viscosity of a fluid can be defined in terms of an experiment based on the observation of a thin but extensive film of fluid that is held between two parallel smooth plates. The lower boundary is fixed and of indefinite extent, whereas the upper one is movable. A steady constant force (F) is applied on the upper boundary. The fluid is set in motion by friction from the moving plate. After an initial transient period of adjustment, the velocity will vary linearly from zero immediately next to the lower boundary to the maximum velocity at the uppermost layer, which is the same as the velocity of the upper boundary (no-slip condition). For straight, parallel and uniform flow, the shear stress (τ) between the layers is proportional to the vertical gradient of velocity (dU/dY) in the direction perpendicular to the layers: $\dfrac{\tau}{dU/dY} = \mu$.

The constant μ is known as the dynamic viscosity. The ratio of dynamic viscosity to mass density is the kinematic viscosity (v): $v = \dfrac{\mu}{\rho}$. Kinematic viscosity is sometimes referred to as diffusivity of momentum, because it has the same dimensions and is comparable to diffusivity of heat and mass. Both dynamic and kinematic viscosity of water and air vary with temperature and pressure (Table 18.1).

Flows: fluids in motion

An important distinction among fluids concerns their response to stress. When a stress is applied to a fluid, it tends to deform. A Newtonian fluid refers to a fluid whose shear rate (rate at which a fluid is deformed) is linearly related to the applied shear stress. The relation passes through the origin and the constant of proportionality (i.e. the slope of the linear relation) is the viscosity.

Table 18.1 Fluid properties of water and air at 1 atm

Air/water	Mass density, ρ (kg m^{-3})	Dynamic viscosity, μ (N s m^{-2})	Kinematic viscosity, v (m^2 s^{-1})
Air			
0°C	1.29	1.72E^{-5}	1.33E^{-5}
20°C	1.20	1.82E^{-5}	1.51E^{-5}
Water			
0°C	1000	1.788E^{-3}	1.788E^{-6}
20°C	998	1.003E^{-3}	1.005E^{-6}

The behavior of Newtonian fluids can be described as a function of temperature and pressure. Both air and water belong to this type of fluid. There are other kinds of fluids for which the relationship between shear stress and shear rate is not linear. These are classified as non-Newtonian fluids that cannot be described by a single constant value of viscosity (e.g. blood). In these cases, viscosity varies as a function of the applied stress.

In fluid flows, the rate at which mass enters a volume in the system is equal to the rate at which mass leaves the volume with the exception of what is being stored or released from the volume. In other words, there is no gain or loss of matter and this corresponds to the principle of continuity.

Important variables to describe fluid flow are velocity and discharge both of which are expressing flow rates. Velocity can be assessed at an instant and at a point or can be an average over a temporal interval and/or on a transect, a cross-section or a volume. Discharge is the volume of fluid passing across a point, transect or cross-section per unit time.

Flow classification

There are numerous ways to classify fluid flows. Two dimensionless quantities are often used to classify flows based on the relative magnitude of gravity, viscous and inertial forces. The Reynolds number is a measure of the ratio of inertial to viscous forces. The Reynolds number is given by $Re = \dfrac{UL}{v}$ where U is the average flow velocity of the fluid and L is a characteristic length (i.e. flow depth, particle size). For instance, it is used to distinguish between the laminar and turbulent flow regimes. In this case, L is a metric of the flow such as flow depth (Y). When Re is less than 500, viscous forces dominate, and the flow is said to be laminar and characterized by parallel layers of moving fluid stacked on top of one another. Any small disturbance in the flow is damped out by fluid viscosity and the parallel laminae are re-established. If the flow is disturbed, it can grow into a turbulent state which tends to produce eddies and flow mixing. At the onset of turbulence, as for instance when a laminar flow encounters a cylinder, regular and periodic vortices may shed in the lee of the cylinder. The flow may be considered to be in transition when Re is between 500 and 2000. When $Re > 2000$, inertial forces dominate and the flow becomes fully turbulent. In nature, most flows are turbulent.

The Froude number is based on the ratio between the inertial and gravitational forces and it is expressed as $Fr = \dfrac{U}{(gY)^{1/2}}$ where Y is the flow depth. The two alternative flow states are called subcritical ($Fr < 1$) and supercritical flow ($Fr > 1$). Subcritical flows are characterized by a water surface that is smooth and on which a wave produced at the water surface is quickly damped out. Subcritical flows are also said to be tranquil. When $Fr = 1$, the flow is critical. At $F > 1$, flow is supercritical and it is usually fast and shallow. Supercritical flows can sustain large standing waves that are formed over obstructions in the river bed or at a constriction. Supercritical flows can be found in rapids and waterfalls, or can be created artificially at weirs.

It is also possible to classify flows with respect to their behavior in space and time. In a uniform flow, velocity or discharge do not show longitudinal variations along the flow path while in a steady flow, velocity and discharge remain constant over time. When deriving fluid flow properties, it is often assumed that the flow is uniform and steady, the simplest case. In nature, however, flow varies in space and time and becomes therefore non-uniform and/or unsteady. Such situations occur when flow accelerates and decelerates.

Forces involved in fluid flow

Fluid flow occurs when the forces that induce motion exceed those resisting motion. The fundamental forces involved in fluid flow of water are mainly gravitational acceleration and resistance motion caused by fluid viscosity and turbulence. In specific cases of water flows, pressure gradients may influence positively or negatively fluid motion as for instance when water accelerates or decelerates in response to changes in flow depth. Pressure gradients are the main driver of air flows. Other forces at right angles to the fluid motion like the centrifugal and the Coriolis forces are involved in particular circumstances but are sometimes neglected. Centrifugal forces are considered when the flow is following a curved path and are critical for many applications in both air and water flows. The Coriolis force is very important in air flows as it acts perpendicular to the flow and it affects the direction of the wind. It is negligible in water flows.

In overland flow and river channel flow, the driving force is the acceleration caused by gravity expressed by:

$$F_g = \gamma wXY \sin\theta$$

or per unit mass

$$\frac{F_g}{M} = \frac{\gamma wXY \sin\theta}{\rho wXY} = \frac{\gamma}{\rho}\sin\theta = g\sin\theta$$

where θ is the slope (Figure 18.2). This force is countered by resistance from the forces arising

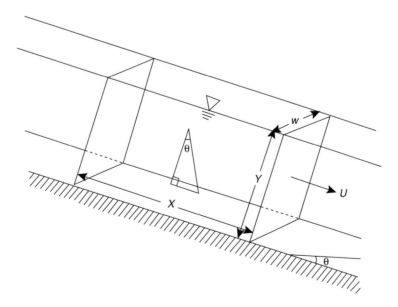

Figure 18.2 Definitions of terms for computing the relative magnitude of forces in open channel flows (Dingman, 1984)

from the viscous effect in fluid flow and the turbulent force associated with fluid motion. The viscous force per unit mass is given by:

$$\frac{Fv}{M} = \frac{\mu UXw}{Y\rho wXY} = \frac{\mu U}{\rho Y^2} = \frac{\upsilon U}{Y^2}$$

while what are called turbulent forces are expressed by:

$$\frac{Ft}{M} = \frac{\rho U^2 wX}{\rho wXY} = \frac{U^2}{Y}$$

Fluid motion occurs when $g\sin\theta$ is greater than $\left(\dfrac{U^2}{Y} + \dfrac{\upsilon U}{Y^2}\right)$.

Other forces may be involved in some particular cases. For instance, the force associated with a pressure-gradient contributes to the acceleration–deceleration of flow due to a spatial pressure difference. Such differences occur at river channel constriction or expansion. In open-channel flow, pressure is proportional to the distance below the water surface. The pressure gradient per unit mass in the flow is

$$\frac{Fp}{M} = \frac{\gamma\Delta YwY}{\rho wXY} = \frac{\gamma\Delta Y}{\rho X} = \frac{g\Delta Y}{X}$$

Air flow is mainly dependent upon pressure gradients caused by differences in heating of the Earth surface that entrains differences in air density. Flow is from high to low pressure zones. The centrifugal force represents the effects of inertia that arise in connection with rotation and which are experienced as an outward force away from the center of rotation. The force depends on the mass of the object, the speed of rotation and the distance from the center. The centrifugal force per unit mass is given by

$$\frac{F_r}{M} = \frac{U^2}{r}$$

where r is the radius of curvature of the path. This force may be important in river meanders. The Coriolis effect is an apparent deflection of moving objects when they are viewed from a rotation reference frame. Per unit of mass, the Coriolis force is given by

$$\frac{F_c}{M} = 2\varpi U\sin\Phi$$

where ω is the angular velocity of the earth and Φ is the latitude. The Coriolis effect does not change the velocity of the flow but affects its direction.

Resistance to flow partly determines routing velocity and is critical for both stream channel flows and overland flow over hillsides (Smith et al., 2007). Frictional resistance is often described by empirical coefficients that are used to estimate flow velocity in runoff and erosion models. One important coefficient is Manning's n as used for decades in engineering. The empirical coefficient is given by

$$n = \frac{Y^{\frac{2}{3}}S^{\frac{1}{2}}}{U}$$

The value of n can also be estimated from photographs (see Barnes, 1967) from typical streams and it can be used to predict the average velocity for a given channel depth and slope. There are some alternatives to the basic equation of Manning. In cases where the value is estimated in a straight channel with uniform bed material, n can be measured using particle size only (Carson and Griffiths, 1987). The Darcy–Weisbach friction factor is based on a more theoretical basis and is dimensionless. It is defined as

$$f = \frac{8gYS}{U^2}$$

Shear stress

A shear stress is defined as a force per unit area which is applied parallel or tangential to a face of a boundary. It is opposed to a normal stress which is applied perpendicularly. For Newtonian fluids the shear stress is proportional to the strain in the fluid where the viscosity is the constant of proportionality. Bed shear stress can be estimated through several methods depending on the type of flows and on the measurements available. For a river reach where flow is assumed to be uniform and steady, bed shear stress is often estimated from $\tau = \rho gYs\sin\theta$.

TURBULENT FLOWS

In general, fluid flow over the surface of the Earth is turbulent. The presence of turbulence is a source of diversity and change in earth surface systems. Turbulence is a complex phenomenon and as a result it is difficult to understand and predict. The turbulent flow regime was originally described as chaotic or stochastic with important changes including low momentum diffusion, high momentum convection and rapid variation of pressure and velocity in space and time. Significant progress has provided evidence that turbulent flow exhibits structures of irregular, but repetitive spatial–temporal flow patterns that can be viewed as coherent turbulent flow structures often responsible for most of the turbulent energy production. Turbulence causes the formation of eddies at many different length scales. They are swirls of fluid with highly irregular shape, a wide range of sizes and in continual state of development and decay (Middleton and Southard, 1984). Each eddy can be characterized by an intensity of rotation referring to vorticity.

The turbulent boundary layer

In geomorphology, fluid motion has the greatest impact when in contact with or near a solid boundary. Thus, boundary layer flow will be emphasized here. In most geomorphological applications, only one boundary is concerned while in some particular contexts two boundaries (e.g. river flow under an ice cover) or pipe flow are involved. The boundary layer is the zone of flow in the immediate vicinity of a solid surface

where the motion of the fluid is affected by the frictional resistance exerted by the boundary (Allen, 1994). The velocities across the boundary layer vary from zero if the boundary is static to the *free-stream* velocity at the outer edge of the boundary layer. The thickness of the boundary layer (δ) corresponds to the distance from the boundary to the point where flow velocity equals 99 per cent the velocity of the free-stream. In a very thin layer close to the boundary, the viscous shear force is sufficient to maintain the fluid in laminar motion. This is the viscous sub-layer whose height is very small with respect to the boundary layer. There is a short transition between the layer dominated by viscous force and the surface layer where the turbulent forces dominate.

There are several ways of classifying the flow above a boundary. The most common is based on the turbulence characteristics of the flow: viscous sub-layer, buffer zone and turbulent zone. Others classify the flow with respect to the position with respect to the boundary either as the inner and outer regions or as the near-bed or wall region, intermediate and free-surface zones as we get further away from the boundary (Figure 18.3).

Boundary layers develop on objects of any shape immersed in a fluid moving relative to the

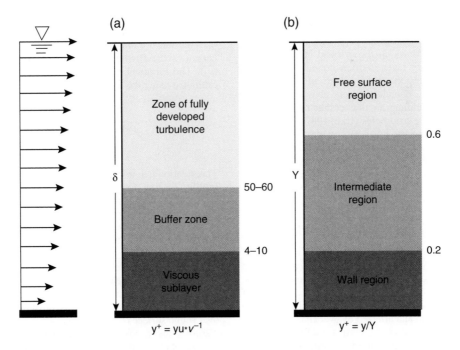

Figure 18.3   Classification of the flow above a boundary ((a) Dingman, 1984; (b) Nezu and Nakagawa, 1993). (Reproduced with the permission of Balkema Publishers)

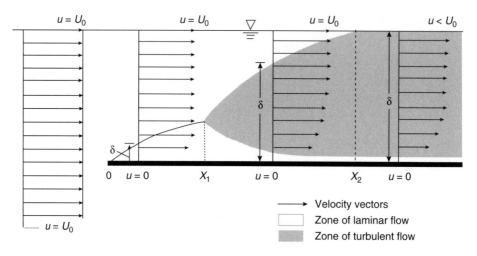

Figure 18.4 **The structure of the turbulent boundary layer (Chow, 1959; Dingman, 1984)**

object. For example, a boundary layer develops on a surface of a stationary flat plate held parallel to a uniform free-stream of fluid (see Middleton and Southard, 1984) (Figure 18.4). Immediately downstream of the leading edge of the plate, the boundary layer is very thin and is lower than 4 to 10 $y^+ = yu_*v^{-1}$ (Figure 18.3). As we move along the plate, the frictional effect of the plate affects a greater thickness of fluid and the boundary layer becomes thicker. The thickness of the boundary layer increases with the roughness of the surface. In shallow flows, as those found in many rivers, the boundary layer is often not fully developed and occupies the entire depth of flow. In the atmosphere, boundary layers above a city with high rise buildings are thicker than rural boundary layers above crop fields (Oke, 1987). The atmospheric boundary layer also varies in thickness with surface heat and the maximal extent of the layer is 1 to 2 km. In colder conditions, the boundary layer may only be 100 m thick.

The logarithmic velocity profile

A first approach to deal with the turbulent structure of the boundary layer is to assume that the flow is uniform and steady. In this approach, the interest lies in the vertical distribution of average flow velocity. Although the flow lines are disorganized and affected by ever changing eddies in a turbulent flow, the vertical profile of average flow velocity measured at different heights above a static bed often behaves predictably (Figure 18.5). Above the viscous sub-layer,

time-averaged flow velocity u varies with height y above the surface according to the law of the wall established by Prandtl and von Karman. The general equation is given by

$$u = \frac{1}{k}(gYS_0)^{1/2} \ln\left(\frac{y}{y_0}\right)$$

where k is von Karman's constant (0.4) and $(gYS_0)^{1/2}$ has the unit of velocity and has become known as the friction velocity or shear velocity (u_*). The turbulent velocity profile can then be estimated using the following equation:

$$u(y) = 2.5u_* \ln\left(\frac{y}{y_0}\right)$$

This relation is known as the logarithmic velocity profile. In this formulation, the average shear stress at the bed is given by ρu_*^2 and is a direct function of the slope of the relation between average velocity and height above the bed. Because $u = 0$ at $y = 0$ due to the no-slip condition, the application of the log law is intractable because ln 0 is indefinite. In order to avoid this problem, it is assumed that flow velocity is null slightly above the bed at y_0. Over smooth surfaces, y_0 is within the viscous sub-layer. The determination of y_0 over rough surfaces where the viscous sub-layer is disrupted by protruding bed elements is often problematic. In general, y_0 may be defined through the roughness height (k_s), which represents a representative roughness from the elements

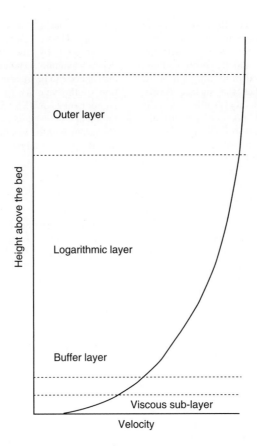

Figure 18.5   Velocity profile and boundary layer for turbulent flow. Thicknesses above the layers are not to scale (Robert, 2003). (Reproducted with the permission of Oxford University Press)

composing the bed and is often a function of the particle size. It is also possible to estimate y_0 from the intercept of the relation between velocity and height above the bed.

The estimate of bed shear stress from the velocity profile ($\tau = \rho u_*^2$) has also been the subject of debate concerning its accuracy and usefulness, especially in shallow flows. Biron et al. (1999) have shown that the estimates of both shear stress and roughness height are highly sensitive to the height of velocity measurements above the bed and have recommended that measurements be taken as close to the bed as possible.

For the last few decades, investigations of the shape of the velocity profiles and of the turbulent flow characteristics in the near-bed region of rivers and of their links with sediment have been central in the study of the turbulent boundary layer (Robert, 2003). For instance, the effect of roughness transitions on the shape of the velocity profile has been studied both in air and water.

Robert et al. (1992, 1996) have indicated the magnitude of the changes occurring downstream from a bed roughness transition in natural coarse-grained river channels from the interpretation of velocity profile shapes. Upstream from a transition between a smooth and very rough section, the velocity profile is generally linear on semi-logarithmic scales and presents a small velocity gradient and correspondent shear stress. Immediately downstream from a roughness transition, the velocities near the bed are considerably reduced and the velocity gradient increased. The linear profile is modified into a more concave shape. The authors also noted an increase of turbulent intensity downstream from the obstacle. Similar results have been observed in airflow at roughness transitions.

Flow over dunes also exhibits a distortion of the velocity profile as the fluid comes across the stoss and the lee sides of the form. As the fluid approaches the form, the vertical velocity profile begins to change its shape with the acceleration of the flow. The flow reaches a maximum acceleration near the crest of the dune before it detaches from the surface. The velocity profile shows signs of a velocity reversal in the through of the dune. The flow reattaches the surface further downstream and the velocity profile gradually reestablishes itself and become logarithmic. Modifications in the shape of the profiles are associated with changes in bed shear stress and in the capacity of the fluid to entrain sediments. Bed shear stress reaches a maximum near the crest.

The log–velocity profile may be applicable as an average model but it is variable and complicated temporally at a point even in a uniform steady flow and in the absence of bedforms. For instance, in gravel-bed rivers, instantaneous velocity profiles show clear phases of near-bed flow deceleration and acceleration and of concomitant changes in bed shear stress (Kirkbride and McLelland, 1994; Buffin-Bélanger et al., 2000). These changes in the profile occur in a preferential manner from the top of the profile to the bottom and are related with the presence of large-scale eddies. Other complications in the form of the velocity profile are associated with the presence of a porous boundary through which flows above and within the substrate are exchanged. Hyporheic flows within river bed sediments are critical for the maintenance of habitats (Buffington and Tonina, 2009; Tonina and Buffington, 2009). They also interact with the flow and modify the form of the velocity profile and especially the location of the 0 velocity point as it can be above or within the sediments whether the flow is subjected to injection of fluid from the substrate or to suction of fluid drawn into the sediments (Lu et al., 2008).

Reynolds decomposition of velocities in turbulent flows

In spite of the usefulness of the velocity profile in geomorphology, the assumptions behind its derivation remain simplistic with respect to the nature of turbulent flows. Turbulence is produced in three dimensions of the Euclidian space and it implies velocity or pressure changes for the three components of the flow: a longitudinal component (X), a vertical component (Y) (normal to the boundary, V) and a lateral component (Z) which is parallel to the boundary and normal to the longitudinal flow (W). Following the approach proposed by Reynolds (1895) for the decomposition of velocity, instantaneous velocities measured at a point (u_t, v_t, w_t) in the flow may be partitioned into a mean (u,v,w) and a fluctuating $(u', v'$ and $w')$

part such that $u_t = u + u'$, $v_t = v + v'$ and $w_t = w + w'$. The turbulent component of the flow is captured by the fluctuating part of the velocity signals. Hence, the intensity of the turbulence can be measured by the standard deviation of the velocity time series of the various components: RMS_u, RMS_v and RMS_w. The turbulent kinetic energy per unit volume combines the variance of the velocity times-series measured of all three components of the flow and is defined by:

$$TKE = 0.5\rho(RMS_u^2 + RMS_v^2 + RMS_w^2)$$

Reynolds turbulent stresses represent the degree of momentum exchange at a given point in the flow and are associated with all combinations of the flow components and of the decomposed velocity. There are six stressors for each of the terms of the velocity decomposition:

$$\rho\begin{bmatrix} u^2 & uv & uw \\ & v^2 & vw \\ & & w^2 \end{bmatrix} \quad \text{and} \quad -\rho\begin{bmatrix} u'^2 & u'v' & u'w' \\ & v'^2 & v'w' \\ & & w' \end{bmatrix}$$

Reynolds turbulent shear stresses are very important quantities not only as measures of momentum exchange but also as forces involved in the mobilization and transport of sediments. The stresses associated with the longitudinal and vertical velocity components are often used in sediment transport studies. The normal stresses ρu^2 and ρv^2 are related to the drag and lift forces and $-\rho u'v'$ to the bed shear stress. Biron et al. (2004) have compared the estimates for bed shear stress in complex water flows and have shown that the best estimate is derived from the turbulent kinetic energy (TKE).

The structure of the turbulent boundary layer: coherent turbulent flow structures

The turbulent boundary layer is also characterized by the presence of well-organized structures. The structures are coherent inasmuch as one of their properties (e.g. velocity, pressure) is correlated with itself or with another property in time and/or space. The main turbulent coherent structures in boundary layers are (Best, 1993):

1 *Low-speed streaks.* Using flow visualization, Kline et al. (1967) have observed the presence of well-organized spatially and temporally dependant motions within the viscous sub-layer that lead

to the formation of narrow and elongated zones of low-speed fluid. These low-speed streaks play an important role in the production of turbulent kinetic energy and in the development of the bursting process within the boundary layer on smooth walls. As they move away from the boundary, the degree of organization and coherence decrease gradually. The lateral spacing of the streaks scales with shear velocity. It has been frequently observed that:

$$\lambda^+ = \lambda u_*/v$$

where λ is the mean spacing between streaks is equal to 100 ± 20 in the near-bed region (Zacksenhouse et al., 2001). Low-speed streaks tend to become disorganized and to disappear as bed roughness increases.

2 *Ejection-sweep cycle.* Ejections are parcels of slow-moving fluid that rise from the bed towards the water surface. A sweep is a wall-directed inrush of higher than the average downstream velocity fluid and has a spanwise width larger that the streaks. Sweeps loose momentum as they impact against the boundary (Best, 1993). These structures are intertwined and they tend to follow one another. They are often associated with the bursting process and with the development of hairpin vortices. Ejections and sweeps are the main mechanisms of turbulence production in the boundary layer.

3 *Large-scale vortices.* Large-scale vortices have been described as recurrent patterns of high-speed fluid wedges interconnected by low-speed regions. They represent advecting regions that occupy a large portion of the flow depth extending into the outer layer of the flow. These flow structures were mostly investigated theoretically or through laboratory experiments but have also been reported in nature. Shvidchenko and Pender (2001) have shown the presence of macroturbulent structures in a laboratory flume. These structures are composed of large scale ejections and sweeps that occupy the entire depth of flow. The size of the large-scale turbulent structures scales with the depth of flow or with the height of the turbulent boundary layer. In gravel-bed rivers, the structures occupy the full depth of flow and they are elongated and narrow with a length that is in the order of 3 to 5 times the flow depth (Roy et al., 2004).

The presence of turbulent coherent structures in geophysical flows is ubiquitous even over rough boundaries. Using detailed velocity measurements in the field, Buffin-Bélanger et al. (2000) confirmed the presence of large-scale turbulent flow structures in gravel-bed rivers. They showed the presence of large wedges of faster fluid joined by regions of slower fluid. The wedges have an average angle of inclination with the bed 36°. The authors also reported the importance of the large scale vortices in bedload sediment transport. Other types of coherent turbulent flow structures are associated with flows around obstacles or with the confluence of flows with different properties (e.g. density, velocity) resulting from Kelvin–Helmholtz instabilities. They often occur in the presence of separation when the flow detaches itself from the boundary as it comes around an obstacle (e.g. boulder, dune, woody debris and groin). Flow separation is very common in nature and it is a major process by which landforms are maintained. Flow separation produces a zone of intense shear away from the solid boundary and is therefore a generator of turbulence as eddies shed into the main flow (Middleton and Wilcock, 1994).

In water flume experiments, Kirkbride (1993, 1994) has shown the presence of shedding motions from the lee of protruding particles. The author suggested that the structures are initiated by mechanisms of vortex coalescence and interactions between the wake interference region and the turbulent flow field. It can also be linked to large scale vortices already present in the flow (Roy et al., 1999). Buffin-Bélanger and Roy (1998) have tested the effects of a pebble cluster on the turbulent flow field in natural gravel-bed rivers. They have shown the existence of complex flow patterns associated with the presence of pebble clusters and particularly a zone of eddy shedding downstream of the obstacle characterized by a band of intense turbulence intensity. Flow separation is also important in the development of dunes both in water and air flows. It is also critical in the dynamics of river channel confluences where it occurs downstream of the tributary as it joins the main flow. These zones are preferential sites for the deposition of sediments and for the formation of bars.

Secondary flows can develop in a straight channel, in river meanders and at channel confluences. Their formation can be linked to the creation of turbulence initiated from the boundary roughness and to the density differences associated with the variations of temperature or fluid density. Secondary flows are defined as currents that occur in the plane normal to the local axis of the primary flow and are brought about interaction of the primary flow (Bathurst et al., 1979). They are analyzed in terms of streamwise velocity, or rotation about the primary axis. In rivers, a simple example of such secondary flow consists of a pair of spiral currents rising at the banks and plunging in mid-channel (Richards, 1982). Keller and Melhorn (1978) have shown that secondary circulations also occur in riffles and pools. They affect the processes of flow resistance, sediment transport and bed and bank erosion, and influence the development of channel morphology. Secondary flows are the main drivers for the maintenance of pools at meander bends (Bathurst et al., 1979) and for scouring at river channel confluences with concordant beds (Ashmore et al., 1992).

FLOWS AS GEOMORPHOLOGICAL AGENTS: FLOW–SEDIMENT INTERACTIONS

Running water is the most important agent of denudation of the Earth surface. It acts in two basic forms: overland and channel flow. Overland flow can be described as the movement of runoff downhill on the ground surface in a more or less disturbed sheet whereas channel flow is associated with the water moving in streams bounded by banks. Running water can also be associated with mass movement with high content of water such as debris and mud flow. A distinct flow path can be recognized in meltwater which is part of the glacial system. The most important source of meltwater is the surface of the glaciers that can occasionally be supplemented by rainfall and runoff valley-side slope in temperate regions.

Aeolian flows are associated with winds that are highly variable in direction and force. The distribution and intensity of winds in combination with climate has much to do with the location of wind erosion (deflation) and windblown deposits on the Earth. The effectiveness of wind as an erosional agent is limited compared to running water flows such as channel flow. This is mostly due to the fact that air has a lower density and viscosity than water. In this case, only fine particles can be transported in suspension or saltation. The effectiveness of wind as an erosional process is also limited by the presence of a vegetation cover that reduces the flow speeds near the ground and the sediment availability. The most intense aeolian activity occurs in arid regions and in coastal zones.

It is when these agents interact and move sediments that geomorphological work is achieved. The grains are entrained from the bed with the flow depending on their size, shape and excess density, and the fluid viscosity, velocity and shear stress. Once the threshold for initial motion has been reached, particles are moving up and down-flow. Although the fundamental principles remain the same in water and air, sediment transport differs because of the density and viscosity of the fluid.

Four modes of transport are recognized in water: traction, saltation, suspension and solution.

Traction occurs when the particles move mainly in contact with the bed surface as they are either rolling or sliding. Sliding motion refers to the particles that do not experience any rotation during transport. Saltating motion describes a series of ballistic jumps characterized by steep-angle ascent from the bed to a height of a few grain diameters which turns into a descent path back to the bed (Leeder, 1982). Rolling, sliding and saltating particles collectively form the bedload. Suspension motion refers to particles moving with an irregular path higher up from the bed than the particles in saltation. It is usually associated with fine sand, silt and clay that only settle at very low flow velocity. Suspended sediment loads are by far the largest contributor of the other major transport modes. Dissolved load varies in magnitude and composition according to the dominant sources, the rates of solute mobilization and the hydrological pathways operating within the catchment (Knighton, 1998).

Fundamental forces on particles

Sediment transport and fluxes can be examined at the particle scale where we are interested in the initiation and maintenance of individual particles in motion. On a static bed, the forces acting on a stationary particle include the fluid drag (F_D) and lift (F_L), the particle-immersed weight (F_W) and the interparticle cohesion (F_C). For sand and small gravel particles in water or air, F_C can normally be ignored (Allen, 1994). F_D refers to forces that oppose the relative motion of an object through a fluid. It acts in the opposite direction to the oncoming flow velocity and is equal to $F_D = \frac{1}{2}\rho u^2 C_D A$, where A is the reference area and C_D is the drag coefficient, a dimensionless constant. F_L is the component that is perpendicular to this direction and is also proportional to $\frac{1}{2}\rho u^2 A$. These forces are associated with pressure gradients around particles.

Particle movement occurs when the lift and drag force are greater than the resisting forces. Thresholds of particle entrainment have been defined in the literature and one of the most used approaches has been developed by Shields (1936). The initial movement of particles is expressed in terms of a critical shear stress. The critical shear stress for spherical grain of diameter (D) resting on a flat bed is given by: $\tau_{cr} = \eta g \left(\rho_d - \rho\right)\frac{\pi}{6}D\tan\phi$ where η = eddy viscosity and ρ_s is the particle density. Shields' diagram is based on the relation between a dimensionless critical shear stress $(\Theta_{cr} = \tau_{cr} /(\rho_s - \rho_w)gD)$ and a particle Reynolds number $(Re_* = u_* D/v)$ at incipient particle motion (Figure 18.6).

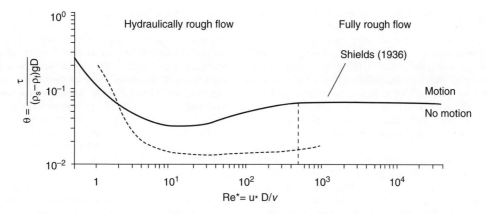

Figure 18.6   Shields diagram. Relation between critical dimensionless shear stress and erosive Reynold number for turbulent flow. Solid line = water (Graf, 1971) and dotted line = air (Mantz, 1977). (Reproduced with the permission of American Society of Civil Engineers)

The plot shows that the threshold for the incipient motion of particles is not as straightforward as a simple resolution of forces would suggest. There is also a major difference between air and water flows that are associated with the differences in density and viscosity. From its initial development, the diagram has been tested in various environments and subsequently complemented and modified. One has to remember that Shields' test conditions were for particles of different densities examined in isolation on an artificial flat bed. Such conditions are not typical of what is encountered in nature and they limit the application of the approach. Also, Shields' relation represents an average behavior of particles and there is much variability even in uniform flow conditions. The forces applied on the particles fluctuate according to the nature of the turbulent flow. It has been shown that turbulent coherent flow structures induce variability in bed shear stress and that they affect sediment transport on the bed. The evidence as to the role of turbulence in sediment transport is not overwhelming, however. The presence of a particle mixture of various sizes, of imbricated clasts and of bedforms on the river bed complicates the application of any entrainment threshold curves. Similar situations occur in air flows when the sediment surface is subject to various moisture conditions or when sediment supply becomes limited.

Sediment and matter fluxes

Because of the interplay of many factors in the initiation and maintenance of sediment transport, the response of fluxes of sediments to changes in the critical flow parameters (velocity, shear stress, discharge or power) is highly variable. Typical relations between bedload and suspended sediment transport rates with peak discharges in two streams. Show that there is a wide range of responses in both the suspended sediment load and the bedload for a given discharge. Solute transport, however, tends to behave well as a function of discharge. The variability in sediment fluxes within a catchment is caused by numerous factors that interact with one another. From the variations of the sediment supply to the particles organization on the bed, these factors entrain severe difficulties in predicting sediment loads for a given event.

The sum of sediment events over a period of time allows for an estimation of the sediment yield of a watershed and of the denudation rate. Denudation rates vary in space and time and are largely controlled by climate and tectonics. There is a debate as to the relative role of these two critical factors, but there may exist a mutual adjustment between the two (Bonnet and Crave, 2003; Whipple, 2009). Climate and tectonics also govern the frequency and magnitude of geophysical flows. The concept of frequency–magnitude is of critical importance when examining the effects of fluid flows on the landscape and in determining the amount of geomorphological work associated with particular flows. The question as to whether frequent low magnitude events contribute more to the overall denudation of the landscape than rare events of very large magnitude is still debated.

Frequency–magnitude analysis, originally developed in the context of fluvial geomorphology and described by Wolman and Miller (1960), has been applied to a wide spectrum of geomorphic processes (e.g. Crozier and Glade, 1999). The product of the frequency and of the magnitude of events gives the amount of work associated with events relative to a recurrence interval. Wolman and Miller (1960) have shown that at least 50% of sediment transport was done by flows of a magnitude that occurred at least once per year. Intermediate magnitude and frequency events seems dominate fluvial sediment transport whereas low frequency and high magnitude events are contributing only a small proportion of the sediment transported in rivers.

Pickup and Warner (1976) have also suggested that more frequent floods dominated bedload transport, but extreme events control the erosion of cohesive banks defining channel capacity and channel adjustments. More recently, Mao et al. (2009) used direct monitoring in two Italian alpine streams characterized by different sediment transport mechanisms: one catchment is bedload dominated while the other is debris flow dominated. For a similar event recurrence interval, the amount of sediment is two to three orders of magnitude larger for the debris flow dominated catchment. They also show that the transport of large volumes of sediments as bedload is unlikely while smaller and more frequent episodes are more effective than high magnitude events. This may be explained by the relatively low sediment supply of the bedload dominated system.

Gonzalez-Hidalgo et al. (2010) have analyzed a large data set of daily events from rivers in the United States. They have calculated the percentage of the total suspended sediment load for the *n* largest events. Their findings show that most of the sediment transport is associated with infrequent events of large magnitude. For example, 50% of the total load is transported by the 25 largest events. Although there is some spatial variability in the findings, these results clearly show the importance of high magnitude events on sedimentary fluxes. This also confirms the highly episodic nature of sediment transport and of erosional and depositional processes (Bergstrom and Schumm, 1981).

GLOBAL MATTER FLUXES OF SEDIMENT, TIMESCALES AND EFFECTS OF HUMAN ACTIVITIES

Estimating the global budget of matter over continental surfaces is a difficult task and there exist various approaches that can be applied at different timescales. Because most denudation at continental scales is associated with the work of running water, sediment budgets are generally based on the products exiting from watersheds and feeding into oceanic basins. Using the work from various sources, Summerfield and Hulton (1994) have estimated the contributions in suspended and solute loads of the major rivers of the world. The data were used to calculate the denudation rates as presented in Table 18.2. The proportion of chemical denudation is highly variable, ranging from 3 to 94 per cent with a relatively low median value of 17.6%. Thus, most denudation is associated with sediment loads and with the suspended load in particular.

It is, however, very difficult to obtain precise estimates of the sedimentary budgets of rivers. This is mainly caused by the lack of reliable and complete data. For short timescales (day, year or decade), sediment budgets are made through measurements at gauging stations while over longer periods (decades and centuries) estimates of sedimentary productions come mainly from accumulation in reservoirs. Such approaches consider the historical timescale or shorter and have often emphasized the role of human activities in the acceleration of erosion. In a classical study, Wolman (1967) has highlighted the effects of changing land use in the Piedmont of the USA over the past two centuries. An increase of several orders of magnitude in sediment yields is shown to be the result of crop agriculture and of urban construction.

Over the very long term, we face the impossibility to have direct measurements of sediment transport rates. Cosmogenic methods, however, are now capable of providing effective denudation rates. Kirchner et al. (2001) have compared the denudation rates obtained through data from reservoirs behind dams with cosmogenic estimates over millennium timescale in a region of California. They reported that long-term denudation rates are one order of magnitude larger than historical rates. Their argument discounted the fact that such a difference could be due to methodological errors or other factors such as climate or tectonics. This difference is also confirmed when using a dating method over an even longer timescale. Their explanation of the increased rates for longer time periods rests on the role of high frequency episodic events that may be detected over 10,000 years but that are not likely to occur in a timescale of a century or less. It is interesting to note that their estimates of denudation rates do not confirm the large impacts of human activities on erosion rates. In contrast to this conclusion, Wilkinson (2005) has shown from the sedimentary rock record that human activities such as construction and agriculture are intense enough to denude ice-free continental surfaces by several hundred meters per million years. This is one order of

Table 18.2 Solid load, solute load and denudation rate data (Summerfield and Hulton, 1994, and sources cited therein)

Basin	Mean annual solid load ($Mt\ y^{-1}$)	Mean annual specific load ($t\ km^{-2}\ y^{-1}$)	Mechanical denudation rate ($mm\ ky^{-1}$)	Mean annual solute load ($Mt\ y^{-1}$)	Mean annual denudation solute load ($Mt\ y^{-1}$)	Mean annual specific denudation solute load ($t\ km^{-2}\ y^{-1}$)	Chemical denudation rate ($t\ km^{-2}\ y^{-1}$)	Total denudation rate ($mm\ ky^{-1}$)	Chemical denudation as proportion of total (%)
Amazon	1320	221	82	275	171	29	11	93	11.6
Amur	57	28	10	22	12	6	2	12	17.6
Brahmaputra	1157	1208	670	51	31	49	18	688	2.6
Chiang Jiang	468	281	104	226	124	72	27	131	20.4
Colorado	167	239	89	16	13	19	7	96	7.4
Columbia	32	48	18	33	21	32	12	30	40.0
Danube	74	94	35	63	35	45	17	52	32.4
Dnepr	1	2	1	11	7	12	4	5	85.7
Ganges	680	694	257	75	41	42	16	273	5.7
Huang He	100	127	47	22	14	18	7	54	12.4
Indus	300	323	120	62	38	42	16	136	11.5
Kolyma	6	9	3	4	2	4	1	4	30.8
La Plata Parana	87	30	11	38	25	9	3	14	23.1
Lena	17	7	3	88	55	22	8	11	75.9
Mackenzie	110	62	23	65	40	23	9	32	27.1
Mekong	176	232	86	47	27	36	13	99	13.4
Mississippi	605	189	70	105	74	20	7	77	9.6
Murray	33	30	11	8	6	6	2	13	9.7
Nelson	—	—	—	31	20	16	6	—	—
Niger	40	19	7	13	8	4	1	8	17.4
Nile	100	28	10	20	11	3	1	11	9.7
Ob	18	6	2	50	31	11	4	6	64.7
Orange	58	65	24	17	10	11	4	28	14.5
Orinoco	165	179	66	29	21	23	9	75	11.4
Rio Grande	30	48	18	3	2	4	1	19	7.7
Sao Francisco	7	11	4	—	—	—	—	—	—
Shatt El Arab	50	56	21	19	13	14	5	26	20.0
St Lawrence	2	2	1	60	35	34	13	14	94.4
Tocantins	—	—	—	—	—	—	—	—	—
Yenisei	14	5	2	73	45	18	7	9	78.3
Yukon	79	94	35	34	19	23	9	44	19.7
Zaire	51	14	5	37	23	6	2	7	30.0
Zambezi	48	34	13	13	9	6	2	15	15.0

magnitude larger than the rates recorded in the absence of humans. Wilkinson argues that humans are a prime geomorphic agent. In a further study, Wilkinson and McElroy (2007) have estimated the erosion rates from the Lower Cambrian to the Pliocene. They have shown an increase in erosion rates up to 53 mm/m.y. (an estimated 16 Gt/yr). This value is close to the current estimate of 62 mm/m.y. (approximately 21 Gt/yr) obtained

from data on large river sediment loads (Syvitski et al., 2005). The breakdown of this amount of sediments into the main erosion processes is 14 per cent as solute, 8 per cent as bedload and 61 per cent as suspended load. The rest is trapped by dams and structures on the continent. Most of recent continental erosion (estimated at 83 per cent) is associated with high elevation watersheds confined to a relatively small proportion of the surface of the Earth (10 per cent). Human activities and predominantly agriculture are concentrated in the lowlands of the Earth. The increase in the sedimentary contributions of these activities is estimated by Wilkinson and McElroy (2007) to be 52 Gt/yr. For the most part, this sediment has yet to reach the oceanic basins and is stored in valley bottoms and floodplains. The authors argue that this alluvium represents the most important sedimentary processes that shape the surface of the Earth. In a context of population growth, human activities are likely to become an even greater influence on local and global sediment fluxes.

In spite of major progress achieved over the years, understanding and predicting fluid flow and the resulting fluxes of matter on the Earth surface remain a formidable challenge. One of the fundamental questions rests on the integration of the scales at which erosional and depositional processes operate from the sources to the sinks and at which forms develop as a result of the movement and storage of sediments. Major efforts have been targeted to understanding processes at the small scales of the spectrum with major advances in our knowledge of the role of turbulent flows in the formation and maintenance of fluvial and aeolian forms. In parallel, concerns about the impacts of environmental changes, including human activities at the regional and global scales, have motivated major investigations of the sediment production system. Reconciling these scales and approaches is critical for the future of geomorphology.

REFERENCES

Allen, J.R.L. (1970) *Physical Processes of Sedimentation*. Allen and Unwin, London.

Allen, J.R.L. (1994) Fundamental properties of fluids and their relation to sediment transport processes, In: K. Pye (ed.). *Sediment Transport and Depositional Processes*. Blackwell Scientific Publication. Cambridge (US). pp. 25–58.

Ashmore, P.E., Ferguson, R.I., Prestegaard, K.L., Ashworth, P.J. and Paola, C. (1992) Secondary flow in anabranch confluences of a braided, gravel-bed stream. *Earth Surface Processes and Landforms* 17, 299–311.

Bagnold, R.A. (1941) *The Physics of Blown Sand and Desert Dunes*. Chapman and Hall, London.

Barnes, H.H. (1967). *Roughness Characteristics of Natural Channels*. U.S. Geological Survey Water-Supply Paper. 1849.

Bathurst, J.C., Thorne, C.R. and Hey, R. (1979) Secondary flow and shear stress at river bends. *Journal of the Hydraulics Division* HY10, 1277–93.

Bergstrom, F.W. and Schumm, S.A. (1981) *Episodic Behaviour in Badlands. Erosion and Sediment Transport in Pacific*

Rim Steeplands. I.A.H.S. Publication, no. 132, Christchurch. pp. 478–92.

Best, J. (1993) Interactions between flow, transport and bedform, In: N.J. Clifford, J.R. French and J. Hardsity (eds) *Turbulence: Perspectives on Flow and Sediment Transport*. Wiley, Chichester. pp. 61–92.

Biron, P.M., Robson, C., Lapointe, M.F. and Gaskin, S.J. (2004) Comparing different methods of bed shear stress estimates in simple and complex flow fields. *Earth Surface Processes and Landforms* 29, 1403–15.

Biron, P.M., Lane, S.N., Roy, A.G., Bradbrook, K.F. and Richards, K.S. (1999) Sensitivity of bed shear stress estimated from vertical velocity profiles: the problem of sampling resolution. *Earth Surface Processes and Landforms* 23, 133–9.

Bonnet, S. and Crave, A. (2003) Landscape response to climate change: insights from experimental modeling and implications for tectonics versus climatic uplift of topography. *Geology* 31, 123–6.

Buffin-Bélanger, T. and Roy, A.G. (1998) Effects of a pebble cluster on the turbulent structures in a gravel-bed river. *Geomorphology* 25, 249–67.

Buffin-Bélanger, T., Roy, A.G. and Kirkbride, A. (2000) On large-scale flow structure in a gravel-bed river. *Geomorphology* 32, 417–35.

Buffington, J.M. and Tonina, D. (2009) Hyporheic exchange in mountain rivers II: effects of channel morphology on mechanics, scales, and rates of exchange. *Geography Compass* 3, 1038–62. doi:10.1111/j.1749-8198.2009.00225.x.

Carson, M.A. (1971) *The Mechanics of Erosion*. Pion. London.

Carson, M.A. and Griffiths, G.A. (1987) Bedload transport in gravel channels. *Journal of Hydrology* 26, 1–151.

Chow, V.T. (1959) Open Channel Hydraulics. New York: McGraw-Hill.

Crozier, M.J. and Glade, T. (1999) Frequency and magnitude of landsliding: fundamental research issues. *Zeitschrift für Geomorphologie* 115, 141–55.

Dingman, S.L. (1984) *Fluvial Hydrology*. W.H. Freeman. New York.

Gilbert, G.K. (1914) *The Transportation of Débris by Running Water*. United States Geological Survey, Professional Paper 86. p. 263.

Gonzalez-Hidalgo, J.C., Batalla, R.J., Cerda, A. and de Luis, M. (2010) Contributions of the largest events to suspended sediment transport across the USA. *Land Degradation and Development* 21, 83–91. doi:10.1002/ldr.897.

Graf, W.H. (1971) *Hydraulics of Sediment Transport*. McGraw Hill, New York.

Keller, E.A. and Melhorn, N. (1978) Rhythmic spacing and origin of pools and riffles. *Geological Society America Bulletin* 82, 753–6.

Kirchner, J.W., Finkel, R.C., Riebe, C.S., Granger, D.E., Clayton, J.L., King, J.G. and Megahan, W.F. (2001) Mountain erosion over 10 yr, 10 k.y., and 10 m.y. time scales. *Geology* 29, 591.

Kirkbride, A.D. (1993) Observation of the influence of bed roughness on turbulence structure in depth limited flows over gravel beds, In: N.J. Clifford, J.R. French, and

J. Hardisty, (eds) *Turbulence: Perspective on Flow and Sediment Transport.* John Wiley and Sons, Chichester, pp. 185–96.

Kirkbride, A.D. and McLelland, S.J. (1994) Visualization of the turbulent-flow structure in a gravel-bed river. *Earth Surface Processes and Landforms* 19, 819–25.

Kline, S.J., Reynolds, W.C., Schraub, F.A. and Runstadler, P.W. (1967) The structure of turbulent boundary layers. *Journal of Fluid Mechanics* 30, 741–73.

Knighton, A.D. (1998) *Fluvial Forms and Processes: A New Perspective.* Arnold, London.

Leeder, M.R. (1982) *Sedimentology: Process and Product.* Unwin Hyman Ltd., London.

Leeder, M.R. (1983) On the interactions between turbulent flow, sediment transport and bedform mechanics in channelized flows, In: J.D. Collinson and J. Lewin. *Modern and Ancient Fluvial Systems.* Special Publication of the International Association of Sedimentologists. 6, 5–18.

Leopold, L.B., Wolman, M.G. and Miller, J.P. (1964) *Fluvial Processes in Geomorphology.* Freeman, San Francisco.

Lu, Y., Chiew, Y.M. and Cheng N.S. (2008) Review of seepage effects on turbulent open-channel flow and sediment entrainment. *Journal of Hydraulic Research* 46, 476–88. doi:10.3826/jhr.2008.2942.

Mantz, P.A. (1977) Incipient transport of fine grains and flakes by fluids – extended shields diagram. *Journal of the Hydraulics Division* ASCE, 103, 601–15.

Mao, L., Cavalli, M., Comiti, F., Marchi, L., Lenzi, M.A. and Arattano, M. (2009) Sediment transfer in two alpine catchments of contrasting morphological settings. *Journal of Hydrology* 364, 88–98.

Middleton, G. and Southard, J.B. (1984) *Mechanics of Sediment Movement.* SEPM, Short Course Number 3.

Middleton, G. and Wilcock, P. (1994) *Mechanics in the Earth and Environmental Sciences.* Cambridge University Press, Cambridge (UK).

Nezu, I. and Nakagawa, H. (1993) *Turbulence in Open-Channel Flows.* IAHR/AIRH, Monograph series. A.A. Balkema, Rotterdam.

Oke, T.R. (1987) *Boundary Layer Climates.* Methuen, London.

Pickup, G. and Warner, R.F. (1976) Effects of hydrologic regime on magnitude and frequency of dominant discharge. *Journal of Hydrology* 29, 51–75.

Reynolds, O. (1895). On the dynamical theory of incompressible viscous fluids and the determination of the criterion. *Philos. Trans. R. Soc.* 186, 123–164

Richards, K.S. (1982) *Rivers, Form and Process in Alluvial Channels.* Methuen, London.

Robert, A. (2003) *River Processes: An Introduction to Fluvial Dynamics.* Oxford University Press, New York. p. 214.

Robert, A., Roy, A. and De Serres, B. (1992) Changes in velocity profiles at roughness transitions in coarse grained channels. *Sedimentology* 39, 725–35.

Robert, A., Roy, A.G. and De Serres, B. (1996) Turbulence at a roughness transition in a depth-limited flow over a gravel-bed river. *Geomorphology* 16, 175–87.

Roy A.G., Biron, P.M., Buffin-Bélanger, T. and Levasseur M. (1999) Combined visual and quantitative techniques in the study of natural turbulent flows. *Water Resources Research* 35, 871–7.

Roy, A.G., Buffin-Bélanger, T., Lamarre, H. and Kirkbride, A.D. (2004) Size, shape and dynamics of large-scale turbulent flow structures in a gravel-bed river. *Journal of Fluid Mechanics* 500, 1–27. doi:10.1017/S0022112003006396.

Schvidchenko, A.B. and Pender, G. (2001) Macroturbulent structure of open-channel flow over gravel beds. *Water Resources Research* 37, 709–19.

Shields. A. (1936) Application of similarity principles and turbulence research to bed-load movement. *Mitteilunger der Preussischen Versuchsanstalt für Wasserbau und Schiffbau* 26, 5–24.

Smith, M.W., Cox, N.J. and Bracken, L.J. (2007) Applying flow resistance equations to overland flows. *Progress in Physical Geography* 31, 363–87.

Summerfield, M.A. and Hulton, N.J. (1994) Natural controls on fluvial denudation rates in major world wide drainage basins. *Journal of Geophysical Research* 99, 13871–83.

Syvitski, J.P.M., Vörösmarty C.J., Kettner, A.J. and Green, P. (2005) Impact of humans on the flux of terrestrial sediment to the global coastal ocean. *Science* 308, 376–80. doi:10.1126/science.1109454.

Tonina, D. and Buffington, J.M. (2009) Hyporheic exchange in mountain rivers I: mechanics and environmental effects. *Geography Compass* 3, 1063–86. doi:10.1111/j.1749-8198.2009.00226.x.

Whipple, K.X. (2009) The influence of climate on the tectonic evolution of mountain belts. *Nature Geoscience* 2, 97–104. doi:10.1038/NGEO413.

Wilkinson, B.H. (2005) Humans as geologic agents: a deep-time perspective. *Geology* 33, 161–4. doi:10.1130/G21108.1.

Wilkinson, B.H. and McElroy, B.J. (2007) The impact of humans on continental erosion and sedimentation. *Geological Society of America Bulletin* 119, 140–56. doi:10.1130/B25899.1

Williams, P.J. (1982) *The Surface of the Earth: An Introduction to Geotechnical Sciences.* Longman, London.

Wolman, M.G. (1967) A cycle of sedimentation and erosion in urban river channels. *Geografiska Annaler* 49A, 385–95.

Wolman, M.G. and Miller, W.P. (1960) Magnitude and frequency of forces in geomorphic processes. *Journal of Geology* 68, 54–74.

Zacksenhouse, M., Abramovich, G. and Hetsroni, G. (2001) Automatic spatial characterization of low-speed streaks from thermal images. *Experiments in Fluids* 31, 229–39.

19

Sediment Transport and Deposition

Jeff Warburton

Sediment transport and deposition are fundamental geomorphological processes that govern the evolution and preservation of landforms (Selby, 1993; Knighton, 1998). The operation of such processes over time acts to fundamentally shape the landscape. A basic understanding of these processes is therefore essential to understanding how landforms are produced under present-day process regimes and how landscapes have evolved in the past in response to variations in the magnitude and disposition of geomorphic drivers (Summerfield, 1991). What is more, natural sediment transport processes impinge on human activity in a multitude of ways. Obvious examples include damage to buildings and infrastructure by landslides and soil erosion of agricultural land (Statham, 1977). Deposition is also highly significant and this is best illustrated by the importance of alluvial floodplains (Walter and Merritts, 2008). Floodplains have historically been the focus of settlement and growth of civilization due to the fertile floodplain soils and ready access to water. They also provide construction materials (sand and gravel); have been exploited for alluvial metal mining and over longer timescales the rocks from ancient floodplains form important aquifers and hydrocarbon reservoirs as well as providing sedimentary archives of environmental change (Knox, 1972; Bridge, 2003).

However, the development of sediment transport theory and empirical studies of erosion and deposition processes have enabled many of these destructive geomorphic events to be controlled, modified or even accommodated in landscape design (Julien, 1998). Inevitably much of the work on sediment transport and deposition has been undertaken by engineers and geologists but more recently the importance of geomorphology is being increasingly recognized as a key component in this work (Thorne et al., 1997).

The aim of this chapter is to provide a concise overview of sediment transport and deposition in earth surface geomorphic systems. The scope is inevitably constrained by available space; hence the emphasis is on describing the key processes of erosion and deposition. Therefore, weathering, chemical processes, solute transport and geological controls are largely neglected and the specific physics not discussed in detail. These assumptions inevitably simplify and restrict detailed consideration of the full complexity of natural geomorphic systems in which the interactions of physical, chemical and biological components are intricately linked. Nevertheless through the selection of examples and case studies an appreciation of the key concepts governing sediment transport and deposition will emerge.

The chapter is structured around four main sections. It begins with a brief consideration of the sediment cascade and then discusses approaches to estimating erosion, sediment transport and deposition. This considers sediment transport by rivers and the assessment of sediment storage in the landscape. This is followed by a discussion of sediment budgets which includes justification for the sediment budget as a unifying framework and then considers a series of four sediment budget case studies as examples of sediment transfer and deposition systems. The step is made from the catchment scale to the global scale in the last substantive section which considers global sediment yields and denudation.

THE SEDIMENT CASCADE

In most landscapes, bedrock is mantled by regolith: a thin layer of stored sediment which is produced by mechanical breakdown and chemical weathering (see Chapter 16). The dissolved products of chemical weathering are flushed away to and in streamflow; solid sediment is progressively transported down hillslopes into river valleys and eventually enters the river system whence it is transported downstream towards the sea. An important conceptual model in understanding the role of erosion and deposition in geomorphic systems is the sediment cascade (Caine, 1974; Reid and Dunne, 1996; Owens, 2005). In simple terms this depicts the sediment system as a network of sediment stores linked by a series of transfer processes. Using this approach, Caine (1974) provided an overview of Alpine geomorphic processes and described these as a series of sediment flux cascades. This consisted of two dynamic geomorphic subsystems, namely the slopes and stream channels, overlain by three sediment subsystems: the geochemical, the fine sediment and the coarse detritus (Caine and Swanson, 1989; Warburton, 2007). The sediment cascade is driven directly or indirectly by gravity but with different dominant processes as gradients reduce downstream. In terms of the fluvial system as a whole headwater mountain catchments are viewed as sediment production zones feeding bed load and wash load sediment downstream.

A simple schematic model describing the linkages between mountain catchments and variations in fluvial form along the river profile from the headwaters to the coastal lowlands was developed by Nevin in 1965 (Mosley and Schumm, 2001) (Figure 19.1). This shows the transition from the headwater catchments which are dominated by steep channels such as cascades, step-pools systems and coarse braided rivers through to meandering lowland channels. River load is deposited, coarsest first then finer at progressively lower velocity or shear stress. Landforms of fluvial deposition include floodplains, alluvial fans, inland and coastal deltas and offshore fans. Generally, the sediment load in the headwater channels is dominated by coarse bed load (50–90 per cent of the total load) whereas downstream fine suspended load dominates (often exceeding 80–90 per cent of the total annual load) (Figure 19.1) (Reid et al., 1997). It is well known that in the absence of significant tributaries grain-size systematically decreases downstream (Frings, 2008). Any downstream change in bed material flux causes aggradation or degradation (incision) of the channel, and these feedback via slopes to the input and output of sediment. Bed material size at any point in the drainage basin is a function of sediment supply and the combined action of sediment sorting and abrasion (Ferguson et al., 1996). This is often expressed as an exponential decline in grain size with distance downstream:

$$D = D_0 e^{-aL} \tag{19.1}$$

where D is the particle size, D_0 is the initial particle size, L is distance downstream and a is a coefficient representing the combined action of sorting and abrasion. Such simple patterns do not hold for all rivers, especially where there are variations in different sediment source rock types or where tributary inputs of sediment become significant (Rice, 1999). Figure 19.1 is useful in highlighting the down gradient continuity of the sediment cascade but also shows, that along the downstream course, there is continual adjustment in process form relationships and transformation of the transported sediment. This association, between (geo)morphology and sediment transport (erosion/deposition) is controlled by the energy available to move sediment which is derived largely from the local catchment slope. In fluvial geomorphology the general form of this relationship can be neatly generalized in the form of a 'trinity' (Best, 1993) which links morphology, flow and sediment transport in a feedback relationship. This is a useful concept because it emphasizes important feedbacks in the geomorphic system which can manifest themselves at a range of scales from the local to the global (e.g. deposition/erosion on a channel bend to continental erosion and uplift); and over short and long timescales (e.g. from changes in an individual storm to evolution over millennia).

These feedbacks mean that sediment is rarely transported in a continuous manner but often will enter temporary storage before resuming transport at a later date. Therefore although sediment moves along the conveyor, it does so in a jerky fashion. Such movements can be better understood by considering in greater detail the pathways sediment can take when moving between the slope, channel and floodplain domains (Figure 19.2). Figure 19.2 provides an overview of erosion, transport and deposition processes acting in the slope, channel and floodplain geomorphic domains. The diagram highlights the links and feedbacks between processes and shows how these major catchment domains are related. In addition, to the key process pathways the

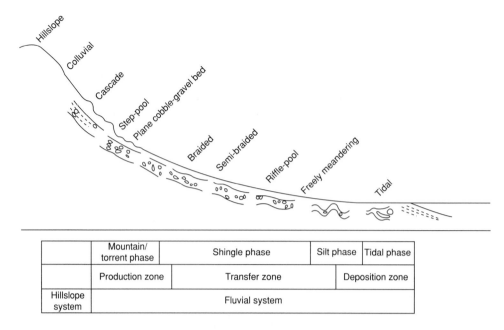

Figure 19.1   Schematic model describing the linkages between mountain catchments and variations in fluvial form along the river profile (Mosley and Schumm, 2001)

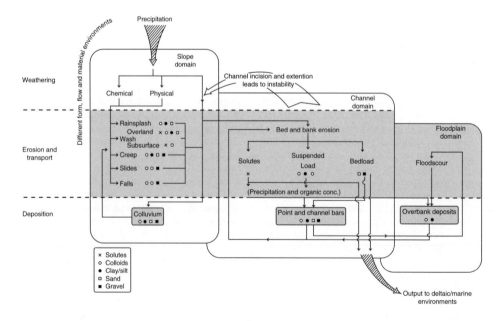

Figure 19.2   Overview of erosion, transport and deposition processes acting in the slope, channel and floodplain geomorphic domains (reproduced with permission from J. Lewin)

dominant grain sizes involved in the sediment transfer are also indicated. These show a general reduction in size from the slope to the channel and floodplain domains. The diagram has many features in common with Figure 19.1 which depicts the fluvial sediment cascade, emphasizing the morphological changes that result from the process links and feedbacks shown in Figure 19.2. Recognizing and quantifying transport processes and sediment storage elements; and identifying the linkages amongst these are the fundamental basis of the sediment budget approach (Reid and Dunne, 1996; Orwin et al., 2010).

ESTIMATING EROSION, SEDIMENT TRANSPORT AND DEPOSITION

Sediment transfer processes are driven by gravity. However, a general distinction can be made between hydrologically related processes and mass movements. Gravity may act directly on the mass being moved (e.g. in a landslide block) or may act indirectly through a fluid such as erosion by flowing water (Statham, 1977). Hence some sediment production is linked to surface runoff pathways (e.g. rainsplash, sheet erosion and rills/gully development) but sediment is also supplied from rockfalls, soil creep and landslides. These processes vary in their magnitude and frequency and can often be classified as widespread, frequent, low magnitude processes (e.g. soil creep, rainsplash, small-scale rockfall) or localized, rare, high magnitude processes (e.g. landslides, debris flows, rock avalanches, etc). Within a particular environment the occurrence of a particular process, and rate of operation, depends on the existing landform assemblage since process is driven either by the slope gradient alone or in combination with upslope contributing area. Processes where the rate depends mainly on the slope gradient alone (e.g. soil creep) are diffusive and lead to typically convex slope profiles. However, where the rate depends on gradient and upslope contributing area, which determines the discharge, then fluvial processes (e.g. slope wash, rills, gullies and streams) dominate, which tend to produce concave slopes.

In the *slope domain* (Figure 19.2) sediment transfer by hillslope processes is often episodic and transfer rates can be either slow in the case of soil creep and solifluction (Ridefelt et al., 2009) or rapid in the case of slope failure (Fort et al., 2009). Magnitude and frequency relationships are, therefore, an important component when assessing mass movement contributions to sediment transfer (Rapp, 1960). There are also distinct differences

in the volume or mass of sediment transferred and subsequent effects on the reworking and mobilization of these deposits by other transfer processes. For example, large slope failures can significantly alter valley morphology by shifting interfluves as well as introducing significant amounts of sediment from high elevations into low elevation fluvial and lacustrine systems (Korup, 2005; Hewitt et al., 2008; Johnson et al., 2010). Mass movement deposits may persist in the landscape for considerable periods of time, elevating sediment transfer fluxes or where deposits act as barriers, interrupting the transfer of sediment downstream (Fort et al., 2009). The variability in both the timing and rate of mass movement sediment transfer by necessity means that monitoring programs need to be long term in order to accurately characterize mass movement transfer, regardless of spatial scale (Warburton, 2007).

In the *channel domain* (Figure 19.2) bed material transport involves erosion and deposition. But, if transport is in equilibrium (no net erosion or deposition) then geomorphic change will be minimal (Church, 2006). However, if a river cannot transport all the coarse sediment supplied to it the excess sediment is deposited and the river will aggrade (Fuller et al., 2003). Conversely if less sediment enters a channel reach than leaves it, the channel will incise (degrade) (Simon and Rinaldi, 2006). Fine sediment (wash load) is less significant in this scenario since there is unlimited capacity for the transport of this material. Aggradation is commonly caused by massive sediment inputs (Korup, 2005) or a reduction in channel slope (e.g. above a dam, Petts and Gurnell, 2005). Degradation, on the other hand is often initiated when the sediment supply is cut off (e.g. below a dam) or the local flow strength is increased possibly by channel shortening (e.g. by straightening meanders).

In the *floodplain domain* (Figure 19.2) deposition occurs through lateral or vertical accretion. Coarse bed load may be laterally accreted onto bars, and bars incorporated into the floodplain. Vertical accretion generally involves fine wash load which is deposited on the floodplain in overbank floods. Sediment may also be deposited in catchments where the transport capacity declines and the flow can no longer transport the excess sediment. This typically occurs in alluvial fans; tributary fans/deltas; above valley blockages (e.g. landslide dams); in lake deltas (including reservoirs); and within alluvial rivers.

Based on these main domains, erosion rates may be estimated at source, in transit or at the site of deposition (sediment sink or store) (Walter and Maerritts, 2008). At-source estimates usually involve the direct measurement of processes or the lowering of a surface over a known period of time

(Cockburn and Summerfield, 1994). Sediment transfer rates involve the measurement of river sediment loads (sediment in transit) (Walling, 1999) and estimates of erosion from deposition are determined from the volume of material accumulated over a known period in depositional areas on land (e.g. alluvial fans, lakes, reservoirs (Morche et al., 2008)) or at and beyond the coast (e.g. deltas, ocean fans) (Islam et al., 1999). Here we focus on sediment transport by rivers and deposition (sediment storage).

Sediment transfer by rivers

Estimating rates of erosion by measuring sediment in transit predominantly relies on the measurement of river sediment loads. These integrate the load from the drainage basin or landscape unit and in principle can be assessed at any scale. However, these only provide estimates of modern rates and are rarely sustained over long measurement intervals (Walling and Webb, 1996; Syvitski et al., 2005). River sediment load consists of three main components relating to the dominant mechanisms of transport. The dissolved load consists of solutes derived mainly from chemical weathering within the catchment (< 1 µm). Wash load/suspended load is mainly composed of fine sediment, which is maintained in suspension by turbulence in the water column and is typically derived from extra-channel sources (hillslopes). Bed load (traction transport) consists of coarse sediment which is transported by rolling, sliding or bouncing along the bed of the river and is transported from sources adjacent to the channel (bed and banks). Saltation is a category of motion in which particles are temporarily launched into the water column but quickly return to the bed following a ballistic trajectory (Church, 2006). This latter category is intermediate between bed load and suspended load. When considering erosion and sedimentation problems (geomorphic change) sediments are most effectively divided into bed material and wash material (Church, 2006). Bed material is coarse material that makes up the bed and lower banks of a river and determines the morphology of the channel. Wash material is fine sediment that once entrained is transported great distances. It is not found in appreciable quantities in the bed and lower banks of the channel and plays only a minor role in channel formation.

During periods of active sediment transport the sediment concentration with depth profile of the river will vary with the competence of the flow, available sediment sizes and turbulence intensity (Gordon et al., 1992). Upward pulses of flow balance the tendency of sediment particles to settle under the influence of gravity. The settling velocity generally increases with larger grain diameters but is very low for silt and clay particles (<0.1 mm). Hence this load can be transported at very low discharge and is referred to as wash load. Wash load is flushed downstream very efficiently and only deposited where the flow is very weak. Thus the transport capacity for very fine sediment is almost unlimited and rivers are efficient gutters for fine sediment eroded from the hillslopes. Rates of wash load transport therefore largely depend on supply from catchment and not the power of local river flow. Concentrations tend to increase with discharge but shows considerable scatter (well below transport capacity). With coarse sediment transport the settling velocity shows greater dependence on sediment grain-size. Sand-sized material is only suspended during floods and gravel hardly ever moves by this mechanism. Hence sand and gravel form the dominant river bed material. Gravel moves by rolling and sliding due to fluid drag (bed load transport) only at high flows and sand moves predominantly as bed load but may be temporarily suspended during floods. In small floods wash load will dominate but as flow strength increases the bed load component will increase and more and more particles will be temporarily suspended and move by saltation.

In alluvial channels bed-material is usually in abundant supply from the channel bed therefore transport rates (the maximum size moved and quantity) is controlled by the flow strength which is usually quantified in terms of shear stress, or sometimes stream power. Shear stress is usually considered the standard measure because it has the strongest empirical and theoretical basis. Shear stress (τ) is the fluid drag per unit area of river bed:

$$\tau = \rho g d s \qquad (19.2)$$

where ρ is the fluid density, g is gravitational constant, d is depth and s is the water surface slope. The relationship between sediment grain size (mm) and shear stress is a threshold process. and transport only occurs when the flow exceeds a critical shear stress which is $\propto$ to the average grain size of the bed. Hence this is much lower for sand-bed than for gravel-bed rivers. Once the threshold is reached transport capacity increases rapidly and sediment may be suspended at very high flows. The thresholds of motion are represented by zones of entrainment which reflect the uncertainty associated with many of the parameters (local slope, relative size, grain shape, etc) that control motion (Bridge, 2003). Generally, fine sand, silt and clay can move in all but the weakest flows. Medium/coarse sand and gravel once entrained will first move as bed load but eventually suspended load as flow strength increases and with a progressive increase in transport rate (Gomez, 1991; Bridge, 2003).The mode of sediment transport largely depends on particle size and strength of fluid flow. Below ~1 μm (0.001 mm) the load is considered as dissolved. From 0.001 to 1 mm this is suspended and coarser than this it is considered as bed load. The maximum size that can be suspended depends on intensity of turbulence, which depends on shear stress and is usually in the sand range (~0.2 to 2 mm).

Fluvial sediment transport is usually measured in accordance with the definition of transport processes and the nature of the concentration depth profile. Consideration of the modes and mechanism is important because this will help determine the most relevant methods for load estimation (Orwin et al., 2010). Dissolved loads and suspended load concentrations are usually directly determined from water samples (Edwards and Glysson, 1988; Walling and Collins, 2000). However, wash load/suspended load concentration time series are often characterized by frequent, non-periodic pulses in suspended sediment (Hammer and Smith, 1983; Stott and Mount, 2007). These suspended sediment pulses are important because they can make up a significant proportion of the fluvial sediment flux (Hodgkins et al., 2003; Lenzi et al., 2003). As a result, temporal changes in suspended sediment concentrations must be adequately characterized in a fluvial sediment-sampling scheme. A solution to capturing these non-periodic changes in suspended sediment concentration is the deployment of continuously recording, datalogger compatible turbidimeters (Orwin et al., 2010). Turbidimeters indirectly monitor suspended sediment concentration through the measurement of turbidity and rely on establishing a statistically significant relationship between turbidity and concentration (Gippel, 1995). The major benefit of using turbidimeters to monitor suspended sediment concentrations is the ability of these instruments to capture non-periodic, transient pulses of sediment. This increase in temporal resolution significantly reduces the amount of error in sediment flux calculations and allows for more detailed examination of suspended sediment dynamics.

When measuring bed load transport, a range of direct and indirect methods are available. Short-term, local bed load transport rates may be determined using portable samplers (<100% efficient for gravel (Sterling and Church, 2002), continuously recording traps (Gomez, 1991), impact sensors or measurements of dune migration. Longer term estimates may be derived from tracer pebbles (Ferguson et al., 1996) or inversely calculated from channel change (morphologic approach). However, due to the ephemeral nature of transport there is great reliance on sediment transport formulae which are mostly semi-empirical and based on flume data. Bed load transport equations use the premise that specific relations exist between the sedimentology and hydraulics of a river channel as a basis for predicting bed load transport rates. Most formulae are based on a correlation between bed load transport and bed shear stress, stream discharge (e.g. the Schoklitsch equation), stochastic functions for sediment movement or stream power. In sand-bed rivers where the bed load threshold is so low that it can be neglected, total load equations such as Engelund–Hansen can be used (Gomez and Church, 1989). Alternatively, in gravel-bed rivers, where a critical transport threshold exists and transport rate increases rapidly with flow strength, one of the most widely used is the Meyer–Peter-Müller equation. But despite the widespread use of these formulae, several recent studies have questioned their value (Gomez and Church, 1989). More recently the morphologic approach to fluvial sediment budgets has grown greatly in popularity following a number of publications outlining the technique (Martin and Church, 1995; Ham and Church, 2000; Fuller et al., 2003) and the increased accuracy and speed of modern topographic survey equipment (Rumsby et al., 2008). Although requiring much less field effort than some of the methods outlined above, Martin and Church (1995) note that there are a number of methodological issues, including the requirement for at least one transport rate measurement and the assumption that there has not been equal erosion and deposition between surveys.

Once sediment transport concentrations and rates have been estimated loads are determined by multiplying the concentration (C, g m^{-3}) by water discharge (Q, m^3 s^{-1}) to obtain instantaneous load ($L = CQ$, g s^{-1}) (Ferguson, 1986). River sediment loads vary enormously being much higher in floods but much lower at normal low flow. This presents a challenge when estimating longer term yields and when attempting to determine annual sediment loads (t y^{-1}) a carefully designed sampling programme is required (Orwin et al., 2010). Annual sediment loads relate to a specific drainage area so if the annual load is divided by catchment area a specific sediment yield (t km^{-2} y^{-1}) can be estimated. This can be converted into an equivalent average denudation rate ('lowering') of the catchment; where 1 t km^{-2} y^{-1} ≈ 0.4 mm ky^{-1}. The bed load component is far more difficult to measure but in many rivers is usually less than suspended load and is often excluded from sediment load calculations.

Assessing sediment storage

In many environments sediment fluxes may be ungauged and the availability of sediment transport records may simply be too short to assess inter-annual variability and trends. An alternative is therefore to evaluate sediment fluxes through the sedimentary record contained in terrestrial sediment stores. Sedimentary deposits cover most of the land surface and are highly variable, consisting of a range of grain sizes derived from rock or biological material. Sediments cover a million-fold range in grain size, <1 μm to >1 m. Information on the grain size of sediment is usually plotted on a logarithmic scale (mm) with descriptors based on a scale with powers of 2 (Knighton, 1998). The most common descriptors include 'boulder' >256 mm, 'gravel' 2–256 mm, 'sand' 0.062–2 mm, 'silt' 0.062–0.002 mm, 'clay' <0.002 mm (2 μm). Depending on the dominant grain size, sediments may be cohesive or non-cohesive. Sand and coarser material is non-cohesive and derives strength through (intergranular) friction at grain contacts. Clay, and some silts, are cohesive and possess additional strength through short-range electrostatic bonding.

Sediment sinks hold the potential for evaluating sediment fluxes over a wide range of timescales, typically between decades to millennia, and potentially provide high resolution sedimentary records (Orwin et al., 2010). Sediment storage volumes in alluvial fans, lakes/reservoirs, floodplains, deltas and deep-sea basins have typically been estimated. These provide time and space integrated estimates of sediment accumulation. Several approaches have been adopted to interpret recent sediment fluxes from sedimentary records (e.g. Lamoureux, 2000; Schiefer and Gilbert, 2007). Dating using ^{210}Pb and ^{14}C methods have been used to infer catchment sediment yield estimates (Foster, 2006) and document changes in sedimentation over intervals of typically centuries or longer. Many such studies point to the importance of anthropogenic processes in controlling fluvial sediment fluxes (Dearing and Jones, 2003).

The ability to assess sediment storage has been greatly advanced in recent years by the application of geophysical survey methods (Hoffman and Schrott, 2002; Schrott and Sass, 2008) and terrestrial laser scanning. A good example of the geophysical approach is demonstrated by Hoffmann and Schrott (2002) who measured the sediment thicknesses in an Alpine valley (Reintal, Bavaria) using two-dimensional seismic refraction techniques (Schrott et al., 2003). They found that geophysical derived sediment thicknesses (3–24 m) were significantly smaller than those estimated from morphometrical analysis. This led to significant differences in rockwall retreat rates based on the revised estimates of sediment storage volumes. Schrott and Sass (2008) provide a useful overview of the relative merits of ground penetrating radar, one-dimensional/two-dimensional resistivity and seismic refraction measurements in characterising sediment storage units. Terrestrial laser scanning provides an effective means of capturing high-resolution data very rapidly (Heritage and Hetherington, 2007; Heritage and Large, 2009). The technique is particularly valuable in geomorphological studies of rapid landscape change as the survey technique is very fast and can collect complex surfaces at high resolution (mm accuracy over a 100 m range); can be executed remotely (e.g. away from a dangerous or hazardous environment) and provides a complete archive of the current scene (Heritage and Large, 2009). Morche et al. (2008), as part of a larger sediment budget study, used terrestrial laser scanning to quantify the change in an actively eroding talus cone in the Riental valley over the summer period June to September 2006. The method proved very useful as the sediment store had complex topography, was inaccessible at high flow and very unstable for direct access.

THE SEDIMENT BUDGET AS A UNIFYING FRAMEWORK

As Anderson (2008) points out there is no general theory of geomorphology and the topic covers an array of disciplines and a multitude of different environmental settings. Yet, there is order in the natural landscape which betrays some commonality of earth surface processes. Anderson (2008) suggests four main features are common in many geomorphic settings:

1 Surface materials tend to move in one dominant direction be it downhill, downstream, down drift or downwind (e.g. sediment cascade, Figure 19.1).
2 Materials are transformed as they move through geomorphic systems often involving the breakdown of bedrock, sediment grains and soils into smaller constituent parts. Significant chemical changes also occur.
3 Motion is 'concentrative' as material gathers into more efficient streams resulting in self-organized branching systems.
4 Over sufficient time the geomorphic system will evolve to a state in which the material flow is adjusted to transport the sediment supplied to it.

Collectively, these features provide some universality within geomorphology which connect elements of the landscape through material flows (sediment transfers) which can be couched in terms of conservation of mass, energy and momentum. Conservation of mass is particularly important because it forms the basis of the sediment mass balance equation which relates sediment inputs (I = from upstream, tributaries and slopes) to outputs (O = downstream sediment movement) and a change in sediment storage (ΔS = deposition and erosion of valley, floodplain and lacustrine sediment stores):

$$I = O \pm \Delta S \qquad (19.3)$$

The sediment mass balance equation is the fundamental concept that underpins sediment budget studies (Reid and Dunne, 1996; Warburton, 2007).

A sediment budget is a quantitative statement of the rates of sediment production, transport and discharge of detritus which accounts for the sources and disposition of sediment as it travels from its point of origin to its eventual exit from a drainage basin (Reid and Dunne, 1996). Sediment budgets provide an understanding of current sediment production and transport regime in a catchment and are usually evaluated in terms of a simple sediment balance (Equation (19.3)). Three elements are needed to construct a sediment budget (Swanson et al., 1982):

1 Recognition and quantification of transport processes
2 Recognition and quantification of storage elements
3 Identification of linkages amongst transport processes.

Hence this is distinct from the sediment yield which is the total mass of particulate material reaching the outlet of a drainage basin or landscape unit. This is usually evaluated on an annual basis and expressed in either tonnes per year (t y^{-1}) or as a specific sediment yield (t km^{-2} y^{-1}).

The sediment delivery ratio (SDR) is a related concept which describes the ratio between sediment output (sediment yield, Y) and sediment production (erosion, E) from a particular landscape unit or drainage area (SDR = Y/E). This provides a gross *estimate* of the sediment balance of a drainage area but does not really convey what is actually happening to the sediment in a river basin. Therefore in terms of the understanding of drainage basin sediment systems these concepts can be ranked: sediment yield → specific sediment yield → sediment delivery ratio → sediment budget, in ascending order of sophistication.

Recognizing and quantifying sediment stores and the processes which link them is the basis of the sediment budget approach. The seminal work of Rapp (1960), working in the Kärkevagge catchment in Sweden, used this methodology and inspired others to follow a similar approach in identifying the significance of sediment storage in regulating the sediment yield from mountain catchments (Church and Ryder, 1972, Church and Slaymaker, 1989).

Dietrich and Dunne (1978) provide an early example of the methodology used to construct a small catchment sediment budget. Their example, constructed for the 16 km^2 Rock Creek basin in the Coast range of Oregon, is based on field mapping of sediment accumulations and from data available in the geomorphic literature; estimates are checked against sediment yields measured in nearby basins. The sediment budget produced using this approach is summarized in Figure 19.3 and shows the significant processes controlling sediment transfer in the Rock Creek basin. In this diagram the rectangles represent storage elements, octagonals indicate transfer processes and circles show outputs. The solid lines indicate transfers of sediment and dotted lines migration of solutes. This type of system diagram has been adopted widely in design sediment budget programmes (Warburton, 1990; Reid and Dunne, 1996). The budget highlights the importance of the linkages that need to be identified before measurements of individual processes can be combined. The definition of key linkages requires intensive fieldwork to estimate the frequency of processes and define the extent and magnitude of sediment stores. The approximate sediment budget which was developed identified key areas for more detailed monitoring. In particular, it was important to monitor soil creep and deep creep associated with soil wedges. In addition the approximate sediment budget also emphasized the important role of valley floor sediment storage (debris fans at tributary mouths) which, if reactivated, would have a large effect on the sediment yield from the basin. However, adequately characterizing the sediment balance equation is notoriously difficult even for small catchments and the results, often measured over the short term, are extremely difficult to extrapolate from year to year and to adjacent areas in similar environments (Lenzi et al., 2003; Johnson et al., 2010).

SEDIMENT BUDGET CASE STUDIES OF SEDIMENT TRANSFER AND DEPOSITION

Sediment budget case studies provide examples of sediment transport and deposition processes acting within landscape units. The examples below include a hillslope sediment budget; a channel sediment budget (Duijsings, 1987); a catchment

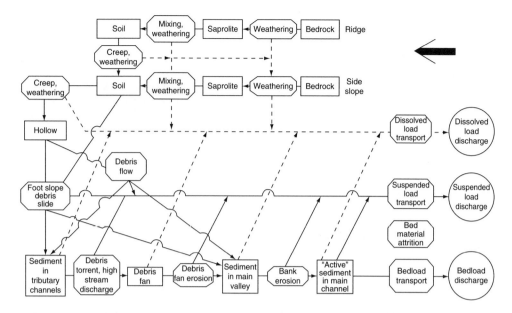

Figure 19.3   Sediment budget model for the Rock Creek basin. Rectangles represent storage systems. Octagonals indicate transfer processes. Circles represent outputs. Solid lines represent the transfer of sediment and dotted lines represent the migration of solutes (Dietrich and Dunne, 1978)

sediment budget (Walling et al., 2001) and a paired sediment budget examining change over time (Trimble, 1981, 1983). Collectively they illustrate the main features of the sediment budget approach.

Hillslope sediment budget

The slope domain is a very active geomorphic zone and the observed activity will depend on the relief, materials and key environmental drivers (precipitation and runoff) (Figure 19.2). Figure 19.4 shows the results of a sediment budget investigation of hillslope erosion of a small 6 km^2 upland catchment in Cumbria, northern England, following a large storm in October 2008. The slope domain in this example is partitioned into general hillslopes, footpaths and areas of mine waste reflecting 'natural' and human impacted slopes. Results clearly show the importance of catchment disturbance in triggering instability in the landscape. Most notable are the high levels of activity (channel erosion, slope undercutting and debris flows) associated with abandoned mine waste. Locally, footpaths also have an impact through small-scale landslides and delivery of sediment from slope wash. The more 'natural' hillslopes are only affected by channel erosion.

However, in terms of sediment delivery to the main fluvial system (channel domain, Figure 19.2) only a small proportion of the sediment is delivered from the mine waste (6 per cent) but footpaths, due to their connectivity with tributary stream course and landslide runout connection to the main stream channel, deliver more of the eroded sediment (48 per cent). In absolute terms, because of the great connectivity of the footpaths with water courses, the main source of sediment are the footpaths which have triggered local slope instability.

Channel sediment budget

Duijsings (1987) constructed a channel-based sediment budget for the forested Schrondweiler-baach catchment on the Luxembourg Keuper area to assess the importance of streambank erosion in controlling sediment output. Over a 2-year study it was shown that stream bank erosion contributed 53 per cent of the total sediment supply, with lateral corrosion (43 per cent) and subsoil fall (39 per cent) being the main sediment producing bank erosion processes. However, differences in the annual sediment budgets between the 2 years of monitoring are very marked with almost twice as much geomorphic activity in the second year.

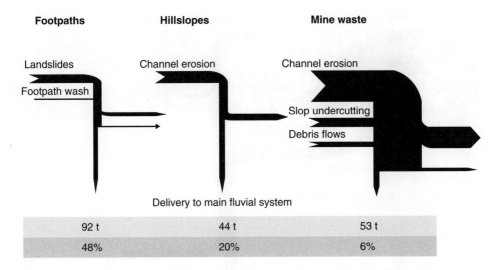

Figure 19.4 Hillslope sediment budget: flood-based upland sediment budget for a small catchment (6 km²) in northern England. Hillslope activity is partitioned into footpath and mine waste disturbance zones alongside more natural areas

This relates to more intense frost action (controlling soil fall and lateral corrosion) and the occurrence of a major storm which evacuated about 43 per cent of the annual sediment output in a 3-day period. Furthermore, the sediment data do not tell the full erosion story becaue dissolved load output for this small forested catchment, despite the actively eroding channel, is twice as large as the suspended load flux. Interpreting comparisons should be treated cautiously because the short record of data collection cannot be easily extrapolated beyond the period of record as not all geomorphic processes may have been adequately characterized (e.g. bank failures).

Catchment sediment budget – the snapshot

Figure 19.5 shows the results of an integrated approach to sediment budget construction for the upper Kaleya catchment, southern Zambia (Walling et al., 2001). The 'integrated' approach uses fallout radionuclides to estimate soil redistribution and floodplain deposition rates, sediment fingerprinting to establish sediment sources, direct sampling to document fine sediment on the channel bed and continuous turbidity monitoring to quantify suspended sediment flux from the catchment (Walling and Collins, 2000, 2008; Owens, 2005). The budget for the 63 km² catchment indicated that areas of communal cultivation and bush grazing were the most important

sediment sources, but a large proportion of this sediment was deposited before reaching the main channel. There was significant sediment storage in reservoirs and on the floodplain. As a consequence, the overall sediment delivery ratio was only about 9 per cent. This is significant from a management point of view because soil erosion control strategies within the catchment are unlikely to have a significant impact on the sediment flux at the catchment outlet (Walling and Collins, 2008). Furthermore, although bank erosion represents a relatively minor component of the overall budget it is supplied directly to the channel (SDR ~ 100%). Hence this component could be targeted in an attempt to reduce fine sediment supply to the lower reaches of the river (Owens, 2005).

Catchment sediment budget – historical comparison

A widely cited example of the catchment sediment budget is the Coon Creek catchment in Wisconsin (Trimble, 1983, 1999). Sediment budgets for this 360 km² catchment were originally established for two periods, 1853–1938 and 1938–1975, based on a range of sedimentological and morphological evidence. During the first period, poor land management resulted in severe soil erosion leading to extensive upland sheet and rill erosion. However, this only resulted in a sediment delivery ratio of 55 per cent because

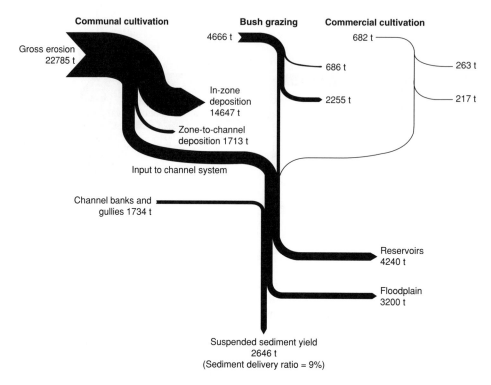

Figure 19.5   Catchment sediment budget: fine sediment budget of the upper Kaleya catchment, southern Zambia (Walling et al., 2001)

much of this sediment remained in the basin, being redeposited on hillslopes and in the lower valleys. The second sediment budget, 1938–1975, records the impact of widespread soil conservation measures within the catchment. These measures proved very effective, reducing upland soil erosion by approximately 25 per cent. However; the catchment sediment yield remained approximately the same due to remobilisation of previously stored sediment. Trimble (1999) published an update to this story by constructing a third sediment budget for the period 1975–1993. This demonstrated a decreased rate of alluvial sediment storage (~ 6% of the 1930s rate); however, sediment yield still remained relatively constant despite significant changes in within-catchment sediment transfer processes. This is an excellent example which demonstrates the utility of the historical sediment budget comparison by identifying the significance of sediment storage in the catchment and the long-term adjustment of the erosion deposition system. In this respect sediment production, transport and storage are characterized by marked spatial and temporal variability which needs to be accounted for when constructing sediment budgets (Walling and Collins, 2008).

GLOBAL SEDIMENT YIELDS AND DENUDATION

Sediment yield is defined as the total output of sediment from a catchment or landscape unit over a particular period of time. This is conventionally expressed in terms of t y^{-1}. It is calculated based on measurements of sediment transfer usually measured as the suspended load component of the fluvial sediment flux but may also includes bed load. Denudation rates are quoted either as a depth per unit time, e.g. mm ky^{-1}, or a mass per unit area per unit time (e.g. t km^{-2} y^{-1}). For typical rock densities, 1 mm ky^{-1} is about 0.4 t km^{-2} y^{-1} or 4 kg ha^{-1} y^{-1}. The geomorphic activity in a particular environment can be estimated by measuring the sediment yield from the catchments draining such areas. There have been several attempts to interpret global patterns of fluvial sediment yield (net erosion) in terms of the factors controlling sediment delivery (Walling and Webb, 1983; Jansson, 1988; Milliman and Syvitski, 1992; Summerfield and Hulton 1994; Walling and Webb, 1996). Walling and Webb (1983) mapped the global distribution of suspended sediment yield showing that although some exceptions exist, such as the low suspended sediment yields from areas of old tablelands and continental shields (Russia, Canada, Australia, Africa), the correspondence between areas of high relief, active tectonics and seismicity and elevated suspended sediment delivery at the global scale is clearly apparent. The importance of drainage basin topography in influencing mechanical denudation rates is clearly demonstrated in the study of Summerfield and Hulton (1994). Their results showed a relatively strong statistical association between basin relief and mechanical denudation; albeit partly a function of other factors related to relief such as seismicity and weak rock structure. Therefore the main factor controlling the broad global patterns in sediment yield is relief. This is distinct from absolute elevation because it is the local topography (height range and mean slope) in an area which is significant. Hence some of the highest sediment yields are from rivers draining young mountain regions in Asia and the Andes (Warburton, 2007).

However, the spatial variation present in maps of suspended sediment yield is very large. Explaining these global geographical patterns is difficult because even correlations with rainfall show hardly any trend (Langbein and Schumm, 1958; Milliman and Meade, 1983) and such patterns are also greatly influenced by variations in surface cover which protects the surface from erosion (splash and surface flow). These may include land use change, which has an important local impact (Walling and Collins, 2008), but these changes rarely affect much of a large catchment. Geology can be important due to its impact on topography and sediment characteristics but much more significant is sediment storage in catchments. As shown in sediment budget studies, the material produced by hillslope erosion is not delivered to the main river system; sediment is stored as colluvium (slope deposits), alluvium (fans, floodplains) and in lakes and reservoirs (which may reduce global sediment delivery to oceans by ~66%, Walling (2008)). For example, in Bangladesh of the ~1000 Mt y^{-1} sediment produced from the Ganges and Brahmaputra river basins only about 51 per cent reaches the sea (Islam et al., 1999). In addition variability also arises due to the incommensurate nature of the measurement of fluvial sediment fluxes caused by differences in the methods used to estimate sediment transfer in catchments (Harbor and Warburton, 1993; Hallet et al., 1996; Warburton, 1999).

These data have also been used to construct global sediment budgets. Estimates of the total load delivered to the oceans vary enormously but recently have been estimated to be ~12 to 36 Gt y^{-1}. This translates into a global average

Table 19.1 Estimates of the major components of the global sediment budget and their modification by human activity (from Syvitski et al., 2005 and Walling, 2008)

Component	Syvitiski et al. (2005)	Walling (2008)
Pre-human land-ocean flux (Gt y⁻¹)	14.0	14
Contemporary land-ocean sediment flux (Gt y⁻¹)	12.6	12.6
Reduction in flux associated with reservoir trapping (Gt y⁻¹)	3.6	24
Contemporary flux in the absence of reservoir trapping (Gt y⁻¹)	16.2	36.6
Increase in pre-human flux due to human activity (%)	16	160
Reduction in contemporary gross flux due to reservoir trapping (%)	22	66

specific sediment yield of ~120 to 360 t km^{-2} y^{-1}. Table 19.1 compares the global sediment budget estimates of Syvitski et al. (2005) with the revised estimates from Walling (2008). The data presented by Syvitski et al. (2005) suggest human activity has increased sediment flux to the oceans by 16 per cent, with reservoirs having an important role in regulating the terrestrial-ocean sediment flux. Walling (2008), using new data and different assumptions regarding to conveyance losses and dam trapping efficiencies, arrives at a very different estimate of the reduction in flux associated with reservoir trapping. The revised estimate suggests human activity has increased the sediment flux by up to 160 per cent. Although highly speculative, these important calculations emphasize our lack of accurate knowledge about the global sediment budget, particularly the extent it has been perturbed by human activity (Walling, 2008).

The significance of these large-scale patterns of sediment yield raise interesting questions about landscape evolution and also the relative efficiency of erosion processes. Over long timescales there has been considerable interest in evaluating the feedbacks between form and denudation processes (Bierman and Nichols, 2004; Slaymaker, 2008). A well-established hypothesis is that actively uplifting mountain belts have high relief which leads to greater erosion and yet exist in a state of dynamic equilibrium where uplift approximately equals erosion. However, as uplift rate declines, erosion begins to dominate and landscape relief is reduced leading to lower erosion rates (Molnar and England, 1990). In order to test these types of hypothesis erosion and deposition rates need to be estimated. This requires the application of a range of geomorphological techniques which include: annual sediment yields (Jansson, 1988), fallout radionuclides (Owens et al., 1997), radiocarbon dating (Macklin and Rumsby, 2007), cosmogenic surface exposure age techniques (Blanckenberg, 2005), and low-temperature thermochronometry (Summerfield, 1991).

This general approach has also been used to assess more specific geomorphic questions such as the relative efficacy of fluvial and glacial erosion over modern and orogenic timescales (Koppes and Montgomery, 2009). This has been a long standing debate in geomorphology (Harbor and Warburton, 1992) but has been more rigorously investigated recently due to improved techniques. Conventionally, estimates of fluvial sediment yield (including measurements of bed load, suspended load and dissolved load flux) have been widely used to infer the relative importance of glacial versus non-glacial processes (Harbor and Warburton, 1992; 1993; Hallet et al., 1996); to assess the impact of climate change on cold climate land systems (Hodgkins et al., 2003; Stott and Mount, 2007) and are used to compare catchments from different cold environments (Warburton et al., 2007).

Koppes and Montgomery (2009) present a recent revised compilation of global erosion rates which questions the conventional wisdom on glaciers and erosion (Figure 19.6). Their overall conclusion is that tectonics controls rates of both fluvial and glacial erosion over millennial and longer timescales and the highest rates of erosion (> 10 mm y^{-1}) result from a transient response to disturbance by volcanic eruptions, climate change and modern agriculture. Figure 19.6a shows erosion rates calculated from measurements of sediment yield over short timescales (1–10 years). This contemporary data compilation clearly shows that volcanic rivers have the greatest erosion rates but, surprisingly, agricultural lands have similar erosion rates to both glaciers and rivers in tectonically active regions (Koppes and Montgomery, 2009). This adds increasing support to the hypothesis that modern sediment yields substantially reflect the impact of human activity (Walling, 2008) (Table 19.1). Comparing erosion rates over longer timescales is problematic due to the incommensurate nature of the data (Harbor and Warburton, 1993). However, with the improvement of methods for estimating erosion rates over differing timescales there are now a suite of

techniques which make such comparisons meaningful. Figure 19.6b shows a comparison between short-term and long-term erosion rates from fluvial and glaciated basins over a variety of timescales from 10^0 to 10^8 years. A variety of methods are used for erosion rate determination. Results demonstrate that over the late Cenozoic erosion rates compared across rivers in the same or proximal regions in five different settings (Himalaya, Taiwan, Italian Apennines, Oregon Coast Range and Australian craton) do not appear to have varied appreciably, although the efficiency of erosion between the tectonically active areas and passive landscapes varies over two orders of magnitude. These detailed analyses support the general patterns shown in and related analyses that show that tectonically active areas of high relief have the highest erosion rates (Summerfield and Hulton, 1994; Dedkov, 2004).

CONCLUSIONS AND PROSPECT

Earth surface sediment transport and deposition are the processes that shape the land surface and are therefore central to any study of geomorphology. The sediment budget approach provides an excellent framework for discussing this topic because it provides a unifying framework for considering sediment erosion, transfer and deposition; it requires conceptualization of landforming processes and hence forces scientists to view the landscape as an ordered system; it emphasizes sediment 'connectivity'; and can also be applied at many scales. Application of the sediment budget framework and integration with long-term sedimentary records will improve understanding of sediment flux and allow better assessment of the sensitivity of sediment transport and deposition systems to environmental change. Long-term, high-resolution sedimentary records offer an opportunity to evaluate magnitude–frequency characteristics of sediment fluxes. These records are important to extend available instrumental records in order to document the range of potential sediment flux behaviour in a given catchment, and to establish a quantitative frequency-magnitude framework to interpret sediment yields.

One of the major barriers to our understanding of sediment flux is the paucity of high temporal resolution, long-term data sets that characterize sediment flux from source to sink. The data available indicate that sediment flux varies significantly in space and time due to differences in sediment sources, transfer processes, the legacy of past glacial activity and changes in the intensity and duration of precipitation events (Gurnell

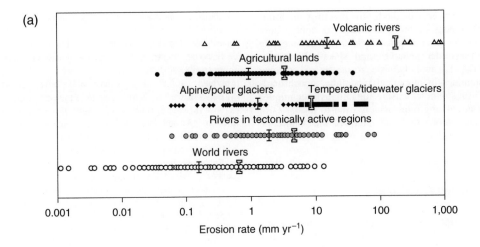

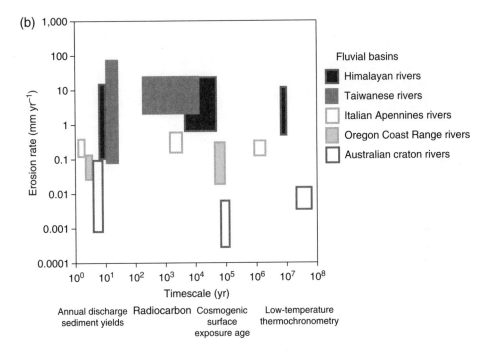

Figure 19.6   Comparison of short-term and long-term erosion rates from glaciated and fluvial basins. (a) Short-term erosion rates calculated from measurements of sediment yield over timescales of 1–10 years. The median of each dataset is shown by black bars, the mean by white bars (Koppes and Montgomery, 2009). (b) Erosion rates measured in the same or adjacent fluvial basins in a range of orogens. Boxes represent errors in estimation (vertical) and timescale of measurement (horizontal)

and Clark, 1987). Unfortunately, accurate characterization of this variability has been limited primarily by data record length and differences in instrumentation and sampling strategies. As a result, fully integrated studies on sediment flux variation from source to sink in cold environments are still relatively rare.

The significance of variations in sediment transport and deposition can only be established if we have good contemporary process understanding;

the necessary methods to carefully monitor such changes; and knowledge to apply these methods at appropriate scales which integrate both sediment sources and sinks (Warburton, 2007). Fortunately, due to advances in new techniques in numerical modelling, geochronology and remote sensing, quantitative techniques are now being more widely used to model earth surface processes and interpret the landscape (Pelletier, 2008).

ACKNOWLEDGEMENTS

This chapter has benefited from a number of research and teaching collaborations. In particular I am grateful to Rob Ferguson, Scott Lamoureux, John Lewin and John Orwin who have allowed me to draw on collaborative material and some unpublished work. I would also like to acknowledge the NERC UK for funding to carry out work on which Figure 19.4 is based (NERC grant NE/G015309/1). Assistance with preparation of the diagrams was provided by the DIU, Department of Geography, and Durham University.

REFERENCES

Anderson, R.S. (2008) *The Little Book of Geomorphology: Exercising the Principle of Conservation.* http://instaar.colorado.edu/~andersrs/publications.html (Accessed, June 2010).

Best, J.L. (1993) On the interactions between turbulent flow, sediment transport and bedform development: some considerations from recent experiment research, in N.J. Clifford, J.R. French and J. Hardisty (eds) *Turbulence: Perspectives on Flow and Sediment Transport.* Hoboken, NJ, John Wiley, 61–92.

Bierman, P.R. and Nichols, K.K. (2004) Rock to sediment – slope to sea with ^{10}Be – rates of landscape change. *Annual Review of Earth and Planetary Science* 32, 215–55.

Bridge, J.S. (2003) *Rivers and Floodplains.* Oxford, Blackwell Science.

Caine, N. (1974) The geomorphic processes of the alpine environment, in J.D. Ives and R.G. Barry, (eds) *Arctic and Alpine Environments.* London, Methuen, pp. 721–48.

Caine, N. and Swanson, F.J. (1989) Geomorphic coupling of hillslope and channel systems in two small mountain basins. *Zeitschrift für Geomorphologie* 33, 189–203.

Church, M. (2006) Bed material transport and the morphology of alluvial river channels, *Annual Reviews Earth and Planetary Science* 34, 325–54.

Church, M. and Ryder, J.M. (1972) Paraglacial sedimentation: a consideration of fluvial processes conditioned by glaciation. *Bulletin Geological Society of America* 83, 3059–71.

Church, M. and Slaymaker, O. (1989) Disequilibrium of Holocene sediment yield in glaciated British Columbia. *Nature* 337, 452–4.

Cockburn, H.A.P. and Summerfield, M.A. (2004) Geomorphological applications of cosmogenic isotope analysis. *Progress in Physical Geography* 28, 1–42.

Dearing, J.A. and Jones, R.T. (2003) Coupling temporal and spatial dimensions of global sediment flux through lake and marine sediment records. *Global and Planetary Change* 39, 147–68.

Dedkov, A. (2004) The relationship between sediment yield and drainage basin area, in V.N. Golosov, V.R. Belyaev and D.E. Waling (eds) *Sediment Transfer Through the Fluvial System.* IAHS. 288, 197–204.

Dietrich, W.E. and Dunne, T. (1978) Sediment budget for a small catchment in mountainous terrain, *Zeitschrift für Geomorphologie* 29(Suppl), 191–206.

Duijsings, J.J.H.M. (1987) A sediment budget for a forested catchment in Luxembourg and its implications for channel development. *Earth Surface Processes and Landforms* 12, 173–184.

Edwards, T.K. and Glysson, G.D. (1988) Field Methods for Measurement of Fluvial Sediment. U.S. Geological Survey Open-File Report 96-531, p. 118.

Ferguson, R.I. (1986). River Loads Underestimated by Rating Curves. *Water Resources Research* 22, 74–76.

Ferguson, R.I., Hoey, T., Wathen, S. and Werritty, A. (1996) Field evidence for rapid downstream fining of river gravels through selective transport, *Geology* 24, 179–82.

Fort, M., Cossart, E., Deline, P., Dzikowski, M., Nicoud, G. and Ravenel, L. (2009) Geomorphic impacts of large and rapid mass movements: a review. *Geomorphologie – Relief Processus Environnement* 1, 47–63.

Foster, I.D.L. (2006) Lakes in the sediment delivery system, in P.N. Owens and A.J. Collins (eds) *Soil Erosion and Sediment Redistribution in River Catchments.* CAB International, Wallingford. pp. 128–42.

Frings, R.M. (2008) Downstream fining in large sand-bed rivers. *Earth-Science Reviews* 87, 39–60.

Fuller, I.C., Large, A.R.G., Charlton, M.E., Heritage, G.L. and Milan, D.J. (2003) Reach-scale sediment transfers: an evaluation of two morphological budgeting approaches. *Earth Surface Processes and Landforms* 28, 889.

Gippel, C.J. (1995) Potential of turbidity monitoring for measuring the transport of suspended solids in streams. *Hydrological Processes* 9, 83–97.

Gomez, B. (1991) Bed load transport. *Earth-Science Reviews* 31, 89–132.

Gomez, B. and Church, M. (1989) An assessment of bed load sediment transport formulae for gravel bed rivers. *Water Resources Research* 25, 161–86.

Gordon, N.D., McMahon, T.A. and Finlayson, B.L. (1992) *Stream Hydrology: An Introduction for Ecologists.* John Wiley & Sons, Chichester.

Gurnell, A.M. and Clark, M.J. (eds) (1987) *Glacio-fluvial Sediment Transfer: An Alpine Perspective.* John Wiley and Sons, London.

Hallet, B., Hunter, L. and Bogen, J. (1996) Rates of erosion and sediment evacuation by glaciers: a review of field data and their implications. *Global and Planetary Change* 12, 213–35.

Ham, D.G. and Church, M. (2000) Bed-material transport estimated from channel morphodynamics: Chilliwack River, British Columbia. *Earth Surface Processes and Landforms* 25, 1123–42.

Hammer, K.M. and Smith, N.D. (1983) Sediment production and transport in a proglacial stream: Hilda Glacier, Alberta, Canada. *Boreas* 12, 91–106.

Harbor, J. and Warburton J. (1992) Glaciation and denudation rates. *Nature* 356, 751.

Harbor, J. and Warburton, J. (1993) Relative rates of glacial and non-glacial erosion processes in Alpine environments. *Arctic and Alpine Research* 25, 1–7.

Heritage, G. and Hetherington, D. (2007) Towards a protocol for laser scanning in fluvial geomorphology. *Earth Surface Processes and Landforms* 32, 66–74.

Heritage, G.L. and Large, A.R.G. (eds) (2009) *Laser Scanning for the Environmental Sciences.* Wiley–Blackwell, Chichester. 288p.

Hewitt, K., Clague, J.J. and Orwin, J.F. (2008) Legacies of rock slope failures in mountain landscapes. *Earth Science Reviews*, 87, 1–38.

Hodgkins, R., Cooper, R., Wadham, J. and Tranter, M. (2003) Suspended sediment fluxes in a high-Arctic glacierised catchment: implications for fluvial storage. *Sedimentary Geology* 162, 105–17.

Hoffman, T. and Schrott, L. (2002) Modelling sediment thickness and rockwall retreat in an Alpine valley using 2D-seimic refraction (Reintal, Bavarian Alps). *Zeitschift für Geomorphologie* 127, 153–73.

Islam, M.R., Begum, S.F., Yamaguchi, Y. and Ogawa, K. (1999) The Ganges and Brahmaputra rivers in Bangladesh: basin denudation and sedimentation. *Hydrological Processes* 13, 2907–23.

Jansson, M.B. (1988) A global survey of sediment yield. *Geografiska Annaler* 70A, 81–98.

Johnson, R.M., Warburton, J., Mills, A.J. and Winter, C. (2010) Evaluating the significance of event and post-event sediment dynamics in a first order tributary using multiple sediment budgets. *Geografiska Annaler* 92, 189–209.

Julien, P.Y. (1998) *Erosion and Sedimentaton.* Cambridge University Press, Cambridge.

Knighton, D.A. (1998) *Fluvial Forms and Processes – A New Perspective.* Arnold, London.

Knox, J.C. (1972) Valley alluviation in south-western Wisconsin. *Annals of the Association of American Geographers* 62, 401–10.

Koppes, M.N. and Montgomery, D.R. (2009) The relative efficiacy of fluvial and glacial erosion over modern to orographic timescales. *Nature Geoscience* 2, 644–7.

Korup, O. (2005) Large landslides and their effect on sediment flux in South Westland, New Zealand. *Earth Surface Processes and Landforms* 30, 305–23.

Lamoureux, S.F. (2000) Five centuries of interannual sediment yield and rainfall-induced erosion in the Canadian High Arctic recorded in lacustrine varves. *Water Resources Research* 36, 309–18.

Langbein, W.B. and Schumm, S.A. (1958) Yield of sediment in relation to mean annual precipitation. *Transactions of the American Geophysical Union* 39, 1076–84.

Lenzi, M.A., Mao, L. and Comiti, F. (2003) Interannual variation of suspended sediment load and sediment yield in an alpine catchment. *Hydrological Sciences Journal* 48, 899–915.

Macklin, M.G. and Rumsby, B.T. (2007) Changing climate and extreme floods in the British uplands. *Transactions Institute of British Geographers* NS 32, 168–86.

Martin, Y. and Church, M. (1995) Bed-material transport estimated from channel surveys: Vedder River, British Columbia. *Earth Surface Processes and Landforms* 20, 347–61.

Milliman, J.D. and Meade, R.H. (1983) World-wide delivery of river sediment to the oceans. *Journal of Geology* 91, 1–21.

Milliman, J.D. and Sivitski, J.P.M. (1992) Geomorphic/tectonic control of sediment discharge to the ocean: the importance of small mountainous rivers. *Journal of Geology* 100, 525–44.

Molnar, P. and England, P. (1990) Late Cenozoic uplift of mountain ranges and global climate change: chicken or egg? *Nature* 346, 29–34.

Morche, D., Schmidt, K.H., Sahling, I., Herkommer, M. and Kutschera, J. (2008) Volume changes of Alpine sediment stores in a state of post-event disequilibrium and the implications for downstream hydrology and bed load transport. *Norsk Geografisk Tidsskrift – Norwegian Journal of Geography* 62, 89–101.

Mosley, M.P., and Schumm, S.A. (2001) Gravel bed rivers - the view from the hills. In: M. Paul Mosley (ed.) *Gravel bed Rivers V*, New Zealand Hydrological Society, Wellington, 479–505.

Orwin, J.F., Lamoureux, S.F., Warburton, J. and Beylich, A. (2010) A framework for characterizing fluvial sediment fluxes from source to sink in cold environments. *Geografiska Annaler* 92, 155–76.

Owens, P.N. (2005) Conceptual models and budgets for sediment management at the river basin scale. *Journal of Soils and Sediments* 5, 201–12.

Owens, P.N., Walling, D.E., He, Q.P., Shanahan, J. and Foster, I.D.L. (1997) The use of caesium-137 measurements to establish a sediment budget for the Start catchment, Devon, UK. *Hydrological Sciences Journal* 42, 405–23.

Pelletier, J.D. (2008) *Quantitative Modelling of Earth Surface Processes.* Cambridge University Press, Cambridge.

Petts, G.E. and Gurnell, A.M. (2005) Dams and geomorphology: research progress and future directions. *Geomorphology* 71, 27–47.

Rapp, A. (1960) Recent developments of mountain slopes in Kärkevagge and surroundings, northern Scandinavia. *Geografiska Annaler* 42A, 71–200.

Reid, L.M. and Dunne, T. (1996) *Rapid Evaluation of Sediment Budgets.* Catena Verlag GMBH, Reiskirchen.

Reid, I., Bathurst, J.C., Carling, P.A., Walling, D.E. and Webb, B.W. (1997) Sediment erosion, transport and deposition, In: C.R. Thorne, R.D. Hey and M.D. Newson (eds) *Applied Fluvial Geomorphology for River Engineers and Managers*. Wiley, pp. 95–135.

Rice, S. (1999) The nature and controls on downstream fining within sedimentary links, *Journal of Sedimentary Research* 69, 32–9.

Ridefelt, H., Boelhouwers, J. and Eiken, T. (2009) Measurement of solifluction rates using multi-temporal aerial photography. *Earth Surface Processes and Landforms* 34, 725–37.

Rumsby, B.T., Brasington, J., Langham, J.A., McLelland, S.J., Middleton, R. and Rollinson, G. (2008) Monitoring and modelling particle and reach-scale morphological change in gravel-bed rivers: applications and challenges. *Geomorphology* 93, 1–2, 40–54.

Schiefer, E. and Gilbert, R. (2007) Proglacial sediment trapping in recently formed Silt Lake, upper Lillooet Valley, Coast Mountains, British Columbia. *Earth Surface Processes and Landforms* 33, 1542–56.

Schrott, L. and Sass, O. (2008) Application of field geophysics in geomorphology: advances and limitations exemplified by case studies. *Geomorphology* 93, 55–73.

Schrott, L., Hufschmidt, G., Hankammer, M., Hoffman, T. and Dikau, R. (2003) Spatial distribution of sediment storage types and quantification of valley fill deposits in an alpine basin, Reintal, Bavarian Alps, Germany. *Geomorphology* 55, 45–63.

Selby, M.J. (1993) *Hillslope Materials and Processes*. 2nd edn. Oxford University Press, Oxford.

Simon, A. and Rinaldi, M. (2006) Disturbance, stream incision, and channel evolution: the roles of excess transport capacity and boundary materials in controlling channel response. *Geomorphology* 79, 361–83.

Slaymaker, O. (2008) Sediment budget and sediment flux studies under accelerating global change in cold environments. *Zeitschrift fur Geomorphologie* 52, 123–48.

Statham, I. (1977) *Earth Surface Sediment Transport*. Clarendon Press, Oxford.

Sterling, S.M. and Church, M. (2002) Sediment trapping characteristics of a pit trap and the Helley-Smith sampler in a cobble gravel bed river. *Water Resources Research*, 38, 19.1–19.11.

Stott, T. and Mount, N. (2007) Alpine proglacial suspended sediment dynamics in warm and cool ablation seasons: implications for global warming. *Journal of Hydrology*, 332, 259–70.

Summerfield, M.A. (1991) *Global Geomorphology; an Introduction to the Study of Landforms*. Longman, Harlow.

Summerfield, M.A. and Hulton, N.J. (1994) Natural controls of fluvial denudation rates in major world drainage basins. *Journal of Geophysical Research B (Solid Earth)* 99, 13871–84.

Swanson, F.J., Janda, R.J., Dunne, T. and Swanston, D.N. (eds) (1982) *Sediment Budgets and Routing in Forested Drainage Basins*. General Technical Report PNW-141. Department of Agriculture, Forest Service, Portland: U.S.

Syvitski, J.P.M., Vörosmarty, C.J., Kettner, A.J. and Green, P. (2005) Impact of humans on the flux of terrestrial sediment to the global coastal ocean. *Science* 303, 376–80.

Thorne, C.R., Hey, R.D. and Newson, M.D. (eds) (1997) *Applied Fluvial Geomorphology for River Management and Engineering*. John Wiley & Sons. Chichester.

Trimble, S.W. (1981) Changes in sediment storage in the Coon Creek basin, Driftless Area, Wisconsin, 1853–1975. *Science* 214, 181–3.

Trimble, S.W. (1983) A sediment budget for Coon Creek in the Driftless Area, Wisconsin, 1853–1977, *American Journal of Science* 283, 454–74.

Trimble, S.W. (1999) Decreased rates of alluvial sediment storage in the Coon Creek Basin, Wisconsin, 1975–93. *Science* 285, 1244–46.

von Blanckenburg, F. (2005) The control mechanisms of erosion and weathering at basin scale from cosmogenic nuclides in river sediment. *Earth and Planetary Science Letters* 237, 462–79.

Walling, D.E. (1999). Linking land use, erosion and sediment yields in river basins. *Hydrobiologia* 419, 223–40.

Walling, D.E. (2008) The changing sediment loads of the world's rivers, in Sediment Dynamics in Changing Environments (proceedings of a symposium held in Christchurch, New Zealand, December 2008). IAHS Publ. 325, 323–8.

Walling, D.E. and Webb, B.W. (1983) Patterns of sediment yield, In: Gregory, K.J. (ed.) *Background to Palaeohydrology*, Wiley, Chichester, pp. 69–100.

Walling, D.E. and Webb, B.W. (1996) Erosion and sediment yield: a global overview. *Erosion and Sediment Yield: Global and Regional Perspectives*. IAHS, 236, 3–19.

Walling, D.E. and Collins, A.L. (2000) *Integrated Assessment of Catchment Sediment Budgets: A Technical Manual*. University of Exeter/DFID, Exeter.

Walling, D.E. and Collins, A.L. (2008) The catchment sediment budget as a management tool. *Environmental Science and Policy* 11, 136–43.

Walling, D.E., Collins, A.L., Sichingbala, H.M. and Leeks, G.J.L. (2001) Integrated assessment of catchment sediment budgets: a Zambian example. *Land Degradation and Development* 12, 387–415.

Walter, R.C. and Merritts, D.J. (2008) Natural streams and the legacy of water-powered mills. *Science* 319, 299–304.

Warburton, J. (1990) An Alpine proglacial fluvial sediment budget. *Geografiska Annaler* 72A, 261–72.

Warburton, J. (1999) Environmental change and sediment yield from glacierized basins: the role of fluvial processes and sediment storage, In: A.G. Brown and T. Quinne (eds) *Fluvial Processes and Environmental Change*. Wiley, Chichester. pp. 363–84.

Warburton, J. (2007) Mountain Environments, In: C. Perry and K. Taylor (eds) *Environmental Sedimentology*. Blackwell, Oxford. pp. 32–74.

20

Hillslopes

David Petley

The majority of the terrestrial surface of the Earth is formed from slopes; indeed plains are essentially the exception, usually formed by comparatively recent depositional processes in fluvial or marine environments. Even areas of low relief generally consist primarily of low-angled slopes. Thus, the understanding of the processes and behaviour of slope systems must form an essential component of geomorphology. However, slopes vary vastly in their character, ranging from gentle slopes formed from young, weak sediments to >1000 m high near-vertical cliffs formed from ancient, strong metamorphic rocks. Thus, it is unsurprising that processes vary greatly across and between slope systems.

In geomorphology the terms slope and hillslope can be used essentially interchangeably when discussing land units. Thus the term *hillslope*, which might appear to indicate those slopes associated with hills rather than mountains, is generally used to refer to all slopes. The current chapter follows this convention, with the term hillslope being used to encompass the full range of slopes in sub-aerial environments, excluding riverbanks, coastal cliffs, and of course submarine slopes. However, the focus of this chapter is not explicitly upon near-vertical cliffs, but instead in general upon lower angled slopes. Similarly, following recent widely accepted conventions the term *landslide* is used to describe all mass movements, even though many do not include any element of sliding.

THE HISTORICAL DEVELOPMENT OF HILLSLOPE STUDIES

Current approaches to the study of hillslopes have evolved over a long period of time, following a number of successive paradigms. In each case the paradigm in question was not all encompassing, but the dominant work tended to lie within that particular framework. On this basis the development of hillslope geomorphology can be grouped into four distinct stages (after Selby, 1993).

Landscape evolution

Landscape evolution theories sought to generate models that explained the way that landscapes, and in particular land surfaces, evolve through time. Slopes have inevitably been a key component of all of these models. Perhaps the best known and most influential, although now essentially discredited, model was that of the hugely influential work of W.M. Davis (1899 and subsequent papers). This is mostly remembered for its focus upon concepts of continental-scale landscape development, which might appear to preclude consideration of individual slopes. However, the Davis model is essentially one of hillslope change, initiated by a short burst of extremely rapid uplift that quickly ceased, and the creation of river valleys with very steep walls as a

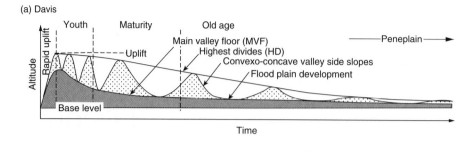

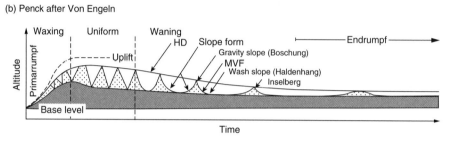

Figure 20.1   Schematic illustration of two of the key landscape evolution models (after Chorley et al., 1984), from a slope perspective. (a) The Davis model, which is essentially one of hillslope change, controlled by initial uplift and the creation of river valleys with very steep walls as a result of fluvial incision. (b) The Penck model, which is also characterized by slope angle reduction with time, albeit in a more complex manner

result of fluvial incision (Figure 20.1a). The model essentially seeks to describe the evolution of the landscape through the development of these slopes through time, hypothesising that the slopes erode back and gradually reduce in gradient, whilst the valley floor widens, such that eventually a peneplain is formed. A peneplain is essentially a surface with slope gradients that are sufficiently low that hillslope processes reduce to near zero. The model of Davis had been hugely influenced by the work of G.K. Gilbert (e.g. Gilbert, 1880), who played a critical role in developing and promoting the idea that slopes evolve through a slow reduction in angle towards a subdued and rounded topography. Whilst from contemporary standpoints the idea that slopes undergo this evolutionary process appears to be surprisingly at odds with simple observations, alternative views of hillslope, mostly notably those of Walther Penck (1924), were slow to gain traction. Interestingly, Penck (1924) is often characterized as providing a description of parallel retreat processes; indeed, Davis (1932) depicted Penck's model as one of parallel slope retreat. This was not, in fact, the case; Penck also described slope evolution as a slow reduction in gradient, although he recognized that this was a

complex process (Figure 20.1b). In fact, although descriptions of the hillslope work of Penck and Davis often portray them as being in opposition to each other, in reality Penck's model was largely in accordance with that of Davis. Penck's three major contributions were (1) recognition of the importance of 'endogenous' (i.e. internally controlled) rather than exogenous (external agencies) in determining slope behaviour, a point that is often missed even today (see below); (2) recognition that uplift is not a short-term process that initiates landscape evolution; and (3) recognition that slope evolution is more complex than a simple relaxation process.

Models of slope-based landscape evolution made a substantial step forward when L.C. King (1951, 1953) recognized that Davis's misrepresentation of Penck's model was, in fact, probably right; that is, that slopes do retreat at an approximately constant slope angle, with that angle being controlled by a combination of rock strength and the rates of external processes. In so doing King rejected the Davisian cycle of erosion, but retained a strong sense in the cyclicity of landform evolution. Once again, landscape was considered to evolve primarily through the development of hillslopes, in this case though with parallel retreat

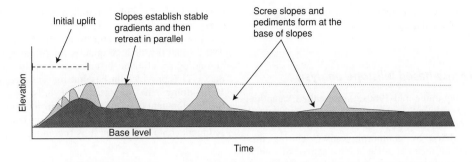

Figure 20.2 Schematic illustration of the King (1951, 1953) model of landscape evolution (after Chorley et al., 1984), from a slope perspective. The landscape was considered to evolve primarily through the development of hillslopes, in this case though with parallel retreat

rather than slope gradient reduction (Figure 20.2). This model for slope, rather than landscape, evolution was developed through observation primarily of escarpments of southern and eastern Africa. The observation of the existence of 'threshold slopes', in which slope angle was observed to be controlled by material properties and climatic drivers (Carson and Petley, 1970), proved to be vindication of this concept, although not of the larger-scale models themselves.

Ultimately, the development of slope-driven landscape evolution models was largely ended by the development of the understanding of tectonic processes after World War 2, as the models quickly proved to be largely incompatible with the nascent plate tectonics theory. In particular, studies in the 1950s and onwards effectively demonstrated that uplift was a complex, continuous process, which did not fit with the early landscape models. It is, however, interesting to note that as our understanding of tectonic chronologies has improved, the modelling of slopes as a landscape-evolving process has become fashionable once again (see below). Whilst it is easy to dismiss these early attempts at the understanding of slope-based landscape evolution – as geomorphologists have had a tendency to do – they represent an important stage in our understanding of geomorphic, and in particular slope, systems.

Slope form measurement

The immediate period after World War 2 was characterized as a phase in which there was a growing sense of discomfort with the models of long-term landscape evolution, but a paucity of clear alternative approaches. Considerable efforts

were placed during this period on the collection of slope information in terms of maps of slope morphometry and the measurement of slope angles (e.g. Fair, 1947, 1948; Savigear, 1952, 1962, 1967). It should be noted that a substantial component of this work was driven by the need to have objective methods that could be used to evaluate and develop the models and theories of Davis and Penck, but in fact they gradually undermined their validity. However, at the same time other disciplines were also developing an understanding of slope behaviour, most notably through the development of an understanding of endogenetic processes in soil slopes via soil mechanics (e.g. Terzaghi, 1943). It is thus unsurprising that these detailed studies of form began to merge into a quantitative consideration of slope processes (e.g. Young, 1971). Thus, in parallel with the form-based studies of hillslopes, a process-based understanding began to emerge.

Paradigm change is always viewed as a sudden revolution, but in reality change can be surprisingly gradual. This is perhaps illustrated in this case by the development of the nine-unit land surface model of Dalrymple et al. (1968), which sought to examine soil processes within a framework of nine defined slope components (the interfluves, the seepage slope, etc.), although not all were considered to be present in any landscape. Evolution of these units was considered to occur through the interplay of pedological and geomorphological processes. Such a model was attractive in that it allowed the landscape to be evaluated on the basis of easily manageable sections, and it provided a transition between large-scale models and detailed process studies. It is also rare in the context of these early (or indeed many recent) models in its explicit consideration of the soil. Such models are clearly rooted in a morphological/morphometrical approach, but start to hint at detailed processes.

Process-based hillslope analysis

Starting in the mid-1960s, the so-called 'quantitative revolution' in geography drove a substantial change in the ways in which hillslopes were considered by geomorphologists. During this time, the focus of slope studies started to switch from studies of morphology, and in particular of hypothetical models of the evolution of slope form, to an understanding of slope processes, underpinned by rigorous and detailed data collection (e.g. Hewlett and Hibbert, 1963; Whipkey, 1965; Weyman, 1970). This general approach is probably exemplified by Carson and Kirkby (1972), which provided a framework for the development of process-based hillslope research, and in particular the use of quantitative analysis of process data. It is this type of work, which continues to this day, that forms the fundamental basis of our understanding of hillslope development and behaviour. It is also important to note that through time this style of research has become increasingly integrated across disciplines, including engineering geology, hydrogeology, geotechnical engineering, geophysics and soil science, primarily as a result of cross-fertilization of technologies and methods (e.g. Rosser et al., 2007). There has been a notable evolution in this type of research through time, most notably exploiting the improvements in data quality that have accompanied technological advances. Thus, it is clear that the research into hillslope processes is increasing 'geophysical' in terms of approach, analysis and presentation. This trend is likely to continue for the foreseeable future.

Modelling revisited

In the last two decades slope evolution modelling has re-emerged as a key theme (e.g. Wicks and Bathurst, 1996; Bronstert and Plate, 1997; Coulthard et al., 2000). In essence there are two parallel elements to this; first, there is an increasing emphasis on the modelling of processes for individual slopes (or sometimes for the slopes within a catchment, in particular to understand complex process responses to external stimuli such as intense precipitation events (e.g. Bronstert and Plate, 1997), climate change (e.g. Coulthard et al., 2000) or seismic excitation (Sepulveda et al., 2005a). Second, especially in North America there has been a trend towards the analysis of long term landscape evolution, particularly on a

continental scale, with recognition that slopes are a key control on many aspects of this complex process (e.g. Tucker and Slingerland, 1997). The availability of high-quality, long-term uplift chronologies, climatic histories and erosion rate datasets has provided the baseline for such models. In so-doing, the importance of hillslopes in the development of a landscape, and in controlling the behaviour of both the fluvial system and the sediment contained within it, is becoming clear.

THE TWO TRIBES IN HILLSLOPE STUDIES – HYDROLOGY AND MASS MOVEMENTS

Although for over 40 years hillslope studies have been based primarily upon the measurement and analysis of processes, and odd schism has developed between those who are interested primarily in hillslope hydrology and those with an interest in mass movements. This schism has not been antagonistic or adversarial to any large extent, but has been based upon the fundamentally different approaches taken by these two research areas. In many ways this divide has been surprising, not least because of the high levels of synergy between these two research themes; for example, the division between debris flows and hyper-concentrated flows is nominal, whilst the stability of slopes is frequently controlled by pore water pressures, which reflect the hillslope hydrology. Whilst more recently there has been increasing evidence that the two aspects are becoming more integrated, a substantial difference remains between the two fields. Thus, in the sections below hydrology is considered separately from mass movements, before a synthesis of the two areas is presented.

Hillslope hydrology

Research into hillslope hydrology has historically been centred on two key approaches: field investigation of hydrological processes, ranging from detailed, slope specific investigations to studies across whole catchments, and modelling-based analyses. Field measurement of hillslope hydrology ranges from the instrumentation of real (but usually small) catchments to the construction of artificial systems that allow greater control on key variables, albeit at the cost of some degree of realistic representation of the complexity of natural systems. The combination of these two approaches reached the point some years ago that practical prediction of hillslope hydrological behaviour, at the catchment scale at least, became viable to the level of reliability that is required in most operational circumstances (Klemes, 1986, 1988; Grayson et al., 1992; Beven, 2002; Kirchener, 2006). However, it is also clear that the current generation of hydrological hillslope models still fails to capture the essential underlying behaviour of natural systems in a satisfactory manner. Kirchener (2006) described the key symptoms that indicate that a number of substantial advances are needed in our understanding of hydrological systems in general. These key aspects apply as much to hillslope hydrology as they do to catchment-scale studies:

1   Conventional networks used for the measurement of hydrological behaviour are generally inappropriate for fine-grained understanding of natural systems, mostly having been designed for a different purpose. At the catchment scale this has often been the measurement of the impact of dam construction and management; at the hillslope scale this has often been the understanding of pore pressure control on slope stability (e.g. Petley, 2004). A key issue is the lack of good quality precipitation and evapo-transpiration data in upland systems (Anders et al., 2006; Smith, 2006). There is a tendency to assume that comparatively large areas receive similar precipitation inputs, whereas in reality there is huge variability in rainfall patterns across small areas due to terrain and convective rainfall effects. Systematic analysis of assumptions that underpin upscaling from small catchments suggest that such approaches can lead to high levels of error (Tetzlaff et al., 2006).

2   The empirical mathematical tools used to analyse hillslope hydrological systems are frequently unrealistic in their handling of system function, assuming for example that hydrologic systems display linear, additive behaviour. In reality, most systems are nonlinear and non-additive, meaning that behaviour in extreme events is poorly represented (e.g. Lehmann et al., 2007; Clark et al., 2009). Thus, most existing models can be characterized to a greater or lesser degree as linear, black-box systems. Clearly, future models need to handle this behaviour rather better.

3   Models based purely upon the application of physical laws for fluid flow etc., such as Darcy's law and Richards' equation, at the superficial level appear to be the ideal modelling approach. However, such models usually perform poorly when applied across a range of spatial scales and catchments (Seibert, 2003), and in systems that are complex (which applies to most real world situations). Thus the models fail in most cases to capture the essence of true behaviour, primarily because although we understand well the physical laws governing behaviour at small

scales, the ways in which complex systems behave are poorly parameterized. In essence, the models currently used require multiple calibrating factors. Unfortunately, the behaviour of most small-scale systems can be represented with a vast range of combinations of particular parameters, meaning that the actual combination used in the model will rarely be correct. Inevitably, as conditions change beyond the bounds of the calibration system behaviour fails to be represented adequately (e.g. Wagener, 2003). This might be taken to imply that better calibration is needed, but actually it is indicative that the modelling approach is inadequate.

4   The linkage between models and field data is often surprisingly poor, with comparison between the two being achieved by simple overlays of modelled and observed behaviour, or two-dimensional regressions of one against the other. This approach tends to mask model deficiencies. Clearly, improved testing of models using field data, ideally using advanced statistical techniques, is required.

In addition to these four key points, Kumar (2007) and Hopp et al. (2009) noted that system co-evolution is a major challenge in hillslope hydrology. This recognizes that the current state of any hydrological system reflects the evolution of the system over time, in which biological and physical factors have interacted to create the current structure. Co-evolution of the system state continues to occur of course. Field measurements can only represent the current state of the system (Brantley et al., 2007). The existing range of models may struggle to represent the complexity that arises from this co-evolution, and will in consequence face challenges in the representation of the impacts of environmental change, which will cause structural changes to the systems that current models will struggle to represent (Sivapalan, 2005).

These key concerns highlight the need for in essence a paradigm change in hillslope hydrology in the form of a unifying model of behaviour, just as a key challenge in physics is to find the unifying model that ties the four fundamental interactions of nature. Unfortunately, finding a unifying model in hillslope hydrology is unlikely to be any more straightforward, and indeed could prove to be impossible. It is difficult to predict where the next key advance will be made, but a number of exciting developments point to interesting opportunities:

1   In recent years there have been substantive improvements in our capabilities to collect comprehensive field data at high spatial and temporal resolutions. High-precision sensors are

now (comparatively) cheap and reliable, allowing high density data to be collected at high frequencies. In addition, the advent of mobile sensor networks (Kienzler and Naef, 2008), allows dynamic system behaviour to be captured (see for example the overview of Soulsby et al., 2008).

2 The development of chemical- and isotope-based monitoring is providing a powerful suite of techniques that allow detailed analysis of system behaviour, presenting new opportunities both to improve and to test models. The greatest potential lies in the automation of these techniques, allowing data collection at sufficiently high frequencies that system behaviour is exposed (Kirchner et al., 2004; Kirchner, 2006).

3 The use of better statistical analyses is allowing improved linkage between measured system parameters and modelled behaviour (Tucker et al., 2001; Kim, 2009).

4 A new wave of large-scale physical models can provide insights into system behaviour in a way that monitoring of the natural system cannot. For example, the BIOSPHERE II project involves the construction of three 33 m × 18 m artificial slopes near Tucson, Arizona, USA. In each case conditions on and in the slope can be carefully controlled and the behaviour of the system measured to a high level of precision. The intention is that the slopes will be monitored for a decade, allowing the coupling of geochemical, biological and physical processes to be observed, during which co-evolution of the system should occur (Dontsova et al., 2009; Hopp et al., 2009).

5 A final, more nebulous but important development has been the move towards interdisciplinary working across the environmental sciences in recent years. For many years research opportunities have been stifled by the failure of funding agencies, researchers and editors/reviewers to recognize the opportunities presented by working across the disciplines. This has undoubtedly changed in recent years, although there remains some considerable resistance to this approach, with research funders in particular being eager to foster interdisciplinarity. The benefits of this change are becoming apparent, for example allowing insights into linkages between morphology, ecosystem behaviour and hydrology (Istanbulluoglu and Bras, 2005; Belyea and Baird, 2006; Bogaart and Troch, 2006).

The essence of the future of hillslope hydrology appears to lie in the concept of process conceptualization, most succinctly summarized at the catchment scale by Tetzaff et al. (2008) as a:

focus on the form of catchments; that is, how and why they are geomorphologically, ecologically and pedologically structured in the way that they

are. Further, there is the need to understand and quantify the ways in which this form determines how and why catchments function hydrologically and behave dynamically with temporal scale (Wagener et al., 2007). In turn, function feeds back into the subsequent evolution of catchment

At the hillslope scale this new approach is already yielding interesting advances, for example through the use of 'virtual experiments' to investigate macropore hydrology hillslopes (Weiler and McDonnell, 2004) and the development flow networks for hillslopes (Spence and Woo, 2003, Tromp-van Meerveld and McDonnell, 2006). This is an area that will develop rapidly in the next few years, and will increasingly allow the development of linkages with the mass movement research.

Hillslope mass movements

Mass movement processes

In recent years, hillslope mass movements have become an increasingly prominent research area, driven primarily by a desire to reduce the economic and social costs of landslides. To a much greater extent than for hillslope hydrology, landslide research sits across a range of academic domains, most notably civil engineering (geotechnics), engineering geology and geomorphology, with less substantial but nonetheless important contributions from, for example, soil science and forestry. Thus, it is perhaps unsurprising that the discipline has been, and remains, quite fragmented. However, within the geotechnical and earth science areas mass movement studies have remained unified by reliance on a single key principle – that of the factor of safety. This principle suggests that instability occurs when the forces driving movement exceed those that restrain movement. Thus, any slope that has failed is assumed to have done so at a factor of safety of one. This principle is of course physically sensible and thus in essence is reasonable. However, as will be demonstrated below, many assumptions have been built upon this principle that have been less helpful in understanding mass movement processes, and the resultant hazards that derive from hillslope failure.

The concept of the factor of safety underpins process-based landslide research in geomorphology principally through the effective stress concept, which provides a mechanism to explain reductions in the restraining forces within a hillslope. Thus, the conditions controlling the development of landslide strain (displacement) can be analysed in terms of the stress state of the slope, which can be determined through the measurement of the landslide depth, material properties and shear surface inclination (to obtain the disturbing forces) and the pore water pressures (to obtain the restraining forces). If the processes operating within a slope can be determined, analysed and modelled, the movement of the landslide should be understandable. As such, for rainfall-induced mass movements, which represent the major type of slope failure occurring worldwide on a daily basis, substantial advances in understanding of slope processes have been rare over the last two decades, which might be taken to suggest that this basis of understanding slope systems is not unreasonable. Indeed, as with hillslope hydrology our understanding of the physical processes occurring within most slopes is sufficient to provide reliability in mitigation approaches. There are exceptions to this – for example during seismic events (see below), and estimates of runout distances/rates remain problematic.

However, there has been a naivety in the understanding of slope stress states that has stifled the development of the discipline and which, fortunately, is now being slowly swept aside. There are two central, inter-related, elements to this.

Consideration of global, rather than local, stress states

The factor of safety approach of course represents the entirety of the portion of the slope that is considered to be potentially unstable, which in this context is defined as the global factor of safety. Thus, a body defined as a landslide is potentially unstable when the global factor of safety (i.e. that of the whole landslide body) reaches unity. However, even where the global factor of safety exceeds unity, the shear stress may exceed the shear resistance of the slope at various points in the system (Figure 20.3). In this area deformation will occur, which allows the system to deform and evolve. Even under static stress states these areas of stress concentration may change in time, allowing evolution of the landslide system.

A lack of appreciation of endogenous processes

There has been a tendency within geomorphology to consider exogenous processes, and thus those components within a slope system that react directly to those processes, as being dynamic whilst endogenous components are considered to be static. Thus, in particular, it is widely recognized that pore water pressures fluctuate

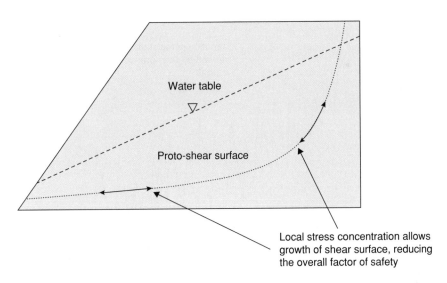

Water table

Proto-shear surface

Local stress concentration allows growth of shear surface, reducing the overall factor of safety

Figure 20.3   A schematic illustration of the ways in which the evolution of slope systems can occur even when the global factor of safety is greater than unity. The local factor of safety can be less than one, which allows the growth of the potential shear surface. This has the effect of reducing the overall factor of safety, which in turn can allow further development of the shear surface. If conditions are right, and enough time, this can allow failure of the slope without external forcing. A proportion of slopes, especially in high mountain areas, appear to show this type of behaviour

continually, primarily in response to precipitation and evapo-transpiration. On the other hand, the strength of the materials within the slope are considered to be a constant except when responding to either weathering, which may be considered to degrade the slope slowly with time (Mottershead et al., 2007), or in the immediate aftermath of failure, when they may transition from peak to residual strength (Petley and Allison, 1997). In consequence, diagrams such as Figure 20.4 are sometimes produced, with the strength of the materials degrading slowly over time whilst the pore pressures fluctuate with a much higher frequency. In this model, failure is considered to occur when the strength of the slope reduces to the point that pore pressures allow a factor of safety of one to be attained. However, there is little evidence to support that view that strengths decline in this way, and indeed there is some evidence that the process is considerably more complex than this (e.g. Fan et al., 1997). In reality, the strength of the materials within a landslide probably vary in a far more complex manner, reacting to changes in pore water chemistry (Moore and Brunsden,

1996), material chemistry (Bogaardi et al., 2007) and, probably most importantly, to weakening caused by evolution of the slope towards failure (Petley et al., 2002, 2005a). Thus collapse of a slope may be caused by changes of properties within a slope associated with the evolution of the putative landslide towards final failure, rather than an external forcing (Petley et al., 2002, 2005a). This is not to discount the importance of external forcing, but rather to note that external forcing is important in the context of internal processes.

In general there has been a tendency to view slope systems in a linear manner, in which the system responds in a simple manner to external forcing. Thus, for example, there has been a wide range of studies that attempt to determine intensity-duration relationships for the initiation of landslides (e.g. Caine, 1980; see the review of multiple subsequent studies Guzetti et al., 2008). Here, rainfall is seen as the driving factor, and the slopes are assumed to start to move when a critical combination of intensity and duration of rainfall is achieved. Such studies are based upon the movement of a single landslide (Sengupta et al., 2010

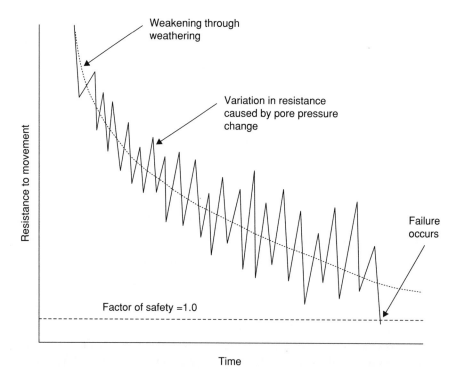

Figure 20.4   The role of weathering in slope failures is often represented in this way. Weathering progressively reduces the resistance of the slope to shear stress. In addition, pore water pressure fluctuations allow the resistance to change with a higher frequency. Failure occurs when the two effects combine to allow the factor of safety to reach unity

for example) or across a wide area (e.g. Frattini et al., 2009). This appears to simplify the mechanical behaviour of slopes to a high degree.

In recent years the analysis of data arising from improvements in techniques for measuring the landslide processes has started to challenge these assumptions. Two advances have been critical in this. First, the availability of reliable data-loggers with large storage capacities and low power demands have allowed high-frequency monitoring of slope parameters (e.g. Coe et al., 2003; Petley et al., 2005b). Thus, for example, pore water pressures can be recorded almost constantly with high levels of precision, allowing for example the rapid response within shallow landslides to high intensity storms to be parameterized (Simoni et al., 2004). Second, the ability to measure landslide displacement has changed profoundly, especially through the development of high resolution, laser-based instrumentation that allow the development of strain within a slope to be resolved in three dimensions at a high frequency of data collection (Rosser et al., 2005). Thus, the accumulation of strain within slopes can be properly related to the stress state of the slope (e.g. Corominas et al., 2005; Schulz et al., 2009a).

Such studies have served to demonstrate the restrictions of simple, non-dynamic factor of safety-based models. These suggest that for a landslide in which the materials are at residual strength the movement rate should be directly related to the pore pressure. Thus, a pore pressure threshold can be defined at which movement should initiate, representing a factor of safety of unity. The rate of movement should increase in inverse proportion to the effective stress state beyond this threshold. This is unlikely to be a linear relationship, but nonetheless should be comparatively simple in a slide with a simple geometry (e.g. a translational slide) with low total strains. As the pore pressures decline the movement should also reduce, with movement ceasing when the same threshold is reached (Figure 20.5).

The recent high-resolution monitoring datasets have demonstrated that such a simple relationship between displacement and pore pressure are not seen in real landslide systems (Gonzalez et al., 2008; Schulz et al., 2009a). For example, Schulz et al. (2009b) demonstrated that increased movement rates were associated with pore pressure decreases along the margin of the Slumgullion landslide, probably due to shear-induced dilation. On the other hand, Gonzalez et al. (2008) showed that there is hysteresis in the relationship between the movement rate and the pore pressure, implying a viscous component to the movement mechanisms.

Similar problems have also arisen with the understanding of first-time slope failures. As described above, the conventional approach to landslide analysis assumes that the system is responding to exogenous processes, which bring the system to a factor of safety of one. Thus, it should be possible to determine a parameter that

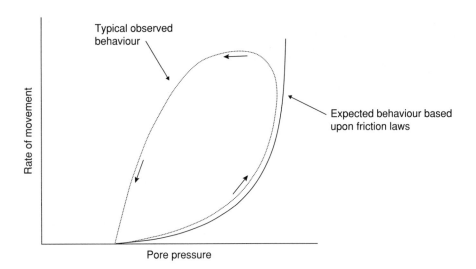

Figure 20.5 Simple friction-based movement laws suggest there should be a direct correlation between movement rate and pore pressure. However, field studies suggest that this relationship is more complex, with strong hysteresis in the relationship. However, the exact form of this hysteresis appears to vary between landslides

has changed in response to an external forcing that has caused the slope to fail; usually this is considered to be the effective stress state. However, it is also clear that many large first-time failures, especially those in hard rock systems, do not have a clearly identifiable trigger process. For example, the 1991 Mount Cook landslide in New Zealand had no identifiable trigger, even though the volume of the slide was large. This is a clear indication of the role of endogenous processes, in this case probably creep rupture or stress corrosion effects that provide a time/strain control on the development of failure. These principles have been long-recognized in rock mechanics (Anderson and Grew, 1977), but are only just being imported into our understanding of large slope systems.

A further paradox persists as well, best illustrated by the behaviour of slopes in the high mountains of Taiwan. The Central Mountain Range of Taiwan is composed of rapidly uplifted sedimentary rocks that are subject to frequent seismic activity, resulting in a topography that is exceptionally steep. Instability is inevitable. Outside of the occasional occurrence of a substantial seismic event, most mass movements are triggered by the precipitation associated with tropical cyclones (typhoons). Taiwan is affected by an average of five typhoons per annum. Each typically brings >400 mm of rainfall to the affected area of the Central Mountains over a 3–4 day period. However, the most intense storms bring >1000 mm over this period. For example, Typhoon Herb in 1997 deposited >1900 mm of rainfall, whist Typhoon Morakot in 2009 deposited >2800 mm. The paradox is that first-time failures occur in both small and large typhoon events, even when the small typhoon occurs some time after the large event, although unsurprisingly the number of landslides is related to the intensity of the storm. Pore pressures are likely to be higher in the large event than in the smaller one, implying that slopes should fail in the large event rather than in the subsequent, less intense rainfall event.

The answer lies in an important, but largely forgotten, set of papers published in the 1960s (Bishop, 1967; Bjerrum, 1967; Rowe, 1969) which introduced the concept of progressive failure. This recognized that first-time failure in a cohesive material requires the development of a shear surface which progressively reduces the strength of the materials in the basal regime of the slide. Thus, the basal processes can be related to those of crack growth, especially in hard rock systems. In essence this can be considered to be the accumulation of damage within the landslide mass. The rate of growth of the crack can be directly related to the stress state of the slide, with lower effective stresses and higher shear stresses promoting higher rates of crack growth. As the

shear surface extends the landslide progressively weakens, but failure cannot occur until the crack is fully developed. In hard rock slopes crack growth may occur through the process of stress corrosion; in softer materials the process may be creep rupture (Singh and Mitchell, 1969).

This model has two key implications that are commonly forgotten within geomorphology. First, the role of a strong storm event may be to progressively develop a shear plane rather than to induce failure. Thus, a large storm will inevitably take many slopes to the critical point at which failure occurs, and hence there are many landslides. However, many other slopes will move towards failure without actually reaching the failure point. Subsequent smaller storms may well take those slopes closest to failure to a state beyond the critical point, and collapse will occur. Thus, it is unsurprising that mass movements continue to occur in small events that follow large ones. Second, slopes may continue to progress towards a failure event when there is no trigger event through the stress corrosion/steady state creep process. Indeed it can be argued that any slope whose stress state means that the factor of safety would be unity or below for the residual strength condition continues to undergo the progressive, albeit very slow, development of failure. Thus, occasionally the point of initiation of final failure will occur without a trigger, although clearly as the rate of accumulation of damage is greatest during large external events, failure will most frequently occur during this time.

The critical point in the development of failure is of course the point at which the failure surface is fully developed, allowing release. This is most simply visualized in a block detaching from a vertical cliff (Figure 20.6), in which failure occurs when the release surface is fully formed. However, there may well be another critical point prior to this detachment, when the cross-sectional area of the intact rock bridge allows the stress concentration to exceed the strength of the rock. At this point failure becomes inevitable, but will only occur when the surface is fully formed. Through time the stress on the unsheared rock will progressively increase as the cross section of unsheared rock reduces, which will generate a progressively increasing rate of strain development. As the cross-sectional area of unsheared rock tends to zero, the stress must tend to infinity, yielding a rapidly increasing rate of movement characteristic of first-time failures.

The complexity of the behaviour of slopes has been illustrated recently in study of the behaviour of the Slumgullion landslide in Colorado, USA (Schulz et al., 2009b). This landslide is a large, continually active earthflow that moves on average about 10 mm per day. Detailed, high

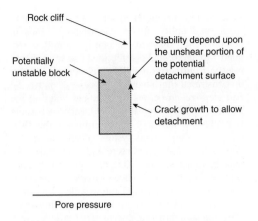

Figure 20.6   Conceptual model of a block detaching from a vertical cliff. Failure occurs when the release surface is fully formed

resolution monitoring of movement patterns has shown that on a daily scale the landslide shows a cyclicity of movement that can be directly related to changes in air pressure associated with atmospheric tides. That such small changes in pressure can cause modifications to the rate of movement of a large landslide is surprising. Slumgullion may well be almost unique in terms of the fact that it permanently resides in a state that is very close to a factor of safety of one. Thus, minor changes in pressure can allow the slide to stabilize and to reactivate. Few other landslides are in this state. The value of the study is to show that small changes in stress state can cause changes in system behaviour; this serves to strengthen the argument that internal changes are under-examined.

Seismic slope behaviour

For many years studies of hillslope mass movement processes have been focussed primarily upon hydrologically triggered mass movement events. This is understandable as in most environments this is the dominant landslide process most of the time. However, Keefer (1984) sought to demonstrate the coupling between large seismic events and the occurrence of landslides, most notably highlighting the link between the surface area of the region affected by high-intensity shaking and the earthquake magnitude. The 1999 Chi-Chi earthquake in Taiwan represented a landmark in the understanding of seismically triggered mass movements, currently representing the most-studied large-scale landslide event of all time (Petley, 2010). These studies, and those in

other key study sites, have highlighted a number of key aspects of the geomorphology of hillslopes affected by seismic shaking:

1   The occurrence of seismically induced landslides is strongly controlled by interactions between seismic waves and the landscape, most notably through the process known as topographic amplification (Murphy et al., 2002; Sepulveda et al., 2005b; Meunier et al., 2007, 2008). This creates a morphological fingerprint of seismically induced failures that differs from that of those induced by pore water pressure effects (Densmore and Hovius, 2000).
2   The spatial distribution of landslides in earthquake affected areas is controlled by a complex and poorly understood combination of topographic, geological and seismic parameters (Meunier et al., 2007, 2008; Parker et al., 2009).
3   The number of post-seismic landslides may exceed the number triggered in the main shock, representing a substantive hazard and playing a key role in both landscape evolution and fluvial system behaviour (Lin et al., 2003, 2006, 2008; Chen and Petley, 2005).
4   The magnitude–frequency distribution of seismically induced landslides is essentially similar across a range of environments and is indistinguishable from that induced by intense precipitation and snowmelt events (Guzetti et al., 2002; Malamud et al., 2004).

More recent earthquake events, most notably the 2005 Kashmir earthquake (Dunning et al., 2007) and the 2008 Wenchuan (Sichuan) earthquake (Parker et al., 2009) have emphasized the need to better understand the controls on seismically induced mass movements. The key mechanisms of seismically-induced landsliding remain poorly understood. This will inevitably be a major research area over the next decade, both through field investigations and, increasingly, modelling based approaches.

LINKING HILLSLOPE HYDROLOGY AND MASS MOVEMENT PROCESSES: RECONCILING THE TWO TRIBES

Given that the majority of mass movement studies are undertaken on slides that are hydrologically induced, the lack of interaction between the hillslope hydrology and mass movement communities is surprising. In the early part of the 21st century, application of hydraulic models to shallow landslide initiation became a key topic. In most cases early studies focussed upon steep, forested catchments in which soil cover was thin. In particular, field measurement of the development of instability during simulated rainfall events allowed the role of water movement, in the form of pressure waves, to propagate through a hillslope, inducing instability (Torres et al., 1998). Such studies have allowed the development of hydrological models for shallow landslide initiation that have provided substantial insight into these processes (e.g. Dietrich, 2001), and which have allowed the development of models that simulate situations such as the impacts of deforestation (Montgomery et al., 2000) and the effects of high intensity rainstorms in tropical environments. However, the application of these new hydrological models to deeper-seated landslide systems remains comparatively rare (e.g. Malet et al., 2005), even though this may well provide at least a partial explanation to some of the more surprising aspects of landslide behaviour. Combined with laboratory-based recent work on the linkage between pore pressure changes and strain development in simulated slope environments (Ng and Petley, 2009, for example), the opportunity for understanding landslide development is within reach.

LANDSCAPE EVOLUTION AND SEDIMENT MOBILITY

It has long been a criticism of hillslope research that studies tend to be excessively focussed on very detailed studies at the individual slope or catchment scale, primarily over short timescales, with comparatively little emphasis on upscaling. This has also been manifested in a compartmentalization of research fields, with few studies attempting to link for example hillslope and channel systems. In part this is a response to the rejection of the early large-scale landscape evolution models, and it has been to the detriment of the advancement of geomorphological knowledge. However, in recent years there has been a growing interest in understanding landscape evolution, and in the mechanisms of sediment production. The core philosophy behind such research is essentially simple. Tectonic processes move material into an orogenic zone. Over geological timescales this material is removed from the mountain chain through the actions of the fluvial system, but in order to do so the material must be released from the hillslopes. Mass movements are the primary mechanism through which this can occur, and thus play a key role in long term landscape evolution. Dadson et al. (2004) examined the coupling between erosion rates, as indicated by the suspended sediment concentration in large river

systems and seismic energy release, precipitation inputs and uplift rate for the island of Taiwan, demonstrating that erosion rate is strongly correlated with the spatial distribution of seismic energy release. Furthermore, Scheurch et al. (2006) demonstrated that landslide processes control the particle size properties of sediment entering the fluvial system. Given that it is clear that particle size plays a key role in determining the flow patterns within fluvial systems (Turowski et al., 2009), the coupling between landslides and rivers may be rather stronger than been previously considered. The high incidence of mass movement in upland areas subjected to large seismic events does generate an interesting paradox, however. Movement on thrusts allows material to advect into mountain chains such that it might be expected that a large seismic event produces a net increase in the volume of rock within the orogenic zone. However, seismic events also induce landslides that release material for removal by the fluvial system. Interestingly, analyses of recent earthquakes suggest that the net erosion on hillslopes is on a similar scale to the advection of material into the mountain chain, although more work is needed. This raises interesting questions about the manner in which mountain chains develop; these can probably only be answered by considering the whole landscape system, of which hillslopes are a key component.

CONCLUSIONS

The majority of the terrestrial surface of the earth is formed of slopes. Thus, it is unsurprising that studies of slope morphology and slope processes have long been key components of geomorphology. However, hillslope studies have evolved in a very particular way because of the historic development of the discipline. Early studies that focussed upon hillslopes as a component of large-scale landscape evolution models dominated research for several decades. The rejection of these theories in the 1960s and 1970s in the face of improved understanding of tectonics led to a strong focus on process studies on individual slopes, with few attempts to upscale to whole landscapes. In the development of this process based approach two parallel research tracks developed, one examining mass movement processes and the other examining hillslope hydrology. Strangely, for two decades or more interaction between these two themes was comparatively rare, despite the obvious synergies between them. Notwithstanding this, considerable advances were made in both fields, although key challenges remain. In recent years there has been an increase in the level of integration between hillslope hydrology and hillslope mass movement studies, in part because of a reawakened interest in sediment routing and long term landscape evolution. In addition, the availability of both high resolution field data collection techniques and high performance computing has allowed a better understanding of micro-scale processes. However, in both hillslope hydrology and mass movement studies there is now a need to exploit this enhanced understanding through a process of reconceptualization, in particular reflecting complex interactions within slopes in response to external forcing. In both cases this will represent a move from simplistic, process-response models to detailed, integrated representations of complex system behaviour. Ideally, these models should be equally applicable across a range of spatial and temporal scales; they should be testable; they should be physically sensible, using parameters that relate to measurable entities; and they should integrate both hydrology and mass movements where appropriate. The development of such models represents a fearsome challenge; the rewards will be substantive.

REFERENCES

Anders, A.M., Roe, G.H., Hallet, B., Montgomery, D.R., Finnegan, N.J. and Putkonen, J. (2006) Spatial patterns of precipitation and topography in the Himalaya. *Geological Society of America Special Papers* 398, 39–53.

Anderson, O.L. and Grew, P.C. (1977) Stress-corrosion theory of crack-propagation with applications to geophysics. *Reviews of Geophysics* 15, 77–104.

Belyea, L.R. and Baird, A.J. (2006) Beyond the "limits to peat bog growth": cross-scale feedback in peatland development. *Ecological Monographs* 76, 299–322.

Beven, K. (2002) Towards a coherent philosophy for modelling the environment, *Proceedings of the Royal Society of London, Series A* 458, 2465–84.

Bishop, A.W. (1967) Progressive failure–with special reference to the mechanism causing it, in *Proceedings of the Geotechnical Engineering Conference, Oslo*, vol. 2. pp. 142–50.

Bjerrum, L. (1967) Progressive failure in slopes of overconsolidated plastic clays and plastic shales. *Journal of Soil Mechanics and Foundations Division, ASCE* 93, 3–49.

Bogaardi, T., Guglielmi, Y., Marc V., Emblanch, C., Bertrand, C. and Mudry, J. (2007) Hydrogeochemistry in landslide research: a review. *Bulletin de la Societe Geologique de France* 178, 113–26.

Bogaart, P.W. and Troch, P.A. (2006) Curvature distribution within hillslopes and catchments and its effect on the hydrological response. *Hydrology and Earth System Sciences* 10, 925–36.

Brantley, S.L., Goldhaber, M.B. and Ragnarsdottir, K.V. (2007) Crossing disciplines and scales to understand the Critical Zone. *Elements*, 3, 307–314.

Bronstert, A. and Plate, E.J. (1997) Modelling of runoff generation and soil moisture dynamics for hillslopes and micro-catchments. *Journal of Hydrology* 198, 177–95.

Caine, N. (1980) The rainfall intensity–duration control of shallow landslides and debris flows. *Geografiska Annaler* Series A62, 23–7.

Carson, M.A. and Petley, D.J. (1970) The existence of threshold hillslopes in the denudation of the landscape. *Transactions of the Institute of British Geographers* 49, 71–95.

Carson, M.A. and Kirkby, M.J. (1972) *Hillslope Form and Process.* Cambridge University Press, Cambridge.

Chen, H. and Petley, D.N. (2005) The impact of landslides and debris flows triggered by Typhoon Mindulle in Taiwan. *Quarterly Journal of Engineering Geology and Hydrogeology* 38, 301–4.

Chorley, R.J., Schumm, S.A. and Sugden, D.E. (1984) *Geomorphology.* Methuen, London, p. 611.

Clark, M.P., Rupp, D.E., Woods, R.A., Tromp-van Meerveld, H.J., Peters, N.E. and Freer, J.E. (2009) Consistency between hydrological models and field observations: linking processes at the hillslope scale to hydrological responses at the watershed scale. *Hydrological Processes* 23, 311–9.

Coe, J.A., Ellis, W.L., Godt, J.W., Savage, W.Z., Savage, J.E., Michael, J.A. et al. (2003) Seasonal movement of the Slumgullion landslide determined from global positioning system surveys and field instrumentation, July 1998–March 2002. *Engineering Geology* 68, 67–101.

Corominas, J., Moya, J., Ledesma, A., Lloret, A. and Gili, J.A. (2005) Prediction of ground displacements and velocities from groundwater level changes at the Vallcebre landslide (Eastern Pyrenees, Spain). *Landslides* 2, 83–96.

Coulthard, T.J., Kirkby, M.J. and Macklin, M.G. (2000) Modelling geomorphic response to environmental change in an upland catchment. *Hydrological Processes* 14, 2031–45.

Dadson, S.J., Hovius, N., Chen, H., Dade, W.B., Lin, J.C., Hsu, M.L. et al. (2004) Earthquake-triggered increase in sediment delivery from an active mountain belt. *Geology* 32, 733–6.

Dalrymple, J.B., Blong, R.J. and Conacher, A.J. (1968) An hypothetical nine unit landsurface model. *Zeitschrift für Geomorphologie* 12, 60–76.

Davis, W.M. (1899) The geographical cycle. *Geographical Journal* 14, 481–505.

Davis, W.M. (1932) Piedmont benchlands and Primarrumpfe, *Bulletin of the Geological Society of America* 43, 399–440.

Densmore, A.L. and Hovius, N. (2000) Topographic fingerprints of bedrock landslides. *Geology* 28, 371–4.

Dietrich, W.E. (2001) Validation of the shallow landslide model, SHALSTAB, for forest management. *Land Use and Watersheds* 2, 195–227.

Dontsova, K., Steefel, C.I., Desilets, S., Thompson, A. and Chorover, J. (2009) Coupled modeling of hydrologic and geochemical fluxes for prediction of solid phase evolution in the Biosphere 2 hillslope experiment. *Hydrological and Earth System Sciences Discussions* 6, 4449–83, http://www.hydrol-earth-syst-sci-discuss.net/6/4449/2009/.

Dunning, S.A., Mitchell, W.A., Rosser, N.J. and Petley, D.N. (2007) The Hattian Bala rock avalanche and associated landslides triggered by the Kashmir Earthquake of 8 October 2005. *Engineering Geology* 93, 30–44.

Fair, T.D.J. (1947) Slope form and development in the interior of Natal. *Transactions of the Geological Society of Africa* 50, 105–20.

Fair, T.D.J. (1948) Slope form and development in the coastal hinterland of Natal. *Transactions of the Geological Society of Africa* 51, 37–53.

Fan, C.H., Allison, R.J. and Jones, M.E. (1997) Weathering effects on the geotechnical properties of argillaceous sediments in tropical environments and their geomorphological implications. *Earth Surface Processes and Landforms* 21, 49–66.

Frattini, P., Crosta, G. and Sosio, R. (2009) Approaches for defining thresholds and return periods for rainfall-triggered shallow landslides. *Hydrological Processes* 23, 1444–60.

Gilbert, G.K. (1880) *Report on the Geology of the Henry Mountains*. US Geographical and Geological Survey on the Rocky Mountain Regions. 108–18.

Gonzalez, D.A., Ledesma, A. and Corominas, J. (2008) The viscous component in slow moving landslides: A practical case, in Z. Chen, J.M. Zhang, Z.K. Li, F.Q. Wu and K. Ho (eds) *Landslides and Engineered Slopes: From the Past to the Future*. vol. 1, pp. 237–42.

Grayson, R.B., Moore, I.D. and McMahon, T.A. (1992) Physically based hydrologic modeling: 2. Is the concept realistic? *Water Resources Research* 28, 2659–66.

Guzzetti, F., Malamud, B.D., Turcotte, D.L. and Reichenbach, P. (2002) Power-law correlations of landslide areas in central Italy. *Earth and Planetary Science Letters* 195, 169–183.

Guzzetti, F., Peruccacci, S., Rossi, M. and Stark, C.P. (2008) The rainfall intensity-duration control of shallow landslides and debris flows: an update. *Landslides* 5, 3–17.

Hewlett, J.D. and Hibbert, A.R. (1963) Moisture and energy conditions within a sloping soil mass during drainage. *Journal of Geophysical Research* 68, 1081–7.

Hopp, L., Harman, C., Desilets, S.L.E., Graham, C.B., McDonnell, J.J. and Troch, P.A. (2009) Hillslope hydrology under glass: confronting fundamental questions of soil-water-biota co-evolution at Biosphere 2. *Hydrology and Earth System Sciences* 13, 2105–18.

Istanbulluoglu, E. and Bras, R.L. (2005) Vegetation-modulated landscape evolution: effects of vegetation on landscape processes, drainage density, and topography. *Journal of Geophysical Research* 110: F02012.

Keefer, D.K. (1984) Landslides caused by earthquakes. *Geological Society of America Bulletin* 95, 406–21.

Kienzler, P.M. and Naef, F. (2008) Subsurface storm flow formation at different hillslopes and implications for the 'old water paradox'. *Hydrological Processes* 22, 104–16.

Kim, S. (2009) Multivariate analysis of soil moisture history for a hillslope. *Journal of Hydrology* 374, 318–28.

King, L.C. (1951) South *African scenery: A textbook of geomorphology*. Oliver and Boyd, Edinburgh.

King, L.C. (1953) Canons of landscape evolution. *Bulletin Geological Society of America* 64, 721–52.

Kirchner, J.W. (2006) Getting the right answers for the right reasons: linking measurements, analyses, and models to advance the science of hydrology. *Water Resources Research* 42, W03S04.

Kirchner, J.W., Feng, X.H., Neal, C. and Robson, A.J. (2004) The fine structure of water-quality dynamics: the (high-frequency) wave of the future. *Hydrological Processes* 18, 1353–9.

Klemes, V. (1986) Delettantism in hydrology: transition or destiny? *Water Resources Research* 22, S177–88.

Klemes, V. (1988) A hydrological perspective. *Journal of Hydology* 100, 3–28.

Kumar, P. (2007) Variability, feedback, and cooperative process dynamics: elements of a unifying hydrologic theory. *Geography Compass* 1, 1338–60.

Lehmann, P., Hinz, C., McGrath, G., Tromp-van Meerveld, H.J. and McDonnell, J.J. (2007) Rainfall threshold for hillslope outflow: an emergent property of flow pathway connectivity. *Hydrology and Earth System Sciences* 11, 1047–63.

Lin, J-C., Petley, D.N., Jen, C-H. and Hsu, M-L. (2006) Slope movements in a dynamic environment – a case study of Tachia river, Central Taiwan. *Quaternary International* 147, 103–12.

Lin, C.W., Chen, H., Chen, Y.H. and Horng, M.J. (2008) Influence of typhoons and earthquakes on rainfall-induced landslides and suspended sediments discharge. *Engineering Geology* 97, 32–41.

Lin, C-W., Shieh, C-L., Yuan, B.D., Shieh, Y.C., Liu, S.H. and Lee, S.Y. (2003) Impact of Chi-Chi earthquake on the occurrence of landslides and debris flows: example from Chenyulan River watershed, Nantou, Taiwan. *Engineering Geology* 71, 49–61.

Malamud, B.D., Turcotte, D.L., Guzzetti, F. and Reichenbach, P. (2004) Landslide inventories and their statistical properties. *Earth Surface Processes and Landforms* 29, 687–711.

Malet, J.-P., van Asch, T.W.J., van Beek, R. and, Maquaire, O. (2005) Forecasting the behaviour of complex landslides with a spatially distributed hydrological model. *Natural Hazards and Earth System Sciences* 5, 71–85.

Meunier, P., Hovius N. and Haines, A.J. (2007) Regional patterns of earthquake-triggered landslides and their relation to ground motion. *Geophysical Research Letters* 34, L20408.

Meunier, P., Hovius, N. and Haines, J.A. (2008) Topographic site effects and the location of earthquake induced landslides. *Earth and Planetary Science Letters* 275, 221–32.

Montgomery, D.R., Schmidt, K.M., Greenberg, H.M. and Dietrich, W.E. (2000) Forest clearing and regional landsliding. *Geology* 28, 311–4.

Moore, R. and Brunsden, D. (1996) Physico-chemical effects on the behaviour of a coastal mudslide. *Geotechnique* 46, 259–78.

Mottershead, D.N., Wright, J.S., Inkpen, R.J. and Duane, W. (2007) Bedrock slope evolution in saltrock terrain. *Zeitschrift fur Geomorphologie* 51, 81–102.

Murphy, W., Petley, D.N., Bommer, J.J. and Mankelow, J.M. (2002) Uncertainty in ground motion estimates for the evaluation of slope stability during earthquakes. *Quarterly Journal of Engineering Geology and Hydrogeology* 35, 71–8.

Ng, K-Y. and Petley, D.N. (2009) A process approach towards landslide risk management in Hong Kong. *Quarterly Journal of Engineering Geology and Hydrogeology* 42, 487–98.

Parker, R., Rosser, N.J., Densmore, A. and Petley, D.N. (2009) Automated landslide detection algorithms to investigate controls on the spatial distribution of landslides triggered by the Wenchuan Earthquake, Sichuan Province, China. Proceedings of the International Conference on Next Generation Research on Earthquake-induced Landslides (invited keynote paper), Taiwan. pp. 27–32.

Penck, W. (1924) Die morphologische Analyse: ein Kapitel der physicalischen Geologie. *Geogr. Abhandlungungen* 2.Reihe, Heft 2, Stuttgart.

Petley, D.N. (2004) The evolution of slope failures: mechanisms of rupture propagation. *Natural Hazards and Earth System Sciences* 4, 147–52.

Petley, D.N. and Allison, R.J. (1997) The mechanics of deep-seated landslides. *Earth Surface Processes and Landforms* 22, 747–58.

Petley, D.N. (2010) Landslide Hazards in I. Alcántara-Ayala and A. Goudie (eds) *Geomorphological Hazards and Disaster Prevention.* Cambridge University Press. pp. 63–74.

Petley, D.N., Bulmer, M.H.K. and Murphy, W. (2002) Patterns of movement in rotational and translational landslides. *Geology* 30, 719–22.

Petley, D.N., Higuchi, T., Petley, D.J., Bulmer, M.H. and Carey, J. (2005a) The development of progressive landslide failure in cohesive materials. *Geology* 33, 201–4.

Petley, D.N., Mantovani, F., Bulmer, M.H.K. and Zannoni, F. (2005b) The interpretation of landslide monitoring data for movement forecasting. *Geomorphology* 66(1–4), 133–47.

Rosser, N.J., Lim, M., Petley, D.N. and Dunning, S.A. (2007) Patterns of precursory rockfall prior to slope failure. *Journal of Geophysical Research (Earth Surface)* 112, F04014.

Rosser, N.J., Petley, D.N., Lim, M., Dunning, S.A. and Allison, R.J. (2005) Terrestrial laser scanning for monitoring the process of hard rock coastal cliff erosion. *Quarterly Journal of Engineering Geology and Hydrogeology* 38, 363–76.

Rowe, P.W. (1969) Progressive failure and strength of a sand mass, in *Proceedings of the 7th International Conference on Soil Mechanics and Foundation Engineering.* vol. 1. pp. 341–9.

Savigear, R.A.G. (1952) Some observations on slope development in South Wales. *Transactions of the Institute of British Geographers* 18, 31–52.

Savigear, R.A.G. (1962) Some observations on slope development in north Devon and north Cornwall. *Transactions of the Institute of British Geographers* 31, 23–42.

Savigear, R.A.G. (1967) On surveying slope profiles, in Study of slope and fluvial processes, *Reviews de Géomorphologie Dynamique* 17, 153–5.

Schuerch, P., Densmore, A.L., McArdell, B.W. and Molnar, P. (2006) The influence of landsliding on sediment supply and channel change in a steep mountain catchment. *Geomorphology* 78, 222–35.

Schulz, W.H., Kean, J.W. and Wang, G.H. (2009a) Landslide movement in southwest Colorado triggered by atmospheric tides. *Nature Geoscience*, 2, 863–6.

Schulz, W.H., McKenna, J.P., Kibler, J.D. and Biavati, G. (2009b) Relations between hydrology and velocity of a continuously moving landslide-evidence of pore-pressure feedback regulating landslide motion? *Landslides* 6, 181–90.

Seibert, J. (2003) Reliability of model predictions outside calibration conditions. *Nordic Hydrology* 34, 477–92.

Selby, M.J. (1993) *Hillslope Processes and Landforms.* Oxford University Press, Oxford. p. 451.

Sengupta, A., Gupta, S. and Anbarasu, K. (2010) Rainfall thresholds for the initiation of landslide at Lanta Khola in north Sikkim, India. *Natural Hazards* 52, 31–42.

Sepulveda, S., Murphy, W. and Petley, D.N. (2005a) Topographic controls on coseismic rock slides during the 1999 Chi-Chi earthquake, Taiwan. *Quarterly Journal of Engineering Geology and Hydrogeology* 38, 189–96.

Sepulveda, S.A., Murphy, W., Jibson, R.W. and Petley, D.N (2005b) Seismically induced rock slope failures resulting from topographic amplification of strong ground motions: the case of Pacoima Canyon, California. *Engineering Geology* 80, 336–48.

Simoni, A., Berti, M., Generali, M., Elmi, C. and Ghirotti, M. (2004) Preliminary result from pore pressure monitoring on an unstable clay slope. *Engineering Geology* 73, 117–28.

Singh, A. and Mitchell, J.K. (1969) Creep Potential and creep rupture of soils. *Proceedings of the 7th International Conference on Soil Mechanics and Foundation Engineering,* vol. 1. pp. 379–84.

Sivapalan, M. (2005) Pattern, process and function: elements of a unified theory of hydrology at the catchment scale, in M.G. Anderson (ed.) *Encyclopedia of Hydrological Sciences.* John Wiley & Sons Ltd, Chichester. pp. 193–219.

Smith, R.B. (2006) Progress on the theory of orographic precipitation. *Geological Society of America Special Papers* 398, 1–16.

Soulsby, C., Neal, C., Laudon, H., Burns, D., Merot, P., Bonell, M. et al. (2008) Catchment data for process conceptualization: simply not enough? *Hydrological Processes* 22, 2057–61.

Spence, C. and Woo, M-K. (2003) Hydrology of sub-arctic Canadian shield: soil-filled valleys. *Journal of Hydrology* 279, 151–66.

Terzaghi, K. (1943) *Theoretical Soil Mechanics.* John Wiley and Sons, New York.

Tetzlaff, D., Soulsby, C., Waldron, S., Malcolm, I.A., Bacon, P.J., Dunn, S.M. et al. (2006) Conceptualization of runoff processes using GIS and tracers in a nested mesoscale catchment. *Hydrological Processes* 21, 1289–307.

Tetzlaff, D., McDonnell, J.J., Uhlenbrook, S., McGuire, K.J., Bogaart, P.W., Naef, F. et al. (2008) Conceptualizing catchment processes: simply too complex? *Hydrological Processes* 22, 1727–30.

Torres, R., Dietrich, W.E., Montgomery, D.R., Anderson, S.P. and Loague, K. (1998) Unsaturated zone processes and the hydrologic response of a steep, unchanneled catchment. *Water Resources Research*, 34, 1865–1879.

Tromp-van Meerveld, H.J. and McDonnell, J.J. (2006) Threshold relations in subsurface stormflow: 2. The fill and spill hypothesis. *Water Resources Research* 42, W02411.

Tucker, G.E. and Slingerland, R. (1997) Drainage basin responses to climate change. *Water Resources Research* 33, 2031–2047.

Tucker, G.E., Catani, F., Rinaldo, A. and Bras R.L. (2001) Statistical analysis of drainage density from digital terrain data. *Geomorphology* 36, 187–202.

Turowski, J.M., Yager, E.M., Badoux, A., Rickenmann, D. and Molnar, P. (2009) The impact of exceptional events on erosion, bedload transport and channel stability in a step-pool channel. *Earth Surface Processes and Landforms* 34, 1661–73.

Wagener, T. (2003) Evaluation of catchment models. *Hydrological Processes* 17, 3375–8.

Wagener, T., Sivapalan, M., Troch, P. and Woods, R. (2007) Catchment classification and hydrologic similarity. *Geography Compass* 1, 901–31.

Weiler, M. and McDonnell, J. (2004) Virtual experiments: a new approach for improving process conceptualization in hillslope hydrology. *Journal of Hydrology* 285, 3–18.

Weyman, D.R. (1970) Throughflow on hillslopes and its relation to the stream hydrograph. *International Association of Scientific Hydrology Bulletin* 15, 25–33.

Whipkey, R.Z. (1965) Subsurface stormflow from forested slopes, *International Association of Scientific Hydrology Bulletin*, 10, 74–85.

Wicks, J.M. and Bathurst, J.C. (1996) SHESED: A physically based, distributed erosion and sediment yield component for the SHE hydrological modelling system. *Journal of Hydrology* 175, 213–38.

Young, A. (1971) Slope profile analysis: the system of best units. IBG Special Paper, 3, 1–14.

21

Riverine Environments

Jim Pizzuto

Fluvial geomorphology is the study of landforms created by rivers. Rivers are channelized flows larger than gullies or rills, while a landform is an 'element of the landscape that can be observed in its entirety, and has consistence of form or regular change in form' (Bloom, 1998). Fluvial landforms cover spatial scales from a few millimeters to more than 1000 km, and they evolve over timescales varying seconds to millions of years (Figure 21.1).

This chapter proceeds from a broad overview of fluvial geomorphology to more specific topics. The first section presents three general categories of knowledge in fluvial geomorphology, followed by discussions of the spatial and temporal scales of interest to fluvial geomorphologists and the research methods currently in use. A classification of rivers and fluvial landforms introduces the features that fluvial geomorphologists strive to understand and explain. Then, several ideas that have dominated fluvial geomorphology during the last 50 years are introduced, beginning with the concept of 'the fluvial system', followed by related ideas of alluvial river channel equilibrium and the development of river channel planforms. To contrast concepts related to equilibrium, steady-state fluvial landforms, models summarizing how rivers evolve through time are discussed next. Finally, topics that have occupied fluvial geomorphologists more recently are presented, including bedrock channels, the influences of plants, animals and humans on

rivers, and the increasingly important topic of river restoration.

THREE CATEGORIES OF KNOWLEDGE IN FLUVIAL GEOMORPHOLOGY

To interpret fluvial landforms, fluvial geomorphologists seek knowledge in three distinct categories (Table 21.1). Description is necessarily the first, a process that involves mapping a landform's morphology and additional observations and measurements to determine its composition. The surficial morphology of landforms is quantified directly by surveying or indirectly using remote sensing. Because landforms are three dimensional, part of a landform extends beneath the Earth's surface; the subsurface nature of landforms is defined using geological techniques such as drilling and coring, or geophysical methods such as ground-penetrating radar or seismic reflection. Describing a landform also requires defining its geological, biological and chemical features. Once a landform has been thoroughly described, geomorphologists seek to understand the processes influencing the form and nature of the landform. This could involve studies of geological, chemical, biological, physical and social (anthropogenic) processes. Finally, geomorphologists seek to understand the temporal context of a landform. This involves determining

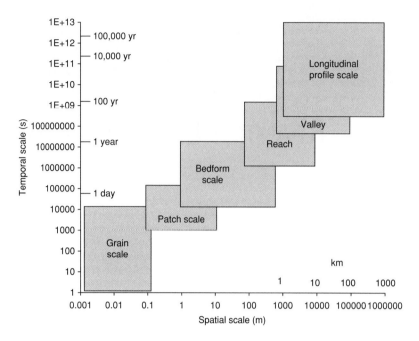

Figure 21.1   A classification of spatial and temporal scales in fluvial geomorphology

Table 21.1 Categories of knowledge in fluvial geomorphology

Knowledge category	Questions answered	Approach, method or category
Description	What is the shape of the landform?	Surveying, LiDAR, remote sensing, geophysical methods
	What is the composition?	Geology, pedology, biology, chemistry
Understanding process	What processes affect the landform?	*Geological* (erosion, deposition, tectonics, faulting), chemical, biological, cultural (anthropogenic)
Temporal context	How often/fast do processes act?	Define past events (historical or geological methods) Monitor contemporary processes. Predict future state (extrapolate past trends or use predictive theory)

how often and over what length of time important processes act.

TEMPORAL AND SPATIAL SCALES IN FLUVIAL GEOMORPHOLOGY

Figure 21.1 defines spatial and temporal scales of fluvial landforms. The 'spatial scale' of a landform largely refers to its size, while the 'temporal scale' refers to the length of time required for the feature to develop or change significantly. 'Grain' scales primarily involve individual sedimentary particles (Table 21.2); landforms at these scales

are influenced by sediment transport processes such as bedload or suspended load transport. Grain scales cover spatial scales equal to the size of the particles involved, and relevant temporal scales range from seconds (or possibly less) to a few hours. Patch scales involve arrangements of groups of particles that cover parts of a streambed from 0.1 to a few meters in length (Paola and Seal, 1995); patches may have temporal dynamics ranging from around 1 h to 1 day or so. Bedform scales are those associated with dunes, pools and riffles, and bars. These features may range from 1 to 1000 m in length, and develop over periods ranging from less than a day to a few years. A river 'reach' is a length of stream on the order

Table 21.2 Morphological elements of rivers at different spatial scales

Scale	Descriptive elements
Watershed	Longitudinal profile
Valley	Floodplain, channel planform, slope, riparian vegetation
Reach	Width, depth, slope, bars (alternate, mid-channel), pools/riffles, step/pools, large woody debris (LWD) jams
Patch	Bedload sheets, ripples, dunes, transverse ribs, pebble clusters, single LWD
Grain	Clay, silt, sand, boulder

of 10–20 channel widths in length; properties of reaches include a river's bankfull depth, width and average slope. Reaches have spatial scales of 100 m to 10 km and temporal scales ranging from less than 1 year to a century. The valley scale obviously refers to river valleys, features that range in size from 1 to more than 100 km. River valleys may form by incision or floodplain development, processes that typically range from 1 to at least 10^5 years in duration. Finally, the form of the longitudinal profile of an entire river can range in length from around 1 km to the size of an entire continent. Longitudinal profiles of small channels

may adjust over periods of a decade or less, while large channels may require more than 10^6 years to adjust their longitudinal profiles.

RESEARCH METHODS OF FLUVIAL GEOMORPHOLOGY

Fluvial geomorphologists rely on diverse methods (Table 21.3) that range from field observations to physical and mathematical modeling. The methods used are often related to the temporal and spatial scale of the landform being studied; contemporary observations are not very useful for studying landforms that require thousands of years to form, and traditional instrument surveys cannot efficiently capture the morphology of huge rivers such as the Amazon. Field methods that capture the morphology of fluvial landforms include traditional surveying and newer, innovative methods based on laser technology such as light detection and ranging (LiDAR). Geomorphologists monitor contemporary fluvial processes by measuring river currents, sediment transport rates and rates of bedrock incision (see Kondolf and Piegay, 2003). Processes with longer timescales require historical methods, geological studies of river deposits and relative and absolute dating methods. Laboratory models are increasingly used to study sediment transport processes

Table 21.3 Examples of research methods and their uses in fluvial geomorphology. Specific examples may be found in Kondolf and Piegay (2003)

General category	Use	Specific category	Example
Field	Define surface morphology	Land surveying	Tripod lidar – erosion surveys Total station, levels – cross-sections
		Remote sensing	Aerial photos – valley/channel morphology Aerial lidar – valley/channel morphology Satellite – valley/channel morphology
	Determine process rates	Water flow measurements	Current meters, discharge gaging
		Sediment movement	Sediment gaging
	Temporal perspective	Stratigraphy	Coring, geophysical methods, soil description
		Dating methods	Historical sources, radiometric, tree rings
Physical modeling	Model calibration, hypothesis testing	Sediment transport	Observations of bedload, suspended load
		Fluvial landforms	Origin of channel form or bedform morphology
Mathematical modeling	Prediction and hypothesis testing	Mathematical theories	One-dimensional modeling of meander migration
		Established 'engineering' models	One-dimensional hydraulic models
		Landscape evolution models	Models of river basin evolution
		Channel evolution models	Cellular automata – braided channels

and development of fluvial landforms under controlled conditions. To address scaling issues, Hooke (1968) considers a scale model as 'a small system in its own right, not as a scale model of a prototype'. As an example, Hooke (1968) describes how studying the slopes of very small alluvial fans in the laboratory provided insights to explain the slopes of alluvial fans observed in the field. Mathematical modeling has also proven fruitful in fluvial geomorphology. For example, mathematical and numerical models have provided a very thorough explanation of the morphology of freely meandering streams (Johanneson and Parker, 1989; Furbish, 1991; Stolum, 1998). Models for computing water surface elevations and mean velocities are widely used and detailed three-dimensional hydraulic models have provided new insights on the origin of fluvial landforms (see Kondolf and Piegay, 2003). Numerical models of fluvial transport, deposition and erosion have provided interesting insights regarding the evolution of entire drainage basins over geological time (Coulthard et al., 2002) and the development of braided channels (Murray and Paola, 1994).

CLASSIFICATION OF RIVERINE LANDFORMS

Rivers are exceedingly diverse, reflecting many different climatic and geological settings. Accordingly, schemes for organizing and classifying observations of rivers are also highly diverse, with different scales of interest and different objectives. Important features of rivers and watersheds with varying spatial and temporal scales are introduced below, so the reader can appreciate the breadth of fluvial geomorphology.

The morphology of entire watersheds represents the largest spatial scale for defining fluvial landforms (Figure 21.1). The geometrical patterns of drainage networks have been classified in a variety of interesting ways (Rodriguez-Iturbe, 1997; Bloom, 1998). Process views of drainage networks highlight the role of rivers in sediment transport and storage: upland mountain settings are considered areas where sediment is supplied to rivers, the middle parts of drainage basins are areas of long-term sediment transport downstream, and the downstream areas of watersheds are areas of sediment deposition and storage (Schumm, 1977). Because of the long time periods required to form an entire drainage basin, defining the temporal evolution of drainage basin networks has proven challenging. Geomorphologists have studied useful small-scale analogues of natural watersheds in quarries (Howard and Kerby, 1983) and in laboratory flumes (Hasbargen and Paola,

2000). Landscapes abandoned by glaciers at different times have presented opportunities to study relatively younger and older watersheds. Recently, computer modeling of drainage basin evolution has clarified the relationships between sediment yield, climate and the internal system dynamics of channel networks (Coulthard et al., 2002; Van de Wiel and Coulthard, 2010).

Interactions between rivers and their valleys are both diverse and profound. Church (1992) notes that nearly all rivers are influenced by underlying lithology and geological structure. If rivers deposit and store some of their sediment load in valleys, floodplains result. Nanson and Croke (1992) present a useful floodplain classification based on stream power.

Rivers display characteristic forms when viewed from above that also provide a useful basis for classification (Schumm 1977; Church, 1992; Nanson and Knighton, 1996). Planform morphologies of rivers include sinuous, meandering, braided and anastomosing channels (Figure 21.2). Sinuous and meandering channels are 'single-thread' channels that differ in sinuosity, which is defined as the ratio of the distance measured along the channel to the distance measured directly downvalley. Sinuous channels have sinuosities less than 1.5, whereas meandering channels have sinuousities greater than 1.5 (Knighton, 1998). In braided channels, the primary flow is divided by unvegetated bars that are submerged during high flows. Anastomosing streams have multiple channels separated by vegetated floodplains that are only submerged during major floods (Nanson and Knighton, 1996).

Reach scale attributes of rivers include measures of channel size and characteristic forms of the channel bed (Table 21.2). The size of rivers is often represented by the channel width or depth, which are typically defined at 'bankfull stage' when the channel is filled with water (Church, 1992). The river's slope provides the primary driving force for moving water and sediment through a reach, and it is therefore another fundamental characteristic of reach scale river morphology. River beds are also often organized into characteristic morphologies known as bedforms; at the reach scale these include pools and riffles that alternate with a characteristic wavelength of five to seven channel widths, and alternate and mid-channel bars. In coarse-grained, boulder channels, pools and cascades alternate downstream with characteristic wavelengths of about 1 channel width (Grant et al., 1990); these are known as 'step-pool' bedforms.

Smaller bedforms are also commonly observed in rivers. In sand bed channels, river beds are often deformed into periodic bedforms (in order of increasing size) such as ripples, dunes and

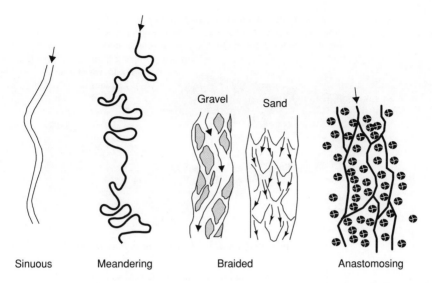

Figure 21.2   Sinuous, meandering, braided, and anastomosing river channel planforms

antidunes (Knighton, 1998). In coarser-grained gravel and boulder bed channels, characteristic bedforms include pebble clusters (Knighton, 1998) and transverse ribs. Bedload sheets (Whiting et al., 1988) and longitudinal streaks (Colombini and Parker, 1995) are common features of mixed sand and gravel bed channels.

'Patches' of characteristic grain size have been noted on streambeds. These have not been widely studied, but they may locally influence sediment transport rates and rates of downstream fining (Paola and Seal, 1995).

The material that comprises the channel perimeter reflects the smallest scale used to describe river channels. Typically, a characteristic grain diameter for bed and bank sediments is defined (though this approach is obviously not very useful where the channel boundaries are composed of bedrock, ice, or plants). The most important grain size classification for river channels simply involves determining if the materials are silt or clay, sand or gravel (these terms are defined in Kondolf and Piegay, 2003).

Many classification schemes have been proposed to organize and interpret different features of rivers. A recent summary is provided by Kondolf and Piegay (2003).

A river reach as a 'system' for transporting water and sediment under 'constraints'

Geomorphologists and engineers often idealize rivers as an 'open system' (see Kondolf and Piegay, 2003, Chapter 5) (Figure 21.3). In this view, a river reach is supplied with water, sediment of specified sizes and energy (from gravity). The 'river reach' is a section of a channel or valley that varies in length depending on the problem of interest. The 'system' is represented by a series of principles that determine how the water and sediment is passed through the channel or valley length of interest; the 'operation' of the system according to these principles: (1) converts potential energy into work (transport of water and sediment) and heat; (2) creates the river's valley and the fluvial landforms contained within it; and (3) passes water and sediment out of the 'system' to reaches downstream.

The principles that govern the fluvial system involve (1) hydraulic principles that govern the transport of water and sediment, (2) constraints imposed by external variables and (3) time. The external variables are of profound importance, because otherwise fluvial geomorphology would have evolved into a sub-discipline of hydraulics or physics. Constraints are imposed by the geological materials that rivers must carve their channels and valleys into, by plants and animals and by the widespread effects of humans on channels and watersheds. The temporal context of a fluvial system is also of paramount importance: the system may be considered at equilibrium (a steady state system), passing through a series of predictable evolutionary states, or responding to historical and current events. All of these possibilities are explored below as examples of ongoing research.

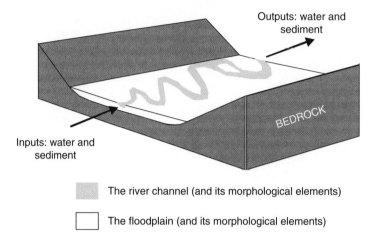

Elements of a reach-scale fluvial system
(the system adjusts its morphological elements given
inputs of water and sediment, subject to 'constraints')

Outputs: water and sediment

BEDROCK

Inputs: water and sediment

The river channel (and its morphological elements)

The floodplain (and its morphological elements)

Figure 21.3   Elements of a 'reach-scale' fluvial system, a mechanism for transporting and storing water and sediment along a channel or valley of specified length. Selected 'morphological elements' are listed in Table 21.2

ALLUVIAL CHANNELS

Alluvial channels are those whose morphology is carved into sediment deposited by the river itself (Figure 21.3 illustrates an alluvial channel). Alluvial channels are often referred to as 'self-formed', which means that the river has created its morphology through time through the transport and deposition of sediment by hydraulic processes. The morphology of alluvial rivers is therefore determined by processes associated with the river itself, without overwhelming control by external factors such as tectonic uplift, faulting, or engineering works (including those imposed by both humans and beavers).

Equilibrium theory of alluvial channels

The elevation of the streambed, plotted as a function of the distance downstream, defines a river's longitudinal profile (Figure 21.4). Studies of longitudinal profiles have greatly influenced fluvial geomorphology in the 20th century. For example, fluvial geomorphologists noticed that when alluvium covered bedrock surfaces in river valleys, longitudinal profiles often assume a characteristic shape, with steeply sloping upstream reaches that smoothly decrease in slope in the downstream direction. Geomorphologists hypothesized that this smooth, concave longitudinal profile represents a special condition referred to as a 'graded stream' (Figure 21.4a).

Geomorphologists also recognized, however, that some longitudinal profiles were interrupted by 'knickpoints', short areas of very steep slopes that might interrupt an otherwise smooth profile (Figure 21.4b to c). Knickpoints could represent the influence of occasional outcrops of resistant bedrock (Figure 21.4b), or they could represent upstream migrating 'pulses' of erosion initiated by tectonic uplift (Figure 21.4c) or a rapid drop in sea level (referred to a fall in the stream's 'base level') (Figure 21.4d).

Geomorphologists have interpreted the smooth, 'graded' longitudinal profile (Figure 21.4a) as representing a state of balance, or equilibrium, created by the river through processes of erosion and deposition of sediment. When the graded state was achieved, the river's slope was believed to be specifically adjusted to ensure that the sediment supplied to the river could be transported downstream without either erosion or deposition. As famously articulated by Mackin (1948):

> A graded river is one in which, over a period of years, slope and channel characteristics are delicately adjusted to provide, with available discharge, just the velocity required for the transportation of the load supplied from the drainage

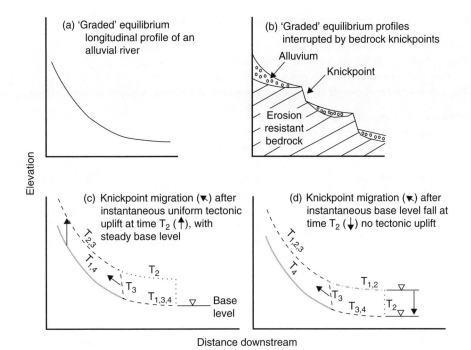

Elevation

Distance downstream

Figure 21.4 Longitudinal profiles of stream channels. (a) Classic 'graded' or equilibrium longitudinal profile for an alluvial river. (b) Equilbrium longitudinal profile interrupted by bedrock-controlled knickpoints. (c) Upstream migrating knickpoint developing from a spatially uniform instantaneous tectonic uplift. (d) Upstream migrating knickpoint developing from a drop in sea level

basin. The graded stream is a system in equilibrium...... any change in any of the controlling factors will cause a displacement of the equilibrium in a direction that will tend to absorb the effect of the change.

The idea that rivers adjust their slopes to achieve equilibrium was extended in important ways in the 1950s, 1960s and early 1970s. First, Wolman and Miller (1960) argued that moderate discharges that re-occur every 1–2 years control river morphology, and that very frequent and very rare flood events are less important. Leopold et al. (1964) further suggested that these moderately recurring flow events typically represent bankfull conditions in alluvial stream channels further suggesting that rivers are adjusted to these discharges. Second, many geomorphologists discovered systematic relationships between water discharge and fluvial morphology, including width, depth, slope, channel roughness (a measure of the friction exerted by the channel on the water flowing within the channel), and even the diameter of the sediment exposed on the streambed. These relationships suggested that river

morphology was 'adjusted' or controlled by the amount of water that flowed in the channel every 1–2 years.

As a result, the concept of the graded stream was expanded considerably (Table 21.4). Streams are referred to as graded when at least three conditions are achieved (Table 21.4), including (1) transporting the 1.5 year water discharge within the bankfull channel, (2) transporting all grain sizes supplied by the watershed without net erosion or deposition and (3) balancing any bank erosion with an equal amount of bank sedimentation (some researchers have also suggested additional conditions, including maximum hydraulic resistance to flow, maximum sediment transport efficiency, or minimum expenditure of energy). To achieve these conditions, the watershed supplies a certain quantity of water and sediment, and through hydraulic processes of erosion and deposition, morphologic variables listed in Table 21.4 are adjusted by the stream until all the conditions are satisfied.

The most important characteristic of graded streams involves their morphology, which is considered to be approximately constant when

Table 21.4 Idealized conditions and variables attributed to 'graded' stream reaches

Idealized conditions satisfied by a graded reach

1   1.5-year discharge is transported within the bankful channel

2   Sediment inputs to a reach and outputs from a reach are equal

3   The reach-averaged morphology of the channel is approximately constant when averaged over a period of years

Independent variables, dependent variables and constraints

Supplied from upstream ('independent variables')	Adjusted by the stream ('dependent variables')	Constraints
1.5-year discharge	Bankful width, depth	Riparian vegetation
Sediment volume and grain sizes	Bed slope	Bedrock exposures
Large woody debris volume and size distribution	Planform (Figure 21.3)	Animals
	Bed surface texture (armouring)	Engineering structures
	Bed grain size patches	
	bedforms (ripples, dunes, bars)	
	Large woody debris architecture	

averaged over a few years. Changes wrought by individual floods, for example, are considered to be short-lived (this concept is explicitly illustrated in Figure 21.6B). Furthermore, the morphology of graded streams should be controlled by contemporary processes that can, at least in principle, be observed today. Graded streams could have been influenced by past events, but most elements of their morphology are adjusted to (and hence dominantly controlled by) present processes.

The conditions and variables listed in Table 21.4 have inspired hundreds of researchers to propose mathematical models of the graded state based on principles of hydraulics and sediment transport. These models are typically suggested for specified idealized conditions that may be approximated by some real rivers; for example, rivers that transport uniform gravel sediment (Parker, 1978b), rivers with sandy beds (Parker, 1978a), or rivers than cannot transport bed material at all (Diplas, 1992). Although these models have been useful conceptually (i.e. Lane, 1955) and in some cases for engineering design (Diplas, 1992), they are not fully accepted by most geomorphologists either as complete scientific descriptions of the graded state or as useful design tools in applied fluvial geomorphology, largely because they are incapable of incorporating the inherent variability of fluvial systems created by differing sediment types, the presence of underlying bedrock and the importance of plants and animals on fluvial processes. An additional, equally vexing problem, is that it has proven nearly impossible in most cases to determine the quantity of sediment supplied to a specific river reach, and without specifying all the independent variables listed in

Table 21.4, the morphology of the stream cannot be determined (even if the equations available are perfectly accurate). Knighton (1998, p. 171) summarizes our theoretical understanding of equilibrium channel geometry by referring to it as 'a problem … which has yet to be satisfactorily solved'.

It is not only the difficulty in quantifying the graded state that presently limits the applicability of this concept: recent research has also emphasized that the morphology of many streams is not always well-adjusted to contemporary processes. To provide a single example: according to Wolman (1955):

> Brandywine Creek [of the mid-Atlantic U.S. Piedmont region] is a graded stream..it..has attained and maintains an equilibrium through the mutual adjustment of the discharge, load, bed material, slope, width, depth, and velocity.

Several decades later, stratigraphic and historical studies (described in greater detail below) demonstrate that the morphology of mid-Atlantic US streams remains strongly influenced by valley alluviation related to European settlement hundreds of years ago, a radical departure from the equilibrium envisioned by Wolman. In an assessment of trends in geomorphic thinking, Phillips (1992) perceives 'the possible end of classic equilibrium studies', and the rise of ideas that include 'non-equilibrium responses and multiple equilibria'. Geomorphologists can no longer assume that particular streams should be graded or close to an equilibrium condition. Rather, careful analysis is required to determine how a stream has changed during its history, and the extent to which historical changes will continue influence channel morphology into the future.

The origin of and development of alluvial river channel planforms

Studies of the origin of alluvial river channel planforms represent a voluminous literature. Theories for the origin of meandering are well-developed, some progress has been made in understanding braided channels, while explanations for anastomosing planforms are in their infancy.

The development of a meandering channel from an initially straight channel with a flat bed has been described in field studies (Lewin, 1976) and theory (Parker, 1976). Flume studies of meandering have proven difficult, though recent progress is encouraging (Braudrick et al., 2009). The first step involves the growth of bars near the banks of the stream on alternating sides of the channel; these bars deflect the flow into adjacent banks, which leads to bank erosion. As the flow becomes curved, the bars are stabilized and they grow into 'point' bars on the insides of meander bends. Once the flow becomes curved, the curves grow as a result of a hydrodynamic instability associated with flow in curved channels (Johanneson and Parker, 1989).

The processes described above have been thoroughly quantified, and the morphology predicted by theory has been compared with observed meander morphology. While researchers in the 1960s believed that meanders could be adequately described by symmetrical 'sine-generated' curves, meanders are know known to be typically asymmetrical (Carson and Lapointe, 1983), without following a single mathematical form (Furbish, 1991). Stolum (1998) has demonstrated that the evolution of meanders illustrates processes of self-organization, and that meander planforms generated by numerical models are statistically indistinguishable from those of real meandering rivers. Quantitative modeling of meandering has become a mature discipline, with useful protocols for model parameterization (Constantine et al., 2009). Several studies demonstrate that predictive models of meander development can be used to evaluate river restoration plans (Larson and Greco, 2002).

Braided channels develop when multiple bars form in wide river channels (Parker, 1976). Many of the specific processes that result in the development of a braided channel from an initially single-thread channel have been documented in flume studies (e.g. Ashmore, 1990). Modeling studies of braided channels using cellular automata (Murray and Paola, 1994) demonstrate that

braids will develop where the flow is unconstrained laterally, and where bedload transport is active.

The origin of anastomosing channels remains controversial. Jerolmack and Mohrig (2007) suggest that anastomosing channels develop where avulsion (a rapid lateral 'jump' of a channel across a floodplain to a new location) is the dominant mechanism of lateral channel migration. They distinguished anastomosing channels from meandering and braided channels using a dimensionless ratio $M = T_a/T_c$, where T_a is the time required for the channel to aggrade one channel depth, and T_c is the time required for the channel to migration laterally 1 channel width. Where $M \gg 1$, channels are typically anastomosing, while where $M \ll 1$, channels are single-thread, suggesting that aggradation and avulsion are the necessary factors that lead to the development of anastomosing rivers. Huang and Nanson (2007) propose that anastomosing channels develop because multiple channels are hydraulically more efficient than single-thread channels in transporting high sediment loads.

Changes in alluvial channel systems with time

Determining the geomorphic changes that have occurred to river systems through time is fundamental (Table 21.1). Depending on the spatial and temporal scales of the landforms of interest and their history, fluvial processes can be viewed as time-invariant (i.e. as representing some sort of equilibrium), as progressing through a series of predictable stages, or as responding to external forcing driven by changes in tectonics, climate or land-use.

While studies of historical maps, monitoring and physical and mathematical modeling can provide useful information on temporal changes to rivers (Table 21.3), geological studies of the evolution of river valleys provide the most useful and comprehensive information on temporal changes to river systems. Valley filling, incision and downcutting are typically recorded by alluvial terraces, which can be dated to determine rates of these processes (e.g. Lave and Avouac, 2001).

Defining and dating stratigraphic units in floodplain deposits can also reveal profound changes in rivers through time. Stratigraphic studies of sinuous rivers of the mid-Atlantic Piedmont of the United States present an instructive example. As described above, geomorphologists initially believed that the morphology of these streams represented a profound equilibrium adjustment to the contemporary hydraulic regime. More recent stratigraphic studies, however, suggest otherwise (Jacobson and Coleman, 1986; Pizzuto, 1987; Walter and Merrits, 2008). Because the floodplains of the region lack the characteristic point bar deposits typical of meandering channels (Figure 21.5a), these rivers were probably not meandering before European settlement.

Rather, they were more likely small, shallow sinuous channels (Figure 21.5b). Although detailed evidence for their planform morphologies is lacking, Walter and Merritts (2008) hypothesize that pre-settlement rivers were anastomosing channels that flowed through extensive wetlands.

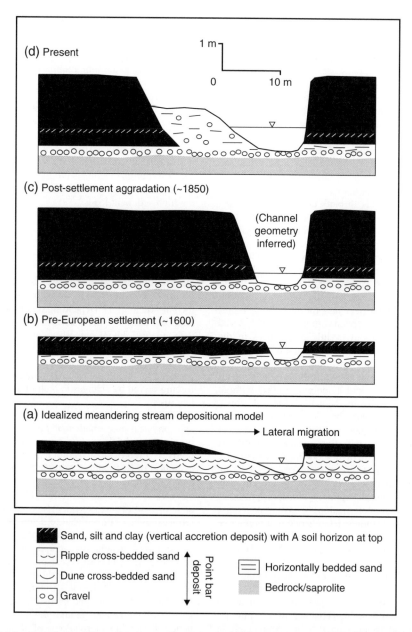

Figure 21.5 Reconstruction of stream channel morphology in the mid-Atlantic Piedmont from stratigraphic data (Jacobson and Coleman, 1986; Pizzuto, 1987; Walter and Merritts, 2008)

Following the growth of agriculture, extensive soil erosion and the construction of colonial mill dams, floodplains accreted vertically in the nineteenth century (Figure 21.5c). By the end of the 20th century, rivers of the region enlarged their channels and locally developed large, gravelly point bars in response to the decline of agriculture, the demise of colonial mill dams (Pizzuto and O'Neal, 2009) and urban development (Hammer, 1973) (Figure 21.5d). Thus, the present morphology (e.g. the elevated cutbank of Figure 21.5d) retains elements created during an early episode of the history of the region, elements that are not being formed by current processes,

but are rather slowly being removed through erosion.

From many studies of the evolution of rivers through time, fluvial geomorphologists have developed several idealized models describing the temporal evolution of fluvial landforms (Figure 21.6). Figure 21.6a illustrates changes to one element of a reach-scale fluvial system (bed elevation) through time following changes in the supply of water and sediment (these are labeled 'forcing' in Figure 21.6a). After each forcing event, the elevation of the bed adjusts towards an equilibrium level at a gradually decreasing rate. However, before a quasi-study bed elevation is

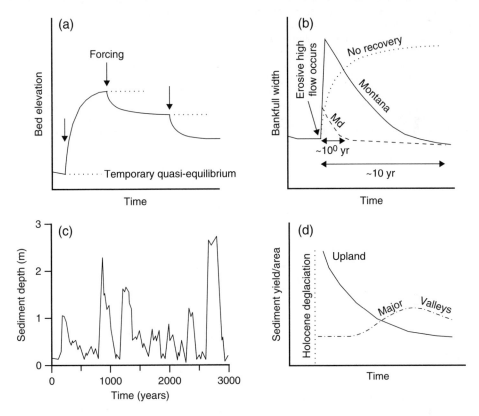

Figure 21.6 Selected models of fluvial channel evolution through time. (a) Response of river bed elevation to periodic forcing. After each forcing event, the stream bed tends to evolve towards a new temporary equilibrium. Before the new equilibrium elevation is reached, another forcing event occurs (after Bull, 1991). (b) Response and recovery of channel width following erosive stormflows in a humid temperate climate (Maryland, Md), a semi-arid climate (Montana), and an arid region where no recovery occurs (after Wolman and Gerson, 1978). (c) Episodic variations in depth of alluvial cover in a Pacific north-west stream channel related to passage of sediment pulses induced by storm events and periodic fires that destroy forest cover (after Benda and Dunne, 1997). (d) Spatial variations in Holocene sediment yield related to glaciation and episodic sediment storage and remobilization in British Columbia (Church and Slaymaker, 1989)

achieved, another forcing event occurs, initiating another gradual adjustment towards a new equilibrium level. In Figure 21.6a, the bed elevation never reaches a steady equilibrium value, but is rather in a state on continual adjustment spurred by ongoing external forcing events.

Figure 21.6b illustrates how channel width responds to individual floods in different climatic settings. In the eastern United States, channels widened in response to Hurricane Agnes, but the width quickly recovered after just a few years (Costa, 1974). This is the kind of response and recovery that is consistent with the notion that equilibrium morphology is 'constant' when averaged 'over a period of years'. In semi-arid and arid regions of the American West, however, some widened channels may require several decades to recover from floods (Pizzuto, 1994). Under these conditions, the concept of equilibrium is only meaningful if the channel morphology is averaged over many decades, possibly centuries. In other arid and semi-arid regions, floods may completely destroy a channel and its floodplain, and recovery may not occur under present conditions (see Knighton, 1998, pages 295–302).

Benda and Dunne (1997) have described pulses of sediment supplied to channels of the Pacific north-west that are driven by stochastic temporal variations in rainfall, colluvial sediment storage and landslides and debris flows. These pulses are then translated downstream by fluvial processes, creating large variations in bed elevation on millennial timescales (Figure 21.6c). Figure 21.6c suggests that channel morphology should change abruptly at intervals varying from decades to centuries as these pulses of sediment pass through the fluvial network. The bed elevation of these streams does not appear to be in a state of long-term equilibrium adjustment.

Figure 21.6d illustrates a transition in the location of sediment supply in a watershed from upland areas to major valleys. This pattern was originally proposed for Holocene changes in sediment supply in British Columbia (Church and Slaymaker, 1989), where glaciation provided abundant sediment sources in upland areas during the early Holocene. This sediment was eroded from upland areas and stored in river valleys downstream, where it is presently being remobilized by fluvial processes. Although specifically developed for a Canadian glaciated landscape over Holocene time, this model can also apply to similar transitions in sediment yield, for example: (1) soil erosion and valley alluviation related to post-settlement deforestation on centennial timescales in the eastern US (also summarized in Figure 21.5) and in New Zealand (Gomez et al., 2003); (2) movement of sediment introduced into rivers through mining (Knighton, 1998: 322–329);

and (3) transfers of sediment from uplands to valleys driven by logging, road construction and storm frequency on decadal timescales in northern California (Knighton, 1998).

The evolutionary models of Figure 21.6 provide a useful summary of how fluvial geomorphologists currently view the concept of grade and equilibrium of alluvial river channels. While the concept of equilibrium is a useful idealization of an ultimate endpoint that rivers could potentially achieve, inputs of water and sediment to a reach and 'constraints' within a reach are rarely constant long enough to achieve a meaningful equilibrium form. As a result, the attainment of equilibrium by a specific stream system should never be assumed, but must be demonstrated by careful field observations (Table 21.4).

BEDROCK CHANNELS

A bedrock channel cannot substantially widen, lower, or shift its bed without eroding bedrock' (Turowski et al., 2008). In alluvial channels, deposition is an important process influencing channel morphology. Bedrock channels, however, primarily evolve through erosional processes.

Bedrock channels erode through a suite of processes that are poorly understood. Bedrock may be removed by plucking along pre-existing joints or fractures, by the forceful collapse of bubbles in turbulent flows known as cavitation, by abrasion, and by dissolution (Whipple et al., 2000). These processes create characteristic features similar to the bedforms of alluvial channels, including flutes, scallops and potholes that are only in their initial stages of characterization and study (e.g. Richardson and Carling, 2005).

Lacking a detailed understanding of the physical process of bedrock incision, geomorphologists initially proposed generalized equations that relate the rate of erosion to 'stream power' (Howard et al., 1994). These incision laws typically take the form $E = U - kA^mS^n$, where E is the incision rate (with units of length/time), U is the rate of tectonic uplift, k is a coefficient, k is a coefficient that mostly represents the erosional resistance of the bedrock, A is the drainage basin area, S is the slope of the channel and m and n are empirically determined exponents that range from 0.3 to 1 and 0.7 to 1 (Howard and Kerby, 1983; Tucker and Whipple, 2002).

The erosion rate E may be represented as the change in elevation, z, of a point with time, t (dz/dt). Furthermore, the slope S can be represented as the change in elevation with distance downstream along the channel, x (dz/dx). Many geomorphological studies have also demonstrated that the drainage basin area, A, varies systematically with distance, x, so that A can be eliminated from the erosion rate law and represented as an empirical function of x. With a few additional manipulations, the erosion rate law is transformed into an equation that represents the elevation of the stream channel, z, as a function of the distance downstream, x, the uplift rate, U, and a variety of coefficients that represent bedrock properties, the delivery of water to the channel (hence climate) and the geomorphic structure of the drainage basin. This essentially provides a theory for predicting the longitudinal profile of bedrock streams, and how it varies with tectonics, climate and lithology.

Studies of bedrock rivers are one of the most exciting current frontiers in fluvial geomorphology. As new cosmogenic radionuclide dating techniques and digital elevation models of mountainous regions have become widely available, modeling analyses of the longitudinal profiles of bedrock rivers have provided profound insights into the relationships between fluvial erosion, climate, lithology and tectonics (Roe et al., 2002; Whipple, 2004). For example, modeling studies have defined patterns of knickpoint migration caused by tectonic activity and base level changes (i.e. Figure 21.4). Field observations defining contemporary processes and rates of bedrock erosion, once believed to be too slow to be directly observed, are providing intriguing opportunities to document how bedrock rivers modify their morphology (Hartshorn et al., 2002). Currently, researchers are developing new, more sophisticated erosion rate laws that account for important variables such as sediment concentration (Sklar and Dietrich, 2004), and that incorporate new erosional mechanisms, such as debris flows (Stock and Dietrich, 2006).

BIOLOGICAL PROCESSES AND RIVERS

Riparian vegetation

Riparian vegetation exerts complex and often poorly understood influences on rivers. Dense riparian vegetation generally reduces rates of bank erosion and lateral channel migration (Pizzuto and Meckelnburg, 1989). Large trees create coherent spatial and temporal patterns of bank erosion rates and processes that control the magnitude and frequency of bank retreat (Pizzuto et al., 2010). Paradoxically, small stream channels with forested riparian zones are often wider than otherwise comparable channels with riparian zones in pasture (Trimble, 1997). Flume studies using alfalfa (*Midicaga sativa*) to represent riparian vegetation (Gran and Paola, 2001) have

demonstrated that dense riparian vegetation can stabilize bar and bank surfaces in braided rivers, resulting in narrower channels that migrate more slowly. Braudrick et al. (2009) argue that bank cohesion (imparted by alfalfa in their laboratory flume) is one of several necessary ingredients required to form stable meander bends. The influence of riparian vegetation, however, varies significantly with scale (Abernethy and Rutherfurd, 1998): the interactions between vegetation and fluvial processes differ as one compares headwater streams to higher order channels. Riparian vegetation probably has little influence on large streams such as the Mississippi River (Nanson and Hickin, 1986).

Large woody debris

Tree branches and trunks in river channels are termed *large woody debris* (LWD). LWD can be supplied by landslides and debris flows in steep, forested watersheds, or by reworking of forested floodplains by laterally migrating channels. LWD can occur as scattered, individual pieces, or as massive log jams (Keller and Swanson, 1979). Before the late 20th century, LWD was commonly removed from rivers to aid flood conveyance and navigation. However, most geomorphologists now view LWD as an essential element of natural stream channels in forested watersheds (Abbe and Montgomery, 2003).

LWD has a variety of important effects on stream channels. LWD can trap sediment (Gurnell et al., 2002; Skalak and Pizzuto, in press), modifying sediment budgets and enhancing floodplain development. LWD can have a dramatic impact on channel morphology, stabilizing bars (Abbe and Montgomery, 1996) and forcing the development of pools and riffles (Montgomery et al., 1995). LWD also can exert a profound influence on river channel hydraulics, greatly increasing the resistance to flow (Wilcox and Wohl, 2006). As with riparian vegetation, the influence of LWD on stream channels varies greatly with channel size.

Animals

Many different animals can influence fluvial processes, but most scientific study has focused on the effects of beavers and grazing animals (particularly cattle). Beaver dams on smaller streams have a profound influence on longitudinal profiles (Persico and Meyer, 2009), sediment routing, channel width and depth and other stream channel characteristics (Naiman et al., 1986). Beaver trapping during the 18th and 19th century reversed many of these effects, causing profound changes to streams (Wohl, 2006). Even burrowing beavers who do not build dams can influence rates of bank erosion and sediment budgets (Meentemeyer et al., 1998). Cattle influence rivers by changing the density of riparian vegetation through grazing, and by trampling riverbanks (Trimble and Mendel, 1995).

Summary: biological influences on stream channels

Although the influence of organisms on stream channels has been widely acknowledged, scientific understanding of this area is rudimentary. Furthermore, the interaction and mutual feedback between organisms and the fluvial environment remains little studied. For example, flooding processes help disperse riparian plants and trees, but the further growth of floodplain plants in turn influences flooding processes. Trees influence bank erosion processes, but the process of bank retreat itself can trigger changes to tree root morphology, which further influences the rate of bank retreat (Pizzuto et al., in press). Beaver dams can influence flood hydrology, which may either aid or inhibit further efforts of beavers to build new dams. These interactions and feedback processes should provide fascinating fruitful areas for future research.

RIVERS AND HUMAN ACTIVITY

As a consequence of increased population and technological prowess, humans now are most effective geomorphic agent on Earth (Hooke, 1994). Not surprisingly, the impact of human activities on rivers is pervasive and profound (Table 21.5).

River engineering projects have had a profound influence on river morphology, hydrology and hydraulics. Large dams now intercept 25–30 per cent of the world's global sediment yield, a mass equal to over 100 billion metric tons (Syvitski et al., 2005). Water impounded by dams worldwide is large enough to influence rates of eustatic sea level change (Gornitz et al., 1997). Even after dams have been breached or removed, their influence on stream morphology and fluvial processes may persist for centuries (Walter and Merrits, 2008; Pizzuto and O'Neal, 2009). Streams in many agricultural watersheds are entirely channelized, often leading to channel incision (Simon, 1989) and changing the distribution of flood discharges. Levee construction, bank stabilization, dredging and removal of large woody debris all have profound effects on channel

Table 21.5 Summary of human activities that influence river channels

Category	Examples (with primary impacts indicated)
River engineering	Dam/weir construction[1], channelization[1,3], removal of large woody debris[1,3], levees[1,3], water diversion/irrigation[1], dredging[1,2,3], bank stabilization[1,3], bridge crossings[1,3]
Land use changes[1,2]	Deforestation and logging, agriculture, urbanization, mining
Other	Instream gravel mining[2,3], global climate changes[1,2]

[1] Directly influences stream hydraulics and hydrology.
[2] Primarily influences sediment caliber and supply.
[3] Influences fluvial landforms (bed elevation, bedforms, floodplain features, bars, etc.).

morphology, flood conveyance and sediment transport processes.

Changing land uses have also caused dramatic changes to stream channels. Logging has increased sediment yield in mountain drainage basins, causing widespread sedimentation during episodic storms (Madej and Ozaki, 1996). Past logging practices involved transporting a season's harvest down stream channels, filling entire river networks with logs (Wohl, 2001). During European colonization of the New World, deforestation and poor agricultural practices led to widespread upland soil erosion and valley alluviation (Trimble, 1964). Wilkinson and McElroy (2007) have compared these processes to natural rates of sediment yield through geological time. They conclude that 'accumulation of post-settlement alluvium on higher-order tributary channels and floodplains is the most important geomorphic process in terms of the erosion and deposition of sediment that is currently shaping the landscape of Earth. It far exceeds even the impact of Pleistocene continental glaciers or the current impact of alpine erosion by glacial and/or fluvial processes' (Figure 21.5 provides a concrete illustration). Urban and suburban development also influences stream channels: the increased runoff associated with soil compaction and paved surfaces often causes widespread channel widening, incision and reduced channel roughness (Hammer, 1973; Trimble, 2003; Chin, 2006; Chin and Gregory, 2009).

Anthropogenic global climate change will also drive important changes to rivers in the future. Only a few studies to date, however, have addressed this issue (Goudie, 2006).

Although it is often difficult to evaluate the relative importance of anthropogenic and non-anthropogenic controls on fluvial processes, there are many examples where natural controls are swamped by the many changes wrought by humans. Rivers often require decades, centuries and even millennia to fully adjust to significant changes in the controlling variables of sediment supply, grain size, water discharge and riparian vegetation (i.e. Figures 21.5 and 21.6). As a result, river morphology observed today may not reflect an equilibrium adjustment to sediment and discharge supplied by natural processes, but rather is best viewed as undergoing a long-term, slow evolution following massive changes wrought by humans. Even apparently pristine streams in mountainous regions have been described as 'dramatically...altered...virtual rivers...{with} the appearance of natural rivers but ...{without} much of a natural river's ecosystem functions' (Wohl, 2001). Establishing the limits of human impacts on rivers and the timescales required for their eventual adjustment to these impacts represents a major research challenge to fluvial geomorphologists.

RIVER RESTORATION

The widespread perception that human activities have significantly degraded the environmental quality of rivers (Gleick, 2003) has led to an exponential increase in the number of river restoration projects (Bernhardt et al., 2005), which are broadly defined as activities designed to 'return ... a degraded stream ecosystem to a close approximation of its remaining natural potential' (Fichenich, 2006). The goals and activities of river restoration are varied (Table 21.6), and include improving water quality, fish passage, in-stream habitat, stabilizing eroding banks, managing riparian zones and removing dams (Graf, 2001).

The goals of river restoration projects rarely involve fluvial geomorphology explicitly, but rather focus on ecological services or societal values such as improved water quality. However, fluvial geomorphology is often central to achieving these goals, as geomorphic processes set the physical template for the biogeochemical processes of primary concern (Graf, 2001; Brierly et al., 2002). As a result, fluvial geomorphology has become an important component of river restoration practice.

A few examples illustrate the diverse contributions of fluvial geomorphology to river restoration projects. Along the Colorado River downstream of Glen Canyon Dam, sandbar river habitats utilized by tourists and endangered fish and bird species have been gradually disappearing. An effort has been under way for several decades to restore

Table 21.6 Goals of river restoration projects and examples of common restoration activities (modified from Bernhardt et al., 2005)

Restoration goal category	Examples of common activities
Aesthetics/recreation/ education	Trash removal
Bank stabilization	Revegetation, bank grading
Channel reconfiguration	Bank or channel reshaping
Dam removal/retrofit	Revegetation
Fish passage	Fish ladder installation, dam removal
Flow modification	Flow regime enhancement
Floodplain reconnection	Bank or channel reshaping
Instream habitat improvement	Boulders/woody debris added
Land acquisition	(Self explanatory)
Riparian management	Livestock exclusion
Stormwater management	Wetland construction
Water quality management	Riparian buffer creation/ maintenance

these habitats; much of the work has focused on designing artificial floods that can resupply sand from the center of the channel to marginal sandbars (Schmidt et al., 2001). While the goals of the restoration involve recreation and ecological services, geomorphologists have played critical roles in (1) establishing the geomorphic setting in which the sandbar habitats are created, (2) defining the processes by which the Colorado River moves sand down its channel and (3) establishing how much sand is available for rebuilding beaches. Despite several decades of study, the prognosis for success is not encouraging (Dalton, 2005). Other restoration projects of similar scope include the CALFED project (to restore the Sacramento–San Joaquin River Delta) (www.calwaterca.gov, accessed 17 November 2009) and the Penobscot Bay restoration project (to revitalize native Atlantic Salmon and their habitat) (www.penobscotriver.org, accessed 17 November 2009).

Most river restoration projects, however, are small projects focused on short river reaches (Bernhardt et al., 2005). Smith and Prestegaard (2005), for example, describe how several hundred meters of Deep Run, Maryland, were reconfigured in an effort to reduce sediment loading to a wetland immediately downstream of the restored reach. The originally irregular, sinuous, single-thread channel was reconfigured into a series of symmetrical meanders.

Despite the burgeoning efforts to restore rivers, river restoration is a new, relatively immature discipline. Many projects are designed without well-defined goals (Bernhardt et al., 2005) and 'are conducted with minimal scientific context' (Wohl et al., 2005). There is no standard training or certification of practitioners, and the relationship between river restoration and more traditional disciplines such as river engineering is poorly defined (Slate et al., 2007). It is not surprising, therefore, that the proper use of fluvial geomorphology in river restoration practice is currently being debated. The utility of stream classification schemes for aiding restoration practice has been promoted by Rosgen (1994), while others remain unconvinced (Miller and Ritter, 1996; Simon et al., 2009). A watershed approach is advocated by Brierly and Fryirs (2005), while others combine geomorphic principles with more traditional river engineering (Federal Interagency Stream Restoration Working Group, 1998; Shields et al., 2003). It will be an important goal for fluvial geomorphologists and others interested in river restoration to work together in the near future to better define this new discipline. Urgent priorities include establishing measureable goals for restoration projects and working to better define the scientific foundations for restoration practice (Wohl et al., 2005; Palmer, 2009).

SUMMARY

Fluvial geomorphology is a vital, healthy discipline. Exciting field, laboratory and theoretical studies of alluvial rivers, long the focus of mainstream fluvial geomorphology, have lead to importance recent advances. However, during the past 20 years, new areas are providing important contributions. The development of cosmogenic isotopic dating, combined with new theoretical and experimental tools, has enhanced our understanding of the interactions between tectonics, geological structure and climate over geological timescales. Increased appreciation of the degraded nature of riverine ecosystems has also lead to rapid developments in applied fluvial geomorphology, where geomorphic principles help manage watersheds and restore ecological services.

In the future, these trends will continue to develop and mature, but fluvial geomorphologists will also increasingly focus on understanding and managing the impacts of humans on rivers. This will require fluvial geomorphology to become more predictive and interdisciplinary. It will be an exciting time.

REFERENCES

Abbe, T.B. and Montgomery, D.R. (1996) Large woody debris jams, channel hydraulics and habitat formation in large rivers. *Regulated Rivers: Research and Management*, 12: 201–21.

Abbe, T.B. and Montgomery, D.R. (2003) Patterns and processes of wood debris accumulation in the Queets River basin, Washington. *Geomorphology*, 51: 81–107.

Abernethy, B. and Rutherfurd, I.D. (1998) Where along a river's length will vegetation most effectively stabilise stream banks. *Geomorphology*, 23: 55–75.

Ashmore, P.E. (1990) How do gravel rivers braid? *Canadian Journal of Earth Sciences*, 28: 326–41.

Benda, L. and Dunne, T. (1997) Stochastic forcing of sediment routing and storage in channel networks. *Water Resources Research*, 33: 2865–80.

Palmer, M.A., Allan, J.D., Alexander, G., Barnas, K., Brooks, S., Carr, J., Clayton, S., Dahm, C., Follstad-Shah, J., Galat, D., Gloss, S., Goodwin, P., Hart, D.Hassett, B., Jenkinson, R., Katz, S. Kondolf, G.M., Lake, P.S., Lave, R., Meyer, J.L., O'Donnell, T.K., Pagano, L., Pwell, B. and Sudduth, E. (2005) Synthesizing U.S. river restoration efforts. *Science*, 308: 636–7.

Bloom, A.L. (1998) *Geomorphology: A Systematic Analysis of late Cenozoic Landforms*. Prentice Hall, New Jersey, p. 482.

Braudrick, C.A., Dietrich, W.E., Leverich, G.T. and Sklar, L.S. (2009) Experimental evidence for the conditions necessary to sustain meandering in coarse-bedded rivers. *Proceedings of the National Academy of Science*, 106. doi:10.1073/pnas.0909417106.

Brierly, G.J. and Fryirs, K.A. (2005) *Geomorphology and River Management: Applications of the River Styles Framework*. Blackwell, Oxford. p. 398.

Brierly, G., Fryirs, K., Outhet, D. and Massey, C. (2002) Application of the River Styles framework as a basis for river management in New South Wales, Australia. *Applied Geography*, 22: 91–122.

Bull, W.B. (1991) *Geomorphic Responses to Climate Change*. Oxford University Press, New York. p. 326.

Carson, M.A. and LaPointe, M.F. (1983) The inherent asymmetry of river meander planform. *Journal of Geology*, 91: 41–55.

Chin, A. 2006. Urban transformation of river landscapes in a global context. *Geomorphology*, 79: 460–87.

Chin, A. and Gregory, K.J. (2009) From research to application: management implications from studies of urban river channel adjustment. *Geography Compass*, 3/1: 297–328.

Church, M.A. (1992) Channel morphology and typology, in P. Calow, and G.E. Petts, (eds.) *The Rivers Handbook*. Blackwell Scientific, Oxford. pp. 126–143, 511.

Church, M.A. and Slaymaker, O. (1989) Disequilibrium in Holocene sediment yield in glaciated British Columbia. *Nature*, 337: 452–4.

Colombini, M. and Parker, G. (1995) Longitudinal streaks. *Journal of Fluid Mechanics*, 304: 161–83.

Constantine, C.R., Dunne, T. and Hanson, G.J. (2009) Examining the physical meaning of the bank erosion coefficient used in meander migration modeling. *Geomorphology*, 106: 242–52.

Costa, J.E. (1974) Response and recovery of a piedmont watershed from tropical storm Agnes, June 1972. *Water Resources Research*, 10, 106–112.

Coulthard, T.J., Macklin, M.G. and Kirkby, M.J. (2002) A cellular model of Holocene upland river basin and alluvial fan evolution. *Earth Surface Processes and Landforms*, 27: 269–88.

Dalton, R. (2005) Floods fail to save canyon beaches. *Nature*, 438: 10.

Diplas, P. (1992) Hydraulic geometry of threshold channels. *Journal of Hydraulic Engineering*, 118: 597–614.

Federal Interagency Stream Restoration Working Group (FISRWG) (1998) *Stream Corridor Restoration: Principles, Processes, and Practices.* US Department of Agriculture.

Fichenich, J.C. (2006) Functional objectives for stream restoration. U.S. Army Corps of Engineers, ERDC TN-EMRRP SR-52, p.18.

Furbish, D.J. (1991) Spatial autoregressive structure in meander evolution. *Geological Society of America Bulleting*, 103: 1576–89.

Gleick, P.H. (2003) Global freshwater resources: soft-path solutions for the 21st century. *Science*, 302, 1524–8.

Gomez, B., Trustrum, N.A., Hicks, D.M., Page, M.J., Rogers, K.M. and Tate, K.R. (2003) Production, storage, and output of particulate organic carbon: Waipaoa River Basin, New Zealand. *Water Resources Research*, 39. doi:10.10029/2002WR00169.

Gornitz, V., Rosenzweig, C. and Hillel, D. (1997) Effects of anthropogenic intervention in the land hydrological cycle on global sea level rise. *Global and Planetary Change*, 14: 147–61.

Goudie, A.S. (2006) Global warming and fluvial geomorphology. *Geomorphology*, 79: 384–94.

Graf, W.L. (2001) *Dam*age control: restoring the physical integrity of America's rivers. *Annals of the Association of American Geographers*, 91: 1–27.

Gran, K. and Paola, C. (2001) Riparian vegetation controls on braided stream dynamics. *Water Resources Research*, 37: 3275–83.

Grant, G.E., Swanson, F.J. and Wolman, M.G. (1990) Pattern and origin of stepped-bed morphology in high-gradient streams, western Cascades, Oregon. *Geological Society of America Bulletin*, 102: 340–52.

Gurnell, A.M., Iegay, H., Swanson, F.J. and Gregory, S.V. (2002) Large wood and fluvial processes. *Freshwater Biology*, 47: 601–19.

Hammer, T.R. (1973) Stream channel enlargement due to urbanization. *Water Resources Research*, 8: 1530–40.

Hartshorn, K., Hovius, N., Dade, W.B. and Slingerland, R.L. (2002) Climate-driven bedrock incision in an active mountain belt. *Science*, 297: 2036–8.

Hasbargen, L.E. and Paola, C. (2000) Landscape instability in an experimental drainage basin. *Geology*, 28: 1067–70.

Hooke, R. LeB. (1968) Model geology: prototype and laboratory streams: discussion. *Geological Society of America Bulletin*, 79: 391–4.

Hooke, R. LeB. (1994) On the efficacy of humans as geomorphic agents. *GSA Today*, 4: 217, 224–5.

Howard, A.D. and Kerby, G. (1983) Channel changes in badlands. *Geological Society of America Bulletin*, 94: 739–52.

Howard, A.D., Dietrich, W.E. and Siedl, M.A. (1994) Modeling fluvial erosion on regional to continental scales. *Journal of Geophysical Research*, 99: 13971–86.

Huang, H.Q. and Nanson, G.C. (2007) Why some alluvial rivers develop and anabranching pattern. *Water Resources Research*, 43. doi:10.1029/2006WR005223.

Jacobson, R.J. and Coleman, D.J. (1986) Stratigraphy and recent evolution of Maryland Piedmont floodplains. *American Journal of Science*, 286: 617–37.

Jerolmack, D.J. and Mohrig, D. (2007) Conditions for branching in depositional rivers. *Geology*, 35: 463–6.

Johanneson, H. and Parker, G. (1989) Linear theory of river meanders, in Ikeda, S. and Parker, G. (eds) *River Meandering*. AGU Water Resources Monograph 12, American Geophysical Union, Washington D.C. pp. 181–214. 485 p.

Keller, E.A. and Swanson, F.J. (1979) Effects of large organic material on channel form and fluvial processes. *Earth Surface Processes*, 4: 361–80.

Knighton, D. (1998) *Fluvial Forms and Processes – A New Perspective*. Hodder Arnold, London. p. 383.

Kondolf, G.M. and Piegay, H. (2003) *Tools in Fluvial Geomorphology*. John Wiley and Sons, Chichester. p. 688.

Lane, E.W. (1955) Design of stable channels. *Transactions of the American Society of Civil Engineers*, 120: 1234–60.

Larson, E.W. and Greco, S.E. (2002) Modeling channel management impacts on river migration: a case study of Woodson Bridge State Recreation Area, Sacramento River, California. *Environmental Management*, 30: 209–24.

Lave, J. and Avouac, J.P. (2001) Fluvial incision and tectonic uplift across the Himalayas of central Nepal. *Journal of Geophysical Research*, 106: 26561–91.

Leopold, L.B., Wolman, M.G. and Miller, J.P. (1964) *Fluvial Processes in Geomorphology*. W.H. Freeman, San Francisco. p. 522.

Lewin, J. (1976) Initiation of bed forms and meanders in coarse-grained sediment. *Geological Society of America Bulletin*, 87: 281–5.

Mackin, J.H. (1948) Concept of the graded river. *Geological Society of America Bulletin*, 59: 463–512.

Madej, M.A. and Ozaki, V. (1996) Channel response to sediment wave propagation and movement, Redwood Creek, California, USA. *Earth Surface Processes and Landforms*, 21: 911–27.

Meentemeyer, R., Vogler, J.B. and Butler, D.R. (1998) The geomorphic influences of burrowing beavers on streambanks, Bolin Creek, North Carolina. *Z. Geomorph. N.*, 42: 453–68.

Miller, J.R. and Ritter, J.B. (1996) Discussion: an examination of the Rosgen classification of natural rivers. *Catena*, 27: 295–9.

Montgomery, D.R., Buffington, J.M., Smith, R.D., Schmidt, K.M. and Pess, G. (1995) Pool spacing in forest channels. *Water Resources Research*, 31: 1097–105.

Murray, A.B. and Paola, C. (1994) A cellular model of braided rivers. *Nature*, 371: 54–7.

Naiman, R.J., Melillo, J.M. and Hobbie, J.E. (1986) Ecosystem alteration of boreal forest streams by beaver (Castor Canadensis). *Ecology*, 67: 1254–69.

Nanson, G.C. and Hickin, E.J. (1986) A statistical analysis of bank erosion and channel migration in western Canada. *Geological Society of America Bulletin*, 97: 497–504.

Nanson, G.C. and Croke, J.C. (1992) A genetic classification of floodplains. *Geomorphology*, 4: 459–86.

Nanson, G.C. and Knighton, A.D. (1996) Anabranching rivers: their cause, character, and classification. *Earth Surface Processes and Landforms*, 21: 217–39.

Palmer, M.A. (2009) Restoring watershed restoration: science in need of application and applications in need of science. *Estuaries and Coasts*, 32: 1–17.

Paola, C. and Seal, R. (1995) Grain size patchiness as a cause of selective deposition and downstream fining. *Water Resources Research*, 31: 1395–407.

Parker, G. (1976) On the cause and characteristic scales of meandering and braiding in rivers. *Journal of Fluid Mechanics*, 76: 457–80.

Parker, G. (1978a) Self-formed straight rivers with equilibrium banks and mobile bed: Pt. 1: the sand-silt river. *Journal of Fluid Mechanics*, 89: 109–25.

Parker, G. (1978b) Self-formed straight rivers with equilibrium banks and mobile bed: Pt. 2: the gravel river. *Journal of Fluid Mechanics*, 89: 127–46.

Persico, L. and Meyer, G. (2009) Holocene beaver damming, fluvial geomorphology, and climate in Yellowstone National Park, Wyoming. *Quaternary Research*, 71: 340–53.

Phillips, J.D. (1992) The end of equilibrium? *Geomorphology*, 5: 195–201.

Pizzuto, J.E. (1987) Sediment diffusion during overbank flows. *Sedimentology*, 34: 301–17.

Pizzuto, J.E. (1994) Channel adjustments to changing discharges, Powder River between Moorhead and Broadus, Montana. *Geological Society of America Bulletin*, 106: 1494–501.

Pizzuto, J.E. and Meckelnburg, T.S. (1989) Evaluation of a linear bank erosion equation. *Water Resources Research*, 25: 1005–13.

Pizzuto, J. and O'Neal, M. (2009) Increased mid-twentieth century riverbank erosion rates related to the demise of mill dams, South River, Virginia. *Geology*, 37: 19–22.

Pizzuto, J.E., O'Neal, M. and Stotts, S. (2010). On the retreat of forested, cohesive riverbanks. *Geomorphology*, 116: 341–352.

Richardson, K. and Carling, P.A. (2005) *The Typology of Sculptured Forms in Open Bedrock Channels*. Geological Society of America Special Paper 392. Boulder, CO, USA. p. 108.

Rodriguez-Iturbe, I. (1997) *Fractal River Basins: Chance and Self-Organization*. Cambridge, New York. p. 547.

Roe, G.H., Montgomery, D.R. and Hallet, B. (2002) Effects of orographic precipitation on the concavity of steady-state river profiles. *Geology*, 30: 143–6.

Rosgen, D.L. (1994) A classification of natural rivers. *Catena*, 22: 169–99.

Schmidt, J.C., Parnell, R.A., Grams, P.E., Hazel, J.E., Kaplinski, M.A., Stevens, L.E. et al. (2001) The 1996 controlled flood in Grand Canyon: flow, sediment transport, and geomorphic change. *Ecological Applications*, 11: 657–71.

Schumm, S.A. (1977) *The Fluvial System*. John Wiley, New York. 338 p.

Shields, F.D., Copeland, R.R., Klingeman, P.C., Doyle, M.W. and Simon, A. (2003) Design for stream restoration. *Journal of Hydraulic Engineering*, 129: 676–84.

Simon, A. (1989) A model of channel response in disturbed alluvial channels. *Earth Surface Processes and Landforms*, 14: 11–26.

Simon, A., Doyle, M., Kondolf, M., Shields, F.D., Rhoads, B. and McPhillips, M. (2009) Critical evaluation of how the Rosgen Classification and associated "natural channel design" methods fail to integrate and quantify fluvial processes and channel response. *Journal of the American Water Resources Association*, 43: 1117–31.

Skalak, K. and Pizzuto, J. (2010) The distribution and residence time of suspended sediment stored within the channel margins of a gravel–bed bedrock river. *Earth Surface Processes and Landforms* 35, 435–446.

Sklar, L. and Dietrich, W.E. (2004) A mechanistic model for river incision into bedrock by saltating bed load. *Water Resources Research*, 40: W06301. doi:10.1029/2003WR002496.

Slate, L.O., Shields, F.D., Schwartz, J.S., Carpenter, D.D. and Freeman, G.E. (2007) Engineering design standards and liability for stream channel engineering. *Journal of Hydraulic Engineering*, 133: 1099–102.

Smith, S.M. and Prestegaard, K.L. (2005) Hydraulic performance of a morphology-based stream channel design. *Water Resources Research*, 41. doi:10.1029/2004WR003926.

Stock, J.D. and Dietrich, W.E. (2006) Erosion of steepland valleys by debris flows. *Geological Society of America Bulletin*, 118: 1125–48.

Stolum, H. (1998) Planform geometry and dynamics of meandering rivers. *Geological Society of America Bulletin*, 110: 1485–98.

Syvitski, J.P.M., Vorosmarty, C.J., Kettner, A.J. and Green, P. (2005) Impact of humans on the flux of terrestrial sediment to the global coastal ocean. *Science*, 308: 376–80.

Trimble, S.W. (1964) Culturally accelerated sedimentation. *Bulletin of the Georgia Academy of Sciences*, 28: 131–41.

Trimble, S.W. (1997) Stream channel erosion and change resulting from riparian forests. *Geology*, 25: 467–9.

Trimble, S.W. (2003) Contribution of stream channel erosion to sediment yield from an urbanizing watershed. *Science*, 278: 1442–4.

Trimble, S.W. and Mendel, A.C. (1995) The cow as a geomorphic agent – a critical review. *Geomorphology*, 13: 233–53.

Tucker, G.E. and Whipple, K.X. (2002) Topographic outcomes predicted by stream erosion models: sensitivity analysis and intermodel comparision. *Journal of Geophysical Research*, 107. doi:10.1029/2001JB000162.

Turowski J.M., Hovius N., Wilson A. and Horng M.-J. (2008) Hydraulic geometry, river sediment and the definition of bedrock channels. *Geomorphology*, 99(1–4), 26–38.

Van de Wiel, M.J. and Coulthard, T.J. (2010) Self-organized criticality in river basins: challenging sedimentary records of environmental change. *Geology*, 38: 87–90.

Walter, R.C. and Merritts, D.J. (2008) Natural streams and the legacy of water-powered mills. *Science*, 399: 299–304.

Whipple, K.X. (2004) Bedrock rivers and the geomorphology of active orogens. *Annual Review of Earth and Planetary Sciences*, 32: 151–85.

Whipple, K.X., Hancock, G.S. and Anderson, R.S. (2000) River incision into bedrock: mechanics and relative efficacy of plucking. *Geological Society of America Bulletin*, 112: 490–503.

Whiting, P.J., Dietrich, W.E., Leopold, L.B., Drake, T.G. and Shreve, R.L. (1988) Bedload sheets in heterogeneous sediment. *Geology*, 16: 105–8.

Wilcox, A.C. and Wohl, E.E. (2006) Flow resistance dynamics in step-pool stream channels: 1. Large woody debris and controls on total resistance. *Water Resources Research*, doi:10.1029/2005WR004277.

Wilkinson, B.H. and McElroy, B.J. (2007) The impact of humans on continental scale erosion and sedimentation. *Geological Society of America Bulletin*, 119: 140–56.

Wohl, E.E. (2001) *Virtual Rivers: Lessons from the Mountain Rivers of the Colorado Front Range.* Yale University Press, New Haven, CT. p. 210.

Wohl, E. (2006) Human impacts to mountain streams. *Geomorphology*, 79: 217–48.

Wohl, E., Angermeier, P.L., Bledsoe, B., Kondolf, G.M., MacDonnell, L., Merritt, D.M. et al. (2005) River restoration. *Water Resources Research*, 41. doi:10.1029/2005WR003985.

Wolman, M.G. (1955) *The Natural Channel of Brandywine Creek.* U.S. Geological Survey Professional Paper 271.

Wolman, M.G. and Miller, J.P. (1960) Magnitude and frequency of forces in geomorphic processes. *Journal of Geology*, 68: 54–74.

Wolman, M.G. and Gerson, R. (1978) Relative scales of time and effectiveness of climate in watershed geomorphology. *Earth Surface Processes and Landforms*, 3: 189–208.

22

Glacial Geomorphology

John Menzies

THE 'ORIGINS' OF GLACIAL GEOMORPHOLOGY

It is only a mere 10,000 years ago or, in some places at higher altitudes and latitudes, even less when much of the mid-continental and poleward landmasses of the northern hemisphere, the southern regions of South America and large parts of the Southern Alps, New Zealand were ice covered. In the almost interminably vast stretch of geological time, the interval between that period of global glaciation and today is the merest 'flick of an eyelash'. As a consequence the impacts and after-effects of glaciation are clearly visible today in these landscapes. The influence of glaciation is felt in innumerable ways in the daily lives of people in these once glaciated lands and far beyond. At the height of the Last Ice Age over 30 per cent of the Earth's land masses were ice covered and, if the ocean basins are included, then at least 60 per cent of the Earth's surface was impacted by ice in the form of vast ice sheets, mountain glaciers, piedmont glaciers and ice shelves.

It is during this most recent period of global glaciation that humans seem to have 'emerged', and with deglaciation most of the once glaciated areas has since been re-colonized by plant and animal species; thus, not only the nature of the landscape has been drastically altered but likewise the composition and diversity of fauna and flora that now populate these areas has been changed as a consequence of glaciation. When these vast tracts of ice covered the continents, ocean margins and shallow seas, the effect of the huge uptake of water into the ice mass systems caused general sea-level to fall by as much as 120 m below today's level, land bridges appeared and the general global climatic circulation patterns were diverted and drastically altered.

From the Middle Ages to the early part of the 19th century amongst local populations in Iceland, Norway and Switzerland and in most mountainous regions of the world there was a realization that the nearby glaciers had repeatedly advanced and retreated. In Europe, for example, during what became known as the Little Ice Age, glaciers advanced into pastureland. In the mountainous areas of south-eastern France a close relationship had been noted and documented by local parish priests, relating the timing of grape blossoming, the sometimes lower and at other times higher temperatures of the 15th and 16th centuries and at the same time the advance and retreat of nearby valley glaciers.

It was a Swiss paleontologist, Louis Agassiz in 1840, who is credited with advocating an Ice Age affecting large areas of the Earth (cf. Imbrie and Imbrie, 2005). Many others in Europe had essentially arrived at a similar conclusion by the beginning of the 19th century (Nilsson, 1983). Such a revolutionary idea, of course, met with almost immediate repudiation in Europe and North America. The idea was seen to verge upon the heretical with the strongly Christian–Judiac theosophy of the time. However, the idea could not, nor would not, be repressed and quickly advocates across Western Europe and North America, recognizing the typical characteristics of glaciated landscapes, began to support the concept of a global glaciation. In Britain, for example, Agassiz embarked upon a sojourn across England and Scotland and quickly pointed out the evidence supportive of extensive glaciation. Initially, Lyell (1830–1833), in an attempt to reconcile the religious with the scientific, advocated that evidence such as far-traveled sediments and perched boulders could be explained and rationalized if a vast flood, as described in the Bible, had swept these sediments and rocks encased within icebergs across the land surface. As the icebergs 'drifted' across this flooded land so with melting the sediments and rocks fell to the ground surface. That the term 'drift' is still occasionally found in glacial literature is testament to this once postulated explanation.

Within a short time following Agassiz's visit to Scotland, Archibald Geikie published what is perhaps the first glacial map containing glacial geomorphological information on the Glasgow area of Scotland (Geikie, 1863). From then until the beginning of the 20th century there followed a flurry of activity in Europe and North America where large tracts of terrain were mapped illustrating glacial geomorphological features. The scientific approach adopted by 'glacialists' at this time was a combination of sedimentological and geomorphological expertise brought to bear upon the problem of understanding glacial features and sediments. The early work by G.K. Gilbert (1906), for example, in the western United States illustrates a strong process-based bias that remains even today as a superb example of a thoroughly scientific approach to understanding glacial features. By the early decades of the 20th century there began a movement toward a more morphological or 'form-based' understanding of glacial features and an increasing de-emphasis of sedimentological and process viewpoints. By the 1950s this de-emphasis had reached a level of imbalance such that glacial geomorphology perhaps overly relied upon the purely morphological analysis of glacial terrains. For example, in tracing the history of drumlin research from the early 20th century there is a strong emphasis on the internal sediments found within drumlins related to glacial processes and to drumlin form (cf. Slater, 1926), yet by the late 1950s drumlins were almost universally viewed as morphological features whose origin was considered almost exclusively as a function of drumlin shape (Chorley, 1959) while internal sedimentology was all but ignored. How this state of affairs should be arrived at cannot readily be explained. Perhaps the demise of glacial studies within the larger field of geology, where the advent of plate tectonic theory, for example, gained an increasing ascendancy and the fact that glacial studies were left almost entirely to geographers who had a much greater and ingrained sense of spatial patterns and morphological appreciation, may be a partial explanation.

Sometime in the 1970s a process-based approach to understanding glacial geomorphology began to be reasserted. It is perhaps a mark of the lack of explanatory power that was missing in the purely morphological approach that the need to understand the processes involved in the glacial system and its subsystems has re-emerged (cf. Spedding, 1997). However this philosophical/methodological dichotomy should not be construed as implying that either the morphological or the process-form approach can be regarded as superior to one another.

Beginning with scientific exploration after World War 2 and the advent of the Cold War increasing interest in polar lands heralded a new period of understanding in studies of all form of ice masses. The rush to plant a foothold on the Antarctic continent quickly led by the mid-1960s to a vast outpouring of scientific data that provide knowledge of glaciology, sea-ice, ice sheets and ice shelves that still continues today. Increasingly, knowledge of present-day ice masses and ice physics and related glacial sediments began to be translated into an appreciation of past glacial sediments and landscapes. Research in Iceland, Svalbard, Greenland, Alaska and Canada, for example, led to a greater understanding of how glaciers 'operate', their impact on terrain that includes not only subglacial deposition but the effects of meltwater and associated sediments across the landscape. Likewise appreciation of glacial lake sedimentation and subaquatic processes, including glaciomarine processes and forms, has since emerged. One of the consequences of increased polar exploration, especially in Antarctica, Greenland, Alaska, the Canadian North and Iceland, has been the immense growth of glaciology as a distinct science devoted to understanding the physics, chemistry and climatology of ice masses. Glaciology has made an enormous impact upon glacial geomorphology. Although often the two sciences seem to have moved along separate paths, the link between both has had a great benefit for the comprehension of glacial processes and landforms. This collaboration has been especially true in Antarctica.

GLACIAL LANDSCAPES: TODAY

During the Pleistocene, vast ice sheets covered the Earth's surface beginning with a relatively slow build-up to repeated maximum extensions into the mid-latitudes (Figure 22.1). These ice sheets had an enormous impact in moving across terrain, and eroding, transporting and depositing vast quantities of sediments both on land and in the ocean basins. The surface topography of the continents, today, for example, in the northern states of USA and throughout Canada and northern Europe, was totally modified with distinctive landscapes and landforms repeatedly overprinted until the terrain was finally exposed from beneath the ice some 10,000 years ago. With the rapid demise of the ice sheets, sea level very quickly rose and, since the ice sheets had isostatically depressed the continental land masses, in many instances sea level rose above present levels, drowning large coastal areas. Much later continental land masses readjusted to the loss of the ice sheet load and slowly rose and raised shorelines and cliffs began to re-appear above sea level. These raised beaches and strandlines, for example, in Europe, became the colonization pathways for early humans as they moved north and westward.

Glacial erosion can be seen on bedrock surfaces in the form of glacial striae and chattermarks, and the effects of high pressure meltwater scour. Evidence of glacial transport can be observed in the form of boulder trains and isolated erratic boulders left strewn across glaciated landscapes. Finally, the effects of glacial deposition can be seen as thick glacial sediments and landforms such as moraines, drumlins and eskers, glacial lake sediments, and immense thicknesses of glacial sediments within marine environments.

GLACIAL ENVIRONMENTS

Glacial geomorphology attempts to understand and interpret glaciated landscapes and their associated sediments and landforms by trying to comprehend the myriad of processes that occur within the many sub-environments that make up any glacial system. To untangle complex sets of often spatially and temporally interlinked glacial sub-environments it is critical that patterns and distinctive signatures characteristic of these sub-environments be detected. As Plate 9(a,b) (see colour plate section, page 594) illustrates, there are many inter-related sub-environments; supraglacial, englacial, subglacial, terrestrial proglacial, subaquatic proglacial, and each of these environments can often be further subdivided. For example, the terrestrial proglacial subdivides into proximal and distal and within each can be isolated glaciolacustrine and glaciofluvial subunits.

Perhaps one of the greatest problems in glacial geomorphology is in the partition and identification of glacial sub-environments since all too often an equifinality of processes and forms emerges. The differentiation, for example, between englacial and supraglacial environments is all too often virtually impossible. Likewise, following glacier and ice sheet advance and retreat, intense overprinting led to sediment reworking and complete or, in some places, partial removal and/or destruction of previous sediments and landforms. It is therefore a glacial geomorphologist's task to attempt to untangle the patterns and attributes of a glaciated landscape. One approach to this problem of overriding has been the development of a glacial landsystem approach whereby specific terrain 'types' can be 'isolated' and clearly demarcated (cf. Boulton and Eyles, 1979; Evans, 2003). Much as this method has its merits in terrains

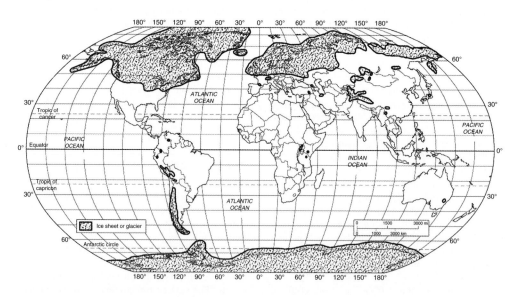

Figure 22.1   Extent of global glaciation at approx. 18,000 years B.P. (Modified from Tarbuck and Lutgens, 2003)

such as Iceland, where often essentially primary glaciation has occurred in the relatively recent past, in other terrains such as the Prairies of North America or the North German Plain in Europe where repeated glacial over-riding has taken place, a landsystem approach is difficult to apply (cf. Kaszycki, 1989; Kemmis, 1996).

As our understanding of glacial environments has increased, many old and strongly held ideas and paradigms have been abandoned or modified. The concept that held sway for decades regarding glacial unconsolidated sediments (tills) was that they could be simply subdivided into ablation and basal tills. Later a new nomenclature appeared in which, for example, basal tills were divided into lodgement tills, melt-out tills and flow tills. Later again this tripartite subdivision has fallen into question (van der Meer et al., 2003). With the acknowledgment of subglacial sediment deformation, the usage of the term 'till' has now often to be qualified in its application. A new paradigm has emerged that may demand the re-interpretation of long-held glacial stratigraphies, origins of landforms and associated processes (cf. Boulton, 1986; Murray, 1997; Menzies, 2002; Stokes and Clark, 2003; Evans et al., 2006; Menzies et al., 2006).

The mechanics of glacier and ice sheet movement and underlying glaciodynamics

It was perhaps James Forbes, Professor of Natural Philosophy at the University of Edinburgh, who in 1859 (Forbes, 1859) first proposed that glacier ice moves much in the same fashion as 'treacle' due to what he termed viscous or plastic motion by its own weight (cf. Geikie, 1888: p. 33). Subsequently, Tyndall and Faraday noted the pressure melting phenomena that can be attributed to ice at 0°C when subject to pressure. These early scientists were well aware that glacier ice moves, in part, by processes of internal deformation or dislocation climb.

Over the next century inquiry into how glacier ice moves advanced relatively little. Even when Ahlmann (1948) noted that glacier ice may be at or below pressure melting, thus introducing the idea of temperate (at or above the pressure melting point of ice) or polar (below pressure melting), there still remained a remarkably limited grasp of glacier physics (Paterson, 1994). However, when in 1955 John Glen experimented with the stress–strain relationships of polycrystalline glacier ice did a significant advance was made. What has become known as Glen's law was empirically formulated. Glen subjected a block of ice to triaxial strain under conditions of −60°C over a 24 h

period, observed the change in shape of the block over time and under strain, and from that formulated a 'best-fit' equation that fitted the data. The equation known as Glen's law is $\dot{e} = A\tau^n$, where $\dot{e}$ is the effective longitudinal strain rate, A is an ice stiffness parameter, n is an empirical constant and τ is the effective basal shear stress. This fundamental equation established the essential parameters in explaining glacier ice motion by internal deformation. Significant alterations and variations have since appeared in the literature but the key elements remain. The value of τ can be defined as $\tau = \rho_i \times h_i \times \sin \alpha$, where τ is the basal shear stress, ρ_i is the density of glacier ice, g is the acceleration due to gravity, h_i is the glacier ice thickness and $\sin \alpha$ is the surface angle of slope of the ice mass. Even though this equation led to a major advance in understanding ice movement, when actual observations of ice mass velocities were made ice mass velocity by internal deformation underestimates the actual velocities by several magnitudes. This inability to predict rates of ice deformation led to further attempts to find a more exact solution (Paterson, 1994; Talalay and Hooke, 2007).

By the mid-1960s the concept of basal slip was introduced whereby a temperate, warm-based ice mass was thought to slip across its bed due to bed decoupling either on a very fine film of water or by the coalescence of multiple cavities beneath the ice mass (Weertman, 1964; Lliboutry, 1968). It was hypothesized that under warm-based ice masses as meltwater was released from the glacier sole an incompressible film or a series of water filled cavities would form and begin to lift the ice mass from its bed, thereby reducing basal friction and increasing forward movement. This process of glacier movement does, in addition to internal deformation, account for more of the observed ice motion of temperate ice masses but still actual measurements of ice velocity were significantly underestimated and the so-called 'glacier ice unsolved sliding problem' remained (Weertman, 1979). As early as 1979 new hypothetical ideas on glacier sliding emerged in terms of another possible mechanism to account for the higher than predicted ice velocities (Boulton and Jones, 1979). This mechanism, commonly referred to as subglacial soft sediment deformation, permits glacier ice to slide across its bed on a saturated substrate of water soaked sediment. Under warm-based ice masses where instead of water behaving as a film or a series of interlinked cavities the underlying sediment becomes saturated, effective stress levels fall and the ice moves across this already moving sediment layer in the manner of a person slipping on a banana peel (cf. Alley et al., 1986; Boulton, 1996). More recently, with subglacial investigations in many parts of the world including the West Antarctic, such layers of deforming sediment have been discovered and appear to be the final 'piece' in the puzzle to explain the glacier sliding bed problem and forward ice motion. Even under sub-polar ice masses, where large areas of the glacier bed are frozen, such sediment movement seems to account for a large portion of forward ice motion (Echelmeyer and Wang, 1987). In the meantime, as well as the glacier sliding bed problems being seemingly almost resolved, it has become apparent that most glaciers are not simply temperate or polar but are more typically poly-thermal (Blatter and Hutter, 1991). It has become clear that under most ice masses, subglacial bed conditions are a key to many processes of glacial erosion, transport and deposition. As a clearer picture of these bed conditions emerges in terms of fluctuating temperatures, sediment rheology, stress and strain conditions so understanding of glacial processes is vastly improving.

In recent decades there have been vast improvements in our understanding of how ice sheets and glaciers advance and margins retreat, and down-waste along varying types of margins whether on land or into the ocean. The complexity, especially, of ice sheet modeling and subglacial conditions during processes of advance and retreat, are well established (Alley et al., 2005; Joughin et al., 2006; Bamber et al., 2007). Likewise, the very close links between subglacial groundwater flow and ice sheet stability for various past ice sheets have been fairly successfully modeled (Boulton et al., 1995; Sugden et al., 2006). What remains to still be more closely modeled and fully understood is the close relationship between ice sheet stability, subglacial conditions, subglacial hydrology, and the role played by ice streams and fringing ice shelves (cf. Echelmeyer et al., 1993; Stokes and Clark, 1999; Oerlemans, 2002; Hindmarsh, 2006).

Recent research, for example Hart's work beneath Briksdalsbreen in Norway, where subglacial conditions and ice mass activity have been remotely monitored, should greatly advance our grasp of subglacial conditions of stress/strain and thermal conditions (Hart, 2006). Likewise, the work by Anderson, Smith and others off the shore of the West Antarctic ice sheet and by Tulaczyk and others on subglacial ice stream activity has greatly expanded our knowledge (cf. Anderson et al., 2002; Tulaczyk et al., 2000). Ice sheet modeling not only helps understand past glaciated terrains in terms of landforms, landform/bedform assemblages, sediments and the interplay between ice movement and various landsystems both at continental and marine margins (Evans et al., 2003) but has enormous implications for future trends in relationship with the consequences of global warming.

Everyone is familiar with the 'doomsday' dire warnings that may result from sea-level rise, the potential socio-economic and political consequences of which cannot be ignored or in any way diminished. Many countries will be very badly affected if not virtually wiped out as in the case of the Maldives. Other countries such as Bangladesh will suffer untold tragic consequences as will large maritime fringing tracts of all ocean-marginal countries (cf. Church and White, 2006).

However, following this scenario of global ocean-rise and increased evaporation from expanding oceans, one reaches the somewhat paradoxical situation when with increasing precipitation polar-ward of continental regions, the possibility, if not inevitability, of a new ice age developing cannot be ignored!

Glaciology and glacial processes – ice sheets and valley glaciers – a question of scale and functionality

Several strands of disparate knowledge concerning ice physics (glaciology) and glacial geomorphology have begun to merge over the past few years. Where in the past both sciences seemed to almost ignore each other's findings and 'issues'; today there has been amalgamation of concepts and a sense of mutual benefit in sharing many basic ideas concerning glaciology and geomorphology. In brief, research from such locations as Greenland and Antarctica has provided new insights into glaciodynamics and glacial processes. With continuing concern for global warming and the detection of 'signatures' indicative of what has now been realized as very rapid environmental change, a vast bank of data has accumulated that pertains to glacial processes (cf. Anderson, 1999; Escutia et al., 2005; Jamieson and Sugden, 2008; Siegert, 2008).

It is critical to appreciate the fundamental differences between ice sheets and valley glaciers. Ice sheet movement is unconfined, traveling across all manner of terrains ignoring up-ice sloping and facing gradients such as escarpments and tributary valley floors; moving largely as a function of the upper ice mass slope (sin a). In contrast, valley glaciers are confined ice masses, restricted by terrain and only, where such an ice mass becomes sufficiently thick to overtop valley interfluves forming diffluent lobes, do valley glaciers begin to move and usually coalesce into small ice caps or ice fields. The landsystems developed by these two fundamentally different forms of ice mass are largely unique to each other (Evans et al., 2003). Until relatively recently (1970s) this fundamental difference was largely overlooked, but even now newspapers in North America continue to refer to 'glaciers' covering the northern half of the continent rather than an ice sheet. Similarly, only recently have textbooks on glacial geomorphology begun to emphasize this fundamental difference. A further division within ice sheet types exists between continental and marine-based ice sheets and is vital to the understanding of ice sheet style glaciation. A continental ice sheet is an ice mass covering continental land masses whose base is above sea level even following total ice retreat. Thus it is inherently stable in term of its underpinning. In contrast, a marine-based ice sheet lies below sea level and thus is potentially unstable. The vast Laurentide Ice Sheet that covered the largest part of northern North America contained both marine and continental margins. Thus the forms and styles of erosion and deposition varied along the margins both in place and over time.

Where lateral moraines and kame terraces abound in valley glacier systems, such landforms are largely absent from any ice sheet system. Of course, many subglacial forms, such as drumlins and eskers, are to be found within both systems, but the scale of form and extent of development under valley glaciation is much reduced. Drumlins and eskers, for example, form within valley glacier systems but are rare and at the small scale (cf. van der Meer, 1983) whereas within ice sheet systems drumlin fields of tens of thousands of individual forms occur as do extensive esker networks (Menzies and Shilts, 2002). Where valley glaciers may ultimately form fjord and U-shaped valleys of intense localized, linear erosion so too do ice streams within ice sheets (Stokes and Clark, 1999). These latter forms may not only produce similar localized bedrock incision but also may act as the fundamental constraints on large areas of ice sheet stability as possibly witnessed today in the West Antarctic Ice Sheet (Vaughan and Spouge, 2002; Hindmarsh, 2006).

Much of what has been learned about glacial geomorphology has until the 1970s been largely gleaned from valley glacier systems in the mountainous parts of the world. However, with intensive research on the Greenland and Antarctic Ice Sheets a more profound and better understanding of the major glaciations to envelop the northern hemisphere during the Quaternary period has been achieved.

An essential aspect of much glacial geomorphology today centres on the impact and effects of global climate change. It has become well recognized that ice sheet stability linked to subglacial sole conditions and grounding line fluctuations hold an important signal to ice sheet stability which is, in turn, linked to sea-level rise, climate warming and ocean circulation patterns. Grounding line locations, potential fluctuations in the past and in the future, hold a crucial and fundamental role in the understanding of ice sheet stability both in the past and into the future. Based upon geomorphological mapping, sediment analyses through drilling and standard sedimentological investigations coupled to geochronology, the positioning of these grounding lines and the general geomorphology of past subglacial surfaces are key elements in this forward looking scientific research.

GLACIAL EROSION: PROCESSES, LANDFORMS AND LANDSCAPE SYSTEMS

The processes of glacial erosion produce sediments for transport and later glacial deposition. Processes of erosion occur in all glacial environments. Although rather limited, supraglacial erosion in the form of abrasion and meltwater action occurs on the surfaces of ice masses, especially during the summer months when melted ice and snow move across glacier surfaces. In mountainous areas, considerable wind-blown debris, and avalanched and mass movement debris often end up in the supraglacial environment where debris may experience abrasion and meltwater action. Even more limited is the amount of erosion that occurs within ice masses within englacial tunnels where glacial debris is transported spasmodically depending on meltwater activity within an ice mass. Proglacial erosion is considerable since significant volumes of debris are released from ice masses either along lateral and frontal margins in valley glaciers, or along the front margins of large ice sheets, or into oceanic basins where ice margins are floating or have ice shelves attached and thus debris is exited via a grounding-line. In all these locations debris exiting into the proglacial environment is subject to substantial erosion largely due to meltwater transport and mass movement activity. Most glacial erosion, however, occurs in the subglacial environment where high stress levels, meltwater under hydrostatic pressures and vast sources of erodable material are available.

Processes

Glacial erosional processes can be subdivided under five main headings: abrasion, plucking/quarrying, meltwater action, chemical action and freeze–thaw processes. In general all of these processes may operate on the same rock surfaces such that examples of glacial erosion typically possess all the characteristic 'marks' of all these processes.

Abrasion is the wearing down of bedrock surfaces and rock fragment surfaces by the scouring processes of debris-laden ice and meltwater. In both instances, as the debris within the base of the ice or debris moving within high pressure meltwater passes over the surface of bedrock or other rock surfaces, scratching and wearing down of the surface occur. The evidence is in the form of minute scratches or striae on rock surfaces, or extremely smooth rock faces with smooth sculpted geometries illustrate rapid wear. These latter forms of erosion are termed P-forms. To be effective such abrasion processes demand high basal ice debris concentrations and/or high debris content in high pressure meltwater streams, debris that is sharp, angular and harder than the rock surfaces to be cut, sufficiently high basal ice pressures and finally an effective means of evacuation of the abraded debris. This latter requirement is essential if the abrasion process is to continue and not become 'clogged'. The production of immense volumes of abraded debris is apparent in the down-stream 'milky' nature of glacial streams that exhibit a blue or greenish-blue color due to the high fine debris content referred to as glacial milk or rock flour.

Plucking or quarrying is a set of processes that remove fractured, jointed or disaggregated rock fragments from bedrock or the surface of rocks. Typically, rocks fracture under tensile or compressive stress levels produced by the overall stress levels created by the overlying moving ice mass, or by percussive processes where rock fragments crash against rock surfaces and generate flakes and shards of rock materials. Rock plucking can produce enormous boulders and minuscule rock flakes. Where large bedrock knobs have been overridden, roches moutoneés may be produced as distinctive erosional landforms with smooth up-ice (stoss) sides and steep craggy down-ice (lee) sides. Where percussive plucking occurs on rock surfaces, tiny chattermarks, troughs and fractures are developed.

Meltwater, in association with high debris content and high pressure discharges, creates other forms of abrasive wear on rock surfaces. A little understood process is the effects of chemical weathering beneath ice masses where hydrostatically pressured meltwater, high stress levels and fluctuating temperatures lead to carbonate, silicate and iron solution and re-precipitation down-ice. Characteristically, such chemical processes can usually be observed as distinctive stains or 'trails' on rock surfaces, often in the lee-side of bedrock protrusions. Finally, in glacial environments, freeze–thaw processes are active, seasonally and diurnally, producing large volumes of frost shattered rock materials.

Landforms

Glacial erosional landforms can be subdivided into areal and linear forms. The range of scale of these landforms can be immense, from centimeter-sized 'rats-tails' and flutes on rock surfaces to roches moutoneés, tens of meters in height. Vast areas of the Canadian and Fenno-Scandian Shields best portray widespread areal scour by glacial erosion. Linear forms such as fjords, trough (U-shaped) valleys (Plate 10, page 595), finger lakes, tunnel valleys, and, at the much smaller scale, fluted bedrock knobs (p-forms), occur in all glaciated regions. All of these landforms reflect the erosive power of glaciers, ice sheets and associated meltwater action.

Landscape systems

Perhaps the best known example of a glaciated landscape system is that of high mountain glaciation where pyramidal peaks (horns), arêtes, cirques, tarn lakes, rock steps (riegels) and hanging valleys are found in close association (Plate 11a,b, page 596). No similar landscape system exists for ice sheet erosive landscapes except for the areal scour, bedrock trough, fluted bedrock knob landscapes, such as those of the Canadian and Fenno-Scandian Shields (Plate 10, page 595).

GLACIAL TRANSPORT

Glaciers and ice sheets transport vast quantities of sediment from fine-grained clays and silts to huge boulders. Sediment is transported on the surface of the ice as supraglacial debris, within the ice as englacial debris, beneath the ice as subglacial debris and beyond the ice margins as proglacial sediment. Each environment affixes the transported sediment with a potentially distinctive signature that, although complex due to overprinting and re-transportation, can be used, at times, to differentiate glacial sediment types (Boulton, 1978; Benn et al., 2003). Sediment is transported encased within the basal layers of the ice, on the surface of the ice and beneath the ice within deforming soft sediment beds.

GLACIAL DEPOSITION: PROCESSES, LANDFORMS AND LANDSCAPE SYSTEMS

The deposition of glacial sediment is largely a function of the means of sediment transportation.

Since ice masses scavenge sediments from their glaciated basins, and in the case of ice sheets this may be continental landmasses, the provenance of glacial sediment is typically immense, containing far-traveled sediment and boulders as well as dominantly local source rock materials. In fact most glacial sediments reflect a provenance only a few tens of kilometers up-ice from any point of deposition. However, repeated glaciation may increase the original distance from the origin to the final deposition site.

Glacial deposits occur as unsorted sediments with a wide range of particle sizes, and as sorted sediments deposited by meltwater on land and in lacustrine and marine environments. Unsorted sediments are typically referred to as till or glacial diamicton. These sediments are exceedingly complex sedimentologically and stratigraphically. They occur in all subenvironments of the glacier system and range from subglacial clay-rich, consolidated lodgement tills to coarse supraglacial flow tills, to subglacial and submarginal areas of basal ice melting leading to melt-out tills. In recent years this classification with reference to subglacial tills is probably inaccurate. It would be more rigorous to term these tills, 'tectomics', products of the complex subglacial environment (van der Meer et al., 2003).

The sorted or fluvioglacial sediments range from fine-grained clays deposited in glacial lakes (glaciolacustrine) to coarser sands and gravels deposited in front of ice masses as outwash fans or sandur, to rain-out sediments deposited in marine settings (glaciomarine).

Processes

The processes of glacial deposition range from direct smearing on of subglacial debris as lodgement tills; to mass movement of sediments from the frontal and lateral margins of glaciers as flow tills; to direct ablation of ice forming melt-out tills in front of ice masses in proglacial areas and within subglacial cavities.

Landforms

Glacial landforms range from those formed (1) transverse to ice motion, (2) parallel to ice motion, (3) unoriented, non-linear forms, (4) ice marginal forms, to (5) fluvioglacial landforms. Although a complex myriad of glacial depositionary forms occur in most glaciated areas the dominant landforms, discussed below, include (1) Rogen moraines, (2) drumlins and fluted moraines, (3) hummocky moraines, (4) end moraines and (5) eskers and kames.

Rogen or ribbed moraines are a series of conspicuous ridges formed transverse to ice movement. These ridges typically rise 10–20 m in height, are 50–100 m in width and are spaced 100–300 m apart. They tend to occur in large numbers as fields as, for example, in the Mistassini area of Quebec. They are composed of a range of subglacial sediments and are thought to be formed by the basal deformation of underlying sediment, possibly at times in association with a floating ice margin. Drumlins are perhaps one of the most known glacial landforms. A considerable literature exists on their formation and debate continues on their mechanic(s) of formation. Drumlins are streamlined, roughly elliptical or ovoid-shaped hills with a steep stoss side facing up-ice and a gentler lee-side facing down-ice (Figure 22.2). Drumlins range in height from a few meters to over 250 m, and may be from 100 m to several kilometers in length. They tend to be found in vast swarms or fields of many thousands. In central New York State, for example, over 70,000 occur, and likewise in Finland, Poland, Scotland and Canada vast fields exist. Drumlins are composed of a wide range of subglacial sediments and may contain boulder cores, bedrock cores, sand dykes and 'rafted' non-glacial sediments. It seems likely these landforms are developed below relatively fast moving ice beneath which a deforming layer of sediment developed inequalities and preferentially stiffer units become nuclei around which sediment plasters and the characteristic shape evolves (Boulton, 1987; Menzies and Shilts, 2002; King et al., 2009). Other hypotheses of formation range from the streamlining by erosion of pre-existing glacial sediments, to changes in the dilatancy of glacial sediments leading to stiffer nuclei at the ice-bed interface, to fluctuations in porewater content and pressure within subglacial sediments again at the ice-bed interface and to the infilling of subglacial cavities by massive subglacial floods (Menzies and Shilts, 2002: 222–223).

Fluted moraines are regarded as subglacial streamlined bedforms akin to drumlins. These landforms are typically linked in formation with drumlins and Rogen moraines. They tend to be much smaller in height and width as compared to drumlins but may stretch for tens of kilometers in length. They are typically composed of subglacial sediments. Fluted moraines have been found to develop in the lee of large boulders but may also be attenuated drumlin forms, the result of high basal ice shear stress and high ice velocities. In central New York State fluted moraines occur beside and amongst the large drumlin fields (cf. Benn, 2006; Hess and Briner, 2009).

Hummocky moraine is a term used to denote an area of terrain in which a somewhat chaotic

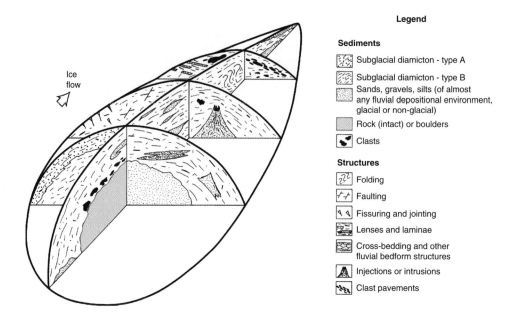

Figure 22.2   General model of drumlin plan and variability of internal composition

deposition pattern occurs similar to a series of small sediment dumps. These moraines rarely rise above a few meters but may exist over considerable areas. Hummocky moraine often marks locations where massive down-wasting of an ice mass may have occurred (Hambrey et al., 1997; Menzies and Shilts, 2002; Lukas, 2008).

End moraines or terminal, retreat or frontal moraines occur at the front margins of ice masses. Typically, these landforms, transverse to ice motion, are a few meters to tens of meters in height and, if built up over several years, may be even higher and of some considerable width (1–5 km). Moraines contain subglacial and supraglacial debris, and often have an arcuate shape closely mirroring the shape of an ice front. Since these moraines mark the edge of an ice mass at any given time, the longer ice remains at that location the higher and large the moraines become. However, as an ice mass retreats, with periodic stationary periods (stillstands), a series or sequence of moraines develop that mark the retreat stages of the ice (Figure 22.3). Sequences of end moraines can be observed in the mid-west states of Illinois, Ohio and Michigan.

Eskers are products of subglacial meltwater streams in which the meltwater channel has become blocked by fluvioglacial sediments (Clark and Walder, 1994; Boulton et al., 2007a, 2007b; Boulton et al., 2009). Eskers are long ridges of sands and gravels that form anastomosing patterns across the landscape. They range in height from

a few meters to >50 m and may run for tens of kilometers. In Canada some eskers cross the Canadian Shield for hundreds of kilometers. These landforms often have branching ridges and a dendritic morphology. Eskers are dominantly composed of fluvioglacial sands and gravels with distinctive faulted strata along the edges where the ice tunnel wall melted and collapsed. In many instances eskers are observed 'running' uphill, or obliquely crossing over drumlins, a testament to their formation within high pressure subglacial meltwater tunnels.

In many glaciated areas during ice retreat large, roughly, circular dumps of fluvioglacial sands and gravels occur. These forms, termed kames, appear to develop when infilled crevasses or buried ice melts, leaving the sediment behind in a complex but chaotic series of mounds. The term, kame or kame-delta, has been applied to fluvioglacial sediments that have formed temporary deltas into long gone glacial lakes. In Finland a long line of such kame-deltas formed into a moraine-like series of linear ridges transverse to ice front retreat (Salpuasselkä).

Glacial depositional landscape systems

Distinctive suites of glacial depositional landforms can be found in valley glacier and ice sheet

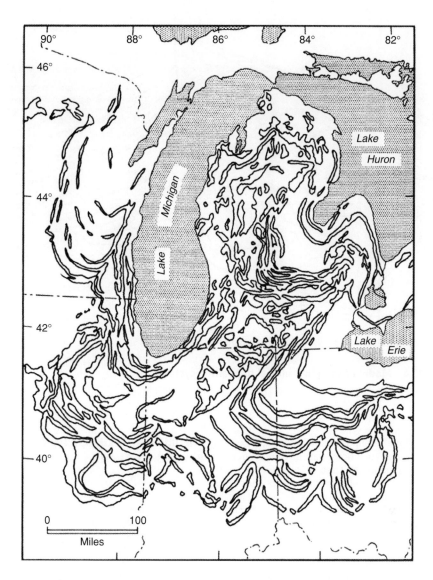

Figure 22.3   End moraine complexes south of the Great Lakes in the Mid-West USA (Modified from Strahler, 1968)

settings (Boulton and Eyles, 1979; Eyles et al., 1983; Evans, 2003; Golledge, 2007). Typically these sequences of landforms are often overprinted due to subsequent glaciations, but in the mid-west of the United States such suites of depositional landforms from the last glaciation (Late Wisconsinan) to affect the area can be clearly discerned. Distinctive landscapes systems attributable to the marginal areas of ice sheets also carry a unique supraglacial suite of landforms as can be observed in parts of the mid-west USA (Figure 22.4) and around modern ice masses.

Subglacial soft sediment deformation

Perhaps one of the greatest discoveries in glacier environments in recent decades has been the realization that soft sediments rather than bedrock beds underlie most temperate ice masses (Alley et al., 1986; Boulton and Hindmarsh, 1987; van der Meer, 1993; Van der Wateren, 1995; Le Heron et al., 2005). Coupled to this has been the detection that the subglacial thermal regime in many ice masses is best described as polythermal

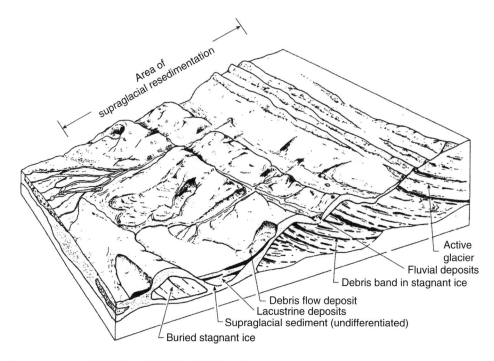

Figure 22.4 Model of marginal glacial deposition systems

(Paterson, 1994). Both these soft deformable beds and polythermal conditions, has, in combination, led to a recognition that ice mass velocity, ice surface profiles and, subglacial processes must be thoroughly reappraised (cf. Murray, 1997; Menzies and Shilts, 2002). Not only are glacial sediments formed under these conditions likely to carry distinctive sedimentological signatures and structures but many landforming processes are likely to evolve landforms in a differing manner than previously perceived. For example, explanations of the nature of Rogen Moraine and drumlin styles of formation need to take such rapidly changing subglacial conditions into account where previously such conditions had never been contemplated. Where multiple till sequences had previously been considered evidence of multiple ice advances, it is now likely that one advance under deforming bed conditions might account for a similar stratigraphic package (cf. Menzies and van der Meer, 1998). However, much remains unknown concerning the impact and influence of subglacial soft beds in terms of the spatial and temporal variability of such beds and how this might affect and generate landform patterns (cf. Stokes and Clark, 2003) and distinctive sediment packages (van der Meer et al., 2003; Menzies et al., 2006).

THE IMPACT OF GLACIATION BEYOND ICE LIMITS

Areas beyond the ice limits are greatly influenced by glaciation due to deteriorating climatic conditions, the deposition of wind-blown dust (loess), the divergence of river systems where headwaters or partial drainage basins may be intersected by advancing ice, the impact of outburst floods from ice fronts (jökulhlaups) and major faunal and floral changes due to encroaching ice. In terms of the impact of ice masses on human life and society, the only evidence is somewhat anecdotal and at the close of the Late Wisconsinan may have acted to spur human migration as, for example, through the short-lived ice-free corridor between the Cordilleran and Laurentian Ice Sheets in central Alberta, Canada, into the prairies to the south.

Within a few hundreds of kilometers of the vast ice sheets that covered North America and Europe, climatic conditions must have been severe, with strong katabatic cold winds descending from the ice (Orme, 2002). Associated with these poor climates, periglacial conditions must have prevailed in which the ground became permanently frozen to a considerable depth. Periglacial activity produced frozen ground phenomena such as solifluction sediments down slopes, ice and sand wedges and localized ice lenses and pingo formation. Dramatic changes in vegetation types and animal life occurred, with the southern migration, in the northern hemisphere, of many species. For example, unlike today, central southern Texas would have supported hardwood forests of the type now found in Ohio and Pennsylvania.

Due to the katabatic winds, fine sediments were picked up from the proglacial areas along the margins of the ice sheets and the dust was transported away from the ice and deposited as thick, massive loess (glacioaeolian sediments). Considerable thicknesses of these occur in the mid-west states especially in Iowa and Kansas, and in central Europe, in Hungary and in central Asia and China (Zhang, 1980; Muhs and Bettis, 2000; Frechnen et al., 2003).

In some instances, vast outpourings of meltwater (jökulhlaups) flooded from the ice fronts leading to distinctive heavily dissected terrains being formed (badlands). Such an occurrence is recorded in the Columbia River Badlands of Washington and Oregon States where a huge lake formed and then dramatically drained (the Lake Missoula floods) (Bretz, 1969; Baker, 1985; Zuffa et al., 2000).

FUTURE PROSPECTS

With such intensive study, especially over the past 50 years and the resultant accumulation of a vast data base of geomorphological terrain maps, landform studies and glacial sediment inventories, glacial geomorphology remains a rich and vibrant science. Much remains to be discovered, reassessed and new avenues of research pursued. It is no easy task to 'cherry-pick' particular areas within glacial geomorphology; however, certain distinct lines of enquiry seem to reveal great future potential.

Over the past decade there have been major advances in understanding large-scale megageomorphology especially in relation to glacial lineations over the beds of large former ice sheets (cf. Stokes and Clark, 2003). At around the same time the development of micromorphology in glacial sediment research has heralded a new phase in glacial sedimentology (cf. van der Meer et al., 2003). It is increasingly critical that a vertical integration of findings be developed that will permit a scalar incorporation of processes at differing scales, thus permitting an assimilation of process mechanisms to be clearly understood from the micro to the megascale.

A key element to understanding the connectivity between ice basal motion, surging, subglacial hydraulics and therefore subglacial processes of erosion, transport and deposition is the spatial, temporal and geotechnical aspects of subglacial deforming sediment (soft) beds. Much remains to be investigated and understood about soft subglacial beds in terms of landforming processes and resultant forms as a consequence of soft bed deformation.

Glacial erosion mechanics has been an avenue of limited research up to the present. Much can be learned from other sciences where wear processes (tribology) have been extensively studied (cf. Bhushan, 2002). In geomorphic terms perhaps too much heed has been paid to morphometric considerations regarding glacial erosional forms and not enough to the processes involved in form evolution.

A persistent and nagging problem all too often in glacial geomorphology has been the acquisition of suitable datable materials or methods of accurate dating. Where in the past datable material in the form of organics was all too often lacking in certain terrains or with reference to specific landforms such as drumlins; now with the advent of optical luminescence and cosmogenic dating finding organics is unnecessary (cf. Preusser, 1999; Owen et al., 2001).

It is hardly surprising that so much glacial geomorphology is Earth continental surface-based yet so much glacial activity in terms of deposits and forms occurs within the marine environment. It is roughly estimated that over 60% of glacial sediments covering the Earth's surface lie in ocean basins. Of especial future research interest are the interface areas in grounding line zones where ice masses enter the marine environment from the subglacial, whether as floating or tidewater ice masses. The dynamics of grounding line sediment expulsion and deposition/emplacement is a key factor in understanding ice sheet stability and sensitivity to climatic and other forcing factors that impact on major ice sheets. So much remains within glaciomarine environments to be researched that a vast data source remains to be tapped.

As meltwater emanates from subglacial tunnels, from englacial moulins and surface streams, in addition to enormous amounts of suspended sediments, there is often a considerable dissolved load within the ice-cold meltwater. Many substances are dissolved at these low temperatures, especially calcium carbonates but also silicates and ferruginous oxides. The nature of the low temperature geochemistry that permits rapid dissolution followed by re-precipitation requires further research. In the case of calcium carbonates it is not uncommon in ice contact glaciofluvial sediments to find the sands and gravels heavily cemented as a result of sudden calcium carbonate re-precipitation probably following a combination of pressure reduction and rapid meltwater temperature rise (cf. Menzies and Brand, 2007).

Glacial geomorphology continues to be vibrant and dynamic area of geomorphology. In only the past decade alone it has made many major advances in our understanding of glacial sediments and landforms, yet much remains to be fully understood and, in the case of glaciomarine environments especially, much to be discovered. It was Eiju Yatsu who, in 1966, in his seminal text *Rock Control in Geomorphology*, noted that although the 'where', 'when' and 'how' of geomorphology was fairly well understood, the 'why' remained a burning and, at times, open question – so too is the case in glacial geomorphology.

REFERENCES

Ahlmann, H.W. (1948) Glaciological research on the North Atlantic coasts. Royal Geographical Society, Research Series No. 1, p. 83.

Alley, R.B., Blankenship, D.D., Bentley, C.R. and Rooney, S.T. (1986) Deformation of till beneath ice stream B, West Antarctica. *Nature*, 322, 57–9.

Alley, R.B., Clark, P.U., Huybrechts, P. and Joughin, I. (2005) Ice-sheet and sea-level changes. *Science*, 21, 456–60.

Anderson, J.B. (1999) *Antarctic Marine Geology*. Cambridge University Press, Cambridge. p. 292.

Anderson, J.B., Shipp, S.S., Lowe, A.L., Smith Wellner, J. and Mosola, A.B. (2002) The Antarctic ice sheet during the last glacial maximum and its subsequent retreat history: a review. *Quaternary Science Reviews*, 21, 49–70.

Baker, V. (1985) Cataclysmic Late Pleistocene flooding from glacial Lake Missoula: a review. *Quaternary Science Reviews*, 4, 1–41.

Bamber, J.L., Alley, R.B. and Joughin, I. (2007) Rapid response of modern day ice sheets to external forcing. *Earth and Planetary Letters*, 257, 1–13.

Benn, D.I. (2006) Fluted moraine formation and till genesis below a temperate valley glacier: Slettmarkbreen, Jotunheimen, southern Norway. *Sedimentology*, 41, 279–92.

Benn, D.I., Kirkbride, M.P., Owen, L.A. and Brazier, V. (2003) Glaciated valley landsystems, in Evans, D.J.A. (ed.) *Glacial Landsystems*. Arnold, London. pp. 372–406.

Bhushan, B. (2002) *Introduction to Tribology*. John Wiley, New York. p. 708.

Blatter, H. and Hutter, K. (1991) Polythermal conditions in Arctic glaciers. *Journal of Glaciology*, 37, 261–9.

Boulton, G.S. (1978) Boulder shapes and grain-size distributions of debris as indicators of transport paths through a glacier and till genesis. *Sedimentology*, 25, 773–99.

Boulton, G.S. (1986) Geophysics – a paradigm shift in glaciology. *Nature*, 322, 18.

Boulton, G.S. (1987) A theory of drumlin formation by subglacial sediment deformation, in Menzies, J. and Rose, J. (eds) *Drumlin Symposium*. A.A. Balkema, Rotterdam, pp. 25–80.

Boulton, G.S. (1996) The origin of till sequences by subglacial sediment deformation beneath mid-latitude ice sheets. *Annals of Glaciology*, 22, 75–84.

Boulton, G.S. and Eyles, N. (1979) Sedimentation by valley glaciers: a model and genetic classification, in Ch. Schlüchter, (ed.) *Moraines and Varves; Origin, Genesis, Classification*. A.A. Balkema, Rotterdam. pp. 11–23.

Boulton, G.S. and Jones, A.S. (1979) Stability of temperate ice caps and ice sheets resting on beds of deformable sediment. *Journal of Glaciology*, 24, 29–43.

Boulton, G.S. and Hindmarsh, R.C.A. (1987) Sediment deformation beneath glaciers: rheology and geological consequences. *Journal of Geophysical Research*, 92(B9), 9059–82.

Boulton, G.S., Caban, P.E. and Van Gijssel, K. (1995) Groundwater flow beneath ice sheets: Part I – large scale patterns. *Quaternary Science Reviews*, 14, 545–62.

Boulton, G.S., Lunn, R., Vidstrand, P. and Zatsepin, S. (2007a) Subglacial drainage by groundwater–channel coupling, and the origin of esker systems: Part I – glaciological observations. *Quaternary Science Reviews*, 26, 1067–90.

Boulton, G.S., Lunn, R., Vidstrand, P. and Zatsepin, S. (2007b). Subglacial drainage by groundwater–channel coupling, and the origin of esker systems: Part II – theory and simulation of a modern system. *Quaternary Science Reviews*, 26, 1091–105.

Boulton, G.S., Hagdorn, M., Maillot, P.B. and Zatsepin, S. (2009) Drainage beneath ice sheets: groundwater–channel coupling, and the origin of esker systems from former ice sheets. *Quaternary Science Reviews*, 28, 621–38.

Bretz, J.H. (1969) The Lake Missoula floods and the channeled Scabland. *Journal of Geology*, 77, 505–43.

Chorley, R.J. (1959) The shape of drumlins. *Journal of Glaciology*, 3, 339–44.

Church, J.A. and White, N.J. (2006) A 20th century acceleration in global sea-level rise. *Geophysical Research Letters*, 33, L01602. p. 4.

Clark, P.U. and Walder, J.S. (1994) Subglacial drainage, eskers, and deforming beds beneath the Laurentide and Eurasian ice sheets. *Geological Society of America* Bulletin, 106, 304–14.

Echelmeyer, K.A. and Wang Z. (1987) Direct observation of basal sliding and deformation of basal drift at sub-freezing temperatures. *Journal of Glaciology*, 33, 83–98.

Echelmeyer, K.A., Harrison, W.D., Larsen, C. and Mitchell, J.E. (1993) The role of the margins in the dynamics of an active ice stream. *Journal of Glaciology*, 40, 527–38.

Escutia, C., De Santis, L., Donda, F., Dunbar, R.B., Cooper, A.K., Brancolini, G. et al. (2005) Cenozoic ice sheet history from East Antarctic Wilkes Land continental margin sediments. *Global and Planetary Change*, 45, 51–81.

Evans, D. J. A. (ed.) (2003) *Glacial Landsystems*. Hodder Arnold, London. p. 544.

Evans, D.J.A., Phillips, E.R., Hiemstra, J.F. and Auton, C.A. (2006) Subglacial till: formation, sedimentary characteristics and classification. *Earth-Science Reviews*, 78, 115–76.

Eyles, N., Dearman, W.R. and Douglas, T.D. (1983) The distribution of glacial landsystems in Britain and North America, in Eyles, N. (ed.) *Glacial Geology*. Pergamon, Oxford. pp. 213–28.

Forbes, J. (1859) *Occasional Papers on the Theory of Glaciers*. Adam and Charles Black, Edinburgh. p. 278.

Frechnen, M., Oches, E.A. and Kohfeld, K.E. (2003) Loess in Europe – mass accumulation rates during the Last Glacial Period. *Quaternary Science Reviews*, 22, 1835–57.

Geikie, A. (1863) On the phenomena of the glacial drift of Scotland. *Transactions of the Geological Society of Glasgow*, 1, 1–190.

Geikie, J. (1888) *The Great Ice Age*. (3rd edn.) Appleton and Co., New York.

Gilbert, G.K. (1906) Crescentic gouges on glaciated surfaces. *Geological Society of America Bulletin*, 17, 303–16.

Golledge, N.R. (2007) An ice cap landsystem for palaeoglaciological reconstructions: characterizing the Younger Dryas in western Scotland. *Quaternary Science Reviews*, 26, 213–29.

Hambrey, M., Huddart, D., Bennett, M.R. and Glasser, N.F. (1997) Genesis of 'hummocky moraines' by thrusting in glacier ice: evidence from Svalbard and Britain. *Journal of the Geological Society*, 154, 623–32.

Hart, J.K. (2006) An investigation of subglacial processes at the microscale from Briksdalsbreen, Norway. *Sedimentology*, 53, 125–46.

Hess, D.P. and Briner, J.P. (2009) Geospatial analysis of controls on subglacial bedform morphometry in the New York Drumlin Field – implications for Laurentide Ice Sheet dynamics. *Earth Surface Processes and Landforms*, 34, 1126–35.

Hindmarsh, R.C. (2006) The role of membrane-like stresses in determining the stability and sensitivity of the Antarctic ice sheets: back pressure and grounding line motion. *Phil. Trans. Royal Soc. (A)*, 364, 1733–67.

Imbrie, J. and Imbrie, K.P. (2005) *Ice Ages: Solving the Mystery*. (Reprint edn.) Harvard University Press; Cambridge, Mass. p. 224.

Jamieson, S.S.R. and Sugden, D.E. (2008) Landscape evolution of Antarctica, in A.K., Cooper, P.J., Barrett, H., Stagg, B., Storey, E., Stump, W. Wise, and the 10th ISAES editorial team, (eds). *Antarctica: A Keystone in a Changing World*. Procs. of the 10th International Symposium on Antarctic Earth Sciences. Washington, DC. The National Academies Press, pp. 39–54.

Joughin, I., Bamber, J.L., Scambos, T., Tulaczyk, S., Fahnestock, M. and MacAyeal, D.R. (2006) Integrating satellite observations with modeling: basal shear stress of the Filchner-Ronne ice streams, Antarctica. *Phil. Trans. Royal Soc. (A)*, 364, 1795–814.

Kaszycki, C.A. (1989) Quaternary Geology and Glacial History of the Haliburton Region, South Central Ontario, Canada: A Model for Glacial and Proglacial Sedimentation. Unpublished Ph.D. dissertation. University of Illinois at Urbana-Champaign. p. 219.

Kemmis, T.J. (1996) Lithofacies associations for terrestrial glacigenic successions, in Menzies, J. (ed.) *Past Glacial Environments*. Butterworth-Heinemann, Oxford. pp. 285–300.

King, E.C., Hindmarsh, R.C.A. and Stokes, C.R. (2009) Formation of mega-scale glacial lineations observed beneath a West Antarctic ice stream. *Nature Geoscience*, 2, 585–8.

Le Heron, D.P., Sutcliffe, O.E., Wittington, R.J. and Craig, J. (2005) The origins of glacially related soft-sediment deformation structures in Upper Ordovician glaciogenic rocks: implication for ice-sheet dynamics. *Palaeogeography, Palaeoclimatology, Palaeoecology*, 218, 75–103.

Lliboutry, L. (1968) General theory of subglacial cavitation and sliding of temperate glaciers. *Journal of Glaciology*, 7, 21–58.

Lukas, S. (2008) A test of the englacial thrusting hypothesis of 'hummocky' moraine formation: case studies from the northwest Highlands, Scotland. *Boreas*, 34, 287–307.

Lyell, C. (1830–1833) *Principles of Geology* (3 Volumes). John Murray, London.

Menzies, J. (ed.) (2002) *Modern and Past Glacial Environments*, Butterworth-Heinemann, Oxford. p. 543.

Menzies, J. and van der Meer, J.J.M. (1998) Sedimentological and micromorphological examination of Late Devensian multiple diamicton sequence near Moneydie, Perthshire, east-central Scotland. *Scottish Journal of Geology*, 34, 15–21.

Menzies, J. and Shilts, W.W. (2002) Subglacial environments, in Menzies, J. (ed.) *Modern and Past Glacial Environments*. Butterworth-Heinemann, Oxford. pp. 183–278.

Menzies, J. and Brand, U. (2007) The internal sediment architecture of a drumlin, Port Byron, New York State, USA. *Quaternary Science Reviews*, 26, 322–35.

Menzies, J., van der Meer, J.J.M. and Rose, J. (2006) Till – as a Glacial 'Tectomict', its internal architecture, and the development of a 'typing' method for till differentiation. *Geomorphology*, 75, 172–200.

Muhs, D.R. and Bettis, E.A. (2000) Geochemical variations in Peoria Loess of western Iowa indicate paleowinds of midcontinental North America during last glaciation. *Quaternary Research*, 53, 49–61.

Murray, T. (1997) Assessing the paradigm shift; deformable glacier beds. *Quaternary Science Reviews*, 16, 995–1016.

Nilsson, T. (1983) *The Pleistocene; Geology and Life in the Quaternary Ice Age*. D. Reidel, Dordrecht. p. 651.

Oerlemans, J. (2002) Global dynamics of the Antarctic ice sheet. *Climate Dynamics*, 19, 85–93.

Orme, A.R. (2002) The Pleistocene legacy – beyond the ice front, in Orme, A.R. (ed.) *The Physical Geography of North America*. Oxford University Press, New York and Oxford. pp. 55–85.

Owen, L.A., Gualtieri, L., Finkel, R.C., Caffe, M.W., Benn, D.I. and Sharma, M.P. (2001) Cosmogenic radionuclide dating of glacial landforms in the Lahul Himalaya, northern India: defining the timing of Late Quaternary glaciation. *Journal of Quaternary Science*, 16, 555–63.

Paterson, W.S.B. (1994) *The Physics of Glaciers*. 3rd edn. Pergamon Press, Oxford. p. 480.

Preusser, F. (1999) Luminescence dating of fluvial sediments and overbank deposits from Gossau, Switzerland: fine grain dating. *Quaternary Science Reviews*, 18, 217–22.

Siegert, M.J. (2008) Antarctic subglacial topography and ice-sheet evolution. *Earth Surface Processes and Landforms*, 33, 646–60.

Slater, G. (1926) Glacial tectonics as reflected in disturbed drift deposits. *Proceedings of the Geologists' Association*, 37, 392–400.

Spedding, N. (1997) On growth and forms in geomorphology. *Earth Surface Processes and Landforms*, 22, 261–5.

Stokes, C.R. and Clark, C.D. (1999) Geomorphological criteria for identifying Pleistocene ice streams. *Annals of Glaciology*, 28, 67–74.

Stokes, C.R. and Clark, C.D. (2003) Laurentide ice streaming on the Canadian Shield: a conflict with the soft-bedded ice stream paradigm? *Geology*, 31, 347–50.

Strahler, A.N. (1966) *The Earth Sciences*. Harper Row, New York.

Sugden, D.E., Bentley, M.J. and O'Cofaigh, C. (2006) Geological and geomorphological insights into Antarctic ice sheet evolution. *Phil. Trans. Royal Soc. (A)*, 364, 1607–25.

Talalay, P.G. and Hooke, R. LeB. (2007) Closure of deep boreholes in ice sheets: a discussion. *Annals of Glaciology*, 47, 125–33.

Tarbuck, E.J., Lutgens, F.K. and Tasa, D. (2009) *Earth Science*. Prentice Hall. NJ.

Tulaczyk, S., Kamb, W.B. and Engelhardt, H.F. (2000) Basal mechanics of Ice Stream B, West Antarctica 1. Till mechanics. *Journal of Geophysical Research*, 105(B1), 463–81.

van der Meer, J.J.M. (1983) A recent drumlin with fluted surface in the Swiss Alps, in Evenson, E.B. Schlüchter, Ch. and Rabassa, J. (eds) *Tills and Related Deposits: Genesis, Petrology, Application, Stratigraphy*. A.A. Balkema, Rotterdam. pp. 105–10.

van der Meer, J.J.M. (1993) Microscopic evidence of subglacial deformation. *Quaternary Science Reviews*, 12, 553–87.

van der Meer, J.J.M., Menzies, J. and Rose, J. (2003) Subglacial till: the deforming glacier bed. *Quaternary Science Reviews*, 22, 1659–85.

Van der Wateren, F.M. (1995). Structural geology and sedimentology of push moraines – processes of soft sediment deformation in a glacial environment and the distribution of glaciotectonic styles. *Mededelingen Rijks Geologische Dienst*, 54, 1–168.

Vaughan, D.S. and Spouge, J.R. (2002) Risk estimation of collapse of the West Antarctic Ice Sheet. *Climatic Change*, 52, 65–91.

Weertman, J. (1964) The theory of glacier sliding. *Journal of Glaciology*, 5, 287–303.

Weertman, J. (1979) The unsolved general glacier sliding problem. *Journal of Glaciology*, 23, 97–115.

Yatsu, E. (1966) *Rock Control in Geomorphology*, Sozosha, Tokyo. 135 pp.

Zhang, Z. (1980) Loess in China. *Geojournal*, 4, 525–40.

Zuffa, G.G., Normark, W.R., Serra, F. and Brunner, C.A. (2000) Turbidite megabeds in an oceanic rift valley recording Jökulhlaups of Late Pleistocene glacial lakes of the western United States. *Journal of Geology*, 108, 253–74.

23

Periglacial Environments

Hugh French

Periglacial geomorphology is the sub-discipline of geomorphology concerned with the landforms and processes of the cold non-glacial regions of the world. Fundamental to the discipline is the freezing of water and its associated frost heaving and ice segregation. Permafrost is a central, but not defining element. Other components include the impact of seasonal freezing, the role of seasonal snow and the relevance of fluvial, lacustrine and sea-ice covers. The study of those azonal processes that exhibit distinct behavioural and/or magnitude and frequency distributions under cold-climate conditions is also regarded as falling within the sphere of periglacial geomorphology.

Periglacial environments occupy approximately 20 per cent of the world land area. But their human population is only 7–9 million, mostly living in Russia, or only 0.3 per cent of the world's population. Thus, the larger importance of periglacial environments lies not only in their spatial extent but also in their natural resources. For example, the Precambrian basement rocks that outcrop as huge tablelands in both Canada and Siberia contain precious minerals, such as gold and diamonds, and sizable deposits of lead, zinc and copper. Equally important are the sedimentary basins of western Siberia, northern Alaska, and the Canadian High Arctic that contain large hydrocarbon reserves. A second reason why periglacial environments are of significance is their place within the cryosphere (the main components of which are snow, glacier, lake and river ice, frozen ground and sea ice) and the critical role which the cryosphere plays in global climate change.

THE HISTORICAL CONTEXT

The term 'periglacial' was first used by a Polish geologist, Walery von Łozinski, in the context of the mechanical disintegration of sandstones in the Gorgany Range of the southern Carpathian Mountains, a region now part of central Romania. Łozinski described the angular rock-rubble surfaces that characterize the mountain summits as so called 'periglacial facies' formed by the previous action of intense frost (von Łozinski, 1909). Following the XI Geological Congress in Stockholm in 1910 and the subsequent field excursion to Svalbard in 1911 (von Łozinski, 1912), the concept of a 'periglacial zone' was introduced to refer to the climatic and geomorphic conditions of areas peripheral to Pleistocene ice sheets and glaciers. Theoretically, this was a tundra zone that extended as far south as the treeline. In the mountains, it was a zone between timberline and snow line (Figure 23.1).

Today, Łozinski's original definition is regarded as unnecessarily restricting; few, if any, modern analogs exist (French, 2000). There are two main reasons. First, frost-action phenomena are known to occur at great distances from both present-day and Pleistocene ice margins. In fact, frost-action phenomena can be completely unrelated to ice-marginal conditions. For example, parts of central and eastern Siberia and north-western Arctic North America were not glaciated during the Pleistocene. Yet these are regions in which frost action was, and currently is, very important. Second, although Łozinski used the term to refer primarily to areas rather than processes, the term

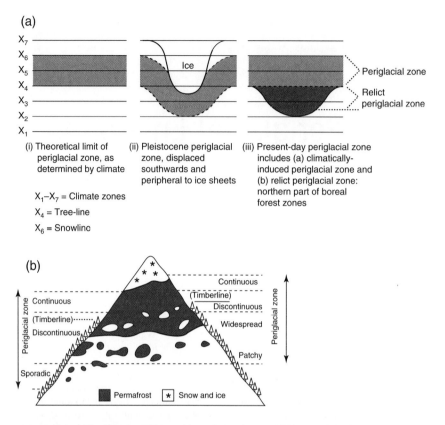

Figure 23.1   Schematic diagram illustrating the concept of the periglacial zone in (a) high-latitude and (b) high-altitude (alpine) areas (From French, 2007)

has increasingly been understood to refer to a complex of cold-dominated geomorphic processes. These include not only unique frost action and permafrost-related processes but also a range of azonal processes, such as those associated with snow, running water and wind, which demand neither a peripheral ice-marginal location nor excessive cold. Instead, these processes assume distinct or extreme characteristics under cold non-glacial conditions.

THE WEAKNESS OF TRADITIONAL PERIGLACIAL GEOMORPHOLOGY

Following the Geological Congress in Stockholm in 1910 the periglacial concept was enthusiastically embraced and several influential benchmark papers were subsequently published. For example, cold-climate patterned ground was described (Meinardus, 1912) and the importance of frost shattering highlighted (Hogbom, 1914). The 1920s

and 1930s witnessed a slow development of periglacial geomorphology. Most studies were Pleistocene-oriented (e.g. Marr, 1919; Bryan, 1928; Losche, 1930; Mortensen, 1932; Denny, 1936; Edelman et al., 1936) and relatively few described present-day phenomena (e.g. Jorre, 1933; Sorensen, 1935; Poser, 1932; Paterson, 1940). The real growth of periglacial geomorphology occurred in Europe in the two decades following 1945 (French, 2003). Most dominant were Pleistocene studies dealing with either paleogeographic reconstruction in the mid-latitudes or terminology (e.g. Bryan, 1946, 1949; Poser, 1948; Budel, 1951, 1953; Peltier, 1950). At that time, a so-called 'periglacial fever' (Andre, 2003) fostered a trendy sub-discipline termed climatic geomorphology (e.g. Troll, 1948; Budel, 1963, 1977).

Two widely held assumptions or interpretations fuelled this intense disciplinary growth. First was the uncritical acceptance of mechanical (frost) weathering and of rapid cold-climate landscape modification. In Europe, a sequence of influential texts by J. Tricart (1963; Tricart and Cailleux, 1967) and A. Cailleux (Cailleux and Taylor, 1954) promoted these ideas. At the same time, an IGU Periglacial Commission under the leadership of J. Dylik was especially active between 1952 and 1972 and an international journal, *Biuletyn Peryglacjalny*, was started in Łódź, Poland.

As early as the mid-1970s, this first assumption was being seriously challenged. Initially, air climates were shown to be poor indicators of the relevant ground climates. Observations in both high latitudes and at high elevations failed to record the numerous freeze–thaw cycles thought responsible, a shortcoming compounded by a lack of moisture in many regions. It was quickly realized that mechanical weathering involved not only frost action but also other mechanisms such as thermal stress and hydration shattering. It is now understood that a variety of chemical, biochemical, physical and mechanical processes operate, and often interact, in cold regions.

The second weakness of traditional periglacial geomorphology was that insufficient consideration was given to, first, the influence of lithology upon so-called 'periglacial' landscapes and, second, the variability, duration and efficacy of cold-climatic conditions, both today and during the Quaternary. Both of these weaknesses were reflected in the standard periglacial texts that became available in the 1960s.

In hindsight, it can be seen that the process assumptions underpinning traditional Pleistocene periglacial geomorphology were erroneous. Not only were the necessary observational data lacking but the conceptual or theoretical framework was flawed.

PERIGLACIAL ENVIRONMENTS, PERIGLACIAL ECOSYSTEMS

Periglacial environments are simple to define because they are characterized by intense frost and restricted to areas that experience cold, but essentially non-glacial, climates (French, 2007). They occur not only as tundra zones in the high latitudes, as defined by Łozinski's concept, but also as forested areas south of the tree-line and in the high elevation regions of the mid latitudes. They include (1) the polar deserts and semi-deserts of the High Arctic, (2) the extensive tundra zones of high northern latitudes, (3) the northern parts of the boreal forests of North America and Eurasia and (4) the alpine zones that lie above timberline and below the snowline in mid- and low-latitude mountains. To these must be added (1) the ice-free areas of Antarctica, (2) the

high-elevation montane environments of central Asia, the largest of which is the Qinghai-Xizang (Tibet) Plateau of China, and (3) small oceanic islands in the high latitudes of both Polar Regions. The freezing and thawing conditions experienced by these different periglacial environments are summarized in Figure 23.2.

In climatic terms, periglacial environments characterize areas where the mean annual air temperature is less than +3°C. This temperature range can be subdivided by the −2°C mean annual air temperature to define environments in which frost action dominates (mean annual air temperature less than −2°C) and those in which frost action occurs but does not necessarily dominate (mean annual air temperature between −2°C and +3°C).

In terms of periglacial landscape dynamics, ground temperature is probably more important than air temperature. Typically, the depth of ground freezing varies from as little as 10–20 cm beneath organic materials to over 500 cm in areas of exposed bedrock. It is important to stress that relatively few freeze–thaw cycles occur at depths in excess of 30 cm; there, only the annual temperature cycle usually occurs. It is also important to differentiate between the mean annual air temperature (MAAT) and the mean annual ground surface temperature (MAGST). This gives rise to the so-called 'surface offset'. The difference between the mean annual ground surface temperature (MAGST) and the temperature at the top of permafrost (TTOP) results in the so-called 'thermal offset' (Figure 23.3) (Smith and Riseborough, 2002). The surface offset reflects primarily the influence of snow cover and vegetation while the thermal offset is conditioned largely by the physical properties of the active layer (thermal conductivity and moisture content).

Periglacial geomorphology must include an understanding of the ecosystems that characterize these climatic environments. The relationship is often intimate and perhaps no more obvious than in the tundra and polar deserts of the high latitudes (Sorensen, 1935; Porsild, 1938, 1955, 1957; Tyrtikov, 1959; Raup, 1965).

The most extensive periglacial environments can be regarded as either *arctic* or *sub-arctic* in nature. The boundary between the two approximates the northern limit of trees, the so-called *tree-line*, a zone, 30–150 km in extent, north of which trees are no longer able to survive. North of tree-line, the terrain is perennially frozen and the surface thaws for a period of only 2–3 months each summer. Ecologists refer to the vegetated but treeless arctic as *tundra*. Where Precambrian basement rocks occur, as in the tablelands of northern Canada and Siberia, the tundra is barren. At higher latitudes, the tundra progressively

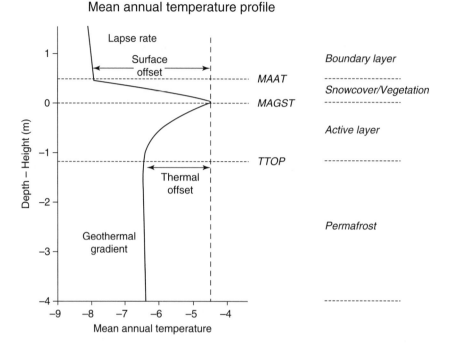

Figure 23.2 Schematic graph that shows the mean annual temperature profile through the surface boundary layer in a periglacial region underlain by permafrost. It illustrates the surface and thermal offsets. (From Smith and Riseborough (2002); reproduced by permission of John Wiley and Sons Ltd.)

changes into semi-desert and, ultimately, into polar desert terrain (a 'frost-rubble' zone) in the high latitudes of Arctic Canada, north-east Greenland, Svalbard and Novaya Zemblya. In Antarctica, the relatively small ice-free areas are true polar deserts or rock-rubble surfaces that are kept free of snow and ice by sublimation from strong katabatic winds that flow outwards from the Antarctic ice sheet.

In the sub-arctic, two major ecological zones can be recognized. Near the tree-line is a zone of transition from tundra to forest consisting of either open woodland or forest–tundra. Here, the trees are stunted and deformed, often being less than 3–4 m high. This zone grades into the boreal forest, or *taiga*, an immense zone of almost continuous coniferous forest extending across both North America and Eurasia. In North America, the dominant species is spruce (*Picea glauca* and *Picea mariana*). In Siberia the dominant species are pine (*Pinus silvestris*) and tamarack (*Larix dahurica*). In northern Scandinavia, it is largely birch (*Betula tortuosa*). The southern boundary of the sub-arctic is less clearly defined than its northern boundary; typically, coniferous species begin to be replaced by others of either local or temperate distribution, such as oak, hemlock and beech, or by steppe, grassland and semi-arid woodland in more continental areas. These cool-climate ecosystems, which experience deep seasonal frost, represent the outer spatial extent of the periglacial realm.

Alpine periglacial environments are spatially less extensive and are dominated by both diurnal and seasonal temperature effects and by much higher solar radiation. In such environments, the timberline constitutes the boundary between the *alpine* and *sub-alpine*. The alpine environments are dominated by steep slopes, tundra (alpine) plants, rocky outcrops, and snow and ice. The *montane* environments of central Asia differ from alpine environments in that they are more extensive, far more arid, and consist of steppe grasslands and intervening desert-like uplands.

PERIGLACIAL LANDSCAPES

It is generally accepted that the geomorphic footprint of periglacial environments is not always achieved because most periglacial landscapes

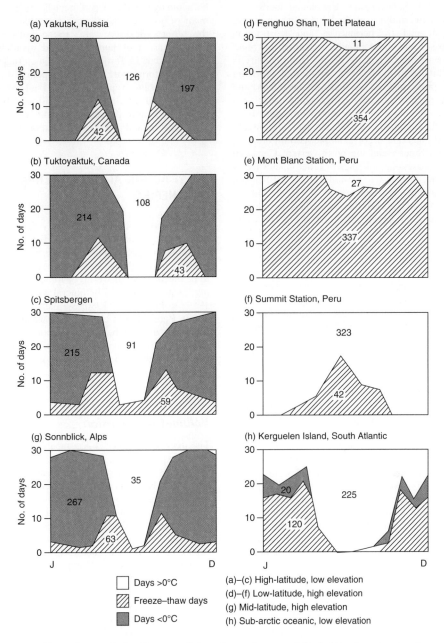

Figure 23.3 Freezing and thawing conditions in various periglacial environments of the world. (a) Yakutsk (lat. 62°N; 108 m asl), Siberia, Russia; (b) Tuktoyaktuk (lat. 69°N; 10 m asl), Mackenzie Delta, NWT, Canada; (c) Green Harbour (lat. 78°N, 7 m asl), Spitsbergen; (d) Fenghuo Shan (lat. 34°N, 4800 m asl), Qinghai-Xizang (Tibet) Plateau, China; (e) Mont Blanc Station, El Misti (lat.16°S, 4760 m asl), Peru; (f) Summit Station, El Misti (lat.16°S, 5850 m asl), Peru; (g) Sonnblick (lat. 47°N, 3060 m asl), Austria; (h) Kerguelen Island (lat. 49°S, sea level), southern Indian ocean. (From French, 2007)

possess some degree of inherited paraglacial or proglacial characteristics. Periglacial landscapes range between those in which the entire landscape is fashioned by permafrost and frost-action processes and those in which frost-action processes are subservient to others. This diversity is accentuated by the fact that certain rock types are more prone to frost weathering than others, and many regions currently experiencing periglacial conditions have only recently emerged from beneath continental ice sheets and are largely glacial landscapes. For example, certain areas of western Siberia and the north-western Canadian Arctic possess large bodies of relict glacier ice of Late-Pleistocene age, partially preserved beneath ablation till (e.g. Astakhov et al., 1996; Murton et al., 2005). It is clear these so-called 'periglacial' landscapes are largely relict and of glacial origin, and that periglacial processes are slowly modifying the landscape. Accordingly, the adjectival term 'periglaciation' is sometimes used to describe the degree or intensity to which periglacial conditions either dominate or affect a specific landscape or environment. The term 'paraglacial' is also relevant; this refers to the disequilibrium that occurs as one geomorphic environment moves from one equilibrium condition to another. In the case of the periglacial environment, the transition is usually to, or from, either a glacial or temperate environment.

Several English-language texts, published over the last 40 years, provide either overviews of periglacial landscapes or summarize the degree of periglaciation that has affected essentially non-periglacial landscapes (Washburn, 1979; Ballantyne and Harris, 1994; French, 2007). Therefore, there is no need for detail at this point.

The only periglacial landscapes that are probably in true geomorphic equilibrium are those that have protracted histories of cold non-glacial conditions. In the northern hemisphere, these include (1) parts of central Alaska and interior Yukon, (2) much of central and eastern Siberia, and (3) much of the montane and steppe environments on, and surrounding, the Qinghai-Xizang (Tibet) Plateau. In the southern hemisphere, some of the ice-free areas of Victoria Land are thought to have been ice-free for several million years.

In all these areas, it is clear that geological structure and lithology largely control the macro-scale periglacial landscape. For example, in areas of resistant igneous, metamorphic and sedimentary bedrock, the higher elevations consist of structurally controlled rock outcrops. Everywhere, the upland surfaces and upper valley-side slopes are covered by angular rock-rubble accumulations (variously termed 'mountain-top detritus', block-fields, or kurums). Bedrock is frequently disrupted by joint and fissure widening, the frost-jacking of

blocks, and by brecciation. Typically, uplands are bordered by low-angle, pediment-like, surfaces. In many ways, these landscapes resemble those of the hot deserts of the world. By contrast, areas of poorly lithified bedrock and unconsolidated Tertiary and Quaternary-age sediments form more undulating, poorly drained, lowland terrain. Typically, these landscapes are characterized by large-scale tundra polygons, thaw lakes and depressions, and widespread mass-wasting and patterned-ground phenomena.

FROST ACTION AND COLD-CLIMATE WEATHERING

It is the action of frost that clearly differentiates periglacial geomorphology from the other sub-disciplines of geomorphology. The weathering of bedrock in periglacial areas is generally assumed to be mechanical, the result of freezing and thawing of water within rock or mineral soil (Tsytovitch, 1959; Romanovskii, 1980; Williams and Smith, 1989). Rates of cold-climate rock weathering are usually assumed to be as great, if not greater, than those in warmer environments and are generally assumed to result from either volumetric expansion of ice or ice segregation.

Volumetric expansion

The freezing of water is accompanied by a volumetric expansion of approximately 9 per cent. In theory, this can generate pressures as high as 270 MPa inside cracks in a rock strong enough to withstand such pressures. While volumetric expansion was probably the mechanism that Łozinski envisaged when he talked of 'periglacial facies', the dominant role attributed to simple volumetric expansion is probably incorrect. This is because the conditions necessary for frost weathering by volumetric expansion are somewhat unusual. Not only must the rock be water-saturated but also freezing must occur rapidly from all side. On the other hand, there is no doubt that volumetric expansion of water within existing joints and other lines of weakness within bedrock outcrops can lead to bedrock heave and joint widening (e.g. Matsuoka, 2001a), and that near-surface frost wedging in fissile sedimentary rocks is a common occurrence.

A related mechanism is hydrofracturing, in which rock disintegration results from pressures generated by pore-water expulsion. For this to happen, the water-saturated rock must possess large interconnected pores, the expelled pore water is unable to drain away as quickly as it is expelled, and pore-water pressures must rise sufficiently to deform or 'hydrofracture' the rock. Rapid inward freezing is the ideal circumstance; for example, there have been occasional instances where Arctic field observations (e.g. Mackay, 1999, 132–133) indicate that boulders or rocks on the ground surface have burst or 'exploded' during periods of rapid temperature drop during mid-winter.

A third possible mechanism relates to the break-up of individual mineral particles by either the wedging effect of ice formed in micro-cracks or by the freezing of water within gas–liquid inclusions at cryogenic (i.e. sub-zero) temperatures. For example, certain Russian laboratory experiments (Konishchev and Rogov, 1993) indicate that, during repeated freeze–thaw cycles, quartz sand particles break down more readily than feldspar minerals and produce finer particles; this may also reflect the increasing brittleness of quartz at very low temperatures when compared to other minerals.

Ice segregation

There is increasing acceptance that the progressive growth of ice lenses as liquid water migrates to the freezing plane is the most likely cause of the widespread fracture of moist porous rocks (Walder and Hallet, 1986). This is because moisture migrates within freezing or frozen ground, the result of a temperature gradient-induced suction (dPw) that affects the unfrozen water held in capillaries and adsorbed on the surfaces of mineral particles. In theory, a temperature drop of 1°C induces a cryo-suction of 1.2 MPa (12 atmospheres). According to Williams and Smith (1989, 190):

$$dPw = \frac{dT\,l}{VT} \qquad (23.1)$$

where dT is the lowering of the freezing point, l is the latent heat of fusion, V is the specific volume of water and T is temperature.

The conditions needed for ice segregation are slow rates of freezing and sustained sub-zero temperatures. These are relatively common in most periglacial environments. In frost-susceptible sedimentary bedrock, long-continued ice segregation can lead to the brecciation of bedrock to a depth of several metres. Ice segregation and rock fracture have also been verified in laboratory experiments (Matsuoka, 2001b; Murton et al., 2006) that simulate natural uni- and bidirectional freezing; the most susceptible rock types appear to be fine-grained porous rocks such as chalk and shale.

Other mechanisms

A number of other weathering mechanisms are also thought to operate in periglacial environments. These are briefly summarized below.

First, insolation weathering refers to cracking in bedrock caused by temperature-induced volume changes such as expansion and contraction. For many years these thermally induced stresses were regarded as more appropriate for hot arid, rather than cold arid, regions. However, laboratory studies suggest the threshold value for thermal shock approximates to a rate of temperature change of 2°C/min and experimental studies, using cold-room facilities, have established that different minerals have varying coefficients of linear thermal expansion in the range +10°C to −10°C. These parameters certainly apply to a number of periglacial environments. For example, in parts of Antarctica, field studies document daily temperature ranges in excess of 40°C and rates of heating and cooling of 0.8°C/min and of 15°C to 20°C/h. These measurements suggest thermal stress, or fatigue, may be a viable rock-weathering process (Hall, 1999; Hall and Andre, 2001). Unfortunately, until further field, laboratory and experimental studies are undertaken, this important mechanism is still largely speculative.

Second, equally perplexing is the relationship between salt, present either in snow in areas adjacent to marine environments or the result of evaporation/sublimation in arid regions, and the granular disintegration of coarse-grained igneous rocks. This is suggested to be the partial explanation for the unusual cavernous-weathering (taffoni) present in the ice-free polar deserts of Antarctica (French and Guglielmin, 2000; Strini et al., 2008). Third, the efficacy of chemical weathering at low temperatures is unclear and the biological and biochemical weathering processes associated with rock-colonising organisms are poorly understood (Etienne, 2002; Dixon and Thorn, 2005; Guglielmin et al., 2005).

It would appear that cold-climate weathering is complex; frost action takes many forms. Certain processes act alone, others in combination, and some may be physico-chemical in nature. It is clear that the study of frost action and cold-climate weathering will continue to be a defining theme in periglacial geomorphology (Matsuoka and Murton, 2008).

FROZEN GROUND

All periglacial environments experience, to varying degrees, either seasonally frozen or perennially frozen ground. The latter, if it persists for more than 2 years, is termed *permafrost*

(Muller, 1943). In areas underlain by permafrost, the *active layer* refers to the near-surface layer of ground which thaws during summer. Where discontinuous permafrost is present, the active layer may be separated from underlying permafrost by either an unfrozen layer (tálik) or a residual thaw layer if permafrost is relict. If no permafrost is present, the active layer no longer exists and the near-surface layer is one of seasonal freezing and thawing.

The typical ground thermal regime of an area underlain by permafrost is illustrated in Figure 23.4. Thawing begins in early summer and the depth of thaw reaches a maximum in late summer at which point, freeze-back occurs. The freeze-back is slower than the thaw because the release of latent heat offsets the temperature drop. This gives rise to the so-called 'zero curtain' effect, in which near-isothermal conditions persist in the active layer for several weeks. Both thawing and freezing are one-sided processes, from the surface downwards. However, if permafrost were present, freezing is a two-sided process, occurring both downwards from the surface and upwards from the perennially frozen ground.

The Stefan equation is sometimes used to approximate the thickness of the active layer:

$$Z = \frac{\sqrt{2TKt}}{Qi} \qquad (23.2)$$

where Z is the thickness of the active layer (meters or centimeters), T is the ground surface temperature during thaw season (°C), K is the thermal conductivity of unfrozen soil (W/m K or kcal/m °C h), t is the duration of the thawing season (day, hour, second), and Qi is the volumetric latent heat of fusion (kJ/m³). The Stefan equation can also be used to calculate the depth of seasonal frost penetration. In this case, time t is the duration of the freezing season ($T < $ °C) and K represents the thermal conductivity of frozen soil.

The base of the active layer represents an unconformity between frozen and unfrozen earth material. Because the annual depth of thaw may vary from year to year depending upon the variability of summer climate, the concept of a *transient layer* recognizes the different periodicities at which near-surface permafrost cycles through 0°C (Shur et al., 2005). The active-layer permafrost interface is commonly ice-rich. This is because, in summer, as the active layer thaws, moisture migrates downwards and refreezes at the base while, during winter, unfrozen water migrates upwards in response to the colder temperatures at the surface. Thus, the active layer not only indicates the depth to which freeze–thaw action occurs but its base acts as a slip plane for gravity-induced near-surface mass movements such as solifluction, active-layer detachments, and for slope instability.

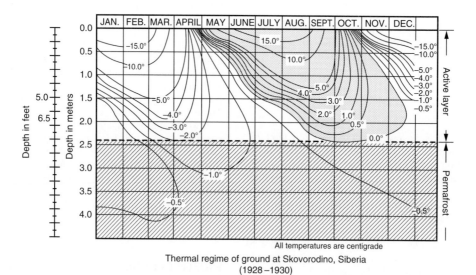

All temperatures are centigrade

Thermal regime of ground at Skovorodino, Siberia
(1928–1930)

Figure 23.4   Diagram illustrating the typical ground thermal regime of a permafrost area, Skovorodino, Siberia, 1928–1930 (From Muller, 1943)

PERIGLACIAL PROCESSES AND LANDFORMS

Like other sub-disciplines of geomorphology, the study of processes is central to modern periglacial geomorphology. Processes that are clearly unique to periglacial environments are related to perennially frozen ground. Others, not necessarily restricted to periglacial environments, are important either on account of their high magnitude or frequency or assume special characteristics in cold environments.

Permafrost-related processes

Direct permafrost-related processes of interest to periglacial geomorphology include aggradation of permafrost and the formation of ground-ice bodies, thermal-contraction-cracking of frozen ground, thawing of permafrost (thermokarst) and the creep of ice-rich permafrost. An understanding of ground ice is also relevant; in general, fine-grained materials are ice-rich and frost-susceptible whereas coarse-grained materials are ice-poor and generally regarded as non-frost-susceptible.

The most widespread and characteristic surface feature of permafrost terrain is the network of thermal-contraction cracks that divides the ground surface up into orthogonal or random-orthogonal patterns or polygons (Romanovskii, 1973, 1977). This phenomenon has attracted considerable attention in terms of the nature and controls over the cracking process. Likewise, the study of frost mounds, reflecting the growth of ground-ice bodies, constitutes another attractive, albeit specialised, aspect of periglacial geomorphology.

Notwithstanding important early Russian research, the study of permafrost-related processes in recent years has been dominated by the Canadian geographer J. Ross Mackay. Over a 50 year period starting in 1951, Mackay conducted quantitative field studies in the western Arctic of Canada (e.g. Mackay, 1963, 1997, 1998, 2000; Mackay and Burn, 2002, 2005). One can argue that his studies brought periglacial geomorphology perilously close to being a sub-branch of geocryology.

Although the nature and origin of ground ice is primarily the concern of geocryology (e.g. see Shumskii, 1959; Popov, 1962; Vtyurin, 1975), it is also highly relevant to periglacial geomorphology. This is because the thaw of ice-rich permafrost gives rise to a complex of processes and forms that are generally termed 'thermokarst'. A large Russian literature exists on thermokarst (e.g. Popov, 1956; Shur, 1977, 1988) and its study has become a major component of periglacial geomorphology in both North America and Russia.

In many instances, the desire to minimize human induced terrain disturbance has been the stimulus. While thermokarst is best developed in lowland terrain underlain by unconsolidated or weakly-lithified sediments (e.g. Hopkins, 1949; Romanovskii et al., 2000), even in mountainous terrain developed on igneous and high-grade metamorphic rocks, the thaw of ice within joints can lead to instability and enhanced rockfall activity. It follows that an overlap exists between periglacial geomorphology and the study of cryostratigraphy as practiced in Russia (e.g. Popov et al., 1985; Melnikov and Spesivtsev, 2000).

Frost-induced near-surface movements give rise to a variety of patterned ground phenomena. These intrigued the early explorers and scientists of the Polar Regions. As a consequence, a voluminous periglacial literature now describes these visually interesting 'decorations' to the ground surface. However, the relatively few experimental or process studies that have been undertaken suggest universal explanations (e.g. see Gleason et al., 1986; Mackay, 1980; Hallet et al., 1988; Werner and Hallet, 1993). Even in non-permafrost environments, frost action gives rise to the formation of small-scale sorted and non-sorted patterned ground (circles nets, stripes), frost heaving of bedrock and small hummocks or frost mounds (thufurs; earth hummocks).

Azonal processes

Mass wasting processes are not unique to cold environments but their study constitutes an important component of periglacial geomorphology. This is because these processes can assume distinctive characteristics in cold environments. For example, solifluction gives rise to the widespread heterogeneous surficial slope deposits, or diamicts, that mantle undulating and lower valley-side slopes in the same way that colluvium mantles slopes in temperate environments. The process is called gelifluction in permafrost terrain. More rapid movements such as rockfalls, debris flows and slush avalanches are common in the alpine periglacial environments. The many slope process studies that have been undertaken in nearly all the major periglacial environments are too numerous to cite individually. However, mention must be made of the early pioneering studies by A. Jahn (1960, 1961), A. Rapp (1960a,b) and A. L.Washburn (1967, 1969) that initiated this component of periglacial geomorphology.

The action of wind, although central to aeolian geomorphology, is also a legitimate topic within periglacial geomorphology (Seppala, 2004). This is especially the case for the tundra and polar desert environments where, typically, upland surfaces are blown clear of snow while lee slopes and lower valley-side slopes are sites of snow-bank accumulation. In the absence of vegetation, deflation and wind erosion assume local importance especially in Antarctica and alpine (montane) environments (Fristrup, 1953; Matsuoka et al., 1996). Therefore, a close relationship exists between aspects of periglacial geomorphology and aeolian (desert) geomorphology.

Certain aspects of cold-climate coastal geomorphology are also distinctive; for example, the presence of sea ice restricts the duration of open water conditions, there is ice-push activity on the beaches, and coastal bluffs developed in frozen, ice-rich unconsolidated sediments may experience fluvio-thermal undercutting and dramatic block collapse along the lines of ice wedges.

In spite of the apparent aridity of many periglacial environments, fluvial studies are another important aspect of periglacial geomorphology. This is because low temperatures and frozen ground minimize losses through evaporation and infiltration. Although a case can be made for treating ice-marginal and proglacial drainage systems as being reasonably distinct (Church, 1972), the fluvial dynamics of periglacial terrain are no different to other environments. Here again, there is an overlap between periglacial geomorphology and mainstream geomorphology.

DISCIPLINARY CONSIDERATIONS

There are two emergent issues within periglacial geomorphology. The more fundamental is the relationship between the older sub-discipline of periglacial geomorphology, with its roots in climatic geomorphology, and more broadly, geography, and the younger discipline of geocryology, here defined simply as permafrost science. The second is the relationship between periglacial geomorphology and Quaternary science, a relationship implicit in Łozinski's original definition of a 'periglacial zone'. These overlapping interests can be viewed as an illustration of Church's (2005) recent observation of the drift of geomorphology away from geography.

The growth of geocryology

The early development of geocryology occurred in Russia where, as early as 1939, an Institute of Permafrost was established at Yakutsk, central Siberia, by the Soviet Academy of Sciences. By 1940, the first edition of what was to become a standard text in the Soviet Union, *Obshchceye Merzlotovedeniya* (*General Permafrostology*) had been published (Sumgin et al., 1940) and by the mid 1960s the first of many undergraduate textbooks had emerged. By comparison, North American geocryology is of relatively recent origin with interest in permafrost becoming important only during and immediately after World War 2 (Muller, 1943). In China, geocryology developed even more recently but in a Soviet-style context (Academia Sinica, 1975). Expansion of permafrost studies into alpine regions is also relatively new and has centred largely upon rock glaciers and the creep and stability of frozen rock masses, especially in the mid-latitude mountains of Europe (Haeberli, 1985; Gruber and Haeberli, 2007) and central Asia (Gorbunov, 1988; Gorbunov and Tytkov, 1989).

For several reasons, the relations between geocryology and geomorphology are complex. First, for many years, permafrost studies in North America and the Soviet Union were conducted not only in relative isolation to each other but also in isolation from mainstream (geographical) geomorphology. Second, both Russian and Chinese geocryology adopt a holistic, all-encompassing approach whereas North American permafrost studies are usually characterized as being either 'science' or 'engineering'. Thus, there is no North American text that equals the breadth and depth presented by the most recent Russian and Chinese texts, *General Geocryolgy* (Yershov, 1990) and *Geocryology in China* (Zhou et al., 2000). Third, permafrost studies sit awkwardly between the disciplines of geology and geography. For example, in North America, periglacial geomorphology is taught usually in geography departments while permafrost is within geology, geophysics or earth sciences departments. Similar fractionation occurs in Russia.

It follows that, while geocryologists prefer to focus upon the thermal implications of the terrain and the presence of ice within the ground, periglacial geomorphologists emphasize the associated landforms and their growth and modification through time. Obviously, there is considerable overlap between the two. For example, in the case of J.R. Mackay, the recognition of anti-syngenetic wedges on hillslopes, pingo growth rates, the controls over thermal-contraction cracking, and the Illisarvik drained-lake experiment, to name just a few of his contributions, all illustrate the intimate connection between landscape evolution (geomorphology) and permafrost-related processes (geocryology).

The changing nature of geomorphology and Quaternary science

As Church (2005) laments, geographical geomorphology is slowly losing its discipline to geophysics and to programs located in refocused Earth science departments. This trend, ongoing since the 1960s, came first as the result of quantification, then of increasingly rigorous process studies founded on Newtonian principles, and finally, as the inevitable product of the all-embracing theory of plate tectonics which ultimately led geophysicists to be interested in topics previously held to be largely geomorphic. Church (2005) goes even further to suggest that the Newtonian approach of geographic geomorphologists is actually a shriven one in comparison to that undertaken by those entering geomorphology from other disciplines. Within periglacial geomorphology this approach produced a sub-discipline focused largely upon quantitative process studies. By the early 1990s process measurement had clearly demonstrated not only the serious shortcomings inherent to traditional versions of periglacial processes (and especially of weathering processes) but also that the azonal processes operating in cold environments, such as running water, wind, waves, and gravity-controlled mass movements, differ little if at all from similar processes in other climatic environments.

A pervasive theme in periglacial geomorphology is that a distinct sub-branch of geomorphology devoted specifically to the operation of 'mainstream' processes in cold climates is unnecessary. This is open to debate. For example, in a volume of collected papers recently published, French (2004) explicitly characterized modern periglacial geomorphology as a branch of geocryology, leaving the common azonal processes associated with running water and wind action to the companion volumes in the series. However, this operational definition is not without criticism because it neglects the important and unusual role played by snow in controlling ground temperatures and in influencing soil moisture conditions and slope runoff regimes. It also fails to consider the enhanced action of wind in high latitudes and the role of sea ice in Arctic and Antarctic coastal processes. An earlier answer to such a concern was to emphasize the role of snow as a unifying concept within periglacial geomorphology (e.g. see Thorn, 1988) and to subsequently attempt a definition of periglacial geomorphology in purely process terms (Thorn, 1992). Another recent trend recognizes the importance of the 'non-periglacial' contributions to cold region landscape development (Andre, 1999, 2003) serves to further question the legitimacy and scope of modern periglacial geomorphology.

There is also overlap between periglacial geomorphology and Quaternary science. The traditional Łozinski concept of Pleistocene periglacial geomorphology has been put in question by the rapid growth of Quaternary science in the last 50 years, due largely to the expansion and proliferation of sophisticated dating techniques. For example, traditional Pleistocene studies involving paleoenvironmental reconstruction based upon morphological features have been largely replaced by detailed Quaternary stratigraphic studies. This trend is best illustrated by T.L. Péwé's studies of the loess-like deposits of central Alaska (Péwé, 1955, 1975; Péwé et al., 1997) (Figure 23.5), central Yakutia (Péwé and Journaux, 1983) and the Qinghai-Xizang (Tibet) Plateau of China (Péwé et al., 1995). More generally, the problematic origin and palaeoenvironmental significance of loess has been a recurrent theme in Pleistocene studies in western Europe, southern Russia and central China (e.g. Soergel, 1936; Vandenberghe and Nugteren, 2001; Jary, 2008). Likewise, the silty 'yadoma' or 'Ice Complex' deposits of northern Siberia have also received much attention over the years (e.g. Tomirdiaro, 1980; Andreev et al., 2008; Schirrmeister et al., 2009).

It is clear that cryostratigraphic principles allow one to infer past permafrost history. It is now common to see cryostratigraphic techniques applied to problems within the more traditional fields of both geomorphology and Quaternary science (e.g. Burn, 1997; Shur and Jorgenson, 1998; Murton et al., 2004, 2005; French and Shur, 2010). Today, studies involving cold-climate paleo-environmental and paleo-geographical reconstruction rely upon a broader range of features and organisms than before. In fact, traditional Pleistocene periglacial geomorphology has been largely replaced by modern Quaternary science and cryostratigraphy.

THE SCOPE OF PERIGLACIAL GEOMORPHOLOGY

Periglacial geomorphology must be viewed as one of the cryospheric sciences. The cryosphere is that part of the Earth's crust, hydrosphere and atmosphere subject to cryotic temperatures (i.e. below 0°C) for at least part of the year. Obviously, the cryolithosphere (perennially and seasonally frozen ground) is central, and the cryohydrosphere (snow cover, glaciers, and river, lake and sea ice) slightly less central, to periglacial geomorphology.

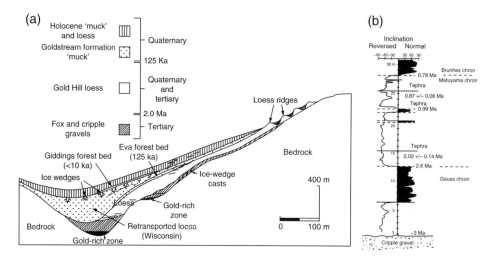

Figure 23.5   Schematic diagram summarizing the Quaternary stratigraphy of organic-rich loess-like silt deposits in central Alaska. (a) Valley cross section illustrating surficial materials and presence of ice wedges and ice-wedge casts. (b) Magneto-stratigraphy of the Gold Hill loess deposits, Fairbanks. (Modified from Péwé et al., 1997; Preece et al., 1999)

Figure 23.6 illustrates the various interactions between geomorphology, periglacial geomorphology, geocryology and the cryospheric sciences.

Several review papers summarize the nature of modern periglacial geomorphology (e.g. Pissart, 1990; Thorn, 1992; Barsch, 1993). Here, the scope of periglacial geomorphology is summarized under a number of headings (French and Thorn, 2006).

Understanding the nature of permafrost-related processes, ground ice and associated landforms

Processes that are clearly unique to periglacial environments relate to ground freezing. These include the growth of segregated ice and associated frost heaving, the formation of permafrost, the development of cryostructures and cryotextures in perennially frozen soil and/or rock, the occurrence of thermal-contraction cracking, and the growth of frost mounds of various sorts. The stratigraphic study of frozen earth material, especially the amount, distribution, and origin of the ice that is contained within it, constitutes the geocryological sub-discipline of cryolithology. Although not strictly geomorphological in nature, cryolithology is relevant to understanding permafrost history, to the interpretation of permafrost-related landforms, to describing periglacial sediments and to when

one undertakes inferences as to past climates or Pleistocene paleo-geographic reconstruction.

A number of frost-action processes operate in (1) the near-surface layer subject to seasonal thaw (the active layer), (2) the near-surface permafrost located above the depth of zero annual amplitude, and (3) the zone of seasonal freezing and thawing in non-permafrost regions. These processes include moisture migration within frozen ground and numerous processes associated with repeated freezing and thawing (e.g. soil churning or cryoturbation, frost creep, solifluction and gelifluction, the upfreezing of stones and particle-size sorting). Many of these processes give rise to distinct small-scale forms of patterned ground that complement the large-scale polygons that result from thermal contraction cracking.

Slopes that are frozen, or in the process of thawing, experience relatively unusual conditions associated with pore water expulsion and thaw consolidation. These processes promote rapid mass failures that are relatively distinct from other failures that might occur on slopes that evolve under non-frozen conditions in either temperate or warm climates.

It is by the study of all these processes, either in the field, the laboratory, or by modeling and simulation, that periglacial geomorphology has become a sub-branch of the broader discipline of geocryology. An understanding of all these processes constitutes the essential underpinning to modern periglacial geomorphology.

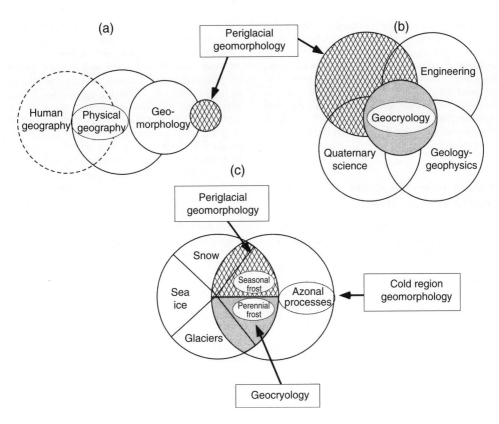

Figure 23.6 Schematic diagram illustrating the disciplinary interacts and overlaps of periglacial geomorphology. (a) Relations between physical geography, geomorphology and periglacial geomorphology. (b) Relations between periglacial geomorphology, geocryology and their interactions with Quaternary science and other natural sciences. (c) Periglacial geomorphology and its overlap with the cryospheric earth sciences

Understanding the azonal processes that operate in cold non-glacial environments

Some processes, not necessarily restricted to environments traditionally labelled as periglacial, are important on account of either their high magnitude or frequency, or their widespread occurrence. These often centre upon the seasonal freezing of soil and bedrock, including the disintegration of exposed rock by either mechanical (frost) wedging or the poorly understood complex of cryotic physical, biochemical or physico-chemical processes. Slightly better understood are the azonal processes associated with running water, wind, snow, and waves. These may assume distinctive characteristics under periglacial conditions but require neither an excessively cold climate nor a peripheral ice-marginal location for their effective operation. Nevertheless, their study complements many other aspects of geomorphology, and permits comparison and better comprehension of the process involved.

The study of all these processes constitute part of mainstream geomorphology, irrespective of the direction in which geomorphology is currently moving.

Understanding the ice-marginal (proglacial) environment and associated paraglacial transitions

While the majority of periglacial environments do not demand an ice-marginal location, those areas that are immediately adjacent to major ice sheets and/or glaciers experience cold-climate conditions. These fluctuate in time and space. In the early beginnings of periglacial geomorphology, the ice-marginal environments of Svalbard and Greenland were regarded as the typical high-latitude analogs for the periglacial conditions that probably existed in the mid-latitudes of Europe during the cold periods of the Pleistocene. This is clearly not the case, and Łozinski's so-called 'periglacial realm' finds no modern counterpart. Nevertheless, ice-marginal environments, by virtue of their proximity to glaciers and the cold-climate conditions that they experience, provide an instructive example of specific and specialized periglacial terrain undergoing paraglacial transition.

Understanding the alpine (montane) environment

Like the proglacial environment, the alpine environments that occur above timberline were initially regarded as typical 'periglacial' environments. However, they also represent site-specific and specialized terrains in which steep slopes and gravity-controlled processes play a critical role together with cold-climate processes aided by wind and snow. As such, they are also instructive for periglacial geomorphology.

Many recent studies focus upon the occurrence of mountain permafrost, the controls over its distribution, the creep of ice-rich debris (rock glaciers), and the instability of bedrock as temperatures warm (e.g. Haeberli, 1992; Harris and Haeberli, 2003). These examples of alpine geomorphology are equally integral components of geocryology.

Undertaking Pleistocene cold-climate paleo-environmental reconstructions

The growth of cryostratigraphy, together with the increasing sophistication of Quaternary science, means that traditional Pleistocene periglacial paleo-geographic studies have become obsolete. Morphological and stratigraphical evidence must now be interpreted within the context of a more realistic appreciation of permafrost and the climatic controls over cold-climate processes, augmented by isotopic and other dating techniques. In other words, advances in cryostratigraphy and Quaternary science have encroached upon, and now largely outgrown, traditional Pleistocene periglacial paleo-climatic studies.

Undertaking environmental and geotechnical studies associated with frozen ground, ground freezing and global climate change

All disciplines thrive best when they have societal relevance. Periglacial geomorphology is no exception. Here, the growth opportunities are

many. For example, the most obvious applications lie in the development and utilization of the natural resources of the northern regions of North America and Eurasia. This will require sound geotechnical engineering and infrastructure design in both urban and remote settings. Roads, buildings, bridges, pipelines and mining operations are just some of the examples where an understanding of periglacial geomorphology is essential. Man-induced thermokarst and other disturbances need to be minimized through management and regulatory practice. The fact that global climate change may first become apparent at high latitudes, and that permafrost can be regarded as a temperature archive, both as regards past and future temperatures, has promoted international monitoring studies such as CALM (Brown et al., 2000) and PACE (Harris et al., 2009). In alpine regions, the increased utilization of upper slopes for recreation activities, and the potential for slope instability consequent upon permafrost thaw, has promoted the study of mountain permafrost. Likewise, the thinning of sea ice and the potential expansion of arctic shipping lanes has prompted international attention towards cold-climate coasts, as illustrated by the Arctic Coastal Dynamics (ACD) initiative of IASC (Rachold et al., 2003). There are many other examples where periglacial geomorphology can play a useful societal role. It must be anticipated that the applied aspects of periglacial geomorphology will become more dominant in the years to come.

THE FUTURE

A major aim of geomorphology is to create models of landscape evolution. Such models embody assumptions as to the processes involved, their speed of operation and their associated rates of erosion, transport and deposition, and the manner in which surface morphology changes through time. While various cyclic and non-cyclic (equilibrium) models characterize the history of geomorphology, only the descriptive and non-quantitative cyclic models of S.G. Boch and I.I. Krashnov (1943) on cryoplanation, D.M. Hopkins (1949) on thaw lakes, L.C. Peltier (1950) on landscape evolution, and P.A. Soloviev (1973) on alas thermokarst relief explicitly apply to cold-climate terrain. The reality is that landscape modification under cold non-glacial conditions remains largely neglected. Notable exceptions are the application of thaw-consolidation theory to the creep and deformation of permafrost (McRoberts and Morgenstern, 1974), the evolution and stability of thawing slopes (Harris and Lewkowicz, 2000), and recent laboratory studies that simulate solifluction process and rock disintegration (Harris et al.,

1993, 2000; Murton et al., 2006). An increased understanding of the geomorphic processes occurring in the exceptionally cold and arid part of ice-free Antarctica may also lead to insights into the nature of planetary permafrost and the geomorphology of Mars.

One must reluctantly conclude that much geographically based periglacial geomorphology currently lacks a rigorous theoretical base. It remains characterized by flawed thinking that reflects its qualitative and climatic heritage. Unless a research agenda that rectifies this weakness is put in place, so-called 'periglacial geomorphology', as presently configured, runs the risk of falling by the wayside. Currently, more progress is being made in geocryology, Quaternary science and geomorphology, even as the latter is losing its geographical base to 'reductionist' science, than in periglacial geomorphology. This is because those disciplines are seen to be more 'scientific' and to have more pragmatic and pressing research agendas. In short, periglacial geomorphology needs to sharpen its scientific rigor before it is either consumed or by-passed, as the case may be, by geocryology, Quaternary science or the other sub-disciplines of geomorphology.

Inevitably, geocryologists have a vested interest in periglacial geomorphology; equally, periglacial geomorphologists are committed to understanding the role of permafrost and frozen ground in shaping the terrestrial landscape. Quaternary scientists also have a legitimate interest in periglacial geomorphology because the fluctuation environments of the Quaternary included significant periods that experienced cold-climate conditions. Therefore, the challenge for periglacial geomorphology is to maintain a bridging position between the changing nature of geomorphology, the emerging discipline of geocryology, and the increasing sophistication of Quaternary science.

REFERENCES

Academia Sinica (1975) *Permafrost*. Research Institute of Glaciology, Cryopedology and Desert Research, Lanzhou, China (in Chinese). English translation, National Research Council of Canada, Technical Translation 2006, 1981, 224 pp.

André, M.-F. (1999) La livrée périglaciaire des paysages polaires: l'arbre qui cache la forêt? *Géomorphologie; Relief, Processus, Environnement*, 1999, 3, 231–52.

André, M.-F. (2003) Do periglacial landscapes evolve under periglacial conditions? *Geomorphology*, 52, 149–64.

Andreev, A.A., Grosse, G., Schirrmeister, L., Kuznetsova, T.V., Kuzmina S.A., Bobrov, A.A., Tarasov, P.E., Novenko, E.Y., Meyer, H.Y. Derevyagin, A.Y., Kienast F., Bryantseva, A. And Kunitsky, V.V. (2008) Weichselian and Holocene palaeoenvironmental history of the Bol'shoy Lyakhovsky Island, New Siberian Archipelago, Arctic Siberia. *Boreas*, 38, 72–110.

Astakhov, V.I., Kaplyanskaya, F.A. and Tarnogradsky, V.D. (1996) Pleistocene permafrost of West Siberia as a deformable glacier bed. *Permafrost and Periglacial Processes*, 7, 165–91.

Ballantyne, C.K. and Harris, C. (1994) *The Periglaciation of Great Britain*. Cambridge University Press, Cambridge. p. 330.

Barsch, D. (1993) Periglacial geomorphology in the 21st century. *Geomorphology*, 7, 141–63.

Boch, S.G. and Krasnov, I.I. (1943) O Nagornykh terraskh i drevnikh poverkhnostyakh vyravnivaniya na Urale isvyazannykh s nimi problemakh. *Vsesoyuznogo Geograficheskogo obshchestva, Izvestiya*, 75, 14–25. (English translation by A. Gladunova, 1994, On altiplanation terraces and ancient surfaces of leveling in the Urals and associated problems, in D.J.A. Evans (ed.), *Cold Climate Landforms*. J. Wiley and Sons, Chichester. pp. 177–86).

Brown, J., Hinkel, K.M. and Nelson, F.E. (2000) The Circumpolar Active Layer Monitoring (CALM) program. Research design and initial results. *Polar Geography*, 24, 165–258.

Bryan, K. (1928) Glacial climate in non-glaciated regions. *American Journal of Science*, 16, 162–4.

Bryan, K. (1946) Cryopedology – the study of frozen ground and intensive frost-action with suggestions on nomenclature. *American Journal of Science*, 244, 622–42.

Bryan, K. (1949) The geologic implications of cryopedology. *Journal of Geology*, 57, 101–4.

Büdel, J. (1951) Die klimazonen des Eiszeitalters. *Eiszeitalter und Gegenwart*, 1, 16–26. (English translation in 1959: The climate of the Ice Age. *International Geology Review*, 1, 72–9).

Büdel, J. (1953) Die 'periglaziale' morphologischen Wirkungen des Eiszeitklimas auf der Ganzen Erde. *Erdkunde*, 7, 249–66.

Büdel, J. (1963) Klimatische geomorphologie. *Geographische Rundschau*, 15, 269–85. (English translation by Perry, J.M. and Derbyshire, E, 1973, Climato-genetic geomorphology, in E. Derbyshire (ed.) *Climatic Geomorphology*, Macmillan. pp. 202–27).

Büdel, J. (1977) *Klima-geomorphologie*. Berlin and Stuttgart, Gebruder Borntraeger, p. 304. English translation by Fischer, L. and Busche, D, 1982, *Climatic Geomorphology*, Princeton University Press, New Jersey. p. 443.

Burn, C.R. (1997) Cryostratigraphy, paleogeography, and climate change during the early Holocene warm interval, western Arctic coast, Canada. *Canadian Journal of Earth Sciences*, 34, 912–25.

Cailleux, A. and Taylor, G. (1954) *Cryopédologie, études des sols gelées: Expéditions Polaires Françaises, Missions Paul-Emile Victor IV*. Hermann & Cie, Actualités Scientifiques et Industrielles 1203, Paris. p. 218.

Church, M. (1972) Baffin Island sandurs: A study in Arctic fluvial processes. *Geological Survey of Canada Bulletin* 216, p. 208.

Church, M. (2005) Continental drift. *Earth Surface Processes and Landforms*, 30, 129–30.

Denny, S. (1936) Periglacial phenomena in southern Connecticut. *American Journal of Science,* 32, 322–42.

Dixon, J.C. and Thorn, C.E. (2005) Chemical weathering and landscape development in mid-latitude alpine environments. *Geomorphology,* 67, 85–106.

Edelman, H., Florshutz, F. and Jeswiet, J. (1936) Uber spatpleistozäne und fruhholozäne kryoturbate Ablagerungen in den ostlichen Niederlanden. *Geologische Mijnbouwkundig Genootschap voor Nederland en Kolonien,* Verhandlingen Geologische Series 11, 301–60.

Etienne, S. (2002) The role of biological weathering in periglacial areas: a study of weathering rind in south Iceland. *Geomorphology,* 47, 75–86.

French, H.M. (2000) Does Lozinski's periglacial realm exist today? A discussion relevant to modern usage of the term 'periglacial'. *Permafrost and Periglacial Processes,* 11, 35–42.

French, H.M. (2003) The development of periglacial geomorphology; 1 – up to 1965. *Permafrost and Periglacial Processes,* 14, 29–60.

French, H.M. (2004) Introduction, in H.M. French, (ed.) *Volume V, Periglacial Geomorphology. Critical Concepts in Geography.* Evans, D.J. A General Editor, Routledge, London and New York. pp. 1–40.

French, H.M. (2007) *The Periglacial Environment.* 3rd. edn. John Wiley and Sons, Chichester. p. 458.

French, H.M. and Guglielmin, M. (2000) Cryogenic weathering of granite, Northern Victoria Land, Antarctica. *Permafrost and Periglacial Processes,* 11, 305–14.

French. H.M. and Thorn, C.E. (2006) The changing nature of periglacial geomorphology. *Géomorphologie: Relief, Processus, Environnement,* 3, 165–74.

French, H.M. and Shur,Y. (2010) The principles of cryostratigraphy. *Earth Science Reviews,* 101, 190–206.

Fristrup, B. (1953) Wind erosion within the Arctic desert. *Geografisk Tidsskrift,* 52, 51–65.

Gleason, K.J., Krantz, W.B., Caine, N., George, J.H. and Gunn, R.D. (1986) Geometric aspects of sorted patterned ground in recurrently frozen soil. *Science,* 232, 216–20.

Gorbunov, A.P. (1988) *Rock Glaciers.* (in Russian). Russian Academy of Sciences, Siberian Division, Novosibirsk. p. 108.

Gorbunov, A.P. and Tytkov, C.H. (1989) *Rock Glaciers in the Mid-Asian Mountains.* (in Russian). Russian Academy of Sciences, Siberian Division, Yakutsk. p. 164.

Gruber, S. and Haeberli, W. (2007) Permafrost in steep bedrock slopes and its temperature-related destabilization following climate change. *Journal of Geophysical Research,* 112, F02S18, doi:10.1029/2006JF000547.

Guglielmin, M., Cannone, N., Strini. A. and Lewkowicz, A.G. (2005) Biotic and abiotic processes on granite weathering landforms in a cryotic environment, Northern Victoria Land, Antarctica. *Permafrost and Periglacial Processes,* 16, 69–85.

Haeberli, W. (1985) Creep of mountain permafrost: internal structure and flow of alpine rock glaciers. *Mittgeilungen der Versuchanstalt fur Wassenbau und Glaziologie,* 77, 142.

Haeberli, W. (1992) Construction, environmental problems and natural hazards in periglacial mountain belts. *Permafrost and Periglacial Processes,* 3, 111–24.

Hall, K. (1999) The role of thermal stress fatigue in the breakdown of rock in cold regions. *Geomorphology,* 31, 47–63.

Hall, K. and Andre, M-F. (2001) New insights into rock weathering from high frequency rock temperature data: an Antarctic study of weathering by thermal stress. *Geomorphology,* 41, 23–35.

Hallet, B., Anderson, S.P., Stubbs, C.W. and Gregory, E.C. (1988) Surface soil displacement in sorted circles, Western Spitzbergen, in K. Senneset, (ed.) *Permafrost, Proceedings of the Fifth International Conference on Permafrost, 2–5 August, 1988.* volume 1. Tapir, Trondheim. pp. 770–5.

Harris, C. and Lewkowicz, A.G. (2000) An analysis of the stability of thawing slopes, Ellesmere Island, Nunavut, Canada. *Canadian Geotechnical Journal,* 37, 449–62.

Harris, C. and Haeberli, W. (2003) Warming permafrost in the mountains of Europe. *World Meteorological Organization,* Bulletin, 52, 1–5.

Harris, C., Gallop, M. and Coutard, J.-P. (1993) Physical modelling of gelifluction and frost creep: some results of a large scale laboratory experiment. *Earth Surface Processes and Landforms,* 18, 383–98.

Harris, C., Rea, B.R. and Davies, M.R.C. (2000) Geotechnical centrifuge modelling of geliflcution processes: validation of a new approach to periglacial slope studies. *Annals of Glaciology,* 31, 263–8.

Harris, C. et al. (2009) Permafrost and climate in Europe: monitoring and modelling thermal, geomorphological and geotechnical responses. *Earth Science Reviews,* 92, 117–171.

Hogbom, B. (1914) Uber die geologische Bedeutung des Frostes. *Uppsala Universiteit, Geological Institute Bulletin,* 12, 257–389.

Hopkins, D.M. (1949) Thaw lakes and thaw sinks in the Imuruk Lake area, Seward Peninsula, Alaska. *Journal of Geology,* 57, 119–30.

Jahn, A. (1960) Some remarks on the evolution of slopes on Spitsbergen. *Zeitschrift fur geomoprhologie, Supplementband,* 1, 49–58.

Jahn, A. (1961) Quantitative analysis of some periglacial processes on Spitsbergen. *Nauka O Zeimi, II,* seria B, 5, 3–34.

Jary, Z. (2008) Periglacial markers within the Late Pleistocene loess-paleosol sequences in Poland and Western Ukraine. *Quaternary International,* doi:10.10.1016/j.quaint. 2008.01.008.

Jorré, G. (1933) Problèmes des 'terrasse goletz' sibériennes. *Revue de Géographie Alpine,* 21, 347–71.

Konischchev, V.N. and Rogov, V.V. (1993) Investigations of cryogenic weathering in Europe and Northern Asia. *Permafrost and Periglacial Processes,* 4, 49–64.

Losche, H. (1930) Lassen sich die diluvialen breitenkreise aus klimabedingten vorzeitformen rekonstruieren? *Archiv der Deutschen Seewarte,* 48, 39.

Mackay, J.R. (1963) The Mackenzie Delta area. *Geographical Branch Memoir* no 8, Ottawa, p. 202.

Mackay, J.R. (1980) The origin of hummocks, western Arctic coast. *Canadian Journal of Earth Sciences,* 17, 996–1006.

Mackay, J.R. (1997) A full-scale field experiment (1978–1995) on the growth of permafrost by means of lake drainage, western Arctic coast: a discussion of the method and some results. *Canadian Journal of Earth Sciences*, 34, 17–33.

Mackay, J.R. (1998) Pingo growth and collapse, Tuktoyaktuk Peninsula area, western Arctic coast, Canada: a long-term field study. *Géographie physique et Quaternaire*, 52, 271–323.

Mackay, J.R. (1999) Cold climate shattering (1974–1993) of 200 glacial erratic on the exposed bottom of a recently drained Arctic lake, western Arctic coast, Canada. *Permafrost and Periglacial Processes*, 10, 125–36.

Mackay, J.R. (2000) Thermally-induced movements in ice-wedge polygons, western Arctic coast. *Géographie physique et Quaternaire*, 54, 41–68.

Mackay, J.R. and Burn, C.R. (2002) The first 20 years (1978–1979 to 1998–1999) of active-layer development, Illisarvik experimental drained lake-site, western Arctic coast, Canada. *Canadian Journal of Earth Sciences*, 39, 1657–74.

Mackay, J.R. and Burn, C.R. (2005) A long-term study (1951–2003) of ventifacts formed by katabatic winds at Paulatuk, western Arctic coast, Canada. *Canadian Journal of Earth Sciences*, 42, 1615–35.

Marr, J.E. (1919) The Pleistocene deposits around Cambridge. *Quarterly Journal of the Geological Society*, 75, 204–44.

Matsuoka, N. (2001a) Direct observation of frost wedging in alpine bedrock. *Earth Surface Processes and Landforms*, 26, 601–14.

Matsuoka, N. (2001b) Microgelivation versus macrogelivation: towards bridging the gap between laboratory and field frost weathering. *Permafrost and Periglacial Processes* 12, 299–313.

Matsuoka, N. and Murton, J.B. (2008) Frost weathering: recent advances and future directions. *Permafrost and Periglacial Processes*, 19, 195–210.

Matsuoka, N., Morikawa, K. and Hirakawa, K. (1996) Field experiments on physical weathering and wind erosion in an Antarctic cold desert. *Earth Surface Processes and Landforms*, 21, 687–99.

McRoberts, E.C. and Morgenstern N.R. (1974) The stability of thawing slopes. *Canadian Geotechnical Journal*, 11, 447–69.

Meinardus, W. (1912) Beobachtungen uber Detritussortierung und Strukturboden uaf Spitzbergen. *Zeitschrift de Gesellschaft für Erdkunde zu Berlin*. pp. 250–59.

Mel'nikov, V.P. and Spesivtsev, V.I. (2000) *Cryogenic formations in the Earth's lithosphere*. Novosibirsk Scientific Publishing Center UIGGM, Siberian Branch, Russian Academy of Sciences Publishing House. pp. 1–172 in Russian; pp. 173–343 in English.

Mortensen, H. (1932) Blockmere und felsburgen in den deutschen Mittelgebirge. *Zeitschrift der Gesellschaft für Erdkunde zu Berlin*. pp. 279–87.

Muller, S.W. (1943) *Permafrost or Permanently Frozen Ground and Related Engineering Problems*. Special Report, Strategic Engineering Study, Intelligence Branch Chief of Engineers, no 62, p.136. Second printing, 1945, p. 230. Reprinted in 1947, J.W. Edwards, Ann Arbor, Michigan. p. 231.

Murton, J.B., Peterson, R. and Ozouf, J-C. (2006) Bedrock fracture by ice segregation in cold regions. *Science*, 314, 1127–9.

Murton, J.B., Waller, R.I., Hart, J.K., Whiteman, C.A., Pollard, W.H. and Clark, I.D. (2004) Stratigraphy and glaciotectonic structures of permafrost deformed beneath the northwest margin of the Laurentide ice sheet, Tuktoyaktuk Coastlands, Canada. *Journal of Glaciology*, 50, 399–412.

Murton, J.B., Whiteman, C.A., Waller, R.I., Pollard, W.H., Clark, I.D. and Dallimore, S.R. (2005) Basal ice facies and supraglacial melt-out till of the Laurentide ice sheet, Tuktoyaktuk Coastlands, western Arctic Canada. *Quaternary Science Reviews*, 24, 681–708.

Paterson, T.T. (1940) The effects of frost action and solifluxion around Baffin Bay and in the Cambridge District. *Quarterly Journal of the Geological Society*, London, 96, 99–130.

Peltier, L.C. (1950) The geographical cycle in periglacial regions as it is related to climatic geomorphology. *Annals, Association of American Geographers*, 40, 214–36.

Péwé, T.L. (1955) Origin of the upland silt near Fairbanks, Alaska. *Geological Society of America Bulletin*, 66, 699–724.

Péwé, T.L. (1975) Quaternary geology of Alaska. *United States Geological Survey Professional Paper* 835, p.145.

Péwé, T.L. and Journaux, A. (1983) Origin and character of loesslike silt in unglaciated south-central Yakutia, Siberia, U.S.S.R. *United States Geological Survey Professional Paper* 1262, p. 46.

Péwé, T.L., Tungsheng, L., Slatt, R.M. and Bingyuan, L. (1995) Origin and character of loess like silt in the southern Qinghai-Xizang (Tibet) Plateau, China. *United States Geological Survey Professional Paper* 1549, p. 55.

Péwé, T.L., Berger, G.W., Westgate, J.A., Brown, P.M. and Leavitt, S.W. (1997) Eva Interglaciation Forest Bed, unglaciated east-central Alaska: global warming 125,000 years ago. *Geological Society of America, Special Paper* 319, p. 54.

Pissart, A. (1990) Advances in periglacial geomorphology. *Zeitschrift fur Geomorphologie, Supplementband*, 79, 119–31.

Popov, A. I. (1956) Le thermokarst. *Biuletyn Peryglacjalny*, 4, 319–30.

Popov, A.I. (1962) *The Origin and Development of Massive Fossil Ice, Issue II*. Moscow, USSR: V.A. Obruchev Institute of Permafrost Studies, Academy of Sciences. National Research Council of Canada, Ottawa, Technical translation No 1006, 1962, pp. 5–24.

Popov, A.I., Rozenbaum G.E. and Tumel, N.V. (1985) *Cryolithology*. (in Russian). Moscow State University Press, Moscow. p. 238.

Porsild, A.E. (1938) Earth mounds in unglaciated arctic northwestern America. *Geographical Review*, 28, 46–58.

Porsild, A.E. (1955) The vascular plants of the Western Canadian Arctic archipelago. *Bulletin National Museum of Canada*, 135, 226.

Porsild, A.E. (1957) Illustrated flora of the Canadian Arctic Archipelago. *Bulletin National Museum of Canada* 146, 209.

Poser, H. (1932) Einege untersuchungen zur morphologie Ostgrönlands. *Meddellelser om Gronland*, 94, 55.

Poser, H. (1948) Böden und klimaverhalnisse in Mittel and Westeuropa der Würmeiszeit. *Erdkunde*, 2, 53–68.

(English translation by Wright, H.E., Schmautz, J. and Pokopovitch, N. 1957. Soil and climate relations in Central and Western Europe during the Wurm glaciation, in D.J.A. Evans, (ed.) 1994. *Cold Climate Landforms*. John Wiley and Sons, Chichester. pp. 3–22).

Preece, S.J., Westgate, J.A., Stemper, B.A. and Péwé, T.L. (1999) Tephrochronology of late Cenozoic loess at Fairbanks, central Alaska. *Geological Society of America, Bulletin*, 111, 71–80.

Rachold, V., Brown, J., Solomon, S. and Sollid, J-L. (2003) Arctic Coastal Dynamics; Report of the 3rd International Workshop, University of Oslo, Norway, 2-5 December 2002. *Beriche zur Polar- und Meeresforschung*, 443, 127.

Rapp, A. (1960a) Recent development of mountain slopes in Karkevagge and surroundings, northern Sweden. *Geografiska Annaler*, 42, 71–200.

Rapp, A. (1960b) Talus slopes and mountain walls at Tempelfjorden, Spitsbergen. *Norsk Polarinstitutt Skrifter*, 119, 96.

Raup, H.M. (1965) The structure and development of turf hummocks in the Mesters Vig District, Northeast Greenland. *Meddelelser om Grønland*, 166, 3, 112.

Romanovskii, N.N. (1973) Regularities in the formation of frost-fissures and development of frost-fissure polygons. *Biuletyn Peryglacjalny*, 23, 237–77.

Romaonovskii, N.N. (1977) *Formation of Polygonal Wedge Structures.* (in Russian). Academia Nauka SSSR, Novosibirsk, p. 212.

Romanovskii, N.N. (1980) *The Frozen Earth*. (in Russian). Moscow University Press, Moscow, 188pp.

Romanovskii, N.N., Hubberten, H.-W., Gavrilov, A.V., Tumskoy, V.E., Tipenko, G.S. and Grigoriev, M.N. (2000) Thermokarst and land-ocean interactions, Laptev Sea region. *Permafrost and Periglacial Processes*, 11, 137–52.

Schirrmeister, L. et al. (2009) Periglacial landscape evolution and environmental changes of Arctic lowland areas for the last 60,000 years (western Laptev Sea coast, Cape Mamontov Klyk). *Polar Research*, 27, 249–72.

Seppala, M. (2004) *Wind as a Geomorphic Agent in Cold Climates*. Cambridge University Press, Cambridge, p. 368.

Shumskii, P.A. (1959) Ground (subsurface) ice, in *Principles of Geocryology*, Part 1, *General Geocryology* (in Russian). Academy of Sciences of the USSR, Moscow, Chapter IX, pp. 274–327. English translation, National Research Council of Canada, technical translation 1130, p. 118.

Shur, Y.L. (1977) *Thermo-Physical Principles Behind the Thermokarst Process*. (in Russian). Nedra, Moscow. p. 80.

Shur, Y.L. (1988) *The Upper Horizon of Permafrost and Thermokarst*. (in Russian). Akademia Moscow, Nauka. p. 210.

Shur, Y. and Jorgenson, T. (1998) Cryostructure development on the floodplain of the Colville River Delta, Northern Alaska, in *Permafrost*. Seventh International Conference, June 23–27, 1998, Proceedings, Yellowknife, Canada, Centre d'études Nordiques, Université Laval, Collection Nordicana, 57, 993–9.

Shur, Y., Hinkel, K.M. and Nelson, F.E. (2005) The transient layer: implications for geocryology and climate-change science. *Permafrost and Periglacial Processes*, 16, 5–18.

Smith, M.W. and Riseborough, D.W. (2002) Climate and the limits of permafrost: a zonal analysis. *Permafrost and Periglacial Processes*, 13, 1–15.

Soergel, W. (1936) Diluviale Eiskeile. *Zeitschrift Deutsche Geologische Gessellschaft*, 88, 223–47.

Soloviev, P.A. (1973) Thermokarst phenomena and landforms due to frost heaving in Central Yakutia. *Biuletyn Peryglacjalny*, 23, 135–55.

Sorensen, T. (1935) Bödenformen und pflanzendecke in Nordostgrönland. *Meddelelser om Gronland*, 93, p. 69. (English translation by C. Halstead, 1993, Ground form and plant cover in North East Greenland, in Evans D.J.A. (ed.), *Cold Climate Landforms*, 1994, Chichester, John Wiley and Sons, Chichester. pp. 135– 75).

Strini, A., Guglielmin, M. and Hall, K. (2008) Tafoni development in a cryotic environment: an example from Northern Victoria Land, Antarctica. *Earth Surface Processes and Landforms*, 33, 1502–19.

Sumgin, M.I., Kachurin, S.P., Tolstikhin, N.I. and Tumel, V.F. (1940) *Obshcheye merzlotovedeniya (General permafrostology)* (in Russian). Akademia Nauka SSSR, Moscow - Leningrad, p. 340.

Thorn, C.E. (1988) Nivation: a geomorphic chimera, in M.J. Clark, (ed.) *Advances in Periglacial Geomorphology*. Wiley, Chichester. pp. 3–31.

Thorn, C.E. (1992) Periglacial geomorphology. What? Where? When?, in J.C. Dixon, and A.D. Abrahams, (eds.) *Periglacial Geomorphology*. John Wiley and Sons, Chichester. pp. 1–30.

Tomirdiaro, S.V. (1980) *Loess-Glacial Formation of East Siberia During Late Pleistocene and Holocene*. (in Russian). Moscow, Nauka, p. 184

Tricart, J. (1963) *Géomorphologie des Régions Froides*. Presses Universitaires de France, Paris. p. 389. English translation by E. Watson, 1970. *The Geomorphology of Cold Regions*. Macmillan, London. p. 320.

Tricart, J. and Cailleux, A. (1967) *Le Modélé des Régions Périglaciaires*. Traité de Géomorphologie, 2, Paris, SEDES. p. 512.

Troll, C. (1948) Der subnival oder periglaziale der denudation. *Erdkunde*, 2, 1–21. (English translation by Wright, H.E. and Thomson, C. 1956. The subnival or periglacial cycle of denudation, in D.J.A. Evans, (ed.), *Cold Climate Landforms*. 1994. Chichester, John Wiley and Sons, Chichester. pp. 23–46).

Tsytovitch, N.A. (1959) Physical phenomena and processes in freezing, frozen and thawing soils, in *General Geocryology*. Moscow, USSR: V.A. Obruchev Institute of Permafrost Studies, Academy of Sciences, Part 1, Chapter V, pp. 108–152. National Research Council of Canada, Ottawa, Technical Translation 1164. 1964. 109 pp.

Tyrtikov, A.P. (1959) Perennially frozen ground and vegetation, in *General Geocryology*. Moscow, USSR: V.A. Obruchev Institute of Permafrost Studies, Academy of Sciences, Part 1, Chapter XII, pp. 399–421. National Research Council of Canada, Ottawa, Technical Translation 1163. 1964. p. 34.

Vandenberghe, J and Nugteren, G. (2001) Rapid climate changes recorded in loess successions. *Global and Planetary Change*, 28, 1–9.

von Łozinski, W. (1909) Uber die mechanische Verwitterung der Sandsteine im gemässigten klima. *Bulletin International de l'Academie des Sciences de Cracovie, Classe des Sciences Mathematiques et Naturelles*, 1, 1–25. (English translation by T. Mrozek, 1992. On the mechanical weathering of sandstones in temperate climates, in D.J.A. Evans, (ed.) *Cold Climate Landforms*. John Wiley and Sons. pp. 119–34).

von Łozinski, W. (1912) Die periglaziale fazies der mechanischen Verwitterung. *Comptes Rendus, XI Congrès Internationale Geologie*, Stockholm. 1910, 1-39-1053.

Vtyurin, B.I. (1975) *Underground Ice in the USSR*. (in Russian). Nauka, Moscow, p. 212.

Walder, J.S. and Hallet, B. (1986) The physical basis of frost weathering: toward a more fundamental and unified perspective. *Arctic and Alpine Research*, 18, 27–32.

Washburn, A.L. (1967) Instrumental observation of mass wasting in the Mesters Vig District, Northeast Greenland. *Meddelelser om Gronland*, 166, p. 318.

Washburn, A.L. (1969) Weathering, frost action and patterned ground in the Mesters Vig District, Northeast Greenland. *Meddelelser om Gronland*, 176, p. 3030.

Washburn, A.L. (1979) *Geocryology: A Survey of Periglacial Processes and Environments*. Edward Arnold, London. p. 406.

Werner, B.T. and Hallet, B. (1993) Numerical simulation of self-organized stone stripes. *Nature*, 361, 142–5.

Williams, P.J. and Smith, M.W. (1989) *The frozen Earth. Fundamentals of Geocryology*. Cambridge University Press, Cambridge. p. 306.

Yershov, E.D. (1990) *Obshcheya Geokriologiya*. Nedra, Moscow. English translation by Williams, P.J., 1998, *General Geocryology*, Cambridge University Press, Cambridge. p. 580.

Zhou Y., Dongxin, G., Guodong, C., and Shude, L. (2000) *Geocryology in China*. (in Chinese). Cold and Arid regions Environmental and Engineering Research Institute (CAREERI), Chinese Academy of Sciences. p. 450.

24

Coastal Environments

Colin D. Woodroffe, Peter J. Cowell and Mark E. Dickson

Coastal geomorphologists study one of the most dynamic and changeable parts of the earth, the coastal zone. Coasts evolve, in profile and planform, as a result of the depositional or erosional outcome of predominantly marine processes acting on the shoreline. The operative processes often leave a vivid imprint and even a casual observer can postulate why a beach may be undergoing erosion, speculate on the pattern of future deposition along an accreting shore, or hypothesize the next stage in the retreat of an undercut cliff. Understanding the way that the shoreline changes is of considerable importance to society because so many people live along the coast, and it is the site of intensive agricultural, aquacultural, industrial, residential and recreational land use. Consequently, there is a wide range of disciplines that study the beaches, dunes, deltas and estuaries, cliffs and reefs of the coast. Studies are driven by scientific curiosity, coastal management or an engineering need to protect infrastructure, transportation and other human activities. Coastal geomorphology has evolved in parallel with engineering, oceanographic and geological disciplines that have also focused on coastal processes or coastal landforms. In some cases the contribution of the geomorphologist has been distinctive, and in other instances geomorphological outcomes have been the product of complex hydrodynamic or engineering models. In particular, coastal geomorphologists have recognized the interactions and mutual co-adjustments between coastal landforms and the processes that shape them, which they have termed coastal morphodynamics, and which is the focus of this chapter (Wright and Thom, 1977; Woodroffe, 2003; Dronkers, 2005).

Many of the largest settlements in the world, including huge megacities of several million people, occur on the coast and are subject to the vagaries of ongoing coastal processes. They are also subject to the impact of infrequent, but devastating, extreme events, as tragically demonstrated by the Indian Ocean tsunami that demolished Aceh in 2004 and Hurricane Katrina that caused such havoc in New Orleans in 2005. Coasts are particularly vulnerable to climate change and individual extreme events, like Katrina, are likely to exact more devastating impacts in the future, especially because the proportion of the population living on the coast is increasing (Small and Nicholls, 2003), placing more people and assets at risk from future coastal hazards.

The morphodynamic concept has become a central tenet of coastal geomorphology, recognizing that a coastal landform, such as a beach, responds as the wave processes that affect it alter, and that in turn those changes in shape further modify the processes. This morphodynamic approach eventually spread to other areas of geomorphology, after ruminations about the need for 'unified theory', and growing awareness of the importance of understanding the nonlinear dynamics of geomorphological systems (Phillips, 1996). For example, many of the papers in the 1996 Binghampton review of 'The Scientific Nature of Geomorphology' tentatively alluded to the morphodynamic essence of geomorphic systems without specifically referring to the existing morphodynamic concept despite it having been widely formalized in the physical coastal sciences. That the morphodynamic approach has now been embraced more widely in geomorphology, as well as other interdisciplinary fields, is evident with the running of the inaugural *IAHR Symposium on River, Coastal and Estuarine Morphodynamics* at Genoa in 1999, with the 6th Symposium run in 2009 at Santa Fe City, Argentina. This unified approach is scale independent and thus applicable across all scales.

This chapter therefore explores the historical context which enabled geomorphologists studying the coast to frame their science in the context of morphodynamics. It traces how this progress has been underpinned by technological developments in computing, field instrumentation, remote sensing, and radiometric dating, before considering the directions in which coastal geomorphology is going, and new approaches that will become more central in the future.

THEORETICAL CONTEXT

Geomorphological study of the coast has progressed at a differing pace from that of the other principal branches of geomorphology, at times lagging behind, and at others setting new directions and developing new approaches through the links that coastal geomorphologists have forged with other scientists studying the coast. For example, coastal geomorphology was relatively slow to embrace the concept of the geographical cycle as advocated by William Morris Davis at the beginning of the 20th century, but was at the forefront of the adoption of the concept of morphostratigraphy, driven by a combination of stratigraphy of coastal sediments and the physical understanding behind processes operating on the coast across a broad range of timescales. Coastal morphodynamics was firmly established as a conceptual basis for coastal geomorphology in the 1970s, and the research directions identified in a key review by Wright and Thom (1977) have remained at the forefront of the discipline over the subsequent three decades.

Initially, the discipline of geomorphology focused primarily on the land. There was considerable evidence that landforms sought to adopt a balance with the processes that operated on them. Paradoxically, although he worked in the arid American West, far from the coast, the great geomorphologist Grove Karl Gilbert derived some of his most enduring concepts in relation to dynamic equilibrium in the context of paleoshorelines of former glacial lakes, particularly Lake Bonneville in the western US (Gilbert, 1885). His ideas concerning dynamic equilibrium remain widely applied in coastal geomorphology today, but owe much to changes in shoreline planform that Gilbert inferred were driven by variability in longshore sediment transport and input of sediment from tributary creeks, with consequent variation in the processes of erosion and deposition. He recognized that:

> if the shore drift receives locally a small increment from stream drift, this increment by adding to the shore contour, encroaches on the margin of the littoral current and produces a local acceleration, which acceleration leads to the removal of the obstruction. Similarly, if from some temporary cause there is a local defect of shore drift, the indentation of the shore contour slackens the littoral current and causes deposition, whereby the equilibrium is restored.
>
> (Gilbert, 1885)

In providing this description, Gilbert outlined a process of dynamic equilibrium involving change in coastal form driven by variability in sediment transport, a concept that was to become widely adopted in many branches of geomorphology. At about the same time, an Italian, Paolo Cornaglia (1889) proposed the concept of an equilibrium profile that implied that sandy open-ocean shorelines adopted a regular concave-up shoreface topography in balance with wave processes.

The concept of the geographical cycle proposed by William Morris Davis suggested that landscape denudation went through a series of stages culminating in peneplanation, with subsequent rejuvenation by uplift (Davis, 1909). Base level, the level down to which the landscape would ultimately be planed, is a major element of this cycle of erosion, and is often equated to sea level. Ironically it was recognized at an early stage in the 20th century that this level had not remained constant but that there had been fluctuations of sea level associated with the Quaternary glaciations, leading to submergence or emergence of coasts (Daly, 1925). Davis himself wrote prolifically, but he did not focus on the coast until late in his career, when he undertook two quite remarkable studies; first a very detailed appraisal of the study of coral reefs (Davis, 1928), and second a study of the former shorelines on the uplifting coast of the Santa Monica mountains in California (Davis, 1933). By contrast Davis' students adopted many of his approaches; Gulliver (1899) recognized progressive infill of estuaries, based on a concept of initial, sequential and ultimate stages, and Douglas Johnson developed this framework of a 'life cycle' from youthful to mature, which became deeply embedded in coastal geomorphology (Johnson, 1919, 1925). He inferred that an indented shoreline would erode and become straighter and smoother with time, which became a basis for the more detailed models of estuary infill described below.

Up until the second half of the 20th Century, geomorphology was dominated by qualitative observation and explanatory description (Walker, 2008). Coastal studies expanded after the Second World War because there had been a series of situations where landing craft and military personnel had been lost as a result of the poor understanding of the dynamics of coastal environments (Williams, 1960). Several substantial programs were funded by the Geography Branch of the US Office of Naval Research, and the Beach Erosion Board of the US Corps of Engineers. Similar programs were also instigated through Delft Hydraulics in the Netherlands and Wallingford Hydraulics Research laboratory in the UK, both of which focused on response to the extensive North Sea flooding that occurred in 1953. While this was also a time in which coastal engineers applied equilibrium concepts to nearshore profiles (Bruun, 1954) and tidal inlets (O'Brien, 1966), Per Bruun proposed a more daring application of the concept of an equilibrium profile. He hypothesized that under a higher sea level the equilibrium profile would be displaced upwards and landwards, and that, assuming conservation of mass, one could estimate the amount of shoreline retreat, with erosion of sediment from the beachface and its deposition in the nearshore. The Bruun rule (Bruun, 1954, 1962), as it became known, remains a simple heuristic approach to assessing the impact of sea-level rise that has been both widely employed and vigorously contested as discussed below.

In the mid 20th century mathematical frameworks were provided for the description of physical processes of longshore sediment transport (Inman and Bagnold, 1963), the longshore diffusivity of sand (LeMéhauté and Soldate, 1977) and the rate of accretion or erosion as a function of wave angle and longshore transport rate (Pelnard-Considere, 1956). Following field experiments using sand tracers and measurements of sand impoundment behind structures, it became possible to calibrate coefficients within equations and provide quantitative predictions of longshore sand transport rates (Komar and Inman, 1970). Coastal geomorphologists developed empirical field and laboratory studies in which largely black-box models of cause and effect were developed through statistical inference. For example, associations between beach gradient, sediment size and wave characteristics were demonstrated (King, 1972), an early computer simulation reconstructed the evolution of Hurst Castle spit (King and McCullach, 1971) and a Markov-chain approach was adopted to illustrate a sequence to changes in beach morphology (Sonu and James, 1973). Increasingly these studies not only involved morphometric measurements, but instrumentation of sedimentation processes, including flows of water or sediment.

The middle part of the 20th century also saw two fundamental advances in associated disciplines that had far-reaching consequences for coastal geomorphology. First, coastal geomorphology, as with other earth sciences, went through a major paradigm shift with the development of the ideas of plate tectonics in the 1960s. For instance, the discrimination of plate margin coasts, termed collision coasts, from those in mid plate settings, termed trailing edge coasts, formed the basis for a revised classification of shorelines (Inman and Nordstrom, 1971) and encapsulated the distinction between cliffed and predominantly uplifting Pacific coasts of the Americas and the largely sedimentary Atlantic coasts. Previously, several attempts had been made to develop a classification of coasts following the taxonomic paradigm of classical biology and geology (Shepard, 1948; Valentin, 1952; Cotton, 1954). Classification of shorelines has received little attention since (Finkl, 2004), but has recently been revisited in the context of geographical information systems (GIS) which enable the association of numerous attributes along a stretch of shoreline and its assessment in terms of susceptibility to erosion or inundation (Manson et al., 2005; Sharples, 2006; Vafeidis et al., 2008).

Second, the dating of organic material using the decay of radiocarbon marked the beginning of several decades of chronostratigraphic studies. It became possible to get a much more precise idea of when Holocene landforms might have been deposited, and hence to calibrate paleoenvironmental reconstructions with rates of process operation. This had a particular impact on determining the pattern of Holocene sea-level change. Initially, this involved plotting radiocarbon ages for coral and other shoreline indicators from shorelines that were thought to be tectonically stable onto one age-depth diagram (Fairbridge, 1961). However, the plots showed geographical disparity; for example, mangrove sediments from Florida and the West Indies implied that sea level over recent millennia had been below present, while other evidence, such as shoreline deposits in Australia, implied that it had been above present. It became increasingly clear that the discrepancies could not be accounted for by a series of oscillations in sea level over the past few millennia, and that the pattern of sea-level rise varied as a consequence of global isostatic adjustments to changes in the ice-ocean masses (Thom and Chappell, 1975; Adey, 1978).

Geophysics provided the framework within which to interpret the isostatic response of the earth to altered loading of the mass of ice and water, and geomorphologists played an important role in compiling an atlas of sea-level curves, especially during a sequence of International Geological Correlation Program (IGCP) projects (Pirazzoli, 1991, 1996). Geomorphological studies of sea-level change at millennial and decadal timescales provided observational data to test geophysical models (Shennan and Andrews, 2000; Lambeck and Chappell, 2001). One of the exciting aspects of this collaboration was that it appeared possible to span such a wide range of timescales.

SPACE AND TIMESCALES

The timescale at which the geomorphologist works is important, and geomorphology as a whole was invigorated by the seminal paper on time and causality by Schumm and Lichty (1965). This had greater impact on terrestrial geomorphology than it did on coastal geomorphology, because in the latter case it depends upon the coastal type in which interest lies. Adoption of a schematic representation of space and timescales proposed by Cowell and Thom (1994) has extended these ideas into coastal evolution (Figure 24.1). At the smallest scale, geomorphology is concerned with 'instantaneous' processes following the principles of fluid dynamics, sediment entrainment and deposition. The 'event' timescale involves individual storm events or an aggregation of several lesser events. The mechanistic relationships from instantaneous time are scaled up in a deterministic or empirical way to understand the operation of coasts at larger spatial and temporal scales; for example, beach erosion during a single storm (reaction) and its recovery over subsequent weeks (relaxation). Coastal managers involved with planning need to consider the 'engineering' timescale of several decades. The longest timescales are 'geological' timescales, concerned with thousands to millions of years. This is the timescale over which deltas prograde and reefs accrete; evolutionary history is generally inferred from the stratigraphy and age of sedimentary sequences. The 'engineering' timescale is of particular significance as it is the scale at which planning decisions need to be made, anticipating behaviour of the shoreline, but it is perhaps the most difficult to understand because it falls between the knowledge based on observations of processes and events over the past few decades, and the long-term behaviour recorded in the sedimentary record.

The concepts of a hierarchy of spatial and temporal scales, and their aggregation, have received attention from various groups of researchers for different types of coastal systems (e.g. sandy shoreface: Swift and Thorne, 1991;

Swift et al., 1991; Roy et al., 1994; muddy coasts: Boyd et al., 1992). More recently, the concept of a coastal tract cascade has been advanced as a template for modelling coastal behaviour over a range of scales (Cowell et al., 2003a, 2003b; Stive et al., 2009).

The nature of the coast, however, is not simply a function of the modern processes that are acting on it. It is a much more complex outcome of changes over time. Many coasts bear the clear imprint of previous landforms; there is a strong geological inheritance. In the case of cliffs, former erosional landforms may be preserved as notches and marine terraces, whereas the morphology of gently sloping shore platforms may be re-sculpted many times through sea-level transgressions (Trenhaile, 1987). The morphology of cliffs is a direct product of erosion, such that there is always an incomplete record of the past. Depositional environments (such as, marsh, coral reefs, mudflat and delta sediments) comprise sedimentary deposits that provide a stratigraphic record from which it is possible to study process and response at longer timescales (Woodroffe, 2003). Inheritance can be important at shorter timescales also; the morphology of a beach is only partly a response to incident wave conditions, its shape is also dependent on antecedent conditions, the previous shape of the beach, which itself was in the process of adjusting to wave conditions incident at that time (Short, 1999).

TECHNOLOGICAL DEVELOPMENTS AND MORPHODYNAMIC MODELS

Coastal geomorphology today, with its theoretical foundations from the geomorphological tradition, combines a suite of new techniques that have resulted from the technological and computer revolution. Requirements for coastal management, and in the case of engineering, coastal 'protection', often direct the type of research that is undertaken, but it is also the case that advances in technology have preconditioned conceptual and theoretical perspectives.

Advances in dating

Radiocarbon dating provided the first absolute radiometric dating technique. It had immediate application in the coastal zone, particularly along the coast of the Gulf of Mexico. Existing conceptual models of the development of the Mississippi delta were based on historical and archaeological evidence, such as potsherds, and Indian mounds. Radiocarbon dating enabled age determinations of

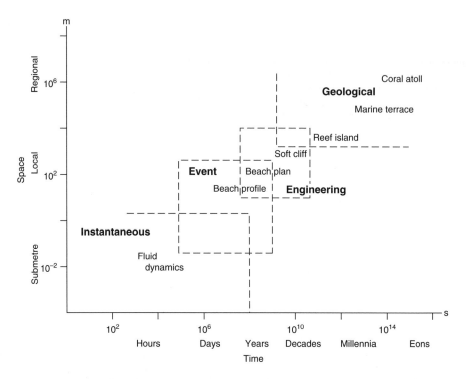

Figure 24.1   The recognition of instantaneous, event, engineering and geological space and timescales in coastal geomorphology (based on Cowell and Thom, 1994), and the identification of the broad domain in which some of the key morphodynamic models operate. Fluid dynamics applies only at the smallest and shortest timescales. Beach profile models such as SBEACH apply at event scales, whereas sand barrier models such as the Shoreline Translation Model (STM) are scaled up to longer timescales, as also are SCAPE (Walkden and Hall, 2005) and reef island models (Barry et al., 2007). Models for marine terraces, such as that generated by Anderson et al. (1999); and atoll formation, such as the subsidence theory of coral atoll evolution proposed by Darwin (1842), operate at geological timescales. Details of several of these models are discussed in the text

shells that provided time control, first on the chenier ridges of the Louisiana coast (Gould and McFarlan, 1959), and then on the chronology of delta distributary progradation and abandonment across the Mississippi delta region (Frazier, 1967). Dating remains central to interpretation of the evolutionary history of that coast (Blum and Roberts, 2009).

The coast abounds in organic materials suitable for radiocarbon dating, and suddenly the opportunity to determine the dates at which organic remains had died provided a means to calibrate models of sediment accumulation that had otherwise been largely unconstrained in time. Coral, amenable both to radiocarbon dating and to dating by uranium-series disequilibrium techniques, became particularly useful as a fossil

for the reconstruction of former shorelines (Bloom et al., 1974). Dating of coral has continued to provide one of the most fundamental controls on the timing of former ice sheet advances, and a constraint on ice volume, which is inversely correlated with ocean volume, and hence sea level (Thompson and Goldstein, 2005).

Other dating techniques, such as luminescence dating, have also been widely applied to coastal environments (Jacobs, 2008). Refinements, discussed elsewhere in this volume, outline how radiocarbon through AMS, and U-series through TIMS, have developed to be able to date smaller samples to greater precision, and how cross-correlation of techniques enables greater certainty about absolute age, increasing the temporal resolution possible for coastal stratigraphy.

Advances in photogrammetry and aerial survey

Understanding coastal behaviour involves methods to acquire topographical data on the morphology of coastal landforms taken at different moments in time. In the past, this has meant tedious survey of often poorly accessible coastal sites, but increasingly sophisticated techniques have been developed that overcome many of the previous obstacles. In rare instances sufficient foresight has seen the instigation of systematic repeat surveys; for example, detailed surveys were initiated at Duck in North Carolina (Holman and Sallenger, 1993), and the beach at Moruya has been surveyed several times a year for more than 30 years (Thom and Hall, 1991; McLean and Shen, 2006). Elsewhere, longer-term perspectives on change have been provided through systematic analysis of historical maps and survey data, but photogrammetry has played a particularly significant role (Dixon et al., 1998). Initially, aerial photography became important in wartime, and early attempts at inferring water depth from wave refraction patterns developed into post-war research programmes that looked at links between process and form. Satellite imagery supplemented the photographic archive, and there are now a range of satellite and airborne techniques which provide unprecedented detail at a range of spatial and temporal resolutions (Monmonier, 2008).

Detailed shallow seismic records have been available offshore for many decades using acoustic methods, but in the past decade ground-penetrating radar (GPR) has enabled the same type of remote sensing capability for investigation of sub-surface onshore coastal deposits (Roy et al., 1994). The belated availability of GPR provided data that has generally validated prevailing models of coastal evolution, such as the progradation of strandplains, for which GPR frequently outlines former depositional surfaces (Bristow et al., 2000, Neal and Roberts, 2000).

Early observations of the subaerial beach were collected by traditional survey using a level, or a series of simpler profiling techniques (Larson et al., 2003; Romanczyk et al., 2005). The availability of total stations and differential GPS, particularly real-time kinematic GPS, has facilitated increasing accuracy and resolution in surveys of beachface shape (e.g. Norcross et al., 2002). Moreover, three-dimensional laser scanner systems now enable rapid capture of wide swaths of beachface as well as steep cliff-face morphology that cannot be readily studied using aerial photogrammetric methods (McCaffrey et al., 2005). However, a more significant advance over recent decades, with extensive application in coastal geomorphology, has been airborne and satellite altimetry (laser and radar). Until recently high vertical resolution (decimetre or better) and high horizontal resolution (submetre) data have typically been unavailable for coastal terrain, except in those areas where photogrammetric surveys had been commissioned. This has limited a detailed knowledge of coastal morphology more than in other aspects of geomorphology because depositional coastal terrain has such subdued relief. High-resolution DTMs can now be acquired using airborne laser surveying (ALS, LIDAR) along long stretches of coast to map regional-scale morphology (e.g. Sallenger et al., 2003; Gesch, 2009).

Photogrammetric technology has been progressively refined, including the introduction of colour and infrared photography flown on low-altitude missions (e.g. 1:4000). Beach conditions can be monitored from aerial photography and satellite imagery, with data extraction possible on wave parameters (such as breaker height, direction), water depth, bar position, and rip current conditions (Boak and Turner, 2005). Imaging of beach morphology can be undertaken using video-based coastal imaging systems such as ARGUS, enabling real-time beach monitoring (Turner et al., 2006). Beach profile surveys can also be extended into the nearshore at low tide, or by use of CRAB (as at Duck), but the development of sidescan sonar in the 1970s and more recently multi-frequency and multi-beam swath mapping, facilitate high-resolution mapping of topography and sediments along wide swaths of the nearshore. Developments in laser technology have resulted in marine laser that penetrates through water (SHOALS, LADS MkII), unless turbid, and enables mapping of the coastal zone from 50 m elevation to 50 m depth (Finkl et al., 2005).

Advances in computing power, modelling and visualisation

The revolution in computer and information technology made possible more sophisticated analysis of detailed time series from environmental sensors, making it possible to obtain measurements of flows and sediment fluxes. Increasing hard-disk space has enabled storage of high-resolution photographs and images, as well as dense point clouds of topographic data enabling generation and manipulation of digital terrain models with a sophistication never previously envisaged. Increasing processing power has facilitated modelling and display of spatial data, and mapping and visualisation, particularly using geographical information systems (GIS), has provided a much more accessible interface with other advances in computing and developments in numerical modelling.

The rapid increase in availability and power of computers from the 1980s underpinned developments in the level of sophistication of morphodynamic models for geomorphic evolution of deltas, estuaries, beaches, shorefaces, dunes and cliffs (Hanson et al., 2003). These models synthesize and integrate principles of coastal geomorphology and related sedimentary processes painstakingly established over decades of theoretical, laboratory and field research. Morphodynamic models have been shown to have application at very different scales. For instance, shoreline engineering projects often use highly detailed hydrodynamic models that have timescales of days to years (e.g. Delft3D, MIKE21), whereas at geological timescales (millions of years) basin-margin models for the evolution of coastal lowlands and continental-shelf sediment bodies generally have time steps involving thousands of years (Syvitski and Daughney, 1992). The need to simulate long timescales means that coastal geomorphic models often have low dimensionality. Despite this, relatively simple models have been successfully used to investigate some fundamental aspects of coastal evolution, such as the growth of small-perturbations, driven by high-angle waves, into large-scale features that resemble forms such as sandy cusps and spits (Ashton et al., 2001, 2009). One aspect of simple geomorphic models is the consideration of nearshore morphodynamics as non-linear dynamical systems that may evolve and self-organize through critical states, which helps understanding of coastal features such as beach cusps (Coco et al., 2001, 2003). Models such as these help operationalize concepts on self organising systems (Cowell and Thom, 1994; Phillips, 1995). That is, the time dependence of these models, and interaction of processes within them, yields evolution of coherent broad patterns in the morphology as seen in nature, but not inherent in the primary dynamics. Examples of such patterns are channel bifurcations in deltas (Seybold et al., 2007), and the formation of reef islands on the rim of coral atolls (Barry et al., 2007, 2008).

Coastal evolution can be explored with models of low dimensionality, and shoreline engineering can be informed by highly parameterized hydrodynamic models. However, this leaves a gap on timescales of decades to millennia in which it has been useful to aggregate representation of processes in models of coastal morphodynamics (de Vriend et al., 1993; Niedoroda et al., 1995). On eroding soft-rock shorelines, decadal-scale coastal change has been addressed using a broad systems model (SCAPE, Soft Cliff and Platform Erosion model) that includes relatively simple representations of many processes, including wave shoaling and refraction, alongshore sediment transport, beach formation, and erosion of the rock shore profile and lower cliff (Walkden and Hall, 2005; Walkden and Dickson, 2008). In this approach, detailed representation of processes is subordinate to representation of the interaction between system components, because these regulate shore evolution in response to hydrodynamic drivers (Figure 24.2). The SCAPE model has been used to simulate the evolution and possible response of tens of kilometres of the eroding Norfolk coast to multiple scenarios of management and climate change (Dickson et al., 2007). The planform response of the shoreline was shown to be a highly complex morphodynamic issue. Sea-level rise resulted in more rapid erosion of some sections of shoreline, but in turn that erosion released sediment that increased the volume of some beaches down-drift. While not initially intuitive, the modelling suggested that some eroding coasts are likely to erode less rapidly under sea-level rise.

As development pressure on coastal land has increased in recent years, so too has the requirement for coastal models to provide outputs that can be readily used by coastal policymakers and planners. GIS offers an opportunity to add value in this respect by overlaying outputs from models on other data sets relevant to society and biotic environments. By the 1990s, the advent of computer-based GIS became a means of managing and integrating the burgeoning datasets flowing from environmental sensors and the harmonisation of previously disparate data kept in unconnected archives (Ricketts, 1992). The integration of morphodynamic models into a GIS would make their output readily available for application in broader environmental analyses and as overarching decision-support models for use in coastal management. GIS also provided a practical means of operationalizing spatial analysis and models that had previously been limited to interesting but impracticable concepts (Bartlett and Smith, 2005; Green, 2010).

One such example is provided by SCAPEGIS, a tool developed to include the outputs of two cliff-erosion models within a GIS, thereby providing users with an opportunity to visualize erosion predictions and analyse their implications by integrating with other spatial datasets (Koukoulas et al., 2005). The system employed an offline protocol in which models and the GIS tool could be developed and operated independently. Simulated future cliff-toe positions, shore platform levels and beach volumes (Dickson et al., 2005), as well as cliff-top recession distances output from a probabilistic model (Hall et al., 2002), were incorporated within the GIS to generate areas of cliff-top land considered to be lost,

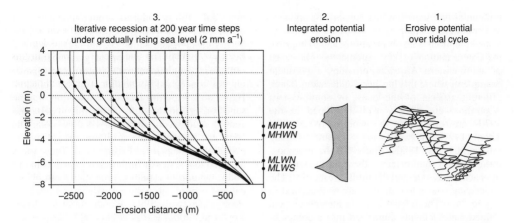

Figure 24.2 A schematization of the cliff model SCAPE and an illustration of how a shore platform evolves from a vertical cliff, over time, using SCAPE (after Walkden and Hall, 2005, Walkden and Dickson, 2008). Stage 1 shows the distribution of potential retreat over a tidal cycle, stage 2 the integration of this erosion potential and stage 3 the pattern of recession, with each line representing a successive 200-year period, superimposed on a gradually rising sea level

at risk, and safe. In a recent extension to this work, outputs from the cliff-erosion modelling, conducted under varied management and climate-change scenarios, were used to evaluate the changing risk of flooding in an adjacent low-lying section of coast (Dawson et al., 2009). In this modelling scheme, the risk of coastal flooding reduced as the size of sandy barriers increased. Management decisions to remove coastal defences and allow greater coastal erosion, thereby supplying the system with greater amounts of sand, resulted in a high financial benefit in terms of the lowered risk of coastal flooding.

The example above illustrates that linked models offer considerable benefit in terms of objectively weighing the significance of various factors under a complex array of scenarios. There is considerable further scope for GIS applications that intersect physical risks with socio-economic impacts (e.g. Hennecke et al., 2004). However, incorporating outputs from numerical morpho-logical models within GIS carries with it the challenge of communicating model assumptions and limitations. This is not necessarily straightfor-ward, as outputs from an already abstracted representation of reality, such as predicted shore-line positions, are given apparent precision when represented as a solid line on a map. There is clearly a need for more explicit incorporation of error terms, as well as scope for the application of fuzzy logic and probabilistic approaches (Cowell et al., 2006; Ranasinghe and Stive, 2009), and visualisation techniques that provide tools for

coastal managers to examine public reaction to alternative scenarios (Jude et al., 2005).

An overview of the development, use and future of morphodynamic models can be recog-nized through a description of the models that were developed to describe the depositional coastal environments of south-eastern Australia where it can be seen that conceptual models were formalized into computational models, which have provided the capacity to further explore the behaviour of the coastal systems, and to forecast future trends.

MODELLING THE COASTAL ENVIRONMENTS OF SOUTH-EASTERN AUSTRALIA AND IMPLICATIONS

There have been significant advances in the devel-opment of conceptual models, following the key review of morphodynamic concepts by Wright and Thom (1977). Models can provide a link between the knowledge of morphostratigraphic character of the coast and the shorter timescale of understanding drivers of coastal change. Models tend to be simple formulations of the principal interrelationships between the main factors that capture a small number of representative con-ditions. Such models ignore much of the detail, but the mark of a good model is that it embodies the major variations that a coastal system can undergo. In order to build models across annual to millennial timescales, it is necessary to organize and simplify the complexity of coastal systems, to test assumptions with observations, and to improve the determination of key parameters. The coast of south-eastern Australia provides a pertinent example of where this has been undertaken. Three different aspects of the coast of south-eastern Australia were already well-developed in the 1970s, are reviewed in the paper by Wright and Thom (1977), and were widely researched along this coast in the 1980s. These are the morphostratigraphy of sand barriers, the morphodynamics of beach behaviour, and the infill of estuaries. The underlying conceptual models are summarized in Figure 24.3. These have been further developed, refined, and elements combined into a computer simulation model, termed the Shoreface Translation Model (STM), and incorporated into an approach termed the coastal tract (Cowell et al., 2003a, 2003b).

Preliminary modelling

The coast of south-eastern Australia has served as a natural laboratory; it consists of a series of embayments which are flanked by headlands, dividing the coast into discrete coastal compartments (Davies, 1974). The coastal compartments have experienced a similar history of climate and sea-level change, but each has reached a different stage in the evolutionary pathway, driven by local variations in topography, catchment and wave characteristics, with different degrees of closure or interchange of sediment.

The morphostratigraphy of sand barriers along this coast was comprehensively investigated in the 1970s and 1980s through a programme of drilling and radiocarbon dating (summarized in Thom, 1984a). These studies indicate that the majority of sand barriers along the NSW coast first assumed their present form as a consequence of sea-level stabilization towards the end of the postglacial marine transgression, around 6000 years ago (Roy and Thom, 1981). A number of different barrier types can be discriminated, ranging from receded barriers which are still experiencing shoreline recession, to prograded barriers, that have experienced continual progradation during the Holocene, to more complex dune barriers (Figure 24.3).

South-eastern Australia has also provided an appropriate field laboratory for the development of models of estuary infill. The simultaneous formation of sand barriers along most of the New South Wales coast around 6000 years ago as sea level stabilized at a level close to its present, served to occlude a series of estuaries and coastal lagoons. The stratigraphy and infill of these estuarine embayments has been examined in detail by Roy et al. (1980), and a conceptual model of each of the three principal types, drowned river valley, barrier estuary, and saline coastal lake, described by Roy (1984). The notion of gradual infill of estuaries through immature to mature stages builds on the conceptual foundation laid by Johnson. Based on a tripartite facies classification, marine sand barrier and flood tide delta, fluvial delta and central mud basin, different systems along the coast can be seen to be at different stages in the evolutionary sequence (Figure 24.3). These ideas have led to a series of further developments in our understanding of estuarine processes, both physical and ecological, within Australia and overseas (Dalrymple et al., 1992; Woodroffe et al., 1993; Roy et al., 2001; Harris et al., 2002, Heap et al., 2004; Dalrymple and Choi, 2007).

At the same time that the Holocene history of barrier formation and estuarine infill was being examined, detailed studies of the morphology and dominant processes on wave-dominated beaches were being undertaken in the Sydney region (Wright and Short, 1984). Beaches change in response to wave energy, but it is possible to recognize a modal beach state, the morphology that a particular beach adopts for most of the time (Short, 1999). Six beach states were identified spanning a range of energy conditions from high-energy, dissipative beaches, which are broad and flat with fine sand and a wide surf zone, to low-energy, steep reflective beaches on which waves surge, often forming beach cusps from the coarser sediments (Figure 24.3). Four intermediate beach types have been identified with various types of bar and rip currents. This classification of wave-dominated beaches has proved remarkably robust; it has been extended to 15 beach types around the entire Australian coastline (Short, 2006), and these ideas have also been incorporated into beach models that have been applied in a broader, international context (Lippman and Holman, 1990).

The morphostratigraphy of barriers represents a study at the geological timescale, whereas beach state is an instantaneous condition of the beach, with a response time of several weeks at the event scale following high-energy wave conditions. Studies along this part of the Australian coastline have also addressed the decadal engineering timescale that is so important in a planning context. The impacts of short but extreme events (storm cut) as opposed to long-term mean adjustments, have been examined, in particular, by regular surveys across several decades of the foreshore at Moruya, demonstrating the significance of major erosional episodes. Storms in 1974 resulted in an erosion-dominated period during which the volume of sediment removed was comparable to about 200 years of mean-trend

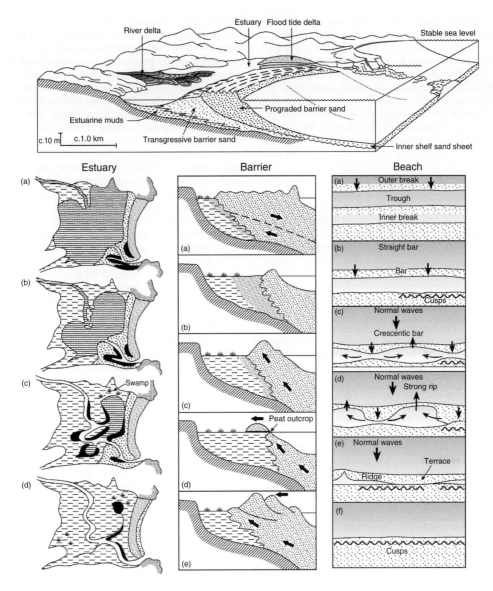

Figure 24.3   A schematic representation of a section of the coast of south-eastern Australia showing a sand barrier that partially occludes an estuary and the relationship of three morphodynamic models of different components of the coastal zone. The estuary model is based on the conceptualization by Roy (1984); it models the successive stages of estuary infill ((a) initial stages of infill of prior embayment, (b) fluvial delta begins to infill central basin, (c) infill nearly complete, residual cut-off embayments, (d) mature riverine system with river discharging to the coast and extensive alluvial plains). The characterization of sand barriers is based on Chapman et al. (1982), and describes the different type of barrier at locations along the coast ((a) prograded barrier, (b) stationary barrier with low foredune, (c) stationary barrier with high foredune, (c) receded barrier, (d) episodic transgressive barrier). The description of beach morphodynamics is based on Wright and Short (1984) recognizing the response of beach state to incident wave energy ((a) dissipative, (b) longshore bar and trough, (c) rhythmic bar and beach, (d) transverse bar and rip, (e) low-tide terrace, (f) reflective)

change as determined from radiocarbon-dated morphostratigraphic studies of this barrier. Erosion has been followed by a subsequent accretion-dominated period during which the development of the foredune has become evident (McLean and Shen, 2006). Similar continual monitoring on Narrabeen Beach in Sydney has revealed a pattern of beach rotation associated with wind changes that accompany the El Niño-Southern Oscillation (ENSO) phenomenon (Ranasinghe et al., 2004).

Simulation modelling

The lessons from a consideration at a longer time frame have been incorporated into the Shoreline Translation Model (STM) which builds upon both the Quaternary perspective and the process models of beach behaviour (Short, 1999). This coast lends itself to such modelling because the history of natural processes is relatively accessible, largely without interference by tectonic or anthropogenic complications. The STM is therefore consistent with established principles regarding the underlying processes, but also grounded in the longer term behaviour of the coast, and involves the computational capacity to undertake conventional forward simulations to obtain estimates of morphological responses purely to varying input parameters (Cowell et al., 1995). Testing of this model approach includes use of inverse-simulation methods to hindcast details of coastal evolution as reconstructed for the transgressive and strandplain deposits in south-eastern Australia where it was shown that the model produced morphostratigraphic outputs with high fidelity (Thom and Cowell, 2005).

Although geomorphic change broadly involves modifications to the topography of the coastal deposits and abrasion surfaces, most attention is commonly placed on changes in location of shorelines. Shoreline change has particular significance to society, through land loss and effects of erosion on property and infrastructure. Shoreline movements are a manifestation of horizontal and vertical translations in the shoreface due to changes in sea level, sediment availability, or geometry of the shorface (Cowell et al., 2003a). In this sense, the shoreface comprises the beach and bed of any coastal water body, extending from the landward limit of sedimentary processes associated with wave runup, out to the offshore limit of sediment transport (Cowell et al., 1999). Erosion, deposition and sediment transport rates generally decrease in deeper water due to the weakening of wave motions with water depth.

Figure 24.4 captures the progression from hypothesized simple equilibrial morphology to flexible simulation within a computational framework, as encapsulated in the STM, in which the geometric detail is time-dependent: that is, it evolves through time (Cowell et al., 2003b). In the context of sandy shorelines, the premise that there might be some equilibrium profile was initially conceived by Cornaglia, and since adopted in various forms. The mechanisms that control that morphology are complex and remain incompletely understood, and the fact that, in individual cases, it is often difficult to parameterize the profile, and to accommodate the vagaries exerted by exogenous factors, such as bedrock outcrops, mean the concept is still regarded as contentious by some researchers (Thieler et al., 2000). Nevertheless, the idea of an equilibrium profile has become widely utilized. For instance, the Bruun rule that relates the rate of retreat of a shoreline to re-adjustment of its equilibrium profile in response to sea-level rise, has been widely applied by engineers in the context of anticipated future greenhouse-related climatic change, despite continued criticism (e.g. Cooper and Pilkey, 2004), and recent model results that demonstrate the complexity of shoreline planform response to climate change (Dickson et al., 2007; Dawson et al., 2009).

The fact that the Bruun rule has been widely applied probably reflects its simplicity rather than its appropriateness. Obsession with the Bruun rule may be a useful simplistic illustration for novices, but this obsolete concept hinders a deeper understanding of processes involved in coastal change. At a more general level, equilibrium considerations are not strictly relevant to understanding changes in coastal morphology. Rather, changes in coastal morphology involve transfers of sediments between sources and sinks within sediment-sharing systems (Cowell et al., 2003a). Over the long term, sources and sinks are signified by the positive (or negative) capacity for the morphology to accommodate deposits (Sloss, 1962) for a given period of time characterized by a prevailing hydrological regime that is itself subject to morphodynamic feedback (Cowell and Thom, 1994). The sediment transfers and changes in accommodation occur at rates that are constrained by the sediment-budget continuity principle that can now be explored using new computational methods (Cowell et al., 1995, 2006; Storms et al., 2002; Stolper et al., 2005). These methods also take into account resistance of substrates to erosion. Recently, the resulting morphological behaviour has been formalized through analytical solutions of the sediment-mass-continuity equation (the Exner equation), under a wide range of conditions (Wolinsky, 2009; Wolinsky and Murray, 2009).

On this basis, coastal change can be characterized by the rate of shoreline advance or retreat, c_S, that Stive (2004) demonstrated can be generalized

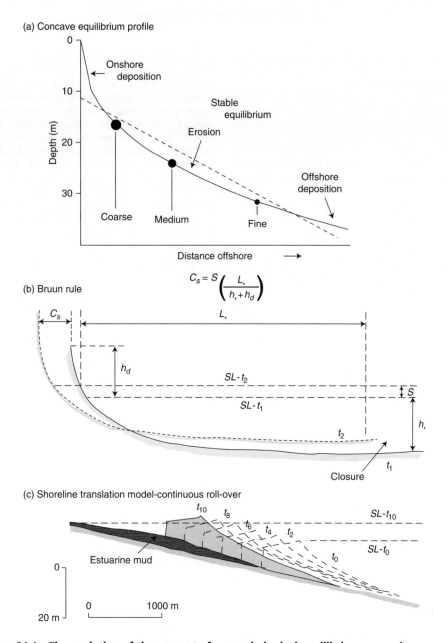

(a) Concave equilibrium profile

Onshore deposition

Stable equilibrium

Erosion

Offshore deposition

Coarse   Medium   Fine

Depth (m)

Distance offshore

$$C_s = S\left(\frac{L_*}{h_* + h_d}\right)$$

(b) Bruun rule

C_s

L_*

h_d

SL-t_2

SL-t_1

S

h_*

t_2

t_1

Closure

(c) Shoreline translation model-continuous roll-over

t_{10} t_8 t_6 t_4 t_2

SL-t_{10}

SL-t_0

t_0

Estuarine mud

0

20 m

0   1000 m

Figure 24.4   The evolution of the concept of a morphological equilibrium on sandy shorefaces (based on Woodroffe, 2003). (a) the concept of an equilibrium profile as a concave-up shoreface, proposed by Cornaglia, producing a graded profile on the basis that wave energy increases towards the shore as waves become increasingly asymmetrical in comparison to gravitational forces which operate to move sediment offshore. A null point exists for any particular grain size where the two forces are equal, with that null point occurring further seaward for finer grains. Gravity also increases onshore if the profile is concave, and the balance between onshore and offshore movement was considered by Cornaglia to represent a stable equilibrium; (b) representation of an equilibrium profile in a simple rule, the Bruun rule, with parameters defined as in equation (24.3). As sea level rises, there is a translation of the equilibrium profile landwards; and (c) the formalization of these concepts into a simulation model, the Shoreline Translation Model (STM) in which the shoreface is parameterized and simulations can be run hindcasting Holocene paleoshoreline conditions that can be partially validated by morphostratigraphic studies, and providing a tool for forward modelling

from integration of the sediment-mass continuity condition across the active zone of the shoreface to an offshore distance L_*:

$$c_S\left(h_* + h_d\right) = \frac{\partial S}{\partial t}L_* + \frac{\partial Q_y}{\partial y} - Q_x - V \qquad (24.1)$$

where S is the rate of sea-level rise, L_* and h_* are the distance and water depth defining the offshore limits of the active zone, h_d is the height of the dune or friable cliff, Q_x is the time-averaged shoreward sediment flux into the active zone from the lower shoreface, Q_y is the alongshore sediment flux integrated over L_*, V is a local sediment source or sink (e.g. delta or estuary), and x and y signify across-shore and alongshore dimensions. The *active zone* is the upper-shoreface region across which rates of fluctuation in bed elevation, due to cycles of erosion and deposition, are larger than the rate of sea-level rise (Cowell et al., 1999). Thus, time-averaged shoreface geometry within this zone ($0 \leq x \leq L_*$) can be used in representation of shoreface kinematics from which equation (24.1) is obtained.

Further aggregation can be attained by integration of equation (24.1) alongshore, provided homogeneity can be assumed for time-averaged processes and morphology along a tract of coastline (Cowell et al., 2003a). Such conditions, for example, are often applicable to a tract of coast between headlands. Alongshore integration thus allows aggregation of all submarine and subaerial line and point sources and sinks for sediments into two variables, Q_x and V, within a given coastal tract:

$$c_S\left(h_* + h_d\right) = \frac{\partial S}{\partial t}L_* - Q_x - V \qquad (24.2)$$

Equation (24.2) is perfectly general, especially if the source/sink term, V, incorporates any changes in the maximum transient-erosion volume due to storms. Such changes can be expected if systematic variation occurs over time in storm magnitudes and frequencies. For special cases in which sources, sinks, net littoral transport, and shoreface supply of sediments can all be ignored, equation (24.1) reduces to the Bruun rule (Bruun, 1962):

$$c_S = \frac{\partial S}{\partial t}\left(\frac{L_*}{h_* + h_d}\right) \qquad (24.3)$$

Unfortunately the special-case conditions underpinning equation (24.3) (Figure 24.4b) seldom arise in nature, so application of the Bruun rule in isolation is generally invalid. One of the few field studies to have validated equation (24.1) demonstrated that applicability existed only along coastal tracts free of tidal-inlet effects and littoral transport gradients, which comprised little more than 30 per cent of the US Delaware, Maryland and Virginia coastline studied (Zhang et al., 2004). Another study, in which validity of equation (24.3) was evident, involved dominance by strong relative sea-level rise (Mimura and Nobuoka, 1995). More generally, under conditions of weak sea-level rise, the other terms in equation (24.1) tend to dominate coastal change (Roy et al., 1994).

Equation (24.3) is widely applied because of its practical simplicity, whereas equation (24.1) demands use of elaborate numerical models that do not depend on assumptions about equilibrium to obtain solutions (Cowell et al., 2006). Caution about misuse of principles that rely on equilibrium concepts must also be exercised if such principles are to be extended to other types of shorelines, such as estuaries or soft-cliffs (Pethick, 2001; Dickson et al., 2007). In south-eastern Australia, the STM provides one approach that avoids reliance on the concept of equilibrial profile morphology in the numerical experimentation on the response of coastal geomorphology to sea-level rise, sediment budgets and variable substrate resistance to erosion. Informed by the history of former shorelines derived from stratigraphic studies and consistent with the observed conditions on modern shorelines, the model allows computation of successive states simulated in response to a suite of possible parameters (Cowell et al., 1995). Such models can be further extended by the incorporation of fuzzy or probabilistic principles to extend the power of such simulations in a 'what if' context (Cowell et al., 2006).

Models similar to those developed for southeastern Australia have been adopted elsewhere, for example in the United States (Stolper et al., 2005) and north-western Europe (Storms et al., 2002), but the Australian examples are particularly appropriate in our review in that they can be seen to have been derived simultaneously with the adoption of morphodynamic concepts in coastal geomorphology, and because they overlap geographically and provide a rounded view of the course of coastal geomorphology in Australia (see Thom, 1984b; Thom and Short, 2006).

In most cases equilibrium is a moving target towards which the shoreline is continually adjusting (Wright, 1995). Nevertheless, limits can be set to the range of shoreface geometries, the domain within which the coast behaves, based on morphological field research for a particular region. The uncertainty that exists, both about shoreface geometry, and boundary conditions such as changes in climate or sea level, could be readily represented probabilistically or through fuzzy membership functions developed empirically or on the basis of available data. Environmental variables such as alongshore sand supply and losses from a coastal cell, are notoriously difficult to estimate with the range of available empirical formulae, and cannot be forecast given the additional uncertainties associated with future rates of sea-level rise. However, probability distributions can be broadened to reflect these high levels of uncertainty, to provide scenarios that at least guide the coastal manager. Although deterministic models may offer unrealistically precise answers, the exciting developments in the broad field of coastal modelling indicate the range of tools and the diversity of approaches (Lakhan and Trenhaile, 1989; Lakhan, 2003, 2010). Coastal geomorphology will continue to stimulate research and generate outcomes that can guide coastal managers.

PROSPECTS

The geomorphology of coastal environments has played a key role in the development of geomorphology as a discipline, and is an exciting and vibrant field of study. It is addressing real issues, such as sea-level rise, of concern to several complementary fields, with important implications for sustainability and management of our coasts. The healthy status of coastal geomorphology appears related to three factors: first, the early recognition of the value of models as a basis for conceptualisation; second, the adoption of a morphodynamic framework in which to develop those models; and, third, a multi-/trans-/inter-disciplinary approach to coastal behaviour. Perhaps the very unusual nature of the coast also explains why coastal geomorphology has been at the forefront of the development and adoption of morphodynamics and modelling when compared with other branches of geomorphology, and why in many instances coastal geomorphologists have looked to coastal scientists in other fields, rather than to their geomorphological colleagues who study terrestrial environments, for their inspiration. The shoreline is such an unambiguous discontinuity between land and sea that its changing vista begs interpretation by specialist or non-expert, and the urgency necessitated by extreme events at this uneasy boundary demand the attention of planners, engineers and coastal managers alike.

The shoreline appears simple, but in practice coastal systems are complex, involving a myriad of complicating factors. Geomorphological study of coasts has progressed through a series of stages from the simple descriptive, empirical and experimental, to the theoretical and conceptual, and most recently, computational. A series of models has been developed and refined to describe coastal systems, and these models are now being used to simulate behaviour and explore the likely response of coasts to future scenarios, particularly of climate and sea level, but with the potential to incorporate other factors, such as human use. Nevertheless, it is important that coastal geomorphology maintain its strong emphasis on fieldwork, including empirical hypothesis testing and shoreline monitoring, in order that the models be validated and verified to the extent possible.

Morphodynamic concepts are centrally embedded in coastal geomorphology, providing a rich framework for development of new models, as well as promoting dialogue outside the traditional disciplines of geography and geology. In the latter part of the 20th century considerable emphasis was placed on two broad approaches to coastal geomorphology: process geomorphology, measuring rates of process operation and morphological responses; and, changes in sea level and their impact on coastal evolution. The convergence of these approaches led to morphodynamic models, built on theoretical and empirical approaches, offering the prospect of a formalisation of coastal geomorphological principles. Perhaps for the first time, there is the prospect of a unified theory for the discipline of coastal geomorphology founded on morphodynamics and using models that are firmly grounded on field experience.

The study of coasts has benefited greatly from interaction with other disciplines that also study the coast. Coastal geomorphologists have at times played a central role in the broader discipline of geomorphology better prepared by their trans-disciplinary experiences, and at other times, have ventured into new areas in conjunction with specialists in other fields. Coastal geomorphologists are already actively involved in development of simulation models that address the inevitable impact of future changes on the coast, whether anticipated variations in sea-level change, extreme events such as storms and tsunami, or management issues arising from accentuated human use of the coast. Coasts are vulnerable to each of these impacts, and the discipline of coastal geomorphology can play a central role in ensuring that human utilisation of coasts does not endanger the longer-term sustainability of these environments.

REFERENCES

Adey, W.H. (1978) Coral reef morphogenesis: a multi-dimensional model. *Science*, 202: 831–7.

Anderson, R.S., Densmore, A.L. and Ellis, M.A. (1999) The generation and degradation of marine terraces. *Basin Research*, 11: 7–19.

Ashton, A., Murray, A.B. and Arnault, O. (2001) Formation of coastline features by large-scale instabilities induced by high-angle waves. *Nature*, 414: 296–300.

Ashton, A., Murray, A.B., Littlewood, R., Lewis, D.A. and Hong, P. (2009) Fetch-limited self-organization of elongate water bodies. *Geology*, 37: 187–90.

Barry, S.J., Cowell, P.J. and Woodroffe, C.D. (2007) A morphodynamic model of reef-island development on atolls. *Sedimentary Geology*, 197: 47–63.

Barry, S.J., Cowell, P.J. and Woodroffe, C.D. (2008) Growth-limited size of atoll-islets: morphodynamics in nature. *Marine Geology*, 247: 159–77.

Bartlett, D. and Smith, J.L. (eds) (2005) *GIS for Coastal Zone Management*. CRC Press, Boca Raton, 310 pp.

Bloom, A.L., Broecker, W.S., Chappell, J.M.A., Matthews, R.K. and Mesolella, K.J. (1974) Quaternary sea level fluctuations on a tectonic coast, new $^{230}Th/^{234}U$ dates for the Huon Peninsula, New Guinea. *Quaternary Research*, 4: 185–205.

Blum, M.D. and H.H. Roberts. (2009) Drowning of the Mississippi Delta due to insufficient sediment supply and global sea-level rise. *Nature Geoscience*, 2: 488–91. doi:10.1038/ngeo553.

Boak, E.H. and Turner, I.L. (2005) Shoreline definition and detection: a review. *Journal of Coastal Research*, 21: 688–703.

Boyd, R., Dalrymple, R. and Zaitlin, B.A. (1992) Classification of clastic coastal depositional environments. *Sedimentary Geology*, 80: 139–50.

Bristow, C.S., Chrostan, P.N. and Bailey, S.D. (2000) The structure and development of foredunes on a locally prograding coast: insights from ground penetrating radar surveys, Norfolk, United Kingdom. *Sedimentology*, 47: 923–44.

Bruun, P. (1954) Coast erosion and the development of beach profiles. Beach Erosion Board, U.S. Army Corps of Engineers, Technical Memorandum, 44: 1–79.

Bruun, P. (1962) Sea-level rise as a cause of shore erosion. American Society Civil Engineering Proceedings, *Journal of Waterways and Harbors Division*, 88: 117–30.

Chapman, D.M., Geary, M., Roy, P.S. and Thom, B.G. (1982) *Coastal Evolution and Coastal Erosion in New South Wales*. Coastal Council of New South Wales, Sydney, p. 340.

Coco, G., Huntley, D.A. and O'Hare, T.J. (2001) Regularity and randomness in the formation of beach cusps. *Marine Geology*, 178: 1–9.

Coco, G., Burnet, T.K., Werner, B.T. and Elder, S. (2003) Test of self-organization in beach cusp formation. *Journal of Geophysical Research,* 108(C3): 3101. doi 10.1029/2002JC001496.

Cooper, J.A.G. and Pilkey, O.H. (2004) Sea-level rise and shoreline retreat: time to abandon the Bruun Rule. *Global and Planetary Change*, 43: 157–71.

Cornaglia, P. (1889) Delle Spiaggie. Accademia Nazionale dei Lincei, Atti. Cl. Sci. Fis., Mat.e Nat. Mem., 5: 284–304.

Cotton, C.A. (1954) Deductive morphology and genetic classification of coasts. *Science Monthly*, 78: 163–81.

Cowell, P.J. and Thom, B.G. (1994) Morphodynamics of coastal evolution, in R.W.G. Carter, and C.D. Woodroffe, (eds) *Coastal Evolution, Late Quaternary Shoreline Morphodynamics*. Cambridge University Press, Cambridge. pp. 33–86.

Cowell, P.J., Roy, P.S. and Jones, R.A. (1995) Simulation of large-scale coastal change using a morphological behaviour model. *Marine Geology*, 126: 45–61.

Cowell, P.J., Hanslow, D.J. and Meleo, J.F. (1999) The Shoreface, in A.D. Short, (ed.) *Handbook of Beach and Shoreface Morphodynamics*. Wiley, Chichester. pp. 37–71.

Cowell, P.J., Thom, B.G., Jones, R.A., Everts, C.H. and Simanovic, D. (2006) Management of uncertainty in predicting climate-change impacts on beaches. *Journal of Coastal Research*, 22: 232.

Cowell, P.J., Stive, M.J.F., Niedoroda, A.W., de Vriend, H.J., Swift, D.J.P. and Kaminsky, G.M. (2003a) The coastal-tract (part 1): A conceptual approach to aggregated modeling of low-order coastal change. *Journal of Coastal Research*, 19: 812–27.

Cowell, P.J., Stive, M.J.F., Niedoroda, A.W., Swift, D.J.P., de Vriend, H.J. and Buijsman, M.C. (2003b) The coastal-tract (part 2): Applications of aggregated modeling of lower-order coastal change. *Journal of Coastal Research*, 19: 828–48.

Dalrymple, R.W. and Choi, K. (2007) Morphologic and facies trends through the fluvial-marine transition in tide-dominated depositional systems: a schematic framework for environmental and sequence-stratigraphic interpretation. *Earth-Science Reviews*, 81: 135–74.

Dalrymple, R.W., Zaitlin, B.A. and Boyd, R. (1992) Estuarine facies models: conceptual basis and stratigraphic implications. *Journal of Sedimentary Petrology*, 62: 1130–46.

Daly, R.A. (1925) Pleistocene changes of level. *American Journal of Science*, 10: 281–313.

Darwin, C. (1842) *The Structure and Distribution of Coral Reefs*. Smith, Elder and Co., London. p. 214.

Davies, J.L. (1974) The coastal sediment compartment. *Australian Geographical Studies*, 12: 139–51.

Davis, W.M. (1909) *Geographical Essays*. Ginn and Co., Boston, p. 777.

Davis, W.M. (1928) *The Coral Reef Problem*. Special Publication, 9. American Geographical Society, p. 596.

Davis, W.M. (1933) Glacial episodes of the Santa Monica Mountains, California. *Geological Society of America Bulletin*, 44: 1041–133.

Dawson, R.J., Dickson, M.E., Nicholls, R.J., Hall, J.W., Walkden, M.J.A. and Stansby, P. (2009) Integrated analysis of risks of coastal flooding and cliff erosion under scenarios of long term change. *Climatic Change*, 95: 249–88.

de Vriend, H.J., Capobianco, M., Chesher, T., de Swart, H.E., Latteux, B. and Stive, M.J.F. (1993) Approaches to long-term modelling of coastal morphology: a review. *Coastal Engineering*, 21: 225–69.

Dickson, M.E., Walkden, M.J.A. and Hall, J.W. (2007) Systemic impacts of climate change on an eroding coastal region over the 21st century. *Climatic Change,* 84: 141–66.

Dickson, M.E., Walkden, M.J.A., Hall, J.W., Pearson, S.G. and Rees, J. (2005) Numerical modelling of potential climate-change impacts on rates of soft-cliff recession, northeast Norfolk, UK. Proceedings of Coastal Dynamics. Barcelona, Spain.

Dixon, L.F.J., Barker, R., Bray, M., Farres, P., Hooke, J., Inkpen, R. et al. (1998) Analytical photogrammetry for geomorpohlogical research, in S.N. Lane, K.S. Richards, and J.H. Chandler, (eds) *Landform Monitoring, Modelling and Analysis.* Wiley.

Dronkers, J. (2005) *Dynamics Of Coastal Systems.* World Scientific, Hackensack, N.J.

Fairbridge, R.W. (1961) Eustatic changes in sea level, physics and chemistry of the earth. Pergamon Press, New York. pp. 99–185.

Finkl, C.W. (2004) Coastal classification: systematic approaches to consider in the development of a comprehensive scheme. *Journal of Coastal Research,* 20: 166–213.

Finkl, C.W., Benedet, L. and Andrews, J.L. (2005) Submarine geomorphology of the continental shelf off Southwest Florida based on interpretation of airborne laser bathymetry. *Journal of Coastal Research,* 21: 1178–94.

Frazier, D.E. (1967) Recent deltaic deposits of the Mississippi River: their development and chronology. *Transactions of the Gulf Coast Association of Geological Societies,* 17: 287–315.

Gesch, D.B. (2009) Analysis of Lidar elevation data for improved identification and delineation of lands vulnerable to sea-level rise. *Journal of Coastal Research, Special Issue,* 53: 49–58.

Gilbert, G.K. (1885) The topographic features of lake shores. *U.S. Geological Survey Annual Report,* 5: 75–123.

Gould, H.R. and McFarlan, E. (1959) Geologic history of the chenier plain, Southwestern Louisiana. *Transactions-Gulf Coast Association of Geological Societies,* 9: 261–70.

Green, D.R. (ed.) (2010) *Coastal and Marine Geospatial Technologies.* Springer, Dordrecht. p. 451.

Gulliver, F. (1899) Shoreline topography. *Proceedings of the American Academy of Arts and Sciences,* 34: 151–258.

Hall, J.W., Meadowcroft, I.C., Lee, E.M. and van Gelder, P. (2002) Stochastic simulation of episodic soft coastal cliff recession. *Coastal Engineering,* 46: 159–74.

Hanson, H., Aarninkof, S., Capobianco, M., Jiménez, J.A., Larson, M. and Nicholls, R.J. (2003) Modelling of coastal evolution on yearly to decadal time scales. *Journal of Coastal Research,* 19: 790–811.

Harris, P.T., Heap, A.D., Bryce, S.M., Porter-Smith, R., Ryan, D.A. and Heggie, D.T. (2002) Classification of Australian clastic coastal depositional environments based upon a quantitative analysis of wave, tidal, and river power. *Journal of Sedimentary Research,* 72: 858–70.

Heap, A.D., Bryce, S. and Ryan, D.A. (2004) Facies evolution of Holocene estuaries and deltas: a large-sample statistical study from Australia. *Sedimentary Geology,* 168: 1–17.

Hennecke, W.G., Greve, C.A., Cowell, P.J. and Thom, B.G. (2004) GIS-based coastal behavior modeling and simulation of potential land and property loss: implications of sea-level rise at Collaroy/Narrabeen Beach, Sydney (Australia). *Coastal Management,* 32: 449–70.

Holman, R.A. and Sallenger, A.H. (1993) Sand bar generation: a discussion of the Duck experiment series. *Journal of Coastal Research,* 115: 76–92.

Inman, D.L. and Bagnold, R.A. (1963) Littoral processes, in D. Hanes, and B. Le Méhauté, (eds) *The Sea.* Interscience, New York. pp. 529–43.

Inman, D.L. and Nordstrom, C.E. (1971) On the tectonic and morphologic classification of coasts. *Journal of Geology,* 79: 1–21.

Jacobs, Z. (2008) Luminescence chronologies for coastal and marine sediments. *Boreas,* 37: 508–35.

Johnson, D.W. (1919) *Shore Processes and Shoreline Development.* Prentice Hall, New York.

Johnson, D.W. (1925) *The New England - Acadian Shoreline.* Wiley, New York. p. 608.

Jude, S.R., Jones, A.P. and Andrews, J.E. (2005) Visualisation for coastal zone management, in D. Bartlett, and J. Smith, (eds) *GIS for Coastal Zone Management.* CRC Press, Boca Raton. pp. 95–107.

King, C.A.M. (1972) *Beaches and Coasts.* Arnold, London.

King, C.A.M. and McCullach, M.J. (1971) A simulation model of a complex recurved spit. *Journal of Geology,* 79: 22–36.

Komar, P.D. and Inman, D.L. (1970) Longshore sand transport on beaches. *Journal of Geophysical Research,* 75, 5514–27.

Koukoulas, S., Nicholls, R.J., Dickson, M.E., Walkden, M.J.A., Hall, J.W. and Pearson, S.G. (2005) A GIS tool for analysis and interpretation of coastal erosion model outputs (SCAPEGIS). *Proceedings of Coastal Dynamics.* Barcelona, Spain.

Lakhan, V.C. (2003) *Advances in Coastal Modelling.* Elsevier Oceanography Series.

Lakhan, V.C. (2010) Modelling the coastal system, in D.R. Green, (ed.) *Coastal Zone Management.* Thomas Telford, London. pp. 185–205.

Lakhan, V.C. and Trenhaile, A.S. (1989) *Applications in Coastal Modeling.* Elsevier Oceanography Series, 49. Elsevier, Amsterdam. p. 387.

Lambeck, K. and Chappell, J. (2001) Sea level change through the last glacial cycle. *Science,* 292: 679–86.

Larson, M., Hanson, H. and Kraus, N.C. (2003) Numerical modelling of beach topography change, in V.C. Lakhan, (ed.) 2003. *Advances in Coastal Modelling.* Elsevier Oceanography Series. pp. 337–66.

LeMéhauté, E.M. and Soldate, M. (1977) *Mathematical Modeling of Shoreline Evolution.* U.S. Army Corps of Engineers, Coastal Engineering Research Centre Misc. Report No.77-10.

Lippman, T.C. and Holman, R.A. (1990) The spatial and temporal variability of sand bar morphology. *Journal of Geophysical Research,* 95: 11575–90.

Manson, G.K., Solomon, S.M., Forbes, D.L., Atkinson, D.E. and Craymer, M. (2005) Spatial variability of factors influencing coastal change in the western Canadian Arctic. *Geo-Marine Letters,* 25: 138–45.

McCaffrey, K.J.W., Jones, R.R., Holdsworth, R.E., Wilson, R.W., Clegg, P. and Imber, J. (2005) Unlocking the spatial dimension: digital technologies and the future of geoscience fieldwork. *Journal of the Geological Society*, 162, 927–38.

McLean, R. and Shen, J.-S. (2006) From foreshore to foredune: foredune development over the last 30 years at Moruya Beach, New South Wales, Australia. *Journal of Coastal Research*, 22: 28–36.

Mimura, N. and Nobuoka, H. (1995) Verification of Bruun rule for the estimate of shoreline retreat caused by sea-level rise, in W.R. Dally, and R.B. Zeidler, (eds) *Coastal Dynamics 95*. American Society of Civil Engineers, New York. pp. 607–16.

Monmonier, M. (2008) *Coastlines: How Mapmakers Frame the World and Chart Environmental Change*. University of Chicago Press, Chicago. p. 228.

Neal, A. and Roberts, C.L. (2000) Applications of ground-penetrating radar (GPR) to sedimentological, geomorphological and geoarchaeological studies in coastal environments, in K. Pye, and J.R.L. Allen, (eds) *Coastal and Estuarine Environments: Sedimentology, Geomorphology And Geoarchaeology*. Geological Society, Special Publication, London. pp. 139–71.

Niedoroda, A.W., Reed, C.W., Swift, D.J.P., Arato, H. and Hoyanagi, K. (1995) Modeling shore-normal large-scale coastal evolution. *Marine Geology*, 126: 181–99.

Norcross, Z.M., Fletcher, C.H. and Merrifield, M. (2002) Annual and interannual changes on a reef-fringed pocket beach: Kailua Bay, Hawaii. *Marine Geology*, 190: 553–80.

O'Brien, M.P. (1966) Equilibrium flow areas of tidal inlets on sandy coasts. Proceedings of the 10th Coastal Engineering Conference, 1, 676–86.

Pelnard-Considere, R. (1956) Essai de theorie de l'evolution des forms de ravage en plage de sable et de galets. 4th Journees de l'Hydraulique, Les Energies de la Mer. Question III. pp. 792–808.

Pethick, J. (2001) Coastal management and sea-level rise. *Catena*, 42: 307–22.

Phillips, J.D. (1995) Self-organization and landscape evolution. *Progress in Physical Geography*, 19: 309–21.

Phillips, J.D. (1996) Deterministic complexity, explanation, and predictability in geomorphic systems, in B.L. Rhoadds and C.E. Thorn, (eds) *The Scientific Nature of Geomorphology*. Wiley, Chichester. pp. 315–35.

Pirazzoli, P.A. (1991) *World Atlas of Holocene Sea-Level Changes*. Elsevier Oceanography Series, 58. Elsevier, Amsterdam. p. 300.

Pirazzoli, P.A. (1996) *Sea-level changes: the last 20,000 years*. Wiley, Chichester p. 211.

Ranasinghe, R. and Stive, M.J.F. (2009) Rising seas and retreating coastlines. *Climatic Change*, 97: 465-468. doi:10.1007/s10584-009-9593-3.

Ranasinghe, R., McLoughlin, R., Short, A.D. and Symonds, G. (2004) The Southern Oscillation Index, wave climate, and beach rotation. *Marine Geology*, 204: 273–87.

Ricketts, P.J. (1992) Current approaches in geographic information systems for coastal management. *Marine Pollution Bulletin*, 25: 82–7.

Romanczyk, W., Boczar-Karakiewicz, B. and Bona, J.L. (2005) Extended equilibrium beach profiles. *Coastal Engineering*, 52: 727–44.

Roy, P.S. (1984) New South Wales estuaries: their origin and evolution, in B.G. Thom, (ed.) *Coastal Geomorphology in Australia*. Academic Press, Sydney. pp. 99–121.

Roy, P.S. and Thom. B.G. (1981) Late Quaternary marine deposition in New South Wales and southern Queensland – an evolutionary model. *Journal of Geology Society of Australia*, 28: 471–89.

Roy, P.S., Thom, B.G. and Wright, L.D. (1980) Holocene sequences on an embayed high-energy coast: an evolutionary model. *Sedimentary Geology*, 16: 1–9.

Roy, P.S., Cowell, P.J., Ferland, M.A. and Thom, B.G. (1994) Wave-dominated coasts, in R.W.G. Carter, and C.D. Woodroffe, (eds) *Coastal Evolution: Late Quaternary Shoreline Morphodynamics*. Cambridge University Press, Cambridge. pp. 121–86.

Roy, P.S., Williams, R.J., Jones, A.R., Yassini, I., Gibbs, P.J. and Coates, B. (2001) Structure and function of south-east Australian estuaries. *Estuarine Coastal and Shelf Science*, 53: 351–84.

Sallenger, A.H., Krabill, W.B., Swift, R.N., Brock, J., List, J., Hansen, M. et al. (2003) Evaluation of airborne topographic lidar for quantifying beach changes. *Journal of Coastal Research*, 19: 125–33.

Schumm, S.A. and Lichty, R.W. (1965) Time, space, and causality in geomorphology. *American Journal of Science*, 263: 110–19.

Seybold, H., Andrade, J.S. and Herrmann, H.J. (2007) Modeling river delta formation. *Proceedings of the National Academy of Sciences*, 104: 16804–9.

Sharples, C. (2006) Indicative Mapping of Tasmanian Coastal Vulnerability to Climate Change and Sea-Level Rise: Explanatory Report 2nd Edition. Report to Department of Primary Industries & Water, Tasmania, p. 173 plus accompanying mapping.

Shennan, I. and J. Andrews, J. (eds) (2000) *Holocene Land-Ocean Interaction and Environmental Change Around the North Sea*. Geological Society, Special Publication, London.

Shepard, F.P. (1948) Evidence of world-wide submergence. *Journal of Marine Research*, 7: 661–78.

Short, A.D. (ed.) (1999) *Handbook of Beach and Shoreface Morphodynamics*. Wiley, Chichester. p. 379.

Short, A.D. (2006) Australian beach systems-nature and distribution. *Journal of Coastal Research*, 22: 11–27.

Sloss, L.L. (1962) Stratigraphic models in exploration. *Journal of Sedimentary Petrology*, 32: 415–22.

Small, C. and Nicholls, R.J. (2003) A global analysis of human settlement in coastal zones. *Journal of Coastal Research*, 19: 584–99.

Sonu, C.J. and James, W.R. (1973) A Markov model for beach profile changes. *Journal of Geophysical Research*, 78: 1462–71.

Stive, M.J.F. (2004) How important is global warming for coastal erosion? *Climatic Change*, 64: 27–39.

Stive, M.J.F., Cowell, P.J. and Nicholls, R.J. (2009) Impacts of global environmental change on beaches, cliffs and deltas, in O., Slaymaker, T. Spencer, and C. Embleton-Howman, (eds) *Geomorphology and Global Environmental Change*. Cambridge University Press, Cambridge. pp. 158–79.

Stolper, D., List, J.H. and Thieler, R.E. (2005) Simulating the evolution of coastal morphology and stratigraphy with a new morphological-behaviour model (GEOMBEST), *Marine Geology*, 218: 17–36.

Storms J.E.A., Weltje G.J., Van Dijke J.J., Geel C.R. and Kroonenberg S.B. (2002) Process-response modelling of wave-dominated coastal systems: simulating evolution and stratigraphy on geological timescales. *Journal of Sedimentary Research*, 72: 226–39.

Swift, D.J.P. and Thorne, J.A. (1991) Sedimentation on continental margins. I. A general model for shelf sedimentation, in D.J.P., Swift, G.F., Oertel, R.W. Tillman, and J.A. Thorne, (eds) *Shelf Sand and Sandstone Bodies: Geometry, Facies and Sequence Stratigraphy*. Special Publication of the International Association of Sedimentologists. Blackwell Scientific Publications, Oxford. pp. 3–31.

Swift, D.J.P., Phillips, S. and Thorne, J.A. (1991) Sedimentation on continental margins, IV: lithofacies and depositional systems. *Special Publication of the International Association of Sedimentologists*, 14: 89–152.

Syvitski, J.P.M. and Daughney, S. (1992) DELTA2: Delta progradation and basin filling. *Computers and Geosciences*, 18: 839–97.

Thieler, E.R., Pilkey, O.H., Young, R.S., Bush, D.M. and Chai, F. (2000) The use of mathematical models to predict beach behavior for U.S. Coastal Engineering: a critical review. *Journal of Coastal Research*, 16: 48–70.

Thom, B.G. (1984a) Transgressive and regressive stratigraphies of coastal sand barriers in eastern Australia. *Marine Geology*, 56: 137–58.

Thom, B.G. (ed.) (1984b) *Coastal Geomorphology in Australia*. Academic Press, Sydney. p. 349.

Thom, B.G. and Chappell, J. (1975) Holocene sea levels relative to Australia. *Search*, 6: 90–3.

Thom, B.G. and Hall, W. (1991) Behaviour of beach profiles during accretion and erosion dominated phases. *Earth Surface Processes and Landforms*, 16: 113–27.

Thom, B.G. and Cowell, P.J. (2005) Coastal changes, gradual, in Schwartz, M.L. (ed.) *Encyclopedia of Coastal Science*. Springer, Dordrecht. pp. 251–3.

Thom, B.G. and Short, A.D. (2006) Introduction: Australian Coastal Geomorphology, 1984–2004. *Journal of Coastal Research*, 22: 1–10.

Thompson, W.G. and Goldstein, S.L. (2005) Open-system coral ages reveal persistent suborbital sea-level cycles. *Science*, 308: 401–4.

Trenhaile, A.S. (1987) *The Geomorphology of Rock Coasts*. Clarendon Press, Oxford. p. 384.

Turner, I.L., Aarninkhof, S.G.J. and Holman, R.A. (2006) Coastal imaging applications and research in Australia. *Journal of Coastal Research*, 22: 37–48.

Vafeidis, A.T., Nicholls, R.J., McFadden, L., Tol, R.S.J., Hinkel, J. and Spencer, T. (2008) A new global coastal database for impact and vulnerability analysis to sea-level rise. *Journal of Coastal Research*, 24: 917–24.

Valentin, H. (1952) *Die Küsten der Erde*. Petermanns Geographische Mitteilungen, 246. J. Perthes, Gotha. p. 118.

Walkden, M.J.A. and Hall, J.W. (2005) A predictive mesoscale model of the erosion and profile development of soft rock shores. *Coastal Engineering*, 52: 535–63.

Walkden M.J.A. and Dickson, M.E. (2008) Equilibrium erosion of soft rock shores with a shallow or absent beach under increased sea level rise. *Marine Geology*, 251: 75–84.

Walker, H.J. (2008) Coastal landforms, in T.P., Burt, R.J., Chorley, D., Brunsden, N.J. Cox, and A.S. Goudie, (eds) *The History of the Study of Landforms, or the Development of Geomorphology*. Vol 4. *Quaternary and Recent Processes and Forms (1890–1965) and the Mid-Century Revolutions*. The Geological Society of London. pp. 805–62.

Williams, W.W. (1960) *Coastal Changes*. Routledge & Kegan Paul, London. p. 220.

Wolinsky, M.A. (2009) A unifying framework for shoreline migration: 1. Multiscale shoreline evolution on sedimentary coasts. *Journal of Geophysical Research*, 114: F01009. doi:10.1029/2007JF000855.

Wolinsky, M.A. and Murray, A.B. (2009). A unifying framework for shoreline migration: 2. Application to wave-dominated coasts. *Journal of Geophysical Research*, 114: F01009. doi:10.1029/2007JF000856.

Woodroffe, C.D. (2003) *Coasts: Form, Process and Evolution*. Cambridge University Press, Cambridge. p. 623.

Woodroffe, C.D., Mulrennan, M.E. and Chappell, J. (1993) Estuarine infill and coastal progradation, southern van Diemen Gulf, northern Australia. *Sedimentary Geology*, 83: 257–75.

Wright, L.D. (1995) *Morphodynamics of Inner Continental Shelves*. CRC Press, Boca Raton. p. 241.

Wright, L.D. and Thom, B.G. (1977) Coastal depositional landforms: a morphodynamic approach. *Progress in Physical Geography*, 1: 412–59.

Wright, L.D. and Short, A.D. (1984) Morphodynamic variability of surf zones and beaches: a synthesis. *Marine Geology*, 56: 93–118.

Zhang, K., Douglas, B.C. and Leatherman, S.P. (2004) Global warming and coastal erosion. *Climatic Change*, 64: 41–58.

25

Aeolian Environments

Joanna E. Bullard

The contemporary scope of aeolian geomorphology encompasses the erosion, transport and deposition of sand and dust-sized particles, and the landforms and landscapes produced by these processes, including dunes, sand seas, loess deposits as well as erosional features such as yardangs and ventifacts (which are not considered here). Many other landforms are formed by the interaction of a range of geomorphic processes of which aeolian activity is only one, including sand ramps, pans or playas and stone pavements, and the aeolian modification of landforms such as alluvial fans and glacial moraines has also been observed. There is evidence of aeolian landforms (whether active, stabilized or relict) in most terrestrial environments, but the majority are associated with either arid conditions or coastal environments. Although, with the exception of some polar regions, wind energy availability is usually no greater in arid than in humid areas, aeolian processes often affect large regions of arid areas due to the widespread availability of suitable erodible sediment and a low density vegetation cover which increases the erosive potential of the wind. Similarly, coastal zones often feature the required strong winds and abundant sediment supply for aeolian landforms to develop. The aeolian processes operating and landforms produced in arid and coastal environments (not themselves mutually exclusive), are similar and it is noted by Bauer and Sherman (1999: 73) that '… the only distinctive coastal dune is the foredune, because it is … integrally coupled to the complex suite of nearshore processes fronting the beach-dune system'. This chapter focuses on aeolian processes and landforms in arid environments, but much of what is discussed here also applies to coastal locations.

Notwithstanding some early works, such as Cornish's (1914) *Waves of Sand and Snow*, aeolian geomorphology as a sub-discipline developed following the publication of Bagnold's (1941) *The Physics of Blown Sand and Desert Dunes*. Whilst a few highly influential papers were produced over the following decades, this classic book remained the only reference text in the field for nearly 40 years. A resurgence of interest in aeolian research in the past 30 years has been triggered by a range of factors and marked by a number of significant publications. These factors include space missions, which have highlighted the importance of understanding aeolian processes and landforms on Earth to use as analogues for landscape development on other planets (Greeley and Iverson, 1985); the successful launches of increasingly sophisticated satellite-based remote sensing devices around both Earth and Mars which have allowed data acquisition over scales appropriate for major sand seas (McKee, 1979); fears about desertification and land degradation in arid areas (Middleton and Thomas, 1992; UNEP, 1997); and an increasing recognition of the role of dust in atmospheric processes and associated global climate change (Goudie and Middleton, 2006).

A recent review of publications in aeolian research from 1646 to 2005 by Stout et al. (2008) documents the impacts of political and social events on the field. These include the two world wars which were marked by declines in the numbers of papers published and the 1930s Dust Bowl years in the USA which heralded an increase in research on wind erosion. More recently, following the end of the Cultural Revolution in China in 1976, the number of aeolian research papers published by Chinese authors (in all languages and locations) has risen from fewer than ten per year to over 350 per year, accounting for 44 per cent of all aeolian research publications in 2005. This reflects both the increase in support for scientific research by the Chinese government and the importance of aeolian processes throughout China, which has extensive dunefields, loess deposits and frequent dust storms (Stout et al., 2008).

Technology that can now be applied to aeolian activity at scales ranging from fractions of a second (e.g. sonic anemometers that sample at >20 Hz) to over 300,000 years (e.g. luminescence dating), and from single-grain studies (using computer modelling) to planetary-scale studies (using satellite remote sensing) is rapidly increasing the resolution of our understanding of aeolian processes and landforms. However, many of the basic principles laid down in early volumes remain sound and have been spatially and temporally refined rather than rejected and replaced. This is illustrated by the sustained positive reference to classic works in aeolian geomorphology, that have, for example, led Bagnold's (1941) book to be the second most widely cited volume in papers published in the journal *Geomorphology* (1242 times from February 1995 to August 2004; Doyle and Julian, 2005).

ARID ENVIRONMENTS

Arid environments can be differentiated from humid areas using various criteria including surface processes, biomes, rainfall and habitability. One of the most widely used schemes is the aridity index (UNEP, 1997) based on moisture balance, or water stress. The aridity index, calculated as the mean annual ratio of precipitation (P) to potential evapo-transpiration (PET), enables the differentiation of four arid zones – dry sub-humid, semi-arid, arid and hyper-arid – that together cover 47 per cent of the terrestrial globe, or 62 million km^2 (Table 25.1, Figure 25.1). The UNEP (1997) map of arid areas largely excludes

cold climate zones but some high latitude or high altitude regions are very dry and have similar geomorphological characteristics to warm deserts. These include parts of the Andean altiplano, the Dry Valleys of Antarctica and some ice-free regions of the Arctic. Very cold conditions result in low mean precipitation, for example, the mean annual precipitation in the Dry Valleys of Antarctica is less than 60 mm (rainfall equivalent), although low potential evapo-transpiration values can mean soils are waterlogged or characterized by permafrost. Seppälä (2004) estimates that cold arid areas – defined as those where the mean temperature of the warmest month is less than 10°C and mean annual precipitation is less than 250 mm – cover around 5 million km^2.

Four main factors contribute to aridity. First, at the global scale, descending, stable warm air associated with the tropical high-pressure belts and cold, dry subsiding air at the North and South Poles lead to low, and highly variable, rainfall in these regions. Second, as air masses move over land they lose their moisture as rainfall in maritime regions and consequently continentality, or distance from the ocean, is also a major cause of aridity, and particularly important in central Asia. Third, at the regional scale, rain-shadow caused by high mountain ranges, such as the southern Andes, Sierra Nevada and the Australian Great Dividing Range can exacerbate aridity. Finally, winds blowing across the cold ocean currents flowing north from southern polar latitudes, such as the Humboldt (South America) and Benguela (southern Africa) currents, are stable, cold and have a low moisture-bearing capacity. Evaporation from the sea surface in these areas is low and coastal areas receive little precipitation; however, the coincidence of warmer air over the land mass and the cool ocean air can result in coastal fog which is an important moisture source in deserts such as the Atacama and Namib.

Tectonic setting and history exert important controls on large-scale relief and the long-term climate of arid regions, including the relative positions of land masses and orogeny. The major deserts form two main groups in terms of their

Table 25.1 Classification and extent of arid environments

Descriptor	Aridity Index (UNEP, 1997) P/PET	Area	Approximate mean annual precipitation
Hyper-arid	<0.05	10 million km^2	<50 mm
Arid	0.05–<0.2	16 million km^2	50–250 mm
Semi-arid	0.2–<0.5	23 million km^2	250–500 mm
Dry sub-humid	0.5–<0.65	13 million km^2	—
Cold arid	—	5 million km^2	<250 mm

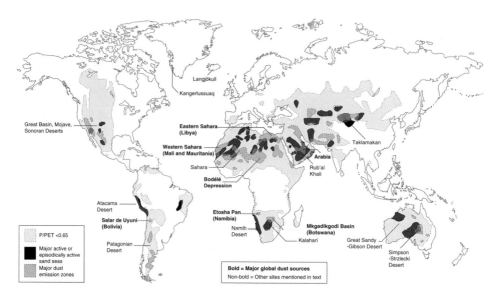

Figure 25.1 Global distribution of arid environments, active sand seas and major dust sources

structural and tectonic controls – the shield and platform deserts (such as the central Sahara, Kalahari and central Australia) and the mountain and basin deserts (such as the Atacama, central Asia and Great Basin, USA). The predominant climate in shield and platform deserts is arid, warm, continental and largely uniform or gently grading across the area, whilst mountain and basin deserts tend to have a highly variable climate ranging from cool, semi-arid to hyper-arid. This, in part, reflects contrasts in relief between the two types. Shield and platform deserts are tectonically stable with low mean elevation, low relative relief and slow rates of uplift which result in the formation of extensive interior basins and sporadic or seasonal (often exotic) fluvial discharge. The stability of these landscapes and availability of accommodation space means sand deposits are often large and extensive. In contrast, the mountain and basin deserts have steep and high relative relief and are often tectonically active. Local erosion rates are high and interior drainage basins can be small. In such active environments, relict landforms are uncommon and sand deposits are confined to small, localized and often discrete areas.

Most geomorphic processes operate to some degree in arid environments, but what differentiates arid from humid regions are the direct and indirect impacts of limited water availability. In arid and hyper-arid areas this is most evident in restricting the quantity and permanence of vegetation. The lack of vegetation cover, combined

with high diurnal temperature ranges, exposes sediments to extremes of temperature, for example, ground surface temperatures can exceed 70°C with diurnal ranges in excess of 50°C (Goudie, 1997). This exposure can mean that local geology is a more important control of weathering and sediment production than in humid environments. Extensive areas of unconsolidated and unprotected fine sediments are vulnerable to wind erosion and this lack of protection can also mean that when rainfall events, which can be very intense, do occur they result in rapid erosion leading to high sediment loads in some dryland rivers. More vegetation is present in semi-arid and sub-humid areas and can play an important geomorphological role. The development of some dunes such as nebkha or parabolic dunes is dependent on vegetation; the localized destruction of vegetation can lead to the formation of erosional features such as blowouts (Bullard, 1997). In cold climate arid zones, strong ice sheet-driven winds and rapid and persistent production of fine sediments can result in intense seasonal dust storms and the development of proglacial dunes and niveo-aeolian processes may be locally significant. As noted earlier, aeolian processes are not confined to arid environments nor do they dominate them in terms of surface area – in fact sand dunes cover only 20–25 per cent of arid lands and whilst Quaternary loess deposits are estimated to cover 10 per cent of the terrestrial globe (Pécsi, 1990), they extend far beyond contemporary dryland boundaries. Spatial and temporal patterns of dust emissions are often closely controlled by desert characteristics; however, dust plumes disperse as they travel and, although there are broadly defined contemporary arid zone dust sources (Figure 25.1), these finer particles are transported and deposited worldwide from the Amazon to Antarctica.

PRINCIPLES OF AEOLIAN SEDIMENT TRANSPORT

Aeolian sediment entrainment is determined by the interaction between meteorological variables and the forces acting to transport or stabilize particles. The relative importance of cohesive, gravitation and aerodynamic forces is dependent upon the size of the particles – below 10 μm cohesive forces tend to dominate, for particles in the range 10–300 μm aerodynamic forces dominate, and greater than 300 μm gravitational force dominates. For this reason the majority of wind-transported sediments are in the 10–300 μm category. Particles less than 10 μm in diameter can be transported if the cohesive forces enable them to aggregate to form a larger, usually less dense, unit, which is entrained, rather than the individual grains. Climate and meteorological variables can modify these boundaries and are important in determining the minimum wind velocities at which particle entrainment is initiated.

Sediment entrainment and transport

When air moves over a surface it exerts both lift and drag forces and if these forces exceed those of gravity and cohesion the particle will start to move. There are three main types of particle movement: creep, saltation and suspension. During creep, particles on the surface move by rolling or sliding and remain in contact with the surface. Saltating particles are those which travel via a series of short jumps or hops but remain within ≈3 m of the surface. Suspended particles travel longer distances and their trajectories are governed by regional atmospheric pressure systems. Of these, saltation is the most important process because it largely drives the other two modes of transport and is to an extent self-perpetuating. Impacts from saltating grains can dislodge particles on the surface causing creep, and can eject particles from the surface into the airstream where they become part of the saltation load or, in the case of finer particles they may be lifted higher into suspension. Commonly, sand-sized material

(63–2000 μm) travels by saltation and creep within the lowest levels of the atmospheric boundary layer and is transported short distances. In contrast, dust-sized material (<63 μm) is transported in suspension at a wide range of heights above the surface and can rapidly travel extensive distances.

The way in which a wind-borne particle travels is governed by the balance between the forces holding it aloft in the atmosphere and those pulling it towards the ground surface. If the terminal velocity (w_t, the rate at which gravitational forces pull the particle towards the surface) is the same as, or less than the mean velocity at which the air parcel within which the particle is contained is moving upwards by atmospheric turbulence (the Lagrangian vertical velocity), then the particle is carried in suspension. In a thermally neutral atmospheric surface layer, the Lagrangian vertical velocity is κu_*, where κ is von Karman's constant (0.4) and u_* is the friction velocity (Hunt and Weber, 1979). Particles travel in suspension when $|w_t|/\kappa u_* \ll 1$ and travel by saltation or creep when $|w_t|/\kappa u_* \gg 1$. This balance between upward and downward forces means that larger particles can be carried in suspension in stronger and more turbulent airflows whilst in weaker or less turbulent airflows the same sized particle might travel by saltation or by creep.

Overall wind speed has a fundamental impact on the bulk rate of sediment transport, however initial particle entrainment is determined by airflow turbulence and the transport rate generally increases as gust frequency increases. In reality, at the point where sediment is being entrained, assuming a range of particle sizes are available, the load transported by the wind is likely to comprise a mix of saltating, creeping and suspended particles and also to include particles travelling in a mode part way between pure suspension and pure saltation in what has been called modified suspension (Nalpanis, 1985) where particle trajectories are influenced by both their inertia and settling velocity. Relatively large particles (>300 μm) have been observed travelling up to 4 m on the lee side of dunes in this temporary, turbulent mode of suspension, which is a transport distance far longer than that predicted by numerical simulation of true saltation (Nickling et al., 2002). In addition, the proportion of the total load moving by each mode varies continually according to wind speed, turbulence and particle characteristics. For example, wind tunnel experiments by Dong and Qian (2007) show the proportion of aeolian sediment flux moving as creep varies from 0.04 to 0.29 depending on wind speed and grain size. Saltation not only causes creep and suspension, but impacting particles can also cause a 'splash' type ejection of grains from the bed in a process known as reptation. Initial movement of reptating particles can be in any direction, including upwind, and the trajectory is both low and short. Reptation can be important in affecting the consolidation of the bed surface and may be a key component in determining aeolian ripple formation (Anderson, 1987). Finally, it is important to note that saltating particles rarely travel as a uniform body of sediment but instead exhibit marked spatial and temporal patterns, primarily forming streamer patterns (Baas and Sherman, 2005).

As discussed earlier, the relative importance of the different forces acting upon particles varies with grain size and the strengths of these different forces in turn affects the wind speeds required to entrain particles. The main inorganic variables determining the strength of cohesive forces binding particles to each other are temperature and moisture availability (or humidity). Wind tunnel experiments by McKenna Neuman (2004) have demonstrated that cold air can transport more sediment than warm air for a given wind speed. At the extremes, wind blowing in the Dry Valleys of Antarctica during winter at temperatures of around −40°C can transport up to 70 per cent more sediment than wind of the same velocity blowing in a hot desert at +40°C. This is attributed to the impact of temperature on air density and turbulence, and the lower humidity that is associated with low temperatures and reduces interparticle cohesion (McKenna Neuman, 2004).

Numerous studies have tried to determine a threshold or critical moisture content above which aeolian sediment transport will cease due to inter-particle cohesion. Suggested values of such a threshold range from 5 to 25 per cent and vary with wind speed, turbulence intensity and grain characteristics. Wiggs et al. (2004) demonstrated that aeolian sediment transport is highly sensitive to changes in moisture content and that the main control on aeolian saltation can rapidly switch between wind velocity and moisture content. In their study (conducted in a temperate environment), up to 2 per cent moisture content had no impact on the rate of sand transport and the threshold was found to be 4 to 6 per cent. In warm arid environments moisture sources include precipitation, flooding, fog and groundwater but the persistence of moisture in surface sediments is low due to high evaporation rates and desiccation by wind. Particle cohesion can also be affected by the development of physical or biological crusts that bind or seal the surface reducing erodibility (Thomas and Dougill, 2007).

Few, if any, natural surfaces are exactly horizontal. Dune slopes range from zero to the angle of repose of sand (usually 32–34°),

undulating aeolian ripples are common on sand sheets and superimposed on dunes and in addition the underlying topography on which aeolian material has been deposited can have considerable relief. Hardisty and Whitehouse (1988) used field experiments to demonstrate that the mass flux of sediment transport increased when particles travelled downslope and decreased as they travelled upslope. A recent study of wind erosion in Langjökull, Iceland, found that slopes exceeding 10 per cent were sufficient to stop sand movement, although finer particles travelling in suspension were not affected (Gisladottir et al., 2005).

Most dune sediments are dominated by quartz or feldspar, although carbonate, gypsum and volcanic sands can accumulate under some circumstances, and are broadly sub-rounded to sub-angular. The roundness of particles reflects both the lithological origins of the grains and their geomorphological history which may include not only aeolian processes but also fluvial, glacial and weathering processes; unsurprisingly therefore, different dunefields exhibit distinct grain roundness characteristics. The effect of sand grain shape (both roundness and sphericity) on entrainment is poorly understood but likely to be less important than the overall size and sorting of the sediment population. In contrast, the shape of dust particles can affect distance travelled due to the role of shape (and mass) in determining terminal velocity (w_t). Natural dust particles are rarely perfectly spherical and it is more common for particles to be platy or 'flattened'. As with sand-sized material, the shape of dust particles is linked to lithology and also to the mechanism by which it was formed; for example, crushing or grinding results in blade-shaped particles (Assallay et al., 1998). For clay-sized sediments the effective particle shape is determined by aggregation. As the number of particles comprising the aggregate increases, its overall shape will become more irregular and there will be a concomitant decrease in density (Goossens, 2005).

Although dryland environments are often associated with a lack of vegetation, few arid areas are entirely devoid of vegetation or biological material (such as cyanobacterial crusts). Although it is often spatially and temporally discontinuous, vegetation can have a profound impact on aeolian processes, protecting the surface immediately underneath the plant from erosion, increasing surface friction, reducing wind erosivity and promoting the deposition of particles either as an obstacle to movement or by creating zones of reduced wind velocity in their lee (Bullard, 1997). As with moisture content, researchers have attempted to determine thresholds of vegetation cover which will prevent aeolian sediment transport (and are particularly of interest for dune management and the prevention of wind erosion). Values of 15 to 20 per cent areal cover are typical but factors such as vegetation spacing, flexibility and shape cause this to vary considerably (Wolfe and Nickling, 1993).

Aeolian deposition

Following aeolian entrainment, particles can be transported distances varying from less than 1 grain diameter (creep) to greater than 10,000 km (suspension) depending on their size and mode of travel. Deposition occurs when wind velocity decreases below that required to keep particles in motion. This can be caused both by natural fluctuations in the wind, such as gustiness, but also by changes in the surface over which the particles are travelling. For example, moving particles transfer some energy to the ground surface on impact and it is this energy exchange that initiates creep, impact saltation, suspension and reptation by surface particles as well as some reorganization of the bed. It therefore takes more wind energy to transport saltating particles over an unconsolidated (e.g. loose sand) bed than over consolidated material (e.g. bedrock) where the energy exchange on impact is much lower. Consequently, for a given wind speed, particles travelling over a consolidated surface may be deposited if the surface changes, for example, they encounter a sand patch. Wind speed also changes with topography and roughness; flow compression (acceleration) occurs as air flows over or around an obstacle and expansion (deceleration) occurs in the lee of the obstacle. Such obstacles may be static, such as boulders, hills and escarpments; dynamic, such as existing aeolian bedforms; or semi-permeable or flexible in the case of vegetation. Particles are typically deposited where wind speed is reduced on the downwind side of such obstacles or when the steepness of a slope up which they are being transported becomes too great. Very fine material carried in suspension is typically removed from the atmosphere by rainfall (wet deposition), although particles travelling near the ground surface can become trapped by the leaves of vegetation.

Aeolian sand deposits include ripples, dunes and sand sheets and dust deposits can accumulate to form loess. Sand ripples are the smallest aeolian bedform and can be superimposed on larger scale features such as dunes and sand sheets. Ripple morphometry (wavelength, height, sinuosity) is determined by wind speed and grain size and ripple orientation responds rapidly to changes in the direction of sand transporting winds. Dunes have been classified in many different ways (Tables 25.2 and 25.3), but perhaps the most useful basic distinction is between free dunes – the morphology of which is a direct reflection of the relationship between wind directional variability and sand supply – and impeded or anchored dunes, where there are additional controlling factors such as vegetation (e.g. nebkha), topography (e.g. climbing, falling or echo dunes) or sediment size (e.g. zibar). Three basic modes of dune 'activity' can also be discerned. Migrating dunes, such as barchans and transverse dunes, advance downwind whilst maintaining their basic shape; extending dunes, typified by linear dunes, elongate rather than migrate and only their downwind end advances; accumulating dunes exhibit little or no migration but gain volume by basal expansion and increasing height and are exemplified by star dunes (Thomas, 1992; Tsoar et al., 2004). McKee (1979) introduced the idea of simple dunes, a single, discernible free dune form, compound dunes, where two dunes of the same type are superimposed, and complex dunes, where two dunes of different types are superimposed. The development of compound and complex dunes can be triggered by secondary air flow on large dunes, for example, resulting in the formation of small star dunes or crescentic dunes on the flanks of large linear dunes, increase or decrease in sediment supply over short or long time scales and changes in wind direction. Sand dunes, particularly free dunes, often, but not always, occur in groups forming dunefields or sand seas.

APPROACHES TO STUDYING AEOLIAN PROCESSES AND LANDFORMS

The study of aeolian processes and landforms has a long pedigree and the methods and approaches employed by researchers have evolved both as knowledge of aeolian systems has increased, and as technology, particularly computers and data loggers have improved and become more portable and robust. Aeolian research questions are approached in a variety of ways and draw upon field measurement and monitoring, controlled experiments using wind tunnels and other laboratory equipment, remote sensing from aeroplanes and satellites and numerical modelling. Whilst one data source might be used, it is more common for researchers to draw upon more than one of these approaches.

For sand and dust-sized particles characterization of the physical and chemical properties provides information pertinent to the erodibility of the sediments, their transport and geomorphological history. The size and sorting of sediment

Table 25.2 Summary of major dune classification schemes (Bullard and Nash, 2000)

Basis of classification	Author(s)	First classificatory division	Subsequent divisions
Extent to which sand accumulation is controlled by vegetation or topography	Pye and Tsoar (1990)	Sand accumulation related to topographic obstacles	Windward accumulations (echo, climbing), cliff-top accumulations, or leeward accumulations (lee, falling)
		Sand accumulation related to bed roughness changes or aerodynamic fluctuations	Self-accumulated dunes formed from fine sand (barchan, transverse, unvegetated linear, dome, star) or from poorly sorted/bimodal sand (sand sheet, zibar)
		Sand accumulation related to vegetation	Parabolic, hummock and vegetated linear
	Cooke et al. (1993)	Stabilized	Cemented (clay dunes, aeolianite) or vegetated (nabkha, beach/stream-side dune ridges, randwallen, vegetated parabolic, blowout)
		Anchored	Echo, climbing, flank and lee
		Mobile	Transverse (dome, barchan, transverse), linear (sandridge), network, star, sand sheet and zibar
	Livingstone and Warren (1996)	Free	Transverse (transverse, barchan, dome, reversing), linear (sandridge, seif) and star dunes (star, network), sand sheets (zibar, streaks).
		Anchored	Vegetation-anchored (nebkha, parabolic, coastal and blowout) or topographically anchored dunes (echo, climbing, cliff-top, falling, lee, lunette)
Dune morphology	McKee (1979)	Dune shape, number of slip faces/internal structure	Sand sheets, stringers, dome, barchan, barchanoid ridge, transverse ridge, blowout, parabolic, linear (seif), reversing and star dunes [supplemented by Fryberger (1979) who examined wind environments for the various dune types, and modified by Greeley and Iversen (1985) to include relationship between dune axis orientation and wind direction]
Major environmental factors influencing dune formation	Wasson and Hyde (1983)	Wind directional variability, equivalent sand thickness	Star, linear, transverse, barchan [modified by Livingstone and Warren (1996) using additional data from Lancaster (1994) to include networks]
	Lancaster (1995)	Basic dune types: Wind regime complexity, Relative sediment thickness	Crescentic (barchans, crescentic ridges, compound crescentic dunes), linear (simple, compound, complex), reversing and star dunes.
		Non-basic dune types: Relative sediment thickness, sand texture, presence of vegetation, proximity to sediment source, topographic influences	Sand sheets, zibars, nebkhas, parabolics, lunettes, falling and climbing dunes
Dune morphology and major environmental factors influencing dune formation	Bagnold (1941)	Presence of obstructions, shape (including number of slip faces)	Deposits caused directly by fixed obstructions (sand shadows, sand drifts), true dunes (barchan, longitudinal), coarse-grained remnants of seif dunes (whalebacks), large scale undulations and sand sheets
	Thomas (1997)	Dune shape and number of slip faces, formative wind regime and nature of dune movement	Zibar, dome, blowout, parabolic, transverse (barchan, barchanoid, transverse), linear (linear ridge, seif dune), reversing and star dune

Table 25.3 Summary of the main characteristics of the most widely studied simple free dune types

Dune type	Description	Wind regime and sediment supply	Dynamics	Stratigraphy	Location
Barchan	Ellipsoidal/crescentic plan; single concave slipface; 'horns' extend downwind; stoss slope 2–10°, lee slope (slipface) 32–34°; interdunes gravelly, irregular dune spacing; horn to horn width of simple barchans up to 250 m; height can exceed 10 m	Unidirectional winds; directional variability <15° about the mean; low sediment supply	Migrate downwind in direction of sand transport; rates decrease with increasing dune size, e.g. 5–7 m yr⁻¹ (Taklamakan), 12.8 m yr⁻¹ (Egypt), 80 m yr⁻¹ (Peru)	Dominated by grainfall + grain flow cross-strata deposited on lee slope; cross-beds can be truncated by low-angle erosion surfaces; wind ripple laminae on stoss slope	Margins of sand seas; sand transport corridors Barchan occupy 40% of sand seas worldwide
Transverse	Straight or scalloped plan view; subdued or absent horns; oriented perpendicular to resultant wind direction; one slipface on lee slope	Unidirectional winds, but more variable than barchans; medium sediment supply	Migrate downwind in direction of sand transport but at a slower rate than barchans; 3.4–7.7 m yr⁻¹ (Inner Mongolia)	Dominated by grainfall + grain flow cross-strata deposited on lee slope; cross-beds can be truncated by low-angle erosion surfaces; wind ripple laminae on stoss slope	Transverse dunes cover 30–40% Sahara desert and are the dominant dune type in USA (70% all dunes)
Linear	Sand transport parallel to dune crest; dunes up to 2 km wide and 200 km long; steep-sided crest (5–20°) incorporates slipface, shallow plinth (2–5°); asymmetric profile; frequently associated with vegetation; merge to create y-junctions	Non-unidirectional winds, several models propose bidirectional winds; medium sediment supply	Predominantly extending forms; some possible lateral migration. 12–14 m high Sinai dunes extend up to 14.5 m yr⁻¹ 50 m high Namib dunes extend up to 2 m yr⁻¹	Combination of wind ripple and grainflow deposits; interpretation depends partly on model for dune formation adopted	Always occur as dunefields which can be extensive. 99% Kalahari, 60% Namib and 99% Australian desert dunes are linear
Star	Three or more radial arms; central high point up to 400 m; arms have slipfaces over all or part of slope which move seasonally; can occur as linked chains of dunes	Multidirectional (complex) winds; high sediment supply	Accumulating forms; individual arms may move 10–20 m yr⁻¹, but centre of dune only migrates very slowly if at all	Complex sedimentary structures; evidence of avalanche slipfaces in multiple directions	Centres of sand seas; local areas of complex wind regimes. Star dunes comprise 40% NE Sahara and 18% Namib sand sea

populations less than ≈2 mm diameter can be quantified in a variety of ways – sieving at 0.25 φ intervals differentiates 36 size classes between 2000 μm (upper size limit of sand) and 39 μm (lower size limit of silt) and is useful for describing sand-sized populations; however, higher resolution techniques such as the Coulter Multisizer, which can resolve 256 size classes in the range 2 to 75 μm (McTainsh et al., 1997), are more appropriate for detailed analysis of finer fractions. Few techniques are ideal for very mixed sediment populations and there are often problems comparing particle-size data obtained using different techniques due to the range of criteria by which grain size is defined. Newer approaches exploiting improvements in digital technology include the use of calibrated digital image analysis which can be applied rapidly in situ in the field (Rubin, 2004) and also in the laboratory where it has been applied to the detection of loess material in sediment cores (Seelos and Sirocko, 2005). Other tools such as automated scanning electron microscopy can quantify grain size, shape and texture as well as creating mineralogical maps of samples (Pirrie et al., 2004).

The ability to quantify accurately the physical characteristics of a sediment population and its chemical and organic components can be used to 'fingerprint' the source area from which the sediment was derived. Muhs (2004) demonstrated that mineralogical maturity of sands determined using X-ray diffractometry and other methods could provide insights into the evolution and origins of dunefields, whilst Marx et al. (2005) used ultra-trace-element composition to demonstrate the source of New Zealand dust deposits. Higher-resolution particle size distribution analyses have enabled more sophisticated approaches to describing and deconstructing populations than were available to earlier researchers (e.g. Folk and Ward, 1957); however, there remains some methodological inertia, particularly in the employment of statistical descriptors, to enable comparison of new research with that which has gone before.

The processes of aeolian sand transport – particularly interactions amongst saltating, creeping, suspended and reptating particles and an unconsolidated bed – are highly complex and difficult to observe. High-speed video of such interactions has proven a useful research tool as have other techniques such as particle-imaging velocimetry and modelling. For example, discrete element modelling (DEM), in which the movement of individual particles is calculated from equations of motion taking into account contact force and wind drag, has been used to investigate dynamic behaviour amongst sand grains, particularly interactions between moving grains and an unconsolidated bed.

between moving grains and the surface. Sun et al. (2001) used DEM to simulate the motion of 9000 sand grains and were able to quantify the energy exchanges between saltating and creeping particles and to generate ripples on the bed. In the field and wind tunnels, spatial and temporal patterns of aeolian sediment transport can be elucidated using sensitive particle counters that indicate whether or not saltation is occurring. Synchronized with wind velocity or turbulence measurements, these can be used to determine variations in the threshold wind speed and turbulence intensity required to transport sediment (Figure 25.2). Using this technique Stout (1998) demonstrated that the actual wind speed at which saltation was initiated was considerably higher than that implied when wind speed is averaged over up to 60 s, a typical averaging period used in many field studies. Recognition of the importance of small-scale temporal and spatial variations in sediment transport caused by rapid fluctuations in turbulence intensity and sediment availability, has been facilitated by the advent of equipment such as sonic anemometers and particle counters and for some studies it has become practical to deploy high-density instrumentation.

Much of what is known about the development and dynamics of individual sand dunes has been based on the systematic collection of empirical field data. Studies of dune morphology, wind fields and sedimentology and their interactions over barchans (e.g. Wiggs, 1993), linear dunes (e.g. Livingstone, 1989; Tsoar, 1983), star dunes (e.g. Lancaster, 1989) and other dune types have provided important baseline data about such landforms and set the agenda for subsequent more detailed field campaigns and modelling studies. Insight into dune stratigraphy, and consequently dune dynamics, has also been provided by the application of techniques such as ground-penetrating radar (e.g. Bristow et al., 2005). The data and conclusions drawn from these studies have been vital in developing understanding of dune dynamics and behaviour; however, as related by Livingstone et al. (2007), by the end of the 1990s it had became clear that for understanding of dune dynamics and development to progress further, more sophisticated approaches needed to be developed. These approaches largely focus around numerical modelling and include computational fluid dynamics, which has been successful in predicting upwind and downwind flow structures over two-dimensional transverse dunes (e.g. Parsons et al., 2004), and linear expansion models (e.g. Hersen, 2004).

Whilst single dunes can occur in arid environments, most free dunes are not isolated and occur as part of a dunefield or sand sea (≤ 30,000 km² and >30,000 km², respectively, although the terms

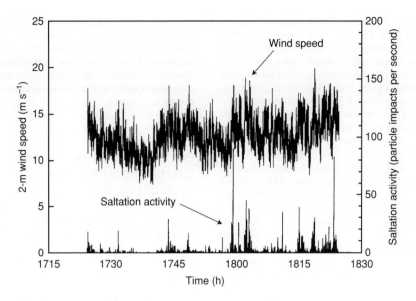

Figure 25.2 Wind speed (measured at 2 m) and saltation activity measured during a 1-h sampling period (After Stout, 1998)

are often used interchangeably). These groups of dunes can cover immense areas – the largest active sand sea is the 560,000 km² Rub'al Khali, Saudi Arabia – and whilst some early researchers did offer insights at the dunefield scale, systematic studies have largely been facilitated by the advent of satellite remote sensing data. Breed et al. (1979) mapped the distribution of dune morphology and sandsheets in eight desert regions and this, and subsequent studies, demonstrated the variability of dune forms both within individual deserts and across different arid regions. Some contain a mix of dune types, for example, the Namib Sand Sea comprises 60 per cent linear dunes, 22 per cent transverse dunes and 18 per cent star dunes, whereas others, such as those in the south-west Kalahari and Australia, where more than 99 per cent of the desert dunes are linear, are dominated by a single dune type. An important factor contributing to the variety and complexity of dunes is the evolution of a dunefield through time. The development and continued application of luminescence dating has revolutionized understanding of long-term development of dunefields and the ways in which multiple phases of dune development lead to distinct dune generations (for a recent review see Lancaster, 2008). Remote sensing data remain the keystone of dunefield scale morphological studies (e.g. Fitzsimmons, 2007); however, researchers are also starting to apply computer and mathematical modelling to try and understand interactions

amongst groups of dunes and the development of dune patterns (e.g. Kocurek and Ewing, 2005; Baas and Nield, 2007). Some of these models have successfully reproduced dune patterns and behaviour observed in the field; however, others have suggested dune dynamics for which there is little or no field-based geomorphological evidence (for a discussion see Livingstone et al., 2007).

Ascertaining rates of sand and dust transport yields insights into erodibility, soil loss and bedform dynamics. To obtain mass flux data this often entails trapping of sediments, which has the added advantage that they are available for further analysis. Goossens and Offer (2000) and Goossens et al. (2000) have reviewed the performance of various sand and dust samplers and new devices are continually being developed. Passive deposition traps are usually low technology, robust and appropriate for deployment in remote or hostile environments but rarely provide high-resolution temporal data. Active samplers, such as laser photometers, can provide a continuous detailed temporal record. Whilst it is possible to relate measurements of some different parameters of aeolian transport (e.g. horizontal and vertical dust fluxes; Goossens, 2008) it is increasingly the case that, where possible, researchers use complimentary techniques.

Satellite-based sensors have not only enabled the detailed mapping of dunefields but they have revolutionized the mapping of dust emissions at the global scale. Whilst data from meteorological stations provide ground-based information about dust emissions, these are spatially discontinuous and there are large arid regions for which little information is available. Satellite data either alone (Prospero et al., 2002), or in combination with surface observations (Washington et al., 2003) can be used to produce a more complete picture of the spatial distribution of global dust sources and their seasonal variation. Data from the total ozone monitoring spectrometer (TOMS) and ozone monitoring instrument (OMI) have been widely used for dust studies, however there are limitations to this and data from most other satellite-based sensors (e.g. Baddock et al., 2009). Lee et al. (2008) have demonstrated that the use of complementary satellite sensors, such as TOMS, MODIS (moderate- resolution imaging spectroradiometer), GOES (geostationary operational environmental satellite) and SeaWIFS (sea-viewing wide field of view sensor), each of which has strengths and limitations, can enable very precise mapping of dust emissions.

As with sand transport and dune development, considerable effort is currently being expended on modelling global dust emissions at a range of scales. This agenda is driven predominantly by the recognition that atmospheric dust is capable of driving or affecting global climate change. Different models predict varying amounts of dust with the majority suggesting total global dust emissions in the range 1000 to 2000 Tg yr^{-1} (Table 25.4) but considerably higher and lower estimates have been reported due largely to the different underlying assumptions incorporated into the models. Such models require high quality geomorphological data concerning surface roughness, topography and hydrology and a good understanding of how landscape components interact to produce realistic simulations of contemporary dust emissions as well as to inform simulations of past and future dust emission scenarios (Callot et al., 2000; Zender et al., 2003).

Although modelling is proving useful for studying aeolian processes and landforms at a variety of scales, the data input to models and model testing still rely on empirical field studies. Similarly, remotely sensed data can not always be interpreted without recourse to ground-truthing. Early empirical field and wind tunnel studies provided aeolian geomorphology with a sound foundation and computer modelling and remote sensing have enabled further progress, although should not be relied upon exclusively. For example, a recent comparison of eight dust emission/transport models concluded that 'reliable surface land use conditions and soil/surface information are more important than the complexity of the dust emission scheme or model horizontal resolution' (Uno et al., 2006, p. 19). Models must be continually tested using field data and as the models become more sophisticated, in many cases so too must the methods and approaches used in the field. Feedback and iteration between field studies and modelling will drive forward the next decade of research.

CONSENSUS, CONFLICTS AND CHALLENGES IN AEOLIAN GEOMORPHOLOGY

This final section outlines just a few of the challenges that have been tackled by aeolian geomorphologists working in arid environments that have largely been resolved, and highlights others that are likely to be key areas of concern in the future.

Considerable prolonged debate has surrounded the origin of silt-sized material that comprises loess deposits. Many major loess deposits are associated with glacial outwash and for years it was argued that loess was a Quaternary phenomenon generated by glacial and possibly periglacial processes (e.g. Smalley and Derbyshire, 1990). There are, however, loess deposits in Africa, Arabia and Australia which are unlikely to be the product of cold climate conditions, and extensive laboratory experiments have now demonstrated that other processes such as fluvial comminution, aeolian abrasion and hot climate weathering are effective at producing silt-sized material (Figure 25.3). Wright (2001) suggests that in most desert regions several different mechanisms of silt formation have contributed to the production of loess sediments and, perhaps more importantly, emphasizes that although mechanisms of sediment production need to be considered, loess deposits result from the complex interaction of environmental, lithological, tectonic, geomorphic and topographic variables and these all play an important role in the formation of loess sequences.

Over the past few decades, studies of the morphology of free dunes and of the relationships among groups of dunes, combined with local and regional meteorological and climate data analyses have enabled the identification of characteristic conditions for the main free dune types. Wasson and Hyde's (1983) model of the relationship between sediment supply and wind directional variability provided a good working hypothesis for differentiating free dunes using a combination of wind directional variability and sand supply. Whilst their model has engendered some debate, particularly concerning the role of sediment

Table 25.4 Comparison of the regional annual mean dust flux (Tg yr⁻¹) from selected global dust models

Study	Africa		Asia			America		Australia	Global
	North	South	Arabia	Central	East	North	South		
Ginoux et al. (2001)	1430 (69%)	63 (3.4%)	221 (11.8%)	140 (7.5%)	214 (11.4%)	2 (0.1%)	44 (2.3%)	106 (5.7%)	1877
Luo et al. (2003)	1114 (67%)	—	119 (7.2%)	—	54 (3.2%)	—	—	132 (8.0%)	1654
Miller et al. (2004)	517 (51%)	—	43 (4.2%)	163 (16%)	50 (4.9%)	53 (5.2%)	—	148 (15%)	1019
Tanaka and Chiba (2006)	1087 (57.9%)	63 (3.4%)	221 (11.8%)	140 (7.5%)	214 (11.4%)	2 (0.1%)	44 (2.3%)	106 (5.7%)	1988
Werner et al. (2002)	693 (65%)	—	101 (9.5%)	96 (9.0%)	—	—	—	52 (4.9%)	1060
Zender et al. (2003)	980 (66%)	—	415 (28%)	—	—	8 (0.5%)	35 (2.3%)	37 (2.5%)	1490

Numbers in brackets indicate the percentages for the annual mean global emission flux predicted by each model (From Tanaka and Chiba, 2006).

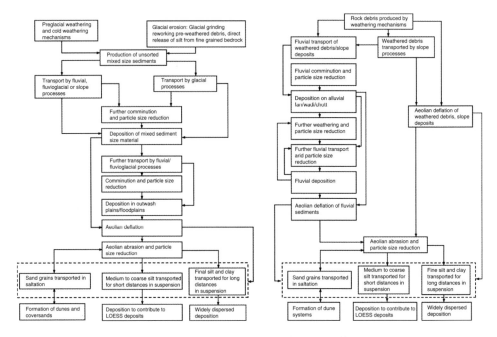

Figure 25.3   Events in the formation of loess deposits. Hypothetical pathways to explain the formation of loess deposits associated with (a) cold environments and (b) hot environments (Wright, 2001)

supply (e.g. Mabbutt, 1984; Rubin, 1984; Wasson and Hyde, 1984; Lancaster, 1994), the broad domains of different dune types are widely accepted (Figure 25.4).

Considerably different levels of understanding have so far been reached for different dune types. Transverse dunes (particularly barchans) have undergone the most scrutiny due probably to their apparent simple relationship to wind regime, single slip face and tendency to be unvegetated. However, Livingstone et al. (2007) discuss a range of uncertainties concerning the dynamics of these dunes that have yet to be resolved. Linear dunes are the most widespread dune type and have also been intensively studied to the extent that over ten different, often highly conflicting, models for their formation have been proposed and over 15 different linear dune classification schemes have been devised. Two types of model have gained most attention: the first attributes the dunes to a bi-directional wind regime; the second proposes a unidirectional wind regime (roll–vortex theory). Empirical field data have been presented to support both model types (e.g. Tsoar, 1983; Livingstone, 1989; Tseo, 1993); however, it has proven difficult to distinguish formative wind regimes from secondary airflow patterns generated by the presence of the dune. Some researchers have proposed different linear dune

formation processes in different dunefields (e.g. Tseo, 1993), and the lack of any consensus has implications for the interpretation of dune stratigraphy and chronology (Munyikwa, 2005). Whilst linear dune activity is predominantly by extension, a recurring debate concerns whether or not linear dunes can also migrate laterally (Hesp et al., 1989; Rubin, 1990; Tsoar et al., 2004; Bristow et al., 2005, 2007). It is argued that if linear dunes form in a bi-directional wind regime where one wind component is stronger than another, then the dunes may move sideways. There is morphological and sedimentological evidence from some, but by no means all, linear dune studies to support this and it is likely that the existence and extent of linear dune lateral migration reflects local and regional conditions. Star dunes are the least well-studied of the four main dune types possibly reflecting the fact that only around 8 per cent of all dunes are of this type.

Bedform dynamics reflect the ongoing interaction of the bedform, the flow-field and sediment transport (Leeder, 1983). Individual bedforms have been well-described morphologically; however, establishing their response to and effect on airflow and sediment transport is still challenging researchers. Studies combining field data, wind tunnel experiments and mathematical modelling have improved our knowledge of

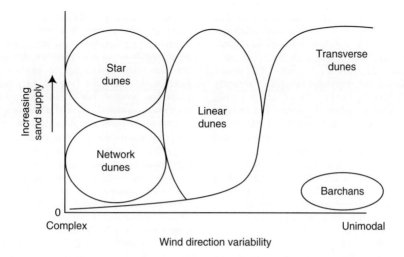

Figure 25.4 Free dune types differentiated using wind directional variability and sand supply. This figure, adapted from Wasson and Hyde (1983) by Livingstone and Warren (1996), to include network dunes, expands the domains of individual dunes beyond the original study

airflow structures over dunes; however, these are still not fully understood. Predicting the nature and rate of sediment transport over dunes is problematic not only because of relating high frequency variations in airflow and turbulence to simultaneous sediment transport, but also because the mode of transport (e.g. creep, saltation, suspension) of individual particles varies temporally and spatially across the bedform. The initiation and development of sand dunes, particularly groups of dunes, is also poorly understood. For example, the regularity of pattern and spacing exhibited by linear sand dunes over extensive areas has been comprehensively described but until recently there have been few convincing attempts to explain how or why these extensive, regular patterns occur.

Early approaches to landscape evolution such as that of Schumm and Lichty (1965) suggest that whether a variable is dependent or independent depends on the temporal and spatial scale at which it is being considered. However, Leeder's (1983) flow–form–sediment transport 'trinity' model, whilst it can only be applied at a single timescale, demonstrated that the dependency/independency relationship between form and flow can change without the need for a change in scale (see Walker and Nickling, 2002 or Lancaster, 2009 for suggested application of this model to aeolian landforms). This contradiction emerges from an increased knowledge of the more complex dynamics of smaller scale geomorphological systems, including feedbacks and interdependency,

and advances in this can be viewed as a product of the increasingly multi-disciplinary approaches and influences within geomorphology. A long-recognized problem in aeolian geomorphology is that scaling-up from what is known about aeolian sediment transport to sand dune development and from the behaviour of individual sand dunes to the accumulation of dunefields and sand seas is rarely successful. The main areas of difficulty (which are not unique to aeolian systems) are first, identifying correctly the processes involved; second, being able to determine the magnitude of the effect of the errors as they inevitably propagate from smaller to larger scales in the hierarchy; and third, perhaps most importantly, the inherent nonlinearity of the physical relationships involved. These limitations challenge the usefulness of 'reductionist' studies for explaining dune form and pattern and are also problematic for deterministic modellers (Sherman, 1995). Progress in understanding aeolian geomorphology at the grain, dune and dunefield scales is however being made following the recognition of aeolian bedforms as the result of self-organizing complex systems (Werner, 1995).

Werner (1995) demonstrated that it was possible to model dunefield patterns without recourse to small-scale airflow and sand transport dynamics. He cited dunefields as a good example of a complex system; in other words, a system that can not be fully explained by an understanding of its component parts. Since Werner's (1995) study, researchers have examined the onset of aeolian sediment transport (McMenamin et al., 2002), sand ripples (Yizhaq et al., 2004), sand streamers (Baas and Sherman, 2005) and dunefield organisation (Kocurek and Ewing, 2005) using such a framework. The overall concept of complex systems in aeolian geomorphology has recently been reviewed by Baas (2007) and viewing aeolian systems as complex systems is undoubtedly enabling progress to be made in understanding the behaviour of aeolian processes and development of dunefield patterns. The impact of the complex systems approach on the discipline is as yet uncertain; acknowledging that aeolian systems are inherently non-linear raises questions concerning whether or not an understanding of interactions amongst different scales of aeolian geomorphology can be achieved.

Although this chapter focuses on the action of wind in the arid realm, it is unrealistic to consider aeolian processes independently of other geomorphological activity. The pioneers of dryland geomorphology may have been divided into 'aeolianists' and 'fluvialists', each of which fervently promoted the dominance of their chosen medium as the key geomorphological agent in deserts (Bullard and McTainsh, 2003), but most contemporary geomorphologists recognize that understanding landform and landscape development requires consideration of the interaction between wind and water to explain sediment production, transport, supply, deposition and landscape preservation. These interactions have both a temporal and a spatial dimension.

One of the most useful models for considering temporal interactions is that of Kocurek (1998) (Figure 25.5) which associated sediment production peaks with humid periods during which rates of weathering, fluvial sorting and transport are enhanced. During and immediately following these periods, sediments are unavailable to the aeolian system due to higher moisture content and vegetation cover. As conditions become more arid, sediment production (and fluvial delivery) declines but availability of sediment to the aeolian system increases. Although intended for long timescales, the model is also applicable to shorter periods such as alternating wet–dry seasons or single events where rivers deliver an influx of sediment to a floodplain or ephemeral lake which subsequently desiccates and is then deflated. The spatial

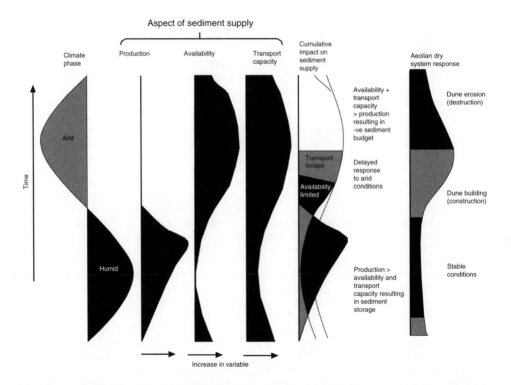

Figure 25.5 Model of the impact of humid-arid phases on sediment production/availability and transport and the response of the aeolian dry system (Bullard and McTainsh, 2003, simplified from Kocurek, 1998)

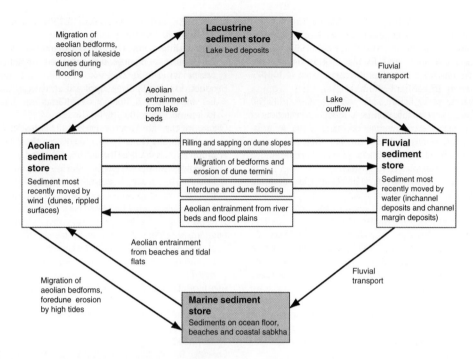

Figure 25.6   Links between four sediment storage areas for sand-sized material in arid environments (Bullard and Livingstone, 2002)

dimension to interactions also occurs at local and global scales and is defined by sediment storage zones and the ways in which sediment is transferred from one sediment storage area to another (Figure 25.6). For example, the inland basins hypothesis, which places contemporary dust sources in topographic lows at the distal reaches of internally draining basins, reflects the transport of material by ephemeral rivers to the lower sections of these catchments where they accumulate and are later deflated. Examples of such catchments include the Murray–Darling, Lake Eyre and Tigris–Euphrates basins, although more detailed studies of some major dust sources suggest that many other factors can come into play (Mahowald et al., 2003; Washington et al., 2006).

Aeolian–fluvial interactions in arid environments on a Quaternary timescale are widely reported in the literature and luminescence dating has contributed significantly to efforts in identifying transitions between humid and arid phases (Lancaster, 2008). Some caution has to be applied to the interpretation of such dates because, as illustrated in Figure 25.5, aeolian sediment deposition may not be a simple reflection of heightened aridity, it may instead reflect a change in sediment supply, transport capacity and/or

sediment availability instigated by shifts in either the aeolian or the fluvial system. Bullard and Livingstone (2002) highlighted the lack of contemporary process data for quantifying the rates and nature of sediment exchange between aeolian and fluvial systems for use as modern analogues to Quaternary studies and, despite some efforts, this paucity of data remains. There is also a poor understanding of the ways in which aeolian and fluvial processes interact during phases of climate transition and this limits understanding not only of past and present relationships between the aeolian and fluvial sediments and landforms, but will also impact on the ability to assess the implications of future climate change and land use strategies in both hot and cold arid regions.

REFERENCES

Anderson, R.S. (1987) A theoretical model for aeolian impact ripples. *Sedimentology* 34, 943–56.

Assallay, A.M., Rogers, C.D.F., Smalley, I.J. and Jefferson, I.F. (1998) Silt: 2–62 μm, 9–4 φ. *Earth Science Reviews* 45, 61–88.

Baas, A.C.W. (2007) Complex systems in aeolian geomorphology. *Geomorphology*, doi:10.1016/j.geomorph.2007.04.012.

Baas, A.C.W. and Sherman, D.J. (2005) The formation and behavior of aeolian streamers. *Journal of Geophysical Research* 110, F03011, doi:10.1029/2004JF000270.

Baas, A.C.W. and Nield, J.M. (2007) Modelling vegetated dune landscapes. *Geophysical Research Letters* 34, L06405. doi:10.1029/2006GL029152.

Baddock, M.C., Bullard, J.E. and Bryant, R.G. (2009) Dust source identification using MODIS: a comparison of techniques applied to the Lake Eyre Basin, Australia. *Remote Sensing of Environments* 113, 1522–28.

Bagnold, R.A. (1941) *The Physics of Blown Sand and Desert Dunes*. Chapman and Hall, London.

Bauer, B.O. and Sherman, D.J. (1999) Coastal dune dynamics: problems and prospects, in A.S. Goudie, I. Livingstone, S. Stokes (eds) *Aeolian Environments, Sediments and Landforms*. Wiley, Chichester, 71–104.

Breed, C.S., Fryberger, S.C., Andrews, S., McCauley, C., Lennartz, F. and Gebel, D. (1979) Regional studies of sand seas using Landsat (ERTS) imagery, in E.D. McKee (ed.) *A Study of Global Sand Seas*. USGS Professional Paper, 1052, 305–98.

Bristow, C.S., Lancaster, N. and Duller, G.A.T. (2005) Combining ground penetrating radar surveys and optical dating to determine dune migration in Namibia. *Journal of the Geological Society of London* 161, 315–61.

Bristow, C.S., Duller, G.A.T. and Lancaster, N. (2007) Age and dynamics of linear dunes in the Namib Desert. *Geology* 35, 555–558.

Bullard, J.E. (1997) Vegetation and dryland geomorphology, in D.S.G. Thomas (ed.) *Arid Zone Geomorphology*. Wiley, Chichester, 109–31.

Bullard, J.E. and Nash, D.J. (2000) Valley-marginal sand dunes in the south-west Kalahari: their nature, classification and possible origins. *Journal of Arid Environments* 45, 369–83.

Bullard, J.E. and Livingstone, I. (2002) Interactions between aeolian and fluvial systems in dryland environments. *Area* 34, 8–16.

Bullard, J.E. and McTainsh, G.H. (2003) Aeolian-fluvial interactions in dryland environments: scales, concepts and Australia case study. *Progress in Physical Geography* 27, 471–501.

Callot, Y., Marticorena, B. and Bergametti, G. (2000) Geomorphologic approach for modelling the surface features of arid environments in a model of dust emissions: application to the Sahara Desert. *Geodinamica Acta* 13, 245–70.

Cooke, R.U., Warren, A. and Goudie, A. (1993) *Desert Geomorphology*. UCL Press, London.

Cornish, V. (1914) *Waves of Sand and Snow*. Unwin-Fisher, London.

Dong, Z. and Qian, G. (2007) Characterizing the height profile of the flux of wind eroded sediment. *Environmental Geology* 51, 835–45.

Doyle, M.W. and Julian, J.P. (2005) The most-cited works in Geomorphology. *Geomorphology* 72, 238–49.

Fitzsimmons, K.E. (2007) Morphological variability in the linear dunefields of the Strzelecki and Tirari Deserts,

Australia. *Geomorphology*, doi:10.1016/j.geomorph. 2007.02.004.

Folk, R.L. and Ward, W.C. (1957) Brazos River bar: a study in the significance of grain size parameters. *Journal of Sedimentary Petrology* 27, 3–26.

Fryberger, S.G. (1979) Dune forms and wind regimes, in E.D. McKee (ed.) *A Study of Global Sand Seas*. USGS Professional Paper, 1052, 305–97.

Ginoux, P., Chin, M., Tegen, I., Prospero, J.M., Holben, B., Dubovik, O. and Lin, S.J. (2001) Sources and distributions of dust aerosols simulated with the GOCART model. *Journal of Geophysical Research* 106, D17, 20255–73.

Gisladottir, F.O., Arnalds, O. and Gisladottir, G. (2005) The effect of landscape and retreating glaciers on wind erosion in south Iceland. *Land Degradation and Development* 16, 177–87.

Goossens, D. (2005) Quantification of the dry deposition of dust on horizontal surfaces: an experimental comparison of theory and measurements. *Sedimentology* 52, 859–73.

Goossens, D. (2008) Relationships between horizontal transport flux and vertical deposition flux during dry deposition of atmospheric dust particles. *Journal of Geophysical Research – Earth Surface* 113, art. no. F02S13.

Goossens, D. and Offer, Z.Y. (2000) Wind tunnel and field calibration of six aeolian dust samplers. *Atmospheric Environment* 34, 1043–57.

Goossens, D., Offer, Z. and London, G. (2000) Wind tunnel and field calibration of five aeolian sand traps. *Geomorphology* 35, 233–52.

Goudie, A.S. (1997) Weathering Processes, in D.S.G. Thomas (ed.) *Arid Zone Geomorphology*. Wiley, Chichester, 25–39.

Goudie, A.S. and Middleton, N.J. (2006) *Desert Dust in the Global System*. Springer, Berlin.

Greeley, R. and Iverson, J.D. (1985) *Wind as a Geological Process*. Cambridge University Press, Cambridge.

Hardisty, J. and Whitehouse, R.J.S. (1988) Evidence for a new sand transport process from experiments on Saharan dunes. *Nature* 332, 532–4.

Hersen, P. (2004) On the crescentic shape of barchan dunes. *European Physical Journal B* 37, 507–14.

Hesp, P., Hyde, R., Hesp, V. and Zhengyu, Q. (1989) Longitudinal dunes can move sideways. *Earth Surface Processes and Landforms* 14, 447–51.

Hunt, J.C.R. and Weber, A.H. (1979) A Lagrangian statistical analysis of diffusion from a ground-level source in a turbulent boundary layer. *Quarterly Journal of the Royal Meteorological Society* 105, 423–43.

Kocurek, G. (1998) Aeolian system response to external forcing factors – a sequence stratigraphic view of the Saharan region, in A.S. Alsharhan, K.W. Glennie, G.L. Whittle and C.G.St.C. Kendall (eds) *Quaternary Deserts and Climate Change*. Balkema, Rotterdam, 327–49.

Kocurek, G. and Ewing, R.C. (2005) Aeolian dunefield self-organization – implications for the formation of simple versus complex dunefield patterns. *Geomorphology* 72, 94–105.

Lancaster, N. (1989) The dynamics of star dunes: an example from the Gran Desierto, Mexico. *Sedimentology* 36, 273–89.

Lancaster, N. (1994) Dune morphology and dynamics, in A.D. Abrahams and A.J. Parsons (ed.) *Geomorphology of Desert Environments*. Chapman and Hall, London, 475–505.

Lancaster, N. (1995) *Geomorphology of Desert Dunes.* Routledge, London.

Lancaster, N. (2008) Desert dune dynamics and development: insights from luminescence dating. *Boreas* 37, 559–73.

Lancaster, N. (2009) Dune morphology and dynamics, in Parsons, A.J. and Abrahams, A.D. (eds) *Geomorphology of Desert Environments*. 2nd edn. Chapmans and Hall, London. pp. 557–96.

Lee, J.A., Gill, T.E., Mulligan, K.R., Acosta, M.D. and Perez, A.E. (2008) Land use/land cover and point sources of the 15 December 2003 dust storm in southwestern North America. *Geomorphology*, doi:10.1016/j.geomorph.2007.12.016.

Leeder, M.J. (1983) On the interactions between turbulent flow, sediment transport and beform mechanics in channelised flows, in J.D. Collinson and J. Lewin (eds) *Modern and Ancient Fluvial Systems.* Blackwell Scientific, Oxford. pp. 5–18.

Livingstone, I. (1989) Monitoring surface change on a Namib linear dune. *Earth Surface Processes and Landforms* 14, 317–32.

Livingstone, I. and Warren, A. (1996) *Aeolian Geomorphology, An Introduction.* Longman, Essex.

Livingstone, I., Wiggs, G.F.S. and Weaver, C.M. (2007) Geomorphology of desert sand dunes: a review of recent progress. *Earth-Science Reviews* 80, 239–57.

Luo, C., Mahowald, N.M. and del Correl, J. (2003) Sensitivity study of meterological parameters on mineral aerosol mobilisation, transport and distribution. *Journal of Geophysical Research* 108, D15, 4447 doi:10.1029/2003JD003483.

Mabbutt, J.A. (1984) Factors determining dune type – comment. *Nature* 309, 5963–92.

Mahowald, N.M., Bryant, R.G., del Corral, J. and Steinberger, L. (2003) Ephemeral lakes and desert dust sources. *Geophysical Research Letters* 30, art. No. 1074.

Marx, S.K., Kamber, B.S. and McGowan, H.A. (2005) Provenance of long-travelled dust determined with ultra-trace-element composition: a pilot study with examples from New Zealand glaciers. *Earth Surface Processes and Landforms* 33, 699–716.

McKee, E.D. (1979) *A Study of Global Sand Seas.* USGS Professional Paper 1052.

McKenna Neuman, C. (2004) Effects of temperature and humidity on the transport of sedimentary particles by wind. *Sedimentology* 51, 1–17.

McMenamin, R., Cassidy, R. and McCloskey, J. (2002) Self-organised criticality at the onset of aeolian sediment transport. *Journal of Coastal Research* SI36, 498–505.

McTainsh, G.H., Lynch, A.W. and Hales, R. (1997) Particle size analysis of aeolian dusts, soils and sediments in very small quantities using a Coulter Multisizer. *Earth Surface Processes and Landforms* 22, 147–58.

Middleton, N.J. and Thomas, D.S.G. (1992) *World Atlas of Desertification.* Edward Arnold, London.

Miller, R.L., Tegen, I. and Perlwitz, J. (2004) Surface radiative forcing by soil dust aerosols and the hydrologic cycle. *Journal of Geophysical Research* 106, D16 18193, doi: 10.1029/2003JD004085.

Muhs, D.R. (2004) Mineralogical maturity in dunefields of North America, Africa and Australia. *Geomorphology* 59, 247–69.

Munyikwa, K. (2005) The role of dune morphogenetic history in the interpretation of linear dune luminescence chronologies: a review of linear dune dynamics. *Progress in Physical Geography* 29, 317–36.

Nalpanis, P. (1985) Saltating and suspended particles over flat and sloping surfaces. II. Experiments and numerical simulations, in O.E. Barndorff-Nielsen, J.T. Møller, W.G. Nickling, C. McKenna Neuman and N. Lancaster 2002. Grainfall processes in the lee of transverse dunes, Silver Peak, Nevada. *Sedimentology* 49, 191–209.

Parsons, D.R., Walker, I.J. and Wiggs, G.F.S. (2004) Numerical modelling of flow structures are idealised transverse aeolian dunes of varying geometry. *Geomorphology* 59, 149–64.

Pécsi, M. (1990) Loess is not just the accumulation of dust. *Quaternary International* 7/8, 1–21.

Pirrie, D., Butcher, A.R., Power, M.R., Gottlieb, P. and Miller, G.L. (2004) Rapid quantitative mineral and phase analysis using automated SEM (QemSCAN): potential applications in forensic geoscience, in K. Pye and D.J. Croft (eds) *Forensic Geoscience.* Geological Society Special Publication 232, 123–36.

Prospero, J.M., Ginoux. P., Torres, O., Nicholson, S.E. and Gill, T.E. (2002) Environmental characterization of global sources of atmospheric soil dust identified with the Nimbus-7 Total Ozone Mapping Spectrometer (TOMS) absorbing aerosol product. *Reviews of Geophysics* 40, doi:10.1029/2000RG000095.

Pye, K. and Tsoar, H. (1990) *Aeolian Sand and Sand Deposits.* Unwin Hyman, London.

Romer-Rasmussen, K. and Willetts, B.B. (eds) *Proceedings of an International Workshop on the Physics of Blown Sand.* Department of Theoretical Statistics, Institute of Mathematics, University of Aarhus Memoir 8, 1, 37–66.

Rubin, D.M. (1984) Factors determining desert dune type (discussion). *Nature* 309, 91–2.

Rubin, D.M. (1990) Lateral migration of linear dunes in the Strzelecki Desert, Australia. *Earth Surface Processes and Landforms* 15, 1–14.

Rubin, D.M. (2004) A simple autocorrelation algorithm for determining grain size from digital images of sediment. *Journal of Sedimentary Research* 74, 160–5.

Schumm, S.A. and Lichty, R.W. (1965) Time, space and causality in geomorphology. *American Journal of Science* 263, 110–9.

Seelos, K. and Sirocko, F. (2005) RADIUS- rapid particle analysis of digital images by ultra-high resolution scanning of thin sections. *Sedimentology* 52, 669–81.

Seppälä, M. (2004) *Wind as a Geomorphic Agent in Cold Climates.* Cambridge University Press, Cambridge.

Sherman, D.J. (1995) Problems of scale in the modelling and interpretation of coastal dunes. *Marine Geology* 124, 339–49.

Smalley, I. and Derbyshire, E. (1990) The definition of 'ice sheet' and 'mountain' loess. *Area* 22, 300–1.

Stout, J.E. (1998) Effect of averaging time on the apparent threshold for aeolian transport. *Journal of Arid Environments* 39, 395–401.

Stout, J.E., Warren, A. and Gill, T.E. (2008) Publication trends in aeolian research: an analysis of the Bibliography of Aeolian Research. *Geomorphology* doi:10.1016/j.geomorph.2008.02.015.

Sun, Q., Wang, G. and Xu, Y. (2001) DEM applications to aeolian sediment transport and impact process in saltation. *Particulate Science and Technology* 19, 339–53.

Tanaka, T.Y. and Chiba, M. (2006) A numerical study of the contributions of dust source regions to the global dust budget. *Global Planetary Change* 52, 88–104.

Thomas, D.S.G. (1992) Desert dune activity: concepts and significance. *Journal of Arid Environments* 22, 31–8.

Thomas, D.S.G. (1997) Sand seas and aeolian bedforms, in Thomas, D.S.G. (ed.) *Arid Zone Geomorphology*. Belhaven Press, London. pp. 373–412.

Thomas, A.D. and Dougill, A.J. (2007) Spatial and temporal distribution of cyanobacterial soil crusts in the Kalahari: implications for soil surface properties. *Geomorphology* 85, 17–29.

Tseo, G. (1993) Two types of longitudinal dune fields and possible mechanisms for their development. *Earth Surface Processes and Landforms* 18, 627–43.

Tsoar, H. (1983) Dynamic processes operating on a longitudinal (seif) dune. *Sedimentology* 30, 356–78.

Tsoar, H., Blumberg, D.G. and Stoler, Y. (2004) Elongation and migration of sand dunes. *Geomorphology* 57, 293–302.

UNEP (1997) *World Atlas of Desertification*. 2nd edn. Arnold, London.

Uno, I., Wang, Z., Chiba, M., Chun, Y.S., Gong, S.L., Hara, Y. et al. (2006) Dust model intercomparison (DMIP) study over Asia: overview. *Journal of Geophysical Research* 111, D12213. doi:10.1029/2005JD006575.

Walker, I.J. and Nickling, W.G. (2002) Dynamics of secondary airflow and sediment transport over and in the lee of transverse dunes. *Progress in Physical Geography* 26, 47–75.

Washington, R., Todd, M.C., Middleton, N.J. and Goudie, A.S. (2003) Dust storm source areas determined by the Total Ozone Mapping Spectrometer and surface observations. *Annals of the Association of American Geographers* 93, 297–313.

Washington, R., Todd, M.C., Lizcano, G., Tegen, I., Flamant, C. and Koren, I. (2006) Links between topography, wind, deflation, lakes and dust; the case of the Bodélé Depression, Chad. *Geophysical Research Letters* 33, L09401. doi:10.1029/2006GL025827.

Wasson, R.J. and Hyde, R. (1983) Factors determining desert dune type. *Nature* 304, 337–9.

Wasson, R.J. and Hyde, R. (1984) Factors determining desert dune type (reply). *Nature* 209, 92.

Werner, B.T. (1995) Complexity in natural landform patterns. *Science* 284, 102–4.

Werner, M., Tegen, I., Harrison, S.P., Kohfeld, K.E., Prentice, I.C., Balkanski, Y., Rodhe, H. and Roelandt, C. (2002) Seasonal and interannual variability of the mineral dust cycle under present and glacial climate conditions. *Journal of Geophysical Research* 107, D24 4744 doi:10.1029/2002JD002365.

Wiggs, G.F.S. (1993) Desert dune dynamics and the evaluation of shear velocity: an integrated approach, in K. Pye (ed.) *The Dynamics and Environmental Context of Aeolian Sedimentary Systems*. Geological Society of London Special Publication 72, 32–46.

Wiggs, G.F.S., Baird, A.J. and Atherton, R.J. (2004) The dynamic effects of moisture on the entrainment and transport of sand by wind. *Geomorphology* 59, 13–30.

Wolfe, S.A. and Nickling, W.G. (1993) The protective role of sparse vegetation in wind erosion. *Progress in Physical Geography* 17, 50–68.

Wright, J.S. (2001) 'Desert' loess versus 'glacial' loess: quartz silt formation, source areas and sediment pathways in the formation of loess deposits. *Geomorphology* 36, 231–56.

Yizhaq, H., Balmforth, N.J. and Provenzale, A. (2004) Blown by wind: nonlinear dynamics of aeolian sand ripples. *Physica* D195, 207–28.

Zender, C.S., Newman, D. and Torres, O. (2003) Spatial heterogeneity in aeolian erodibility: uniform, topographic, geomorphic and hydrologic hypotheses. *Journal of Geophysical Research* 108. doi: 10.1029/2002JD003039.

26

Tropical Environments

Michael Thomas and Vishwas Kale

THE NATURE OF TROPICAL ENVIRONMENTS

'The tropics' encompass environments ranging from the per-humid to the semi-arid, but for most purposes the latter areas are considered separately or with the arid zone. The humid and sub-humid tropics include both the rainforests (P >1600 mm y^{-1}; dry season 2–4 months) and the wetter savannas (P 800–1600 mm y^{-1}; dry season 5–7 months). Some highly seasonal (usually monsoon) climates that receive high rainfalls (>2000 mm y^{-1}) also support luxuriant forests, while 'tropical' conditions extend to 30° north and south of the equator along east coasts of the continents.

Many tropical environments experience extremes of rainfall, wind and drought, and all of these have major impacts on the geomorphology. The high moisture capacity of warm oceanic air can lead to high intensity convectional rainfall, and persistent heavy rainfall can result from major atmospheric perturbations such as tropical cyclones, and the effects of upper air troughs that produce disturbance lines across West Africa and prolonged rainfalls in eastern Brazil, and within monsoon-driven rains of India and south-east Asia. Extreme falls also arise from orographic effects, particularly in the path of on-shore winds. Charrapunji, India is famous for its high rainfall, which averages 11,430 mm y^{-1}, with an extreme value of 22,987 mm (1860–1861). Similar values of 9000–11,000 mm y^{-1} occur in other tropical mountain areas. Very intense (24 h) rainfall is also experienced widely, the world record standing at 1,825 mm at Foc-foc, La Réunion (21° 14′ S) (World Meteorological Organisation/Arizona State University website).

GEOMORPHOLOGY IN THE TROPICS

Early descriptions of tropical landforms by visitors from Europe (see Thomas, 2008a) tended to emphasize the exotic: vast plains punctured by granite domes (inselberge); huge rivers enclosed by impenetratable forests; towering karst hills, and confirmed the prevailing paradigm of a humid temperate normality in geomorphology (Davis, 1899). The classification of world climates by Köppen in 1923 also effectively subdivided the world into environmental compartments for separate study. 'Climatic geomorphology' became formalized (Büdel, 1948, 1982), and despite sceptics and objectors (Stoddart, 1969), it has persisted in both theoretical (this volume) and applied (Fookes et al., 2005) contexts. Arguments against the separation of the tropics for specific study are cogent (see Thomas, 2006), but they frequently come up against a solid pragmatism. It is quite simply easier to subdivide the globe geographically for studies in the field sciences, and the clear links between climate, vegetation, soils and landforms almost dictate that climatic environments should be used for this purpose.

Pioneer studies in the tropics were often pursued in the so-called wet–dry regions of open, savanna vegetation, where access was easier and intervisibility better than in the wetter, rainforest areas. Thus Büdel (1982) devoted 64 pages of his *Climatic Geomorphology* to discussion of the landforms of the seasonal tropics, but only 15 pages to the inner, forested areas. Detailed research in the Amazon and Congo basins is comparatively recent, and still sparse. Much early work concentrated on the ancient continental fragments of former Gondwanaland. Debate centred on the long-term evolution of the continental relief, and became strongly influenced by the concept of pediplanation, applied worldwide by Lester King (King, 1950, 1953, 1962), a geologist working in South Africa. King's nomenclature for continent-wide planation surfaces was widely adopted. Recent work by Burke and Gunnell (2008) illustrates a modern approach to key issues in this model.

Although Branner published observations on processes operating in Brazil in 1896 few institutions existed to enable the systematic study of surface processes in the humid tropics until the mid 20th century. Almost by default the supposed 'normality' of landform evolution in temperate climates became adopted by the advocates for a dynamic geomorphology, following Horton (1945) and Strahler (1952). This led Tricart (1972) to argue that the study of tropical geomorphology had 'been delayed by uniformist views of "normal erosion", and consequently the "tropics were systematically neglected"' (p. xiii).

It is now understood that, while the processes of mineral decay and the dynamics of water flow may be universal, the natural systems which contain them exhibit chaotic behaviour and produce very different outputs over time (emergent structures) (Phillips, 1999; Aziz-Alaoui and Bertelle, 2006). In the tropics some weathering phenomena are more frequently encountered and require greater emphasis than in colder or more arid environments. Similarly, seasonal contrasts in discharge and high flood peaks in many tropical rivers require specific research and a different approach to stream behaviour (Baker, 1978).

Geomorphic outcomes are also influenced by the geological setting. There are important contrasts between the largely cratonized Gondwana continents, which dominate large areas of the humid tropics and the mobile, orogenic areas, including the island-arc terrains of the western Pacific. No single entity that can be called 'tropical geomorphology' really exists. Modern texts on geomorphology in the tropics are few (Thomas, 1994; Wirthmann, 1994), the study of Indonesia by Verstappen (2000) is valuable, and there are modern studies on regolith (Taylor and Eggleton, 2001) and residual tropical soils (Fookes, 1997).

CLIMATE CHANGE IN THE TROPICS

The tropics, no less than other regions of the globe, have experienced important climate changes over extended timescales, as the continental plates drifted northward during the last 100 Ma, and world climates responded to mountain building.

There is evidence for a mid-Miocene deterioration of climate in South America probably associated with the growth of the Antarctic ice-cap, while desert sands of pre-Quaternary age are found in the rainforests in Angola, and these probably indicate Mio-Pliocene aridity. Throughout the tropics, below ~2500 m altitude, climate change has meant temperature depression by perhaps 4–6°C over land and 2–3°C in sea surface temperatures (SSTs) during the Quaternary ice ages, and this reduced moisture flux from the oceans to the land, leading to widespread reductions in rainfall during glacial maxima. In regions close to the mid-latitude deserts aridification was intense and dune systems or *acacia* thorn scrub replaced the savannas. Elsewhere forests became degraded to savanna or woodland–savanna mosaics. Rivers were transformed by these changes: in the extreme case drying up, with their channels blocked by dune sand. Sediments are absent from some alluvial and lacustrine sequences at glacial maxima. In other instances braided rivers developed or fans accumulated, where meandering rivers in swampy floodplains had previously flowed and have now returned (see Douglas and Spencer, 1985; Thomas, 1994; Thomas and Thorp, 1995; Gregory and Benito, 2003 – Section III, Low Latitude Regions). As climates recovered and intense rainfalls became more frequent in the early Holocene, a lag in vegetation response led to instances of widespread erosion and colluviation. The fall in global sea levels by more than 100 m during glacial maxima also meant that stream courses became extended over areas of shallow continental shelf. The islands of Sumatra, Singapore and Borneo became linked by land bridges with rivers flowing north to the South China Sea. With deep water off-shore deep channels were eroded as in the lower Amazon.

WEATHERING PROCESSES AND PROFILES

Deep weathering in the tropics

The processes of chemical weathering and the formation of regolith have been considered in Chapters 16 and 17. In the humid and sub-humid tropics, weathering processes and products play central roles in modelling the landscape. Deep rock decay leads to the separation of the landsurface from the unaltered rock by tens of metres over wide areas, and in the process of formation these weathering profiles penetrate unevenly along fractures and susceptible rock formations. The products of weathering are dominantly clays (especially kaolinite), oxides (mainly of Fe, Al,

Si), residual quartz sand (plus heavy minerals), and solutes carried away in groundwater (carbonates, sulphates, hydroxides). These materials can be rapidly eroded if exposed to intense rainfall.

Details of tropical weathering products and profiles (*saprolite*) can be found in several studies (Thomas, 1994; Fookes, 1997; Taylor and Eggleton, 2001). In the humid tropics *ferralitic* soils are formed (ISRIC, 2006), which means that almost all the mobile cations have been removed to leave a mixture of kaolinite clay, Fe and Al oxides. In equatorial climates, free draining saprolites are increasingly dominated by gibbsite, as Fe compounds are removed, and economic bauxites are common. In the seasonal tropics more iron (as haematite and goethite) is retained in the saprolite, giving the distinct red colour to many tropical soils. Iron oxides tend to be mobile within the soil profile under reducing conditions that require either organic complexing (*chelation*) or saturation, and where the reduced Fe^{2+} becomes oxidized, usually due to drying out, ferric iron (Fe^{3+}) oxides are precipitated. These commonly occur in the form of small segregations that can harden to form nodules and become integrated to form resistant *duricrust*. The residual red soils of the tropics are often called *laterites*, but they are correctly called 'ferralsols' (also oxisols, latossolos, ferralitic soils – see ISRIC, 2006).

Many deep weathering profiles have been shown to be of considerable geological age, often dating to the Cretaceous or before. The great continental plates of former Gondwanaland, which had been grouped around the southern pole in South Africa, today straddle tropical latitudes, though Australia only edges into the tropics at its northern extremities. Prolonged geological stability has allowed weathering processes to operate over wide areas, sometimes without truncation of ancient profiles. The past 100 Ma of plate migration was accompanied by significant cooling in higher latitudes. On the 'old tropical Earth' warm moist conditions penetrated to 60° N and S latitudes, and affected the landscapes of North America and northern Europe. Traces of these conditions are found in the red inter-basaltic soils, at the Giant's Causeway in Northern Ireland, dated to 62 Ma. On the other hand deep profiles exhibiting advanced weathering have formed in Neogene rocks in western Kalimantan (Indonesia) (Thomas et al., 1999).

The remnants of thick weathering profiles were seen by early geologists to coexist with residual, rocky hills, and authors such as Falconer (1911) drew the conclusion that extensive, deeply weathered landscapes had been stripped over wide areas to expose the underlying rock boulders and domes. Early accounts often emphasized the need for conditions of extreme planation to form a widespread, deep mantle of weathered material, but most weathering processes are advanced by down-profile and lateral movements of water, which evacuate the solutes produced by weathering and maintain the systems in disequilibrium. This requires a hydraulic gradient and it has been shown that deeper weathering often occurs around plateau edges in dissected landscapes.

The weathering profile

The description and understanding of weathered rocks (saprolites) involves several different criteria:

- Morphological zonation of the weathering profile.
- Geochemistry and mineral composition and their transitions down profile.
- Texture (particle size distribution) and fabric (micro-morphology).
- Position and catenary relationship of each vertical profile to slope, local relief and drainage.
- Age determinations for neo-formed minerals using isotopic ratios and paleomagnetism.

The weathered mantle can be very thick and reach to depths of more than 100 m, commonly undulating in a pattern of domes and basins over granitic rocks. At the base of the profiles there may be a sharp transition to fresh rock at the *weathering front* (or 'basal surface of weathering') where fundamental chemical changes to the rocks are taking place. In fissile sedimentary and metamorphic rocks (phyllites, schists) the transition towards fresh rock can be gradual and a distinct weathering front may be difficult to identify. Descriptions of weathering profiles are commonly based on a model developed for granite weathering in Hong Kong (Ruxton and Berry, 1957) and adapted for wider use (Fookes, 1997) (Figure 26.1). A generic regolith profile is illustrated in Chapter 16, Figure 16.3.

Weathering patterns and landforms

Saprolite-covered convex hills often show a transition to dome-shaped hills of rounded boulders, and actual granite domes, often called inselbergs (or *bornhardts*, after a German geologist of that name) (Plate 12 and Figure 26.2). (See the colour plate section, page 596, for Plate 12.) In savanna areas low shield or whaleback exposures of granite can often be seen. Many sections display residual corestones and occasionally convex arches of solid rock, while buttresses and domes

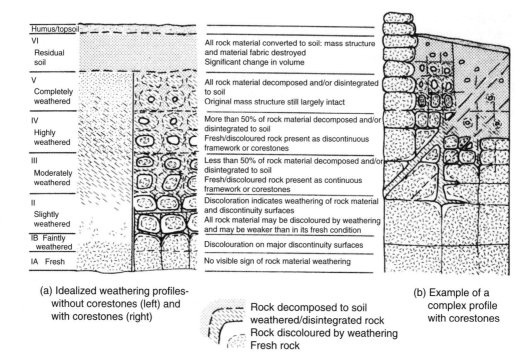

Humus/topsoil		
VI Residual soil	All rock material converted to soil: mass structure and material fabric destroyed Significant change in volume	
V Completely weathered	All rock material decomposed and/or disintegrated to soil Original mass structure still largely intact	
IV Highly weathered	More than 50% of rock material decomposed and/or disintegrated to soil Fresh/discoloured rock present as discontinuous framework or corestones	
III Moderately weathered	Less than 50% of rock material decomposed and/or disintegrated to soil Fresh/discoloured rock present as continuous framework or corestones	
II Slightly weathered	Discoloration indicates weathering of rock material and discontinuity surfaces All rock material may be discoloured by weathering and may be weaker than in its fresh condition	
IB Faintly weathered	Discolouration on major discontinuity surfaces	
IA Fresh	No visible sign of rock material weathering	

(a) Idealized weathering profiles-without corestones (left) and with corestones (right)

Rock decomposed to soil
weathered/disintegrated rock
Rock discoloured by weathering
Fresh rock

(b) Example of a complex profile with corestones

Figure 26.1 Characteristic weathering profiles using an engineering-based classification (Zones I–VI) (Compiled by the author for Fookes (1997))

(a)

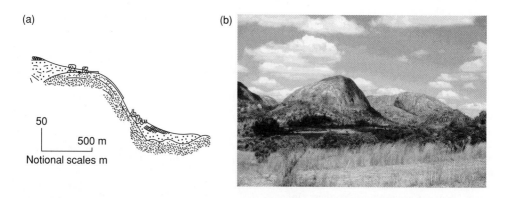

(b)

50
500 m
Notional scales m

Figure 26.2 Characteristic granite domes (inselbergs, borhnhardts) illustrated from Zimbabwe. (a) Diagram to show association of duricrust cores, sheeted granite exposures and footslope colluvium. (b) Domes at Dombashawa, Zimbabwe

of fresh granite occur along major escarpments and margins of intrusive bodies of granite. The association of deep basins of weathering with interspersed rock domes and boulders has led many writers to see a causal link between the two: the rock forms being fashioned by weathering, and then exposed by excavation from the saprolite by increased rates of erosion. The rock forms range from boulder fields, through *tors* (kopjes) to large domes and complex massifs.

One of the most eloquent exponents of this hypothesis was Willis (1936), while Büdel (1957) contended that the widespread weathered mantle led to a separation of the erosional or wash-surface from the unweathered bedrock by many tens of metres over large areas, forming double surfaces of levelling. The rocky inselbergs were rock masses that had resisted weathering and their shapes were controlled by rock structures and the effects of deep rock decay. Other research has shown that most granite domes are not located on major watersheds, but are defined by geological factors, and two recent studies from Zimbabwe by Römer (2005, 2006) used modelling to predict the persistent control by geology as the landscape evolves. King (1953) had argued that these landforms developed by slope retreat and the extension of pediments. But, while measurable slope retreat is found around inselbergs (Ahnert, 1982), this is not the decisive factor in their evolution.

Ages of saprolites and rates of weathering

Until recently, ages of weathering profiles and rates of chemical weathering were mainly estimates from geological evidence. Rates have been calculated using mass–balance studies in small catchments but these apply to present-day conditions. Isotopic studies of clays and oxides have proved more important. The results from all methods range from <4 m Ma^{-1} to >20 m Ma^{-1}, though higher figures come from materials such as volcanic ash (see Thomas, 1994). The age of formation of manganese oxides (cryptomelane and hollandite), produced by the weathering of manganese-rich rocks, has been determined using K/Ar and ^{40}Ar/^{39}Ar methods. Results from cratonic areas in Brazil and West Africa suggested that many profiles were pre-Miocene (Vasconcelos et al., 1994), and that propagation of the weathering front was slow: ranging from 3.8 m Ma^{-1} (Vasconcelos and Conroy, 2003) to 8.9 m Ma^{-1} (Carmo and Vasconcelos, 2006). But Feng and Vasconcelos (2007) have recently shown that, on Mary Island (26° S latitude), south-east Queensland, in a tropical forest environment, rates of 15–17 m Ma^{-1} occurred during the last 1 Ma. Within this period peak rates were attained at 1000–800 ka, 630–510 ka, and 400–50 ka (peaking at 313 ± 4 ka and 213 ± 7 ka). The authors point out that these figures show the sensitivity of weathering rates to climate and climate changes as determined by oxygen isotopes and pollen analysis of ocean sediments. The apparent correlation of periods of rapid decay with episodes of 'greenhouse Earth' (high CO$_2$) conditions in the Quaternary, draws attention to the importance of rock decay as a component in the global carbon cycle (reviewed by Kump et al., 2000).

Weathering extremes

Under aggressive conditions the breakdown of clay minerals, such as kaolinite, occurs. Both Fe^{2+} and Al^{3+} ions may be lost in solution (*congruent dissolution*) leaving only quartz sand and heavy minerals, and this depends on either saturated (*hydromorphic*) conditions or the presence of organic acids (*chelation*). If Fe^{2+} ions are partly exported but the less mobile Al^{3+} remains as an oxide (*incongruent* dissolution), then yellow–red *bauxite* (mainly gibbsite) remains. In profiles that experience strong periodic evaporation most of the Fe^{2+} ions mobilized by weathering become oxidized to form insoluble Fe^{3+} sesquioxides (goethite, haematite).

Congruent dissolution of kaolinite is also accompanied by removal of iron and takes place in hydromorphic environments, especially swamps (Figure 26.3). It can also result from chelation under rainforest, where perched water-tables also occur after heavy rain, and it has been argued that ferralsols may evolve towards podzols under these conditions (Lucas et al., 1987). The Rio Negro basin (Amazonas, Brazil) has become a focus for this research, but evidence for past climate changes (river-bordering dunes, duricrust fragments) in this area suggests this may also be a factor. Stallard and Edmond (1987) pointed out that the formation of podzolized 'white sands' (>98% quartz) was only one step away from the total dissolution of rocks, which emphasizes the geomorphic significance of these processes.

The oxides of Al and Fe, when concentrated in definite horizons form *bauxite* and *laterite* which may initially be soft and earthy, and their distributions broadly correspond with the tropics. The term *laterite* was first used by Buchanan in 1807 to describe a soft, clay-rich material in India that hardened on exposure and is used for building, and confusion about its use has led many authors to prefer the term *ferricrete* to describe indurated deposits. *Ferricrete*, *silcrete* and *calcrete* are all duricrusts, which frequently form a protective carapace (fr. *cuirasse*) over the terrain. A popular view, that duricrusts can form rapidly by exposure of laterite in cuttings, mistakes the formation of surface crusts for a more fundamental change in the crystallinity of the iron oxides throughout a deposit.

Iron-rich rocks (volcanics, metavolcanics, ultramafic intrusions, some sandstones) can form laterite (ferricrete) by relative enrichment of Fe,

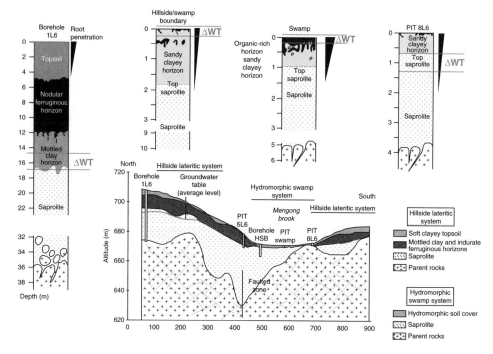

Figure 26.3   The hillside and swamp soil system of the Mengong Brook catchment (L6 catena), Cameroun. WT: variation of the groundwater level. (From Braun et al., 2005)

as Si is leached from elevated sites. But most ferricretes form where lateral movement of iron in solution leads to absolute enrichment of saprolite. Where periodic drying of the material occurs, oxidation of the Fe^{2+} ions leads to precipitation of goethite, which may become dehydrated to form haematite. This takes place below seasonally dry valley floors and at breaks of slope, and was described from West Africa by Maignien (1966) (Figure 26.2a and Figure 26.4). Some ferricretes are cemented accumulations of iron oxide segregations (pisoliths); others are vesicular or vermicular, containing many cavities or tubes within an indurated skeleton. Ollier and Galloway (1990) have argued that most occurrences of ferricrete have no genetic connection to an underlying profile and that there is effectively an unconformity between the duricrust and the saprolite. Some accounts make a clear distinction between upper (or primary) and lower (or secondary) laterites, which implies that the duricrust caps a complete profile on hill summits, but is formed by detrital and chemical accumulation on lower benches and footslopes. A problem with this concept is that many ferricretes in elevated positions also contain detrital material, even pebbles and cobbles, and these can be interpreted as resulting from the inversion of relief, having first

been deposited in valley floors or on the lower slopes of a now denuded landscape (for general accounts see: Goudie, 1973; McFarlane, 1976; Aleva, 1994; Boulangé et al., 1997).

Calcretes and *silcretes* are both found in highly seasonal, and often semi-arid areas but their climatic and landscape relationships can be complex, involving long-term climate changes and relief inversion (see Chapter 16). Calcium goes into solution as the bicarbonate and comes not only from limestone rocks but also from calcic plagioclase in crystalline rocks. It is readily precipitated by evaporation in both soil and stream environments, in addition to forming speleothems in solution tunnels and caverns (karstic weathering).

SLOPE PROCESSES AND LAND INSTABILITY

Weathering can have a pivotal role in influencing land instability in humid tropical areas. Deep-seated slides generally occur as bedrock failures, but in deeply weathered terrains slopes may fail by the stripping of saprolite from smooth rock surfaces that form the weathering front

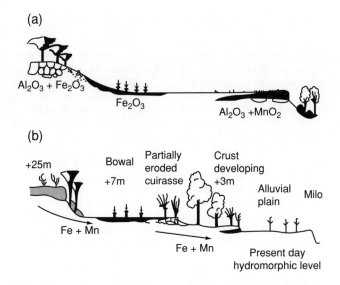

Figure 26.4 **Association of duricrusts (ferricrete) with relief features. (a) accumulation of Fe$_2$O$_3$ on the terrace near Labé, Guinea; (b) Catena showing levels of crust formation on the terraces of the Milo River, Guinea. (After Maignein, 1966)**

(Lacerda, 2007). Thick saprolite is also liable to rotational failure, and is susceptible to the formation of deep canyon gullies. But debris flows are probably the most common, and the most destructive, form of mass transport downslope. They usually originate as sudden floods in small catchments during intense rainfall. Debris from the slope is entrained and a turbulent mass of water and regolith flows downslope, often at great speed, causing a serious hazard to people and structures.

The influence of slope angle and form

Most studies of landslides emphasize relationships to slope angle and form. The frequency of slides increases steeply on slopes above ~20°. In the Serra do Mar, near Rio de Janeiro in eastern Brazil, Fernandes et al. (2004) found slopes classes 18.6–37.0° and 37.1–55.5° were most prone to failure. The former had the greater frequency in the landscape, but the latter had the higher landslide potential. Concave slopes failed three times more often than convex slopes in this area. Hillslope concavities are part of the drainage system, and they receive convergent water flow and accumulate colluvium, leading to slope instability and repeated failure (Dietrich et al., 1986). In Hong Kong the encroachment of headwater channels in an area of high relief has led to

multiple steep concavities and Ng (2006) has shown a clear relationship between landslide density and first-order drainage basins. Colluvium also accumulates towards the base of hillslopes, and a catastrophic landslide that occurred on Hong Kong Island in 1972 involved a weathered volcanic colluvium towards the base of the island's main mountain ridge.

The prediction of where landslides will occur remains imprecise, but identifying the most likely sites of future slides is important for the avoidance of risk. This task includes (1) identification of naturally failing bedrock structures, (2) identification of old landslides (palaeoslides) which are often subject to further failures, (3) mapping the most susceptible geomorphic sites, and (4) understanding substrate materials and discontinuities within soils and slope deposits. Fernandes et al. (2004) also showed from Brazil that a reduction in hydraulic conductivity often occurs at depths between 0.8 and 3.0 m corresponding with the pedologic B horizon, and that below this conductivity rises again within the more-sandy saprolite. This factor can lead to loss of cohesion and failure at shallow depths and is probably a widespread phenomenon.

The impact of tropical rainfall

The 1972 event in Hong Kong stimulated new research into the prediction of landslides, where the principal cause is rainfall. This factor has different aspects: (1) high intensity (>100 mm h^{-1}) which produces copious runoff and can lead to debris flows; (2) duration of intense rainfall influences the severity of debris flows and frequency of shallow failures; and (3) antecedent rainfall beyond the 24 h period may be measured over weeks or months, and influences deep-seated failures. High winds cause tree fall downslope exposing the root mat and the domino effect of multiple tree collapse can have catastrophic effects on local slope stability.

Tropical cyclones (hurricanes, typhoons) combine these factors in a unique combination of intense rainfall and high winds that cause havoc along coastlines and on inland slopes. They are associated with widespread flooding, accompanied by frequent debris flows and slides, which can be extremely destructive. Mountainous islands, such as Puerto Rico and the Philippines, which have short, steep catchments are particularly vulnerable. Other areas, such as eastern Brazil, Natal (RSA) and Mozambique are subject to slow-moving troughs and frontal systems that produce copious amounts of rainfall. The city of Rio de Janeiro and its hinterland are especially prone to these events. More organized rains take place within the Asian and African monsoon circulation, and heavy, orographic rain falls in mountainous regions and along major escarpments (in north-east Queensland 2000–3000 mm y^{-1} falls near sea level, rising to 8000–11,000 mm y^{-1} on summits at 1600–1800 m asl). All these mechanisms can lead to slope failures under natural forest vegetation, and frequencies of landslides where forest clearance has occurred are not always higher.

QUATERNARY SLOPE DEPOSITS

In tropical areas (except in high mountains) the impacts of Quaternary climate change have been related to variations in effective rainfall and vegetation cover. Widespread cooling during glacial advances was frequently accompanied by reduced rainfall. In some areas pollen evidence shows that the vegetation cover was depleted and savanna or open sclerophyll woodland replaced rainforests. But in other areas changes may have been restricted to species loss within surviving forests and the invasion of grasses in the savannas. The geomorphic impacts of these periods indicate both long term and abrupt climate changes. Many valleys in eastern Brazil contain great thicknesses of colluvium derived from local hillslopes, and Coelho-Netto (1997) demonstrated that most of these deposits accumulated rapidly following the

last glacial termination, post 15 ka (12 ky ^{14}C). Much older palaeo-landslides are common and have been mapped in Hong Kong, Zambia (Thomas and Murray, 2001) and in north-east Queensland, Australia, where several were successfully dated to moist phases of the last glacial cycle (Thomas et al., 2007; Thomas, 2008b).

Layered slope deposits can also result from fluvial activity, fan deposition (often debris flows) and more diffuse sheet flow. Many sequences consist of sands intercalated with coarse flow deposits: gravels, pebbles and cobbles, which denote pulses of higher energy affecting the system. The widespread occurrence of these materials and their susceptibility to gullying has been described from Brazil (De Oliveira, 1990). Fans fronting the east-facing escarpment of the Great Dividing Range in north-east Queensland have been dated to dry climates associated with the Last Glacial Maximum, and they were dissected when climates recovered around 14 ka (Thomas et al., 2007) (Figure 26.5).

Stone-lines

Many soil profiles in the seasonal tropics contain stony horizons, usually at depths of 1–3 m, and their origins have been disputed. Pedologists have argued that stone-lines are formed when soil fauna such as termites and earthworms bring fine particles to the surface, leaving larger rock fragments (usually quartz or quartzite) at depth. This is the process of *bioturbation*. On the other hand many geomorphologists view stone-lines as *lag deposits* that developed as fine sediment was removed by surface water flows, and were later covered by bioturbation.

Because runoff is most erosive in the absence of a protective plant cover many stone-lines have been interpreted as results of climate change. Some stone-lines contain smooth pebbles which denote fluvial activity and are remnants of river terrace sediments. Many deposits contain *duricrust fragments* derived from benches or cappings of ferricrete or silcrete that now form higher parts of the landscape. Ultimately, many stonelines represent the residues of long-term weathering and erosion.

FLUVIAL GEOMORPHOLOGY

Fluvial hydrology and flow regime

The position of the Inter Tropical Convergence Zone (ITCZ), the inter-annual variability of

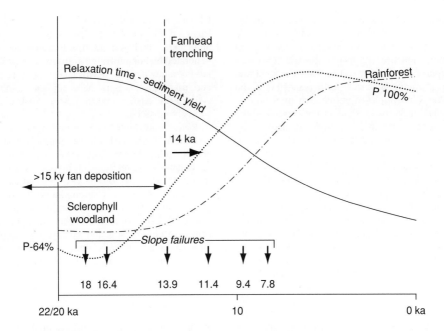

Figure 26.5 Geomorphic system responses to global climate warming after the termination of the last glacial. Data from north-east Queensland, showing 64% reduction in P at the glacial maximum leading to deposition of fans, which were dissected as P increased (c 14 ka). A series of slope failures took place, many during the humid period of the early Holocene. (From Thomas, 2008a,b)

rainfall linked to ENSO, and the influence of tropical cyclones all contribute to the characteristic flow regimes of the tropical rivers. The equatorial zone experiences abundant rainfall year round, with two peaks coinciding with the movement of the ITCZ across the equator, but it is rarely affected by tropical cyclones. Rivers draining this region reflect a bimodal discharge pattern with the annual discharge peaks suppressed and no marked low flow season.

The bimodal distribution of rainfall gives way to a more unimodal pattern with distance from the equator. Consequently, the rivers experience a distinct low flow season and a single period of high peak. All but a few large tropical rivers under composite climatic influences, such as the Amazon, Paraná, Congo and Orinoco, display such an annual discharge pattern. The seasonality and peakedness varies, depending upon the basin size and the influence of altitude, tropical storms, and the position of the ITCZ. The seasonality index (R), which is the ratio between maximum and minimum average monthly discharges, is low for large equatorial–tropical rivers such as the Amazon ($R = 2.1$, at Obidos), the Congo, ($R = 1.8$, at Kinshasha), the Parana ($R = 1.8$, at Corrientes) and the White Nile ($R = 1.5$, at Mogren), but

is higher for the Orinoco River ($R = 9.2$, at Puente Angostura), probably because the basin is situated closer to the northern limit of the ITCZ migration.

Discharge regimes in monsoon Asia are known for their high peakedness. The south-west or summer monsoon initiates the rainy season in south, south-east and east Asia, when the ITCZ moves northward in early boreal summer. The common feature of the larger rivers, such as the Ganga and Mekong is that there is just a single annual flood hydrograph in response to the south-west monsoon, and a severe low flow season. This is reflected by a high seasonality index (R). The value for the Ganga ($R = 24.6$, at Farakka) is the highest, followed by other rivers. The Niger ($R = 11.1$, at Malanville) and Volta ($R = 11.3$, at Senchi-Halcrow), which drain the monsoon impacted region of central and West Africa have similar regimes. In the savanna tropics, the non-rainy season is characterized by very low or no flow. The Senegal ($R = 129$, at Dagana), Atbara (Ethiopia–Sudan), and Luni (western India) rivers, all belong to this category.

The hydrological characteristics of small- and intermediate-sized tropical basins closely reflect the regional rainfall regime, but the flow regime conditions of large basins are more complex because they traverse different hydro-climatic regions. For instance, the Nile River discharge in the lower reaches (Egypt) is derived from two principal sources – the White Nile (Uganda and Sudan), which is under the influence of ITCZ and the Blue Nile, which is dominated by monsoons in Ethiopia. Similarly, the distinct flow regime of the lower Congo River is strongly controlled by the integration of discharges of large tributaries heading south (Kassai and upper Congo) and north (Ubangui, Sangha and Likouala) of the equator (examples of hydrographs may be found at: http://www.grdc.sr.unh.edu).

Significant inter-annual anomalies in rainfall and annual discharge are associated with the El-Niño/Southern Oscillation (ENSO) that roughly occurs once in 3–7 years and is characterized by anomalous warming of the sea surface waters in the tropical eastern Pacific Ocean and lower surface pressure over the eastern than the western tropical Pacific Ocean. Floods, droughts and other weather disturbances in many regions of the world have been ascribed to ENSO. El Niño events are associated with widespread tropical droughts, while La Niña, the opposite phase of the oscillation, generally yields wetter than normal conditions. The Nile historical records are among the longest hydrological records that have near-annual resolution (Kondrashov et al., 2005) and they show a strong negative correlation between the Nile discharge and El-Niño severity (Amarasekera et al., 1997). Similarly, in south Asia, below-normal (above-normal) rainfall and discharge conditions are associated with El-Niño years (La Niña years). The two large equatorial tropical rivers, Amazon and Congo, are weakly correlated with the ENSO events (Amarasekera et al., 1997). Epochal variability in annual rainfall also has an effect on annual discharges as well as flood peaks. The Senegal River, for instance, like other West African rivers, experienced wet periods, separated by three periods of shortage, in 1911–1915, 1940–1944 and 1972–1992 (Albergel et al., 1997).

Flood hydrology

Floods are common hydrological events in the tropics. The largest documented flood, with a peak discharge >300,000 m^3 s^{-1}, occurred in 1953 on the Amazon River at Obidos, Brazil (O'Connor and Costa, 2004). Four other rivers have registered flood magnitudes of ~100,000 m^3 s^{-1}: 102,534 m^3 s^{-1} on the Brahmaputra River at Bahadurabad, Bangladesh (1998); 99,400 m^{3}s^{-1} on the Godavari River at Dowleswaram, India (1986); 110,000 m^3 s^{-1} on the Changjiang River,

China (1870) and 98,120 m^3 s^{-1} on the Orinoco at Puente Angostura (1905) (Mirza, 2003; O'Connor and Costa, 2004; Kale, 2007). Extreme floods are generated by copious rains from tropical cyclones. All but a few tropical storms take a westerly track, and many river basins in south and south-east Asia, east Africa and northern Australia are severely impacted.

Rivers with high flood regimes generally correspond to high discharge variability (Latrubesse et al., 2005). Figure 26.6 shows that there is a reasonably good relationship between seasonality index (R) and peak discharges, indexed by mean discharge (Q_{max}/Q_m). Cyclone-induced floods in small- and intermediate-sized basins are generally 1–2 orders of magnitude higher than the mean flows. In comparison, the average discharges of large rivers are generally higher and, therefore, the Q_{max}/Q_m ratio may not appear to be high.

Sediment load

Rivers with exceptionally high annual suspended sediment load occupy the tropics. The Amazon River, with 1.2 × 10^9 t y^{-1} tops the list. It is followed by the Ganga–Brahmaputra (1.0 × 10^9 t y^{-1}), Changjiyan (0.48 × 10^9 t y^{-1}), Irrawaddy (0.26 × 10^9 t y^{-1}) and Indus (0.25 × 10^9 t y^{-1}) rivers (Milliman and Syvitski, 1992). All these rivers are impacted by monsoons and have their headwaters in tectonically active mountains (Andes and Himalay) or the Tibetan Plateau, which are the main sources of sediment. Nearly half of the total suspended load of the Amazon is delivered by one tributary, the Rio Madeira, for this reason.

Rivers that drain stable cratonic/shield areas, in comparison, deliver modest amounts of sediment to the ocean. The Zambezi and the Congo rivers as well as many rivers of the Indian Peninsula belong to this category. The contribution of Guayana and Brazilian shields to the total suspended sediment load of the Orinoco and Amazon respectively is far lower than the Andean tributaries. The Rio Negro, which delivers a large quantity of water to the Amazon, contributes only a fraction of the sediment load brought by the Rio Madeira.

Suspended sediment load averaged over the drainage area, or sediment yield, is a better measure of the amount of material annually eroded from a unit surface area. Sediment yields are generally higher in tropical environments and in tectonically active belts in particular. Highest are the Fly, Papua New Guinea (1087 t km^{-2} y^{-1}) and Brahmaputra rivers (852 t km^{-2} y^{-1}), (Hovius, 1998). The specific yields of these rivers are 1–2 orders of magnitude higher than for the Congo or

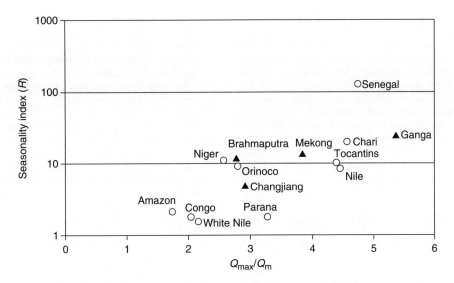

Figure 26.6   Plot of seasonality index (R) versus the ratio between peak discharge on record and mean discharge (Q_{max}/Q_m). Solid triangles refer to monsoon-fed rivers

Zambezi. In general, the sediment yields are low for large rivers and highly variable for intermediate-sized basins (Latrubesse et al., 2005).

Dissolved load is another important constituent of the total sediment load, especially where chemical weathering is pervasive. The solute load of the Amazon (223×10^6 t yr^{-1}) is substantially higher than for any other river (Hovius, 1998). The annual solute load of other rivers discussed here ranges between 58 and 92×10^6 tyr^{-1} (Hovius, 1998). The Zambezi and Congo rivers carry by far the most dilute waters, in spite of their equatorial–tropical location. This is probably due to the dominance of siliceous sediments and ancient crystalline rocks in their catchments.

Most tropical lowland rivers are sand-bedded. Sand is relatively mobile and, therefore, bedload transport is equally important in tropical environments. It is notoriously difficult to calculate; consequently there are no accurate estimates of bedload transport available for comparison. The proportions of bedload for the Ganga–Brahmaputra (in Bangladesh), Brahmaputra (in Assam, India) and Paraná are ~38%, ~17% and ~15% respectively (Goswami, 1985; Goswamy, 1989; Ramonell et al., 2002; Wasson, 2003; Sarma, 2005), while the estimated figure for the Amazon at Obidos is 10% (Milliman and Meade, 1983) and for the lower Nile (pre-Aswan High Dam) it is only 1–2% (Shahin, 1985).

Many earlier workers, such as Tricart (1972), believed that the rivers of humid tropics lack coarse bedload because of intense weathering.

These views have changed over the years. Many rivers in their headwaters have gravel bars and riffles, including the Sungai Kuamut (Sabah, Malaysia) and the Rio Chagres (Panama), which are both located in tropical rainforests (Fletcher and Muda, 2003; Rengers and Wohl, 2007). The Strickland River, a tributary of the Fly River, is widely cited as an example of humid–tropical river with a distinct gravel–sand transition (Lauer, et al., 2008).

Not all the sediment produced within a basin reaches the ocean. A significant portion of the load accumulates in foreland basins (e.g. Andes and Himalay), inland deltas, floodplains, midstream bars and deltas of large rivers. Figures range from 39 to 71 per cent for the Ganga–Brahmaputra (Allison et al., 1998), and from 17 to 50 per cent for the inland delta of the Niger in Mali (Dembele, 2008). The sediments are not stored forever, but are reworked, and recycled on different timescales, and may be exchanged between channels and floodplains. The residence time for sediments on floodplains and other sinks can vary from 10^1 to 10^3 years. Mertes et al. (1996) have demonstrated that the Amazon floodplain sediments have been recycled in <5000 y, and the recycling process is more rapid in the upstream than the downstream reaches.

Construction of large dams has dramatically altered the hydrologic regimes of many tropical rivers during the last 100 years (Nilsson et al., 2005), and their sediment dynamics have changed dramatically. Major examples include large dams on the Nile (Aswan Dam), the Changjiang (Three Gorges Dam), the Zambezi (Kariba and Cahora Bassa Dams), and the Narmada (Sardar Sarovar and Indirasagar Dams). In these and other cases the reduction in sediment load has been estimated at 75–100 per cent (Vörösmarty et al., 2003). The world's major high-discharge rivers (Amazon, Orinoco, Congo, Brahmaputra) have not been seriously affected by flow regulation, while others such as upper Mekong, Rio Magdalena and Fly have recorded increased suspended sediment load due to substantial land use changes in their catchments (Walling, 2006).

Channel morphology and patterns

The channel cross-section and reach morphology are largely governed by the water and sediment discharges, the recurrence interval of extreme events and the nature of bed and bank material. Nearly all the rivers of tropics reveal the effect of bankfull discharges and/or large floods (Gupta, 1988). Given the enormous variability in the channel morphology on account of varied geology, structure, tectonics, hydrological regime and base level conditions, it is difficult to generalize. Regarding river channels in the seasonal tropics (Gupta, 1988; Kale, 2002):

- Geomorphic work is confined to the wet season.
- Due to large variations in flood magnitude seasonal rivers are adjusted to a range of discharges.

- High-magnitude floods are significant geomorphic events because they are able to exceed the bank resistance threshold and are capable of transporting sediments (including coarse debris) on an enormous scale.
- Inter-season fluvial transport rate can range over 2–3 orders of magnitude.

The amount of geomorphic work that a river can accomplish at the bankfull stage and during peak floods is generally evaluated in terms of specific stream power. River reaches with lower gradient and higher width–depth ratio are characterized by lower specific stream power (Table 26.1), and values for large tropical rivers are typically less than 25 W m^{-2} (Latrubesse, 2008). Large monsoonal floods on intermediate-size rivers, such as the Narmada and Godavari, generate high unit stream power, close to 300 W m^{-2} (Kale, 2007). Such floods are capable of coarse sediment transport and perhaps bedrock erosion. The Narmada River, analysed by Kale (2008), reveals high total energy expenditure within the monsoon season (43–380 MJ) as well as during large floods (40–145 MJ). Extraordinary floods have a 20–60 per cent share in the total monsoon energy expenditure, indicating significant role of floods in landscape evolution.

Although meandering and braided patterns are widespread, all but a few large tropical rivers anabranch (Table 26.1) (Latrubesse, 2008). The tendency to anabranch has been attributed to very low channel gradient, low variability of grain size and high width–depth ratio. Also such rivers

Table 26.1 Channel morphological properties of some tropical rivers (mostly after Latrubesse, 2008)

River	Site	Channel pattern	Bankfull discharge in m³ s⁻¹	Width/depth ratio	Gradient	Specific stream power in W m⁻²
Amazon	Jatuarana	A	161,330	72	0.00002	12
Solimões	Manacapuru	A	120,000	120	0.00002	7
Yangtze	Datong	A	87,204	130	0.00009	26
Orinoco	Musinacio	A	64,600	101	0.00006	20
Brahmaputra	Bahadurabad	A	60,000	200	0.00007	10
Madeira	Fazenda Vista Alegre	A	57,000	64	0.00006	23
Ganga	Patna	A	32,000	134	0.00016	43
Parana	Corrientes	A	27,330	110	0.00005	9
Kosi	Kurshela	A	11,338	117	0.00038	69
Paraguay	Porto Murtinho	M	3333	20	0.00004	4
Fly	Kuambjt	M	3018	26	0.00005	15

A, anabranching; M, meandering.

are primarily bedload-dominant. In comparison, suspended load-dominant rivers with lower width–depth ratio are highly sinuous (Latrubesse, 2008). Due to intermittent control by geology and structure, some tropical rivers flow alternately through bedrock and alluvial reaches. Consequently, the channels undergo dramatic changes in morphology, gradient and pattern, as in courses of Orinoco and Mekong and Narmada Rivers. Bedrock reaches are often characterized by scabland-like topography, with gorges, cascades and cataracts. A multi-channel pattern is not uncommon in bedrock reaches, as in the Rio Negro, Zambezi (upstream of Victoria Falls), Mekong (*Si Phan Don* – Four Thousand Islands) and Narmada rivers. In comparison, the alluvial stretches are marked by primarily aggradational features.

The Congo River displays a very complex channel adjustment along its course due to varied lithological and tectonic controls. The river profile is stepped in appearance with several steeper segments separated by low gradient reaches (Runge, 2007). The Congo River has to cross a series of rapids and waterfalls to reach the Atlantic Ocean, and the Zambezi River and Kaveri River (south India), both have a significantly steeper middle section with waterfalls, rapids and bedrock gorges.

Tropical floodplains

Many tropical rivers are storehouses of enormous quantities of sediments, and a significant proportion of these sediments are stacked on vast floodplains that fringe the river channels, and in coastal deltaic plains. With an area of ~0.9×10^6 km^2 the floodplain developed by Amazon is the largest, others are shown in Table 26.2. In Indonesia, the combined area of floodplains, primarily in Kalimantan and Irian Jaya, is in excess of 100,000 km^2 (Tockner and Stanford, 2002).

By definition, floodplains are lowlands adjacent to the river that are subjected to frequent inundation. Floodplains are built by meandering, braided and anabranching rivers, by lateral

Table 26.2 Major fringe and deltaic floodplains of tropics (modified after Tockner and Stanford, 2002)

River system	Approximate area (km²) of floodplain during maximum inundation	Remarks
Amazon River	890,000	Floodplain known as *várzea* Affected by structural arches across the river; inland delta
Pantanal	130,000	The 'largest' single wetland complex
Ganges and Brahmaputra	120,000	Dynamic plain with frequent changes in channel size and position, incised rivers Kosi inland delta/megafan
Nile	93,000	Sudd, Kagera basin
Congo	70,000	Middle Congo depression, Kamulondo, Malagarasi
Fly River	45,000	Unusually large floodplain
Niger-Benue	38,900	Niger central delta, Benue River Large inland delta
Orinoco fringing floodplain and delta	37,000	Extensive alluvial plain in Andean forelands known as *llanos* Meta inland delta
Irrawaddy	31,000	Wide floodplain and numerous sandbars
Magdalena floodplain and deltic plain	20,000	Consists of alluvial plains, beach ridges, and marginal lagoon systems. Floodplain ~ 35 km wide
Zambezi	19,000	Kafue flats, Barotse plain, Liuwa plain Pan and dambos common
Mekong	11,000	Combined Tonle Sap lake/river and lower Mekong floodplain widest Flow reversals in Tonle Sap River during monsoon floods
Paraná	20,000	Pantanal sedimentary basin with largest alluvial fans and vast wetlands Floodplain ~ 15–40 km wide

accretion during channel migration and by overbank deposition during high flows. During peak floods the inundated floodplain area runs into several tens of thousands of square kilometres as in case of Sudd (Nile), Okavango and Niger inland deltas. About 60 per cent of the catchment area (76,500 km²) of the Fly River is seasonally flooded.

Tropical floodplains are generally categorized into three types: fringing, inland delta-like and coastal deltaic floodplains. Fringing floodplains extend, roughly parallel and along major rivers. The average floodplain width along large tropical rivers such as the Amazon and Orinoco ranges between 10 and 40 km (Hamilton and Lewis, 1990). Even in the case of modest-sized rivers, such as the Fly, the average floodplain width of the lower 800 km is more than 40 km (Tockner and Stanford, 2002). Generally speaking, all floodplains contain large numbers of oxbow lakes, scroll-bar and levee complexes and permanent or seasonal lakes. Frequently, old channel belts occur, as in the case of Ganga (Singh, 2007) and Amazon floodplains (Mertes et al., 1996). Headwater reaches of small tributaries draining elevated plateaus in Africa, and other cratonic areas in the seasonal tropics are characterized by shallow, often channel-less valleys known as *Dambos*. These valleys can be dry for most of the year, but become inundated after heavy rains.

A special feature of some of tropical rivers, such as Nile, Niger, Okavango, Kosi (tributary of the Ganga), Meta (tributary of the Orinoco) and Taquar (Paraná Basin), is that they have all developed large inland, delta-like floodplains, sometimes described as megafans. Such features usually develop when the river gradient suddenly drops and the river is forced to shed most of its sediment load. The result is a vast region of multi-thread or anabranching rivers, labyrinthine watercourses, abandoned channels, extensive wetlands, a multitude of lakes and frequent avulsion. The plains are extensively flooded during the wet season. Consequently, some of the largest wetlands are associated with these inland deltas. A classic example is the Okavango River in Botswana. Its vast, delta-like floodplain (area ~20,000 km²) marks the terminus of the Okavango River within the Kalahari Basin. The White Nile flows through an enormous area of swampland in southern Sudan known as the *Sudd*. This region of low gradient, clayey soils and multiple channels covers an area of about 30,000–40,000 km² (Mohamed et al., 2006). A similar topographic situation occurs in the Sahel region of Mali, where the Niger River has developed a large inland delta. Several Himalayan rivers also have developed multiple megafans at the foot of the lofty mountain ranges. Amongst these, the migrating Kosi River has built an inland delta or megafan with an area of ~15,000 km². The river has shifted by over 100 km westward across its gigantic fan by successive avulsions during the last 250 years or so (Gole and Chitale, 1966).

However, by far the largest megafan in the world occupies the *el Gran* Pantanal region of west-central Brazil (Table 26.2). The Pantanal is a tectonically active interior sedimentary basin with diverse geomorphic features, including several coalescing and overlapping alluvial fans, numerous lakes, multiple meandering and multi-thread channels, extensive floodplains and large wetlands (Assine, 2005). The Pantanal is considered as one of the largest wetland complexes in the world. However, the most outstanding feature of the Pantanal region is the Taquari megafan. The fan is drained by the Taquari River, a tributary of the Paraguay, covers an area of about 50,000 km² and is characterized by an entrenched meander belt, intricate network of paleochannels and channel–levée complexes (Assine, 2005). Avulsion is frequent and may take place every few years and recurs on the decadal scale.

Apart from inland floodplains, coastal deltaic plains sequester a significant proportion of the sediment load. As rivers are the primary source of sediment to the deltas, rivers with larger catchments and/or higher sediment load develop large deltas. Seasonal tropical rivers are subjected to great fluctuations in water and sediment discharge. Seasonal flooding tends to deposit huge amounts of sediment over the deltaic plains through multiple distributaries, adding to the levees and silting up the shallow basins. Accumulation of sediment induces channel shifting and avulsion. Marshes or swamplands or lagoons occur between distributaries. The Amazon River, with the highest annual sediment load, has formed the largest delta in the world. In monsoon Asia, the mangrove-dominated *Sunderban* delta of the Ganga–Brahmaputra is an example of a large composite delta formed by three rivers: Ganga, Brahmaputra and Meghna. Details of other notable tropical deltas are shown in Table 26.3.

Flood risks and mitigation

Irrespective of the basin size and location, all tropical rivers experience large floods. Some of them are extreme even by world standards. Heavy rainfall is the fundamental reason for flooding and widespread inundation (Table 26.4).

Tropical floods are of four types – flash floods, wet season floods, cyclone-induced floods and storm surges. Short duration, intense rainfall produces flash floods in small basins and in mountainous areas. Flash floods frequently generate debris flows and sometimes landslides. They are a

Table 26.3 Delta type and area of some major tropical rivers

Name of the river	Approximate area in × 10³ km²	Coastal environment type
Amazon	470	Tide-dominated
Ganga-Brahmaputra	105	Tide-dominated
Mekong	94	Mixed wave- and tide-dominated
Changjiang	67	Tide-dominated
Indus	30	Mixed wave- and fluvial-dominated
Orinoco	20	Mixed wave- and tide-dominated
Irrawaddy	20	Tide-dominated
Niger	19	Mixed tide- and wave-dominated
Nile	12	Wave-dominated
Song Hong	12	Mixed wave- and tide-dominated
Chao Phraya	11	Tide-dominated (low-energy)
Fly	7	Tide-dominated
Senegal	5	Wave-dominated
Magdalena	2	Wave-dominated

major hazard particularly in steep terrains and more semi-arid areas. Intermediate and large rivers overflow their banks and inundate large areas when the bankfull capacity is exceeded. Such floods are the results of rainfall over an extended period and an extended area during the wet season and reflect catchment characteristics and basin-wide climatic conditions. Cyclone-generated floods are most severe because they usually exceed normal wet season peak discharges. Exceptional floods occur when the wet season and tropical storms coincide, as in the case of the rivers of south Asia and south-east Asia during monsoon season. Notable floods are also linked to El-Niño and La Niña phenomena, and the impact is particularly severe if an El Niño year is followed by La Niña year (or vice versa).

Many of the coastal plains, particularly in south Asia, are also susceptible to storm surges, induced by tropical storms. This hazard is particularly severe in Bangladesh and eastern India. During the 1970 storm surge in Bangladesh, between 0.3 and 0.5 million people perished. Similarly, the 1999 storm surge caused by the super-cyclone in Orissa and the 1977 surge in Krishna delta in eastern India resulted in large

numbers of casualities. Other areas affected by this storm hazard include China, central America and the Caribbean islands; the Philippines and north-east Queensland in Australia.

Floods are one of the most prevalent natural hazards in the tropics and account for most damages and casualties (Table 26.4). Many large cities and densely populated regions in Africa, Asia and South America lie in inland and coastal plains and hence are generally vulnerable to flooding. Flood hazard is connected to the magnitude of damaging event and the human vulnerability. Human vulnerability results from physical, socio-economic and environmental conditions, which determine the likelihood and scale of impact (UNDP, 2004). The plot given in Figure 26.7 reveals that the average number of people exposed to the flood hazard is highest in four populous Asian countries, namely, India, China, Indonesia and Bangladesh (113, 103, 49 and 42 million respectively). This is 1–2 orders of magnitude higher than other tropical countries, except Brazil (18 million). It is also apparent from the plot that the density of population in areas prone to floods is highest in Bangladesh (912/km²) followed by Puerto Rico, and countries having a high rural population density and a low Human Development Index, where the scale and impact of flood risk is very high.

Traditionally, structural measures (dams, weirs and embankments) and non-structural measures (flood forecasting and warning systems) have been adopted all over the world, and the tropics are no exception. Numerous large dams and thousands of kilometres long embankments have been built to control floods and to reduce damage. However, failure of embankments and smaller dams has only exacerbated the situation, as in the case of the August 2008 extraordinary flood on the Kosi that uprooted three million people.

Several governmental and international agencies (such as UNDP, USAID/OFDA, World Bank, Asian Development Bank), along with non-government organizations (NGOs) have initiated or supported flood risk mitigation projects in the flood-affected countries. Greater emphasis is now being laid on managing the risks of floods rather than on protection against floods and on strengthening the flood forecasting and warning capabilities. To help minimize exposure and vulnerability, flood hazard zone mapping and flood risk mapping have received greatest priority in all flood risk mitigation programs. The former provides information on the flood extent and water depths for low, medium and high probability events, and the latter gives spatial information regarding the number of inhabitants and the type of economic activity potentially affected. Remote sensing and GIS technologies have proved to be

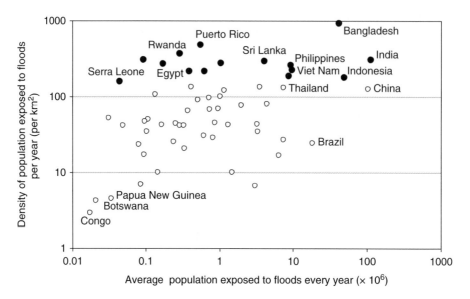

Figure 26.7 Plot of density of population in flood prone area versus the average population in flood-affected area. Solid circles represent countries with high population density (>150/km²). (UNDP (2004))

vital tools in delineating flood-prone areas and in combining information on hazard, exposure and vulnerability. The paradigm shift in the approach is primarily due to the growing realization that floods are natural events and cannot be avoided or completely eliminated.

CONCLUSIONS

Geomorphology in tropical environments is no longer justified by the study of exotic or particular landforms, but distinctive features of the process systems and their geomorphic outcomes require recognition. In the humid tropics, pervasive rock-mass weathering and the impacts of intense rainfall and violent storms indicate priorities seldom emphasized in temperate geomorphology. The importance of the weathered mantle was recognized early, in the search for mineral resources, and was integrated into many accounts of landform evolution (Branner, 1896; Falconer, 1911; Willis, 1936), before King (1950, 1953, 1962) argued persuasively for universal slope retreat and the formation of worldwide pediplains. Büdel (1957, 1982), however, re-emphasized the role of weathering within his model of double surfaces of levelling, and this stimulated enquiry into patterns of weathering (Thomas, 1966). Increasing interest in process-based studies and in climate change is reflected in the work of Tricart

(1972) and is discussed by several authors in Douglas and Spencer (1985). Thomas (1994) and Wirthmann (1994) reviewed progress until the early 1990s. Radiometric techniques have now expanded to enable the dating of regolith minerals and sediments, the estimation of rates of denudation, and the duration of rock exposure. These data are leading to reappraisal of the timescales of change and landform evolution. In parallel, the expansion of hydrological recording and observations on tropical rivers, their channels and floodplains has provided new tools for geomorphological analysis of stream behaviour.

The hazards faced by both rural and urban communities in the tropics often arise from a combination of floods, landslides and destructive coastal erosion. Coastal processes are discussed elsewhere, but certain areas are particularly susceptible to storm surges during tropical cyclones. Features that can be related to tropical conditions are mainly driven by biological processes, particularly the growth of mangroves in sediment-rich environments and corals in clear-water environments away from river estuaries and deltas. On a global view humid tropical coasts are dominated by muddy sediments (50 per cent) and corals (about 20 per cent), the remainder being sandy environments. Both mangroves and corals can afford some protection against coastal erosion but are particularly sensitive to environmental and sea level changes. The patterns and frequencies of hazards from

Table 26.4 Extreme flood events (recurrence interval >100 y) between 1985 and 2009

Country	Date	Duration in days	Number of deaths	Number of persons displaced	Area affected, km^2	FM
Zambia, Angola	27 Feb 09	58	150	54,000	1,148,000	8.1
Thailand	1 Aug 95	101	260	4,220,000	444,500	8.0
Argentina, Uruguay	1 Oct 01	62	2	14,500	627,900	7.9
Somalia, Kenya, Ethiopia	19 Oct 97	41	2000	230,000	713,800	7.8
Democratic Republic of Congo, Congo Republic	29 Nov 99	33	2	75,000	738,800	7.7
China	8 Aug 98	38	89	2,160,000	620,800	7.7
China	9 Jun 94	21	719	—	1,015,000	7.6
Brazil	22 Apr 09	23	39	408,000	524,400	7.4
China	25 Jul 95	35	123	3,000,000	325,400	7.4
Peru	1 Feb 86	90	—	300,000	94,880	7.2
Ecuador	30 Jan 08	93	51	265,000	75,830	7.1
China	28 Jul 08	12	—	93,000	533,900	7.1
Pakistan	12 Aug 97	23	165	836,300	276,900	7.1
India, Nepal	18 Aug 08	38	400	10,000,000	163,700	7.1
India, Bangladesh	20 Jul 08	20	142	225,000	283,800	7.1
India, Bangladesh	18 Sep 00	34	1468	24,000,000	155,000	7.0
Philippines	1 Jul 01	19	178	800,000	205,200	6.9
Vietnam, Thailand, Cambodia, Malaysia	25 Oct 99	16	632	1,114,000	232,600	6.9
India	22 Sep 08	8	2400	—	464,300	6.9
Venezuela	1 Jul 96	20	—	15,000	179,100	6.9
Pakistan	2 Aug 08	100	35	200,000	32,490	6.8
India	29 Oct 99	15	9803	10,000,000	191,900	6.8
China	8 Jun 02	11	326	300,000	252,000	6.7
China	19 Jun 05	12	200	1,500,000	226,600	6.7
Argentina	28 Apr 03	13	24	161,500	202,800	6.7
China	3 May 97	14	135	360,000	175,000	6.7
Honduras	19 Oct 08	25	110	320,000	96,390	6.7
China	31 May 00	20	94	8,725	114,000	6.7
Thailand	11 Aug 08	10	130	4,000	219,500	6.6
Venezuela, Colombia	15 Dec 99	6	20,006	400,000	328,200	6.6
Sri Lanka	22 Nov 08	13	61	171,000	103,400	6.4
China	7 Jun 08	16	176	1,600,000	79,370	6.4
Brazil	22 Nov 08	12	117	80,000	103,400	6.4

Source: Dartmouth Flood Observatory (2009). http://www.dartmouth.edu/~floods/index.html

FM, Flood magnitude = log (duration × severity × affected area).

floods, landslides and coastal erosion are likely to be influenced by global climate changes, and geomorphology can play a central role in understanding the impacts of extreme events and change on different timescales (Nott, 2006).

REFERENCES

Ahnert, F. (1982) Investigations on the morphoclimate and on the morphology of the inselberg region of Machakos, Kenia. Beiträge zur Geomorphologie der Tropen (Ostafrika, Brasilien, Zentral- und Westafrika). *Catena Supplement*, 2, 1–72.

Albergel, J., Bader, J-C. and Lamagat, J-P. (1997) Flood and drought: application to the Senegal River management. Sustainability of Water Resources under Increasing Uncertainty. Proceedings of Rabat Symposium S1, April 1997. IAHS Publication no. 240.

Aleva, G.J.J. (compiler) (1994) *Laterites. Concepts, Geology, Morphology and Chemistry*. ISRIC, Wageningen, Netherlands. p. 169. ISBN 90-6672-053-0.

Allison, M.A., Kuehl, S.A., Martin, T.C. and Hassan, A. (1998) Importance of flood-plain sedimentation for river sediment budgets and terrigenous input to the oceans: insights from the Brahmaputra-Jamuna River. *Geology*, 26, 175–8.

Amarasekera, K.N., Lee, R.F., Williams, E.R. and Eltahir, E.A.B. (1997) ENSO and the natural variability in the flow of tropical rivers. *Journal of Hydrology*, 200, 24–39.

Assine, M.L. (2005) River avulsions on the Taquari megafan, Pantanal wetland, Brazil. *Geomorphology*, 70, 357–71.

Aziz-Alaoui, M.A. and Bertelle, C. (eds) (2006) *Emergent Properties in Natural and Artificial Dynamical Systems*. Springer, p. 281.

Baker, V.R. (1978) Adjustment of fluvial systems to climate and source terrain in tropical and sub-tropical environments, in A.D. Miall (ed.), *Fluvial Sedimentology*. Canadian Society Petroleum Geologists, Memoir 5, pp. 211–30.

Benito, G. and Gregory, K.J. (eds) (2003) *Palaeohydrology: Understanding Global Change*. John Wiley and Sons, Chichester. p. 396.

Boulangé, B., Ambrosi, J-P. and Nahon, D. (1997) Laterites and bauxites, in H. Paquet and N. Clauer (eds), *Soils and Sediments – Mineralogy and Geochemistry*. Springer, Berlin. pp. 49–65.

Branner, J.C. (1896) Decomposition of rocks in Brazil. *Bulletin of the Geological Society of America*, 7, 255–314.

Braun, J-J., Ngoupayou, J.R.N., Viers, J., Dupre, B., Bedimo, J-P., Boeglin, J-L. et al. (2005) Present weathering rates in a humid tropical watershed: Nsimi, South Cameroon. *Geochimica et Cosmochimica Acta*, 69, 357–87.

Büdel, J. (1948) *Das System de Klimatischen Morphologie*. Deutscher Geographen-Tag, Munchen. pp. 65–100.

Büdel, J. (1957) Die 'doppelten Einebnungsflächen' in den feuchten Tropen. *Zeitschrift für Geomorphologie*, N.F.1, 201–88.

Büdel, J. (1982) *Climatic Geomorphology*. (translation L. Fischer and D. Busche). Princeton University Press, Princeton. p. 443.

Burke K. and Gunnell, Y. (2008) *The African Erosion Surface*. The Geological Society of America, Memoir 201. p. 66.

Carmo, I de Oliveira and Vasconcelos, P.M. (2006) ^{40}Ar/^{39}Ar geochronology constraints on late Miocene weathering rates in Minas Gerais, Brazil. *Earth and Planetary Science Letters*, 241, 80–94.

Coelho-Netto, A.L. (1997) Catastrophic landscape evolution in a humid region (SE Brazil): inheritances from tectonic, climatic and land use induced changes. Geogr. Fisi. Dinam. Quat., Supplement 111 T.3, 21–48.

Dartmouth Flood Observatory. (2009) Global active archive of large flood events. http://www.dartmouth.edu/%7Efloods/Archives/index.html (accessed 14 October 2009).

Davis, W.M. (1899) The geographical cycle. *Geographical Journal*, 14, 481–504.

Dembele, N.J. (2008) An evaluation of sediment storage in the Niger inland floodplain in Mali. *Online Journal of Earth Sciences*, 2, 64–9.

De Oliveira, M.A.T. (1990) Slope geometry and gully erosion development: Bananal, Sao Paulo, Brazil. *Zeitschrift für Geomorphologie*, N.F., 34, 423–34.

Dietrich, W.E., Wilson, C.J. and Reneau, S.L. (1986) Hollows, colluvium and landslides in soil-mantled slopes, in A.D. Abrahams (ed.), *Hillslope Processes*. The Binghamton Symposia in Geomorphology: International Series, no. 16. pp. 361–88.

Douglas, I. and Spencer, T. (eds) (1985) *Environmental Change and Tropical Geomorphology*. Allen and Unwin, London. p. 378.

Falconer, J.D. (1911) *The Geology and Geography of Northern Nigeria*. Macmillan, London. p. 295.

Feng, Y.-X. and Vasconcelos, P. (2007) Chronology of Pleistocene weathering processes, southeast Queensland, Australia. *Earth and Planetary Science Letters* 263, 275–87.

Fernandes, N.A., Guimarães, R.F., Gomes, R.A.T., Vieira, B.C., Montgomery, D.R. and Greenberg, H. (2004) Topographic controls of landslides in Rio de Janeiro: field evidence and modeling. *Catena*, 55, 163–81.

Fletcher, W.K. and Muda, J. (2003) Dispersion of gold in stream sediments in the Sungai Kuli region, Sabah, Malaysia. *Geochemistry: Exploration, Environment, Analysis*, 3, 51–6.

Fookes, P.G. (ed) (1997) *Tropical Residual Soils*. Geological Society Professional Handbook. The Geological Society, London. p. 184.

Fookes, P.G., Lee, E.M. and Milligan, G. (2005) *Geomorphology for Engineers*. Whittles Publishing, CRC Press. p. 851.

Gole, C.V. and Chitale, S.V. (1966) Inland delta-building activity of the Kosi River. *Journal Hydraulic Division, American Society Civil Engineering*, HY2, 111–26.

Goswami, D.C. (1985) Brahmaputra River, Assam, India: Physiography, basin denudation, and channel aggradation. *Water Resources Research*, 21, 959–78.

Goswamy, D. (1989) Estimation of bedload transport in the Brahmaputra River, Assam. *Journal of Earth Science*, 15, 14–26.

Goudie, A. (1973) *Duricrusts in Tropical and Sub-Tropical Landscapes*. Clarendon Press, Oxford. p. 174.

Gregory, K.J. and Benito, G. (eds) (2003) *Palaeohydrology–Understanding Global Change*. John Wiley and Sons, Chichester. p. 396.

Gupta, A. (1988) Large floods as geomorphic events in the humid tropics, in Baker, V.R., Kochel, R.C. and Patton, P.C. (eds), *Flood Geomorphology*. John Wiley and Sons, Chichester. pp. 301–315.

Hamilton, S.K. and Lewis Jr., W.M. (1990) Physical characteristics of the fringing floodplain of the Orinoco River, Venezuela. *Intersciencia*, 15, 491–500.

Horton, R.E. (1945) Erosional development of streams and their drainage basins: hydrophysical approach to quantitative morphology. *Bulletin Geological Society of America*, 56, 275–370.

Hovius, N. (1998) Controls on sediment supply by large rivers, in K.W. Shanley and P.J. McCabe (eds), *Relative Role of Eustasy, Climate, and Tectonics in Continental Rocks*. SEPM Special Publication 59, Tulsa. pp. 3–16.

ISRIC (International Soil Reference Information Centre) (2006) *World Reference Base for Soil Resources* 2006. FAO, Rome. p. 128.

Kale, V.S. (2002) Fluvial geomorphology of Indian rivers: an overview. *Progress in Physical Geography*, 26, 400–33.

Kale, V.S. (2007) Geomorphic effectiveness of extraordinary floods on three large rivers of the Indian Peninsula. *Geomorphology*, 85, 306–16.

Kale, V.S. (2008) A half-a-century record of annual energy expenditure and geomorphic effectiveness of the monsoon-fed Narmada River, central India. *Catena*, 75, 154–63.

King, L.C. (1950) The study of the world's plainlands. *Quarterly Journal Geological Society, London*, 106, 101–31.

King, L.C. (1953) Canons of landscape evolution. *Bulletin Geological Society of America*, 64, 721–52.

King, L.C. (1962) *The Morphology of the Earth*. (2nd edn. 1967). Oliver and Boyd, Edinburgh. p. 726.

Kondrashov, D., Feliks, Y. and Ghil, M. (2005) Oscillatory modes of extended Nile River records (A.D. 622–1922). *Geophysical Research Letters*, 32, L10702. doi:10.1029/2004GL022156.

Köppen, W. (1923) Die Klimate der Erde; Grundriss der Klimakunde. Griuyter, Berlin. p. 369.

Kump, L.R., Brantley, S.L. and Arthur, M.A. (2000) Chemical weathering, atmospheric CO_2, and climate. *Annual Review of Earth and Planetary Sciences*, 28, 1–67.

Lacerda, W. (2007) Landslide initiation in saprolite and colluvium in southern Brazil: Field and laboratory observations. *Geomorphology*, 87, 104–19.

Latrubesse, E.M. (2008) Patterns of anabranching channels: the ultimate end-member adjustment of mega-rivers. *Geomorphology*, 101, 130–45.

Latrubesse, E.M., Stevaux, T, J.C. and Sinha, R. (2005) Tropical rivers. *Geomorphology*, 70, 187–206.

Lauer, J.W., Parker, G. and Dietrich, W.E. (2008) Response of the Strickland and Fly River confluence to postglacial sea level rise, *Journal of Geophysical Research*, 113, F01S06. doi:10.1029/2006JF000626, 2008.

Lucas, Y., Boulet, R., Chauvel, A. and Veillon, L. (1987) Systèmes sols ferrallitiques – podzols en région Amazonienne. In D. Righi and A. Chauvel (eds), Podzols et Podzolisation, *Comptes Rendus de la Table Ronde Internationale, Association française pour l'Etude du Sol*, ORSTOM/INRA, 53-65.

Maignien, R. (1966) Review of Research on Laterites. UNESCO, Natural Resources Research 4, Paris. p. 148.

McFarlane, M.J. (1976) *Laterite and Landscape*. Academic Press, London. p. 151.

Mertes, L.A.K., Dunne, T. and Martinelli, L.A. (1996) Channel-floodplain geomorphology along the Solimões-Amazon River, Brazil. *Bulletin Geological Society of America*, 108, 1089–107.

Milliman, J.D. and Meade, R.H. (1983) Worldwide delivery of river sediment to the oceans. *The Journal of Geology*, 91, 1–21.

Milliman, J.D. and Syvitski, P.M. (1992) Geomorphic/tectonic control of sediment discharge to the ocean: the importance of small mountain rivers. *Journal of Geology* 100, 525–544.

Mirza, M.M.Q. (2003) Three recent extreme floods in Bangladesh: a hydro-meteorological analysis. *Natural Hazards*, 28, 35–64.

Mohamed, Y.A., Savenije, H.H.G., Bastiaanssen, W.G.M. and van den Hurk, B.J.J.M. (2006) New lessons on the Sudd hydrology learned from remote sensing and climate modelling. *Hydrology Earth System Science*, 10, 507–18.

Nilsson, C., Reidy, C.A., Dynesius, M. and Revenga, C. (2005) Fragmentation and flow regulation of the world's large river systems. *Science*, 308, 405–8.

Nott, J.F. (2006) *Extreme Events – A Physical Re-Construction and Risk Assessment*. Cambridge University Press, Cambridge, U.K. p. 300.

O'Connor, J.E. and. Costa, J.E. (2004) *The World's Largest Floods, Past and Present: Their Causes and Magnitudes*. Circular 1254. U.S. Department of the Interior, U.S. Geological Survey, Washington, D.C.

Ollier, C.D. and Galloway, R.W. (1990) The laterite profile, ferricrete and unconformity. *Catena*, 17, 97–109.

Phillips, J.D. (1999). *Earth Surface Systems*. Blackwell, Oxford. p. 180.

Ramonell, C., Amsler, M. and Toniolo, H. (2002) Shifting modes of the Parana River thalweg in its middle-lower reach. *Zeitschrift fur Geomorphologie* 129, 129–42.

Rengers, F. and Wohl, E. (2007) Trends of grain sizes on gravel bars in the Rio Chagres, Panama. *Geomorphology*, 83, 282–93.

Römer, W. (2005) The distribution of inselbergs and their relationship to geomorphological, structural and lithological controls in Southern Zimbabwe. *Geomorphology*, 72, 156–76.

Römer, W. (2006) Differential weathering and erosion in an inselberg landscape in southern Zimbabwe: a morphometric study and some notes on factors influencing the long-term development of inselbergs. *Geomorphology*, 86, 349–68.

Runge, J. (2007) The Congo River, Central Africa, in A. Gupta (ed.), *Large Rivers: Geomorphology and Management*. John Wiley and Sons, Chichester. pp. 293–309.

Ruxton, B.P. and Berry, L. (1957) Weathering of granite and associated erosional features in Hong Kong. *Geological Society of America Bulletin*, 68, 1263–92.

Sarma, J.N. (2005) Fluvial processes and morphology of the Brahmaputra River in Assam, India. *Geomorphology*, 70, 226–56.

Shahin, M. (1985) *Hydrology of the Nile Basin*. Elsevier, Amsterdam. p. 575.

Singh, I.B. (2007) Ganga River, in A. Gupta (ed.), *Large Rivers: Geomorphology and Management*. John Wiley and Sons, Chichester. pp. 347–71.

Stallard, R.F. and Edmond, J.M. (1987) Geochemistry of the Amazon. 3. Weathering chemistry and limits to dissolved inputs. *Journal of Geophysical Research*, 92, 8293–302.

Stoddart, D.R. (1969). Climatic geomorphology: review and assessment. *Progress in Physical Geography*, 1, 160–222.

Strahler, A.N. (1952) Dynamic basis of geomorphology. *Bulletin Geological Society America*, 63, 923–38.

Taylor, G.A. and Eggleton, R.A. (2001) *Regolith Geology and Geomorphology*. John Wiley and Sons, Chichester. p. 375.

Thomas, M.F. (1966) Some geomorphological implications of deep weathering patterns in crystalline rocks in Nigeria. *Transactions of the Institute of British Geographers*, 40, 173–93.

Thomas, M.F. (1994) *Geomorphology in the Tropics*. John Wiley and Sons, Chichester. p. 460.

Thomas, M.F. (2006) Lessons from the tropics for a global geomorphology. *Singapore Journal of Tropical Geography* 27, 111–27.

Thomas, M.F. (2008a) Geomorphology in the humid tropics, 1890–1965, in T.B. Burt, R.J. Chorley, D. Brunsden, N.J. Cox and A.S. Goudie (eds), *The History of the Study of Landforms or The Development of Geomorphology*. 4, 679–727.

Thomas, M.F. (2008b) Understanding the impacts of late Quaternary climate change in tropical and sub-tropical regions. *Geomorphology*, 101, 146–58.

Thomas, M.F. and Thorp, M.B. (1995) Geomorphic response to rapid climatic and hydrologic change during the Late Pleistocene and Early Holocene in the humid and sub-humid tropics. *Quaternary Science Reviews*, 14, 193–207.

Thomas, M.F. and Murray, A.S. (2001) On the age and significance of Quaternary colluvium in eastern Zambia. *Palaeoecology of Africa*, 27, 117–33.

Thomas, M.F., Thorp, M.B. and McAlister, J. (1999) Equatorial weathering, landform development and the formation of white sands in north western Kalimantan, Indonesia. *Catena*, 36, 205–32.

Thomas, M.F., Nott, J., Murray, A.S. and Price, D.M. (2007) Fluvial response to late Quaternary climate change in NE Queensland, Australia. *Palaeogeography, Palaeoclimatology, Palaeoecology*, 251, 119–36.

Tockner, K. and Stanford, J.A. (2002) Riverine flood plains: present state and future trends. *Environmental Conservation*, 28, 308–30.

Tricart, J. (1972) *The Landforms of the Humid Tropics, Forests and Savannas*. (translation C.J.K. De Jonge). Longman, London. p. 306.

UNDP (2004) *Reducing Disaster Risk: A Challenge for Development*. United Nations Development Programme, Bureau for Crisis Prevention and Recovery, New York.

Vasconcelos P.M. and Conroy, M. (2003) Geochronology of weathering and landscape evolution, Dugald River valley NW Queensland, Australia. *Geochimica Gosmochimica Acta*, 67, 2913–30.

Vasconcelos, P.M., Renne, P.R., Brimhall, G.H. and Becker, T.A. (1994) Direct dating of weathering phenomena by $^{40}Ar/^{39}Ar$ and K-Ar analysis of supergene K-Mn oxides. *Geochimica et Cosmochimica Acta*, 58, 1635–65.

Verstappen, H.Th. (2000) *Outline of the Geomorphology of Indonesia*. ITC Publication 79, Enschede, Netherlands. p. 212.

Vörösmarty, C., Meybeck, M., Fekete, B., Sharma, K., Green, P. and Syvytski, J. (2003) Anthropogenic sediments retention: major global impact from registered river impoundments. *Global and Planetary Change*, 39, 169–90.

Walling, D.E. (2006) Human impact on land–ocean sediment transfer by the world's rivers. *Geomorphology*, 79, 192–216.

Wasson, R.J. (2003) A sediment budget for the Ganga–Brahmaputra catchment. *Current Science*, 84, 1041–7.

Willis, B. (1936) *East African Plateaus and Rift Valleys: Studies in Comparative Seismology*. Carnegie Institute, Publication 470, Washington. p. 358.

Wirthmann, A. (1994) *Geomorphology of the Tropics*. (English translation D. Busche 1999). Springer, p. 314.

World Meteorological Organisation/Arizona State University. Global weather and climate extremes at: http://wmo.asu.edu/#continental.

27

Geomorphology Underground: The Study of Karst and Karst Processes

D.C. Ford and P.W. Williams

The central focus of karst research is on understanding karst landforms above and below ground. This requires understanding of the nature and controls of dissolution processes in the comparatively soluble karst rocks and the underground flow systems that may develop as consequences of these processes. Karst studies thus are rock-specific. The principal host rocks are limestone, marble and dolomite (carbonates), and gypsum, anhydrite and salt (evaporites). Individual beds, members or even entire formations of these rocks can be *monominerallic* or nearly so, with the result that there is little insoluble residuum to obstruct the propagation of initially tiny solution conduits along fractures in them. Pure siliceous sandstones and quartzites are another example of the monominerallic that may exhibit karstic evolution but, because of their lesser solubility, development is limited in form and scale and mingled with conventional fluvial or other geomorphic forms. A bulk insoluble fraction of 10 per cent or more will similarly inhibit wholly karstic development in many carbonate strata but in the evaporites, because they have greater solubility, karst is often found where the insolubles exceed 50 per cent.

Where karst rocks outcrop in humid regions, normal runoff into networks of surface erosional channels is inhibited and may be absent altogether. The channels are replaced by a hierarchy of solutional landforms that direct runoff underground, from which it eventually emerges in springs at regional base levels or at contacts with insoluble rocks. *Karren* is the general term for a large variety of small pit, groove and runnel forms that may intensely dissect the surface of the rock, whether exposed or subsoil. Particularly intense dissolution near the surface widens joints and fissures to depths as great as several metres, creating an *epikarst* zone of high effective (interconnected) porosity. Located within the epikarst are *dolines* (or *sinkholes*), topographically enclosed depressions with funnel or bowl shapes, 10–100 m or more in diameter and proportionally deep. They may occur as scattered individuals, in groups determined by local geological or hydrological conditions (e.g. where streams flow onto limestone from adjoining insoluble rocks) or can be regularly packed to fully occupy a surface – *polygonal karst*. They are often considered to be the diagnostic karst landforms. At larger scales there are *dry valleys* and *gorges* where rivers have entrenched into karst rocks. Large marginal or internal plains of corrosional or depositional origin that discharge water underground are termed *poljes* (Figure 27.1).

KARST STUDIES

The early karst studies

The significance of dissolution of limestone by rainwater was recognized by James Hutton (Hutton, 1795: 219) and the first quantitative

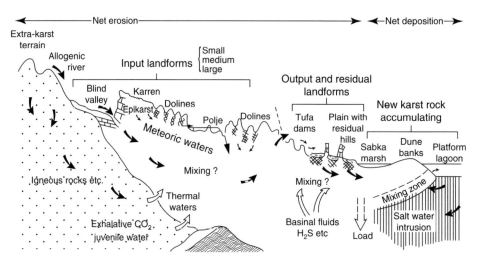

Figure 27.1   The comprehensive karst system: a composite diagram illustrating the major phenomena encountered in active karst terrains (From Ford and Williams, 1989)

studies of solution appeared early in the 19th century. Groundbreaking works that established karst as a quasi-separate branch of geomorphology and of hydrogeology came in the later 19th and early 20th centuries. They were centred in Slovenia and adjoining areas known since classical times as The Karst (Gams, 2003). Similar terrain extends down the Adriatic coast to Albania and the entire region is known as the Dinaric Karst, the type site of karst. By extension, landscapes possessing similar features elsewhere in the world are also known as karst, another particularly extensive karst region being found in southern China (Yuan et al., 1991; Sweeting, 1995). Karst is best developed on carbonate rocks, which cover about 14 percent of the ice-free continental area but, as noted, is also found on other especially soluble rocks such as gypsum and rock salt.

Jovan Cvijić (1893) published a classification and discussion of the principal surface landforms of the Dinaric Karst and their roles in the groundwater systems that came to guide much of the future development of ideas. It included the earliest morphometric analysis to be found anywhere in geomorphology. This work was his doctoral thesis, supervised by Albrecht Penck who shortly after invited W.M. Davis to tour the region with them. Perhaps predictably, attempts to develop a cycle of karst erosion soon followed, evolving landscapes from initial fluvial or scattered doline stages to poljes or corrosional end plains (Ford and Williams, 2007, pp. 391–5). The early efforts were not successful, but in a 1918

study and later work published posthumously Cvijić analysed situations where, first, the karst massif extends below the general base level of erosion (sea level) and, second, the underlying insoluble rocks occur above general base level and geological structures are complex like those of The Karst (Cvijić, 1918, 1960; Ford, 2005). His proposals were formative and many remain pertinent today.

In the following decades karst specialists continued to focus on the surface landforms but with a new emphasis, however, that of climato-genesis (Lehmann, 1936, 1954, 1956). The proposition that contrasting morphologies develop in different climatic zones was, perhaps, pursued further in karst research than elsewhere in geomorphology (Jakucs, 1977; Salomon and Maire, 1992). It stimulated renewed attention to measurable properties of karst that objectively compared morphology in order to test the climato-genetic assertions. The research that followed (e.g. Panos and Stelcl, 1968; Williams, 1971, 1972, 1978; Drake and Ford, 1972; Day, 1976, 1979; White and White, 1979) found that almost as much morphological variety occurred within a single climatic zone as between climatic zones, and that lithology and structure play a larger role than previously appreciated. These evolving views on karst development were captured in textbooks by Jennings (1971, 1985), Sweeting (1972) and Trudgill (1985). At present the climato-genetic approach is still emphasized rather more in Europe than elsewhere (e.g. Salomon, 2000,

2006), but it is now generally recognized that while broad differences in karst landscape styles exist between climatic zones, overlap occurs and differentiation at too fine a scale is unjustified (Ford and Williams, 1989, 2007; White, 1988).

Speleology: the study of caves

In work that was largely independent of the karst geomorphology of the period, others sought to explain development of the solution caves that they were able to explore. Necessarily, these were in the *vadose zone* (above the contemporary water tables), so that models for enlargement (and even for initiation) in vadose conditions were the first to be advanced. These assumed that a water table stable at significant depth in the rock must precede most or all cave development. The flow is *unconfined* (not trapped beneath impermeable strata in an artesian situation). In a remarkable intervention, W.M. Davis then turned matters on their head by analysing a large collection of cave maps and descriptions to suggest that most caves (including many large ones) must have evolved beneath the water table (in the *phreatic zone*) by dissolution in slowly flowing ground waters circulating to random depths (Davis, 1930). He and Bretz (1942) ingeniously adapted this model to fit into the mature and old stages of the Davisian landform cycle. In response to such ideas, Swinnerton (1932) and Rhoades and Sinacori (1941) argued that solutional cave systems should develop principally *along* a water table rather than above or below it. Meanwhile in France, Trombe (1952) developed ideas on speleogenesis with a focus on discrete channel networks that were quite independent of those proposed elsewhere. The Anglo-American and European schools of speleogenesis were yet to meet. The establishment of the International Union of Speleology in 1965 considerably assisted this.

EXPANDING THE SCOPE OF KARST STUDIES

It is most important to emphasize that, following this fragmented beginning to studies of processes and morphology with ideas developing almost independently in different parts of the world, the scope and theoretical basis of karst research has increased and matured considerably in recent decades, in the physical extent of the flow systems being considered, the range of physical, chemical and biochemical processes affecting them, and the evolution of landforms above and below ground.

This point is underscored in Figure 27.1, where the suite of surface karst landforms discussed earlier are seen to be perched upon a welter of subsurface flow and dissolution settings that may operate on the karst rocks to considerable depths or from the moment of their first precipitation on a sea floor.

The importance of 'rock control' in karst evolution is firmly established. The number of initially distinct facies of limestone and dolomite exceeds that of all other sedimentary facies combined (Scholle et al., 1983). During their subsequent compaction and cementation in sedimentary basins (their *diagenesis*) widely differing conditions alter given facies in different ways, increasing the variety of morphologic structures within the rock pile and their effective permeability and solubility. To this must be added the profound effects of concurrent or later folding, faulting and other fracturing. All can have significant impacts upon the types of groundwater flow systems, caves and surface landforms that eventually arise. They explain much of the variation in surface karst morphology which earlier students sought to attribute to differing climates. From studies in the Mendip Hills, England, Ford (1971) resolved the vadose or phreatic or water table cave debate, showing that each type or mixtures of them can be created, depending upon factors of lithology, fissure (fracture and bedding plane) density, and the hydrologic setting. It is an irony that the karst geomorphologist is concerned with a much smaller number of rock types than are other geomorphologists, but must pay much closer attention to many of their properties!

It is convenient to consider the significance of karst facies, diagenesis and structural deformation in the context of the three diagenetic stages that all sedimentary rocks may experience (James and Choquette, 1984). Our analysis here applies primarily to carbonate rocks, which are predominant in karst globally: evaporites are less affected by facies and diagenetic variations because of their greater bulk solubility. *Eogenesis* is the superficial cementation of an accumulating layer or pile of loose sediments. *Mesogenesis* takes place as compaction by burial plus precipitation of cements from expelled formation waters converts the sediments into consolidated rocks. Other fluids such as waters and gases of volcanic origin may also pass through them (Figure 27.1). When erosion of the overburden exposes the rocks to the circulation of meteoric waters, they enter the *telogenetic* stage: the hydrological setting will be *confined* where the karst strata are in artesian structures, *unconfined* where there is simple gravity-driven flow through them.

Diagenesis and karstification may begin during the *eogenetic* stage of carbonate particles accumulating on the sea floor, along beaches and in dunes, or in biostructures such as patch reefs. If exposed sub-aerially, dissolution in rain followed by evaporation can quickly produce skin-deep crusts of consolidated rock. Colonizing plants punch holes through them along root paths (a first generation of karren forms), while inter-tidal or other streams may pierce them to mechanically wash out the unconsolidated sand, etc. of the interior in bulk, so creating a cave. Jennings (1968) first described such features, terming them *syngenetic karst*. Such vadose zone cementation can rapidly consolidate limestone to significant depths beneath the crusts. By dating speleothems, Mylroie and Carew (1990) showed that carbonate sand dunes of Last Interglacial age (~125,000 years BP) in the Bahamas were fully consolidated and subject to karstification by marine mixing zone dissolution within 30,000 years or so.

Compaction and cementation are the dominant processes during mesogenesis. Pressure solution can dissolve as much as 40 per cent of the remaining thickness of an already compacted limestone mass beneath as little as 500 m of overburden, and continue to act to depths of many kilometres. Despite such effects, however, delicate features of individual facies structure or eogenetic alteration are retained in most limestones and many dolomites, and can greatly affect karst development during the telogenetic stage. When and where the rocks become sufficiently consolidated to mechanically support voids of centimetres or more in diameter, fluid flow may then dissolve vugs or larger caves. These can remain as open voids, suffer roof and wall failure that converts them into breccia bodies, fill or partly fill with phreatic encrustations of calcite or silica, host Mississippi Valley Type (MVT) deposits of massive sulphides (pyrite, galena, sphalerite), or exhibit combinations of two or more of these alternatives. For example, the Nanisivik zinc/lead deposit, Baffin Island, is a sinuous, flat-roofed body ~3.5 km in length, ~100 m in width, filling a syngenetic cavity in Proterozoic dolomites. It appears to have developed along a deep gas–water interface (Ford, 2009) and cannot be younger than Lower Paleozoic in age (Ghazban et al., 1992). It is now 300 m above sea level, overlooking a fiord. Goongewa zinc and lead sulphide deposits in Devonian reef limestones in the Kimberley of north-west Australia occupied a cavern 600 m long and 70 m across and were emplaced in the early Carboniferous following dissolution of the cavity by hydrothermal fluids when the carbonates were buried to a depth of at least 2 km (Wallace et al., 2002). The sulphides are now at the surface.

Mesogenetic cavities such as Nanisivik and Goongewa are almost always created by fluid flow that ascends or flows laterally into them from other, non-karstic, formations and then passes out of them by further ascending or lateral flow, all under deep phreatic conditions. Klimchouk (2000, 2003, 2007) terms such flow *transformational* and classifies the caves and other karst that is created (e.g. breccias) as *hypogene*. In contrast, cave systems and surface karst created by unconfined meteoric waters belong to the *hypergene* (or *epigene*) zone and can develop only in the eogenetic or telogenetic stages. A significant complication is that meteoric waters that become confined by artesian trapping within a telogenetic setting may also function in a transformational manner and thus create caves with hypogene, rather than hypergene, morphology; in fact, these are quite common and include some of the longest (measured by aggregate passage length) that are known (Ford and Williams, 2007; Klimchouk, 2007; Stafford et al., 2009).

Figure 27.2 illustrates the many permutations and combinations that can occur in karst analysis because of the potential for dissolution at all three stages, eo-, meso- and telogenetic, especially where the soluble strata are underlain and overlain by insoluble rocks that are likely to have radically different effective permeabilities. Most limestones experience some karstification in the eogenetic stage, chalk and other weakly consolidated rocks being the exception. They may then be buried and fully consolidated in the mesogenetic setting and experience no further karstification until uplift and erosion expose them at the surface ('open karst' of Figure 27.2). They may be subject to 'deep-seated' transformational karst solution beneath cover strata. Alternatively, first karstification may occur under telogenetic but confined flow conditions where ascending transformational waters create solutional caves *en route* to springs that have opened in eroding cover rocks, or where descending meteoric waters can leak through thicker cover into the karst strata *en route* to springs at lower elevation. Both situations are classified as *subjacent karst* and may not be distinguishable morphologically. The cave morphology, ground water flow patterns, and any consequent surface karst (chiefly, collapse dolines) may be much more complex where such subjacent karst is superimposed upon pre-existing deep karst. Similarly, where rivers initiated on cover strata entrench through karst rocks and begin to drain them (unconfined flow), there may be (i) no prior karst history, or (ii) superimposition onto deep karst, (iii) onto subjacent karst or (iv) onto both in combination. Complete removal of the cover rocks will expose these karsts, complex or otherwise, to the open surface and thus to formation of the 'classic' suite of landforms (epikarst, dolines, erosional poljes, etc.) in denuded karst. Open and denuded karsts can be

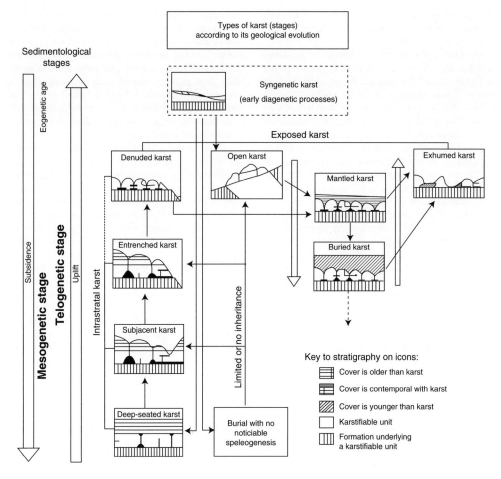

Figure 27.2 Evolutionary types of karst (From Klimchouk and Ford, 2000)

pene-contemporaneously *mantled* by their insoluble residua or by transported sediments so that their surface morphology is concealed but the karstic ground water circulation is maintained (usually creating *suffosion dolines* where mantle sediments are piped into the caves below), or rapidly *buried* by transported sediments so that they become hydrologically inert and so are classified as *paleokarst* features. Mantled and buried karsts may later be *exhumed*.

Other fields of geomorphology are usually concerned only with (1) unconsolidated rocks in their earliest, mechanically accumulative, sub-stages of eogenesis as fans, dunes, beaches, etc. before diagenetic effects exert any significant influence on process or form or (2) consolidated rocks in their final telogenetic sub-stages where the available erosional energy so predominates in the system of forces *versus* resistances that details

of lithology or geologic structure can be (and often are) ignored. Karst studies embrace all stages of diagenesis and consider both deep (hypogene) and surficial (hypergene) conditions. It can be difficult to evaluate the role of inheritance of morphology in such circumstances, which is one of the big problems facing future karst specialists.

Aeolian speleomorphology!

It is a further irony that, following our emphasis on the exceptional importance of deep sub-surface diagenetic and other processes in karst geomorphology, we must conclude this opening review of the modern field with mention of some distinctive effects of air flow in caves that are known collectively as *condensation corrosion*. Their role was recognized and explained by Trombe (1952: 120), but has been the subject of more research in the past 20–30 years.

In relict (drained) caves diurnal or seasonal wetting and drying can delicately weather and etch the walls of the entrance zones, extending far inside where there is through airflow (Zupan Hajna, 2003). The strongest or steadiest flows create 'air scallops' (spoon-shaped corrosional facets) up to several metres in length along walls and ceilings and may reduce obstructions such as stalagmites to aerofoil cross-sections with only 10–20 per cent of the original mass (e.g. Tarhule-Lips and Ford, 1998).

Condensation corrosion is quantitatively more important (and visually striking) immediately above water tables in caves of hypogene origin. Over pools of thermal water, condensation of CO_2-rich steam may create semi-spherical chambers of beautiful regularity; Szunyogh (1989) offers comprehensive analysis of their geometric development. Very large galleries and chambers can develop where there is release of deep-source H_2S into humid atmospheres in limestone caves. Sulphuric acid condenses onto the walls, converting the calcite to gypsum crusts that then fall away under their own weight, exposing fresh limestone to attack (i.e. it is a process of sulphation similar to that attacking limestone buildings in many automobile-clogged cities today). It is now recognized that H_2S (or partially H_2S) caves form a rare but significant sub-class of caverns, including such magnificent examples as Carlsbad Caverns and Lechuguilla Cave in New Mexico, Novy Afonskaya in Georgia and Grotta di Frasassi in Italy (Ford and Williams, 2007: 562): for example, the Big Room in Carlsbad Caverns has a volume of 2 million m^3, almost entirely attributable to condensation corrosion processes (Hill, 1987, 1995).

MODERN MODELLING

The nature and development of karst aquifers

The key to karst resides in understanding the processes that develop its distinctive groundwater systems. Despite our emphasis on the morphogenetic significance of deep karst, the bulk of karst groundwater circulating at a given time is hypergene, derived from rainfall, and flows unconfined or locally confined in artesian traps. Although some direct observation of this is possible in vadose caves, most water flow is inaccessible, and so its behaviour must be investigated by modelling. As was the case in early studies where

students of surface landforms and of cave development were separate schools, scarcely intercommunicating after Cvijić's time, so in recent decades there is a marked division between hydrogeologists trained generally and believing that their training equips them to tackle flow in cavernous rocks, and karst scientists who find conventional groundwater models inapplicable and capable of grossly misleading managers of karst groundwater resources.

Most karst rocks in hand specimen are almost impervious, usually with <2 per cent porosity. However, in young carbonate rocks such as coral and calcareous aeolianite, and also chalk, porosities of 20–50 per cent can be found. This porosity arises from the interstitial spaces between the grains comprising the rock. It is high when the sediment is first deposited, but decreases during diagenesis. It is termed *matrix porosity*. Fissures running through the rock, such as bedding planes, joints and faults, impart *fracture porosity*. Once groundwater saturates the rock, dissolution occurs along fissures penetrable by moving water. This initiates conduits and leads to considerable enhancement of porosity – *conduit porosity*. Interconnected pores give rise to rock *permeability*. Although many rocks are porous and permeable, karst rocks are unique in that the very act of water circulation promotes dissolution and so progressively increases porosity and permeability over time.

The basic principles of groundwater movement are explained in Ford and Williams (2007: 562). A critical point is that the general law describing groundwater flow in porous media, Darcy's law, usually does not apply in karst rocks because it assumes laminar flow and isotropic conditions, both of which seldom apply to karst where flow is frequently turbulent and conditions anisotropic. Consequently, the application of traditional groundwater exploration and management practices to karst aquifers is often misguided, sometimes with serious practical consequences.

Four major conceptual approaches have been used to model groundwater flow through karst aquifers: (1) the *equivalent porous medium*, which assumes that the rock matrix is similar to a uniformly porous sandstone; (2) the *discrete fracture network*, which assumes that an otherwise dense rock is traversed by an intersecting series of permeable fissures; (3) the *double porosity continuum*, which assumes the rock matrix is porous and is traversed by permeable fissures or conduits and (4) the *triple porosity* approach – matrix, fracture and conduit – which assumes that the rock is porous and is traversed by permeable fissures and conduits. The latter is of most significance to karst, because the bulk of water being stored in a given volume of rock is often predominantly in the rock matrix and fissure system, whereas flow is achieved mainly through conduits (caves). In a triple porosity model flow is assumed to be laminar in the pores and fissures of the rock matrix, but turbulent in conduits (or macrofissures). The development of realistic triple porosity models for karst aquifers represents the frontier in karst hydrogeologic modelling. It is the ideal approach but, currently, is generally impracticable because the data are unavailable and the level of detail required beyond current computer capability. Nevertheless, progress is being made. Some steps in the development of modelling approaches can be seen from the stage of hardware modelling of conduit development by Ewers (1982), through computer modelling of conduits by Dreybrodt (1988, 1996), Hanna and Rajaram (1998), Palmer et al. (1999), Dreybrodt and Gabrovšek (2000), Annable (2003), Dreybrodt et al. (2005). A recent advance by Kiraly (2002) involves coupling the epikarst to the main karst aquifer and uses a combined discrete channel and continuum approach (Figure 27.3).

A different approach has been used to understand processes operating *within* karst aquifers, as opposed to flow *through* them. In this case emphasis is on the karst spring, the output of which is an integration of flow from the entire basin upstream. Spring hydrograph (Mangin, 1973, 1975, 1998) and chemograph (Lakey and Krothe, 1996; Sauter, 1997) analyses have been used successfully to interpret the dynamics of the groundwater system, to understand recharge–response processes, pulse-through and flow-through times, and to measure groundwater storage volume.

However, theoretical modelling is one thing; factual evidence of water flow and input to output connections proven by conventional water tracing techniques remains essential and irreplaceable in providing 'ground truth' of aquifer limits, structure and behaviour (e.g. see Käss, 1998 and US EPA, 2002).

Surface landform morphometry and modelling

The first models of karst landscape evolution were conceptual. They attempted to generalize and encapsulate the current state of understanding of karst landform evolution. As noted, those developed early in the 20th century were strongly influenced by the prevailing Davisian ideas of landscape evolution through youth, maturity and old age. A model by Grund (1914) constituted an intellectual collation of ideas drawn from contemporary descriptions of karst landforms from Europe to the humid tropics. He imagined cone

karst (kegelkarst) to represent the end stage of an evolution that commenced with doline karst. A later model by Cvijić (1918), on the other hand, was based strongly on his personal experience of karst in the Dinaric region, where the thick limestones are strongly folded and faulted. He presented a sequence rather than a cycle of evolution, because he recognized that karst development ceases when the underlying clastic rocks are revealed.

Later models of this type were developed to portray landscape evolution in specific regions. Thus Lehmann's (1936) model of the Gunung Sewu karst in Java showed the development of cone karst from a landscape that was initially crossed and incised by rivers that were subsequently dismembered by karst processes and captured underground. Corbel's (1959) model of evolution in the humid tropics suggested 100 million years is needed to develop tower karst, reflecting his hypothesis that karst dissolution – and therefore landscape evolution – proceeds more slowly in the tropics than in cold temperate zones. Zhu's (1988) model of *fengcong* to *fenglin* ('peak-cluster depression' to 'peak forest plain') landscape development in humid subtropical southern China took particular account of the role of structure, relief and hydrology in the evolution of these styles of relief. They are broadly equivalent to 'cockpit' and 'tower' karst as depicted by Williams (1987). Conceptual models such as these are representations of reality that are intended to portray the essential elements of the landscape; detail assumed irrelevant being ignored. It is therefore significant that the ideas conveyed in the independently developed models by Zhu and Williams are essentially in agreement.

The limitations of conceptual models include their subjectivity and qualitative nature. A second set of models – morphometric models – was therefore developed to achieve greater objectivity and quantitative reproducibility. Examples of two- and three-dimensional morphometric models are by Williams (1971, 1972), Day (1976, 1979, 1983), Brook (1981), Fang (1984), Vincent (1987), Ferrarese et al. (1997), Telbisz (2001) and Šušteršić (2006). This morphometric approach is used to define landform attributes and to describe statistical inter-relationships, but evolution and process controls, for example spatial competition, steady-state interaction, etc., can only be inferred. These models do not permit exploration of the factors that guide landscape evolution over time. For this, process–response models are required.

Computer models designed deductively from general principles based on prior empirical research are better equipped to provide

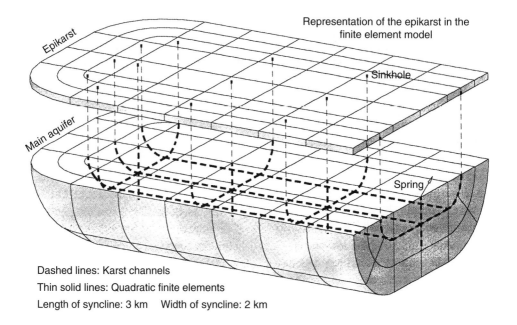

Dashed lines: Karst channels

Thin solid lines: Quadratic finite elements

Length of syncline: 3 km   Width of syncline: 2 km

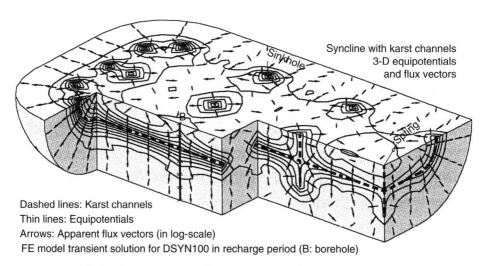

Dashed lines: Karst channels

Thin lines: Equipotentials

Arrows: Apparent flux vectors (in log-scale)

FE model transient solution for DSYN100 in recharge period (B: borehole)

Figure 27.3   *Upper*. The coupling of the epikarst to the main aquifer in a finite element model of a karstified syncline. *Lower*. Variation of hydraulic head in a karstified syncline following recharge by concentrated infiltration through the epikarst (From Kiraly, 2002)

process–response models, because they enable investigation of variables that influence karst landscape evolution. The first three-dimensional process–response model, by Ahnert and Williams (1997), applied to a simple autogenic karst system represented by an uplifted limestone block with already established input to output hydrologic connections. For simplicity, dissolution rate is proportional to rainfall and all rain passes through

the system without storage or overflow. The initial X–Y grid surface is of very low relief with randomly variable elevation. Runoff follows the steepest slope, rainfall at each input point having eight grid-point neighbours towards which it could flow. Water sinks underground at the lowest points of topographic hollows. The analysis set out to assess (1) the minimum requirements for karst landscape evolution, (2) the effect of different starting conditions on end-stage land-forms, and (3) whether different landform types are the result of different environments or merely represent successive stages under unchanging conditions. Variables tested that influence the rate and location of dissolution included topographic control, structural control, flow divergence and slope of the initial surface.

Runs of the model suggested that locally higher dissolution caused by flow convergence is sufficient to explain the formation of solution dolines and polygonal karst, but insufficient by itself to lead to the development of cone or tower karst, which also requires dissolution to be reduced at points of flow divergence. In different runs of the model, cones (or towers) were never found to develop directly, but were always derived from inter-doline residual hills. To explain the model landscapes developed, there was no need to invoke any climatic factors except sufficient rainfall to permit dissolution. Figure 27.4 illustrates results of a run of the model. In this case the initial surface at time zero ($T = 0$) is horizontal, but has small random irregularities in elevation.

An arbitrary base level is shown by a dashed line at $Z = 450$. By eleven model time units ($T = 11$) dissolution by water converging on topographic low points has caused solution depressions to develop. Doline karst at $T = 20$ evolves into a cockpit karst with conical inter-depression hills by $T = 59$. Also by $T = 59$ the bottoms of some depressions have reached base level. Cockpit floors converge at water table level producing a sloping corrosion plain (that follows the hydraulic gradient), the beginning of which is evident by $T = 98$. Further development sees the isolation of cones on the corrosion plain ($T = 150$). These results clearly point to the importance of runoff divergence and convergence effects in generating strong spatial variation in solutional denudation, which gains topographic expression in the evolution of cockpit karst (fengcong depression). The model also shows that tower karst (fenglin) can develop from a prior polygonal cone karst stage.

Although still in its infancy, landscape modelling can already be seen to provide insights into key factors that control landform evolution and can generate plausible sequences of landscape

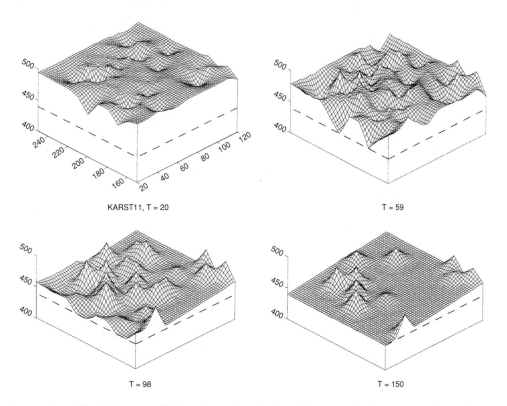

Figure 27.4   Block diagrams illustrating selected time unit stages in the course of running a process–response model (*KARST11*) of karst landscape development. The sloping corrosion plain at *T* = 150 follows the hydraulic gradient (From Ahnert and Williams, 1997)

development. Radiometric dating in karst, to which we now turn, will then provide the time frames.

SPELEOCHRONOLOGY: THE KEY TO DATING EROSIONAL LANDSCAPES IN GEOMORPHOLOGY

Dissolutional caves survive in eroding terrains for greater spans of time than any other landforms because they extend deep into the bedrock itself. When erosional base levels are lowered, older hypergene caves are progressively abandoned by their formative waters as younger systems develop to new spring positions below them. The relict caves retain fluvial deposits of their abandonment phases and, afterwards, may slowly accrete particulate and dissolved mineral and organic matter representative of environmental conditions in the soils and epikarst overhead. For example, in the Calcareous Alps, the Pyrenees, the Rocky Mountains and elsewhere innumerable fragments of relict phreatic caves survive in summit rocks, recording the existence of karst groundwater systems that once discharged into valley floors now above the modern peaks (Ford et al., 1981; Maire, 1990).

Most karst terrains adjoin regular fluvial or arid, glacial, coastal, even volcanic, terrains and so share with them specific local histories of deformation, uplift and denudation. Dating relict caves and their contents thus can provide a framework for dating of all types of geomorphic regions and, further, supply details of the paleo-temperatures, precipitation, soils and vegetation prevailing at specific times. For example, the existence of a major floodplain, first reduced to a terrace and then totally destroyed at the surface, might be reconstructed from cave evidence within a regional chronology. Absolute dating and paleo-environmental analysis of cave deposits is a rapidly growing area of science at present, something of a bandwagon in fact. Space limitations permit us to present only a brief summary here; for details see Ford and Williams (2007: 562).

Dating methods and prospects

It is a seeming truism of geomorphology that the time of actual removal of bedrock from an eroding surface cannot be measured because, by definition, the evidence is being destroyed. A breakthrough was achieved by Polyak et al. (1998) when they showed that, in some caves being created by the H_2S processes noted above, clay minerals such as montmorillonite, illite and kaolinite present in small quantities in the limestone reacted with the acid to form hydrated sulphate minerals. In particular, alunite $[KAl_3(SO_4)_2(OH)_6]$ is produced as small, impervious crystals that may be dated by the long-established $^{40}Ar/^{39}Ar$ method. Figure 27.5 shows their exciting finding that the ~1500 m of modern relief in the Guadalupe Mountains of New Mexico has developed over the past 12–13 Ma.

There are silts, sands and gravels from exterior, non-karst, sources in most relict stream caves, stranded during abandonment. Safely underground and thus out of range of cosmic radiation, the residence time of quartz grains may be dated by the differential decay of ^{10}Be and ^{26}Al cosmogenic isotopes. The dating range is 0–5 Ma or a little more at present. Granger et al. (1997, 2001) and Anthony and Granger (2004) have applied the method to date river gorge entrenchment in Virginia and to confirm that the previously speculative pre-Quaternary history of Mammoth Cave and caves of the Cumberland Plateau, Kentucky and Tennessee, begins at ~5 Ma. Sediments in older levels of the Siebenhengste–Bärenshacht cave systems overlooking Lake Thun, Switzerland, may be 4.4 Ma in age (Häuselmann and Granger, 2004). Caution is needed in the interpretations, however. The individual dates obtained represent the *maximum* residence times underground; before being carried into a cave, a given quartz pebble may have buried deep in a river terrace and thus also out of range of some or all cosmic radiation.

Calcite and aragonite speleothems are precipitated from seepage waters in perhaps the majority of limestone caves, beginning during solutional enlargement and potentially lasting throughout all relict phases thereafter. Uranium series dating methods exploiting the decay of trace quantities of ^{234}U and ^{238}U co-precipitated with the calcite, etc. are of principal importance. Atomic ratios of ^{234}U to daughter ^{230}Th can be measured by thermal ionization or induction-coupled plasma mass spectrometers to a limit of ~550 ka, with a 2σ uncertainty of only ~1 per cent back to 300 ka or beyond, rising to ~10 per cent at the limit. Mass spectrometry replaces older alpha-particle counting that, as early as 1972, allowed estimates of maximum entrenchment rates for a series of spectacular antecedent river canyons in the Northwest Territories, Canada (Ford, 1973). There have been many other applications since (e.g. Ford et al., 1981, cited above in the Canadian Rockies; Williams, 1982, on raised coastal terraces in New Zealand) but most Th–U measurements have been made to provide the chronology for paleotemperature and other reconstructions utilizing the O and C stable isotope ratios in the calcite.

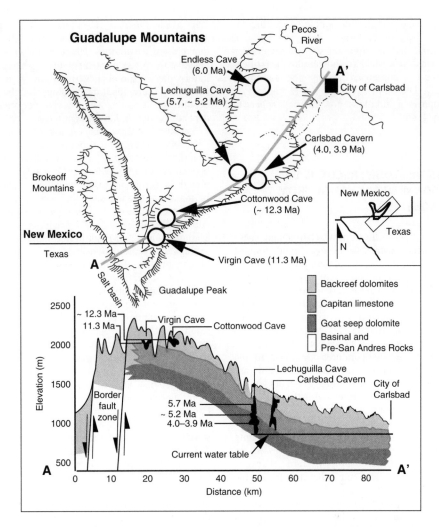

Figure 27.5 ^{40}Ar/^{39}Ar **ages of alunite crystals from H$_2$S caves of the Guadalupe Mountains, New Mexico, and reconstruction of the orogenic history (From Polyak et al., 1998)**

The Uranium series time range can be expanded using the ^{234}U:^{238}U ratio in samples >500 ka in age, to a limit of ~1.5 Ma. It has seen little use because of difficulties in accurately estimating this ratio at time of deposition, but can be adopted in favourable circumstances ('RUBE dating'; Gascoyne et al., 1983). Ludwig et al. (1992) used it to confirm the basal ages of DH II, a Nevada thermal cave calcite deposit that at the time of writing has yielded the longest precisely dated published record of Quaternary climate change (Coplen et al., 1994). Ford et al. (1993) applied it to determine the rate of fall of the water table over the past 500 ka in Wind Cave, South Dakota, another thermal water cave.

The decay of uranium to stable lead in principle permits dating of speleothems and other calcites over all of geomorphic time on Earth. There is a great deal of background lead in the environment; however, it also accumulates in trace amounts in the precipitate. Modern mass spectrometric techniques are beginning to permit discrimination using non-radiogenic ^{204}Pb as the control in isochrons based on differing concentrations of the isotopes in adjoining calcite samples (Richards et al., 1996). At time of writing, most published results have been for speleothems within the U series range, but Woodhead et al. (2006) report reasonable late Tertiary ages for relict speleothems in desert caves of the Nullarbor Plain, Australia, while Lundberg et al. (2000) obtained a range of 90–96 ± 7 Ma for the earliest, hypogene, phase of dissolution and precipitation in the Guadalupe Mountains, New Mexico, showing that karstification there commenced during the Laramide Orogeny. Difficulties remain, but technology is fast improving and the future prospects appear to be very good.

'TREES OF STONE': RECORDS OF ENVIRONMENTAL CHANGE FROM STALAGMITES

Caves are superb sites from which to obtain paleo-environmental data. Just as trees yield climatic information from analysis of their growth rings, so stalagmites and flowstones (collectively known as *speleothems*) yield information from analysis of their calcite. By providing a terrestrial paleo-environmental record, speleothem sequences complement the marine and polar records from deep sea cores and glacier ice cores.

Cave interior temperatures are close to the mean annual temperature of the external environment. As environmental change occurs at the surface, cave temperatures adjust correspondingly. Percolation water also transmits evidence of environmental change underground, and speleothems record the detail in their accumulating calcite. The information can be accessed by analysing for stable isotopes, trace elements, luminescence, pollen and other larger organic matter. Depending on growth rate and method used, data can be obtained to one year resolution, or better, in some cases. Details are available in Ford and Williams (2007: 562).

Many of the shortcomings of location and dating in current Quaternary science are overcome by speleothems, because their environmental data relates directly to the land above the cave, they occur across all latitudes on ice-free continents, and they can be closely and precisely dated back to ~0.55 Ma BP using thorium ages and with the prospect of millions of years as U/Pb dating improves. McDermott (2004) emphasizes the importance of speleothems in providing precise estimates for the timing and duration of major oxygen isotope-defined climatic events. Analysis of the stable isotopes ^{16}O, ^{18}O, ^{12}C and ^{13}C is the most widely used approach (Yonge et al, 1985). The $\delta^{18}O$ value of speleothems growing under isotopic equilibrium is determined by two sets of characteristics: the first represents the thermodynamic fractionation between calcite and water, the *cave temperature effect*, and the second relates to the combination of factors that influence the isotopic composition of the feed water, termed the *drip water function* (Lauritzen and Lundberg, 1999). These characteristics have different temperature sensitivities. The former always has a negative response to temperature, whereas the latter may respond negatively or positively, depending on regional meteorology and the scale of climatic change. The overall net outcome depends on their relative magnitudes.

Because the *cave temperature effect* and the *drip water function* may oppose each other, the isotopic characteristics in each karst region must be studied carefully before $\delta^{18}O_c$ trends are interpreted in terms of paleo-temperature changes. In many continental sites the cave temperature effect predominates, for example in China and Austria (Wang et al., 2001; Yuan et al., 2004; Mangini et al., 2005); but in mid-latitude oceanic settings and some continental sites the influence of precipitation dominates (Dorale et al., 1992, Gascoyne 1992; Goede et al., 1986; Paulsen et al., 2003; Williams et al., 2005). The over-riding lesson is that no assumptions should be made about the direction of a temperature relationship: results for each speleothem must be checked individually. Furthermore, variations in $\delta^{18}O$ cannot always be assumed to be determined by temperature changes, because changes in rainfall source and amount can sometimes be more important (Bar-Matthews and Ayalon, 1997; Cruz et al., 2005a, 2005b).

Five factors are especially important in understanding the reasons for change in speleothem $\delta^{13}C$ values: the concentration and isotopic composition of CO_2 in the atmosphere; production of biogenic CO_2 by plant and soil processes; carbon sourced from karst bedrock and aging soil; the ratio of open to closed system dissolution; and degassing of CO_2. Wet conditions tend to give rise to relatively low $\delta^{13}C$ values and cold, near-freezing conditions to high values.

Good examples of interpretation of O and C isotopic information from speleothems are from caves in Israel (Bar-Matthews and Ayalon et al., 1997; Frumkin et al., 1999), while the Devil's Hole record from Nevada (Coplen et al., 1994) tracks the climatic cycles of most of the past 500,000 years. The ratios $^{18}O/^{16}O$ and $^{13}C/^{12}C$ are expressed in delta (δ) units, which are parts per thousand (per mil, ‰) deviations of the isotopic ratio from standards.

Regionally representative stable isotope time series can be obtained by merging the individual records of several speleothems. Examples are available from New Zealand and Israel, where a composite 185 ka series was compiled from 21 overlapping speleothem records (Ayalon et al., 2002).

When sectioned some speleothems display fine banding in their calcite texture or colour, although it does not always occur: Baker et al. (1993)

reported it in only a few of more than 40 samples studied. When present, banding can be resolved under the optical microscope to layers only 1–100 μm in thickness. Shopov (1987) and Shopov et al. (1994) measured variations in the luminescence intensity at a still finer scale, this being attributable to varying concentrations of fulvic and humic acids in the calcite lattice. Particulate organic matter (van Beynen et al., 2001) and lipids (Blyth et al., 2006) have also been identified, offering promise of reconstruction of vegetation associations. The correlation of fine scale optical or luminescence banding with the hydrological year at particular cave sites has been firmly established in recent years (Genty, 1992; Genty and Quinif, 1996). Trace element ratios such as Mg/Ca, Sr/Ca and Ba/Ca have also been used as proxies showing a relationship to precipitation (Treble et al., 2003, Fairchild et al., 2006).

One of the interesting attributes of speleothem records is the periodicity often shown by the data. In pioneering work, Shopov (e.g. 1987) attributed this directly to solar cycles, but others emphasized changes in the terrestrial environment (McDermott et al., 2001; Spötl and Mangini, 2002; Genty et al., 2003; Cruz et al., 2005a). The periodicity can be analysed using the program SPECTRUM (Schulz and Statteger, 1997), which was specially created for unevenly spaced paleoclimatic time series. Wavelet analysis has also been used to identify variations in $\delta^{18}O$ and $\delta^{13}C$. For instance, Paulsen et al. (2003) detected cycles of 33, 22, 11, 9.6 and 7.2 years in a 1270 year high resolution record from China. It is therefore clear that the 'trees of stone' we find in caves provide some of the best available records of environmental change on the continents. Research here, however, must be strictly constrained by the ethical imperative to preserve fragile speleothem formations for the enjoyment of future generations. With available narrow diameter drills for sampling and minute sample size requirements for analysis, this is becoming increasing possible.

CONCLUSIONS: THE WAY AHEAD

Karst 'punches above its weight' in the science of geomorphology for three very good reasons: first, caves contain deposits that can now be radiometrically dated back for many millions of years; second, the dated geomorphic history of a karst region provides a time scale for geomorphic evolution in the wider non-karst region around it; and third, the palaeoclimatic record contained within speleothems is applicable to the entire region around it, and is more accurately and precisely dated and of higher resolution than comparable palaeoclimatic records obtained from deep sea cores or ice cores. Karst and caves therefore provide an almost unparalleled insight into the geomorphic and palaeo-environmental evolution of the continents. Much effort is being made at present to build high-resolution precisely dated records of terrestrial climatic change that reveal not only changes in temperature and precipitation but also give insight into changes in general circulation. A 550,000 year record has already been published from Devil's Hole, Nevada, and records of more than a million years duration are being worked on elsewhere.

For several decades uranium series dating of speleothems has provided reliable information back to 350,000 years BP, and with improved instrumentation this has been extended to ~550,000 years BP. After a slow start U/Pb dating is proving to be viable, and plausible dates on speleothems have now been achieved for 5 million years or more. This is the present frontier of dating technology, because not all speleothems analysed have proven to be datable by this technique.

Cave sediments dated by palaeomagnetism in Slovenia, for example, indicate that some caves in the Classical Karst may have formed at the time of the Messinian crisis in the Mediterranean. Reliable U/Pb dates on interbedded speleothems would help confirm this hypothesis, but at present techniques are not sufficiently sensitive on speleothems with low uranium concentrations to resolve U/Pb ages. The challenge is to improve U/Pb dating of speleothems and to make it more widely applicable.

Given our greater understanding of the antiquity of relief, it is clear that many landscapes must have evolved when climatic circumstances were different and even when tectonic settings were different. Thus the landscapes we see today must to a greater or lesser degree have traits inherited from the past. In the classical karst of Slovenia, for example, we now know that many caves formed millions of years ago in the phreatic zone are now unroofed and exposed at the surface, and so are partly responsible for the present topography. It is also probable that many towers in the spectacular fenglin karst of southern China have an inheritance as conical hills between enclosed depressions in an earlier fengcong landscape. One of the great challenges for karst geomorphologists is in understanding the role of inheritance and its imprint/direction on modern landscapes.

One way to tackle the issue of inheritance will be with the aid of process-response modelling. At the moment we have groundwater models that link processes in the epikarst with processes in the phreatic zone. We also have less sophisticated process–response models that give insight into the development of surface relief. The challenge for the future is to develop linked process-response models above and below ground, so that it is possible to explore the relationship between the development of surface landforms and subterranean networks. By varying initial conditions at the start of such a model, it will be possible to gain insight into the role of inheritance.

Karst groundwater models are used for two reasons: first to characterize and understand the groundwater system, mainly for water resource management purposes; and second to try to reproduce the evolution of its characteristics, which is important for understanding speleogenesis. The most pressing challenge at present is perfecting triple porosity hydrogeological modelling for water resources management. Beyond that will be the adaptation of current two-dimensional computer models of solution conduit evolution to the full three dimensions of reality.

In common with other geomorphologists, karst specialists also face the problem of encouraging the conservation of the areas that they study. In every sphere of geomorphology there is a pressing need to ensure conservation of the most important representative landforms and landscapes, including the 'classical' sites that were the subject of pioneering investigations that helped lay the theoretical foundations of our discipline. In karst we are making progress with this, because several great karst landscapes and caves are now protected as World Heritage properties, and a thematic study has recently been conducted for the IUCN on World Heritage caves and karst that has identified gaps in conservation coverage (Williams, 2008). Efforts are now being made to fill them. In addition, the International Speleological Union is proactively aware of issues associated with conservation of caves and karst terrains (Williams, 2009). Thus karst specialists may arguably be amongst the leaders of landscape conservation, but there is a glaring omission: the value of the classical karst of Europe, the region that gave Karst its name, remains unrecognized so far as conservation protection is concerned. Only small parts of it are in the World Heritage estate. So an international effort is required to secure a trans-national serial World Heritage Park that truly represents the natural values of the Dinaric Karst.

REFERENCES

Ahnert, F. and Williams, P.W. (1997) Karst landform development in a three-dimensional theoretical model. *Zeitschrift für Geomorphologie, Supplement Band* 108, 63–80.

Annable, W.K. (2003) *Numerical Analysis of Conduit Evolution in Karstic Aquifers.* University of Waterloo, PhD thesis. p. 139.

Anthony, D.M. and Granger, D.E. (2004) A Tertiary origin for multilevel caves along the western escarpment of the Cumberland Plateau, Tennessee and Kentucky, established by cosmogenic 26Al and 10Be. *Journal of Cave and Karst Studies* 66, 46–55.

Ayalon, A., Bar-Matthews, M. and Kaufman, A. (2002) Climatic conditions during marine oxygen isotope stage 6 in the eastern Mediterranean region from the isotopic composition of speleothems of Soreq Cave, Israel. *Geology* 30, 303–6.

Baker, A., Smart, P.L., Edwards, R.L. and Richards, D.A. (1993) Annual growth banding in a cave stalagmite. *Nature* 364, 518–20.

Bar-Matthews, M. and Ayalon, A. (1997) Late Quaternary paleoclimate in the eastern Mediterranean region from stable isotope analysis of speleothems at Soreq Cave, Israel. *Quaternary Research* 47, 155–68.

Blyth, A. J., Farrimond, P. and Jones, M. (2006) An optimized method for the extraction and analysis of lipid biomarkers from stalagmites. Organic Geochemistry, 37, 882–90.

Bretz, J.H. (1942) Vadose and phreatic features of limestone caves. *Journal of Geology* 50, 675–811.

Brook, G.A. (1981) An approach to modelling karst landscapes. *South African Geography Journal* 63, 60–76.

Coplen, T.B., Winograd, I.J., Landwehr, J.M. and Riggs, A.C. (1994) 500,000 year stable carbon isotopic record from Devils Hole, Nevada. *Science*, 263, 361–5.

Corbel, J. (1959) Erosion en terrain calcaire. *Annales de Géographie* 68, 97–120.

Cruz, F.W., Burns, S.J. and Karmann, I. (2005a) Insolation-driven changes in atmospheric circulation over the past 116,000 years in subtropical Brazil. *Nature* 434, 63–6.

Cruz, F.W., Karmann, I., Viana, O., Burns, S.J., Ferrari, J.A. and Vuille, M. (2005b) Stable isotope study of cave percolation waters in subtropical Brazil: implications for palaeoclimate inferences from speleothems. *Chemical Geology* 220, 245–62.

Cvijić, J. (1893) *Das Karstphanomen.* Versuch einer morphologischen Monographie. *Geographische Abhandlung, Wien.* 5, 218–329.

Cvijić, J. (1901) Morphologische und glaciale Studien aus Bosnien, der Hercegovina und Montenegro: die Karst-Poljen. *Abhandlung Geographisches Geschichte, Wien.* 3, 1–85.

Cvijić, J. (1918) Hydrographie souterraine et evolution morphologique du karst. *Receuil des Travaux de l' Institute de Géographie Alpine* 6, 375–426.

Cvijić, J. (1960) *La Géographie des Terrains Calcaires.* Academie Serb des Sciénces et des Arts, Beograd; Monographies, t.CCCXLI, p. 212.

Davis, W.M. (1930) Origin of limestone caves. *Geological Society of America, Bulletin* 41, 475–628.

Day, M.J. (1976) The morphology and hydrology of some Jamaican karst depressions. *Earth Surface Processes* 1, 111–29.

Day, M.J. (1979) Surface roughness as a discriminator of tropical karst styles. *Zeitschrif fur Geomorpholige. Supplement Band* 32, 1–8.

Day, M.J. (1983) Doline morphology and development in Barbados. *Annals of the Association of American Geographers*, 73, 206–19.

Dorale, J.A., Gonzalez, L.A. and Reagan, M.K. (1992) A high resolution record of Holocene climate change in speleothem calcite from Cold Water Cave, northeast Iowa. *Science* 258, 1626–30.

Drake, J.J. and Ford, D.C. (1972) The analysis of growth patterns of two generations: the example of karst sinkholes. *Canadian Geographer* 16, 381–4.

Dreybrodt, W. (1988) *Processes in Karst Systems*. Physics, Chemistry and Geology Series, Springer Verlag, Berlin. p. 288.

Dreybrodt, W. (1996) Principles of early development of karst conduits under natural and man-made conditions revealed by mathematical analysis of numerical models. *Water Resources Research* 32, 2923–35.

Dreybrodt, W. and Gabrovsek, F. (2000) Dynamics of the evolution of single karst conduits, in A.V. Klimchouk, D.C. Ford, A.N. Palmer and W. Dreybrodt (eds), *Speleogenesis: Evolution of Karst Aquifers*. National Speleological Society of America, Huntsville, AL. pp. 184–93.

Dreybrodt, W., Gabrovsek, F. and Romanov, D. (2005) *Processes of Speleogenesis: A Modeling Approach*. ZRC Publishing, Carsologica, Lubljana. p. 376.

Ewers, R.O. (1982) *Cavern Development in the Dimensions of Length and Breadth*. PhD thesis, McMaster University, p. 398.

Fairchild, I.J., Smith, C.L. and Baker, A. (2006) Modification and preservation of environmental signals in speleothems. *Earth Science Reviews* 75, 105–15.

Fang Lingchang (1984) Application of distances between nearest neighbours to the study of karst. *Carsologica Sinica* 3, 97–101.

Ferrarese, F., Sauro, U. and Tonello, C. (1997) The Montello Plateau: evolution of an alpine neotectonic morphostructure. *Zeitschrift für Geomorphologie, Supplement Band* 109, 41–62.

Ford, D.C. (1971) Geologic structure and a new explanation of limestone cavern genesis. *Cave Research Group of Great Britain, Transactions* 13, 81–94.

Ford, D.C. (1973) Development of the canyons of the South Nahanni River, N.W.T. *Canadian Journal of Earth Sciences* 10, 366–78.

Ford, D.C. (2005) Jovan Cvijić and the founding of karst geomorphology, in Z. Stevanovic and B. Mijatovic (eds), *Cvijić and Karst*. Serbian Academy of Science and Arts, Belgrade. pp. 305–21.

Ford, D.C. (2009) Carbonate-hosted massive sulfide deposits and hypogene speleogenesis: a case study from Nanisivik zinc/lead mine, Baffin Island, Canada, in K.W. Stafford, L. Land and G. Veni (eds), *Advances in Hypogene Karst Studies*. National Cave and Karst Research Institute, Symposium 1, 136–48.

Ford, D.C. and Ewers, R.O. (1978) The development of limestone cave systems in the dimensions of length and breadth. *Canadian Journal of Earth Science* 15, 1783–98.

Ford, D.C. and Williams, P.W. (1989) *Karst Geomorphology and Hydrology*. Unwin Hyman, London. p. 601.

Ford, D.C. and Williams, P.W. (2007) *Karst Hydrogeology and Geomorphology*. Wiley, Chichester. p. 562.

Ford, D.C., Schwarcz, H.P., Drake, J.J., Gascoyne, M., Harmon, R.S. and Latham, A.G. (1981) On the age of existing relief in the southern Rocky Mountains in Canada, *Arctic and Alpine Research* 13, 1–10.

Ford D.C., Palmer A.N., Palmer M.V., Dreybrodt W., Lundberg J. and Schwarcz H.P. (1993) Uranium series dating of the draining of an aquifer: the example of Wind Cave, Black Hills, South Dakota. *Geological Society of America Bulletin* 105, 241–50.

Frumkin, A., Ford, D.C. and Schwarcz, H.P. (1999) Continental oxygen isotope record of the last 170,000 years in Jerusalem. *Quaternary Research* 51, 317–27.

Gams, I. (2003) *Kras v Sloveniji v prostoru in èasu*, Založba ZRC, ZRC SAZU, Ljubljana. p. 516.

Gascoyne, M. (1992) Palaeoclimate determination from cave calcite deposits. *Quaternary Science Reviews* 11, 609–32.

Gascoyne, M., Ford, D.C. and Schwarcz, H.P. (1983) Rate of cave and landform development in the Yorkshire Dales from speleothem age data. *Earth Surface Processes and Landforms* 8, 557–68.

Genty, D. (1992) Les spéléothemes du tunnel de Godarville (Belgique) – un exemple exceptionnel de concrétionnement moderne. *Spéléochronos* 4, 3–29.

Genty, D. and Quinif, Y. (1996) Annually laminated sequences in the internal structure of some Belgian stalagmites – importance for paleoclimatology. *Journal of Sedimentary Research* 66, 275–88.

Genty, D., Blamart, D., Ouahdi, R. et al. (2003) Precise dating of Dansgaard–Oeschger climate oscillations in western Europe from stalagmite data. *Nature* 421, 833–7.

Ghazban, F., Schwarcz, H.P, and Ford, D.C. (1992) Correlated strontium, carbon and oxygen isotopes in carbonate gangue at the Nanisivik zinc-lead deposits, northern Baffin Island, Canada. *Chemical Geology (Isotope Geoscience Section)* 87, 137–46.

Goede, A., Green, D.C. and Harmon, R.S. (1986) Late Pleistocene paleotemperature record from a Tasmanian speleothem. *Australian Journal of Earth Science* 33, 333–42.

Granger, D.E., Kirchner, J. and Finkel, R. (1997) Quaternary downcutting rate of the New River, Virginia, measured from differential decay of cosmogenic ^{26}Al and ^{10}Be in cave-deposited alluvium. *Geology* 25, 107–10.

Granger, D.E., Fabel, D. and Palmer, A.N. (2001) Plio-Pleistocene incision of the Green River, KY, from radioactive decay of cosmogenic ^{26}Al and ^{10}Be in Mammoth Cave sediments. *Bulletin, Geological Society of America* 113, 825–36.

Grund, A. (1914) Der geographische Zyklus im Karst. (English translation: Sweeting, 1981). *Geschichte Erdkunde* 52, 621–40.

Hanna, R.B. and Rajaram, H. (1998) Influence of aperture variability on dissolutional growth of fissures in karst formations. *Water Resources Research* 34, 2843–53.

Häuselmann, P. and Granger, D.E. (2004) Dating caves with cosmogenic nuclides: methods, possibilities and the Siebenhengste example, in A. Mihevc and Zupan N. Hajna (eds), *Dating of Cave Sediments*. SAZU, Postojna. p. 50.

Hill, C.A. (1987) Geology of Carlsbad Caverns and other caves of the Guadalupe Mountains. *New Mexico Bureau Mines and Minerals Bulletin* 117, p. 150.

Hill, C.A. (1995) Sulfur redox reactions: hydrocarbons, native sulfur, Mississippi Valley-type deposits, and sulfuric acid karst in the Delaware Basin, New Mexico and Texas. *Environmental Geology* 25, 16–23.

Hutton, J. (1795) *Theory of the Earth*. vol. II. Creech, Edinburgh.

Jakucs, L. (1977) *Morphogenetics of Karst Regions: Variants of Karst Evolution*. Akademiai Kiado, Budapest, and Hilger, Bristol. p. 284.

James, N.P. and Choquette, P.W. (1984) Diagenesis. (9). Limestones – the meteoric diagenetic environment. *Geoscience Canada* 11, 161–94.

Jennings, J.N. (1968) Syngenetic karst in Australia, in P.W. Williams and J.N. Jennings (eds), *Contributions to the Study of Karst*. Publication G5, Research School for Pacific Studies, Australian National University, Canberra. pp. 41–110.

Jennings, J.N. (1971) *Karst*. MIT Press, Cambridge, MA.

Jennings, J.N. (1985) *Karst Geomorphology*. Basil Blackwell, Oxford and New York.

Kass, W. (1998) *Tracing Techniques in Geohydrology*. Balkema, Rotterdam. p. 581.

Kiraly, L. (2002) Karstification and groundwater flow, in F. Gabrovsek (ed.), *Evolution of Karst: From Prekarst to Cessation*. Institut za raziskovanje krasa, ZRC SAZU, Postojna-Ljubljana. pp. 155–90.

Klimchouk, A.B. (2000) Speleogenesis of the Great Gypsum Mazes in the Western Ukraine, in A.B. Klimchouk, D.C. Ford, A.N. Palmer and W. Dreybrodt (eds), *Speleogenesis; Evolution of Karst Aquifers*. National Speleological Society Press, Huntsville, AL. pp. 261–73.

Klimchouk, A.B. (2003) Conceptualisation of speleogenesis in multi-storey artesian systems: a transverse speleogenesis. *Speleogenesis* 1(2), 21.

Klimchouk, A.B. (2007) *Hypogene Speleogenesis: Hydrogeological and Morphological Perspectives*. Special Paper Series No. 1. National Cave and Karst Research Institute, USA. p. 163.

Klimchouk, A.B. and Ford, D.C. (2000) Types of karst and evolution of hydrogeologic settings, in A.B. Klimchouk, D.C. Ford, A.N. Palmer and W. Dreybrodt (eds), *Speleogenesis; Evolution of Karst Aquifers*. National Speleological Society Press, Huntsville, AL. pp. 45–53.

Lakey, B. and Krothe N.C. (1996) Stable isotopic variation of storm discharge from a perennial karst spring, Indiana. *Water Resources Research* 32, 721–31.

Lauritzen, S.-E. and Lundberg, J. (1999) Speleothems and climate: a special issue of The Holocene. *The Holocene* 9, 643–7.

Lehmann, H. (1936) *Morphologische Studien auf Java*. Series 3, No. 9, Geographische. Abhandlungen, Stuttgart. p. 114.

Lehmann, H. (1954) Das Karstphänomen in den verschiedenen Klimazonen. *Erdkunde* 8, 112–22.

Lehmann, H. (1956) Der Einfluss des Klimas auf die Morphologische entwicklung des Karstes. 18 International Geographical Congress (Rio de Janeiro), *Report on the Commission on Karst Phenomena*. pp. 3–7.

Ludwig, K.R., Simmons, K.R., Szabo, B.J. et al. (1992) Mass-spectrometric ^{230}Th–^{234}U–^{238}U dating of the Devils Hole calcite vein. *Science* 258, 284–7.

Lundberg, J., Ford, D.C. and Hill, C.A. (2000) A preliminary U–Pb date on cave spar, Big Canyon, Guadalupe Mountains, New Mexico, USA. *Journal of Cave and Karst Studies* 62, 144–8.

Maire, R. (1990) La Haute Montagne Calcaire. *Karstologia-Mémoires* 3, p. 731.

Mangin, A. (1973) Sur la dynamiques des transferts en aquifere karstique. *6 International Congress of Speleology Proceedings*, Olomouc 6, 157–62.

Mangin, A. (1975) *Contribution a l'etude hydrodynamique des aquiferes karstiques*. These Doctorale de Science, Universite de Dijon, France. *Annales de Speleologie* 1974, 29, 283–332; 495–601; and 1975, 30, 21–124.

Mangin, A. (1998) L'approche hydrogéologique des karsts. *Spéléochronos* 9, 3–26.

Mangini, A., Spotl, C. and Verdes, P. (2005) Reconstruction of temperature in the central Alps during the past 2000 yr from a δ^{18}O stalagmite record. *Earth and Planetary Science Letters* 235, 741–51.

McDermott, F. (2004) Palaeo-climate reconstruction from stable isotope variations in speleothems: a review. *Quaternary Science Reviews* 23, 901–18.

McDermott, F., Mattey, D.P. and Hawkesworth, C. (2001) Centennial-scale Holocene climate variability revealed by a high-resolution speleothem δ^{18}O record from SW Ireland. *Science* 294, 1328–31.

Mylroie, J.E. and Carew, J.L. (1990) The flank margin model for dissolution cave development in carbonate platforms. *Earth Surface Processes and Landforms* 15, 413–24.

Palmer, A. N., Palmer, M.V. and Sasowsky, I.D. (1999) *Karst Modeling*. Special Publication 5. Karst Waters Institute, Charles Town, WV. p. 265.

Panos, V. and Stelcl , O. (1968) Physiographic and geologic control in development of Cuban mogotes. *Zeitschrift fur Geomorphologie* 12, 117–73.

Paulsen, D.E., Li, H.-C. and Ku, T.-L. (2003) Climate variability in central China over the last 1270 years revealed by high-resolution stalagmite records. *Quaternary Science Reviews* 22, 691–701.

Polyak, V.J., McIntosh, W.C., Güven, N. and Provencio, P. (1998) Age and origin of Carlsbad Cavern and related caves from ^{40}Ar/^{39}Ar of Alunite. *Science* 279, 1919–22.

Rhoades, R. and Sinacori, N.M. (1941) Patterns of groundwater flow and solution. *Journal of Geology* 49, 785–94.

Richards, D.A., Bottrell, S.H., Cliff, R.A. and Strohle, K.D. (1996) U–Pb dating of Quaternary age speleothems, in S.-E. Lauritzen (ed.), *Climate Change: The Karst Record*. Special Publication 2. Karst Waters Institute, Charles Town, WV. pp. 136–7.

Salomon, J.-N. (2000) *Précis de Karstologie*. Presses Universitaires, Bordeaux. p. 250.

Salomon, J.-N. (2006) *Précis de Karstologie*. 2nd edn. Presses Universitaires, Bordeaux. p. 289.

Salomon, J.-N. and Maire, R. (eds) (1992) *Karst et Evolutions Climatiques*. Presses Universitaires, Bordeaux. p. 520.

Sauter, M. (1997) Differentiation of flow components in a karst aquifer using the $\delta^{18}O$ signature, in Kranjc, A. (ed.), Tracer Hydrology 97. Balkema, Rotterdam. pp. 435–41.

Scholle P.A., Bebout, D.G. and Moore, C.H. (eds) (1983) *Carbonate Depositional Environments*. Memoir 33, American Association of Petroleum Geologists, Tulsa, OK, p. 761.

Schulz, M. and Statteger, K. (1997) SPECTRUM: spectral analysis of unevenly spaced paleoclimate time series. *Computers and Geosciences* 23, 929–45.

Shopov, Y.Y. (1987) Laser luminescent micro-zonal analysis: a new method for investigation of the alterations of the climate and solar activity during the Quaternary, in T. Kiknadze (ed.), *Problems of Karst Studies of Mountainous Countries*. Metsniereba, Tbilisi, Georgia. pp. 104–8.

Shopov, Y.Y., Ford, D.C. and Schwarcz, H.P. (1994) Luminescent microbanding in speleothems: high resolution chronology and paleoclimate. *Geology* 22, 407–10.

Spötl, C. and Mangini, A. (2002) Stalagmite from the Austrian Alps reveals Dansgaard-Oeschger events during isotope stage 3: implications for the absolute chronology of Greenland ice cores. *Earth and Plantary Science Letter* 203, 507–18.

Stafford, K.W, Land, L. and Veni, G. (eds) (2009) *Advances in Hypogene Karst Studies*. National Cave and Karst Research Institute, Symposium 1. p. 182.

Šušteršić, F. (2006) A power function model for the basic geometry of solution dolines: considerations from the classical karst of south-central Slovenia. *Earth Surface Processes and Landforms* 31, 293–302.

Sweeting, M.M. (1972) *Karst Landforms*. Macmillan, London. p. 362.

Sweeting, M.M. (1995) *Karst in China: Its Geomorphology and Environment*. Springer-Verlag, Berlin.

Swinnerton, A.C. (1932) Origin of limestone caverns. *Geological Society of America, Bulletin* 43, 662–93.

Szunyogh, G. (1989) Theoretical investigation of the development of spheroidal niches of thermal water origin. *Proceedings 10th International Congress of Speleology*. v. III. pp. 766–8.

Tarhule-Lips, R. and Ford, D.C. (1998) Condensation corrosion in caves on Cayman Brac and Isla de Mona, P.R. *Journal of Caves and Karst Studies* 60, 84–95.

Telbisz, T. (2001) Töbrös felszínfejlödes számítógépes modellezése. *Karsztfejlödés* (Szombathely), VI, 27–43.

Treble, P., Shelley, J.M.J. and Chappell, J. (2003) Comparison of high resolution sub-annual records of trace elements in a modern (1911–1992) speleothem with instrumental climate data from southwest Australia. *Earth and Planetary Science Letters* 216, 141–53.

Trombe, F. (1952) *Traité de Spéléologie*. Payot, Paris. p. 376.

Trudgill, S. (1985) *Limestone Geomorphology*. Longman, London.

US EPA (2002) *The QTRACER2 Program for Tracer-Breakthrough Curve Analysis for Tracer Tests in Karst Aquifers and Other Hydrologic Systems*. EPA/600/R-02/001, US Environmental Protection Agency. p. 179 plus diskette.

Van Beynen, P.E., Bourbonniere, R., Ford, D.C. and Schwarcz, H.P. (2001) Causes of colour and fluorescence in speleothems. *Chemical Geology* 175, 319–41.

Vincent, P.J. (1987) Spatial dispersion of polygonal karst sinks. *Zeitschrift fur Geomorphologie* N.F. 31, 65–72.

Wallace, M.W., Moxham, H., Johns, B. and Marshallsea, S. (2002) Hydrocarbons and Mississippi Valley-type sulfides in the Devonian reef complexes of the eastern Lennard Shelf, Canning Basin, Western Australia, in M. Keep and S.J. Moss (eds), *The Sedimentary Basins of Western Australia 3*. Proceedings of the Petroleum Exploration Society of Australia Symposium, Perth, Western Australia. pp. 795–815.

Wang, Y.J., Chen, H. and Edwards, R.L. (2001) A high-resolution absolute-dated late Pleistocene monsoon record from Hulu Cave, China. *Science* 294, 2345–8.

White, E.L. and White, W.B. (1979) Quantitative morphology of landforms in carbonate rock basins in the Appalachian Highlands. *Geological Society of America, Bulletin* 90, 385–96.

White, W.B. (1988) *Geomorphology and Hydrology of Karst Terrains*. Oxford University Press, Oxford. p. 464.

Williams, P.W. (1971) Illustrating morphometric analysis of karst with examples from New Guinea. *Zeischrift fur Geomorphologie* 15, 40–61.

Williams, P.W. (1972) Morphometric analysis of polygonal karst in New Guinea. *Geological Society of America, Bulletin* 83, 761–96.

Williams, P.W. (1978) Interpretations of Australasian karsts, in J.L. Davies and M.A.J. Williams (eds), *Landform Evolution in Australasia*. ANU Press, Canberra. pp. 259–86.

Williams, P.W. (1982) Speleothem dates, Quaternary terraces and uplift rates in New Zealand. *Nature* 298, 257–60.

Williams, P.W. (1987) Geomorphic inheritance and the development of tower karst. *Earth Surface Processes and Landform* 12, 453–65.

Williams, P.W. (2008) World Heritage caves and karst: a thematic study. IUCN World Heritage Studies, No. 2, Gland, Switzerland. p. 50.

Williams, P.W. (2009) UNESCO *World Heritage Caves and Karst: Present Situation, Future Prospects and Management Requirements*. 15th International Congress of Speleology Proceedings. 1, 38–44.

Williams, P.W., King, D.N.T., Zhao, J.-X. and Collerson, K.D. (2005) Late Pleistocene to Holocene composite speleothem chronologies from South Island, New Zealand: did a global Younger Dryas really exist? *Earth and Planetary Science Letters* 230, 301–17.

Woodhead, J.D., Hellstrom, J., Maas, R., Drysdale, R., Zanchetta, G. and Devine, P. (2006) U-Pb geochronology of speloethems by MC-ICPMS. *Quaternary Geochronology* 1, 208–21.

Yonge, C.J., Ford, D.C., Gray, J. and Schwarcz, H.P. (1985) Stable isotope studies of cave seepage water. *Chemical Geology* 58, 97–105.

Yuan, D. et al. (1991) *Karst of China.* Geological Publishing House, Beijing. p. 224.

Yuan, D., Cheng, H. and Edwards, R.L. (2004) Timing, duration transitions of the last interglacial Asian monsoon. *Science* 304, 575–8.

Zhu X. (1988) *Guilin Karst.* Shanghai Scientific and Technical Publishers, p. 188.

Zupan Hajna, N. (2003) Incomplete solution: weathering of cave walls and the production, transport and deposition of carbonate fines. *Carsologica,* p. 167.

PART 4

Environmental Change

28

Landscape Evolution and Tectonics

Paul Bishop

It has long been realized that the landscape evolves and that it is not a static unchanging feature of the Earth's surface. Chorley et al. (1964) noted that Leonardo da Vinci clearly grasped the notion that the landscape evolves, and so we can reasonably say that landscape evolution has been central to geomorphology from the dawn of the discipline. Indeed, landscape evolution is at the heart of the geosciences: the great realization by James Hutton (1788), who provided the permanent foundation of the geosciences (Chorley et al., 1964), was that landscapes evolve over time. One of Hutton's great contributions was the proposition that riverine (fluvial) processes remove material at one point on the Earth's surface and deposit it elsewhere as a sedimentary deposit. Fluvial incision modifies the landscape through time – the landscape evolves – and, in certain favourable situations, the rate, as well as the nature, of landscape evolution can be specified (Figure 28.1). Following these foundations, 'modern' geomorphology can be thought of as starting with William Morris Davis and he, almost more than anyone else, was fundamentally concerned with landscape evolution.

In this chapter, we review the major (and often competing) models of landscape and then we examine the approaches to understanding landscape evolution that have been made possible by recent advances in dating of landscapes and in understanding rates of landscape evolution. Finally, we look forward to emerging issues in landscape evolution.

THE 'BIG' QUESTIONS

'Landscape evolution' is the study of how the landscape of the Earth's surface evolves over long timescales, that is, over timescales of hundreds of thousands to millions to tens of millions of years. The study entails considerations of the rates of landscape evolution and the sequence(s) of the forms that the landscape passes through as it evolves. These two broad issues constitute the 'big' questions of the discipline, with a set of secondary questions that concern relationships between landscape morphology and lithology and climate (including past climates), and between landscape morphology and tectonics; we here take the latter to mean the vertical and, perhaps less importantly, lateral movement of the rock of the Earth's crust. These tectonic movements are currently understood to reflect the operation of plate tectonics, itself driven by the interactions of the deep Earth, the asthenosphere and the lithosphere; see the review by Bishop (2007).

The sequence of forms (the various morphologies) through which the landscape passes as it evolves can be considered simply, at first order, in three broad categories built around the way the relief of the landscape – the height difference between the highest and lowest points in a given landscape area – changes as the landscape evolves. Thus, relief can either: case 1, decline as the landscape evolves; case 2, increase during landscape evolution; or case 3, stay constant through time. Declining relief (case 1) must mean that

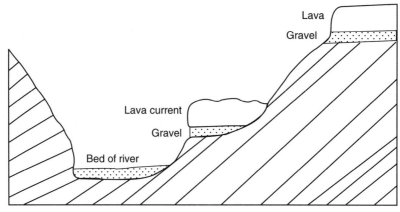

Lavas of auvergne resting on alluviums of different ages

Figure 28.1   Charles Lyell's illustration of how successive volcanic lavas, flowing into and preserving progressively lower river beds containing sediments of progressively younger ages, provide clear evidence of the rate of landscape evolution and perhaps of the forms of the landscape as it evolves. (Lyell, 1833: Figure 61, p. 267)

overall slope angle is lowering (the landscape is being lowered), whereas in case 2 slope angle must be increasing; in case 3, slope angles stay constant.

Several of the broad schemes of landscape evolution are cyclical, in that they propose that the landscape evolves from a high-elevation block to, ultimately, a low elevation landscape approximating a type of plain; the cycle starts again with uplift and renewed erosion. Figure 28.2 diagrammatically summarizes these various cyclical schemes. The landscape evolution scheme of Hack is not shown on Figure 28.2 as it proposes essentially constant landscape morphology through time; nor is the scheme of Crickmay who emphasized increasing relief with time (case 2). We now look at each of these in turn; the reader may refer to the original texts themselves for more detail and Palmquist (1975), Summerfield (1991) and Phillips (1995) have provided useful summaries of some of the various models.

Declining relief and landscape evolution: W.M. Davis

The name of William Morris Davis and his 'Geographical Cycle' (Davis, 1899) are inextricably associated with slope decline during landscape evolution; for more detail see the exhaustive review of Davis' life and work by Chorley et al. (1973). Davis's scheme of landscape evolution is a deductive scheme (Chorley et al., 1973, 160): he deduced logically what happens when the land surface in an area with a humid climate is tectonically uplifted as a block and then the processes of rock weathering and soil formation, hillslope processes of creep, overland flow and mass movement, and fluvial processes of channel erosion and sediment transport and deposition operate on that block of uplifted crust. A key part of this scheme is that the tectonic processes of rock uplift cease after the initial uplift and so the only processes acting to 'fashion' the landscape are the surficial, geomorphological processes listed in the previous sentence. As these processes operate on the uplifted block, the landscape soon develops a rugged incised character, corresponding to Davis' so-called 'Youthful' stage in the Cycle, characterized by incised gorges, steep bedrock channels and high levels of geomorphic activity. At some point in this evolution, the landscape will develop the maximum local relief between valley bottoms and ridge lines/interfluves, signalling the onset of the Mature stage in the Cycle, and thereafter relief declines through Maturity as hillslope angles decline and the elevations of divides and hillcrests are lowered. This decline in slope angle is asserted by Davis to result from the fact that rivers effectively cease to incise as the cycle progresses (the rivers have approached close to base level) whereas surface processes, including weathering, slow downslope mass movement and overland flow, continue to act

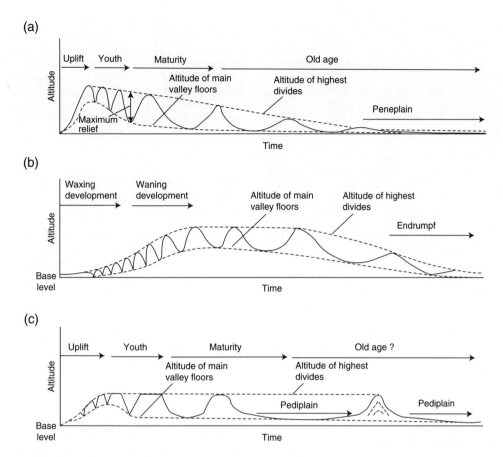

Figure 28.2   Diagram illustrating the development of relief in three of the principal schemes of long-term landscape evolution: (a) William Morris Davis; (b) Walther Penck; and (c) Lester King. Note that in (a) maximum relief marks the transition from the Youthful stage to the Mature, and that relief declines thereafter through the stages of Maturity and Old Age, culminating in the low-relief plain, the peneplain. King's scheme of landscape evolution in (c) is characterized by early river incision down to base level and valley widening by parallel retreat of slopes thereafter; relief stays essentially constant. The end stage, the pediplain, is reached when the last residual hills are consumed by parallel retreat of slopes and relief only starts to decrease once hillslopes on opposite sides of a remnant hill intersect at the hill crest, which then starts to lower in elevation (dotted lines beneath summit of right-hand hill). (Summerfield, 1991: Figure 18.1)

on the sloping parts of the landscape, with the result that these landscape elements continue to be lowered. The end stage – Old Age – consists of a plain of low relief, the peneplain, with a deep weathered regolith and sluggish meandering streams. Another pulse of surface uplift starts a new Cycle.

Davis emphasized many times that the sequence through which the landscape passes as it evolves is a function of what he called the trio of geographic controls, namely, Structure (i.e. lithology and geological structure), Process (the geomorphological processes that act on the Earth's crust to produce a landscape), and Time (the passage of time) (Davis, 1899, 481). He argued that the importance and influence of Structure decline throughout the Cycle – as Davis said: the peneplain, 'the almost featureless plain … show[s] little sympathy with structure' (Davis, 1899, 488) – and that the key control is the passage of time or Stage. Stage was at the heart of the Cycle and, if uninterrupted, landscapes evolved through the Cycle's fixed sequence of landforms; see Bishop's (1980) further discussion of this point.

Perhaps the most enduring element of the application of the Cycle is the peneplain, the low relief surface that Davis argued characterizes the Cycle's end stage ('old age'). High-elevation, low-relief plateau surfaces have been widely recognized and the Davisian interpretation is that these are uplifted peneplains. If a high-elevation plateau surface is indeed a peneplain, it is implied that landscape evolution in that area must have been via an extended period of tectonic stability for the peneplain to form close to base level (generally taken to be sea level), followed by surface uplift and the initiation of a new Geographical Cycle, eroding back into the uplifted peneplain (see Bishop's (1998) fuller discussion of the use of the Geographical Cycle in this context). Such an interpretation has been invoked many times in the literature, but, for some, the issue of the origin of high-elevation plateau surfaces remains at least unresolved and often contentious. Many such surfaces have been reconstructed, often by the visual 'joining-up' of ridge crests by projecting from one ridge crest or peak to another to produce an uplifted surface.

Criticisms of the Cycle

A wide range of criticisms has been levelled at the Davisian Cycle, from the time of the Cycle itself (e.g. Tarr, 1898; Shaler, 1899) through to the mid and late 20th century (Hettner, 1928; Strahler, 1952; Chorley, 1965; Chorley et al., 1973; Bishop, 1980). We now understand that it is unlikely that tectonic uplift would be restricted to a short sharp episode followed by the tectonic quiescence required for the Cycle to run its full course. Throughout much of the 20th century, however, it was believed that tectonic activity was a highly episodic, worldwide event. In fact, in the context of the major early 20th century debate about whether orogenesis was episodic, Davis' Cycle was taken as 'proof' that tectonic activity was indeed episodic Chamberlin, 1914; see Bishop's (2008) discussion of this point. Episodicity of tectonics was a particularly American viewpoint and the German geomorphologist/geologist Walther Penck (1924) pointed out that Davis' methodology (and especially his deductive approach) meant that Davis in effect 'imposed' such episodicity on the Earth's crust (rather than establishing empirically that episodic tectonic activity was a 'reality' – see also the discussion by Beckinsale and Chorley, 1991).

A key element of the Geographical Cycle is Davis' disregard of the effects of denudational isostasy, whereby unloading of the Earth's crust by denudation causes the crust to float up, in effect generating surface uplift and concomitant base-level fall. We return to this issue below when we consider the applicability of the Davisian scheme (or, more generally, a scheme of landscape evolution in which relief declines over time).

Increasing relief and landscape evolution: C.H. Crickmay and C.R. Twidale

A more poorly known, but nonetheless important, scheme of landscape evolution is that of Colin H. Crickmay, a somewhat enigmatic figure who was not a 'mainstream' researcher and published relatively little (e.g. Twidale, 2004). Crickmay echoed W. Penck (1924) in highlighting the illogicality of the preservation of former peneplains as high-elevation (i.e. uplifted), low-relief plateau surfaces (Crickmay, 1968): quite simply, if Davis' Cycle of Erosion operates via downwasting (Figure 28.2), then the preservation of an uplifted peneplain is logically impossible – the surface of the uplifted peneplain would soon disappear as a result of downwasting. Crickmay observed that much of the Earth's surface, at all elevations, is relatively flat because the processes of landscape evolution are concentrated, as it were, in river valleys, leaving broad areas of the landscape subject to geomorphic processes that are, in effect, 'dead quiet' (Crickmay, 1975, 105).

Crickmay's work has received increased attention of late (e.g. Nott and Roberts, 1996) and C.R. Twidale has been a major champion of Crickmay's (1974, 1975) 'hypothesis of unequal activity'. This hypothesis proposes that fluvial erosional energy is progressively concentrated in the valley bottoms of larger rivers and that denudation of upland areas slows because of lower erosion rates in the smaller rivers that drain these areas. Major channels incise, becoming in effect de-coupled from slopes, which evolve more slowly. In other words, as rivers continue to incise and upstream areas erode more slowly, landscape evolution must involve increasing relief amplitude (Crickmay, 1975; Twidale, 1976). Twidale (1991) formalized this approach in a model of landscape evolution involving increased and increasing relief amplitude, highlighting the role of resistant lithologies, structure and different groundwater conditions throughout a drainage basin as the important factors leading to increasing relief (rather than simply the relative sizes of trunk and tributary streams, as proposed by

Crickmay) (see also Twidale, 1999). Twidale has argued that these factors act to concentrate erosion in valleys and to reduce rates in upland areas:

> water ... is concentrated in and near major channels, for, once a master stream develops, not only surface water but subsurface drainage too gravitates towards it ... [U]plift induces stream incision and water-table lowering, leaving high plains and plateaux perched and dry.
>
> (Twidale, 1998: 663)

Constant relief and landscape evolution: J.T. Hack, L.C. King and W. Penck

A range of workers have proposed that relief remains relatively constant as the landscape evolves over time. The schemes of two of these – King and Penck – are essentially cyclical and developments of, or reactions to, Davis' scheme; these are considered first. Hack's scheme was also a reaction to Davis but was substantially different in the way in which it considered landscape evolution.

Before considering these schemes of landscape evolution, the statement that relief remains constant must be qualified. Figure 28.2a shows how the relief between the highest points in the landscape (the divides) and the valley bottoms declines through time in the Cycle of Erosion – this is a key element of the Cycle. In the King scheme, which is discussed below,

relief stays relatively constant (Figure 28.2c). Likewise, the Penck scheme, also discussed below (Figure 28.2b), involves essentially constant relief, but slope angles decline through time by replacement from below of steeper slopes by gentler slopes. Once the ridge crests in the King and Penck schemes are consumed, however, relief must decline and constant relief can only be maintained by ongoing surface uplift or rock uplift, itself an important distinction that is explored in more detail when the Hack scheme is discussed below.

Lester King's scheme of landscape evolution is cyclical, with the major difference from the Davisian scheme being that King proposed that slopes do not decline with time but retreat parallel to each other (King, 1962, 154). The primacy of parallel slope retreat in this scheme of landscape evolution reflects King's life-long experience of landscape evolution in semi-arid southern Africa. Based both on first principles and on the character of those semi-arid landscapes, King argued that landscapes in general consist of four morphological elements: a convex *crest* zone, passing laterally and often abruptly into a more-or-less rectilinear *scarp* zone, with a *debris slope* at its foot, and a gently-sloping, concave-up *pediment* between the foot of the debris slope and the drainage line (King, 1962) (Chapter 5) (Figure 28.3). The final plain of low relief, resulting from the coalescence of pediments, is termed a pediplain (Figure 28.2c). The processes that characterize humid temperate terrains, such as weathering, soil creep and channelized flow, are less common in semi-arid landscapes (where, for example,

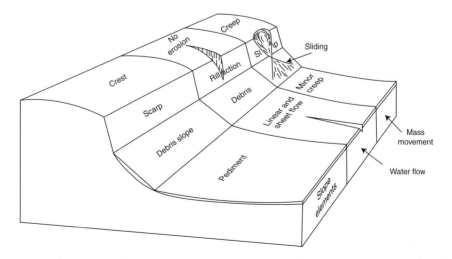

Figure 28.3   King's illustration of the four morphological elements of all landscapes (King, 1962: Figure 53)

overland (sheet) flow across pediments is much more the rule (e.g. King, 1962: 138)) but King was at pains to emphasize that pediments are not restricted to semi-arid environments and that the ensemble of the four morphological elements (Figure 28.3) is found in humid and arid environments and 'form[s] naturally under the normal flow of meteoric water across a landscape' (King, 1962: 139).

In King's scheme, resistant lithologies are associated with well-developed scarp faces which persist as they evolve by parallel retreat; on weaker lithologies the scarp face and debris slope may disappear, especially if there is only low relief between the crest and valley bottom, and/or closely-spaced streams. The primacy of the four-element model is strongly asserted by King (1962: 139, 140, 142) and finds its optimum form in bornhardts in regions of highly resistant lithology and high relief. Where there is no scarp face and debris slope, slopes are convexo-concave, with a convex upper slope (reflecting the convexity of the crest) and a concave pediment forming the lower slope. This adjustment of the four-element landscape model to account for landscapes consisting dominantly of convexo-concave hillslopes is a logical, but untested, rationalization of the four-element model.

W. Penck's scheme of landscape evolution, *Morphological Analysis of Landforms*, was published in English in 1953 and is a translation

of his *Die morphologische Analyse* of nearly three decades earlier (Penck, 1924). The title provides the clues to Penck's objective – as Davis himself expressed it:

> [Penck's 'morphological analysis' may be defined as] a procedure by which the nature of crustal movement is inferred from external processes and surface forms
>
> (W.M. Davis quoted by Chorley et al., 1973: 699).

Thus, Penck used slope characteristics to infer the nature of tectonic activity and, specifically, surface uplift, on the assumption that streams are able to respond relatively rapidly to changes in this surface uplift (i.e. to changes in the rate of change of base level). An increase in the intensity of stream incision (in response, say, to an increase in the rate of surface uplift) will lead, Penck argued, to convex lower hillslopes, in a phase that Penck termed waxing development. A period of decreased rate of base-level fall (i.e. a slowing in surface uplift and stream incision) will result in concave slopes in Penck's phase of waning development (Figure 28.4a). Thus, whereas Penck's morphological analysis has often been cast as a viewpoint in opposition to that of Davis (an opposition that Davis, in private, seems to have promoted (e.g. Chorley et al., 1973: Chapter 28), the two schemes of landscape evolution strictly

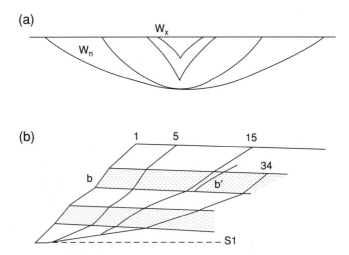

Figure 28.4 **Diagrammatic representations of Penck's models of hillslope and landscape development. (a) Slope morphology as a landscape under waxing development (Wx) and waning development (Wn). (b) Illustration of the way in which slope replacement operates, with steeper slopes replaced from below by lower-angled slopes. Resistant (stippled) lithologies are associated with steeper slopes within the overall scheme of slope replacement and declining slope angle. (Palmquist, 1975: Figure 2)**

had different objectives: Davis' aim was to bring order to and explain landscape morphology in a sequence of landscapes that reflect the interaction of structure, process and time (and especially the latter), whereas Penck aimed to deduce from the landscape the intensity of tectonic movements that have affected the Earth's surface. A further important difference between the two schemes is that Davis' Cycle of Erosion was specifically formulated for a humid climate (with variants for glacial and arid environments developed subsequently by Davis) whereas Penck's scheme is aclimatic and focused on the landscape expression of different rates of tectonic activity.

In one area, namely the evolution of hillslopes, Davis misrepresented Penck's views when he reported that Penck argued that slopes retreat parallel to each other (e.g. Simons, 1962; Chorley et al., 1973; Summerfield, 1991). In fact, Penck argued that as denudation slows, slopes will flatten from the bottom up, with the process of slope replacement leading to the progressive upslope replacement of steeper slopes by lower-angled slopes (Summerfield, 1991) (Figure 28.4b). It is also noteworthy that Penck placed greater emphasis than did Davis on the role of lithology in landscape evolution (Palmquist, 1975), and the persistence of lithological influences (Figure 28.4b).

John Hack's scheme of landscape evolution was articulated in relatively few papers (e.g. Hack, 1960, 1975) that built on his earlier empirical work on the morphological characteristics of bedrock rivers (e.g. Hack, 1957). Following Arthur Strahler's (1952) call to abandon the Davisian approach (and indeed, by implication, to move away more generally from grand schemes of landscape evolution), fluvial geomorphology during the third quarter of the twentieth century largely consisted of the quantitative characterization of alluvial rivers, which Clifford (2008) has reviewed. Hack (1957) explored relationships between morphometric variables in bedrock rivers, including channel width, gradient, mean grain size, downstream distance, and so on, as Leopold and colleagues were doing in alluvial systems (e.g. Leopold et al., 1964).

One of Hack's key findings was that bedrock rivers on resistant lithologies exhibit steeper gradients, which Hack explained by the observation that such lithologies generate coarser sediment that requires steeper channel gradients to be transported. This equilibrium relationship between lithological resistance and gradient means that for given lithology and other factors (e.g. climate and tectonic setting), river gradient must be essentially constant and cannot decline with the passage of time, as Davis (1899) had argued it does. Hack (1960) then made the next logical step

and argued that slopes as well as channels adjusted their gradients to transport the material available to them (as a result of weathering in the case of slopes). In other words, Hack (1960, 1975) argued that landscapes attain a dynamic equilibrium between process, lithology and gradient, and that each example of dynamic equilibrium (and its associated landscape morphology) remains essentially constant through time. The implication of this dynamic equilibrium viewpoint – which derived from G.K. Gilbert's late 19th century thinking about, and experiments in, geomorphological processes and his laws of erosion (Hack, 1975) – is that landscapes do not evolve through a sequence of time-dependent morphological stages of Youthful, Mature and Old age forms, as Davis (1899) had argued. Once the adjustment (dynamic equilibrium) between lithology and slope is attained for the particular process domain of interest (i.e. slope processes or fluvial channel processes) the adjustment persists: landforms do not evolve, their form is constant and 'time independent', to use Adams' (1985) term. Such a landscape consists of an orderly mosaic of elements, each exhibiting a dynamic equilibrium between lithology, process and slope, with the land surface downwasting at the same rate across the whole landscape. In such a landscape 'every slope and every channel in an erosional system is adjusted to every other' (Hack, 1960: 80). Such 'catchment-wide downwasting' persists until perturbed by, for example, the land surface encountering a change in lithology or experiencing a change of climate that shifts the process regime.

An important, but implicit and generally unstated, component of the constant relief and morphologies of Hack's catchment-side, dynamic equilibrium model must be ongoing rock uplift. The terms 'rock uplift' and 'surface uplift' do not have the same meanings. Rock uplift means uplift – upward movement – of the crustal rock column outwards, away from the centre of the Earth in a whole-of-Earth frame of reference. This uplift corresponds to surface uplift (i.e. upward movement of the Earth's surface) to the extent that surface processes of weathering and erosion do not 'keep pace' with that rock uplift (Summerfield, 1991). Zero erosion of the uplifting rock column means that the surface is uplifted by the amount of rock uplift, whereas if the rate of surface processes is the same as the rate of rock uplift, there is no surface uplift (see Summerfield, 1991 Figure 15.2). In a Hack-type dynamic equilibrium scheme of landscape 'evolution', catchment-wide downwasting can only continue for as long as river valley bottoms (the lowest parts of the landscape) are sufficiently above base level for river incision to continue. Once rivers reach base level (or at least the small distance above base level required for rivers to continue to flow), then those valley bottoms cannot continue to incise and they must begin to widen by lateral erosion. In other words, when that point is reached, relief must begin to decline: the valley sides will continue to downwaste at the rate set by the equilibrium between process, lithology and slope, but the valley bottoms are no longer being lowered. Thus, rock uplift is necessary for the persistence of Hack-type landscape dynamic equilibrium, a point that Hack does not seem to have addressed. We return to this point below.

The key 'controls' of landscape evolution in various schemes

Davis considered that of his trio of geographic controls of landscape evolution – namely, Structure (lithology and geological structure), Process and Time (Stage) – the most important is Stage, the control 'of the most frequent application and of a most practical value in geographical description' (Davis, 1899: 481). The influence or effect of lithology (Davis' 'Structure') declines throughout the Cycle so that by the late Mature and Old age Stages, expressions of lithological difference are muted (or even non-existent) in the landscape, with the 'peneplain … showing little sympathy with structure' (Davis, 1899: 498). Process, especially the processes of humid weathering and fluvial erosion, is central to Davis's scheme but only to the extent that process provides the 'engine', as it were, for the passage of the Cycle. To that extent, climate (and its accompanying processes) were relegated by Davis to a somewhat secondary role in his Cycle of Erosion: 'the wind of arid deserts and the ice of frigid deserts will be considered as climatic modifications of the norm' [of the humid temperate Cycle], and set apart for particular discussion' (Davis, 1899: 482). The passage of Time is the key element and so the Geographical Cycle in an arid climate is built around the same sequences of Stages as the 'normal' Geographical Cycle (Davis, 1905), and Davis (1900) likewise presented a cycle of glacial denudation.

All of the other schemes of landscape evolution – those of Penck, King, Hack, Crickmay and Twidale – incorporate key roles for lithology and rock structure in determining the form of the Earth's surface and its evolution, but not acting in the same ways for all schemes. Penck and King argued that the influence of rock type is felt throughout their version of the cycle of landscape development, as slopes adjust to the rocks on which they are formed. Indeed, as we have seen, King emphasized that the best development of

his four landscape elements occurs on resistant lithologies in areas of substantial relief. Likewise, Hack gave lithology a central and fundamental role in the evolution of landscape morphology, but in this case via the way that lithology is expressed in the calibre of the weathering products of that rock type. For Hack (1960), the interaction between process and lithology (and the ways in which interaction determines slope angle) is via the calibre of the breakdown products of the lithology, the grain size of the regolith formed on that lithology: coarser weathering products require steeper slopes for those products to be transported away.

For the schemes that involve increasing relief during landscape evolution (Crickmay and Twidale), lithology and structure play different roles, especially in Twidale's elaboration and development of Crickmay's ideas. For Twidale, structure is an important element in focusing river incision and concentrating landscape lowering in valley bottoms, especially in the valleys of larger rivers. This incision, as we have seen, is more rapid than that in smaller channels higher up the drainage net and leads to increasing relief. Resistant lithologies act to slow the upstream propagation of base-level falls. Resistant lithologies thus also separate the landscape into downstream segments that are incising/lowering and upper segments that are lowering at very much lower rates because of low stream power in the upstream areas and/or isolation by lithological barriers from the catchments downstream, leaving these upper catchments isolated and lowering at very low rates, if at all.

All schemes we have considered have a climatic element, for it is climate, along with slope, that 'drives' the processes of landscape evolution. As we have seen, Davis's Geographical Cycle was formulated for humid temperate climate, as was Hack's, in a direct response to Davis: Hack's 1960 paper is entitled 'Interpretations of erosional topography in humid temperate regions'. Penck's concern was inferring the endogenetic factors (those internal to the Earth, the crustal movements) from the landscape morphology. Penck claimed that the exogenetic processes of rock weathering and surface processes, driven by climate, can be observed, as can the morphologies that result from the interactions of the endo- and exogenetic processes, which means in turn that the endogenetic processes can then be inferred (Penck, 1924: Section 1.2). Thus, for Penck the endogenetic processes (and hence the climate) are observable 'givens' in his quest to understand the endogenetic (tectonic) processes and he says very little explicitly about a particular role for climate. King argued that his model of landscape evolution is general but it is usually agreed that it strongly reflected his background and extensive experience of landscape evolution in semi-arid southern Africa.

Before moving to an assessment of how these various models of landscape evolution can be tested, we make a brief digression to consider climatic geomorphology

CLIMATIC GEOMORPHOLOGY

A series of models of landscape evolution, or at least of the development of characteristic suites of landforms, can be grouped under the heading of climatic geomorphology; this branch of geomorphology self-evidently attributes to climate a much greater role than the models of landscape evolution considered so far. Climatic geomorphology is founded on the premise that the geomorphic processes associated with the world's climatic belts act in such a way as to produce characteristic and distinctive landform assemblages in these climatic belts (Derbyshire, 1973; Beckinsale and Chorley, 1991). One of the principal applications of climatic geomorphology has been in the identification of relict landforms characteristic of former, now-changed climatic regimes and hence in the documenting of these climate changes. English language publications on climatic geomorphology include those of Birot (1960), Tricart et al. (1972) and Büdel (1982).

Beckinsale and Chorley (1991) have noted how, unlike in the United States and United Kingdom, where Davis' historical geomorphological analysis embodied in the Geographical Cycle ruled, German geomorphology in the late 19th and early 20th centuries was strongly concerned with exogenetic and endogenetic processes; we have seen the legacy of that interest in Walther Penck's (1924) work. The publication in the early 20th century of Köppen's classification of world climate zones, with its basis partly in temperature and seasonality, prompted the search for corresponding sets of geomorphological regions with assemblages of characteristic landforms.

Research in climatic geomorphology was thus initially dominated by German language publications, with the approach becoming more widely accepted from the 1940s onwards (e.g. Cotton, 1942; see Beckinsale and Chorley, 1991). Peltier (1950) produced a series of diagrams that classify geomorphological processes by temperature and rainfall, attempting thereby to characterize intensity of these processes. Nine morphogenetic regions were identified by combining these six plots, comprising glacial, periglacial, boreal, maritime, moderate, selva, savanna, semi-arid and arid. Whether all of these are associated with distinctive suites of landforms, however, is the key issue.

Acceptance of the climatic geomorphology approach has not been particularly widespread among Anglophone researchers and there has been some stout opposition to it. Twidale and Lageat's (1994) critique includes the following issues. Some suites of landforms are unambiguously indicative of the climatic zones that they characterize, namely arid, periglacial and glacial landforms. That said, one of the most striking features of arid systems is, in fact, the importance of running water in generating desert landscapes. Fluvial landforms abound in many deserts, but they are only very infrequently active. And even in those areas exhibiting 'climatically distinctive' landform suites, structurally and lithologically controlled elements remain significant. Perhaps most importantly, Twidale and Lageat (1994) pointed out that many geomorphologically significant processes, such as the fluvial processes in arid (and even hyper-arid) environments that we have just noted, transgress boundaries between climatic zones.

RATES OF LANDSCAPE EVOLUTION AND TESTING THE MODELS

Rates

The two issues central to an understanding of landscape evolution are the rates at which whole landscapes develop (i.e. the rate of geomorphological processes) and the sequences, if any, through which landscapes pass as they evolve. The latter has been our focus so far and we now turn to rates. Charles Darwin's (1881) calculation of rates of soil turnover by careful monitoring of how much earth is cast by worms on the ground surface each year or every few years is an elegant demonstration of how rates of geomorphological processes can be monitored and quantified. Thus, early attempts to estimate rates of geomorphological processes were made using the volumes of sediment carried annually by rivers; this volume of fluvially transported sediment was taken to be a measure of the rate at which the land is being eroded (e.g. Schumm, 1963; Judson and Ritter, 1964). It was not clear whether these sediment loads do actually result from land-surface erosion (as opposed to, say, fluvial remobilization of sediment from sediment storage sites such as floodplains and in-stream deposits) nor, if the sediment loads are indeed derived from erosion of the land surface, whether they represent anthropogenically accelerated erosion rates. The pitfalls of the approach are apparent in Thornbury's (1969) conclusion, following his simplistic dividing of the average elevation of the Earth's surface by measured rates of surface processes, that landscapes cannot be much older than the Tertiary (i.e. no older than ~65 million years [Myrs]) and are probably no older than the Pleistocene (i.e. about 2 Myrs old). Even when the calculation used long-term erosion rates that were not artificially increased by human land-use, the conclusion was that the Davisian Cycle required between 10 and 25 Myrs to run its full course (e.g. Gilluly, 1955; Schumm, 1963; Judson and Ritter, 1964; Melhorn and Edgar, 1975). Given that we now know that landscapes may be many tens of millions of years old (e.g. Twidale, 1998), one or more key factors were being missed in even these more accurate analyses (see below).

Several ways of avoiding anthropogenically enhanced erosion rates in these calculations include using demonstrably ancient geological features to determine erosion rates. Thus, if the ages of the basaltic lava flows shown in Figure 28.1 are known (e.g. based on fossils contained in the 'Gravel' capped by each lava), the rates of river incision can be calculated for two time intervals: the time between the eruption of the higher and lower lavas, and the time since the eruption of the lower lava. The same approach can use any features and/or materials demonstrably marking a former land surface or river bed, such as a fluvial terrace. The absolute ages of fossils were, however, not known accurately until absolute radiometric dating was developed, initially by Ernest Rutherford, following the discovery of radioactivity in the early 20th century, and then in more detail and more thoroughly by Arthur Holmes (1913). Once these absolute ages were known – and such radiometric ages did not become routinely available until the second half of the 20th century – 'geological' rates of landscape evolution (i.e. rates of landscape evolution prior to human disturbance of the landscape) could be determined accurately. Approaches using ancient lavas that mark old river beds, as in Figure 28.1, have been notably useful in establishing rates of landscape evolution (e.g. Young and McDougall, 1982, 1993; Bishop, 1985; Nott et al., 1996).

In the last two decades or so, new methods have been developed to quantify landscape evolution. Whereas ancient lavas are used to determine the rate of incision below a dated surface, these new techniques, including low-temperature thermochronology and the analysis of terrestrial (or *in situ*) cosmogenic nuclides, quantify the rates at which rocks have been brought to the surface by denudation. Low-temperature thermochronology (LTT) provides data on when rocks cooled below certain temperatures, which correspond to depths

in the crust according to the geothermal gradient (Gleadow and Brown, 2000). The known (or assumed) geothermal gradient can be converted to depth, which, combined with the age, corresponds to a rate of landscape evolution (in effect, the rate of denudation required to bring the rock to the surface over the time interval that is provided by the age from the low-temperature thermochronometer). The commonest LTT techniques for geomorphological analysis include apatite fission-track analysis, which provides the time since the rock cooled below ~120–60°C (say ~100°C or ~4 km depth in the crust) as denudation brings the rock to the Earth's surface, and (U–Th)/He analysis in apatite, which tracks rocks from temperatures of ~80–40°C (say ~60°C or ~2–3 km depth). These crustal depths might be thought irrelevant to geomorphology but it can be seen that the same thermochronological ages for the fission-track and apatite-helium systems must mean very rapid rock uplift and denudation (because the rocks have passed through the ~120–60°C and ~80–40°C isotherms at essentially the same times; for example, Zeitler et al., 1993, 2001). The low-temperature systems can also be used to date the longevity of major, long-wavelength topography. This is done by exploiting the way in which long-wavelength topography perturbs the thermal structure of the shallow crust, deforming the shallow isotherms (House et al., 1998).

Analysis of terrestrial (or *in situ*) cosmogenic nuclides (TCNs) is introduced only briefly here and more information is in the reviews by Bierman (1994), Cerling and Craig (1994), Bierman and Nichols (2004), Gosse and Phillips (2001), Niedermann (2002) and Cockburn and Summerfield (2004). Just as LTT 'tracks' rocks from kilometres depth in the Earth's crust, TCN analysis provides the rate at which rocks reach the surface from ~2m depth. The technique exploits the fact that cosmic radiation from outer space interacts with elements, such as oxygen and silicon, in the crystal lattices of minerals, producing small amounts of distinctive nuclides that can be measured using conventional and accelerator mass spectrometry. For our purposes here, it can be assumed that these cosmogenic nuclide production reactions are restricted essentially to the upper 2 m of the Earth's surface, and so the TCN content of a mineral at the Earth's surface is a measure of the rate at which that mineral (and its host rock) were brought through the top 2 m of the crust (a measure, in other words, of the rate of erosion at the spot where the rock was sampled). Analysis of the TCN content of a sample of sediment collected from a river basin outlet provides an average rate of erosion for that river basin (because the TCN content of each grain records the rate of erosion of that grain's source rock). If the rock is not eroding (if it remains, say, glacially striated after the last period of glaciation), the TCN content corresponds to the age of exposure of that rock surface. These techniques have revolutionized geomorphology because, for the first time, it has become possible to sample a rock and determine directly the rates of erosion that brought it to the surface over several time intervals, from depths in the crust ranging from several kilometres to several metres.

Testing the schemes of landscape evolution

The Earth's surface and the landscapes that comprise it are highly complex systems that evolve by an almost infinite number of processes and interactions, including the physical processes of rock uplift, the physical and chemical processes of rock weathering and soil formation (including the impacts on the Earth's surface of the growth of vegetation), the physical and chemical processes of movement of materials across the landscape and down through landscape systems by fluvial, groundwater, glacial, aeolian and coastal processes. The various schemes of landscape evolution are difficult to test, and perhaps impossible, especially in the case of the Davisian scheme (Bishop, 1980).

The literature contains relatively few examples of tests of these schemes; historically the schemes have been used more to interpret landscape evolution in particular areas. Bishop et al. (1985) argued that evolution under conditions of Hack-type dynamic equilibrium would be demonstrated by fulfilment of two conditions: (1) the evolution of an area under conditions of catchment-wide downwasting, with (2) close association between resistant lithologies and steeper slopes, especially river gradients. These conditions were apparently satisfied in a 20,000 km^2 drainage basin in south-east Australia, where Neogene rates of river incision and Neogene rates of sediment flux to basins are broadly equivalent (i.e. catchment-wide downwasting), and there is an apparently close adjustment between long profile gradient and lithology (i.e. steeper bedrock channels associated with resistant lithologies) (Bishop et al., 1985). Subsequent analyses have shown, however, that the steeper reaches on more resistant lithologies are in fact disequilibrium reaches (knickpoints) which are 'pinned' to (i.e. have not propagated through) these lithologies (Bishop and Goldrick, 2000, 2010; Goldrick and Bishop, 2007). Landscape evolution in that situation therefore corresponds more to the Crickmay–Twidale model of increasing relief.

Tests of some of the models of landscape evolution may be provided by numerical models (computer simulations) of landscape evolution, which are reviewed in some detail by Beaumont et al. (2000), Coulthard (2001), Willgoose (2005) and Codilean et al. (2006). These models attempt to simulate in a linked way surface (geomorphological) processes and lithospheric (tectonic) processes in order to simulate landscape evolution over millions to hundreds of millions of years. The most common form of these models of long-term landscape evolution involves the setting-up of a synthetic landscape in a regularly or irregularly gridded digital elevation model and, at each node of this DEM, the representation of the processes relevant to long-term landscape evolution. These processes are conventionally: fluvial processes of bedrock channel incision and sediment transport; slope processes of bedrock weathering and mass movement by slow (creep) and rapid (landsliding) processes; and tectonic processes of rock uplift by active tectonic processes and passive processes of isostatic rebound and flexure. It is impossible at present to implement full, physics-based formulations of these processes and so they are represented by empirical 'rules' expressed as relatively simple algorithms. Thus, fluvial and slope processes are represented by simple rules based essentially around measures of channel slope and hill slope, respectively. Simply put, where these slopes are greater, process rates and erosion are greater.

Kooi and Beaumont (1996) used such a numerical simulation to test the 'classical' models of landscape evolution considered above (Davis, Penck, King and Hack). Their simulations show that all forms of landscape evolution can be simulated, depending on the relationship between the tectonic 'forcing' of the model and the timescale of response of the geomorphological processes. When the tectonic forcing (rock uplift) is constant and at a rate to which the geomorphological processes can respond to remove the amount of rock uplift, a steady-state adjustment between rock uplift and denudation is established and time-independent landforms develop. The landscape is akin to the dynamic equilibrium landscape proposed by Hack (1960). If the rate of rock uplift exceeds the capacity of the surface processes to respond, some elements of a Penckian-type landscape evolution are produced by the Kooi and Beaumont (1996) modelling but the phases of waxing and waning development are not. A short, sharp rock uplift event ('impulsive tectonic forcing'), followed by tectonic stability, results in simulated landscape evolution similar to the Davisian Cycle, with progressive slope lowering and ultimate development of a plain of low relief similar to a peneplain.

RECENT VIEWS

Tectonics and landscape evolution

Kooi and Beaumont's (1996) simulations of landscape evolution models show that all of the 'traditional' geomorphological models of landscape evolution might be judged to be valid, depending on the relative magnitudes of rates of rock uplift and rates of surface processes that respond to that tectonic activity. In recent decades, the focus of understanding landscape evolution has shifted towards models that seek more closely to integrate tectonics and landscape response, thereby circumventing to some extent the issues of whether landscapes evolve via downwasting (i.e. as Davis envisaged) or by backwearing, as envisaged by, say, King. Uplift figures in all the various theories of landscape evolution but this uplift is generally either simply taken as a given, as the starting point of a Davisian Cycle, for example, or is itself to be interpreted from the landscape evolution, as in the Penckian scheme. The issue of mechanisms of uplift was generally ignored in landscape evolution studies, to the extent that, as we have seen above, Davisian theory ended up being used to impose tectonic and orogenic histories on the Earth's crust (Chamberlin, 1914; Bishop, 2008). The theory of plate tectonics has provided a rock uplift mechanism that lies outside the theories of landscape evolution and hence may be one tool in interpreting that landscape evolution. We have already noted that the key tectonic process is rock uplift, and that this translates into surface uplift to the extent that surface processes cannot keep pace with the rock uplift. In the plate tectonic framework, rock uplift results from either (1) crustal thickening as a result of plate convergence, or (2) crustal buoyancy (due to heating of the crust), or (3) crustal flexure (due to loading or unloading of the crust). All of these responses are fundamentally isostatic responses: (1) thicker crust floats higher (rock is uplifted), (2) warmer, less dense crust floats higher, and (3) loading of the crust causes vertical crustal movement: positive loading by ice or sediment results in depression of the crust; negative loading as a result of the melting of an ice sheet ice (glacial unloading) or denudation of the land (denudational unloading) result in rock uplift. Flexural isostasy means that loading by ice or sediment in one place may be transmitted to adjacent areas as rock uplift, depending on the flexural properties of the lithosphere (the lithosphere's mechanical strength) (Figure 28.5). A stiff, relatively inflexible lithosphere (corresponding to high values of the effective elastic thickness of the lithosphere) transmits such loading over long distances) whereas a weaker, more flexible lithosphere, with low values for the effective elastic thickness of the lithosphere, accommodates positive and negative loading locally and does not transmit that loading to adjacent areas. Such flexural effects are implicated in many aspects of the landscape evolution of passive continental margins (Figure 28.5), a class of landscapes that has received substantial attention over the last 20 years or so from geomorphologists, geologists, geophysicists and geodynamicists (Bishop, 2007). This attention reflects the potential to provide a coherent integration of landscape evolution, global-scale (plate tectonic) geodynamics, volcanic history, denudation, and offshore and onshore sedimentation (e.g. Summerfield, 2000). Thus, recent attention has focused more on landscape history in particular tectonic settings rather than on theories and schema of landscape evolution *per se*.

A second major focus of such integrated studies has been areas of crustal convergence and orogenic uplift, such as the Himalayas, European Alps, Southern Alps and Taiwan. In these areas, rock is being fed into convergence zones at rates of, for example, 38 ± 3 mm a^{-1} for the Southern Alps (translating into a rate of ~11 mm a^{-1} normal to the Alpine Fault (Tippett and Hovius, 2000)) and ~70 mm a^{-1} for Taiwan (Lin, 2000). These rates of convergence translate into rates of rock uplift of 5–10 mm a^{-1} and higher, with the mountains of many of these convergence zones projecting up into very moist air masses, thus generating very high rainfalls on the mountain belts [e.g. annual rainfalls of 5 m in the Central Range of Taiwan and with very high intensity rainfall during typhoons (Lin, 2000), 11–12 m in the foothills of the Himalayas, and an astonishing 15 m at the crest of the Southern Alps (Tippett and Hovius, 2000)]. These convergence zones are also characterized by intense seismic shaking and it has been increasingly recognized that a dynamic equilibrium will develop between rock strength and hillslope and channel slope angles in areas of high rates of rock uplift, seismic shaking (which destabilizes slopes and supplies abundant sediment to channels), and precipitation. High rates of all three processes are closely associated with extreme rainfalls in seismically active areas at actively uplifting plate convergence zones, such as Taiwan (Hartshorn et al., 2002; Dadson et al., 2003, 2004), the Himalayas (Burbank et al., 1996; Hancock et al., 1998) and the Southern Alps (Adams, 1985; Hovius et al., 1997). These steady-state landscapes are composed of essentially time-independent landforms, and such landscapes are also developed, as we have seen, in numerical models. They are also developed in physical models of landscape development under conditions of continuous rock uplift and sufficient

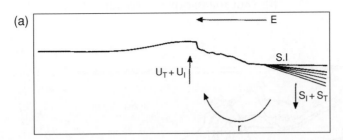

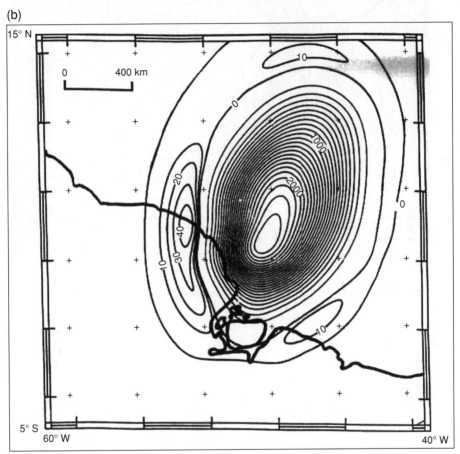

Figure 28.5   (a) Flexural uplift of the onshore region of a passive continental margin as a result of continental shelf subsidence due to sediment loading (SI) and thermal subsidence (cooling; ST) of the continental margin after continental breakup. The isostatically driven subsidence of the shelf drives isostatic flexural uplift (UI) of the onshore (via the rotation 'arm', u) which may also experience rock uplift as a result of thermal effects (UT). An escarpment retreats (E) into the flexurally uplifted hinterland. Note how the mechanical strength of the lithosphere (the strength of the lithospheric 'lever arm') will determine the inland extent and amplitude of the flexural rock uplift (from Summerfield, 1991: Figure 4.20). Pazzaglia and Gardner (2000) have attributed the formation of the Fall Zone on the Atlantic continental margin of North America to erosion into a flexural bulge formed as in this diagram. (b) Calculated depression resulting from post-Middle Miocene sediment loading of the Amazon fan (offshore; contour interval 100 m) and onshore flexural uplift resulting ('peripheral bulge') from that loading (contour interval 10 m). (c) Projection of the peripheral bulge onto the drainage net of coastal Amazon highlighting the way in which small tributaries have their headwaters on that peripheral bulge. (Driscoll and Karner, 1994: Figures 3 and 4)

(c)

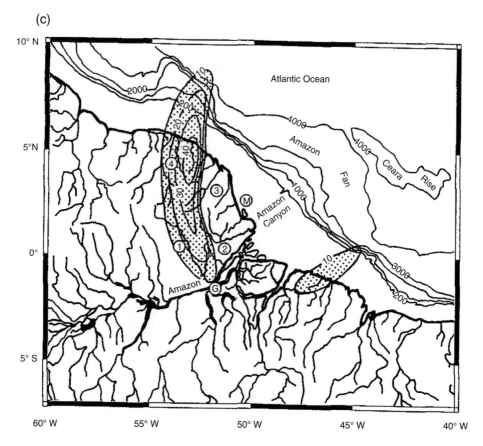

Figure 28.5   Cont'd

precipitation for surface processes to match
that rock uplift (e.g. Bonnet and Crave, 2003)
(Figure 28.6). An important difference from the
Hack dynamic equilibrium model, however, is
that the hillslopes in tectonically active areas, such
as Taiwan, the Himalayas and the Southern Alps,
are adjusted to the strength characteristics of the
rock rather than to the characteristics of the regolith, as Hack (1960) had proposed (Burbank et al.,
1996). These steep slopes, standing on average at
the limiting slope angle for the strength of the
rock(s) of which they are composed, are closely
coupled to channels, feeding sediment directly to
the channels. This sediment constitutes the fluvial
'tools' (Sklar and Dietrich, 2001), which, coupled
with the extreme discharges of these monsoonal
and typhoon- and/or storm-fed rivers, enable the
rivers to incise at the rate of rock uplift and to
maintain steady-state landscapes.

The processes of landscape evolution in
tectonically active areas that result in steady-
state landscapes can be considered 'top-down'

processes (Bishop, 2007): the high water and
sediment discharges enable rivers to incise, and
slopes to track this incision, at the rate of rock
uplift. Perhaps it is semantically incorrect to
state that such steady-state, time-independent
landscapes constitute 'landscape evolution'. Be
that as it may, it is clear that many (and perhaps
most) landscapes do evolve, with the landscape
morphology changing over time. That is because
most parts of the Earth's surface do not experience
the high rates of rock uplift, rainfall and seismic
shaking that characterize convergence zones.
In the absence of such high water and sediment
discharges, the response to base-level fall is
accomplished by 'bottom-up' fluvial processes
of knickpoint retreat (Bishop, 2007). Such knickpoints, triggered and propagating headwards in
response to base-level fall, communicate the
rejuvenation up through the drainage net and
thence to the hillslopes. Such bottom-up responses
are found in tectonically active settings (e.g.
Bishop et al., 2005; Reinhardt et al., 2007), where

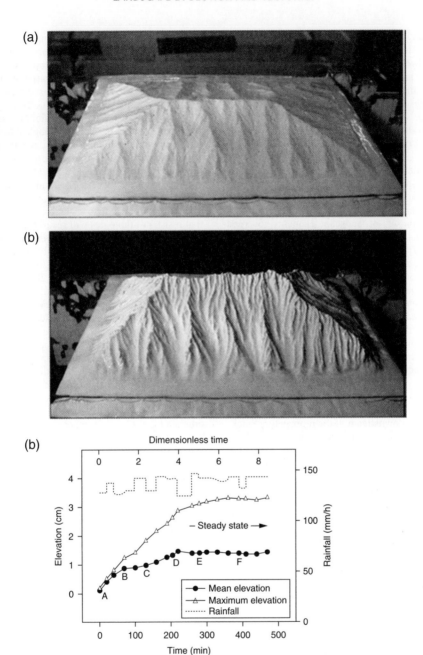

Figure 28.6 **(a) Development of steady-state topography in Bonnet and Crave's (2003) physical model of landscape development with rock uplift rate of 1.5 cm/h and under (top) high rainfall rate conditions (mean rainfall rate 166 ± 5 mm/h), and (bottom) low rainfall rate conditions (mean rainfall rate 98 ± 7 mm/h) (Bonnet and Crave, 2003: Figure 3). (b) Time sequence of development of steady-state topography in Bonnet and Crave's (2003) physical model of landscape development with rock uplift rate 1.5 cm/h and mean rainfall rate of 137 ±7 mm/h. The model evolves for ~200 min of model run, by which time the mean elevation of the model becomes constant. From that time, maximum elevation (the peaks in the model) asymptotically approach a constant value which they attain at ~350 min of model run. Attainment of constant mean elevation can be considered topographic steady-state. (Bonnet and Crave, 2003: Figure 1)**

water and sediment discharges are insufficient to maintain steady-state incision and hillslope lowering, or where tectonic activity, and the corresponding base-level falls, are intermittent. In these settings, the rate at which the hillslopes become fully adjusted to the base-level fall determines the landscape's overall response time for that base-level fall (Reinhardt et al., 2007).

Thus, orogenic settings may exhibit both top-down and bottom-up modes of response to rock uplift. In the former case, landscapes are in steady-state with gradients of hillslopes and channels adjusted to rock strength and rate of rock uplift, whereas the latter case is characterized by a complex mosaic of landscape elements governed by the rates at which (1) rejuvenation heads propagate through the drainage net and (2) the hillslopes respond to the passage of knickpoints in the channels. The lack of high rainfalls and seismic shaking and landslides in the latter case is the fundamental cause for the lack of landscape-wide steady-state topography: there is simply insufficient water and sediment to achieve bedrock channel incision at the rate of rock uplift and insufficient landsliding for hillslopes to be maintained at the critical angle corresponding to their rock strength. An episode of rock uplift (base-level fall) and the corresponding rejuvenation may lead to the development of steady-state topography downstream of the propagating knickpoint (Reinhardt et al., 2007), but the lack of continuous tectonic activity will likely mean that that steady-state landscape will not be maintained, until it is 'refreshed' by the next base-level fall and episode of knickpoint propagation.

That second type of orogenic setting, characterized by lower water and sediment discharges and/or more intermittent base-level fall, can be thought of as transitional to a completely post-orogenic setting, examples of which are widespread across the Earth's surface (Bishop, 2007). Such post-orogenic settings, which superficially resemble the classic Davisian situation (landscape evolution following the cessation of tectonic activity), are not experiencing active rock uplift and the landscape evolves by the interaction of surface processes and base-level falls driven by isostatic rebound in response to this denudation (e.g. Bishop and Brown, 1992). The recognition of the importance of such denudational isostasy as providing the ongoing base-level falls necessary for continued landscape evolution has been of fundamental importance in accounting for the extended landscape longevity that has been increasingly accepted over the last half-century or so (e.g. Twidale, 1998). The numerical modelling by Baldwin et al. (2003) points to several other factors, additional to denudational isostasy, as important in prolonging the timescales of

post-orogenic landscape evolution. These include a progressive change from detachment-limited to transport-limited conditions, as slopes and stream power decline, and changing magnitude-frequency characteristics of erosional and sediment transporting events. And we have already noted Crickmay's and Twidale's highlighting of how important are structural and lithological factors in focussing and/or slowing erosion and thus compartmentalizing the post-orogenic landscape into actively eroding elements, those that are slowly eroding and areas that are essentially dead. Under typical post-orogenic conditions, of low stream gradients and low hillslope angles, it is very likely that resistant lithologies will retard the passage of the signals of base-level fall (rejuvenation) through the drainage net, thus substantially slowing landscape evolution and prolonging the 'life' of the landscape (Bishop and Goldrick, 2010).

As we have already noted, denudational isostasy – the vertical uplift of rock in response to the unloading of the crust by denudation – is a key element in prolonging landscape evolution and was ignored by Davis and many subsequent workers. For lithosphere in isostatic equilibrium and free to respond to loading, the denudational removal of 1 m of rock across the landscape generates 0.8 m of lithospheric rebound (rock uplift) as a result of the density contrasts between the crust and the mantle. In other words, the denudation of 1 m across a landscape results in the lowering of that landscape's elevation by only 0.2 m. A further effect of such denudational isostasy is highly relevant to understanding landscape evolution, namely, the way in which localized denudational unloading by valley incision can drive the uplift of mountain peaks (Figure 28.7). Of course the latter effect cannot continue indefinitely because the mountain peaks will become too high and steep for the strength of the crust to support them and they will fail by major deep-seated sliding of crustal blocks of kilometre-scale thicknesses, in the process of tectonic denudation. Notwithstanding that point, the links between valley incision and peak uplift are a powerful way in which denudational isostasy drives landscape evolution. More generally, denudational isostasy is a key mechanism determining landscape longevity.

Bedrock rivers and tectonics

If landscapes in areas of high rates of rock uplift are time-independent with slopes standing on average at the limiting slope angle for the strength of the rock(s) of which they are composed, then higher rates of rock uplift cannot be expressed in a change in the morphology of hillslopes (because

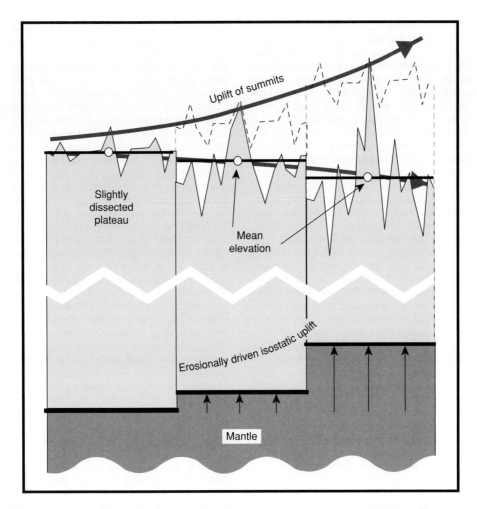

Figure 28.7   Concentration of erosional unloading of the lithosphere in valleys and limited erosion of the adjacent peaks can lead to uplift of those peaks. Note that any surface erosion must lead to an overall decline in mean surface elevation of an area in isostatic equilibrium and free to respond isostatically to the erosional unloading. In the case illustrated here, the mechanical strength of the lithosphere means that localized unloading leads to more regional isostatic response and so the peaks may rise as the lithosphere floats up regionally by 80 per cent of the regional unloading, which in this case is concentrated in the valleys. (Burbank and Anderson, 2001: Figure 10.26)

slopes are already at their limiting angle) (Burbank et al., 1996; Densmore et al., 1998). Higher rates of rock uplift *are* expressed in the geomorphology, however, in the long profile characteristics of the bedrock rivers of such areas. Bedrock rivers have been intensively studied over the last quarter of a century (e.g. Whipple, 2004). W.M. Davis (1899) had argued that river long profile gradient declines with the passage of his Cycle of Erosion and Hack argued that steeper rivers are associated with more resistant lithologies, but the focus of

recent work has been on the ways in which rock uplift is expressed in long profile characteristics. That recent work has been built around the well-known relationship between channel gradient and upstream drainage area known as Flint's law: $S = k_s A^{-\theta}$, where S is channel slope, A is catchment area (a surrogate for discharge), k_s is the steepness index and θ the concavity index (Howard and Kerby, 1983; Whipple, 2004). Whipple (2004) has summarized many bedrock channel studies and reported that: low concavities (<0.4) are associated with short, steep debris flow-influenced channels or with downstream increases in incision rate or rock strength; moderate concavities (0.4–0.7) typify bedrock channels being actively and uniformly uplifted; and high concavity bedrock channels (0.7–1.0) are associated with downstream decreases in rock uplift rate or rock strength and with disequilibria associated with a temporal decline in rock uplift rate. Thus, and notwithstanding the caveats presented by Whipple (2004), it is not unreasonable to seek a tectonic influence in bedrock river concavity.

Likewise, rock uplift rate is a determinant of bedrock river steepness (k_s in Flint's law, above) (Whipple and Tucker, 1999). Higher rock uplift rates are associated with steeper bedrock channels, as has been elegantly demonstrated by data from areas with a range of rates of rock uplift, such as the Mendocino triple junction region of northern California (Snyder et al., 2000) and the Siwalik Hills (Kirby and Whipple, 2001). Notwithstanding these relationships, Whipple (2004) has emphasized that there is still some way to go before a tectonic signature can be unequivocally read from bedrock river characteristics, with confounding factors including the range of ways in which channels can respond to changing rates of rock uplift (e.g. via changing channel width), and the role of lithology in setting bedrock river long profile concavity and steepness. Nonetheless, successes to date provide promising signs that aspects of rates of rock uplift will be discernible from long profile characteristics, thereby fulfilling an objective set many decades ago by Walther Penck.

Climate and tectonics

Much of the foregoing discussion of landscape evolution, and especially that on the research from the last few decades, has been built implicitly and explicitly around the broad notions of tectonics and climate. From Davis's time and earlier, the interaction of, or competition between, tectonics and climate has been seen as the key determinant of landscape evolution. Tectonics (Penck's exogenetic factor) is essentially rock uplift, which elevates the Earth's surface with respect to base level and thereby provides the potential energy for water (and regolith and ice) to flow downhill and do work in lowering the landscape. And it is climate that drives those landscape lowering processes (Penck's exogenetic factors). Steady-state landscapes are taken to reflect a balance between rock uplift and surface processes, and it is clear that tectonics strongly influences climate to the extent that tectonic activity is ultimately responsible for upland areas, which themselves are key elements in the generation of the world's

highest rates of precipitation. An intriguing current debate concerns the reverse interaction, the extent to which climate can be said to affect tectonics, via impacts on rock uplift. In areas of very high precipitation and very high rates of denudation, such as in monsoonal areas characterized by steady-state topography, the very high rates of denudation must be associated with substantial rates of denudational isostatic rebound (i.e. rock uplift) as the crust floats up isostatically in response to the denudation. In effect, climate is generating rock uplift as a result of the isostatic response to high rates of denudation (Figure 28.8). Erosionally controlled rock uplift (i.e. a strong link between climate, discharge and rock uplift) has been proposed by, for example, Wobus et al. (2003), Thiede et al. (2005) and Grujic et al. (2006). Gilchrist et al. (1994) assessed the degree to which the elevation of peaks in the European Alps can be explained by isostatic uplift in response to the excavation of Alpine valleys (as envisaged in Figure 28.7). For maximum isostatic response, the volume of the valleys must be a maximum: deep (excavated to base level) and wide (wider than the spacing of the peaks). The maximum valley cross-sections are at the edges of Alps; in other words, where the valleys are close to base-level denudational isostatic uplift is maximum. Valley cross-sections are smaller in the interior of the Alps, and overall Gilchrist et al. (1994) found that denudational isostatic uplift in response to the excavation of Alpine valleys can account for ~1000 m of the elevation of the highest Alpine peaks, and even less if the crust has significant flexural rigidity (mechanical strength). Others (e.g. Burbank et al., 2003) are likewise unconvinced that surface processes (i.e. erosion) exert a fundamental control on rock uplift rates, but this remains an intriguing area of research. Of course, in such areas of extreme rates of tectonic and surface processes, it is perhaps unexpected that steady-state landscapes on average experience little overall morphological change (or evolution). In effect, very high rates of rock uplift feed the crust through a landscape of essentially constant morphology that is being eroded by climate-driven forces as rapidly as the rock is uplifted.

Climate versus tectonics as major controls on rates of landscape processes

The debate summarized in the previous section, namely, whether climate can significantly affect rock uplift via the operation of the processes of denudation and denudational rebound of rock, is paralleled by a similar debate around whether

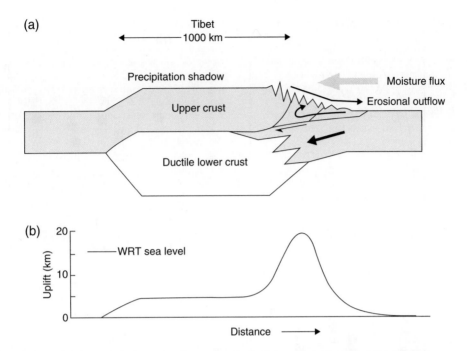

Figure 28.8 Upper panel shows a diagrammatic crustal section through the Himalayas, with the Himalayan front with peak mountain heights on the middle right and the Tibetan plateau at centre and centre left. The overall elevation of the Tibetan plateau and its fronting Himalayan mountain range is due to the double crustal thickness resulting from the collision of the Indian-Australian plate from the right with the Asian plate from the left – the thicker crust floats higher due to isostasy. That ~5 km of rock uplift is indicated in the lower plot. The high monsoonal precipitation on the Himalayan mountain front denudes that mountain front at very high rates leading to the second isostatic response, namely, very high rates of rock uplift in response to that denudational unloading, as indicated by the upward flow of crust in the upper right of the top diagram. (Adapted from Fielding, 2000: Figure 10.11)

climate or tectonics are major controls of the rates of surface processes. A range of data has been used to assess this issue, including compilations of rates of denudation for different tectonic and climatic settings, and cosmogenic nuclide analysis. Summaries of extensive data were presented by Summerfield (1991) to show that relief (i.e. slope angle) is a major control on rates of erosion, a conclusion reiterated by Summerfield and Hulton (1994) based on their compilation of modern rates of mechanical and chemical denudation (i.e. stream load data) for different settings. They found that basin relief is most strongly associated with both mechanical and chemical denudation rates, as is runoff, but less so. Denudation rates were found to be poorly associated with climatic measures (acknowledging, of course, that runoff is an expression of climate). Relief is an expression of tectonic setting and so

these results point strongly to the importance of tectonic setting as a control of denudation rate (see also Binnie et al., 2007). Von Blanckenburg (2005) likewise did not find an association between denudation and mean annual temperature or precipitation, nor, unlike the various findings noted above, with relief alone. The major association found by von Blanckenburg (2005) is between erosion rate and landscape rejuvenation (that is, rock uplift). As we have already seen, many studies report very high rates of denudation in the tectonically most active areas (namely, plate convergence zones) and the data noted here are consistent with that finding. It must be remembered, however, that those high rates of denudation, as we have also seen, reflect the combination of high rates of rock uplift (tectonics), high to extreme precipitation rates (climate) and seismic shaking (tectonics). The compilation of various

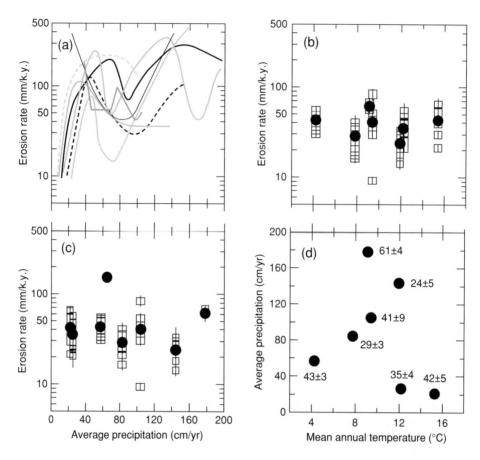

Figure 28.9 Plots from a test for climatic control of erosion rates according to Riebe et al. (2001). (a) Compilation of published relationships between erosion rate and mean annual precipitation; (b–d) cosmogenic nuclide-based erosion rate data plotted against various parameters of climate (Riebe et al., 2001). In (d) the measured rate is given beside each data point which is plotted against that site's mean annual rainfall and temperature

published relationships between erosion rate and mean annual precipitation (Figure 28.9a) by Riebe et al. (2001) confirms further the apparent lack of a meaningful relationship between erosion rate and climate. Their measured rates of erosion, based on cosmogenic nuclide analyses, likewise show little relationship with mean annual precipitation (Figures 28.9b–d).

FINAL WORD

This survey of the evolution of landforms, and of the approaches used to elucidate them, has highlighted the range of models and theoretical positions that have been adopted in understanding landscape evolution. The models remain difficult

to test but new techniques and approaches for understanding and quantifying landscape evolution bode well for future research. These recent techniques include numerical modelling and cosmogenic nuclide analysis, the use of high-resolution DEMs and ever-increasing computer power to characterize, quantitatively, the Earth's surface, ever more detailed palaeoclimatic data that extends deeper into geological time, and increasingly sophisticated understanding of Earth processes in the deep and shallow subsurface. It is easy to see how the plate tectonics revolution re-stimulated research into landscape evolution, because plate tectonics provides an elegant framework that brings together tectonics and climate, the two key factors in landscape evolution.

It is apparent from the foregoing that the different models of landscape evolution may all be applicable, depending on the climatic, tectonic and lithological character of the landscape in question. Teasing out the detail of which models are appropriate in which settings remains a substantial challenge, as does answering some of the major questions that underpinned the work of Davis, Penck, King and Crickmay. The big question that Penck posed – can the history of crustal movements be read from landscape morphology? – remains to be answered, but recent work has made useful advances in that area. Perhaps most curiously, the nature and evolution of the upland surfaces of low relief that so exercised Davis – his uplifted peneplains – remain enigmatic. They apparently preserve the morphology of an ancient low relief plain that has been uplifted, but several lines of evidence and analysis prompt caution with such an interpretation. Thus, the supposedly ancient (Tertiary-aged) upland surfaces of Scotland have relatively young cosmogenic nuclide exposure ages (all <~650 kyr) and rates of lowering of at least 30 m Myr^{-1} (Phillips et al., 2005). In other words, the apparently 'ancient' morphology of an apparently relict upland surface may not match the age of the surface, and an ancient plateau morphology may be retained as the surface is lowered (by, for example, solution, etch and other chemical weathering processes, as well as some glacial erosion). The 'age' of the 'present' surface may be much younger than the age of formation of the original surface the morphology of which is retained in the present surface. Weathering processes are lowering such a landscape – the landscape is evolving but its morphology is not changing (or is changing in very subtle ways).

Is it necessary for such surfaces, which so intrigued Davis and a host of subsequent workers, to have formed at low elevations (close to base level) and to have been subsequently uplifted, as is the almost universal interpretation? Young (1977) proposed that such a surface might be a planation surface that has been formed well above sea level, graded to resistant lithologies in the downstream portions of the drainage net that drains the surface.

Crucially, in modern geomorphology we have the techniques and over-arching paradigms (e.g. plate tectonics) that enable us to re-focus attention on landscape evolution. It is an exciting time for landscape evolution.

REFERENCES

Adams, J. (1985) Large-scale tectonic geomorphology of the Southern Alps, New Zealand, in M. Morisawa and J.T. Hack (eds) *Tectonic Geomorphology*. Allen and Unwin, Massachusetts. pp. 105–28.

Baldwin, J.A., Whipple, K.X. and Tucker, G.E. (2003) Implications of the shear stress river incision model for the timescale of postorogenic decay of topography. *Journal of Geophysical Research*, 108, 2158. doi:10.1029/2001JB000550.

Beaumont, C., Kooi, H. and Willett, S. (2000) Coupled tectonic-surface process models with applications to rifted margins and collisional orogens, in Summerfield M.A. (ed.) *Geomorphology and Global Tectonics*. John Wiley, Chichester. pp. 29–55.

Beckinsale, R.P. and Chorley, R.J. (1991) *The History of the Study of Landforms*. Vol. 3. Routledge, London.

Bierman, P.R. (1994) Using in situ produced cosmogenic isotopes to estimate rates of landscape evolution: A review from the geomorphic perspective. *Journal of Geophysical Research*, 99, 13885–96.

Bierman, P.R. and Nichols, K.K. (2004) Rock to sediment – slope to sea with ^{10}Be – rates of landscape change. *Annual Review of Earth and Planetary Science*, 32, 215–55.

Binnie, S.A., Phillips, W.M., Summerfield, M.A. and Fifield, L.K. (2007) Tectonic uplift, threshold hillslopes, and denudation rates in a developing mountain range. *Geology*, 35, 743–6. doi:10.1130/G23641A.1.

Birot, P. (1960) *The Cycle of Erosion in Different Climates*. (Translated Jackson, C.I. and Clayton, K.M. 1968). Batsford, London.

Bishop, P. (1980) Popper's principle of falsifiability and the irrefutability of the Davisian Cycle. *Professional Geographer*, 32, 310–315.

Bishop, P. (1985) Southeast Australian late Mesozoic and Cenozoic denudation rates: a test for late Tertiary increases in continental denudation. *Geology*, 13, 479–82.

Bishop, P. (1998) Griffith Taylor and the SE Australian highlands: testability of models of longterm drainage history and landscape evolution. *Australian Geographer*, 29, 7–29.

Bishop, P. (2007) Long-term landscape evolution: Linking tectonics and surface processes. *Earth Surface Processes and Landforms*, 32, 329–65. doi:10.1002/esp.1493.

Bishop, P. (2008) Tectonic and related landforms, in T.P. Burt, R.J. Chorley, D. Brunsden, N.J. Cox and A.S. Goudie (eds) *The History of the Study of Landforms*. Vol. 4. The Geological Society, London. pp. 55–105.

Bishop, P. and Brown, R. (1992) Denudational isostatic rebound of intraplate highlands: The Lachlan River valley, Australia. *Earth Surface Processes and Landforms*. 17, 345–60.

Bishop, P. and Goldrick, G. (2000) Geomorphological evolution of the East Australian continental margin, in M.A. Summerfield (ed.) *Geomorphology and Global Tectonics*, John Wiley, Chichester. pp. 227–55.

Bishop, P. and Goldrick, G. (2010) Lithology and the evolution of bedrock rivers in post-orogenic settings: Constraints from the high elevation passive continental margin of SE Australia, in P. Bishop and B. Pillans (eds) *Australian Landscapes*. Geological Society of London Special Publications 346, 267–287.

Bishop, P., Young, R.W. and McDougall, I. (1985) Stream profile change and long-term landscape evolution: Early Miocene and modern rivers of the east Australian highland crest, central New South Wales, Australia. *Journal of Geology*, 93, 455–74.

Bishop, P., Hoey. T.B., Jansen, J.D. and Artza, I.L. (2005) Knickpoint recession rates and catchment area: The case of uplifted rivers in Eastern Scotland. *Earth Surface Processes and Landforms*, 30, 767–78.

Bonnet, S. and Crave, A. (2003) Landscape response to climate change: Insights from experimental modeling and implications for tectonic versus climatic uplift of topography. *Geology*, 31, 123–6.

Büdel, J. (1982) *Climatic Geomorphology*. Princeton University Press, Princeton.

Burbank, D.W. and Anderson, R.S. (2001) *Tectonic Geomorphology*. Blackwell, Malden, MA.

Burbank, D.W., Leland, J., Fielding, E., Anderson, R.S., Brozovic, N. and Reid, M.R. (1996) Bedrock incision, rock uplift and threshold hillslopes in the northwestern Himalayas. *Nature*, 379, 505–10.

Burbank, D.W., Blythe, A.E., Putkonen, J., Pratt-Sitaula, B., Gabet, E., Oskin, M., Barros, A. and Ojha, T.P. (2003) Decoupling of erosion and precipitation in the Himalayas. *Nature* 426, 652–5.

Cerling, T.E. and Craig, H. (1994) Geomorphology and in-situ cosmogenic isotopes. *Annual Reviews of Earth and Planetary Science*, 22, 273–317.

Chorley. R.J. (1965) A re-evaluation of the geomorphic system of W.M. Davis, in Chorley, R.J. and Haggett, P. (eds) *Frontiers in Geographical Teaching*. Methuen, London. pp. 21–38.

Chorley, R.J., Dunn, A.J. and Beckinsale, R.P. (1964) *The History of the Study of Landforms*. Vol. 1. Methuen, London.

Chorley, R.J., Beckinsale, R.P. and Dunn, A.J. (1973) *The History of the Study of Landforms*. Vol. 2. Methuen, London.

Clifford, N. (2008) River channel processes and forms, in T.P. Burt, R.J. Chorley, D. Brunsden, N.J. Cox and A.S. Goudie (eds) *The History of the Study of Landforms*. Vol. 4. The Geological Society, London. pp. 217–324.

Cockburn, H.A.P. and Summerfield, M.A. (2004) Geomorphological applications of cosmogenic isotope analysis. *Progress in Physical Geography*, 28, 1–42.

Codilean, A., Bishop, P. and Hoey, T.B. (2006) Surface Process Models and the links between tectonics and topography. *Progress in Physical Geography*, 30, 307–33.

Cotton, C.A. (1942) *Climatic Accidents in Landscape Making*. 2nd edn. Whitcombe and Tombs, Christchurch.

Coulthard, T.J. (2001) Landscape evolution models: A software review. *Hydrological Processes*, 15, 165–73.

Crickmay, C.H. (1968) *Some Central Aspects of the Scientific Study of Scenery*. Author publication, Calgary.

Crickmay, C.H. (1974) *The Work of the River*. Macmillan, London.

Crickmay, C.H. (1975) The hypothesis of unequal activity, in W.M. Melhorn and R.C. Flemal (eds) *Theories of Landform Development*. Publications in Geomorphology, State University of New York, Binghamton. pp. 103–9.

Dadson, S.J., Hovius, N., Chen, H., Dade, W.B., Hsieh, M-L. and Willett, S.D. (2003) Links between erosion, runoff variability and seismicity in the Taiwan orogen. *Nature*, 426, 648–51.

Dadson, S.J., Hovius, N., Chen, H., Dade, W.B., Lin, J-C. and Hsu, M-L. (2004) Earthquake-triggered increase in sediment delivery from an active mountain belt. *Geology*, 32, 733–6.

Darwin, C. (1881) *The Formation of Vegetable Mould through the Action of Worms with Observations on Their Habits*. John Murray, London.

Davis, W.M. (1899) The Geographical Cycle. *Geographical Journal*, 14, 481–504.

Davis, W.M. (1900) Glacial erosion in France, Switzerland, and Norway. *Proceedings of the Boston Society of Natural History*, 29, 273–322.

Davis, W.M. (1905) The Geographical Cycle in an arid climate. *Journal of Geology*, 13, 381–407.

Densmore, A.L., Ellis, M.A. and Anderson, R.S. (1998) Landsliding and the evolution of normal fault-bounded mountains. *Journal of Geophysical Research: Solid Earth Planets*, 103, 15203–19.

Derbyshire, E. (ed.) (1973) *Climatic Geomorphology*. Macmillan, London.

Driscoll, N.W. and Karner, G.D. (1994) Flexural deformation due to Amazon fan loading: a feedback mechanism affecting sediment delivery to margins. *Geology*, 22, 1015–8.

Fielding, E.J. (2000) Morphotectonic evolution of the Himalayas and Tibetan PLateau, in M.A. Summerfield (ed.) *Geomorphology and Global Tectonics*, John Wiley, Chichester. pp. 201–22.

Gilchrist, A.R., Summerfield, M.A. and Cockburn, H.A. (1994) Landscape dissection, isostatic uplift, and the morphologic development of orogens. *Geology*, 22, 963–6.

Gilluly, J. (1955) Geologic contrasts between continents and ocean basins. *Geological Society of America*, *Special Paper*, 62, 7–18.

Gleadow, A.J.W. and Brown, R.W. (2000) Fission-track thermochronology and the long-term denudational response to tectonics, in M.A. Summerfield (ed.) *Geomorphology and Global Tectonics*. John Wiley, Chichester. pp. 57–75.

Goldrick, G. and Bishop, P. (2007) Regional analysis of bedrock stream long profiles: Evaluation of Hack's SL form, and formulation and assessment of an alternative (the DS form). *Earth Surface Processes and Landforms*. 32, 649–71.

Gosse, J.C. and Phillips, F.M. (2001) Terrestrial in situ cosmogenic nuclides: Theory and application. *Quaternary Science Reviews*, 20, 1475–560.

Grujic, D., Coutand, I., Bookhagen, B., Bonnet, S., Blythe, A. and Duncan, C. (2006) Climatic forcing of erosion, landscape, and tectonics in the Bhutan Himalayas. *Geology* 34, 801–4.

Hack, J.T. (1957) Studies of longitudinal stream profiles in Virginia and Maryland. *US Geological Survey Professional Paper* 294-B, 45–97.

Hack. J.T. (1960) Interpretations of erosional topography in humid temperate regions. *American Journal of Science*, 258-A, 80–97.

Hack. J.T. (1975) Dynamic equilibrium and landscape evolution, in W.M. Melhorn and R.C. Flemal (eds) *Theories of Landform Development*. Publications in Geomorphology, State University of New York, Binghamton. pp. 87–102.

Hancock, G.S., Anderson, R.S. and Whipple, K.X. (1998) Beyond power: Bedrock river incision process and form, in K.J. Tinkler and E.E. Wohl (eds) *Rivers Over Rock: Fluvial Processes in Bedrock Channels*. American Geophysical Union, Washington D.C. *Geophysical Monograph*, 107, 35–60.

Hartshorn, K., Hovius, N., Dade, W.B. and Slingerland, R.L. (2002) Climate-drive bedrock incision in an active mountain belt. *Science*, 297 2036–8.

Hettner, A. (1928) *The Surface Features of the Earth.* (Translated Tilley, P. 1972). Macmillan, London.

Holmes, A. (1913) *The Age of the Earth.* Harper, London.

House, M., Wernicke, B.P. and Farley, K.A. 1998. Dating topography of the Sierra Nevada, California, using apatite (U-Th)/He ages. *Nature*, 396, 66–9.

Hovius, N., Stark, C. and Allen, P. (1997) Sediment flux from a mountain belt derived by landslide mapping. *Geology*, 25, 231–4.

Howard, A.D. and Kerby, G. (1983) Channel changes in badlands. *Geological Society of America Bulletin*, 94, 739–52.

Hutton, J. (1788) Theory of the Earth; or an investigation of the laws observable in the composition, dissolution, and restoration of land upon the globe. *Transactions of the Royal Society of Edinburgh*, 1, 209–304.

Judson, S. and Ritter, D.F. (1964) Rates of regional denudation in the United States. *Journal of Geophysical Research*, 69, 3395–401.

King, L.C. (1962) *The Morphology of the Earth*, Oliver and Boyd, Edinburgh.

Kirby, E. and Whipple, K. (2001) Quantifying differential rock-uplift rates via stream profile analysis. *Geology*, 29, 415–8.

Kooi, H. and Beaumont, C. (1996) Large-scale geomorphology: Classical concepts reconciled and integrated with contemporary ideas via a surface process model. *Journal of Geophysical Research*, 101, 3361–86.

Leopold, L.B., Wolman, M.G. and Miller, J.P. (1964) *Fluvial Processes in Geomorphology*. Freeman, San Francisco.

Lin, J-C. (2000) Morphotectonic evolution of Taiwan, in Summerfield, M.A. (ed.) *Geomorphology and Global Tectonics*. John Wiley, Chichester. pp. 135–46.

Lyell, C. (1833) *Principles of Geology*. Vol. 3. John Murray, London.

Melhorn, W.N. and Edgar, D.E. (1975) The case for episodic, continental scale erosion surfaces: a tentative geodynamic model, in W.M. Melhorn and R.C. Flemal (eds) *Theories of Landform Development*. Publications in Geomorphology, State University of New York, Binghamton. pp. 243–72.

Niedermann, S. (2002) Cosmic-ray-produced noble gases in terrestrial rocks: Dating tools for surface processes. *Reviews in Mineralogy and Geochemistry*, 47, 731–84.

Nott, J. and Roberts, R.G. (1996) Time and process rates over the past 100 Myr: A case for dramatically increased landscape denudation rates during the late Quaternary in northern Australia. *Geology*, 24, 883–7.

Nott, J., Young, R. and McDougall, I. (1996) Wearing down, wearing back, and gorge extension in the long-term denudation of a highland mass: Quantitative evidence from the Shoalhaven catchment, southeast Australia. *Journal of Geology*, 104, 224–32.

Palmquist, R.C. (1975) The compatibility of structure, lithology and geomorphic models, in W.M. Melhorn and R.C. Flemal (eds) *Theories of Landform Development*. Publications in Geomorphology, State University of New York, Binghamton. pp. 145–67.

Pazzaglia, F.J. and Gardner, T.W. (2000) Late Cenozoic landscape evolution of the US Atlantic passive margin: Insights into a North American Great Escarpment, in M.A. Summerfield (ed.) *Geomorphology and Global Tectonics*. John Wiley, Chichester. pp. 283–302.

Peltier, L.C. (1950) The geographic cycle in periglacial regions as it is related to climatic geomorphology. *Annals of the Association of American Geographers*, 40, 214–36.

Penck, W. (1924) *Morphological Analysis of Landforms.* (Translated H. Czech and K.C. Boswell 1953). Macmillan, London.

Phillips, J. (1995) Nonlinear dynamics and the evolution of relief. *Geomorphology*, 14, 57–64.

Reinhardt, L.J., Bishop, P., Hoey, T.B., Dempster, T.J. and Sanderson, D.C.W. (2007) Quantification of the transient response to base-level fall in a small mountain catchment: Sierra Nevada, southern Spain. *Journal of Geophysical Research–Earth Surface*, 112, F03S05. doi:10.1029/ 2006JF000524.

Riebe, C.S., Kirchner, J.W., Granger, D.E. and Finkel, R.C. (2001) Minimal climatic control on erosion rates in the Sierra Nevada, California. *Geology*, 29, 447–50.

Schumm, S.A. (1963) The disparity between present rates of denudation and orogeny. *US Geological Survey Professional Paper*, 454-H, 1–13.

Shaler, N.S. (1899) Spacing of rivers with respect to the hypothesis of base-leveling. *Geological Society of America Bulletin*, 19, 263–76.

Simons, M. (1962) The morphological analysis of landforms, a new review of the work of Albrecht Penck (1888–1923). *Transactions of the Institute of British Geographers*, 31, 1–14.

Sklar, L. and Dietrich, W.E. (2001) Sediment supply, grain size and rock strength controls on rates of river incision into bedrock. *Geology*, 29, 1087–90.

Snyder, N., Whipple, K., Tucker, G. and Merritts, D. (2000) Landscape response to tectonic forcing: digital elevation model analysis of stream profiles in the Mendocino triple junction region, northern California. *Geological Society of America Bulletin*, 112, 1250–63.

Strahler, A.N. (1952) Dynamic basis of geomorphology. *Geological Society of America Bulletin*, 63, 923–38.

Summerfield, M.A. (1991) *Global Geomorphology*. Longman, Harlow.

Summerfield, M.A. (ed.) (2000) *Geomorphology and Global Tectonics*. John Wiley, Chichester.

Summerfield, M.A and Hulton, N.J. (1994) Natural controls of fluvial denudation rates in major world drainage basins. *Journal of Geophysical Research*, 99, 13871–83.

Tarr, R.S. (1898) The peneplain. *American Geologist*, 21, 351–70.

Thiede, R.C., Bookhagen, B., Arrowsmith, J.R., Sobel, E.R. and Strecker, M.R. (2004) Climatic control on rapid exhumation along the Southern Himalayan Front. *Earth and Planetary Science Letters*, 222, 791–806.

Thornbury, W.D. (1969) *Principles of Geomorphology*. 2nd edn. John Wiley, New York.

Tippett, J.M. and Hovius, N. (2000) Geodynamic processes in the Southern Alps, New Zealand, in Summerfield, M.A. (ed.) *Geomorphology and Global Tectonics*. John Wiley, Chichester. pp. 109–34.

Tricart, J., Cailleux, A. and Kiewietdejonge, C.J. (1972) *Introduction to Climatic Geomorphology*. (Translated Kiewiet de Jonge, C.J.) Longman, Harlow.

Twidale, C.R. (1976) On the survival of palaeoforms. *American Journal of Science*, 276, 77–94.

Twidale, C.R. (1991) A model of landscape evolution involving increased and increasing relief amplitude. *Zeitschrift für Geomorphologie*, 35, 85–109.

Twidale, C.R. (1998) Antiquity of landforms: an 'extremely unlikely' concept vindicated. *Australian Journal of Earth Sciences*, 45, 657–68.

Twidale, C.R. (1999) Landforms ancient and recent: the paradox. *Geografiska Annaler*, 81A, 431–41.

Twidale, C.R. (2004) Reinventing the wheel: Recurrent conception in geomorphology. *Earth Sciences History*, 23, 297–313.

Twidale, C.R. and Lageat, Y. (1994) Climatic geomorphology: a critique. *Progress in Physical Geography*, 3, 319–34.

von Blanckenburg, F. (2005) The control mechanisms of erosion and weathering at basin scale from cosmogenic nuclides in river sediment. *Earth and Planetary Science Letters*, 237, 462–79. doi:10.1016/j.epsl.2005.06.030.

Whipple, K.X. (2004) Bedrock rivers and the geomorphology of active orogens. *Annual Review of Earth and Planetary Sciences*, 32, 151–85. doi 10.1146/annurev.earth.32.101802.120356.

Whipple, K.X. and Tucker, G.E. (1999) Dynamics of the stream-power incision model: implications for height limits of mountain ranges, landscape response timescales, and research needs. *Journal of Geophysical Research*, 104, 17661–74.

Willgoose, G. (2005) Mathematical modeling of whole landscape evolution. *Annual Review of Earth and Planetary Sciences*, 33, 443–59.

Wobus, C.W., Hodges, K.V. and Whipple, K.X. (2003) Has focused denudation sustained active thrusting at the Himalayan topographic front? *Geology*, 31, 861–4.

Young, R.W. (1977) Landscape development in the Shoalhaven River catchment of southern New South Wales. *Zeitschrift für Geomorphologie*, 21, 262–83.

Young, R.W. and McDougall, I. (1982) Basalts and silcretes on the coast near Ulladulla, southern New South Wales. *Journal of the Geological Society of Australia*, 29, 425–30.

Young, R.W. and McDougall, I. (1993) Long-term landscape evolution: Early Miocene and modern rivers in southern New South Wales, Australia. *Journal of Geology*, 101, 35–49.

Zeitler, P.K., Chamberlain, C.P. and Smith, H.A. (1993) Synchronous anatexis, metamorphism, and rapid denudation at Nanga Parbat (Pakistan Himalaya). *Geology*, 21, 347–50.

Zeitler, P.K., Meltzer, A.S., Koons, P.O., Craw, D., Hallet, B. and Chamberlain C.P. (2001) Erosion, Himalayan geodynamics, and the geomorphology of metamorphism. *GSA Today*, 11, 4–9.

29

Interpreting Quaternary Environments

Anne Mather

Deposits and associated landforms of the late Cenozoic (the last 2.6 Ma, also termed the Quaternary, see below) potentially provide a high-resolution database of environmental change if we can successfully interpret them. These records are locked in both terrestrial (ice cores, glacial moraines; river terraces, raised beaches; organic deposits; tufa/travertine deposits; speleothems, cave deposits) and subaqueous (marine and lacustrine) sequences. Problems arise, however, particularly with the land-based sequences, in that the record is laterally and vertically discontinuous. This poses problems for placing any palaeoenvironmental data into a suitable temporal and spatial scale for interpretation beyond individual sites. Thus a reliable stratigraphic approach is essential to any palaeoenvironmental interpretation. More recently the development of new dating methods has helped resolve some of the issues by facilitating chronostratigraphic correlations over wider areas. There have also been major steps forward in interrogating the sedimentary record in terms of its fauna, flora and sedimentology. These new approaches offer quantitative reconstructions of the palaeoenvironment based on analysis of, for example, stable isotopes or transfer functions. Many of these approaches to interpreting the geological record are unique to Quaternary studies as they have been adopted to maximize the high-resolution record of the Quaternary, and particularly the Holocene timescale of the deposits. These methodologies are thus suitable for other studies operating on similar, high-resolution records such as archaeological studies.

It is beyond the scope of this chapter to include interpretation of environments based on geomorphological (landform) approaches and sedimentological (facies) analysis of field sections. Elements of these areas are dealt with in other chapters within this handbook. Instead this chapter focuses on the laboratory-based approaches used to facilitate environmental interpretation of Quaternary sediments from field section and core. This chapter examines the stratigraphic approaches that have been used to provide a temporal and spatial template from which to interpret Quaternary environmental change on regional and global scales. This will be followed by an examination of some of the key sedimentological and palae-oecological approaches and techniques which provide the backbone to detailed environmental reconstructions.

STRATIGRAPHIC APPROACHES: PROVIDING THE TEMPORAL AND SPATIAL FRAMEWORK

Terminology

In 1839 Lyell proposed the term 'Pleistocene' to describe the unconsolidated deposits overlying bedrock, and 'Recent' to describe the younger material overlying the Pleistocene. The term 'Holocene' (meaning 'wholly recent') was coined by the third Geological Congress in London in 1885 as the more official name for the 'Recent' deposits (Gibbard and Van Kolfschoten, 2004).

The stratigraphic usage of the term 'Quaternary' and its chronostratigraphic status has recently been revised (Gibbard et al., 2010). In the latest revision the term Quaternary is used to refer to materials deposited over the last 2.58 Ma, embracing the interval of oscillating glacial and interglacial episodes. It is this revised definition which will be used within this chapter.

Approaches

The study of late Cenozoic environments necessitates a temporal as well as spatial awareness of environmental change. The ordering of the evidence at a locality provides the temporal stratigraphic framework. How that relates to adjacent localities and its spatial relationship involves correlation. Stratigraphy lends itself well to the more continuous older geological record. However, the Quaternary record, particularly the continental record, is typically fragmentary in nature leading to problems in erecting stratigraphies and correlating them. Stratigraphies are thus most successfully applied to the more continuous Quaternary records such as marine cores but these can be correlated with terrestrial records using marker horizons (e.g. magnetic reversals, radiometric dating, tephra layers). Initially stratigraphy relied on describing the visible attributes of the sections (e.g. colour, grain size, bedding) that is, 'lithostratigraphy'. With the advent of new techniques, however, new methods of stratigraphic reconstruction have developed. Whilst there are codes of practice to aid in stratigraphy, most of these are based on material older and of lower temporal resolution than the Quaternary. The challenge for the Quaternary is that the record is often fragmentary and lateral continuity poor, for example, for terraces in terrestrial sequences, the sediments are preserved at a far greater resolution than older geological sequences. Boundaries between strata in older geological units are taken in areas of lithological change. However, in many Late Cenozoic sequences these stratal definitions are not concordant with fossil ranges or other stratal definitions (e.g. biostratigraphic) and may be time transgressive. As a result the standard use of stratotypes applied to older geological sequences has generally not been applied to Quaternary sequences. A brief resumé of the different stratigraphic approaches is outlined below and summarized in Table 29.1.

Lithostratigraphy

Lithostratigraphic units should be defined on the basis of sediment properties such as grain size, grain shape and Munsell colour. Unfortunately,

however, the Quaternary has suffered from the classification of sedimentary units by their inferred mode of origin (e.g. till, fluvial etc.) rather than relying on primary outcrop observations. Lithostratigraphy based on inference rather than observation is open to interpretation error. In the more recent literature this practice has become less common and has employed sedimentological terms without genetic inference. The genetic interpretation comes after careful examination of the sedimentology, often through facies analysis. Lithostratigraphic approaches are often strengthened by incorporating other dating techniques and approaches such as event stratigraphy, magnetostratigraphy or biostratigraphy.

Biostratigraphy

Biostratigraphy within a Quaternary framework has traditionally involved the use of pollen and molluscs. Pollen as a stratigraphic tool was pioneered in the United Kingdom in the 1950s and 1960s. These pioneering works used high-resolution records in lacustrine sediments, but it has proved more difficult to extrapolate this success to more incomplete records, for example, in Quaternary river environments (Schreve and Thomas, 2001). Although generally rarer in Quaternary sequences than microfauna and flora, vertebrates have had a rapid evolutionary turnover making them valuable stratigraphic indicators (Lister, 1992). Mollusca, in contrast, have undergone limited change over Quaternary timescales. As a result mollusca have been mostly used as 'assemblage zones' which reflect the ecological response of organisms to environmental change. The potential stratigraphic problems with this approach lie in the fact that these environmental changes are reversible and may be repeated so that the same assemblage may appear at different stratigraphic levels. Also, because biozones are time-transgressive, they can cut across other stratigraphic units. Other forms of biostratigraphy use an event approach and are based around climatostratigraphy (see below).

Morphostratigraphy

A morphostratigaphic unit is a body of sediment that can be associated with a particular surface form (Willman and Frye, 1970). Initially, this term was developed for geological surveyors who had problems applying stratigraphy to glacial sediments using lithostratigraphic approaches alone; mainly due to the often lateral stratigraphy presented by moraines as a result of ice retreat. Typically the morphostratigraphy has to be integrated with the lithostratigraphy and/or biostratigraphic approaches to fully interpret a sequence.

Table 29.1 Stratigraphic approaches applied to the Quaternary as discussed in this chapter

Approach	Stratigraphic subdivision within Quaternary studies	Comments
Lithostratigraphy	Lithology: sediment properties such as colour, grain size, grain shape, bedding	Environmental interpretations should be based on detailed sedimentology (e.g. facies analysis) rather than inferred origin
Biostratigraphy	Flora/fauna. Evolutionary changes or assemblage zones	Stratigraphies most successfully erected using planktonic foraminifera and calcareous nannoplankton. Also uses pollen, molluscs and vertebrates. Molluscs tend to be used as 'assemblage zones'. Problems arise as biozones are time-transgressive and assemblages may be repeated
Morphostratigraphy	Surface form	Needs to be integrated with other approaches (lithostratigraphy/biostratigraphy)
Pedostratigraphy and magnetostratigraphy	Soils and magnetic properties	Soils are of local to regional significance in correlations but combined with megnetostratigraphy can give global scale stratigraphies. Polygenetic soils may be problematic in a stratigraphic context
Marine isotope stratigraphy	Marine oxygen isotope stages	A form of climatostratigraphy and chemical stratigraphy. Uses the value and pattern of oxygen isotope values to infer climate. Rapid changes (terminations) can be used as marker horizons
Chronostratigraphy	Inferred age	Correlated by isochronous surfaces. Chronozones used where biozones dated by radiometric techniques. Problems with dating methods (age and error range)
Event stratigraphy	Isochronous events	Mainly used to correlate regional stratigraphies. Uses a specific recognizable event so typically used with tephra, biozones, isotope spikes etc.

Soil/pedo-stratigraphy and magnetostratigraphy

Palaeosols have been used to define soil stratigraphic units/pedostratigraphic units (North American Commission on Stratigraphic Nomenclature, 1983) and act as time–stratigraphic markers (Kraus, 1999). Commonly, much of the stratigraphy is established by integrating soil and magnetic data. Magnetostratigraphy uses stratigraphical variations in the magnetic properties of rocks for correlation. These properties enable global-scale correlations within and between marine and terrestrial sequences (Maher, 1998). However, it is not a very-high-resolution technique and is restricted to specific rock types or depositional environments. The soils are often identified using physical description, chemical, magnetic (Table 29.2) and micromorphological characteristics. Generally, soils tend to be used on a regional scale as the degree of soil development varies locally to regionally depending on parent material, climate, slope etc. Soils usually represent a depositional hiatus, and as such

Table 29.2 Iron oxides commonly found in soils and their magnetic susceptibilities (from Maher, 1998)

Mineral	Magnetic susceptibility $(10^{-8}\ m^3\ kg^{-1})$
Hematite	40
Goethite	70
Maghemite	26,000
Lepidocrocite	70
Ferrihydrite	40
Magnetite	56,500

may contain polygenetic soils developing under different environments. Whilst this may make a good marker horizon it also raises issues over the use of the soil as a stratigraphic marker. For example, the Sangamon soil of North America was considered to be of the last interglacial age. However, more recent work revealed it developed during the last interglacial but also in preceding and succeeding cold stages (Follmer, 1983).

Climatostratigraphy

Climatostratigraphy is recognized by the American Commision on Stratigraphic Nomenclature (1970), as 'a widespread climatic episode defined from a subdivision of Quaternary rocks'. Originally, it was envisaged that the boundaries for the climatic change would follow biostratigraphic changes. However, the higher resolution of Quaternary sediments compared to older geological successions has meant that this does not occur. There are a number of issues in using this type of stratigraphy for regional and continental scales. First, it assumes that the cause of difference between units is climatically driven. This makes the stratigraphy reliant on interpretations (and thus potential errors in interpretation) not observations. Also climate may not be the primary driver of changes within the local stratigraphy. In south-east Spain, for example, significant differences in sedimentology within fluvial deposits have been attributed to the impact of river capture (Mather, 2000) driven by the regional tectonics. Also the boundaries will be diachronous, making correlation with chronostratigraphic units unrealistic, although this has often formally been done. There is also the added complication that continental records such as glaciogenic sediments, represent only sediments deposited during deglaciation whereas in more continuous records, such as lake sediments, a more complete record may be present. As a result of the above the use of climatic units has been abandoned by the North American Commission on Stratigraphic Nomenclature (1983) but remains in the form of oxygen isotope stages (see below).

Marine isotope stratigraphy

Emiliani (1955) provided the pioneering work on isotopic profiles in deep-ocean cores and divided the cores into marine isotope stages (MISs). Formal stratigraphy now takes the MISs to 116, dating back to 7.73 Ma (Patience and Kroon, 1991). The MIS for the Quaternary can be seen in Figure 29.1. The warm stages are associated with lighter $\delta^{18}O$ and odd numbers, and the heavier $\delta^{18}O$ is associated with cold stages and even numbers. These numbers are counted from the top. Since the original code was established substages have been identified that represent colder/ warmer periods within a general interglacial/ glacial. More recent developments have given these excursions odd and even numbers depending on whether they are a positive or negative excursion. The MISs form a signal that can

be recognized in a wide range of ocean cores, underlining their significance as a proxy climate signal. The rapid changes at the end of a glacial to interglacial transition are referred to as 'terminations'. These events, numbered from the top down, now constitute key marker horizons.

Within ice cores the reciprocal arrangement of oxygen isotopes can facilitate correlation. In Vostok Station (Antarctica) the record extends back some 420 ka (Figure 29.1). Correlation between marine and terrestrial records is further strengthened by dust horizons in cores (Petit et al., 1990) which can also be found in cores from the subpolar Indian Ocean. Evidence from the Greenland summit ice-core research programmes (GRIP and GISP2) suggests strong correlation between ice and marine cores. Bond et al. (1992) showed changes in the Greenland ice sheet matched changes in the North Atlantic sea surface temperatures over the last 90 ka. This correlation is further strengthened by marker horizons ('Dansgaard-Oeschger cycles' in the ice cores and 'Heinrich layers' in the ocean sediment) (Figure 29.2).

Chronostratigraphy

Chronostratigraphy relies on classifying the strata in terms of time. The chronostratigraphic units are correlated by isochronous surfaces. These can be biozone or lithological boundaries. Like climatostratigraphy, chronostratigraphy is inferred from the original sediments rather than being directly based on field observations. Most commonly chronozones have been used where biozones have been dated by radiometric research. Most problems have arisen with the absolute dating of such units. Many methods are only applicable for limited parts of the timescale, and sometimes the error range on the date may exceed the stratigraphic interval. Such issues can lead to problems in correlation.

Event stratigraphy

Features relating to a specific event (e.g. stable isotope spike; tsunami, flood, volcanic eruption, appearance/disappearance of climatically indicative fauna/flora) are valuable in stratigraphic reconstruction. The key use of event stratigraphy is in correlating regional stratigraphies. Often techniques may be integrated, for example, tephrochronology and biostratigraphy. Tephra layers are isochronous layers which can be recognized in a wide variety of environments making them good marker horizons on land and in lakes and marine environments, facilitating correlation between land and marine based records (Einarsson, 1986). However, the range for an individual

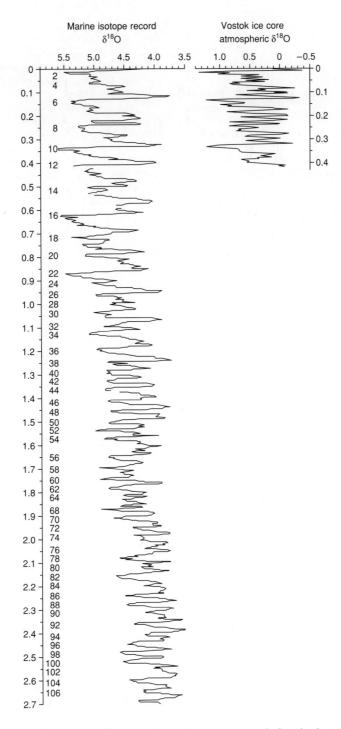

Figure 29.1   Marine isotope stratigraphy from the Late Cenozoic for the last 2.7 million years and the continental Vostok ice core record. The continental record is being extended and used to refine timing of the isotope stages using carbonate landbased records which can be more accurately dated by uranium-series methods such as speleothems. (Simplified from Gibbard and Van Kolfschoten, 2004)

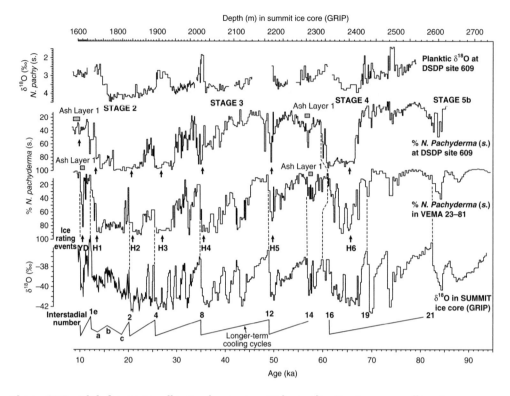

Figure 29.2   High-frequency climate change events in marine Quaternary sedimentary records (Heinrich events) and the GRIP Summit ice core (Dansgaard–Oeschger events). The lowermost plot demonstrates the sawtooth nature of 'Bond cycles'. (Adapted from Bond et al., 1993)

tephra is related to the proximity to the eruptive source, hence it has mainly been used on a local or regional scale. To correlate on a larger scale we need to use global events such as magnetic reversals in sediments or 'Heinrich events' (Figure 29.2). The latter are layers of carbonate-rich debris located in the North Atlantic ocean sediments which reflect major ice rafting events (Heinrich Events) during the last cold stage (Bond et al., 1992). Such global events facilitate correlation between off-shore and on-shore records. Other correlations include stable isotopes of oxygen, carbon, sulfur and strontium as stratigraphic markers (Odin et al., 1982; Walker et al., 1999). Biostratigraphic markers may also be important. For example, the decline of *Ulnus* in Holocene pollen records around 5 ka in western Europe (Huntley and Birks, 1983).

It is clear that on regional and continental scales lithostratigraphy, biostratigraphy and climatostratigraphy are not sufficiently sensitive tools with which to develop meaningful time–stratigraphic correlations. A geochronological

basis is required. However, the range of dating tools available to the researcher depends very much upon the nature of the sediments and each method is applicable over different timescales (see Chapter 11). The adoption of multi-disciplinary techniques and approaches together with the development of new dating techniques such as cosmogenic dating which cover much wider time-ranges and are applicable over a range of different materials may help overcome some of these issues.

LITHOLOGICALLY BASED APPROACHES AND THE QUATERNARY

Sedimentary sections in the field can yield valuable information on the rates and styles of processes operating at the time of deposition and may facilitate reconstruction of former flow events. This section will focus on inorganic, clastic and authigenic deposits (Table 29.3).

Table 29.3 Approaches and techniques commonly used in interpreting the environments of Quaternary deposits as dealt with in this chapter

Approach	Quaternary archive	Techniques
Lithological	Provenance • Continental and marine clastic sediments	• Heavy minerals • Magnetic susceptibility • Clay minerals and geochemistry
	Depositional process • Continental and marine clastic sediments	• Particle characteristics (shape, size, surface texture) • Geochemistry
	Proxy climate • Lacustrine carbonates • Cave deposits	• Stable isotopes
Palaeoecological	Fauna • Insects (beetles, midges) • Ostracods	Proxy climate indicators • Transfer functions • Mutual temperature range (MTR) • Stable isotopes • Trace elements
	Protists • Foraminifera • Testate amoebae	
	Flora • Diatoms • Playnology • Phytoliths • Fossil wood/charcoal	

The interpretation of sediments associated with specific sedimentary environments (fluvial, glacial etc.) will not be dealt with individually within this section as these are dealt with in earlier chapters.

Provenance indicators

In reconstructing the regional environmental context of field sections and regional palaeogeographies the provenance of sediments becomes important. Traditionally this was done by examining the lithology of larger clasts such as glacial erratics in the field. The term 'erratic' derives from the French term 'terrain erratique' which was used initially by the geologist de Saussure in the late 18th century to describe areas where exotic material overlie local bedrock. On a smaller scale lithological analysis of courser clasts within sedimentary sequences is still a good way of indicating provenance, particularly where combined with other sedimentary evidence such as palaeoflow indicators. There are, however, potential dangers of sediment reworking so that the sediments being analysed may not reflect their true provenance. Thus most analyses of provenance will use more than one line of supporting evidence.

Heavy minerals

Heavy minerals have a specific gravity greater than 2.85. The presence of heavy minerals in both core and field outcrop within late Cenozoic sediments has been used to determine the provenance of fluvial deposits, marine turbidites, lake deposits, glacial till and aeolian sediments. The presence of heavy minerals and their association with particular lithofacies within a sedimentary environment has led to their use as environmental indicators. For example Frihy et al. (1998) linked particular suites of heavy minerals to their source area and with their associated environments and then successfully tested the use of the mineral assemblages on their own, in core, to infer sedimentary environment. Clearly there are weaknesses in this approach as the heavy minerals alone are not environmental indicators of the sediments in which they are located and may be reworked. Other researchers have combined heavy mineral analysis with examination of grain surfaces to strengthen interpretations of transport and provenance rather than relying solely on the presence of the heavy minerals. Heavy mineral presence and their source area information have been successfully applied to linking climate controls with drainage development (Foucault and Stanley, 1989).

Magnetic susceptibility

Magnetic susceptibility of minerals is easily measured in the core and in the field and has been successfully used, typically in conjunction with other basic sedimentological analysis such as grain size, as a provenance indicator. Thus mineral magnetics have played an important role in identifying the source of Quaternary marine terrigenous debris, particularly associated with glacial origins. Commonly, the magnetics are used in conjunction with other analyses such as Sr and Nd isotopes (Walter et al., 2000) and core composition. Magnetic susceptibility has also been used on lake cores to indicate periods of enhanced soil erosion into lake environments. Some of the pioneering research in this area was undertaken by Mackereth (1965) and Pennington et al. (1972). They demonstrated how changes in the chemical composition of lake sediments could be better understood if the sediments were regarded as sequences of catchment derived soils. Increased soil erosion transfers large amounts of relatively unweathered material into lake basins, and the mineral fraction is dominated by elements derived from the bedrock and exposed sediment (Na, K, Mg, Ca, Fe, Mn). The quantity, size and mineralogy of associated ferromagnetic particles associated with such erosion are reflected in magnetic susceptibility, measurements which are easily measured in core. During periods of lower erosion more organic carbon would be present and relatively fewer metal elements.

Mineralogy and geochemistry

Clay mineralogy can provide important information on provenance. It has been used to identify environment of the provenance in coastal areas (Stanley et al., 1998) and how this changes with changing sea level (Wang et al., 2006). The research by Stanley et al. showed that clay assemblages are not uniformly distributed in the south-east Mediterranean and could be tied into specific source areas. Smectite-rich clay assemblages east of the Nile delta were derived from storm wave and coastal current erosion of the Nile delta, reworking of Quaternary deposits along the coast and on the seafloor east of the delta, and from specific Israeli rivers between Tel Aviv and Atlit. Alternatively, kaolinite and illite at offshore sites were supplied from erosion of coastal cliff sections, river input between Wadi El Arish in Sinai and the Lebanon–Israel border, and from wind-borne dust from African and Middle East deserts released seaward of the coast. Clay minerals have also been used successfully in identifying the sources of glacial tills and loess (Haldorsen et al., 1989; Dilli and Pant, 1994). Most commonly the clay minerals are used in conjunction

with other geochemistry such as Sr–Nd isotopes or Al, Na and K (Huisman and Kiden, 1998) to add strength to provenance identification in cores. Indeed combined with geochemistry the clay minerals can act as a powerful tool. Underwood and Pickering (1996) used this approach to identify the source (marine sediments and volcanic inputs) of clays comprising the late Cenozoic trench-wedge and associated sediment dispersal patterns of an accretionary prism off south-west Japan.

Mineralogy has aided in the interrogation of lacustrine sediments. In a maar (volcanic crater) lake in Turkey, for example, Roberts et al. (2001) used mineralogy to identify hydroclimatic controls on lake geochemistry and episodic volcanic/geothermal activity. Changes from aragonite to high magnesium calcite to dolomite are interpreted as mirroring increasing magnesium concentration in the water due to increasing aridity and impact of evaporation. The presence of quartz (non-biogenic silica) which is not available in the immediate vicinity points towards aeolian input into the lake. Associations between peaks in talc are related to increases in geothermal activity within the crater, which in turn may have led to dissolution of diatoms releasing peaks in biogenic silica.

Depositional process indicators

The sedimentology of a section and the bedforms therein can yield key data on the depositional processes and thus aid environmental interpretation. It is beyond the scope of this chapter to examine the use of sedimentary structures in detail. This section will focus on the analyses most commonly carried out on core, which is typically the finer grain size fraction (sands, silts and clays). To produce fine grain sizes such as silt in nature typically requires glacial grinding, or intense weathering processes in high, cold, tectonically active mountain regions (e.g. Asia), although some silt and loess production occurs in hot deserts. Supplies of dust were thus prolific in the Quaternary.

Particle characteristics

The most commonly used particle characteristics to infer depositional process are grain size, grain shape, and grain surface texture. Grain size distribution can reflect the energy of deposition. However, care should be taken in such interpretations as the source area characteristics are a key control on grain size and may thus limit the available grain size within geomorphic systems (e.g. Mather and Hartley, 2005).

Records of dust can be obtained from ice cores, marine sediments, and terrestrial (loess) deposits. These records document a variety of processes (source, transport and deposition) in the dust cycle, stored in each archive as changes in clay mineralogy, isotopes, grain size, and concentration of terrigenous materials. Changes in the dust flux and particle size are typically controlled by a combination of source area extent, wind speed, atmospheric transport, atmospheric residence time and/or relative contributions of dry settling and rainout of dust (see, for example, Ding et al., 2000). Vriend and Prins (2005) used grain size distribution and a modeling algorithm to 'unmix' sediment from multiple dust sources and sand fraction abundance has been used to track the expansion and contraction of deserts (e.g. the Tengger Desert; Xu et al., 2007). Where these data are related to specific geochemistry, provenance can be clearly identified, such as the desiccated and shrinking palaeolakes which supplied aeolian dust to the central Colorado Plateau (Reynolds et al., 2006).

The combination of these geological data with process-based, forward-modelling schemes in global earth system models provides an excellent means of achieving a comprehensive picture of the global pattern of dust accumulation rates, their controlling mechanisms, and how those mechanisms may vary regionally (Kohfeld and Harrison, 2003). Sediment fluxes (in $g/m^2/yr$) for terrestrial sites obtained from loess are typically higher than those obtained from ocean and ice cores (Derbyshire, 2003). This suggests that future climatic models designed to study the role of global atmospheric dust will have to take full account of the present and past dust records over the continents.

In other Quaternary sediments grain size has also been used to indicate the nature of the depositional environment. For example, in moving pyroclastic systems where particles are sorted as a function of their sizes, densities and shapes the analysis of these variables can supply valuable information on the flow process (Taddeucci and Palladino, 2002). The analysis of the distribution of these characteristics in particle populations of pyroclastic deposits is a major tool in evaluating properties and regimes of parent transport systems. In the Weddell Sea Pudsey (1992) used the proportion of silt and well-sorted clay sand in surface sediments to indicate Antarctic bottom water velocity (with increases in these elements associated with increasing current activity).

There have also been attempts to use the coarser elements of the depositional systems at outcrop level to reconstruct late Cenozoic environments in a semi-quantitative way based on pebble size. For example, Adams (2003) estimated palaeowind strength using particle size and fetch length to calculate wave height and thus wind velocity. In fluvial geomorphology clast size and palaeochannel size have been used to reconstruct palaeoflows (e.g. Mather and Hartley, 2005).

Pebble shape and surface texture have also been used as indicators of palaeoenvironment. Plakht (1998) even went as far as to use pebble shape as an indicator of different climate in Quaternary terraces in Israel, although this is a dangerous practice as source lithology will play a major role here. More traditionally, grain surface texture has been used to distinguish between glacial and aeolian-sourced quartz grains. Helland et al. (1997) used scanning electron microscopy (SEM) examination on the grains of a sedimentary deposit in China formerly interpreted as glacial to indicate an alluvial/colluvial origin.

Proxy climate indicators: stable isotopes

The key approach in the area of quantitative reconstruction of climate has been through the use of stable isotopes of $\delta^{18}O$, $\delta^{13}C$ and $\delta^{15}N$. Stable isotopes are chemically stable and although chemically the same, their physical properties are different with the heavier isotopes having lower mobility. Thus differences in the abundance of the heavier/lighter isotopes ($^{18}O/^{16}O$, $^{13}C/^{12}C$, $^{15}N/^{14}N$) record the palaeoenvironment at the time of precipitation of the ice or carbonate. Stable isotopes have been used in ice cores and on the tests of microfossils. Current research is focussing on correlating the marine isotope records with those of continental sequences. The continental sequence of stable isotopes is recorded within lacustrine environments (Leng and Marshall, 2004), cave deposits (principally speleothems; McDermott, 2004) and more rarely, riverine tufas (Andrews, 2006), This section will focus on their use within lacustrine and cave environments as the marine environment has in part been examined in the first section.

Lacustrine environments

In lacustrine environments the stratigraphic changes in $\delta^{18}O$ are commonly linked to changes in temperature or precipitation/evaporation ratio, whereas $\delta^{13}C$ and $\delta^{15}N$ are used to demonstrate changes in carbon nutrient cycling and productivity within the lake catchment which may be climatically driven or relate to other environmental controls within the lake catchment. Identifying the cause of observed change typically necessitates a multi-disciplinary approach, as whilst palaeoclimate may induce changes in isotope values (Figure 29.3), so too may other factors such as tephra falls or deforestation within the catchment (see Leng and Marshall, 2004, and references therein). Most commonly the isotope composition is measured from authigenic and biogenic carbonates and diatom silica. When analysing biogenic sources the type of shell should be considered (Table 29.4). Jones et al. (2002) noted that two species of freshwater snail extracted from the same lacustrine sampling intervals yielded different isotope ratios, probably reflecting different microhabitats of the snails.

Cave environments

Within cave environments speleothems have been one of the key sources of material for stable

isotope analysis (McDermott, 2004). They can grow continuously on timescales of 10^3 to 10^5 years and can be accurately dated with uranium-series approaches. Speleothems are multiproxy palaeoclimate records and have been used to provide data on mean annual temperatures, rainfall variability, atmospheric circulation changes and vegetation response. These data are recorded in stable isotope ratios, thickness of growth laminae, variations in trace element ratios, and trapped pollen grains amongst other sources. However, there have been problems in interpreting the stable isotope data in terms of palaeotemperature reconstructions and more recent studies have focused on providing precise estimates of the timing of major oxygen isotope-defined climatic events (glacial interglacial transitions etc.).

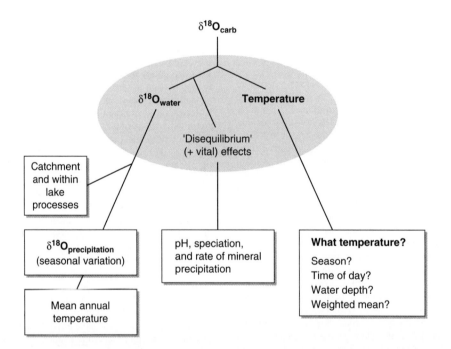

Figure 29.3 An example of how oxygen isotopes are affected by their environment for lacustrine carbonates ($\delta^{18}O_{carb}$). If the carbonate is precipitated in isotopic equilibrium, the lacustrine carbonate depends entirely on temperature and the isotopic composition of the lake water ($\delta^{18}O_{water}$). Disequilibrium effects ('vital effects') in biogenic precipitates, caused by local changes in microenvironment or rate of precipitation can induce systematic or non-systematic offsets in the lacustrine carbonates. Thus factors such as time of year in which a particular type of authigenic or biogenic carbonate forms is important. In lakes with optimum hydrology (size and precipitation/evaporation regime) there is a simple relationship to $\delta^{18}O_{precipitation}$ but in others the water composition is strongly influenced by processes such as evaporation within the catchment and within the lake itself. $\delta^{18}O_{precipitation}$ is increasingly being shown to be an important indicator of climate change: it typically changes with mean annual temperature. (After Leng and Marshall, 2004)

Table 29.4 Selection of biogenic carbonate materials for stable isotope analysis of lacustrine systems (based on Leng and Marshall 2004). Note that biogenic silica may be the only remaining source of biogenic material for isotope analysis in acid lakes. In this case diatoms and sponges have been used successfully for palaeoenvironmental reconstructions

Biogenic source	Advantages	Disadvantages
Ostracod shells	• Ostracod multiple moults means individuals can be used to examine seasonal variation, where from the same population • Short duration of calcification after each moult • Shell surface morphology (indicates preservation) • Ease of location	• May be washed into the lake environment from streams • Carbonate may not precipitate in equilibrium with its environment ('vital effects') • Very small species may not provide the required size of sample
Mollusc shells (bivalves and gastropods)	• Widespread occurrence • Some species of snail form carbonate in a regular manner over a fixed, short time-period offering analysis of continuous inter-seasonal changes	• Often composed of thermodynamically unstable aragonite which can convert to calcite and 'reset' the isotope signal • Differences may arise between species reflecting local habitats

PALAEOECOLOGICAL APPROACHES AND THE QUATERNARY

Quaternary palaeoecological studies involve the use of a plethora of evidence ranging from microfossils of plants, animals and protists to vertebrates. Vertebrates have been used mainly for stratigraphic context, commonly using the teeth of mammalian fauna such as rats or mammoths. Many of the vertebrates, because of their relatively rapid evolutionary turnover, are more restricted in terms of palaeoenvironmental interpretation than other organisms as they may not have living examples from which to gain an accurate picture of preferred habitats. Thus most quantitative palaeoenvironmental reconstructions rely on simpler organisms that exist today and throughout the Quaternary. This section cannot examine all of the available archives due to constraints of space so will focus on the key ones commonly used in quantitative palaeoenvironmental reconstructions of Quaternary archives (Table 29.3). The use of vertebrates, dinoflagellate tests and molluscs, amongst others, are not included. Information on these can be found in standard texts such as Lowe and Walker (1997). The approaches have involved the use of geochemistry of tests (mainly stable isotopes and trace elements) to identify changes in palaeotemperature and salinities recorded in the faunal and floral remains. Other approaches use the mutual climatic range (MCR) approach which was originally designed for beetles and pollen. With this technique modern species are calibrated by plotting their distributions in 'climate space',

usually defined by two parameters (e.g. mean temperature of the warmest month and the range between that and the mean temperature of the coldest month) to establish their climate envelopes. The overlapping climate envelopes of species in a fossil assemblage can then be used to determine these variables in the palaeo-assemblage. Alternatively, the use of transfer functions has also been used for quantitative reconstructions of environmental parameters such as temperature or hydrologic conditions from some species. Transfer functions are variants of multiple regression models. The approach assumes that a particular assemblage of organisms, with the original data taken from a modern 'training set', is related to particular environmental parameters such as temperature by complex functions (transfer functions). Thus these transfer functions, derived from modern data sets, can be used to quantitatively reconstruct past environments with the same assemblages. Both these approaches (MCR and transfer functions) require the use of large modern data sets to be effective.

Fauna: insects

Although a range of insects have been found preserved in Quaternary sequences this section will focus on the most commonly used insects in Quaternary studies. It is also worth noting that trace fossils (ichnofacies) of insects have also been used successfully to provide useful palaeoenvironmental data (e.g. De, 2002) and through this environmental information to identify sea level and climate change (e.g. Alonso-Zara and Silva, 2002; De, 2002).

Beetles (Coleoptera)

Beetles (Coleoptera) have been found well preserved in environments such as glacier ice, permafrost, glacial kettle hole fill, wetlands and packrat middens. Since the early 1980s MCR has been used to estimate palaeotemperatures from assemblages. Beetles have been used to demonstrate that desert zones were currently larger than today and to establish the varying discharge of rivers (Smith and Howard, 2004). Fossil beetles can provide significant information on factors such as tree cover, health of trees, forest composition etc. that is difficult to obtain using other macrofossils (Lavoie, 2001).

Midges (Chironomids)

Chironomids (non-biting midges) have been used in palaeoenvironmental studies in Eurasia since the 1920s. Initially, changes in chironomid assemblages were largely interpreted as a response to changes in trophic status or water depth. The development of chironomid–temperature inference models in the early 1990s was the springboard for their palaeoclimatic exploitation. Since then better taxonomic resolution of fossil midges, expansion of modern training sets, use of air temperature data derived from meteorological stations rather than surface-water temperature data, and Bayesian statistical approaches have led to improvements in the performance of chironomid–temperature transfer functions (Figure 29.4). Chironomids have thus been used in a range of studies from reconstructing palaeotemperatures to detecting the significance of stream influences on lacustrine environments (e.g. Ruck et al., 1998) and have been found to be more sensitive to environmental change than vegetation reconstructions using pollen. They have most extensively been used in the northern hemisphere. Work in the southern hemisphere has suffered from a lack of traditional surface-sediment calibration data sets. New approaches may reveal further information. For example, Walker and Cwynar (2006) have suggested that the isotopic signature of chitin in midge head capsules may provide additional information on palaeotemperatures and improve existing applications.

Fauna: ostracods

Ostracods are found in a range of environments from open-water bodies in marine to brackish and freshwater environments to hot and freshwater springs and the deposits of craters and sinkhole lakes. Their faunal assemblages, use of indicator species and variations in valve morphology have been used for biostratigraphy and palaeoenvironmental reconstruction and to interpret lake sediments in terms of productivity, oxygen availability and water temperature. Although there are many assumptions that have to be made about the climatic tolerances of living and fossil ostracod species in interpreting past records, as well as many existing complicating factors (such as the relationship between water temperature and air temperature, habitat preferences and taxonomic errors) ostracod shells can still provide valuable information on water salinity, temperature, chemistry, hydrodynamic conditions, and substrate characteristics. Freshwater ostracods have been known to respond to physical properties of their host environment and nutrients within that environment. They have been used to identify lake isolation from marine influences, to indicate fluctuations of groundwater discharge in lakes, to identify periods of droughts and flooding and to identify changes in regional and global climate circulations. New approaches to provide more quantitative estimates of palaeotemperatures include the mutual ostracod temperature range (MOTR) method which uses a non-analogue approach based on the presence/absence of species in a fossil assemblage but further testing and refinement are needed (Horne, 2007). Careful use of ostracod data, with other proxies such as pollen can provide valuable insights into how past eco-systems and environments have responded to environmental change. For example, Lamb et al. (1995) demonstrated that whilst vegetation was not affected by a given climate change, the ostracods were. This suggests that although overall water availability for plant growth was little different, seasonality was enhanced to produce dry events in the lake record. Broader similarities in climate change (e.g. between the Caribbean and Africa) have been used to suggest orbitally induced (Milankovitch) variations in climate which are reflected both in the biogeography and human cultural developments (Hodell et al., 1991).

Protists

Protists are simple organisms with an external hard shell or 'test' that can be preserved within Quaternary sequences. The two most commonly used protists for reconstructing Quaternary environments are foraminifers (identified as far back as the 16th century and used more systematically since the late 1800s) and testate amoebae (a more recent approach).

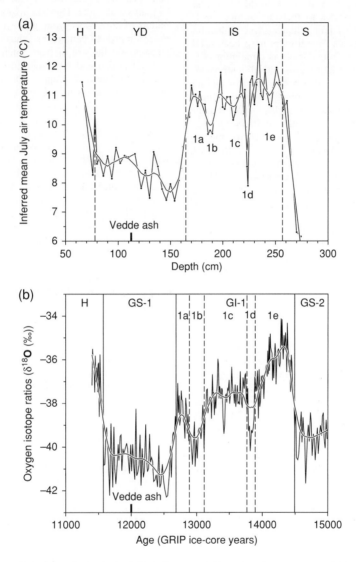

Figure 29.4 A comparison of (a) chironomid-inferred temperatures from Whitrig Bog, south-east Scotland and (b) the oxygen isotope record from the GRIP ice core during the Late Glacial. The reconstruction suggested that the thermal maximum occurred early in the last Interstadial with temperatures reaching about 12–13°C. Thereafter, there was a gradual downward trend to about 11°C, punctuated by four distinct cold oscillations of varying intensity. At the beginning of the Younger Dryas, summer temperatures fell to about 7.5°C but gradually increased to about 9°C before a rapid rise at the onset of the Holocene. The chironomid-inferred temperature curve agrees closely with the GRIP ice-core oxygen-isotope curve from Greenland. (After Brooks, 2006)

Foraminifera

Foraminifera have been used extensively in marine micropalaeontology. Recently the advances in stable isotope analysis, have led to the use of the shells of foraminifera to recreate surface water temperatures and salinity changes. In conjunction with trace elements foraminifera have been used to record changes in major ocean circulation patterns (over the last four major glacial–interglacial changes). They have also been used to examine sediment dispersal in relation to sea-level change, tsunamis and climate variability. Most recent studies have been typically multi-proxy and use transfer functions to derive environmental information (e.g. Gehrels, 2000). In the last decade there has also been a growing use of foraminifera from saltmarsh environments where they are very close to the highest astronomical tide level (Gehrels et al., 2001). The indicative meaning of fossil assemblages in this latter environment has been shown to be most accurately predicted by a training set that incorporates live and dead foraminifera and that uses flooding duration, not height, as the predictor variable (Gehrels, 2000).

Testate amoebae (rhizopods/thecamoebians/arcellaceans)

Testate amoebae inhabit a wide range of terrestrial and aquatic habitats, including wet soils, lakes and saltmarshes (Charman, 2001). They can be influenced by summer precipitation variability and summer temperature (Charman et al., 2004) thus they can be used as proxy climate indicators in environmental and climate reconstructions although there can be problems with preservation and subsequent analysis (Wilmshurst et al., 2003). Reconstructions are typically based on transfer functions, relating modern assemblage composition to water table and moisture content, applied to fossil sequences. Existing transfer functions in Europe and elsewhere are limited geographically and there are often problems with missing or poor analogues (Charman et al., 2007). Assemblages of testate amoebae in UK saltmarshes are strongly correlated with elevation and flooding duration, suggesting that if adequately preserved in sediments they may be used as accurate sea-level indicators (Roe et al., 2002). Although their vertical distribution in saltmarshes is small (their lower tolerance limit in modern saltmarshes occurs where tides cover the marsh less than 2% of the year), they are important sea-level indicators as information for sea-level reconstruction is best derived from sediments that originate in the highest part of the intertidal zone (Gehrels et al., 2001).

Flora

The analysis of Quaternary environments has benefited from the use of the fossil remains of plants and algae. This has involved the examination of tree rings from plant macrofossils to ascertain growth patterns and thus environmental stresses and the use of plant microfossils in the form of pollen and phytoliths to reconstruct the vegetation. These latter approaches have also been adopted extensively in archaeology.

Diatoms (Bacillariophyceae)

Diatoms are particularly useful environmental indicators as they are found in a range of environments from marine to saltmarsh and freshwater. They are unicellular algae with a shell (frustule) composed of silica and have been used to reconstruct pH, salinity, water depth and productivity. There can be problems in their interpretation due to differential preservation potential (Battarbee et al., 2005) and significance of local factors such as the presence of aquatic peat (Sarmaja-Korjonen and Alhonen, 1999). Diatoms, along with other proxies, such as isotopes, have been used to reconstruct palaeoclimate. They have also been used to identify catastrophic sea-water incursions as a result of tsunami, aid in the identification of the major collapse of ice sheets in Antarctica and to identify changes in runoff from landmasses through examination of off-shore freshwater versus marine diatoms. Where used with other proxies such as foraminifera they have been found to be the best representation of flooding duration (Gehrels et al., 2001).

Palynology (pollen analysis)

The value of pollen in reconstructing Quaternary environments has long been understood. It has been used for ecological and Biome reconstruction (Marchant et al., 2002; Anhuf et al., 2006), environmental change (e.g. Kiage and Liu, 2006; Grimley et al., 2009); and chronostratigraphy (e.g. Kuhlman et al., 2006). It is typically used alongside other approaches described in this chapter (see Huntley, 1996, for a review). Pollen has been found in a range of environments including fossil rodent middens, tufa/travertine, alluvial sequences, lacustrine sequences, speleothems and ice cores. The use of pollen in speleothems is a recent and valuable approach due to the dating resolution and length of record offered by dating of carbonate (TIMS U–Th; McGarry and Caseldine, 2004). Pollen has most commonly been used to recreate vegetation characteristics and thus palaeoenvironmental conditions although the cause of the vegetation change may not always be clear to identify. The pollen record provides a useful proxy for climate which can be used in conjunction with other records. For example, Thorndycraft and Benito (2006) use sedimentological evidence of flooding from Spanish river sections and relate these to pollen records of climate change to identify wetter, flood prone periods in the Holocene (Figure 29.5). Additional records from Poland and Britain suggest these may have been Europe-wide flooding events during periods of increased moisture balance.

In the North Island of New Zealand pollen has been used to demonstrate the spread of *Pteridium* (bracken) after the 1314 AD Kaharoa eruption (Newnham et al., 1998). The spread of bracken indicated at around the time of the Kaharoa eruption is palaeoenvironmental evidence for the first discernible impact of human settlement by Polynesians in New Zealand.

Phytoliths

Plants secrete opaline silica bodies which assume the shape of the cell in which they were deposited. It is these forms that are known as phytoliths. They are useful in vegetational reconstructions as they are characteristic of the plants in which they originally formed. Most commonly phytoliths are used to reconstruct past vegetation patterns, and through this local environmental conditions. They have been found in palaeosols, where they can often be more sensitive than the soils in picking out shorter term periods of aridity that are

effectively overprinted by the soils (Sedov et al., 2003). They have also been located in the deposits of crater lakes, loess and sand dunes. As soils on aeolian landforms are often short-lived and produce low phytolith inheritance, such assemblages may lack the temporal and spatial integration necessary to be clear indicators of regional vegetative succession and thus climate change. However, as phytoliths are more robust than pollen, and typically more localized in origin, they can provide evidence of climate periods not recorded in fossil pollen alone. Phytolith assemblages have been used in the Late Cenozoic record to identify climate change, marine transgression, impact of climate on human populations, coastal environmental change and the impact of tephra falls. Transfer functions have been used to determine proxies on soil weathering and to aid in quantitative palaeoenvironmental reconstruction (Prebble and Schulmeister, 2002).

Fossil wood/charcoal

Charcoal has long been used as an indicator of human interaction with the environment by archaeologists (Figueiral and Mosbrugger, 2000). However there is good evidence in the geological record that fires were a part of the natural ecosystem well before humans. Charcoal and fossil wood is often used with other floral records such as pollen to reconstruct vegetation and from this, climate. Fossil wood has been used to indicate the presence of mangroves hence sea level. Dicotyledonous wood (wood that has growth

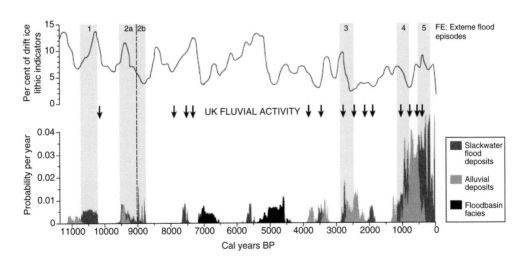

Figure 29.5   Holocene fluvial chronology of Spain in relation to the North Atlantic drift-ice record (Bond et al., 2001) (From Thorndycraft and Benito (2006))

rings etc.) that is preserved in Quaternary sequences can be used to indicate periods of drought etc. which have stressed the macro flora but its use in this context is limited (Wheeler and Baas, 1993).

GEOARCHAEOLOGY, HISTORIC RECONSTRUCTIONS AND THE QUATERNARY

Geoarchaeology focuses on the interface between earth sciences and past human activity. Geoarchaeology includes the study of sediments both at archaeological sites, and in a wider geographical context to understand local and regional environmental histories. The techniques developed by the geosciences that are applicable to the Quaternary are generally the most appropriate geoscience techniques for archaeologists as they are adapted to the highest resolution timescale. As a result stratigraphy and sedimentology have influenced archaeology since the 19th century. Indeed, Lyell used archaeology (the borings on the three pillars of the Roman Market of Pozuolli) as the frontispiece to his book *Principles of Geology* in 1830. Archaeology has played an important role in establishing sea-level fluctuations including the use of Roman fish tanks to identify sea-level change over the last 2000 years (Schmiedt, 1972) and tectonic activity (e.g. Neev et al., 1987). Techniques used in these studies include geomorphological mapping, geophysical survey, biostratigraphy, sedimentology, and geochemistry.

Historical records have facilitated the reconstruction of Holocene environments. Metereological records, historical land-use maps, ethnographic data, historical literature have all been used to understand Holocene environmental interactions when combined with sedimentary evidence (e.g. Dotterweich, 2005; Kraft et al., 2007). Other approaches involve using the historical records to reconstruct events such as floods (Benito et al., 2004), reinforce interpretation of pollen records (Nielsen and Odgaard, 2004) and examine the impact of historic eruptions (Grattan and Brayshay, 1995).

Off-shore studies

Within port environments the identification of the presence of trace metals such as lead has helped understand human–environment interactions. The trace metals can be provenanced from stable isotopes (Marriner and Morhange, 2007) to aid in understanding how the environment was being used by humans. Similarly, other Quaternary geological approaches have magnified the value of archaeological studies through analysis of environmental information from sediments and their fauna (e.g. Reindhardt et al., 1994). The artefacts combined with the geoarchaeology can help date the sequences and identify their palaeoclimatic significance.

Archaeology has also encompassed recent geological approaches to stratigraphy such as sequence stratigraphy in coastal settings, developing, for example, the Ancient Harbour Parasequence with enhanced sedimentation rates against the natural background of a rise in sea level. The bio-sedimentary data and archaeological structures can be used to determine the relative rise in sea level.

Marine microfossils (principally foraminifera and diatom assemblages) have been used widely to indicate human activity. This may be by examining the water quality of areas surrounding former human settlement through agricultural or industrial changes.

On-shore studies

The interdisciplinary nature of geoarchaeology means that it can include many different techniques and can also use regional stratigraphy to correlate between archaeological sites. Of particular interest in this field has been the study of soil loss as a direct consequence of land-use changes. This often involves the use of magnetic susceptibility and magnetic remanence to pinpoint the provenance of eroded materials. Integrated archaeological and geomorphological studies have taught us much about the environment. For example, badland sites are typically considered to be actively eroding sites. However, research by Wise et al. (1982) revealed that Bronze Age sites lay fairly undisturbed on badland slopes, indicating that the badlands pre-dated the settlements which have been subsequently little affected by changing agricultural practices in the region.

Charcoal analysis has been extensively used in archaeological studies. Charcoal scattered through archaeological horizons tends to reflect local, long-term environmental conditions, facilitating the recreation of archaeological vegetation histories (see Figueiral and Mosbrugger 2000 and references therein). Palaeoecological approaches have also proved valuable. Beetle populations, for example, will change with changing vegetation brought about by land clearance. Some beetles are associated with stored grain products whilst others are concentrated around tanneries, sewers etc. There is, however considerable noise in the data due to contamination of local assemblages with beetles that have flown in or been deposited through dung etc. This means large death assemblages have to be examined to provide reliable data.

INTERPRETING QUATERNARY ENVIRONMENTS: A SUMMARY AND THE FUTURE

Our ability to interpret the Quaternary environment has progressed rapidly in the last few decades. This progress has been largely a function of (1) the scientific communities increased understanding of geomorphological systems and how they operate, and (2) the improvement and development of new techniques (e.g. dating methods, stable isotopes and trace element analysis). Many of these approaches are summarized in key texts such as Brown, 1997; Roberts, 1998; MacKay et al., 2005; Oldfield, 2005 and Anderson et al., 2007.

The successful application of the approaches described in this chapter relies on a detailed understanding of the geomorphological processes which constructed the Quaternary record being examined, and which may have modified it subsequently (for example, see discussions on the application of cosmogenic isotope techniques in Cockburn and Summerfield 2004 and optically stimulated luminescence in Thrasher et al., 2009). An appreciation of the geomorphic stratigraphy enables us to build up a framework on which to base the sampling techniques utilized for the study. It is due to the nature of the Quaternary geomorphological record and its deposits (often fragmentary and of higher resolution in comparison to longer geological records) that stratigraphic approaches adopted from geology have had to be modified. Some stratigraphic techniques have translated successfully e.g. the use of sequence stratigraphic concepts from geology to geoarchaeological studies of harbours. Other approaches such as stable isotope stratigraphy applied to deep ocean cores (marine isotope stratigraphy) have been embraced by Quaternary geomorphologists and adapted to use in terrestrial geomorphological sequences (e.g. lacustrine and cave environments) where they can be used for climate reconstruction and may lead to successful correlation with the marine records in the future.

Within Quaternary studies integrating the above techniques with palaeoecological approaches has further strengthened our interpretation of the Quaternary record by enabling quantitative reconstructions of the palaeoenvironment to be achieved

(e.g. undertaking isotope analysis of the tests of foraminifera). The development of transfer functions has meant that modern environmental ranges of particular organisms can be used to recreate, quantitatively, the palaeoenvironments of Quaternary populations. Multiproxy approaches can be used to pinpoint more accurately the causes of identified changes in the Quaternary. For example, pollen flora records can be used in conjunction with faunal records such as foraminifera to identify the most likely causal factor for identified vegetation change. Similarly Phytoliths have proved valuable when used in conjunction with charcoal or pollen.

Despite the fact that some multi-proxy methods apparently give contradictory findings we can use these data to better understand the subtleties of past environments. For example for the same sedimentary records beetles and macro remains of plants, pollen etc. appear to give contradictory messages in response to the same environmental change. This in part reflects the sensitivity of the individual records (Brooks and Birks, 2001) and the fact that seasonal changes in temperature and precipitation which do not result in much mean annual change may affect vegetation more than the insect fauna. Stable-isotope data and pollen data from Gölhisar lake sediments in south-west Turkey show a divergence between pollen inferred and stable-isotope palaeoclimate data, suggesting vegetation may take several millennia to reach a climatic equilibrium (Eastwood et al., 2007). Likewise adjacent environments may show different response to the same environmental impact. For example also at this location, work by Eastwood et al. (2002) demonstrates that tephra fall out from the mid-second millennium bc 'Minoan' eruption of Santorini in the Aegean impacted on the lake environment but had little discernible impact on the terrestrial pollen record from the site.

As a caution, we must be aware that we are generally only able to determine mean records from Quaternary archives. Most hydrological climate–proxy indicators, for example, reflect 'effective moisture availability', not rainfall *per se*. Palaeodata integration is typically achieved by time–series regression against instrumental data such as precipitation or derived measures of hydrological change (e.g. effective moisture) or though transfer functions. Times–series calibrations are often considered the most reliable approach, although where, for example, local instrumental data is lacking this may not be the case (Jones et al., 1998). Thus other approaches using a regional climate field reconstruction approach, using multi-variate statistical calibration of proxy data against spatial networks of instrumental dating (e.g. Zhang et al., 2004) may be more appropriate. In addition limited age constraints mean correlations between proxies and regions is limited although possible. For example there is strong correspondence between peat surface wetness, lake levels and glacier mass-balance in Europe (Figure 29.6) over the last 4000 years which probably reflects warm season

water balance. High precipitation and low temperatures during summer months would increase glacier mass balance and outflows from peatland and lakes. The speleothem record, however, is governed more by winter rainfall conditions and annual temperature to precipitation records (see discussion in Verschuren and

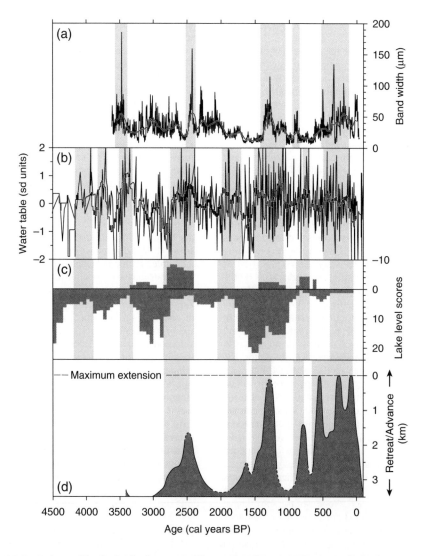

Figure 29.6 Selected hydrologic data records resolved to century scale for temperate Europe. Shaded areas indicate inferred high moisture availability. (a) Mean annual band thickness, speleothems, north-west Scotland; (b) mean standardized peatland surface wetness, northern Britain; (c) summed lake-level scores from sediment stratigraphy, French pre-Alps; (d) mass balance fluctuation (advance and retreat), Great Alettsch glacier, Switzerland. (Based on original data from Proctor et al., 2002, Magny, 2004; Charman et al., 2006, Holzhauser et al., 2005, summarized in Verschuren and Charman, 2008)

Charman, 2008, and references therein) so compares less favorably.

The future for the study of Quaternary studies offers exciting possibilities. As the development of new techniques and our understanding of our contemporary environment increases so our ability to interrogate Quaternary sequences improves, especially where these new techniques form part of multiproxy, multidisciplinary approaches.

REFERENCES

Adams, K.D. (2003) Estimating palaeowind strength from beach deposits. *Sedimentology*, 50, 565–77.

Alonso-Zarza, A.M. and Silva, P.G. (2002) Quaternary laminar calcretes with bee nests: evidences of small-scale climatic fluctuations, Eastern Canary Islands, Spain. *Palaeogeography, Palaeoclimatology and Palaeoecology*, 178, 119–35.

American Commission on Stratigraphic Nomenclature (1970) Code of stratigraphic nomenclature. *American Association of Petroleum Geologists Bulletin*, 60, 1–45.

Anderson, D.E., Goudie, A.S. and Parker A.G. (2007) *Global Environments through the Quaternary: Exploring Environmental Change*. Oxford University Press. p. 372.

Andrews, J.E. (2006) Palaeoclimate records from stable isotopes in riverine tufas: synthesis and review. *Earth Science Reviews*, 75, 85–104.

Anhuf, D., Ledru, M.P., Behling, H., Da Cruz, F.W., Cordeiro, R.C., Van der Hammen, T. et al. (2006) Palaeo-environmental change in Amazonian and African rainforest during the LGM. *Palaeogeography, Palaeoclimatology and Palaeoecology*, 239, 510–27.

Battarbee, R.W., Mackay, A.W., Jewson, D.H., Ryves, D.B. and Sturm, M. (2005) Differential dissolution of Lake Baikal diatoms: correction factors and implications for palaeoclimatic reconstruction. *Global and Planetary Change*, 46, 75–86.

Benito, G., Lang, M., Barriendos, M., Llasat, M.C., Frances, F., Ouarda, T. et al. (2004) Use of systematic, palaeoflood and historical data for the improvement of flood risk estimation: review of scientific methods. *Natural Hazards*, 31, 623–43.

Bond, G., Heinrich, H., Broecker, W., Labeyrie, L., McManus, J., Andrews, J. et al. (1992) Evidence for massive discharges of icebergs into the North Atlantic during the last glacial period. *Nature*, 360, 245–9.

Bond, G., Broecker, W.S., Johnson, S., McManus, J., Labeyrie, L., Jouzel, J. et al. (1993) Correlations between climate records from North Atlantic sediments and Greenland Ice. *Nature*, 365, 143–7.

Brooks, S.J. (2006) Fossil midges (Diptera : Chironomidae) as palaeoclimatic indicators for the Eurasian region. *Quaternary Science Reviews*, 25, 1894–910.

Brooks, S.J. and Birks, H.J.B. (2001) Chironomid-inferred air temperatures from Lateglacial and Holocene sites in north-west Europe: progress and problems. *Quaternary Science Reviews*, 20, 1723–41.

Brown, A.G. (1997) *Alluvial Geoarchaeology: Floodplain Archaeology and Environmental Change*. Cambridge Manuals in Archaeology. Cambridge University Press. p. 404.

Charman, D.J. (2001) Biostratigraphic and palaeoenvironmental applications of testate amoebae. *Quaternary Science Reviews*, 20, 1753–64.

Charman, D.J., Blundell, A. and Members, A. (2007) A new European testate amoebae transfer function for palaeohydrological reconstruction on ombrotrophic peatlands. *Journal of Quaternary Science*, 22, 209–21.

Charman, D.J., Brown, A.D., Hendon, D. and Karofeld, E. (2004) Testing the relationship between Holocene peatland palaeoclimate reconstructions and instrumental data at two European sites. *Quaternary Science Reviews*, 23, 137–43.

Charman, D.J., Blundell, A., Chiverrell, R.C., Hendon, D. and Langdon, P.G. (2006) Compilation of non-annually resolved Holocene proxy climate records: stacked Holocene peatland palaeo-water table reconstructions from northern Britain. *Quaternary Science Reviews*, 25, 336–50.

De, C. (2002) Continental mayfly burrows within relict-ground in inter-tidal beach profile of Bay of Bengal coast: a new ichnological evidence of Holocene marine transgression. *Current Science*, 83, 64–7.

Derbyshire, E. (2003) Loess, and the dust indicators and records of terrestrial and marine palaeoenvironments (DIRTMAP) database. *Quaternary Science Reviews*, 22, 1813–19.

Dilli, K. and Pant, R.K. (1994) Clay-minerals as indicators of the provenance and paleoclimatic record of the Kashmir-loess. *Journal of the Geological Society of India*, 44, 563–74.

Ding, Z.L., Rutter, N.W., Sun, J.M., Yang, S.L. and Liu, T.S. (2000) Re-arrangement of atmospheric circulation at about 2.6 Ma over northern China: evidence from grain size records of loess-palaeosol and red clay sequences. *Quaternary Science Reviews*, 19, 547–58.

Dotterweich, M. (2005) High-resolution reconstruction of a 1300 year old gully system in northern Bavaria, Germany: a basis for modeling long-term human-induced landscape evolution. *The Holocene*, 15, 994–1005.

Eastwood, W.J., Leng, M.J., Roberts, N. and Davis, B. (2007) Holocene climate change in the eastern Mediterranean region: a comparison of stable isotope and pollen data from lake Golhisar, southwest Turkey. *Journal of Quaternary Science*, 22, 327–41.

Eastwood, W.J., Tibby, J., Roberts, N., Birks, H.J.B. and Lamb, H.F. (2002) The environmental impact of the Minoan eruption of Antorini (Thera): statistical analysis of palaeoecological data from Golhisar, southwest Turkey. *The Holocene*, 12, 431–44.

Einarsson, T. (1986) Tephrochronology, in B.E. Berglund (ed.) *Handbook of Holocene Palaeoecology and Palaeohydrology*. John Wiley, Chichester and New York. pp. 329–42.

Emiliani, C. (1955) Pleistocene temperatures. *Journal of Geology*, 63, 538–75.

Figueiral, I. and Mosbrugger, V. (2000) A review of charcoal analysis as a tool for assessing Quaternary and Tertiary environments: achievements and limits. *Palaeogeography, Palaeoclimatology and Palaeoecology*, 164, 397–407.

Follmer, L.R. (1983) Sangamon and Wisconsinian pedogenesis in the Midwestern United States, in H.E. Wright Jr. (ed.) *Late Quaternary Environments of the United States*. Vol. 1. The Late Pleistocene. Longman, London. pp. 138–44.

Foucault, A. and Stanley, D.J. (1989) Late Quaternary paleoclimatic oscillations in East-Africa recorded by heavy minerals in the Nile delta. *Nature*, 339, 44–6.

Frihy, O.E., El Askary, M.A., Deghidy, E.M. and Moufaddal W.M. (1998) Distinguishing fluvio-marine environments in the Nile delta using heavy minerals. *Journal of Coastal Research*, 14, 970–80.

Gehrels, W.R. (2000) Using foraminiferal transfer functions to produce high-resolution sea-level records from salt-marsh deposits, Maine, USA. *The Holocene*, 10, 367–76.

Gehrels, W.R., Roe, H.M. and Charman, D.J. (2001) Foraminifera, testate amoebae and diatoms as sea-level indicators in UK saltmarshes: a quantitative multiproxy approach. *Journal of Quaternary Science*, 16, 201–20.

Gibbard, P.L. and Van Kolfschoten, T. (2004) The Pleistocene and Holocene Epochs in F.M. Gradstein, J.G. Ogg and A.G. Smith (eds) *A Geologic Time Scale*. Cambridge University Press. pp. 441–52.

Gibbard, P.L., Head, M.J., Walker, M.J.C. and The Subcomission on Quaternary Stratigraphy (2010) Formal ratification of the Quaternary System/Period and the Pleistocene Series/Epoch with a base at 2.58 Ma. *Journal of Quaternary Science*, 25, 96–102.

Grattan, J. and Brayshay, M. (1995) An amazing and portentous summer: environmental and social responses in Britain to the 1783 eruption of an Iceland Volcano. *Geographical Journal*, 161, 125–34.

Grimley, D.A., Larsen, D., Kaplan, S.W., Yansa, C.H., Curry, B.B. and Oches, E.A. (2009) A multi-proxy palaeoecological and palaeoclimatic record within full glacial lacustrine deposits, western Tennessee, USA. *Journal of Quaternary Science*, 24, 960–81.

Haldorsen, S., Jørgensen, P., Rappol, M. and Riezebos, P.A. (1989) Composition and source of the clay-sized fraction of Saalian till in the Netherlands. *Boreas*, 18, 89–98.

Helland, P.E., Huang, P.H. and Diffendal, R.F. (1997) SEM analysis of quartz sand grain surface textures indicates alluvial/colluvial origin of the Quaternary 'glacia' boulder clays at Huangshan (Yellow Mountain), east-central China. *Quaternary Research*, 48, 177–86.

Hodell, D.A., Curtis, J.H., Jones, G.A., Higueragundy, A., Brenner, M., Binford, M.W. et al. (1991) Reconstruction of Caribbean climate change over the past 10,500 years. *Nature*, 352, 790–3.

Holzhauser H., Magny, M. and Zumbuhl, H.J. (2005) Glacier and lake-level variations in west-central Europe over the last 3500 years. *The Holocene*, 15, 789–801.

Horne, D.J. (2007) A mutual temperature range method for Quaternary palaeoclimatic analysis using European nonmarine Ostracoda. *Quaternary Science Reviews*. 26, 1398–415.

Huisman, D.J. and Kiden, P. (1998) A geochemical record of late Cenozoic sedimentation history in the southern Netherlands. *Geologie En Mijnbouw-Netherlands Journal of Geosciences* 76, 277–92.

Huntley, B. (1996) Quaternary palaeoecology and ecology. *Quaternary Science Reviews*, 15, 591–606.

Huntley, B. and Birks, H.J.B. (1983) *An Atlas of Past and Present Pollen Maps for Europe: 0-13,000 Years Ago*. Cambridge University Press, Cambridge.

Jones, P.D., Briffa, K.R., Barnett, T.P. and Tett, S.F.B. (1998) High-resolution palaeoclimatic records for the past millennium: interpretation, integration and comparison with general circulation model control-run temperatures. *The Holocene*, 8, 455–71.

Jones, M.D., Leng, M.J., Eastwood, W.J., Keen, D.H. and Turney, C.S.M. (2002) Interpreting stable-isotope records from freshwater snail-shell carbonate: a Holocene case study from Lake Golhisar, Turkey. *The Holocene*, 12, 629–34.

Kiage, L.M. and Liu, K.B. (2006) Late Quaternary paleoenvironmental changes in East Africa: a review of multiproxy evidence from palynology, lake sediments, and associated records. *Progress in Physical Geography*, 30, 633–58.

Kohfeld, K.E. and Harrison, S.P. (2003) Glacial-interglacial changes in dust deposition on the Chinese Loess Plateau. *Quaternary Science Reviews*, 22, 1859–78.

Kraft, J.C., Bruckner, H., Kayan, I. and Engelmann, H. (2007) The geographies of ancient Ephesus and the Artemision in Anatolia. *Geoarchaeology International Journal*, 22, 121–49.

Kraus, M.J. (1999) Paleosols in clastic sedimentary rocks: their geologic applications. *Earth-Science Reviews*, 47, 41–70.

Kuhlmann, G., Langereis, C., Munsterman, D., van Leeuwen, R.J., Verreussel, R., Meulenkamp, J. et al. (2006) Chronostratigraphy of late Neogene sediments in the southern North Sea Basin and paleoenvironmental interpretations. *Palaeogeography, Palaeoclimatology and Palaeoecology*, 239, 426–55.

Lamb, H.F., Gasse, F., Benkaddour, A., Elhamouti, N., Vanderkaars, S., Perkins, W.T. et al. (1995) Relation between century-scale Holocene arid intervals in tropical and temperate zones. *Nature*, 373, 134–7.

Lavoie, C. (2001) Reconstructing the late-Holocene history of a subalpine environment (Charlevoix, Quebec) using fossil insects. *The Holocene* 11, 89–99.

Leng, M.J. and Marshall, J.D. (2004) Palaeoclimate interpretation of stable isotope data from lake sediment archives. *Quaternary Science Reviews*, 23, 811–31.

Lister, A.M. (1992) Mammalian fossils and Quaternary Biostratigraphy. *Quaternary Science Reviews*, 11, 329–44.

Lowe, J.J. and Walker, M.J.C. (1997) *Reconstructing Quaternary Environments*. 2nd edn. Prentice Hall.

MacKay, A., Battarbee, R.W., Birks, J. and Oldfield, F. (2005) *Global Change in the Holocene*. Hodder Arnold. p. 544.

Mackereth, F.J.H. (1965) Chemical investigation of lake sediments and their interpretation. *Proceedings of the Royal Society of London*. B161, 295–309.

Magny, M. (2004) Holocene climate variability as reflected by mid-European lak-level fluctuations and its probable impact on prehistoric human settlements. *Quaternary International*, 113, 65–79.

Maher, B.A. (1998) Magnetic properties of modern soils and Quaternary loessic paleosols: paleoclimatic implications. *Palaeogeography, Palaeoclimatology and Palaeoecology*, 137, 25–54.

Marchant, R., Behling, H., Berrio, J.C., Cleef, A., Duivenvoorden, J., Hooghiemstra, H. et al. (2002) Pollen-based biome reconstructions for Colombia at 3000, 6000, 9000, 12 000, 15 000 and 18 000 (14)Cyr ago: Late Quaternary tropical vegetation dynamics. *Journal of Quaternary Science*, 17, 113–29.

Marriner, N. and Morhange, C. (2007) Geoscience of ancient Mediterranean harbours. *Earth-Science Reviews*, 80, 137–94.

Mather, A.E. (2000) Impact of headwater river capture on alluvial system development. *Journal of the Geological Society of London*, 157, 957–66.

Mather, A. and Hartley, A.J. (2005) Flow events on a hyper-arid alluvial fan: Quedrada Tambores, Salar de Atacama, northern Chile, in A.M. Harvey, A.E. Mather and M. Stokes (eds) Alluvial Fans: Geomorphology, Sedimentology, Dynamics. Geology Society Special Publication 251, London. pp. 9–29.

McDermott, F. (2004) Palaeo-climate reconstruction from stable isotope variations in speleothems: a review. *Quaternary Science Reviews*, 12, 901–18.

McGarry, S.F. and Caseldine, C. (2004) Speleothem palynology: an undervalued tool in Quaternary studies. *Quaternary Science Reviews*, 23, 2389–404.

Neev, D., Bakler, N. and Emery, K.O. (eds) (1987) *Mediterranean Coasts of Israel and Sinai. Holocene Tectonism from Geology, Geophysics and Archaeology.* Taylor and Francis, London.

Newnham, R.M., Lowe, D.J., McGlone, M.S., Wilmshurst, J.M. and Higham, T.F.G. (1998) The Kaharoa Tephra as a critical datum for earliest human impact in northern New Zealand. *Journal of Archaeological Science*, 25, 533–44.

Nielsen, A.B. and Odgaard, B.V. (2004) The use of historical analogues for interpreting fossil pollen records. *Vegetation History and Archaeobotany*, 13, 33–43.

North American Commission on Stratigraphic Nomenclature (1983) North American Stratigraphic Code. *American Association of Petroleum Geologists Bulletin* 67, 841–75.

Odin, G.S., Renard, M. and Grazinni, C.V. (1982) Geochemical events as a means of correlation, in G.S. Odin (ed.) *Numerical Dating in Stratigraphy*. John Wiley, Chichester and New York. pp. 37–71.

Oldfield, F. (2005) *Environmental Change: Key Issues and Alternative Perspectives.* Cambridge University Press. p. 386.

Patience, A.J. and Kroon, D. (1991) Oxygen Isotope chronostratigraphy, in P.L. Smart and P.D. Frances (eds) *Quaternary Dating Methods – a User's Guide.* Technical Guide 4, Quaternary Research Association, Cambridge. pp. 199–228.

Pennington, W., Haworth, E.Y., Bonny, A.P. and Lishman, J.P. (1972) Lake sediments in northern Scotland. *Philosophical Transactions of the Royal Society of London.* B264. 191–294.

Petit, J.R., Mournier, L., Jouzel, J., Korotevitch, Y.S., Kotlyakov, V.I. and Lorius, C. (1990) Palaeoclimatological and chronological implications of the Vostok core dust record. *Nature*, 343, 56–8.

Plakht, J. (1998) Shape of pebbles as an indicator of climatic changes during the Quaternary in Makhtesh Ramon, Negev, Israel. *Zeitschrift Für Geomorphologie*, 42, 221–31.

Prebble, M. and Shulmeister, J. (2002) An analysis of phytolith assemblages for the quantitative reconstruction of late Quaternary environments of the Lower Taieri Plain, otago, South Island, New Zealand II. Paleoenvironmental reconstruction. *Journal of Paleolimnology*, 27, 415–27.

Proctor, C.J., Baker, A. and Barnes, W.L. (2002) A three thousand year record of North Atlantic climate. *Climate Dynamics*, 19, 449–54.

Pudsey, C.J. (1992) Late Quaternary changes in Antarctic bottom water velocity inferred from sediment grain-size in the Northern Weddell Sea. *Marine Geology*, 107, 9–33.

Reindhardt, E.G., Patterson, R.T. and Schröder-Adams, C.J. (1994) Geoarchaeology of the ancient harbour site of Caesarea Maritima, Israel: evidence from sedimentology and paleoecology of benthic foraminifera. *Journal of Foraminiferal Research*, 21, 37–48.

Reynolds, R.L., Reheis, M.C., Neff, J.C., Goldstein, H. and Yount, J. (2006) Late Quaternary eolian dust in surficial deposits of a Colorado Plateau grassland: controls on distribution and ecologic effects. *Catena*, 66, 251–66.

Roberts, N. (1998) *The Holocene: An Environmental History.* 2nd edn. Wiley/Blackwell, p. 344.

Roberts, N., Reed, J.M., Leng, M.J., Kuzucuoğlu, C., Fontugne, M., Bertaux, J. et al. (2001) The tempo of Holocene climatic change in the eastern Mediterranean region: new high resolution crater-lake sediment data from central Turkey. *The Holocene* 2001, 11, 721–736.

Roe, H.M., Charman, D.J. and Gehrels, W.R. (2002) Fossil testate amoebae in coastal deposits in the UK: implications for studies of sea-level change. *Journal of Quaternary Science*, 17, 411–29.

Ruck, A., Walker, I.R. and Hebda, R. (1998) A palaeolimnological study of Tugulnuit Lake, British Columbia, Canada, with special emphasis on river influence as recorded by chironomids in the lake's sediment. *Journal of Paleolimnology*, 19, 63–75.

Sarmaja-Korjonen, K. and Alhonen, P. (1999) Cladoceran and diatom evidence of lake-level fluctuations from a Finnish lake and the effect of aquatic-moss layers on microfossil assemblages. *Journal of Paleolimnology*, 22, 277–90.

Schmiedt, G. (1972) *Il livello antico del mar Tirreno: Testimonianze dei resti Archeologici.* Leo S. Olschki Editore, Florence.

Schreve, D.C. and Thomas, G.N. (2001) Critical issues in European Quaternary Biostratigraphy. *Quaternary Science Reviews*, 20, 1577–82.

Sedov, S., Solleiro-Rebolledo, E., Morales-Puente, P., Arias-Herreia, A., Vallejo-Gomez, E. and Jasso-Castaneda, C. (2003) Mineral and organic components of the buried paleosols of the Nevado de Toluca, Central Mexico as indicators of paleoenvironments and soil evolution. *Quaternary International*, 106, 169–84.

Smith, D.N. and Howard, A.J. (2004) Identifying changing fluvial conditions in low gradient alluvial archaeological landscapes: can coleoptera provide insights into changing discharge rates and floodplain evolution? *Journal of Archaeological Science*, 31, 109–20.

Stanley, D.J., Nir, Y. and Galili, E. (1998) Clay mineral distributions to interpret Nile cell provenance and dispersal: III. Offshore margin between Nile delta and northern Israel. *Journal Of Coastal Research*, 14, 196–217.

Iaddeucci, J. and Palladino, D.M. (2002) Particle size-density relationships in pyroclastic deposits: inferences for emplacement processes. *Bulletin of Volcanology*, 64, 273–84.

Thorndycraft, V.R. and Benito, G. (2006) The Holocene fluvial chronology of Spain: evidence from a newly compiled radiocarbon database. *Quaternary Science Reviews*, 25, 223–34.

Underwood, M.B. and Pickering, K.T. (1996) Clay-mineral provenance, sediment dispersal patterns, and mudrock diagenesis in the Nankai accretionary prism, southwest Japan. *Clays and Clay Minerals*, 44, 339–56.

Verschuren, D. and Charman, D.J. (2008) Latitudinal linkages in late Holocene moisture-balance variation, in H. Binney and R. Battabee (eds) *Natural Climate Variability and Global Warming*. Blackwells. pp. 189–231.

Vriend, M. and Prins, M.A. (2005) Calibration of modelled mixing patterns in loess grain-size distributions: an example from the north-eastern margin of the Tibetan Plateau, China. *Sedimentology*, 52, 1361–74.

Walker, I.R. and Cwynar, L.C. (2006) Midges and palaeotemperature reconstruction – the North American experience. *Quaternary Science Reviews*, 25, 1911–25.

Walker, M.J.C., Bjorck, S., Lowe, J.J., Cwynar, L.C., Johnsen, S., Knudsen, K.L. et al. (1999) Isotopic 'events' in the GRIP ice core: a stratotype for the Late Pleistocene. *Quaternary Science Reviews*, 18, 1143–50.

Walter, H.J., Hegner, E., Diekmann, B., Kuhn, G. and van der Loeff, M.M.R. (2000) Provenance and transport of terrigenous sediment in the South Atlantic Ocean and their relations to glacial and interglacial cycles: Nd and Sr isotopic evidence. *Geochimica et Cosmochimica Acta*, 64, 3813–27.

Wang, Z.H., Chen, Z.Y. and Tao, J. (2006) Clay mineral analysis of sediments in the Changjiang Delta Plain and its application to the Late Quaternary variations of sea level and sediment provenance. *Journal of Coastal Research*, 22, 683–91.

Wheeler, E.A. and Baas, P. (1993) The potentials and limitations of dicotyledonous wood anatomy for climatic reconstructions. *Palaeobiology*, 19, 487–98.

Willman, H.B. and Frye, J.C. (1970) *Pleistocene Stratigraphy of Illinois*. Illinois Geological Survey Bulletin 94.

Wilmshurst, J.M., Wiser, S.K. and Charman, D.J. (2003) Reconstructing Holocene water tables in New Zealand using testate amoebae: differential preservation of tests and implications for the use of transfer functions. *The Holocene*, 13, 61–72.

Wise, S.M., Thornes, J.B. and Gilman, A. (1982) How old are the badlands? A case study from south-east Spain, in R. Ryan and A. Yair (eds) *Badland Geomorphology and Piping*. Geobooks, Norwich. pp. 259–78.

Xu, S.J., Pan, B.T., Gao, H.S., Cao, G.J. and Su, H. (2007) Changes in sand fractions of Binggou section and the expansion and contraction of the Tengger Desert during 50–30 ka. *Earth Surface Processes and Landforms*, 32, 475–80.

Zhang, Z.H., Mann, M.E. and Cook, E.R. (2004) Alternative methods of proxy-based climate field reconstruction: application to summer drought over the conterminous United States back to AD 1700 from tree-ring data. *The Holocene*, 14, 502–16.

30

Environmental Change

Martin Williams

Geomorphology is the study of landscape origins and evolution. Every landscape is a legacy of past tectonic and climatic events and of present-day geomorphic and anthropogenic processes. Certain landforms, such as volcanoes, limestone caves or granite domes, reflect a strong geological control, while others, such as desert dunes or glacial moraines, require for their formation a specific set of geomorphic processes associated with a particular type of climate. Most landscapes are polygenic and contain landforms fashioned under ever changing climatic conditions, so that it is unwise to rely on purely geomorphic evidence to reconstruct past environmental change. In fact, geomorphology is as much a consumer as a producer of data relating to environmental change. Reconstruction of environmental change relies on a wide array of evidence from many disciplines, all of which are appropriate at particular scales in time and space so that an eclectic approach to the reconstruction of past, present and future environmental change is essential.

Rivers have always played a central theme in geomorphology. The first serious attempt at quantification date from the time of the late 18th century French and Italian engineers Guglielmini and Fabre (1797), who were responsible for flood mitigation in the valleys of the Alps (see Baulig, 1950, for a comprehensive review). Then followed a reversal to qualitative or descriptive studies, epitomized in the work of W.M. Davis (1909) and his *Explanatory Description of Landforms (Die Erklärende Beschreibung der Landformen*: 1912, written by him in English, translated by Alfred Rühl and published only in German) and what he termed the 'normal cycle of erosion'. Davis's ideas about valley development were refuted at length by Albrecht Penck in his *Morphologische*

Analyse (1924), published in the year after his death, but not available in English until 30 years later (Penck, 1953). This rebuttal was followed in more recent times by a widespread rejection of Davisian concepts (Beckinsale and Chorley, 1991: Chapter 4) and by more rigorous, process-based studies of fluvial hydraulics and palaeohydrology (Gregory et al., 1995).

Understanding the effects of past climatic changes on the world's great river systems, the life-blood of civilizations, ancient and modern, can also help us to appreciate how human societies have coped with environmental changes in the past, and is also essential to an understanding of how these systems may respond to future change.

Throughout the Quaternary the climate has oscillated between cold glacial and warm interglacial conditions, with the duration of each glacial–interglacial cycle lasting about 100,000 years (100 ka) during about the past million years, with each cycle showing a long, slow, fluctuating build-up to extreme glacial conditions (Imbrie and Imbrie, 1979; Williams et al., 1998; Bintanja and van de Wal, 2008; Raymo and Huybers, 2008). The relatively short-lived interglacial intervals were characterized by globally warmer climates and by enhanced summer monsoon activity, in contrast to the cold glacial maxima that were dry and windy in many parts of the world (Williams, 1975, 1985; Rognon and Williams, 1977; Williams et al., 1998). Prehistoric human societies have responded in one of three ways to the occasional abrupt climatic events and periodic extremes of Quaternary rainfall or temperature: adaptation, migration (Sharma, 1973), or extinction. Fundamental to our reconstruction of past human adaptation is our ability to find, date and analyse prehistoric occupation sites in primary context. In the decades before and after World War 2, influenced by the work of Zeuner (1958) and his contemporaries, many archaeologists working in Eurasia and Africa were preoccupied with discoveries of stone artefacts and fossil bones associated with river gravels, leading to some over simplified equations between alluvial gravels and their associated stone tools, and over-reliance on transported and often mixed assemblages. The involvement in archaeological research of earth scientists trained in geomorphology led to a rapid change in focus. For example, investigations of the Quaternary geology of big river valleys in India (Williams and Royce, 1982, 1983; Williams and Clarke, 1984, 1995; Williams et al., 2004, 2006a; Gibling et al., 2008; Tandon et al., 2008) have involved a more rigorous approach to identifying and dating alluvial formations and their spatial and temporal limits, based upon a battery of new techniques, provoking some spirited and much-needed debate that continues to this day (Tandon et al., 2008). Archaeologists have sometimes been slow to capitalize on these advances in fluvial geomorphology and chronostratigraphy, and have perhaps not always paid enough attention to taphonomic issues and questions of differential preservation and sorting of bone, stone and other archaeological remains. With some exceptions, Quaternary geologists and fluvial geomorphologists working in India have been equally slow to engage with their archaeologist counterparts in genuine interdisciplinary investigation, a tendency evident elsewhere and deplored by Butzer (1971) over three decades ago.

An interesting question which has arisen from this interdisciplinary research to which geomorphology has made a valuable contribution concerns the degree to which the artefact assemblages preserved in alluvial and other sediments are representative of particular types of activity associated with particular geomorphic elements in the landscape, such as flood plain, back-swamp, channel bank, or particular ecotones such as the interface between grassy flood plain and wooded hillside. Another issue is how much of the archaeological record is preserved and how much is missing, which immediately raises the question of how long are the time intervals characterized by erosion of the Quaternary sedimentary record. In seeking some answers to these questions we need first to consider the Quaternary alluvial record preserved in river valleys large and small, since a reliable supply of water is essential to all forms of life. Before dealing with these questions in greater detail it is useful to begin by reviewing some of the methods used to reconstruct past environments, using the 65 million years of the Cenozoic era as a convenient starting point.

LATE CENOZOIC ENVIRONMENTAL CHANGE

The types of evidence used in the reconstruction of Cenozoic environmental change are summarized in Table 30.1. In every case it is important to have a reliable chronology. Age control should always be accurate, meaning that the ages provided must relate directly to the event or material being dated. Ideally, the ages should also be precise, but in many cases the error terms are unavoidably large.

Table 30.2 provides an overview of some of the more important late Cenozoic tectonic and climatic events. Major environmental changes were not always globally synchronous. For example, the Eocene was a time of progressive cooling in high latitudes but of stable warm tropical climate in low latitudes (Pearson et al., 2007).

Events in both hemispheres contributed to late Cenozoic desiccation. After separating from Antarctica about 45 Ma ago, Australia has been moving north into dry subtropical latitudes at a mean rate of 6 cm/yr (Wellman and McDougall, 1974). This northward movement across one or more hot spots was associated with volcanic activity and renewed uplift along the Eastern Highlands of Australia, accentuating the rain shadow west of the divide. Within Australia, forest gave way to woodland, and woodland gave way to savanna. The net effects for Australia were climatic desiccation, progressive disruption of the drainage network, expansion of the desert, and successive plant and animal extinctions.

The Drake Passage between South America and Antarctica opened about 34 Ma ago, leading to the establishment of the circum-Antarctic ocean current driven by the prevailing westerly winds (McGowran et al., 2004). As a result, Antarctica became thermally isolated from warmer ocean waters to the north, and rapid cooling ensued. In the Southern Ocean, the changing isotopic composition of both planktonic and benthic foraminifera indicates major cooling of deep ocean water as well as surface water (Shackleton and Kennett, 1975; McGowran et al., 2004). Cumulative ice build-up in Antarctica saw the creation of mountain glaciers followed by the growth of a major ice cap, first in East Antarctica and later in West Antarctica.

Uplift of the Tibetan plateau as a result of the collision of India and Asia ~ 45 Ma ago caused a major change in the distribution of land and sea and was followed by severe desiccation of the region to the north and east of the plateau. Widespread sedimentation in playa lakes at the north-eastern edge of the Tibetan plateau persisted during the Eocene and ended abruptly at the Eocene–Oligocene

Table 30.1 Evidence used to reconstruct environmental change (Williams et al., 1998)

Proxy data source	Variable measured
Geology and geomorphology – continental	
Relict soils	Soil types
Closed-basin lakes	Lake level
Lake sediments	Varve thickness
Eolian sediments – loess, desert dust, dunes, sand plains	Mineralogical composition and surface texture; geochemistry
Lacustrine deposits and erosional features	Mineralogical composition and surface texture; geochemistry
	Age
Evaporites, tufas	Stable isotope composition
Speleothems	
Geology and geomorphology – marine	
Ocean sediments	Accumulation rates
	Fossil plankton composition
	Isotopic composition of planktonic and benthic fossils
Continental dust; fluviatile inputs	Mineralogical composition and surface texture; geochemistry
Biogenic dust: pollen, diatoms, phytoliths	Provenance
Marine shorelines	Coastal features, reef growth
Glaciology	
Mountain glaciers, ice sheets	Terminal positions
Glacial deposits and features of glacial erosion	Equilibrium snowline
Periglacial features	Distribution and age
Glacio-eustatic features	Shorelines
Layered ice-cores	Oxygen isotope concentration; physical properties (e.g. ice fabric); trace element and micro-particle concentrations
Biology and biogeography – continental	
Tree rings	Ring-width anomaly, density; isotopic composition
Fossil pollen and spores; plant macrofossils and microfossils; vertebrate fossils; invertebrate fossils: mollusca,ostracods	Type, relative abundance and/or absolute concentrations; age; distribution
Diatoms	Type, assemblage, abundance
Insects	
Modern population distributions	Refuges; relict populations of plants and animals
Molecular biology and genetics	
Biology and biogeography – marine	
Diatoms; foraminifera; coral reefs	Abundance, assemblage, trace element geochemistry, oxygen isotopic composition
Archaeology	—
Written records; plant remains; animal remains, including hominids; rock art; hearths, dwellings, workshops; artefacts: bone, stone, wood, shell, leather	

transition, coincident with the global cooling 33–34 Ma ago associated with the inception of permanent Antarctic ice sheets at that time (Dupont-Nivet et al., 2007). A major drop in temperature is also evident 33–34 Ma ago in the Great Plains of North America (Zanazzi et al., 2007).

We do not yet know when the northern hemisphere ice caps began to grow. Drop-stones from ice-rafted debris laid down in the Norwegian–Greenland Sea between 38 and 30 Ma ago and apparently derived from East Greenland suggest that northern high latitude ice accumulation may

Table 30.2 Late Cenozoic tectonic and climatic events (Sources cited in text)

Period (Ma)	Event
50–45	Separation of Australia from Antarctica
	Northward movement of Australia into dry subtropical latitudes
45	Collision of Greater India with Asia
	Progressive uplift of the Tibetan Plateau and Himalayas
38–30	Dropstones in Norwegian–Greenland Sea
	Ice present in Greenland?
34–33	Opening of the Drake Passage between Antarctica and South America
	Creation of the circum-Antarctic current
	Major ice accumulation in Antarctica
	Major global cooling
	Severe desiccation in central Asia
6–5	Miocene salinity crisis
	Mediterranean salt desert
	Incision of Nile canyon
	Genetic isolation of Africa from Eurasia
	Emergence of bipedal hominids
4–3	Closure of the Indonesian seaway
	Diversion of cool ocean water towards East Africa
	Desiccation in East Africa
	Closure of the Panama Isthmus
2.7–2.5	Rapid accumulation of ice over North America
	Drying out of East Africa and the Sahara
	First appearance of stone tool-making in East Africa
	Enhanced aridity in central Australia
2.4–0.9	High frequency low amplitude 41 ka glacial–interglacial cycles
0.9–0	Low frequency, high amplitude 100 ka glacial–interglacial cycles

be far older than previously envisaged (Eldrett et al., 2007).

In East Africa, uplift and rifting created the Neogene sedimentary basins with their unrivalled record of Pliocene and Pleistocene hominid evolution. The emergence in this region of the early Pliocene hominids may be linked to the Messinian salinity crisis of 6 to 5 Ma ago, during which the Mediterranean Sea dried out, refilled and dried out repeatedly, resulting in the creation of a salt desert

and the genetic isolation of Africa from Eurasia (Williams et al., 1998). The late Miocene Nile responded to this change in base level by cutting a gorge over 1000 km long and up to 2 km deep at its northern end (Said, 1993).

It is often hard to separate out the precise causes of environmental change. For example, closure of the Indonesian seaway 3–4 Ma ago as a result of northward displacement of New Guinea in the early Pliocene may have triggered a change in the source of water flowing through Indonesia into the Indian Ocean from previously warm South Pacific water to cooler North Pacific water, leading to reduced rainfall over East Africa (Cane and Molnar, 2001). Just as plausibly, the late Pliocene increase in aridity evident in East Africa and Ethiopia 3–4 Ma ago (Feakins et al., 2005) may have arisen from closure of the Panama Isthmus and northward diversion of the warm equatorial water which until then had flowed westwards from the Atlantic into the Pacific Ocean. The presence of warm moist air over the North Atlantic, allied to changes in solar insolation, resulted in widespread and persistent snow accumulation over North America (Williams et al., 1998). The rapid accumulation of ice over North America at 3.0–2.5 Ma was accompanied by global cooling in high latitudes and intertropical desiccation, revealed in the drying out of the large ate Pliocene tropical lakes of the Sahara and East Africa. The emergence of stone toolmaking at this time in East Africa may have been an adaptation by our ancestors to the increase in seasonality and the need to diversify their sources of food protein (Williams et al., 1998: 329). This was also a time of widespread loess accumulation in central China (Heller and Liu, 1982) and of the first appearance of stony desert plains in central Australia (Fujioka et al., 2005). The region around the Mediterranean also developed its now characteristic dry summer, wet winter climatic regime (Suc, 1984).

The net effects of these late Cenozoic environmental changes were an increase in the temperature gradients between high and low latitudes, a more seasonal rainfall regime, a reduction in forest and the replacement of woodlands by deserts in North Africa, Arabia and Australia, and the emergence in Africa some 2.5 Ma ago of ancestral human who walked upright and made stone tools.

QUATERNARY GLACIAL–INTERGLACIAL CYCLES AND THEIR LEGACY

The rapid accumulation of ice over North America at 2.7–2.5 Ma was followed by repeated oscillations from glacial to interglacial conditions

(Williams et al., 1998). These glacial–interglacial cycles were accompanied by glacio-eustatic sea level oscillations and by alternating morphogenetic systems (Reid, 2009; Williams, 2009a). Initial interest was focussed almost entirely upon the often fragmentary glacial, glaci-fluvial and glaci-aeolian landforms and their associated fossils, with the result that both the chronology and extent of the more recent glacial deposits are now reasonably well established in Europe and North America, but much less so in South America, Asia and Africa (Ehlers and Gibberd, 2004, 2007). In contrast, areas outside the northern mid-latitudes were comparatively neglected, particularly the hot deserts. In China the magnificent sequence of alternating loess (wind-blown desert dust) and fossil soils was to receive world recognition thanks to the efforts of Professor Liu Tungsheng and his successors (Liu, 1985, 1987, 1991; An and Porter, 1997). The hot wet tropics received somewhat belated attention (Flenley, 1978). Within the tropics, the cold and dry glacial maxima were times of desert expansion and forest retreat. Over the past 0.9 Ma, each cycle lasted 100 ka, reflecting the influence of the 100-ka orbital eccentricity cycle (Williams et al., 1998; Raymo and Huybers, 2008; Bintanja and van de Wal, 2008). In those parts of North America beyond the limits of the Laurentide ice cap, lake levels were high as a result of locally higher precipitation and runoff, allied to reduced evaporation (see Orme, 2008, for an historical survey of Pleistocene pluvial lakes in the American West). The synchronism between glacial phases and mid-latitude high lake levels led to the notion of glacial pluvials, a concept eagerly but erroneously applied to lakes and rivers in Africa (Nilsson, 1931, 1940; Butzer and Hansen, 1968; Büdel, 1977) until radiocarbon dating of high lake shorelines in the Sahara (Faure et al., 1963; Faure, 1969) and East Africa showed them to be of Holocene age (Butzer et al., 1972). Equally understandable, albeit misguided, was the development of a glacial–pluvial chronology for prehistoric sites in Africa, cogently criticized by Cooke (1958) and by Flint (1959a, 1959b). Grove (2008) has provided a scholarly account of what he termed the 'revolution in palaeoclimatology' initiated, *inter alia*, by the widespread use of radiocarbon dating of lake shorelines across the world from about 1970 onwards (Street and Grove, 1976, 1979; Street-Perrott et al., 1985).

The impact of these rapid climatic fluctuations may be illustrated with reference to Australia and India. In the case of Australia we will focus primarily on the Last Glacial Maximum (LGM) and on the very diverse array of evidence used in reconstructing the LGM environments. Bard (1999) defined the LGM (i.e. the time of maximum global ice volume, determined from the marine oxygen isotope record) as the 6000-year interval between 24,000 and 18,000 calendar years ago (24–18 ka). Mix et al. (2001) later narrowed down this age to 21 ± 2 ka. The LGM represents the time of most extreme global climate since the dominantly warm, wet Last Interglacial some 125,000 years ago. For example, the area now covered by humid rain forest in the Amazon was probably reduced by ~54 per cent as a result of a 20–40 per cent decrease in LGM precipitation accompanied by a temperature drop of 4.5–5°C; corresponding figures for the Congo humid forest are 84 per cent, 30–40 per cent and 5°C (Anhuf et al., 2006). Elsewhere in Africa and Asia, lake levels fell and desert areas expanded while woodland areas contracted.

THE LAST GLACIAL MAXIMUM IN AUSTRALIA

The evidence used to reconstruct the LGM environment in Australia includes marine sediment cores and coral reefs; glacial and periglacial deposits; desert dunes and loess; lakes, lunettes and wetlands; river alluvium; pollen; isotope geochemistry; cave deposits; and vertebrate and invertebrate fossils (Williams et al., 2009). The northern third of the continent lies within the tropical summer rainfall zone while the temperate south receives most of its rain in winter. The arid centre lies in the zone of mainly summer rainfall with hot summers and cold dry winters. Desert dunes and sand sheets cover nearly two fifths of the continent and today are vegetated and stable except along the dune crests.

During the LGM sea level was ~120 m lower than today (Yokoyama et al., 2000) and the land area was increased by ~25 per cent. The now submerged continental shelf is narrow in the east, wide in the north (200–300 km), south (up to 200 km) and north-west, where longitudinal desert dunes now lie below modern sea level. The late Pleistocene land bridge linking mainland Australia to Papua New Guinea (PNG) diverted the warm South Equatorial Current northwards around PNG and southwards outside the then exposed Great Barrier Reef, enhancing aridity in the tropical north (Figure 30.1).

During the LGM, sea surface temperatures (SSTs) were 1–2°C cooler to the north-west and north-east of Australia and 2–6°C cooler to the west, south and east in both August and February. SSTs during the LGM in the Indo-Pacific Warm Pool were ~3°C cooler (Gagan et al., 2004). The tropics were up to 4°C cooler in the eastern Indian Ocean and 0–3°C cooler elsewhere along the

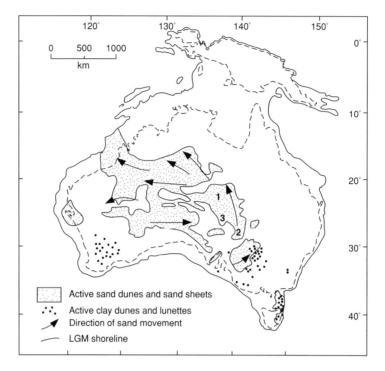

Figure 30.1   Australia during the Last Glacial Maximum when sea level was 120 m lower and the desert dunes were active. Note the land bridges connecting mainland Australia to Papua New Guinea and to Tasmania. Arrows show direction of sand flow. Black dots represent crescent-shaped clay dunes or lunettes. 1, Simpson Desert; 2, Strzelecki Desert; 3, Tirari Desert

equator (Barrows and Juggins, 2005). Greatest cooling was in high latitudes, with a maximum of 7–9°C in the south-west Pacific. The time of coldest SST was 20.5 ± 1.4 ka. Glacial and periglacial landforms are confined to a small area of the south-eastern highlands and larger areas of Tasmania. The three most recent glacial advances in Snowy Mountains have been dated using the cosmogenic nuclide [10]Be to 32 ± 2.5 ka, 19.1 ± 1.6 ka and 16.8 ± 1.4 ka (Barrows et al., 2001, 2002). These glacial advances were broadly synchronous with phases of blockstream and rock glacier activity dated by cosmogenic [36]Cl. Periglacial activity was concentrated during the time interval 23–16 ka (Barrows et al., 2004). In the Snowy Mountains the orographic snowline was 600–700 m lower, the lower limit of periglacial solifluction was at least 975 m lower and the temperature in the warmest month was at least 9°C cooler (Galloway, 1965). Blue Lake in the Snowy Mountains was free of ice by 15.8 ka (Barrows et al., 2001).

During and after the LGM dunes were active over 40 per cent of the continent, including north-eastern Tasmania. OSL age estimates for

dunes in the Strzelecki and Tirari deserts show well-separated peaks at ~14 ka, 20 ka (LGM), and 34 ka (Fitzsimmons et al., 2007a, 2007b). The event with the most samples is that at 20 ka when glaciers were at their maximum global extent and sea level was ~120 m lower.

The marine record shows a three-fold increase in dust flux during the LGM relative to the Holocene in temperate and tropical Australia (Hesse and McTainsh, 2003). The causes appear to be weakened Australian monsoon rains (tropical north) and drier westerly circulation (temperate south). During 33–16 ka, there was enhanced aeolian dust flux to the east and south over the southern half of the continent. The northern limit of the dust plume was 350 km or 3 degree north of the present limit during 22–18 ka (Hesse et al., 2004). Clay dunes and gypseous lunettes were active on the downwind margins of seasonally fluctuating lakes in the south-east and south-west of the continent immediately before and during 21–19 ka. Aeolian dust began to accumulate in the lunettes on the eastern side of Pleistocene Lake Mungo and adjacent lakes from

~35 ka until ~16 ka with a peak centred on the LGM (Bowler, 1998). Gingele and De Deckker (2005) have recorded aeolian dust in two cores located on the continental margin of South Australia immediately south of Kangaroo Island in the Murray Canyons area. During periods of minimum insolation at this latitude, strong northerly winds blew dust from the continental interior, with one peak at ~20 ka.

During the LGM the tropical northern lakes were mostly dry, except for Lake Carpentaria. Lake Eyre in central Australia was dry and its bed was being lowered by wind erosion at this time (Magee et al., 1995, 2004; Magee and Miller, 1998). In presently semi-arid western New South Wales, the Willandra Lake levels were declining progressively during 34–26 ka, dried out towards 25 ka, were high during 24–22 ka, dried out during 22–19 ka, rose at 18 ka and have remained dry thereafter (Bowler, 1998; Bowler and Price, 1998).

During and before the LGM, many rivers draining westwards from the Eastern Highlands transported coarse sand and gravel. In the lowlands there was widespread deposition until ~15 ka by all major rivers west of the Eastern Highlands, with deposition concentrated towards 30–25 ka, 20–18 ka and 18–14 ka. Meander wavelengths and channel widths indicate greater fluvial discharge, at least seasonally, before and during the LGM (Bowler, 1978; Page et al., 1996).

The pollen record from Lynch's Crater on the Atherton Tableland of north-east Queensland shows a prolonged interval of very low precipitation culminating at the LGM, with replacement of tropical rainforest by sclerophyll eucalyptus woodland. Kershaw (1976, 1994) estimated that at the LGM the annual rainfall at Lynch's Crater amounted to ~500 mm; today it is ~2500 mm. In the south-eastern uplands and Tasmania, alpine shrubland, grassland and herbland expanded during the LGM, while woodland gave way to grassland in less elevated areas in the south and south-east (Kershaw, 1995).

The record of cyclic peat humification from Lynch's Crater in north-east Queensland (Turney et al., 2004) adds significant new detail to Kershaw's earlier reconstructions of precipitation based on pollen spectra (Kershaw, 1976, 1994). Increased peat humification reflects microbial activity under aerobic conditions and drier surface conditions. The new record spans the last 45,000 years and is consistent with drying in northern Australia centred on 40 ka, 25 ka and 15 ka with cooler events centred on 30 ka and 21 ka (Turney et al., 2004). To sum up, pollen spectra from tropical Australia are consistent with a weaker summer monsoon during the LGM and with northward displacement of the ITCZ. Pollen spectra from Tasmania and mainland south-east Australia are likewise consistent with lower winter rainfall during the LGM.

In addition to the increasingly well-documented impacts of these geologically recent extreme climates upon the flora, there has also been speculation as to the possible impacts upon the fauna, with claims of widespread faunal extinctions resulting from climatic desiccation during the LGM. Recent advances in molecular biology have brought these issues into sharper relief by providing independent age estimates for times of species divergence and radiation in Europe and North America (Hewitt, 2000). In Australia many such speciation events extend well back into the Pliocene or Miocene and often appear to relate to pulses of climatic desiccation (Byrne et al., 2008). However, a substantial number of changes in the Australian biota appear to be of relatively recent origin, possibly reflecting the influence of rapidly changing climates linked to Quaternary glacial–interglacial cycles, or even the impact of the severe cold and aridity that afflicted the continent during the LGM (Byrne et al., 2008).

Johnson et al. (1999) analysed carbon isotopes in fossil emu eggshell from around Lake Eyre in central Australia. They found significant changes in the proportions of C4 grasses over the last 65,000 years. The data imply that the Australian monsoon was most effective between 65 and 45 ka, least effective during the LGM, and moderately effective during the Holocene. They noted that in arid inland Australia the effectiveness of the summer monsoon decreased at about the time that the megafauna became extinct and at about the time that humans arrived on the continent.

Widespread faunal extinctions took place in a concentric zone around the arid core that appears to have supported more abundant woodlands during wetter intervals before the LGM. Some of these bone beds occur around former spring sites that dried out during arid phases. Most of the extinct megafauna were browsers rather than grazers. Debate continues as to the relative role of climate and human impact, both direct (through hunting) and indirect (through modification of vegetation by fire) on late Quaternary faunal extinctions in Australia (Miller et al., 1999, 2005). Similar debates in North America are polarized between climate and prehistoric humans as the agents of faunal extinction.

We will now return to the questions raised in the introduction concerning the links between river responses to climatic change and the impact of such responses to prehistoric human occupation of river valleys, using two of the better studied rivers in north-central India (the Son and the Belan) as examples.

FLUVIAL GEOMORPHOLOGY, THE TOBA SUPER-ERUPTION AND PREHISTORIC OCCUPATION IN NORTH-CENTRAL INDIA

In the Son valley the least abraded Lower Palaeolithic Acheulian assemblages are associated with local alluvial fan sediments that are overlain by archaeologically sterile grey silty clays. These latter deposits are very well sorted and have been interpreted as wind-blown silts or loess (Williams and Royce, 1982; Williams and Clarke, 1984), although the possibility remains that they may have been re-worked by overland flow. The composite formation is known informally as the Sihawal Formation (Figure 30.2), after the village of that name on the left bank of the river close to the type section (Williams and Royce, 1982; Williams and Clarke, 1995; Williams et al., 2006a). The deposits are found at varying elevations in the valley and, in places, may crop out close to present low water river level. In contrast to later alluvial formations, the lower member of the Sihawal Formation consists of locally derived sandstone clasts. In the Belan valley, the Acheulian artefacts recovered by the teams led by the late Professors G.R. Sharma and J.D. Clark are more abraded and do not seem to be in primary context. Geomorphic considerations suggest that the search for undisturbed Acheulian sites is likely to prove most fruitful in sandstone rock shelters in the Kaimur Hills and in the transitional zone between the Kaimur Hills and the innermost margin of the former flood plain (Williams and Royce, 1982; Clark and Williams, 1986, 1990).

An abrupt change in depositional regime succeeded the deposition of the Sihawal Formation in the Son valley. Two younger alluvial formations have now been identified, each associated with Middle Palaeolithic artefacts. Some of the stone tools are abraded and transported, while others retain their sharp cutting edges and appear to lie close to their original place of deposition. The Patpara Formation (Figure 30.2) is a fining-upward alluvial deposit with abundant agate, chalcedony and jasper pebbles in the lower gravel members, indicating a source of alluvium in the Deccan Traps close to the interfluve (Williams and Royce, 1982). These raw materials were used by the Middle Palaeolithic stone toolmakers and some show signs of heat treatment (Clark and Williams, 1990). The sediments have undergone at least one phase of prolonged weathering and many of the finer sediments are coated with hydrated iron oxides and are characteristically red in colour.

The Toba volcanic eruption and its environmental impact

Conditions changed abruptly immediately after the 73 ± 2 ka eruption of Toba volcano in northern Sumatra (Chesner et al., 1991), which produced

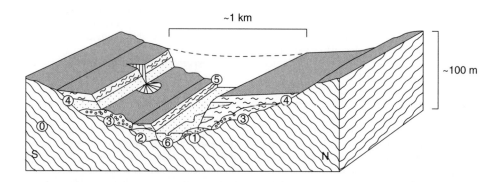

⓪ Lower Proterozoic metasediments
① Middle Pleistocene gravels and clays of the Sihawal Formation
② Upper Pleistocene sands and gravels of the Khunteli Formation
③ Upper Pleistocene gravels, sands and clays of the Patpara Formation
④ Terminal Pleistocene sands and clays of the Baghor Formation
⑤ Late Holocene clays, silts and sands of the Khetaunhi Formation
⑥ Present-day channel sands and point-bars of the River Son

Figure 30.2 Block diagram showing the five major alluvial formations investigated in the middle Son valley (After Williams et al., 2006a)

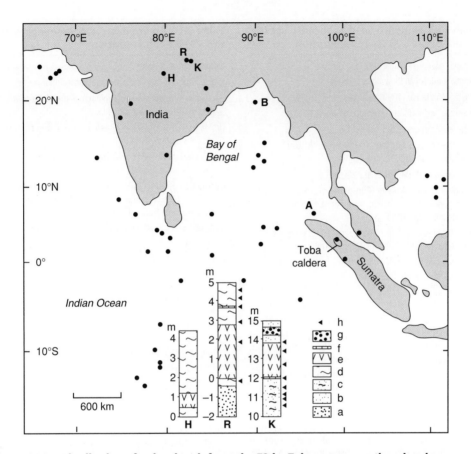

Figure 30.3 Distribution of volcanic ash from the 73 ka Toba super-eruption showing location of marine cores and sections sampled in India. Black dots represent Toba tephra occurrences on land and in marine cores. R is site of first Toba ash discovery at Son-Rehi confluence. B is marine core SO188-342KL in the Bay of Bengal; K is Khunteli; R is Rehi; H is Hirapur. Key to stratigraphic sections in India: a is coarse sand; b is medium/fine sand; c is silt loam/sandy loam/interstratified sand and loam; d is clay; e is Toba volcanic ash; f is massive carbonate; g is gravel; h is sampled pedogenic carbonate horizon (After Williams et al., 2009)

2000–3000 km³ of ejecta, and covered all of peninsular India in a layer of volcanic ash ~10 cm thick termed the Youngest Toba Tephra or YTT (Figure 30.3) (Acharyya and Basu, 1993; Shane et al., 1995, 1996; Westgate et al., 1998). By way of comparison, the eruption of Krakatoa in 1883 produced no more than 20 km³ of ejecta. The YTT has been recovered from marine cores in the Bay of Bengal (Ninkovich et al., 1978a, b; Ninkovich, 1979), the Indian Ocean to 12 degree south of the equator, the Arabian Sea and the South China Sea (Bühring and Sarnthein, 2000) (Figure 30.3).

Discovery of the Toba ash in India nicely illustrates one very human aspect of scientific

endeavour: the tendency to only notice what we expect to see and to ignore or fail to recognize the unexpected or out of place. On the afternoon of February 5th, 1980, the author noticed a buried channel (3.5 m by 10 m) filled with ash in a cliff section on the left bank of the Son slightly downstream from its confluence with the Rehi (Williams and Royce, 1982). Having worked extensively in Ethiopia, the author was familiar with the appearance of volcanic ash in outcrop, but was nonetheless surprised to see it here. This was the first time that any Quaternary volcanic ash had ever been recorded in India so that he decided to sample it at close vertical intervals. The sequel is well known (Williams and Clarke, 1995).

Table 30.3 Evidence used to reconstruct the impact of the ~73 ka Toba eruption

Historic eruptions (Williams et al., 2009)

Climate models (Williams et al., 2009)

Genetics (Ambrose, 1998; Williams et al., 2009)

Prehistoric archaeology (Petraglia et al., 2007)

Ice cores (Williams et al., 2009)

Stable isotope geochemistry (Ambrose et al., 2007; Williams et al., 2009)

Pollen analysis (Williams et al., 2009)

Geology and geochemistry (Chesner et al., 1991; Acharyya and Basu, 1993; Shane et al., 1995, 1996; Westgate et al., 1998)

Geomorphology (Williams and Royce, 1972; Williams and Clarke, 1995; Williams et al., 2009)

Marine cores (Ninkovich et al., 1978a, b; Ninkovich, 1979; Buhring and Sarnthein, 2000)

Table 30.4 Environmental consequences of the ~73 ka Toba eruption

Definite consequences

Eruption of ~3 000 km³ of ejecta and sulphur particles.

Formation of a volcanic ash veil across the globe that persisted for at least six years.

Deposition of sulphates in Greenland ice during at least six consecutive years.

Accumulation of wind blown dust in Greenland ice for several centuries after the eruption.

Destruction of forest in Sumatra as a result of the blast of the explosion and of fires triggered by very high temperatures from the erupted ash flow tuffs.

Deposition of a layer of ash over adjacent lands and oceans.

Deposition of a 10–15 cm thick ash mantle across peninsular India, leading to interference with respiration, transpiration and photo synthesis and associated widespread damage to the plant cover. Rivers and lakes choked with ash transported down slope by mass movement and surface runoff.

Change from widespread forest across central India before the eruption to open woodland and grassland for many centuries after the eruption.

Sharp drop in temperature recorded in Greenland ice cores.

Probable consequences

Acid rain resulting from oxidation of the sulphur to sulphur dioxide and its conversion to sulphuric acid on reacting with water vapour.

Possible consequences

Possible sharp drop in global temperature caused by the volcanic dust veil and sulphate aerosols.

Pollution of rivers, lakes and other wetlands as a result of toxins leached from the ash.

Possible abrupt decline in animal and prehistoric human populations linked to widespread ecosystem damage caused directly and indirectly by the eruption.

YTT ash has been found across India (Westgate et al., 1998) and isolated outcrops reworked soon after the eruption (cf. Collins and Dunne, 1986) continue to be discovered.

At all events, it seems *a priori* unlikely that the 73 ka Toba eruption had little impact on environment and people in South Asia and India, despite claims to the contrary (Gathorne-Hardy and Harcourt-Smith, 2003; Petraglia et al., 2007). It is not impossible, on the basis of mitochondrial DNA evidence (Ambrose, 1998), that the Middle Palaeolithic peoples of South Asia were decimated by this event, but this remains a working hypothesis.

Late Pleistocene valley aggradation and post-LGM incision

In north-central India the Late Pleistocene was a time of climatic extremes. Radiocarbon and IRSL dating of alluvial sediments in the Son and Belan valleys shows that the interval from ca. 39 ± 9 ka to ca. 16 ± 3 ka was one of widespread and prolonged alluvial sedimentation in these valleys (Pal et al., 2004). The time of most extreme climate coincided with the Last Glacial Maximum or LGM (21 ± 2 ka) (Mix et al., 2001), which was cold, dry and windy in much of the inter-tropical zone, including peninsular India (Williams, 1985), with sparse vegetation in the catchment headwaters, high sediment loads of coarse sands and gravels, highly seasonal flow regimes, and periodic influxes of wind-blown silts or loess (Williams and Clarke, 1984; Tandon et al., 2008). In the Son valley this interval of valley aggradation resulted in deposition of the Baghor Formation (Figure 30.2), which in the type section consists of a lower coarse member and an upper

fine member. The coarse member is comprised of 10–15 m of cross-bedded coarse sands and fine gravels with abundant transported bones of large mammals and abraded Upper Palaeolithic artefacts, and is often cemented with calcium carbonate, indicating dry climatic conditions. The upper fine member consists of alluvial clays and silts 10–15 m thick, and is overlain locally by Terminal Pleistocene to Early Holocene clays with Mesolithic artefacts and Sambur deer tracks (Sharma and Clark, 1983). The long interval of Late Pleistocene valley aggradation in this part of India ended with rapid and sustained vertical incision after 16 ± 3 ka, coincident with the onset of a warmer and wetter climate associated with a return of the summer monsoon regime so familiar today.

A similar response to these climatic extremes (cold, dry LGM; warm, wet Postglacial) is evident in other big tropical river systems, including the Nile (Adamson et al., 1980; Williams et al., 2006b). In the case of the Nile basin a simple model of river response to climate change was advanced nearly three decades ago (Adamson et al., 1980; Williams and Adamson, 1980; Woodward et al., 2007), with late Pleistocene river aggradation reflecting a high sediment load to discharge ratio caused by lower and more erratic river discharge linked to a weaker summer monsoon, and high sediment yields from relatively bare and easily eroded hill slopes. In this model river incision was brought about by a return to higher and less erratic discharge (a function of high summer monsoon rainfall) and an influx of finer sediment from the now vegetated and stable valley sides. River channels in which the bed and banks were lined with clay and the transported load were either a mixed load of sand, silt and clay (as in the Nile today) or a suspension load of silt and clay (as in the Huang Ho/Yellow River early last century, before upstream water abstraction had curtailed its flow) often have a surplus of kinetic energy available for vertical incision. This very simple model is in good general accord with the dated alluvial records from the Son and Belan valleys (Williams et al., 2006a; Gibling et al., 2008) and needs to be rigorously tested against the more detailed studies that are now beginning to emerge from different regions of India.

Terminal Pleistocene and Holocene aggradation

A younger phase of aggradation began towards ~5.5 ka and is especially well displayed in the Son valley in the form of an inset terrace consisting of inter-bedded silts, clays and fine sands. This terrace contains Neolithic artefacts, some in primary context (Sharma and Clark, 1983; Clark and Williams, 1986, 1990) and is on occasion submerged during present-day Son floods. On the right bank of the river, opposite the village of Sihawal, beds dipping 10–15° towards the present river with very well preserved sedimentary structures, indicative of recent rapid deposition, were observed in January 2004. These deposits may well prove to be ephemeral features at the mercy of subsequent big floods. The alternation of silts, clays and sands is indicative of variations in flow that may reflect variations in summer monsoon. There are several present-day climatic phenomena responsible for inter-annual and inter-decadal variations in monsoon strength and duration, including El Niño–Southern Oscillation events, which are discussed later in this chapter. The

essential point here is that there was an increase in the frequency of El Niño events towards 4.8 ka (Moy et al., 2002), resulting in an increase in climatic variability in widely scattered regions of the world, including India, eastern China, Indonesia, eastern Australia and the Ethiopian headwaters of the Nile (Adamson et al., 1987; Whetton et al., 1990). This period of enhanced climatic variability seems to coincide in time with the emergence of agriculture in north-central India (Sharma et al., 1980; Sharma and Clark, 1983), although the inception of plant and animal domestication had begun over five millennia earlier in the Fertile Crescent region of the Near East, during a return to a warmer, wetter postglacial climate (Williams et al., 1998).

RAPID LATE PLEISTOCENE AND HOLOCENE CLIMATIC FLUCTUATIONS

The Holocene spans the past 11.7 ka (Walker et al., 2009). Initial ideas as to the climatic stability of the Holocene have proven as unfounded as early ideas about the ability of the tropical rain forests to endure the climatic vicissitudes of the Quaternary unchanged (see Flenley, 1978 for an overview). Indeed, the ice core records from Greenland have shown that the last glacial–interglacial cycle was marked by frequent and rapid temperature changes, reflected in the oxygen isotope record. Dansgaard et al. (1993) documented 24 brief, warm 'interstadial' events, each lasting ~500 to ~2000 years, characterized by a rapid temperature rise of up to 7°C within a few decades, followed by a slower cooling, attaining 12–13°C below modern levels (cold 'stadial' events). Each packet, comprising a warm interstadial and a cold stadial, is termed a Dansgaard–Oeschger (D–O) event, lasting 1000–3000 years (Williams et al., 1998: Chapters 3 and 4). These warm and cold climatic phases were initially identified in the Scandinavian pollen record. The Bölling–Allerød (B–A) interstadial event, which began abruptly at ~14.6 ka, is not evident in the Antarctic ice core records, which often show a temperature signal opposite to that in high northern latitudes. Likewise, the Younger Dryas (YD) cold phase (12.8–11.7 ka) has long been evident in the northern hemisphere, but not, on present evidence, in Australia or New Zealand (Williams et al., 2005; Barrows et al., 2007; Williams et al., 2009). The abrupt onset and end of both the B–A interstadial and the YD stadial are also characteristic of shorter-term climatic fluctuations operating at centennial to decadal scales. Among the best known examples of these short, sharp climatic interludes are the periodic influxes of ice-rafted debris into the North Atlantic reported by Heinrich (1988) and now termed 'Heinrich events' and the Holocene glacial advances discussed by Denton and Karlén (1973) and comprehensively documented by the late Jean Grove in her magisterial compendium (Grove, 2004). These 'neoglacial' advances culminated in the Little Ice Age immortalized by Flemish painters and described in graphic detail by Le Roy Ladurie (1971). Other abrupt climatic perturbations include the 8.2 ka cooling event identified in Greenland and also evident as far away as equatorial Africa and China (Alley et al., 1997; Kendall et al., 2008); and the 4.2 ka severe drought which led to the demise of the Akkadian Empire in Mesopotamia (Weiss, 2000) and the disintegration of the Old Kingdom in Egypt (Figure 30.4).

THE LAST THOUSAND YEARS

The climatic fluctuations of the last 1000 years have been well documented for certain parts of the northern hemisphere (Le Roy Ladurie, 1971; Lamb, 1977; Jones et al., 2001; Grove, 2004), and

the cultural responses to these fluctuations have seen a renewal of scholarly interest (deMenocal, 2001; Diamond, 2005). In the northern hemisphere, the most recent of the Holocene Little Ice Ages is bracketed between about 1550 and 1900 AD, following what is widely termed the Mediaeval Warm Period of 99 to 1200 AD (Jones et al., 2001). Later research has revealed that the advance and retreat of mountain glaciers during the LIA was not always synchronous, and masked a high degree of local and regional variability (Grove, 2004). Furthermore, a growing body of evidence from the equatorial Pacific region strongly suggests that the terms Little Ice Age and Mediaeval Warm Period are misnomers for this region and should perhaps be replaced by 'Little Warm Age' and 'Mediaeval Cold Period' (Allen, 2006). Given the size of India and the range of regional climates within the sub-continent, it will be useful to obtain high-resolution records of this important time interval and to examine with care the response of Indian rivers to external forcing factors and the cultural responses to any documented changes in river behaviour, after taking into account any tectonic influences on

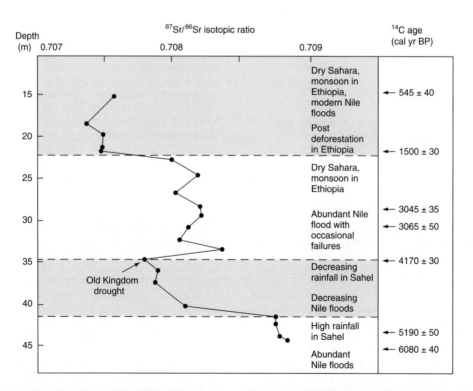

Figure 30.4   Depth profile of ^{87}Sr/^{86}Sr from coastal core S-21 in the Nile delta east of the Suez Canal, showing that the 4.2 ka drought coincided with the demise of the Old Kingdom in Egypt (From Williams, 2009b, after Krom et al., 2002 and Stanley et al., 2003)

river systems. To this end, it is illuminating to consider the historic records of floods and droughts in India and to place these within a wider synoptic context.

HUMAN IMPACT: PAST, PRESENT, FUTURE

With the advent of forest clearing from Neolithic times onwards, and the exponential increase in the human population as a result of plant and animal domestication, there is evidence of accelerated soil erosion. In Ethiopia the long-term rate of erosion in the forested highlands amounted to 10–15 m/Ma (McDougall et al., 1975). The present rate is one to two orders of magnitude higher (Hurni, 1999; Nyssen et al., 2004). A similar bleak picture is true of many of the once densely forested uplands of Africa, Asia and South America. Five large rivers in Asia provide nearly two thirds of the total sediment load of some 20 billion tonnes deposited each year in the ocean. Change in land use now outweighs climate as a geomorphic agent.

There is particular concern over the geomorphic impact of human activities in the arid, semi-arid and dry sub-humid regions that comprise 40 per cent of our total land area. Table 30.5 shows the extent and severity of desertification processes grouped according to continent. Following the 1992 Earth Summit Conference in Rio de Janeiro, desertification was re-defined as 'land degradation in arid, semi-arid and dry sub-humid areas resulting from various factors, including climatic variations and human activities' (UNCED, 1992). Before then, the United Nations Environment Programme (UNEP) definition specified 'adverse human activities' as the sole cause (UNEP, 1990).

The difficulty in defining desertification arose partly from uncertainty over sifting out the impact of humans from that of natural climatic variability in arid and semi-arid areas (Hare and Ogallo, 1993). Seventeen years of satellite monitoring of vegetation cover along the southern margins of the Sahara revealed considerable variation from year to year in response to annual variations in rainfall (Tucker and Nicholson, 1999). An earlier study by Tucker et al. (1991) extending from 1980 to 1990 concluded that at least 10 years of observations were needed to detect any possible trends in the vegetation cover in this region. This is true of other dry regions of the world where the annual rainfall is highly variable.

Dregne et al. (1991) indicated that roughly three quarters of all rangelands on each of the six continents were to some degree degraded. From 15 to 30 per cent of the land in irrigated areas was degraded. Over half of the rain-fed croplands in Africa, Asia and Europe were considered degraded, over a third in Australia and South America, and almost a fifth in North America.

In an effort to improve the quality of these data, UNEP approached the International Soil Reference Centre (ISRIC) in Holland and in 1987 UNEP and ISRIC began the Global Assessment of Soil Degradation (GLASOD) project. Table 30.5 is based on the GLASOD data published in the 1997 edition of the *World Atlas of Desertification* (UNEP, 1997).

Williams and Balling (1996) reviewed the worldwide evidence for desertification. Some of the more obvious signs of dryland degradation include accelerated soil erosion by wind and water; salt accumulation in the surface horizons of dryland soils; a decline in soil structural stability with an attendant increase in surface crusting and surface runoff and a concomitant reduction in soil infiltration capacity and soil

Table 30.5 Extent of soil degradation in susceptible drylands, grouped by continent, in millions of hectares (UNEP, 1997)

Region	Type of soil degradation				Total
	Water erosion	Wind erosion	Chemical deterioration	Physical deterioration	
Africa	119.1	159.9	26.5	13.9	319.4
Asia	157.5	153.2	50.2	9.6	370.5
Australasia	69.6	16.0	0.6	1.2	87.4
Europe	48.1	38.6	4.1	8.6	99.4
North America	38.4	37.8	2.2	1.0	79.4
South America	34.7	26.9	17.0	0.4	79.0
Total	467.4	432.4	100.7	34.7	1035.2

moisture storage; replacement of forest or woodland by secondary savanna grassland or scrub; an increase in the flow variability of dryland rivers and streams; an increase in the salt content of previously freshwater lakes, wetlands and rivers; and a reduction in species diversity and plant biomass in dryland ecosystems (Williams and Balling, 1996).

We now recognize the role of El Niño Southern Oscillation (ENSO) events and of regional anomalies in sea surface temperature in controlling the incidence of major floods and droughts in widely scattered parts of the world. Sir Gilbert Walker defined the Southern Oscillation in 1924 (Walker, 1924). The Southern Oscillation Index (SOI) is a measure of the surface atmospheric pressure difference between Darwin and Tahiti and is widely used today in predicting wet and dry years. Time series analysis shows a statistically significant correlation between years of low Nile flow, drought in Indonesia and years of weak summer monsoon rainfall in India and eastern China (Williams and Balling, 1996). The converse is equally true, with years of extreme flooding synchronous in each of these regions.

Despite earlier claims to the contrary, overgrazing was not responsible for the prolonged drought that began in the Sahel in 1968 (Williams and Balling, 1996). However, drought can lead to local overgrazing which can in turn accelerate soil erosion by wind and water and re-activate previously stable vegetated dunes. The southern margins of the Sahara, the eastern and southern flanks of the Gobi desert and the eastern borders of the Rajasthan desert of north-western India are all covered in sand dunes that have remained vegetated and stable for the last 10,000 years. Recent severe droughts combined with tree clearing and over-stocking have led to remobilization of the once stable desert dunes in each of these regions. In the Alashan region of Inner Mongolia in northern China, dunes are now advancing from the north-west at rates of up to 10 m/yr, and many millions of tons of sand are now blowing into the Yellow River. Local estimates suggest that some 30,000 km² of land are now severely degraded in Alashan and that the rate of desertification is increasing by about 1000 km² each year. Control measures to arrest dune movement include tree-planting and extensive use of the straw mulch chequer-board technique, especially along the main railway line linking Lanzhou to Baotou (Williams, 2000, 2002).

Aubréville (1949) documented the replacement of tropical African rain forest by secondary savanna and scrub as a result of tree clearing and burning associated with shifting cultivation and coined the term desertification to encapsulate this process. He concluded that humanly induced deserts are forming today in Africa in areas receiving 750–1500 mm of rain a year.

Removal of forest in tropical uplands can alter the local hydrological balance through increased runoff and reduced infiltration. Accelerated loss of soil from highland catchments can lead to sedimentation in reservoirs far downstream and not always within the same country. By 1996, the capacity of the Roseires reservoir on the Blue Nile had been reduced by almost 60 per cent through silt accumulation (Ayoub, 1999). The silt was from the Ethiopian highlands.

Hurni (1999) has monitored accelerated soil erosion in the Ethiopian uplands. In one region in Gojjam Province the area cultivated increased from 40 per cent in 1957 to 77 per cent in 1995, while the area under natural forest decreased from 27 per cent to 0.3 per cent. Annual rates of soil loss amount to about 2 mm/yr on mountain slopes, but attain rates of over 15 mm/yr during cultivation years, or some five to ten times more than in non-mountainous areas (Hurni, 1999).

Further examples of adverse human impacts upon our landscapes include the progressive accumulation of salt in agricultural soils as a result of clearing of the native vegetation, poor canal maintenance and inadequate soil drainage. Salt accumulation is now a major problem in northern China, the Indus basin and the Murray–Darling basin in Australia (Williams, 2000).

CONCLUDING REMARKS: HISTORIC FLOODS AND DROUGHTS

During his time as Director of the Indian Meteorological Department, Sir Gilbert Walker puzzled over the causes of inter-annual variations in the strength of the Indian summer monsoon. His statistical researches led him to recognize the Southern Oscillation (Walker, 1924). He noted that when the surface atmospheric pressure in the equatorial Pacific region off the coast of Peru was below average (El Niño years), the atmospheric pressure centred over northern Australia and Indonesia was above average, resulting in reduced precipitation in that region, and conversely. The Southern Oscillation Index (SOI) is a measure of the atmospheric pressure difference between Darwin and Tahiti. In those years when the SOI was strongly negative, as for instance in 1877, 1899, 1902, 1941, 1965 to 1966, 1972, and 1982 to 1983, extreme droughts were synchronous in north-eastern China, India, Ethiopia and eastern Australia (Whetton et al., 1990; Williams and Balling, 1996: 106–109 and 155–161). Years of synchronous floods in these same regions coincided with years of strongly positive SOI, as for example in 1887, 1889 to 1890, 1894, 1916 to 1917, 1955 to 1956, 1964 and 1975.

The year 1999 was an exceptionally wet year, marked by severe floods in eastern China, India and eastern Australia and very high flow in the Nile. The prolonged drought that had afflicted the Alashan region of Inner Mongolia ended abruptly in August that year, an event witnessed by the author. Exceptionally heavy rain in central Sudan at that time caused a major irrigation canal to breach its banks, an event captured on satellite imagery (Williams and Nottage, 2006). The extreme rainfall events over north-eastern Africa and India were associated with a warm equatorial Indian Ocean, a shift of the Inter-tropical Convergence Zone earlier and further north than usual, and the presence of deep, well developed westerly air masses accompanied by a strong Tropical Easterly Jet that allowed more moisture transport into Africa from the South Atlantic via the Congo basin (Williams and Nottage, 2006).

Understanding the synoptic conditions that give rise to extreme floods and droughts today can provide us with modern analogues of past events (Gasse et al., 2008), although the geomorphic response to past climatic change is often complex and sometimes counter-intuitive (Williams et al., 2006c). In order to achieve the full potential of this approach, existing stratigraphic work needs to be combined with stable isotope analysis (carbon, oxygen, strontium) and trace element geochemistry in order to quantify past changes in temperature, precipitation/evaporation and salinity. Such an approach has been used with good effect in the Nile basin and adjoining area (Abell and Williams, 1989; De Deckker and Williams, 1993; Ayliffe et al., 1996; Talbot et al., 2000). In addition, there is ample scope in India and elsewhere for using a wide variety of dating techniques, including the use of cosmogenic isotopes and palaeomagnetic dating (Pillans et al., 2005). Another promising approach that has yet to be widely attempted is the integration of palaeoenvironmental data with phylogenetic and phylogeographic evidence derived from molecular biology (Byrne et al., 2008). Finally, we need to use our growing appreciation of past environmental change to bolster our awareness of possible future change, since, ultimately, the past is our only proven guide to the present and future (Williams, 1993a, b, c; Williams, 2004).

With a world population of ~ 6.5 billion, and an exponential increase in greenhouse gases, it is highly probable that current global warming trends are primarily anthropogenic (IPCC, 2007). The predicted outcomes are for more extreme rainfall events and an increase in tropical summer rainfall. The geomorphic consequences are likely to be greater runoff and sediment yield from tropical rivers. It is hard to predict the climatic consequences of the albedo changes linked to the decrease in polar ice, since the response may be nonlinear. We will need to learn to live with uncertainty in a time of rapid environmental change, just as our ancestors did.

REFERENCES

Abell, P.I. and Williams, M.A.J. (1989) Oxygen and carbon isotope ratios in gastropod shells as indicators of palaeoenvironments in the Afar region of Ethiopia. *Palaeogeography, Palaeoclimatology, Palaeoecology*, 74, 265–78.

Acharyya, S.K. and Basu, P.K. (1993) Toba ash on the Indian subcontinent and its implications for correlation of late Pleistocene alluvium. *Quaternary Research*, 40, 10–19.

Adamson, D.A., Gasse, F., Street, F.A. and Williams, M.A.J. (1980) Late Quaternary history of the Nile. *Nature*, 287, 50–5.

Adamson, D.A., Williams, M.A.J. and Baxter, J.T. (1987) Complex late Quaternary alluvial history in the Nile, Murray–Darling and Ganges basins: three river systems presently linked to the Southern Oscillation, in V. Gardiner (ed.) *Proceedings of the First International Conference on Geomorphology*. Part II. Manchester, September 1985, John Wiley & Sons, Chichester. pp. 875–87.

Allen, M.S. (2006) New ideas about Late Holocene climate variability in the central Pacific. *Current Anthropology*, 47, 521–35.

Alley, R.B., Mayewski, P.A., Sowers, T., Stuiver, M., Taylor, K.C. and Clark, P.U. (1997) Holocene climatic instability: a prominent, widespread event 8200 yr ago. *Geology*, 25, 483–6.

Ambrose, S.H. (1998) Late Pleistocene human population bottlenecks, volcanic winter, and differentiation of modern humans. *Journal of Human Evolution*, 34, 623–51.

Ambrose, S.H., Williams, M.A.J., Chattopadhyaya, U., Pal, J.N. and Chauhan, P. (2007) Environmental impact of the 73 ka Toba eruption reflected by paleosol carbonate carbon isotope ratios in central India. INQUA 2007 Abstracts. *Quaternary International*, 8, 167–8.

An, Z. and Porter, S. (1997) Millennial-scale climatic oscillations during the last interglacial in central China. *Geology*, 25, 603–6.

Anhuf, D., Ledru, M.-P., Behling, H., Da Cruz Jr., F.W., Cordeiro, R.C., Van der Hammen, T. et al. (2006) Paleoenvironmental change in the Amazonian and African rainforest during the LGM. *Palaeogeography, Palaeoclimatology, Palaeoecology*, 239, 510–27.

Aubréville, A. (1949) *Climats, Forêts et Désertification de l'Afrique Tropicale*. Société d'Editions Géographiques, Maritimes et Coloniales, Paris.

Ayliffe, D., Williams, M.A.J. and Sheldon, F. (1996) Stable carbon and oxygen isotopic composition of early Holocene gastropods from Wadi Mansurub, north-central Sudan. *The Holocene*, 6, 157–69.

Ayoub, A.T. (1999) Land degradation, rainfall variability and food production in the Sahelian zone of the Sudan. *Land Degradation and Development*, 10, 489–500.

Bard, E. (1999) Ice Age temperatures and geochemistry. *Science*, 284, 1133–4.

Barrows, T.T. and Juggins, S. (2005) Sea-surface temperatures around the Australian margin and Indian Ocean during the Last Glacial Maximum. *Quaternary Science Reviews*, 24, 1017–47.

Barrows, T.T., Stone, J.O. and Fifield, L.K. (2004) Exposure ages for Pleistocene periglacial deposits in Australia. *Quaternary Science Reviews*, 23, 697–708.

Barrows, T.T., Stone, J.O., Fifield, L.K. and Cresswell, R.G. (2001) Late Pleistocene glaciation of the Kosciuszko Massif, Snowy Mountains, Australia. *Quaternary Research*, 55, 179–89.

Barrows, T.T., Stone, J.O., Fifield, L.K. and Cresswell, R.G. (2002) The timing of the Last Glacial Maximum in Australia. *Quaternary Science Reviews*, 21, 159–73.

Barrows, T.T., Lehman, S.J., Fifield, L.K. and De Deckker, P. (2007) Absence of cooling in New Zealand and the adjacent ocean during the Younger Dryas chronozone. *Science*, 318, 86–9

Baulig, H. (1950) La notion de profil d'équilibre: histoire et critique, in H. Baulig (ed.) *Essais de Géomorphologie*. Publications de la Faculté des Lettres de l'Université de Strasbourg, Fascicule 114. Société d'Édition, Les Belles Lettres, Paris. pp. 43–86.

Beckinsale, R.P. and Chorley, R.J. (1991) *The History of the Study of Landforms or the Development of Geomorphology*. Vol. 3: *Historical and Regional Geomorphology 1890–1950*. Routledge, London and New York.

Bintanja, R. and van de Wal, R.S.W. (2008) North American ice-sheet dynamics and the onset of 100, 000-year glacial cycles. *Nature*, 454, 869–72.

Bowler, J.M. (1978) Quaternary climate and tectonics in the evolution of the Riverine Plain, southeastern Australia, in J.L. Davies and M.A.J. Williams (eds) *Landform Evolution in Australasia*. Australian National University Press, Canberra. pp. 70–112.

Bowler, J.M. (1998) Willandra Lakes revisited: environmental framework for human occupation. *Archaeology in Oceania*, 33, 120–55.

Bowler, J.M. and Price, D.M. (1998) Luminescence dates and stratigraphic analyses at Lake Mungo: review and new perspectives. *Archaeology in Oceania*, 33, 156–68.

Büdel, J. (1977) *Klima-Geomorphologie*. Borntraeger, Berlin and Stuttgart.

Bühring, C. and Sarnthein, M. (2000) Toba ash layers in the South China Sea: evidence of contrasting wind directions during eruption ca. 74ka. *Geology*, 28, 275–8.

Butzer, K.W. (1971) *Environment and Archaeology*. 2nd. edn. Methuen, London.

Butzer, K.W. and Hansen, C.L. (1968) *Desert and River in Nubia*. University of Wisconsin Press, Madison.

Butzer, K.W., Isaac, G.L., Richardson, J.L. and Washbourn-Kamau, C. (1972) Radiocarbon dating of East African lake levels. *Science*, 175, 1069–76.

Byrne, M., Yeates, D.K., Joseph, L., Kearney, M., Bowler, J., Williams, M.A.J. et al. (2008) Birth of a biome: insights into the assembly and maintenance of the Australian arid zone biota. *Molecular Ecology*, 17, 4398–417.

Cane, M.A. and Molnar, P. (2001) Closing of the Indonesian seaway as a precursor to east African aridification around 3-4 million years ago. *Nature*, 411, 157–62.

Chesner, C.A., Rose, W.I., Deino, A., Drake, R. and Westgate, J.A. (1991) Eruptive history of earth's largest Quaternary caldera (Toba, Indonesia) clarified. *Geology*, 19, 200–3.

Clark, J.D. and Williams, M.A.J. (1986) Palaeoenvironments and prehistory in north central India: a preliminary report, in J. Jacobson (ed.) *Studies in the Archaeology of India and Pakistan*. New Delhi, Oxford. pp. 18–41.

Clark, J.D. and Williams, M.A.J. (1990) Prehistoric ecology, resource strategies and culture change in the Son valley, northern Madhya Pradesh, central India. *Man and Environment*, 15, 13–24.

Collins, B.D. and Dunne, T. (1986) Erosion of tephra from the 1980 eruption of Mount St. Helens. *Geological Society of America Bulletin*, 97, 896–905.

Cooke, H.B.S. (1958) Observations relating to Quaternary environments in East and Southern Africa. Alex du Toit Memorial Lecture no. 5, Annex to vol. LX. *Geological Society of South Africa Bulletin*.

Dansgaard, W., Johnsen, S.J., Clausen, H.B., Dahl-Jensen, D., Gundestrup, N.S., Hammer, C.U., et al. (1993) Evidence for general instability of past climate from a 250-kyr ice-core record. *Nature*, 364, 218–20.

Davis, W.M. (1909) *Geographical Essays*. Ginn, Boston (reprinted 1954, Dover).

Davis, W.M. (1912) *Die erklärende Beschreibung der Landformen*. B.G. Teubner, Leipzig.

De Deckker, P. and Williams, M.A.J. (1993) Lacustrine paleoenvironments of the area of Bir Tarfawi-Bir Sahara reconstructed from fossil ostracods and the chemistry of their shells, in F. Wendorf, R. Schild, A.E. Close and Associates. *Egypt During the Last Interglacial: The Middle Palaeolithic of Bir Tarfawi and Bir Sahara East*. Plenum, New York. pp. 115–9.

de Menocal, P.B. (2001) Cultural responses to climate change during the Late Holocene. *Science*, 292, 667–73.

Denton, G.H. and Karlén, W. (1973) Holocene climatic variations: their pattern and possible cause. *Quaternary Research*, 3, 155–205.

Diamond, J. (2005) *Collapse: How Societies Choose to Fail or Survive*. Penguin, Allen Lane, New York. p. 575.

Dregne, H.E., Kassas, M. and Rozanov, B. (1991) A new assessment of the world status of desertification. *Desertification Control Bulletin*, 19, 6–18.

Dupont-Nivet, G., Krijgsman, W., Langereis, C.G., Abels, H.A., Dai, S. and Fang, X. (2007) Tibetan plateau aridification linked to global cooling at the Eocene-Oligocene transition. *Nature*, 445, 635–8.

Ehlers, J. and Gibbard, P. (2004) *Quaternary Glaciations – Extent and Chronolgy. Part I: Europe. Part II: North America. Part III: South America, Asia, Africa, Australasia, Antarctica*. Elsevier, Amsterdam.

Ehlers, J. and Gibbard, P. (2007) The extent and chronolgy of Cenozoic global glaciation. *Quaternary International*, 164–5, 6–20.

Eldrett, J.S., Harding, I.C., Wilson, P.A., Butler, E. and Roberts, A.P. (2007) Continental ice in Greenland during the Eocene and Oligocene. *Nature*, 446, 176–9.

Fabre (1797) *Essai sur la Théorie des Torrens et des Rivières*. Paris. Cited in H. Baulig, *Essais de Géomorphologie*, 1950. pp. 52–3.

Faure, H. (1969) Lacs quaternaires du Sahara. *Internationale Vereiningung für theoretische und angewandte Limnologie*, 17, 131–46.

Faure, H., Manguin, E. and Nydal, R. (1963) Formations lacustres du Quaternaire supérieur du Niger oriental: diatomites et âges absolus. *Bulletin du Bureau de Recherches Géologiques et Minières (Dakar)*, 3, 41–63.

Feakins, S.J., deMenocal, P.B. and Eglinton, T.I. (2005) Biomarker records of late Neogene changes in northeast African vegetation. *Geology*, 33, 977–80.

Fitzsimmons, K., Bowler, J.M., Rhodes, E.J. and Magee, J.M. (2007a) Relationships between desert dunes during the late Quaternary in the Lake Frome region, Strzelecki Desert, Australia. *Journal of Quaternary Science*, 22, 549–48.

Fitzsimmons, K., Rhodes, E.J., Magee, J.W. and Barrows, T.T. (2007b) The timing of linear dune activity in the Strzelecki and Tirari Deserts, Australia. *Quaternary Science Reviews*, 26, 2598–616.

Flenley, J.R. (1978) *The Equatorial Rainforest: A Geological History*. Butterworth, London.

Flint, R.F. (1959a) On the basis of Pleistocene correlation in East Africa. *Geological Magazine*, 96, 265–84.

Flint, R.F. (1959b) Pleistocene climates in eastern and southern Africa. *Bulletin Geological Society of America*, 70, 343–74.

Fujioka, T., Chappell, J., Honda, M., Yatsevich, I., Fifield, K. and Fabel, D. (2005) Global cooling initiated stony deserts in central Australia 2-4 Ma, dated by cosmogenic ^{21}Ne-^{10}Be. *Geology*, 33, 993–6.

Gagan, M.K., Hendy, E.J., Haberle, S.G. and Hantoro, W.S. (2004) Post-glacial evolution of the Indo-Pacific Warm Pool and El Niño-Southern Oscillation. *Quaternary International*, 118–9, 127–43.

Galloway, R.W. (1965) Late Quaternary climates in Australia. *Journal of Geology*, 73, 603–18.

Gasse, F., Chalié, F., Vincens, A., Williams, M.A.J. and Williamson, D. (2008) Climatic patterns in equatorial and southern Africa from 30 000 to 10 000 years ago reconstructed from terrestrial and near-shore proxy data. *Quaternary Science Reviews*, 27, 2316–40.

Gathorne-Hardy, F.J. and Harcourt-Smith, W.E.H. (2003) The super-eruption of Toba, did it cause a human bottleneck? *Journal of Human Evolution*, 45, 227–30.

Gibling, M.R., Sinha, R., Roy, N.G., Tandon, S.K. and Jain, M. (2008) Quaternary fluvial and eolian deposits on the Belan River, India: paleoclimatic setting of Paleolithic to Neolithic archeological sites over the past 85,000 years. *Quaternary Science Reviews*, 27, 392–411.

Gingele, F.X. and De Deckker, P. (2005) Late Quaternary fluctuations of palaeoproductivity in the Murray Canyons area, South Australian continental margin. *Palaeogeography, Palaeoclimatology, Palaeoecology*, 220, 361–73.

Gregory, K.J., Starkel, L. and Baker, V.R. (eds) (1995) *Global Continental Palaeohydrology*. Wiley, Chichester.

Grove, A.T. (2008) The revolution in palaeoclimatology around 1970, in T.P. Burt, R.J. Chorley, D. Brunsden, N.J. Cox and A.S. Goudie (eds), *The History of the Study of Landforms*. Vol. 4. The Geological Society, London. pp. 961–1004.

Grove, J. (2004) *Little Ice Ages: Ancient and Modern*. Vols. I and II. Routledge, London.

Hare, F.K. and Ogallo, L.A.J. (1993) *Climate Variations, Drought and Desertification*. World Meteorological Organisation, Geneva.

Heinrich, H. (1988) Origin and consequences of cyclic ice-rafting in the northeast Atlantic Ocean during the last 130,000 years. *Quaternary Research*, 29, 142–52.

Heller, F. and Liu, T.-S. (1982) Magnetostratigraphical dating of loess deposits in China. *Nature*, 300, 431–3.

Hesse, P.P. and McTainsh, G.H. (2003) Australian dust deposits: modern processes and the Quaternary record. *Quaternary Science Reviews*, 22, 2007–35.

Hesse, P.P., Magee, J.W. and van der Kaars, S. (2004) Late Quaternary climates of the Australian arid zone: a review. *Quaternary International*, 118–19, 87–102.

Hewitt, G. (2000) The genetic legacy of the Quaternary ice ages. *Nature*, 405, 907–13.

Hurni, H. (1999) Sustainable management of natural resources in African and Asian mountains. *Ambio*, 28, 382–9.

Imbrie, J. and Imbrie, K.P. (1979) *Ice Ages: Solving the Mystery*. Macmillan, London.

IPCC (2007) Contribution of working group I to the fourth assessment report of the intergovernmental panel on climate change, in S. Solomon, S. Qin, M. Manning, Z. Chen, M. Marquis and K.B. Averyt (eds) *Climate Change 2007: The Physical Science Basis*. Cambridge University Press, Cambridge.

Johnson, B.J., Miller, G.H., Fogel, M.L., Gagan, M.K. and Chivas, A.R. (1999) 65,000 years of vegetation change in Central Australia and the Australian summer monsoon. *Science*, 284, 1150–2.

Jones, P.D., Osbourn, T.J. and Briffa, K.R. (2001) The evolution of climate over the last millennium. *Science*, 292, 662–7.

Kendall, R.A., Mitrovica, J.X., Milne, G.A., Törnqvist, T.E. and Li, Y. (2008) The sea-level fingerprint of the 8.2 ka climate event. *Geology*, 36, 423–6.

Kershaw, A.P. (1976) A late Pleistocene and Holocene pollen diagram from Lynchs Crater, north-eastern Queensland, Australia, *New Phytologist*, 77: 469–98.

Kershaw, A. P. (1994) Pleistocene vegetation of the humid tropics of northeastern Queensland, Australia. *Palaeogeography, Palaeoclimatology, Palaeoecology*, 109, 339–412.

Kershaw, A.P. (1995) Environmental change in Greater Australia, in J. Allen and J.F. O'Connell (eds.) Transitions: Pleistocene to Holocene in Australia and Papua New Guinea. *Antiquity*, 69: 656–75.

Krom, M.D., Stanley, D., Cliff, R.A. and Woodward, J.C. (2002) River Nile sediment fluctuations over the past 7000 yr and their key role in sapropel development. *Geology*, 30, 71–4.

Lamb, H.H. (1977) *Climate: Present, Past and Future*. Vol. 2. *Climatic History and the Future*. Methuen, London.

Le Roy Ladurie, E. (1971) *Times of Feast, Times of Famine: A History of Climate since the Year 1000*. (Translated Barbara Bray from French). George Allen & Unwin, London.

Liu, T.S. (ed.) (1985) *Quaternary Geology and Environment of China*. China Ocean Press, Beijing.

Liu, T.S. (ed.) (1987) *Aspects of Loess Research*. China Ocean Press, Beijing.

Liu, T.S. (ed.) (1991) *Loess, Environment and Global Change*. Science Press, Beijing.

Magee, J.W. and Miller, G.H. (1998) Lake Eyre palaeohydrology from 60 ka to the present: beach ridges and glacial maximum aridity. *Palaeogeography, Palaeoclimatology, Palaeoecology*, 144, 307–29.

Magee, J.W., Bowler, J.M., Miller, G.H. and Williams, D.L.G. (1995) Stratigraphy, sedimentology, chronology and palaeohydrology of Quaternary lacustrine deposits at Madigan Gulf, Lake Eyre, South Australia. *Palaeogeography, Palaeoclimatology, Palaeoecology*, 113, 3–42.

Magee, J.W., Miller, G.H., Spooner, N. and Questiaux, D. (2004) Continuous 150 k.y. monsoon record from Lake Eyre, Australia: Insolation-forcing implications and unexpected Holocene failure. *Geology*, 32, 885–8.

McDougall, I., Morton, W.H. and Williams, M.A.J. (1975) Age and rates of denudation of Trap Series basalts at Blue Nile gorge, Ethiopia. *Nature*, 254, 207–9.

McGowran, B., Holdgate, G.R., Li, Q. and Gallagher, S.J. (2004) Cenozoic stratigraphic succession in southeastern Australia. *Australian Journal of Earth Sciences*, 51, 459–96.

Miller, G.H., Magee, J.W., Johnson, B.J., Fogel, M.L., Spooner, N.A., McCulloch, M.T., et al. (1999) Pleistocene extinction of *Genyornis newtoni*: human impact on Australian megafauna. *Science*, 283, 205–8.

Miller, G.H., Mangan, J., Pollard, D., Thompson, S.L., Felzer, B.S. and Magee, J.W. (2005) Sensitivity of the Australian Monsoon to insolation and vegetation: implications for human impact on continental moisture balance. *Geology*, 33, 65–8.

Mix, A.C., Bard, E. and Schneider, R. (2001) Environmental processes of the ice age: land, oceans, glaciers (EPILOG). *Quaternary Science Reviews*, 20, 627–57.

Moy, C.M., Seltzer, G.O., Rodbell, D.T. and Anderson, D.M. (2002) Variability of El Niño-Southern Oscillation activity at millennial timescales during the Holocene epoch. *Nature*, 420, 162–5.

Nilsson, E. (1931) Quaternary glaciations and pluvial lakes in British East Africa. *Geografiska Annaler A*, 13, 249–349.

Nilsson, E. (1940) Ancient changes of climate in British East Africa and Abyssinia: A study of ancient lakes and glaciers. *Geografiska Annaler A*, 22, 1–79.

Ninkovich, D. (1979) Distribution, age and chemical composition of tephra layers in deep-sea sediments off western Indonesia. *Journal of Volcanology and Geothermal Research*, 5, 67–86.

Ninkovich, D., Sparks, R.S.J. and Ledbetter, M.T. (1978a) The exceptional magnitude and intensity of the Toba

eruption, Sumatra: an example of the use of deep-sea tephra layers as a geological tool. *Bulletin Volcanologique*, 41, 1–13.

Ninkovich, D., Shackleton, N.J., Abdel-Monem, A.A., Obradovich, J.D. and Izett, G. (1978b) K-Ar age of the late Pleistocene eruption of Toba, north Sumatra. *Nature*, 276, 574–7.

Nyssen, J., Poesen, J., Moeyersons, J., Deckers, J., Haile, M. and Lang, A. (2004) Human impact on the environment in the Ethiopian and Eritrean highlands – a state of the art. *Earth-Science Reviews*, 64, 273–320.

Orme, A.R. (2008) Pleistocene pluvial lakes of the American West: a short history of research, in R.H. Grapes, D. Oldroyd and A. Grigelis (eds.) *History of Geomorphology and Quaternary Geology*. Geological Society Special Publication, 301, 51–78.

Page, K., Nanson, G. and Price, D.M. (1996) Thermoluminescence chronology of Murrumbidgee paleochannels on the Riverine Plain, south-eastern Australia. *Journal of Quaternary Science*, 11, 311–26.

Pal, J.N., Williams, M.A.J., Jaiswal, M. and Singhvi, A.K. (2004) Infra Red Stimulated Luminescence ages for prehistoric cultures in the Son and Belan valleys, north central India. *Journal of Interdisciplinary Studies in History and Archaeology*, 1, 51–62.

Pearson, P.N., van Dongen, B.E., Nicholas, C.J., Pancost, R.D., Schouten, S., Singano, J.M., et al. (2007) Stable warm tropical climate through the Eocene Epoch. *Geology*, 35, 211–4.

Penck, W. (1924) Die morphologische Analyse. Ein Kapitel der physikalischen Geologie. *Geographische Abhandlungen*, 2, 1–283.

Penck, W. (1953) *Morphological Analysis of Landforms: A Contribution to Physical Geology*. (Translated Czech, H. and Boswell. K.C. from German). Macmillan, London.

Petraglia, M., Korisettar, R., Boivin, N., Clarkson, C., Ditchfield, P., Jones, S. et al. (2007) Middle Pleistocene assemblages from the Indian subcontinent before and after the Toba super-eruption. *Science*, 317, 114–6.

Pillans, B., Williams, M., Cameron, D., Patnaik, R., Hogarth, J., Sahni, A. et al. (2005) Revised correlation of the Haritalyangar magnetostratigraphy, Indian Siwaliks: implications for the age of the Miocene hominids Indopithecus and Sivapithecus, with a note on a new hominid tooth. *Journal of Human Evolution*, 48, 507–15.

Raymo, M.E. and Huybers, P. (2008) Unlocking the mysteries of the ice ages. *Nature*, 451, 284–5.

Reid, I. (2009) River landforms and sediments: evidence of climatic change, in A.J. Parsons and A.D. Abrahams (eds) *Geomorphology of Desert Environments*. 2nd. edn. Springer, Berlin and New York. pp. 695–721.

Rognon, P. and Williams, M.A.J. (1977) Late Quaternary climatic changes in Australia and North Africa: a preliminary interpretation. *Palaeogeography, Palaeoclimatology, Palaeoecology*, 21, 285–327.

Said, R. (1993) *The River Nile: Geology, Hydrology and Utilization*. Pergamon, Oxford.

Shackleton, N.J. and Kennett, J.P. (1975) Paleotemperature history of the Cenozoic and the initiation of Antarctic glaciation: oxygen and carbon isotope analyses in DSDP sites 277, 279 and 281, in J.P. Kennett, R.E. Houtz, P.B. Andrews and A.R. Edwards (eds) *Initial Reports of the Deep Sea Drilling Project No. 29*. U.S. Government Printing Office, Washington DC. pp. 743–55.

Shane, P., Westgate, J., Williams, M. and Korisettar, R. (1995) New geochemical evidence for the Youngest Toba Tuff in India. *Quaternary Research*, 44, 200–4.

Shane, P., Westgate, J., Williams, M. and Korisettar, R. (1996) Reply to comments by S. Mishra and S.N. Rajaguru on 'New Geochemical Evidence for the Youngest Toba Tuff in India'. *Quaternary Research*, 46, 342–3.

Sharma, G.R. (1973) Stone Age in the Vindhyas and the Ganga Valley, in D.P. Agrawal and A. Ghosh (eds) *Radiocarbon and Indian Archaeology*. TIFR, Bombay.

Sharma, G.R. and Clark, J.D. (eds) (1983) *Palaeoenvironments and Prehistory in the Middle Son Valley, Madhya Pradesh, North Central India*. Abinash Prakashan, Allahabad. p. 320.

Sharma, G.R., Misra, V.D., Mandal, D., Misra, B.B. and Pal, J.N. (1980) *Beginnings of Agriculture*. Abinash Prakashan, Allahabad. p. 200.

Stanley, J.-D., Krom, M.D., Cliff, R.A. and Woodward, J.A. (2003) Nile flow failure at the end of the Old Kingdom, Egypt: strontium isotopic and petrologic evidence. *Geoarchaeology*, 18, 395–402.

Street, F.A. and Grove, A.T. (1976) Environmental and climatic implications of late Quaternary lake-level fluctuations in Africa. *Nature*, 261, 385–90.

Street, F.A. and Grove, A.T. (1979) Global maps of lake-level fluctuations since 20 000 yr BP. *Quaternary Research*, 12, 83–118.

Street-Perrott, F.A., Roberts, N. and Metcalfe, S. (1985) Geomorphic implications of late Quaternary hydrological and climatic changes in the Northern Hemisphere tropics, in I. Douglas and T. Spencer (eds) *Environmental Change and Tropical Geomorphology*. George Allen & Unwin, London. pp. 165–83.

Suc, J.-P. (1984) Origin and evolution of the Mediterranean vegetation and climate in Europe. *Nature*, 307, 429–32.

Talbot, M.R., Williams, M.A.J. and Adamson, D.A. (2000) Strontium isotopic evidence for Late Pleistocene re-establishment of an integrated Nile drainage network. *Geology*, 28, 343–6.

Tandon, S.K., Sinha, R., Gibling, M.R., Dasgupta, A.S. and Ghazanfari, P. (2008) Late Quaternary evolution of the ganga Plains: myths and misconceptions, recent developments and future directions. *Memoir Geological Society of India*, 2008, 1–40.

Tucker, C.J. and Nicholson, S.E. (1999) Variations in the size of the Sahara Desert from 1980 to 1997. *Ambio*, 28, 587–91.

Tucker, C.J., Dregne, H.E. and Newcomb, W.W. (1991) Expansion and contraction of the Sahara Desert from 1980 to 1990. *Science*, 253, 299–301.

Turney, C.S.M., Kershaw, A.P., Clemens, S.C., Branch, N., Moss, P.T. and Fifield, L.K. (2004) Millennial and orbital variations of Le Niño/Southern Oscillation and high-latitude climate in the last glacial period. *Nature*, 428, 306–10.

UNCED (1992) *Earth Summit Agenda 21: Programme of Action for Sustainable Development*. United Nations Environment Programme, New York.

UNEP (1990) *Report of the ad-hoc consultative meeting on the assessment of desertification*. United Nations Environment Programme, Nairobi.

UNEP (1997) Middleton, N and Thomas, D.S.G. (eds) *World Atlas of Desertification*, 2nd. edn. Arnold, London.

Walker, G.T. (1924) Correlation in seasonal variations of weather. IX: a further study of world weather. *Memoirs of the Indian Meteorological Department*, 24, 275–332.

Walker, M., Johnsen, S., Rasmussen, S.O., Popp, T., Steffensen, J.-P. and Gibbard, P. (2009) Formal definition and dating of the GSSP (Global Stratotype Section and Point) for the base of the Holocene using the Greenland NGRIP ice core, and selected auxiliary records. *Journal of Quaternary Science*, 24, 3–17.

Weiss, H. (2000) Beyond the Younger Dryas: collapse as adaptation to abrupt climate change in ancient West Asia and the Eastern Mediterranean, in G. Bawden and R.M. Reycraft (eds) *Environmental Disaster and the Archaeology of Human Response*. Maxwell Museum of Anthropology, University of New Mexico, Albuquerque, Anthropological Papers No.7. pp. 75–98.

Wellman, P. and McDougall, I. (1974) Cainozoic igneous activity in Eastern Australia. *Tectonophysics*, 23, 49–65.

Westgate, J.A., Shane, P.A.R., Pearce, N.J.G., Perkins, W.T., Korisettar, R. and Chesner, C.A. (1998) All Toba tephra occurrences across peninsula India belong to 75 ka eruption. *Quaternary Research*, 50, 107–12.

Whetton, P., Adamson, D.A. and Williams, M.A.J. (1990) Rainfall and river flow variability in Africa, Australia and East Asia linked to El Niño–Southern Oscillation events, in P. Bishop (ed.) *Lessons for Human Survival: Nature's record from the Quaternary*. Geological Society of Australia Symposium Proceedings, 1, 71–82.

Williams, M., Cook, E., van der Kaars, S., Barrows, T., Shulmeister, J. and Kershaw, P. (2009) Glacial and deglacial climatic patterns in Australia and surrounding regions from 35 000 to 10 000 years ago reconstructed from terrestrial and near-shore proxy data. *Quaternary Science Reviews*, 28, 2398–419.

Williams, M., Dunkerley, D., De Deckker, P., Kershaw, P. and Chappell, J. (1998) *Quaternary Environments*. 2nd. edn. Arnold, London. p. 329.

Williams, M., Nitschke, N. and Chor, C. (2006c) Complex geomorphic response to late Pleistocene climatic changes in the arid Flinders Ranges of South Australia. *Géomorphologie: Relief, Processus, Environnement*, 4, 249–58.

Williams, M. and Nottage, J. (2006) Impact of extreme rainfall in the central Sudan during 1999 as a partial analogue for reconstructing early Holocene prehistoric environments. *Quaternary International*, 150, 82–94.

Williams, M., Talbot, M., Aharon, P., Abdl Salaam, Y., Williams, F. and Brendeland, K.I. (2006b) Abrupt return of the summer monsoon 15,000 years ago: new supporting evidence from the lower White Nile valley and Lake Albert. *Quaternary Science Reviews*, 25, 2651–65.

Williams, M.A.J. (1975) Late Pleistocene tropical aridity synchronous in both hemispheres? *Nature*, 253, 617–8.

Williams, M.A.J. (1985) Pleistocene aridity in tropical Africa, Australia and Asia, in I. Douglas and T. Spencer (eds) *Environmental Change and Tropical Geomorphology*. George Allen and Unwin, London. pp. 219–33.

Williams, M.A.J. (1993a) Late Quaternary desert margin systems in Africa, Australia and Asia: can a study of the past help us to predict possible future change? *Abstracts, International Scientific Conference on the Taklimakan Desert*. Urumqi, China, September 1993. pp. 141–2.

Williams, M.A.J. (1993b) Drought, desertification and climatic change. *Abstracts, International Scientific Conference on the Taklimakan Desert*. Urumqi, China, September 1993. pp. 151–2.

Williams, M.A.J. (1993c) The response of big rivers in semi-arid areas to environmental change. Three examples from Africa, Australia and India: the Nile, the Darling and the Son', *Abstracts, International Scientific Conference on the Taklimakan Desert*. Urumqi, China, September 1993. p. 319.

Williams, M.A.J. (2000) Desertification: general debates explored through local studies. *Progress in Environmental Science*, 2, 229–51.

Williams, M.A.J. (2002) Desertification, in I. Douglas (ed.) *Encyclopedia of Global Environmental Change*. Vol. 3: *Causes and Consequences of Global Environmental Change*. Wiley, Chichester. pp. 282–90.

Williams, M.A.J. (2004) Desertification in Africa, Asia and Australia: human impact or climatic variability? *Annals of Arid Zone*, 42, 213–30.

Williams, M.A.J. (2009a) Cenozoic climates in deserts, in A.J. Parsons and A.D. Abrahams (eds) *Geomorphology of Desert Environments*. 2nd. edn. Springer, Berlin and New York. pp. 799–824.

Williams, M.A.J. (2009b) Late Pleistocene and Holocene environments in the Nile basin. *Global and Planetary Change*. 69, 1–15.

Williams, M.A.J. and Adamson, D.A. (1980) Late Quaternary depositional history of the Blue and White Nile rivers in central Sudan, in M.A.J. Williams and H. Faure (eds) *The Sahara and The Nile: Quaternary Environments and Prehistoric Occupation in Northern Africa*. Balkema, Rotterdam. pp. 281–304.

Williams, M.A.J. and Royce, K. (1982) Quaternary geology of the middle Son valley, north central India: implications for prehistoric archaeology. *Palaeogeography, Palaeoclimatology, Palaeoecology*, 38, 139–62.

Williams, M.A.J. and Royce, K. (1983) Alluvial history of the middle Son valley, north central India, in G.R. Sharma and J.D. Clark (eds), *Palaeoenvironments and prehistory in the middle Son valley, Madhya Pradesh, north central India*. University of Allahabad, 9–21.

Williams, M.A.J. and Clarke, M.F. (1984) Late Quaternary environments in north central India. *Nature*, 308, 633–5.

Williams, M.A.J. and Clarke, M.F. (1995) Quaternary geology and prehistoric environments in the Son and Belan Valleys, north-central India. *Geological Society of India Memoir*, 32, 282–308.

Williams, M.A.J. and Balling Jr., R.C. (1996) *Interactions of Desertification and Climate*. Arnold, London. p. 270.

Williams, M.A.J., Pal, J.N., Jaiswal, M. and Singhvi, A.K. (2004) Infra Red Stimulated Luminescence ages for prehistoric cultures in the Son and Belan valleys, north central India. *Journal of Interdisciplinary Studies in History and Archaeology* 1, 51–62.

Williams, M.A.J., Pal, J.N., Jaiswal, M. and Singhvi, A.K. (2006a) River response to Quaternary climatic fluctuations: evidence from the Son and Belan valleys, north central India. *Quaternary Science Reviews*, 25, 2619–31.

Williams, M.A.J., Ambrose, S.H., van der Kaars, S., Chattopadhyaya, U., Pal, J. and Chauhan, P.R. (2009) Environmental impact of the 73 ka Toba super-eruption in South Asia. *Palaeogeography, Palaeoclimatology, Palaeoecology* 284, 295–314.

Williams, P., King, D., Zhao, J. and Collerson, K. (2005) Late Pleistocene to Holocene composite speleothem ^{18}O and ^{13}C chronologies from South Island, New Zealand: did a global Younger Dryas really exist? *Earth and Planetary Science Letters*, 230, 301–17.

Woodward, J.C., Macklin, M.G., Krom, M.D. and Williams, M.A.J. (2007) The Nile: evolution, Quaternary river environments and material fluxes, in A. Gupta (ed.) *Large Rivers: Geomorphology and Management*. John Wiley & Sons, Chichester. pp. 261–92.

Yokoyama, Y., Lambeck, K., De Deckker, P., Johnston, P. and Fifield, L.K. (2000) Timing of the Last Glacial Maximum from observed sea-level minima. *Nature*, 406, 713–6.

Zanazzi, A., Kohn, M.J, MacFadden, B.J. and Terry, D.O. (2007) Large temperature drop across the Eocene-Oligocene transition in central North America. *Nature*, 445, 639–42.

Zeuner, F.E. (1958) *Dating the Past: An Introduction to Geochronology*. Methuen, London.

31

Disturbance and Responses in Geomorphic Systems

Jonathan D. Phillips

INTRODUCTION

For the average person landforms and landscapes are immutable; 'solid as a rock' or 'old as dirt' or at most 'moving at a glacial pace'. Geomorphic change is, to most laypersons, too slow to notice, except when landscapes are palpably disturbed or shaped by humans or in the case of events such landslides, floods or storm erosion, which are seen as cataclysmic or catastrophic.

Of course, landforms and landscapes are actually in constant, though sometimes slow and subtle, flux. Processes operating at spatial scales from the atomic to the cosmic, at temporal scales from instantaneous to billions of years, and at rates and magnitudes ranging from almost inconceivably large and rapid to equally unimaginably gradual and protracted are constantly influencing Earth's surface. The study of earth surface processes and landforms is fundamentally the study of constantly changing features and phenomena.

This chapter deals with changes to and perturbations of geomorphic systems, and the responses to those disturbances. The focus will be on geomorphic responses; that is, how landforms and landscapes respond to externally imposed changes. However, we must bear in mind that geomorphic, biotic, hydrological, climatic, geological and pedological components often respond in related and interconnected ways. Furthermore, geomorphic change may itself trigger responses in other aspects of the environment.

Some general perceptions and concepts of change in geomorphology are first reviewed,

leading to a critical discussion of the idea of normative conditions. Finally, some general approaches to assessments of geomorphic change and response are outlined.

Geomorphic disturbance: concepts and perceptions

The term disturbance has connotations of negativity and abnormality, and indeed this was mostly intended in the early days of studies of externally driven changes to ecosystems and landscapes. Disturbance is used here in a general sense to refer to any externally driven perturbation to geomorphic systems, and relatively persistent changes in boundary conditions such as climate and sea level change are not excluded from consideration. The term disturbance has persisted in the environmental sciences, but without the value-laden implications, as it became clear that in many earth surface systems factors such as fire, tropical cyclones, droughts, floods and earthquakes are common, not abnormal and critical to system function, maintenance and development.

Perceptions of disturbance in geomorphology are necessarily scale contingent; what appears as a discrete event or episode at one resolution, for instance, may be treated as continuous at another. Fluctuations that seem abrupt over long timescales may be seen as gradual viewed over more restricted time frames. For any given timescale or temporal frame of reference, disturbances of geomorphic systems fall into two general categories: persistent or semi-permanent changes in boundary conditions or environmental context (e.g. glacial–interglacial climate cycles) or singular, discrete disturbances (e.g. a volcanic eruption). Due to scale contingency the same phenomenon may sometimes be treated as either a disturbance (e.g. a fire) or a persistent change (a shift to a new fire regime).

Geomorphic forcings may have inherent tempos and cycles that influence our perception of them (e.g. tides, diurnal and seasonal climate fluctuations). So, too, are perceptions affected by intrinsic properties of the frequency of occurrence, particularly relative to human lifespans. In addition to these factors intrinsic to geomorphic disturbances themselves, human diversity of world views, methodologies, metaphors and personal preferences shape perceptions of change. Concepts of disturbances and responses are therefore linked to a combination of the intrinsic nature of the changes themselves, the scale of the analysis or problem, and the philosophical, methodological and psychological characteristics of the observer.

Approaches to geomorphology tend to focus either on process mechanics and process–response relationships, or on histories and trajectories of landscape evolution. Process-based approaches lend themselves to a stimulus–response or disturbance–recovery perspective of geomorphic change. Historical approaches are more likely to be associated with a chronological, sequential view. In either case, geomorphic changes may be intrinsic, due to thresholds and interactions within a geomorphic system, or extrinsic and external to the system itself. Sometimes the intrinsic–extrinsic nature of change may be unambiguous; in other cases the distinction is unclear. Human impacts, for example, have traditionally been treated as exogenous, external disturbances, but some recent work (recognizing the ubiquity of human impacts and the role of *Homo sapiens* as a keystone species) has treated humans as intrinsic components of geomorphic systems (e.g. Nordstrom, 1987; Roberge, 2002; Riegert and Turkington, 2003; Craghan, 2005).

Conceptualizations of geomorphic responses to persistent changes are closely tied to general theories of landscape evolution, covered in detail elsewhere in this volume. The key considerations with respect to disturbance are whether the model or theoretical construct is based on single or multiple potential pathways of development or evolution, and on single or multiple possible outcomes. Davis' erosion cycle, for example, postulates a singular developmental pathway toward a single outcome, in common with traditional soil zonalism concepts and Clementsian ecological succession, arising as they do from the same general intellectual milieu (Osterkamp and

Hupp, 1996; Stoddart, 1966; Reynolds, 2006). Dynamic (steady-state) equilibrium and self-organized criticality-based models and schools of thought, however, are based on a single outcome (steady-state equilibrium form or a critical state), but except in a general phenomenological sense do not imply any specific pathway (Sapozhinikov and Foufoula-Georgiou, 1996; Stolum, 1998; Phillips, 1999b).

Dynamic instability and deterministic chaos, by contrast, lead to conceptions of geomorphic and environmental change based on or allowing for multiple outcomes (Scheidegger, 1983, 1991; Huggett, 1995, 1997; Phillips, 1999a,b). In dynamically unstable and chaotic systems minor variations in initial conditions can lead to much larger differences later on, so that effects of small disturbances are likely to grow and persist. Other more-or-less related conceptual frameworks that emphasize the importance of historical contingency and path-dependence are also in the multiple-outcome category (Lane and Richards, 1997; Harrison, 1999; Thomas, 2001; Phillips, 2006).

Equifinality describes phenomena whereby different pathways or processes lead to similar results; for instance, models of channel network evolution based on a wide variety of mechanisms and assumptions all converge on networks of similar geometry and topology. Equifinality is consistent with single-outcome multiple-pathway frameworks, since the dynamic steady-state equilibrium or the critical state is seen as an attractor towards which the system will always move. However, equifinality may also be consistent with instability and chaos, in that a given outcome does not necessarily imply any particular pathway of development (Phillips, 1997a).

Single-outcome models of change are based on the notion that for a given set of environmental controls there exists a best-fit state or condition towards which the system moves, though disturbances or insufficient time may prevent full realization of this ideal state. These ideal states are typically based on some concept of balance, goal functions, or a critical state. In the first case, the geomorphic system is conceptualized as making adjustments to achieve some sort of approximate balance or steady state, such as hillslope or channel profiles adjusted so that the available sediment can just be transported, with little or no persistent net removal or accumulation (e.g. Strahler, 1952; Hack, 1960; Chorley, 1962; Brunsden, 1990). Goal functions infer that the system adjusts so as to achieve some goal, generally related to energy use and dissipation. An example is the theory that channel networks evolve to achieve a balance between maximum energy dissipation (least work) at any given point and equal energy dissipation throughout the network (Woldenberg, 1969; Rodtriguez-Iturbe and Rigon, 1997). Critical-state concepts postulate evolution towards a threshold, a condition of incipient instability, or a condition of self-organized criticality (Schumm, 1979; Begin and Schumm, 1984; Hergarten, 2002).

The language of adaptation and fit makes it tempting to draw analogies between single-outcome theories and biological evolution. This analogy is misleading, however, as one principal tenet of evolutionary theory is multiple degrees of freedom in adaptation; for example, the variety of adaptations of desert plants to moisture scarcity. Evolution theory also places great emphasis on path dependence, contingency and random events. Huggett (1995, 1997) contrasts an evolutionary view of environmental change and geoecosystems based on multiple possible outcomes, path dependency and historical contingency with a developmental view based on single or few pathways and outcomes.

While single-path, single-outcome viewpoints remain quite common in geomorphology, the general theoretical, methodological and epistemological trend is toward inclusion of various multi-path, multi-outcome perspectives (cf. Scheidegger, 1983, 1991; Huggett, 1995, 1997; Lane and Richards, 1997; Harrison, 1999; Phillips, 1999a, 2006; Thomas, 2001). This trend does not necessarily entail rejection of single-path or single-outcome scenarios, as in general the pluralistic approaches include the singular approaches as special cases.

Reversibility and repeatability

Some geomorphic changes are reversible – sediment infilling and flushing of a channel, for example, or the erosion and rebuilding of sandy beaches during and after storms. Other changes are irreversible: once moved downslope, weathered debris does not climb back up; and once disintegrated, weathered rock does not reconstitute. Geomorphic changes can also sometimes be repeatable. Back-and-forth migration of a dune in response to opposing winds, for instance, could conceivably be repeated indefinitely, as could aggradation–degradation episodes in river channels. And while regolith stripping is irreversible in most instances, a sequence of regolith accumulation and removal might be repeated multiple times. However, assessments of geomorphic responses to change should distinguish between the *possibility* of reversal or repetition, as opposed to the likelihood.

Conditioning and triggering

Concepts of change in nature are often assumed to be either gradual and more-or-less continuous, analogous to the growth and development of an organism, or discrete, as in a stimulus–response sequence. Geomorphic change, however, is often a combination of relatively gradual, continuous or chronic modifications and more abrupt changes. The latter often depend on the system being close to a threshold, and the occurrence of a specific event. This type of change can be conceptualized as a two-phase sequence of preconditioning followed by a trigger. On hillslopes, for instance, gradual weathering and hydrological processes may bring the shear strength–shear stress ratio near a threshold, and then a trigger such as a large rainstorm results in slope failure. Explicit use of the conditioning-and-triggering framework can be found in the study of river avulsions (Slingerland and Smith, 2004; Aslan et al., 2005).

NORMATIVE STATES

Kennedy (1992) contrasted the views of some key figures in the development of geomorphology on whether the contemporary landscape is of some particular significance (e.g. all previous Earth history was destined to bring the planet to the current state), or just a snapshot along a continuing pathway of change. The former is admittedly an extreme view in contemporary geomorphology, but Kennedy (1992) showed that this perspective has been prevalent and influential in the history of the earth sciences. Even if one attaches no particular theoretical significance to contemporary conditions, a human tendency is to at least tacitly identify a 'normal' baseline condition. This normative condition might be based on theory, on ubiquity, on perceived representativeness, on personal experience and familiarity, or on an arbitrary designation of a reference condition. Informally, due to long experience, I (for instance) tend to view features of humid subtropical environments as my baseline, and to evaluate other environments by comparison to those standards. As discussed earlier, single-outcome theories imply a normal condition.

Though a 'snapshot' view is prevalent nowadays, geomorphologists have long tended to view some types of change as 'normal' and others as exceptional. Davis (1905, 1930) contrasted 'normal', fluvially dominated downwasting in humid climates with, for example, arid and karst cycles. Fluvial geomorphologists have implicitly treated alluvial channels as the standard, and bedrock streams as an exception (Whipple, 2004).

Likewise, the existence of subfields and specialties in tropical, arid, and arctic–alpine geomorphology – but not humid temperate or subtropical – indicates that the latter is the implicit norm.

The most common theoretically derived or arbitrarily designated type of normative state in earth and environmental sciences is some form of equilibrium. Equilibrium concepts play a major role in geomorphology, with implications too broad to fully cover here. However, equilibrium is unavoidable in any discussion of disturbances and responses in geomorphology, as notions of equilibrium influence not just the analysis of geomorphic responses, but also perceptions and value judgments.

Equilibrium concepts in geomorphology may be grouped into three progressively restrictive and rigorous categories. *Relaxation time* equilibrium merely implies that a landform or landscape has completed its response to a disturbance. If a stream aggrades because of base level change, for instance, this weakest concept of equilibrium is achieved when aggradation ceases or declines to negligible rates. The stronger notion of *characteristic form* implies relaxation time equilibrium, and additionally that the system achieves (or at least clearly moves toward) a form or state which is adjusted to its environmental constraints and context. Characteristic form equilibrium has traditionally been conceptualized as a more-or-less universal characteristic form, such as a strongly concave longitudinal river profile, or a zonal soil, but could also be interpreted in a local or regional context. *Steady-state* equilibrium, the strongest and most restrictive variety, implies that characteristic forms and states are stable in response to all but the largest perturbations, and self-maintaining. Steady-state soil thickness, where erosion and weathering feedbacks maintain a consistent soil depth, or river channels which continually adjust to maintain steady-state sediment transport, are both examples.

Relaxation time equilibrium concepts have little theoretical baggage, and are in fact an inevitable result of the fact that geomorphic response and relaxation times are finite (and often decelerate over time). Characteristic form may imply a preferred or normal state, but may also simply reflect broad geographical relationships between landforms and climate, lithology and other factors. The extent to which steady-state and characteristic form equilibrium states are considered normative relates to another dialectic in views of equilibrium: privileged versus potential.

Privileged concepts are based on some notion of equilibrium as a normal and preferred condition. At least three variants exist. First, and

probably less common in earth science than in ecology and in popular views of the environment, is a balance-of-nature view, that self-maintaining steady states (be they river channels, salt marshes or climax forests) are the normal, expected, natural – and implicitly or explicitly preferred – condition. Disequilibrium is considered a temporary or aberrant condition. A second, weaker variant of this viewpoint sees steady state as the inevitable goal of a geomorphic system, such that adjustments to disturbances or boundary condition fluctuations operate so as to move the system toward equilibrium. However, it is recognized that disturbances are common enough in some landscapes so that disequilibrium may be relatively common. A third privileged concept is steady-state equilibrium as a reference standard, against which other conditions (disequilibrium or non-equilibrium) are to be judged. The reference standard may in fact be a matter of analytical convenience or simply tradition. However, any reference standard is implicitly privileged, having greater ontological and epistemological significance relative to other conditions.

Potential concepts of equilibrium, by contrast, acknowledge the existence of steady states, but the latter are not considered to be necessarily any more common, normal, desirable, or important than other states. Steady-state equilibrium is seen (along with dis- and non-equilibrium) as one possible state for a geomorphic system. A more extreme view – generally applied to specific issues, such as a balance between stream sediment supply and transport capacity – sees steady states as rare, transient and unlikely.

We turn now to some conceptual frameworks useful in addressing geomorphic responses to disturbances.

THE HAZARDS MATRIX

Geomorphic hazards are a particularly important type of geomorphic disturbance–response problems, as they are by definition directly relevant to humans. Gilbert White (1974) devised a conceptual framework (the hazards matrix) for analyzing natural hazards, and Gares et al. (1994) specifically adapted it to geomorphic hazards. While Gares et al. (1994) applied the hazards matrix to responses (i.e. geomorphic hazards generated by various disturbances), it is equally applicable to the disturbances themselves.

The hazards matrix (White, 1974) framework is based on seven criteria. Four of these are related to temporal characteristics: (1) frequency, generally expressed as the probability or recurrence interval of an event; (2) duration; (3) speed

Table 31.1 Qualitative assessment of disturbance parameters for selected geomorphic changes and disturbances

Parameter	Disturbance				
	Fire	Drought	Volcanic eruption	Eustatic sea-level change	Mining
Frequency	Frequent to rare	Frequent to rare	Rare	Rare	Singular
Magnitude	Low to moderate	Low to moderate	Low to extreme	Moderate to extreme	Extreme
Duration	Short	Short to moderate	Short	Long	Short to moderate
Areal extent	Moderate to extensive	Extensive	Local to extensive	Extensive to global	Local
Speed of onset	Rapid	Slow	Rapid	Slow	Rapid
Spatial dispersion	Diffuse	Diffuse	Concentrated	Diffuse	Concentrated
Temporal spacing	Random	Random to cyclical	Random	Cyclical	Singular

of onset; and (4) temporal spacing (random vs. regular). Two criteria are related to spatial characteristics: the areal extent and spatial dispersion (diffuse vs. concentrated). The final criterion is magnitude, which is often some measure of force, power, energy or mass flux.

Gares et al. (1994: Table 7) compared soil, coastal, river and aeolian erosion, slope failures and land subsidence according the seven hazard parameters. Table 31.1 is structured similarly, but with respect to disturbances which may trigger or influence geomorphic responses.

LANDSCAPE SENSITIVITY

Landscape sensitivity addresses the probability that a given change in boundary conditions or forcings of a geomorphic system will 'produce a sensible, recognizable and persistent response' (Brunsden, 2001: 99). The concept in a formal sense was introduced by Brunsden and Thornes (1979), and was the subject of a book (Thomas and Allison, 1993) and of a special issue of *Catena* in 2001. Examples of explicit applications of the landscape sensitivity framework can be found in that issue (Thomas and Stirling, 2001), and are also provided by Sauchyn (2001), Marker (2003), Thomas (2004) and Marker and Holmes (2005).

Four prevailing themes in the landscape sensitivity literature provide a framework for approaching disturbances to and responses of geomorphic systems: (1) force and resistance; (2) frequencies, durations and relaxation times; (3) stability; and (4) contingency.

Consideration of the factors (e.g. energy, force, power) driving change versus the factors resisting change is fundamental to geomorphology. Assessing disturbances and potential responses thus involves some measures or estimates of the magnitude of the forcing or disturbance, such as the maximum sustained winds, storm surge or Saffir–Simpson category of a tropical cyclone. The resisting factors include resistance *per se*, (e.g. particle size, rock hardness or shear strength), and the ability to absorb change (e.g. buffering capacity or sediment accommodation space). Resistance in the broadest sense is also related to resilience (the ability of a system to recover from disturbance), and the degrees of freedom available to respond to disturbances.

With respect to temporal aspects of landscape sensitivity to disturbances, one useful approach comes from Chappell's (1983) assessment of whether climate variations constitute changes or perturbations to geomorphic systems, based on comparing the duration of the climate fluctuations to the geomorphic relaxation. This can be generalized to other types of geomorphic changes and responses. If the fluctuation is at least as long as the relaxation time (e.g. if a shift to more arid conditions lasts long enough for landforms adjusted to aridity to develop) then a change in boundary conditions has occurred. Otherwise (for instance, a drought influences erosion rates but does not persist long enough to result in climatically adjusted landforms), the shift may be considered a disturbance or perturbation. Phillips (1997b) presented a formal version of this analysis with respect to human impacts.

The transient form ratio devised by Brunsden and Thornes (1979) is closely related and compares the relaxation time of the landform with the recurrence interval of a landform-changing event. A ratio greater than unity represents a sensitive landscape dominated by transient forms, while a ratio <1 indicates insensitivity and a prevalence of permanent forms.

The role of stability (the ability to resist change) in the most general sense in sensitivity is straightforward. Dynamic stability is directly related to resilience. Geomorphic systems may be intrinsically dynamical stable or unstable. Dynamical stable systems are likely to be less sensitive and more resilient than unstable ones, and the response to relatively small disturbances is likely to be recovery toward the pre-disturbance condition. Besides implications for landscape sensitivity, dynamic instability in geomorphic systems also implies that responses to perturbations may be disproportionately large and long-lived compared to the disturbance. Relaxation times, resilience and stability are considered further in the next section.

Contingency refers to the dependence of geomorphic responses on the particulars of place and time. Studies of geomorphic impacts of events such as storms and floods, for example, has shown that event magnitude and the resisting framework do not always predict the response. Time- and place-contingent factors such as the timing, sequence and initial conditions may play comparable or even greater roles in determining the response (e.g. Magilligan et al., 1997; Phillips, 1999c). The soil erosion induced by an intense thunderstorm, for instance, may depend as much on whether the storm occurs before or after a crop has emerged as on the intensity of the rainfall and the properties of the soil surface.

AMPLIFIERS AND FILTERS

External forcings of geomorphic systems may sometimes be exaggerated and reinforced, such that the geomorphic response is disproportionately large relative to the forcing. This can occur due to knock-on effects of external changes, threshold exceedences, dynamic instabilities or a high degree of landscape sensitivity. These phenomena act as amplifiers of disturbances. For example, the effects of climate cooling on fluvial systems may be amplified by blockage of drainage systems by ice sheets or outwash. Other phenomena – filters – reduce, mitigate or obscure geomorphic responses. Filters may be related to high levels of resistance or distance from thresholds, as for example when a climatically controlled increase of discharge and stream power in a bedrock-controlled channel fails to result in

morphological change because force:resistance thresholds are not exceeded. Filters can also be attributed to dynamic stability, and to factors that directly offset the effects of the disturbance. Responses to the same disturbance may be dominated by amplifiers or filters with respect to different geomorphic phenomena (Figure 31.1).

Amplifiers and filters may be factors intermediary between the forcing and the geomorphic system. Effects of climate change on erosion, for instance, may be amplified or filtered via climate effects on vegetation cover. Factors internal to the geomorphic system may also enhance disturbances via intrinsic thresholds and dynamical instabilities, as for example when the effects of individual trees on regolith development grow unstably over time (see Phillips and Marion, 2005; Phillips et al., 2008). Or, intrinsic stability may erase disturbance effects; for example, when sandy beaches recover their pre-storm profile after storm erosion events. Finally, other external factors may offset or reinforce the disturbance. Land uplift or subsidence, for instance, may counteract or exacerbate effects of eustatic sea-level change.

Several levels of amplifiers and/or filters may exist between a disturbance and the geomorphic response, such that no simple relationships exist between magnitudes of disturbance and response (Figure 31.2). Large, catastrophic change may sometimes result from minor disturbances, and large disturbances may produce unnoticeable geomorphic impacts. For example, some mega-gullies in the southeastern US, such as Providence Canyon, Georgia, are known to have originated via the amplification of small, localized soil erosion features (e.g. Magilligan and Stamp, 1997). On the other hand, in the same region, clearcutting of forests may have negligible lasting geomorphic impacts, despite large short-term increases in runoff and erosion, because vegetation cover recovers so rapidly in the humid subtropical climate (Phillips, 1997b).

A FRAMEWORK FOR CHANGE ASSESSMENT

An approach to assessing geomorphic change should incorporate the principles inherent in the landscape sensitivity and hazards frameworks. It should be based on recognizing the possibility of multiple response pathways or trajectories, of amplifier and filter effects, and of multiple outcomes. The framework should also not assume *a priori* the existence of any single normative state for the system, except perhaps in applied projects where management goals, for whatever reason, are based on some particular desired or target state

Figure 31.1   Site of a tornado forest blowdown in the Ouachita Mountains, Arkansas, USA

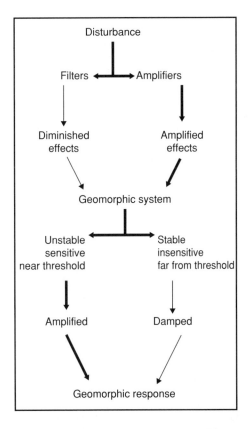

Figure 31.2   Conceptual diagram showing how potential amplifiers and filters intrinsic and extrinsic to geomorphic systems may enhance or reduce disturbance impacts

or condition. Even then, it should be recognized that such goals are often based on engineering, social, cultural, economic or aesthetic criteria and may or may not correspond to a normal or natural condition.

The hazard parameters (magnitude, frequency, duration, speed of onset, tempo, areal extent and spatial concentration) for a given disturbance should be identified if possible. Landscape sensitivity and potential amplifiers and filters should also be evaluated. Then (potential) geomorphic repercussions can then be assessed as a function of the four Rs outlined below: response, resistance, resilience and recursion (feedback).

Response

Here response is used as a shorthand for the timing and rate of the geomorphic response to disturbances. Two key components are reaction time, the period required for the system to begin responding and relaxation time. The latter is the time needed to complete the response. Reaction time is often closely related to the frequency and timing of trigger events. For instance, the reaction time of a hillslope to vegetation removal may depend on how soon afterwards a precipitation event occurs. Relaxation time was discussed above in connection with landscape sensitivity.

Letting R_i represent the relaxation time for disturbance i, the ratio of R_i to the frequency of the event (F_i) gives the transient form ratio (TFR) of Brunsden and Thornes (1979):

$$TFR = R_i/F_i \qquad (31.1)$$

A value of TFR > 1 suggests a landscape dominated by transient forms. For changes treated as continuous rather than discrete, by denoting the duration of an event Δt_i, a relaxation/duration ratio (RDR) could also be defined:

$$RDR = R_i/\Delta t_i \qquad (31.2)$$

If RDR > 1, there is insufficient time for relaxation to occur.

Resistance

The ability of a system to avoid or minimize responses to externally imposed changes – resistance has two main aspects. First, is some indication or measure of strength, chemical or mechanical stability, or susceptibility to modification. Examples include indices of shear strength, erodibility, hardness, mineral stability

and cohesion. These must be evaluated in the context of the relationship to a related measure of the magnitude of the disturbance. Classic examples include the shear strength–shear stress ratio used to assess slope stability, and the critical shear stress for boundary erosion in stream channels.

The second major aspect of resistance is absorption, generally assessed as some measure of capacity. Effects of sediment delivered to a channel by a landslide or a construction site, for example, will depend in part on the sediment transport capacity of the stream. Capacities are typically defined in terms of transport or conveyance (as above), storage or buffering. The impact of acid deposition on weathering and stream chemistry, for instance, may be mitigated by buffering capacity of soils and rocks. An example of storage effects is the influence of accommodation space on the response to a pulse of sediment inputs into an alluvial valley.

Resilience

Resilience – the ability to recover toward a predisturbance state – is directly related to dynamic stability. At least four general approaches for assessment of resilience exist: direct assessments, and three variants of stability analysis. The most direct methods are observations, experiments, historical reconstructions and models that examine the response and recovery of geomorphic systems to disturbance. Using S_a, S_b, S_c, etc. to denote different system states or conditions, if S_a represents a pre-disturbance state, then any of the following sequences would represent resilience (alphabetic rather than numeric subscripts are used to emphasize that the states are not necessarily sequential and irreversible):

$$S_a \rightarrow S_b \rightarrow S_a$$
$$S_a \rightarrow S_c \rightarrow S_a$$
$$S_a \rightarrow S_b \rightarrow S_c \rightarrow S_a$$

In other words, if the pre-disturbance state is restored, the system is resilient. Any sequence whereby S_a is not restored indicates a non-resilient or low-resilience system.

Forested landscapes of the humid subtropics, for instance, are generally resilient to effects of deforestation. Vegetation removal triggers erosion, but restoration of plant cover is rapid, and the increases in runoff and erosion resulting from deforestation are quickly damped. By contrast, many dryland landscapes have low resilience. Erosion following vegetation removal may quickly remove thin topsoils, exposing hard, infertile subsoils that inhibit or prevent vegetation and soil recovery.

Often the empirical observations necessary to directly assess resilience are absent or equivocal. In these cases resilience can be determined from structural characteristics of the system via stability analysis. Three variants of this are quantitative stability analysis, qualitative asymptotic stability analysis and convergence–divergence assessment.

If a geomorphic system can be represented quantitatively in terms of a dynamical equation system or interaction matrix, standard methods of stability analysis can be employed, as outlined in any matrix algebra text. Even nonlinear systems are amenable to these methods, as linearization (e.g. via Taylor series) does not affect the stability properties of the nonlinear parent system. Stability can be evaluated, however, even if the system is only partly specified (only the qualitative nature or direction of influences of system components on each other is known). In this case dynamical stability (or the conditions under which stability or instability occur) can be determined with qualitative asymptotic stability analysis. These methods are outlined in general terms by Puccia and Levins (1985), and in a geomorphic context by Phillips (1999a).

One can often characterize geomorphic systems as n-dimensional systems with components x_i, $i = 1, 2, ..., n$

$$dx_i/dt = f(dx/dt) \quad (31.3)$$

where x indicates the vector of all x_i. Thus the components of the system potentially affect, and are potentially affected by, each other. The system state at time t following a disturbance is given by:

$$x(t) = C\, x(0)\, e^{\lambda t} \quad (31.4)$$

where $x(0)$ is the initial state at the time of disturbance and C is a vector constant related to the initial conditions. The λ are the n Lyapunov exponents of the system (equivalent to the real parts of the complex eigenvalues of a Jacobian interaction matrix of the system), where $\lambda_1 > \lambda_2 > \ldots \lambda_n$ Equation (31.4) indicates that perturbations are exponentially damped if $\lambda < 0$, and grow exponentially if the exponent is positive.

Equation systems, box-and-arrow diagrams or signed, directed graphs and interaction matrices can all be analyzed using qualitative stability analysis, as outlined by Phillips (1999a), to determine whether there are any positive λ values. This theoretical and methodological framework can also be directly linked to empirical geomorphology via convergence–divergence analysis (Phillips, 2006). This is applicable if observations or inferences are sufficient to detect general trends in landscape response, but not to determine the entire response trajectory (otherwise a direct assessment is possible).

The formal theoretical justification is based on random pairs of locations in a landscape compared with respect to some indicator of system state (elevation or regolith thickness, for example). The mean difference at time t is:

$$\delta(t) = k e^{\lambda_1 t} \quad (31.5)$$

where k normalizes the initial separation and λ_1 is the largest Lyapunov exponent. The presence of at least one positive Lyapunov exponent indicates dynamical instability and chaos, while stability requires that all $\lambda < 0$.

Rewriting Equation (31.5) as:

$$\lambda_1 = \ln \delta(t) - \ln k \quad (31.6)$$

shows that λ_1 can be calculated from mean divergence or convergence; that is, whether on average, initial differences are increasing or decreasing. Evidence of a convergent, stable response indicates a resilient system, while divergence and instability indicate low resilience.

Convergence–divergence analysis can be quantified via a relevant metric of spatial variability or richness of the phenomenon of interest for successive time periods based on chronosequences, historical reconstructions, longitudinal studies, etc. Generally,

$$DI = V_t - V_{t-x} \quad (31.7)$$

where DI is a divergence index, and V is some measure of variability measured at two time increments t and $t - x$. DI > 0 indicates divergent and DI < 0 convergent behavior.

Measurements of Shannon and Kolmogorov entropy as indicators of variability may be especially useful in this regard, because entropy can be directly related to nonlinear dynamics (Culling, 1988; Ibanez, 1994; Phillips, 2006), and because spatial entropy analysis is commonly used in quantitative geography, landscape ecology, pedometrics and related fields.

Resilience is particularly relevant to theoretical constructs based on steady-state equilibrium, whether the latter is considered a normative, reference or potential state. Steady states are by definition resilient to small perturbations. The distinction between small or large perturbations is not always obvious, in which case a perturbation magnitude index (PMI) (Phillips, 2006) can be used for guidance:

$$PMI = M_t/M \quad (31.8)$$

where M_i is a measure of the magnitude or extent of disturbance i, and M indicates the extent or capacity of the affected system. For instance, PMI = A_i/A is based on the areal extent of disturbance compared to the total area. Or, for the example of a change in stream discharge (Q), PMI = Q_i/Q, where Q_i is the change in flow, and Q the mean or some other reference discharge. PMI ≥ 1 reflects a large perturbation, while PMI $\ll 1$ is a small perturbation.

Recursion

Disturbance responses often feed back upon themselves. These recursive feedbacks may be positive, reinforcing and thus perpetuating or even accelerating the change. Feedbacks may also be negative, slowing or even negating the change. The initiation of a weathering cavity, for instance, often involves positive feedback as the depression tends to collect more moisture and enlarge further. Negative feedback is exhibited by oversteepened slopes, which fail to the angle of repose.

Formal quantitative or qualitative stability analyses require the specification of feedbacks between system components. One of the advantages of stability analysis is that for systems with multiple feedbacks the cumulative effects of both positive and negative feedback can be assessed. In any case, however, change-and-response studies should seek to identify potential feedbacks, determine their signs and assess their relative importance. Uncertainty and complexity in geomorphic systems make this more easily said than done, but potential recursive effects should at least be considered in studies of geomorphic responses to disturbance.

CONCLUSIONS

In geomorphology, and in other environmental sciences such as ecology and pedology, perceptions and conceptions of change, disturbance, response and recovery have broadened. Multiple pathway, multi-outcome perspectives have supplemented (but not replaced) conceptual models emphasizing single-path, single-outcome trajectories of change. Geomorphology has also seen a transition from the idea of normative standards such as characteristic (steady-state), equilibrium, zonal and mature forms to the recognition that some systems may have multiple potential characteristic or equilibrium forms, and that some may have no particular normative state at all.

Trends such as these are often portrayed as replacements of outmoded ideas, but the trends described above are more accurately seen as a broadening of approaches to the study of changes and responses. The single-path single-outcome frameworks are, in essence, special cases of the broader pluralistic analytical structures. As a case in point, nonlinear dynamical systems approaches that recognize and even emphasize disequilibrium, non-equilibrium and multi-equilibrium, also include the possibility of stable steady-state equilibria.

In the final analysis, geomorphology is all about how landforms and landscapes respond to variable boundary conditions and to disturbances, and how geomorphic systems co-evolve with climate, ecosystems, soils and other environmental systems. In this respect conceptual frameworks or methods for the study of geomorphic disturbances and responses are those of geomorphology as a whole. However, at least three perspectives lend themselves particularly well to studies of recent and contemporary changes in earth surface systems.

The *hazards matrix* originally designed for analysis of natural hazards is also suitable for application to landscape disturbances and geomorphic responses themselves, independently of any hazards to humans. This matrix provides a useful template or checklist for disturbance studies, as it considers the magnitude of landscape-changing events, temporal aspects (frequency, duration, speed of onset and randomness or regularity of temporal spacing) and spatial aspects (areal extent and spatial concentration). *Landscape sensitivity* is explicitly concerned with the potential for landscape change, and is based on four key considerations: force versus resistance, frequency and duration of disturbances relative to geomorphic relaxation times, the stability or resilience of landscapes to perturbations and historical and geographical contingency. The concept of *amplifiers and filters* recognizes the potential presence of several levels of factors that may either exaggerate or (partially) negate the effects of disturbances, thus possibly producing disproportionately large or small geomorphic responses.

These perspectives can be synthesized into a framework for the assessment of geomorphic changes and responses based on the four Rs: response (reaction and relaxation times), resistance (relative to the drivers of change), resilience (recovery ability, based on dynamical stability) and recursion (positive and/or negative feedbacks).

ACKNOWLEDGEMENTS

This chapter is based on a paper published in *Progress in Physical Geography:* Phillips, J.D.

2009. Changes, Perturbations and Responses in Geomorphic Systems, 33(1): 17–30. The author is grateful to the editor and publisher of the journal for granting permission to adapt the article for this volume.

REFERENCES

Aslan, A., Autin, W.J. and Blum, M.D. (2005) Causes of river avulsion: insights from the late Holocene avulsion history of the Mississippi River, USA. *Journal of Sedimentary Research*, 75, 650–64.

Begin, Z.B. and Schumm, S.A. (1984) Gradational thresholds and landform singularity – significance for Quaternary studies. *Quaternary Research*, 21, 267–274.

Brunsden, D. (1990) Tablets of stone: towards the ten commandments of geomorphology. *Zeitschrift fur Geomorphologie* (supplement), 79, 1–37.

Brunsden, D. (2001) A critical assessment of the sensitivity concept in geomorphology. *Catena*, 42, 99–123.

Brunsden, D. and Thornes, J.B. (1979) Landscape sensitivity and change. *Transactions of the Institute of British Geographers*, 4, 463–84.

Chappell, J. (1983) Thresholds and lags in geomorphologic changes. *Australian Geographer*, 15, 358–66.

Chorley, R.J. (1962) Geomorphology and general systems theory. *U.S. Geological Survey Professional Paper*, 500-B, 1–10.

Craghan, M. (2005) Sediment delivery and accumulation in a developed area during coastal floods. *Geomorphology*, 69, 57–75.

Culling, W.E.H. (1988) Dimension and entropy in the soil-covered landscape. *Earth Surface Processes and Landforms*, 13, 619–48.

Davis, W.M. (1905) The geographical cycle in an arid climate. *Journal of Geology*, 13, 381–407.

Davis, W.M. (1930) Origin of limestone caverns. *Geological Society of America Bulletin*, 41, 475–628.

Gares, P.A., Sherman, D.J. and Nordstrom, K.F. (1994) Geomorphology and natural hazards. *Geomorphology*, 10, 1–18.

Hack, J.T. (1960) Interpretation of erosional topography in humid temperate regions. *American Journal of Science*, 258, 80–97.

Harrison, S. (1999) The problem with landscape. Some philosophical and practical questions. *Geography*, 84, 355–63.

Hergarten, S. (2002) *Self-Organized Criticality in Earth Systems*. Springer, Berlin.

Huggett, R.J. (1995) *Geoecology: An Evolutionary Approach.* Routledge, London.

Huggett, R.J. (1997) *Environmental Change. The Evolving Ecosphere*. Routledge, London.

Ibanez, J.J. (1994) Evolution of fluvial dissection landscapes in Mediterranean environments: quantitative estimates and geomorphic, pedologic, and phytocenotic repercussions. *Zeitschrift fur Geomorphologie*, 38, 105–19.

Kennedy, B.A. (1992) Hutton to Horton: views of sequence, progression, and equilibrium in geomorphology. *Geomorphology*, 5, 231–50.

Lane, S.N. and Richards, K.S. (1997) Linking river channel form and process: time, space, and causality revisited. *Earth Surface Processes and Landforms*, 22, 249–60.

Magilligan, F.J. and Stamp, M.L. (1997) Historical land-cover changes and hydrogeomorphic adjustment in a small Georgia watershed. *Annals, Association of American Geographers*, 87, 614–35.

Magilligan, F.J., Phillips, J.D., Gomez, B. and James, L.A. (1997) Geomorphic and sedimentologcal controls on the effectiveness of an extreme flood. *Journal of Geology*, 106, 87–95.

Marker, M.E. (2003) The Knysna Basin, South Africa: geomorphology, landscape sensitivity and sustainability. *Geographical Journal*, 169, 32–42.

Marker, M.E. and Holmes, P.J. (2005) Landscape evolution and landscape sensitivity: the case of the southern Cape. *South African Journal of Science*, 101, 53–60.

Nordstrom, K.F. (1987) Predicting shoreline changes at tidal inlets on a developed coast. *Professional Geographer*, 39, 457–65.

Osterkamp, W.R. and Hupp, C.R. (1996) The evolution of geomorphology, ecology, and other composite sciences, in B.L. Rhoads and C.E. Thorn (eds) *The Scientific Nature of Geomorphology*. John Wiley, New York. pp. 415–41.

Phillips, J.D. (1997a) Simplexity and the reinvention of equifinality. *Geographical Analysis*, 29, 1–15.

Phillips, J.D. (1997b) Humans as geological agents and the question of scale. *American Journal of Science*, 297, 98–115.

Phillips, J.D. (1999a) *Earth Surface Systems. Complexity, Order, and Scale.* Basil Blackwell, Oxford.

Phillips, J.D. (1999b) Divergence, convergence, and self-organization in landscapes. *Annals of the Association of American Geographers*, 89, 466–88.

Phillips, J.D. (1999c) Event timing and sequence in coastal shoreline erosion: hurricanes Bertha and Fran and the Neuse estuary. *Journal of Coastal Research,* 15, 616–23.

Phillips, J.D. (2006) Deterministic chaos and historical geomorphology: a review and look forward. *Geomorphology*, 76, 109–21.

Phillips, J.D. and Marion, D.A. (2005) Biomechanical effects, lithological variations, and local pedodiversity in some forest soils of Arkansas. *Geoderma* 124, 73–89.

Phillips, J.D., Marion, D.A. and Turkington, A.V. (2008) Pedologic and geomorphic impacts of a tornado blowdown event in a mixed pine–hardwood forest. *Catena*, 75, 278–287.

Puccia, C.J. and Levins, R. (1985) *Qualitative Modeling of Complex Systems.* Harvard University Press, Cambridge, MA.

Reynolds, V.N. (2006) Life Cycles, Systems, and Chaos: The Adoption and Application of Metaphors in Geomorphology. PhD dissertation, University of Kentucky, Lexington.

Riegert, M.A. and Turkington, A.V. (2003) Setting stone decay in a cultural context: conservation at African Cemetery No. 2, Lexington, Kentucky, USA. *Building and Environment*, 38, 1105–11.

Roberge, M. (2002) Human modification of the geomorphically unstable Salt River in metropolitan Phoenix. *Professional Geographer*, 54, 175–89.

Rodriguez-Iturbe, I. and Rigon, R. (1997) *Fractal River Basins*. Cambridge University Press, New York.

Sapozhinikov, V.B. and Foufoula-Georgiou, E. (1996) Do the current landscape evolution models show self-organized criticality? *Water Resources Research*, 32, 525–35.

Sauchyn, D.J. (2001) Modeling the hydroclimatic disturbance of soil landscapes in the southern Canadian plains: the problems of scale and place. *Environmental Monitoring and Assessment*, 67, 277–91.

Scheidegger, A.E. (1983) The instability principle in geomorphic equilibrium. *Zeitschrift fur Geomorphologie*, 27, 1–19.

Scheidegger, A.E. (1991) *Theoretical Geomorphology*. 3rd. edn. Springer, Berlin.

Schumm, S.A. (1979) Geomorphic thresholds: the concept and its applications. *Transactions of the Institute of British Geographers*, 4, 485–515.

Slingerland, R. and Smith, N.D. (2004) River avulsions and their deposits. *Annual Review of Earth and Planetary Science*, 32, 257–85.

Stoddart, D.R. (1966) Darwin's impact on geography. *Annals of the Association of American Geographers*, 56, 683–98.

Stolum, H.-H. (1998) Planform geometry and dynamics of meandering rivers. *Geological Society of America Bulletin*, 110, 1485–98.

Strahler, A.N. (1952) Dynamic basis of geomorphology. *Geological Society of America Bulletin*, 63, 923–38.

Thomas, D.S.G. and Allison, R.J. (1993) *Landscape Sensitivity*. John Wiley, Chichester.

Thomas, M.F. (2001) Landscape sensitivity in time and space: an introduction. *Catena*, 42, 83–98.

Thomas, M.F. (2004) Landscape sensitivity to rapid environmental change – a Quaternary perspective with examples from tropical areas. *Catena*, 55, 107–24.

Thomas, M.F. and Stirling, I.A. (eds) (2001) Landscape sensitivity. (Special issue). *Catena*, 42, 81–383.

Whipple, K.X. (2004) Bedrock rivers and the geomorphology of active orogens. *Annual Review of Earth and Planetary Sciences*, 32, 151–85.

White, G.F. (1974) Natural hazards research: concepts, methods, and policy implications, in G.F. White (ed.) *Natural Hazards: Local, National, Global*. Oxford University Press, New York. pp. 3–16.

Woldenberg, M.J. (1969) Spatial order in fluvial systems: Horton's laws derived from mixed hexagonal hierarchies of drainage basin areas. *Geological Society of America Bulletin*, 80, 97–112.

PART 5

Conclusion

32

Challenges and Perspectives

Mike Crozier, Paul Bierman, Andreas Lang
and Victor R. Baker

Prior to the conclusion it seemed pragmatic to invite several short challenge statements giving thought-provoking contributions focusing on future directions and trends to provide a perspective for geomorphology. These were invited to be relatively short, to focus on complementary themes and to be written by senior established geomorphologists who currently hold, or have recently held, major offices in geomorphological societies. The four statements are written by Michael Crozier who is President of the International Association of Geomorphology (2009–2013); Paul Bierman who is Chair, Quaternary Geology and Geomorphology Division, Geological Society of America (2009–2010); Andreas Lang who is Chair of the British Society for Geomorphology (2009–2010); and Vic Baker who was President of the Geological Society of America (1998).

GEOMORPHOLOGY AND SOCIETY

Michael Crozier

A defining trait of mankind is its thirst for knowledge and its ability to communicate and record its findings. Originally, those endeavours were directed at the day-to-day needs of survival – shelter, and the means to locate, gain and protect resources; no doubt aspects of what we now know as geomorphology formed an important part of that accumulated knowledge. But the prosaic needs of survival, as today, were

also accompanied by the bigger 'supernatural' questions on the origin of mankind and the origin of the Earth, its home. From the earliest times, gathering information about the terrain and its processes was an essential and integral part of social survival.

As the ability to travel great distances and explore the world became widespread in the 18th and 19th centuries, the natural sciences became part of a rapidly expanding frontier, represented by scientific expeditions such as those of Cook, Agassiz, Humboldt, Darwin and Wallace, to name a few. As with other philosophies, those concerned with geomorphology compiled detailed observations, descriptions and classifications. New theories emerged, many of which sorely tested strongly held conventional beliefs of the time. Scientific discoveries were therefore not only a source of wonder and social advancement but also a challenge to the established beliefs of society.

As the flood of new information became increasingly difficult to master, the age of the polymath, represented by the early global explorer, inevitably gave way to a plethora of specialists. For instance, the geographer ultimately became either a human or physical geographer, the physical geographer became a climatologist, biogeographer, hydrologist or geomorphologist. In turn, geomorphologists began to identify themselves with components of their discipline such as rivers, tectonics, coasts, glaciers and the Quaternary, and more recently, by subdivisions such as, modelling, dating, paleoenvironment, etc. In many cases, this sectorization spawned an esoteric technical language that obscured research findings from society's view.

This increase is specialization has demanded and been driven by the replacement of observation by measurement and the consequential quantification which allowed new and sophisticated ways of establishing and verifying relationships. In geomorphology, from the 1950s, measuring devices diversified and multiplied. Initially, manpower requirements and expense meant that research was confined to small catchments and specific processes, rather than the large-scale features which had occupied the energies of earlier geomorphologists. The advent of automatic electronic recording (e.g. pressure transducers and data loggers) extended the duration and intensity of both laboratory and field studies. Processing of increasingly large amounts of data was facilitated by expanding computer capacity. The perspective of geomorphology has also been influenced by capabilities to digitize, rectify and geo-reference aerial photography, increasing precision of global positioning systems, the advent of remote sensing capabilities, such as satellite multi-band imagery, surface and airborne radar and laser technology. These advances, together with the expanding use of geographic information system (GIS) platforms, have seen a re-widening of the geographic scale of interest. The shift from small catchment and plot-scaled studies to GIS- driven regional studies has been mirrored by the increased focus on the temporal/historic spectrum. In particular, this has been stimulated by the increasing range and precision of dating techniques and techniques to unravel paleoenvironmental conditions (e.g. exposure dating, and isotope analysis).

This increased specialization and technological sophistication has placed much of geomorphology beyond the reach of society – science has in some quarters been seen as self-indulgent and only marginally relevant to society. Government research support in a number of countries has reacted by defining the category of 'public good science' and funding it to a greater extent than 'blue sky' (curiosity-driven) research. The adaptive response of some researchers has been to re-orient their research goals or at least re-label their endeavours to make their research more socially relevant. Polar research and paleoenvironmental research have found relevance by informing the climate change debate while process studies are readily adapted to hazard issues. Scientific ethics are also being promoted – even my current subscription to the Hydrological Society contains an amount directed to 'water for survival'.

There is a rubric that says 'there is not such a thing as "applied geomorphology", only geomorphology to be applied.' Whatever construct you

place on this statement, geomorphologists, as with other scientists, are currently confronted with increasing opportunities, if not an ethical prerogative to apply their science. The reasons for this are the dramatic contemporary changes in two of the most important drivers within our environment, climate and human population. Geomorphologists have long studied the role of climate and humans in controlling process behaviour and landform development. Furthermore, increases in population, urbanization, social infrastructure and economy are accelerating the risk and impacts on the human condition, irrespective of the threats from accelerated climate change.

In the period between 1650 AD and 1850 AD, world population doubled from 550 million to 1200 million. But in half that time, from 1900 AD to 2000 AD the world population increased four-fold to almost 6 billion and economic activity increased 40-fold. While population increase is placing huge demands on food production, the capacity to produce is being severely limited by anthropogenic and climate-driven soil degradation and erosion. Recent research indicates that during the 40 years between 1955 and 1995, nearly one third of the world's arable land had been lost by erosion and losses continue at a rate of 10 million hectares per year.

Population increases, together with increased urbanization and increased standard of living, have created unprecedented levels of intervention within the geomorphic system. Spiralling demands for resources such as aggregates are impacting river and coastal systems, while demands for hydropower and irrigation affect the quality and quantity of surface and subsurface hydrological systems. Mineral, timber and soil resources are also being exploited at unprecedented rates, in many cases with unanticipated and disastrous results.

It is against this background that the geomorphologist has an important social role to play. Collectively and individually we can inform society of how systems will respond to human intervention and enhanced climate activity. We can identify the outcomes and risks and communicate them effectively to resource managers, planners, industry, government and the public. In jurisdictions where resource management legislation exists there are frameworks such as environmental impact assessment procedures that are well suited to geomorphological input.

Environmental and resource law has also inadvertently spawned what might be referred to as 'forensic geomorphology'. Some examples from New Zealand illustrate the nature of the geomorphic challenge. Recently the agency responsible for the national hazard insurance scheme refused a multi-million dollar payout to a community because it deemed that the damage was caused by a flood (their schedule excludes flood damage but not mass movement damage). Expert geomorphological opinion was sought and resolved that the damage was in fact caused by a debris flow – the decision was reversed. Similar issues are confronted by coastal geomorphologists. In some jurisdictions, foredunes are protected from development, resulting in protracted legal debates as to what actually constitutes the real foredune. In property law, ambulatory property boundaries, often defined by the centre line of a river, allow conjoint property owners to gain or lose valuable land as the river shifts course. But this rule applies only if the migration is 'natural' (slow and imperceptible) and does not apply if the migration is rapid and event-related. Robust resolution of such issues is the realm of forensic geomorphology and requires expert geomorphic evidence to resolve.

The effectiveness of the geomorphological contribution, however, ultimately relies on the integrity and rigour of our science and continued exercise of curiosity. My conclusion therefore is that there is 'geomorphology to be applied' and that its application is desperately needed by society. Given the immensity of current global change and its impact on the well-being of mankind, perhaps we should learn from our hydrology cousins and institute an initiative of 'geomorphology for survival'.

GEOMORPHOLOGY: LINKING PAST AND PRESENT

Paul Bierman

In an increasingly crowded world, where population pressures and climate change juxtapose people and natural hazards with ever-increasing frequency, understanding the rate, distribution and magnitude of Earth surface processes is critical to the health and well-being of billions. Geomorphologists, with their intellectual focus on and near Earth's surface, have the skills and tools to understand our planet's behaviour on timescales relevant to society.

Over the past several decades, a suite of technological advances has radically transformed the field of geomorphology. LIDAR now maps Earth's surface at sub-metre scales, allowing detection of previously unknown fault scarps, monitoring the movement of river channels and clearly delineating the location of landslides. Exquisitely sensitive mass spectrometers and new means of chemical analysis estimate rates of erosion, explain processes of sediment generation and constrain the rate of chemical weathering.

Today, we can measure precisely what the last generation of geomorphologists could barely, if at all, estimate.

The ability to characterize in detail the morphology of Earth's surface, the age of sedimentary deposits, and the rate of erosion and sediment generation has changed geomorphology from a descriptive science with little predicative capability to one with the potential to serve society. Our science can link Earth's past, present, and future. The question is – will we as a field use the tools we have to better the world in which we, and many billions of other people, live?

As geologists, looking back into the past is a familiar means of learning what the future might hold. There are many ways in which understanding the past behaviour of our planet can guide and inform the future. For example, we can use past local and planetary responses to examine what happens when the hydrological cycle intensifies, drying some parts of the planet and moistening others. Past climate changes, and the records they leave on the landscape, serve as analogues for a changing future.

In a time of rapid, human-induced climate change triggered by changing levels of CO_2 and other atmospheric gasses, geomorphology's view back in time is crucial. The climate system is complex, with numerous feedbacks limiting the certainty of model results. In contrast, the geological record is tangible and decipherable in terms of the planet's past reaction to a changing climate. Glacial terminations provide at least a partial analogue for rapid warming, particularly the end of the Younger Dryas cooling at the dawn of the Holocene.

Deciphering the planet's reaction to past climate change is by no means simple; the resolution of many dating techniques, often limited by uncertainties in geological assumptions, remains insufficient to determine leads and lags – critical to interpreting which processes drive climate change and which processes respond to changing climate. Reliable, unambiguous paleo-climate archives are few and far between. Clearly, we as a community need to continue refining analytical techniques but that is not enough. We need detailed field studies to place geochemical and isotopic records in context. We need creative approaches to find and then exploit new and different paleoclimatic archives around the world.

With the tools to estimate background rates of surface processes, geomorphologists can be an integral part of setting regulatory goals. For example, the US Environmental Protection Agency has classified suspended sediment as a pollutant and in many watersheds has determined that a total maximum daily load (TMDL) must be established as a regulatory endpoint. TMDLs can be set by examining sediment yields from reference watersheds, those that seem close to their natural conditions, or by determining loads based on biological impacts downstream. Yet, we can do better. Using cosmogenic nuclides and other means of sediment budgeting, geomorphologists can estimate long-terms rate of sediment generation, transport, and delivery – rates more representative of the natural system than even a few decades of suspended sediment data could ever deliver.

With the world population still growing at a tremendous rate, the pressure is on to use marginal lands for housing, agriculture and resource extraction. Such intensive land use pushes people into areas where geologic hazards are common – the flanks of volcanoes, active flood plains and eroding coastal zones. It is here that geomorphology and geomorphologists can make a difference, guiding zoning decisions, producing hazard maps, participating in emergency response missions. Although developing nations are most typically considered at risk from such natural hazards, one needs look no farther than New Orleans, Louisiana, to understand the importance of geomorphology to society. The loss of coastal marsh land, the re-routing of Mississippi River sediment and levée building all contributed to a situation that is at best non-sustainable and at worst, catastrophic.

Understanding Earth surface processes and the history of our planet are critical in a future that includes many billions of people inhabiting an ever more crowded and warming planet. Geomorphologists and geomorphology as a discipline have the knowledge to reduce society's exposure to both natural and human-induced hazards. By understanding how quickly and where our planet's surface changes, we can adapt to rising sea levels, a warming climate, erupting volcanoes and flooding rivers. The question is: Are we as a profession up to this task of understanding coupled natural/human systems and how will we ensure that our relevant knowledge is used by those managing Earth's dynamic surface?

BRINGING PROCESS, SEDIMENT, AND LANDFORM EVOLUTION CLOSER TOGETHER: A CALL FOR QUANTIFICATION AND PREDICTIVE MODELLING IN GEOMORPHOLOGY

Andreas Lang

Geomorphology is in a prime position to help tackle the great challenges for mankind, including natural hazards or climate change impacts. The Earth's surface, the focus of our discipline, is where change is happening, provides the ground on which we are living and supports the resources that sustain us. Challenges for geomorphology are plentiful: climate warming, sea-level rise, changing rainfall patterns and, especially, human transformation of landscapes through direct engineering and unintended land-use impacts. The question which society poses to us quite rightly is: How will landscapes react to environmental change and what is the sensitivity of the landscape to change of a given magnitude?

The usual approach to answering such questions would be to utilize monitoring data where geomorphic response to environmental change has been determined in detail over the short term. But monitoring data are sparse for many geomorphic systems and, if available, the timescales covered rarely extend beyond a decade. The geomorphology toolbox allows for another approach: we can carefully reconstruct landscape evolution to deduce how Earth surface systems reacted to changing constraints in the past, whether climatic or anthropogenic or both, and thus provide analogues for future environmental changes. In practice both approaches are flawed: geomorphic response depends strongly on how a system has evolved, and understanding the system trajectory is a necessary prerequisite to predict systems response. Also, we may now approach a situation in which we experience changes of a 'non-analogue' state unmatched by past environmental changes. What else can we do?

Much of my geomorphic upbringing was dominated by the idea that erosion, transport and deposition processes can be unravelled by carefully studying sediments, supposedly the processes correlate with deposits. At that time – and despite conceptual ideas like those of Schumm (e.g. Schumm, 1991) having been around for some time already – the study of sedimentary deposits largely ignored issues like equifinality, contingency or complexity, and argued in a rather linear and reductionist fashion. 'Contemporary' was mistaken as 'causal' (still commonplace in some communities) and a process-based understanding was rarely attempted. Since that time much progress has been achieved in geomorphological research. We now have a much better understanding of geomorphic process and also much more sophisticated ideas on longer term landscape evolution. Recent progress can be grouped roughly in two domains: (1) technology-driven and (2) concept-driven.

Technology-driven progress is plentiful and often based on advances made in cognate fields that are then utilized in geomorphology. Examples include techniques for dating, modelling and monitoring like LIDAR, shallow geophysical sounding, sediment analyses, radar, low-temperature thermochronometry, *in situ* produced cosmogenic isotopes, three-dimensional morphometric datasets, SAR interferometry and many others. This progress has allowed the testing of many long-standing hypotheses on small and large scales and on short as well as long timescales.

Concept-driven progress is much rarer. Even so, geomorphology has for many years been at the forefront of developing conceptual frameworks (e.g. models of complex system response to external forcing as in Schumm's geomorphic work are only now considered in cognate disciplines like ecology). Still, geomorphological contributions, for example to complexity science, are significant and often most convincing.

The way ahead will need to ensure that we continue to drive progress in all fields: first, we need further technological development, but we have to ensure that geomorphological questions are the drivers rather than just adopting new technologies developed elsewhere. Geomorphology is asking the questions and it needs to develop the right tools to solve these. We cannot just be opportunistic, exploit cognate fields and hope the perfect tool will come along at some point. Second, we have to continue refining conceptual ideas and developing new ones. We can build on a rich body of theory and should try not to reinvent the wheel but ensure the progress integrates existing knowledge. Third, and probably most challenging, we need to accept uncertainty and have to learn to live with it. We now have much more precise and accurate estimates of process rates and many other parameters but we still ignore uncertainty and error propagation. Most values can be determined only to a limited precision and any further calculation based on a value needs to propagate its uncertainty. Besides the technical uncertainties we need also to consider conceptual uncertainty: how to deal with gaps (temporally as well as spatially) in observations and records; how to deal with equifinality and the possibility that we may derive a correct answer for the wrong reason; and how can we identify complex behaviour like self-organized criticality and deal with it?

Recent geomorphic progress is, however, mainly achieved within sub-disciplines focusing on specific processes or landforms at specific spatial and temporal scales. This leads to diverging developments where, for example, process understanding unravels more and more minute detail and evolutionary studies are applied to longer timescales and larger areas but are based only on rudimentary process understanding. Also, much of the technological advance has not been used to test existing hypotheses but new and often rather naive ideas of landscape functioning have been developed to explain the new data that suddenly become available. The desire to follow the

attraction of new and sexy technologies has often resulted in brushing aside the wealth of knowledge that exists already. How can we learn to make the developments converge, integrate approaches and develop more holistic methodologies across process domains, spatial and temporal scales, and that also include people?

In my view we need a more holistic approach and to marry process understanding and evolutionary information. I am dreaming of a computer modelling framework, an 'Earth surface simulator', to provide a unifying platform: something like GCM technology, representing dynamic process interactions, and including interfaces to the lithosphere, biosphere and atmosphere. To represent different spatial and temporal scales a nested hierarchy of model components could be introduced. Our excellent process understanding needs to be coded at the core of the model (as is already done for some process domains) and sedimentary bodies could be represented as dynamic storage elements. Sediment flux could form the link between system components; internal feedbacks, thresholds and complex nonlinear dynamics need to be implemented; and capturing the emergent character of many geomorphic systems will be essential. To calibrate and validate model scenarios we can then use sedimentary records, essential to enable prediction of future landform response to environmental change. Besides prediction, such a modelling tool will also greatly benefit our understanding of system evolution as we will be enabled to run different scenarios for hypotheses testing for past events. In cases where we are currently limited to state 'contemporary' between an inferred cause and its potential effect, the modelling framework will allow exploring different scenarios in detail: we should test evolutionary hypotheses and unravel cause and consequence by considering several sets of initial conditions, different starting points along the system trajectory, and different process interactions.

However, such an approach will work only if we accept uncertainty. Besides the need to quantify uncertainties in observations and modelling, we also need to be prepared to lose resolution spatially and temporally. For example, we may be able to determine the sediment flux from a catchment under a certain scenario but we may well lose the spatial information about which part of a gully is active at what time. We need to be prepared to look beyond the single landform (slope, channel cross-profile, etc.) and take a more generalist viewpoint. We can find guidance in other disciplines that have faced similar problems. Techniques to treat uncertainty in data and in prediction are commonplace in climate research, for example. Probabilistic approaches could be utilized comparable to those used in weather forecasting: for example, with an increase in precipitation by 50 mm p.a. with a probability of 80 per cent the increase in sediment flux will be 20 t p.a.

Will this remain a dream or can we work together to make it happen? Community approaches currently used in Earth System Modelling (like CSDM; http://csdms.colorado.edu) offer the opportunity to progress towards such a system. There is a lot of work still to be done but some of the necessary tools seem to be available. A major challenge will be incorporation of human activity. In many landscapes over recent centuries human interference (intentional and unintentional) with the Earth's surface has outperformed 'natural' rates of change; this has to be included in our reasoning and geomorphological models, otherwise our prediction will fail. Given the increasing population pressure this is very unlikely to change.

FUTURE SCOPE OF GEOMORPHOLOGY

Victor R. Baker

There are at least 10 billion stars in our moderate-sized Milky Way galaxy (one of at least 200 billion galaxies in the known universe). A high proportion of these stars have orbiting planets, at least according to recent discoveries, which, in the past 15 years, have identified (measured) more than 400 planets outside our own solar system. Although this accelerating pace of exoplanet discovery was initially biased towards finding huge, gas-ball objects (so-called 'hot Jupiters'), orbiting close to their respective stars, future missions, including the recently launched Kepler space observatory, will be more focused on the discovery of Earth-like planets. Given that our own solar system has four rocky planets with Earth-like attributes (Mercury, Venus, Earth and Mars), prospects from other star systems foretell an astronomical expansion in the sample size from which to develop a generalized understanding of Earth-like landscapes, that is the science of geomorphology.

Of course, it may be many decades before adequate resolution is achieved to analyse surface details on the immense number of potential planets on which to pursue future geomorphology. In the interim, the science of Earth-like planetary surfaces (Sharp, 1980; Baker, 2008) will continue to be focused on Earth's surface and that of nearby neighbours, particularly Mars. Its scientific scope will involve both the range of landscape phenomena to be studied and also the range of approaches that will be taken in that study. Of course, these two elements are interwoven, in that the chosen methods of study can limit the range of phenomena deemed to be of interest, and vice versa.

Geomorphology is undergoing rapid change. Musing on the general pace of all scientific change, one easily comes to the view that scientific concerns and methods 40 or 50 years from now will be even more surprising to us than the concerns and methods of today would have been to those viewing the field from a vantage 40 to 50 years ago. Geomorphology of the 1950s to 1970s was characterized by the so-called 'quantitative revolution' and related process studies (Chorley, 2008). This 'bandwagon' (Jennings, 1973) developed into the 'normal science' (Kuhn, 1962), or 'fashion' (Sherman, 1996), of its day. From that vantage point George H. Dury (1978) expressed concern for the long-standing split between geography and geology, while highlighting three areas future development: (1) continuing studies of stream nets, (2) advances in tropical geomorphology, and (3) advances in palaeomorphology and palaeoclimatology. Though geography and geology locally continue to persist in arcane philosophical discord, the palaeoclimatology related to global change and Earth system science has certainly assumed fashionable, bandwagon status, and this popularity can probably be safely extrapolated to the future. The prospects for drainage network studies evolved towards emphases on theoretical understanding of optimal network configurations and fractal models of self-organized criticality (e.g. Rodriguez-Iturbe and Rinaldo, 1997). While understanding of the tropics certainly advanced considerably, the space programme with its associated technological developments from remote sensing has proven to be a more significant development.

In projecting past the 1990s, to our present-day and beyond, Dury (1978) expected the major themes of his day to continue, among them quantification, data processing, computer modelling, systems analysis, climatic geomorphology and the relationship of plate tectonics to geomorphology. In a similar vein, at this time it is probably safe to infer a continued emphasis on tectonic geomorphology (e.g. Burbank and Anderson, 2001). Moreover, in much the same way that the applied mathematics of Scheidegger's (1963) *Principles of Geodynamics* became mainstream in geophysics (Turcotte and Schubert, 1982), the vision of Scheidegger's (1961) *Theoretical Geomorphology* is now being realized through advances in computer modelling (e.g. Wilcock and Iverson, 2003; Pelletier, 2008), a trend that is assured of continuance because of the exponential growth of computational power (Moore's law). The challenges for this new modelling paradigm are (1) to avoid having predictive capabilities outrun any substantive means of testing the predictions against meaningful observations; and, more critically, (2) an issue raised years ago by Chorley (1978): that a functionalism that merely matches predictions to observations will need to be replaced by a realism that can satisfactorily identify the actual causal mechanisms of phenomena.

In the conclusion to his article, it was clear that Dury (1978) was not satisfied with his prognostications. He wanted more from his chosen science (Dury, 1978: p. 273), as follows, 'Let us hope that, 25 years from now, those of us who survive will be doing something very different from what we are doing today'. But how can such a hope be realized? Following reasoning made famous by Thomas Kuhn (1962), Dury (1978: 273) answered this question by advocating a role for, 'recognized and stubborn anomalies, which cannot be assimilated into existing paradigms.' He concluded (Dury, 1978: 273), 'We are, I suggest obliged to hope that surprises are in fact in prospect, and that a forecast of surprise in the long term will not be falsified by events'.

One likely source of such surprises will arise from continued advances in technology. Thus, spectacular advances in geochronology are being generated through studies of terrestrial cosmogenic nuclides, optically stimulated luminescence dating, and thermochronology. Equally spectacular advances have occurred with interferometric radar (InSAR) for measuring Earth surface topography and its deformation. Other important sources of data are the remotely sensed digital terrain models at very high resolution generated by terrestrial laser scanning (TLS) and airborne laser swath mapping (SLSM) using light detection and ranging (LIDAR) technology. These produce sub-meter resolution digital terrain models (DTMs) and high-quality land cover digital surface models (DSMs) of geomorphic phenomena.

A second important source of surprise can arise from a change in methodological approach to problems. One such trend is the replacement of reductionistic methodologies, so successful for the progress of physics in the last century, with a new holism. This affords a prospect for transcending the long-standing divide between historical and process studies, as recently highlighted in the text by Huggett (2007). Here, computation is used in regard to nonlinear dynamic theory (Phillips, 1999), such that timebound elements of the landscape can be treated in a holistic manner.

Finally, a more radical source of surprise can occur when the entire viewpoint of a science is transformed (Kuhn, 1962), and participants come to see the world in a different way. The currently prevailing viewpoint in science, much of geomorphology included, is one that can be labelled 'Aquarian' (Baker, 2001). This is the scientific view that sees nature in detachment, from the outside, such that a presumed objectivity is maintained. The contrasting viewpoint can be labelled 'Piscean' (Baker, 2001), in that it views nature from the inside (as something of which the science is a part). Just as the act of measurement cannot be separated from the reality of sub-atomic particles (at least according to the 'Copenhagen Interpretation' of quantum theory), and just as the 'Anthropic Principle' is arguably a critical element to modern cosmological theory, so the geomorphologist may come to be seen as less than detached from his/her object of study. Current trends in artificial intelligence, again propelled by computer technology, are making advances possible in this sphere by capturing aspects of the abductive mode of scientific inference that has long been a traditional part of some geomorphological inquiry (Baker, 1996). The prospect that this viewpoint holds is for the same power of reasoning that can make it possible for humankind to understand the billions of other 'Earths' to be brought to bear in resolving critical habitability questions in regard to their own.

REFERENCES

Baker, V.R. (1996) Hypotheses and geomorphological reasoning, in B.L. Rhoads and C.E. Thorn (eds) *The Scientific Nature of Geomorphology*. Wiley, New York. pp. 57–85.

Baker, V.R. (2001) Water science: Aquarian and Piscean viewpoints. *Geological Society of America Abstracts with Programs*, 33(7), 429.

Baker, V.R. (2008) Planetary landscape systems. *Earth Surface Processes and Landforms*, 33, 1341–53.

Burbank, D.W. and Anderson, R.S. (2001) *Tectonic Geomorphology*. Blackwell, Oxford.

Chorley, R.J. (1978) Bases for theory in geomorphology, in C. Embleton, D. Brunsden and D.K.C. Jones (eds) *Geomorphology: Present Problems, and Future Prospects*. Oxford University Press, Oxford. pp. 1–13.

Chorley, R.J. (2008) The mid-century revolution in fluvial geomorphology, in T. Burt, R.J. Chorley, D. Brunsden, J.J. Cox and A.S. Goudie, A.S. (eds) *The History of the Study of Landforms, or the Development of Geomorphology*. Vol. 4. *Quaternary and Recent Processes and Forms (1890–1965) and the Mid-Century Revolutions*. The Geological Society, London. pp. 926–960.

Dury, G.H. (1978) The future of geomorphology, in C. Embleton, D. Brunsden and D.K.C. Jones (eds), *Geomorphology: Present Problems, and Future Prospects*. Oxford University Press, Oxford. pp. 263–74.

Huggett, R.J. (2007) *Fundamentals of Geomorphology.* Routledge, London. p. 458.

Jennings, J.N. (1973) 'Any millenniums today, lady?' The geomorphic bandwagon parade. *Australian Geographical Studies,* 11, 115–33.

Kuhn, T. (1962) *The Structure of Scientific Revolutions.* University of Chicago Press, Chicago. p. 172.

Pelletier, J.P. (2008) *Quantitative Modeling of Earth Surface Processes.* Cambridge University Press, Cambridge.

Phillips, J.D. (1999) *Earth Surface Systems: Complexity, Order and Scale.* Blackwell, Malden.

Rodriguez-Iturbe, I. and Rinaldo, A. (1997) *Fractal River Basins: Chance and Self-Organization.* Cambridge University Press, Cambridge.

Scheidegger, A.W. (1961) *Theoretical Geomorphology.* Springer Verlag, Berlin. p. 331.

Scheidegger, A.W. (1963) *Principles of Geodynamics.* Academic Press, New York. p. 362.

Schumm, S.A. (1991) *To Interpret the Earth. Ten ways to be wrong.* Cambridge University Press. Cambridge, New York, Port Chester, Melbourne, Sydney.

Sharp, R.P. (1980) Geomorphological processes on planetary surfaces. *Annual Reviews of Earth and Planetary Sciences,* 8, 231–61.

Sherman, D.J. (1996) Fashion in geomorphology, in B.L. Rhoads and C.E. Thorn (eds) *The Scientific Nature of Geomorphology.* Wiley, New York. pp. 87–114.

Turcotte, D.L. and Schubert, G. (1982) *Geodynamics, Applications of Continuum Mechanics to Geological Problems.* Cambridge University Press, Cambridge. p. 456.

Wilcock, P.R. and Iverson, R.M. (eds) (2003) *Prediction in Geomorphology.* American Geophysical Union Geophysical Monograph, p. 135.

33

Conclusion

Kenneth J. Gregory and Andrew Goudie

We have noted ways in which geomorphology is now poised to develop in the future in preceding chapters; this conclusion combines some of these proposals, considers a rejuvenated discipline and indicates new issues which geomorphology can pursue.

THEMES IDENTIFIED

Enormous *changes in geomorphology* over the last century include plate tectonics, the revolution in Quaternary science, quantitative process-oriented geomorphology (Goudie, Chapter 2), and automated landform classification (Oguchi and Wasklewicz, Chapter 13), as new instruments have produced an increasingly complex picture of the world (Church, Chapter 7). Such changes mean that models cannot easily be extended from one environment to another , such as temperate to tropical environments (Thomas and Kale, Chapter 26); links with, and awareness of, other disciplines are increasingly significant as in regolith study (Taylor, Chapter 16), and some geomorphologists incline toward the mainstream of geophysical science (Church, Chapter 7). Links with other disciplines occur because geomorphologists are increasingly part of a team in applied projects (Downs and Booth, Chapter 5), results of the increasing ability to propose estimates of dates now have implications well beyond geomorphology (Brown, Chapter 11), and geomorphological research can lead to feedbacks to other disciplines in the way that stratigraphic approaches adopted from geology have had to be modified (Mather, Chapter 29). Geomorphologists now offer environmental problem solutions that for most of the 20th century were left to engineers (Kondolf and Piegay, Chapter 6).

Many branches of geomorphology identify *themes which merit further consideration.* Although technology alone will not replace the need for conceptual ingenuity (Rhoads and Thorn, Chapter 4), data innovations can be central to informing ways in which geomorphological models will develop in the future (Odoni and Lane, Chapter 9), and it is necessary to increase the fields in geomorphology to which GIS analyses can be applied (Oguchi and Wasklewicz, Chapter 13). Further applications of nonlinear dynamics to geomorphic systems should illuminate the process–history problem (Huggett, Chapter 10) and a major future challenge is in increasing the scale of experiments to more nearly simulate real-life conditions and of upscaling the results of experiments to the field scale and integrating them into models of landscape evolution (Robinson and Moses, Chapter 17).

New emphases recently highlighted include the unity of geomorphology (Richards and Clifford, Chapter 3), with the prospect of a unified theory in particular branches such as coastal geomorphology founded on morphodynamics and using models firmly based on field experience (Woodroffe, Cowell and Dickson, Chapter 24). Plate tectonics re-stimulated research, providing an elegant framework to bring together tectonics and climate, as two key factors in landscape evolution (Bishop, Chapter 28). Geomorphological mapping is now seeing something of a renaissance and the ability to handle, manipulate and analyse large, parallel, datasets will begin to be realized (Smith and Pain, Chapter 8). Applied geomorphology, or more broadly geomorphology to be applied (Crozier, Bierman, Lang and Baker, Chapter 32), can support development programmes, in the tropics for example (Thomas and Kale, Chapter 26), although there has been greater awareness of the cultural bias pervading applications of geomorphic research (Kondolf and Piegay, Chapter 6) which can now be improved by expanding the role of geomorphology in environmental problem-solving: the goal is not the control or manipulation of the natural environment, but rather maximizing of ecosystem services (Downs and Booth, Chapter 5). Reconceptualization, necessary in both hillslope hydrology and mass movement studies, representing a move from simplistic, process–response models to detailed, integrated representations of complex system behaviour (Petley, Chapter 20), could provide a more integrated view of the interlinkages between inorganic processes and microorganism, plant and animal activity (Viles, Chapter 14); focus on understanding and managing the impacts of humans on rivers (Pizzuto, Chapter 21); and enable glacial geomorphologists to accomplish a vertical integration of findings to permit a scalar incorporation of processes at differing scales, so that process mechanisms can be clearly understood from the micro to the megascale (Menzies, Chapter 22). Although use of transfer functions has meant that modern environmental ranges of particular organisms can be used to quantitatively recreate the palaeoenvironments of Quaternary populations, multi-proxy approaches can be used to pinpoint more accurately the causes of identified changes in the Quaternary (Mather, Chapter 29).

Further *exploration and development of some contemporary themes* includes the extent to which landforms are in equilibrium with geophysical factors including both climate and tectonics, the degree to which landforms are inherited, whether humans have significantly altered the geomorphic and related cycles and to what extent this mitigates or exaggerates climatic forcing (Brown, Chapter 11). Human impacts, of which global warming is but one example, have become a major focus for the international geomorphological community (Slaymaker et al., 2009). Combining existing stratigraphic research with stable isotope analysis (carbon, oxygen, strontium) and trace element geochemistry, can quantify past changes in temperature, precipitation/evaporation and salinity (Williams, Chapter 30); improved uranium/lead dating of speleothems shows

that the role of inheritance may be aided by process-response modelling (Ford and Williams, Chapter 27), and enhanced quantification of the rates and nature of sediment exchange between aeolian and fluvial systems for modern analogues could elaborate how aeolian and fluvial processes interact during phases of climate transition, useful for understanding past and present relationships and for assessing implications of future climate change in both hot and cold arid regions (Bullard, Chapter 25). Advances in numerical modelling, geochronology and remote sensing quantitative techniques are now being more widely used to model earth surface process and interpret the landscape (Warburton, Chapter 19).

Exciting prospects suggested for the future of geomorphology include an integrative view of landscapes and landforms as circumstantial and contingent outcomes of deterministic laws operating in a specific environmental and historical context, with several outcomes possible for each set of processes and boundary conditions (Huggett, Chapter 10). More specifically, there are prospects of global DEMs at better than 10 m resolution and complete image coverage of the Earth at better than 50 cm (Farr, Chapter 12), of investigating how biogeomorphological interactions influence global carbon and nitrogen cycling (Viles, Chapter 14), of focus on glaciomarine environments (Menzies, Chapter 22), and exploring nonlinear dynamical systems approaches that recognize disequilibrium, non-equilibrium, and multi-equilibrium, as well as the possibility of stable steady-state equilibria (Phillips, Chapter 31). The discipline is ready to provide a more substantial contribution to the solution of environmental problems at all scales, from local to global, possibly employing resilience engineering (Loczy and Suto, Chapter 15), which could involve conservation of representative landforms and landscapes, including the 'classical' sites that were the subject of pioneering geomorphological investigations (Ford and Williams, Chapter 27, The identification, explanation and conservation of geomorphological features as World Heritage sites, geoparks, etc. is a burgeoning area of interest (Migon, 2010), and geomorphologists need to be more involved with public outreach by stressing the value and appeal of great landscapes.

We now have the techniques and over-arching paradigms, including plate tectonics, that enable us to re-focus attention on landscape evolution (Bishop, Chapter 28), and to use appreciation of past environmental change to bolster awareness of possible future change, so that we can learn to live with uncertainty in a time of rapid environmental change, just as our ancestors did in the past (Williams, Chapter 30). In some ways geomorphology is an 'interface' discipline (Smith and

Pain, Chapter 8), with a future which has the potential to reconcile scales and approaches of process and environmental change (Roy and Lamarre, Chapter 18), but at the same time maintaining a clear bridging position in the way that periglacial geomorphology relates to the emerging discipline of geocryology, and the increasing sophistication of Quaternary science (French Chapter 23). Concluding statements (Chapter 32) include suggestions of an initiative of 'geomorphology for survival' (Crozier, Chapter 32), and ways in which geomorphology and geomorphologists can to reduce society's exposure to both natural and human-induced hazards (Biermann, Chapter 32), so that scientific concerns and methods 40 or 50 years from now will cause even more surprise than the concerns and methods of today would have caused to those viewing the field 40 to 50 years ago (Baker, 2001).

CONSEQUENCES OF THEMES IDENTIFIED

There is certainly potential for a rejuvenated discipline and geomorphology in the future could be structured in a way which is distinct from the arrangement in this Handbook. This could be accomplished in a situation where geomorphology has been characterized as simultaneously developing in diverse directions: on one hand towards more rigorous geophysical science on another as more concerned with human social and economic values, with environmental change, conservation ethics, with the human impact on environment, and with issues of social justice and equity (Church, 2010). However, in the light of the themes identified, three particular trends arise: (1) the need for more multidisciplinary research and investigations, (2) the question about how far geomorphology can extend, and (3) the potential to progress further than previously in relation to management and design of environments.

Recent decades have seen an increasing amount of geomorphological research contributing to, or undertaken as part of, *multi-disciplinary research* investigations (Goudie and Viles, 2010) and the requirement for better opportunities for interdisciplinary collaboration and more sophisticated interaction between natural and social scientists are needs identified by Dadson (2010). One way in which this is reflected (see Chapter 1, Table 1.5), is in the emergence of hybrid branches such as ecogeomorphology and hydrogeomorphology. Such hybrids are responses to particular opportunities and an example is illustrated by palaeohydrology which was first explicitly identified in 1954 (e.g. Gregory, 1983) but grew as

methods were required for elaborating hydrological records prior to the periods of continuous monitoring, very often the last 100 years or less. As decision-making requires knowledge of events on timescales longer than those provided by existing records, one of the aims of palaeohydrology was to extend knowledge of past hydrological events so that flood frequency analysis used for management decisions in environmental planning could be as informed as possible (e.g. Gregory and Benito, 2003). This was just one area where a multi-disciplinary approach was required and where contributions from geomorphologists, hydrologists, ecologists, geologists, and archaeologists, among others, were necessary to identify the past hydrological record as fully as possible. At least two alternatives are generally available: one to adapt geomorphology to be increasingly appropriate in a multidisciplinary framework, the other to foster new hybrid subdisciplines as necessary.

The recent report focusing on earth surface processes (NRC, 2009) identified nine grand challenges and proposed four high-priority research initiatives (see Chapter 1, Table 1.6). That report identified earth surface processes as a new field which has emerged over the last 20 years as a result of multi-disciplinary science needed to answer questions not easily dealt with by any single discipline. This therefore assumes that disciplines including geomorphology but also ecology, geology, hydrology and tectonics, together with hybrids such as ecogeomorphology (much cited in the NRC report), will all contribute, but that the new area of Earth Surface Processes is the field which will become pre-eminent. The report suggests that (p. 95) the field of earth surface processes has reached a point at which the imminent research questions cross multiple disciplinary boundaries and require new intellectual collaborations and approaches. Is geomorphology about to be replaced or subsumed or should it evolve to become more adapted to the new challenges and questions that confront the world today? The situation of geomorphology with regard to its two parent disciplines – geography and geology – is a challenging one. On the one hand there seems to be a great gulf between the concerns of much cultural geography and of physical geography, while in geology there appears to be an increasing interest in current processes and in environmental issues. Another fascinating issue is the extent to which national schools of geomorphology are becoming less distinct, possibly because of the wide dissemination of research results in truly international journals but also because of the establishment of bodies that promote international collaboration, such as the International Association of Geomorphologists.

At the end of the introduction (Chapter 1) it was proposed that the establishment of geomorphology might be thought of, slightly tongue in cheek, according to headings borrowed from Davis' geographical cycle of landscape development (Table 1.7) and that geomorphology is now poised, after more than a century of development, to enter a new revitalized stage characteristic of a vibrant, holistic and resilient discipline. Should a new focus for geomorphology be conceived, a rejuvenated discipline, and could this involve rediscovering landscape, particularly as it is now of so much public interest and concern?

Pertinent to this is the question about *how far geomorphology can and should extend.* Adaptation of the discipline has occurred before as exemplified by the impact of advances in plate tectonics (see Chapter 28). Whereas geomorphology up to the mid 20th century embraced some basic global tectonics and continental drift, it was only when geophysicists and geologists had demonstrated the significance of plate tectonic theory that it was appreciated how geomorphology might contribute, hence releasing the debate between small-scale process geomorphology contrasting with macroscale geomorphology (Summerfield, 2005a); potential for links between these two groups of researchers provides enormous scope to advance geomorphology as a whole probably at its most exciting time since it emerged as a discipline (Summerfield, 2005b).

Two obvious extensions for geomorphology beyond the land surface include submarine environments and the geomorphology of other planets and satellites. To advance the subject in these ways is now feasible as a result of techniques available, including the use of multibeam sensing which allows offshore geomorphology to be extended from onshore geomorphology. Investigation of submarine environments is necessarily associated with terrestrial environments because the present boundary between the two is not static and techniques are now available to furnish information from present submarine environments of palaeofeatures that can assist reconstruction of earlier geomorphological systems. Submarine geomorphology has a major contribution to make to hydrocarbon exploration and exploitation in such areas as the Angola Fan, the Nile Delta and the Norwegian Continental Shelf, not least because of the hazards associated with subsea landslides, salt dissolution and turbidity currents. Justification for investigations of other planets is that, for geomorphology to be, a complete science of landforms and landscapes, terrestrial systems of landforms and their generative processes need to be understood in a planetary context so that excluding extraterrestrial landscapes from geomorphology is illogical

(Baker, 2008). There is now a working group of the IAG (http://www.psi.edu/pgwg/reading_list. html), and research, for example in arid environments, can provide analogues for planets and satellites. Increased understanding of geomorphic processes in Antarctica may also lead to insights into the nature of planetary permafrost and the geomorphology of Mars (French, Chapter 23). Equally, other planets may provide environments that do not feature on Earth. The scope of planetary geomorphology requires further exploration but the substantial opportunity has to be balanced against the way in which extending the subject to encompass the geomorphology of other planets necessarily greatly increases the dimensions of the discipline; indeed planetary landscape systems provide a limitless frontier (Baker, 2008).

Applications of geomorphology might go much further than previously in relation to land surface impacts and *management and design of environments.* Preceding chapters have indicated potential applications of geomorphology, and the range of ways in which applications are achievable was mentioned in Chapter 1 and have been enumerated for fluvial geomorphology (Gregory et al., 2008).

Whereas many applications have been area- or problem-specific, further developments are now possible concerned with events or impacts including:

- Solutions which involve working with nature, and soft rather than hard engineering methods, involving a range of restoration approaches (e.g. Gregory, 2000: 265; Downs and Gregory, 2004: 240) applied to river channels.
- Solutions designed to manage environmental hazards have to identify how hazards may be misinterpreted, in the way that Schumm (1994) drew attention to three types of misperceptions of fluvial hazards which were:

  1 stability – the idea that any change is not natural
  2 instability – that change will not cease
  3 excessive response – that change is always major

  He concluded that such misperceptions can lead to litigation and may be the reason for unnecessary engineering works.
- Adaptive management solutions which involve monitoring consequences of implementation of particular projects, expressed as adaptive science by Graf (2003).
- Employing a holistic view has been undertaken in the sense of managing a specific area (e.g. Bahrain Surface Materials Resources Survey, Brunsden et al., 1979; 1980) but extended to consider the environmental setting so that

impacts and consequences elsewhere are considered, and may also require awareness of globalization (Clifford, 2009).

- Concern with landscape in its entirety: a geomorphologist is able to place a particular environment in its temporal and spatial setting. The spatial setting requires consideration of culture as a particular form or stage of civilisation including the pattern of human knowledge, belief and behaviour, embracing language, ideas, beliefs, customs, codes, institutions, tools, techniques and works of art, and so should be reflected in legislation and in ethical values (Gregory, 2010: 299). Thus, what is called nature in one culture at one time may be viewed very differently in a culture affected by different political, historical, and social factors (Palmer, 2003: 33). In the context of river management Hillman (2009) referred to what Aristotle called phronesis – contextualized and place-based wisdom built on experience and incorporation of cultural values, thus providing a key to geomorphological perspectives. The temporal setting requires consideration of inheritance of past conditions and in recent decades conceptual frameworks emphasizing single-path, single-outcome trajectories of change have been supplemented by multi-path, multi-outcome perspectives (Phillips, 2009). Landscape memory has been developed by Brierley (2010) by differentiating the imprint of the past upon contemporary landscape forms and processes in terms of geologic, climatic and anthropogenic memory, although the importance of the land surface and geomorphology should not be assigned to a geological category.
- Design is an inevitable consequence of adopting a holistic perspective with a design science proposed to support the profession (Rhoads and Thorn, 1996: 135) producing impressive examples of ways in which geomorphologists have contributed to the design of restoration projects (e.g. Brookes et al., 2004; Gregory, 2010).

As pointed out by Bierman (Chapter 32), understanding earth surface processes and the history of our planet are critical in a future that includes many billions of people inhabiting an ever more crowded and warming planet; geomorphology as a discipline has the knowledge to reduce society's exposure to both natural and human-induced hazards. He poses the question as to whether we as a profession are up to the task of understanding coupled natural/human systems and ensuring that our relevant knowledge is employed by those managing Earth's dynamic surface; a rejuvenated discipline could respond to the challenge. The study of geomorphological hazards and the contribution of geomorphology to

disaster reduction is indeed a developing theme (Alcántara-Ayala and Goudie, 2010).

IMPACTS OF FUTURE CHANGE AND LANDSCAPE RESILIENCE

New issues which geomorphology can pursue include major potential future applications for the subject in relation to the implications of a high CO_2 world. It is arguable that, in the past, geomorphology was somewhat reticent in addressing the applications of its science but surely is now poised to respond to the challenge by contributing to the changes likely to occur in the next phase of development of the Anthropocene. Relatively little attention has been devoted to potential geomorphological consequences, probably reflecting the dominant, but misinformed, view that geomorphological changes are exceedingly slow and therefore of little relevance to human-centred management systems concerned with contemporary environmental change; in fact environmental change rather than stability has characterized the most recent 2 million years of Earth's history (Anderson et al., 2007) and human influences are for the most part merely modifying the rates of change (Jones, 1993). More recently, when attention was drawn to global warming and global geomorphology Goudie (2006) noted that the major impacts attracting attention, to date, are sea-level rise, the extent of glaciers, the hydrological cycle, especially through changes in precipitation characteristics and runoff, permafrost characteristics, but observed that remarkably few scenarios for future geomorphological changes have been developed. In more general terms it has been suggested that we might develop 'Earthcasts' analogous to weather forecasts, of both gradual changes and extreme landscape-changing events (Murray et al., 2009). When considering potential geomorphological contributions Tooth (2007) reminds us that, because geomorphology is the integrated outcome of interactions between the atmosphere, hydrosphere, biosphere and geosphere, the inherent uncertainties in future scenarios for each of these component parts are compounded when trying to develop geomorphological scenarios. In particular, the complexity of environmental change makes effective modelling and prediction of geomorphological responses extremely difficult. For instance, landscape response to climate change is highly nonlinear, and characterized by numerous feedbacks between different variables.

Global warming is not new, as shown by the IGPCC review of the history of global warming over the last 900,000 years; records for the last

2000 years of coastal change in Britain were used to demonstrate flood risk areas in England and Wales, and coastal erosion vulnerability (Cracknell, 2005). Although comparatively few, there is a growing number of other specific examples of geomorphological research which address the issue of the effects of global warming (Table 33.1). Research over the last three decades, much driven by a concern to understand the global carbon cycle, indicates that there are many factors, both geological and environmental, determining the rates of silicate weathering and the controlling factors offering further opportunities for geomorphological research. Geomorphologists need, however, to assess whether future climate change is the most important factor in global change or whether other factors (such as land cover change brought about by deforestation) are of greater significance (Slaymaker et al., 2009).

Using the achievements of geomorphology, many recounted in the previous chapters, it should now be possible to address broad groups of issues to which geomorphologists are particularly able to contribute. These include:

- Evaluating the consequences of outputs from GCMs for earth surface processes. Establishing how dynamics will be affected.
- How do these consequences translate into environmental hazards? What additional risks arise and what uncertainties surround expectations?
- What consequences will changed processes have for the landsurface, including alterations in the frequency of land-forming events, as well as new consequences from changing events?
- What new process domains could be created? How will landscapes have different degrees of sensitivity and resilience?
- Will new scenarios occur? How can geomorphologists contribute to the design of the Anthropocene under new conditions?

Table 33.1 Published examples of the geomorphological impacts of global warming on the land surface and surface processes

Area studied	Suggested impacts	Reference
Sub-Arctic landscape	Under the predicted global warming, the 'greening' of Arctic and sub-Arctic regions may decrease and also increase the activity of the periglacial processes in sparsely vegetated terrain.	Hjort and Luoto, 2009
Rockglacier, central Andes	Significant change in the active layer and suprapermafrost of this rockglacier of the Cordón del Plata is registered. The observed changes imply direct consequences for the cryogenic environment and the Andean creeping permafrost. Variations of the cryogenic structure of the rockglaciers of the Cordón del Plata caused by warming processes, will have direct consequences for the volume of frozen sediments and therefore for the hydrology of the entire region.	Trombotto and Borzotta, 2009
Andhra Pradesh coast, India	A coastal vulnerability index was prepared which showed that about 43% of the AP coast (a length of 1030 km) is under very high risk.	Nageswara et al., 2009
Future eustatic sea-level rise on the deltaic coasts of Inner Thermaikos Gulf (Aegean Sea) and Kyparissiakos Gulf (Ionian Sea), Greece	Predictions of responses to future sea-level rise, of 0.5 to 1 m, are made. In the case of the deltaic coast of the Alfios, exposed to high waves and protected by dune fields, shoreline retreat of up to 700 m is predicted. The low-lying bird-foot type deltas of the Axios and Aliakmon are expected to retreat by more than 2 km.	Poulos et al., 2009
Alluvial channel patterns in North-European Russia	Results from three geomorphological models indicate that climatic warming in combination with other environmental changes may lead to transformation of the river channel types at selected locations in north-western Russia.	Anisimov et al., 2008
Coastal plains of Greece	Predictions of responses to future sea-level rise, of 0.5 to 1 m, are made. In the case of the deltaic coast of the Alfios, exposed to high waves and protected by dune fields, shoreline retreat of up to 700 m is predicted. The low-lying bird-foot type deltas of the Axios and Aliakmon are expected to retreat by more than 2 km.	Poulos et al., 2009

Continued

Table 33.1 Cont'd

Area studied	Suggested impacts	Reference
California's Sacramento–San Joaquin watershed	Geomorphic processes in rivers are likely to be influenced by global warming through alterations of flood, erosion, and sedimentation processes and rates. Future increases in magnitudes and durations (and changes in timing) of floods. Because the scenarios of warming are developed at coarse scales, only an understanding of the relations between large-scale hydrology and climate on the one hand, and the incidence of levee breaches on the other, will make it possible to project likely geomorphic responses to future warming and flooding. Preliminary results suggest strong relations between levee breaches and discharge, but poor relations to ENSO. Further investigation will help inform models and river management policy that addresses rates and magnitudes of erosion and sedimentation.	Florsheim et al., 2006
UK coast	At many locations maintaining a resilient coast demands the availability of space and sediments, but both are in short supply. Future management should incorporate adaptive measures, including strategic large-scale coastal floodplain reactivation, to support a resilient geomorphic response to sea-level rise. Management decisions will have major implications for coastal waterbirds.	Crooks, 2004
Coastal management and sea-level rise	Spatial adjustments that coastal landforms will exhibit in response to changing energy gradients both normal to and parallel to the shore in many cases, will take the form of the migration of landforms in order that they maintain their position within the coastal energy gradient. Prediction of the rates of such migration will be fundamental to the future management of the changing coastal environment. In estuaries and on the open coast rates of landform migration under an accelerated sea-level rise are predicted and compared with existing rates using examples from the east coast of Britain. Assuming a sea-level rise of 6 mm y^{-1}, the paper predicts that estuaries will migrate landwards at rates of around 10 m y^{-1}, open-coast landforms can exhibit long-shore migration rates of 50 m y^{-1}, while ebb-tidal deltas may extend laterally along the shore at rates of 300 m y^{-1}.	Pethick, 2001
Permafrost in Eurasia	Tries to evaluate the probable catastrophic implications of **global** climatic change in the cold regions of Eurasia. The impact on human activities in permafrost regions is discussed.	Demek, 1994
Sea-level change	For each coastal lowland the rate of rise in sea level given in the scenario to AD 2050 is greater than that experienced for any long-term average during the last 7000 years; coastal management involves a wide range of activities which will be affected to different degrees by enhanced sea level rise. Short-term variations in rate of sea level change occur, and an increase in the intensity and frequency of storm surges could compound this.	Shennan, 1993
Slope instability	Ways of dealing with consequences include: extending and refining National Landslide data base; expanding public information programme already begun by DOE; placing further emphasis upon landslide hazard assessment; extending recently expanded remit of building codes.	Jones, 1993

Investigation of such changes afford great opportunities for geomorphology, using the foundations already established that have been reviewed in the preceding chapters.

REFERENCES

Alcántara-Ayala, I. and Goudie, A.S. (eds) (2010) *Geomorphological Hazards and Disaster Prevention.* Cambridge University Press, Cambridge.

Anderson, D.E., Goudie, A.S. and Parker, A.G. (2007) *Global Environments through the Quaternary.* Oxford University Press, Oxford.

Anisimov, O., Vandenberghe, J., Lobanov, V. and Kondratiev, K. (2008) Predicting changes in alluvial channel patterns in North-European Russia under conditions of global warming. *Geomorphology,* 98, 262–74.

Baker, V.R. (2001) Water science: Aquarian and Piscean Viewpoints. *Geological Society of America, Abstracts with Programs,* 33(7), 429.

Baker, V.R. (2008) Planetary landscape systems: a limitless frontier. *Earth Surface Processes and Landforms,* 33, 1341–53.

Brierley, G.J. (2010) Landscape memory: the imprint of the past on contemporary landscape forms and processes. *Area,* 42, 76–85.

Brookes, A., Chalmers, A. and Vivash, R. (2004) Solving an urban river erosion problem on the Tilmore Brook, Hampshire, (UK). *Journal of the Chartered Institution of Water and Environmental Management,* 19, 199–206.

Brunsden, D., Doornkamp, J.C. and Jones, D.K.C. (1979) The Bahrain surface materials resources survey and its application to regional planning. *Geographical Journal,* 145, 1–35.

Brunsden, D., Doornkamp, J.C. and Jones, D.K.C. (eds) (1980) *Geology, Geomorphology and Pedology of Bahrain.* Geo Books, Norwich.

Church, M. (2010) The trajectory of geomorphology. *Progress in Physical Geography,* 34, 265–86.

Clifford, N.J. (2009) Globalization: a physical geography perspective. *Progress in Physical Geography,* 33, 5–16.

Cracknell, B.E. (2005) *Outrageous Waves: Global Warming and Coastal Change in Britain Through Two Thousand Years.* Phillimore, Chichester.

Crooks, S. (2004) The effect of sea-level rise on coastal geomorphology. *Ibis,* 146, 18–20.

Dadson, S. (2010) Geomorphology and Earth system science. *Progress in Physical Geography,* 34, 385–98.

Demek, J. (1994) Global warming and permafrost in Eurasia: a catastrophic scenario. *Geomorphology,* 10, 317–29.

Downs, P.W. and Gregory, K.J. (2004) *River Channel Management.* Arnold, London. p. 240.

Florsheim, J.L., Dettinger, M., Malamud-Roam, F., Ingram, B. and Mount, J. (2006) Geomorphic response to global warming in the Anthropocene: levee breaches in California's Sacramento-San Joaquin watershed. *American Geophysical Union,* Fall Meeting 2006, abstract #H11B–1246.

Goudie, A.S. (2006) Global warming and fluvial geomorphology. *Geomorphology,* 79, 384–94.

Goudie, A.S. and Viles, H.A. (2010) *Landscapes and Geomorphology: A Very Short Introduction.* Oxford University Press, Oxford.

Graf, W.L. (2003) Summary and perspective, in Graf, W.L. (ed.) *Dam Removal Research: Status and Prospects.* The H. John Heinz III Center for Science, Economics and Environment, Washington DC. pp. 1–21.

Gregory, K.J. (ed.) (1983) *Background to Palaeohydrology.* Wiley, Chichester.

Gregory, K.J. (2000) *The Changing Nature of Physical Geography.* Arnold, London.

Gregory, K.J. (2010) *The Earth's Land Surface.* Sage, London.

Gregory, K.J. and Benito, G. (eds) 2003. *Palaeohydrology: Understanding Global Change.* Wiley, Chichester.

Gregory, K.J., Benito, G. and Downs, P.W. (2008) Applying fluvial geomorphology to river channel management: background for progress towards a palaeohydrology protocol. *Geomorphology,* 98, 153–72.

Hillman, M. (2009) Integrating knowledge: the key challenge for a new paradigm in river management. *Geography Compass* 3, 1988–2010.

Hjort, J. and Luoto, M. (2009) Interaction of Geomorphic and Ecologic features across ecological zones in a subarctic landscape. *Geomorphology,* 112, 324–33.

Jones, D.K.C. (1993) Global warming and geomorphology. *Geographical Journal,* 159, 124–30.

Migoń, P. (ed.) (2010) *Geomorphological Landscapes of the World.* Springer, Berlin.

Murray, B., Lazarus, E., Ashton, A., Baas, A., Coco, G., Coulthard, T. et al. (2009) Geomorphology, complexity, and the emerging science of the Earth's surface. *Geomorphology,* 103, 496–505.

Nageswara Rao, K., Subraelu, P., Venkateswara Rao, T., Hema Malini, B., Ratheesh, R., Bhattacharya, S. et al. (2009) Sea-level rise and coastal vulnerability: an assessment of Andhra Pradesh coast, India through remote sensing and GIS. *Journal of Coastal Conservation,* 12, 1–13.

NRC (National Research Council) (2009) *Landscapes on the Edge: New Horizons for Research in Earth Surface Processes.* Committee on Challenges and Opportunities in Earth Surface Processes. National Research Council, Washington.

Palmer, C. (2003) An overview of environmental ethics, in Light, A. and Rolston, H. (eds) *Environmental Ethics.* Blackwell, Oxford. pp. 15–37.

Pethick, J. (2001) Coastal management and sea-level rise. *Catena,* 42, 307–22.

Phillips, J.D. (2009) Changes, perturbations, and responses in geomorphic systems. *Progress in Physical Geography,* 33, 1–14.

Poulos, S.E., Ghionis, G. and Maroukian, H. (2009) The consequences of a future eustatic sea-level rise on the deltaic coasts of Inner Thermaikos Gulf (Aegean Sea) and Kyparissiakos Gulf (Ionian Sea), Greece. *Geomorphology,* 107, 18–24.

Rhoads, B.L. and Thorn, C.E. (1996) Towards a philosophy of geomorphology, in Rhoads, B.L. and Thorn, C.E. (eds) *The Scientific Nature of Geomorphology*. Wiley, Chichester. pp. 115–43.

Schumm, S.A. (1994) Erroneous perception of fluvial hazards. *Geomorphology*, 10, 129–38.

Shennan, I. (1993) Sea-level change and the threat of coastal inundation. *Geographical Journal*, 159, 148–56.

Slaymaker, O., Spencer, T. and Embleton-Hamann, C. (eds) (2009) *Geomorphology and Global Environmental Change*. Cambridge University Press, Cambridge.

Summerfield, M.A. (2005a) The changing landscape of geomorphology. *Earth Surface Processes and Landforms*, 30, 779–81.

Summerfield, M.A. (2005b) A tale of two scales, or the two geomorphologies. *Transactions of the Institute of British Geographers*, 30, 402–15.

Tooth, S. (2007) Arid geomorphology: investigating past, present and future changes. *Progress in Physical Geography*, 33, 319–35.

Trombotto, D. and Borzotta, E. (2009) Indicators of present global warming through changes in active layer-thickness, estimation of thermal diffusivity and geomorphological observations in the Morenas Coloradas rockglacier, Central Andes of Mendoza, Argentina. *Cold Regions Science and Technology*, 55, 321–30.

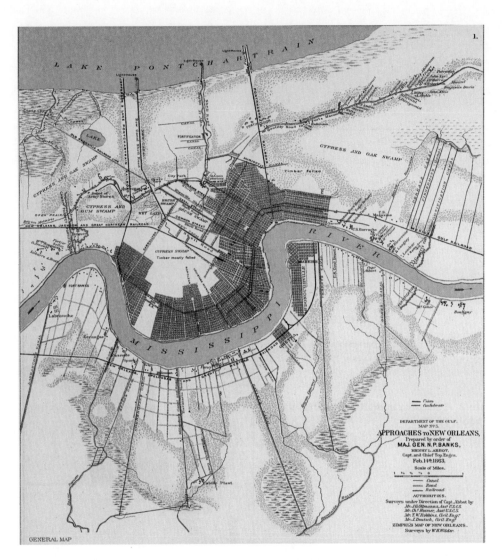

Plate 1   Map of New Orleans in 1895, showing urban development on the natural levees of Mississippi River main channel (e.g. French Quarter), and of former distributary channels, such as Metarie and Gentily ridges, both occupied by roads of the same name. (US Department of War, 1895)

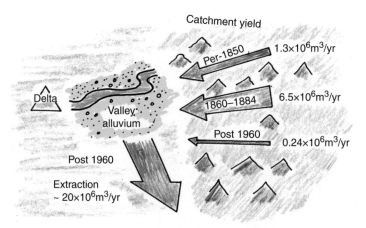

Plate 2   Sediment budget for Central Valley of California, 1850 to present. (Adapted from Kondolf, 2001)

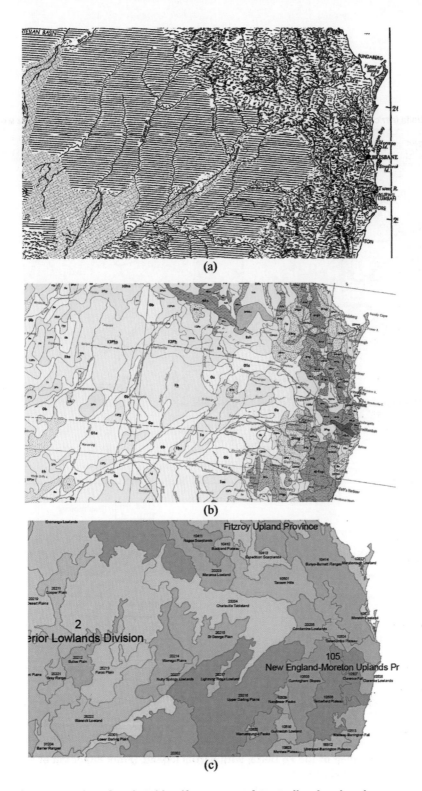

(a)

(b)

(c)

Plate 3   Three examples of regional landform maps of Australia, showing the same part of the eastern part of the continent. (a) extract from Lobeck (1951) showing the use of hachures. (b) extract from Löffler and Ruxton (1969), showing a polygon map derived from land system mapping. (c) extract from Pain et al. (in press) showing a polygon map compiled from the SRTM 30′ DEM (see Pain, 2008)

Po delta (F.77 Comacchio)   M.Bondesan & M.C.Turrini

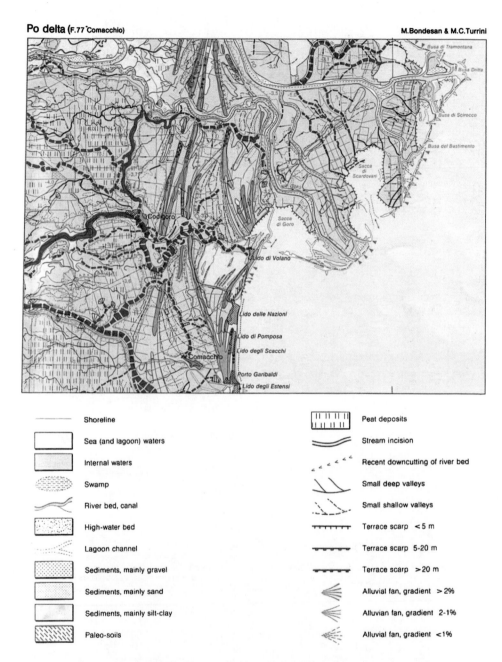

——— Shoreline	Peat deposits
Sea (and lagoon) waters	Stream incision
Internal waters	Recent downcutting of river bed
Swamp	Small deep valleys
River bed, canal	Small shallow valleys
High-water bed	Terrace scarp < 5 m
Lagoon channel	Terrace scarp 5-20 m
Sediments, mainly gravel	Terrace scarp > 20 m
Sediments, mainly sand	Alluvial fan, gradient > 2%
Sediments, mainly silt-clay	Alluvian fan, gradient 2-1%
Paleo-soils	Alluvial fan, gradient < 1%

Plate 4   Landform map of the Po Delta, with partial legend. (From Bondesan et al., 1989; see also Castiglioni et al., 1999)

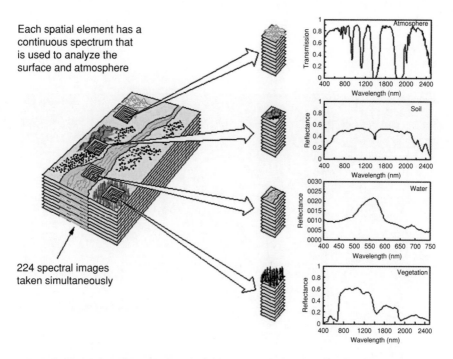

Each spatial element has a continuous spectrum that is used to analyze the surface and atmosphere

224 spectral images taken simultaneously

Plate 5   Principle of imaging spectroscopy. Each pixel in the 224 bands samples nearly continuously the VNIR spectrum of the terrain

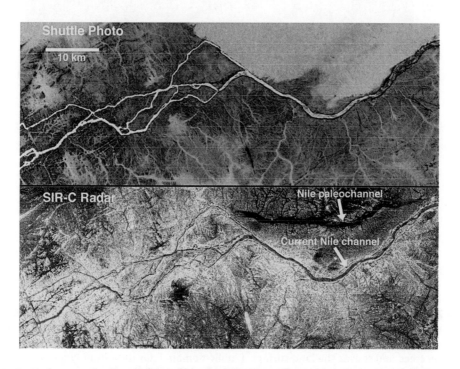

Plate 6   Radar penetration of dry sediments: Nile River. The top image is a hand-held photograph from the Space Shuttle in November 1995 and shows the Nile River in Sudan. The river is brownish due to silt. The lower image is from SIR-C and is a color composite of L-band (25 cm wavelength) and C-band images. Note the old channel of the Nile shows up in the lower radar image, but not the hand-held image, where sand is seen to cover the area

(a)

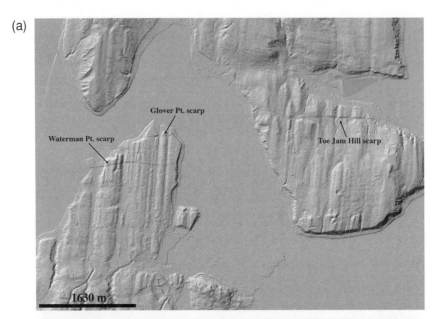

(b)

Plate 7 Lidar DEM showing a forested fault scarp in the state of Washington. (a) Bare-earth lidar DEM showing prominent north–south lineations due to glaciation and east–west fault scarps. (b) Google Earth image of same area showing extensive forest cover; lidar penetrates between trees and senses the land surface. (Public-domain lidar data from Puget Sound Lidar Consortium; http://pugetsoundlidar.ess.washington.edu/index.html)

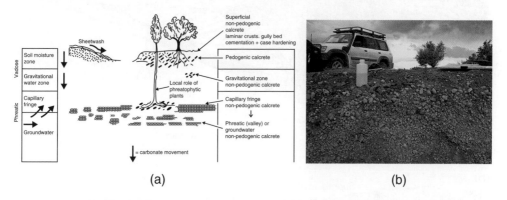

(a)

(b)

Plate 8 **(a) A schematic showing their interpretation of calcrete formation. (From Wright and Tucker, 1991) (b) A nodular calcrete from Broken Hill, New South Wales, formed within the pedogenic zone of and alluvial/aeolian regolith. Here much of the carbonate is thought to be derived from dry lakes and lacustrine environments to the west of Broken Hill**

594

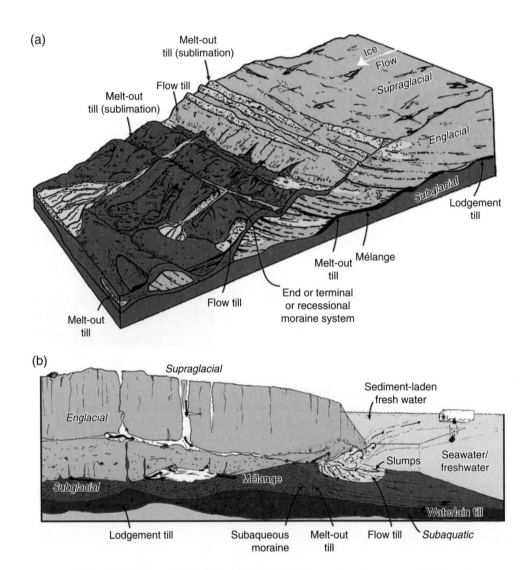

Plate 9 (a) Model of ice sheet/valley glacier frontal terrestrial margin. (b) Model of ice sheet/valley glacier floating subaquatic margin

Plate 10   Glacial valley incision producing a U-shaped valley, the Korok River Valley, Torngat Mountains, Labrador, Canada. (Photograph courtesy John Gosse)

596

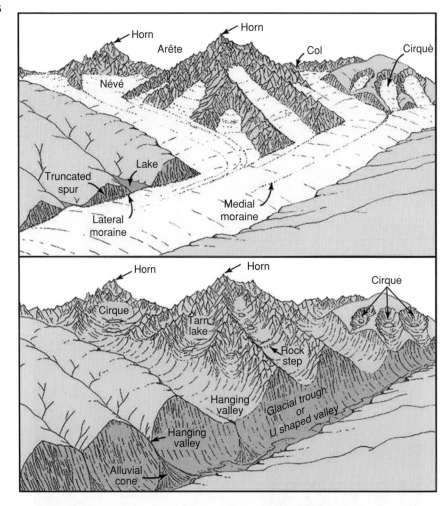

Plate 11 Erosional landscape system due to alpine glaciation. (Modified from Tarbuck and Lutgens, 2003)

(a)　　　　　　　　　　　　　(b)

Plate 12 (a) Tabular hills in Panchagani, India. Weathered lavas at this site are capped by a thick ferricrete duricrust. (b) Deeply decomposed granite-gneiss below convex hill in eastern Brazil, now dissected by canyon gullies (height ~ 50–100 m)

Index

Figures in *italics*; tables in **bold**